Unproblem JEE

Coordinate Geometry

for JEE Main & Advanced

1500+

Problems with Solutions...

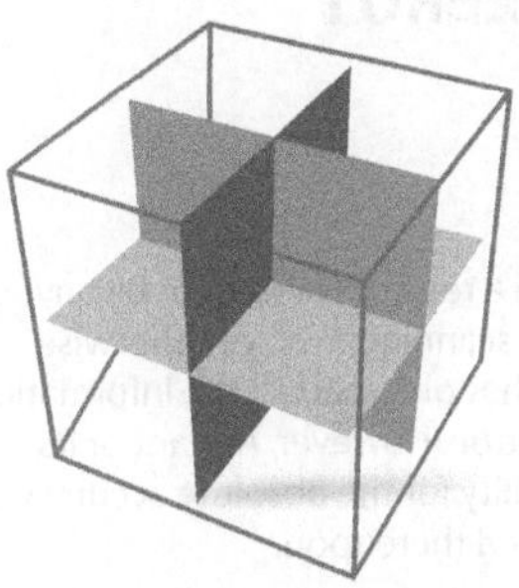

Er. Girish Kumar Jain
ME, MSc, BE

arihant
HANT PRAKASHAN (Series), MEERUT

arihant

ARIHANT PRAKASHAN (Series), MEERUT

卐 **Administrative & Production Offices**

Regd. Office

'Ramchhaya' 4577/15, Agarwal Road, Darya Ganj, New Delhi -110002
Tele: 011- 47630600, 43518550

卐 **Head Office**

Kalindi, TP Nagar, Meerut (UP) - 250002
Tel: 0121-7156203, 7156204

卐 **Sales & Support Offices**

Agra, Ahmedabad, Bengaluru, Bareilly, Chennai, Delhi, Guwahati, Hyderabad, Jaipur, Jhansi, Kolkata, Lucknow, Nagpur & Pune.

卐 **ISBN** : 978-93-26199-96-4

卐 **PRICE** : ₹425.00

PO No : TXT-XX-XXXXXXX-X-XX

Published by Arihant Publications (India) Ltd.

For further information about the books published by Arihant, log on to www.arihantbooks.com or e-mail at info@arihantbooks.com

Follow us on

Preface

Unproblem JEE Series is an attempt to enrich and enliven the Mathematical Problem Skills in JEE & other engineering entrances aspirants. Most of the Problem Books available in the segment are not substantially capable to serve the purpose for which an engineering aspirant goes for the problem books. They are just a compilation of problems packed in different chapters without any logical arrangement & purposeful grading. Thus, they fail to provide the readers with the approach required to solve complex & structured problems generally asked in JEE.

The Third book of Unproblem JEE Series, **'Problems in Coordinate Geometry'** is an effort to strengthen the concepts & problem solving skiil of coordinate geometry in engineering aspirants. Coordinate Geometry is a stream where geometry meets algebra, through the problems given in the book, students will develop the approach to solve Coordinate Geometry Problems and also their applications to solve other mathematical problems.

Problems in each chapter are arranged to promote problem-solving ability and make the students understand that concepts & techniques are logically connected. In this way, students will become comfortable with concepts & will know how and when to use them while solving a problem.

The chapters in the book are divided into three levels and each level is a complete treatment of the chapter but at different levels of difficulty.

- Each chapter is divided into three major heads; **Level 1** 'Starter Level', **Level 2** 'JEE Main Level' & **Level 3** 'JEE Advanced Level'.
- **Level 1 'Starter Level'** consists of preliminary questions that test basic concepts. After completing this level student will have a solid understanding of fundamentals of the chapter necessary to solve the problems.
- **Level 2 'JEE Main Level'** has single choice correct questions and numerical value questions aimed especially for JEE Main.
- **Level 3 'JEE Advanced Level'** consists of various types of questions; MCQ with one or more than One Correct Answer, Comprehension Based Questions, Match the Columns and Numerical Value problems as per JEE Advanced Pattern.
- The problems given in JEE Advanced Level are made to be conceptually challenging resulting in practical and relevant practice for JEE Advanced.
- Detailed solutions to all the problems have been made available online just to help the students in case they stuck somewhere while solving some questions.

At the end, I can say confidently that if you rigorously practice the problems given in this book then surely you will get an edge in Mathematics and will increase your chance of getting good JEE Score.

However, I have taken the utmost care while writing this book but if any error or whatsoever has skipped my attention then please bring it to my notice so that it can be rectified. I would also like to appeal all my fellow Mathematics Teachers to have a look at the book and share their valuable suggestions for its further improvement.

Er. Girish Kumar Jain

ME, MSc, BE

Contents

CHAPTER

01

CARTESIAN COORDINATE SYSTEM

LEVEL 01 \ Starter Level

1.1 Introduction of Coordinates, Distance Formula and Section Formula

1 (i) Find the cartesian coordinates of the following points whose polar coordinates are

(a) $\left(5, \frac{\pi}{3}\right)$ (b) $\left(2, -\frac{\pi}{4}\right)$ (c) $\left(3, \frac{3\pi}{4}\right)$ (d) $\left(6, -\frac{2\pi}{3}\right)$

(e) $\left(1, \pi - \tan^{-1}\frac{4}{3}\right)$

(ii) Find the polar coordinates of the following points whose cartesian coordinates are

(a) (3, 4) (b) (−1,1)

(c) $(\sqrt{6}, -\sqrt{2})$ (d) $(-1, -\sqrt{3})$

2 (i) Transform the following equations into polar form.

(a) $\frac{x^2}{a^2} + \frac{y^2}{b^2} = 1$ (b) $2x^2 + 3xy + 2y^2 = 1$

(c) $(x^2 + y^2)^2 = a^2(x^2 - y^2)$ (d) $x^2 + y^2 = 2ax$

(ii) Transform the following equations into cartesian form.

(a) $r = 2a\cos\theta$ (b) $r^2 = \frac{6}{4 + 3\sin 2\theta}$

(c) $r^2 = a^2 \cos 2\theta$ (d) $r = \frac{3\sin\theta + 4\cos\theta}{7}$

3 Find the distance between following points.

(i) $(am_1^2, 2am_1)$ and $(am_2^2, 2am_2)$ (ii) $(at^2, 2at)$ and $\left(\frac{a}{t^2}, -\frac{2a}{t}\right)$

(iii) $\left(3, \frac{\pi}{4}\right)$ and $\left(7, \frac{5\pi}{4}\right)$ (iv) $(17, 105°)$ and $(5\sqrt{2}, 60°)$

4 Find the value of λ, if

(i) distance between (3, 7) and $(-2, \lambda)$ is 7.

(ii) $(\lambda, 0)$ is equidistant from the points (7, 6) and (3, 4).

5 Find the values of a and b, if

(i) $(-3, 4)$, $(-2, 3)$ and (a, b) are vertices of an equilateral triangle.

(ii) $(-4,-1)$, $(-2,-4)$, $(4,0)$ and $(3a,-b)$ are vertices of a rectangle.

(iii) $P(1, 2)$, $Q(4, 6)$, $R(5, 7)$ and $S(a, b)$ are vertices of a parallelogram.

(iv) $(-3,-4)$, $(2, 6)$ and (a, b) are vertices of a right angled triangle.

6 Find the coordinate of point

(i) which trisects the line segment joining $(1,-2)$ and $(-3,4)$.

(ii) lies on the line $x - y = 1$ and also divides the line segment joining (3,4) and (−7,2) in $k:1$ internally.

(iii) lies on the curve, $r = \frac{3\sin\theta + 4\cos\theta}{5}$ and divides the line segment joining (1,0) and $\left(3, \frac{\pi}{2}\right)$ in $k:1$ internally, also find k.

1.2 Area of Polygon

7 Find the area of

(i) triangle whose vertices are $(4,1)$, $(-3,7)$ and $(5, 8)$.

(ii) quadrilateral whose vertices are $(7, 1)$, $(5,-3)$, $(3,3)$ and $(-3,-7)$.

(iii) pentagon whose vertices are $(4,3)$, $(-5,6)$, $(0,7)$, $(3,-6)$ and $(-7,-2)$.

8 Find the value of λ, if

(i) area of triangle formed by the vertices $(-1,-4)$, $(\lambda,1)$ and $(\lambda,-4)$ is 12.5 sq units.

(ii) $A(-2, 3)$, $B(1, 2)$ and $C(3\lambda, 0)$ are collinear.

(iii) $(\lambda, \lambda+1)$ is an interior point of ΔABC, where $A(0,3)$, $B(-2, 0)$ and $C(6, 1)$.

1.3 Some Special Points of Triangle (Centroid, Incentre, Excentre, Orthocentre, Circumcentre)

9 (i) Find the circumcentre and circumradius of the triangle whose vertices are $A(1,1)$, $B(2,-1)$ and $C(3,2)$.

(ii) Find the incentre of the triangle whose vertices are $A(2,3)$, $B(-2,-5)$ and $C(-4,6)$.

(iii) The circumcentre of a triangle lies at the origin and the centroid is the mid-point of the line segment joining (2, 2) and $(+2,-2)$, then find the orthocentre.

(iv) Two vertices of a triangle are $(5, -1)$ and $(2, 3)$, if origin is the centroid, then find the third vertex of triangle.

(v) Find the orthocentre and nine-point centre of triangle having vertices (0, 0), (3, 0) and (0, 4).

(vi) If a triangle has its orthocentre (2, 3) and circumcentre at $\left(\frac{3}{2}, \frac{4}{5}\right)$, then find the centroid and nine-point centre.

(vii) Find the all excentre of the triangle whose vertices are $(-2, 2)$, $(1, 6)$ and $(4, 2)$.

1.4 Locus of Point

10 Find the locus of the point

(i) which moves such that it is equidistant from the points (1, 2) and (−2,3).

(ii) which moves such that the area of the triangle formed with the points (1, 2) and (2, − 1) is equal to 3 sq units.

(iii) which moves such that the sum of the squares of its distances from the coordinate axes is 5 units.

(iv) which moves such that $PA^2 = 3PB^2$, where $A = (5,0)$ and $B(-5,0)$.

(v) which moves such that the line joining (2, 0) and (−2,0) subtend a right angle at it.

(vi) which moves such that its distance from the point (2, 3) is twice its distance from (−2, 2).

1.5 Transformation of Axes

11 (i) Suppose the origin is shifted to (3, 4), then find the new coordinates of the point (4,5), (3,− 1), (−2,4) and (−3,− 4).

(ii) When the origin is shifted to (−1,4) the axes remaining parallel, the new coordinates of three points are (3,2), (4,− 3), (−1,7). Find their old coordinates.

(iii) On shifting the origin to a point (h,k), axes remaining parallel, the coordinates of (−2,1) becomes (4,− 5). Find h and k.

(iv) When the axes are rotated through 30° in the anti-clockwise sense without changing the origin, find the new coordinates of the point (−2,4).

12 (i) If the axes are rotated through 45° in the clockwise sense, then find the transformed form of the equation $x^2 - y^2 = a^2$.

(ii) The equation of locus is $y^2 + 8y - 4x + 24 = 0$. If the origin in shifted to (2,− 4), the axes remaining parallel. Find the new equation of the locus.

(iii) Find the point to which the origin should be shifted so that the equation $y^2 - 6y - 4x + 13 = 0$ is transformed to form $y^2 + ax = 0$.

(iv) Reduce the equation $x^2 + y^2 + 8x - 6y - 25 = 0$ to the form $AX^2 + BY^2 = k$, by shifting the origin to a suitable point.

LEVEL 02 \\ JEE Main Level

Single Choice Correct Questions

1.1 Introduction of Coordinates, Distance Formula and Section Formula

1 If the perpendicular bisector of the line segment joining $P(1,4)$ and $Q(\alpha,3)$ cut Y-axis at −4, then possible value of α is

(a) −2 (b) −4 (c) 1 (d) 2

2 A line segment AB of length 10 units and $A = (2,-3)$. If the abscissa of B is positive root of the equation $x^2 - 8x - 20 = 0$, then the sum of all possible ordinates of B is

(a) 3 (b) −3 (c) 6 (d) −6

3 If P divides OA internally in the ratio $\lambda_1 : \lambda_2$ and Q divides OA externally in the ratio $\lambda_1 : \lambda_2$, then OP, OA and OQ are in

(a) AP (b) GP (c) HP (d) None of these

4 The distance between the point $\left(\frac{21}{5},-\frac{8}{5}\right)$ to point which lies on the line $3x+y=20$ and also lies on line segment joining (3,–1) and (7, 2), is

(a) $\sqrt{10}$ units
(b) $\sqrt{11}$ units
(c) $\sqrt{13}$ units
(d) None of these

5 The ratio in which the line $r=\frac{4}{\sin\theta-\cos\theta}$ divides the line joining the points $\left(3,\frac{\pi}{2}\right)$ and $\left(3,\frac{\pi}{6}\right)$ is

(a) $5+3\sqrt{3}:2$ (externally)
(b) 5:3 (externally)
(c) 1 : 2 (internally)
(d) None of these

6 The point which divides the line segment in four equal parts joining the points (3, 1) and (–2,5) is

(a) $\left(\frac{-3}{4},4\right)$
(b) $\left(\frac{1}{2},3\right)$
(c) $\left(\frac{7}{4},2\right)$
(d) All of these

1.2 Area of Polygon

7 The area (in sq units) of the triangle formed by the points with polar coordinates (1, 0), $\left(2,\frac{\pi}{3}\right)$ and $\left(3,\frac{2\pi}{3}\right)$ is

(a) $\frac{11\sqrt{3}}{5}$ sq units
(b) $\frac{5\sqrt{3}}{4}$ sq units
(c) $\frac{5}{4}$ sq units
(d) $\frac{11}{4}$ sq units

8 Let $A(1, k)$, $B(1, 1)$ and $C(2, 1)$ be the vertices of a right angled triangle with AC as its hypotenuse. If the area of the triangle is 1 sq unit, then the set of values which k can take, is given by

(a) {–1,3}
(b) {–3,–2}
(c) {1,3}
(d) {0,2}

9 Let $O(0,0)$, $P(3,4)$, $Q(6,0)$ be the vertices of the ΔOPQ. The point R inside the ΔOPQ such that ΔOPR, ΔPQR and ΔOQR are of equal area. The coordinates of R is

(a) $\left(\frac{4}{3},3\right)$
(b) $\left(3,\frac{2}{3}\right)$
(c) $\left(3,\frac{4}{3}\right)$
(d) $\left(\frac{4}{3},\frac{2}{3}\right)$

10 The line $x=c$ cuts the triangle with corners (0,0), (1,1) and (9,1) into two regions. For the area of the two regions to be the same c must be equal to

(a) $\frac{5}{2}$
(b) 3
(c) $\frac{7}{2}$
(d) 3 or 15

11 Let A be the set of all points (α,β) such that the area of triangle formed by the points $A(5, 6)$, $B(3, 2)$ and $C(\alpha, \beta)$ is 12 sq units. Then, the least possible length of a line segment joining the origin to a point in c is **[JEE Main 2021]**

(a) $\frac{4}{\sqrt{5}}$ units
(b) $\frac{16}{\sqrt{5}}$ units
(c) $\frac{8}{\sqrt{5}}$ units
(d) $\frac{12}{\sqrt{5}}$ units

12 In a ΔABC, coordinates of A is (1,2) and the equations of the medians through B and C are $x+y=5$ and $x=4$, respectively. Then, area of ΔABC is **[JEE Main 2018]**

(a) 12 sq units
(b) 4 sq units
(c) 5 sq units
(d) 9 sq units

13 If m_1 and m_2 are roots of the equation $x^2+(\sqrt{3}+2)x+(\sqrt{3}-1)=0$, then the area of the triangle formed by the lines $y=m_1x$, $y=m_2x$ and $y=c$, is

(a) $\left(\frac{\sqrt{33}-\sqrt{11}}{4}\right)c^2$ (b) $\left(\frac{\sqrt{33}+\sqrt{11}}{4}\right)c^2$

(c) $\left(\frac{\sqrt{33}+\sqrt{11}}{2}\right)c^2$ (d) $\left(\frac{\sqrt{33}-\sqrt{11}}{2}\right)c^2$

14 If the vertices of a triangle have integral coordinates, then the triangle is

(a) equilateral (b) never equilateral

(c) isosceles (d) None of these

15 If Δ_1 is the area of the triangle with vertices (0,0), $(a\tan\alpha, b\cot\alpha)$, $(a\sin\alpha, b\cos\alpha)$, Δ_2 is the area of triangle with vertices (a, b), $(a\sec^2\alpha, b\operatorname{cosec}^2\alpha)$, $(a+a\sin^2\alpha, b+b\cos^2\alpha)$ and Δ_3 is the area of the triangle with vertices (0, 0) $(a\tan\alpha, -b\cot\alpha)$, $(a\sin\alpha, b\cos\alpha)$. The value of α, if Δ_1, Δ_2 and Δ_3 are in GP, is

(a) 1 (b) −3

(c) No value of α (d) None of these

1.3 Some Special Points of Triangle (Centroid, Incentre, Excentre, Orthocentre, Circumcentre)

16 If the coordinates of the feet of perpendiculars from the vertices of a triangle on the opposite sides are (20, 25), (8, 16) and (8, 9), then orthocentre of the triangle lies at the point

(a) (5, 10) (b) (10, 15) (c) (15, 30) (d) None of these

17 If α, β and γ are the roots of the equation $x^3-6x^2+11x-6=0$, then the centroid of the triangle whose vertices are $\left(\alpha\beta, \frac{1}{\alpha\beta}\right)$, $\left(\beta\gamma, \frac{1}{\beta\gamma}\right)$, $\left(\gamma\alpha, \frac{1}{\gamma\alpha}\right)$ is

(a) $\left(\frac{11}{3}, \frac{1}{3}\right)$ (b) $\left(\frac{1}{3}, \frac{11}{3}\right)$ (c) $\left(\frac{11}{3}, -\frac{1}{3}\right)$ (d) None of these

18 Harmonic conjugate of the point $C(5,1)$ with respect to the point $A(2, 10)$, $B(6, -2)$ is circumcentre and (3,−4) is the orthocentre of triangle, then its centroid is

(a) $\left(\frac{19}{3}, \frac{20}{3}\right)$ (b) $\left(\frac{19}{3}, -\frac{20}{3}\right)$ (c) $\left(\frac{20}{3}, \frac{-19}{3}\right)$ (d) None of these

19 Let the centroid of an equilateral ΔABC be at origin. Let one of the sides of equilateral triangle be along the straight line $x+y=3$. If R and r be the radius of circumcircle and incircle, respectively of ΔABC, then $R+r$ is equal to **[JEE Main 2021]**

(a) $7\sqrt{2}$ units (b) $\frac{9}{\sqrt{2}}$ units (c) $2\sqrt{2}$ units (d) $3\sqrt{2}$ units

20 The vertices of ΔABC are (−1,−7), (5,1) and (1,4). The bisector of $\angle ABC$ meets the opposite side in the point D. Then, the length of angle bisector BD is

(a) $\frac{10}{3}\sqrt{2}$ units (b) $\frac{5}{3}\sqrt{2}$ units (c) $\frac{8}{3}\sqrt{2}$ units (d) None of these

21 Three vertices of triangle are $A(4, 3)$, $B(1, -1)$ and $C(7, k)$. The value of k for which centroid, orthocentre, incentre and circumcentre of ΔABC lies on the same straight line, is

(a) 7 (b) −1 (c) $\frac{5}{4}$ (d) None of these

22 Area of triangle formed by incentre, centroid and circumcentre of triangle having vertices $(1, 0), (1, \sqrt{3}), (2, 0)$ is

(a) $\frac{1}{2\sqrt{3}}$ sq unit (b) $\frac{1}{3\sqrt{3}}$ sq unit (c) $\frac{5}{2\sqrt{3}}$ sq units (d) None of these

23 The vertices of a triangle are $A(a, a\tan\alpha)$, $B(b, b\tan\beta)$ and $C(c, c\tan\gamma)$. If the circumcentre of ΔABC coincides with the origin and $H(\bar{x}, \bar{y})$ is the orthocentre, then the value of $\frac{\bar{x}}{\bar{y}}$ is

(a) $\frac{\sin\alpha+\sin\beta+\sin\gamma}{\cos\alpha+\cos\beta+\cos\gamma}$ (b) $\frac{\cos\alpha+\cos\beta+\cos\gamma}{\sin\alpha+\sin\beta+\sin\gamma}$

(c) $\frac{\tan\alpha+\tan\beta+\tan\gamma}{\cos\alpha+\cos\beta+\cos\gamma}$ (d) None of these

24 If $P = x_1, x_2, x_3 \ldots\ldots x_i$ and $q = y_1, y_2, y_3 \ldots y_i$ are in AP with the common difference a and b respectively, then the centre of mean position of the points $A_i(x_i, y_i), i = 1,2,3\ldots. n$ lies on the line

(a) $ax - by = aq - bq$ (b) $bx - ay = ap - bq$ (c) $bx - ay = bp - aq$ (d) $ax - by = bq - ap$

Note Centre of mean position $\left(\frac{\Sigma x_i}{n}, \frac{\Sigma y_i}{n}\right)$

25 $A_1, A_2, A_3 \ldots. A_n$ are points on the line $y = x$ lying in the positive quadrant such that $OA_n = nOA_{n-1}$, O being the origin. If $OA_1 = 1$ and the coordinates of A_n are $(2520\sqrt{2}, 2520\sqrt{2})$, then n is equal to

(a) 5 (b) 6 (c) 7 (d) 8

1.4 Locus of Point

26 A and B are two fixed points of ΔABC with side $AB = 2$. If its vertex C moves in such a way that $\cot A + \cot B = \lambda$, where λ is a constant. Then, the locus of the point C is

(a) $y\lambda = 2a$ (b) $ya = 2\lambda$ (c) $y = \lambda a$ (d) None of these

27 The coordinates of the points O, A and B are (0, 0), (0, 4) and (6, 0), respectively. If a point P moves such that the area of ΔPOA is twice the area of a ΔPOB, then equation of both parts of the locus is

(a) $(x-3y)(x+3y)=0$ (b) $(x-3y)(x+y)=0$ (c) $(3x-y)(3x+y)=0$ (d) None of these

28 If $A(\cos\alpha, \sin\alpha)$, $B(\sin\alpha, -\cos\alpha)$ and $C(1, 2)$ are the vertices of a ΔABC. Then, as α varies, the locus of its centroid is

(a) $x^2+y^2-2x-4y+1=0$ (b) $3(x^2+y^2)-2x-4y+1=0$

(c) $x^2+y^2-2x-4y+3=0$ (d) None of these

29 Let B(0, 0) and A(0, 1) be two fixed points. Then, the locus of a point P such that the perimeter of ΔABP is 4 units, is **[JEE Main 2019]**

(a) $9x^2+8y^2-8y=16$ (b) $8x^2-9y^2+9y=18$

(c) $8x^2+9y^2-9y=18$ (d) $9x^2-8y^2+8y=16$

30 Let A be a fixed point (0, 6) and B be a moving point (2t, 0). Let M be the mid-point of AB and the perpendicular bisector of AB meets the Y-axis at C. The locus of the mid-point P at MC is **[JEE Main 2021]**

(a) $3x^2-2y-6=0$ (b) $3x^2+2y-6=0$ (c) $2x^2+3y-9=0$ (d) $2x^2-3y+9=0$

31 If A(−4, 0) and B(4, 0). M and N are the variable points on the Y-axis such that M lies below N and MN = 4. Line joining AM and BN intersect at P, then the locus of P is

(a) $2xy-16-x^2=0$ (b) $2xy+16-x^2=0$ (c) $2xy+16+x^2=0$ (d) $2xy-16+x^2=0$

1.5 Transformation of Axes

32 If in the curve $2x^2+3\sqrt{3}xy+3y^2=2$ the axes are inclined at 30° to rectangular axis (X-axis remaining unchanged), then equation of new curve is

(a) $x^2+\sqrt{3}xy=2$ (b) $x^2+\sqrt{3}xy=1$

(c) $x^2+2\sqrt{3}xy=1$ (d) None of these

33 Reflecting the point (2,–1) about Y-axis. Coordinates axis are rotated at 45° angle in negative direction without shifting the origin. The new coordinates of the point is

(a) $\left(\frac{-1}{\sqrt{2}},\frac{-3}{\sqrt{2}}\right)$ (b) $\left(\frac{-3}{\sqrt{2}},-\sqrt{2}\right)$ (c) $\left(\frac{1}{\sqrt{2}},\frac{3}{\sqrt{2}}\right)$ (d) None of these

34 If in the equation $2x^2+4xy-5y^2+20x-22y-14=0$, when if referred to rectangular axes through the point (–2,–3) and the new equation formed which inclined to an angle of 45° with the old axes, is

(a) $x^2-14xy-7y^2+2=0$ (b) $x^2-14xy-7y^2-2=0$

(c) $x^2-14xy+7y^2-14=0$ (d) None of these

35 If a point P (1, 2) is translated itself 2 units along the positive direction of X-axis and then it is rotated about the origin in anti-clockwise sense through an angle of 90°, then the new position of P is

(a) (–2,–3) (b) (–2, 3)

(c) (2,–3) (d) None of these

Numerical Value Type Questions

36 If points A(3, 4), B(7, 12) and P(x, y) are such that $PA^2>PB^2>AB^2$, then sum of integral value of x less than 99 and divisible by 4, is ....

37 The length of angle bisector of ∠B of ΔABC having vertices A(–1, –7), B(5, 1) and C(1, 4) is l, then the value of $\left[\frac{l}{4}\right]$, {[·] = greatest integer function}, is ....

38 One of the diameters of the circle circumscribing the rectangle ABCD is $4y=x+7$. If A and B are the points (–3, 4) and (5, 4) respectively, then the area of rectangle ABCD, is ....

39 A square ABCD has all its vertices on the curve $x^2y^2=1$. The mid-points of its sides also lie on the same curve. Then, the square root of area of ABCD is (upto 3 decimal) .... [JEE Main 2021]

40 P(2, 1), Q(3, 2) and R(4, –1) are the vertices of a triangle and it through P and R lines parallel to opposite sides are drawn to intersect in S. If the area of PQRS is A, then the value of 0.32A is ....

41 The coordinates of A, B, C are (6, 3), (–3, 5), (4, –2), respectively and P is any point (x, y), if ratio of area of ΔPBC to ΔABC is $\frac{|x+y-a|}{b}$, then the value of $\frac{a+b}{10}$ is ....

42 If P be the mid-point of the straight line joining the points A(1, 2) and Q, where Q is a variable point on the curve $x^2+y^2+x+y=0$ and locus of point P is $p(x^2+y^2)+qx+ry+s$, then the value of $\frac{4p+q+r+s}{25}$ is ....

LEVEL 03 \ JEE Advanced Level

Single Choice Correct Questions

1 The distance between the points $(t_1^3, \sqrt{2}t_1)$ and $(t_2^3, -\sqrt{2}t_2)$, if t_1 and t_2 are the roots of $x^2 - 2\sqrt{2}x + 1 = 0$, is

(a) 3 units (b) $6\sqrt{6}$ units (c) $5\sqrt{2}$ units (d) None of these

2 The coordinates of three points $A(-4,0)$, $B(2,1)$ and $C(3,1)$ determine the vertices of an equilateral trapezium $ABCD$. The coordinates of vertex D is

(a) (6, 0) (b) (−3, 0) (c) (−5, 0) (d) (9, 0)

3 Consider three points $P\{-\sin(\beta-\alpha), -\cos\beta\}$, $Q\{\cos(\beta-\alpha), \sin\beta\}$ and $R\{\cos(\beta-\alpha+\theta), \sin(\beta-\theta)\}$, where $0 < \alpha, \beta, \theta < \frac{\pi}{4}$. Then,

(a) P lies on the line segment RQ (b) Q lies on the line segment PR

(c) R lies on the line segment QP (d) P, Q, R are non-collinear

4 An integral point means that both coordinates of the point are integers. The number of integral points exactly in the interior of the triangle with vertices (0, 0), (0, 21) and (21, 0) is

(a) 133 (b) 190 (c) 233 (d) 105

5 A particle beings at the origin and moves successively in the following manner as shown 1 unit to the right, $\frac{1}{2}$ unit up, $\frac{1}{4}$ unit to the left, $\frac{1}{8}$ unit down, $\frac{1}{16}$ unit to the right, $\frac{1}{32}$ unit up etc. The length of each move is half the length of previous move and movement continuous indefinitely. The coordinates of the point to which the movement converges is

(a) $\left(\frac{4}{5}, \frac{2}{5}\right)$ (b) $\left(\frac{2}{5}, \frac{4}{5}\right)$ (c) $\left(\frac{4}{3}, \frac{2}{3}\right)$ (d) $\left(\frac{2}{3}, \frac{4}{3}\right)$

6 m and n are integer with $o < n < m$. A is the point (m, n) on the cartesian plane. B is the reflection of A in the line $y = x$. C is the reflection of B in the Y-axis. D is the reflection of C in the X-axis and E is the reflection of D in the Y-axis. The area (sq units) of pentagon $ABCDE$ is

(a) $2m(m+n)$ (b) $m(m+3n)$ (c) $m(2m+3n)$ (d) $2m(m+3n)$

7 Each side of a square is of length 4 units. The centre of the square is (3,7) and one of its diagonal is parallel to $y = x$, then the coordinates of its vertices are

(a) (1, 5) (1, 9) (5, 9) (5, 5) (b) (−1, 5) (−1, 9) (−2, 3) (−5, 5)

(c) both (a) and (b) (d) None of these

8 Suppose the bisector AD of the interior angle A of ΔABC divides the sides BC into segments $BD = 4$, $DC = 2$, then the maximum value of k is

(a) 3 (b) 4 (c) 5 (d) None of these

9 A rod of length k slides in a vertical plane, its ends touching the coordinate axes, then the locus of the foot of perpendicular from origin to rod is

(a) $(x^2+y^2)^3 = kx^2y^2$ (b) $(x^2+y^2)^3 = k^2x^2y^2$

(c) $(x^2+y^2)^2 = k^2x^2y^2$ (d) None of these

10 Find the number of triangles with sides of integer length whose perimeter is 10 units, is

(a) 4 (b) 3 (c) 2 (d) 0

11 If x_1, y_1 are the roots of $x^2 + 8x - 20 = 0$, x_2, y_2 are the roots of $4x^2 + 32x - 57 = 0$ and x_3, y_3 are the roots of $9x^2 + 72x - 112 = 0$, then the points (x_1, y_1), (x_2, y_2) and (x_3, y_3) are

(a) collinear (b) form an equilateral triangle

(c) form a right angled isosceles triangle (d) concylic

More than One Correct

12 Coordinates of a point P are (a, b), where a is root of the equation $x^2 + x - 42 = 0$ and b is an integral root of the equation $x^2 + ax + a^2 - 37 = 0$. The coordinates of P can be

(a) (6, 4) (b) (−7, 4) (c) (−7, 3) (d) (6, −3)

13 If the points $\left(\frac{a^3}{a-1}, \frac{a^2-3}{a-1}\right), \left(\frac{b^3}{b-1}, \frac{b^2-3}{b-1}\right)$ and $\left(\frac{c^3}{c-1}, \frac{c^2-3}{c-1}\right)$, where $a, b, c \neq 1$, lie on the line $lx + my + n = 0$, then

(a) $a+b+c = \frac{-m}{l}$ (b) $ab+bc+ca = \frac{n}{l}$

(c) $abc = \frac{-(m+n)}{l}$ (d) $abc-(bc+ca+ab)+3(a+b+c)=0$

14 Two sides of a rhombus $OABC$ (lying entirely in I quadrant or III quadrant) of area equal to 2 sq units are $y = \frac{x}{\sqrt{3}}, y = \sqrt{3}x$. Then, possible coordinate of B is/are (O being origin)

(a) $(1+\sqrt{3}, 1+\sqrt{3})$ (b) $(-1-\sqrt{3}, -1-\sqrt{3})$

(c) $(3+\sqrt{3}, 3+\sqrt{3})$ (d) $(\sqrt{3}-1, \sqrt{3}-1)$

15 The points $O(0, 0)$, $A(\cos\alpha, \sin\alpha)$ and $B(\cos\beta, \sin\beta)$ are the vertices of a right angle triangle, if

(a) $\sin\left(\frac{\alpha-\beta}{2}\right) = \frac{1}{\sqrt{2}}$ (b) $\cos\left(\frac{\alpha-\beta}{2}\right) = \frac{-1}{\sqrt{2}}$ (c) $\cos\left(\frac{\alpha-\beta}{2}\right) = \frac{1}{\sqrt{2}}$ (d) $\sin\left(\frac{\alpha-\beta}{2}\right) = \frac{-1}{\sqrt{2}}$

16 The straight line $3x + 4y - 12 = 0$ meets the coordinate axes at A and B. An equilateral triangle ABC is constructed. The possible coordinates of vertex C are

(a) $\left\{2\left(1-\frac{3\sqrt{3}}{4}\right), \frac{3}{2}\left(1-\frac{4}{\sqrt{3}}\right)\right\}$ (b) $\left\{-2(1+\sqrt{3}), \frac{3}{2}(1-\sqrt{3})\right\}$

(c) $\left\{2(1+\sqrt{3}), \frac{3}{2}(1+\sqrt{3})\right\}$ (d) $\left\{2\left(1+\frac{3\sqrt{3}}{4}\right), \frac{3}{2}\left(1+\frac{4}{\sqrt{3}}\right)\right\}$

Linked Comprehension Type

Paragraph I Directions (Q.Nos. 17-19)

$A_1, A_2, A_3 \ldots\ldots, A_n$ are n points in a plane whose coordinates are $(x_1, y_1), (x_2, y_2) \ldots\ldots (x_n, y_n)$, respectively.

17 A_1A_2 is bisected at the point G_1, G_1A_3 is divided in the ratio 1 : 2 at G_2, G_2A_4 is divided in the ratio 1:3 at G_3, G_3A_5 is divided in the ratio 1 : 4 at G_4 and so on until all n points are exhausted. The coordinates of the final point G so obtain are

(a) $\{n(x_1 + x_2 + \ldots\ldots x_n), n(y_1 + y_2 + \ldots\ldots y_n)\}$

(b) $\left(\frac{x_1 + x_2 + \ldots . x_n}{n}, \frac{y_1 + y_2 + \ldots . y_n}{n}\right)$

(c) $\left(\frac{x_1 + 2x_2 + 3x_3 + \ldots . nx_n}{n}, \frac{y_1 + 2y_2 + 3y_3 + \ldots ny_n}{n}\right)$

(d) None of the above

18 If $x_1 = a, y_1 = b, x_1, x_2, \ldots., x_n$ and $y_1, y_2 \ldots y_n$ form an ascending AP with common difference 2 and 4, respectively. Then, the coordinates of G is

(a) $\{a+n-1, b+2(n-1)\}$ (b) $(a+2b-2, b+n-1)$

(c) $(a+n-1, b+n+1)$ (d) $(a+n+1, b+n+2)$

19 If P_1 is the centroid of the triangle A_1, A_2, A_3, P_2 is the centroid of the triangle A_2, A_3, A_4 and P_3 is the centroid of the triangle A_3, A_4, A_5 and so on. P_{n-2} is the centroid of the triangle A_{n-2}, A_{n-1}, A_n. Then, the centre of mean position of the points $P_1, P_2, \ldots P_{n-2}$ is

(a) $\left(\frac{x_1+2x_2+\ldots nx_n}{n-2}, \frac{y_1+2y_2+\ldots ny_n}{n-2}\right)$

(b) $\left(\frac{x_1+2x_2+3x_3+\ldots 3x_{n-2}+2x_{n-1}+x_n}{n-2}, \frac{y_1+2y_2+3y_3+\ldots 3y_{n-2}+2y_{n-1}+y_n}{n-2}\right)$

(c) $\left(\frac{3(x_1+x_2+\ldots x_n)}{n-2}, \frac{3(y_1+y_2+\ldots y_n)}{n-2}\right)$

(d) at the centre of mean position of the points.

Paragraph II Directions (Q.Nos. 20-22)

$A(P,0)$, $B(4,0)$, $C(5,6)$ and $D(1,4)$ are the vertices of a quadrilateral $ABCD$.

20 If $\angle ADC$ is obtuse, then the maximum integral value of P is

(a) 1 (b) 2 (c) 3 (d) 0

21 The value of P for which $\angle ADC$ is obtuse and $\angle BAD$ is a right angle, is

(a) 1 (b) 2 (c) -3 (d) -4

22 If two sides of the quadrilateral AB and AD are equal, then area of the quadrilateral is

(a) $\frac{25}{3}$ sq units (b) 11 sq units

(c) $\frac{58}{3}$ sq units (d) None of these

Matrix Match Type Questions

23 $ABCD$ is a rectangle in the clockwise direction. The coordinates of A are (1, 3) and of C are (5, 1). Coordinates of vertices B and D satisfy $y = 2x + c$, the coordinates of

	Column I		Column II
(i)	middle point of BD	A.	(2,0)
(ii)	middle point of AB	B.	(3,2)
(iii)	middle point of BC	C.	$\left(\frac{5}{2}, \frac{7}{2}\right)$
(iv)	the point D	D.	$\left(\frac{9}{2}, \frac{5}{2}\right)$

24 If $P(3, 7)$, $Q(8, 9)$ and $R(11, 23)$ are three vertices of ΔPQR. S divides PQ in the ratio 2 : 3 and T divides QR in the ratio 3 : 2.

	Column I		Column II
(i)	$5(ST)^2$	A.	576
(ii)	PR^2+QR^2+51	B.	464
(iii)	$100(PS)^2$	C.	$2PR^2-64$
(iv)	$PQ^2+2(QR)^2+25$	D.	$50QS^2-2PQ^2$
		E.	$5QT^2+QR^2+2$

Numerical Value Type Questions

Question Stem Directions (Q.Nos. 25-26) The answer to each question is a **NUMERICAL VALUE.**

Let $ABCD$ is a square with sides of unit length. Points E and F are taken on sides AB and AD respectively, so that $AE = AF$. Let P be any point inside the square $ABCD$.

25 The maximum possible area (in sq units) of quadrilateral $CDFE$ is ......

26 The value of $PA^2 - PB^2 + PC^2 - PD^2$ is ......

Question Stem Directions (Q.Nos. 27-28) The answer to each question is a **NUMERICAL VALUE.**

Consider three non-collinear points $A(9,3)$, $B(7,-1)$ and $C(1,-1)$. Let $P(a,b)$ be the centre and R is the radius of circle 'S' passing through points A, B and C. Also, $H(\bar{x},\bar{y})$ are the coordinates of the orthocentre of ΔABC whose area be denoted by Δ.

27 If D, E and F are the middle points of BC, CA and AB respectively, then the area of ΔDEF is ......

28 The value of $a + b + R$ is ......

29 $P(3, 1)$, $Q(6, 5)$ and $R(x, y)$ are three points such that $\angle PRQ$ is right angle and the area of ΔPRQ is 7 sq units, then number of such points R is ......

30 Number of values of b for which in an acute ΔABC, if the coordinates of orthocentre H are $(4, b)$, centroid G are $(b, 2b - 8)$ and circumcentre S are $(-4, 8)$, is ......

31 ΔABC with $AB = 13$, $BC = 5$ and $AC = 12$ slides on the coordinates axis with A and B on the positive X-axis and positive Y-axis, respectively. The locus of vertex C is $ax - by$, then value of $a + b$ is ......

32 In a ΔABC, the vertex A is $(1,1)$ and orthocentre is $(2,4)$. If the sides AB and BC are members of the family of straight lines $ax + by + c = 0$, where a, b, c are in AP, then the coordinates of vertex C are (h,k). The value of $2h + 12k$ is ......

33 Each side of a square is of length 4 units. The centre of the square is at (3, 7) and one of the diagonal is parallel to the line $y = x$. If the vertices of the square be (x_1, y_1), (x_2, y_2), (x_3, y_3) and (x_4, y_4), then the value of $\max(y_1, y_2, y_3, y_4) - \min(x_1, x_2, x_3, x_4)$ is ......

ANSWER KEY

Level 1

1. (i) (a) $\left(\frac{5}{2}, \frac{5\sqrt{3}}{2}\right)$, (b) $(\sqrt{2}, -\sqrt{2})$, (c) $\left(\frac{-3}{\sqrt{2}}, \frac{3}{\sqrt{2}}\right)$, (d) $(-3, -3\sqrt{3})$, (e) $\left(\frac{-4}{5}, \frac{3}{5}\right)$

(ii) (a) $\left(5, \tan^{-1}\frac{4}{3}\right)$, (b) $\left(\sqrt{2}, \frac{3\pi}{4}\right)$ (c) $\left(2\sqrt{2}, \frac{-\pi}{6}\right)$, (d) $\left(2, \frac{-2\pi}{3}\right)$

2. (i) (a) $\frac{\cos^2\theta}{a^2} + \frac{\sin^2\theta}{b^2} = \frac{1}{r^2}$, (b) $r^2 = \frac{2}{4 + 3\sin 2\theta}$ (c) $r^2 = a^2\cos 2\theta$ (d) $r = 2a\cos\theta$

(ii) (a) $x^2 + y^2 - 2ax = 0$, (b) $2x^2 + 2y^2 + 3xy - 3 = 0$

(c) $(x^2 + y^2)^2 = a^2(x^2 - y^2)$ (d) $7x^2 + 7y^2 - 4x - 3y = 0$

3. (i) $|a(m_2 - m_1)|\sqrt{m_1^2 + m_2^2 + 2m_1m_2 + 4}$ units (ii) $|a|\left(t + \frac{1}{t}\right)^2$ units (iii) $\sqrt{58}$ units (iv) 13 units

4. (i) $7 \pm 2\sqrt{6}$ (ii) $\frac{15}{2}$

5. (i) $\left(\frac{-5 \pm \sqrt{3}}{2}, \frac{7 \pm \sqrt{3}}{2}\right)$ (ii) $\frac{2}{3}, -3$

(iii) 2, 3 (iv) Infinite possible values of a and b.

6. (i) $\left(\frac{-5}{3}, 2\right)$ (ii) $\left(\frac{11}{2}, \frac{9}{2}\right)$ (iii) At $k = 9$, $\left(\frac{1}{10}, \frac{27}{10}\right)$ or $k = 4$, $\left(\frac{1}{5}, \frac{12}{15}\right)$

7. (i) $\frac{55}{2}$ sq units (ii) 14 sq units (iii) $\frac{63}{2}$ sq units

8. (i) 4 or -6 (ii) $\frac{7}{3}$ (iii) $\lambda \in \left(\frac{-6}{7}, \frac{3}{2}\right)$

9. (i) $\left(\frac{5}{2}, \frac{1}{2}\right)$ and $\sqrt{\frac{5}{2}}$ (ii) $(-1, 2)$ (iii) $(6, 0)$ (iv) $(-7, -2)$ (v) $(0, 0)$ and $\left(\frac{3}{4}, 1\right)$

(vi) $\left(\frac{5}{3}, \frac{23}{15}\right)$ and $\left(\frac{7}{4}, \frac{19}{10}\right)$ (vii) $(6, 6)$, $(1, -4)$ and $(-4, 6)$

10. (i) $3x - y + 4 = 0$ (ii) $3x + y - 11 = 0$ or $3x + y + 1 = 0$ (iii) $x^2 + y^2 = 5$

(iv) $x^2 + y^2 + 20x + 25 = 0$ (v) $x^2 + y^2 = 4$ (vi) $3x^2 + 3y^2 + 20x - 10y + 19 = 0$

11. (i) $(1, 1)$, $(0, -5)$, $(-5, 0)$ and $(-6, -8)$ (ii) $(2, 6)$, $(3, 1)$ and $(-2, 11)$

(iii) $(-6, 6)$ (iv) $(2 - \sqrt{3}, 1 + 2\sqrt{3})$

12. (i) $XY = \frac{a^2}{2}$ (ii) $y^2 = 4x$ (iii) $(1, 3)$ (iv) $x^2 + y^2 = 50$

Level 2

1. (b)	**2.** (d)	**3.** (c)	**4.** (c)	**5.** (a)	**6.** (d)	**7.** (b)
8. (a)	**9.** (c)	**10.** (b)	**11.** (c)	**12.** (d)	**13.** (b)	**14.** (b)
15. (c)	**16.** (b)	**17.** (a)	**18.** (b)	**19.** (b)	**20.** (a)	**21.** (b)
22. (d)	**23.** (b)	**24.** (c)	**25.** (c)	**26.** (a)	**27.** (a)	**28.** (b)
29. (a)	**30.** (c)	**31.** (d)	**32.** (b)	**33.** (a)	**34.** (b)	**35.** (b)
36. (1176)	**37.** (4)	**38.** (32)	**39.** (6.324)	**40.** (1.28)	**41.** (0.9)	**42.** (0.2)

Level 3

1. (d)	**2.** (b)	**3.** (d)	**4.** (b)	**5.** (a)	**6.** (b)	**7.** (a)
8. (b)	**9.** (b)	**10.** (c)	**11.** (a)	**12.** (b,c)	**13.** (a,b,d)	**14.** (a,b)
15. (a,c)	**16.** (a,d)	**17.** (b)	**18.** (a)	**19.** (b)	**20.** (b)	**21.** (a)
22. (c)	**23.** (i) → B (ii) → C (iii) → D (iv) → A			**24.** (i) → A,C,E (ii) → A,C,E (iii) → B,D (iv) → B,D		
25. (0.625)	**26.** (0)	**27.** (3)	**28.** (12)	**29.** (0)	**30.** (0)	**31.** (17)
32. (14)	**33.** (8)					

SOLUTIONS

Level 1

1. (i) (a) Given, $r = 5$ and $\theta = \frac{\pi}{3}$

Now, $x = r\cos\theta = 5\cos\frac{\pi}{3} = 5 \times \frac{1}{2} = \frac{5}{2}$

$y = r\sin\theta = 5\sin\frac{\pi}{3} = 5 \times \frac{\sqrt{3}}{2} = \frac{5\sqrt{3}}{2}$

$\therefore$ Cartesian coordinates of the given point will be $\left(\frac{5}{2}, \frac{5\sqrt{3}}{2}\right)$.

(b) Given, $r = 2$ and $\theta = \frac{-\pi}{4}$

Now, $x = r\cos\theta = 2\cos\left(-\frac{\pi}{4}\right)$

$= 2 \times \frac{1}{\sqrt{2}} = \sqrt{2}$

$y = r\sin\theta = 2\sin\left(-\frac{\pi}{4}\right) = 2 \times \frac{-1}{\sqrt{2}} = -\sqrt{2}$

$\therefore$ Cartesian coordinates of the given point will be $(\sqrt{2}, -\sqrt{2})$

(c) Given, $r = 3$ and $\theta = \frac{3\pi}{4}$

Now, $x = r\cos\theta = 3\cos\frac{3\pi}{4}$

$= -3\cos\frac{\pi}{4} = \frac{-3}{\sqrt{2}}$

$y = r\sin\theta = 3\sin\frac{3\pi}{4}$

$= 3\sin\frac{\pi}{4} = \frac{3}{\sqrt{2}}$

$\therefore$ Cartesian coordinates of the given point will be $\left(\frac{-3}{\sqrt{2}}, \frac{3}{\sqrt{2}}\right)$.

(d) Given, $r = 6$ and $\theta = \frac{-2\pi}{3}$

Now, $x = r\cos\theta = 6\cos\left(\frac{-2\pi}{3}\right)$

$= -6\cos\frac{\pi}{3} = -3$

$y = r\sin\theta = 6\sin\left(-\frac{2\pi}{3}\right)$

$= -6\sin\frac{\pi}{3} = -3\sqrt{3}$

$\therefore$ Cartesian coordinates of the given point will be $(-3, -3\sqrt{3})$.

(e) Given, $r = 1$ and $\theta = \pi - \tan^{-1}\left(\frac{4}{3}\right)$

$\Rightarrow \quad \tan^{-1}\left(\frac{4}{3}\right) = \pi - \theta$

$\Rightarrow \quad \tan(\pi - \theta) = \frac{4}{3}$

$\Rightarrow \quad -\tan\theta = \frac{4}{3}$

$\Rightarrow \quad \tan\theta = -\frac{4}{3}$

$x = r\cos\theta$

$= 1\left(-\frac{4}{5}\right) = -\frac{4}{5}$

$y = r\sin\theta = 1\left(\frac{3}{5}\right) = \frac{3}{5}$

$\therefore$ Cartersian coordinates of the given point will be $\left(-\frac{4}{5}, \frac{3}{5}\right)$.

(ii) (a) Given, $x = 3$ and $y = 4$

$\therefore \quad r = \sqrt{x^2 + y^2} = \sqrt{3^2 + 4^2} = 5$

$\therefore \quad \alpha = \tan^{-1}\left|\frac{y}{x}\right| = \tan^{-1}\frac{4}{3}.$

$\because$ Point (3,4) lies in I quadrant.

$\therefore \quad \theta = \alpha$

Hence, polar coordinates of the given points will be $\left(5, \tan^{-1}\frac{4}{3}\right)$.

(b) Given, $x = -1$ and $y = 1$

$\therefore \quad r = \sqrt{x^2 + y^2} = \sqrt{(-1)^2 + (1)^2} = \sqrt{2}$

$\therefore \quad \alpha = \tan^{-1}\left|\frac{y}{x}\right| = \tan^{-1}(1) = \frac{\pi}{4}$

$\because$ Point $(-1,1)$ lies in II quadrant.

$\therefore \quad \theta = \pi - \alpha = \pi - \frac{\pi}{4} = \frac{3\pi}{4}.$

Hence, polar coordinates of the given point will be $\left(\sqrt{2}, \frac{3\pi}{4}\right)$.

(c) Given, $x = \sqrt{6}$ and $y = -\sqrt{2}$

$\therefore \quad r = \sqrt{x^2 + y^2} = \sqrt{6 + 2} = \sqrt{8} = 2\sqrt{2}$

$\therefore \quad \alpha = \tan^{-1}\left|\frac{y}{x}\right| = \tan^{-1}\left|\frac{-\sqrt{2}}{\sqrt{6}}\right|$

$= \tan^{-1}\left(\frac{1}{\sqrt{3}}\right) = \frac{\pi}{6}$

$\because$ Point $(\sqrt{6},-\sqrt{2})$ lies in IV quadrant.

$$\therefore \quad \theta=-\alpha=-\frac{\pi}{6}$$

Hence, polar coordinates of the given point will be $\left(2\sqrt{2},-\frac{\pi}{6}\right)$.

(d) Given, $x=-1$ and $y=-\sqrt{3}$

$$\therefore \quad r=\sqrt{x^2+y^2}=\sqrt{(-1)^2+(-\sqrt{3})^2}=\sqrt{4}=2$$

$$\therefore \quad \alpha=\tan^{-1}\left|\frac{y}{x}\right|=\tan^{-1}\sqrt{3}=\frac{\pi}{3}$$

$\because$ Point $(-1,-\sqrt{3})$ lies in III quadrant.

$$\therefore \quad \theta=-\pi+\alpha=-\pi+\frac{\pi}{3}=-\frac{2\pi}{3}$$

Hence, polar coordinates of the given point will be $\left(2,-\frac{2\pi}{3}\right)$.

2. (i) (a) We have,

$$\frac{x^2}{a^2}+\frac{y^2}{b^2}=1$$

$$\Rightarrow \quad b^2x^2+a^2y^2=a^2b^2$$

$$\Rightarrow \quad b^2r^2\cos^2\theta+a^2r^2\sin^2\theta=a^2b^2$$

$$\Rightarrow \quad \frac{r^2\cos^2\theta}{a^2}+\frac{r^2\sin^2\theta}{b^2}=1$$

$$\Rightarrow \quad \frac{\cos^2\theta}{a^2}+\frac{\sin^2\theta}{b^2}=\frac{1}{r^2}$$

(b) We have,

$$2x^2+3xy+2y^2=1$$

$$\Rightarrow 2r^2\cos^2\theta+3r^2\sin\theta\cos\theta+2r^2\sin^2\theta=1$$

$$\Rightarrow \quad r^2(2\cos^2\theta+3\sin\theta\cos\theta+2\sin^2\theta)=1$$

$$\Rightarrow \quad r^2\left(2+\frac{3}{2}\sin 2\theta\right)=1$$

$$\Rightarrow \quad r^2=\frac{2}{4+3\sin 2\theta}$$

(c) We have,

$$(x^2+y^2)^2=a^2(x^2-y^2)$$

$$\Rightarrow \quad (r^2\cos^2\theta+r^2\sin^2\theta)^2=a^2(r^2\cos^2\theta-r^2\sin^2\theta)$$

$$\Rightarrow \quad r^4=a^2r^2(\cos^2\theta-\sin^2\theta)$$

$$\Rightarrow \quad r^2=a^2(\cos^2-\sin^2\theta)$$

$$\Rightarrow \quad r^2=a^2\cos 2\theta$$

(d) We have,

$$x^2+y^2=2ax$$

$$\Rightarrow \quad r^2\cos^2\theta+r^2\sin^2\theta=2ar\cos\theta$$

$$\Rightarrow \quad r^2(\cos^2\theta+\sin^2\theta)=2ar\cos\theta$$

$$\Rightarrow \quad r=2a\cos\theta$$

(ii) (a) Where, $r=2a\cos\theta$

$$\therefore \quad \cos\theta=\frac{x}{r}$$

$$\Rightarrow \quad r=2a\left(\frac{x}{r}\right)$$

$$\Rightarrow \quad r^2=2ax$$

$$\because \quad r^2=x^2+y^2 \text{ (in polar form)}$$

$$\Rightarrow \quad x^2+y^2-2ax=0$$

(b) $$r^2=\frac{6}{4+3\sin 2\theta}$$

We know that

$$\sin 2\theta=2\sin\theta\cos\theta$$

$$=2\cdot\left(\frac{y}{r}\right)\left(\frac{x}{r}\right)$$

[$\because$ in polar form $x=r\cos\theta$, $y=r\sin\theta$]

$$=\frac{2xy}{r^2}$$

$$\Rightarrow \quad r^2=\frac{6}{4+\frac{6xy}{r^2}}$$

$$\Rightarrow \quad r^2=\frac{6r^2}{4r^2+6xy}$$

$$\Rightarrow \quad 4r^2+6xy=6$$

$$\Rightarrow \quad 4(x^2+y^2)+6xy=6$$

$$\Rightarrow \quad 2x^2+2y^2+3xy-3=0$$

(c) We have,

$$r^2=a^2\cos 2\theta$$

$$\Rightarrow \quad x^2+y^2=a^2(\cos^2\theta-\sin^2\theta)$$

$$\Rightarrow \quad (x^2+y^2)=a^2\left(\frac{x^2-y^2}{r^2}\right)$$

[in polar form $x=r\cos\theta$, $y=r\sin\theta$]

$$\Rightarrow (x^2+y^2)r^2=a^2(x^2-y^2)$$

$$\Rightarrow (x^2+y^2)^2=a^2(x^2-y^2)$$

(d) We have,

$$7r=3\sin\theta+4\cos\theta$$

$$\Rightarrow \quad 7r=\frac{3y}{r}+\frac{4x}{r}$$

$$\Rightarrow \quad 7r^2=3y+4x$$

$$\Rightarrow 7(x^2+y^2)=4x+3y$$

$$\Rightarrow 7x^2+7y^2-4x-3y=0$$

3. (i) $A\equiv(am_1^2, 2am_1)$ and $B\equiv(am_2^2, 2am_2)$

$$\therefore \quad AB=\sqrt{(am_2^2-am_1^2)^2+(2am_2-2am_1)^2}$$

$$=\sqrt{a^2(m_2-m_1)^2[(m_2+m_1)^2+(2)^2]}$$

$$=|a(m_2-m_1)|\sqrt{m_1^2+m_2^2+2m_1m_2+4} \text{ units}$$

(ii) $A=(at^2,2at)$ and $B=\left(\frac{a}{t^2},-\frac{2a}{t}\right)$

$$\therefore\ AB=\sqrt{\left(at^2-\frac{a}{t^2}\right)^2+\left(2at+\frac{2a}{t}\right)^2}$$

$$=\sqrt{a^2\left(t-\frac{1}{t}\right)^2\left(t+\frac{1}{t}\right)^2+4a^2\left(t+\frac{1}{t}\right)^2}$$

$$=\left|a\left(t+\frac{1}{t}\right)\right|\sqrt{\left(t-\frac{1}{t}\right)^2+4}$$

$$=\left|a\left(t+\frac{1}{t}\right)\right|\sqrt{\left(t+\frac{1}{t}\right)^2}$$

$$=|a|\left(t+\frac{1}{t}\right)^2 \text{ units, if } a>0$$

(iii) $A=\left(3,\frac{\pi}{4}\right)$ and $B=\left(7,\frac{5\pi}{4}\right)$

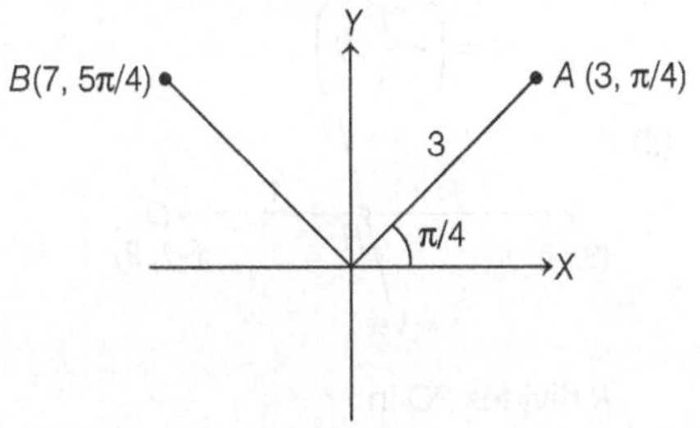

$$A\Rightarrow x=r\cos\theta=3\cos\frac{\pi}{4}=\frac{3}{\sqrt{2}}$$

$$y=r\sin\theta=3\sin\frac{\pi}{4}=\frac{3}{\sqrt{2}}$$

$$B\Rightarrow x=7\cos\frac{5\pi}{4}=7\left(-\frac{1}{\sqrt{2}}\right)=\frac{-7}{\sqrt{2}}$$

$$y=7\sin\frac{5\pi}{4}=7\left(\frac{1}{\sqrt{2}}\right)=\frac{7}{\sqrt{2}}$$

$$\therefore\ AB=\sqrt{\left(\frac{3}{\sqrt{2}}+\frac{7}{\sqrt{2}}\right)^2+\left(\frac{3}{\sqrt{2}}-\frac{7}{\sqrt{2}}\right)^2}$$

$$=\sqrt{\frac{100}{2}+\frac{16}{2}}=\sqrt{58}\text{ units}$$

Alternate Method

$$\therefore\ AB=\sqrt{r_1^2+r_2^2-2r_1r_2\cos(\theta_1-\theta_2)}$$

$$=\sqrt{3^2+7^2-2(3)(7)\cos\left(\frac{\pi}{4}-\frac{5\pi}{4}\right)}$$

$$=\sqrt{9+49-0}=\sqrt{58}\text{ units}$$

(iv) $A=(17,105°)$ and $B=(5\sqrt{2},60°)$

$$\therefore AB=\sqrt{r_1^2+r_2^2-2r_1r_2\cos(\theta_1-\theta_2)}$$

$$=\sqrt{17^2+(5\sqrt{2})^2-2(17)(5\sqrt{2})\cos(105°-60°)}$$

$$=\sqrt{289+50-2(17)(5\sqrt{2})\times\frac{1}{\sqrt{2}}}$$

$$=\sqrt{339-170}=\sqrt{169}=13\text{ units}$$

4. (i) Let $A=(3,7)$ and $B=(-2,\lambda)$, then $AB=7$

$$\sqrt{(3+2)^2+(7-\lambda)^2}=7$$

$$\Rightarrow\quad 25+(\lambda-7)^2=49$$

$$\Rightarrow\quad (\lambda-7)^2=49-25$$

$$\Rightarrow\quad (\lambda-7)^2=24$$

$$\Rightarrow\quad \lambda-7=\pm\sqrt{24}$$

$$\Rightarrow\quad \lambda=\pm2\sqrt{6}+7$$

$$\Rightarrow\quad \lambda=7\pm2\sqrt{6}$$

(ii) Let $A\,(7,6)$, $B\,(3,4)$ and $C\,(\lambda,0)$ are points.
Now, according to question

$$AC=BC$$

$$\Rightarrow \sqrt{(\lambda-7)^2+(6-0)^2}=\sqrt{(\lambda-3)^2+(0-4)^2}$$

$$\Rightarrow\quad (\lambda-7)^2+36=(\lambda-3)^2+16$$

$$\Rightarrow\quad \lambda^2-14\lambda+49+36=\lambda^2-6\lambda+9+16$$

$$\Rightarrow\quad -14\lambda+6\lambda=25-49-36$$

$$\Rightarrow\quad -8\lambda=-60$$

$$\Rightarrow\quad \lambda=\frac{60}{8}=\frac{15}{2}$$

5. (i) ABC is equilateral triangle

$$\Rightarrow\quad AB=BC=AC$$

$$\Rightarrow\quad AB^2=BC^2=AC^2$$

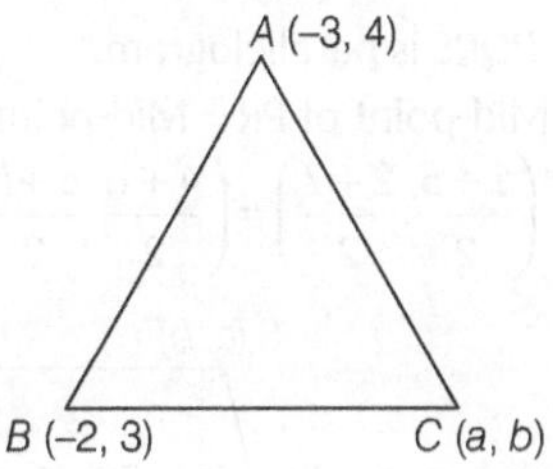

$$\Rightarrow(-2+3)^2+(3-4)^2=(a+2)^2+(b-3)^2$$

$$=(a+3)^2+(b-4)^2$$

$$\Rightarrow\quad 2=a^2+b^2+4a-6b+13$$

$$=a^2+b^2+6a-8b+25$$

From II and III part, we get

$$4a-6b+13=6a-8b+25$$

$$\Rightarrow\quad -2a+2b=12$$

$$\Rightarrow\quad b-a=6$$

$$\Rightarrow\quad b=a+6 \qquad \text{...(i)}$$

From I and II part, we get

$$2=a^2+b^2+4a-6b+13$$

$$\Rightarrow\quad a^2+(a+6)^2+4a-6(a+6)+11=0$$

$$\Rightarrow a^2+a^2+12a+36+4a-6a-36+11=0$$

$$\Rightarrow \quad 2a^2+10a+11=0$$

$$\Rightarrow \quad a=\frac{-10\pm\sqrt{100-88}}{2\times 2}$$

$$=\frac{-10\pm\sqrt{12}}{4}=2\frac{(-5\pm\sqrt{3})}{4}$$

$$=\frac{-5\pm\sqrt{3}}{2}$$

$$\therefore \quad b=a+6$$

$$=\frac{-5\pm\sqrt{3}}{2}+6=\frac{7\pm\sqrt{3}}{2}$$

$$\therefore \quad (a,b)\equiv\left(\frac{-5\pm\sqrt{3}}{2},\frac{7\pm\sqrt{3}}{2}\right)$$

(ii) $A(-4,-1), B(-2,-4), C(4,0)$ and $D(3a,-b)$ are vertices of rectangle.

Mid-point of AC = Mid-point of BD

$$\left(\frac{-4+4}{2},\frac{-1+0}{2}\right)=\left(\frac{-2+3a}{2},\frac{-4-b}{2}\right)$$

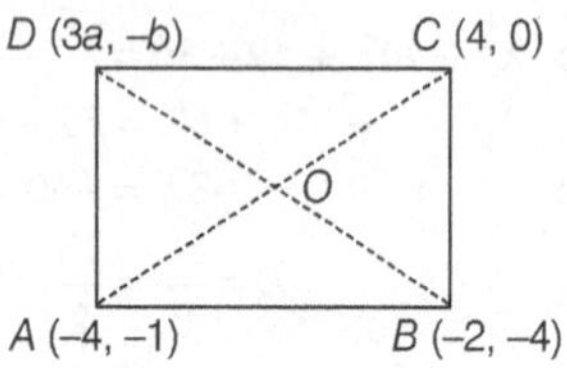

$$0=-2+3a\Rightarrow a=\frac{2}{3}$$

$$-1=-4-b\Rightarrow b=-3$$

(iii) $\because$ $PQRS$ is parallelogram.

$\therefore$ Mid-point of PR = Mid-point of QS

$$\left(\frac{1+5}{2},\frac{2+7}{2}\right)=\left(\frac{4+a}{2},\frac{6+b}{2}\right)$$

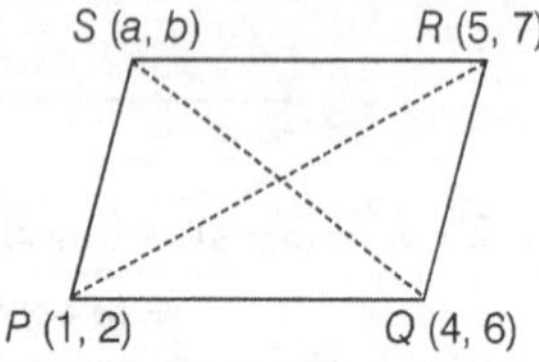

$$6=4+a\Rightarrow a=2$$

$$9=6+b\Rightarrow b=3$$

(iv) $\therefore AB^2+BC^2=AC^2$

$$\Rightarrow (25+100)+(a-2)^2+(b-6)^2$$

$$=(a+3)^2+(b+4)^2$$

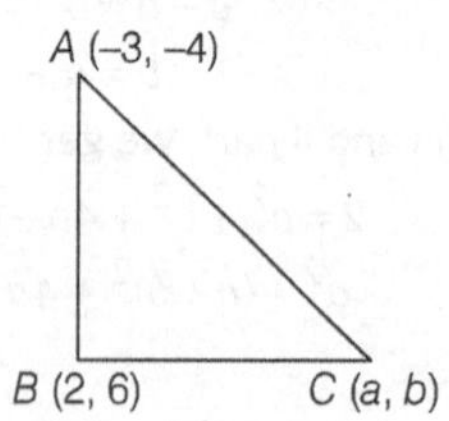

$$\Rightarrow 125+a^2+b^2-4a-12b+4+36$$

$$=a^2+6a+9+b^2+8b+16$$

$$\Rightarrow 10a+20b=140$$

$$a+2b=14$$

So, there are infinite possible values of a and b.

6. (i)

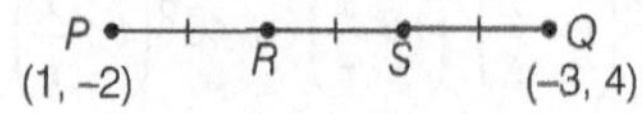

R and S trisects the PQ.

For S ($\because$ S divides PQ internally $2:1$)

$$S=\left(\frac{2\times(-3)+1\times 1}{2+1},\frac{2\times 4+1\times(-2)}{2+1}\right)$$

$$=\left(\frac{-5}{3},2\right)$$

For R ($\because$ R divides PQ internally $1:2$)

$$R=\left(\frac{1\times(-3)+2\times 1}{1+2},\frac{1\times 4+2\times(-2)}{1+2}\right)$$

$$=\left(-\frac{1}{3},0\right)$$

(ii)

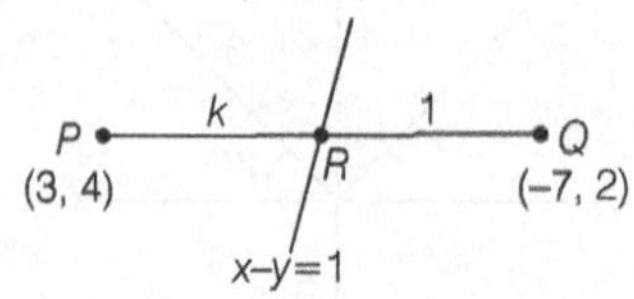

R divides PQ in

$k:1$

$$\therefore \quad R=\left(\frac{-7k+3}{k+1},\frac{2k+4}{k+1}\right)$$

and R lies on line $x-y=1$

$$\Rightarrow \quad \frac{-7k+3}{k+1}-\frac{2k+4}{k+1}=1$$

$$\Rightarrow \quad -7k+3-2k-4=k+1$$

$$\Rightarrow \quad -10k=2\Rightarrow k=\frac{-1}{5}$$

$$\therefore \quad R=\left(\frac{\frac{7}{5}+3}{\frac{-1}{5}+1},\frac{-\frac{2}{5}+4}{\frac{-1}{5}+1}\right)$$

$$=\left(\frac{22}{4},\frac{18}{4}\right)=\left(\frac{11}{2},\frac{9}{2}\right)$$

(iiii) Let $P(1,0)$ and $Q\left(3,\frac{\pi}{2}\right)$ are two points and a point R on the curve.

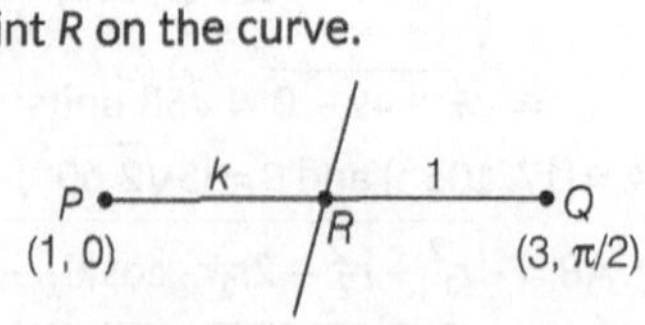

$$r=\frac{3\sin\theta+4\cos\theta}{5}$$ divides PQ in $k:1$

$P(1,0) \Rightarrow r = 1$ and $\theta = 0°$

$\Rightarrow \quad x = r\cos\theta$ and $y = r\sin\theta$

$\Rightarrow \quad x = 1\cos 0°$ and $y = 1\sin 0°$

$\Rightarrow \quad x = 1$ and $y = 0$

$\therefore \quad P = (1, 0)$

$Q\left(3, \frac{\pi}{2}\right) \Rightarrow r = 3$ and $\theta = \frac{\pi}{2}$

$\Rightarrow \quad x = 3\cos\frac{\pi}{2}$ and $y = 3\sin\frac{\pi}{2}$

$\Rightarrow \quad x = 0$ and $y = 3$

$\therefore \quad Q = (0, 3)$

$\therefore$ R is $\left(\frac{1}{k+1}, \frac{3k}{k+1}\right)$ lies on curve

$$r = \frac{3\sin\theta + 4\cos\theta}{5}$$

$\Rightarrow \quad 5r = \frac{3y}{r} + \frac{4x}{r}$ (in cartesian form)

$\Rightarrow \quad 5r^2 = 3y + 4x$

$5(x^2 + y^2) = 4x + 3y$

Now, $5\left(\frac{1}{(k+1)^2} + \frac{9k^2}{(k+1)^2}\right) = \frac{4(1)}{k+1} + \frac{3(3k)}{k+1}$

$$\frac{5(1+9k^2)}{(k+1)^2} = \frac{4+9k}{k+1}$$

$\Rightarrow \quad 5 + 45k^2 = (4+9k)(k+1)$

$\Rightarrow \quad 45k^2 + 5 = 4k + 4 + 9k^2 + 9k$

$\Rightarrow \quad 36k^2 - 13k + 1 = 0$

$\Rightarrow \quad k = \frac{13 \pm \sqrt{169 - 4(36)}}{2}$

$= \frac{13 \pm 5}{2} = 9, 4$

For $k = 9$, point is $\left(\frac{1}{9+1}, \frac{3(9)}{9+1}\right)\left(\frac{1}{10}, \frac{27}{10}\right)$ and ratio is $9:1$.

For $k = 4$, point is $\left(\frac{1}{4+1}, \frac{3(4)}{4+1}\right) = \left(\frac{1}{5}, \frac{12}{5}\right)$ and ratio is $4:1$.

7. (i) $A(4,1)$, $B(-3,7)$ and $C(5,8)$ are vertices of triangle.

$\therefore$ Area of $\Delta ABC = \frac{1}{2}$ mod of $\begin{vmatrix} x_1 & y_1 & 1 \\ x_2 & y_2 & 1 \\ x_3 & y_3 & 1 \end{vmatrix}$

$= \frac{1}{2} \times$ mod of $\begin{vmatrix} 4 & 1 & 1 \\ -3 & 7 & 1 \\ 5 & 8 & 1 \end{vmatrix}$

$= \frac{1}{2}|4(7-8) - 1(-3-5) + 1(-24-35)|$

$= \frac{1}{2}(-4 + 8 - 59)$

$= \frac{55}{2}$ sq units

Or

Alternate Method

$\therefore$ Area of $\Delta ABC = \frac{1}{2}|x_1(y_2 - y_3) + x_2(y_3 - y_1) + x_3(y_1 - y_2)|$

$= \frac{1}{2}|4(7-8) + (-3)(8-1) + 5(1-7)|$

$= \frac{1}{2}|-4 - 21 - 30|$

$= \frac{55}{2}$ sq units

(ii) $A(7,1)$, $B(5,-3)$, $C(3,3)$ and $D(-3,-7)$ are vertices of quadrilateral.

$\therefore$ Area of quadrilateral $ABCD$

$= \frac{1}{2}|x_1(y_2 - y_3) + x_2(y_3 - y_4) + x_3(y_4 - y_1) + x_4(y_1 - y_2)|$

$= \frac{1}{2}|7(-3-3) + 5(3+7) + 3(-7-1) + (-3)(1+3)|$

$= \frac{1}{2}|-42 + 50 - 24 - 12|$

$= \frac{1}{2}|-28| = 14$ sq units

(iii) $A(4,3)$, $B(-5,6)$, $C(0,7)$, $D(3,-6)$ and $E(-7,-2)$.

$\therefore$ Area of pentagon $ABCDE$

$= \frac{1}{2}|x_1(y_2 - y_3) + x_2(y_3 - y_4) + x_3(y_4 - y_5) + x_4(y_5 - y_1) + x_5(y_1 - y_2)|$

$= \frac{1}{2}|4(6-7) + (-5)(7+6) + 0(-6+2) + 3(-2-3) + (-7)(3-6)|$

$= \frac{1}{2}|-4 - 65 + 0 - 15 + 21|$

$= \frac{63}{2}$ sq units

8. (i) Let $A(-1,-4)$, $B(\lambda, 1)$ and $C(\lambda, -4)$ are 3 points

Now, according to question,

Area of $\Delta ABC = 12.5$ sq units

$\Rightarrow \quad \frac{1}{2}$ mod $\begin{vmatrix} -1 & -4 & 1 \\ \lambda & 1 & 1 \\ \lambda & -4 & 1 \end{vmatrix} = 12.5$

$\Rightarrow$ mod of $|-1(1+4) + 4(\lambda - \lambda) + 1(-4\lambda - \lambda)| = 25$

$|-5 - 5\lambda| = 25$

$5\lambda + 5 = \pm 25$

$\Rightarrow \quad \lambda + 1 = \pm 5$

$$\Rightarrow \qquad \lambda = \pm 5 - 1$$
$$\Rightarrow \qquad \lambda = 4, -6$$

(ii) $A(-2,3)$, $B(1,2)$ and $C(3\lambda, 0)$ are collinear.

$\therefore$ A, B, C are collinear, then $\begin{vmatrix} x_1 & y_1 & 1 \\ x_2 & y_2 & 1 \\ x_3 & y_3 & 1 \end{vmatrix} = 0$

$$\Rightarrow \qquad \begin{vmatrix} -2 & 3 & 1 \\ 1 & 2 & 1 \\ 3\lambda & 0 & 1 \end{vmatrix} = 0$$

$$\Rightarrow \qquad -2(2) - 3(1 - 3\lambda) + 1(-6\lambda) = 0$$
$$\Rightarrow \qquad -4 - 3 + 9\lambda - 6\lambda = 0$$
$$\Rightarrow \qquad 3\lambda = 7 \Rightarrow \lambda = \frac{7}{3}$$

(iii) The point $P(\lambda, \lambda + 1)$ will be inside the ΔABC, then Area of ΔPBC + Area of ΔPCA + Area of ΔPAB
= Area of ΔABC

$$\Rightarrow \frac{1}{2} \text{mod} \begin{vmatrix} \lambda & \lambda+1 & 1 \\ -2 & 0 & 1 \\ 6 & 1 & 1 \end{vmatrix} + \frac{1}{2} \text{mod} \begin{vmatrix} \lambda & \lambda+1 & 1 \\ 6 & 1 & 1 \\ 0 & 3 & 1 \end{vmatrix}$$
$$+ \frac{1}{2} \text{mod} \begin{vmatrix} \lambda & \lambda+1 & 1 \\ 0 & 3 & 1 \\ -2 & 0 & 1 \end{vmatrix}$$
$$= \frac{1}{2} \text{mod} \begin{vmatrix} 0 & 3 & 1 \\ -2 & 0 & 1 \\ 6 & 1 & 1 \end{vmatrix}$$

$$\Rightarrow \quad |7\lambda + 6| + |8\lambda - 12| + |\lambda + 4| = 22 \qquad \text{...(i)}$$
$$\lambda = -\frac{6}{7};\ \lambda = \frac{12}{8} = \frac{3}{2};\ \lambda = -4$$

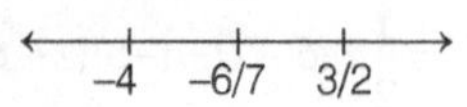

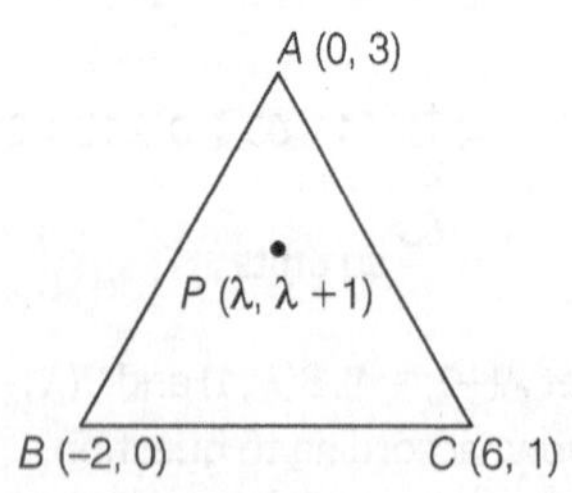

Eq (i) can be written as

$$-7\lambda - 6 - 8\lambda + 12 - \lambda - 4 = 22,\ \lambda \le -4$$
$$\Rightarrow -7\lambda - 6 - 8\lambda + 12 + \lambda + 4 = 22,\ -4 < \lambda \le -\frac{6}{7}$$
$$\Rightarrow 7\lambda + 6 - 8\lambda + 12 + \lambda + 4 = 22,\ \frac{-6}{7} < \lambda \le \frac{3}{2}$$
$$\Rightarrow 7\lambda + 6 + 8\lambda - 12 + \lambda + 4 = 22,\ \lambda > \frac{3}{2}$$

$$\Rightarrow \lambda = \frac{-5}{4}, \text{ when } \lambda \le -4$$
$$\Rightarrow \lambda = \frac{-6}{7}, \text{ when } -4 < \lambda \le \frac{-6}{7}$$
$$\Rightarrow 22 = 22, \text{ when } \frac{-6}{7} < \lambda \le \frac{3}{2} \Rightarrow \lambda \in \left(\frac{-6}{7}, \frac{3}{2}\right)$$
$$\Rightarrow \lambda = \frac{3}{2}, \text{ when } \lambda > \frac{3}{2}$$

Area of $\Delta PCA = 0$

At $\lambda = -\dfrac{6}{7}$

Area of $\Delta PBC = 0$

So, value of λ is $\lambda \in \left(-\dfrac{6}{7}, \dfrac{3}{2}\right)$.

9. (i) Let O be the circumentre of ΔABC, so

$$OA = OB = OC,$$

where $A(1,1), B(2,-1)$ and $C(3,2)$

$OA = OB$	$OB = OC$	$OC = OA$
$(x-1)^2 + (y-1)^2$	$(x-2)^2 + (y+1)^2$	$(x-3)^2 + (y-2)^2$
$= (x-2)^2 + (y+1)^2$	$= (x-3)^2 + (y-2)^2$	$= (x-1)^2 + (y-1)^2$
$\Rightarrow 2x - 4y = 3$ ...(i)	$\Rightarrow 2x + 6y = 8$ ...(ii)	$\Rightarrow 4x + 2y = 11$ ...(iii)

By Eqs. (i), (ii) and (iii), we get

$$x = \frac{5}{2} \text{ and } y = \frac{1}{2}$$

Circumcentre $= \left(\dfrac{5}{2}, \dfrac{1}{2}\right)$

and circumradius $= OA = \sqrt{\left(\dfrac{5}{2} - 1\right)^2 + \left(\dfrac{1}{2} - 1\right)^2}$

$$= \sqrt{\frac{9}{4} + \frac{1}{4}} = \sqrt{\frac{10}{4}} = \sqrt{\frac{5}{2}}$$

(ii) $a = BC = \sqrt{(-2+4)^2 + (-5-6)^2}$

$$= \sqrt{4 + 121} = 5\sqrt{5}$$

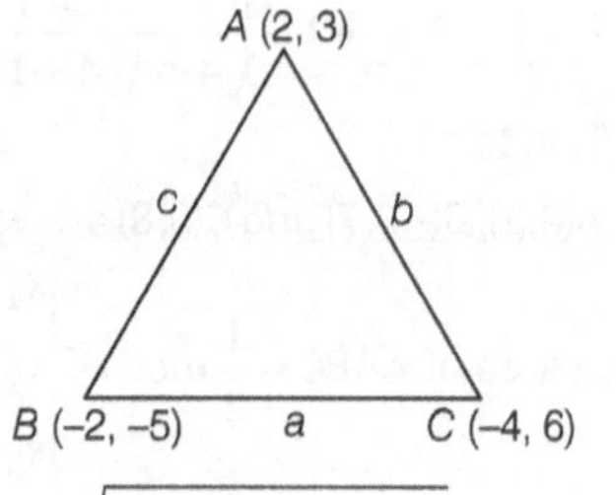

$$b = CA = \sqrt{(2+4)^2 + (3-6)^2}$$
$$= \sqrt{36 + 9} = 3\sqrt{5}$$
$$c = AB = \sqrt{(2+2)^2 + (3+5)^2}$$
$$= \sqrt{16 + 64} = 4\sqrt{5}$$

Now, incentre

$$\left(\frac{ax_1+bx_2+cx_3}{a+b+c}, \frac{ay_1+by_2+cy_3}{a+b+c}\right)$$

$$=\left(\frac{(5\sqrt{5})(2)+3\sqrt{5}(-2)+4\sqrt{5}(-4)}{12\sqrt{5}}, \frac{5\sqrt{5}(3)+3\sqrt{5}(-5)+4\sqrt{5}(6)}{12\sqrt{5}}\right)$$

$$=(-1,2)$$

(iii) In a triangle, circumcentre, centroid and orthocentre are collinear and the centroid divides the line joining the circumcentre and orthocentre in the ratio 1 : 2.

Suppose, $H(x,y)$ is the orthocentre.

Then, $OG:GH=1:2$

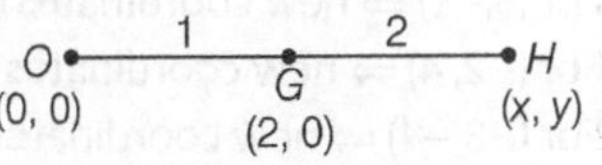

Now, $O(0,0)$, $G(2,0)$ and $H(x,y)$

$$\frac{x+0}{3}=2 \text{ and } \frac{y+0}{3}=0$$

$$x=6 \text{ and } y=0$$

$\therefore$ Orthocentre is (6, 0).

(iv) Let $C(\alpha,\beta)$ be the third vertex.

$$\frac{5+2+\alpha}{3}=0, \frac{-1+3+\beta}{3}=0$$

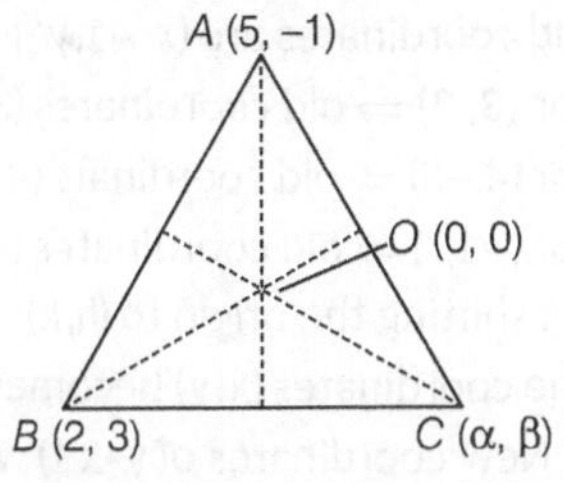

$$\alpha=-7 \text{ and } \beta=-2$$

$\therefore$ $C(-7,-2)$

(v) (0,0), (3, 0) and (0, 4) and the vertices of a right angle triangle

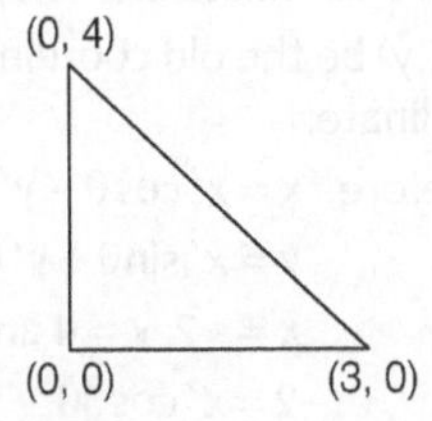

Orthocentre is (0, 0)

Circumcentre = mid-point of (0, 4) and (3, 0)

$$=\left(\frac{3}{2},\frac{4}{2}\right)$$

Nine-point centre is mid-point of orthocentre and circumcentre $=\left(\frac{3}{4},1\right)$.

(vi) $\therefore$ Centroid divides the orthocentre and circumcentre in the ratio 2 : 1 (internally) and if centroid $G(x,y)$, then

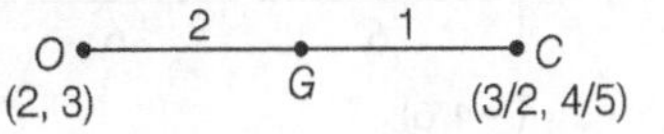

$$\therefore \quad G=\left(\frac{2\times\frac{3}{2}+1\times 2}{2+1}, \frac{2\times\frac{4}{5}+1\times 3}{2+1}\right)$$

$$=\left(\frac{5}{3},\frac{23}{15}\right)$$

$\therefore$ Centroid is $\left(\frac{5}{3},\frac{23}{15}\right)$.

and nine-point centre is mid-point of orthocentre and circumcentre.

$$\therefore \text{ Nine-point centre is}=\left(\frac{2+\frac{3}{2}}{2}, \frac{3+\frac{4}{5}}{2}\right)$$

$$=\left(\frac{7}{4},\frac{19}{10}\right)$$

(vii) Let $A(-2,2)$, $B(1,6)$ and $C(4,2)$ are the vertices of the triangle

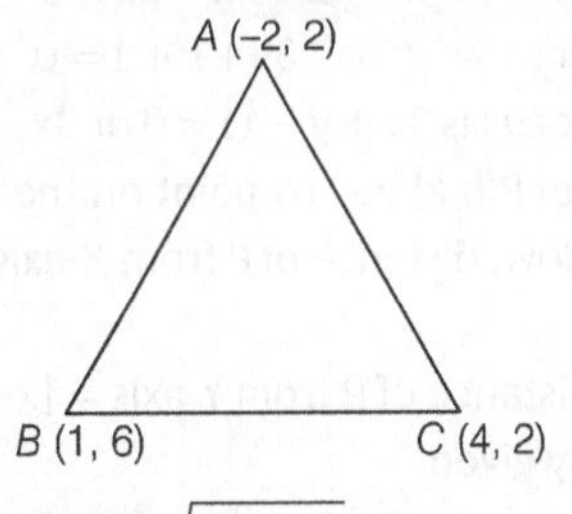

$$a=BC=\sqrt{3^2+4^2}=5$$

$$b=AC=\sqrt{6^2+0^2}=6$$

$$c=AB=\sqrt{3^2+4^2}=5$$

Excentre I_1 w.r.t. A

$$=\left(\frac{-ax_1+bx_2+cx_3}{-a+b+c}, \frac{-ay_1+by_2+cy_3}{-a+b+c}\right)$$

$$=\left(\frac{-5(-2)+6(1)+5(4)}{-5+6+5}, \frac{-5(2)+6(6)+5(2)}{-5+6+5}\right)$$

$$=\left(\frac{36}{6},\frac{36}{6}\right)=(6,6)$$

Similarly, I_2 w.r.t. B

$$=\left(\frac{ax_1-bx_2+cx_3}{a-b+c}, \frac{ay_1-by_2+cy_3}{a-b+c}\right)$$

$$=\left(\frac{-10-6+20}{4}, \frac{10-36+10}{4}\right)$$

$$=(1,-4)$$

and I_3 w.r.t. C

$$=\left(\frac{ax_1+bx_2-cx_3}{a+b-c},\frac{ay_1+by_2-cy_3}{a+b-c}\right)$$

$$=\left(\frac{-10+6-20}{6},\frac{10+36-10}{6}\right)$$

$$=(-4,6)$$

10. (i) Let $A(1,2)$ and $B(-2,3)$

Let $P(h,k)$ be any point on the locus. Thus, we have

$$PA=PB$$

$$\Rightarrow \quad PA^2=PB^2$$

$$\Rightarrow (h-1)^2+(k-2)^2=(h+2)^2+(k-3)^2$$

$$\Rightarrow h^2-2h+1+k^2-4k+4$$

$$=h^2+4h+4+k^2-6k+9$$

$$\Rightarrow \quad 6h-2k+8=0$$

$$\Rightarrow \quad 3h-k+4=0$$

$\therefore$ Equation of the locus is $3x-y+4=0$

(ii) Let $P(h,k)$ be any point on the locus and $A(1,2)$ and $B=(2,-1)$

$\because$ Area $(\Delta PAB)=3$

$$\Rightarrow \frac{1}{2}|h(2+1)+1(-1-k)+2(k-2)|=3$$

$$\Rightarrow \quad |3h-1-k+2k-4|=6$$

$$\Rightarrow \quad 3h+k-5=\pm 6$$

$$\Rightarrow \quad 3h+k-11=0$$

or $\quad 3h+k+1=0$

Locus is $3x+y-11=0$ or $3x+y+1=0$

(iii) Let $P(h,k)$ be any point on the locus.

Now, distance of P from X-axis $=|y\text{-coordinate of } P|$

$=|k|$

distance of P from Y-axis $=|x\text{-coordinate of } P|=|h|$

By given

$$h^2+k^2=5$$

Equation of the locus is $x^2+y^2=5$

(iv) Let $P(h,k)$ be any point

$$PA^2=3PB^2$$

$$\Rightarrow \quad (h-5)^2+k^2=3[(h+5)^2+k^2]$$

$$\Rightarrow h^2+k^2+20h+25=0$$

$\therefore$ Equation of locus is

$$x^2+y^2+20x+25=0$$

(v) Let $A(2,0)$ and $B(-2,0)$

and $P(h,k)$ be any point on the locus

By given condition

AB subtends a right angle at P.

$$\Rightarrow \quad \angle APB=90^\circ$$

$$\Rightarrow \quad AP^2+PB^2=AB^2$$

$$\Rightarrow (h-2)^2+k^2+(h+2)^2+k^2=(2+2)^2$$

$$\Rightarrow \quad h^2+k^2=4$$

$\therefore$ Equation of locus is $x^2+y^2=4$

(vi) Let $P(h,k)$ be point and $A(2,3)$ and $B(-2,2)$

$$PA=2PB$$

$$PA^2=4PB^2$$

$$\Rightarrow (h-2)^2+(k-3)^2=4[(h+2)^2+(k-2)^2]$$

$$\Rightarrow \quad 3h^2+3k^2+20h-10k+19=0$$

$\therefore$ Equation of locus is $3x^2+3y^2+20x-10y+19=0$

11. (i) Here, $O'(h,k)=(3,4)$

$$h=3 \text{ and } k=4$$

On shifting the origin to (3, 4), the new coordinates (x',y') of (x, y) will be given by

$x'=x-h=x-3$ and $y'=y-k=y-4$

Hence, new coordinates are $(x-3, y-4)$.

For $(4,5) \Rightarrow$ new coordinates $(4-3, 5-4)=(1,1)$

For $(3,-1) \Rightarrow$ new coordinates $(3-3,-1-4)=(0,-5)$

For $(-2,4) \Rightarrow$ new coordinates $(-2-3,4-4)=(-5,0)$

For $(-3,-4) \Rightarrow$ new coordinates $(-3-3,-4,-4)$

$=(-6,-8)$

(ii) The new origin $O'(h,k)=(-1,4)$

When origin is shifting to $O'(-1,4)$ the new coordinates of (x,y) be (x',y').

$\therefore \quad x'=x-h$ and $y'=y-k$

$\therefore \quad x'=x-(-1)$ and $y'=y-4$

$=x+1$

$\Rightarrow \quad x=x'-1$ and $y=y'+4$

Old coordinates are $(x'-1, y'+4)$.

For $(3,2) \Rightarrow$ old coordinates $(3-1,2+4)=(2,6)$

For $(4,-3) \Rightarrow$ old coordinats $(4-1,-3+4)=(3,1)$

For $(-1,7) \Rightarrow$ old coordinates $(-1-1,7+4)=(-2,11)$

(iii) On shifting the origin to (h,k).

The coordinates (x,y) becomes $(x-h, y-k)$

$\therefore$ New coordinates of $(-2,1)$ will be $(-2-h, 1-k)$

But given $(4,-5)$

$\therefore \quad -2-h=4$ and $-5=1-k$

$h=-6$ and $k=6$.

Origin was shifted to $(-6,6)$.

(iv) Let (x,y) be the old coordinates and (x',y') be new coordinate.

Therefore, $x=x'\cos\theta-y'\sin\theta$

$y=x'\sin\theta+y'\cos\theta$

$\therefore \quad x=-2, y=4$ and $\theta=30^\circ$

$$-2=x'\cos 30^\circ-y'\sin 30^\circ$$

$$\Rightarrow \quad \sqrt{3}x'-y'=-4 \quad \text{...(i)}$$

and $\quad 4=x'\sin 30^\circ+y'\cos 30^\circ$

$$\Rightarrow \quad x'+\sqrt{3}y'=8 \quad \text{...(ii)}$$

By Eqs. (i) and (ii),

$$x'=2-\sqrt{3} \text{ and } y'=1+2\sqrt{3}$$

New coordinates of $(-2,4)$ is $(2-\sqrt{3}, 1+2\sqrt{3})$.

12. (i) Let (x, y) and (X, Y) be the old and the new coordinates, respectively. Therefore,

$$x = X\cos(-45°) - Y\sin(-45°)$$
$$x = \frac{X+Y}{\sqrt{2}}$$

and $$y = X\sin(-45°) + Y\cos 45°$$
$$y = \frac{-X+Y}{\sqrt{2}}$$

Now, $x^2 - y^2 = a^2$

$$\Rightarrow \left(\frac{X+Y}{\sqrt{2}}\right)^2 - \left(\frac{-X+Y}{\sqrt{2}}\right)^2 = a^2$$

$$\Rightarrow \quad 4XY = 2a^2 \Rightarrow XY = \frac{a^2}{2}$$

(ii) Suppose, (x, y) changes to (X, Y).
When origin is changed to $(2, -4)$.
Here, $h = 2$, $k = -4$
Now, $x = X + h$ and $y = Y + k$

$$x = X + 2 \text{ and } y = Y - 4$$

Putting these values in $y^2 + 8y - 4x + 24 = 0$, we get

$$(Y-4)^2 + 8(Y-4) - 4(X+2) + 24 = 0$$
$$\Rightarrow \quad Y^2 - 8Y + 16 + 8Y - 32 - 4X - 8 + 24 = 0$$
$$\Rightarrow \quad Y^2 - 4X = 0$$

New equation is $y^2 = 4x$.

(iii) Given equation is $y^2 - 6y - 4x + 13 = 0$ ...(i)
Let the origin be shifted to point (α, β).
Now, new Eq. (i) becomes

$$(y+\beta)^2 - 6(y+\beta) - 4(x+\alpha) + 13 = 0$$

or $y^2 + (2\beta - 6)y - 4x + \beta^2 - 6\beta - 4\alpha + 13 = 0$ ...(ii)
Eq. (ii) to be of the form $y^2 + ax = 0$
Then, $2\beta - 6 = 0$, $a = -4$

$$\beta^2 - 6\beta - 4\alpha + 13 = 0 \quad \text{...(iii)}$$
$$\beta = 3 \text{ and } a = -4$$

From Eq. (iii), $9 - 18 - 4\alpha + 13 = 0$

$$\alpha = 1$$

(iv) Let (h, k) be the point to which the origin should be shifted, the equation.
In new coordinates (X, Y) becomes

$$(X+h)^2 + (Y+k)^2 + 8(X+h) - 6(Y+k) - 25 = 0$$
$$\Rightarrow X^2 + Y^2 + (2h+8)X + (2k-6)Y + (h^2 + k^2 + 8h - 6k - 25) = 0$$

Then, cofficient of X and Y vanish, when

$$2h + 8 = 0 \text{ and } 2k - 6 = 0$$
$$h = -4 \text{ and } k = 3$$

and then $h^2 + k^2 + 8h - 6k - 25 = -50$
Hence, on shifting the origin to the point $(-4, 3)$ the given equation reduces to $x^2 + y^2 = 50$ is the required form.

Level 2

1. (b) Every point (x, y) on the perpendicular bisector of PQ is equidistant from both P and Q, we have

$$(x-1)^2 + (y-4)^2 = (x-\alpha)^2 + (y-3)^2$$
$$\Rightarrow \quad -2x - 8y + 17 = -2\alpha x + \alpha^2 - 6y + 9$$

$\therefore$ It cut X-axis at -4
At $x = 0$, $y = -4$

$$\Rightarrow \quad -8(-4) + 17 = \alpha^2 - 6(-4) + 9$$
$$\Rightarrow \quad \alpha^2 = 16 \Rightarrow \alpha = \pm 4$$

2. (d) $AB = 10$ where $A(2, -3)$ and let $B(x, y)$.
Roots of equation

$$x^2 - 8x - 20 = 0$$
$$\Rightarrow \quad x^2 - 10x + 2x - 20 = 0$$
$$\Rightarrow \quad x(x-10) + 2(x-10) = 0$$
$$\Rightarrow \quad (x-10)(x+2) = 0$$
$$\Rightarrow \quad x = 10, -2$$

$\therefore$ x is positive only when $x = 10$

$$\therefore \quad B \Rightarrow (10, y)$$

Now, $$AB = \sqrt{8^2 + (y+3)^2}$$
$$\Rightarrow \quad 100 - 64 = (y+3)^2$$
$$\Rightarrow \quad y + 3 = \pm 6$$
$$\Rightarrow \quad y = \pm 6 - 3$$
$$\Rightarrow \quad y = 3, -9$$

$\therefore$ Required sum $= 3 - 9 = -6$

3. (c) $\because \dfrac{OA}{OP} = \dfrac{\lambda_1 + \lambda_2}{\lambda_1}$

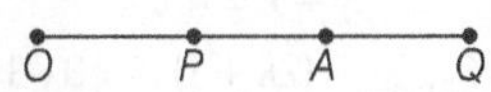

Or $$\frac{OA}{OQ} = \frac{\lambda_1 - \lambda_2}{\lambda_1}$$
$$\Rightarrow \quad \frac{1}{OP} + \frac{1}{OQ} = \frac{2}{OA}$$

OP, OA and OQ are in HP.

4. (c) Let the point divides the line segment joining the points $(3, -1)$ and $(7, 2)$ in $k : 1$.

$$\therefore \quad R = \left(\frac{7k+3}{k+1}, \frac{2k-1}{k+1}\right)$$

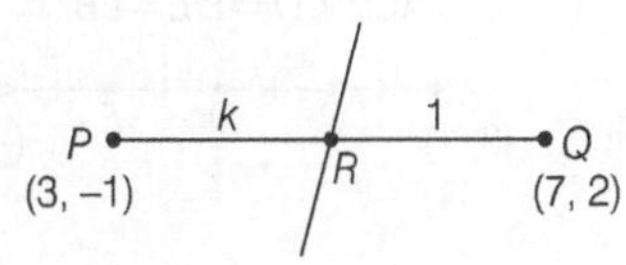

$\because$ R lies on $3x + y = 20$

$$\frac{21k+9}{k+1} + \frac{2k-1}{k+1} = 20$$
$$\Rightarrow \quad 23k + 8 = 20k + 20$$
$$\Rightarrow \quad 3k = 12$$
$$\Rightarrow \quad k = 4$$

$$\Rightarrow \qquad R=\left(\frac{31}{5},\frac{7}{5}\right)$$

$$\therefore \text{ Required distance} = \sqrt{\left(\frac{31}{5}-\frac{21}{5}\right)^2+\left(\frac{7}{5}+\frac{8}{5}\right)^2}$$

$$=\sqrt{2^2+3^2}=\sqrt{13}\text{ units}$$

5. (a) Let the line $r=\dfrac{4}{\sin\theta-\cos\theta}$ divides the line joining the points. $\left(3,\dfrac{\pi}{2}\right)$ and $\left(3,\dfrac{\pi}{6}\right)$ in $k:1$ and point is C

Now, $A\left(3,\dfrac{\pi}{2}\right)=(0,3)$ $\qquad \left\{\begin{array}{l}\because r=3,\theta=\dfrac{\pi}{2}\\ x=r\cos\theta, y=r\sin\theta\end{array}\right\}$

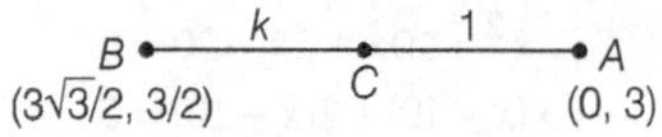

$$B\left(3,\frac{\pi}{6}\right)=\left(\frac{3\sqrt{3}}{2},\frac{3}{2}\right)$$

$$\therefore \qquad C=\left(\frac{\frac{3\sqrt{3}}{2}}{k+1},\frac{3k+\frac{3}{2}}{k+1}\right)$$

$$\Rightarrow \qquad \left[\frac{3\sqrt{3}}{2(k+1)},\frac{3k+3(2k+1)}{2(k+1)}\right]$$

Now, equation of line, $r=\dfrac{4}{\sin\theta-\cos\theta}$

$$r\sin\theta-r\cos\theta=4$$

$$y-x=4$$

and C lies on line $=\dfrac{(2k+1)}{2(k+1)}-\dfrac{3\sqrt{3}}{2(k+1)}=4$

$\Rightarrow\ 6k+3-3\sqrt{3}=8k+8$

$\Rightarrow \qquad 2k=3-8-3\sqrt{3}$

$$k=-\left(\frac{5+3\sqrt{3}}{2}\right)$$

$\therefore$ C divides AB externally in $5+3\sqrt{3}:2$. [externally]

6. (d) Let C,D,E are the points which divides AB in four equal parts.

$\therefore \qquad AC=CD=DE=EB$

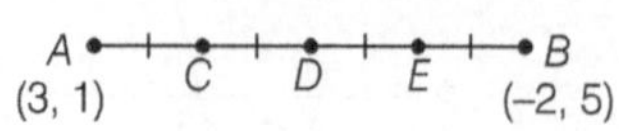

For C

$$C=\left(\frac{-2+9}{1+3},\frac{5+3}{1+3}\right)=\left(\frac{7}{4},2\right)$$

{C divides as in 1 : 3}

Similarly, D divides AB in 2 : 2 i.e. 1 : 1.

$$\therefore \qquad D=\left(\frac{3-2}{1+1},\frac{1+5}{2}\right)=\left(\frac{1}{2},3\right)$$

Similarly, E divides AB in 3 : 1.

$$\therefore \qquad E=\left(\frac{-6+3}{1+3},\frac{15+1}{1+3}\right)=\left(\frac{-3}{4},4\right)$$

7. (b) Area of the triangle formed by $(1,0)$, $\left(2,\dfrac{\pi}{3}\right)$ and $\left(3,\dfrac{2\pi}{3}\right)$ is $\dfrac{1}{2}|\Sigma r_1r_2\sin(\theta_1-\theta_2)|$

$$=\frac{1}{2}\left|2\sin\left(\frac{-\pi}{3}\right)+6\sin\left(\frac{-\pi}{3}\right)+3\sin\left(\frac{2\pi}{3}\right)\right|$$

$$=\frac{1}{2}\left|-2\times\frac{\sqrt{3}}{2}-6\times\frac{\sqrt{3}}{2}+3\times\frac{\sqrt{3}}{2}\right|$$

$$=\frac{1}{2}\times\frac{5\sqrt{3}}{2}=\frac{5\sqrt{3}}{4}\text{ sq units}$$

8. (a) Area of $\Delta ABC=1$ sq unit

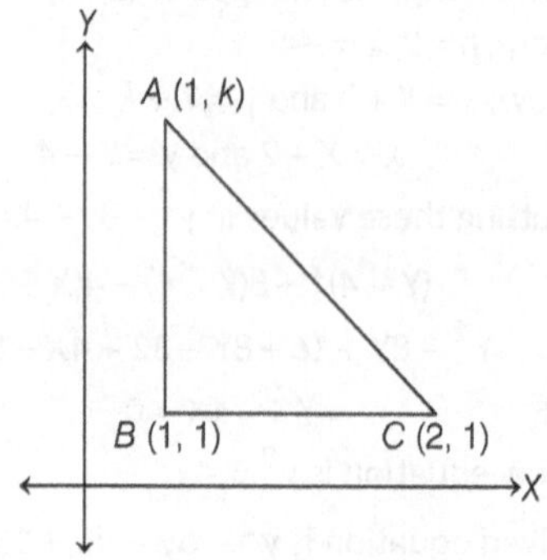

$$\Rightarrow \qquad \frac{1}{2}\times AB\times BC=1$$

$$|k-1|\times1=2$$

$$k-1=\pm2$$

$$k=\pm2+1$$

$$k=3,-1$$

9. (c) $\therefore\ OP=\sqrt{(3-0)^2+(4-0)^2}=5$

$$PQ=\sqrt{(6-3)^2+(4-0)^2}=5$$

$\therefore \qquad OP=PQ$

$\Rightarrow \Delta OPQ$ is an isosceles triangle.

$\because\ \Delta OPR, \Delta PQR$ and ΔOQR are equal in area.

Let $G(h,k)$ be the required point.

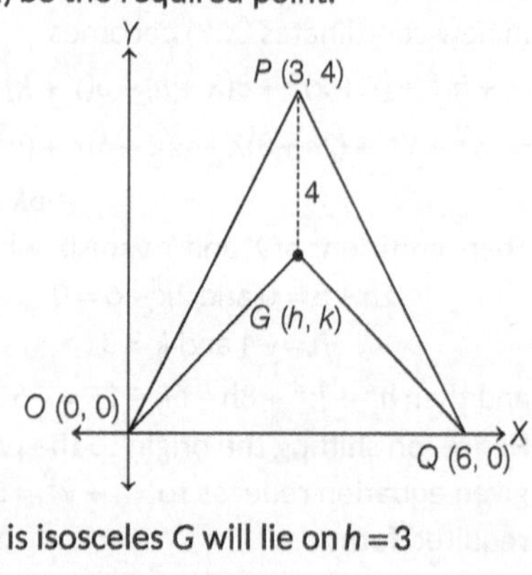

$\therefore\ \Delta OPQ$ is isosceles G will lie on $h=3$

$\therefore$ Area of $\Delta OGQ = \frac{1}{3} \times$ area of $\Delta OPQ = \frac{1}{3} \times 6 \times \frac{4}{2} = 4$

$\Rightarrow \quad \frac{1}{2} \times 6 \times k = 4 \Rightarrow k = \frac{4}{3}$

$\therefore$ Required point is $G\left(3, \frac{4}{3}\right)$ i.e. centroid is triangle

Also, area of (ΔOGP) = area of (ΔGPQ)

10. (b) $\frac{SM}{QN} = \frac{OM}{ON}$

$\frac{y}{1} = \frac{c}{9} \Rightarrow y = \frac{c}{9}$

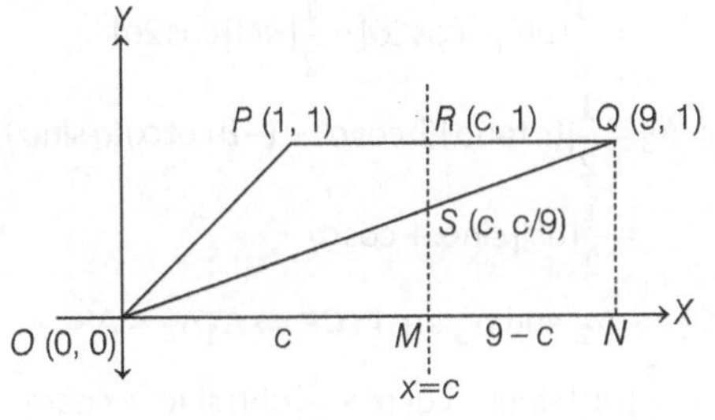

$\therefore$ Area of $\Delta RSQ = \frac{1}{2}$ area of ΔOPQ

$$\Rightarrow \frac{1}{2}\begin{vmatrix} 9 & 1 & 1 \\ c & 1 & 1 \\ c & c/9 & 1 \end{vmatrix} = \frac{1}{2} \times \frac{1}{2}\begin{vmatrix} 0 & 0 & 1 \\ 1 & 1 & 1 \\ 9 & 1 & 1 \end{vmatrix} \rightarrow \text{only + ve}$$

$$\Rightarrow 2\left|\left[9\left(1-\frac{c}{9}\right) - 1(c-c) + 1\left(\frac{c^2}{9} - c\right)\right]\right| = 8$$

$$\Rightarrow 2\left|9 - c + \frac{c^2}{9} - c\right| = 8$$

$$\Rightarrow \quad 9 - 2c + \frac{c^2}{9} = 4$$

$\Rightarrow \quad c^2 - 18c + 45 = 0$

$\Rightarrow \quad (c-3)(c-15) = 0$

$c = 3$ or 15 (not possible)

11. (c)

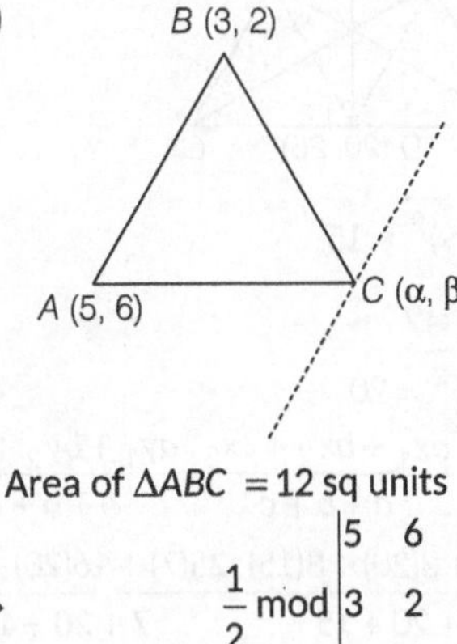

$\therefore$ Area of $\Delta ABC = 12$ sq units

$$\Rightarrow \frac{1}{2} \text{mod} \begin{vmatrix} 5 & 6 & 1 \\ 3 & 2 & 1 \\ \alpha & \beta & 1 \end{vmatrix} = 12$$

$\Rightarrow \quad |5(2-\beta) - 6(3-\alpha) + 1(3\beta - 2\alpha)| = 24$

$\Rightarrow \quad |10 - 5\beta - 18 + 6\alpha + 3\beta - 2\alpha| = 24$

$\Rightarrow \quad 4\alpha - 2\beta - 8 = \pm 24$

$\Rightarrow \quad 4\alpha - 2\beta = \pm 24 + 8$

$\Rightarrow \quad 2\alpha - \beta = \pm 12 + 4$

$\Rightarrow \quad 2\alpha - \beta = 16, -8$

Now, $2\alpha - \beta = 16$ and $2\alpha - \beta = -8$

$\Rightarrow \quad 2x - y = 16$ and $2x - y = -8$

$2x - y - 16 = 0$ and $2x - y + 8 = 0$

$\therefore$ Least distance with O is perpendicular distance

$$= \left|\frac{ax_1 + by_1 + c}{\sqrt{a^2 + b^2}}\right|$$

$$p_1 = \left|\frac{-16}{\sqrt{5}}\right| = \frac{16}{\sqrt{5}}$$

$$p_2 = \left|\frac{8}{\sqrt{5}}\right| = \frac{8}{\sqrt{5}}$$

12. (d) Equation of median through B and C are $x + y = 5$ and $x = 4$

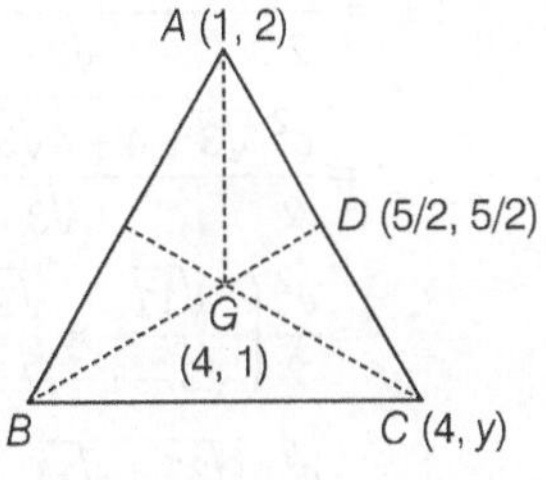

Let $C(4, y)$

Now, mid-point of $AC = \left(\frac{1+4}{2}, \frac{2+y}{2}\right)$

$= \left(\frac{5}{2}, \frac{2+y}{2}\right)$

It lies on BD

$\frac{5}{2} + \frac{2+y}{2} = 5$

$2 + y = 5$

$y = 3 \Rightarrow C = (4,3)$

and centroid G is point of intersection of median.

$\therefore \quad G = (4,1)$

$\therefore$ Area of $\Delta AGC = \frac{1}{3} \times$ Area of (ΔABC)

$$\therefore \text{Area of } \Delta ABC = 3 \times \frac{1}{2}\begin{vmatrix} 1 & 2 & 1 \\ 4 & 3 & 1 \\ 4 & 1 & 1 \end{vmatrix}$$

$= \frac{3}{2}\{|2-8|\} = \frac{3}{2} \times 6 = 9$ sq units

13. (b) m_1 and m_2 are roots of $x^2 + (\sqrt{3}+2)x + (\sqrt{3}-1) = 0$

$m_1 + m_2 = -(\sqrt{3}+2)$

$m_1 m_2 = \sqrt{3} - 1$

$\therefore \quad \text{Area} = \frac{1}{2} \text{mod} \begin{vmatrix} 0 & c/m_1 & c/m_2 \\ 0 & c & c \\ 1 & 1 & 1 \end{vmatrix}$

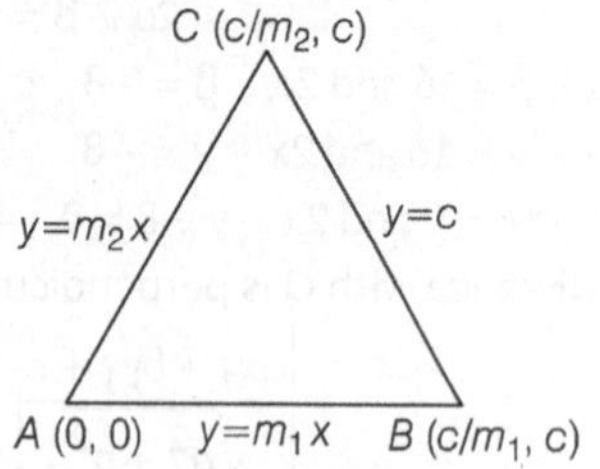

$$\Rightarrow \frac{1}{2}\left|\frac{c^2}{m_1} - \frac{c^2}{m_2}\right| = \frac{c^2}{2}\left|\frac{m_2 - m_1}{m_1 m_2}\right|$$

$$= \frac{c^2}{2} \frac{\sqrt{(m_1+m_2)^2 - 4m_1m_2}}{m_1m_2}$$

$$= \frac{c^2}{2} \frac{\sqrt{(2+\sqrt{3})^2 - 4(\sqrt{3}-1)}}{\sqrt{3}-1}$$

$$= \frac{c^2}{2} \frac{\sqrt{3+4+4\sqrt{3}-4\sqrt{3}+4}}{\sqrt{3}-1}$$

$$= \frac{c^2}{2}\left(\frac{\sqrt{11}}{\sqrt{3}-1} \times \frac{\sqrt{3}+1}{\sqrt{3}+1}\right)$$

$$= \frac{c^2}{2} \cdot \frac{\sqrt{33}+\sqrt{11}}{2}$$

$$= \frac{c^2}{4}(\sqrt{33}+\sqrt{11})$$

14. (b) Let the vertex be (x_i, y_i)

$\Rightarrow i = 1, 2, 3,$

where, x_i and y_i both are integers.

Then, area of $\Delta = \frac{1}{2}\Sigma x_1(y_2 - y_3)$ = rational number ...(i)

Also, if a be side of the triangle.

Then, $a^2 = (x_1 - x_2)^2 + (y_1 - y_2)^2$ = a positive integer.

$\therefore$ Area of equilateral triangle $= \frac{\sqrt{3}}{4}a^2$

$\Rightarrow$ which is irrational.

$\therefore a^2$ is positive integer. ...(ii)

Thus, Eqs. (i) and (ii) are contradict each other.

The triangle is never equilateral.

15. (c) $\therefore \Delta_1 = \frac{1}{2}|(a\tan\alpha)(b\cos\alpha) - (a\sin\alpha)(b\cot\alpha)|$

[$\because$ one vertex is (0,0)]

$= \frac{1}{2}|ab||\sin\alpha - \cos\alpha|$

$\therefore \Delta_2 = \frac{1}{2}$ mod of

$$\begin{vmatrix} a-(a+\sin^2\alpha) & a\sec^2\alpha-(a+a\sin^2\alpha) \\ b-(b+\cos^2\alpha) & b\text{cosec}^2\alpha-(b+b\cos^2\alpha) \end{vmatrix}$$

$$= \frac{1}{2}|ab|\begin{vmatrix} -\sin^2\alpha & \sin^2\alpha\tan^2\alpha \\ -\cos^2\alpha & \cos^2\alpha\cot^2\alpha \end{vmatrix}$$

$$= \frac{1}{2}|ab||-\sin^2\alpha\cos^2\alpha\cot^2\alpha + \sin^2\alpha\cos^2\alpha\tan^2\alpha|$$

$$= \frac{1}{2}|ab||-\cos^4\alpha + \sin^4\alpha|$$

$$= \frac{1}{2}|ab||-\cos 2\alpha| = \frac{1}{2}|ab||\cos 2\alpha|$$

$$\therefore \Delta_3 = \frac{1}{2}|(a\tan\alpha)(b\cos\alpha) - (-b\cot\alpha)(a\sin\alpha)|$$

$$= \frac{1}{2}|ab||\sin\alpha + \cos\alpha|$$

$\because \Delta_1, \Delta_2$ and Δ_3 are in GP $\Rightarrow \Delta_1\Delta_3 = \Delta^2$

$$\Rightarrow \frac{1}{2}|ab||\sin\alpha - \cos\alpha| \times \frac{1}{2}|ab||\sin\alpha + \cos\alpha|$$

$$= \frac{1}{4}|ab|^2|\cos 2\alpha|^2$$

$\Rightarrow |\sin^2\alpha - \cos^2\alpha| = |\cos 2\alpha|^2$

$\Rightarrow |\cos 2\alpha| = |\cos 2\alpha|^2$

$\Rightarrow |\cos 2\alpha|\{1 - |\cos 2\alpha|\} = 0$ $\quad \{\because |\cos 2\alpha| \neq 0\}$

$\Rightarrow 1 - |\cos 2\alpha| = 0$

$\Rightarrow |\cos 2\alpha| = 1$

$\Rightarrow \cos 2\alpha = \pm 1$

$\Rightarrow \alpha = n\pi, n \in I$

For these values of α no triangle is formed.

$\therefore$ No value of α.

16. (b) Orthocentre of ΔABC is incentre of its pedal ΔDEF

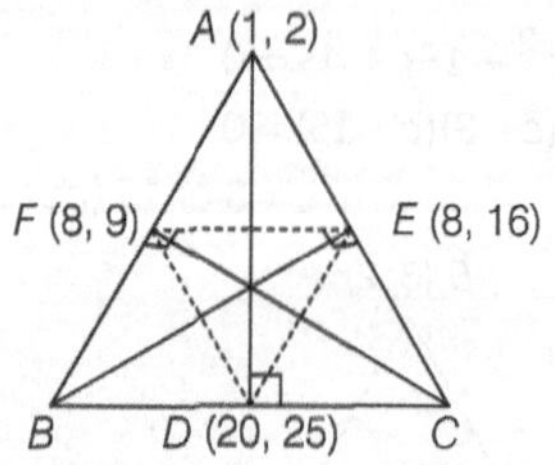

Now, $DE = \sqrt{(12)^2 + (9)^2} = 15$

$EF = \sqrt{0^2 + 7^2} = 7$

$FD = \sqrt{12^2 + 16^2} = 20$

$$\therefore \text{Required point} = \left(\frac{ax_1 + bx_2 + cx_3}{a+b+c}, \frac{ay_1 + by_2 + cy_3}{a+b+c}\right)$$

$$= \left(\frac{20(7) + 8(20) + 8(15)}{7+20+15}, \frac{25(7) + 16(20) + 9(15)}{7+20+15}\right)$$

$= (10, 15)$

17. (a) $x^3 - 6x^2 + 11x - 6 = 0$

$$\therefore \quad \alpha + \beta + \gamma = \frac{-b}{a} = 6$$

$$\therefore \quad \alpha\beta + \beta\gamma + \gamma\alpha = \frac{c}{a} = 11$$

$$\therefore \quad \alpha\beta\gamma = \frac{-d}{a} = 6$$

$$\therefore \text{ Centroid of triangle is} = \left(\frac{x_1 + x_2 + x_3}{3}, \frac{y_1 + y_2 + y_3}{3}\right)$$

$$= \left(\frac{\alpha\beta + \beta\gamma + \gamma\alpha}{3}, \frac{\frac{1}{\alpha\beta} + \frac{1}{\beta\gamma} + \frac{1}{\alpha\gamma}}{3}\right)$$

$$= \left(\frac{11}{3}, \frac{\frac{\alpha + \beta + \gamma}{\alpha\beta\gamma}}{3}\right)$$

$$= \left(\frac{11}{3}, \frac{6/6}{3}\right) = \left(\frac{11}{3}, \frac{1}{3}\right)$$

18. (b) Harmonic conjugate of $C(5,1)$ w.r.t. $A(2,10)$ and $B(6,-2)$.

Let C divides AB in $k:1$

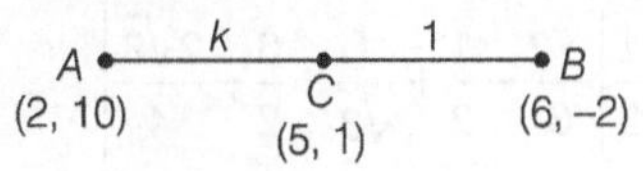

$$\Rightarrow \quad \frac{-2k + 10}{k + 1} = 1$$

$$\Rightarrow \quad -2k + 10 = k + 1$$

$$\Rightarrow \quad 3k = 9 \Rightarrow k = 3$$

Harmonic conjugate is

$$\left[\frac{3 \times 6 - 1 \times 2}{3 - 1}, \left(\frac{3 \times (-2) - 1 \times 10}{3 - 1}\right)\right]$$

$$= \left(\frac{16}{2}, \frac{-16}{2}\right) = (8, -8)$$

$\therefore$ Circumcentre $O = (8, -8)$

and orthocentre $H = (3, -4)$

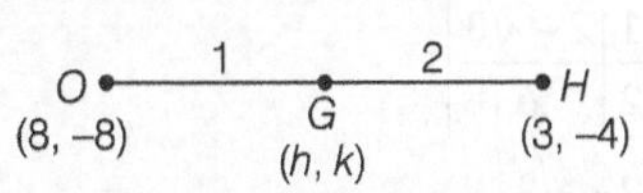

$$\therefore \quad G = \left(\frac{3 + 16}{2 + 1}, \frac{-4 - 16}{2 + 1}\right) = \left(\frac{19}{3}, \frac{-20}{3}\right)$$

19. (b) $\therefore r$ be the inradius of ΔABC

and centroid and incentre of equilateral triangle is same.

$$r = \left|\frac{0 + 0 - 3}{\sqrt{2}}\right| = \frac{3}{\sqrt{2}}$$

[r is perpendicular distance between (0,0) and line $x + y = 3$]

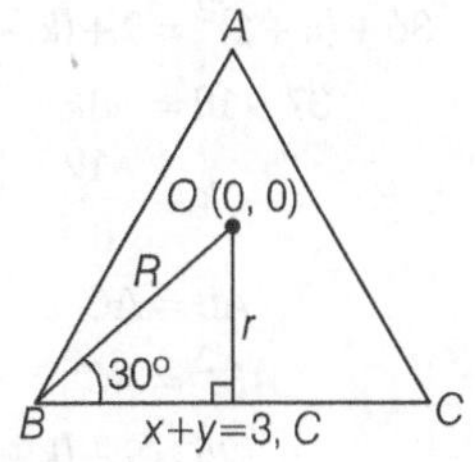

$$\therefore \quad \sin 30° = \frac{r}{R} = \frac{1}{2}$$

$$\Rightarrow \quad R = 2r$$

$$\Rightarrow \quad R + r = 3r = 3 \times \frac{3}{\sqrt{2}} = \frac{9}{\sqrt{2}} \text{ units}$$

20. (a) $AB = 10$ and $BC = 5$

$\because$ Angle bisector divides the opposite sides in the ratio of its containing sides.

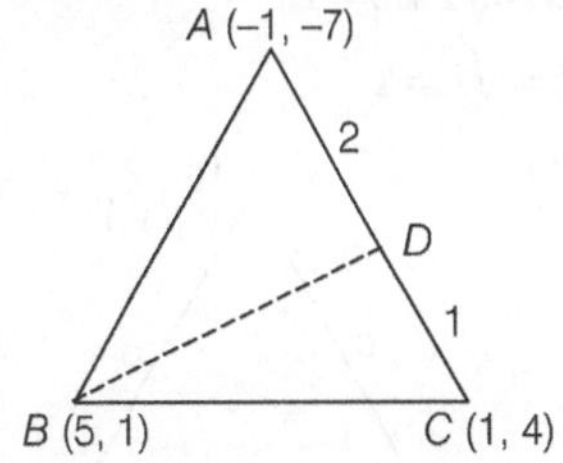

$$\therefore \quad \frac{AD}{CD} = \frac{AB}{CB} = \frac{10}{5} = 2$$

$$\therefore \text{ Coordinates of } D = \left(\frac{2 - 1}{2 + 1}, \frac{8 - 7}{2 + 1}\right) = \left(\frac{1}{3}, \frac{1}{3}\right)$$

$$\therefore \quad BD = \sqrt{\left(5 - \frac{1}{3}\right)^2 + \left(1 - \frac{1}{3}\right)^2} = \sqrt{\frac{196}{9} + \frac{4}{9}}$$

$$= \sqrt{\frac{200}{9}} = \frac{10}{3}\sqrt{2} \text{ units}$$

21. (b) Centroid, orthocentre, incentre and circumcentre of ΔABC lies on the same line.

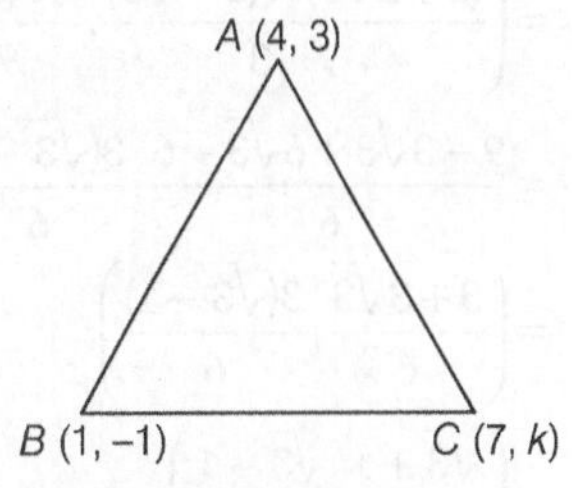

ΔABC must be isosceles. Therefore,

$$AB = BC$$

$$\Rightarrow \quad AB^2 = BC^2$$

$$\Rightarrow \quad 9 + 16 = 36 + (k + 1)^2$$

$$\Rightarrow \quad (k + 1)^2 = -11 \text{ (not possible)}$$

Similarly, $BC = CA$

$$\Rightarrow \quad BC^2 = CA^2$$

$$\Rightarrow \quad 36+(k+1)^2=9+(k-3)^2$$
$$\Rightarrow \quad 37-18=-8k$$
$$k=\frac{-19}{8}$$
Similarly, $\quad AB=AC$
$$\Rightarrow \quad AB^2=AC^2$$
$$\Rightarrow \quad 25=9+(k-3)^2$$
$$\Rightarrow \quad k-3=\pm 4$$
$$\Rightarrow \quad k=\pm 4+3$$
$$\Rightarrow \quad k=-1,+7$$
But $k=7$

A,B,C becomes collinear
$$k=-1$$

22. (d) $AB=c=\sqrt{0+3}=\sqrt{3}$

$BC=a=\sqrt{1+3}=2$

$AC=b=\sqrt{1}=1$

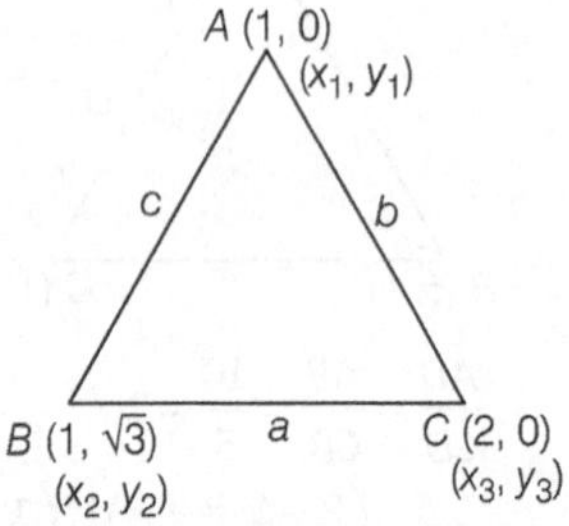

Now, centroid $=\left(\frac{4}{3},\frac{1}{\sqrt{3}}\right)$

In centre $=\left(\frac{2(1)+(1)(1)+\sqrt{3}(2)}{1+2+\sqrt{3}},\frac{2(0)+1(\sqrt{3})+\sqrt{3}(0)}{1+2+\sqrt{3}}\right)$

$$=\left(\frac{3+2\sqrt{3}}{3+\sqrt{3}},\frac{\sqrt{3}}{3+\sqrt{3}}\right)$$
$$=\left(\frac{(3+2\sqrt{3})\times(3-\sqrt{3})}{9-3},\frac{\sqrt{3}(3-\sqrt{3})}{9-3}\right)$$
$$=\frac{9-3\sqrt{3}+6\sqrt{3}-6}{6},\frac{3(\sqrt{3}-1)}{6}$$
$$=\left(\frac{3+3\sqrt{3}}{6},\frac{3(\sqrt{3}-1)}{6}\right)$$
$$=\left(\frac{\sqrt{3}+1}{2},\frac{\sqrt{3}-1}{2}\right)$$

Circumcentre $(O)\rightarrow(x,y)$
$$OA^2=(x-1)^2+y^2$$
$$OB^2=(x-1)^2+(y-\sqrt{3})^2$$
and $\quad OC^2=(x-2)^2+y^2$
$$OA^2=OB^2$$
$$\Rightarrow \quad y=\pm(y-\sqrt{3})$$
$$\Rightarrow \quad 2y=\sqrt{3}$$
$$\Rightarrow \quad y=\frac{\sqrt{3}}{2}$$
$$OA^2=OC^2$$
$$\Rightarrow \quad x-1=-(x-2)$$
$$\Rightarrow \quad 2x=3$$
$$\Rightarrow \quad x=\frac{3}{2}$$
$$\therefore \quad O=\left(\frac{3}{2},\frac{\sqrt{3}}{2}\right)$$

$$\text{Area}=\frac{1}{2}\text{ mod}\begin{vmatrix} 4/3 & 1/\sqrt{3} & 1 \\ \frac{\sqrt{3}+1}{2} & \frac{\sqrt{3}-1}{2} & 1 \\ 3/2 & \sqrt{3}/2 & 1 \end{vmatrix}$$

$$=\frac{1}{2}\left|\frac{4}{3}\left(\frac{\sqrt{3}-1}{2}-\frac{\sqrt{3}}{2}\right)-\frac{1}{\sqrt{3}}\left(\frac{\sqrt{3}+1}{2}-\frac{3}{2}\right)+1\left(\frac{\sqrt{3}}{2}\left(\frac{\sqrt{3}+1}{2}\right)-\frac{3}{2}\left(\frac{\sqrt{3}-1}{2}\right)\right)\right|$$
$$=\frac{1}{2}\left|\frac{4}{3}\left(-\frac{1}{2}\right)-\frac{1}{\sqrt{3}}\left(\frac{\sqrt{3}-2}{2}\right)+\left(\frac{3}{4}+\frac{\sqrt{3}}{4}-\frac{3\sqrt{3}}{4}+\frac{3}{4}\right)\right|$$
$$=\frac{1}{2}\left|\frac{-2}{3}-\frac{1}{2}+\frac{1}{\sqrt{3}}+\frac{3}{2}-\frac{2\sqrt{3}}{4}\right|$$
$$=\frac{1}{2}\left|-\frac{2}{3}-\frac{1}{2}+\frac{3}{2}+\frac{1}{\sqrt{3}}-\frac{2\sqrt{3}}{4}\right|$$
$$=\frac{1}{2}\left|\frac{-4-3+9}{6}+\frac{4-6}{4\sqrt{3}}\right|$$
$$=\frac{1}{2}\left|\frac{2}{6}-\frac{2}{4\sqrt{3}}\right|$$
$$=\frac{1}{2}\left|\frac{2}{6}-\frac{1}{2\sqrt{3}}\right|$$
$$=\frac{1}{2}\left|\frac{2\sqrt{3}-3}{6\sqrt{3}}\right|$$
$$=\frac{1}{2}\left|\frac{2-\sqrt{3}}{6}\right|$$
$$=\frac{2-\sqrt{3}}{12}$$

23. (b) If R is the circumradius and O be the circumcentre
$$OA=OB=OC=R$$
or $\quad \sqrt{a^2+a^2\tan^2\alpha}=\sqrt{b^2+b^2\tan^2\beta}$
$$=\sqrt{c^2+c^2\tan^2\gamma}=R$$
$$\Rightarrow \quad a\sec\alpha=b\sec\beta=c\sec\gamma=R$$
$$\Rightarrow \quad a=R\cos\alpha, b=R\cos\beta \text{ and } c=R\cos\gamma$$

Then, $a\tan\alpha = R\cos\alpha \cdot \dfrac{\sin\alpha}{\cos\alpha} = R\sin\alpha$

$$b\tan\beta = R\cos\beta \frac{\sin\beta}{\cos\beta} = R\sin\beta$$

$$c\tan\gamma = R\cos\gamma \cdot \frac{\sin\gamma}{\cos\gamma} = R\sin\gamma$$

$\therefore$ Coordinates of vertices are

$A(R\cos\alpha, R\sin\alpha), B(R\cos\beta, R\sin\beta)$ and $C(R\cos\gamma, R\sin\gamma)$

$\therefore$ Centroid $\Rightarrow$

$$G\left(\frac{R\cos\alpha + R\cos\beta + R\cos\gamma}{3}, \frac{R\sin\alpha + R\sin\beta + R\sin\gamma}{3}\right)$$

Centroid divides H (orthocentre) and O (circumcentre) in the ratio 2 : 1 internally, then

$$\frac{R}{3}(\cos\alpha + \cos\beta + \cos\gamma) = \frac{2\cdot 0 + 1\cdot \bar{x}}{2+1}$$

$$\bar{x} = (\cos\alpha + \cos\beta + \cos\gamma)R$$

and $\quad \dfrac{R}{3}(\sin\alpha + \sin\beta + \sin\gamma) = \dfrac{2\cdot 0 + 1\cdot \bar{y}}{2+1}$

$\Rightarrow \quad \bar{y} = (\sin\alpha + \sin\beta + \sin\gamma)R$

$\therefore \quad \dfrac{\bar{x}}{\bar{y}} = \dfrac{\cos\alpha + \cos\beta + \cos\gamma}{\sin\alpha + \sin\beta + \sin\gamma}$

24. (c) Let the coordinates of the centre of mean position of the points

$A_i \Rightarrow i = 1, 2, 3, \ldots n$ be (x, y), then

$$x = \frac{x_1 + x_2 + \ldots x_n}{n}$$

$$y = \frac{y_1 + y_2 + \ldots y_n}{n}$$

$$x = \frac{np + a(1+2+3+\ldots n)}{n}$$

$$y = \frac{nq + b(1+2+3+\ldots+n)}{n}$$

$$x = p + \frac{n(n+1)}{2n}a$$

and $\quad y = q + \dfrac{n(n+1)}{2n}b$

$\Rightarrow \quad \dfrac{2(x-p)}{a} = \dfrac{2(y-q)}{b}$

$\Rightarrow \quad bx - ay = bp - aq$

25. (c) We have,

$$OA_n = nOA_{n-1}$$
$$= n(n-1)OA_{n-2}$$
$$= (OA_1) = n!(OA_1)$$
$$= n!$$

Then, $\sqrt{2}(2520\sqrt{2}) = n!$

$7 \times 6 \times 5 \times 4 \times 3 \times 2 = n!$

$\Rightarrow \quad n! = 7!$

$n = 7$

26. (a) Let point C is (h, k) from figure

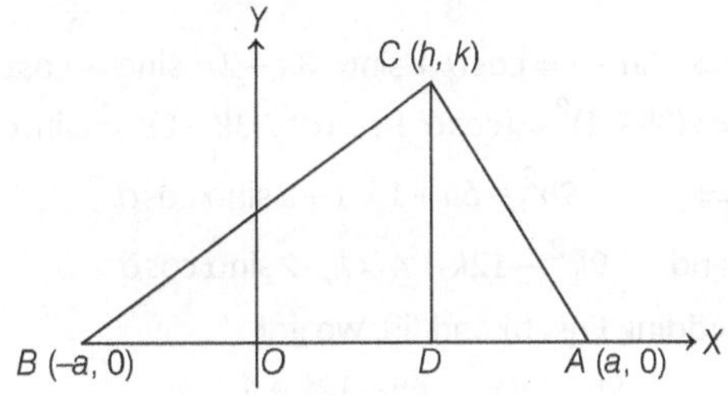

$$\cot A = \frac{DA}{CD} = \frac{a-h}{k}$$

$$\cot B = \frac{BD}{CD} = \frac{a+h}{k}$$

But $\cot A + \cot B = \lambda$

$\Rightarrow \quad \dfrac{a-h}{k} + \dfrac{a+h}{k} = \lambda$

$$\lambda = \frac{2a}{k}$$

$$k\lambda = 2a$$

$\therefore \quad$ Locus is $y\lambda = 2a$

27. (a) $O(0, 0)$, $A(0, 4)$ and $B(6, 0)$.

Let $P(h, k)$

$\therefore A_1 = \text{Area of } \Delta POA = \dfrac{1}{2}\begin{vmatrix} h & k & 1 \\ 0 & 0 & 1 \\ 0 & 4 & 1 \end{vmatrix}$

$$= \frac{1}{2}|-4h| = |-2h|$$

$\therefore A_2 = \text{Area of } \Delta POB = \dfrac{1}{2}\begin{vmatrix} h & k & 1 \\ 0 & 0 & 1 \\ 6 & 0 & 1 \end{vmatrix}$

$$= \frac{1}{2}|6k| = |3k|$$

Given, $\quad A_1 = 2A_2$

$\Rightarrow \quad |-2h| = 2|3k|$

$\Rightarrow \quad h = 3k$ or $h = -3k$

$\therefore$ Locus is $x = 3y$ or $x = -3y$

Combined $\Rightarrow (x - 3y)(x + 3y) = 0$

28. (b) Let the centroid of ΔABC is (h, k)

$\therefore$ Centroid $= \left(\dfrac{x_1 + x_2 + x_3}{3}, \dfrac{y_1 + y_2 + y_3}{3}\right)$

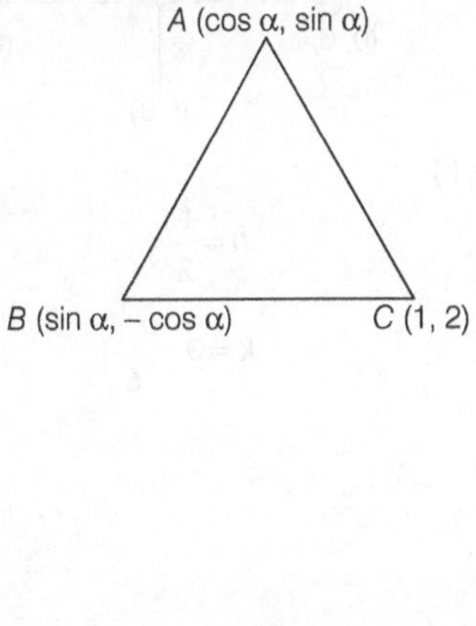

Now, $h=\dfrac{\cos\alpha+\sin\alpha+1}{3}$ and $k=\dfrac{\sin\alpha-\cos\alpha+2}{3}$

$\Rightarrow\quad 3h-1=\cos\alpha+\sin\alpha,\ 3k-2=\sin\alpha-\cos\alpha$

$\Rightarrow (3h-1)^2=(\cos\alpha+\sin\alpha)^2,\ (3k-2)^2=(\sin\alpha-\cos\alpha)^2$

$$\Rightarrow\quad 9h^2-6h+1=1+2\sin\alpha\cos\alpha \quad ...(i)$$

and $$9k^2-12k+4=1-2\sin\alpha\cos\alpha \quad ...(ii)$$

Adding Eqs. (i) and (ii), we get

$$9h^2+9k^2-6h-12k+1+4=2$$

$$\Rightarrow\quad 3h^2+3k^2-2h-4k+1=0$$

$\therefore$ Locus is $3(x^2+y^2)-2x-4y+1=0$

29. (a) Let $P(h,k)$

Given, $$AB+BP+AP=4$$

$$1+\sqrt{h^2+(k-1)^2}+\sqrt{h^2+k^2}=4$$

$$\sqrt{h^2+(k-1)^2}+\sqrt{h^2+k^2}=3$$

$$\sqrt{h^2+k^2}=3-\sqrt{h^2+(k-1)^2}$$

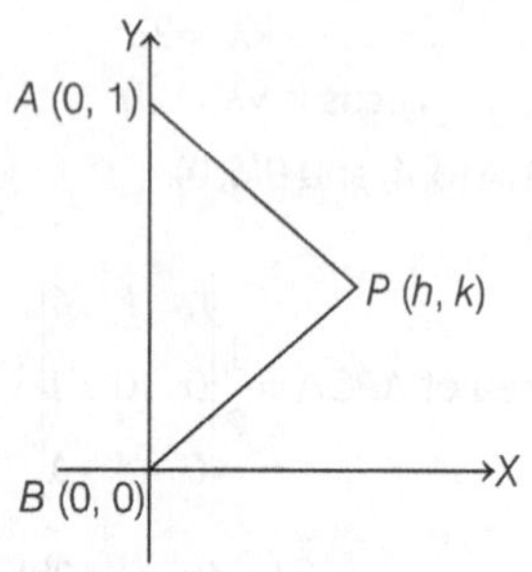

Squaring both sides

$$h^2+k^2=9+h^2+k^2-2k+1-6\sqrt{h^2+(k-1)^2}$$

$$\Rightarrow\quad 2k-10=-6\sqrt{h^2+k^2-2k+1}$$

$$\Rightarrow\quad (k-5)^2=9(h^2+k^2-2k+1)$$

$$\Rightarrow\quad 9h^2+8k^2-8k-16=0$$

$\therefore\quad$ Locus is $9x^2+8y^2-8y=16$

30. (c) Perpendicular bisector of AB is

$$y-3=\frac{t}{3}(x-t)$$

$$C\Rightarrow\left(0,3-\frac{t^2}{3}\right)$$

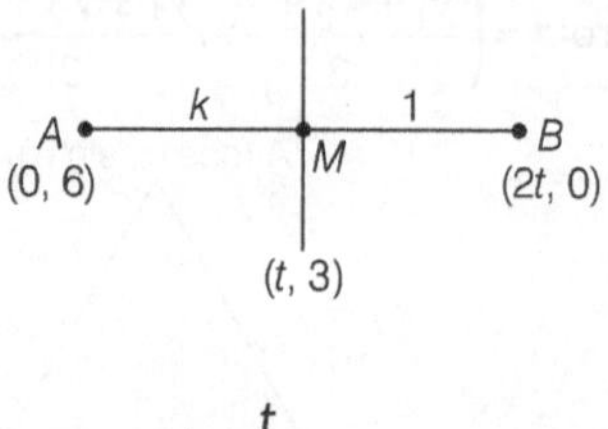

Let $P(h,k)$

$$\Rightarrow\quad h=\frac{t}{2}$$

$$\Rightarrow\quad k=3-\frac{t^2}{6}$$

$$\Rightarrow\quad k=3-\frac{4h^2}{6}$$

$\therefore$ Locus is $2x^2+3y-9=0$

31. (d) $M(0,h)$ and $N(0,h+4)$

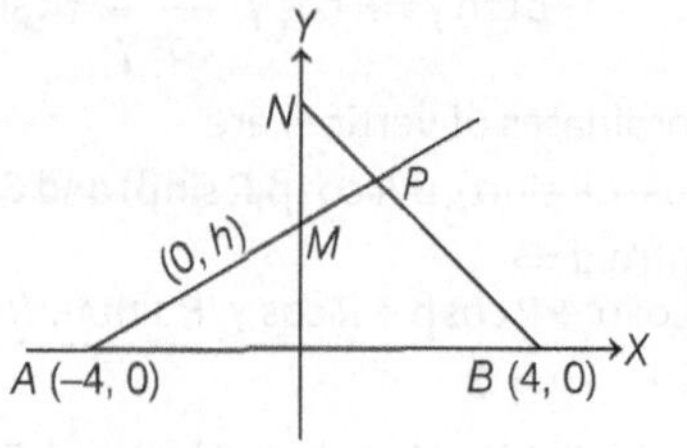

Equation of AM is $\dfrac{x}{-4}+\dfrac{y}{h}=1$

$$\frac{y}{h}=\frac{4+x}{4}$$

$$\Rightarrow\quad h=\frac{4y}{4+x} \quad ...(i)$$

Equation of BN is $\dfrac{x}{4}+\dfrac{y}{h+4}=1$

$$h+4=\frac{4y}{4-x}$$

$$h=\frac{4y-16+4x}{4-x} \quad ...(ii)$$

From Eqs. (i) and (ii),

$$\frac{4(y-4+x)}{4-x}=\frac{4y}{4+x}$$

$$\Rightarrow\quad 2xy-16+x^2=0$$

32. (b) Given equation

$$2x^2+3\sqrt{3}xy+3y^2=2 \quad ...(i)$$

Now, axes is rotated 30°.

Let the new coordinates be (x',y')

$$\omega=30°,\ \omega'=90° \text{ and } \theta=0°$$

$$x\sin 30°=x'\sin 30°+y'\sin(-60°)$$

$$\Rightarrow\quad \frac{x}{2}=\frac{x'}{2}-\frac{\sqrt{3}}{2}y'$$

$$\Rightarrow\quad x=x'-\sqrt{3}y'$$

$$y\sin 30°=x'\sin 0°+y'\sin 90°$$

$$\frac{y}{2}=0+y'$$

$$\Rightarrow\quad y=2y'$$

By Eq. (i),

$$2(x'-\sqrt{3}y')^2+3\sqrt{3}(x'-\sqrt{3}y')2y'+3(2y')^2=2$$

$$\Rightarrow 2(x')^2+6y'^2-4\sqrt{3}x'y'+6\sqrt{3}y'(x'-\sqrt{3}y')+12y'^2=2$$

$$\Rightarrow\quad 2x'^2+6y'^2-4\sqrt{3}x'y'+6\sqrt{3}x'y'-18y'^2+12y'^2=2$$

$$\Rightarrow\quad 2x'^2+2\sqrt{3}x'y'=2$$

$$\Rightarrow\quad x'^2+\sqrt{3}x'y'=1$$

$\therefore$ Required equation $x^2+\sqrt{3}xy=1$

33. (a) After reflection about Y-axis the point $(2,-1)$ will be transformed in $(-2,-1)$.

Now, $45°$ rotation in negative direction

$$x' = x\cos\alpha + y\sin\alpha$$
$$x' = -2\cos 45° + \sin 45°$$
$$= \frac{-2}{\sqrt{2}} + \frac{1}{\sqrt{2}} = \frac{-1}{\sqrt{2}}$$

Similarly, $y' = -x\sin\alpha + y\cos\alpha$

$$y' = -2\sin 45° - \cos 45°$$
$$y' = \frac{-2}{\sqrt{2}} - \frac{1}{\sqrt{2}} = \frac{-3}{\sqrt{2}}$$

$\therefore \left(-\frac{1}{\sqrt{2}}, \frac{-3}{\sqrt{2}}\right)$

34. (b) Let O' be $(-2,-3)$. Since, the axes are rotated about O' by an angle of $45°$ in anti-clockwise direction. Let (x', y') be the new axes and (x, y) be the coordinates with respect to old axes, then

$$x = -2 + x'\cos 45° - y'\sin 45°$$
$$= -2 + \left(\frac{x' - y'}{\sqrt{2}}\right)$$

and $y = -3 + x'\sin 45° + y'\cos 45°$

$$y = -3 + \left(\frac{x' + y'}{\sqrt{2}}\right)$$

The new equation will be

$$2\left\{-2 + \left(\frac{x'-y'}{\sqrt{2}}\right)\right\}^2 + 4\left\{\left(-2 + \frac{(x'-y')}{\sqrt{2}}\right)\left(-3 + \frac{(x'+y')}{\sqrt{2}}\right)\right\} - 5\left\{-3 + \left(\frac{x'+y'}{\sqrt{2}}\right)\right\}^2$$
$$+ 20\left\{-2 + \left(\frac{x'-y'}{\sqrt{2}}\right)\right\} - 22\left\{-3 + \frac{(x'+y')}{\sqrt{2}}\right\} - 14 = 0$$

$\Rightarrow$ $$(x')^2 - 14x'y' - 7(y')^2 - 2 = 0$$

New equation of curve is

$$x^2 - 14xy - 7y^2 - 2 = 0$$

35. (b) If the point $P(1,2)$ is translated itself 2 units along the positive direction of X-axis, then P becomes $(3, 2)$.

Let $Z = (3,2) = 3 + 2i$

Thus, the new position of $P = iZ = i(3 + 2i)$

$$= -2 + 3i = (-2, 3)$$

36. (1176) $PA^2 > PB^2$

$$(x-3)^2 + (x-4)^2 > (x-7)^2 + (x-12)^2$$

$\Rightarrow$ $24x - 168 > 0 \Rightarrow x > 7$ ...(i)

$$PB^2 > AB^2$$
$$(x-7)^2 + (x-12)^2 > (3-7)^2 + (4-12)^2$$

$\Rightarrow$ $2x^2 - 38x + 49 + 144 > 16 + 64$

$\Rightarrow$ $2x^2 - 38x + 113 > 0$

$\Rightarrow$ $x = \frac{38 \pm \sqrt{1444 - 904}}{2 \times 2} = \frac{38 \pm \sqrt{540}}{4}$

$$x \in \left(-\infty, \frac{38 - \sqrt{540}}{4}\right) \cup \left(\frac{38 + \sqrt{540}}{4}, \infty\right) \quad \text{...(ii)}$$

By Eqs. (i) and (ii), $x \in \left(\frac{19 + \sqrt{135}}{2}, \infty\right)$

$\therefore$ Possible integral values are 16,17,18....

Now, required series

$\Rightarrow$ $16 + 20 + ... 96$

$\therefore$ $96 = 16 + (n-1)4$

$$\frac{80}{4} = n - 1 \Rightarrow n = 21$$

$\therefore$ $S_n = \frac{n}{2}(a + l) = \frac{21}{2}(16 + 96)$

$$= 21 \times 56 = 1176$$

37. (4) $AB = \sqrt{6^2 + 8^2} = 10$

and $BC = \sqrt{4^2 + 3^2} = 5$

$\because$ We know that

$\therefore$ $\frac{DA}{DC} = \frac{BA}{BC}$

$\Rightarrow$ $\frac{AD}{CD} = \frac{10}{5} = \frac{2}{1}$

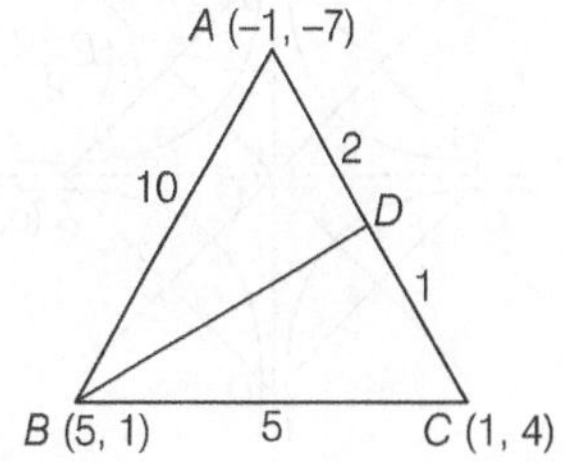

$\therefore$ $D = \left(\frac{2(1) + 1(-1)}{2+1}, \frac{4 \times 2 + (1)(-7)}{2+1}\right) = \left(\frac{1}{3}, \frac{1}{3}\right)$

$\therefore$ Length of angle bisector

$$BD = \sqrt{\left(5 - \frac{1}{3}\right)^2 + \left(1 - \frac{1}{3}\right)^2} = \sqrt{\frac{196}{9} + \frac{4}{9}}$$
$$= \sqrt{\frac{200}{9}} = \frac{10}{3}\sqrt{2}$$
$$l = \frac{10\sqrt{2}}{3}$$
$$[l] = 4$$

38. (32) Ordinates of A and B are same

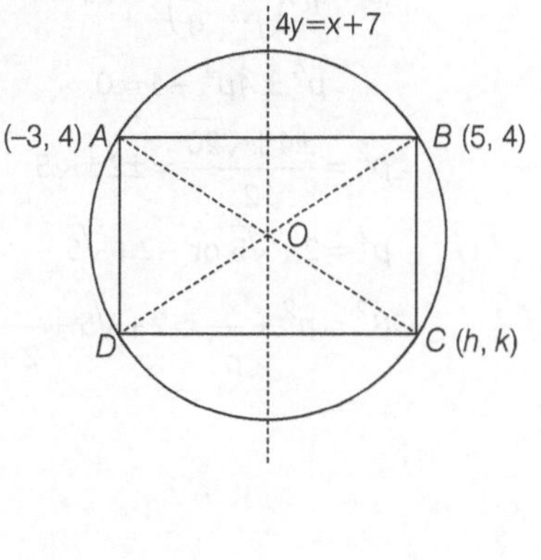

∴ This is parallel to X-axis and $AB = 8$

Now, BC is parallel to Y-axis [∵ AB perpendicular on BC]

x-coordinate will same

$\Rightarrow \quad 5 = h$

Mid-point of $AC \equiv O$

$$\left(\frac{-3+h}{2},\frac{4+k}{2}\right)=\left(1,\frac{4+k}{2}\right)$$

O lies on $4y = x + 7$

$$\Rightarrow \quad 4\left(\frac{4+k}{2}\right)=1+7$$

$$\Rightarrow \quad 4+k=4 \Rightarrow k=0$$

$$\therefore \quad C=(5,0)$$

$$\therefore \quad BC=4$$

∴ Area of rectangle $ABCD = AB \times BC$

$$= 8 \times 4 = 32 \text{ sq units}$$

39. (6.324)

$$x^2y^2=1$$

$$y=\pm\frac{1}{x}$$

$$OA \perp OB$$

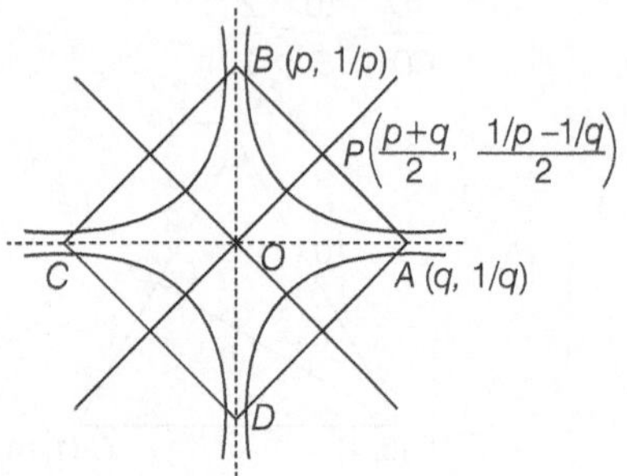

$$\Rightarrow \quad \frac{1}{p^2}\cdot\left(\frac{-1}{q^2}\right)=-1$$

$$\Rightarrow \quad p^2q^2=1 \Rightarrow pq=1$$

Mid-point of $AB = P$

$$P\left(\frac{p+q}{2},\frac{\frac{1}{p}-\frac{1}{q}}{2}\right)$$

lies on $x^2y^2=1$

$$\Rightarrow \quad (p+q)^2\left(\frac{1}{p}-\frac{1}{q}\right)^2=16$$

$$\Rightarrow \quad p^4 \pm 4p^2-1=0 \qquad [\because pq=1]$$

$$\Rightarrow \quad p^2=\frac{\pm4\pm\sqrt{20}}{2}=\pm2\pm\sqrt{5}$$

$$\Rightarrow \quad p^2=2+\sqrt{5} \text{ or } -2+\sqrt{5}$$

$$\therefore \quad OB^2=p^2+\frac{1}{p^2}=2+\sqrt{5}+\frac{1}{2+\sqrt{5}}=2\sqrt{5}$$

or $$-2+\sqrt{5}+\frac{1}{-2+\sqrt{5}}=2\sqrt{5}$$

$$\therefore \quad \text{Area}=4\times\frac{1}{2}\times OA\times OB=2OB^2$$

$$=2(2\sqrt{5})^2=2\times4\times5=40 \text{ sq units}$$

Square root $=\sqrt{40}=2\sqrt{10}=6.324$ (upto 3 decimal)

40. (1.28) Here, $PQRS$ is parallelogram.

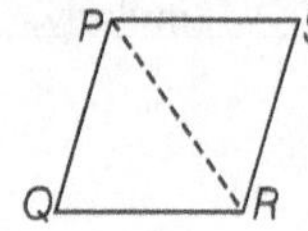

$$\therefore \quad \text{Area}=2 \text{ Area of } (\Delta PQR)$$

$$=2\times\frac{1}{2}\text{mod}\begin{vmatrix}2 & 1 & 1\\3 & 2 & 1\\4 & -1 & 1\end{vmatrix}$$

$$A=4$$

∴ Value of $0.32\,A=0.32\times4=1.28$

41. (0.9)

$$\therefore \text{Area of } \Delta ABC=\frac{1}{2}\text{ mod of}\begin{vmatrix}6 & 3 & 1\\-3 & 5 & 1\\4 & -2 & 1\end{vmatrix}=\frac{49}{2}\text{ sq units}$$

$$\therefore \text{Area of } \Delta PBC=\frac{1}{2}\text{mod}\begin{vmatrix}x & y & 1\\-3 & 5 & 1\\4 & -2 & 1\end{vmatrix}=\frac{1}{2}|7x+7y-14|$$

$$=\frac{7}{2}|x+y-2| \text{ sq units}$$

$$\therefore \text{Required ratio}=\frac{\frac{7}{2}|x+y-2|}{\frac{49}{2}}=\frac{|x+y-2|}{7}$$

$$\therefore \; a=2 \text{ and } b=7 \Rightarrow \frac{a+b}{10}=\frac{9}{10}=0.9$$

42. (0.2)

Let $P(h, k)$

P is mid-point of AQ

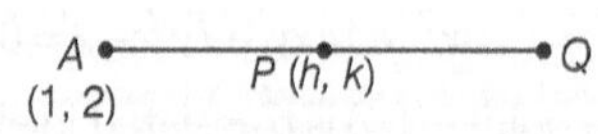

∴ Q is $(2h-1, 2k-2)$

∵ It lies on the curve

$$x^2+y^2+x+y=0$$

∴ It will satisfy the equation

$$(2h-1)^2+(2k-2)^2+(2h-1)+(2k-2)=0$$

$$\Rightarrow 4h^2-4h+1+4k^2+4-8k+2h-1+2k-2=0$$

$$\Rightarrow \quad 4h^2+4k^2-2h-6k+2=0$$

$$\Rightarrow \quad 2(h^2+k^2)-h-3k+1=0$$

$$p=2, q=-1, r=-3 \text{ and } s=1$$

$$\therefore \quad \frac{4p+q+r+s}{25}=\frac{8-1-3+1}{25}=\frac{5}{25}=\frac{1}{5}=0.2$$

Level 3

1. (d) $P(t_1^3, \sqrt{2}t_1)$ and $Q(t_2^3, -\sqrt{2}t_2)$

$\because$ t_1 and t_2 are the the roots of $x^2 - 2\sqrt{2}x + 1 = 0$

$\therefore$ $t_1 + t_2 = 2\sqrt{2}$ and $t_1 t_2 = 1$

Now, $PQ = \sqrt{(t_2^3 - t_1^3)^2 + (-\sqrt{2}t_2 - \sqrt{2}t_1)^2}$

$= \sqrt{(t_2 - t_1)^2 (t_2^2 + t_1^2 + t_1 t_2)^2 + 2(t_2 + t_1)^2}$

$\because$ $t_2 - t_1 = \sqrt{(t_1 + t_2)^2 - 4t_1 t_2} = \sqrt{8 - 4} = \sqrt{4} = 2$

$\because$ $t_2^2 + t_1^2 = (t_1 + t_2)^2 - 2t_1 t_2 = 8 - 2 = 6$

$\therefore$ $PQ = \sqrt{4(6+1)^2 + 2(2\sqrt{2})^2}$

$= \sqrt{196 + 16} = \sqrt{212} = 2\sqrt{53}$ units

2. (b) $ABCD$ is equilateral trapezium

$\Rightarrow$ $AB \parallel CD$ and $AD = BC$

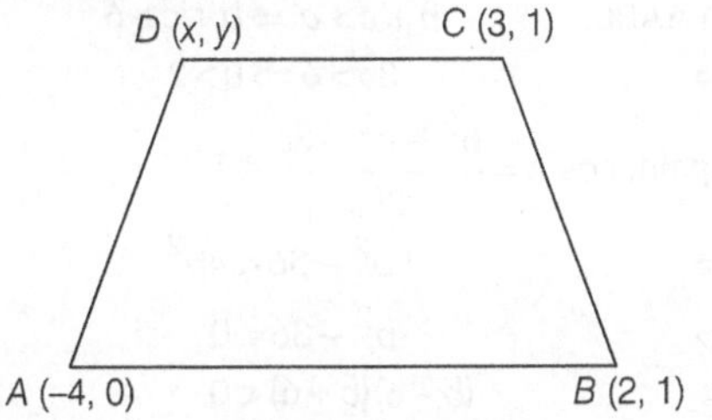

Now, $AD = BC$

$AD^2 = BC^2$

$(x+4)^2 + y^2 = 1$ ...(i)

$\because AB \parallel CD$

Slope of AB = Slope of CD

$\Rightarrow$ $\dfrac{1-0}{2+4} = \dfrac{y-1}{x-3}$ $\left[\because \text{slope } = \dfrac{y_2 - y_1}{x_2 - x_1}\right]$

$\Rightarrow$ $\dfrac{1}{6} = \dfrac{y-1}{x-3}$

$\Rightarrow$ $6y - 6 = x - 3$

$\Rightarrow$ $x - 6y = -3$

$x = 6y - 3$

From Eq. (i),

$(6y+1)^2 + y^2 = 1$

$\Rightarrow$ $36y^2 + 12y + 1 + y^2 = 1$

$\Rightarrow$ $37y^2 + 12y = 0$

$\Rightarrow$ $y(37y + 12) = 0$

$\Rightarrow$ $y = 0$ and $x = -3$ $[\because (-3, 0)]$

$\Rightarrow$ $y = \dfrac{-12}{37}$ and $x = \dfrac{-72 - 111}{37} = \dfrac{-183}{37}$

3. (d) We have,

$P = \{-\sin(\beta - \alpha), -\cos\beta\} \equiv (x_1, y_1)$

$Q = (\cos(\beta - \alpha), \sin\beta) \equiv (x_2, y_2)$

$R = \{\cos(\beta - \alpha + \theta), \sin(\beta - \theta)\}$

The x-coordinate and y-coordinate of R are

$X_R = x_2 \cos\theta + x_1 \sin\theta$

$Y_R = y_2 \cos\theta + y_1 \sin\theta$

The points

$$\left(\frac{x_2 \cos\theta + x_1 \sin\theta}{\cos\theta + \sin\theta}, \frac{y_2 \cos\theta + y_1 \sin\theta}{\cos\theta + \sin\theta}\right)$$

lies on PQ. But R doesn't lies on PQ. For it to lie on PQ. $\cos\theta$ and $\sin\theta$ must be related by $\cos\theta + \sin\theta = 1$ which is not given.

$\therefore$ P, Q, R are non-collinear.

4. (b) The integral points must be on the vertical lines $x = 1, 2, 3 \ldots 20$. The number of integral points on $x = 1$ inside the triangle are

(1, 1), (1, 2), (1, 3) ... (1, 19)

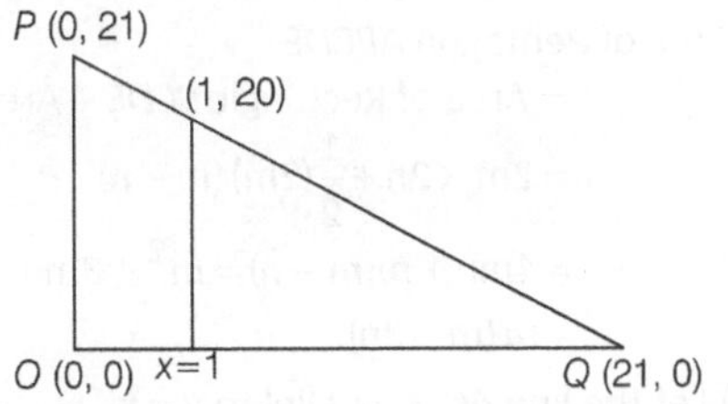

$\Rightarrow$ Total number is 19.

Similarly, the number of points on $x = 2$ is 18.

For $x = 3$ it is $= 17$

and for $x = 20$ it is 0.

$\therefore$ Total number of integral points inside the triangle is

$= 19 + 18 + 17 + \ldots + 0$

$= \dfrac{19}{2}(19 + 1)$

$= \dfrac{19}{2} \times 20$

$= 190$

5. (a) x-coordinate $= 1 - \dfrac{1}{4} + \dfrac{1}{16} - \dfrac{1}{64} + \ldots..$

$= \dfrac{a}{1-r} = \dfrac{1}{1 + \frac{1}{4}} = \dfrac{1}{\frac{5}{4}} = \dfrac{4}{5}$

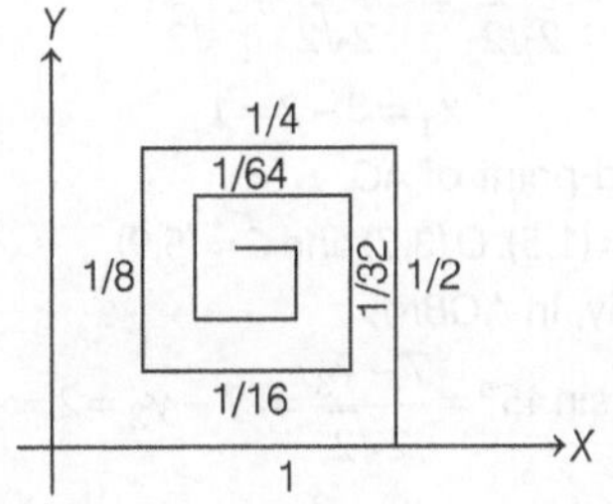

y-coordinate $= \dfrac{1}{2} - \dfrac{1}{8} + \dfrac{1}{32} - \ldots$

$= \dfrac{\frac{1}{2}}{1 + \frac{1}{4}} = \dfrac{2}{5}$

6. (b) $A = (m, n)$

B = reflection of A w.r.t. $y = x = (n, m)$

C = reflection of B in Y-axis $= (-n, m)$

D = reflection of C in X-axis $= (-n, -m)$

E = reflection of D in Y-axis $= (n, -m)$

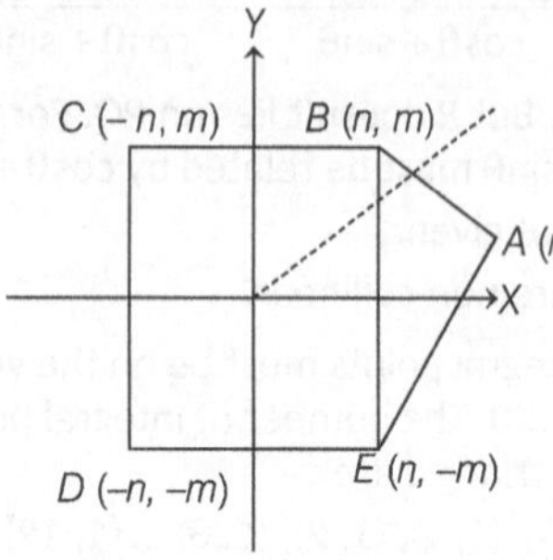

$\therefore$ Area of Pentagon $ABCDE$

$= \text{Area of Rectangle } BCDE + \text{Area of } \Delta ABC$

$= 2m \times 2n + \frac{1}{2}(2m)(m-n)$

$= 4mn + m(m-n) = m^2 + 3mn$

$= m(m+3n)$

7. (a) Let the line AC is parallel to $y = x$

$\therefore$ Side of square $= a = 4$

$\Rightarrow$ Diagonal $= a\sqrt{2} = 4\sqrt{2}$

$OA = OC = 2\sqrt{2}$

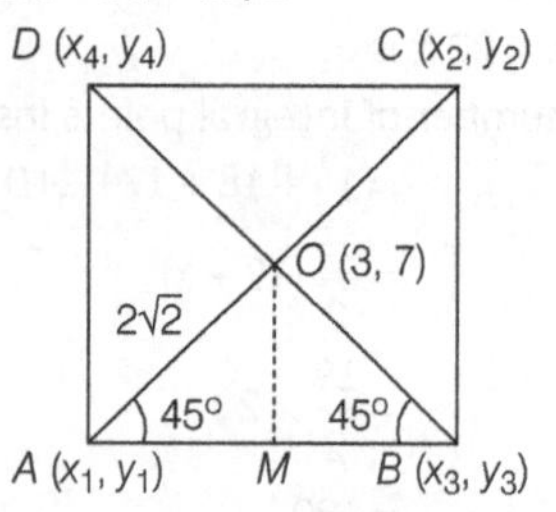

In ΔOAM, $\sin 45° = \frac{7-y_1}{2\sqrt{2}} \Rightarrow \frac{7-y_1}{2} = 1$

$\Rightarrow \quad 7 - y_1 = 2$

$\Rightarrow \quad y_1 = 5$

$\cos 45° = \frac{AM}{2\sqrt{2}} \Rightarrow \frac{3-x_1}{2\sqrt{2}} = \frac{1}{\sqrt{2}}$

$x_1 = 3 - 2 = 1$

O is mid-point of AC.

$\therefore \quad A = (1, 5), O(3, 7)$ and $C = (5, 9)$

Similarly, In ΔOBM,

$\sin 45° = \frac{7-y_3}{2\sqrt{2}} \Rightarrow 7 - y_3 = 2 \Rightarrow y_3 = 5$

$\cos 45° = \frac{x_3-3}{2\sqrt{2}} \Rightarrow x_3 - 3 = 2 \Rightarrow x_3 = 5$

$\therefore \quad B \Rightarrow (5, 5)$

O is mid-point of BD,

$B(5,5), O(3,7)$ and $D = (1,9)$

$\therefore$ Vertices are $A(1,5), B(5,5), C(5,9)$ and $D(1,9)$.

8. (b) Taking B as origin and BC as X-axis. Let A be (h, k)

Since, AD is angle bisector of $\angle A$

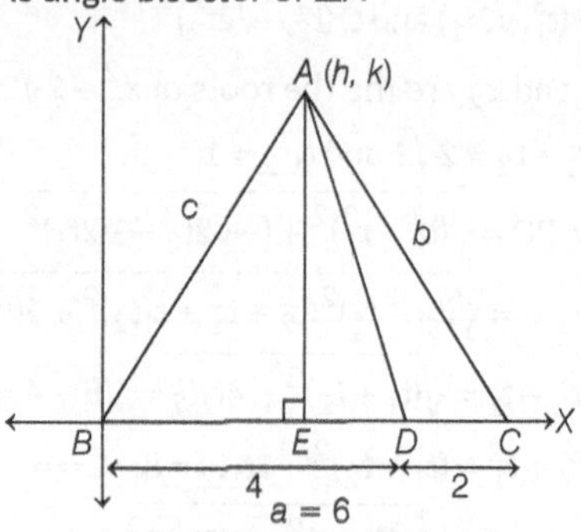

$\therefore \quad \frac{AB}{AC} = \frac{BD}{CD} \Rightarrow c = 2b$

Now, we know that in any triangle.

Sum of two sides > third side

In ΔABC, $\quad b + c > a \Rightarrow b + c > 6$

$\Rightarrow \quad 3b > 6 \Rightarrow b > 2$

Again, $\cos A = \frac{b^2 + c^2 - 36}{2bc} < 1$

$\Rightarrow \quad 5b^2 - 36 < 4b^2 \quad [\because c = 2b]$

$\Rightarrow \quad b^2 - 36 < 0$

$\Rightarrow \quad (b-6)(b+6) < 0$

$\Rightarrow \quad 0 < b < 6 \quad [\therefore b \text{ is '+' ve}]$

$\therefore \quad 2 < b < 6$

Now, in ΔABE $\quad h^2 + k^2 = c^2$

$\Rightarrow \quad h^2 + k^2 = 4b^2 \quad$ ...(i)

In ΔAEC, $\quad (h-6)^2 + k^2 = b^2 \quad$ ...(ii)

By Eqs. (i) and (ii), $3b^2 = h^2 - (h-6)^2$

$\Rightarrow \quad k^2 = 16 - (h-8)^2$

$\therefore$ Maximum value of $k = 4$

9. (b) AB = length of rod $= k$

$AB = \sqrt{a^2 + b^2} = k$

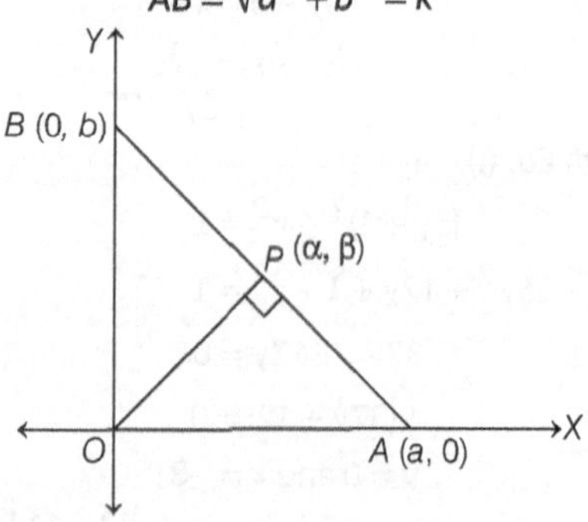

In ΔOPA, $\quad OP^2 + AP^2 = OA^2$

$\Rightarrow \alpha^2 + \beta^2 + \alpha^2 - 2a\alpha + a^2 + \beta^2 = a^2$

$\Rightarrow \quad 2\alpha^2 + 2\beta^2 - 2a\alpha = 0$

$\Rightarrow \quad a = \frac{\alpha^2 + \beta^2}{\alpha}$

In ΔOPB,

$$OP^2 + PB^2 = OB^2$$

$$\Rightarrow \alpha^2 + \beta^2 + \alpha^2 + \beta^2 - 2b\beta + b^2 = b^2$$

$$\Rightarrow \quad 2\alpha^2 + 2\beta^2 - 2b\beta = 0$$

$$\Rightarrow \quad b = \frac{\alpha^2 + \beta^2}{\beta}$$

Now, $\quad a^2 + b^2 = k^2$

$$\Rightarrow \quad \frac{(\alpha^2+\beta^2)^2}{\alpha^2} + \frac{(\alpha^2+\beta^2)^2}{\beta^2} = k^2$$

$$\Rightarrow \quad (\alpha^2+\beta^2)^3 = k^2\alpha^2\beta^2$$

$\therefore$ Locus is $(x^2+y^2)^3 = k^2x^2y^2$

10. (c) Let the sides are $x, y, 10-(x+y)$ and $x, y \in N$

Now, $\quad x + y > 10 - (x+y)$

$\Rightarrow \quad x + y > 5 \quad$ ...(i)

$\Rightarrow \quad x + 10 - (x+y) > y$

$\Rightarrow \quad 10 - y > y$

$\Rightarrow \quad y < 5$

$\Rightarrow \quad 0 < y < 5$

and $\quad y + 10 - (x+y) > x$

$\Rightarrow \quad x < 5$

$\Rightarrow \quad 0 < x < 5$

Put $\quad x = 1 \Rightarrow x + y > 5$

From Eq. (i),

$\Rightarrow \quad y > 4$

$\therefore \quad y = 5, 6 \dots$

But $\quad y < 5$ (not possible)

Put $x = 2 \Rightarrow y > 3 \Rightarrow y = 4 \Rightarrow z = 10 - (x+y) = 4$

Put $x = 3 \Rightarrow y > 2 \Rightarrow y = 3, 4 \Rightarrow z = 4, 3$

Put $x = 4 \Rightarrow y > 1 \Rightarrow y = 2, 3, 4 \Rightarrow z = 4, 3, 2$

Only 2 different types of triangle having sides (2,4,4) and (3,3,4).

$\therefore$ Number of triangle is $= 2$

11. (a) $x^2 + 8x - 20 = 0 \begin{smallmatrix} x_1 \\ y_1 \end{smallmatrix}, \quad 4x^2 + 32x - 57 = 0 \begin{smallmatrix} x_2 \\ y_2 \end{smallmatrix}$

$$\Rightarrow \quad x_1 + y_1 = -8 \Rightarrow x_2 + y_2 = -\frac{32}{4} = -8$$

$$\Rightarrow \quad y_1 = -8 - x_1 \Rightarrow y_2 = -8 - x_2$$

$$\Rightarrow \quad 9x^2 + 72x - 112 = 0 \begin{smallmatrix} x_3 \\ y_3 \end{smallmatrix}$$

$$\Rightarrow \quad x_3 + y_3 = \frac{-72}{9} = -8$$

$$\Rightarrow \quad y_3 = -8 - x_3$$

$$\therefore \quad \Delta = \begin{vmatrix} x_1 & y_1 & 1 \\ x_2 & y_2 & 1 \\ x_3 & y_3 & 1 \end{vmatrix} = \begin{vmatrix} x_1 & -8-x_1 & 1 \\ x_2 & -8-x_2 & 1 \\ x_3 & -8-x_3 & 1 \end{vmatrix}$$

Applying $C_1 \to C_1 + C_2$, we get

$$\begin{vmatrix} -8 & -8-x_1 & 1 \\ -8 & -8-x_2 & 1 \\ -8 & -8-x_3 & 1 \end{vmatrix} = 0 \quad [C_1 \text{ and } C_3 \text{ are same}]$$

$\therefore$ Points are collinear.

12. (b,c) $\because a$ is root of $x^2 + x - 42 = 0$

$\Rightarrow \quad a^2 + a - 42 = 0$

$\Rightarrow \quad (a+7)(a-6) = 0$

$\quad a = -7, 6$

$\because b$ is a root of $x^2 + ax + a^2 - 37 = 0$

For $a = -7$, $x^2 - 7x + 12 = 0$

$\Rightarrow (x-4)(x-3) = 0 \Rightarrow x = 3, 4 \Rightarrow b = 3, 4$

$a = 6 \Rightarrow x^2 + 6x - 1 = 0$

$$\Rightarrow \quad x = \frac{-6 \pm \sqrt{36+4}}{2} = \frac{-6 \pm \sqrt{40}}{2} \quad \text{(not integral)}$$

$a = -7$ and $b = 3$, then $(-7, 3)$

$a = -7$ and $b = 4$, then $(-7, 4)$

13. (a,b,d)

$\because$ The given points lies on the line $lx + my + n = 0$

So, a, b, c are the roots of the equation.

$$l\left(\frac{t^3}{t-1}\right) + m\left(\frac{t^2-3}{t-1}\right) + n = 0$$

$$\Rightarrow \quad lt^3 + mt^2 + nt - (3m+n) = 0$$

$$\therefore \quad a + b + c = \frac{-m}{l} \quad \text{...(i)}$$

$$ab + bc + ca = \frac{n}{l} \quad \text{...(ii)}$$

$$abc = \frac{3m+n}{l} \quad \text{...(iii)}$$

$$\Rightarrow \quad abc - \left(\frac{3m+n}{l}\right) = 0 \quad \text{[from Eq. (iii)]}$$

By Eqs. (i), (ii) and (iii),

$abc - (bc + ca + ab) + 3(a+b+c) = 0$

$$\Rightarrow \quad \frac{3m+n}{l} - \frac{n}{l} + 3\left(\frac{-m}{l}\right) = 0$$

$$\Rightarrow \quad \frac{3m+n-n-3m}{l} = 0$$

14. (a,b) $\because \angle COA = \angle COX - \angle AOX$

$= 60° - 30° = 30°$

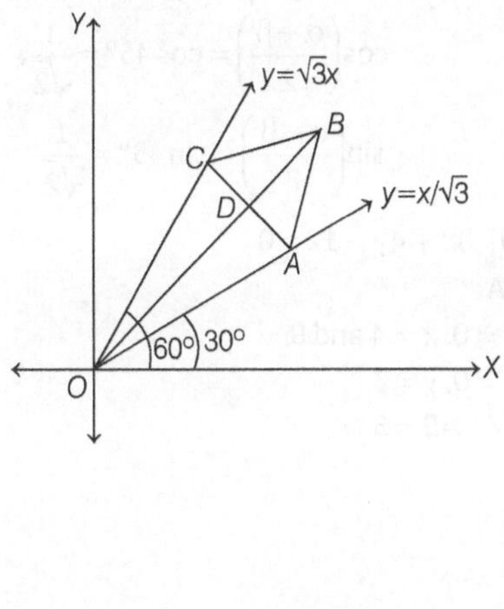

$\therefore$ Area of rhombus $= 2\times\frac{1}{2}\times OA\times OC\sin 30^\circ$

$$2=\frac{1}{2}x^2$$

$$OA=OC=2$$

Also, $\angle OAB = 180^\circ - 30^\circ = 150^\circ$

$$\cos 150^\circ = \frac{OA^2+AB^2-OB^2}{2\times OA\times AB}$$

$\Rightarrow$ $OB^2 = 8+4\sqrt{3}$

$\Rightarrow$ $OB=\sqrt{2}(\sqrt{3}+1)$

$[\because 8+4\sqrt{3}=(\sqrt{6}+\sqrt{2})^2]$

Coordinates of B are

$\{\pm\sqrt{2}(\sqrt{3}+1)\cos 45^\circ, \pm\sqrt{2}(\sqrt{3}+1)\sin 45^\circ\}$

$\Rightarrow$ $\{\pm(\sqrt{3}+1), \pm(\sqrt{3}+1)\}$

$\Rightarrow$ $(\sqrt{3}+1, \sqrt{3}+1)$

and $(-\sqrt{3}-1, -\sqrt{3}-1)$

15. (a,c) It is not possible because $OA=OB=1$

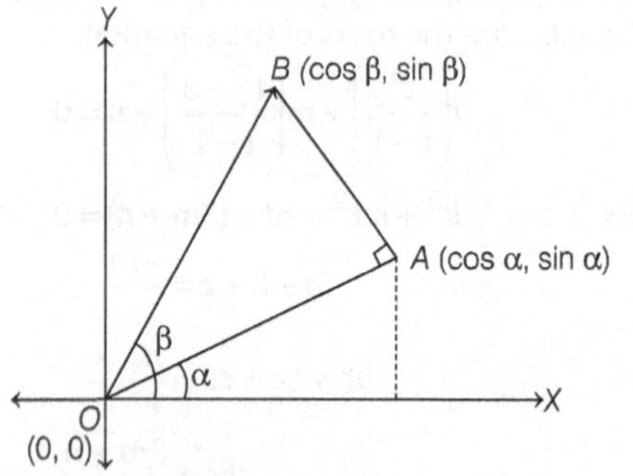

$\therefore$ $OA=OB$ and ΔOAB is right angle triangle

i.e. $\angle O = 90^\circ$

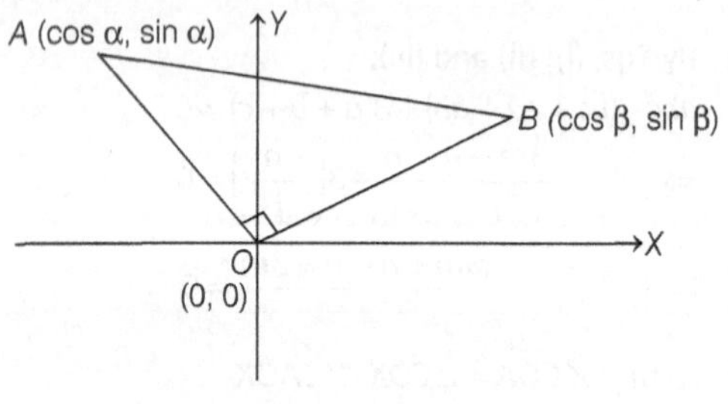

$\therefore$ $\alpha-\beta=90^\circ$

$$\cos\left(\frac{\alpha-\beta}{2}\right)=\cos 45^\circ=\frac{1}{\sqrt{2}},$$

and $$\sin\left(\frac{\alpha-\beta}{2}\right)=\sin 45^\circ=\frac{1}{\sqrt{2}}$$

16. (a, d) $3x+4y-12=0$

For A

$\Rightarrow y=0, x=4$ and B

$\Rightarrow x=0, y=3$

$AB=5$

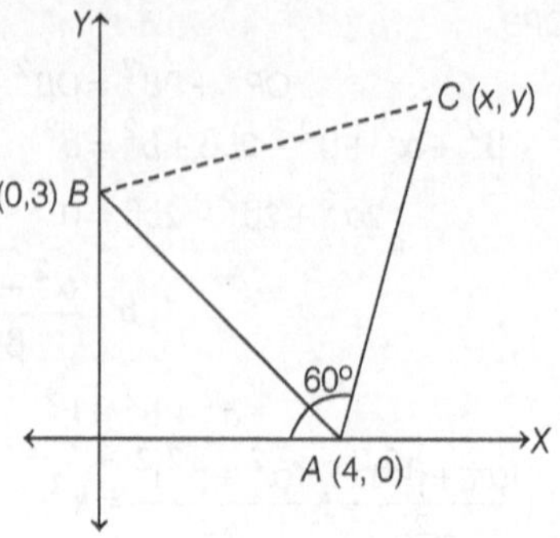

$\therefore$ $AB=BC=AC$

$AC=5=\sqrt{(x-4)^2+y^2}$

$\Rightarrow$ $(x-4)^2+y^2=25$ ...(i)

$\therefore$ $BC\Rightarrow\sqrt{x^2+(y-3)^2}=5$

$\Rightarrow$ $x^2+(y-3)^2=25$ ...(ii)

On subtracting Eq. (ii) from Eq. (i), we get

$$x^2+16-8x-x^2+y^2-y^2+6y-9=0$$

$\Rightarrow$ $-8x+6y=-7$

$\Rightarrow$ $8x-6y=7$

$\Rightarrow$ $6y=8x-7$

$\Rightarrow$ $y=\frac{8x}{6}-\frac{7}{6}=\frac{4x}{3}-\frac{7}{6}$

From Eq. (i), $x^2+16-8x+\frac{16x^2}{9}+\frac{49}{36}-\frac{56x}{18}=25$

$\Rightarrow$ $\frac{25x^2}{9}-\frac{72x+23x}{9}+\frac{49}{36}-9=0$

$\Rightarrow$ $\frac{25x^2}{9}-\frac{95}{9}x+\frac{49-324}{36}=0$

$\Rightarrow$ $100x^2-380x-275=0$

$\Rightarrow$ $x=2\left(1-\frac{3\sqrt{3}}{4}\right), 2\left(1+\frac{3\sqrt{3}}{4}\right)$

Then, $y=\frac{3}{2}\left(1-\frac{4}{\sqrt{3}}\right), \frac{3}{2}\left(1+\frac{4}{\sqrt{3}}\right)$

Sol. (Q.Nos. 17-19)

17. (b) $G_1=\left(\frac{x_1+x_2}{2}, \frac{y_1+y_2}{2}\right)$

G_2 divides G_1A_3 in 1:2

$$G_2=\left(\frac{1}{3}\left(\frac{2(x_1+x_2)}{2}+x_3\right), \frac{1}{3}\left(\frac{2(y_1+y_2)}{2}+y_3\right)\right)$$

$$=\left(\frac{x_1+x_2+x_3}{3}, \frac{y_1+y_2+y_3}{3}\right)$$

Similarly, $G_3=\left(\frac{x_1+x_2+x_3+x_4}{4}, \frac{y_1+y_2+y_3+y_4}{4}\right)$

$\vdots \qquad \vdots$

$\therefore$ $$G=\left(\frac{x_1+x_2+x_3+...x_n}{n}, \frac{y_1+y_2+y_3+...y_n}{n}\right)$$

18. (a) Now, if $x_1 = a$,

$$x_2 = a+2$$
$$x_3 = a+4$$
$$\vdots$$
and $$x_n = a+(n-1)2$$
$$y_1 = b$$
$$y_2 = b+4$$
$$y_3 = b+8$$
$$\vdots$$
$$y_n = b+(n-1)4$$

Coordinates of G

$$= \left[\frac{1}{n}\cdot\frac{n}{2}(2a+(n-1)2), \frac{1}{n}\times\frac{n}{2}(2b+(n-1)4)\right]$$
$$= (a+n-1, b+2(n-1))$$

19. (b) $P_1\left(\frac{x_1+x_2+x_3}{3}, \frac{y_1+y_2+y_3}{3}\right)$

and $P_2\left(\frac{x_2+x_3+x_4}{3}, \frac{y_2+y_3+y_4}{3}\right)$

$$\vdots$$

$P_{n-2}\left(\frac{x_{n-2}+x_{n-1}+x_n}{3}, \frac{y_{n-2}+y_{n-1}+y_n}{3}\right)$

$\therefore$ Coordinates of the centre of mean position of $P_1, P_2 .... P_{n-2}$ is $(\bar{x}, \bar{y})$.

$$n(\bar{x}, \bar{y}) = \left(\frac{x_1+2x_2+3x_3+....3x_{n-2}+2x_{n-1}+x_n}{n-2}, \frac{y_1+2y_2+3y_3+....3y_{n-2}+2y_{n-1}+y_n}{n-2}\right)$$

Sol. (Q.Nos. 20-22)

20. (b) $\cos\angle ADC = \frac{AD^2+CD^2-AC^2}{2AD\cdot CD} < 0$

$\Rightarrow (P-1)^2+4^2+(5-1)^2+(6-4)^2 < (P-5)^2+6^2$

$\Rightarrow 8P < 24 \Rightarrow P < 3$

and maximum value of P is 2.

21. (a) $AB^2+AD^2 = BD^2$

$$(P-4)^2+(P-1)^2+4^2 = 3^2+4^2$$
$\Rightarrow$ $$2P^2-10P+8=0$$
$\Rightarrow$ $$P = 1, 4$$

22. (c) $AB^2 = AD^2$

$\Rightarrow$ $$(P-4)^2 = (P-1)^2+4^2$$
$$P = -\frac{1}{6}$$

$\therefore$ Area of quadrilateral $ABCD$

= Area of ΔABD + Area of ΔBDC

$$= \frac{1}{2}\text{mod}\begin{vmatrix} -1/6 & 0 & 1 \\ 4 & 0 & 1 \\ 1 & 4 & 1 \end{vmatrix} + \frac{1}{2}\text{mod}\begin{vmatrix} 4 & 0 & 1 \\ 1 & 4 & 1 \\ 5 & 6 & 1 \end{vmatrix}$$

$$= \frac{25}{3}+11 = \frac{58}{3} \text{ sq units}$$

23. (i) $\to$ B, (ii) $\to$ C, (iii) $\to$ D, (iv) $\to$ A

(i) Let M be the point of intersection of the diagonals in bisects each of the diagonal.

$\therefore$ The coordinates of the middle point M of BD

$$= \left(\frac{5+1}{2}, \frac{3+1}{2}\right)$$
$$= (3, 2)$$

(ii) If $x = 4$

i.e. $B = (4, 4)$

$\therefore$ Mid-point of $AB = \left(\frac{4+1}{2}, \frac{4+3}{2}\right) = \left(\frac{5}{2}, \frac{7}{2}\right)$

(iii) $\therefore$ Mid-point of $BC = \left(\frac{5+4}{2}, \frac{1+4}{2}\right) = \left(\frac{9}{2}, \frac{5}{2}\right)$

(iv) M satisfy $y = 2x+c$

$\Rightarrow$ $$2 = 6+c$$
$$c = -4$$

Let the coordinates of D which satisfies

$$y = 2x-4 \text{ i.e. } (x, 2x-4)$$
$\Rightarrow$ $$AD^2+DC^2 = AC^2$$
$\Rightarrow (x-1)^2+(2x-4-3)^2+(x-5)^2+(2x-4-1)^2$
$$= (5-1)^2+(1-3)^2$$
$\Rightarrow$ $$x^2-6x+8=0$$
$\Rightarrow$ $$x = 2 \text{ or } 4$$
$\therefore$ $$D = (2, 0)$$

24. (i) $\to$ A, C, E, (ii) $\to$ A, C, E, (iii) $\to$ B, D (iv) $\to$ B, D

$\therefore$ Coordinates of $S = \left(\frac{16+9}{5}, \frac{18+21}{5}\right) = \left(5, \frac{39}{5}\right)$

$\therefore$ Coordinates of $T = \left(\frac{33+16}{5}, \frac{69+18}{5}\right) = \left(\frac{49}{5}, \frac{87}{5}\right)$

$\therefore$ $$ST^2 = \left(5-\frac{49}{5}\right)^2 + \left(\frac{39}{5}-\frac{87}{5}\right)^2 = \left(\frac{576}{5}\right)$$
$\Rightarrow$ $$5ST^2 = 576$$
$\therefore$ $$PQ^2 = 29, QR^2 = 205 \text{ and } PR^2 = 320$$
$$PS^2 = \left(\frac{2}{5}PQ\right)^2 = \frac{116}{25}$$
$$QS^2 = \left(\frac{3}{5}PQ\right)^2 = \frac{261}{25}$$
and $$QT^2 = \left(\frac{3}{2}QR\right)^2 = \frac{369}{5}$$

So that $PR^2+QR^2+51 = 320+205+51 = 576$

$$100PS^2 = 4\times116 = 464$$
and $$PQ^2+2QR^2+25 = 29+410+25 = 464$$
$$2PR^2-64 = 2\times320-64 = 576$$
$$5QT^2+QR^2+2 = 369+205+2 = 576$$
$$50QS^2-2PQ^2 = 522-58 = 464$$

Sol. (Q.Nos. 25-26)

25. (0.625) $\therefore$ Area of $CDFE$

$$A(x) = 1 - \frac{1}{2}x^2 - \frac{1}{2}(1-x) = \frac{1+x-x^2}{2}$$

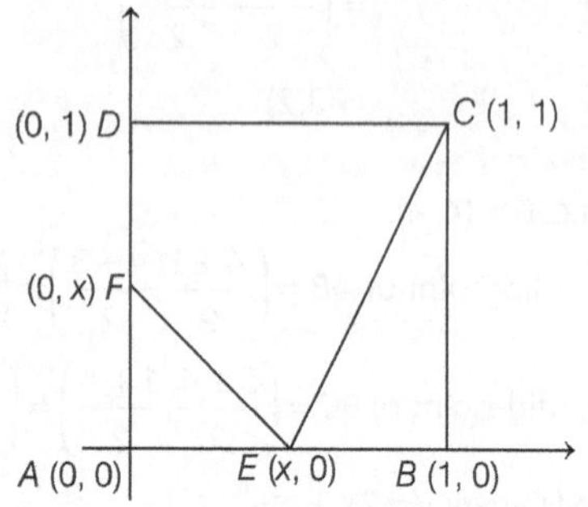

$$A(x) = \frac{1}{2}\left[-\left(x-\frac{1}{2}\right)^2 + 1 + \frac{1}{4}\right]$$

$$= \frac{5}{8} - \frac{1}{2}\left(x-\frac{1}{2}\right)^2$$

$$A_{max} = \frac{5}{8} = 0.625 \text{ at } x = \frac{1}{2}$$

26. (0) $\therefore PA^2 - (PB)^2 + (PC)^2 - (PD)^2$

$$\Rightarrow \alpha^2 + \gamma^2 - (\alpha^2 + \delta^2) + (\beta^2 + \delta^2) - (\gamma^2 + \beta^2) = 0$$

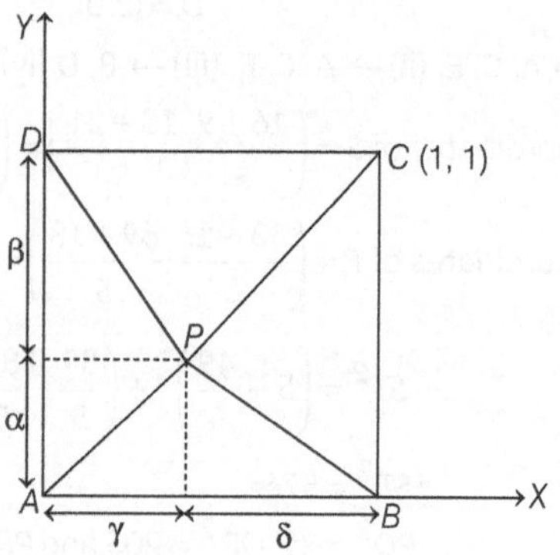

Sol. (Q.Nos. 27-28)

27. (3) $\therefore$ Area of $\Delta DEF = \frac{1}{4}$ Area of ΔABC

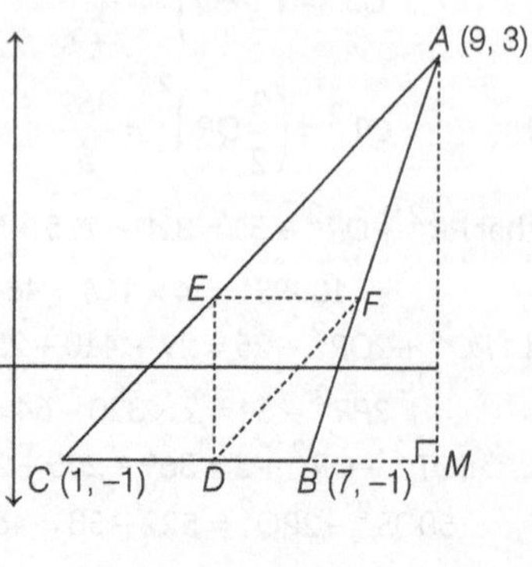

$$= \frac{1}{4} \times \left(\frac{1}{2} \times \text{base} \times \text{height}\right)$$

$$= \frac{1}{4} \times \frac{1}{2} \times BC \times AM$$

$$= \frac{1}{4} \times \frac{1}{2} \times 6 \times 4 = 3 \text{ sq units}$$

28. (12) D is mid-point BC, E is mid point AC for circumcentre

$$OA = OB = OC$$

$$OA = OB$$

$$(a-9)^2 + (b-3)^2 = (a-7)^2 + (b+1)^2$$

$$a + 2b = 10$$

$$OB = OC$$

$$(a-1)^2 + (b+1)^2 = (a-7)^2 + (b+1)^2$$

$$a = 4$$

$\therefore b = 3$

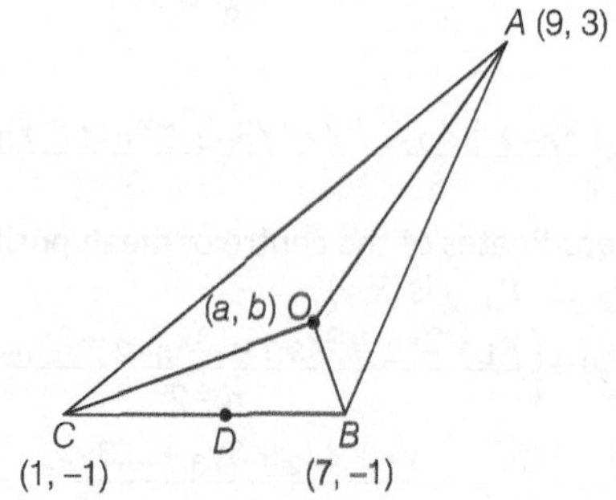

Circumcentre $= (a, b) = (4,3)$

$\Rightarrow$ $$R = OA = \sqrt{(4-1)^2 + (3+1)^2} = 5$$

$$a + b + R = 4 + 3 + 5 = 12$$

29. (0) $PQ = \sqrt{(6-3)^2 + (5-1)^2} = \sqrt{3^2 + 4^2} = 5$

Area of $\Delta PQR = 7$ sq units

$\Rightarrow$ $$\frac{1}{2} \times PQ \times h = 7$$

$$h = \frac{2 \times 7}{5} = \frac{14}{5}$$

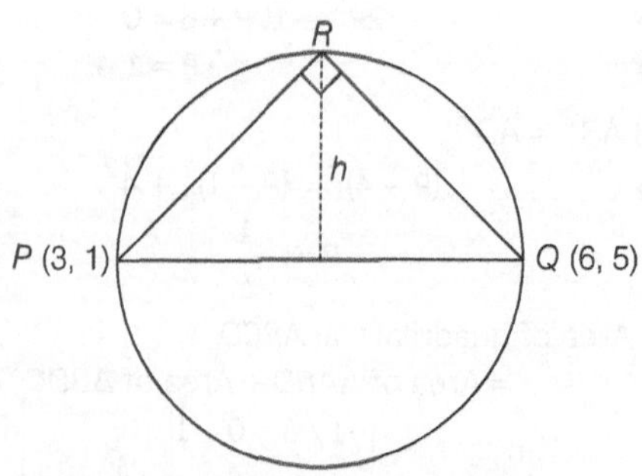

But h_{max} = Radius of circle with diameter PQ

$$= \frac{5}{2} < \frac{14}{5}$$

$\Rightarrow$ No such triangle exists.

30. (0) We know that

H, G and S are collinear

$$\begin{vmatrix} 4 & b & 1 \\ b & 2b-8 & 1 \\ -4 & 8 & 1 \end{vmatrix} = 0$$

Applying $R_2 \to R_2 - R_1$ and $R_3 \to R_3 - R_2$, we get

$$\begin{vmatrix} 4 & b & 1 \\ b-4 & b-8 & 0 \\ -(b+4) & 16-2b & 0 \end{vmatrix} = 0$$

$\therefore \quad (b-4)(16-2b) + (b-8)(b+4) = 0$

$\Rightarrow \quad (b-8)(b-12) = 0$

$\therefore \quad b = 8$ or 12

But $\quad \dfrac{-8+4}{3} = b$

$$b = \frac{-4}{3}$$

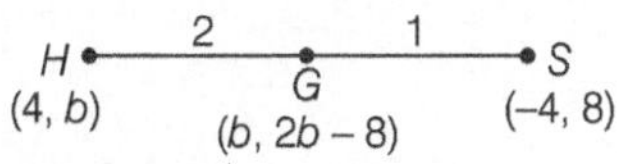

and $\quad \dfrac{16+b}{3} = 2b - 8$

$$b = 8$$

But no common value of b is possible.

31. (17) $\angle BCA = 90°$

Points A, O, B, C are concyclic.

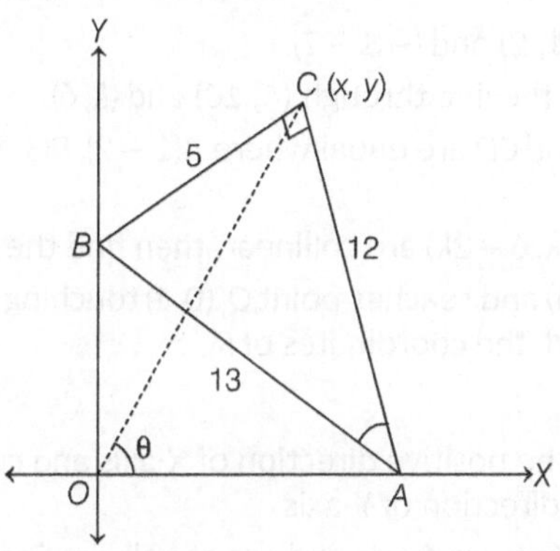

Let $\angle AOC = \theta$

$\therefore \quad \angle BOC = \angle BAC$

$\Rightarrow \quad \tan\left(\dfrac{\pi}{2} - \theta\right) = \dfrac{5}{12} \Rightarrow \dfrac{x}{y} = \dfrac{5}{12}$

$$12x - 5y = 0$$

vertex $ax - by = 0$

$a = 12$ and $b = 5$

$a + b = 12 + 5 = 17$

32. (14) a, b, c are in AP

$$2b = a + c$$

$$a + c - 2b = 0$$

$\Rightarrow \quad ax + by + c = 0$

Passes through a fixed point $(1, -2)$

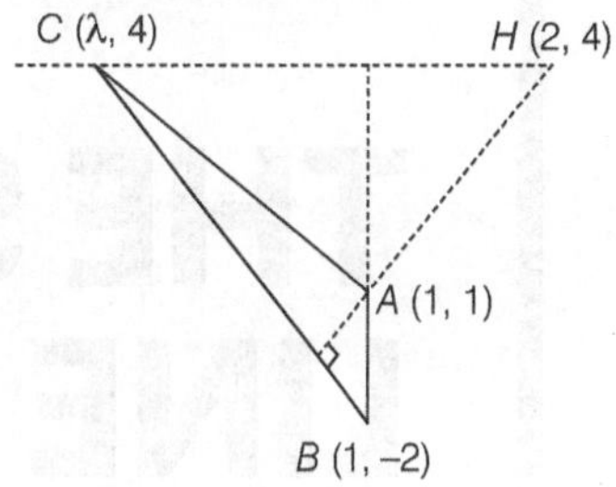

Slope of BC $\dfrac{-(2-1)}{4-1} = \dfrac{4+2}{\lambda - 1}$

$$\lambda = -17$$

$\therefore \quad C \equiv (-17, 4) = (h, k)$

$2h + 12k = -34 + 48 = 14$

33. (8) Vertices of square are

$$\left(3 \pm 2\sqrt{2}\left(\frac{1}{\sqrt{2}}\right), 7 \pm 2\sqrt{2}\left(\frac{1}{\sqrt{2}}\right)\right),$$

$$\left(3 \pm 2\sqrt{2}\left(\frac{-1}{\sqrt{2}}\right), 7 \pm 2\sqrt{2}\left(\frac{1}{\sqrt{2}}\right)\right)$$

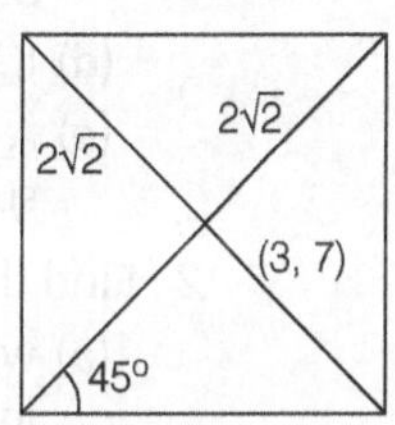

i.e. $(5,9), (1,5), (1,9), (5,5)$

$\therefore \max(y_1, y_2, y_3, y_4) - \min(x_1, x_2, x_3, x_4) = 9 - 1 = 8$

CHAPTER

02

THE STRAIGHT LINE

LEVEL 01 \\ Starter Level

2.1 Slope and Equation of a Line

1 (a) Find the slope of joining of points (1, 2) and (– 3, – 7).

(b) Find value of k, so that 4 is slope of the line through (4, 10) and $(k, 6)$.

(c) Find the value of λ, if slope of AB and CD are equal where $A(1, -2)$, $B(3, \lambda)$, $C(\lambda, 3)$ and $D(4, 7)$.

(d) If $(k, 2-2k)$, $(-k+1, 2k)$ and $(-4-k, 6-2k)$ are collinear, then find the value of k.

(e) A man starts from the point $P(-3, 4)$ and reaches point $Q'(0, 1)$ touching X-axis at R such that $PR + RQ'$ is minimum. Find the coordinates of R.

2 Find the equation of a straight line,

(a) which make an angle of 135° with the positive direction of X-axis and cuts an intercept of 5 units on the positive direction of Y-axis.

(b) which cuts off an intercept of 4 units from Y-axis and are equally inclined with the axes.

(c) which passes through (3, – 2) and (– 5, 4).

(d) which is median of ΔABC from A, where $A(10, 4)$, $B(-9, 3)$ and $C(5, 7)$.

(e) which has y-intercept 4 and perpendicular to line joining (2, – 3) and (4, 2).

(f) which is right bisector of the line joining the points (1, 2) and (5, 7).

(g) which cut off intercept on the axes whose sum and product are 1 and – 6, respectively.

(h) which passes through, the point (3, 4) and the sum of its intercepts on the axes is 14.

(i) if PS is a median of the triangle with vertices $P(2, 2)$, $Q(6, -1)$ and $R(7, 3)$, find the equation of the line passing through (1, – 1) and parallel to PS.

3 (a) Find the equation of line parallel to $3x - 2y + 5 = 0$ and passes through $(5, -6)$.

(b) Find the equation of a line passing through the point $(2, 1)$ and perpendicular to the line $2x - 4y + 7 = 0$.

(c) Find the equation of altitude through A of a ΔABC, where $A(10, 4)$, $B(-4, 9)$ and $C(-2, -1)$.

(d) Find the equation of the internal bisector of $\angle BAC$ of the ΔABC whose vertices A, B, C are $(5, 2)$, $(2, 3)$ and $(6, 5)$, respectively.

4 Find the slope, x-intercept and y-intercept of following lines

(a) $3x - 5y + 7 = 0$ (b) $4\sqrt{3}x - 7\sqrt{3}y + 8 = 0$

(c) $15x + 3y - 4 = 0$ (d) $9x + 8y + 11 = 0$

5 Reduce $2x - 3y + 4 = 0$ in the slope intercept form, intercept form, normal form.

2.2 Point and Line

6 Is the point $(4, -9)$ lies on the origin side of the line $3x + 4y + 5 = 0$?

7 Are the points $(3, 2)$ and $(4, -6)$ on the same or opposite sides of the line $4x - 5y = 9$?

8 Find the distance between the point and line given below

(a) Point $(1, -2)$ and line $y = 2x + 1$

(b) Point $(3, 0)$ and line $5x - 2y + 3 = 0$

(c) Point $(3, 4)$ and line $3x + 4y - 2 = 0$

9 The line $2x + y = 4$ meet X-axis at A and Y-axis at B. The perpendicular bisector of AB meets horizontal line through $(0, -1)$ at C. Let G be the centroid of ΔABC, find the perpendicular distance from G to AB.

10 Find the foot of perpendicular and image of given point w.r.t. given line

(a) point $(1, 2)$ and line $3x + 4y - 1 = 0$

(b) point $(-2, 3)$ and line $x + 3y = 4$

2.3 Family of Lines

11 (a) Find the distance between the parallel lines $l_1 : 4x + 3y + 12 = 0$ and $l_2 : -8x - 6y = 10$.

(b) Find the area of parallelogram formed by lines $x + 3y \pm a = 0, 3x - 2y \pm 3a = 0$.

(c) Find the distance of the point $(2, 5)$ from the line $3x + y + 4 = 0$ measured parallel to a line having slope $3/4$.

(d) If the angle between AB and CD is $90°$, where $A(-k, 1)$, $B(7, 2)$, $C(k - 3, 4)$ and $D(4, 0)$, then find k.

(e) Find the angle between the lines joining the points $(1, 2)$, $(3, 5)$, $(5, 7)$ and $(4, 1)$.

(f) The angle between two lines is $\pi/4$ and the slope of one of them is 3, then find the slope of other line.

(g) Find the equation of a line through $(1, 4)$ which makes an angle of $45°$ with the line $2x + y - 3 = 0$.

12 Find the equation of the straight line which passes through the point of intersection of lines $3x - 4y - 7 = 0$ and $12x - 5y - 13 = 0$ and is perpendicular to the line $2x - 3y + 5 = 0$.

13 The equation of two sides of a triangle are $3x - 2y + 6 = 0$ and $4x + 5y = 20$ and the orthocentre is $(1, 1)$. Find the equation of third side.

14 Two vertices of a triangle are $(3, -1)$ and $(-2, 3)$ and its orthocentre is origin. Find the coordinates of its third vertex.

15 The equation of perpendicular bisectors of sides AB and AC of a ΔABC are $x - y + 5 = 0$ and $x + 2y = 0$, respectively. If the coordinates of vertex A is $(1, -2)$. Find the equation of BC.

16. A straight line through the point (2, 2) intersects the lines $\sqrt{3}x + y = 0$ and $\sqrt{3}x - y = 0$ at the points A and B. Find the equation of the line AB such that ΔOAB is equilateral triangle, where O is origin.

17 (a) Find the value of m so that the lines $y = x + 1$, $2x + y = 16$ and $y = mx - 4$ may be concurrent.

(b) If $4a^2 + 9b^2 - c^2 + 12ab = 0$, the family of the straight lines $ax + by + c = 0$ is either concurrent at (m, n) or (p, q). Find the value of $m + n + p + q + 10$.

(c) Find the equation of a line passing through the point of intersection of $2x + 3y + 1 = 0$, $3x - 5y - 5 = 0$ and equally inclined to the axes.

(d) Two equal sides of an isosceles triangle have the equations $7x - y + 3 = 0$ and $x + y = 3$ and its third side passes through the point $(1, -10)$. Find the equation of the third side.

(e) Find the circumcentre and orthocentre of triangle formed by the lines $x - y + 2 = 0$, $y - 3 = 0$ and $x - 3y + 5 = 0$.

18 (a) Find the equation of the bisectors bisecting the angle containing the origin of the straight lines $4x + 3y = 6$ and $5x + 12y + 9 = 0$.

(b) Find the coordinates of the incentre of the triangle whose sides are $x + 1 = 0$, $3x - 4y = 5$ and $5x + 12y = 27$.

(c) The equations of the perpendicular bisectors of the sides AB and AC of ΔABC are $x - y + 5 = 0$ and $x + 2y = 0$, respectively. If the point A is $(1, -2)$, find the equation of line BC.

(d) Find the acute angle and obtuse angle bisector between the lines $x + y - 3 = 0$ and $7x - y + 5 = 0$.

(e) Straight lines $3x + 4y = 5$ and $4x - 3y = 15$ intersects at the point A. Points B and C are chosen on these two lines such that $AB = AC$. Determine the possible equations of the line BC passing through the point (1, 2).

(f) Find the equations of straight lines passing through $(-2, -7)$ and having an intercept of length 3 units between the straight line $4x + 3y = 12$ and $4x + 3y = 3$.

(g) In a ΔABC, if $A(2, -1)$ and $7x - 10y + 1 = 0$ and $3x - 2y + 5 = 0$ are equations of altitude and angle bisector, respectively drawn from B, then find the equation of side BC.

(h) A straight line L through the point $(3, -2)$ is inclined at an angle of $\pi/3$ to the line $\sqrt{3}x + y = 1$. If line L also intersect the X-axis, then find the equation of line L.

19 (a) A ray of light is sent along the line $x - 2y = 3$ upon reaching the line $3x - 2y = 5$, the ray is reflected from it. Find the equation of the line containing the reflected ray.

(b) A ray of light passing through the point (1, 2) is reflected on the X-axis at a point P and passes through the point (5, 3). Find the abscissa of the point P.

(c) A ray of light is sent along the line $x - 6y = 8$. After refracting across the line $x + y = 1$, it enters the opposite side after turning by 15° away from the line $x + y = 1$. Find the equation of line along with the reflected ray travels.

LEVEL 02 \ JEE Main Level

Single Choice Correct Questions

2.1 Slope and Equation of Line

1 $A(2, -3)$ and $B(-2, 1)$ are vertices of ΔABC. If its centroid lies on the line $2x + 3y = 1$, then locus of the vertex C, is

(a) $3x + 2y = 5$ (b) $2x - 3y = 7$

(c) $2x + 3y = 9$ (d) $3x - 2y = 3$

2 The equation of the line passing through $(-a, 0)$ which makes with the axes a triangle of area T is

(a) $2Tx + a^2y + 2aT = 0$ (b) $2Tx - a^2y + 2aT = 0$

(c) $2Tx - a^2y - 2aT = 0$ (d) None of these

3 An equation of a line through the point (1, 2) whose distance from the point (3, 1) has the greater value is

(a) $y = 2x$ (b) $y = x + 1$ (c) $x + 2y = 5$ (d) $y = 3x - 1$

4 The centroid of an equilateral triangle is (0, 0). If two vertices of triangle lie on $x + y = 2\sqrt{2}$, then one of them will have its coordinates.

(a) $(\sqrt{2}+\sqrt{6}, \sqrt{2}-\sqrt{6})$ (b) $(\sqrt{2}+\sqrt{3}, \sqrt{2}-\sqrt{3})$

(c) $(\sqrt{2}+\sqrt{5}, \sqrt{2}-\sqrt{5})$ (d) None of these

5 If the equation of base of an equilateral triangle is $2x - y = 1$ and vertex is $(-1, 2)$, then the length of the side of the triangle is

(a) $\sqrt{\frac{20}{3}}$ units (b) $\frac{2}{\sqrt{15}}$ units (c) $\sqrt{\frac{8}{15}}$ units (d) $\sqrt{\frac{15}{2}}$ units

6 The line $\frac{x}{3} + \frac{y}{4} = 1$ meets the Y-axis and X-axis at A and B, respectively. A square $ABCD$ is constructed on the line segment AB away from the origin. The coordinates of the vertex of the square farthest from the origin, are

(a) (7, 3) (b) (4, 7) (c) (6, 4) (d) (3, 8)

7 Two sides of a parallelogram are along the lines $4x + 5y = 0$ and $7x + 2y = 0$. If the equation of one of the diagonals of the parallelogram is $11x + 7y = 9$, then other diagonal passes through the point. **[JEE Main 2021]**

(a) (1, 2) (b) (2, 2) (c) (2, 1) (d) (1, 3)

8 Let $A(-1, 1)$, $B(3, 4)$ and $C(2, 0)$ be given three points. A line $y = mx$, $m > 0$ intersects lines AC and BC at point P and Q, respectively. Let A_1 and A_2 be the areas of ΔABC and ΔPQC respectively, such that $A_1 = 3A_2$, then the value of m is equal to **[JEE Main 2021]**

(a) 1 (b) 3 (c) 2 (d) $\frac{4}{15}$

9 The line PQ whose equation is $x - y = 2$ cuts the X-axis and Y-axis at P and Q, respectively. The line PQ is rotated about P through 45° in the anticlockwise direction. The equation of the line PQ, in the new position is

(a) $y = -\sqrt{2}$ (b) $y = 2$ (c) $x = 2$ (d) $x = -2$

2.2 Point and Line

10 A line of fixed length 2 units moves so that its ends are on the positive X-axis and that part of the line $x + y = 0$, which lies in the second quadrant. Then, the locus of the mid-point of the line has the equation.

(a) $x^2 + 5y^2 + 4xy - 1 = 0$ (b) $x^2 + 5y^2 + 4xy + 1 = 0$

(c) $x^2 + 5y^2 - 4xy - 1 = 0$ (d) $4x^2 + 5y^2 + 4xy + 1 = 0$

11 Consider the points $A(0, 1)$ and $B(2, 0)$ and P be a point on the line $4x + 3y + 9 = 0$. Coordinates of P such that $|PA - PB|$ is maximum are

(a) $\left(-\frac{12}{5}, \frac{17}{5}\right)$ (b) $\left(-\frac{84}{5}, \frac{13}{5}\right)$ (c) $\left(-\frac{6}{5}, \frac{17}{5}\right)$ (d) $(0, -3)$

12 In a ΔABC, $A=(\alpha,\beta)$, $B=(2,3)$ and $C=(1,3)$ and point A lies on line $y=2x+3$ where $\alpha \in I$. Area of ΔABC, Δ is such that $[\Delta]=5$. Possible coordinate of A are (where [.] represents greatest integer function)

(a) (2, 3) (b) (5, 13) (c) (– 5, 8) (d) (– 3, – 5)

13 If (α, α^2) falls inside the angle made by the lines $y=\frac{x}{2}$, $x>0$ and $y=3x$, $x>0$, then α belongs to

(a) $(3, \infty)$ (b) $\left(\frac{1}{2}, 3\right)$ (c) $\left(-3, -\frac{1}{2}\right)$ (d) $\left(0, \frac{1}{2}\right)$

14 If p and p' are the distance of origin from the lines $x \sec\alpha + y \operatorname{cosec}\alpha = k$ and $x\cos\alpha - y\sin\alpha = k\cos 2\alpha$, then $4p^2 + p'^2$ is equal to

(a) k (b) $2k$ (c) k^2 (d) $2k^2$

15 The equation of the base of an equilateral triangle is $x+y=2$ and the vertex is (2, – 1). The length of the sides of the triangle is

(a) $\sqrt{\frac{3}{2}}$ units (b) $\sqrt{2}$ units

(c) $\sqrt{\frac{2}{3}}$ units (d) None of these

16 Three distinct point A, B and C are given in the 2-dimensional coordinates such that the ratio of the distance of any one of them from the point (1, 0) to the distance from the point (–1, 0) is equal to $\frac{1}{3}$. Then, the circumcentre of the ΔABC is at the point.

(a) $\left(\frac{5}{4}, 0\right)$ (b) $\left(\frac{5}{2}, 0\right)$ (c) $\left(\frac{5}{3}, 0\right)$ (d) (0, 0)

17 If the perpendicular bisector of the line segment joining $P(1, 4)$ and $Q(\alpha, 3)$ has y-intercept is – 4, then a possible value of α is

(a) – 2 (b) – 4 (c) 1 (d) 2

18 The image of the point (3, 5) in the line $x-y+1=0$ lies on **[JEE main 2021]**

(a) $(x-4)^2+(y-4)^2=8$ (b) $(x-4)^2+(y+2)^2=16$

(c) $(x-2)^2+(y-2)^2=12$ (d) $(x-2)^2+(y-4)^2=4$

19 A triangle is formed by the lines $2x-3y-6=0$, $3x-y+3=0$ and $3x+4y-12=0$. If the points $P(\alpha, 0)$ and $Q(0, \beta)$ always lie on or inside the ΔABC, then

(a) $\alpha \in [-1, 2]$ and $\beta \in [-2, 3]$ (b) $\alpha \in [-1, 3]$ and $\beta \in [-2, 4]$

(c) $\alpha \in [-2, 4]$ and $\beta \in [-3, 4]$ (d) $\alpha \in [-1, 3]$ and $\beta \in [-2, 3]$

20 Let the algebraic sum of perpendicular distances from the point (3, 0), (0, 3) and (2,2) to a variable straight line be zero, then the line passes through a fixed point whose coordinates are

(a) (3, 2) (b) (2, 3) (c) $\left(\frac{3}{5}, \frac{3}{5}\right)$ (d) $\left(\frac{5}{3}, \frac{5}{3}\right)$

2.3 Line and Line (Family of Line)

21 $P(m, n)$ (where m, n are natural numbers) is any point in the interior of the quadrilateral formed by the pair of lines $xy=0$ and the lines $2x+y-2=0$ and $4x+5y=20$. The possible number of positions of the point P is

(a) 7 (b) 5 (c) 4 (d) 6

22 If the straight line $x+y-2=0, 2x-y+1=0$ and $ax+by-c=0$ are concurrent, then the family of lines $2ax+3by+c=0$ (a, b, c are non-zero) is concurrent at

(a) (2,3) (b) $\left(\frac{1}{2}, \frac{1}{3}\right)$ (c) $\left(-\frac{1}{6}, -\frac{5}{9}\right)$ (d) $\left(\frac{2}{3}, -\frac{7}{5}\right)$

23 If the lines $ax+y+1=0, x+by+1=0$ and $x+y+c=0$ (a, b, c being distinct and different from 1) are concurrent, then $\left(\frac{1}{1-a}\right)+\left(\frac{1}{1-b}\right)+\left(\frac{1}{1-c}\right)$ is equal to

(a) 0 (b) 1

(c) $\frac{1}{a+b+c}$ (d) None of these

24 For the straight lines given by the equation $(2+k)x+(1+k)y=5+7k$ for different values of k, which of the following statements is true.

(a) Lines form a parallelogram
(b) Lines pass through the point $(-2, 9)$
(c) Lines pass through the point $(2, -9)$
(d) None of the above

25 Line L has intercepts a and b on the coordinate axes. When the axes are rotated through a given angle keeping the origin fixed, the same line L has intercepting p and q, then

(a) $a^2+b^2=p^2+q^2$ (b) $\frac{1}{a^2}+\frac{1}{b^2}=\frac{1}{p^2}+\frac{1}{q^2}$

(c) $a^2+p^2=b^2+q^2$ (d) $\frac{1}{a^2}+\frac{1}{p^2}=\frac{1}{b^2}+\frac{1}{q^2}$

26 The line L given by $\frac{x}{5}+\frac{y}{b}=1$ passes through the point (13, 32). The line k is parallel to L and has the equation $\frac{x}{c}+\frac{y}{3}=1$. Then, the distance between L and k is

(a) $\sqrt{17}$ units (b) $\frac{17}{\sqrt{15}}$ units (c) $\frac{23}{\sqrt{17}}$ units (d) $\frac{23}{\sqrt{15}}$ units

27 The number of integral values of m for which the x-coordinate of the point of intersection of the lines $3x+4y=9$ and $y=mx+1$ is also in integer is

(a) 2 (b) 0 (c) 4 (d) 1

28 The locus of the orthocentre of the triangle formed by the lines $y=0$ (i.e X-axis) and the lines $(1+p)x-py+p(1+p)=0$ and $(q+1)x-qy+q(1+q)=0$, where $p \neq q$ is

(a) the line $x+y=0$ (b) the line $x-y=0$

(c) the curve $y^2=4pq(p+q)(-pq)x$ (d) the curve $\frac{x^2}{p(p+q)}+\frac{y^2}{q(p+q)}=1$

29 In ΔABC, $\angle C=90^\circ$ and the side AB is of length 60 units. The equations of the medians AD and BE, respectively are $y=x+3$ and $y=2x+4$. Then, the area of ΔABC (in sq units) is

(a) 200 (b) 300 (c) 400 (d) 500

30 Lines $L_1: y-x=0$ and $L_2: 2x+y=0$ intersect the line $L_3: y+2=0$ at P and Q, respectively. The bisector of the acute angle between L_1 and L_2 intersects the line L_3 at R. Consider the following two statements.

Statement I S_1 : The ratio $PR:RQ=2\sqrt{2}:\sqrt{5}$

Statement II S_2 : In any triangle, bisector of an angle divides the triangle into two similar triangles.

Then,

(a) both S_1 and S_2 are true (b) both S_1 and S_2 are false
(c) S_1 is true and S_2 is false (d) S_2 is true and S_1 is false

31 The straight line $2x+3y+1=0$ bisects the angle between two straight lines of which one line is $3x+2y+4=0$. Then, the equation of the other line is

(a) $9x+26y=48$ (b) $9x+46y=28$ (c) $9x+26y=28$ (d) $3x+26y=48$

32 A ray of light coming from the point $(2, 2\sqrt{3})$ is incident at an angle $30°$ on the line $x=1$ at point A. The ray gets reflected on the line $x=1$ and meets X-axis at point B. Then, the line AB passes through the point **[JEE Main 2020]**

(a) $(3, -\sqrt{3})$ (b) $(4, -\sqrt{3})$ (c) $\left(4, -\frac{\sqrt{3}}{2}\right)$ (d) $\left(3, -\frac{1}{\sqrt{3}}\right)$

33 Let ABC be a triangle with $A(-3, 1)$ and $\angle ACB=\theta$, $0<\theta<\frac{\pi}{2}$. If the equation of the median through B is $2x+y-3=0$ and the equation of angle bisector of C is $7x-4y-1=0$, then $\tan\theta$ is equal to **[JEE Main 2021]**

(a) $\frac{1}{2}$ (b) $\frac{3}{4}$ (c) $\frac{4}{3}$ (d) 2

34 The point $(4, 1)$ undergoes the following three transformation, successively.

(i) Reflection about the line $y=x$.
(ii) Translation through a distance 2 units along the positive direction of X-axis.
(iii) Rotation through an angle $\frac{\pi}{4}$ about the origin in the counter clockwise direction.

The final position of the points is given by the coordinates

(a) $\left(\frac{7}{\sqrt{2}}, -\frac{7}{\sqrt{2}}\right)$ (b) $\left(\frac{7}{\sqrt{2}}, \frac{1}{\sqrt{2}}\right)$ (c) $\left(-\frac{1}{\sqrt{2}}, \frac{7}{\sqrt{2}}\right)$ (d) None of these

35 A light ray emerging from the point source placed at $P(2, 3)$ is reflected at a point Q on the Y-axis and then passes through the point $R(5, 10)$. Then, coordinate of Q are

(a) $(0, 3)$ (b) $(0, 2)$ (c) $(0, 5)$ (d) None of these

36 A beam of light is sent along the line $x-y=1$, which after refracting from the X-axis entres the opposite side by turning through $30°$ towards, then normal at the point of incidence on the X-axis. Then, the equation of refracted ray is

(a) $(2-\sqrt{3})x-y=2+\sqrt{3}$ (b) $(2+\sqrt{3})x-y=2+\sqrt{3}$
(c) $(2-\sqrt{3})x+y=2+\sqrt{3}$ (d) $y=(2-\sqrt{3})(x-1)$

Numerical Value Type Questions

37 The interior angle bisector of $\angle A$ for ΔABC whose coordinates of the vertices are $A(-8, 5)$, $B(-15, -19)$ and $C(1, -7)$ has the equation $ax+2y+c=0$. Find the value of $\frac{a+c}{89}$.

38 A variable line passes through $P(2, 3)$ and cuts the coordinates axes at A and B. If the parallelogram $OACB$ (where O is the origin) is completed, then find number of order pair (x, y) of integers which lie on the locus of point C.

39 Consider a triangle having vertices $A(-2, 3)$, $B(1, 9)$ and $C(3, 8)$. If a line L passing through the circumcentre of ΔABC, bisects line BC and intersects Y-axis at point $\left(0, \frac{\alpha}{2}\right)$, then the value of real number α is **[JEE Main 2021]**

40 Absolute value of the sum of the abscissa of all the points on the line $x+y=4$ that lie at a unit distance from the line $4x+3y-10=0$ is

41. Let the point $M(2, 1)$ be shifted through a distance $3\sqrt{2}$ units measured parallel to the line $L: x+y-1=0$ in the direction of decreasing ordinates, to reach at N. If the image of N in the line L is R and Q. If the distance of R from the line $3x-4y+25=0$ is d. Find $0.14d$.

42 If the line $2x-y+3=0$ is at a distance $\frac{1}{\sqrt{5}}$ and $\frac{2}{\sqrt{5}}$ units from the lines $4x-2y+\alpha=0$ and $6x-3y+\beta=0$ respectively, then the sum of all possible values of α and β is **[JEE Main 2020]**

43 The equation of adjacent sides AB and AD of a parallelogram $ABCD$ are $3x-4y+21=0$ and $4x-3y-14=0$, respectively. The perpendicular distances between the parallel sides AB, DC and AD, BC are $\frac{14}{5}$ and $\frac{7}{5}$, respectively. If area of parallelogram is $2k$ sq units, then find the value of k.

44 If the distance of the point $(2, 3)$ from the line $2x-3y+9=0$ measured along the line $2x-2y+5=0$ is $d\sqrt{2}$, then find $d+2$.

45 A line $L_1=3y-2x-6=0$ is rotated about its point of intersection with Y-axis in clockwise direction to make it L_2 such that the area formed by L_1, L_2, X-axis and line $x=5$ is $\frac{49}{3}$ sq units. If its point of intersection with $x=5$ lies below X-axis and equation of line L_2 is $ax+by+c=0$, then find the value of $\frac{3a+4b-c}{10}$.

46 Let $2x+3y=6$ be a line meeting the coordinate axes at A and B, respectively. A variable line $\frac{x}{a}+\frac{y}{b}=1$ meets the axes at P and Q respectively, in such a way that the line BP and AQ always meet at right angle at R. If the locus of the orthocentre of the ΔARB is $ax^2+by^2+cx+dy=0$, then find $\sqrt{(a+b-c-d)}$ upto 3 decimal.

47. The equation of a line through the mid-point of the sides AB and AP of rhombus $ABCD$, whose one diagonal is $3x-4y+5=0$ and one vertex is $A(3, 1)$ is $ax+by+c=0$. Find the absolute values of $a+b+c$, where a, b, c are integers expressed in lowest form.

48. Let $A=(-1, 0)$, $B(3, 0)$ and PQ be any line passing through $(4, 1)$. The range of the slope of PQ for which there are two points on PQ at which AB subtends a right angle is (λ_1, λ_2), then $5(\lambda_1+\lambda_2)$ is equal to

LEVEL 03 \ JEE Advanced Level

Single Choice Correct Questions

1 Given, $A(0, 0)$ and $B(x, y)$ with $x \in (0, 1)$ and $y>0$. Let the slope of the line AB equal to m_1. Point C lies on the line $x=1$ such that the slope of BC equal to m_2, where $0<m_2<m_1$. If the area of ΔABC can be expressed as $(m_1-m_2)\cdot f(x)$, then the largest possible value of x is

(a) 1 (b) 1/2

(c) 1/4 (d) 1/8

2 The line $L_1 : 4x + 3y - 12 = 0$ interesects the X-axis and the Y-axis at A and B. A variable line perpendicular to L_1 intersects the X and the Y-axes at P and Q, respectively. Then, the locus of circumcentre of ΔABQ is

(a) $3x - 4y + 2 = 0$
(b) $4x + 3y + 7 = 0$
(c) $6x - 8y + 7 = 0$
(d) None of these

3 The vertices B and C of ΔABC lie on the lines $4y - 3x = 0$ and $y = 0$, respectively and the side BC passes through the point $P(0, 5)$. If $ABOC$ is a rhombus, where O is the origin and P is an internal point to the rhombus, then the vertex A is

(a) (3, 9)
(b) (9, 3)
(c) (3, – 9)
(d) (– 3, 9)

4 In an acute angle ΔABC, equation of the side BC is $4x - 3y + 3 = 0$, orthocentre is (1, 2) and the circumcentre O is (2, 3). Then, the value of the circumradius R is

(a) $\sqrt{28}/5$
(b) $\sqrt{58}/5$
(c) 23
(d) $\sqrt{23}/5$

5 If P_1, P_2, P_3 are the length of the perpendicular from the points $(2m, m^2)$ $(m + m', mm')$ and $(2m', m'^2)$ respectively, on the line $x\sin\alpha + y\cos\alpha + \dfrac{\sin^2\alpha}{\cos\alpha} = 0$, then P_1, P_2, P_3 are in

(a) AP
(b) GP
(c) HP
(d) none of these

6 If A_1 is area of the triangle with the vertices (0, 0), $(a\tan\theta, b\cot\theta)$ and $(a\sin\theta, b\cos\theta)$, A_2 is area of the triangle with the vertices (a, b), $(a\sec^2\theta, b\operatorname{cosec}^2\theta)$ and $(a + a\sin^2\theta, b + b\cos^2\theta)$ and A_3 is the area of the triangle with the vertices (0, 0), $(a\tan\theta, -b\cot\theta)$, $(a\sin\theta, b\cos\theta)$, then the values of θ for which A_1, A_2 and A_3 are in GP is

(a) 0
(b) $\pi/2$
(c) $\theta, \theta \in R$
(d) None of these

7 OA and OB are equal sides of an isosceles triangle lying in the first quadrant making angle θ and ϕ respectively, with X-axis the gradient of bisector of acute $\angle AOB$ is

(a) $\operatorname{cosec}\theta - \cot\phi$
(b) $\operatorname{cosec}(\theta + \phi) - \cot(\phi + \theta)$
(c) Both (a) and (b)
(d) None of these

8 The equations of the sides AB and AC of ΔABC, respectively are $2x - y = 0$ and $x + y = 3$ and the centroid G is (2, 3). Then, the slope of the side BC is

(a) 4
(b) 3
(c) 2
(d) 5

9 The vertices of triangle are $A(-8, 5)$, $B(-15, -19)$ and $C(1, -7)$. An internal bisector of $\angle A$ has the equation $px + 2y + r = 0$, then find the image of $\left(\dfrac{p}{11}, \dfrac{r}{78}\right)$ w.r.t. line $3x + 4y - 1 = 0$

(a) $\left(\dfrac{11}{25}, \dfrac{23}{25}\right)$
(b) $\left(\dfrac{-11}{25}, \dfrac{-23}{25}\right)$
(c) $\left(\dfrac{-11}{25}, \dfrac{23}{25}\right)$
(d) None of these

More than One Correct

10 Line $\dfrac{x}{a} + \dfrac{y}{b} = 1$ cuts the coordinate axes at $A(a, 0)$ and $B(0, b)$ and the line $\dfrac{x}{a'} + \dfrac{y}{b'} = -1$ at $A'(-a', 0)$ and $B'(0, -b')$. If points A, B, A', B' are concyclic, then the orthocentre of the $\Delta ABA'$ is

(a) (0, 0)
(b) $(0, b')$
(c) $\left(0, \dfrac{aa'}{b}\right)$
(d) $\left(0, \dfrac{bb'}{a}\right)$

11 Let $L_1: 3x+4y=1$ and $L_2: 5x-12y+2=0$ be two given lines. Let image of every point on L_1 w.r.t. a line L lies on L_2, then possible equation of L can be

(a) $14x+112y-23=0$
(b) $64x-8y-3=0$
(c) $11x-4y=0$
(d) $52y-45x=7$

12 Let x_1 and y_1 be the roots of $x^2+8x-2009=0$, x_2 and y_2 be the roots of $3x^2+24x-2010=0$ and x_3 and y_3 be the roots of $9x^2+72x-2011=0$. The points $A(x_1, y_1)$, $B(x_2, y_2)$ and $C(x_3, y_3)$

(a) can not lie on a circle
(b) form a triangle of area 2 sq units
(c) form a right-angled triangle
(d) are collinear

13 The line $3x+2y=24$ meets the Y-axis at A and the X-axis at B. C is a point on the perpendicular bisector of AB such that the area of the ΔABC is 91 sq. units. The coordinates of C are

(a) $\left(\frac{29}{2}, -1\right)$
(b) $\left(\frac{29}{2}, 13\right)$
(c) $\left(\frac{-13}{2}, -1\right)$
(d) $\left(\frac{-13}{2}, 13\right)$

14 Let $u = ax+by+a\sqrt[3]{b}=0$ and $v = bx-ay+b\sqrt[3]{a}=0$, where $a, b \in R$ be two straight lines. The equation of the bisectors of the angle formed by $k_1u-k_2v=0$ and $k_1u+k_2v=0$ for non-zero real k_1 and k_2, are

(a) $u=0$
(b) $k_2u+k_1v=0$
(c) $k_2u-k_1v=0$
(d) $v=0$

15 The equation of straight line with gradient 1 passing through $\left(\frac{m}{2}, \frac{n}{2}\right)$, where $m, n \in R$ satisfies the equation $\sec^2(n(m+2))+m^2=1$, where $n \in \left[\frac{-\pi}{2}, \frac{\pi}{2}\right]$ can be

(a) $x+y=0$
(b) $x-y=0$
(c) $x-y+\frac{\pi}{4}=0$
(d) $x-y-\frac{\pi}{4}=0$

16 Let A be a variable point on the line $y=4$. If B and C are variable points on the line $y=-4$, such that ΔABC is equilateral, then which of the following statement(s) holds good?

(a) Perimeter of ΔABC is $16\sqrt{3}$ units
(b) Radius of circle inscribed in $\Delta ABC = 4/3$ units
(c) Locus of orthocentre of ΔABC is $3y-16=0$
(d) $r_1 : r_2 : r_3 = 1:1:1$, where r_1, r_2 and r_3 are exradii of ΔABC

17 A line l is drawn through the point (1, 2) so that it intersects the line $x+y-4=0$ at a point whose distance from the point (1, 2) is $\sqrt{2/3}$ units, then the angle made by l with X-axis, is

(a) 15°
(b) 75°
(c) 105°
(d) 60°

18 Let $A(4, 3)$, $B(-4, 3)$ and $C(0, -5)$ be the vertices of a triangle and $P=(5, 0)$. Let L, M and N be the feet of the perpendicular drawn from P onto the sides BC, CA and AB respectively, then

(a) the centroid of ΔLMN does not exist
(b) the area of ΔLMN is 5 sq units
(c) the orthocentre of ΔLMN is the origin
(d) L, M and N are collinear

19 The coordinates of the feet of the perpendiculars from the vertices of a triangle on the opposite sides are (20, 25), (8, 16) and (8, 9). The coordinates of a vertex of triangle are

(a) (5, 10)
(b) (50, –5)
(c) (15, 30)
(d) (10, 15)

20 If the two lines represented by $x^2(\tan^2\theta+\cos^2\theta)-2xy\tan\theta+y^2\sin^2\theta=0$ make angles α, β with the X-axis, then

(a) $\tan\alpha+\tan\beta = 4\operatorname{cosec}2\theta$
(b) $\tan\alpha\cdot\tan\beta = \sec^2\theta+\tan^2\theta$
(c) $\tan\alpha-\tan\beta = \pm 2$
(d) $\frac{\tan\alpha}{\tan\beta} = \frac{2+\sin 2\theta}{2-\sin 2\theta}$

21 The straight line $u = 0$ and $v = 0$ passes through the origin and angle between them is $\tan^{-1}(7/9)$. If the ratio of slopes of $u = 0$ and $v = 0$ is 9/2, then their equation are

(a) $y = 3x$ and $3y = 2x$
(b) $2y = 3x$ and $3y = x$
(c) $y + 3x = 0$ and $3y + 2x = 0$
(d) $2y + 3x = 0$ and $3y + x = 0$

22 Three points $A(0, 0)$, $B(\cos\alpha, \sin\alpha)$ and $C(\cos\beta, \sin\beta)$ are the vertices of a right angled triangle, then

(a) $\sin\left(\frac{\alpha-\beta}{2}\right) = 1/\sqrt{2}$
(b) $\cos\left(\frac{\alpha-\beta}{2}\right) = -1/\sqrt{2}$
(c) $\cos\left(\frac{\alpha-\beta}{2}\right) = 1/\sqrt{2}$
(d) $\sin\left(\frac{\alpha-\beta}{2}\right) = -1/\sqrt{2}$

23 The family of lines $(2\cos\theta + 3\sin\theta)x + (3\cos\theta - 5\sin\theta)y - (5\cos\theta - 2\sin\theta) = 0$ is concurrent at (x_1, y_1) and the family of lines $(a + 2b)x + (a - 3b)y + b - a = 0$ for different values of a and b is concurrent at (x_2, y_2), then

(a) $x_2 + y_2 = x_1$
(b) $x_2 + y_2 = y_1$
(c) $6x_2 + y_2 = 2x_1 + y_1$
(d) $2x_2 + y_2 = 6x_1 + y_1$

24 Let $B(1, -3)$ and $D(0, 4)$ represent two vertices of rhombus $ABCD$ in (x, y) plane, then coordinate of vertex A is $\angle BAP = 60°$ can be equal to

(a) $\left(\frac{1-7\sqrt{3}}{2}, \frac{1-\sqrt{3}}{2}\right)$
(b) $\left(\frac{1+7\sqrt{3}}{2}, \frac{1+\sqrt{3}}{2}\right)$
(c) $\left(\frac{-1+7\sqrt{3}}{2}, \frac{-1+\sqrt{3}}{2}\right)$
(d) $\left(\frac{-1-7\sqrt{3}}{2}, \frac{-1-\sqrt{3}}{2}\right)$

Linked Comprehension Type

Paragraph I Directions (Q.Nos. 25-27)

$A(x_1, y_1)$, $B(x_2, y_2)$, $(y_1 < y_2)$ are two points on the line $x + y = 4$ from which perpendicular AQ and BP are drawn on the line $4x + 3y = 10$, where P and Q are the feet of perpendiculars such that $AQ = BP = 1$. Now, considering AB as diameter a circle is drawn which meets the line $4x + 3y = 10$ at C and D such that C is closer to P.

25 The value $\frac{y_1 + y_2}{x_1 + x_2}$ is equal to

(a) -4
(b) -3
(c) 4
(d) None of these

26 The length PQ is equal to

(a) $3\sqrt{14}$ units
(b) $4\sqrt{5}$ units
(c) 14 units
(d) None of these

27 Length QD is equal to

(a) $(3\sqrt{2} + \sqrt{17})$ units
(b) $(3\sqrt{2} - \sqrt{17})$ units
(c) $(4\sqrt{2} + \sqrt{17})$ units
(d) $(5\sqrt{2} - 7)$ units

Paragraph II Directions (Q.Nos. 28-29)

Consider the lines L_1 and L_2 defined by
$L_1 : x\sqrt{2} + y - 1 = 0$ and $L_2 : x\sqrt{2} - y + 1 = 0$
For a fixed constant λ, let C be locus of a point P such that the product of the distance of P from L_1 and the distance of P from L_2 is λ^2. The line $y = 2x + 1$ meets C at two points R and S, where the distance between R and S is $\sqrt{270}$ units.

Let the perpendicular bisector RS meets C at two distinct points R' and S'. Let D be the square of the distance between R' and S'. **[JEE Advanced 2021]**

28 The value of λ^2 is

(a) 8
(b) 9
(c) 7
(d) 6

29 The value of D is

(a) 77.14 (b) 70.11 (c) 69.14 (d) 82.31

Paragraph III Directions (Q.Nos. 30-32)

Let $M\left(2, \frac{13}{8}\right)$ is the circumcentre of ΔPQR whose sides PQ and PR are represented by the straight line $4x - 3y = 0$ and $4x + y = 16$, respectively.

30 The orthocentre of ΔPQR is

(a) $\left(\frac{7}{3}, \frac{4}{3}\right)$ (b) $\left(\frac{4}{3}, \frac{7}{3}\right)$ (c) $\left(3, \frac{3}{4}\right)$ (d) $\left(\frac{3}{4}, 3\right)$

31 If A, B and C are the mid-points of the sides PQ, QR and PR of ΔPQR respectively, then the area of ΔABC equals

(a) 1 sq unit (b) 2 sq units (c) 3 sq units (d) 4 sq units

32 If PB is the median of ΔPQR, then the equation of the straight line passing through $N(-2, 3)$ and perpendicular to PB is

(a) $4x + y + 5 = 0$ (b) $x - 4y + 14 = 0$

(c) $4x - y + 11 = 0$ (d) $x + 4y - 10 = 0$

Paragraph IV Directions (Q.Nos. 33-35)

Given point $A(6, 30)$ and point $B(24, 6)$, equation of line AB is $4x + 3y = 114$. Point $P(0, \lambda)$ is a point on Y-axis such that $0 < \lambda < 38$ and point $Q(0, k)$ is a point on Y-axis such that $k > 38$.

33 For all positions of point P, $\angle APB$ is maximum when point P is

(a) (0, 11) (b) (0, 15) (c) (0, 18) (d) (0, 21)

34 The maximum value of $\angle ADB$ is

(a) $\pi/3$ (b) $\pi/2$ (c) $2\pi/3$ (d) $3\pi/4$

35 For all position of point Q, $\angle AQB$ is maximum when point Q is

(a) (0, 54) (b) (0, 58) (c) (0, 60) (d) None of these

Paragraph V Directions (Q.Nos. 36-38)

Let the straight lines $x + y - 2 = 0$, $2x - y + 1 = 0$ and $px + qy = r$ be concurrent and l_1 and l_2 be the two members of the family of lines $2px + 2qy + 4r = 0$ which are nearest and farthest from origin.

36 The equation of line l_1 is

(a) $y = 5x$ (b) $y = 3x$ (c) $5y = x$ (d) $3y = x$

37 The equation of line l_2 is

(a) $3x + 15y - 52 = 0$ (b) $3x + 15y + 52 = 0$

(c) $3x - 15y + 52 = 0$ (d) $3x - 15y - 52 = 0$

38 The angle between l_1 and l_2 is

(a) 45° (b) 60° (c) 90° (d) 30°

Paragraph VI Directions (Q.Nos. 39-41)

In the diagram, a line is drawn through the points $A(0, 16)$ and $B(8, 0)$. Point P is chosen in the first quadrant on the line through A and B. Points C and D are chosen on the X and Y-axes respectively, so that $PDOC$ is a rectangle.

39 Perpendicular distance of the line AB from the point (2, 2) is

(a) 2 units (b) $\sqrt{16}$ units (c) $2\sqrt{5}$ units (d) $5\sqrt{2}$ units

40 The sum of the coordinates of the point P, if $PDOC$ is a square, is
(a) 32/3 (b) 16/3 (c) 16 (d) 11

41 Number of possible ordered pairs (x, y) of all positions of a point P on AB, so that area of rectangle $PDOC$ is 30 sq units, is
(a) 3 (b) 2 (c) 1 (d) 0

Paragraph VII Directions (Q.Nos. 42-44)

Consider two points $A = (1, 2)$ and $B = (3, -1)$. Let M be a point on the straight line $L = x + y = 0$

42 If M is a point on line $L = 0$ such that $AM + BM$ is minimum, then the reflection of M in the line $y = x$ is
(a) $(1, -1)$ (b) $(-1, 1)$ (c) $(2, -2)$ (d) $(-2, 2)$

43 If M is a point on the line $L = 0$ such that $|AM - BM|$ is maximum, then the distance of M from $N = (1, 1)$, is
(a) $5\sqrt{2}$ units (b) 7 units (c) $3\sqrt{5}$ units (d) 10 units

44 If M is a point on the line $L = 0$ such that $|AM - BM|$ is minimum, then the area of ΔAMB equals
(a) $\frac{13}{4}$ sq units (b) $\frac{13}{2}$ sq units (c) $\frac{13}{6}$ sq units (d) $\frac{13}{8}$ sq units

Paragraph VIII Directions (Q.Nos. 45-46)

Consider the triangle having vertices $O(0, 0)$, $A(2, 0)$ and $B(1, \sqrt{3})$. Also, $b \le \min(a_1, a_2, a_3, \ldots, a_n)$ means $b \le a_1$, when a_1 is least, $b \le a_2$, when a_2 is least and so on. For this we can say.
$b \le a_1, b \le a_2, \ldots, b \le a_n$.

45 Let R be the region consisting of all those points P inside ΔOAB which satisfy $d(P, OA) \le \min[d(P, OB), d(P, AB)]$, where d denotes the distance from the point to the corresponding line. Then, the area of the region R is
(a) $\sqrt{3}$ sq units (b) $(2 + \sqrt{3})$ sq units (c) $\sqrt{3}/2$ sq units (d) $1/\sqrt{3}$ sq units

46 Let R be the region consisting of all those points P inside ΔOAB which satisfy $PO \le \min[BP, AP]$. Then, the area of the region R is
(a) $\sqrt{3}$ sq units (b) $\frac{1}{2\sqrt{3}}$ sq units (c) $\frac{\sqrt{3}}{2}$ sq units (d) None of these

Matrix Match Type Questions

47

	Column I		Column II
(i)	A straight line with negative slope passing through (1, 4) meets the coordinate axes at A and B. The minimum length of $OA + OB$, O being origin is	A	$5\sqrt{2}$
(ii)	If the point $P(4, -1)$ is symmetric to the point Q with respect to the bisector of the first quadrant, then the length of PQ is	B	$3\sqrt{2}$
(iii)	On the portion of the straight line $x + y = 2$ between the axis a square is constructed away from the origin, with this portion as one of its sides. If 'd' denotes the perpendicular distance of a side of this square from the origin, then maximum value of d is	C	9/2
(iv)	If the parametric equation of a line is given by $x = 4 + \lambda/\sqrt{2}$ and $y = 4 + \sqrt{2}\lambda$, where λ is a parameter, then the intercept made by the line on the X-axis is	D	9

48

	Column I		Column II
(i)	The distance between the lines $(x+7y)^2+4\sqrt{2}(x+7y)-42=0$ is	A	2
(ii)	If the sum of the distances of a point from two perpendicular lines in a plane is 1, then its locus is $\|x\|+\|y\|=k$, where k is equal to	B	7
(iii)	If $6x+6y+m=0$ is acute angle bisector of line $x+2y+4=0$ and $4x+2y-1=0$, then m is equal to	C	3
(iv)	Area of triangle formed by the lines $y^2-9xy+18x^2=0$ and $y=6$ is	D	1

49 Let ABC be a triangle such that the coordinates of A are $(-3, 1)$. Equation of the median through B is $2x+y-3=0$ and equation of the angular bisector of C is $7x-4y-1=0$, then match the following column

	Column I		Column II
(i)	Equation of line AB is	A	$2x+y-3=0$
(ii)	Equation of the line BC is	B	$2x-3y+9=0$
(iii)	Equation of the line CA is	C	$4x+7y+5=0$
		D	$18x-y-49=0$

50

	Column I		Column II
(i)	If lines $3x+y-4=0$, $x-2y-6=0$ and $\lambda x+4y+\lambda^2=0$ are concurrent, then the values of λ is	A	-4
(ii)	If the points $(\lambda+1, 1)$, $(2\lambda+1, 3)$ and $(2\lambda+2, 2\lambda)$ are collinear, then the value of λ is	B	$-1/2$
(iii)	If line $x+y-1-\|\lambda/2\|=0$ passing through the intersection of $x-y+1=0$ and $3x+y-5=0$ is perpendicular to one of them, then the value of λ is	C	4
(iv)	If line $y-x-1+\lambda=0$ is equidistant from the points $(1,-2)$ and $(3, 4)$, then λ is	D	2

51

	Column I		Column II
(i)	The lines $y=0$, $y=1$, $x-6y+4=0$ and $x+6y-9=0$ constitute a figure which is	A	cyclic quadrilateral
(ii)	The points $A(a, 0)$, $B(0, b)$, $C(c, 0)$ and $D(0, d)$ are such that $ac=bd$ and a, b, c, d are all non-zero. The points, A, B, C and D always constitute	B	a rhombus
(iii)	The figure formed by the four line $ax \pm by \pm c=0$, $(a \neq b)$ is	C	a square
(iv)	The line pair $x^2-8x+12=0$ and $y^2-14y+45=0$ constitute a figure which is	D	a trapezium

Numerical Value Type Question

52 Let $O(0, 0)$, $A(6, 0)$ and $B(3, \sqrt{3})$ be the vertices of ΔOAB. Let R be the region consisting of all those points P inside ΔOAB which satisfy $d(P, OA) \leq \text{minimum}\,(d(P, OB), d(P, AB))$, where $d(P, OA)$, $d(P, OB)$ and $d(P, AB)$ represent the distance of P from the sides OA, OB and AB, respectively. If the area of region R is $9(a - \sqrt{b})$ sq units, where a and b are coprime, then find the value of $(a + b)$.

53 Points O, A, B, C, ... are shown in figure, where $OA = 2AB = 4BC$ ... so on.

Let A is the centroid of a triangle whose orthocentre and circumcentre are (2, 4) and $\left(\frac{7}{2}, \frac{5}{2}\right)$, respectively. If an insect starts moving from the point $O(0, 0)$, along the straight line in zig-zag fashion and terminates ultimately at point $P(\alpha, \beta)$, then find the value of $\alpha + \beta$.

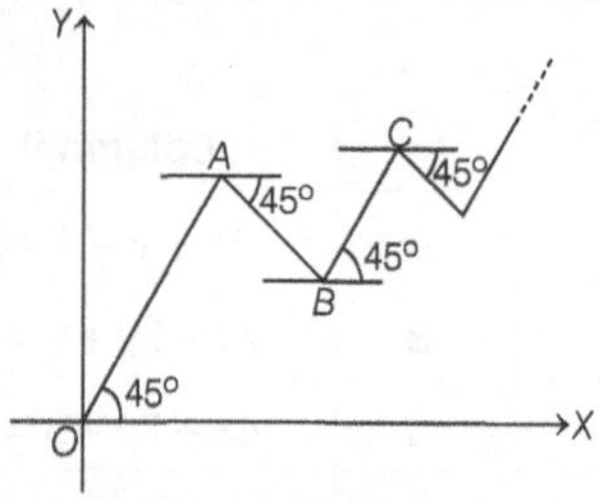

54 The equation of the altitudes AD, BE, CF of a ΔABC are $x + y = 0$, $x - 4y = 0$ and $2x - y = 0$, respectively. The coordinates of A are $(t, -t)$. If t varies, then locus of centroid of ΔABC is $ax + by + c = 0$, then find $a + b + c$

55 A diagonal of rhombus $ABCD$ is member of both the families of lines $(x + y - 1) + \lambda_1 (2x + 3y - 2) = 0$ and $(x - y + 2) + \lambda_2 (2x - 3y + 5) = 0$, where $\lambda_1, \lambda_2 \in R$ and one of the vertex of the rhombus is (3, 2).
If area of rhombus is $12\sqrt{5}$ sq units, then find the length of longer diagonal of the rhombus.

56 In an acute ΔABC, the base BC has the equation $4x - 3y + 3 = 0$. If the coordinates of the orthocentre (H) and circumcentre (P) of the triangle are (1, 2) and (2, 3) respectively, then the radius of the circle circumscribing the triangle is $\frac{\sqrt{m}}{n}$, where m and n are relatively prime number, find value of $m + n$.

57 The parallelogram is bounded by the lines $y = ax + c$, $y = ax + d$ and $y = bx + c$, $y = bx + d$ and has the area equal to 18 sq units. Also, the parallelogram bounded by the lines $y = ax + c$, $y = ax - d$ and $y = bx + c$, $y = bx - d$ has area 72 sq units. Given, a, b, c and d are positive integers, find the smallest possible value of $a + b + c + d$.

58 Consider two lines L_1 and L_2 given by $x - y = 0$ and $x + y = 0$, respectively and a moving point $P(x, y)$. Let $d(P, L_i) i = 1, 2$ represents the distance of point 'P' from the line L_i. If point P moves in certain region 'R' in such a way that $2 \leq d(P, L_1) + d(P, L_2) \leq 4$, find the area of region R.

59 A line through $A(-5, -4)$ meets the lines $x + 3y + 2 = 0$, $2x + y + 4 = 0$ and $x - y - 5 = 0$ at points B, C and D, respectively. If $\left(\frac{15}{AB}\right)^2 + \left(\frac{10}{AC}\right)^2 = \left(\frac{6}{AD}\right)^2$ and equation of the line is $ax + by + c = 0$, then find the value $25a - c$.

60 The sides of a triangle are $U_r = x \cos\alpha_r + y \sin\alpha_r - p_r = 0$ for $r = 1, 2, 3$, if the orthocentre of the triangle is $U_1 \cos(p\alpha_2 - q\alpha_3) = U_2 \cos(r\alpha_3 + s\alpha_1) = U_3 \cos(t\alpha_2 - u\alpha_1)$, then find the value of $5p + q - r + s - t + u$.

ANSWER KEY

Level 1

1. (a) $\frac{9}{4}$ (b) $k = 3$ (c) $\lambda = 0$ or $\lambda = 2$ (d) $k = -1$ (e) $R\left(-\frac{3}{5}, 0\right)$
2. (a) $x + y - 5 = 0$ (b) $x + y - 4 = 0$ (c) $3x + 4y - 1 = 0$ (d) $x + 12y - 58 = 0$ (e) $2x + 5y - 20 = 0$
   (f) $8x + 10y - 6y = 0$ (g) $\frac{x}{3} + \frac{y}{(-2)} = 1$ or $\frac{x}{(-2)} + \frac{y}{3} = 1$ (h) $4x + 3y = 24$ or $x + y = 7$
   (i) $2x + 9y + 7 = 0$
3. (a) $3x - 2y - 27 = 0$ (b) $2x + y - 5 = 0$ (c) $x - 5y + 10 = 0$ (d) $2x + y - 12 = 0$
4. (a) x-intercept $= -\frac{7}{3}$ and y-intercept $= \frac{7}{5}$ (b) x-intercept $= -\frac{2}{\sqrt{3}}$ and y-intercept $= \frac{8}{7\sqrt{3}}$
   (c) x-intercept $= \frac{4}{15}$ and y-intercept $= \frac{4}{3}$ (d) x-intercept $= -\frac{11}{9}$ and y-intercept $= -\frac{11}{8}$
5. Slope intercept form $\Rightarrow m = \frac{2}{3}$ and $c = \frac{4}{3}$, intercept form $\Rightarrow (a = -2)$ and $b = \frac{4}{3}$
   and normal form $\Rightarrow \alpha = \cos^{-1}\left(-\frac{2}{\sqrt{13}}\right)$ and $\alpha = \sin^{-1}\left(\frac{3}{\sqrt{13}}\right)$
6. Non-origin
7. Opposite side
8. (a) $\sqrt{5}$ units (b) $\frac{18}{\sqrt{29}}$ units (c) $\frac{23}{5}$ units
9. $\sqrt{5}$ units
10. (a) Foot $= \left(-\frac{1}{5}, \frac{2}{5}\right)$ and image $= \left(-\frac{7}{5}, -\frac{6}{5}\right)$
    (b) Foot $= \left(-\frac{23}{10}, \frac{21}{10}\right)$ and image $\left(-\frac{13}{5}, \frac{6}{2}\right)$
11. (a) $\frac{7}{5}$ units (b) $\frac{12}{5}a^2$ sq units (c) 5 units (d) $\pm\sqrt{45}$
    (e) $\tan^{-1}\left(\frac{9}{20}\right)$ (f) $m = \frac{1}{2}$ and -2 (g) $x + 3y = 13$ and $3x - y + 1 = 0$
12. $33x + 22y + 1 = 0$
13. $26x - 122y - 1675 = 0$
14. $A = \left(-\frac{36}{7}, -\frac{45}{7}\right)$
15. $14x + 23y - 40 = 0$
16. $y - 2 = 0$
17. (a) $m = 2$ (b) 10 (c) $19x + 19y + 3 = 0$
    (d) $3x + y + 7 = 0$ and $x - 3y = 31$ (e) Circumcentre $= \left(\frac{5}{2}, 0\right)$ and orthocentre $\left(-\frac{1}{2}, \frac{15}{2}\right)$
18. (a) $7x + 9y = 3$ (b) $\left(\frac{1}{3}, \frac{2}{3}\right)$ (c) $14x + 23y = 40$ (d) $6x + 2y - 5 = 0$ (e) $7x + y - 9 = 0$
    (f) $x + 2 = 0$ (g) $5x + y + 17 = 0$ (h) $\sqrt{3}x - y = 2 + 3\sqrt{3}$
19. (a) $29x - 2y = 31$ (b) $\frac{13}{5}$ (c) $(70 - 37\sqrt{3})x - 13y - 153 + 74\sqrt{3} = 0$

Level 2

1. (c)	**2.** (a)	**3.** (a)	**4.** (a)	**5.** (a)	**6.** (b)	**7.** (b)
8. (a)	**9.** (c)	**10.** (a)	**11.** (b)	**12.** (b)	**13.** (b)	**14.** (c)
15. (c)	**16.** (a)	**17.** (b)	**18.** (d)	**19.** (d)	**20.** (d)	**21.** (b)
22. (c)	**23.** (b)	**24.** (b)	**25.** (b)	**26.** (c)	**27.** (a)	**28.** (a)
29. (c)	**30.** (c)	**31.** (b)	**32.** (a)	**33.** (c)	**34.** (c)	**35.** (c)
36. (d)	**37.** (1)	**38.** (8)	**39.** (9)	**40.** (4)	**41.** (1.4)	**42.** (30)
43. (7)	**44.** (6)	**45.** (0.9)	**46.** (2.646)	**47.** (1)	**48.** (6)	

Level 3

1. (d)	**2.** (c)	**3.** (d)	**4.** (b)	**5.** (b)	**6.** (d)	**7.** (b)
8. (d)	**9.** (b)	**10.** (b, c)	**11.** (a, b)	**12.** (a, d)	**13.** (b, c)	**14.** (a, d)
15. (b, c, d)	**16.** (a, d)	**17.** (a, b)	**18.** (a, d)	**19.** (a, b, c)	**20.** (a, c, d)	**21.** (a, b, c, d)
22. (a, b, c, d)	**23.** (a, b, c)	**24.** (a, b)	**25.** (b)	**26.** (c)	**27.** (d)	**28.** (b)
29. (a)	**30.** (c)	**31.** (b)	**32.** (d)	**33.** (c)	**34.** (b)	**35.** (b)
36. (a)	**37.** (b)	**38.** (c)	**39.** (c)	**40.** (a)	**41.** (b)	**42.** (b)
43. (d)	**44.** (a)	**45.** (d)	**46.** (d)			

47. (i)–D, (ii)–A, (iii)–B, (iv)–C
48. (i)–A, (ii)–D, (iii)–B, (iv)–C
49. (i)–C, (ii)–D, (iii)–B
50. (i)–(A, D) (ii)–(B, D) (iii)–(A, C) (iv)–D
51. (i)–(A, D) (ii)–(A) (iii)–(B) (iv)–C

52. (5)	**53.** (8)	**54.** (6)	**55.** (10)	**56.** (63)	**57.** (16)	**58.** (24)
59. (28)	**60.** (4)					

SOLUTIONS

Level 1

1. (a) Given, $A(1,2)$ and $B(-3,-7)$

Slope of $AB \Rightarrow m = \dfrac{y_2 - y_1}{x_2 - x_1}$

$$= \frac{-7-2}{-3-1} = \frac{-9}{-4} = \frac{9}{4}$$

(b) Slope of the line joining (4, 10) and $(k,6)$

$$m = \frac{6-10}{k-4}$$

$$= \frac{-4}{k-4} = 4 \qquad \text{(given)}$$

$\Rightarrow \quad k-4 = -1$

$\Rightarrow \quad k = 3$

(c) Given,

$A(1,-2), B(3,\lambda), C(\lambda,3)$ and $D(4,7)$

$\therefore$ Slope of AB = Slope of CD

$\Rightarrow \quad \dfrac{\lambda+2}{3-1} = \dfrac{7-3}{4-\lambda}$

$\Rightarrow \quad (\lambda+2)(4-\lambda) = (4)(2)$

$\Rightarrow 4\lambda - \lambda^2 + 8 - 2\lambda = 8$

$\Rightarrow \quad \lambda^2 - 2\lambda = 0$

$\Rightarrow \quad \lambda(\lambda-2) = 0$

$\Rightarrow \quad \lambda = 0$ or $\lambda = 2$

(d) $A(k, 2-2k), B(-k+1, 2k)$ and $C(-4-k, 6-2k)$ are three points.

$\because$ A, B and C are collinear.

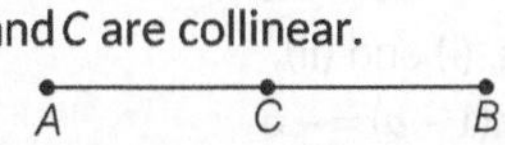

$\Rightarrow \quad m_{AB} = m_{BC}$

$\Rightarrow \left(\dfrac{2k-2+2k}{-k+1-k}\right) = \left(\dfrac{6-2k-2k}{-4-k+k-1}\right)$

$\Rightarrow \quad \dfrac{4k-2}{-2k+1} = \dfrac{6-4k}{-5}$

$\Rightarrow \quad \dfrac{2(2k-1)}{-(2k-1)} = \dfrac{-2(2k-3)}{-5}$

$\Rightarrow \quad -1 = \dfrac{1}{5}(2k-3)$

$\Rightarrow \quad -5 = 2k-3$

$\Rightarrow \quad 2k = -5+3$

$\therefore \quad k = \dfrac{-2}{2} = -1$

(e) $PR + RQ'$ is minimum

$= PR + RQ$ is minimum.

$= PQ$

$\therefore$ PRQ is a line.

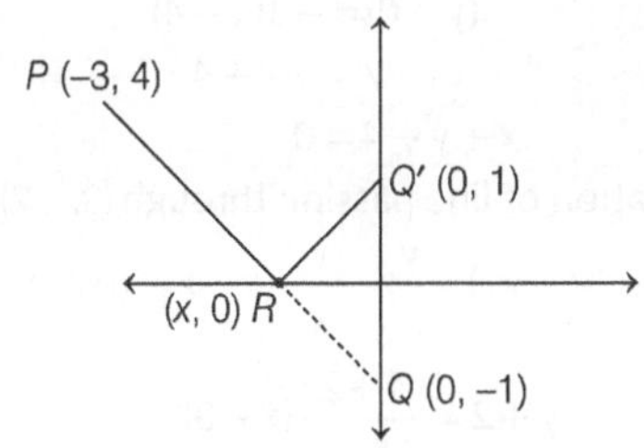

Slope of PR = Slope of PQ

$\Rightarrow \quad \dfrac{0-4}{x+3} = \dfrac{-1-4}{0+3}$

$\Rightarrow \quad \dfrac{4}{x+3} = \dfrac{5}{3}$

$\Rightarrow \quad 12 = 5x + 15$

$\Rightarrow \quad 5x = -3$

$\Rightarrow \quad x = -3/5$

$\therefore$ Coordinate of $R\left(-\dfrac{3}{5}, 0\right)$.

2. (a) Slope of line $m = \tan\theta$

$= \tan 135° = -1$

y-intercept $\Rightarrow c = 5$

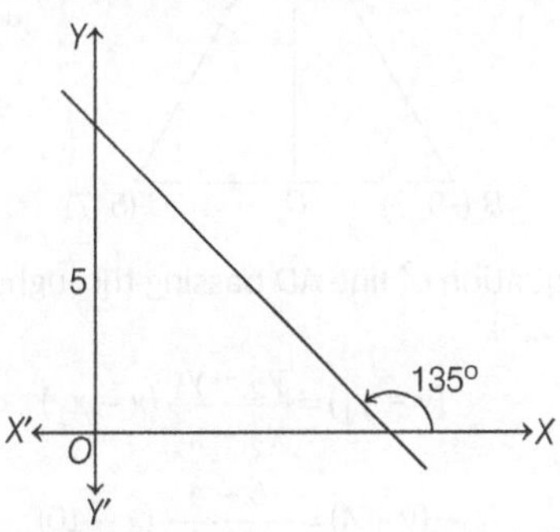

Using slope intercept form, the equation of a line is

$$y = mx + c$$

$$y = -x + 5$$

$\Rightarrow \quad x + y - 5 = 0$

(b) $\theta = 135°$

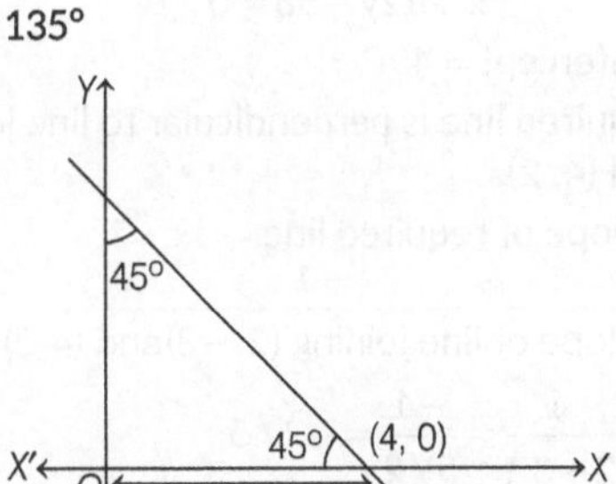

$\Rightarrow$ Slope, $m = \tan\theta$

$= \tan 135°$

$= -1$

Point (4, 0)

$(y - 0) = -1(x - 4)$

$\Rightarrow \quad y = -x + 4$

$\Rightarrow \quad x + y - 4 = 0$

(c) Equation of line passing through $(3, -2)$ and $(-5, 4)$

$$(y - y_1) = \frac{y_2 - y_1}{x_2 - x_1}(x - x_1) \text{ (two point form)}$$

$$\Rightarrow \quad y + 2 = \frac{4+2}{-5-3}(x - 3)$$

$$\Rightarrow \quad y + 2 = \frac{6}{-8}(x - 3)$$

$\Rightarrow \; -4(y + 2) = 3(x - 3)$

$\Rightarrow \; -4y - 8 = 3x - 9$

$\Rightarrow 3x + 4y - 1 = 0$

(d) For median AD

$D =$ mid-point of BC

$$= \left(\frac{-9+5}{2}, \frac{3+7}{2}\right)$$

$= (-2, 5)$

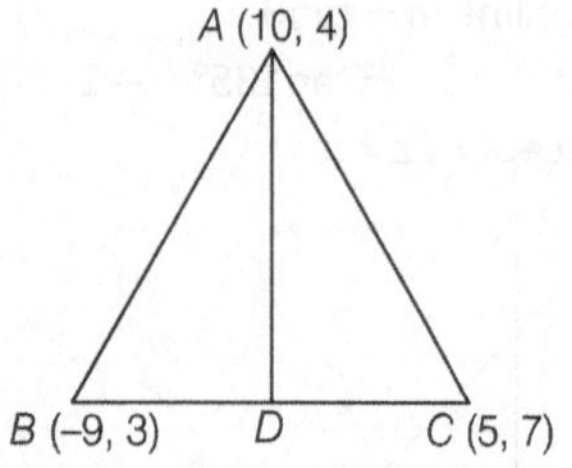

$\therefore$ Equation of line AD passing through $A(10, 4)$ and $D(-2, 5)$.

$$\Rightarrow \quad (y - y_1) = \frac{y_2 - y_1}{x_2 - x_1}(x - x_1)$$

$$\Rightarrow \quad (y - 4) = \frac{5-4}{-2-10}(x - 10)$$

$$\Rightarrow \quad y - 4 = \frac{1}{-12}(x - 10)$$

$\Rightarrow \quad -12y + 48 = x - 10$

$\Rightarrow \quad x + 12y - 58 = 0$

(e) y-intercept $= 4$

Required line is perpendicular to line joining $(2, -3)$ and $(4, 2)$.

$\therefore$ Slope of required line

$$= \frac{-1}{\text{Slope of line joining } (2, -3) \text{ and } (4, 2)}$$

$$= \frac{-1}{\left(\frac{2+3}{4-2}\right)} = \frac{-1}{5/2} = -2/5$$

Using slope intercept form

$y = mx + c$

$$\Rightarrow \quad y = \frac{-2}{5}x + 4 \Rightarrow 2x + 5y - 20 = 0$$

(f) $PQ =$ Right bisector of AB

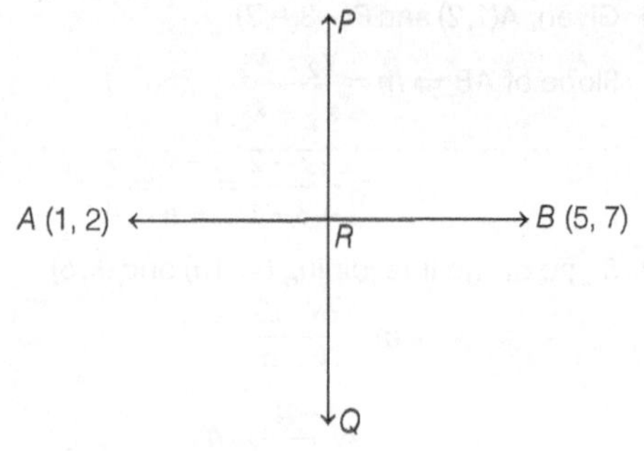

$\therefore R =$ mid-point of AB

$$= \left(\frac{5+1}{2}, \frac{7+2}{2}\right) = \left(3, \frac{9}{2}\right)$$

$$\therefore \quad m_{PQ} = \frac{-1}{m_{AB}} = \frac{-1}{\left(\frac{7-2}{5-1}\right)} = \frac{-1}{5/4} = -4/5$$

$$\therefore \text{ Equation of line } PQ \Rightarrow \left(y - \frac{9}{2}\right) = -\frac{4}{5}(x - 3)$$

$\Rightarrow \quad 10y - 45 = -8x + 24$

$\Rightarrow \quad 8x + 10y - 69 = 0$

(g) Let the equation of line be $\frac{x}{a} + \frac{y}{b} = 1$

Now, using given information,

$a + b = 1 \Rightarrow b = 1 - a$ ... (i)

$ab = -6$ ... (ii)

From Eqs. (i) and (ii),

$a(1 - a) = -6$

$\Rightarrow \quad a - a^2 = -6$

$\Rightarrow \quad a^2 - a - 6 = 0$

$\Rightarrow (a - 3)(a + 2) = 0$

	$a = 3$	$a = -2$
$\Rightarrow$	$b = 1 - a$	$b = 1 + 2$
	$= -2$	$= 3$

Equation of line $\frac{x}{3} + \frac{y}{-2} = 1$ or $\frac{x}{-2} + \frac{y}{3} = 1$

(h) Let equation of line be

$\frac{x}{a} + \frac{y}{b} = 1$ and $a + b = 14$

$\Rightarrow \quad b = 14 - a$ ...(i)

Line passes through the point (3, 4), so

$$\frac{3}{a} + \frac{4}{b} = 1$$

$$\Rightarrow \quad \frac{3b + 4a}{ab} = 1$$

$\Rightarrow \quad 3(14-a)+4a=a(14-a)$

$\Rightarrow \quad 42-3a+4a=14a-a^2$

$\Rightarrow \quad a^2-13a+42=0$

$\Rightarrow \quad (a-6)(a-7)=0$

When $a=6$, then $b=8$

When $a=7$, then $b=7$

$\therefore$ Required equation $\dfrac{x}{6}+\dfrac{y}{8}=1$ | $\dfrac{x}{7}+\dfrac{y}{7}=1$

$4x+3y=24$ | $x+y=7$

(i) Coordinates of $S=\left(\dfrac{7+6}{2},\dfrac{3-1}{2}\right)=\left(\dfrac{13}{2},1\right)$

So, the slope of the line PS is $\dfrac{-2}{9}$.

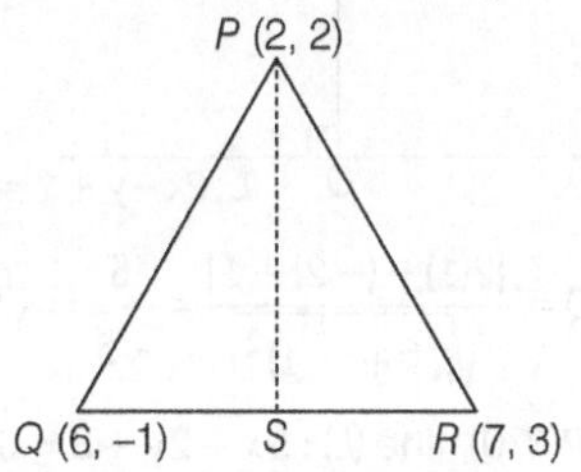

Equation of line passing through (1, – 1) and parallel to PS is

$$y+1=\frac{-2}{9}(x-1)\Rightarrow 2x+9y+7=0$$

3. (a) Equation of line parallel to the line $3x-2y+5=0$ is $3x-2y+\lambda=0$.

$\because$ It passes through $(5,-6)$.

$\therefore \quad 3(5)-2(-6)+\lambda=0$

$\Rightarrow \quad \lambda=-27$

$\therefore$ Equation of required line is

$3x-2y-27=0$

(b) Equation of line perpendicular to line $2x-4y+7=0$ is $4x+2y+k=0$

$\because$ It passes through (2, 1).

$\therefore \quad 4(2)+2(1)+k=0$

$k=-10$

$\therefore$ Equation of required line is $4x+2y-10=0$

$\Rightarrow \quad 2x+y-5=0$

(c) $\because$ AD is altitude.

$\Rightarrow$ Slope of $AD=\dfrac{-1}{m_{BC}}$

$=\dfrac{-1}{10/-2}=\dfrac{2}{10}=\dfrac{1}{5}$

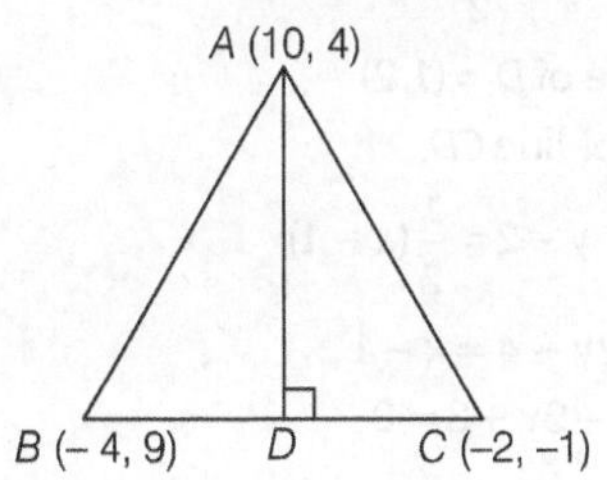

$\therefore$ Equation is $(y-4)=\dfrac{1}{5}(x-10)$

$\Rightarrow \quad 5y-20=x-10 \Rightarrow x-5y+10=0$

(d) Let AD is internal angle bisector of $\angle BAC$.

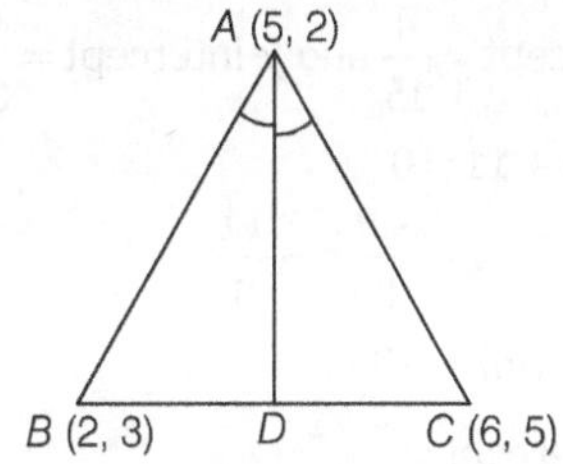

$\therefore \quad \dfrac{AB}{AC}=\dfrac{BD}{CD}$

Now, $AB=\sqrt{3^2+1^2}=\sqrt{10}$

$AC=\sqrt{1^2+3^2}=\sqrt{10}$

$\therefore \quad \dfrac{AB}{AC}=\dfrac{1}{1}\Rightarrow\dfrac{BD}{CD}=1$

$\therefore D$ is mid-point of BC.

$\therefore D\equiv\left(\dfrac{6+2}{2},\dfrac{5+3}{2}\right)=(4,4)$

$\therefore$ Equation of AD is

$\Rightarrow (y-4)=\dfrac{2-4}{5-4}(x-4)$

$\Rightarrow \; y-4=-2(x-4)$

$\Rightarrow \; y-4=-2x+8$

$\Rightarrow \; 2x+y-12=0$

4. (a) $3x-5y+7=0$

For slope $\Rightarrow y=\dfrac{3x}{5}+\dfrac{7}{5}$

$\therefore \quad m=\dfrac{3}{5}$

x-intercept, put $y=0\Rightarrow x=\dfrac{-7}{3}$

$\therefore$ x-intercept $=-7/3$

y-intercept, put $x=0\Rightarrow y=7/5$

$\therefore$ y-intercept $=7/5$

(b) $4\sqrt{3}x-7\sqrt{3}y+8=0$ ... (i)

$\Rightarrow \quad y=\dfrac{4\sqrt{3}x}{7\sqrt{3}}+\dfrac{8}{7\sqrt{3}}$

$\Rightarrow \quad y=\dfrac{4}{7}x+\dfrac{8}{7\sqrt{3}}$

$\therefore$ Slope $(m)=\dfrac{4}{7}$

x-intercept $=\dfrac{-8}{4\sqrt{3}}$

$=\dfrac{-2}{\sqrt{3}}$

and y-intercept $=\dfrac{8}{7\sqrt{3}}$

(c) $15x + 3y - 4 = 0$

$\Rightarrow \quad y = -5x + \frac{4}{3}$

$\therefore$ slope$(m) = -5$

x-intercept $= \frac{4}{15}$ and y-intercept $= \frac{4}{3}$

(d) $9x + 8y + 11 = 0$

$\Rightarrow \quad y = \frac{-9}{8}x - \frac{11}{8}$

$\therefore$ Slope $(m) = -9/8$

$\therefore x$-intercept $= \frac{-11}{9}$

and y-intercept $= \frac{-11}{8}$

5. Given line, $2x - 3y + 4 = 0$

For slope intercept form

$2x - 3y + 4 = 0$

$\Rightarrow \quad 3y = 2x + 4$

$\Rightarrow \quad y = \frac{2}{3}x + \frac{4}{3}$

$\because$ Slope intercept form, $y = mx + c$

$\therefore \quad m = \frac{2}{3}$ and $c = \frac{4}{3}$

For intercept form

$2x - 3y + 4 = 0$

$\Rightarrow \quad 2x - 3y = -4$

$\Rightarrow \quad \frac{2x}{-4} - \frac{3y}{-4} = 1 \Rightarrow \frac{x}{-2} + \frac{y}{4/3} = 1$

Intercept form $\frac{x}{a} + \frac{y}{b} = 1$

$\therefore a = -2$ and $b = 4/3$

Normal form

$2x - 3y + 4 = 0$

$\Rightarrow \quad 2x - 3y = -4$

$\Rightarrow \quad -2x + 3y = 4$

Dividing both the sides by $\sqrt{(-2)^2 + (3)^2} = \sqrt{13}$, we get

$$\left(\frac{-2}{\sqrt{13}}\right)x + \left(\frac{3}{\sqrt{13}}\right)y = \frac{4}{\sqrt{13}}$$

which is the normal for $x\cos\alpha + y\sin\alpha = p$

$\cos\alpha = \frac{-2}{\sqrt{13}}, \sin\alpha = \frac{3}{\sqrt{13}}$ and $p = \frac{4}{\sqrt{13}}$

$\Rightarrow \quad \alpha = \cos^{-1}(-2/\sqrt{13})$ and $\alpha = \sin^{-1}\left(\frac{3}{\sqrt{13}}\right)$

6. Let $f(x, y) = 3x + 4y + 5$

$\therefore f(4, -9) = 3 \times 4 + 4 \times (-9) + 5$

$= 12 - 36 + 5 = -19$

$\therefore f(4, -9) < 0$ and constant $5 > 0$.

Hence, the point $(4, -9)$ lies on non-origin side.

7. Given line, $4x - 5y = 9$

Let $f(x, y) = 4x - 5y - 9$

$\therefore \quad f(3, 2) = 4 \times 3 - 5 \times 2 - 9$

$= 12 - 10 - 9 = -7$

and $f(4, -6) = 4 \times 4 - 5 \times (-6) - 9$

$= 16 + 30 - 9 = 37$

Since, $f(3, 2)$ and $f(4, -6)$ have opposite signs. So, points (3, 2) and (4, −6) lie on opposite sides of the line.

8. (a) Distance between point $P(1, -2)$ and line $L: 2x - y + 1 = 0$ is PQ

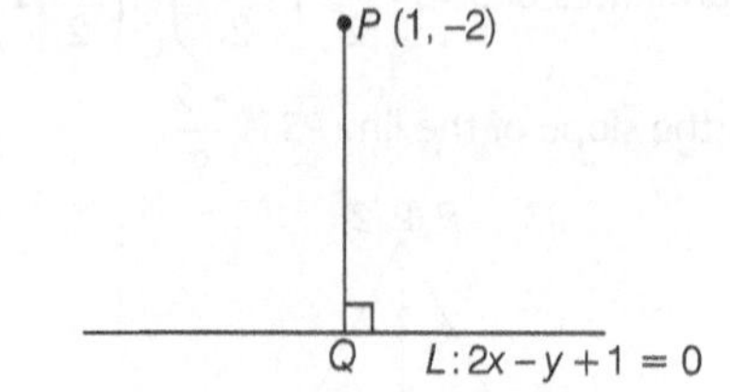

$\Rightarrow PQ = \frac{|2(1) - (-2) + 1|}{\sqrt{2^2 + (-1)^2}} = \frac{5}{\sqrt{5}} = \sqrt{5}$ units

(b) Point $P(3, 0)$, line $(L): 5x - 2y + 3 = 0$

Distance between P and L

$= \left|\frac{5(3) - 2(0) + 3}{\sqrt{5^2 + 2^2}}\right| = \frac{18}{\sqrt{29}}$ units

(c) Point $P(3, 4)$ and line $L: 3x + 4y - 2 = 0$

Distance between P and L $= \frac{|3(3) + 4(4) - 2|}{\sqrt{3^2 + 4^2}}$

$= \frac{|9 + 16 - 2|}{\sqrt{25}} = \frac{23}{5}$ units

9. Slope of $AB = -2$

Slope of line perpendicular to $AB = \frac{1}{2}$

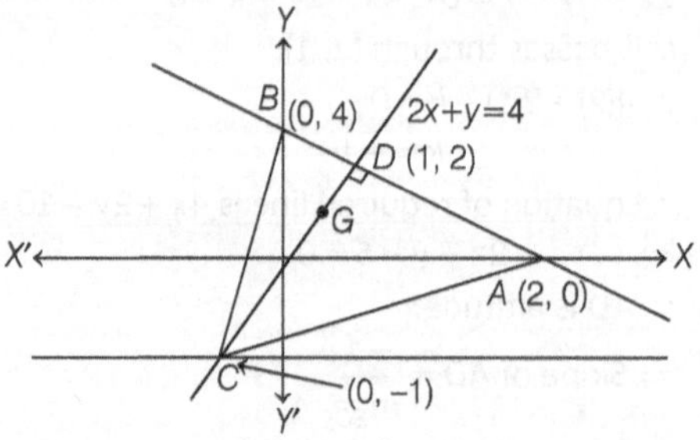

$\therefore$ Slope of $CD = \frac{1}{2}$

Coordinate of $D = (1, 2)$

Equation of line CD

$y - 2 = \frac{1}{2}(x - 1)$

$\Rightarrow \quad 2y - 4 = x - 1$

$\Rightarrow \quad x - 2y + 3 = 0$

Equation of horizontal line through $(0,-1)$ is $y=-1$

$\because$ Coordinate of C is intersection point of line through CD and horizontal line.

$\therefore \quad C=(-5,-1)$

$\therefore$ Perpendicular distance from G to AB

$$=\frac{1}{3}\text{(Perpendicular distance } C \text{ to } AB)$$

$$=\frac{1}{3}\left|\frac{-10-1-4}{\sqrt{4+1}}\right|=\sqrt{5}\text{ units}$$

10. (a) Point (1, 2) and line $3x+4y-1=0$

Let (h,k) and (α,β) are the foot of perpendicular and image of point (1, 2) w.r.t. the line $3x+4y-1=0$.

Then, $\frac{h-1}{3}=\frac{k-2}{4}=-\frac{[3(1)+4(2)-1]}{3^2+4^2}$

$$=\frac{-10}{25}=\frac{-2}{5}$$

$$\frac{h-1}{3}=\frac{-2}{5} \quad \left| \quad \frac{k-2}{4}=-2/5\right.$$

$$\Rightarrow \quad h=1-\frac{6}{5} \quad \left| \quad k=2-\frac{8}{5}\right.$$

$$h=\frac{-1}{5} \quad \left| \quad k=\frac{2}{5}\right.$$

and $\frac{\alpha-1}{3}=\frac{\beta-2}{4}=\frac{-2[3(1)+4(2)-1]}{3^2+4^2}=\frac{-4}{5}$

$$\therefore \alpha=1-\frac{12}{5} \quad \left| \quad \beta=2-\frac{16}{5}\right.$$

$$\alpha=\frac{-7}{5} \quad \left| \quad \beta=\frac{-6}{5}\right.$$

$\therefore$ Foot $=(-1/5, 2/5)$ and image $=\left(-7/5, \frac{-6}{5}\right)$.

(b) Point $(-2,3)$ line $x+3y=4$. Let (h,k) is foot of perpendicular and (α,β) image of point $(-2,3)$ w.r.t. $x+3y=4$

$$\therefore \quad \frac{h+2}{1}=\frac{k-3}{3}=-\frac{[-2+3(3)-4]}{1^2+3^2}=\frac{-3}{10}$$

$$\Rightarrow \quad h=-2-\frac{3}{10} \quad \left| \quad k=3-\frac{9}{10}\right.$$

$$\Rightarrow \quad h=\frac{-23}{10} \quad \left| \quad k=\frac{21}{10}\right.$$

and $\frac{\alpha+2}{1}=\frac{\beta-3}{3}=\frac{-2[-2+3(3)-4]}{1^2+3^2}=\frac{-6}{10}=\frac{-3}{5}$

$$\therefore \quad \alpha=-2-\frac{3}{5};\ \beta=3-\frac{9}{5}$$

$$\alpha=\frac{-13}{5};\quad \beta=\frac{6}{5}$$

Foot of perpendicular,

$$\left(-\frac{23}{10},\frac{21}{10}\right)$$

and Image $\left(\frac{-13}{5},\frac{6}{5}\right)$

11. (a) Given, $l_1: 8x+6y+24=0$ and $l_2: 8x+6y+10=0$

So, $\frac{a_1}{a_2}=\frac{8}{8}=1$ and $\frac{b_1}{b_2}=\frac{6}{6}=1$

and $\frac{a_1}{a_2}=\frac{b_1}{b_2}=1$

$\therefore$ Distance between them $=\left|\frac{c_1-c_2}{\sqrt{a^2+b^2}}\right|$

$$=\left|\frac{24-10}{\sqrt{8^2+6^2}}\right|$$

$$=\frac{14}{10}=\frac{7}{5}\text{ units}$$

(b) $\because$ For parallelogram equation of lines are given by

$$x+3y+a=0,\ 3x-2y+3a=0$$

and $x+3y-a=0,\ 3x-2y-3a=0$

i.e. $c_1=a, c_2=-a, d_1=3a, d_2=-3a$

and $a_1=1, b_1=3, a_2=3, b_2=-2$

$\therefore$ Required area $=\frac{(a+a)(3a+3a)}{(3+2)}$

$$=\frac{2a\times 6a}{5}=\frac{12a^2}{5}\text{ sq units}$$

(c) Let AB be the line passing through the point $A(2,5)$, making an angle θ with the positive direction of X-axis and parallel to another line CD.

$$\Rightarrow \quad AB \parallel CD$$

$\therefore$ Slope of $AB=$ Slope of $CD=\frac{3}{4}$

$\tan\theta=\frac{3}{4} \Rightarrow \sin\theta=\frac{3}{5}$ and $\cos\theta=\frac{4}{5}$

Let the $3x+y+4=0$ intersects BA at $P(x,y)$ such that $AP=r$ ...(i)

$\therefore$ Equation of $AB=\frac{x-2}{\cos\theta}=\frac{y-5}{\sin\theta}=r$

$$\Rightarrow \quad \frac{x-2}{\frac{4}{5}}=\frac{y-5}{\frac{3}{5}}=r$$

$$\therefore \quad P=\left(2+r\cdot\frac{4}{5}, 5+r\cdot\frac{3}{5}\right)$$

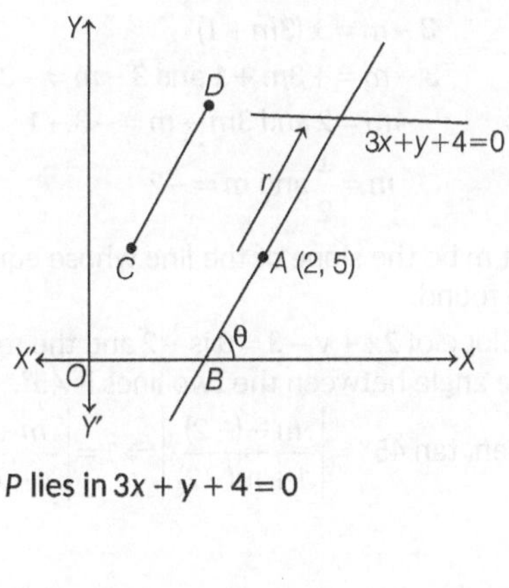

$\because P$ lies in $3x+y+4=0$

$$\Rightarrow 3\left(2+\frac{4r}{5}\right)+\left(5+\frac{3r}{5}\right)+4=0$$

$\Rightarrow \quad 3r=-15 \Rightarrow r=-5$

$\therefore \quad |r|=|-5|=5$

$\therefore$ Required distance = 5 units

(d) $\because$ Angle between AB and $CD = 90°$

$\Rightarrow \quad m_{AB} \times m_{CD} = -1$

$$\Rightarrow \left(\frac{2-1}{7+k}\right)\times\left(\frac{0-4}{4-k+3}\right)=-1$$

$$\Rightarrow \frac{1\times(-4)}{(k+7)(7-k)}=-1$$

$\Rightarrow \quad 4=(k+7)(7-k)$

$\Rightarrow \quad 4=7^2-k^2$

$\Rightarrow \quad k^2=7^2-4$

$\Rightarrow \quad k^2=45 \Rightarrow k=\pm\sqrt{45}$

(e) Let $A(1,2)$, $B(3,5)$, $C(5,7)$ and $D(4,1)$ are four points and angle between AB and CD is θ.

$$\therefore \tan\theta=\left|\frac{m_{AB}-m_{CD}}{1+m_{AB}\cdot m_{CD}}\right|$$

$$=\left|\frac{\frac{3}{2}-6}{1+\frac{3}{2}\cdot 6}\right|$$

$$=\left|\frac{3-12}{2+18}\right|=\frac{9}{20}$$

$$\theta=\tan^{-1}\frac{9}{20}$$

(f) Let slope of the other line is m.

$$\therefore \quad \tan\theta=\left|\frac{m_1-m_2}{1+m_1m_2}\right|$$

$$\Rightarrow \quad \tan\frac{\pi}{4}=\left|\frac{3-m}{1+3m}\right|$$

$$\Rightarrow \quad 1=\left|\frac{3-m}{3m+1}\right|$$

$$\Rightarrow \frac{3-m}{3m+1}=\pm 1$$

$\Rightarrow \quad 3-m=\pm(3m+1)$

$\Rightarrow \quad 3-m=+3m+1$ and $3-m=-3m-1$

$\Rightarrow \quad 4m=2$ and $3m-m=-3-1$

$\Rightarrow \quad m=\frac{1}{2}$ and $m=-2$

(g) Let m be the slope of the line whose equation is to be found.

$\therefore$ Slope of $2x+y-3=0$ is -2 and the measure of the angle between the two lines is 45°.

Then, $\tan 45° = \left|\frac{m-(-2)}{1+m(-2)}\right| \Rightarrow 1=\left|\frac{m+2}{1-2m}\right|$

$$\Rightarrow \quad \frac{m+2}{1-2m}=\pm 1 \Rightarrow m=-\frac{1}{3}, 3$$

$\therefore$ Equation of lines,

$y-4=-\frac{1}{3}(x-1)$ and $y-4=3(x-1)$

$\Rightarrow \quad x+3y=13$ and $3x-y+1=0$

12. Given, lines $3x-4y-7=0$ ...(i)

and $\quad 12x-5y-13=0$ ...(ii)

Equation of line passing through the lines (i) and (ii), we get

$3x-4y-7+\lambda(12x-5y-13)=0$

$\Rightarrow (3+12\lambda)x-(4+5\lambda)y-(7+13\lambda)=0$ ...(iii)

Since, this line is perpendicular to $2x-3y+5=0$

$$\therefore \quad \left(\frac{3+12\lambda}{4+5\lambda}\right)\left(\frac{2}{3}\right)=-1$$

$\Rightarrow \quad 6+24\lambda=-12-15\lambda$

$\Rightarrow \quad \lambda=\frac{-6}{13}$

Putting the value of λ in Eq. (iii), we get

$$\left(3-\frac{72}{13}\right)x-\left(4-\frac{30}{13}\right)y-(7-6)=0$$

$\Rightarrow \quad 33x+22y+1=0$

13. Let the equation of lines AB and AC are

$3x-2y+6=0$ ...(i)

and $\quad 4x+5y-20=0$ ...(ii)

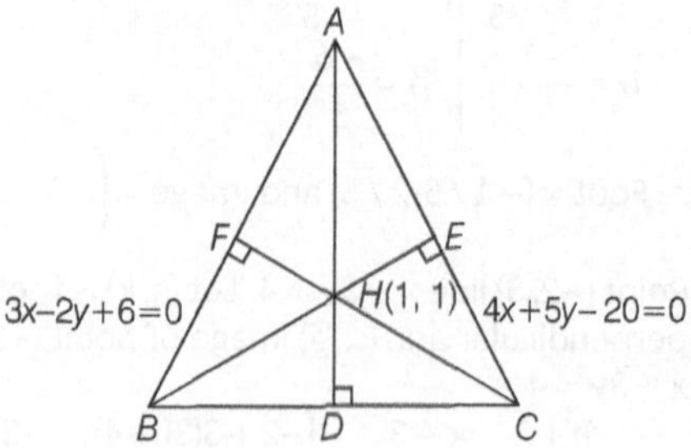

$\because BE$ passes through $H(1,1)$ and perpendicular to AC

$\therefore$ Equation of line BE is

$5x-4y=5(1)-4(1)$

$\Rightarrow \quad 5x-4y=1$ ...(iii)

Similarly, equation of CF is

$2x+3y=5$ ...(iv)

Solving Eqs. (i) and (iii), we get $B=\left(-13,-\frac{33}{2}\right)$

Similarly, solving Eqs. (ii) and (iv), we get $C=\left(\frac{35}{2},-10\right)$

$\therefore$ Equation of BC is

$$y+\frac{33}{2}=\frac{-\frac{33}{2}+10}{-13-\frac{35}{2}}(x+13)$$

$\Rightarrow 26x-122y-1675=0$

14. Let ABC be the given triangle, $B(3,-1)$ and $C(-2,3)$.
Let H be the orthocentre of ΔABC. Given, $H(0,0)$.
$\because$ AB passes through the point $B(3,-1)$ and perpendicular to the line CH.

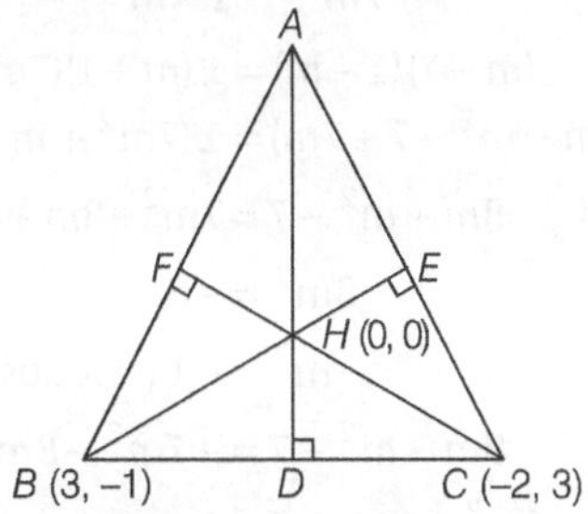

$\therefore$ Equation of AB will be
$$y+1=-\left(\frac{0+2}{0-3}\right)(x-3) \Rightarrow 2x-3y=9 \quad \text{...(i)}$$
$\because$ AC passes through $C(-2,3)$ and perpendicular to BH.
$\therefore$ It's equation will be
$$\Rightarrow \quad y-3=3(x+2) \Rightarrow 3x-y=-9 \quad \text{...(ii)}$$
Solving Eqs. (i) and (ii), we get $x=-\frac{36}{7}$ and $y=-\frac{45}{7}$

$\therefore$ Coordinates of third vertex $A=\left(-\frac{36}{7},-\frac{45}{7}\right)$

15. Let $B(x_1,y_1)$ and $C(x_2,y_2)$ be two vertices of ΔABC.
Given line, $x-y+5=0$ is perpendicular bisector of AB.
$\because$ B is image of $A(1,-2)$ to the line $x-y+5=0$

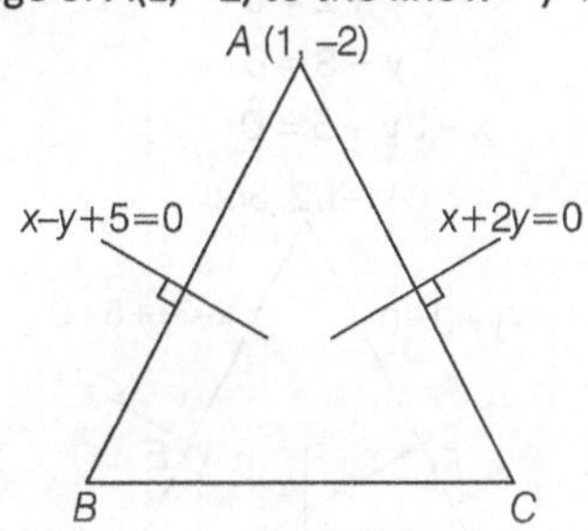

$$\Rightarrow \quad \frac{x_1-1}{1}=\frac{y_1+2}{-1}=-\frac{2(1+2+5)}{2}$$
$$\Rightarrow \quad x_1=-7 \text{ and } y_1=6$$
and line $x+2y=0$ is perpendicular bisector of AC.
$$\therefore \quad \frac{x_2-1}{1}=\frac{y_2+2}{2}=\frac{-2(1-4)}{5}$$
$$\therefore \quad x_2=\frac{11}{5} \text{ and } y_2=\frac{2}{5}$$
$\therefore$ Coordinates of B and C are $(-7,6)$ and $\left(\frac{11}{5},\frac{2}{5}\right)$, respectively.
$\therefore$ Equation of line BC is
$$y-6=\frac{\frac{2}{5}-6}{\frac{11}{5}+7}(x+7) \Rightarrow 14x+23y-40=0$$

16. $\sqrt{3}x+y=0$ makes an angle of 120° with OX.
and $\sqrt{3}x-y=0$ make an angle of 60° with OX.
$\because$ ΔOAB is an equilateral triangle.

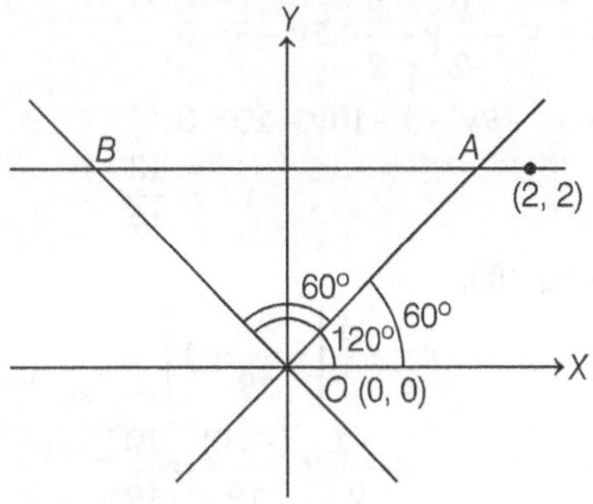

So, the required equation of line is parallel to X-axis and passes through $(2,2)$.
$\therefore$ Equation of line is $y-2=0$

17. (a) On solving $y=x+1$ and $2x+y=16$,
we get $x=5$ and $y=6$
Thus, the point of concurrency is $(5,6)$.
$\because$ The given lines are concurrent, so
$$6=5m-4$$
$$\Rightarrow \quad m=\frac{10}{5}=2$$
(b) We have, $4a^2+9b^2-c^2+12ab=0$
$$\Rightarrow \quad 4a^2+9b^2+12ab-c^2=0$$
$$\Rightarrow \quad (2a+3b)^2-(c)^2=0$$
$$\Rightarrow \quad (2a+3b+c)(2a+3b-c)=0$$
$$\Rightarrow \quad 2a+3b+c=0 \quad \text{...(i)}$$
$$\text{and} \quad 2a+3b-c=0 \quad \text{...(ii)}$$
Given, family of lines are
$$ax+by+c=0 \quad \text{...(iii)}$$
By Eqs. (i) and (iii), we get
$$ax+by-2a-3b=0$$
$$\Rightarrow \quad a(x-2)+b(y-3)=0$$
$$\Rightarrow \quad (x-2)+\frac{b}{a}(y-3)=0$$
$$\Rightarrow \quad x=2 \text{ and } y=3$$
$$\Rightarrow \quad m=2 \text{ and } n=3$$
By Eqs. (ii) and (iii),
$$ax+by+2a+3b=0$$
$$\Rightarrow \quad (x+2)+\frac{b}{a}(y+3)=0$$
$$\Rightarrow \quad x=-2 \text{ and } y=-3$$
$$\Rightarrow \quad p=-2 \text{ and } q=-3$$
$$\therefore \quad m+n+p+q+10=2+3-2-3+10=10$$
(c) Point of intersection
$$2x+3y+1=0 \quad \text{...(i)}$$
$$\text{and} \quad 3x-5y-5=0 \quad \text{...(ii)}$$
From Eq. (i),
$$x=-\frac{1}{2}(3y+1) \quad \text{...(iii)}$$

From Eq. (ii),

$$3\left[-\frac{1}{2}(3y+1)\right]-5y-5=0$$

$$\Rightarrow \quad -\frac{9}{2}y-\frac{3}{2}-5y-5=0$$

$$\Rightarrow \quad -9y-3-10y-10=0$$

$$y=-\frac{13}{19}$$

By Eq. (iii),

$$x=-\frac{1}{2}\left[-\frac{39}{19}+1\right]$$

$$=\frac{-1}{2}\times\frac{(-20)}{19}=\frac{10}{19}$$

$\therefore$ Point is $\left(\frac{10}{19},-\frac{13}{19}\right)$

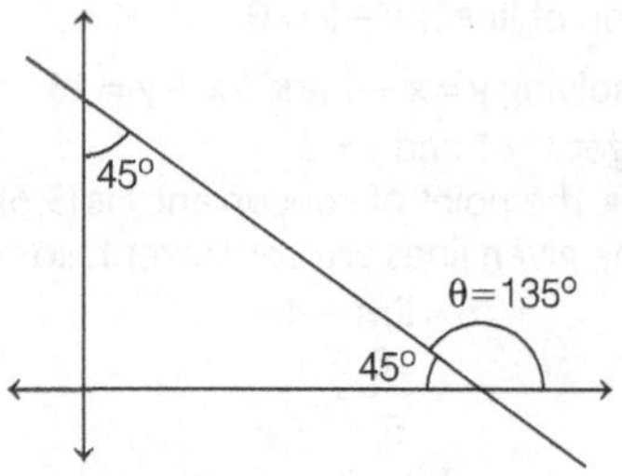

$\because$ Line is equally inclined with axes

$$\theta=135^\circ \Rightarrow \tan\theta=-1,$$

$\therefore$ Equation of line $\left(y+\frac{13}{19}\right)=-\left(x-\frac{10}{19}\right)$

$$\Rightarrow \quad 19y+13=-19x+10$$

$$\Rightarrow \quad 19x+19y+3=0$$

(d) Point $A \Rightarrow$ Point of intersection of AB and AC

$$7x-y+3=0 \quad \text{...(i)}$$

and $\quad x+y-3=0 \quad$ ...(ii)

Solving Eqs. (i) and (ii), we get

$$8x=0$$

$$\Rightarrow \quad x=0$$

Then, $y=3$

$\therefore$ $A(0,3)$

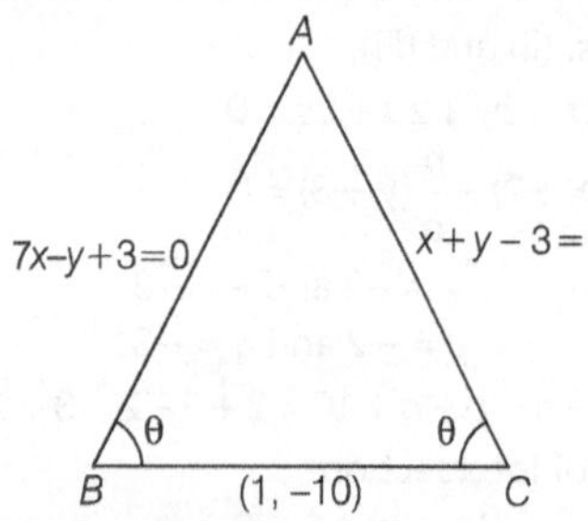

Let slope of $BC=m$

$\therefore$ $m_{AB}=7$ and $m_{AC}=-1$

$\therefore$ $\tan\theta=\left|\frac{m-7}{1+7m}\right|$ and $\tan\theta=\left|\frac{m+1}{1-m}\right|$

$$\Rightarrow \quad \left|\frac{m-7}{1+7m}\right|=\left|\frac{m+1}{1-m}\right|$$

$$\Rightarrow \quad \frac{m-7}{1+7m}=\pm\frac{m+1}{1-m}$$

$$\Rightarrow \quad (m-7)(1-m)=\pm(m+1)(7m+1)$$

$$\Rightarrow \quad (m-m^2-7+7m)=\pm(7m^2+m+7m+1)$$

$$\Rightarrow \quad 8m-m^2-7=7m^2+8m+1$$

$$\Rightarrow \quad 8m^2=-8$$

$$\Rightarrow \quad m^2=-1 \text{ (Not possible)}$$

$$\Rightarrow \quad 8m-m^2-7=-7m^2-8m-1$$

$$\Rightarrow \quad 6m^2+16m-6=0$$

$$\Rightarrow \quad 3m^2+8m-3=0$$

$$\Rightarrow \quad 3m^2+9m-m-3=0$$

$$\Rightarrow \quad 3m(m+3)-1(m+3)=0$$

$$\Rightarrow \quad (m+3)(3m-1)=0$$

$$m=-3, \frac{1}{3}$$

$\therefore$ Equation of third side,

$$\Rightarrow \quad (y+10)=-3(x-1) \text{ and } y+10=\frac{1}{3}(x-1)$$

$$\Rightarrow \quad y+10=-3x+3 \text{ and } 3y+30=x-1$$

$$\Rightarrow \quad 3x+y+7=0 \text{ and } x-3y=31$$

(e) Given line,

$$x-y+2=0 \quad \text{...(i)}$$

$$y-3=0 \quad \text{...(ii)}$$

$$x-3y+5=0 \quad \text{...(iii)}$$

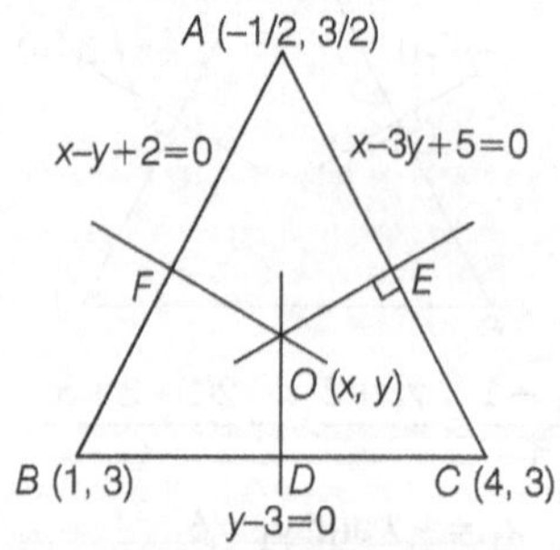

Let $O(x,y)$ be the circumcentre.

$\because$ D, E, F are mid-points of BC, AC and AB, respectively.

And $OD \perp BC$, $OE \perp AC$, $OF \perp AB$

For *B* Solving Eqs. (i) and (ii), we get $y=3$ and $x=1$

For *C* Solving Eqs. (ii) and (iii), we get $y=3$ and $x=4$

For *A* Solving Eqs. (i) and (iii), we get

$$y=\frac{3}{2} \text{ and } x=-\frac{1}{2}$$

Now, $D \Rightarrow \left(\frac{5}{2},3\right)$, $E \Rightarrow \left(\frac{7}{4},\frac{9}{4}\right)$ and $F \Rightarrow \left(\frac{1}{4},\frac{9}{4}\right)$

Now, $\quad m_{OD}\times m_{BC}=-1$

$$\Rightarrow \quad \frac{y-3}{x-\frac{5}{2}} \times 0 = -1$$

$$\Rightarrow \quad x - \frac{5}{2} = 0$$

$$\Rightarrow \quad x = \frac{5}{2}$$

and $\quad m_{OE} \times m_{AC} = -1$

$$\Rightarrow \quad \frac{y-\frac{9}{4}}{x-\frac{7}{4}} \times \frac{1}{3} = -1$$

$$\Rightarrow \quad y - \frac{9}{4} = -3\left(x - \frac{7}{4}\right)$$

Put $x = \frac{5}{2}$

$$\Rightarrow y = \frac{9}{4} - 3\left(\frac{5}{2} - \frac{7}{4}\right)$$

$$= \frac{9}{4} - \frac{9}{4} = 0$$

Similarly, $m_{OF} \times m_{AB} = -1$

$$\Rightarrow \quad \frac{y-\frac{9}{4}}{x-\frac{1}{4}} \times 1 = -1$$

$$\Rightarrow \quad y - \frac{9}{4} = -\left(x - \frac{1}{4}\right)$$

Put $\quad x = \frac{5}{2}$

$$\Rightarrow y - \frac{9}{4} = -\left(\frac{5}{2} - \frac{1}{4}\right)$$

$$y = 0$$

$\therefore$ Circumcentre is $\left(\frac{5}{2}, 0\right)$

Now, for orthocentre

Let $O'(x, y)$ be the orthocentre

$\therefore$ Slope of $AC = \frac{1}{3}$

$\therefore$ Slope of $AB = 1$

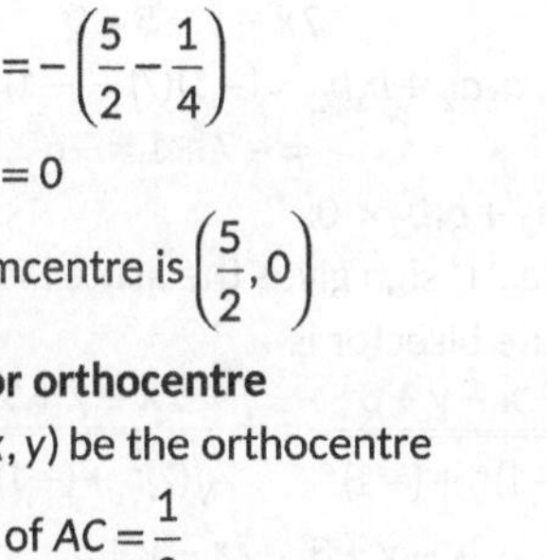

$\therefore$ Slope $O'A \times$ Slope of $BC = -1$

$$\Rightarrow \quad \frac{y-\frac{3}{2}}{x+\frac{1}{2}} \times 0 = -1$$

$$\Rightarrow \quad x + \frac{1}{2} = 0$$

$$\Rightarrow \quad x = -\frac{1}{2}$$

$\therefore$ Slope of $O'B \times$ Slope of $CA = -1$

$$\frac{y-3}{x-1} \times \frac{1}{3} = -1$$

$$y - 3 = -3(x-1)$$

Put $x = -\frac{1}{2}$

$$y - 3 = -3\left(-\frac{3}{2}\right)$$

$$\Rightarrow \quad y = 3 + \frac{9}{2} = \frac{15}{2}$$

$\therefore$ Orthocentre is $\left(-\frac{1}{2}, \frac{15}{2}\right)$.

18. (a) $L_1 : -4x - 3y + 6 = 0$

$L_2 : 5x + 12y + 9 = 0$

Both constant are positive.

$\therefore$ The equation of bisector, containing origin is

$$\frac{-4x-3y+6}{\sqrt{4^2+3^2}} = \frac{5x+12y+9}{\sqrt{5^2+12^2}}$$

$$\Rightarrow \frac{-4x-3y+6}{5} = \frac{5x+12y+9}{13}$$

$$\Rightarrow 13(-4x-3y+6) = 5(5x+12y+9)$$

$$\Rightarrow \quad 7x + 9y = 3$$

(b) Let ABC be a triangle

where, $AB : x + 1 = 0$

$BC : 3x - 4y = 5$

$CA : 5x + 12y = 27$

Case I Acute angle bisector between AB and AC.

Given lines, $x + 1 = 0$ and $-3x + 4y + 5 = 0$

Now, $a_1a_2 + b_1b_2 = 1(-3) + 0(4) = -3 < 0$

Thus, positive one is the acute angle bisector.

$\therefore$ The acute angle bisector is

$$\frac{x+1}{\sqrt{1^2}} = -\frac{3x+4y+5}{\sqrt{3^2+4^2}}$$

$$\Rightarrow \quad \frac{x+1}{1} = \frac{-3x+4y+5}{5}$$

$$\Rightarrow \quad 5x + 5 = -3x + 4y + 5$$

$$\Rightarrow \quad 8x - 4y = 0 \Rightarrow 2x - y = 0$$

Case II Acute angle bisector between BC and AC.

$-3x + 4y + 5 = 0$ and $-5x - 12y + 27 = 0$

Now, $a_1a_2 + b_1b_2 \Rightarrow (-3)(-5) + 4(-12) = -33 < 0$

$\therefore$ The positive one is acute angle bisector.

$\therefore$ Acute angle bisector is

$$\frac{-3x+4y+5}{5}=\frac{-5x-12y+27}{13}$$

$\Rightarrow \quad -39x+52y+65=-25x-60y+135$

$\Rightarrow \quad 14x-112y+70=0$

$\Rightarrow \quad x-8y+5=0$

$\Rightarrow \quad x-8y=-5$

$\therefore$ Incentre is intersection of $2x-y=0$ and $x-8y=-5$

$$x=\frac{-5}{-15}=\frac{1}{3} \text{ and } y=\frac{2}{3}$$

$\therefore$ Incentre is $\left(\frac{1}{3},\frac{2}{3}\right)$.

(c) Point of intersection of L_1 and L_2

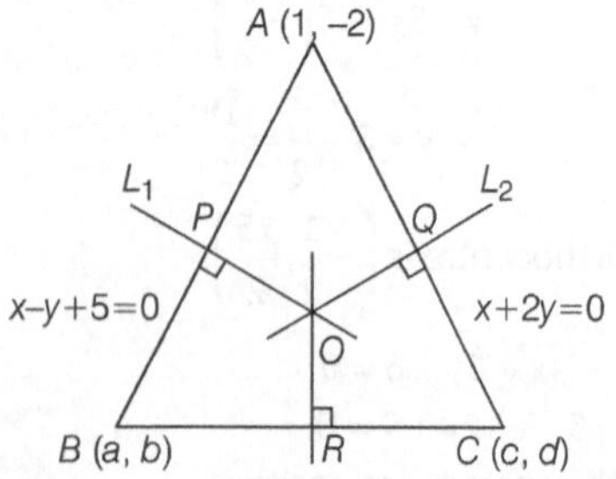

Given line, $x-y+5=0$ ...(i)

$x+2y=0$ ...(ii)

Solving Eqs. (i) and (ii), we get

$-3y=-5$

$\Rightarrow \quad y=\frac{5}{3}$ and $x=-\frac{10}{3}$

$\therefore O\left(-\frac{10}{3},\frac{5}{3}\right)$.

Point $P \Rightarrow \left(\frac{a+1}{2},\frac{b-2}{2}\right)$ lies on $x-y+5=0$

$\Rightarrow \quad \frac{a+1}{2}-\frac{b-2}{2}+5=0$

$\Rightarrow \quad a+1-b+2+10=0$

$a-b=-13$ ...(iii)

$\therefore$ Slope of $L_1=1$

and slope of $AB=\frac{b+2}{a-1}$

$\Rightarrow \quad 1\times\frac{b+2}{a-1}=-1$

$\Rightarrow \quad a-1=-b-2$

$\Rightarrow \quad a+b=-1$ ...(iv)

By Eqs. (iii) and (iv),

$a=-7$ and $b=6$

$\therefore B(-7,6)$

Point $Q\left(\frac{c+1}{2},\frac{d-2}{2}\right)$ lies on $x+2y=0$

$\Rightarrow \quad \frac{c+1}{2}+d-2=0$

$\Rightarrow \quad c+1+2d-4=0$

$\Rightarrow \quad c+2d=3$ ...(v)

$\therefore$ Slope of $L_2=\frac{-1}{2}$

and slope of $AC=\frac{d+2}{c-1}$

$\Rightarrow \quad \frac{d+2}{c-1}\times\left(-\frac{1}{2}\right)=-1$

$\Rightarrow \quad d+2=2c-2$

$\Rightarrow \quad 2c-d=4$ ...(iv)

By Eqs. (v) and (vi),

$6-4d-d=4$

$\Rightarrow \quad -5d=-2$

$\Rightarrow \quad d=\frac{2}{5}$ and $c=3-\frac{4}{5}=\frac{11}{5}$

$\therefore \quad C\left(\frac{11}{5},\frac{2}{5}\right)$.

$\therefore$ Equation of BC is

$$y-6=\frac{\frac{2}{5}-6}{\frac{11}{5}+7}(x+7)$$

$\Rightarrow \quad y-6=-\frac{28}{46}(x+7)$

$\Rightarrow \quad 23y-138=-14x-98$

$\Rightarrow \quad 14x+23y=40$

(d) Given equation is $-x-y+3=0$

and $7x-y+5=0$

Here, $a_1a_2+b_1b_2=(-1)(7)+(-1)(-1)$

$=-7+1=-6$

$\therefore a_1a_2+b_1b_2<0$.

Hence, '+' sign gives the acute bisector.

$\therefore$ Acute bisector is

$$\frac{-x-y+3}{\sqrt{(-1)^2+(-1)^2}}=+\frac{7x-y+5}{\sqrt{(7)^2+(-1)^2}}$$

$\Rightarrow \quad \frac{-x-y+3}{\sqrt{2}}=\frac{7x-y+5}{5\sqrt{2}}$

$\Rightarrow -5x-5y+15=7x-y+5$

$\Rightarrow 12x+4y-10=0$

or $\quad 6x+2y-5=0$

(e) The given straight lines are

$3x+4y=5$

and $\quad 4x-3y=15$

Clearly lines are perpendicular to each other (i.e. $m_1m_2=-1$) and intersect at A. Now, B and C are points on these lines such that $AB=AC$ and BC passes through $(1,2)$.

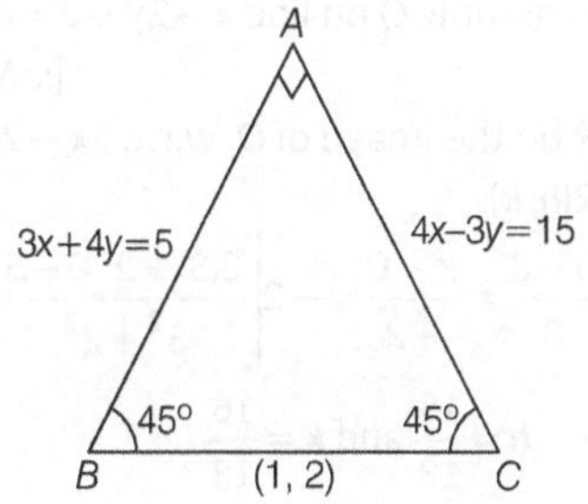

Let slope of BC be m, then

$$\tan B = \left|\frac{m_1 - m_2}{1 + m_1 m_2}\right|$$

$$\tan 45° = \left|\frac{m + \frac{3}{4}}{1 - \frac{3}{4}m}\right|$$

$\Rightarrow \quad 4m + 3 = \pm(4 - 3m)$

$\Rightarrow \quad 4m + 3 = 4 - 3m$

or $\quad 4m + 3 = -4 + 3m$

$\Rightarrow \quad m = \frac{1}{7}$ or $m = -7$

$\therefore$ Equation of BC is

$$y - 2 = \frac{1}{7}(x - 1)$$

or $\quad y - 2 = -7(x - 1)$

$\Rightarrow \quad x - 7y + 13 = 0$

or $\quad 7x + y - 9 = 0$

(f) Given, lines are

$$4x + 3y - 12 = 0$$

and $\quad 4x + 3y - 3 = 0$

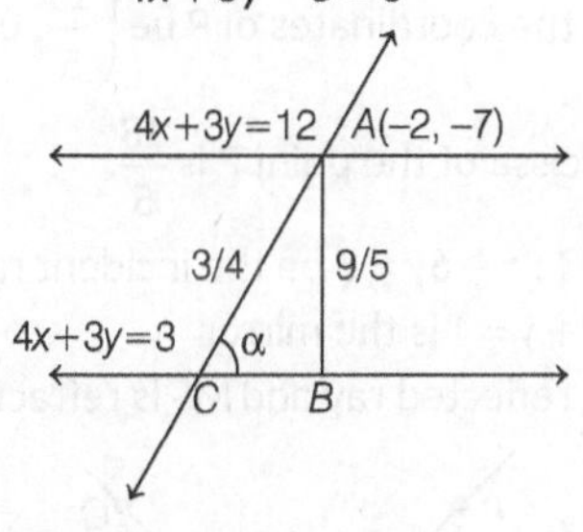

Given, $AC = 3$

AB = Distance between two parallel lines

$$\therefore AB = \left|\frac{12 - 3}{\sqrt{4^2 + 3^2}}\right| = \frac{9}{5}$$

In ΔABC, $\tan\alpha = \frac{AB}{BC} = \frac{\frac{9}{5}}{\frac{12}{5}} = \frac{3}{4}$

$$\left[\because BC = \sqrt{9 - \frac{81}{25}}, BC = \frac{12}{5}\right]$$

Let m be the slope of AC

$\therefore$ Slope of given line $= -\frac{4}{3}$

$$\therefore \quad \tan\alpha = \left|\frac{m + \frac{4}{3}}{1 - \frac{4}{3}m}\right|$$

$$\Rightarrow \quad \frac{3}{4} = \left|\frac{3m + 4}{3 - 4m}\right| \Rightarrow \frac{3m + 4}{3 - 4m} = \pm\frac{3}{4}$$

$\Rightarrow 12m + 16 = 9 - 12m$

or $12m + 16 = -9 + 12m$

$\therefore \quad m = -\frac{7}{24}$ and $m = \frac{25}{0}$ [not possible]

Hence, equation of line passing through $(-2, -7)$ and $m = -\frac{7}{24}$, is

$$y + 7 = -\frac{7}{24}(x + 2)$$

$\Rightarrow \quad 7x + 24y + 182 = 0$

and equation of line passing through $(-2, -7)$ and slope is not defined as

$$x = -2 \Rightarrow x + 2 = 0$$

(g) In ΔABC,

Let BP is altitude and BD is angle bisector respectively, drawn through B.

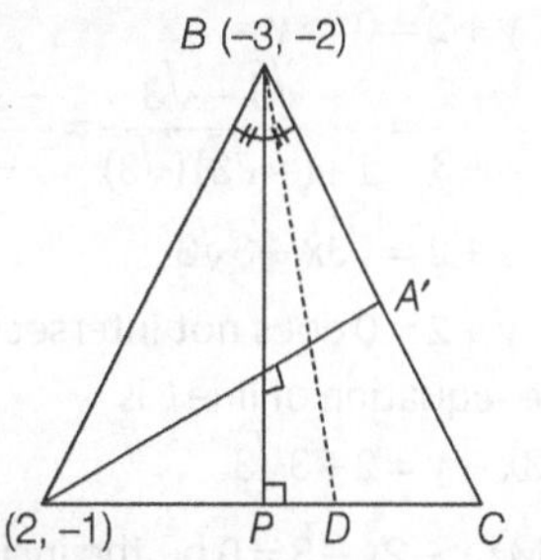

Given, equation of BP

$$7x - 10y + 1 = 0 \quad \text{...(i)}$$

and equation of BD

$$3x - 2y + 5 = 0 \quad \text{...(ii)}$$

Solving Eqs. (i) and (ii), we get

$$x = -3 \text{ and } y = -2$$

$\therefore$ Coordinate of $B(-3, -2)$.

BD is angle bisector.

$\therefore$ Image of A w.r.t. line

$3x - 2y + 5 = 0$ is lie on line BC.

$\because$ Image of A i.e. A' is

$$\frac{x - 2}{3} = \frac{y + 1}{-2}$$

$$= -2\left(\frac{6 + 2 + 5}{13}\right) = -2$$

$$x = -4 \text{ and } y = 3$$

$\therefore \quad A' = (-4, 3)$

Equation of line BC is

$$y-3=\frac{3+2}{-4+3}(x+4)$$

$$\Rightarrow \quad y-3=\frac{5}{-1}(x+4)$$

$$\Rightarrow \quad 5x+y+17=0$$

(h) A straight line passing through P and making an angle of $\alpha=60°$ is given by

$$\frac{y-y_1}{x-x_1}=\tan(\theta\pm\alpha)$$

$$\Rightarrow \quad \sqrt{3}x+y=1 \Rightarrow y=-\sqrt{3}x+1$$

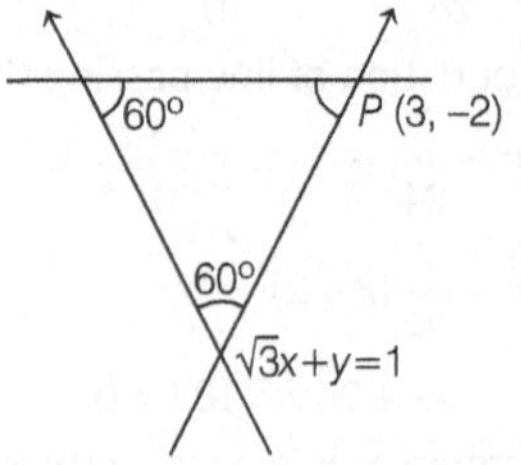

Then, $\tan\theta=-\sqrt{3}$

and $\tan\alpha=\tan 60°=\sqrt{3}$

$$\therefore \quad \frac{y+2}{x-3}=\frac{\tan\theta\pm\tan\alpha}{1\mp\tan\theta\tan\alpha}=\frac{-\sqrt{3}+\sqrt{3}}{1-(-\sqrt{3})(\sqrt{3})}$$

$$\Rightarrow \quad y+2=0 \Rightarrow y=-2$$

and $\dfrac{y+2}{x-3}=\dfrac{-\sqrt{3}-\sqrt{3}}{1+(-\sqrt{3})(\sqrt{3})}=\dfrac{-2\sqrt{3}}{-2}=\sqrt{3}$

$$\Rightarrow \quad y+2=\sqrt{3}x-3\sqrt{3}$$

$\Rightarrow \quad y+2=0$ does not intersect X-axis.

Hence, equation of line L is

$$\sqrt{3}x-y=2+3\sqrt{3}$$

19. (a) Let $PM: x-2y-3=0$ be the incident ray and MR is reflected ray.

Draw MQ such that $PM=MQ$.

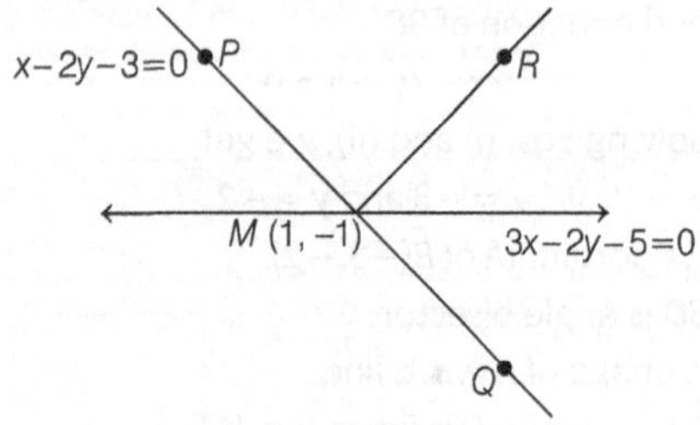

For point M

Given line,

$$x-2y-3=0 \quad \text{...(i)}$$

$$3x-2y-5=0 \quad \text{...(ii)}$$

Solving Eqs. (i) and (ii), we get

$$x=1 \text{ and } y=-1$$

$\therefore M(1,-1)$.

Let us assume P on line $PM \Rightarrow (-1,-2)$.

Let us assume Q on line $x-2y-3=0$ is $(3, 0)$ [$\because M$ is mid-point]

and R be the image of Q, w.r.t. $3x-2y-5=0$ and $R(h,k)$.

$$\therefore \quad \frac{h-3}{3}=\frac{k-0}{-2}=-2\left[\frac{3(3)+2\cdot 0-5}{3^2+2^2}\right]=-\frac{8}{13}$$

$$\Rightarrow \quad h=\frac{15}{13} \text{ and } k=\frac{16}{13}$$

$$\therefore R\left(\frac{15}{13},\frac{16}{13}\right)$$

$\therefore$ Equation of line MR is

$$y+1=\frac{\frac{16}{13}+1}{\frac{15}{13}-1}(x-1) \Rightarrow 29x-2y=31$$

(b) Let N be the image of M w.r.t. X-axis. Thus, N is $(5,-3)$.

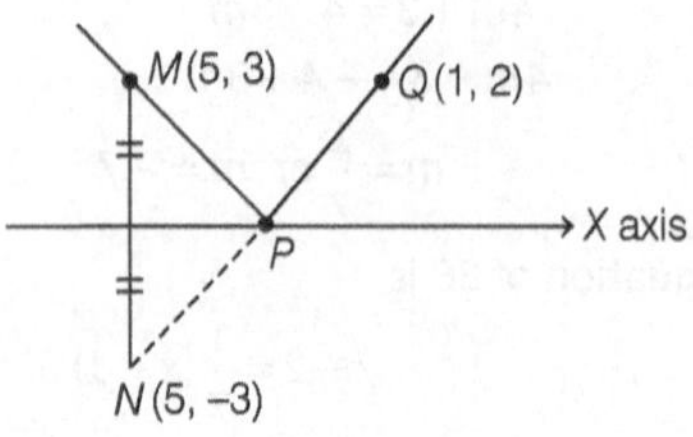

Slope of $MQ=\dfrac{-3-2}{5-1}=-\dfrac{5}{4}$

$\therefore$ Equation of QN is

$$y-2=-\frac{5}{4}(x-1) \Rightarrow 5x+4y=13$$

Thus, the coordinates of P be $\left(\dfrac{13}{5},0\right)$.

$\therefore$ Abscissa of the point P is $\dfrac{13}{5}$.

(c) Let $PM: x-6y=8$ be the incident ray

$AB: x+y=1$ is the mirror

MQ is reflected ray and MS is refracted ray.

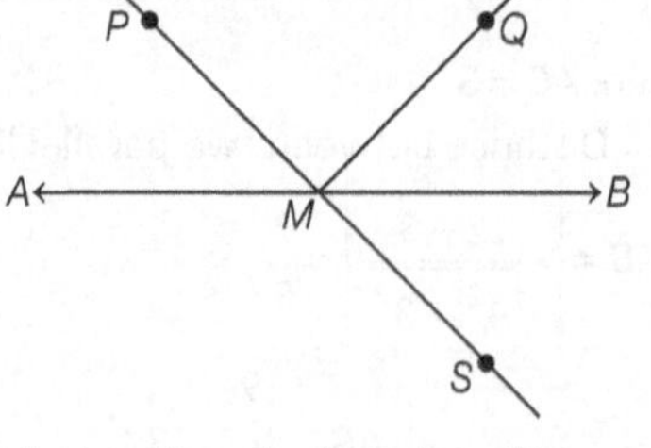

Clearly, $M=(2,-1)$ [$\because$ Point of intersection of lines $x-6y=8$ and $x+y=1$]

$$\text{Now, } \tan 15°=\left|\frac{m-\frac{1}{6}}{1+\frac{m}{6}}\right|$$

$$2-\sqrt{3}=\pm\frac{6m-1}{6m+1}$$

$$m=\frac{70-37\sqrt{3}}{13} \text{ or } \frac{37\sqrt{3}-70}{13}$$

Let the angle between $x+y=1$ and the line through $M(2,-1)$ with the slope

$$m=\frac{70-37\sqrt{3}}{13} \text{ be } \alpha.$$

$$\therefore \tan\alpha=\left|\frac{\frac{70-37\sqrt{3}}{13}+1}{1-\frac{70-37\sqrt{3}}{13}}\right|=\frac{83-37\sqrt{3}}{37\sqrt{3}-57}$$

and the angle between $x+y=1$ and the line through $M(2,-1)$ with slope $m=\frac{37\sqrt{3}-70}{13}$ be β.

$$\text{Then, } \tan\beta=\left|\frac{\frac{37\sqrt{3}-70}{13}+1}{1-\frac{37\sqrt{3}-70}{13}}\right|=\left|\frac{37\sqrt{3}-9}{131-37\sqrt{3}}\right|$$

$$=\frac{37\sqrt{3}-9}{131-37\sqrt{3}}$$

$\because \quad \tan\alpha>\tan\beta$

$\Rightarrow \quad \alpha>\beta$

Therefore, the slope of refracted ray is $\frac{70-37\sqrt{3}}{13}$.

$\therefore$ Equation of refracted ray

$$y+1=\frac{70-37\sqrt{3}}{13}(x-2)$$

$$\Rightarrow (70-37\sqrt{3})x-13y-153+74\sqrt{3}=0$$

Level 2

1. (c) Let $C\equiv(x,y)$ and centroid $G=(h,k)$, then

$$h=\frac{2-2+x}{3}=\frac{x}{3} \text{ and } k=\frac{-3+1+y}{3}=\frac{y-2}{3}.$$

$\because$ G lies on $2x+3y=1$, so $2h+3k=1$

$$\Rightarrow \quad 2\left(\frac{x}{3}\right)+3\left(\frac{y-2}{3}\right)=1$$

$\therefore$ Required equation of locus is

$$2x+3y=9$$

2. (a) Any line passing through $(-a,0)$ is

$$y=m(x+a)$$

$$\Rightarrow \quad -mx+y=ma$$

$$\frac{x}{-a}+\frac{y}{ma}=1 \quad \text{...(i)}$$

Since, it makes with axes a triangle of area T.

So, $\frac{1}{2}(-a)(ma)=T$

$$m=-\frac{2T}{a^2}$$

$\therefore$ Required equation is

$$\Rightarrow \quad 2Tx+a^2y=-2aT \quad \text{[from Eq. (i)]}$$

$$\Rightarrow \quad 2Tx+a^2y+2aT=0$$

3. (a) Obviously line through Q is at greatest distance from point P when it is perpendicular to PQ.

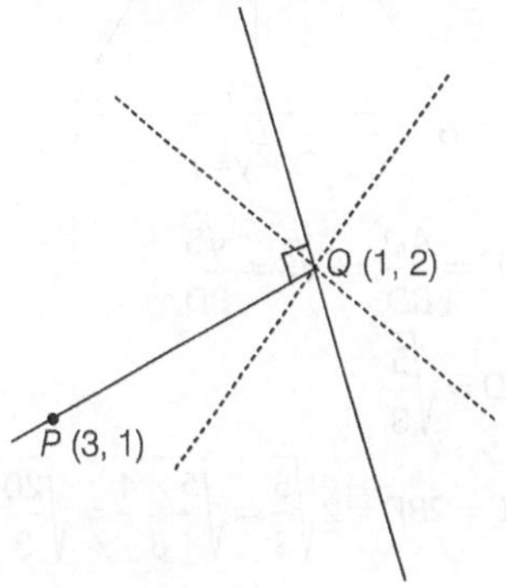

Now, slope of line PQ, is

$$m_{PQ}=-\frac{1}{2}$$

Then, slope is perpendicular line passing through Q, is

$$y-2=2(x-1)$$

$$\Rightarrow \quad 2x-y=0$$

4. (a) Let the vertices B and C lie on the given line.

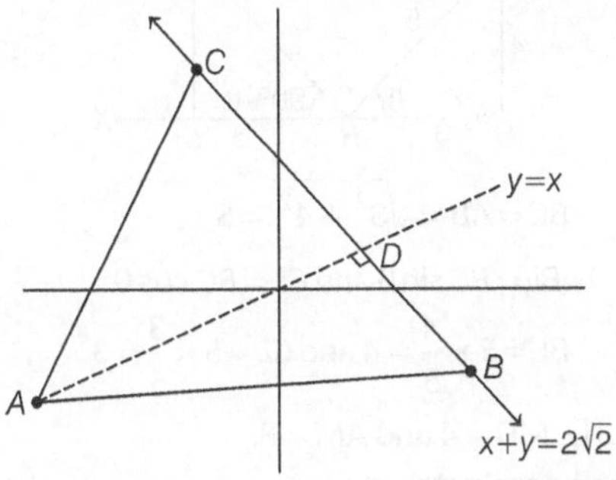

Then, $OD=\frac{2\sqrt{2}}{\sqrt{2}}=2$

$\therefore$ Equation of OD is

$y=x \Rightarrow x=y=\sqrt{2}$ for point D.

Also, $BD=OD\times\tan 60°$

$=2\sqrt{3}$ for the coordinates of B and C.

Using parametric equation of line, we get

$$\frac{x-\sqrt{2}}{-\frac{1}{\sqrt{2}}}=\frac{y-\sqrt{2}}{\frac{1}{\sqrt{2}}}=\pm 2\sqrt{3}$$

$$C=(\sqrt{2}+\sqrt{6},\sqrt{2}-\sqrt{6})$$

and $\quad B=(\sqrt{2}-\sqrt{6},\sqrt{2}+\sqrt{6})$

5. (a) $AD = \left| \dfrac{-2-2-1}{\sqrt{2^2+(-1)^2}} \right| = \left| \dfrac{-5}{\sqrt{5}} \right| = \sqrt{5}$

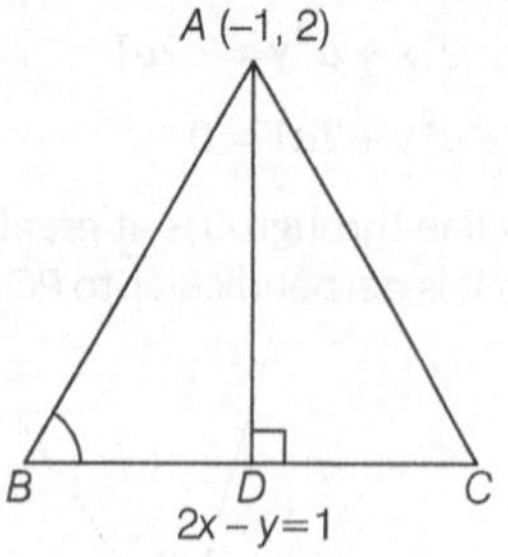

$$\because \quad \tan 60^\circ = \frac{AD}{BD} \Rightarrow \sqrt{3} = \frac{\sqrt{5}}{BD}$$

$$\Rightarrow \quad BD = \sqrt{\frac{5}{3}}$$

$$\Rightarrow \quad BC = 2BD = 2\sqrt{\frac{5}{3}} = \sqrt{\frac{5 \times 4}{3}} = \sqrt{\frac{20}{3}} \text{ units}$$

6. (b) The coordinates of A are $(0, 4)$ and those of B are $(3, 0)$.

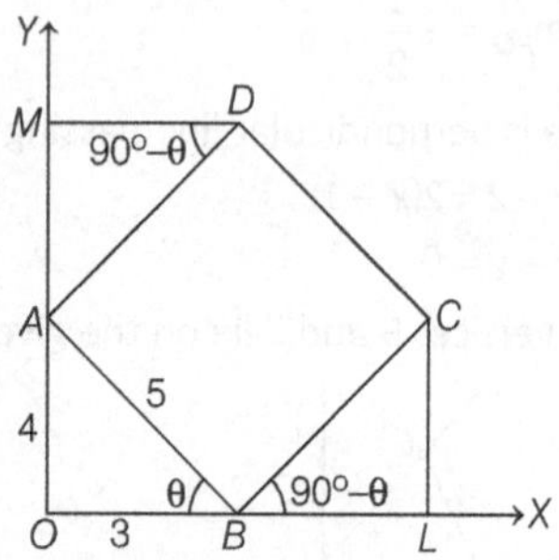

$$BC = AB = \sqrt{3^2+4^2} = 5$$

$$\Rightarrow \quad BL = BC \sin\theta \text{ and } CL = BC \cos\theta$$

$$\Rightarrow \quad BL = 5 \times \frac{4}{5} = 4 \text{ and } CL = 5 \times \frac{3}{5} = 3$$

Similarly, $MD = 4$ and $AM = 3$.

So, the coordinates of C are $(OB + BL,\ CL) = (7, 3)$ and those of D are $(MD, OA + AM) = (4, 7)$

7. (b) Both the lines passes through origin.

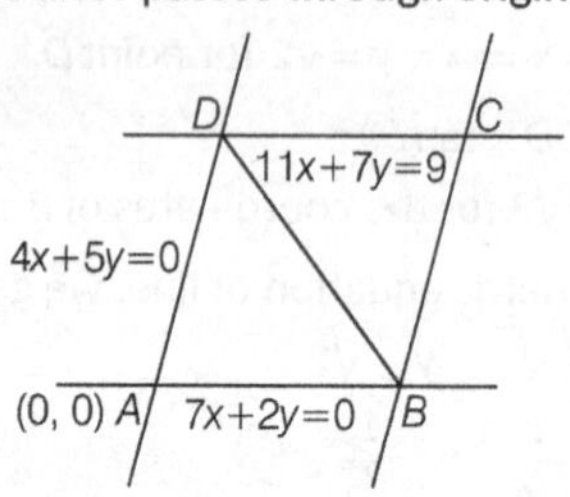

Point D is equal to intersection of $4x + 5y = 0$ and $11x + 7y = 9$.

$$\therefore \ D \text{ is } \left(\frac{5}{3}, -\frac{4}{3}\right)$$

$\because B$ is point of intersection of $7x + 2y = 0$ and $11x + 7y = 9$

$$\therefore \quad B = \left(-\frac{2}{3}, \frac{7}{3}\right)$$

Diagonals of parallelogram intersect at middle.

$$\text{Let middle point of } B, D = \left(\frac{\frac{5}{3} - \frac{2}{3}}{2}, \frac{-\frac{4}{3} + \frac{7}{3}}{2} \right) = \left(\frac{1}{2}, \frac{1}{2}\right)$$

$\therefore$ Equation of diagonal AC is

$$y - 0 = \frac{\frac{1}{2} - 0}{\frac{1}{2} - 0}(x - 0)$$

$$\Rightarrow \quad y = x$$

$\therefore$ Diagonal AC passes through $(2, 2)$.

8. (a) $A_1 = \text{Area}(\Delta ABC)$

$$= \frac{1}{2} \text{ mod of } \begin{vmatrix} -1 & 1 & 1 \\ 2 & 0 & 1 \\ 3 & 4 & 1 \end{vmatrix}$$

$$= \frac{1}{2} |(4 + 1 + 8)| = \frac{13}{2}$$

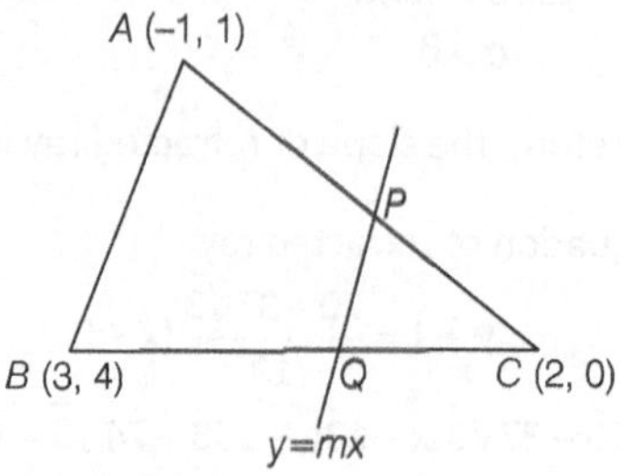

Equation of line AC is

$$y - 1 = -\frac{1}{3}(x + 1)$$

Line AC intersection with the line $y = mx$ at P.

$$\text{Solving, we get } P = \left(\frac{2}{3m+1}, \frac{2m}{3m+1} \right)$$

Equation of line BC, is $y - 0 = 4(x - 2)$

$$\Rightarrow \quad y = 4x - 8$$

$\because Q$ is point of intersection of BC and PQ.

Solving, we get

$$Q = \left(\frac{-8}{m-4}, \frac{-8m}{m-4} \right)$$

$$\therefore \ A_2 = \text{Area}(\Delta PQC) = \frac{1}{2} \text{ mod of } \begin{vmatrix} 2 & 0 & 1 \\ \frac{2}{3m+1} & \frac{2m}{3m+1} & 1 \\ \frac{-8}{m-4} & \frac{-8m}{m-4} & 1 \end{vmatrix}$$

$$A_2 = \frac{A_1}{3} = \frac{13}{6} \text{ (given)}$$

$$= \frac{1}{2}\left\{\begin{bmatrix} 2\left(\frac{2m}{3m+1} + \frac{8m}{m-4}\right) - 1\left(\frac{-16m}{(3m+1)(m-4)} + \frac{16m}{(3m+1)(m-4)}\right)\end{bmatrix}\right\}$$

$$= \frac{+13}{6}$$

$$\Rightarrow \quad \frac{26m^2}{3m^2 - 11m - 4} = \pm\frac{13}{6}$$

$$12m^2 = \pm 3m^2 - 11m - 4$$

Positive	**Negative**
$\Rightarrow 9m^2 + 11m + 4 = 0$	$15m^2 - 11m - 4 = 0$
$\Rightarrow$ (Imaginary roots)	$m = 1, -\frac{4}{15}$

$\because m > 0; m = 1$

9. (c) $x - y = 2 \Rightarrow y = x - 2$

$\therefore \quad m = 1$

$\tan\theta = 1 \Rightarrow \theta = 45°$

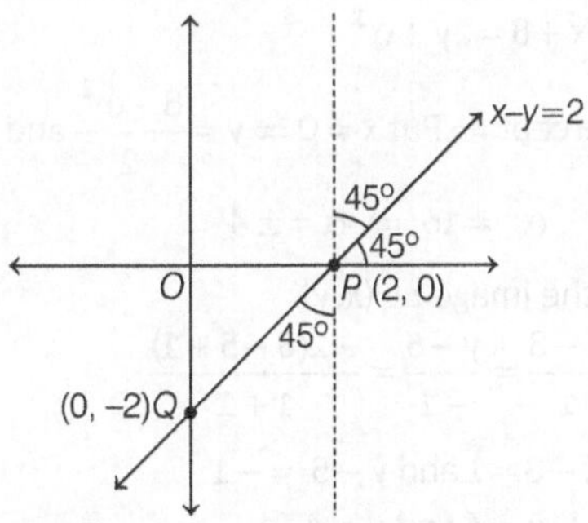

If PQ is rotated 45°, then it clearly seen that the equation of new position of line PQ is $x = 2$.

10. (a) If $\angle BAO = \theta$, then $BM = 2\sin\theta$

and $\quad MO = BM = 2\sin\theta, MA = 2\cos\theta$

Hence, $A = (2\cos\theta - 2\sin\theta, 0)$ and

$B = (-2\sin\theta, 2\sin\theta)$

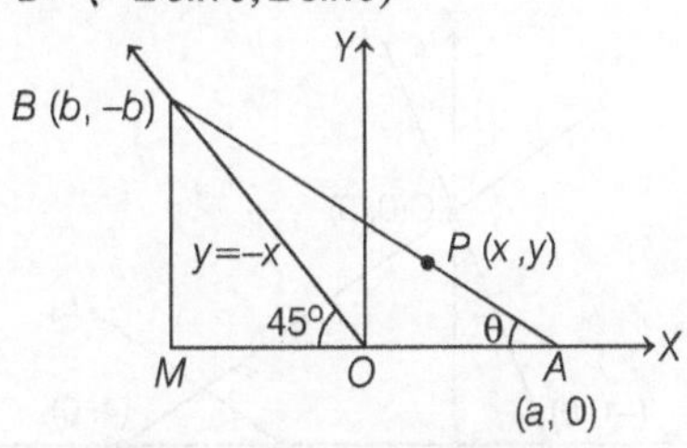

Since, $P(x, y)$ is the mid-point of AB. So

$2x = 2\cos\theta + (-4\sin\theta)$

or $\quad \cos\theta - 2\sin\theta = x$

$2y = 2\sin\theta \Rightarrow \sin\theta = y$

Eliminating θ, we have

$(x + 2y)^2 + y^2 = 1$ or $x^2 + 5y^2 + 4xy - 1 = 0$

11. (b) Equation of AB is

$$y - 1 = \frac{0-1}{2-1}x \Rightarrow x + 2y - 2 = 0$$

$|PA - PB| \leq AB$

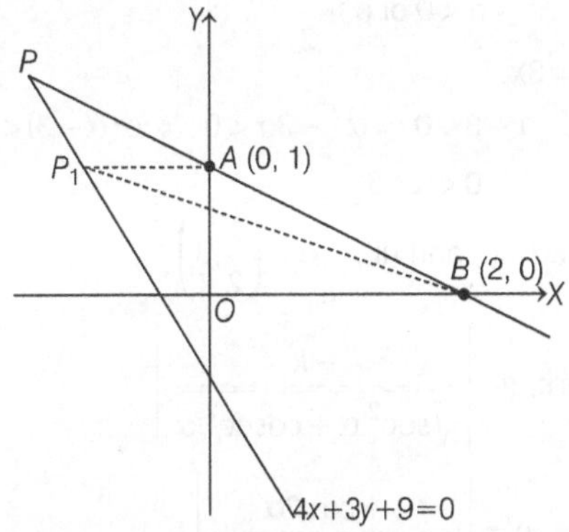

Thus, $|PA - PB|$ is maximum, if points A, B and P are collinear.

Solving $x + 2y - 2 = 0$ and $4x + 3y + 9 = 0$, we get

point $P = \left(\frac{-84}{5}, \frac{13}{5}\right)$

12. (b) In a ΔABC, $A = (\alpha, \beta)$, $B = (2, 3)$, $C = (1, 3)$

$\because$ A lies on the line $y = 2x + 3$

$\Rightarrow \quad \beta = 2\alpha + 3$ ...(i)

$\alpha \in I$

and $BC = \sqrt{(2-1)^2 + 0^2} = 1$

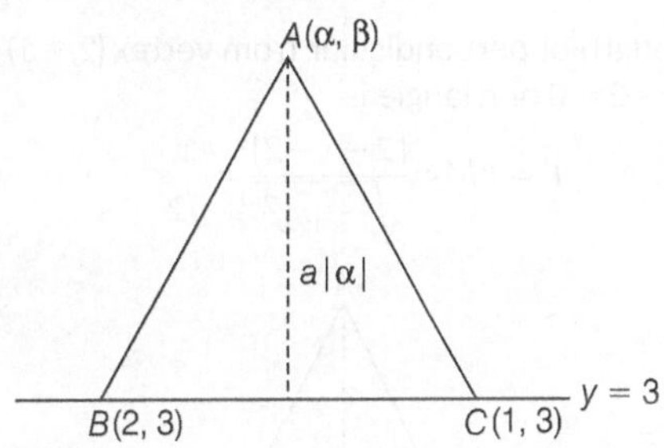

Distance of A from $BC = |2x + 3 - 3| = 2|\alpha|$

$\therefore$ Area of $\Delta ABC = \Delta \Rightarrow \frac{1}{2} \times BC \times 2|\alpha| = \Delta \Rightarrow |\alpha| = \Delta$

$\because [\Delta] = 5$

$\therefore 5 \leq \Delta < 6 \Rightarrow 5 \leq |\alpha| < 6$

$\therefore |\alpha| \geq 5$ and $|\alpha| < 6$

$\alpha \in [-\infty, -5] \cup [5, \infty)$ and $\alpha \in (-6, 6)$

$\therefore \alpha \in (-6, -5] \cup [5, 6)$

$\because \alpha \in I \Rightarrow \alpha = -5, 5$

$\therefore A = (-5, -7)$ and $(5, 13)$

13. (b) Obviously, (1, 1) lies inside the angle made by given lines.

So, (1, 1) and (α, α^2) must lies on the same side of two lines.

Hence, for $y = \frac{x}{2}$

$\Rightarrow \quad 1 - \frac{1}{2} > 0 \Rightarrow \alpha^2 - \frac{\alpha}{2} > 0 \Rightarrow \alpha(2\alpha - 1) > 0$

$\alpha < 0 \text{ or } \alpha > \frac{1}{2}$ ...(i)

For $y = 3x$,

$1 - 3 < 0 \Rightarrow \alpha^2 - 3\alpha < 0 \Rightarrow \alpha(\alpha - 3) < 0$

$\Rightarrow \quad 0 < \alpha < 3$ ...(ii)

From Eqs. (i) and (ii), $\alpha \in \left(\frac{1}{2}, 3\right)$

14. (c) Here, $p = \left|\frac{-k}{\sqrt{\sec^2\alpha + \operatorname{cosec}^2\alpha}}\right|$

and $p' = \left|\frac{-k\cos 2\alpha}{\sqrt{\cos^2\alpha + \sin^2\alpha}}\right|$

$\therefore 4p^2 + p'^2 = \frac{4k^2}{\sec^2\alpha + \operatorname{cosec}^2\alpha} + \frac{k^2\cos^2 2\alpha}{\cos^2\alpha + \sin^2\alpha}$

$\Rightarrow \frac{4k^2}{\sec^2\alpha + \operatorname{cosec}^2\alpha} + \frac{k^2(\cos^2\alpha - \sin^2\alpha)^2}{1}$

$\Rightarrow 4k^2\sin^2\alpha\cos^2\alpha + k^2\cos^4\alpha + k^2\sin^4\alpha - 2k^2\sin^2\alpha\cos^2\alpha$

$\Rightarrow k^2\cos^4\alpha + k^2\sin^4\alpha + 2k^2\sin^2\alpha\cos^2\alpha$

$\Rightarrow k^2(\cos^2\alpha + \sin^2\alpha)^2 = k^2(1)^2 = k^2$

15. (c) Length of perpendicular from vertex $(2, -1)$ to base $x + y - 2 = 0$ of triangle is

$$P = AM = \frac{|2 - 1 - 2|}{\sqrt{1^2 + 1^2}} = \frac{1}{\sqrt{2}}$$

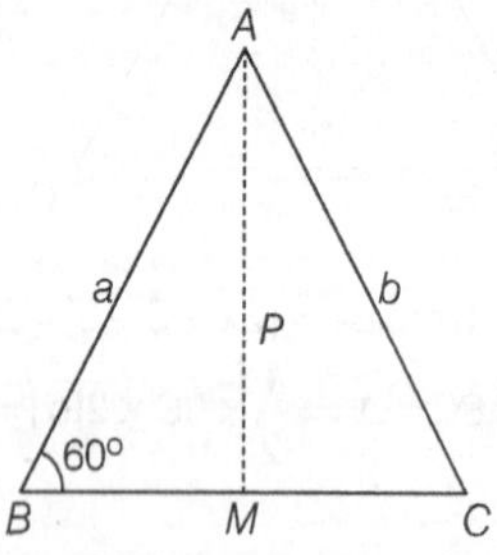

From $\Delta ABM \Rightarrow \frac{AB}{AM} = \operatorname{cosec} 60°$

$\therefore$ Length of side $(a) = \frac{1}{\sqrt{2}} \times \frac{2}{\sqrt{3}} = \sqrt{\frac{2}{3}}$ units

16. (a) Given, that $P(1, 0), Q(-1, 0)$ and $\frac{AP}{AQ} = \frac{BP}{BQ} = \frac{CP}{CQ} = \frac{1}{3}$

$\Rightarrow \quad 3AP = AQ$

Let $A = (x, y)$, then $3AP = AQ$

$\Rightarrow \quad 9AP^2 = AQ^2$

$\Rightarrow 9(x - 1)^2 + 9y^2 = (x + 1)^2 + y^2$

$\Rightarrow 9x^2 - 18x + 9 + 9y^2 = x^2 + 2x + 1 + y^2$

$\Rightarrow 8x^2 - 20x + 8y^2 + 8 = 0$

$\Rightarrow \quad x^2 + y^2 - \frac{5}{2}x + 1 = 0$ ...(i)

$\therefore$ A lies on the circle given by Eq. (i).

As B and C also follow the same condition, they must lie on the same circle.

$\therefore$ Centre of circumcircle of ΔABC

= Centre of circle given by Eq. (i)

$= \left(\frac{5}{4}, 0\right)$

17. (b) $\because$ Every point (x, y) on the perpendicular bisector of $\overline{PQ}$ is equidistant from both P and Q.

We have,

$(x - 1)^2 + (y - 4)^2 = (x - \alpha)^2 + (y - 3)^2$

$\Rightarrow -2x - 8y + 17 = -2\alpha x - 6y + \alpha^2 + 9$

$\therefore$ Equation of perpendicular bisector of PQ is

$2(\alpha - 1)x + 8 = 2y + \alpha^2$

$\therefore$ y-intercept $\Rightarrow$ Put $x = 0 \Rightarrow y = \frac{8 - \alpha^2}{2}$ and $\frac{8 - \alpha^2}{2} = -4$

$\alpha^2 = 16 \Rightarrow \alpha = \pm 4$

18. (d) Let the image be (x, y).

Then, $\frac{x - 3}{1} = \frac{y - 5}{-1} = \frac{-2(3 - 5 + 1)}{1 + 1}$

$\Rightarrow \quad x - 3 = 1$ and $y - 5 = -1$

$x = 4$ and $y = 4$

$\therefore$ Point is (4, 4) which will satify the curve

$(x - 2)^2 + (y - 4)^2 = 4$

19. (d) Given, $L_1 : 2x - 3y - 6 = 0$

$L_2 : 3x - y + 3 = 0$

and $L_3 : 3x + 4y - 12 = 0$

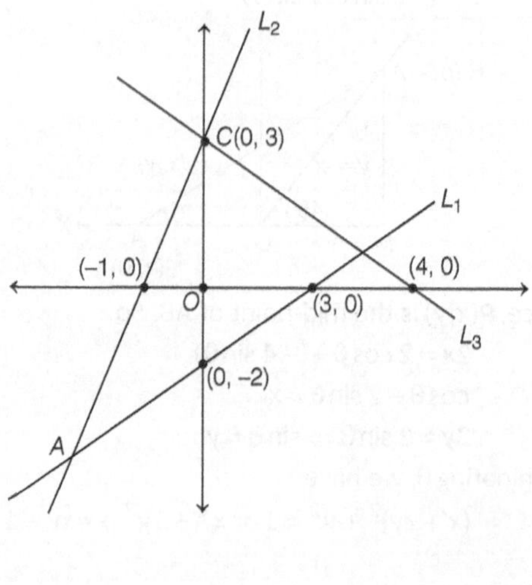

$P(\alpha, 0)$ and $Q(0, \beta)$ lies on the line or inside the ΔABC.

$\because$ By geometry origin also lies in ΔABC

and $\quad L_1(0) \le 0, L_2(0) \ge 0, L_3(0) \le 0$

$\therefore \quad (L_1(P) \le 0), L_2(P) \ge 0, L_3(P) \le 0$

$\Rightarrow \quad \alpha - 3 \le 0, \alpha + 1 \ge 0$ and $\alpha - 4 \le 0$

$\Rightarrow \quad \alpha \le 3, \alpha \ge -1$ and $\alpha \le 0$

$\therefore \quad \alpha \in [-1, 3]$

Similarly, $L_1(Q) \le 0, L_2(Q) \ge 0$ and $L_3(Q) \le 0$

$\Rightarrow \quad -3\beta - 6 \le 0, -\beta + 3 \ge 0$ and $4\beta - 12 \le 0$

$\Rightarrow \quad \beta \ge -2, \beta \le 3$ and $\beta \le 3$

$\therefore \quad \beta \in [-2, 3]$

20. (d) Let a line $ax + by + c = 0$

$\therefore \quad P_1 + P_2 + P_3 = 0$

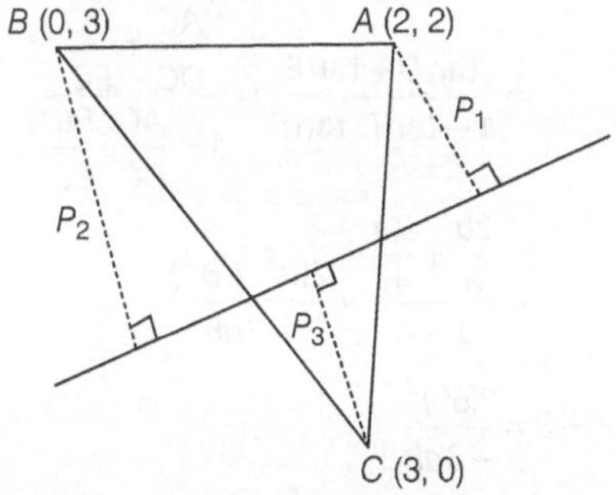

$$\Rightarrow \frac{3a + c}{\sqrt{a^2 + b^2}} + \frac{3b + c}{\sqrt{a^2 + b^2}} + \frac{2a + 2b + c}{\sqrt{a^2 + b^2}} = 0$$

$$\Rightarrow \quad 5a + 5b + 3c = 0$$

$$\Rightarrow \quad a\left(\frac{5}{3}\right) + b\left(\frac{5}{3}\right) + c = 0$$

$\Rightarrow \left(\frac{5}{3}, \frac{5}{3}\right)$ satisfy the given line.

$\Rightarrow$ Fix point $\left(\frac{5}{3}, \frac{5}{3}\right)$ which is centroid of ΔABC.

21. (b)

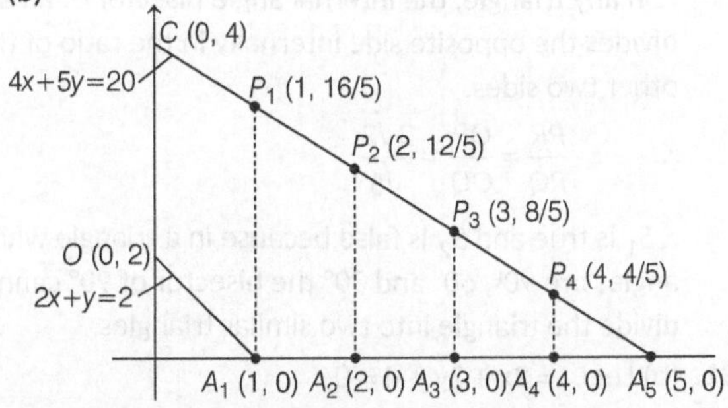

There are clearly 5 points.

22. (c) $\begin{vmatrix} 1 & 1 & -2 \\ 2 & -1 & 1 \\ a & b & -c \end{vmatrix} = 0$

$$a + 5b - 3c = 0 \Rightarrow -\frac{a}{3} - \frac{5}{3}b + c = 0$$

$\therefore \quad 2ax + 3by + c = 0$ is concurrent at

$$2x = -\frac{1}{3} \text{ and } 3y = -\frac{5}{3} \Rightarrow x = -\frac{1}{6} \text{ and } y = -\frac{5}{9}$$

23. (b) $\because$ If the given lines are concurrent, then

$$\begin{vmatrix} a & 1 & 1 \\ 1 & b & 1 \\ 1 & 1 & c \end{vmatrix} = 0$$

Applying $C_2 \to C_2 - C_1$ and $C_3 \to C_3 - C_1$, we get

$$\begin{vmatrix} a & 1-a & 1-a \\ 1 & b-1 & 0 \\ 1 & 0 & c-1 \end{vmatrix} = 0$$

$$\Rightarrow a(b-1)(c-1) - (c-1)(1-a) - (b-1)(1-a) = 0$$

$$\Rightarrow \quad \frac{a}{1-a} + \frac{1}{1-b} + \frac{1}{1-c} = 0$$

By dividing $(1-a)(1-b)(1-c)$

Adding 1 both sides, $\frac{1}{1-a} + \frac{1}{1-b} + \frac{1}{1-c} = 1$

24. (b) Given, $(2 + k)x + (1 + k)y = 5 + 7k$ ...(i)

is true for every value of k.

Putting $k = 1$ and $k = 2$ in Eq. (i), we get

$3x + 2y = 12$ ...(ii)

$4x + 3y = 19$ ...(iii)

Obviously they are not parallel.

Intersection point of lines (ii) and (iii) is $(-2, 9)$.

25. (b) Suppose, we rotate the coordinate axes in the anti-clockwise direction through an $\angle\alpha$.

The equation of the line 'L' w.r.t. old axes is $\frac{x}{a} + \frac{y}{b} = 1$.

In this equation replacing x by $(x \cos\alpha - y \sin\alpha)$ and y by $(x \sin\alpha + y \cos\alpha)$.

The equation of line w.r.t. new axes is

$$\frac{x \cos\alpha - y \sin\alpha}{a} + \frac{x \sin\alpha + y \cos\alpha}{b} = 1$$

$$\Rightarrow x\left(\frac{\cos\alpha}{a} + \frac{\sin\alpha}{b}\right) + y\left(\frac{\cos\alpha}{b} - \frac{\sin\alpha}{a}\right) = 1 \quad \text{...(i)}$$

The intercept made by Eq. (i) on the coordinate axes are given as p and q.

$$\therefore \quad \frac{1}{p} = \frac{\cos\alpha}{a} + \frac{\sin\alpha}{b} \text{ and } \frac{1}{q} = \frac{\cos\alpha}{b} - \frac{\sin\alpha}{a}$$

Both squaring and adding, we get $\frac{1}{p^2} + \frac{1}{q^2} = \frac{1}{a^2} + \frac{1}{b^2}$

26. (c) Given, $\frac{x}{5} + \frac{y}{b} = 1$ ...(i)

By Eq. (i), $\frac{13}{5} + \frac{32}{b} = 1 \Rightarrow \frac{32}{b} = -\frac{8}{5} \Rightarrow b = -20$

$$\frac{x}{5} - \frac{y}{20} = 1 \Rightarrow 4x - y = 20$$

Line k has same slope

$\Rightarrow \quad -\frac{3}{c} = 4 \Rightarrow c = -\frac{3}{4}$

$\Rightarrow \quad 4x - y = -3 \Rightarrow 4x - y + 3 = 0$

$\therefore$ Required distance $= \frac{|4(13) - 32 + 3|}{\sqrt{4^2 + 1^2}} = \frac{23}{\sqrt{17}}$ units

27. (a) Solving $3x + 4y = 9$ and $y = mx + 1$, we get

$x = \frac{5}{3 + 4m}$, x is an integer.

Put $\quad m = -\frac{2}{4}, -\frac{4}{4}, \frac{2}{4}, -\frac{8}{4}$

i.e. $\quad 3 + 4m = 1, -1, 5, -5$

Only two integral values.

28. (a) Given lines, $\quad y = 0 \quad$ ...(i)

$(1+p)x - py + p(1+p) = 0 \quad$ ...(ii)

and $(1+q)x - qy + q(1+q) = 0 \quad$ ...(iii)

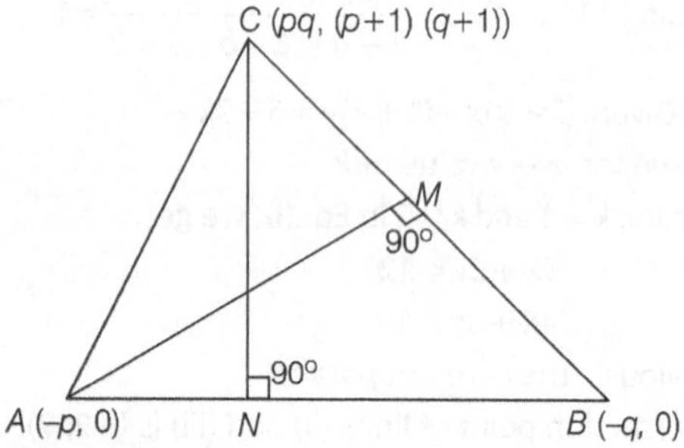

By solving Eqs. (i), (ii) and (iii), we get

$A(-p, 0), B(-q, 0)$ and $C(pq, (p+1)(q+1))$

$\because$ The side AB is along the X-axis, altitude CN, drawn from C to the side is a vertical line so that the equation of CN is $\quad x = pq \quad$ ...(iv)

$\therefore$ Equation of altitude AM is

$y - 0 = \frac{-(pq+q)}{(p+1)(q+1)}(x+p) \Rightarrow y = \frac{-q}{(q+1)}(x+p) \quad$ ...(v)

By Eqs. (iv) and (v), $y = -\frac{q}{(q+1)}(pq+p) = -pq$

If (x, y) is orthocentre, then we have $x = pq$ and $y = -pq$

So, that the locus of the orthocentre is $y = -x$

$\Rightarrow \quad x + y = 0$

29. (c) By pythagoras theorem, $a^2 + b^2 = 60^2 \quad$ ...(i)

Let θ be the acute angle between medians AD and BE.

$\therefore \quad \tan\theta = \left|\frac{2-1}{1+2(1)}\right| = \frac{1}{3} \quad$ ...(ii)

$\because$ In Quadrilateral $DGEC$, we have

$\angle DGE + \angle GDC + 90° + \angle GEC = 360°$

$\Rightarrow \quad \pi - \theta + \angle GDC + \frac{\pi}{2} + \angle GEC = 2\pi$

$\Rightarrow \quad \angle GDC + \angle GEC = \frac{\pi}{2} + \theta$

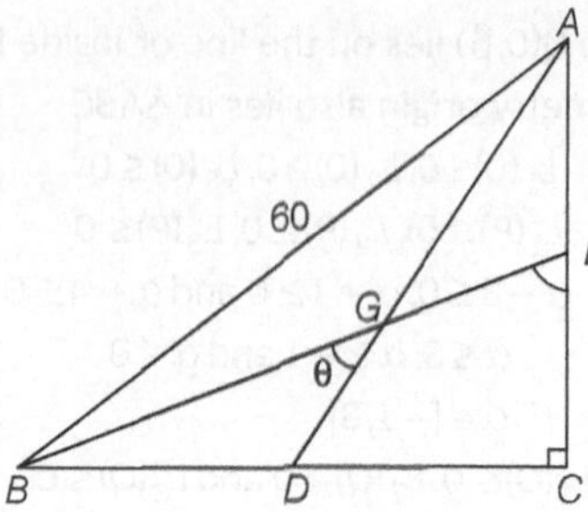

By Eq. (ii),

$-3 = -\cot\theta$

$= \tan\left(\frac{\pi}{2} + \theta\right)$

$= \tan(\angle D + \angle E)$

$= \frac{\tan D + \tan E}{1 - \tan D \tan E} = \frac{\frac{AC}{DC} + \frac{BC}{EC}}{1 - \frac{AC}{DC} \cdot \frac{BC}{EC}}$

$= \frac{\frac{2b}{a} + \frac{2a}{b}}{1 - 4} = \frac{2(a^2 + b^2)}{-3ab}$

$-3 = \frac{2(60)^2}{-3ab}$

$\Rightarrow \quad ab = \frac{2 \times 60^2}{9} \Rightarrow \frac{1}{2}ab = \frac{60^2}{9}$

$\therefore$ Area of $\Delta ABC = \frac{1}{2}ab = \frac{60^2}{9} = \frac{3600}{9} = 400$ sq units

30. (c) Solving $L_1 = 0$ and $L_2 = 0$

$\Rightarrow \quad P = (-2, -2)$

Solving $L_2 = 0$ and $L_3 = 0$

$\Rightarrow \quad Q = (1, -2)$

Therefore, $OP = 2\sqrt{2}$ and $OQ = \sqrt{5}$

$\therefore$ In any triangle, the internal angle bisector of an angle divides the opposite side internally in the ratio of the other two sides.

$\therefore \quad \frac{PR}{RQ} = \frac{OP}{OQ} = \frac{2\sqrt{2}}{\sqrt{5}}$

$\therefore S_1$ is true and S_2 is false because in a triangle whose angles are 90°, 60° and 30° the bisector of 90° cannot divide the triangle into two similar triangles.

31. (b) Let $L_1 = 3x + 2y + 4 = 0$

$L = 2x + 3y + 1 = 0$

and $L_2 = 0$ be the required line

$\because$ The line $L_2 = 0$ passes through point of intersection of the lines $L_1 = 0$ and $L = 0$

$\therefore$ By family of lines

$L_2 = (3x + 2y + 4) + \lambda(2x + 3y + 1) = 0$

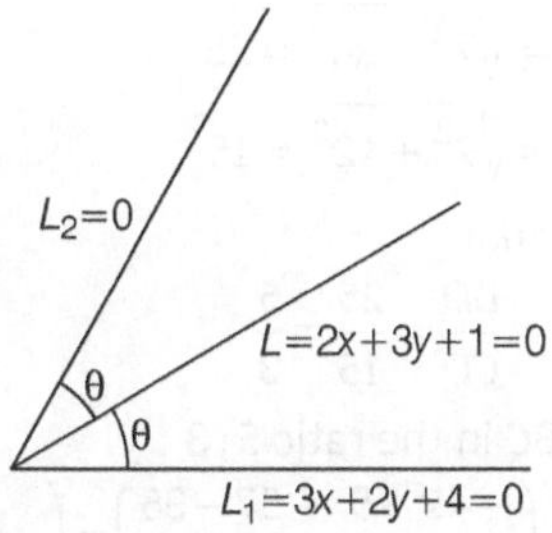

Suppose, (α, β) is intersection of $L_1 = 0$ and $L = 0$ is a point on $L = 0$ so that

$$2\alpha + 3\beta + 1 = 0 \quad \text{...(i)}$$

But $L = 0$ is an angular bisection of $L_1 = 0$ and $L_2 = 0$

$$\Rightarrow \left| \frac{(3\alpha + 2\beta + 4) + \lambda(2\alpha + 3\beta + 1)}{\sqrt{(2\lambda + 3)^2 + (3\lambda + 2)^2}} \right| = \frac{3\alpha + 2\beta + 4}{\sqrt{3^2 + 2^2}}$$

$$\Rightarrow \quad 13\lambda^2 + 24\lambda + 13 = 13 \quad \text{(Eq. (i))}$$

$$\Rightarrow \quad \lambda = 0 \text{ or } \lambda = -\frac{24}{13}$$

$\lambda = 0$ gives $L_1 = 0$

and $\lambda = -\dfrac{24}{13}$ gives the lines

$$L_2 = 3x + 2y + 4 - \frac{24}{13}(2x + 3y + 1) = 0$$

$$\Rightarrow \quad -9x - 46y + 28 = 0 \Rightarrow 9x + 46y = 28$$

32. (a) Equation of reflected ray $P'B$

$$y - 2\sqrt{3} = \tan 120^\circ (x - 0)$$

$$\Rightarrow \quad \sqrt{3}x + y = 2\sqrt{3}$$

and $(3, -\sqrt{3})$ satisfy the line

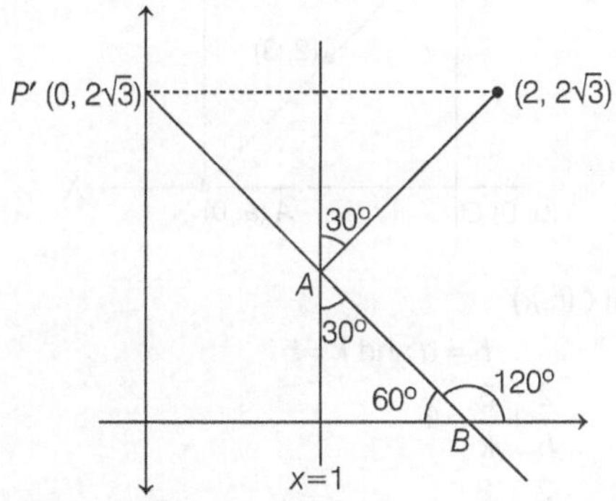

33. (c) $\therefore M\left(\dfrac{a-3}{2}, \dfrac{b+1}{2}\right)$ lies

on $\quad 2x + y - 3 = 0$

$$\Rightarrow \quad 2a + b = 11 \quad \text{...(i)}$$

$\because C$ lies on $7x - 4y = 1$

$$\Rightarrow \quad 7a - 4b = 1 \quad \text{...(ii)}$$

By Eqs. (i) and (ii), we get

$$a = 3 \text{ and } b = 5$$

$$\Rightarrow \quad C(3, 5) \Rightarrow m_{AC} = \frac{2}{3}$$

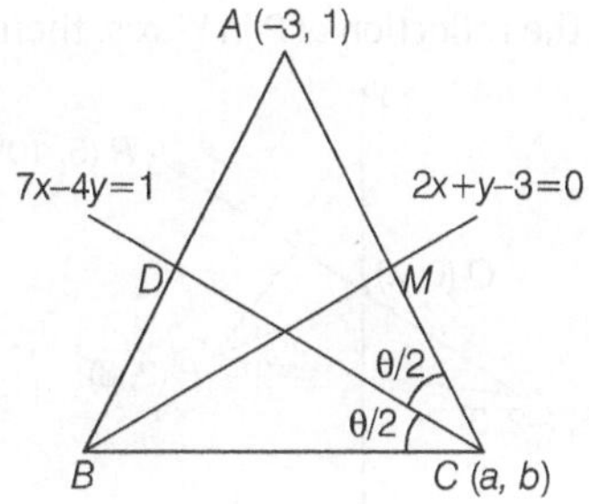

Also, $m_{CD} = \dfrac{7}{4}$

$$\Rightarrow \quad \tan\frac{\theta}{2} = \left| \frac{\frac{2}{3} - \frac{7}{4}}{1 + \frac{2}{3} \cdot \frac{7}{4}} \right| = \frac{1}{2}$$

$$\therefore \quad \tan\theta = \frac{2 \cdot \frac{1}{2}}{1 - \frac{1}{4}} = \frac{4}{3}$$

34. (c)

(i) Reflection about $y = x$ at $(4, 1)$ is $(1, 4)$

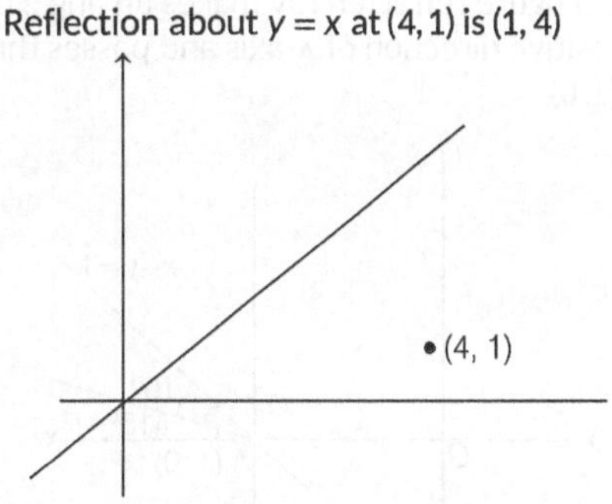

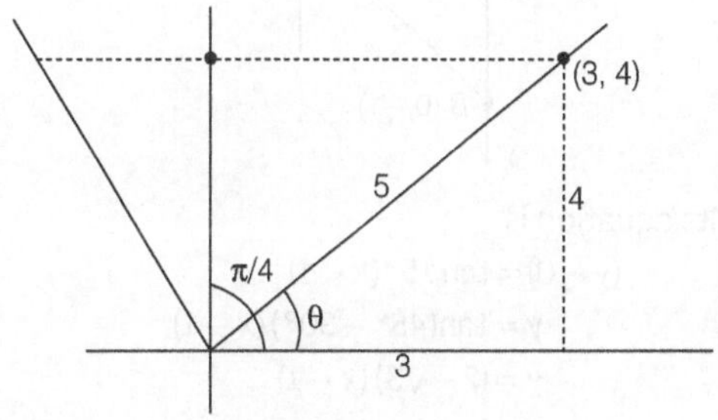

(ii) Now, 2 units along positive x-direction

$$\Rightarrow \quad (1 + 2, 4 + 0) = (3, 4)$$

(iii) We wish to find $\{\because \tan\theta = \frac{4}{3}, \sin\theta = \frac{4}{5}, \cos\theta \frac{3}{5}\}$

$$5\cos\left(\theta + \frac{\pi}{4}\right) - 5\sin\left(\theta + \frac{\pi}{4}\right)$$

$$\Rightarrow \quad x = \frac{5\cos\theta}{\sqrt{2}} - \frac{5\sin\theta}{\sqrt{2}} = \frac{-1}{\sqrt{2}}$$

$$\Rightarrow \quad y = \frac{5\sin\theta}{\sqrt{2}} + \frac{5\cos\theta}{\sqrt{2}} = \frac{7}{\sqrt{2}}$$

$$\Rightarrow (x, y) = \left(-\frac{1}{\sqrt{2}}, \frac{7}{\sqrt{2}}\right)$$

35. (c) If P_1 is the reflection of P in Y-axis, then $P_1(-2,3)$

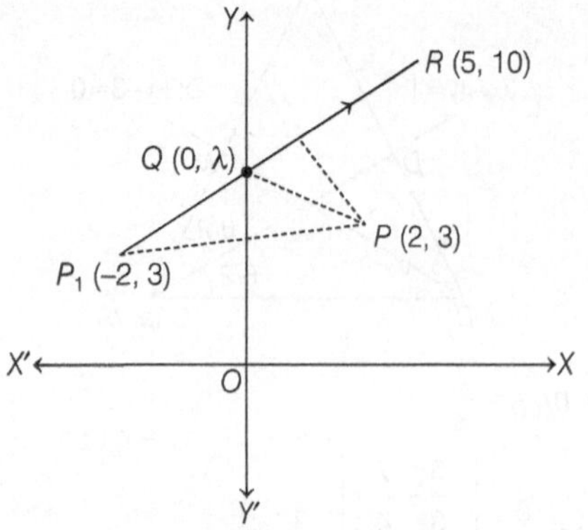

∴ Equation of line P_1R, is

$$y-3=\frac{10-3}{5+2}(x+2)$$

$$y=x+5$$

It meets Y-axis at $(0,5)$

$\Rightarrow \quad Q=(0,5)$

36. (d) From figure refracted ray makes an angle of 15° with positive direction of X-axis and passes through the point (1, 0).

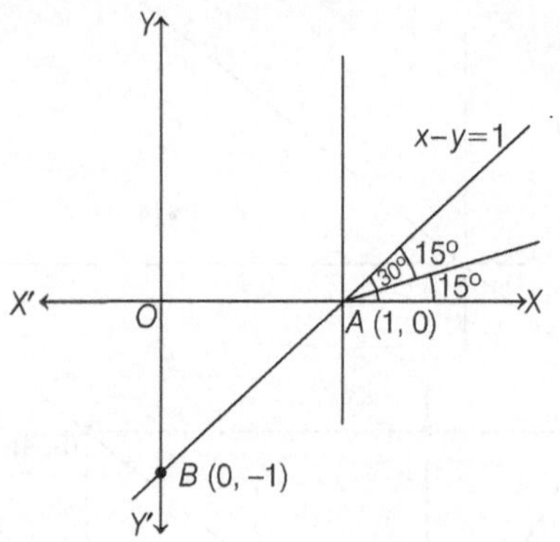

∴ Its equation is

$$(y-0)=\tan 15^\circ(x-1)$$

$$y=\tan(45^\circ-30^\circ)(x-1)$$

$$y=(2-\sqrt{3})(x-1)$$

37. (1) Angle of bisector of A is AD and equation of AD is

$$ax+2y+c=0$$

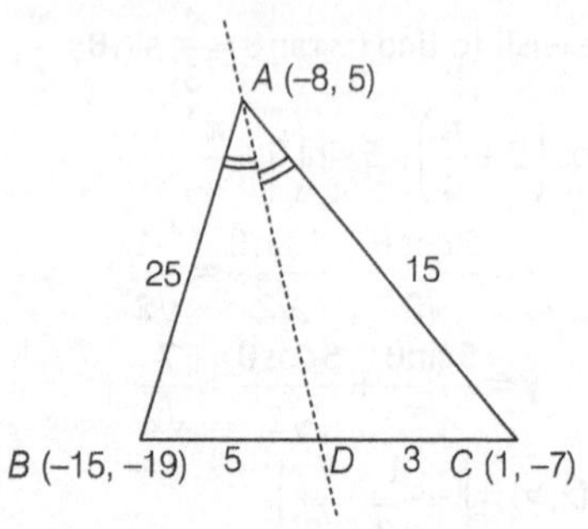

$$AB=\sqrt{7^2+24^2}=25$$

$$AC=\sqrt{9^2+12^2}=15$$

∴ We know that

$$\frac{AB}{AC}=\frac{BD}{CD}=\frac{25}{15}=\frac{5}{3}$$

∵ D divides BC in the ratio 5 : 3

$$\therefore \quad D=\left(\frac{-45+5}{8},\frac{-57-35}{8}\right)=\left(-5,-\frac{23}{2}\right)$$

Line AD is

$$y-5=\frac{-\frac{23}{2}-5}{-5+8}(x+8)$$

$$\Rightarrow \quad y-5=-\frac{33}{2\times 3}(x+8)$$

$$\Rightarrow \quad 2y-10=-11x-88$$

$$\Rightarrow \quad 11x+2y+78=0$$

$$\Rightarrow \quad a=11 \text{ and } c=78$$

$$\therefore \quad \frac{a+c}{89}=\frac{11+78}{89}=\frac{89}{89}=1$$

38. (8) Line AB is $\frac{x}{a}+\frac{y}{b}=1$

Point (2, 3) lies on it.

$$\therefore \quad \frac{2}{a}+\frac{3}{b}=1$$

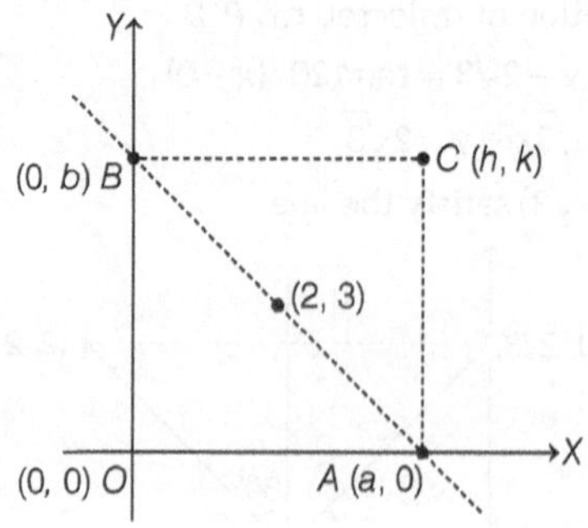

Point $C(h,k)$

$$h=a \text{ and } k=b$$

$$\therefore \quad \frac{2}{h}+\frac{3}{k}=1$$

$$\frac{2}{x}+\frac{3}{y}=1 \text{ which is locus of } C$$

$$\frac{2}{x}=1-\frac{3}{y}\Rightarrow\frac{2}{x}=\frac{y-3}{y}$$

$$x=\frac{2y}{y-3}$$

$\therefore x=\frac{6}{y-3}+2$ for x is to be integer

$y-3=\pm1,\pm2,\pm3,\pm6$

∴ Total 8 points.

39. (9) $\because AB^2 + BC^2 = AC^2$

$\therefore \qquad \angle B = 90°$

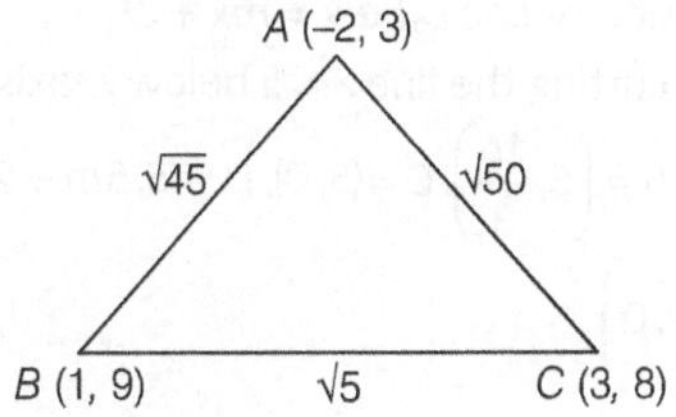

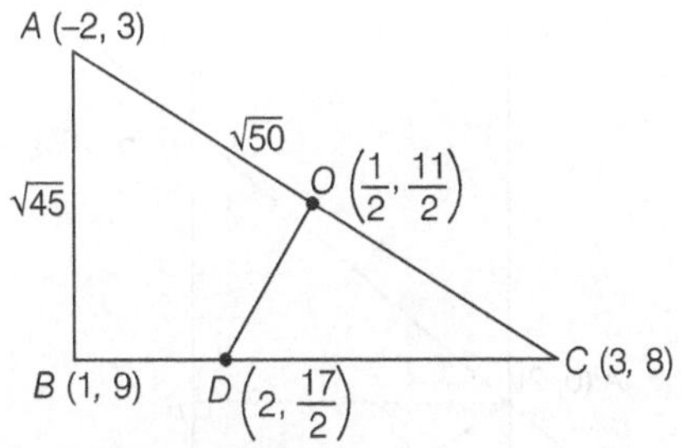

$\therefore$ Circumcentre = mid-point of AC

$$\Rightarrow \qquad O = \left(\frac{1}{2}, \frac{11}{2}\right)$$

Mid-point of BC, $D = \left(2, \frac{17}{2}\right)$

Line OD is

$$y - \frac{11}{2} = 2\left(x - \frac{1}{2}\right)$$

$$\Rightarrow \qquad y = 2x + \frac{9}{2}$$

It will passing through $\left(0, \frac{\alpha}{2}\right)$

$$\Rightarrow \qquad \frac{\alpha}{2} = \frac{9}{2}$$

$$\Rightarrow \qquad \alpha = 9$$

40. (4) Any point on the line $x + y = 4$ is $(t, 4 - t)$, where $t \in R$.

Now, distance of this point from the line $4x + 3y - 10 = 0$ is 1 unit

$$\frac{|4t + 3(4 - t) - 10|}{\sqrt{4^2 + 3^2}} = 1$$

$$\Rightarrow \qquad |t + 2| = 5$$

$$\Rightarrow \qquad t + 2 = \pm 5$$

$$\Rightarrow \qquad t = \pm 5 - 2$$

$$\Rightarrow \qquad t = 3, -7$$

$\therefore$ Sum of values of abscissa $= 3 + (-7) = -4$

$\therefore$ Absolute value = 4

41. (1.4) We have, $MN = 3\sqrt{2}$

$$X_N = 2 - 3\sqrt{2}\left(-\frac{1}{\sqrt{2}}\right) = 5$$

$$Y_N = 1 - 3\sqrt{2}\left(\frac{1}{\sqrt{2}}\right) = -2$$

So, $N = (5, -2)$

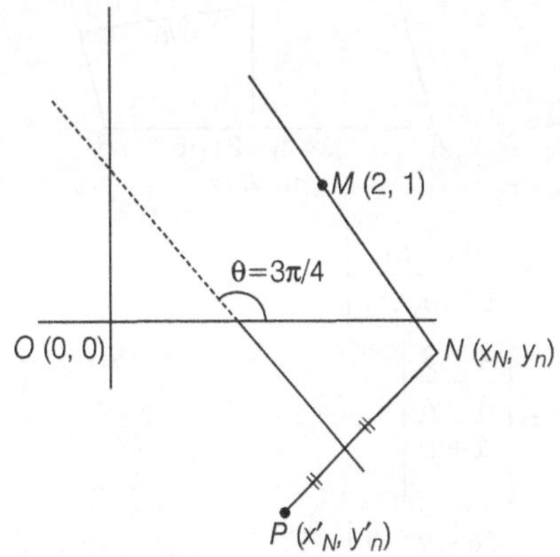

Image of N w.r.t. $x + y - 1 = 0$

$$\Rightarrow \qquad \frac{h-5}{1} = \frac{k+2}{1} = \frac{-2(5-2-1)}{(1^2+1^2)}$$

$$\Rightarrow \qquad h - 5 = -2$$

$$\Rightarrow \qquad h = 3$$

$$k = -4$$

$$\therefore \qquad R = (3, -4)$$

$$\therefore \qquad d = \left|\frac{3(3) - 4(-4) + 25}{\sqrt{3^2 + (-4)^2}}\right| = \left|\frac{50}{5}\right| = 10 \text{ units}$$

and $\quad 0.14\,d \Rightarrow 0.14 \times 10 = 1.4$

42. (30) Apply distance formula between parallel line

$$4x - 2y + \alpha = 0$$

$$4x - 2y + 6 = 0$$

$$\Rightarrow \qquad \frac{|\alpha - 6|}{\sqrt{16+4}} = \frac{1}{\sqrt{5}}$$

$$|\alpha - 6| = \frac{2\sqrt{5}}{\sqrt{5}} = 2$$

$$\alpha = \pm 2 + 6 \Rightarrow \alpha = 8, 4$$

Sum = 12

Again, $6x - 3y + \beta = 0$

$$6x - 3y + 9 = 0$$

$\therefore$ Distance between these lines $\left|\frac{\beta - 9}{\sqrt{36+9}}\right| = \frac{2}{\sqrt{5}}$

$$\Rightarrow \qquad |\beta - 9| = 2 \times \frac{3\sqrt{5}}{\sqrt{5}}$$

$$\Rightarrow \qquad \beta - 9 = \pm 6$$

$$\Rightarrow \qquad \beta = 9 \pm 6 = 15, 3$$

Sum = 18

$\therefore$ Total sum = 12 + 18 = 30

43. (7)

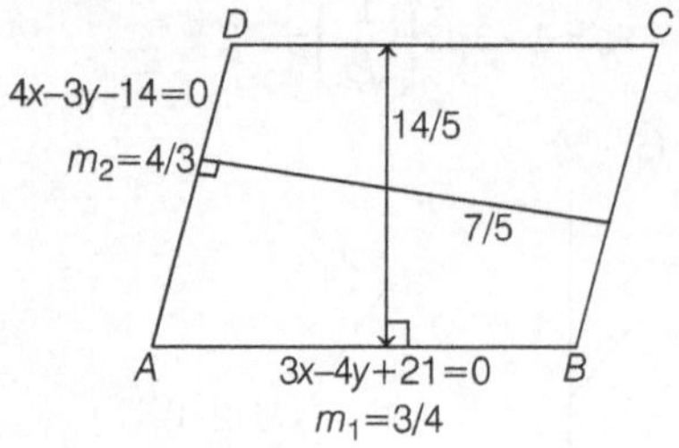

$$\tan\theta = \left|\frac{m_2 - m_1}{1 + m_1 m_2}\right|$$

$$= \left|\frac{\frac{4}{3} - \frac{3}{4}}{1+1}\right|$$

$$= \left|\frac{16-9}{12} \times \frac{1}{2}\right| = \frac{7}{24}$$

$$\therefore \sin\theta = \frac{7}{25}$$

$$\therefore \text{ Area of parallelogram} = \frac{1}{\sin\theta} p_1 p_2$$

$$= \frac{\frac{14}{5} \times \frac{7}{5}}{\frac{7}{25}} = 14$$

$$\Rightarrow \quad 2k = 14 \Rightarrow k = 7$$

44. (6) Clearly, slope of line $2x - 2y + 5 = 0$ is 1

$$\Rightarrow \quad \theta = \frac{\pi}{4}$$

Equation of any line passing through $(2, 3)$ and making an angle $\frac{\pi}{4}$ is

$$\frac{x-2}{\cos\frac{\pi}{4}} = \frac{y-3}{\sin\frac{\pi}{4}} = r$$

$$\Rightarrow \quad x - 2 = y - 3 = \frac{r}{\sqrt{2}}$$

$$\Rightarrow \quad x = 2 + \frac{r}{\sqrt{2}} \text{ and } y = 3 + \frac{r}{\sqrt{2}}$$

which lies on $2x - 3y + 9 = 0$

$$\Rightarrow 2\left(2 + \frac{r}{\sqrt{2}}\right) - 3\left(3 + \frac{r}{\sqrt{2}}\right) + 9 = 0$$

$$\Rightarrow \quad \frac{2r}{\sqrt{2}} - \frac{3r}{\sqrt{2}} + 4 = 0$$

$$\Rightarrow \quad -\frac{r}{\sqrt{2}} = -4 \Rightarrow r = 4\sqrt{2}$$

$$\therefore \quad d\sqrt{2} = 4\sqrt{2}$$

$$d = 4$$

$$d + 2 = 6$$

45. (0.9) $L_1 = 3y - 2x - 6 = 0$

Point about which the line is rotated is $A = (0, 2)$.

Let equation of line L_2 be $y = mx + 2$

As L_2 will cutting the line $x = 5$ below X-axis, so

$A = (0, 2)$, $B = \left(5, \frac{16}{3}\right)$, $C = (5, 0)$, $D = (5, 5m + 2)$ and

$E = \left(-\frac{2}{m}, 0\right)$

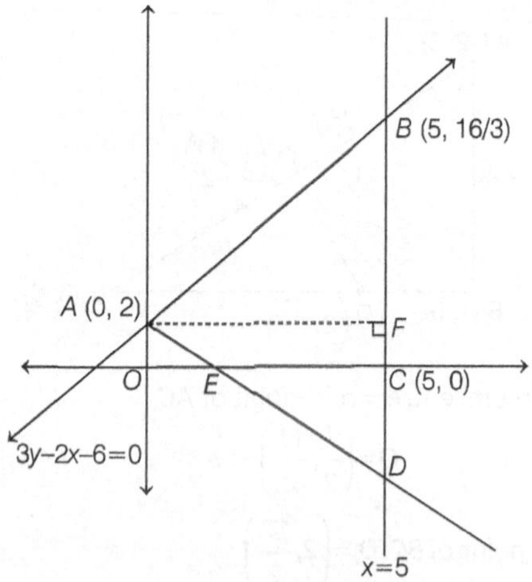

Now, Area of $AECB$ = Area of ΔABD − Area of ΔECD

$$\frac{49}{3} = \frac{1}{2} \times BD \times AF - \frac{1}{2} \times EC \times CD$$

$$= \frac{1}{2}\left[\frac{16}{3} - (5m+2)\right] \times 5 - \frac{1}{2} \times \left(5 + \frac{2}{m}\right)(-5m + 2)$$

$$\frac{49}{3} = \frac{110m + 12}{6m} \Rightarrow m = -1$$

$$\therefore \quad L_2 : y = -x + 2$$

$$x + y - 2 = 0$$

$$a = 1, b = 1 \text{ and } c = -2$$

$$\therefore \frac{3a + 4b - c}{10} \Rightarrow \frac{3 + 4 + 2}{10} = \frac{9}{10}$$

46. (2.646) $\because AB$ is fixed for all position of R and $\angle ARB = 90°$

Locus of R is a circle with diameter as AB.

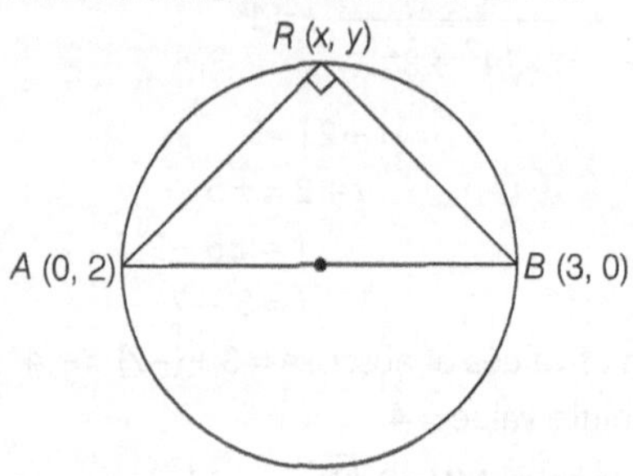

$\therefore$ Required locus is

$$m_{AR} \times m_{BR} = -1$$

$$\Rightarrow \quad \frac{y-2}{x-0} \times \frac{y-0}{x-3} = -1$$

$\Rightarrow \quad (y-2)y = -x(x-3)$

$\Rightarrow x(x-3)+y(y-2)=0$

$\Rightarrow x^2-3x+y^2-2y=0$

$\Rightarrow x^2+y^2-3x-2y=0$

$\Rightarrow \quad a=1, b=1, c=-3$ and $d=-2$

$\therefore a+b-c-d=1+1+3+2=7$

$\therefore \sqrt{a+b-c-d}=\sqrt{7}=2.646$ (upto 3 decimal)

47. (1) $ax+by+c=3x-4y+c$

$a=3$ and $b=-4$

Distance of $3x-4y+c$ from $A(3,1)$ is 1

$$\Rightarrow \left|\frac{9-4+c}{5}\right|=1 \Rightarrow |c+5|=5$$

Also, $3x-4y+c=0$ and $3x-4y+5=0$ lies on the same side of A.

$c+5>0$

$\therefore \quad c+5=5 \Rightarrow c=0$

$\therefore$ Equation is $ax+by+c$ is $3x-4y=0$

$\therefore \quad |a+b+c|=|3-4|=1$

48. (6) Describe a circle whose diameter is AB.

$\therefore$ Centre (1, 0) and radius = 2

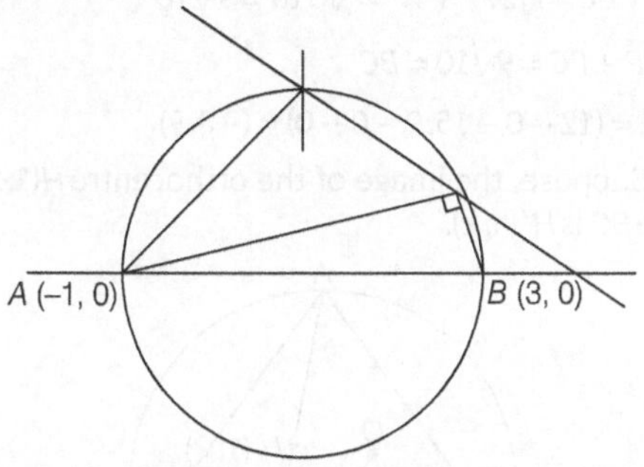

Let m the slope of the line passing through (4,1) $y-1=m(x-4)$ intersect the circle.

$\therefore$ Perpendicular distance from centre < radius of circle

$$\left|\frac{-3m+1}{\sqrt{m^2+1}}\right|<2$$

$\Rightarrow \quad 9m^2-6m+1<4m^2+4$

$\Rightarrow \quad 5m^2-6m-3<0$

$$\Rightarrow \left[m-\left(\frac{6-\sqrt{96}}{10}\right)\right]\left[m-\left(\frac{6+\sqrt{96}}{10}\right)\right]$$

$$m \in \left(\frac{6-\sqrt{96}}{10}, \frac{6+\sqrt{96}}{10}\right)-\left\{\frac{1}{5}, 1\right\}$$

$$\lambda_1+\lambda_2=\frac{12}{10}=\frac{6}{5}$$

$\therefore 5(\lambda_1+\lambda_2)=6$

Level 3

1. (d) Let the coordinates of C be $(1, c)$.

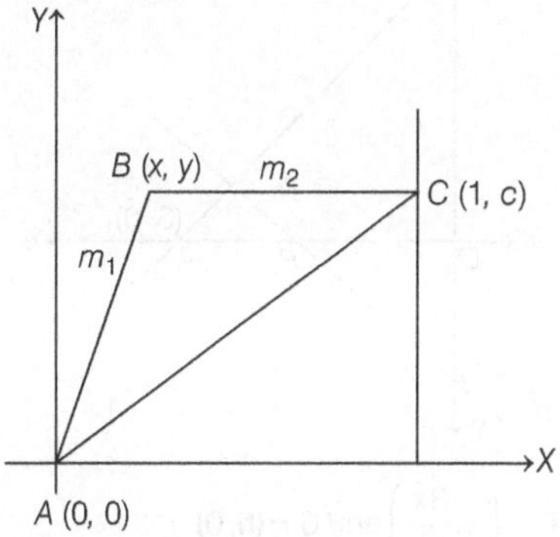

Then, $m_2=\dfrac{c-y}{1-x}$

or $\quad m_2=\dfrac{c-m_1x}{1-x}$

$\Rightarrow \quad m_2-m_2x=c-m_1x$

$\Rightarrow (m_1-m_2)x=c-m_2$

$c=(m_1-m_2)x+m_2 \quad$ ...(i)

Now, Area ΔABC is

$$=\frac{1}{2}\text{mod}\begin{vmatrix}0 & 0 & 1\\ x & m_1x & 1\\ 1 & c & 1\end{vmatrix}$$

$$=\frac{1}{2}|cx-m_1x|$$

$$=\frac{1}{2}|[(m_1-m_2)x+m_2]x-m_1x|$$

$$=\frac{1}{2}|(m_1-m_2)x^2+m_2x-m_1x|$$

$$=\frac{1}{2}|(m_1-m_2)(x-x^2)|$$

$\because x>x^2$ in Eq. (i),

$$f(x)=\frac{1}{2}(x-x^2)$$

$$=1/2\left[\frac{1}{4}-\left(x-\frac{1}{2}\right)^2\right]$$

$\therefore \quad f(4)_{max}=\dfrac{1}{8}$ when $x=1/2$

2. (c) Clearly, circumcentre of ΔABQ will lie on the perpendicular bisector of line AB.

Now, equation of perpendicular bisector of line AB is

$$3x-4y+7/2=0$$

$\therefore$ Locus of circumcentre is $6x-8y+7=0$.

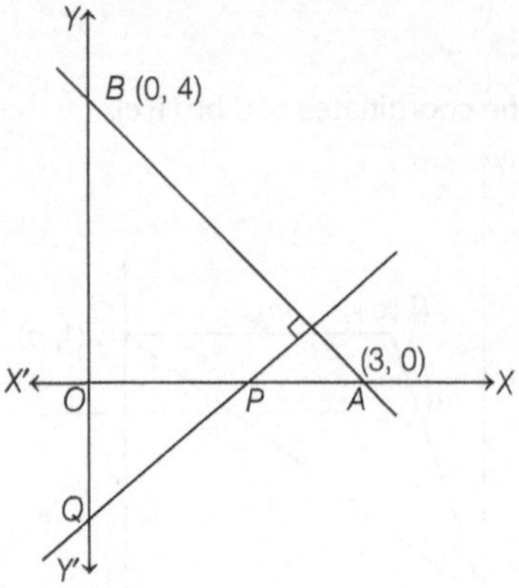

3. (d) Let $B=\left(x, \frac{3x}{4}\right)$ and $C=(h, 0)$

$\because B, P$ and C are collinear.

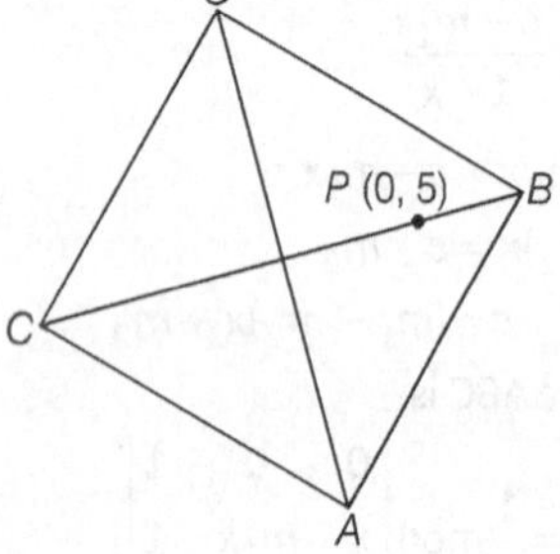

$$\therefore \quad \begin{vmatrix} x & 0 & h \\ \frac{3x}{4} & 5 & 0 \\ 1 & 1 & 1 \end{vmatrix}=0$$

$$\Rightarrow \quad 5x+h\left(\frac{3x}{4}-5\right)=0$$

$$\Rightarrow \quad 20x+3hx-20h=0$$

$$\Rightarrow \quad x(20+3h)=20h$$

$$\Rightarrow \quad x=\frac{20h}{20+3h}$$

Also, $OB^2=OC^2$

$$\Rightarrow \quad x^2+\frac{9x^2}{16}=h^2$$

$$\Rightarrow \quad \frac{25x^2}{16}=h^2$$

$$\Rightarrow \quad \frac{25}{16}\cdot\frac{400h^2}{(20+3h)^2}=h^2$$

$$\Rightarrow \quad 25^2=(20+3h)^2$$

$$\Rightarrow \quad 20+3h=\pm 25$$

$$\Rightarrow \quad 3h=-20\pm 25$$

$$\Rightarrow \quad h=\frac{5}{3}, -15$$

when $h=5/3$	when $h=-15$
$x=\frac{20h}{20+3h}=\frac{4}{3}$	$x=\frac{20(-15)}{20+3(-15)}=12$

$\therefore$ The coordinates of B and C respectively are either $\left(\frac{4}{3}, 1\right)$ and $\left(\frac{5}{3}, 0\right)$ or (12, 9) and (– 15, 0).

Case I : $B=\left(\frac{4}{3}, 1\right), P=(0,5)$ and $C=\left(\frac{5}{3}, 0\right)$

P is an inside point and B, P and C are collinear.

$$\Rightarrow BP+PC=BC$$

$$\Rightarrow BP+PC=\sqrt{\frac{16}{9}+16}+\sqrt{\frac{25}{9}+25}=3\sqrt{10}$$

$$\Rightarrow \quad BC=\sqrt{\frac{1}{9}+1}=\frac{\sqrt{10}}{3}$$

Therefore, $B\neq\left(\frac{4}{3}, 1\right)$ and $C\neq\left(\frac{5}{3}, 0\right)$.

Case II : $B=(12, 9), P=(0, 5)$ and $C=(-15, 0)$

Now, $BP=\sqrt{12^2+4^2}=\sqrt{160}=4\sqrt{10}$

$$PC=\sqrt{225+25}=\sqrt{250}=5\sqrt{10}$$

$$BC=\sqrt{27^2+9^2}=\sqrt{810}=9\sqrt{10}$$

$$\therefore \ BP+PC=9\sqrt{10}=BC$$

$$\therefore \ A=(12-0-15, 9-0-0)=(-3, 9)$$

4. (b) Suppose, the image of the orthocentre $H(1, 2)$ on the side BC is $H'(h, k)$.

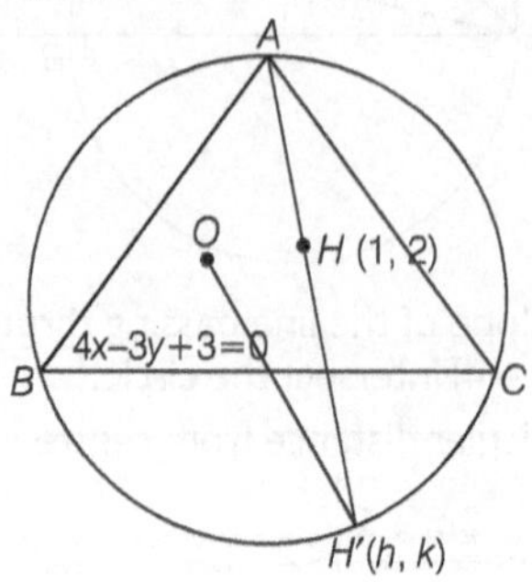

$$\frac{h-1}{4}=\frac{k-2}{-3}=\frac{-2(4-6+3)}{4^2+3^2}=\frac{-2}{25}$$

$$\Rightarrow \quad h=1-\frac{8}{25}=\frac{17}{25}$$

$$\Rightarrow \quad k=2+\frac{6}{25}=\frac{56}{25}$$

$$\therefore \quad H'=\left(\frac{17}{25}, \frac{56}{25}\right)$$

$\therefore$ Circumradius $R=OH'$

$$=\sqrt{\left(2-\frac{17}{25}\right)^2+\left(3-\frac{56}{25}\right)^2}=\frac{\sqrt{1450}}{25}=\frac{\sqrt{58}}{5}$$

5. (b) $P_1 = \dfrac{2m\sin\alpha + m^2\cos\alpha + \dfrac{\sin^2\alpha}{\cos\alpha}}{\sqrt{\sin^2\alpha + \cos^2\alpha}}$

$$= \frac{(m\cos\alpha + \sin\alpha)^2}{\cos\alpha}$$

Similarly, $P_3 = \dfrac{(m'\cos\alpha + \sin\alpha)^2}{\cos\alpha}$

$$P_2 = \frac{(m+m')\sin\alpha + mm'\cos\alpha + \dfrac{\sin^2\alpha}{\cos\alpha}}{\sqrt{\sin^2\alpha + \cos^2\alpha}}$$

$$= \frac{(m\cos\alpha + \sin\alpha)(m'\cos\alpha + \sin\alpha)}{\cos\alpha}$$

$\because \quad P_2^2 = P_1 \times P_3$

$\therefore P_1, P_2, P_3$ are in GP.

6. (d) $A_2^2 = A_1 \times A_3$

$$\begin{vmatrix} a & b & 1 \\ a\sec^2\theta & b\,\text{cosec}^2\theta & 1 \\ a + a\sin^2\theta & b + b\cos^2\theta & 1 \end{vmatrix}^2 = \begin{vmatrix} 0 & 0 & 1 \\ a\tan\theta & b\cot\theta & 1 \\ a\sin\theta & b\cos\theta & 1 \end{vmatrix}$$

$$\begin{vmatrix} 0 & 0 & 1 \\ a\tan\theta & -b\cot\theta & 1 \\ a\sin\theta & b\cos\theta & 1 \end{vmatrix}$$

$$\Rightarrow a^2b^2 \begin{vmatrix} 1 & 1 & 1 \\ \sec^2\theta & \text{cosec}^2\theta & 1 \\ \sin^2\theta + 1 & \cos^2\theta + 1 & 1 \end{vmatrix}^2$$

$$= (ab\tan\theta\cos\theta - ab\sin\theta\cot\theta) \cdot (ab\tan\theta\cos\theta + ab\sin\theta\cot\theta)$$

$$\Rightarrow a^2b^2 \begin{vmatrix} 0 & 0 & 1 \\ \sec^2\theta - \text{cosec}^2\theta & -\cot^2\theta & 1 \\ \sin^2\theta - \cos^2\theta & \cos^2\theta & 1 \end{vmatrix}^2$$

$$= a^2b^2(\sin\theta - \cos\theta)(\sin\theta + \cos\theta)$$

$$= a^2b^2(\sin^2\theta - \cos^2\theta)$$

$$\Rightarrow a^2b^2 \left\{ \left[\frac{\sin^2\theta - \cos^2\theta}{\sin^2\theta \cdot \cos^2\theta} \cdot \cos^2\theta \right] - \frac{\cos^2\theta}{\sin^2\theta}[\sin^2\theta - \cos^2\theta] \right\}$$

$$= a^2b^2(\sin^2\theta - \cos^2\theta)$$

$$\Rightarrow \frac{(\sin^2\theta - \cos^2\theta)}{\sin^2\theta}(1 - \cos^2\theta)$$

$$= (\sin^2\theta - \cos^2\theta)$$

$$\Rightarrow (\sin^2\theta - \cos^2\theta) = \sin^2\theta - \cos^2\theta$$

For $\theta \in R$ but $\theta \neq 0$

7. (b)

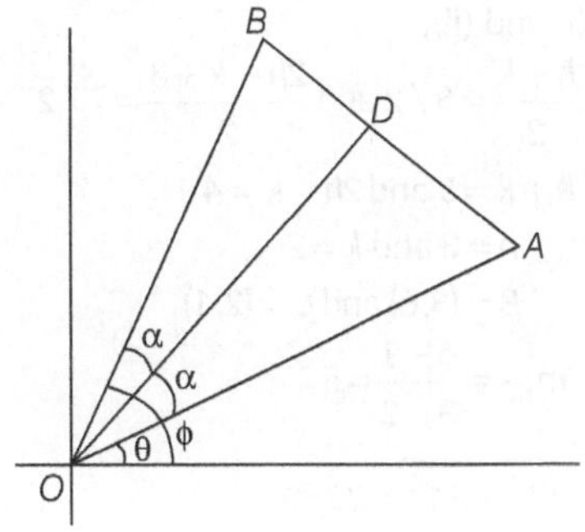

From figure,

$$\angle AOD = \alpha = \angle DOB, \phi - \theta = 2\alpha$$

or $\quad \alpha = \dfrac{\phi - \theta}{2}$ or $\angle DOX = \theta + \alpha = \dfrac{\theta + \phi}{2}$

$\therefore$ Gradiant of $OD = \tan\left(\dfrac{\theta + \phi}{2}\right)$

$$= \frac{\sin\left(\dfrac{\theta+\phi}{2}\right)}{\cos\left(\dfrac{\theta+\phi}{2}\right)}$$

$$= \frac{2\sin^2\left(\dfrac{\theta+\phi}{2}\right)}{2\cos\left(\dfrac{\theta+\phi}{2}\right)\sin\left(\dfrac{\theta+\phi}{2}\right)}$$

$$= \frac{1 - \cos(\theta+\phi)}{\sin(\theta+\phi)}$$

$$= \text{cosec}(\theta+\phi) - \cot(\theta+\phi)$$

8. (d) Solving $2x - y = 0$ and $x + y = 3$

$\therefore$ We get $A = (1, 2)$

Equation of median through A

$$\Rightarrow \quad y - 2 = \frac{2-3}{1-2}(x-1)$$

$$x - y = -1$$

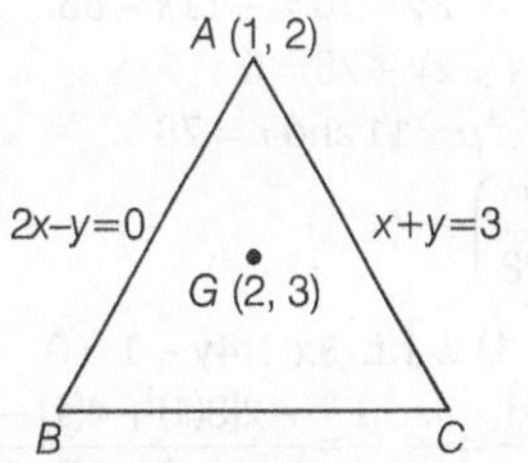

Let $B = (h, 2h)$ and $C(k, 3-k)$

$\therefore$ Mid-point of $BC = \left(\dfrac{h+k}{2}, \dfrac{2h-k+3}{2}\right)$ ...(i)

Let D is mid-point BC

$\because \quad \dfrac{AG}{GD} = \dfrac{2}{1}$

$$D = \left(\frac{5}{2}, \frac{7}{2}\right) \quad \text{...(ii)}$$

By Eqs. (i) and (ii),

$$\frac{h+k}{2}=5/2 \text{ and } \frac{2h-k+3}{2}=7/2$$

$$h+k=5 \text{ and } 2h-k=4$$

$\therefore \quad h=3$ and $k=2$

$\therefore \quad B=(3,6)$ and $C=(2,1)$

$$m_{BC}=\frac{6-1}{3-2}=5$$

9. (b) $AB=\sqrt{7^2+24^2}=25$

$BC=\sqrt{(16)^2+(12)^2}=20$ and $AC=\sqrt{9^2+12^2}=15$

$\because$ We know that internal bisector divide opposite sides in side ratio.

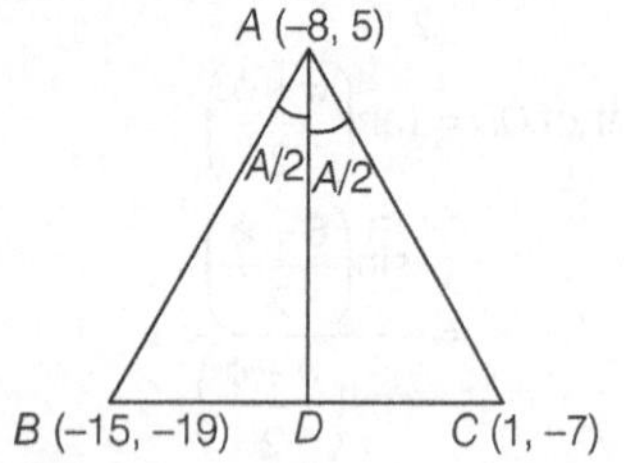

$$\frac{BD}{DC}=\frac{AB}{AC}$$

$$\Rightarrow \quad \frac{BD}{DC}=\frac{AB}{AC}=\frac{25}{15}=\frac{5}{3}$$

$$\Rightarrow \quad D=\left(\frac{-15\times3+5(1)}{5+3},\frac{-19\times3+5\times(-7)}{5+3}\right)$$

$$=\left(-5,\frac{-23}{2}\right)$$

$\therefore$ Equation of AD,

$$y-5=\frac{5+23/2}{-8+5}(x+8)$$

$$\Rightarrow \quad y-5=\frac{-11}{2}(x+8)$$

$$\Rightarrow \quad 2y-10=-11x-88$$

$$\Rightarrow \quad 11x+2y+78=0$$

$$\Rightarrow \quad p=11 \text{ and } r=78$$

Point $\left(\frac{p}{11},\frac{r}{78}\right)\Rightarrow(1,1)$

Image of (1, 1) w.r.t. $3x+4y-1=0$

$$\Rightarrow \quad \frac{h-1}{3}=\frac{k-1}{4}=\frac{-2[3(1)+4(1)-1]}{3^2+4^2}$$

$$\Rightarrow \quad \frac{h-1}{3}=\frac{k-1}{4}=\frac{-12}{25}$$

$$\Rightarrow \quad h=1-\frac{36}{25} \text{ and } k=1-\frac{48}{25}$$

$$\Rightarrow \quad h=\frac{-11}{25} \text{ and } k=\frac{-23}{25}$$

$\therefore$ Required point is $\left(\frac{-11}{25},\frac{-23}{25}\right)$.

10. (b, c) H is image of B'

$$\Rightarrow \quad H=(0,b')$$

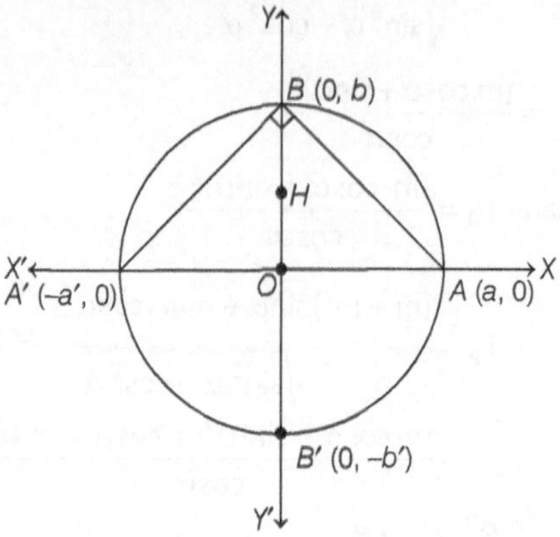

Applying power of 0,

$$0=aa'-bb' \Rightarrow b'=\frac{aa'}{b}$$

$\therefore$ Required points are $(0,b')$ and $\left(0,\frac{aa'}{b}\right)$.

11. (a, b) $\because L$ must be lies on angle bisector of L_1 and L_2.

$\therefore L$ is given by

$$\frac{3x+4y-1}{\sqrt{3^2+4^2}}=\pm\frac{(5x-12y+2)}{\sqrt{5^2+12^2}}$$

$$\Rightarrow \quad \frac{3x+4y-1}{5}=\pm\frac{5x-12y+2}{13}$$

$$\Rightarrow \quad 14x+112y-23=0 \text{ or } 64x-8y-3=0$$

12. (a, d) $x^2+8x-2009=0 <^{x_1}_{y_1}$

$$\Rightarrow \quad x_1+y_1=-8$$

$$3x^2+24x-2010=0 <^{x_2}_{y_2}$$

$$\Rightarrow \quad x_2+y_2=\frac{-24}{3}=-8$$

$$9x^2+72x-2011=0 <^{x_3}_{y_3}$$

$$\Rightarrow \quad x_3+y_3=\frac{-72}{9}=-8$$

Now, value of $\begin{vmatrix} x_1 & y_1 & 1\\ x_2 & y_2 & 1\\ x_3 & y_3 & 1\end{vmatrix}$

Applying $C_1\to C_1+C_2$, we get

$$\begin{vmatrix} x_1+y_1 & y_1 & 1\\ x_2+y_2 & y_2 & 1\\ x_3+y_3 & y_3 & 1\end{vmatrix}=\begin{vmatrix} -8 & y_1 & 1\\ -8 & y_2 & 1\\ -8 & y_3 & 1\end{vmatrix}=0$$

$\therefore A, B, C$ are collinear.

They cannot lie on a circle.

13. (b, c) Equation of perpendicular bisector of AB is

$$2x - 3y + \lambda = 0$$

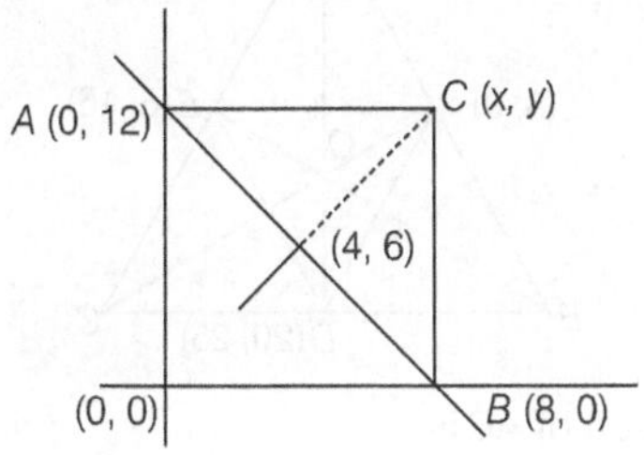

It passes through (4, 6)

$$2(4) - 3(6) + \lambda = 0$$
$$\lambda = 10$$

So, equation is $2x - 3y + 10 = 0$ ...(i)

$\therefore$ Area of $\Delta ABC = 91$

$$\frac{1}{2}\begin{vmatrix} 0 & 8 & x \\ 12 & 0 & y \\ 1 & 1 & 1 \end{vmatrix} = \pm 91$$

$\Rightarrow \quad -8(12 - y) + x(12) = \pm 182$

$\Rightarrow \quad -96 + 8y + 12x = \pm 182$

With '+' sign

$\Rightarrow \quad 6x + 4y - 139 = 0$ ...(i)

Solving Eqs. (i) and (ii),

$$y = 13 \text{ and } x = 29/2$$

$$\therefore C\left(\frac{29}{2}, 13\right)$$

With '–' sign,

$$12x + 8y + 86 = 0$$

$\Rightarrow \quad 6x + 4y + 43 = 0$ ...(iii)

By Eqs. (i) and (iii), we get

$$C\left(\frac{-13}{2}, -1\right)$$

14. (a, d) $\because \quad u = 0 \Rightarrow m_u = \frac{-a}{b}$

$$v = 0 \Rightarrow m_v = \frac{b}{a}$$

$$m_u \times m_v = -1$$

$\therefore$ $u = 0$ and $v = 0$ are perpendicular

Assume the coordinate axes to be directed along $u = 0$ and $v = 0$.

Now, the lines $k_1u - k_2v = 0$ and $k_1u + k_2v = 0$ are equally inclined with UV-axes.

$\therefore$ The bisector of $u = 0$ and $v = 0$

$\Rightarrow \quad \frac{k_1u - k_2v}{\sqrt{k_1^2 + k_2^2}} = \pm \frac{k_1u + k_2v}{\sqrt{k_1^2 + k_2^2}}$

$\Rightarrow \quad k_1u - k_2v = \pm(k_1u + k_2v)$

$\Rightarrow \quad k_2v = 0 \;\big|\; k_1u = 0$

$\qquad v = 0 \;\big|\; u = 0$

15. (b, c, d) $\because \sec^2(n(m+2)) + m^2 = 1$

$$1 + \tan^2(n(m+2)) + m^2 = 1$$

$\Rightarrow \quad \tan^2(m+2)n + m^2 = 0$

$\Rightarrow \quad m = 0 \text{ and } \tan^2(m+2)n = 0$

$\therefore \quad \tan^2 2n = 0 \Rightarrow n = 0, \frac{\pi}{2}, \frac{-\pi}{2}$

$$(m, n) = (0, 0)\,(0, \pi/2)\,(0, -\pi/2)$$

$\therefore$ Equation of straight lines

(i) $y - 0 = 1(x - 0) \Rightarrow y = x \Rightarrow x - y = 0$

(ii) $y - \frac{\pi}{4} = 1(x - 0) \Rightarrow x - y + \frac{\pi}{4} = 0$

(iii) $y + \frac{\pi}{4} = 1(x - 0) \Rightarrow x - y - \frac{\pi}{4} = 0$

16. (a, d) Let length of each side be a.

$\therefore \quad \sin 60° = \frac{h}{a} \Rightarrow a = \frac{16}{\sqrt{3}}$

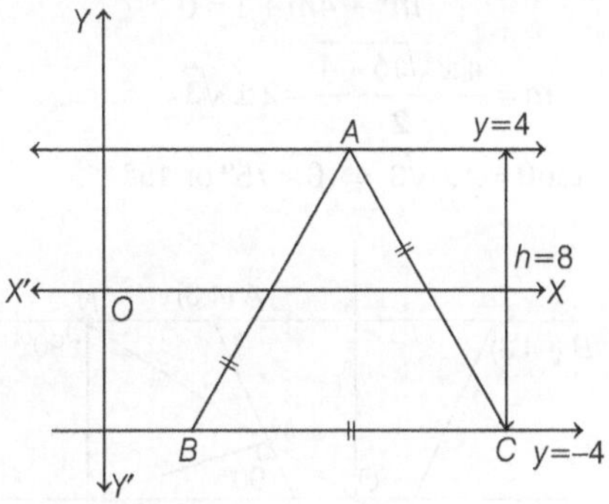

$\therefore$ Perimeter of ΔABC

$$= 3a = \frac{3 \times 16}{\sqrt{3}} = 16\sqrt{3} \text{ units}$$

Also, $\quad \frac{a}{\sin A} = 2R$

$\Rightarrow \quad R = \frac{16/\sqrt{3}}{2 \times \sqrt{3}/2} = \frac{16}{3}$

$\therefore \quad r = \frac{R}{2} = \frac{8}{3}$ ($\because \Delta ABC$ is equilateral triangle)

Now,

$\because \Delta ABC$ is equilateral.

So, locus of orthocentre is same as locus of centroid.

$$\underset{\text{(circumcentre)}}{O} \overset{1}{—} \underset{\text{(centroid)}}{G} \overset{2}{—} \underset{\text{(orthocentre)}}{H}$$

Clearly, distance of centroid from line $y = 4$ is $\frac{16}{3}$ and distance of centroid from line $y = -4$ is $\frac{8}{3}$.

$\therefore$ So, locus of centroid is $y = \frac{-4}{3} \Rightarrow 3y + 4 = 0$

ΔABC is equilateral

So, $\quad r_1 = r_2 = r_3 \Rightarrow r_1 : r_2 : r_3 = 1 : 1 : 1$

17. (a, b) Let $y-2=m(x-1)$ be the equation of the line l

i.e. $y=mx+2-m$ ...(i)

$y=-x+4$ ...(ii)

Solving Eqs. (i) and (ii),

$$x=\frac{m+2}{m+1} \text{ and } y=\frac{3m+2}{m+1}$$

$\because$ Distance between $\left(\frac{m+2}{m+1}, \frac{3m+2}{m+1}\right)$

and (1, 2) is $\sqrt{\frac{2}{3}}$, we have

$$\left(\frac{m+2}{m+1}-1\right)^2+\left(\frac{3m+2}{m+1}-2\right)^2=\frac{2}{3}$$

$$\Rightarrow \left(\frac{1}{m+1}\right)^2+\left(\frac{m^2}{(m+1)^2}\right)=\frac{2}{3}$$

$$\Rightarrow m^2-4m+1=0$$

$$\Rightarrow m=\frac{4\pm\sqrt{16-4}}{2}=2\pm\sqrt{3}$$

$$\Rightarrow \tan\theta=2\pm\sqrt{3} \Rightarrow \theta=75° \text{ or } 15°$$

18. (a,d)

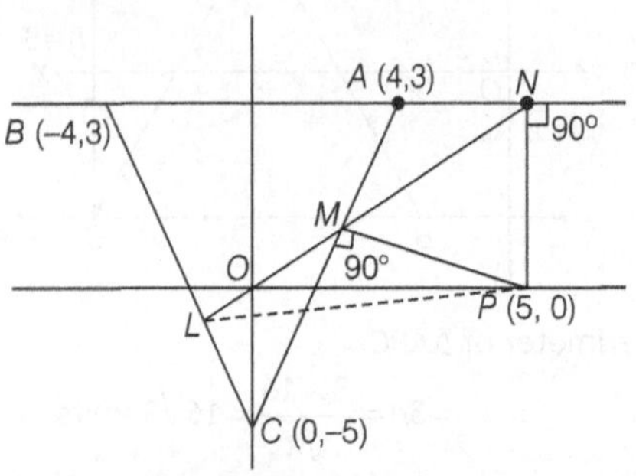

The sides BC, CA and AB are represented by the equation $2x+y=5$, $2x-y-5=0$ and $y-3=0$, respectively.

We can see that $L=(-1,-3)$, $M=(3,1)$ and $N=(5,3)$.

Now, slope of LM = slope of $MN=1$

So, that L, M and N are collinear.

19. (a,b,c) We use the fact that the orthocentre O of the ΔABC is the incentre of the pedal ΔOEF.

Let (h, k) be the coordinates of O.

Now, $ED=\sqrt{(20-8)^2+(25-16)^2}=15$

$FD=20$ and $EF=7$

So, that $h=\frac{7\times20+20\times8+15\times8}{7+20+15}=10$

$$k=\frac{7\times25+20\times16+15\times9}{7+20+15}=15$$

Thus, the coordinates of O or (10, 15).

$\because$ AC is perpendicular to OE. $\left[\because m_{AC}=\frac{-1}{m_{DE}}=\frac{-1}{\frac{(15-16)}{(10-8)}}\right]$

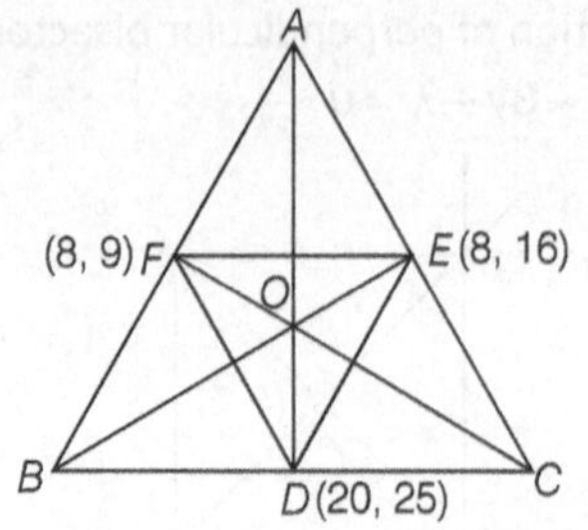

$\therefore$ Equation of AC is

$$y-16=-\left(\frac{10-8}{15-16}\right)(x-8)$$

$\Rightarrow$ $y-16=2(x-8) \Rightarrow y-2x=0$ ...(i)

Similarly, equation of AB is

$$y-9=-\frac{1}{m_{OF}}(x-8) \quad \left[\because m_{OF}=\frac{15-9}{10-8}=\frac{6}{2}=3\right]$$

$$\Rightarrow y-9=\frac{-1}{3}(x-8)$$

$\Rightarrow$ $3y-27=-x+8 \Rightarrow 3y+x=35$ ...(ii)

And equation of BC is

$$y-25=-\frac{1}{m_{OD}}(x-20) \left[\because m_{OD}=\frac{25-15}{20-10}=\frac{10}{10}=y\right]$$

$\Rightarrow y-25=-(x-20) \Rightarrow y+x=45$ ...(iii)

Solving Eqs. (i) and (ii),

$6x+x=35 \Rightarrow 7x=35$ [point (5, 10)]

$\Rightarrow$ $x=5$ and $y=10$

Solving Eqs. (i) and (iii),

$x+2x=45 \Rightarrow 3x=45$ [point (15, 30)]

$\Rightarrow$ $x=15$ and $y=30$

Solving Eqs. (ii) and (iii),

$2y=-10 \Rightarrow y=-5$ [point (50, –5)]

$\Rightarrow$ $x=50$

20. (a,c,d) Let the lines represented by the given equation be $y=x\tan\alpha$ and $y=x\tan\beta$

Then, $\tan\alpha+\tan\beta=\frac{2\tan\theta}{\sin^2\theta}=\frac{2}{\sin\theta\cdot\cos\theta}=4\operatorname{cosec}2\theta$

$$\therefore \quad \tan\theta\cdot\tan\beta=\frac{\tan^2\theta+\cos^2\theta}{\sin^2\theta}$$

$$=\sec^2\theta+\cot^2\theta$$

$$\Rightarrow \tan\alpha-\tan\beta=\pm\sqrt{\frac{4}{\sin^2\theta\cos^2\theta}-4(\sec^2\theta+\cot^2\theta)}$$

$$=\pm2\sqrt{\frac{1-(\sin^2\theta+\cos^4\theta)}{\sin^2\theta\cos^2\theta}}$$

$$=\pm2\sqrt{\frac{\cos^2\theta-\cos^4\theta}{\sin^2\theta\cdot\cos^2\theta}}=\pm2$$

and $\frac{\tan\alpha}{\tan\beta}=\frac{4\operatorname{cosec}2\theta+2}{4\operatorname{cosec}2\theta-2}=\frac{2+\sin2\theta}{2-\sin2\theta}$

21. (a, b, c, d) Let slope of line $u=0$ is m_1 and slope of line $v=0$ is m_2

Given, $\frac{m_1}{m_2}=9/2$ ...(i)

and angle between two lines is $\tan^{-1}7/9$

$$\Rightarrow \quad 7/9=\left|\frac{m_1-m_2}{1+m_1m_2}\right|$$

$$\Rightarrow \quad \frac{m_1-m_2}{1+m_1m_2}=\pm 7/9$$

$$\Rightarrow \quad \frac{m_1-m_2}{1+m_1m_2}=7/9 \quad \Bigg| \quad \frac{m_1-m_2}{1+m_1m_2}=-7/9$$

$$9m_1-9m_2=7+7m_1m_2 \quad \Big| \quad 9m_1-9m_2=-7-7m_1m_2$$

Putting the value of m_1 from Eq. (i) we get

$9m_2^2-9m_2+2=0$ or $9m_2^2+9m_2+2=0$

$$m_2=\frac{2}{3},1/3$$

or $$m_2=\frac{-2}{3},-1/3$$

Case I $m_2=2/3, m_1=3$

$\therefore$ Equation is $y=3x, 3y=2x$

Case II $m_2=1/3, m_1=3/2$

$\therefore$ Equation are $3y=x$ and $2y=3x$

Case III $m_2=-2/3, m_1=-3$

$\therefore$ Equation are $3x+y=0, 2x+3y=0$

Case IV $m_2=-1/3, m_1=-3/2$

$\therefore$ Equation are $x+3y=0, 3x+2y=0$

22. (a,b,c,d) We have,

Vertices of ΔABC are $A(0,0)$, $B(\cos\alpha, \sin\alpha)$ and $C(\cos\beta, \sin\beta)$.

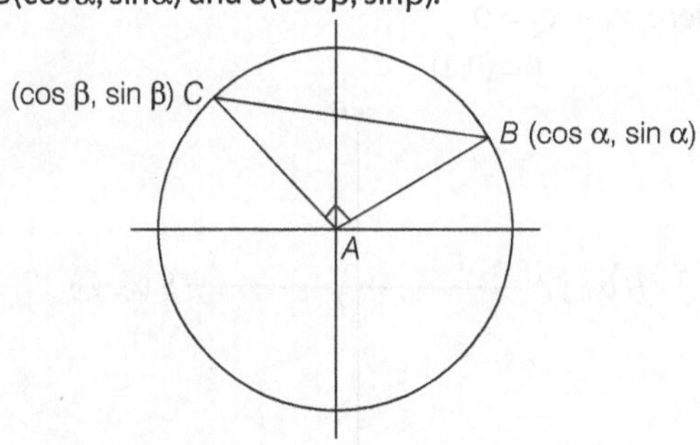

ΔABC is right angle triangle.

It is possible when

$$\alpha-\beta=\pm\pi/2, \pm\frac{3\pi}{2}$$

$$\Rightarrow \quad \frac{\alpha-\beta}{2}=\pm\pi/4, \pm 3\pi/4$$

$$\therefore \cos\left(\frac{\alpha-\beta}{2}\right)=\cos(\pm\pi/4) \text{ or } \cos\left(\pm\frac{3\pi}{4}\right)$$

$$=\frac{1}{\sqrt{2}} \text{ or } \frac{-1}{\sqrt{2}}$$

and $$\sin\frac{(\alpha-\beta)}{2}=\sin(\pm\pi/4) \text{ or } \sin\left(\pm\frac{3\pi}{4}\right)$$

$$=\pm\frac{1}{\sqrt{2}} \text{ or } \pm\frac{1}{\sqrt{2}}$$

$\therefore$ All options are corrects.

23. (a, b, c) Given lines are

$$(2\cos\theta+3\sin\theta)x+(3\cos\theta-5\sin\theta)y-(5\cos\theta-2\sin\theta)=0$$

$$\cos\theta(2x+3y-5)+\sin\theta(3x-5y+2)=0$$

and $$(a+2b)x+(a-3b)y+b-a=0$$

$$a(x+y-1)+b(2x-3y+1)=0$$

Lines are $2x+3y-5=0$ ...(i)

$3x-5y+2=0$ ...(ii)

Solving Eqs. (i) and (ii), we get

$$x_1=1 \text{ and } y_1=1$$

Other lines $x+y-1=0$ ...(iii)

$2x-3y+1=0$ ...(iv)

Solving Eqs. (iii) and (iv), we get

$$x_2=2/5 \text{ and } y_2=\frac{3}{5}$$

(a) $x_2+y_2=\frac{2}{5}+\frac{3}{5}=1=x_1$

(b) $x_2+y_2=1=y_1$

(c) $6x_2+y_2=\frac{12}{5}+\frac{3}{5}=5$ and $2x_1+y_1=2+3=5$

$\therefore 6x_2+y_2=2x_1+y_1$

(d) $2x_2+y_2=\frac{4}{5}+\frac{3}{5}=\frac{7}{5}$ and $6x_1+y_1=7$

$2x_2+y_2\neq 6x_1+y_1$

$\therefore$ a, b and c are correct option.

24. (a, b) We have $ABCD$ is rhombus.

Given, $B(1,-3)$ and $D(0,4)$

$$\therefore \quad BD=\sqrt{(0-1)^2+(4+3)^2}=5\sqrt{2}$$

$$\Rightarrow \quad BP=\frac{BD}{2}=\frac{5\sqrt{2}}{2}$$

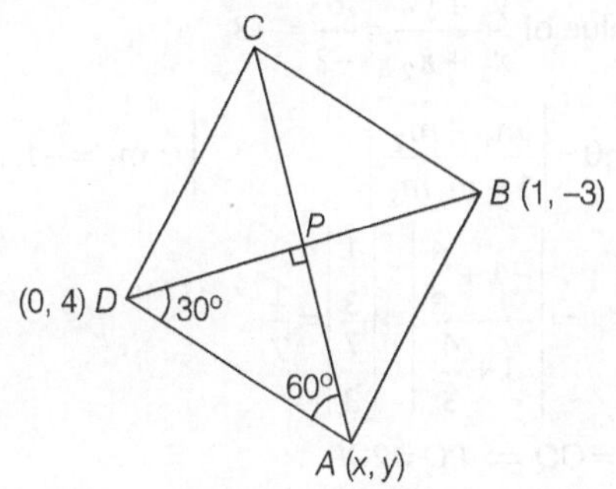

Coordinate of P is $(1/2, 1/2)$.

$$\therefore \text{ Slope of } BD=\frac{4+3}{0-1}=-7$$

$\because$ Slope of $AC = 1/7$

In ΔAPB, $\tan 30° = \dfrac{BP}{AP} \Rightarrow AP = \dfrac{5\sqrt{2}}{2}\cot 30° = \dfrac{5\sqrt{6}}{2}$

Equation line AP in parametric form

$$\frac{x-1/2}{\cos\theta} = \frac{y-1/2}{\sin\theta} = \pm\frac{5\sqrt{6}}{2} \left\{ \begin{aligned} &\therefore \tan\theta = \frac{1}{7}, \\ &\sin\theta = \frac{1}{5\sqrt{2}} \text{ and} \\ &\cos\theta = \frac{7}{5\sqrt{2}} \end{aligned} \right\}$$

$$\Rightarrow \quad x - \frac{1}{2} = \frac{\pm 5\sqrt{6}}{2}\cos\theta \text{ and } y - \frac{1}{2} = \frac{\pm 5\sqrt{6}}{2}\sin\theta$$

$$\Rightarrow \quad x = 1/2 \pm \frac{5\sqrt{6}}{2} \times \frac{7}{5\sqrt{2}} \text{ and } y = \frac{1}{2} \pm \frac{5\sqrt{6}}{2} \times \frac{1}{5\sqrt{2}}$$

$\therefore$ Coordinates are

$$\left(\frac{1-7\sqrt{3}}{2}, \frac{1-\sqrt{3}}{2}\right) \text{ and } \left(\frac{1+7\sqrt{3}}{2}, \frac{1+\sqrt{3}}{2}\right)$$

Sol. (Q.Nos. 25-27)

25. (b) Given, $AQ = BP = 1$, $\angle P = \angle Q = 90°$, $\angle AOP = \angle BOP$

$\therefore \Delta AOP \cong \Delta BOQ \Rightarrow OA = OB$

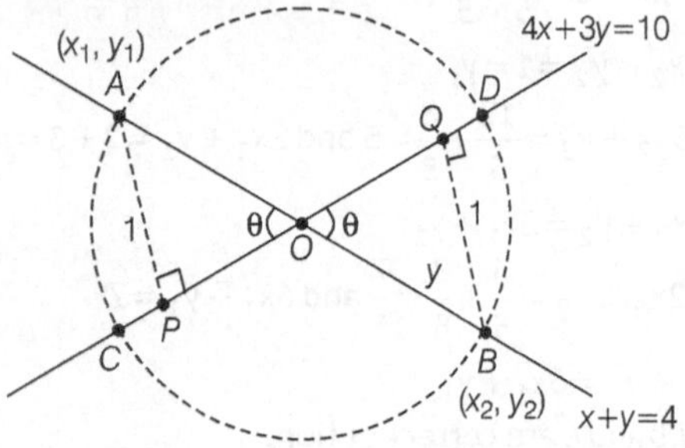

$\Rightarrow O$ is mid-point of AB.

$\Rightarrow$ Point of intersection, $4x + 3y = 10$ and $x + y = 4$

$\Rightarrow O = (-2, 6)$

$$\Rightarrow O\left(\frac{x_1+x_2}{2}, \frac{y_1+y_2}{2}\right) = O(-2, 6)$$

$\therefore$ Value of $\dfrac{y_1+y_2}{x_1+x_2} = \dfrac{6}{-2} = -3$

26. (c) $\tan\theta = \left|\dfrac{m_1 - m_2}{1 + m_1 m_2}\right| \qquad \left[\because m_1 = -1, m_2 = -\dfrac{4}{3}\right]$

$$= \left|\frac{-1+\frac{4}{3}}{1+\frac{4}{3}}\right| = \left|\frac{\frac{1}{3}}{\frac{7}{3}}\right| = \frac{1}{7}$$

$\therefore OP = OQ \Rightarrow PQ = 2OP$

Now, $\dfrac{AP}{OP} = \dfrac{1}{7} \Rightarrow \dfrac{1}{OP} = \dfrac{1}{7} \Rightarrow OP = 7$

$\therefore \quad \phi Q = 2 \times 7 = 14$ units

27. (d) $OD = OA = \sqrt{\left(\dfrac{14}{2}\right)^2 + 1} = \sqrt{50}$

$\Rightarrow OD = 5\sqrt{2}$

and $QD = OD - OQ = 5\sqrt{2} - 7$ units

Sol. (Q.Nos. 28-29)

28. (b) Let $P(x, y)$ be the point.

$$\Rightarrow \left|\frac{\sqrt{2}x + y - 1}{\sqrt{3}}\right| \left|\frac{\sqrt{2}x - y + 1}{\sqrt{3}}\right| = \lambda^2$$

$$\Rightarrow \left|\frac{2x^2 - (y-1)^2}{3}\right| = \lambda^2$$

$$C : |2x^2 - (y-1)^2| = 3\lambda^2$$

Line $y = 2x + 1$

$$RS = \sqrt{(x_1 - x_2)^2 + (y_1 - y_2)^2}$$

$\because \quad y_1 = 2x_1 + 1$ and $y_2 = 2x_2 + 1$

$\Rightarrow y_1 - y_2 = 2(x_1 - x_2)$

Then, $RS = \sqrt{5(x_1 - x_2)^2} = \sqrt{5}\,|x_1 - x_2|$

Solve curve C and line $y = 2x + 1$, we get

$$|2x^2 - (2x)^2| = 3\lambda^2 \Rightarrow x^2 = \frac{3\lambda^2}{2}$$

$$\Rightarrow \qquad RS = \sqrt{5}\left|\frac{2\sqrt{3}\lambda}{\sqrt{2}}\right| = \sqrt{30}\,\lambda = \sqrt{270}$$

$$\Rightarrow \qquad \lambda^2 = \frac{270}{30} \Rightarrow \lambda^2 = 9$$

29. (a) $T = \left(\dfrac{x_1+x_2}{2}, \dfrac{y_1+y_2}{2}\right)$

Here, $x_1 + x_2 = 0$

$$T \equiv (0, 1)$$

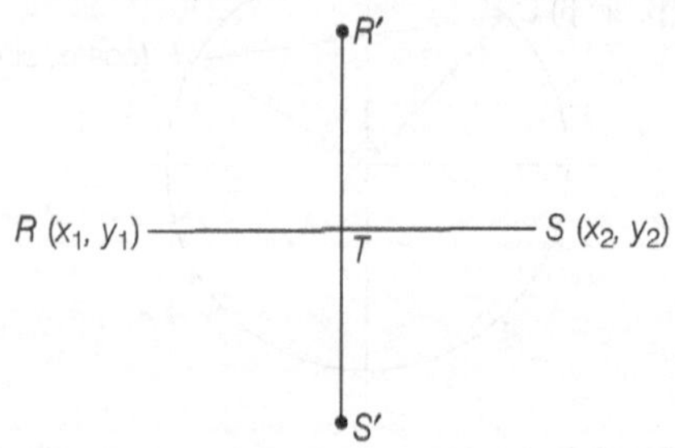

$\therefore$ Equation of $R'S'$ is

$$y - 1 = \frac{1}{2}(x - 0)$$

$$\Rightarrow \qquad x + 2y = 2$$

$$R' = (a_1, b_1) \text{ and } S' = (a_2, b_2)$$

$$D = (a_1 - a_2)^2 + (b_1 - b_2)^2 = 5(b_1 - b_2)^2$$

Solve $x + 2y = 2$ and $|2x^2 - (y-1)^2| = 3\lambda^2$

$$\Rightarrow \qquad |8(y-1)^2 - (y-1)^2| = 3\lambda^2$$

$$\Rightarrow \quad (y-1)^2 = \left(\frac{\sqrt{3}\lambda}{\sqrt{7}}\right)^2$$

$$\Rightarrow \quad y - 1 = \pm \frac{\sqrt{3}}{\sqrt{7}}\lambda$$

$$\Rightarrow \quad b_1 = 1 + \frac{\sqrt{3}}{\sqrt{7}}\lambda \text{ and } b_2 = 1 - \frac{\sqrt{3}}{\sqrt{7}}\lambda$$

$$D = 5\left(\frac{2\sqrt{3}\lambda}{\sqrt{7}}\right)^2$$

$$= \frac{5 \times 4 \times 3\lambda^2}{7} = \frac{5 \times 4 \times 27}{7}$$

$$= 77.14$$

Sol. (Q.Nos. 30-32)

On solving $4x - 3y = 0$ and $4x + y = 16$, we get $P(3, 4)$.

The equation of line perpendicular to PQ is $3x + 4y = k$

As above line perpendicular to PQ is $3x + 4y = k$

As above line passes through $M\left(2, \frac{13}{8}\right)$.

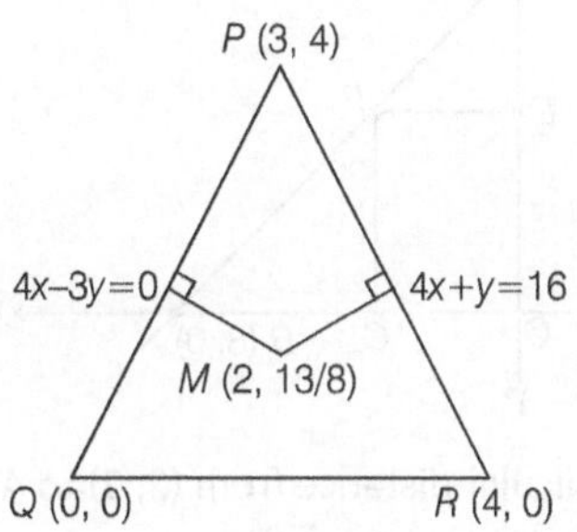

So, $3(2) + 4\left(\frac{13}{8}\right) = k \Rightarrow k = \frac{25}{2}$

$\therefore$ Equation of AM is $3x + 4y = \frac{25}{2}$

Similarly, the equation of line perpendicular to PR is

$$x - 4y = \lambda$$

As above line passes through M is $\left(2, \frac{13}{8}\right)$.

So, $2 - 4\left(\frac{13}{8}\right) = \lambda \Rightarrow \lambda = -9/2$

$\therefore$ Equation of CM is $x - 4y = \frac{-9}{2}$

Now, on solving $4x - 3y = 0$ and $3x + 4y = \frac{25}{2}$,

we get $A(3/2, 2)$

Also, on solving $4x + y = 16$ and $x - 4y = -9/2$, we get $C(7/2, 2)$

As A is the mid-point of PQ, so $Q \equiv (0, 0)$ and C is mid-point of PR so $R = (4, 0)$.

30. (c)

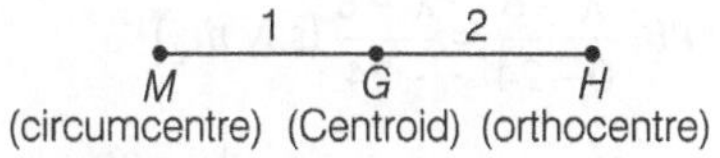

$\left(2, \frac{13}{8}\right)$ $\quad \left(\frac{7}{3}, \frac{4}{3}\right)$ $\quad (\alpha, \beta)$

$\therefore$ Using section formula

$$\frac{1 \times \alpha + 2 \times 2}{1 + 2} = \frac{7}{3}$$

$$\alpha = 7 - 4 = 3$$

and $\dfrac{1 \times \beta + 2 \times \frac{13}{8}}{3} = \dfrac{4}{3} \Rightarrow \beta = 3/4$

$\therefore$ Orthocentre is $\left(3, \frac{3}{4}\right)$.

31. (b) $\therefore$ Area $(\Delta ABC) = 1/2 \text{ mod} \begin{vmatrix} \frac{3}{2} & 2 & 1 \\ 2 & 0 & 1 \\ \frac{7}{2} & 2 & 1 \end{vmatrix} = 2$ sq units

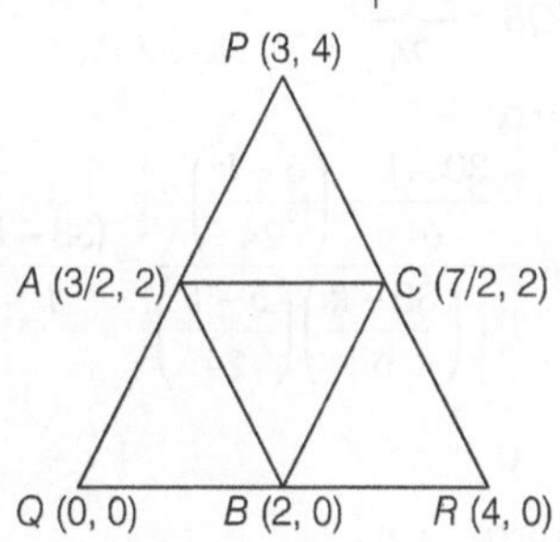

32. (d) $\therefore$ Slope of $PB = \frac{4-0}{3-2} = \frac{4}{1}$

$\therefore$ Equation of straight line passing through $N(-2, 3)$ and perpendicular to PB is

$$y - 3 = \frac{-1}{4}(x + 2) \Rightarrow x + 4y = 10$$

Sol. (Q.Nos. 33-35)

Slope of PA, $\frac{\lambda - 30}{0 - 6} = \frac{\lambda - 30}{-6}$ (say m_1)

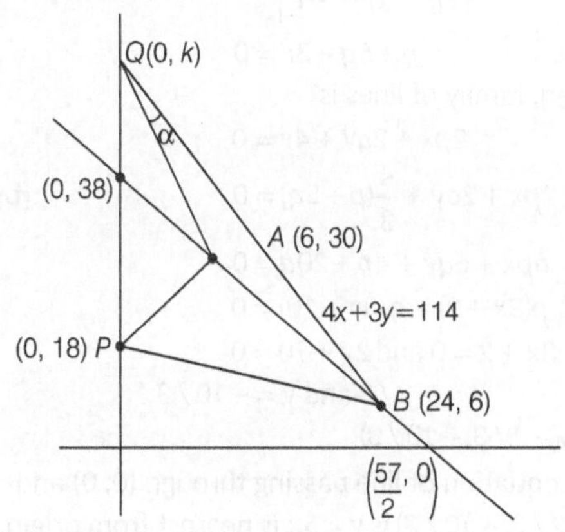

Slope of PB, $\frac{\lambda-6}{0-24}=\frac{\lambda-6}{-24}$ (say m_2)

$$\therefore \quad \tan\theta=\left|\frac{m_1-m_2}{1+m_1m_2}\right|=\left|\frac{\frac{30-\lambda}{6}-\frac{6-\lambda}{24}}{1+\left(\frac{30-\lambda}{6}\right)\left(\frac{6-\lambda}{24}\right)}\right|$$

$$\tan\theta=\frac{(38-\lambda)(18)}{(\lambda-18)^2}$$

Differentiating w.r.t. λ

$$\frac{d}{d\lambda}(\tan\theta)=\frac{18(\lambda-18)(\lambda-58)}{(\lambda-18)^4}$$

33. (c) $\tan\theta$ is maximum at $\lambda=18$

Point $P(0,18)$

34. (b) $\tan\theta$ is maximum at $\lambda=18$

$$\tan\theta=\infty \Rightarrow \theta=\pi/2$$

35. (b) Slope of $QA=\frac{30-k}{6}$

$\therefore$ Slope of $QB=\frac{6-k}{24}$

Let $\angle BQA=\alpha$

$$\tan\alpha=\frac{\frac{30-k}{6}-\left(\frac{6-k}{24}\right)}{1+\left(\frac{30-k}{6}\right)\left(\frac{6-k}{24}\right)}=\frac{(38-k)(18)}{k-18}$$

$$\therefore \frac{d(\tan\alpha)}{dk}=0$$

$$\Rightarrow \frac{18(k-18)(k-58)}{(k-18)^2}=0$$

$\therefore \tan\alpha$ is maximum of $k=58$

$\therefore$ Point is (0, 58).

Sol. (Q.Nos. 36-38)

36. (a) Given, three lines $x+y-2=0$, $2x-y+1=0$, $px+qy-r=0$ are concurrent.

$$\begin{vmatrix}1 & 1 & -2\\ 2 & -1 & 1\\ p & q & -r\end{vmatrix}=0$$

$$\Rightarrow \quad p+5q-3r=0 \quad \text{...(i)}$$

Given, family of lines is

$$2px+2qy+4r=0 \quad \text{...(ii)}$$

$$\Rightarrow 2px+2qy+\frac{4}{3}(p+5q)=0 \quad \text{[by Eq. (i)]}$$

$$\Rightarrow 6px+6qy+4p+20q=0$$

$$\Rightarrow p(3x+2)+q(3y+10)=0$$

$$\Rightarrow 3x+2=0 \text{ and } 2y+10=0$$

$$x=-2/3 \text{ and } y=-10/3$$

$\Rightarrow p(-2/3,-10/3)$

The equation of line passing through (0, 0) and $p(-2/3,-10/3)$ is $y=5x$ is nearest from origin.

37. (b) Line is passing through $\left(\frac{-2}{3},\frac{-10}{3}\right)$ is farthest from origin is perpendicular to OP.

i.e. $\quad y+10/3=-1/5(x+2/3)$

$\Rightarrow \quad 3x+15y+52=0$

$\therefore$ Equation of L_2 is $3x+15y+52=0$

38. (c) Equation of line $L_1 \Rightarrow y=5x$

Equation of line $L_2 \Rightarrow 3x+15y+52=0$

Slope of $l_1=5$ and slope of $l_2=-1/5$

$\therefore \quad m_1m_2=-1$

$\therefore$ Angle between L_1 and L_1 is $\pi/2$.

Sol. (Q.Nos. 39-41)

39. (c) Equation of line AB is

$$\frac{x}{8}+\frac{y}{16}=1$$

$$\Rightarrow \quad 2x+y=16$$

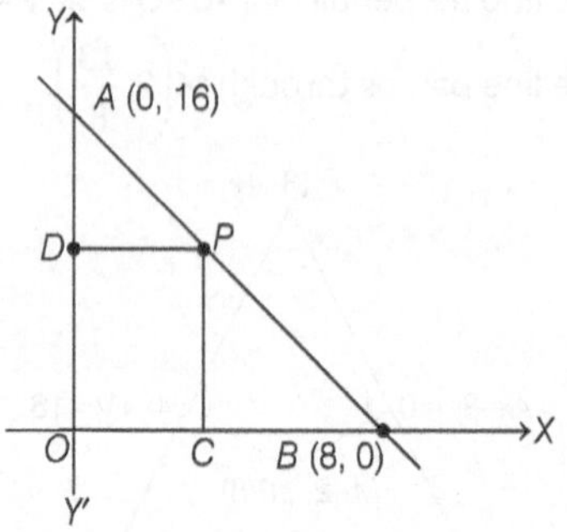

$\therefore$ Perpendicular distance from (2, 2) to AB

$$=\left|\frac{4+2-16}{\sqrt{4+1}}\right|=\frac{10}{\sqrt5}=2\sqrt5 \text{ units}$$

40. (a) Let the point $P(\lambda,\lambda)$ to equation AB the $PDOC$ is a square $2\lambda+\lambda=16$

$\lambda=16/3$

$\therefore$ Sum of coordinate of P is $2\lambda=\frac{32}{3}$

41. (b) Given, point $P(x,y)$ lie on the line

$$2x+y=16 \quad \text{...(i)}$$

$C(x,0), D(0,y)$ and $O(0,0)$

$\therefore$ Area of rectangle $PDOC=30$ sq units

$$xy=30 \quad \text{...(ii)}$$

Put $y=16-2x$ in Eq. (ii)

$$x(16-2x)=30$$

$$\Rightarrow \quad 2x^2-16x+30=0$$

$$\Rightarrow \quad x^2-8x+15=0$$

$$\Rightarrow \quad (x-5)(x-3)=0$$

$$x=3,5$$

$\therefore P(3,10)$ and $P(5,6)$

$\therefore$ Number of possible ordered pairs is 2.

Sol. (Q.Nos. 42-44)

42. (b) Let C be the reflection of A in $L = x + y = 0$

$\Rightarrow \quad C \equiv (-2, -1)$

Now, $AM + BM = LM + BM$

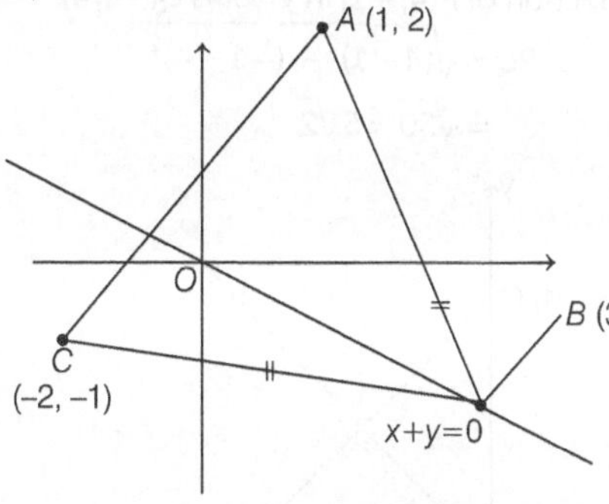

which is minimum, if B, M and C are collinear.

Now, equation of BC is

$$y + 1 = 0(x + 2)$$

$$y + 1 = 0 \quad \text{...(i)}$$

On solving Eq. (i) with $x + y = 0$, we get

$$M = (1, -1)$$

Hence, the reflection of M in the line $x = y$ is $(-1, 1)$.

43. (d) We have, $|AM - BM| \le AB$

$\therefore$ For $AM - BM$ to be maximum.

Points A, B and M must be collinear.

Now, equation of AB is

$$y - 2 = \frac{-3}{2}(x - 1)$$

$$\Rightarrow \quad 3x + 2y = 7 \quad \text{...(i)}$$

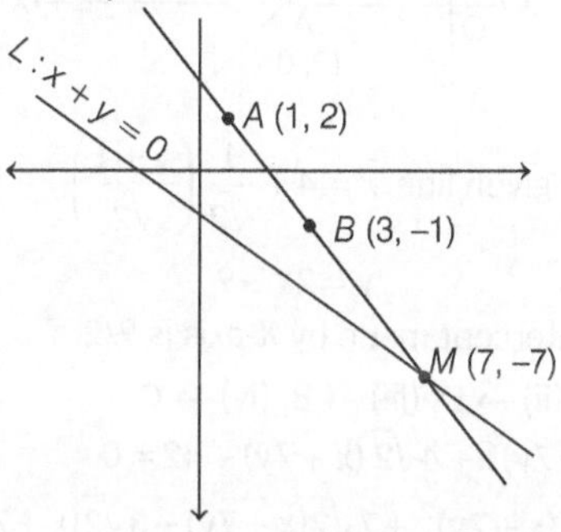

$\therefore$ On solving Eq. (i) with $x + y = 0$, we get $M(7, -7)$

Hence, distance of $M = (7, -7)$ from $N = (1, 1)$

$$MN = \sqrt{(7-1)^2 + (-7-1)^2} = 10 \text{ units}$$

44. (a) For $|AM - BM|$ minimum

$$AM = BM$$

$$\Rightarrow \quad AM^2 = BM^2$$

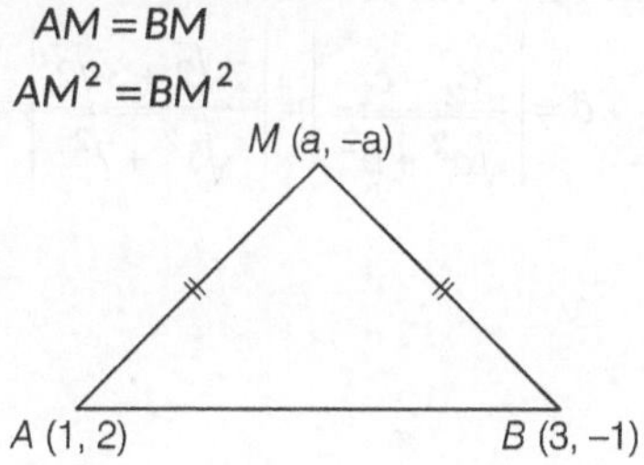

$$(a-1)^2 + (a+2)^2 = (a-3)^2 + (a-1)^2$$

$$\Rightarrow a^2 - 2a + 1 + a^2 + 4a + 4 = a^2 - 6a + 9 + a^2 - 2a + 1$$

$$\Rightarrow \quad 4a + 5 = -6a + 10$$

$$\Rightarrow \quad 10a = 5$$

$$a = 1/2$$

$\therefore M(1/2, -1/2)$

$$\therefore \text{ Area} = 1/2 \bmod \begin{vmatrix} 1 & 2 & 1 \\ 3 & -1 & 1 \\ \frac{1}{2} & -\frac{1}{2} & 1 \end{vmatrix}$$

$$= \frac{1}{2}\left|\left(-1 + \frac{1}{2}\right) - 2\left(3 - \frac{1}{2}\right) + \left(\frac{-3}{2} + \frac{1}{2}\right)\right|$$

$$= \frac{1}{2}\left|\left(-\frac{1}{2}\right) - \frac{10}{2} - \frac{2}{2}\right|$$

$$= \frac{1}{2}\left(\frac{13}{2}\right) = \frac{13}{4} \text{ sq units}$$

Sol. (Q.Nos. 45-46)

45. (d) $d(P, OA) \le \min[d(P, OB), d(P, AB)]$

$$d(P, OA) \le d(P, OB) \text{ and } d(P, OA) \le d(P, AB)$$

When $d(P, OA) \le d(P, OB)$, P is equidistant from OA and OB or P lies on the angle bisector of lines OA and OB.

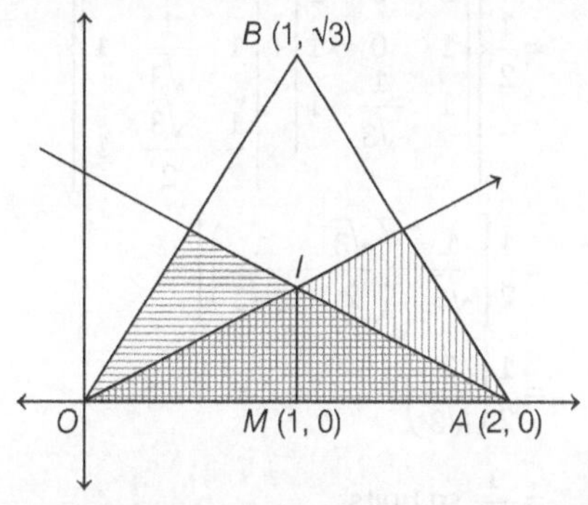

$\therefore$ When $d(P, OA) \le d(P, OB)$ point P is nearest to OA than OB or lies below bisector of OA and OB.

Similarly, when $d(P, OA) \le d(P, AB)$, P is nearest to OA than AB or lies below bisector of OA and AB.

Therefore, the required area is equal to the area of ΔOIA.

Now, $\tan \angle BOA = \dfrac{\sqrt{3}}{1} = \sqrt{3}$

$\Rightarrow \quad \angle BOA = 60°$

$\therefore \Delta$ is equilateral.

Then, I coincides with centroid

$\Rightarrow$ i.e. $\left(1, \dfrac{1}{\sqrt{3}}\right)$

$$\therefore \text{Area of } \Delta OIA = \frac{1}{2} \times OA \times IM$$

$$= \frac{1}{2} \times 2 \times \frac{1}{\sqrt{3}} = \frac{1}{\sqrt{3}} \text{ sq units}$$

46. (d) $OP \le \min[BP, AP]$

$OP \le BP$ (when $BP < AP$)

$OP \le AP$ (when $AP < BP$)

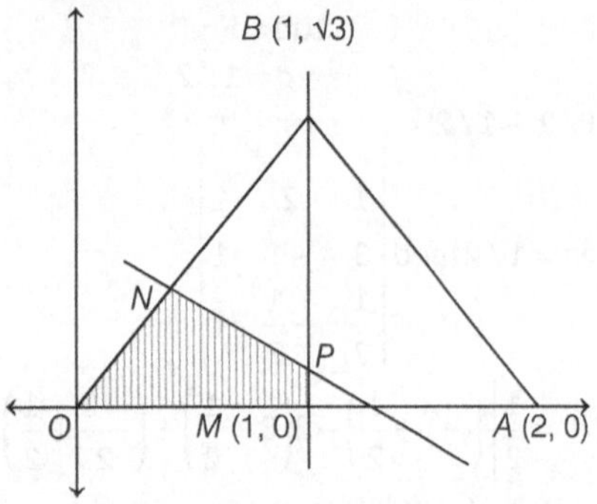

Let $OP = BP$. Then, P lies on perpendicular bisector of OB for $OP = OA$, P lies on the perpendicular bisector of OA.

Then, the required condition, P lies in the region as shown in figure.

$\therefore$ The area of region $OMPN$ is

$$= \frac{1}{2}\left\{\begin{vmatrix} 0 & 0 & 1 \\ 1 & 0 & 1 \\ 1 & \frac{1}{\sqrt{3}} & 1 \end{vmatrix} + \begin{vmatrix} 0 & 0 & 1 \\ 1 & \frac{1}{\sqrt{3}} & 1 \\ \frac{1}{2} & \frac{\sqrt{3}}{2} & 1 \end{vmatrix}\right\}$$

$$= \frac{1}{2}\left[\frac{1}{\sqrt{3}} + \left(\frac{\sqrt{3}}{2} - \frac{1}{2\sqrt{3}}\right)\right]$$

$$= \frac{1}{2}\left(\frac{2}{\sqrt{3}}\right)$$

$$= \frac{1}{\sqrt{3}} \text{ sq units}$$

47. (i) $\rightarrow$ D, (ii) $\rightarrow$ A, (iii) $\rightarrow$ B, (iv) $\rightarrow$ C

(i) $OA = 1 + 4\cot\theta$

$OB = 4 + \tan\theta$

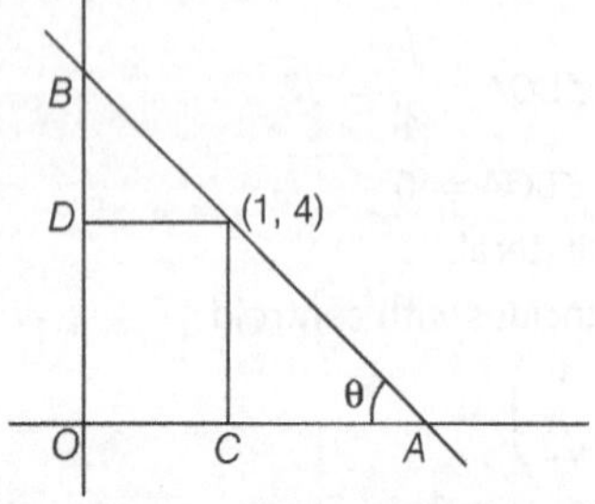

$\therefore OA + OB = 5 + 4\cot\theta + \tan\theta$

$\ge 5 + 2\sqrt{4\cot\theta \tan\theta}$

$= 5 + 2 \times 2 = 9$

(ii) Reflection of $P(4, -1)$ in $y = x$ is $Q(-1, 4)$.

$\therefore \quad PQ = \sqrt{(4+1)^2 + (-1-4)^2}$

$= \sqrt{50} = 5\sqrt{2}$

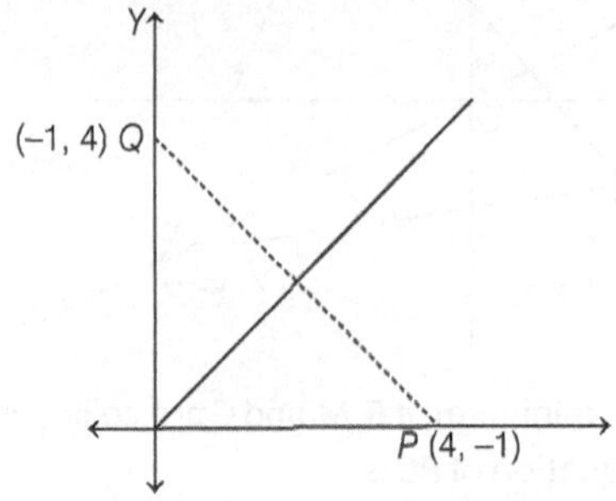

(iii) $AB = 2\sqrt{2}$ and $OC = \sqrt{2}$

Maximum value of d is

$\Rightarrow \quad OF = \sqrt{2} + 2\sqrt{2} = 3\sqrt{2}$

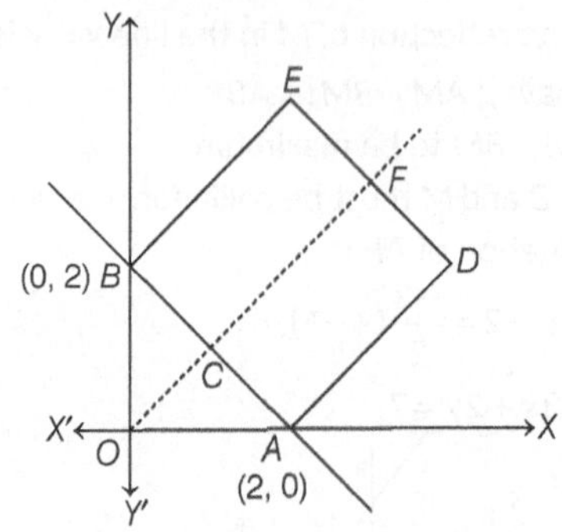

(iv) The given line $x = 4 + \frac{1}{\sqrt{2}}\left(\frac{y+1}{\sqrt{2}}\right)$

$y = 2x - 9$

$\therefore$ Intercept made by X-axis is $9/2$.

48. (i) $\rightarrow$ A, (ii) $\rightarrow$ D, (iii) $\rightarrow$ B, (iv) $\rightarrow$ C

(i) $(x+7y)^2 + 4\sqrt{2}(x+7y) - 42 = 0$

$\Rightarrow (x+7y)^2 + 7\sqrt{2}(x+7y) - 3\sqrt{2}(x+7y) - 42 = 0$

$\Rightarrow (x+7y)[(x+7y) + 7\sqrt{2}] - 3\sqrt{2}[(x+7y) + 7\sqrt{2}] = 0$

$\Rightarrow \quad (x+7y+7\sqrt{2})(x+7y-3\sqrt{2}) = 0$

$\Rightarrow x + 7y + 7\sqrt{2} = 0$ and $x + 7y - 3\sqrt{2} = 0$

$$d = \left|\frac{c_2 - c_1}{\sqrt{a^2+b^2}}\right| = \left|\frac{7\sqrt{2}+3\sqrt{2}}{\sqrt{1^2+7^2}}\right| = \frac{10\sqrt{2}}{\sqrt{50}} = 2$$

(ii) Let two perpendicular lines are coordinate axes

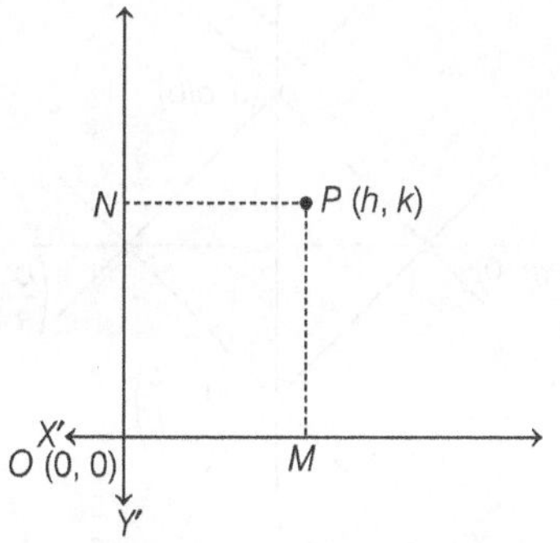

Then, $PM + PN = 1$

$h + k = 1$

Hence, the locus is $x + y = 1$

But if the point lies in other quadrant also then $|x| + |y| = 1$

and $k = 1$

(iii) Angle bisector between the lines

$x + 2y + 4 = 0$ and $4x + 2y - 1 = 0$

$$\frac{x+2y+4}{\sqrt{1+4}} = \pm \frac{(-4x-2y+1)}{\sqrt{16+4}}$$

$$\Rightarrow \quad x+2y+4 = \pm \frac{(-4x-2y+1)}{2}$$

$$\Rightarrow 2(x+2y+4) = \pm(-4x-2y+1)$$

$\because \quad aa' + bb' = 0$

So, the (+ve) sign gives acute angle bisector.

$\therefore \quad 2x + 4y + 8 = -4x - 2y + 1$

$\Rightarrow \quad 6x + 6y + 7 = 0$

$\therefore \quad m = 7$

(iv) We have, $y^2 - 9xy + 18x^2 = 0$

$(y - 3x)(y - 6x) = 0$

$y = 3x$, $y = 6x$ and $y = 6$

$\therefore$ Point of intersection

(0, 0), (1, 6) and (2, 6)

$$\therefore \text{ Area of } \Delta = \frac{1}{2} \text{mod} \begin{vmatrix} 0 & 0 & 1 \\ 2 & 6 & 1 \\ 1 & 6 & 1 \end{vmatrix} = \frac{1}{2} \times 6 = 3 \text{ sq units}$$

49. (i) $\rightarrow$ C, (ii) $\rightarrow$ D, (iii) $\rightarrow$ B

BM is median and *CF* is the angle bisector

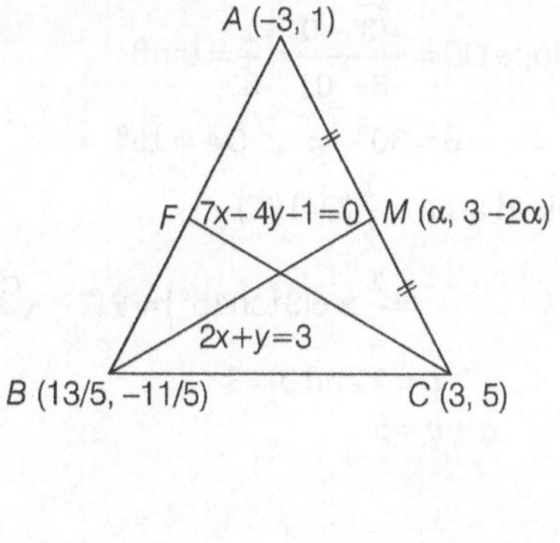

$\because$ *M* lies on the median

Let $M(\alpha, 3 - 2\alpha)$.

Let *C* be (a, b).

$$\therefore \quad \frac{a-3}{2} = \alpha$$

$$\Rightarrow \quad a = 2\alpha + 3$$

$$\text{and} \quad \frac{b+1}{2} = 3 - 2\alpha$$

$$\Rightarrow \quad b = 5 - 4\alpha$$

$\therefore C$ is $(2\alpha + 3, 5 - 4\alpha)$.

As *C* lies on angle bisector $7x - 4y - 1 = 0$

$7(2\alpha + 3) - 4(5 - 4\alpha) - 1 = 0$

$14\alpha + 21 - 20 + 16\alpha - 1 = 0$

$30\alpha = 0$

$\Rightarrow \quad \alpha = 0$

Hence, *C* is (3, 5) and $M = (0, 3)$.

$\therefore$ Equation of *AC* is

$$y - 1 = \frac{4}{6}(x + 3)$$

$\Rightarrow \quad 3y - 3 = 2x + 6$

$\Rightarrow 2x - 3y + 9 = 0 \Rightarrow AC$

Again let equation of *BC* is

$2x - 3y + 9 + \lambda(7x - 4y - 1) = 0$ ...(i)

(As any line through the intersecting of *AC* and *FC*)

Let (x_1, y_1) be a point on $7x - 4y - 1 = 0$

i.e., $7x_1 - 4y_1 - 1 = 0$

$$\therefore \left| \frac{2x_1 - 3y_1 + 9}{\sqrt{13}} \right| = \left| \frac{(2x_1 - 3y_1 + 9) + \lambda(7x_1 - 4y_1 - 1)}{\sqrt{(7\lambda + 2)^2 + (4\lambda + 3)^2}} \right|$$

$\therefore (7\lambda + 2)^2 + (4\lambda + 3)^2 = 13$

Solving $\lambda = \frac{-4}{5}$ or $\lambda = 0$ (not possible)

Hence, equation of *BC*

$10x - 15y + 45 = 28x - 16y - 4$

$\Rightarrow 18x - y - 49 = 0 \rightarrow BC$

Solving $2x + y - 3 = 0$

and $18x - y - 49 = 0$

Point *B* is $\left(\frac{13}{5}, \frac{-11}{5}\right)$

Equation of *AB* is

$$y - 1 = \frac{\frac{-11}{5} - 1}{\frac{13}{5} + 3}(x + 3)$$

$$\Rightarrow \quad y - 1 = \frac{-4}{7}(x + 3)$$

$\Rightarrow \quad 7y - 7 = -4x - 12$

$\Rightarrow 4x + 7y + 5 = 0 \rightarrow AB$

50. (i) $\rightarrow$ (A, D), (ii) $\rightarrow$ (B, D), (iii) $\rightarrow$ (A, C), (iv) $\rightarrow$ (D)

(i) Given, lines are concurrent so

$$\begin{vmatrix} 3 & 1 & -4 \\ 1 & -2 & -6 \\ \lambda & 4 & \lambda^2 \end{vmatrix} = 0$$

$\Rightarrow \quad \lambda^2 + 2\lambda - 8 = 0 \Rightarrow \lambda = 2, -4$

(ii) Points are collinear

i.e. $\begin{vmatrix} \lambda+1 & 1 & 1 \\ 2\lambda+1 & 3 & 1 \\ 2\lambda+2 & 2\lambda & 1 \end{vmatrix} = 0$

$\Rightarrow \quad 2\lambda^2 - 3\lambda - 2 = 0 \Rightarrow \lambda = 2, \dfrac{-1}{2}$

(iii) Point of intersection of $x - y + 1 = 0$ and $3x + y - 5 = 0$ is (1, 2)

It lies on the line $x + y - 1 - |\lambda/2| = 0$

$\Rightarrow \quad \lambda = \pm 4$

(iv) The mid-point of (1, – 2) and (3, 4) will satisfy.

$y - x - 1 + \lambda = 0 \Rightarrow \lambda = 2$

51. (i) $\rightarrow$ (A, D), (ii) $\rightarrow$ (A), (iii) $\rightarrow$ (B), (iv) $\rightarrow$ (C)

(i)

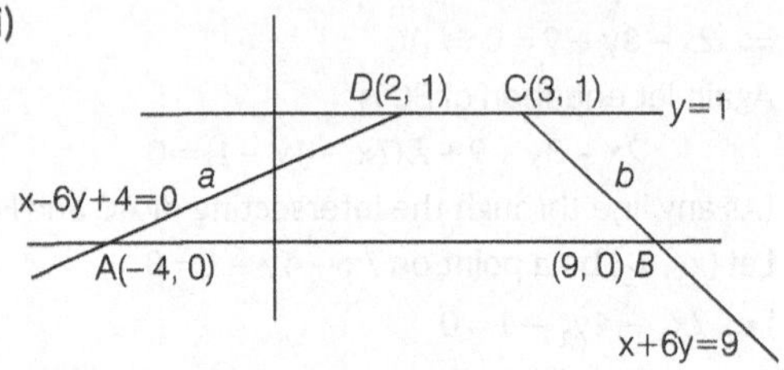

Obviously trapezium

$a = \sqrt{37}$ and $b = \sqrt{37}$

$a = b$

Hence, isosceles trapezium

$\Rightarrow$ a cyclic quadrilateral

(ii) $ac = bd$

$\dfrac{b}{c} = \dfrac{a}{d}$

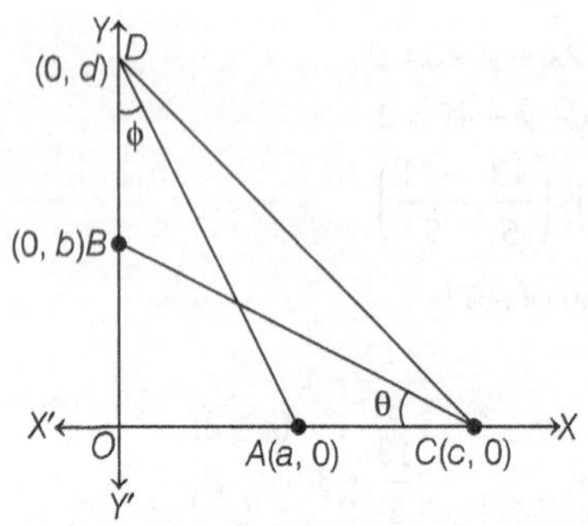

$\therefore \tan\theta = b/c$ and $\tan\phi = \dfrac{a}{d}$

$\Rightarrow \quad \phi = \theta$

$\therefore$ Cyclic quadrilateral.

(iii) Rhombus

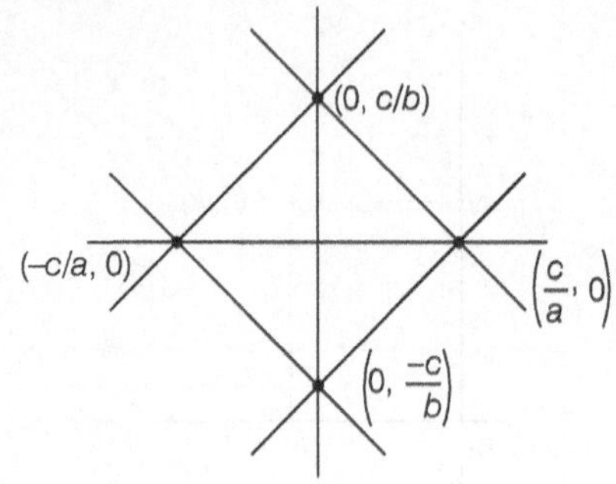

(iv) Given, $x^2 - 8x + 12 = 0$

$\Rightarrow \quad (x-6)(x-2) = 0$

$x = 2$ or $x = 6$ and $y^2 - 14y + 45 = 0$

$\Rightarrow \quad (y-9)(y-5) = 0$

$y = 9$ or $y = 5$

$\Rightarrow$ a square

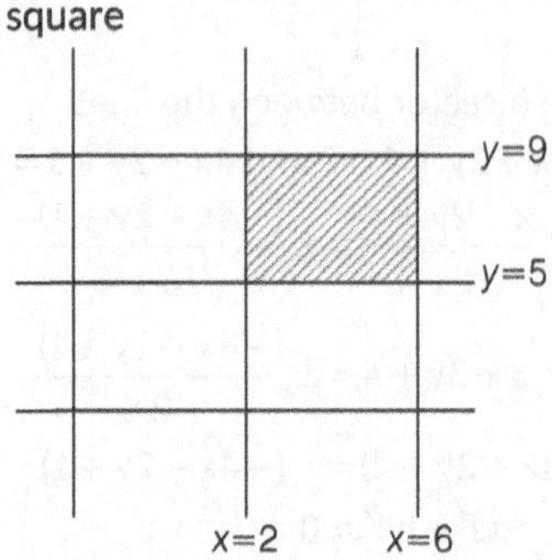

52. (5) $\because OA = \sqrt{6^2} = 6$

$AB = \sqrt{9+3} = \sqrt{12} = 2\sqrt{3}$

$OB = \sqrt{9+3} = \sqrt{12} = 2\sqrt{3}$

$\because \quad OB = AB$

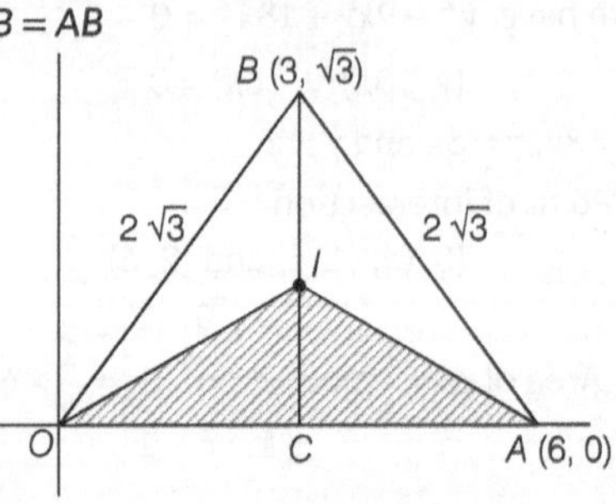

$\therefore \Delta OBA$ is isosceles triangle.

$\therefore$ The region R is the shaded portion.

Now, slope $OB = \dfrac{\sqrt{3} - 0}{3 - 0} = \dfrac{1}{3} = \tan\theta$

$\Rightarrow \quad \theta = 30° \Rightarrow \angle IOA = 15°$

$\therefore$ Required area $= \dfrac{1}{2}(OA)(IC)$

$= \dfrac{1}{2} \times 6(3\tan 15°) = 9(2 - \sqrt{3}) = 9(a - \sqrt{b})$

$\Rightarrow \quad a = 2$ and $b = 3$

$\therefore \quad a + b = 5$

53. (8) $\because$ Centroid, divides line joining orthocentre and circumcentre in the ratio 2 : 1

$$\because A=\left(\frac{2\times\frac{7}{2}+1\times2}{2+1},\frac{2\times5/2+1\times4}{2+1}\right)=(3,3)$$

$$\Rightarrow \quad OA=3\sqrt{2}$$

$\therefore$ x-coordinate of point P

$$=OA\cos45^\circ+AB\cos45^\circ+BC\cos45^\circ+\ldots$$

$$=\cos45^\circ\left(OA+\frac{OA}{2}+\frac{OA}{4}+\ldots\right)$$

$$=\frac{1}{\sqrt{2}}\left(\frac{OA}{1-\frac{1}{2}}\right)=\frac{2}{\sqrt{2}}\times3\sqrt{2}=6$$

$\therefore$ y-coordinate of point P

$$=OA\sin45^\circ-AB\sin45^\circ+BC\sin45^\circ+\ldots$$

$$=\frac{1}{\sqrt{2}}\left(OA-\frac{OA}{2}+\frac{OA}{4}\ldots\right)$$

$$=\frac{1}{\sqrt{2}}\times\frac{OA}{1+1/2}=\frac{2}{3\sqrt{2}}OA$$

$$=\frac{2}{3\sqrt{2}}\times3\sqrt{2}=2$$

$\therefore P(6,2)\Rightarrow\alpha+\beta=8$

54. (6) $\frac{a+t}{4a-t}=\frac{-1}{2}$

$\Rightarrow 2a+2t=-4a+t\Rightarrow 6a=-t$

$$a=\frac{-t}{6}$$

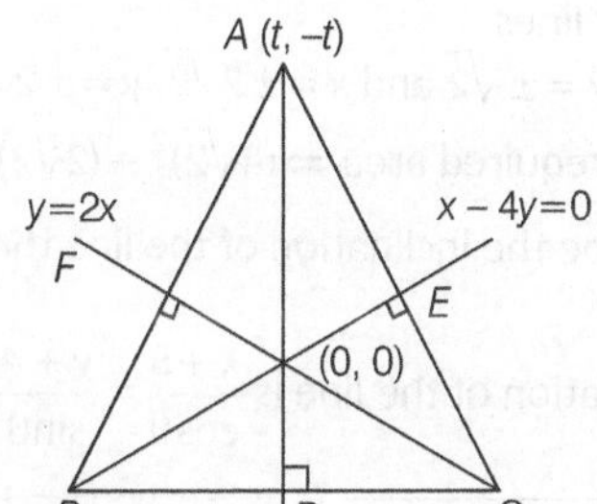

$\Rightarrow B\left(\frac{-2t}{3},\frac{-t}{6}\right)$ and $\frac{2b+t}{b-t}=-4$

$$\Rightarrow \quad 2b+t=-4b+4t\Rightarrow 6b=3t\Rightarrow b=\frac{t}{2}$$

$$\Rightarrow \quad C=\left(\frac{t}{2},t\right)$$

$$\therefore \quad G=\left(\frac{\frac{-2t}{3}+\frac{t}{2}+t}{3},\frac{\frac{-t}{6}+t-t}{3}\right)$$

$$\Rightarrow \quad h=\frac{\frac{3t}{2}-\frac{2t}{3}}{3}=\frac{5t}{18}\text{ and }k=\frac{-t}{18}$$

$$\therefore \quad h=-5k$$

$\therefore$ Locus is $x+5y=0$

$$a=1,b=5\text{ and }c=0$$

$$a+b+c=6$$

55. (10) $\because$ Diagonal is member of both the given families of line, so it will pass through the points (1, 0) and (– 1, 1)

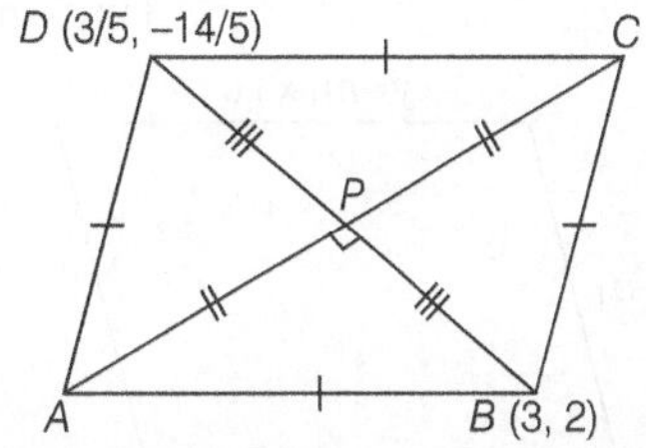

$\Rightarrow$ Equation of diagonal AC is $x+2y-1=0$

$\because$ One of vertices is (3, 2) (which does not lies on AC)

So, equation of diagonal BD is $2x-y=4$

$\Rightarrow$ Point of intersection of diagonals AC and BD is $P\left(\frac{9}{5},-2/5\right)$.

$\Rightarrow$ Vertex D is $\left(\frac{3}{5},\frac{-14}{5}\right)$ ($\because P$ is mid point of BD)

Also, $BD=\frac{12\sqrt{5}}{5}$ units (say d_1)

$\therefore$ Area of Rhombus $=1/2(d_1\times d_2)$

$$=\frac{1}{2}\frac{12\sqrt{5}}{5}d_2=12\sqrt{5}$$

$$d_2=10$$

$\therefore$ Length of longer diagonal is 10 units.

56. (63) Let $HN=P_1$

Now, $PM=P_2$

$$HP^2=R^2(1-8\cos A\cos B\cos C)$$

$$2=R^2-4(R\cos A)(2R\cos B\cos C)$$

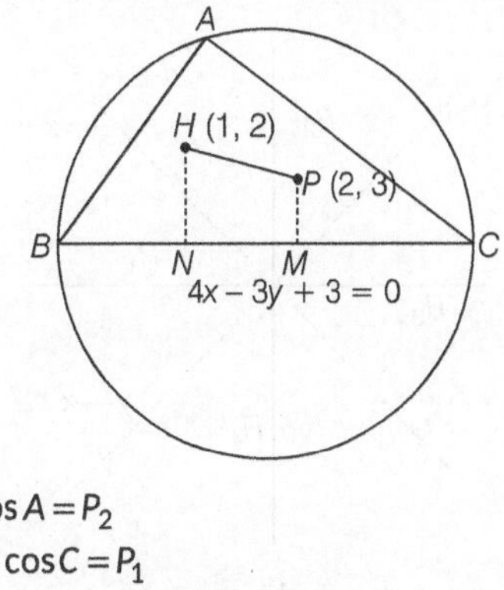

But $R\cos A=P_2$

$2R\cos B\cos C=P_1$

$$2=R^2=4P_1P_2$$

Now, $P_1 = \left|\frac{4-6+3}{5}\right| = \frac{1}{5}$ and $P_2 = \left|\frac{8-9+3}{5}\right| = \frac{2}{5}$

$\therefore \quad R^2 = 2+4\left(\frac{1}{5}\right)(2/5) = \frac{58}{25}$

$\therefore \quad R = \frac{\sqrt{58}}{5} = \frac{\sqrt{m}}{n} \Rightarrow m = 58$ and $n = 5$

$\therefore m+n = 63$

57. (16) $\therefore$ Area of parallelogram $= \frac{|c_1-c_2||d_1-d_2|}{|m_1-m_2|}$

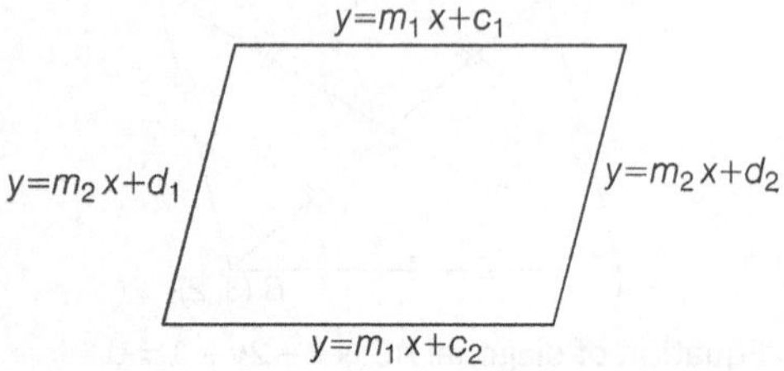

$\therefore$ Area of parallelogram formed by the first set of 4 lines

$$\frac{|c-d|^2}{|a-b|} = 18 \qquad \text{...(i)}$$

and area of parallelogram formed by the second set of 4 lines

$$\frac{|c+d|^2}{|a-b|} = 72 \qquad \text{...(ii)}$$

By Eqs. (i) and (ii),

$$|c-d|^2 = |a-b|(18)$$
$$|c+d|^2 = |a-b|(72)$$

$\because a, b, c, d \in N$

$\therefore$ RHS must be a perfect square for some least value of $(a-b)$ which obviously can be equal to 2

$$|a-b| = 2$$

$\Rightarrow \quad a = 1$ and $b = 3$

$\therefore \quad c-d = 6$ and $c+d = 12$

$c = 9$ and $d = 3$

$\therefore a+b+c+d = 16$

58. (24)

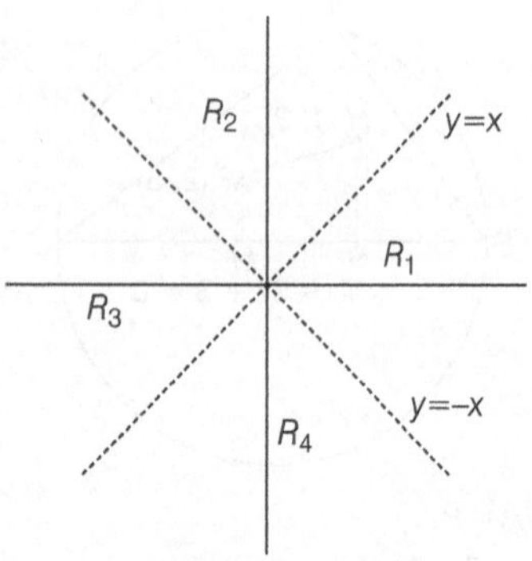

$\therefore \quad d(P, L_1) = \left|\frac{x-y}{\sqrt{2}}\right|$

$\therefore \quad d(P, L_2) = \left|\frac{x+y}{\sqrt{2}}\right|$

Now, we have $2 \le d(P, L_1) + d(P, L_2) \le 4$

$$2\sqrt{2} \le |x-y| + |x+y| \le 4\sqrt{2} \qquad \text{...(i)}$$

Now, let us consider the four region namely R_1, R_2, R_3 and R_4 in the lines L_1 and L_2 dividing the coordinate plane.

In R_1, we have $y < x, y > -x$

In R_2, we have $y > x, y > -x$

In R_3, we have $y > x, y < -x$

In R_4, we have $y < x, y < -x$

Thus in R_1 Eq. (i) becomes

$$2\sqrt{2} \le x-y+x+y \le 4\sqrt{2}$$

$\Rightarrow \quad \sqrt{2} \le x \le 2\sqrt{2}$

Similarly for $R_2 \Rightarrow 2\sqrt{2} \le y-x+x+y \le 4\sqrt{2}$

$\Rightarrow \quad \sqrt{2} \le y \le 2\sqrt{2}$

$R_3 \Rightarrow 2\sqrt{2} \le y-x-x-y \le 4\sqrt{2} \Rightarrow -2\sqrt{2} \le y \le -\sqrt{2}$

$R_4 \Rightarrow 2\sqrt{2} \le x-y-x-y \le 4\sqrt{2} \Rightarrow -2\sqrt{2} \le x < -\sqrt{2}$

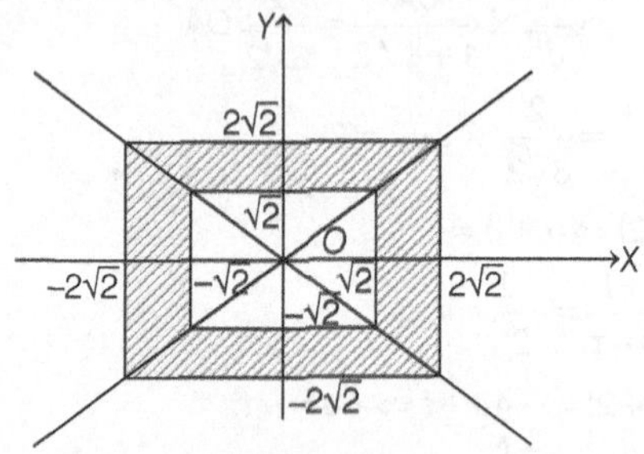

$\therefore$ R will be the region between concentric squares formed by lines

$x = \pm\sqrt{2}, y = \pm\sqrt{2}$ and $x = \pm 2\sqrt{2}, y = \pm 2\sqrt{2}$

Thus, the required area $\Rightarrow (4\sqrt{2})^2 - (2\sqrt{2})^2 = 24$ sq units

59. (28) Let θ be the inclination of the line through $A(-5, -4)$.

Then, equation of the line is $\frac{x+5}{\cos\theta} = \frac{y+4}{\sin\theta} = r$

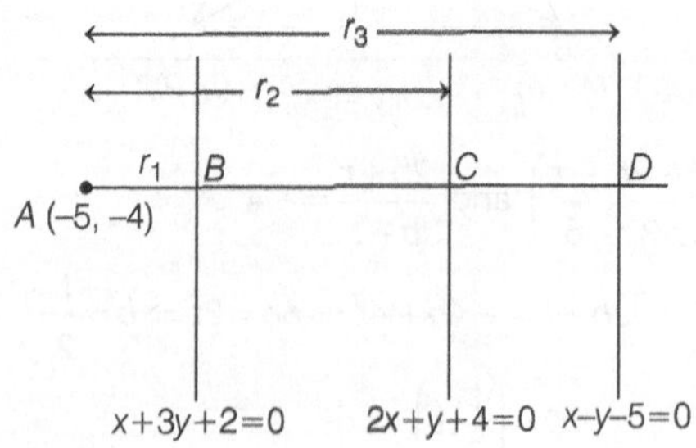

$B = (r_1\cos\theta - 5, r_1\sin\theta - 4)$

$C = (r_2\cos\theta - 5, r_2\sin\theta - 4)$

$D = (r_3\cos\theta - 5, r_3\sin\theta - 4)$

But B lies on $x+3y+2 = 0$

$\therefore r_1\cos\theta - 5 + 3r_1\sin\theta - 12 + 2 = 0$

$$\Rightarrow \quad r_1 = \frac{15}{\cos\theta + 3\sin\theta} = AB$$

$$\Rightarrow \quad \frac{15}{AB} = \cos\theta + 3\sin\theta \qquad \text{...(i)}$$

As C lies on $2x + y + 4 = 0$

$$\because 2(r_2\cos\theta - 5) + (r_2\sin\theta - 4) + 4 = 0$$

$$\frac{10}{r_2} = 2\cos\theta + \sin\theta \Rightarrow \frac{10}{AC} = 2\cos\theta + \sin\theta$$

Similarly, D lies on $x - y - 5 = 0$

$$\therefore r_3\cos\theta - 5 - r_3\sin\theta + 4 - 5 = 0$$

$$\Rightarrow \quad r_3 = \frac{6}{\cos\theta - \sin\theta} = AD$$

$$\Rightarrow \quad \frac{6}{AD} = \cos\theta - \sin\theta \qquad \text{...(iii)}$$

Now, given that

$$\left(\frac{15}{AB}\right)^2 + \left(\frac{10}{AC}\right)^2 = \left(\frac{6}{AD}\right)^2$$

$$\Rightarrow (\cos\theta + 3\sin\theta)^2 + (2\cos\theta + \sin\theta)^2$$

$$= (\cos\theta - \sin\theta)^2 \qquad \text{[by Eqs. (i), (ii) and (iii)]}$$

$$\Rightarrow 4\cos^2\theta + 9\sin^2\theta + 12\sin\theta\cos\theta = 0$$

$$2\cos\theta + 3\sin\theta = 0$$

$$\tan\theta = -2/3$$

$\therefore$ Equation is

$$y + 4 = \frac{-2}{3}(x + 5)$$

$$\Rightarrow \quad 3y + 12 = -2x - 10$$

$$\Rightarrow 2x + 3y + 22 = 0$$

On comparing, we get

$$ax + by + c = 0$$

$$\Rightarrow \quad a = 2 \text{ and } c = 22$$

$$\therefore \quad 25a - c \Rightarrow 50 - 22 = 28$$

60. (4) Let BC be $u_1 = x\cos\alpha_1 + y\sin\alpha_1 - p_1 = 0$
AC be $u_2 = 0$ and AB be $u_3 = 0$

Then, the altitude AB through A is of the form

$$u_2 + \lambda u_3 = 0$$

$$\Rightarrow (x\cos\alpha_2 + y\sin\alpha_2 - p_2) + \lambda(x\cos\alpha_3 + y\sin\alpha_3 - p_3) = 0$$

i.e. AD is given by

$$x(\cos\alpha_2 + \lambda\cos\alpha_3) + y(\sin\alpha_2 + \lambda\sin\alpha_3) - (p_2 + \lambda p_3) = 0$$

$$\therefore \text{Slope of } AD = -\frac{\cos\alpha_2 + \lambda\cos\alpha_3}{\sin\alpha_2 + \lambda\sin\alpha_3}$$

Now, $AD \perp BC$

$$\Rightarrow \quad -\left(\frac{\cos\alpha_2 + \lambda\cos\alpha_3}{\sin\alpha_2 + \lambda\sin\alpha_3}\right)\cdot(-\cot\alpha_1) = -1$$

$$\Rightarrow \quad \cos\alpha_1(\cos\alpha_2 + \lambda\cos\alpha_3) + \sin\alpha_1(\sin\alpha_2 + \lambda\sin\alpha_3) = 0$$

$$\Rightarrow \quad (\cos\alpha_1\cdot\cos\alpha_2 + \sin\alpha_1\cdot\sin\alpha_2) + \lambda(\cos\alpha_1\cos\alpha_3 + \sin\alpha_1\sin\alpha_3) = 0$$

$$\Rightarrow \cos(\alpha_1 - \alpha_2) + \lambda[\cos(\alpha_3 - \alpha_1)] = 0$$

$$\lambda = -\frac{\cos(\alpha_1 - \alpha_2)}{\cos(\alpha_3 - \alpha_1)}$$

$$\therefore \text{Equation } AD \text{ is } u_2 - \frac{\cos(\alpha_1 - \alpha_2)}{\cos(\alpha_3 - \alpha_1)}u_3 = 0$$

or Equivalently $u_2\cos(\alpha_3 - \alpha_1) = u_3\cos(\alpha_1 - \alpha_2)$

Similarly, BE is

$$\Rightarrow u_3\cos(\alpha_1 - \alpha_2) = u_1\cos(\alpha_2 - \alpha_3)$$

$\therefore$ Orthocentre is given by

$$u_1\cos(\alpha_2 - \alpha_3) = u_2\cos(\alpha_3 - \alpha_1) = u_3\cos(\alpha_1 - \alpha_2)$$

but $u_1\cos(p\alpha_2 - q\alpha_3) = u_2\cos(r\alpha_3 + s\alpha_1) = u_3\cos(-u\alpha_1 + t\alpha_2)$

i.e. $p = 1, q = 1, r = 1, s = -1, u = -1$ and $t = -1$

$$5p + q - r + s - t + u$$

$$\Rightarrow 5 + 1 - 1 - 1 + 1 - 1 \Rightarrow 7 - 3 = 4$$

CHAPTER

03

PAIRS OF STRAIGHT LINE

LEVEL 01 \ Pair of Straight Lines

3.1 Homogeneous Equation

1 (i) Find the separate equation of lines for the following Homogeneous equations

(a) $y^2 - 9xy + 18x^2 = 0$ (b) $x^2 - 3xy + 2y^2 = 0$

(c) $x^2 + 5xy + 6y^2 = 0$ (d) $3x^2 + 5xy + 4y^2 = 0$

(e) $2x^2 - 4\sqrt{3}xy + 6y^2 = 0$

(ii) Find the combined equation of the following

(a) $x = 0$ and $y = 0$ (b) $x = 4$ and $y = 3$

(iii) Find the combined equation of straight lines which pass through (1, 2) and perpendicular to the line $3x^2 - 8xy + 5y^2 = 0$

2 (a) Find the area of triangle formed by the lines $12x^2 - 20xy + 7y^2 = 0$ and $2x - 3y + 4 = 0$

(b) Find the centroid of triangle the equation of whose sides are $x^2 - 5xy + 6y^2 = 0$ and $y = 6$.

(c) Find the circumcentre of triangle formed by the lines $xy - x - y + 1 = 0$ and $x + y = 4$.

3 (a) Find the angle between the following lines and also find area of triangle formed by these lines $x^2 - 4xy + y^2 = 0$ and $x + y = 1$

(b) Find the value of a if lines represented by $ax^2 + 5xy + 2y^2 = 0$ are mutually perpendicular.

(c) If sum of gradients of the lines denoted by $x^2 - 2xy \tan A - y^2 = 0$ is 4, then find $\angle A$.

4 (a) Find the equation of the bisectors of the angle between the lines represented by $3x^2+5xy+4y^2=0$.

(b) If $y=mx$ is one of bisectors of the lines $x^2+4xy+y^2=0$ find the value of m.

(c) Find the value of, λ if $9x^2+14xy+16y^2=0$ are equally inclined to the lines $3x^2+2xy+\lambda y^2=0$.

(d) The lines represented by $x^2+2\lambda xy+2y^2=0$ and the lines represented by $(1+\lambda)x^2-8xy+y^2=0$ are equally inclined find the value of λ.

3.2 General Equation of Second Degree

5 Find the value of λ for which given equation represents a pair of straight lines

(a) $\lambda x^2-5xy-6y^2+14x+5y+4=0$
(b) $12x^2-10xy+2y^2+11x-5y+\lambda=0$
(c) $6x^2+5xy-\lambda y^2+7x+13y-3=0$
(d) $6x^2+2\lambda xy-8y^2+23x-10y+7=0$

6 Find the separate equation of lines for the following equation represented by

(a) $x^2-5xy+4y^2+x+2y-2=0$
(b) $12x^2-10xy+2y^2+11x-5y+2=0$
(c) $16x^2+24xy+9y^2+40x+30y-75=0$

7 (a) Find the equation of straight lines passing through $(-1, 2)$ and parallel to the lines represented by $6x^2+8xy-8y^2+23x-10y+7=0$.

(b) Find the distance between the pair of straight lines represented by $x^2+2\sqrt{2}xy+2y^2+4x+4\sqrt{2}y+1=0$

(c) Find the angle between the lines represented by $3x^2-2xy-8y^2-4x+18y-7=0$

(d) If the equation $12x^2+7xy-py^2-18x+qy+6=0$ represents two perpendicular lines, then find the values of p and q.

(e) Find the area of quadrilateral formed by the given four lines $3x^2+8xy-3y^2=0$ and $3x^2+8xy-3y^2+2x-4y-1=0$

3.3 Point of Intersection of Pair of Straight Line and Equation of Bisectors

8 (a) Find the point of intersection of the lines represented by

(i) $x^2-5xy+2y^2+x+2y-2=0$
(ii) $3x^2-2xy-8y^2-4x+18y-7=0$
(iii) $x^2-8xy+16y^2+2x-8y+1=0$

(b) Find the equation of bisectors of the lines represented by

(i) $x^2-5xy+4y^2+3x+2y-1=0$
(ii) $12x^2-10xy+2y^2+9x+2y-12=0$
(iii) If the pair of lines $ax^2+2hxy+by^2+2gx+2fy+c=0$ intersects on the Y-axis, then find the value of bg^2+ch^2 in terms of f, g, h.

3.4 Miscellaneous (Homogenisation)

9 (a) Find the new equation of curve $12x^2+7xy-12y^2-17x-31y-7=0$ after removing the first degree term.

(b) If the equation $x^2+4xy+y^2-2x+2y+4=0$ into the form $\frac{Y'^2}{b^2}-\frac{X'^2}{a^2}=k$, then find a, b and k.

(c) If the pair of straight lines $ax^2+2hxy+by^2=0$ is rotated about the origin through 90°, then find the equation in the new position.

(d) Find the equation of the straight lines joining the origin to the points of intersection of the line $y=mx+c$ and the curve $x^2+y^2=b^2$.

(e) Find the equation of the straight lines joining the origin to the points of intersection of the line $lx+my=1$ and $y^2=4bx$.

(f) Find the angles between the lines joining the origin to the points of intersection of the line $3x-2y=1$ and the curve $3x^2+5xy-3y^2+2x+3y=0$

(g) Find the value of m, if the lines joining the origin and the point of intersection of $y=mx+1$ and $x^2+3y^2=1$ perpendicular to one another.

(h) If the equation $ax^2+2hxy+by^2+2gx+2fy+c=0$ represents two straight lines, find the product of the perpendiculars drawn from the origin to lines.

LEVEL 02\\ JEE Main

3.1 Homogeneous Equation

1 If one of the lines given $6x^2-xy+4cy^2=0$ is $3x+4y=0$, then c equals to

(a) -3 (b) -1 (c) 2 (d) 1

2 If the pair of lines $x^2+2xy+ay^2=0$ and $ax^2+2xy+y^2=0$ have exactly one line in common, then the joint equation of the other two lines is given by

(a) $3x^2+8xy-3y^2=0$ (b) $3x^2+10xy+3y^2=0$

(c) $y^2+2xy-3y^2=0$ (d) $x^2+2xy-3y^2=0$

3 Equation of a line which is parallel to the line common to the pair of lines given by $6x^2-xy-12y^2=0$ and $15x^2+14xy-8y^2=0$ and at a distance 7 from it is

(a) $3x-4y=-35$ (b) $5x-2y=7$

(c) $3x+4y=35$ (d) $2x-3y=7$

4 The condition that one of the straight lines given by the equation $ax^2+2hxy+by^2=0$ may coincide with one of those given by the equation $a'x^2+2h'xy+b'y^2=0$

(a) $(ab'-a'b)^2=4(ha'-h'a)(bh'-b'h)$ (b) $(ab'-a'b)^2=(ha'-h'a)(bh'-b'h)$

(c) $(ha'-h'a)^2=4(ab'-a'b)(bh'-b'h)$ (d) $(bh'-b'h)^2=4(ab'-a'b)(ha'-h'a)$

5 If the sum of the slopes of the lines given by $x^2-2cxy-7y^2=0$ is four times their product, then c has the value

(a) -2 (b) -1 (c) 2 (d) 1

6 The image of the pair of straight lines represented by $ax^2+2hxy+by^2=0$ by the mirror $y=0$ is

(a) $ax^2+2hxy+by^2=0$ (b) $ax^2+2hxy-by^2=0$

(c) $ax^2-2hxy+by^2=0$ (d) $ax^2-2hxy-by^2=0$

7 The line $x-2y=0$ will be a bisector of the angle between the lines represented by the equation $x^2-2hxy-2y^2=0$, if h is equal to

(a) $\frac{1}{2}$ (b) 2 (c) -2 (d) $\left(-\frac{1}{2}\right)$

8 If the bisectors of the angles between the pairs of the lines given by the equation $ax^2+2hxy+by^2=0$ and $ax^2+2hxy+by^2+\lambda(x^2+y^2)=0$ be coincident then λ is equal to

(a) a (b) b (c) h (d) any real number

9 If two of the lines represented by $x^4+x^3y+cx^2y^2-xy^3+y^4=0$ bisects the angle between the other two, then the value of c is

(a) 0 (b) −1 (c) 1 (d) −6

10 If the pair of straight lines given by $Ax^2+2Hxy+By^2=0\ (H^2>AB)$ forms an equilateral triangle with the line $ax+by+c=0$, then $(A+3B)(3A+B)$ is

(a) H^2 (b) $-H^2$ (c) $2H^2$ (d) $4H^2$

3.2 General Equation of Second Degree

11 If the equation $12x^2+7xy-py^2-18x+qy+6=0$ represents two perpendicular lines, the value of $p+q+7$ is

(a) 15 (b) 20 (c) 25 (d) 30

12 The lines represented by the equation $ax^2+2hxy+by^2+2gx+2fy+c=0$ will be equidistant from origin, if

(a) $f^2+g^2=c(b-a)$ (b) $f^4+g^4=c(bf^2+ag^2)$

(c) $f^4-g^4=c(bf^2-ag^2)$ (d) $f^2+g^2=af^2+bg^2$

13 The distance between the two lines represented by the equation $9x^2-24xy+16y^2-12x+16y-12$ is d, then find the pair of straight line passing through origin and $(d, 1)$ and $(1, d)$ is

(a) $40(x^2+y^2)+89xy=0$ (b) $40(x^2+y^2)-89xy=0$

(c) $40x^2+40y^2-59xy$ (d) None of the above

14 The straight lines represented by $(y-mx)^2=a^2(1+m^2)$ and $(y-nx)^2=a^2(1+n^2)$ form a

(a) rectangle (b) rhombus

(c) trapezium (d) None of these

15 Let PQR be a right angled isosceles triangle, right angled at $P(2, 1)$. If the equation of line QR is $2x+y=3$ the equations representing the pair of lines PQ and PR is

(a) $3x^2-3y^2+8xy+20x+10y+25=0$ (b) $3x^2-3y^2-8xy-4x+22y-7=0$

(c) $3x^2-3y^2+8xy+10x+15y+20=0$ (d) $3x^2-3y^2-8xy-10x-15y-20=0$

16 If the distance at two lines passing through origin from the point (x_1, y_1) is d, then the equation of lines is

(a) $(xy_1-yx_1)^2=d^2(x^2+y^2)$ (b) $(x_1y_1-xy)^2=x^2+y^2$

(c) $(xy_1+yx_1)^2=d(x^2-y^2)$ (d) $(x^2-y^2)=d^2(x_1+y_1)$

17 If the equation of the pair of straight lines passing through the point (1, 1), one making an angle θ with the positive direction of Y-axis, is $x^2-(a+2)xy+y^2+a(x+y-1)=0$, $a\neq-2$, then the value of $\sin 2\theta$ is

(a) $a-2$ (b) $a+2$ (c) $\frac{2}{a+2}$ (d) $\frac{2}{a}$

18 The angle between the pair of lines whose equation is $4x^2+10xy+my^2+5x+10y=0$ is

(a) $\tan^{-1}\left(\frac{3}{8}\right)$ (b) $\tan^{-1}\left(\frac{3}{4}\right)$

(c) $\tan^{-1}\left(\frac{2\sqrt{25-4m}}{m+1}\right)$, $m\geq n$ (d) None of these

3.3 Point of Intersection of Pair of Straight Line and Equation of Bisectors

19 If the lines $x^2+2xy-35y^2-4x+44y-12=0$ and $5x+\lambda y-8=0$ are concurrent, then the value of λ is

(a) 0 (b) 1 (c) -1 (d) 2

20 The square of the distance between the point of intersection of the lines represented by the equation $ax^2+2hxy+by^2+2gx+2fy+c=0$ and origin is

(a) $\frac{c(a+b)-f^2-g^2}{ab-h^2}$ (b) $\frac{c(a-b)+f^2+g^2}{\sqrt{ab-h^2}}$ (c) $\frac{c(a+b)-f^2-g^2}{ab+h^2}$ (d) None of these

21 Let the equation of the pair of lines, $y=px$ and $y=qx$ can be written as $(y-px)(y-qx)=0$ Then, the equation of the pair of the angle bisectors of the line $x^2-4xy-5y^2=0$ is

[JEE Main 2021]

(a) $x^2-3xy+y^2=0$ (b) $x^2+4xy-y^2=0$ (c) $x^2+3xy-y^2=0$ (d) $x^2-3xy-y^2=0$

3.4 Miscellaneous

22 The pair of lines $\sqrt{3}y^2-xy-\sqrt{3}xy+x^2=0$ are rotated about the origin by $\frac{\pi}{12}$ in the anti-clockwise sense, then the equation of the pair in the new position is

(a) $y^2-\sqrt{3}xy-xy-\sqrt{3}x^2=0$ (b) $y^2-\sqrt{3}xy-xy+\sqrt{3}x^2=0$

(c) $\sqrt{3}y^2-xy+\sqrt{3}xy+x^2=0$ (d) None of these

23 The equation of the perpendiculars drawn from the origin to the lines represented by the equation $2x^2-10xy+12y^2+5x-16-3=0$ is

(a) $6x^2+5xy+y^2=0$ (b) $6y^2+5xy+x^2=0$ (c) $6x^2-5xy+y^2=0$ (d) None of these

24 If the lines represented by the equation $3y^2-x^2+2\sqrt{3}x-3=0$ are rotated about the point $(\sqrt{3}, 0)$ through an angle 15° one clockwise direction and other in anti-clockwise direction so that they become perpendicular, then the equation of the pair of lines in the new position is

(a) $y^2-x^2+2\sqrt{3}x+3=0$ (b) $y^2-x^2+2\sqrt{3}x-3=0$

(c) $y^2-x^2-2\sqrt{3}x+3=0$ (d) $y^2-x^2+3=0$

25 $3x^2-8xy-3y^2+10x+20y-25=0$ are the bisectors of the angles between two lines of which one line is passing through the origin. Then the equation of the other line is

(a) $x-2y=0$ (b) $2x-y=0$ (c) $x+2y-5=0$ (d) $x-2y-5=0$

26 A straight line is drawn through (1, 0) to the curve $x^2+y^2+6x-10y+1=0$ such that the intercept made on it by the curve subtends a right angle at the origin. Then , the slope of the line is

(a) 1 or $\frac{1}{9}$ (b) -1 or $-\frac{1}{9}$ (c) -1 or $\frac{1}{9}$ (d) 1 or $-\frac{1}{9}$

27 The equation of the line joining the origin to the points of intersection of the curve $x^2+y^2=a^2$ and $x^2+y^2-ax-ay=0$ is

(a) $x^2-y^2=0$ (b) $xy=0$ (c) $xy-x^2=0$ (d) $y^2+xy=0$

28 The angle between the lines joining the points of intersection of line $y=3x+2$ and the curve $x^2+2xy+3y^2+4x+8y-11=0$ to the origin is

(a) $\tan^{-1}\left(\frac{3}{2}\sqrt{2}\right)$ (b) $\tan^{-1}\left(\frac{2\sqrt{2}}{3}\right)$ (c) $\tan^{-1}(\sqrt{3})$ (d) $\tan^{-1}(2\sqrt{2})$

29 The pair of straight lines joining the origin to the points of intersection of the line $y = 2\sqrt{2}x + c$ and the circle $x^2 + y^2 = 2$ are at right angles, if

(a) $c^2 - 4 = 0$ (b) $c^2 - 8 = 0$ (c) $c^2 - 9 = 0$ (d) $c^2 - 10 = 0$

Numerical Type Questions

1 From a point A (1, 1), straight line AL and AM are drawn at right angles to the pair of straight lines $3x^2 + 7xy + 2y^2 = 0$. Find the area of quadrilateral $ALOM$ where O is origin.

2 The base of triangle passes through a fixed point (f, g) and its sides are, respectively, bisected at right angles by the lines $y^2 - 8xy - 9x^2 = 0$, if the locus of its vertex is $a(x^2 + y^2) + (bg + cf)x + (cg - bf)y = 0$, then find $\frac{a+b+c}{5}$.

3 If $ax^3 - 9x^2y - xy^2 + 4y^3 = 0$ represents three straight lines such that two of them are perpendicular, then find the sum of the values of a.

4 A triangle has the lines $ax^2 + 2hxy + by^2 = 0$ for two of its sides and the point (l, m) for its orthocentre then, if the third side has the equation.
$(a + b)(lx + my) = \lambda am^2 - \mu hlm + bl^2$, then find the value of $\sqrt{\lambda^2 + \mu^2}$ upto two decimal.

5 The sides of a triangle have the combined equation $x^2 - 3y^2 - 2xy + 8y - 4 = 0$. The third side, which is variable always passes through the point $(-5, -1)$. If the range of values of the slope of third line is such that the origin is an interior point of the triangle is (a, b), then the value of $\left(a + \frac{1}{b}\right)$ is

6 If area of the triangle formed by the lines $x + y = 3$ and the angle bisectors of the pair of lines $x^2 - y^2 + 4y - 4 = 0$ is A. Then, the value of $16A$ is

7 If (α, β) is the centroid of the triangle whose sides are the lines $ax^2 + 2hxy + by^2 = 0$ and $lx + my = 1$ and a relation $\frac{\alpha}{bl - hm} = \frac{\beta}{am - hl} = \frac{p}{q} \cdot \frac{1}{bl^2 - 2hlm + ah^2}$ follow, then find $\frac{p+q}{10}$, where p, q are coprime.

LEVEL 03\ JEE Advanced

One Option Correct Type Questions

1 A is a point on the X-axis through the point A, a line is drawn parallel to Y-axis so as to meet the lines $ax^2 + 2hxy + by^2 = 0$ in B and C. If $AB = BC$, then

(a) $9x^2 - 8ab$ (b) $8h^2 = 9ab$ (c) $4b^2 = 3ab$ (d) $3x^2 = 8ab$

2 A variable line L passing through the point $B(2, 5)$ intersects the lines $2x^2 - 5xy + 2y^2 = 0$ at P and Q, then the locus of point R on L such that distances BP, BR and BQ are in harmonic progression.

(a) $17x - 10y = 0$ (b) $17x + 10y = 0$ (c) $10x - 17y = 0$ (d) None of these

3 Let a and b be non-zero and real number. Then, the equation $(ax^2 + by^2 + c)(x^2 - 5xy + 6y^2) = 0$ represents

(a) four straight line when $c = 0$ and a, b are of the same sign

(b) two straight lines and a circle when $a = b$ and c is of sign opposite to that of a

(c) two straight lines and a circle when a and b are of the same sign and c is of opposite to that of a
(d) a circle and an pair of straight line when a and b are at the same sign and c is at sign opposite to that of a

4 If the lines $ax^2 + 2hxy + by^2 = 0$ represents the adjacent sides of a parallelogram, then the equation of second diagonal if one is $lx + my = 1$ will be
(a) $(am + hl)x = (bl + hm)y$ (b) $(am - hl)x = (bl - hm)y$
(c) $(am - hl)x = (bl + hm)y$ (d) None of these

5 The angle between the lines $(x^2 + y^2)\sin^2 \alpha = (x\cos\beta - y\sin\beta)^2$ is
(a) 2β (b) 2α (c) α (d) None of these

6 The combined equation of the lines $l_1\, l_2$ is $2x^2 + 6xy + y^2 = 0$ and that of the lines m_1, m_2 is $4x^2 + 18xy + y^2 = 0$. If the angle between l_1 and m_2 be α, then the angle between l_2 and m_1 will be
(a) $\frac{\pi}{2} - \alpha$ (b) 2α (c) $\frac{\pi}{4} + \alpha$ (d) α

7 If the straight lines joining the origin and the points of intersection of curve $5x^2 + 12xy - 6y^2 + 4x - 2y + 3 = 0$ and $x + ky - 1 = 0$ are equally inclined to the coordinate axes, then the value of k
(a) is equal to 1 (b) is equal to -1
(c) is equal to 2 (d) does not exist in the set of real numbers

More Than One Type Questions

8 If $x^2 + 2hxy + y^2 = 0$ represents the equation of the straight lines through the origin which make an angle α with the straight line $y + x = 0$, then
(a) $\sec 2\alpha = h$ (b) $\cos\alpha = \sqrt{\frac{1+h}{2h}}$ (c) $2\sin\alpha = \sqrt{\frac{1+h}{n}}$ (d) $\cot\alpha = \sqrt{\frac{h+1}{h-1}}$

9 The combined equation three sides of a triangle is $(x^2 - y^2)(2x + 3y - 6) = 0$. If $(-2, a)$ is an interior point and $(b, 1)$ is an exterior point of the triangle, then
(a) $2 < a < \frac{10}{3}$ (b) $-2 < a < \frac{10}{3}$ (c) $-1 < b < \frac{9}{2}$ (d) $-1 < b < 1$

10 If one of the lines given by the equation $2x^2 + pxy + 3y^2 = 0$ coincide with one of those given $2x^2 + qxy - 3y^2 = 0$ and the other lines represented by them be perpendicular, then
(a) $p = 5$ (b) $p = -5$ (c) $q = -1$ (d) $q = 1$

11 The lines L_1 and L_2 denoted by $3x^2 + 10xy + 8y^2 + 14x + 22y + 15 = 0$ intersect at the point P and have gradients m_1 and m_2, respectively. The acute angle between them is θ which of following relations hold good?
(a) $m_1 + m_2 = \frac{5}{4}$ (b) $m_1 m_2 = \frac{3}{8}$
(c) $\theta = \sin^{-1}\left(\frac{2}{5\sqrt{5}}\right)$ (d) sum of abscissa and ordinate of point P is -1

12 The lines joining the origin to the point of intersection of $3x^2 + mxy - 4x + 1 = 0$ and $2x + y - 1 = 0$ are at right angles. Then, which of the following is not possible value of m?
(a) -4 (b) 4 (c) 7 (d) 3

Paragraph Type Questions

Paragraph I (Q.Nos. 13-15)

Let the lines represented by $2x^2 - 5xy + 2y^2 = 0$ be the two sides of a parallelogram and the line $5x + 2y = 1$ be one of its diagonal. Then,

13 the one of sides of the parallelogram passing through the origin is
(a) $x - 2y = 0$ (b) $x + 2y = 0$ (c) $x - 3y = 0$ (d) $x + 3y = 0$

14 the equation of other diagonal is
(a) $x - 2y = 0$ (b) $11x - 10y = 0$ (c) $10x - 11y = 0$ (d) None of these

15 The area of the parallelogram is
(a) $\frac{1}{36}$ sq unit (b) $\frac{1}{54}$ sq unit (c) $\frac{1}{72}$ sq unit (d) $\frac{1}{96}$ sq unit

Paragraph II (Q.Nos. 16-18)

Suppose a line $lx + my = 1$ meets a second degree curve $ax^2 + 2hxy + by^2 + 2gx + 2fy + c = 0$ at two points A and B. Then, the combined equation of the pair of lines OA and OB is $ax^2 + 2hxy + by^2 + (2gx + 2fy)(lx + my) + c(lx + my)^2 = 0$ which is a second degree homogeneous equation. Answer the following questions.

16 All chords of the curve $3x^2 - y^2 - 2x + 4y = 0$ which subtend right angle at the origin will pass through a fixed point whose coordinates are
(a) $(-1, 2)$ (b) $(1, -2)$ (c) $(1, 2)$ (d) $(-1, -2)$

17 If the intercept of the line $lx + my = 1$ made by the curve $x^2 + y^2 - a^2 = 0$ subtend right angle at the origin, then $l^2 + m^2 =$
(a) $\frac{2}{a^2}$ (b) $\frac{1}{a^2}$ (c) $2a^2$ (d) $3a^2$

18 The line $y = mx + c$ makes an intercept on the curve $y^2 - 4ax = 0$ which subtends right angle at the origin. Then, the line $y = mx + c$ passes through a fixed point whose coordinates are
(a) $(2a, 0)$ (b) $(9, 0)$ (c) $(3a, 0)$ (d) $(4a, 0)$

Match the Column

19 Match the following.

	Column I		Column II
(a)	If the slope of one of the lines represented by $ax^2 + 2hxy + by^2 = 0$ is square of the other, then $\frac{a+b}{h} + \frac{8h^2}{ab}$ is	p.	odd prime number
(b)	The product of propendiculars drawn from the point (1,2) to the pair of lines $x^2 + 4xy + y^2 = 0$ is λ units, then $[\lambda]$ is a (where [.] denotes the greatest integer function)	q.	composite number
(c)	If the pairs $a_1x^2 + 2h_1xy + b_1y^2 = 0$ and $a_2x^2 + 2h_2xy + b_2y^2 = 0$ have one line common, then $\frac{(a_1b_2 - a_2b_1)^2}{(b_1h_2 - b_2h_1)(a_2h_1 - a_1h_2)}$ is	r.	even prime number
(d)	Distance between two lines represented by the line pair $x^2 - 4xy + 4y^2 + x - 2y - 6 = 0$ is λ unit. Then, $[\lambda]$ is a (where [.] denotes greatest integer function)	s.	multiple of 6

20 Match the items of column I with those of column II.

	Column I		Column II
(a)	If $6x^2+5xy-6y^2+9x+20y+c=0$ represents a pair of lines, then c is equal to	p.	$\frac{\pi}{2}$
(b)	the angle between the lines joining the origin to the points of intersection of the curve $2x^2+6xy+3y^2+4x+2y-36=0$ and line $x-2y-6=0$ is	q.	8
(c)	If the lines joining the origin to the points of intersection of the curve $2x^2-2xy+3y^2+2x-y-1=0$ and the line $x+2y=k$ are at right angles, then the value of k is	r.	1
(d)	if d is the distance between the parallel lines represented by the equation $9x^2-24xy+16y^2-12x+16y-12=0$, then the value of $5d$ is	s.	–6
		t.	–1

Numerical Type Questions

Question Stem (Q.Nos. 21-23) The answer to each question is a **NUMERICAL VALUE.**

Consider a line pair $ax^2+3xy-2y^2-5x+5y+c=0$ representing perpendicular lines intersecting each other at c and forming a ΔABC with the X-axis.

21 If x_1 and x_2 are intercepts on the X-axis and y_1 and y_2 are the intercepts on the Y-axis, then the sum $(x_1+x_2+y_1+y_2)$ is equal to

22 Distance between orthocentre and circumcentre of the ΔABC is

23 If the circle $x^2+y^2-4y+k=0$ is orthogonal with the circumcircle of the ΔABC, then k equals

24 A point moves so that the distance between the feet of perpendiculars drawn from it to the lines $ax^2+2hxy+by^2=0$ is a constant $2n$ and the locus of the point is $(x^2+y^2)(h^2-ab)=h^2((a-b)^2+\lambda h^2)$, then find $\frac{\lambda}{10}$.

25 There exists two ordered triplets (a_1, b_1, c_1) and (a_2, b_2, c_2) for (a, b, c) for which the equation $4x^2-4xy+ay^2+bx+cy+1=0$ represents a pair of identical straight lines in xy-plane. Find the value of $a_1+b_1+c_1+a_2+b_2+c_2$.

26 The equation $9x^3+9x^2y-45x^2=4y^3+4xy^2-20y^3$ represents 3 straight lines, two of which pass through the origin. Find the area of the triangle formed by these lines (in sq units).

ANSWER KEY

Level 1

1. (i) (a) $y - 6x = 0$ and $y - 3x = 0$ (b) $x - 2y = 0$ and $x - y = 0$
(c) $x + 3y = 0$ and $x + 2y = 0$ (d) No separate equation
(e) $x = \sqrt{3}y$ and $x = \sqrt{3}y$
(ii) (a) $xy = 0$ (b) $xy - 3x - 4y + 12 = 0$
(iii) $5x^2 + 8xy + 3y^2 - 26x - 20y + 33 = 0$

2. (a) 4 sq units (b) (10, 4) (c) (2, 2)

3. (a) $\frac{1}{2\sqrt{3}}$ sq units (b) $A = \tan^{-1}(-2)$

4. (a) $5x^2 + 2xy - 5y^2 = 0$ (b) $m = \pm 1$
(c) $\lambda = 4$ (d) No value

5. (a) $\lambda = 6$ (b) $\lambda = 2$ (c) $\lambda = 4$ (d) $\lambda = 4, \frac{143}{7}$

6. (a) $4y - x - 2 = 0$ and $y - x + 1 = 0$ (b) $y - 3x - 2 = 0$ $2y - 2x - 1 = 0$
(c) $3y + 4x - 15 = 0$ and $3y + 4x + 5 = 0$

7. (a) $x^2 + xy - 2y^2 + 9y - 9 = 0$ (b) 2 units (c) $\theta = \tan^{-1}2$
(d) $q = 1, -\frac{23}{2}$ (e) $\frac{1}{10}$ sq units

8. (a) (i) $\left(\frac{14}{17}, \frac{9}{17}\right)$ (ii) (1, 1) (iii) Infinite number
(b) (i) $\dfrac{\left(x - \frac{3y}{9}\right)\left(y - \frac{19}{9}\right)}{-\frac{5}{2}}$ (ii) $20x^2 - 20y^2 + 2269x + 2528y - 82xy - 19721 = 0$
(iii) $bg^2 + ch^2 = 2bg^2 = 2fgh$

9. (a) $12X^2 + 7XY - 12Y^2 = 0$ (b) $a = \pm\sqrt{2}, b = \pm\sqrt{6}, k=1$
(c) $bx^2 + ay^2 - 2hxy = 0$ (d) $(c^2 - b^2m^2)x^2 + 2b^2mxy + (c^2 - b^2)y^2 = 0$
(e) $4blx^2 + 4bmxy - y^2 = 0$ (f) $90°$
(g) $\pm\sqrt{3}$ (h) $\frac{c}{\sqrt{(a-b)^2 + 4h^2}}$

Level 2

1. (a)	**2.** (b)	**3.** (c)	**4.** (a)	**5.** (c)	**6.** (c)	**7.** (c)
8. (d)	**9.** (d)	**10.** (d)	**11.** (b)	**12.** (c)	**13.** (b)	**14.** (b)
15. (b)	**16.** (a)	**17.** (c)	**18.** (b)	**19.** (d)	**20.** (d)	**21.** (c)
22. (b)	**23.** (a)	**24.** (b)	**25.** (c)	**26.** (b)	**27.** (b)	**28.** (b)
29. (c)	**30.** (0.7)	**31.** (2.6)	**32.** (1)	**33.** (2.23)	**34.** (4)	**35.** (8)
36. (0.5)						

Level 3

1. (b)	**2.** (a)	**3.** (b)	**4.** (b)	**5.** (b)	**6.** (d)	**7.** (b)
8. (a, b, d)	**9.** (a, d)	**10.** (a, b, c, d)	**11.** (b, c)	**12.** (a, b, c, d)	**13.** (a)	**14.** (b)
15. (a)	**16.** (b)	**17.** (a)	**18.** (d)			

19. a→(q, s), b→(p), c→(q), d→(r) **20.** a→(s), b→(p), c→(r, t), d→(q)

21. (5)	**22.** (1.75)	**23.** (1.5)	**24.** (0.4)	**25.** (2)	**26.** (30)

SOLUTIONS

Level 1

1. (i) (a) Given equation, $y^2 - 9xy + 18x^2 = 0$

$$\Rightarrow \quad y^2 - 6xy - 3xy + 18x^2 = 0$$
$$\Rightarrow \quad y(y - 6x) - 3x(y - 6x) = 0$$
$$\Rightarrow \quad (y - 6x)(y - 3x) = 0$$

$\therefore$ Separate equations are

$y - 6x = 0$ and $y - 3x = 0$

(b) Given equation, $x^2 - 3xy + 2y^2 = 0$

$\Rightarrow x^2 - 2xy - xy + 2y^2 = 0$

$\Rightarrow x(x - 2y) - y(x - 2y) = 0$

$\Rightarrow (x - 2y)(x - y) = 0$

Separate equations are

$x - 2y = 0$ and $x - y = 0$

(c) Given equation, $x^2 + 5xy + 6y^2 = 0$

$\Rightarrow x^2 + 3xy + 2xy + 6y^2 = 0$

$\Rightarrow x(x + 3y) + 2y(x + 3y) = 0$

$\Rightarrow (x + 3y)(x + 2y) = 0$

Separate equations are

$x + 3y = 0$ and $x + 2y = 0$

(d) Given equation, $3x^2 + 5xy + 4y^2 = 0$

$\Rightarrow$ Consider an quadratic equation in x

$a = 3, b = 5y, c = 4y^2$

$$\therefore \quad x = \frac{-5y \pm \sqrt{(5y)^2 - 4(3)(4)y^2}}{2(3)}$$
$$= \frac{-5y \pm \sqrt{25y^2 - 48y^2}}{6}$$
$$= \frac{-5y \pm \sqrt{-23y^2}}{6}$$

No separate equation.

(e) Given equation,

$2x^2 - 4\sqrt{3}xy + 6y^2 = 0$

Consider quadratic equation in x

$a = 2, b = -4\sqrt{3}y, c = 6y^2$

$$x = \frac{-b \pm \sqrt{b^2 - 4ac}}{2a}$$
$$= \frac{4\sqrt{3}y \pm \sqrt{48y^2 - 4(2)(6y^2)}}{2(2)}$$
$$= \frac{4\sqrt{3}y \pm \sqrt{48y^2 - 48y^2}}{4}$$
$$= \frac{4\sqrt{3}y \pm 0}{4}$$
$$= \sqrt{3}y$$

$\therefore$ Separate equations are $x = \sqrt{3}y$ and $x = \sqrt{3}y$.

(ii) (a) Given separate equations are $x = 0$ and $y = 0$

$\Rightarrow \quad xy = 0$

which is combined equation.

(b) Given separate equations are $x = 4$ and $y = 3$

$x - 4 = 0$ and $y - 3 = 0$

Combined equation is

$$(x - 4)(y - 3) = 0$$
$$\Rightarrow \quad xy - 3x - 4y + 12 = 0$$

(iii) Given equation is

$$3x^2 - 8xy + 5y^2 = 0$$
$$\Rightarrow \quad 3x^2 - 5xy - 3xy + 5y^2 = 0$$
$$\Rightarrow \quad x(3x - 5y) - y(3x - 5y) = 0$$
$$\Rightarrow \quad (3x - 5y)(x - y) = 0$$

$\therefore$ Equation of lines are $3x - 5y = 0$ ...(i)

and $x - y = 0$ ...(ii)

and let perpendicular lines on Eqs. (i) and (ii) be respectively

$$5x + 3y = \lambda_1$$

and $x + y = \lambda_2$

If these lines pass through (1, 2), then

$5 + 3 \times 2 = \lambda_1$ and $1 + 2 = \lambda_2$

$\lambda_1 = 11$ and $\lambda_2 = 3$

$\therefore$ Equation of required lines are $5x + 3y = 11$ and $x + y = 3$

Combined equation is

$(5x + 3y - 11)(x + y - 3) = 0$

$\Rightarrow 5x^2 + 5xy - 15x + 3xy + 3y^2 - 9y$

$-11x - 11y + 33 = 0$

$\Rightarrow \quad 5x^2 + 8xy + 3y^2 - 26x - 20y + 33 = 0$

2. (a) Given lines which are sides of triangle combined equation $12x^2 - 20xy + 7y^2 = 0$ and $2x - 3y + 4 = 0$

$$\Rightarrow \quad 12x^2 - 14xy - 6xy + 7y^2 = 0$$
$$\Rightarrow \quad 2x(6x - 7y) - y(6x - 7y) = 0$$
$$\Rightarrow \quad (2x - y)(6x - 7y) = 0$$
$$\Rightarrow \quad 2x - y = 0$$
$$y = 2x$$
$$6x - 7y = 0$$
$$y = \frac{6x}{7}$$

For C

Solve $\quad y = 2x$ and $2x - 3y + 4 = 0$

$\Rightarrow \quad y - 3y + 4 = 0$

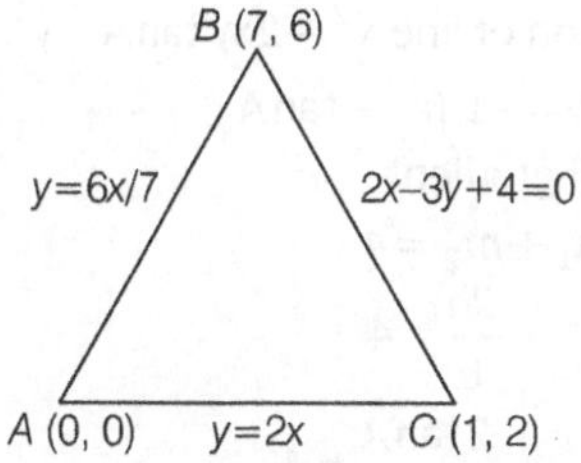

$$y=\frac{4}{2}=2$$

$\therefore \quad x=1$

$\therefore$ Point C is (1, 2)

For B

Solve $y=\dfrac{6x}{7}$ and $2x-3y+4=0$

$$\Rightarrow \quad 2x-3\left(\frac{6x}{7}\right)+4=0$$

$$\Rightarrow \quad \frac{14x-18x}{7}+4=0$$

$$\Rightarrow \quad -\frac{4x}{7}=-4$$

$$x=7$$

$$\Rightarrow \quad y=6$$

Point B is (7, 6)

Area of ΔABC

$$=\frac{1}{2}\text{ mod of}\begin{vmatrix}0 & 0 & 1\\ 7 & 6 & 1\\ 1 & 2 & 1\end{vmatrix}$$

$$=\frac{1}{2}|(14-6)|$$

$$= 4 \text{ sq units}$$

(b) Equation of sides of triangle are

$$x^2-5xy+6y^2=0 \text{ and } y=6$$

$$\Rightarrow \quad x^2-2xy-3xy+6y^2=0 \text{ and } y=6$$

$$\Rightarrow \quad x(x-2y)-3y(x-2y)=0$$

$$(x-2y)(x-3y)=0$$

$\therefore$ Equation of sides are

$$x-2y=0,$$
$$x-3y=0,$$
$$y=6$$

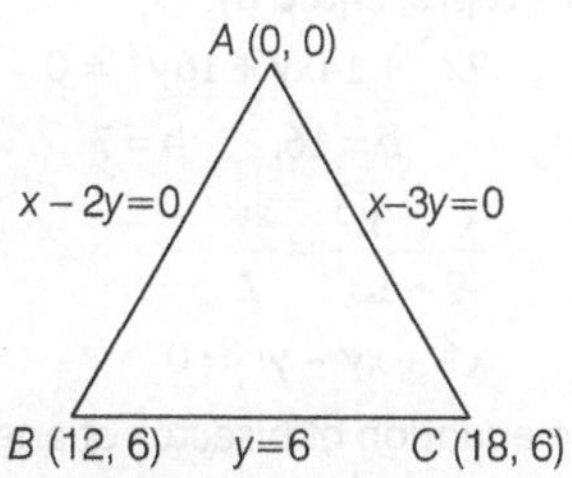

$\therefore$ $A(0,0)$, $B(12,6)$ and $C(18,6)$ are three vertices of ΔABC.

Now, centroid of ΔABC

$$=\left(\frac{x_1+x_2+x_3}{3},\frac{y_1+y_2+y_3}{3}\right)$$

$$=\left(\frac{0+12+18}{3},\frac{0+6+6}{3}\right)$$

$$=(10, 4)$$

(c) The given equations are

$$xy-x-y+1=0$$

and $\quad x+y=4$

$$\Rightarrow \quad x(y-1)-1(y-1)=0$$

$$\Rightarrow \quad (x-1)(y-1)=0$$

$$x=1 \text{ and } y=1$$

and $\quad x+y=4$

By solving line, we get

$$A(1,1), B(3,1) \text{ and } C(1,3)$$

$\because$ $x=1$ and $y=1$ are perpendicular to each other.

$\therefore$ The triangle formed by these lines will be right angled.

Circumcentre is mid-point of hypotenuse BC

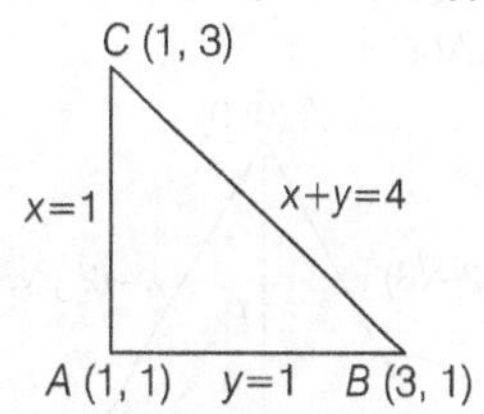

$$\therefore \quad \text{Circumcentre}=\left(\frac{1+3}{2},\frac{3+1}{2}\right)$$

$$=(2, 2)$$

3. (a) Angle between the two lines

$$x^2-4xy+y^2=0 \text{ is}$$

$$\theta=\tan^{-1}\left|\frac{2\sqrt{h^2-ab}}{a+b}\right|$$

$$=\tan^{-1}\left|\frac{2\sqrt{4-1}}{1+1}\right|=\tan^{-1}\sqrt{3}=60°$$

$\therefore \quad \theta=60°$

Now, again $x^2-4xy+y^2=0$

$$x=\frac{4y\pm\sqrt{16y^2-4y^2}}{2}$$

$$=\frac{4y\pm2\sqrt{3}y}{2}=2y\pm\sqrt{3}y$$

$$x=(2\pm\sqrt{3})y$$

Equation of lines represented by

$$x^2-4xy+y^2=0 \text{ is}$$

$$x=(2+\sqrt{3})y \quad \text{...(i)}$$

and $\quad x=(2-\sqrt{3})y \quad$ ...(ii)

Now, the angle between the line $x+y=1$

and line (i) is

$$\theta = \tan^{-1}\left|\frac{m_1 - m_2}{1 + m_1 m_2}\right|$$

$$= \tan^{-1}\left|\frac{-1 - \frac{1}{(2+\sqrt{3})}}{1 - \frac{1}{(2+\sqrt{3})}}\right|$$

$$= \tan^{-1}\left|\frac{-1-(2-\sqrt{3})}{1-(2-\sqrt{3})}\right|$$

$$= \tan^{-1}\left|\frac{-3+\sqrt{3}}{-1+\sqrt{3}}\right|$$

$$= \tan^{-1}\left|\frac{-\sqrt{3}(\sqrt{3}-1)}{\sqrt{3}-1}\right|$$

$$= \tan^{-1}\sqrt{3} = 60°$$

$\therefore$ Similarly angle between $x + y = 1$ and line (ii) is 60°.

$\therefore$ Triangle formed by the lines $x^2 - 4xy + y^2 = 0$ and $x + y = 1$ is equilateral.

Area of ΔABC

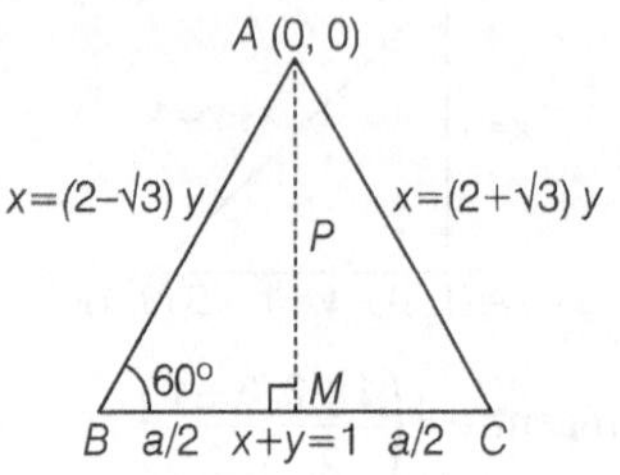

$$= \frac{\sqrt{3}}{4}(a^2)$$

$$= \frac{\sqrt{3}}{4}\left(\frac{2P}{\sqrt{3}}\right)^2 \quad \left\{\because \Delta AMB, \frac{P}{a/2} = \tan 60° \quad a = \frac{2P}{\sqrt{3}}\right\}$$

$\left[\because P = \text{perpendicular distance between } A(0, 0) \text{ and line } x + y = 1 = \left|\frac{0+0-1}{\sqrt{1^2+1^2}}\right| = \frac{1}{\sqrt{2}}\right]$

$$= \frac{\sqrt{3} \times 4P^2}{4 \times 3}$$

$$= \frac{P^2}{\sqrt{3}} = \frac{1}{\sqrt{3}}\left(\frac{1}{\sqrt{2}}\right)^2$$

$$= \frac{1}{\sqrt{3}} \times \frac{1}{2} = \frac{1}{2\sqrt{3}} \text{ sq units}$$

(b) The lines given by

$$ax^2 + 5xy + 2y^2 = 0$$

are mutually perpendicular.

If $\quad a + 2 = 0$

$\Rightarrow \quad a = -2$

(c) Equation of line $x^2 - 2xy \tan A - y^2 = 0$

$a = 1, b = -1, h = -\tan A$

Sum of gradient

$$m_1 + m_2 = 4$$

$$\Rightarrow \quad -\frac{2h}{b} = 4$$

$$\Rightarrow \quad \frac{2\tan A}{-1} = 4$$

$$\Rightarrow \quad \tan A = -2$$

$$\therefore \quad A = \tan^{-1}(-2)$$

4. (a) The given equation is $3x^2 + 5xy + 4y^2 = 0$

Here, $a = 3, b = 4, h = \frac{5}{2}$

$\therefore$ The equation of the bisector is

$$\frac{x^2 - y^2}{a - b} = \frac{xy}{h}$$

$$\Rightarrow \quad \frac{x^2 - y^2}{3 - 4} = \frac{xy}{5/2}$$

$$\Rightarrow \quad 5(x^2 - y^2) = -2xy$$

$$\Rightarrow \quad 5x^2 + 2xy - 5y^2 = 0$$

(b) $\because$ $y = mx$ is one of bisector of the given lines $x^2 + 4xy + y^2 = 0$.

Then, $y = mx$ will satisfies the equation of bisector of given lines.

Bisector $a = 1, b = 1, h = 2$

$$\Rightarrow \quad \frac{x^2 - y^2}{a - b} = \frac{xy}{h}$$

$$\Rightarrow \quad x^2 - y^2 = 0 \quad \text{...(i)}$$

Put $y = mx$ in Eq. (i),

$$x^2 - m^2x^2 = 0$$

$$\Rightarrow \quad 1 - m^2 = 0$$

$$\therefore \quad m = \pm 1$$

(c) The given pair of lines are

$$9x^2 + 14xy + 16y^2 = 0$$

and $\quad 3x^2 + 2xy + \lambda y^2 = 0$

The two pair will be equally inclined, if the two pair of straight lines have same bisector.

$\therefore$ The equation of the bisector of angle between the lines represented by

$$9x^2 + 14xy + 16y^2 = 0$$

$a = 9, \quad b = 16, \quad h = 7$

$$\Rightarrow \quad \frac{x^2 - y^2}{9 - 16} = \frac{xy}{7}$$

$$\Rightarrow \quad x^2 + xy - y^2 = 0 \quad \text{...(i)}$$

and the equation of bisector of angle between the lines represented by

$$3x^2 + 2xy + \lambda y^2 = 0$$

$$\Rightarrow \quad \frac{x^2-y^2}{3-\lambda}=\frac{xy}{1} \qquad \begin{cases} a=3,\ h=1 \\ b=\lambda \end{cases}$$

$$\Rightarrow \quad x^2-y^2=(3-\lambda)xy$$

$$\Rightarrow \quad x^2-y^2+(\lambda-3)xy=0 \qquad \text{...(ii)}$$

On comparing Eqs. (i) and (ii), we get

$$\lambda-3=1$$

$$\Rightarrow \quad \lambda=4$$

(d) The given pair of lines are

$$x^2+2\lambda xy+2y^2=0$$

and $\quad (1+\lambda)x^2-8xy+y^2=0$

both are equally inclined.

$\Rightarrow$ They have same bisectors.

$\therefore$ Equation of bisectors of $x^2+2\lambda xy+2y^2=0$

$$\frac{x^2-y^2}{1-2}=\frac{xy}{\lambda} \qquad [a=1, b=2, h=\lambda]$$

$$\lambda(x^2-y^2)+xy=0 \qquad \text{...(i)}$$

and equation of bisectors of

$(1+\lambda)x^2-8xy+y^2=0$ are

$$\frac{x^2-y^2}{1+\lambda-1}=\frac{xy}{-4} \qquad \{a=1+\lambda, b=1, h=-4\}$$

$$-4(x^2-y^2)=\lambda xy$$

$$4(x^2-y^2)+\lambda xy=0 \qquad \text{...(ii)}$$

By comparing Eqs. (i) and (ii), we get

$$\lambda=4 \text{ and } \lambda=1$$

no same value.

$\therefore$ No value of λ for which they are equally inclined to each other.

5. (a) $\lambda x^2-5xy-6y^2+14x+5y+4=0$

$$a=\lambda, b=-6, h=-\frac{5}{2}, g=\frac{14}{2}=7,$$

$$f=\frac{5}{2}, c=4$$

The equation represents a pair of straight lines.

$$\Delta \equiv abc+2fgh-af^2-bg^2-ch^2=0$$

$$\Rightarrow \lambda(-6)(4)+2\left(\frac{5}{2}\right)(7)\left(-\frac{5}{2}\right)-\lambda\left(\frac{5}{2}\right)^2$$

$$-(-6)(7)^2-4\left(-\frac{5}{2}\right)^2=0$$

$$\Rightarrow -24\lambda+\left(-\frac{175}{2}\right)-\frac{25}{4}\lambda+294-25=0$$

$$\Rightarrow \frac{121}{4}\lambda=\frac{363}{2}$$

$$\lambda=6$$

(b) $12x^2-10xy+2y^2+11x-5y+\lambda=0$

$$a=12, b=2, h=-5, g=\frac{11}{2}, f=-\frac{5}{2}, c=\lambda$$

The equation represents a pair of straight lines

$$\Delta \equiv abc+2fgh-af^2-bg^2-ch^2=0$$

$$\Rightarrow 12(2)(\lambda)+2\left(-\frac{5}{2}\right)\left(\frac{11}{2}\right)(-5)$$

$$-12\left(-\frac{5}{2}\right)^2-2\left(\frac{11}{2}\right)^2-\lambda(-5)^2=0$$

$$\Rightarrow \quad 24\lambda+\frac{275}{2}-75-\frac{121}{2}-25\lambda=0$$

$$\therefore \quad \lambda=\frac{275-150-121}{2}=\frac{275-271}{2}$$

$$=\frac{4}{2}$$

$$\Rightarrow \quad \lambda=2$$

(c) $6x^2+5xy-\lambda y^2+7x+13y-3=0 \qquad \text{...(i)}$

$$a=6, b=-\lambda, h=\frac{5}{2}, g=\frac{7}{2}, f=\frac{13}{2}, c=-3$$

equation (i) represents a pair of straight line

$$\Delta \equiv abc+2fgh-af^2-bg^2-ch^2=0$$

$$\Rightarrow \quad 6(-\lambda)(-3)+2\left(\frac{13}{2}\right)\left(\frac{7}{2}\right)\left(\frac{5}{2}\right)$$

$$-6\left(\frac{13}{2}\right)^2-(-\lambda)\left(\frac{7}{2}\right)^2-(-3)\left(\frac{5}{2}\right)^2=0$$

$$\Rightarrow \quad 18\lambda+\frac{455}{4}-\frac{507}{2}+\frac{49}{4}\lambda+\frac{75}{4}=0$$

$$\Rightarrow \quad 72\lambda+455-1014+49\lambda+75=0$$

$$\Rightarrow \quad 121\lambda=484$$

$$\lambda=\frac{484}{121}=4$$

(d) $6x^2+2\lambda xy-8y^2+23x-10y+7=0 \qquad \text{...(i)}$

$$a=6, b=-8, c=7, h=\lambda, g=\frac{23}{2}, f=-5$$

$\because$ Eq. (i) represents a pair of straight line.

$$\Delta \equiv abc+2fgh-af^2-bg^2-ch^2=0$$

$$\Rightarrow \quad (6)(-8)(7)+2(-5)\left(\frac{23}{2}\right)(\lambda)-6(-5)^2$$

$$-(-8)\left(\frac{23}{2}\right)^2-7(\lambda^2)=0$$

$$\Rightarrow \quad -336-115\lambda-150+1058-7\lambda^2=0$$

$$\Rightarrow \quad 7\lambda^2+115\lambda-572=0$$

$$\therefore \quad \lambda=\frac{-115\pm\sqrt{(115)^2+4(7)(572)}}{2\times 7}$$

$$=\frac{-115\pm 171}{14}$$

$$=\frac{56}{14}, \frac{286}{14}$$

$$\Rightarrow \quad \lambda=4, \frac{143}{7}$$

6. (a) Given combines equation is

$$x^2-5xy+4y^2+x+2y-2=0$$
$$x^2+x(1-5y)+4y^2+2y-2=0$$

Assuming quadratic in x

$$a=1, b=1-5y, c=4y^2+2y-2$$

$$\therefore \quad x=\frac{-b\pm\sqrt{b^2-4ac}}{2a}$$
$$=\frac{-(1-5y)\pm\sqrt{(1-5y)^2-4(1)(4y^2+2y-2)}}{2(1)}$$
$$=\frac{(5y-1)\pm\sqrt{1+25y^2-10y-16y^2-8y+8}}{2}$$
$$=\frac{(5y-1)\pm\sqrt{9y^2-18y+9}}{2}$$
$$=\frac{(5y-1)\pm\sqrt{(3y-3)^2}}{2}$$
$$=\frac{5y-1\pm(3y-3)}{2}$$

Taking positive

$$x=\frac{5y-1+3y-3}{2}$$
$$x=4y-2$$
$$4y-x-2=0$$

Taking negative

$$x=\frac{5y-1-(3y-3)}{2}$$
$$x=\frac{5y-3y-1+3}{2}$$
$$x=y+1$$
$$y-x+1=0$$

$\therefore$ Separate equations are $4y-x-2=0$ and $y-x+1=0$

(b) Given combined equation is

$$12x^2-10xy+2y^2+11x-5y+2=0$$
$$2y^2-y(10x+5)+12x^2+11x+2=0$$

Taking as quadratic in y

$$a=2, b=-(10x+5), c=12x^2+11x+2$$

$$y=\frac{(10x+5)\pm\sqrt{(10x+5)^2-4(2)(12x^2+11x+2)}}{2(2)}$$
$$=\frac{10x+5\pm\sqrt{100x^2+25+100x-96x^2-88x-16}}{4}$$
$$\Rightarrow y=\frac{10x+5\pm\sqrt{4x^2+12x+9}}{4}$$
$$=\frac{(10x+5)\pm\sqrt{(2x+3)^2}}{4}$$
$$=\frac{10x+5\pm(2x+3)}{4}$$

with +ve

$$\therefore \quad y=\frac{10x+5+2x+3}{4}$$
$$y=3x+2$$
$$y-3x-2=0$$

with –ve

$$y=\frac{10x+5-2x-3}{4}$$
$$y=2x+\frac{1}{2}$$
$$2y-2x-1=0$$

$\therefore$ Separate equation are $y-3x-2=0$ and $2y-2x-1=0$

(c) Given combined equation is

$$16x^2+24xy+9y^2+40x+30y-75=0$$
$$9y^2+y(24x+30)+16x^2+40x-75=0$$

Taking quadratic in y

$$a=9, b=24x+30, c=16x^2+40x-75$$

$$\therefore y=\frac{-(24x+30)\pm\sqrt{(24x+30)^2-4(9)(16x^2+40x-75)}}{2(9)}$$
$$\Rightarrow \frac{-6(4x+5)\pm\sqrt{16x^2+25+40x-16x^2-40x+75}}{2(9)}$$
$$y=\frac{-(4x+5)\pm\sqrt{100}}{3}$$
$$3y=-4x+5\pm10$$
$$\therefore \quad 3y=-4x+15$$
and $\quad 3y=-4x-5$
$$3y+4x-15=0$$
and $\quad 3y+4x+5=0$

$\therefore$ Separate equations are

$$3y+4x-15=0$$
and $\quad 3y+4x+5=0$

7. (a) The given combined equation is

$$6x^2+8xy-8y^2+23x-10y+7=0 \quad \text{...(i)}$$
$$6x^2+8xy-8y^2$$
$$=6x^2+12xy-4xy-8y^2$$
$$=6x(x+2y)-4y(x+2y)$$
$$=(x+2y)(6x-4y)$$

Now, Eq. (i) taken as $(x+2y+l)(x-y+m)=0$

Now, the equation of line parallel to the given equation are

$$(x+2y+\lambda)(x-y+\mu)=0$$

$\because$ It passes through $(-1,2)$.

$\therefore$ Separate equations are

$$x+2y+\lambda=0$$

$$-1+4+\lambda=0$$
$$\lambda=-3$$
and $\quad x-y+\mu=0$
$$-1-2+\mu=0$$
$$\mu=3$$
$\therefore$ Equation is $(x+2y-3)(x-y+3)=0$
$\Rightarrow x^2-xy+3x+2xy-2y^2+6y-3x+3y-9=0$
$\Rightarrow x^2+xy-2y^2+9y-9=0$

(b) Given combined equation is
$$x^2+2\sqrt{2}xy+2y^2+4x+4\sqrt{2}y+1=0$$
$$\Rightarrow x^2+x(2\sqrt{2}y+4)+2y^2+4\sqrt{2}y+1=0$$
Taking above as a quadratic in x
$a=1, b=2\sqrt{2}y+4, c=2y^2+4\sqrt{2}y+1$
$$x=\frac{-(2\sqrt{2}y+4)\pm\sqrt{(2\sqrt{2}y+4)^2-4(1)(2y^2+4\sqrt{2}y+1)}}{2(1)}$$
$$=\frac{-(2\sqrt{2}y+4)\pm\sqrt{8y^2+16-16\sqrt{2}y-8y^2-16\sqrt{2}y-4}}{2}$$
$$\Rightarrow \quad x=\frac{-2(\sqrt{2})y-4\pm\sqrt{12}}{2}$$
$$\Rightarrow \quad 2x=-2\sqrt{2}y-4\pm2\sqrt{3}$$
$$\Rightarrow \quad x=-\sqrt{2}y-2\pm\sqrt{3}$$
$\therefore$ Equation are
$$\sqrt{2}y+x+2-\sqrt{3}=0$$
and $\quad \sqrt{2}y+x+2+\sqrt{3}=0$
$\Rightarrow \quad x+\sqrt{2}y+2-\sqrt{3}=0$
and $\quad x+\sqrt{2}y+2+\sqrt{3}=0$
$$\because \frac{b_1}{b_2}=\frac{\sqrt{2}}{\sqrt{2}}=1 \text{ and } \frac{a_1}{a_2}=\frac{1}{1}=1$$
$$\Rightarrow \quad \frac{a_1}{a_2}=\frac{b_1}{b_2}=1$$
Lines are parallel.
$\therefore$ Distance between them
$$=\left|\frac{c_1-c_2}{\sqrt{a^2+b^2}}\right|$$
$$=\left|\frac{2-\sqrt{3}-2-\sqrt{3}}{\sqrt{2+1}}\right|$$
$$=\left|\frac{2\sqrt{3}}{\sqrt{3}}\right|$$
$$=2 \text{ units}$$

(c) Given equation is
$$3x^2-2xy-8y^2-4x+18y-7=0$$
$a=3, b=-8, c=-7, h=-1, g=-2, f=9$

We have,
$$\tan\theta=\left|\frac{2\sqrt{h^2-ab}}{a+b}\right|$$
$$=\left|\frac{2\sqrt{1+24}}{3-8}\right|=\left|\frac{2\times5}{-5}\right|$$
$$\tan\theta=2$$
$$\theta=\tan^{-1}2$$

(d) The equation represents pair of straight lines.
$\therefore \quad \Delta\equiv abc+2fgh-af^2-bg^2-ch^2=0$ ...(i)
and also both are perpendicular.
$\Rightarrow \quad a+b=0$ ...(ii)
Now, given equation is
$$12x^2+7xy-py^2-18x+qy+6=0$$
$$\Rightarrow a=12, h=\frac{7}{2}, b=-p, g=-9, f=\frac{q}{2}, c=6$$
By Eq. (ii),
$$12+(-p)=0$$
$$\Rightarrow \quad p=12$$
By Eq. (i),
$$12(-12)6+2\left(\frac{q}{2}\right)(-9)\left(\frac{7}{2}\right)-12\left(\frac{q}{2}\right)^2-(-12)(-9)^2-6\left(\frac{7}{2}\right)^2=0$$
$$\Rightarrow \quad -864-\frac{63q}{2}-3q^2+972-\frac{147}{2}=0$$
$$\Rightarrow \quad -63q-6q^2+69=0$$
$$\Rightarrow \quad 6q^2+63q-69=0$$
$$\Rightarrow \quad 2q^2+21q-23=0$$
$$\Rightarrow \quad 2q^2+23q-2q-23=0$$
$$q(2q+23)-1(2q+23)=0$$
$$(2q+23)(q-1)=0$$
$$\therefore \quad q=1,-\frac{23}{2}$$

(e) Given, $\quad 3x^2+8xy-3y^2=0$
$$3x^2+9xy-xy-3y^2=0$$
$$3x(x+3y)-y(x+3y)=0$$
$$(x+3y)(3x-y)=0$$
$\therefore \quad y=-\frac{1}{3}x$ and $y=3x$ are two lines
$$m_1=-\frac{1}{3}$$
$$m_2=3$$
$$\Rightarrow \quad m_1m_2=-1$$
$\therefore$ Both are perpendicular.
Now, $3x^2+8xy-3y^2+2x-4y-1=0$
$\Rightarrow (x+3y)(3x-y)+2x-4y-1=0$ ...(i)

Let us assum Eq. (i) is

$$(x+3y+c_1)(3x-y+c_2)=0$$

$$\Rightarrow (x+3y)(3x-y)+(3c_1+c_2)x-(c_1-3c_2)y + c_1c_2=0 \quad \text{...(ii)}$$

By comparing Eqs. (i) and (ii),

$$3c_1+c_2=2,\ c_1-3c_2=4$$

$$c_1c_2=-1$$

Solving, we get

$$c_1=1, c_2=-1$$

∴ Separate equations are

$$(x+3y+1)=0$$

$$y=-\frac{1}{3}x-\frac{1}{3}$$

$$m_3=-\frac{1}{3}$$

and $\quad 3x-y-1=0$

$$y=3x-1$$

$$m_4=3$$

$\Rightarrow \quad m_3\cdot m_4=-1$

⇒ Both are perpendicular

and $m_1=m_3$ and $m_4=m_2$

$y=-\frac{x}{3}$ and $y=-\frac{x}{3}-\frac{1}{3}$

and $y=3x$ and $y=3x-1$ are parallel and

$$AB=AD=\frac{1}{\sqrt{10}}$$

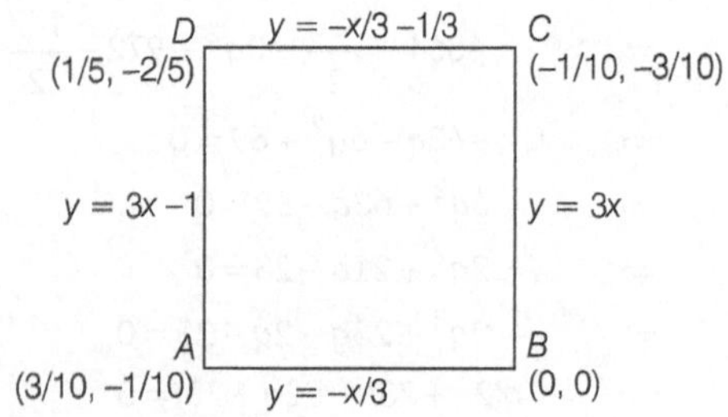

Quadrilateral formed by these four lines is square.

$$\therefore \text{ Area}=(\text{side})^2=\left(\frac{1}{\sqrt{10}}\right)^2=\frac{1}{10}\text{ sq units}$$

8. (a) (i) Given equation is

$$x^2-5xy+2y^2+x+2y-2=0$$

Now, $\quad \frac{\partial f}{\partial x}=0$

$$\Rightarrow \quad 2x-5y+1=0 \quad \text{...(i)}$$

and $\quad \frac{\partial f}{\partial y}=0$

$$\Rightarrow \quad -5x+4y+2=0 \quad \text{...(ii)}$$

Solving Eqs. (i) and (ii), we get

$$x=\frac{14}{17} \text{ and } y=\frac{9}{17}$$

∴ Point of intersection is $\left(\frac{14}{17},\frac{9}{17}\right)$.

(ii) Given equation is

$$3x^2-2xy-8y^2-4x+18y-7=0$$

Now, $\quad \frac{\partial f}{\partial x}=0$

$$\Rightarrow \quad 6x-2y-4=0$$

$$\Rightarrow \quad 3x-y-2=0 \quad \text{...(i)}$$

and $\quad \frac{\partial f}{\partial y}=0$

$$\Rightarrow \quad -2x-16y+18=0$$

$$\Rightarrow \quad -x-8y+9=0 \quad \text{...(ii)}$$

Solving Eqs. (i) and (ii), we get

$$x=1 \text{ and } y=1$$

∴ Point of intersection is (1, 1).

Here, $a=3, h=-1, b=-8$.

$g=-2, f=9$ and $c=-7$

$\therefore\ h^2-ab=1+24=25$

$bg-hf=16+9=25$

$af-gh=27-2=25$

$$\therefore \text{ Point of intersection is } \left(\frac{bg-hf}{h^2-ab},\frac{af-gh}{h^2-ab}\right)$$

$$=\left(\frac{25}{25},\frac{25}{25}\right)=(1,1)$$

(iii) Given equation is

$$x^2-8xy+16y^2+2x-8y+1=0$$

Now, $\quad \frac{\partial f}{\partial x}=0$

$$\Rightarrow \quad 2x-8y+2=0$$

$$\Rightarrow \quad x-4y+1=0 \quad \text{...(i)}$$

$$\frac{\partial f}{\partial y}=0$$

$$\Rightarrow \quad -8x+32y-8=0$$

$$\Rightarrow \quad x-4y+1=0 \quad \text{...(ii)}$$

Since Eqs. (i) and (ii) are same.

∴ Lines are coincident.

⇒ Infinite number of solution.

(b) (i) The given curve is

$$x^2-5xy+4y^2+3x+2y-1=0$$

Now, $\quad \frac{\partial f}{\partial x}=0$

$$\Rightarrow \quad 2x-5y+3=0 \quad \text{...(i)}$$

and $\quad \frac{\partial f}{\partial y}=0$

$$\Rightarrow \quad -5x+8y+2=0 \quad \text{...(ii)}$$

Solving Eqs. (i) and (ii), we get

$$\begin{array}{r} 10x-25y+15=0 \\ -10x+16y+4=0 \\ \hline -9y+19=0 \end{array}$$

$$y = \frac{19}{9}$$

$$x = \frac{5y-3}{2} = \frac{1}{2}\left[\frac{95}{9} - 3\right]$$

$$= \frac{95-27}{2\times 9} = \frac{68}{2\times 9} = \frac{34}{9}$$

$\therefore$ Point of intersection is $\left(\frac{34}{9}, \frac{19}{9}\right)$.

$\therefore$ Equation of the bisectors is

$$\frac{(x-a)^2-(y-b)^2}{a-b} = \frac{(x-a)(y-b)}{h}$$

$$\Rightarrow \frac{\left(x-\frac{34}{9}\right)^2 - \left(y-\frac{19}{9}\right)^2}{\left(\frac{34}{9}-\frac{19}{9}\right)} = \frac{\left(x-\frac{34}{9}\right)\left(y-\frac{19}{9}\right)}{-\frac{5}{2}}$$

(ii) The given curve is

$$12x^2 - 10xy + 2y^2 + 9x + 2y - 12 = 0$$

Now, $\frac{\partial f}{\partial x} = 0$

$\Rightarrow$ $24x - 10y + 9 = 0$

and $\frac{\partial f}{\partial y} = 0$

$\Rightarrow$ $5x - 2y - 1 = 0$

Solving, we get

$$x = 14,\ y = \frac{69}{2}$$

The equation of the bisectors is

$$\frac{(x-a)^2-(y-b)^2}{(a-b)} = \frac{(x-a)(y-b)}{h}$$

$$\Rightarrow \frac{(x-14)^2 - \left(y-\frac{69}{2}\right)^2}{14-\frac{69}{2}} = \frac{(x-14)\left(y-\frac{69}{2}\right)}{-5}$$

$$\Rightarrow \frac{(x^2+196-28x) - \left(y^2+\frac{4761}{4}-69y\right)}{\frac{28-69}{2}} = \frac{(x-14)(2y-69)}{-10}$$

$$\Rightarrow \frac{2\left(x^2-28x-y^2+69y-\frac{3977}{4}\right)}{-41} = \frac{2xy-69x-28y+966}{-10}$$

$$\Rightarrow -20x^2+560x+20y^2-1380y-19885 = -82xy+2829x+1148y-39606$$

$$\Rightarrow -20x^2+20y^2-2269x-2528y+82xy+19721=0$$

$$\Rightarrow 20x^2-20y^2+2269x+2528y-82xy-19721=0$$

(iii) We must have

$$abc + 2fgh - af^2 - bg^2 - ch^2 = 0 \quad \text{...(i)}$$

Let the point of intersection on the Y-axis be $(0, -1)$. Then,

$$\left(\frac{bg-hf}{h^2-ab}, \frac{af-gh}{h^2-ab}\right).$$

$\therefore$ $fh = bg$ ...(ii)

The equation of the Y-axis is $x = 0$

Solving this and the equation of the pair of lines, we get

$$by^2 + 2fy + c = 0$$

which must have equal roots.

$\therefore$ $f^2 = bc$ ...(iii)

Eq. (ii) can be written as

$$fgh = bg^2$$

Put $f^2 = bc$

and $fgh = bg^2$ in Eq. (i), we get

$$bg^2 = ch^2$$

$\therefore$ $bg^2 + ch^2 = 2bg^2 = 2fgh$

9. (a) Let $\phi = 12x^2 + 7xy - 12y^2 - 17x - 31y - 7 = 0$ ...(i)

$\therefore$ $\frac{\partial\phi}{\partial x} = 24x + 7y - 17 = 0$

$\Rightarrow$ $\frac{\partial\phi}{\partial y} = 7x - 24y - 31 = 0$

Their point of intersection is $(x, y) \equiv (1, -1)$.

Here, $\alpha = 1, \beta = -1$

Shift the origin to $(1, -1)$ then replacing $x = X + 1$ and $y = Y - 1$ in Eq. (i) the required equation is

$$12(X+1)^2 + 7(X+1)(Y-1) - 12(Y-1)^2 - 17(X+1) - 31(Y-1) - 7 = 0$$

$$12X^2 + 7XY - 12Y^2 = 0$$

Here, $\alpha = 1$ and $\beta = -1$

and $g = -\frac{17}{2}, f = -\frac{31}{2}, c = -7$

$\therefore$ $g\alpha + f\beta + c = -\frac{17}{2} \times 1 + \left(-\frac{31}{2}\right) \times (-1) - 7 = 0$

Remove equation is

$$aX^2 + 2hXY + bY^2 + (g\alpha + f\beta + c) = 0$$

$$\Rightarrow 12X^2 + 7XY - 12Y^2 = 0$$

(b) To removing the first degree, we shift the origin to the point α, β

Then, $\alpha = \frac{bg - fh}{h^2 - ab} = \frac{1\times(-1) - 1\times 2}{4 - 1\times 1} = -1$

$\beta = \frac{af - gh}{h^2 - ab} = \frac{1\times 1 - (-1)\times 2}{4 - 1\times 1} = 1$

Then, the transformed equation is

$$X^2+4XY+Y^2+(g\alpha+f\beta+c)=0$$

$$\Rightarrow X^2+4XY+Y^2+(-1\times(-1)+1\times1+4)=0$$

$$X^2+4XY+Y^2+6=0 \quad \text{...(i)}$$

Now, to remove the XY term from Eq. (i), we rotate the axes through an angle θ is given by

$$\cot 2\theta=\frac{a-b}{2h}=0$$

$$\cot 2\theta=0$$

$$2\theta=\frac{\pi}{2} \text{ or } \frac{3\pi}{2}$$

$$\theta=\frac{\pi}{4} \text{ or } \frac{3\pi}{4}$$

Taking $\theta=\frac{\pi}{4}$ [we can also take $\theta=\frac{3\pi}{4}$]

$$\Rightarrow \cos\theta=\frac{1}{\sqrt{2}}, \sin\theta=\frac{1}{\sqrt{2}}$$

Now, if (X', Y') be the coordinates of the point when the axes are rotated through $\theta=\frac{\pi}{4}$, we have

$$X=\frac{X'-Y'}{\sqrt{2}}, Y=\frac{X'+Y'}{\sqrt{2}}$$

$\therefore$ Eq. (i) become

$$\frac{Y'^2}{6}-\frac{X'^2}{2}=1$$

$$\Rightarrow a^2=2, b^2=6 \text{ and } k=1$$

$$a=\pm\sqrt{2}, b=\pm\sqrt{6}, k=1$$

(c) Let $y=m_1x$ and $y=m_2x$ be the lines represented by

$$ax^2+2hxy+by^2=0.$$

Then, we have

$$m_1+m_2=\frac{-2h}{b}$$

and $$m_1m_2=\frac{a}{b}$$

Let $y=m_1'x$ and $y=m_2'x$ be new position of $y=m_1x$ and $y=m_2x$ respectively.

Then, $y=m_1x$ is perpendicular to $y=m_1'x$

$$\therefore \quad m_1\cdot m_1'=-1$$

$$m_1'=-\frac{1}{m_1}$$

Similarly $m_2'=-\frac{1}{m_2}$

So, the new lines are

$y=-\frac{1}{m_1}x$ and $y=-\frac{x}{m_2}$ and so their combined equation is

$$(m_1y+x)(m_2y+x)=0$$

$$\Rightarrow m_1m_2y^2+x^2+xy(m_1+m_2)=0$$

$$\Rightarrow \frac{a}{b}y^2+x^2+xy\left(-\frac{2h}{b}\right)=0$$

$$\Rightarrow bx^2+ay^2-2hxy=0$$

(d) The equation of the straight line is

$$y=mx+c$$

$$\Rightarrow \frac{y-mx}{c}=1 \quad \text{...(i)}$$

The equation of the given curve is

$$x^2+y^2=b^2 \quad \text{...(ii)}$$

Making the Eq. (ii) homogeneous of the second degree in x and y by means of Eq. (i), we get

$$\Rightarrow x^2+y^2=b^2\left(\frac{y-mx}{c}\right)^2$$

$$\Rightarrow c^2(x^2+y^2)=b^2(y-mx)^2$$

$$\Rightarrow (c^2-b^2m^2)x^2+2b^2mxy+(c^2-b^2)y^2=0$$

which is the required equation of the straight lines joining the origin to the point of intersection of the lines (i) and (ii).

(e) The given equation is

$$lx+my=1$$

$$\Rightarrow \frac{lx+my}{1}=1 \quad \text{...(i)}$$

The given curve is

$$y^2=4bx \quad \text{...(ii)}$$

Making the equation (ii) homogeneous of the second degree in x and y by means of equation (i), we get

$$y^2=4bx(lx+my)$$

$$\Rightarrow 4blx^2+4bmxy-y^2=0$$

(f) The given curve is

$$3x^2+5xy-3y^2+2x+3y=0 \quad \text{...(i)}$$

The equation of straight line is

$$3x-2y=1 \quad \text{...(ii)}$$

Making in Eq. (i) homogeneous of the second degree in x and y by means of Eq. (ii), we get

$$3x^2+5xy-3y^2+(2x+3y)(3x-2y)=0$$

$$\Rightarrow 9x^2+10xy-9y^2=0 \quad \text{...(ii)}$$

$$\because \quad a=9, b=-9, h=5$$

$$\therefore \quad a+b=9-9=0$$

$\therefore$ It represents that the pair of straight lines by Eq. (iii) are mutually perpendicular to each other.

$\therefore$ Angle between them $=90°$

(g) Given curve is

$$x^2+3y^2=1 \quad \text{...(i)}$$

and given line is

$$y=mx+1 \quad \text{...(ii)}$$

Making the Eq. (i) homogeneous of the second degree in x and y by means of Eq. (ii), we get

$$x^2+3y^2=(y-mx)^2$$

$$\Rightarrow \quad x^2+3y^2=y^2+m^2x^2-2mxy$$

$$\Rightarrow \quad (1-m^2)x^2+2mxy+2y^2=0$$

Clearly, $a+b=0$

$$1-m^2+2=0$$

$$m^2=3$$

$$m=\pm\sqrt{3}$$

$\therefore$ The value of m is $\pm\sqrt{3}$.

(h) Let $ax^2+2hxy+by^2+2gx+2fy+c$

$$=(lx+my+n)(l_1x+m_1y+n_1)$$

$$=ll_1x^2+mm_1y^2+(lm_1+l_1m)xy+(ln_1+l_1n)x+(mn_1+nm_1)y+nn_1$$

Comparing the cofficients, we get

$$ll_1=a,\ mm_1=b,\ nn_1=c,$$

$$lm_1+ml_1=2h$$

$$ln_1+nl_1=2g,\ mn_1+nm_1=2f$$

Let P_1 and P_2 be the length of perpendicular from the origin.

Thus, $P_1\cdot P_2=\dfrac{n}{\sqrt{l^2+m^2}}\times\dfrac{n_1}{\sqrt{l_1^2+m_1^2}}$

$$=\frac{nn_1}{\sqrt{(l^2+m^2)(l_1^2+m_1^2)}}$$

$$=\frac{nn_1}{\sqrt{l^2l_1^2+m^2m_1^2+l^2m_1^2+l_1^2m^2}}$$

$$\Rightarrow \quad \frac{nn_1}{\sqrt{l^2l_1^2+m^2m_1^2+(lm_1+l_1m)^2-2ll_1mm_1}}$$

$$=\frac{c}{\sqrt{a^2+b^2+4h^2-2ab}}$$

$$=\frac{c}{\sqrt{(a-b)^2+4h^2}}$$

Level 2

1. (a) Given, $3x+4y=0$ is the one line of pair of straight line

$$6x^2-xy+4cy^2=0 \quad \text{...(i)}$$

$$3x+4y=0$$

$$y=-\frac{3}{4}x$$

Put $y=-\dfrac{3}{4}x$ in Eq. (i)

$$6x^2-x\left(-\frac{3}{4}x\right)+4\left(-\frac{3}{4}x\right)^2c=0, \forall x$$

$$\Rightarrow \quad 6+\frac{3}{4}+\frac{9c}{4}=0 \Rightarrow c=-3$$

2. (b) Let $y=mx$ be a line common to the given pairs of lines, then

$$am^2+2m+1=0 \text{ and } m^2+2m+a=0$$

$$\Rightarrow \quad \frac{m^2}{2(1-a)}=\frac{m}{a^2-1}=\frac{1}{2(1-a)}$$

$$\Rightarrow \quad m^2=1 \text{ and } m=-\frac{(a+1)}{2}$$

$$(a+1)^2=4$$

$$\Rightarrow \quad a=1 \text{ or } -3$$

But for $a=1$ the two pairs have both the lines common so $a=-3$

$$x^2+2xy+ay^2=x^2+2xy-3y^2=(x-y)(x+3y)$$

and $ax^2+2xy+y^2=-3x^2+2xy+y^2=-(x-y)(3x+y)$

So, the equation of the required lines is

$$(x+3y)(3x+y)=0$$

$$\Rightarrow \quad 3x^2+10xy+3y^2=0$$

3. (c) Given, $6x^2-xy-12y^2=0$

$$\Rightarrow \quad (2x-3y)(3x+4y)=0 \quad \text{...(i)}$$

and $15x^2+14xy-8y^2=0$

$$\Rightarrow \quad (5x-2y)(3x+4y)=0 \quad \text{...(ii)}$$

Equation of common lines to (i) and (ii) is

$$3x+4y=0 \quad \text{...(iii)}$$

Equation of line parallel to $3x+4y=0$ is

$$3x+4y=k \quad \text{...(iv)}$$

Distance between Eqs. (iii) and (iv) is 7.

$$\therefore \quad \left|\frac{c_1-c_2}{\sqrt{a^2+b^2}}\right|=7$$

$$\Rightarrow \quad \left|\frac{k}{\sqrt{3^2+4^2}}\right|=7$$

$$K=\pm 35$$

$\therefore$ Equation is $3x+4y=\pm 35$

4. (a) Let the common lines be $y=mx$. Then, it must satisfy both the equation $ax^2+2hxy+by^2=0$ and $a'x^2+2h'xy+b'y^2=0$

$$\therefore \quad bm^2+2hm+a=0 \quad \text{...(i)}$$

$$b'm^2+2h'm+a'=0 \quad \text{...(ii)}$$

Solving Eqs. (i) and (ii), we get

$$\frac{m^2}{2(ha'-h'a)}=\frac{m}{(ab'-a'b)}=\frac{1}{2(bh'-b'h)}$$

Eliminating m, we get

$$\left[\frac{ab'-a'b}{2(bh'-b'h)}\right]^2=\frac{ha'-h'a}{bh'-b'h}$$

$$\Rightarrow \quad (ab'-a'b)^2=4(ha'-h'a)(bh'-b'h)$$

5. (c) If the slope are m_1 and m_2.

Given, $x^2 - 2cxy - 7y^2 = 0$

Here, $a = 1, b = -7, h = -c$

$$\because \quad m_1 + m_2 = \frac{-2h}{b} = \frac{2c}{-7}$$

$$= \frac{-2c}{7}$$

and $m_1 m_2 = \frac{a}{b} = \frac{1}{-7} = -\frac{1}{7}$

According to the question,

$$m_1 + m_1 = 4m_1 m_2$$

$$\therefore \quad -\frac{2c}{7} = -4\left(\frac{1}{7}\right)$$

$$\Rightarrow \quad c = 2$$

6. (c) Given equation

$$ax^2 + 2hxy + by^2 = 0 \quad \text{...(i)}$$

Let m_1 and m_2 are slope of lines represented by Eq. (i).

$$\therefore \quad m_1 + m_2 = -\frac{2h}{b}$$

and $\quad m_1 m_2 = \frac{a}{b}$

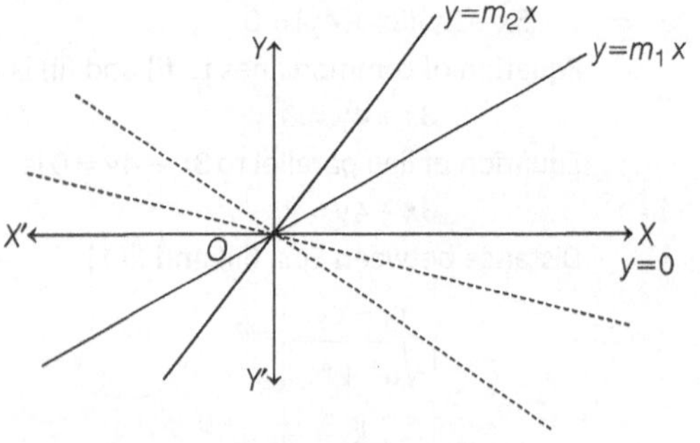

Now, slope of the lines after taking image w.r.t. $y = 0$, i.e. m_1', m_2'

$$m_1' = -m_1 \text{ and } m_2' = -m_2$$

Now, $\quad m_1' + m_2' = -(m_1 + m_2) = \frac{2h}{b}$

$$m_1' \times m_2' = (-m_1)(-m_2) = m_1 m_2 = \frac{a}{b}$$

$\therefore$ The required equation becomes

$$ax^2 - 2hxy + by^2 = 0$$

7. (c) Given, $x^2 - 2hxy - 2y^2 = 0$ be a pair of straight lines whose one bisector is line $x = 2y$

Now, equation of bisectors of given equation

$$\frac{x^2 - y^2}{a - b} = \frac{xy}{h'}$$

Here, $\quad a = 1, b = -2, h' = -h$

$$\therefore \quad \frac{x^2 - y^2}{1 - (-2)} = \frac{xy}{-h}$$

$$\Rightarrow \quad \frac{x^2 - y^2}{3} = \frac{xy}{-h} \quad \text{...(i)}$$

and $x = 2y$ be a bisector.

It will satisfy Eq. (i)

$$\frac{4y^2 - y^2}{3} = \frac{2y^2}{-h}$$

$$\Rightarrow \quad \frac{3y^2}{3} = \frac{2y^2}{-h}$$

$$\Rightarrow \quad -h = 2$$

$$h = -2$$

8. (d) Bisector of $ax^2 + 2hxy + by^2 = 0$ are

$$\frac{x^2 - y^2}{a - b} = \frac{xy}{h} \quad \text{...(i)}$$

and of $ax^2 + 2hxy + by^2 + \lambda(x^2 + y^2) = 0$

i.e. $(a + \lambda)x^2 + 2hxy + (b + \lambda)y^2 = 0$ are

$$\frac{x^2 - y^2}{(a + \lambda) - (b + \lambda)} = \frac{xy}{h}$$

$$\Rightarrow \quad \frac{x^2 - y^2}{a - b} = \frac{xy}{h} \quad \text{...(ii)}$$

which is same equation as Eq. (i).

$\therefore$ For any λ belonging to real number the lines will have same bisectors.

9. (d) Given equation is

$$x^4 + x^3y + cx^2y^2 - xy^3 + y^4 = 0 \quad \text{...(i)}$$

Let m_1, m_2, m_3 and m_4 are slope of 4 lines represented by Eq. (i)

$$m_1 m_2 m_3 m_4 = 1$$

The product of the slope of the 4 lines represented by Eq. (i) and a pair of lines represented the bisectors of the angles between the other two, the product of slope of each pair is -1.

So let the equation of one pair be

$$ax^2 + 2hxy - ay^2 = 0,$$

then equation of its bisector is

$$\frac{x^2 - y^2}{2a} = \frac{xy}{h}$$

$$\Rightarrow \quad hx^2 - hy^2 - 2axy = 0$$

By hypothesis

$$x^4 + x^3y + cx^2y^2 - xy^3 + y^4$$

$$= (ax^2 + 2hxy - ay^2)(hx^2 - 2axy - hy^2)$$

$$= ah(x^4 + y^4) + 2(h^2 - a^2)(x^3y - xy^3) - 6ahx^2y^2$$

By comparing

$$ah = 1, 2(h^2 - a^2) = 1 \text{ and } -6ah = c$$

$$0 \Rightarrow c = -6 \quad \text{[put } ah = 1\text{]}$$

10. (d) We know that the pair of lines

$$(a^2-3b^2)x^2+8abxy+(b^2-3a^2)y^2=0$$

with the lines $ax+by+c=0$ forms an equilateral triangle by comparing with $Ax^2+2Hxy+By^2=0$

then $A=a^2-3b^2, 2H=8ab,$

$B=b^2-3a^2$

Now, $(A+3B)(3A+B)=(-8a^2)(-8b^2)$

$=(8ab)^2=(2H)^2=4H^2$

11. (b) Given equation

$12x^2+7xy-py^2-18x+qy+6=0$ represents a pair of straight line

$a=12, b=-p, c=6$

$h=\frac{7}{2}, \quad g=-9, \quad f=\frac{q}{2}$

$\therefore \quad \Delta=\begin{vmatrix} a & h & g \\ h & b & f \\ g & f & c \end{vmatrix}=0$

$$=\begin{vmatrix} 12 & \frac{7}{2} & -9 \\ \frac{7}{2} & -p & \frac{q}{2} \\ -9 & \frac{q}{2} & 6 \end{vmatrix}=0$$

$\Rightarrow \quad 12\left(-6p-\frac{q^2}{4}\right)-\frac{7}{2}\left(21+\frac{9q}{2}\right)-9\left(\frac{7q}{4}-9p\right)=0$

$\Rightarrow \quad -72p-3q^2-\frac{147}{2}-\frac{63q}{4}-\frac{63q}{4}+81p=0$

$\Rightarrow \quad 9p-3q^2-\frac{63}{2}q-\frac{147}{2}=0$

$\Rightarrow \quad 18p-6q^2-63q-147=0 \quad$ ...(i)

Lines are perpendicular.

$\therefore \quad a+b=0$

$12-p=0$

$p=12$

Put $p=12$ in Eq. (i)

$12\times 18-6q^2-63q-147=0$

$216-147-6q^2-63q=0$

$\Rightarrow \quad 6q^2+63q-69=0$

$\Rightarrow \quad 2q^2+21q-23=0$

$\Rightarrow \quad 2q^2+23q-2q-23=0$

$\Rightarrow \quad q(2q+23)-1(2q+23)=0$

$q=1, q=-\frac{23}{2}$

Hence, at $q=1$

$p+q+7$

$\Rightarrow \quad 12+1+7=20$

At $\quad q=-\frac{23}{2}$

$p+q+7$

$=12-\frac{23}{2}+7$

$\Rightarrow \quad \frac{38-23}{2}=\frac{15}{2}$

12. (c) Let $ax^2+by^2+2hxy+2gx+2fy+c=0$ represents two lines $lx+my+n=0$ and $l'x+m'y+n'=0$

Then, the combined equation represented by these lines is given by

$(lx+my+n)(l'x+m'y+n')=0$

So, it must be similar with the given equation.

On comparing, we get

$ll'=a, \; mm'=b, \; nn'=c$

$lm'+ml'=2n, ln'+l'n=2g,$

$mn'+m'n=2f$

According to the condition, the length of perpendicular drawn from origin to the lines are same.

So, $\frac{n}{\sqrt{l^2+m^2}}=\frac{n'}{\sqrt{l'^2+m'^2}}=\frac{nn'}{(l^2+m^2)(l'^2+m'^2)}$

Now, on eliminating l, m, l', m' n and n', we get the required condition

$f^4-g^4=c(bf^2-ag^2)$

13. (b) Given, $9x^2-24xy+16y^2-12x+16y-12=0$

$\Rightarrow \quad (3x-4y+2)(3x-4y-6)=0$

$\Rightarrow \quad 3x-4y+2=0$ and $3x-4y-6=0$

$\because$ Lines are parallel.

$\therefore \quad d=\left|\frac{(c_1-c_2)}{\sqrt{a^2+b^2}}\right|$

$=\left|\frac{2+6}{\sqrt{3^2+4^2}}\right|=\frac{8}{5}$

$\therefore$ The required points are $\left(\frac{8}{5},1\right)$ and $\left(1,\frac{8}{5}\right)$.

Now, equation of pair of lines passing through origin

$(y-m_1x)(y-m_2x)=0$

$\Rightarrow \quad \left(y-\frac{5}{8}x\right)\left(y-\frac{8}{5}x\right)=0$

$\Rightarrow \quad (8y-5x)(5y-8x)=0$

$\Rightarrow \quad 40y^2+40x^2-89xy=0$

$\left\{\because m_1=\frac{1}{8/5}, m_2=\frac{8}{5}\right\}$

14. (b) The straight lines represented by

$(y-mx)^2=a^2(1+m^2)$ are

$y-mx=\pm a\sqrt{1+m^2}$

$$y - mx = a\sqrt{1+m^2} \quad \text{...(i)}$$

and $$y - mx = -a\sqrt{1+m^2} \quad \text{...(ii)}$$

Similarly, the straight lines represented by

$(y - nx)^2 = a^2(1+n^2)$ are

$$y - nx = a\sqrt{1+n^2} \quad \text{...(iii)}$$

$$y - nx = -a\sqrt{1+n^2} \quad \text{...(iv)}$$

$\because$ The line Eqs. (i) and (ii) are parallel so the distance between Eqs. (i) and (ii) is

$$\Rightarrow \left|\frac{a\sqrt{1+m^2} + a\sqrt{1+m^2}}{\sqrt{1+m^2}}\right| = |2a| = 2a$$

Similarly lines (iii) and (iv) are parallel and distance between them is $|2a|$.

$\therefore$ Lines are parallel and distance between are same

$\therefore$ It forms a rhombus.

15. (b) Clearly the slope of PM is $-\frac{1}{2}$.

Here, PM is equally inclined with PQ and PR.

Let the slope of PQ be m, then the slope of PR will be $-\frac{1}{m}$.

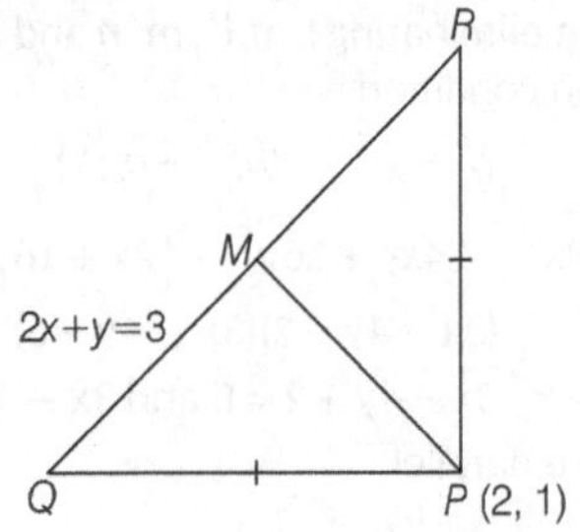

Thus, $$\frac{m - \left(-\frac{1}{2}\right)}{1 + m\left(-\frac{1}{2}\right)} = \frac{-\frac{1}{2} - \left(-\frac{1}{m}\right)}{1 + \left(-\frac{1}{2}\right)\left(-\frac{1}{m}\right)}$$

$$\Rightarrow \frac{2m+1}{2-m} = \frac{2-m}{2m+1}$$

$$\Rightarrow (2m+1)^2 = (2-m)^2$$

$$\Rightarrow 4m^2 + 4m + 1 = 4 - 4m + m^2$$

$$\Rightarrow 3m^2 + 8m - 3 = 0$$

$$\Rightarrow 3m^2 + 9m - m - 3 = 0$$

$$\Rightarrow 3m(m+3) - 1(m+3) = 0$$

$$(m+3)(3m-1) = 0$$

$$\Rightarrow m = -3, \frac{1}{3}$$

Thus, the equations of PQ and PR are

$$(y-1) = -3(x-2)$$

and $$y - 1 = \frac{1}{3}(x-2)$$

$$\Rightarrow y - 1 = -3(x-2),$$

$$3(y-1) = (x-2)$$

Joint equation

$$\{(y-1) + 3(x-2)\}\{3(y-1) - (x-2)\} = 0$$

$$\Rightarrow \{y - 1 + 3x - 6\}\{3y - 3 - x + 2\} = 0$$

$$\Rightarrow (y + 3x - 7)(3y - x - 1) = 0$$

$$\Rightarrow 3y^2 - xy - y + 9xy - 3x^2 - 3x - 21y + 7x + 7 = 0$$

$$\Rightarrow 3y^2 - 3x^2 + 8xy + 4x - 22y + 7 = 0$$

$$3x^2 - 3y^2 - 8xy - 4x + 22y - 7 = 0$$

16. (a) If the equation of lines is $y = mx$ and length of perpendicular drawn on it from the point (x_1, y_1) is d.

Then, $$d = \left|\frac{y_1 - mx_1}{\sqrt{1+m^2}}\right|$$

$$\Rightarrow \left(\frac{y_1 - mx_1}{\sqrt{1+m^2}}\right)^2 = d^2$$

$$\Rightarrow (y_1 - mx_1)^2 = d^2(1+m^2)$$

But $$m = \frac{y}{x}$$

$$\therefore \left(y_1 - \frac{y}{x}x_1\right)^2 = d^2\left(1 + \frac{y^2}{x^2}\right)$$

$$\Rightarrow (xy_1 - yx_1)^2 = d^2(x^2 + y^2)$$

17. (c) The equation of the given lines are

$$y - 1 = \tan\theta(x-1)$$

and $$y - 1 = \cot\theta(x-1)$$

So their joint equation is

$$\{(y-1) - \tan\theta(x-1)\}\{(y-1) - \cot\theta(x-1)\} = 0$$

$$\Rightarrow (y-1)^2 - (\tan\theta + \cot\theta)(x-1)(y-1) + (x-1)^2 = 0$$

$$\Rightarrow x^2 - (\tan\theta + \cot\theta)xy + y^2 + (\tan\theta + \cot\theta - 2)(x + y - 1) = 0 \quad \text{...(i)}$$

The given equation is

$$x^2 - (a+2)xy + y^2 + a(x + y - 1) = 0 \quad \text{...(ii)}$$

$\therefore$ By comparing Eqs. (i) and (ii),

$$\tan\theta + \cot\theta = a + 2$$

$$\Rightarrow \frac{1}{\sin\theta\cos\theta} = a + 2$$

$$\sin 2\theta = \frac{2}{a+2}$$

18. (b) Given equation is

$$4x^2 + 10xy + my^2 + 5x + 10y = 0$$

$$a = 4, b = m, h = 5, g = \frac{5}{2},$$

$$f = 5, \ c = 0$$

$\because$ For pair of straight lines

$$abc + 2fgh - af^2 - bg^2 - ch^2 = 0$$

$$\Rightarrow 4(m)(0) + 2(5)\left(\frac{5}{2}\right)(5) - 4(5)^2 - m\left(\frac{5}{2}\right)^2 - 0(5)^2 = 0$$

$$\Rightarrow \quad 125-100-\frac{25}{4}m=0$$

$$\Rightarrow \quad \frac{25}{4}m=25$$

$$m=4$$

and angle between lines

$$\theta=\tan^{-1}\left\{\frac{2\sqrt{h^2-ab}}{a+b}\right\}$$

$$=\tan^{-1}\left(\frac{2\sqrt{5^2-4m}}{4+m}\right)$$

$$=\tan^{-1}\left(\frac{2\sqrt{25-4m}}{m+4}\right)$$

$$=\tan^{-1}\left(\frac{2\sqrt{25-16}}{4+4}\right)$$

$$=\tan^{-1}\left(\frac{6}{8}\right)=\tan^{-1}\left(\frac{3}{4}\right)$$

19. (d) Given equation is

$$f=x^2+2xy-35y^2-4x+44y-12=0 \qquad \text{...(i)}$$

For point of intersection

$$\frac{\partial f}{\partial x}=0 \text{ and } \frac{\partial f}{\partial y}=0$$

$$2x+2y-4=0$$

and $\quad 2x-70y+44=0$

Solving $\quad 72y=48$

$$\Rightarrow \quad y=\frac{48}{72}$$

$$\Rightarrow \quad y=\frac{2}{3}$$

$$\therefore \quad x=\frac{4}{3}$$

$\therefore$ Point of intersection $\left(\frac{4}{3},\frac{2}{3}\right)$.

$\because$ Line $5x+\lambda y-8=0$ is concurrent with line (i)

$$\therefore \quad 5\left(\frac{4}{3}\right)+\lambda\left(\frac{2}{3}\right)-8=0$$

$$\Rightarrow \quad 20+2\lambda-24=0$$

$$\lambda=2$$

20. (d) Let the lines represented by given equation be

$$y=m_1x+c_1$$

and $\quad y=m_2x+c_2$

Then, $\quad ax^2+2hxy+by^2+2gx+2fy+c=0$

$$\Rightarrow \quad b(y-m_1x-c_1)(y-m_2x-c_2)=0$$

Comparing the coefficients of x^2, xy, x, y and constant term, we get

$$m_1m_2=\frac{a}{b} \text{ and } m_1m_2=-\frac{2h}{b}$$

$$m_1c_2+m_2c_1=\frac{2g}{b}$$

$$c_1+c_2=\frac{2f}{b}$$

and $\quad c_1c_2=\frac{c}{b}$

Also point of intersection of $y=m_1x+c_1$ and $y=m_2x+c_2$ is

$$\left(\frac{c_2-c_1}{m_1-m_2},\frac{m_1c_2-m_2c_1}{m_1-m_2}\right)$$

The square of distance of this point from origin is

$$\frac{(c_2-c_1)^2}{(m_1-m_2)^2}+\frac{(m_1c_2-c_1m_2)^2}{(m_1-m_2)^2}$$

$$=\frac{[(c_1+c_2)^2-4c_1c_2][(m_1c_2+m_2c_1)^2-4m_1m_2-c_1c_2]}{(m_1+m_2)^2-4m_1m_2}$$

Put the values,

$$\frac{-c(a+b)+f^2+g^2}{h^2-ab}$$

21. (c) Equation of angle bisector of homogeneous equation of pair of straight line $ax^2+2hxy+by^2$ is

$$\frac{x^2-y^2}{a-b}=\frac{xy}{h}$$

For $\quad x^2-4xy-5y^2=0$

$$a=1, h=-2, b=-5$$

So, equation of angle bisector is

$$\frac{x^2-y^2}{1-(-5)}=\frac{xy}{-2}$$

$$\Rightarrow \quad x^2-y^2=-3xy$$

$$\Rightarrow \quad x^2+3xy-y^2=0$$

So, combined equation of angle bisector is $x^2+3xy-y^2=0$.

22. (b) Given equation $\sqrt{3}y^2-xy-\sqrt{3}xy+x^2=0$

$$\Rightarrow \quad y(\sqrt{3}y-x)-x(\sqrt{3}y-x)=0$$

$$\Rightarrow \quad (y-x)(\sqrt{3}y-x)=0$$

$$y=x$$

$$m_1=1$$

$$\Rightarrow \quad \tan\theta_1=1$$

$$\theta_1=45°$$

$$y=\frac{1}{\sqrt{3}}x$$

$$m_2=\frac{1}{\sqrt{3}}$$

$$\Rightarrow \quad \tan\theta_2=\frac{1}{\sqrt{3}}$$

$$\theta_2=30°$$

Now, after rotation $\frac{\pi}{12} = 15°$ in anti-clockwise direction the new angles becomes

$$\theta_1' = 45°+15° = 60°$$

and $\theta_2' = 30°+15° = 45°$

$\therefore$ $m_1' = \tan 60° = \sqrt{3}$

and $m_2' = \tan 45° = 1$

$\therefore$ Equation of pair

$$(y-\sqrt{3}x)(y-x) = 0$$

$$\Rightarrow \quad y^2 - \sqrt{3}xy - xy + \sqrt{3}x^2 = 0$$

23. (a) We know that the equation of perpendicular drawn from origin on $ax^2 + 2hxy + by^2 + 2gx + 2fy + c = 0$ is

$$bx^2 - 2hxy + ay^2 = 0$$

$\therefore$ The required equation is given by

$$12x^2 + 10xy + 2y^2 = 0$$

$$\Rightarrow \quad 6x^2 + 5xy + y^2 = 0$$

24. (b) The given equation is

$$3y^2 - x^2 + 2\sqrt{3}x - 3 = 0$$

$$\Rightarrow \quad (\sqrt{3}y - x + \sqrt{3})(\sqrt{3}y + x - \sqrt{3}) = 0$$

$$\sqrt{3}y - x + \sqrt{3} = 0$$

and $\sqrt{3}y + x - \sqrt{3} = 0$

are separate equation

$$y = \frac{1}{\sqrt{3}}x - 1$$

and $y = -\frac{1}{\sqrt{3}}x + 1$

$$y = \tan 30° x - 1$$

and $y = (\tan 150°)x + 1$

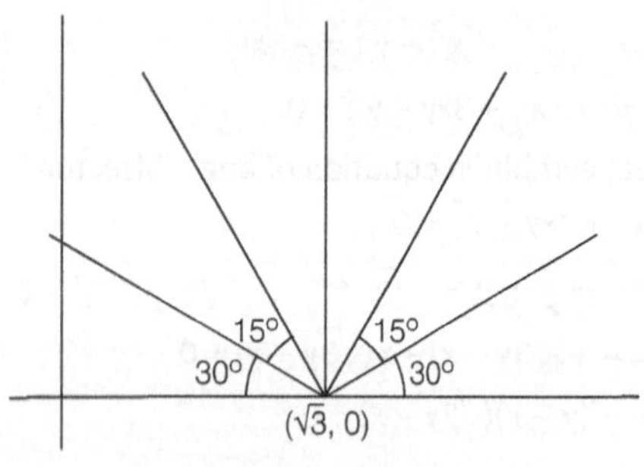

After rotation through an angle of 15° the lines are

$$y - 0 = \tan 45°(x - \sqrt{3})$$

and $y - 0 = \tan 135°(x - \sqrt{3})$

$$y = x - \sqrt{3}$$

and $y = -x + \sqrt{3}$

Combined equation is

$$(y - x + \sqrt{3})(y + x - \sqrt{3}) = 0$$

$$\Rightarrow \quad y^2 - x^2 + 2\sqrt{3}x - 3 = 0$$

25. (c) Given equation of bisector B_1 and B_2 is

$$3x^2 - 8xy - 3y^2 + 10x + 20y - 25 = 0$$

$$\Rightarrow \quad (3x + y - 5)(x - 3y + 5) = 0$$

Now, point of intersection P.

Solve $3x + y - 5 = 0$ and $x - 3y + 5 = 0$, we get $x = 1, y = 2$

$\therefore$ P is (1, 2).

Now, L_1 passes through the origin.

$\therefore$ $y = 2x$ and $m_1 = 2$

Slope of the bisector : L_3 $x - 3y + 5$ is $m_3 = \frac{1}{3}$

Slope of L_2 is m_2 given by

$$\frac{m_1 - m_3}{1 + m_1 m_3} = \frac{m_3 - m_2}{1 + m_2 m_3}$$

$$\Rightarrow \quad \frac{2 - \frac{1}{3}}{1 + \frac{2}{3}} = \frac{\frac{1}{3} - m_2}{1 + \frac{m_2}{3}}$$

$$\Rightarrow \quad 1 - 3m_2 = 3 + m_2$$

$$m_2 = -\frac{1}{2}$$

$\therefore$ Equation of L_2 is $2(y-2) = -1(x-1)$

$$\Rightarrow \quad x + 2y - 5 = 0$$

26. (b) Let $y = m(x-1)$ be the line meeting the curve in the points A and B.

The combined equation of the pair of lines OA and $OB(0)$ is the origin.

$$\Rightarrow \quad x^2 + y^2 + (6x - 10y)\frac{(mx - y)}{m} + 1 \cdot \frac{(mx-y)^2}{m^2} = 0$$

$\because \angle AOB = 90°$, from the above equation, we have

Coefficient of x^2 + Coefficient of $y^2 = 0$

$$\Rightarrow \quad (1 + 6 + 1) + \left(1 + \frac{10}{m} + \frac{1}{m^2}\right) = 0$$

$$\Rightarrow \quad 9m^2 + 10m + 1 = 0$$

$$\Rightarrow \quad (9m + 1)(m + 1) = 0$$

$$\Rightarrow \quad m = -\frac{1}{9} \text{ or } -1$$

27. (b) $x^2 + y^2 = a^2$

represent a circle with centre (0, 0) and radius a and $x^2 + y^2 - ax - ay = 0$ also represents a circle with centre $\left(\frac{a}{2}, \frac{a}{2}\right)$ and radius $\frac{a}{\sqrt{2}}$.

From to the figure, we can directly say the lines joining origin and points of intersection are $x = 0$ and $y = 0$

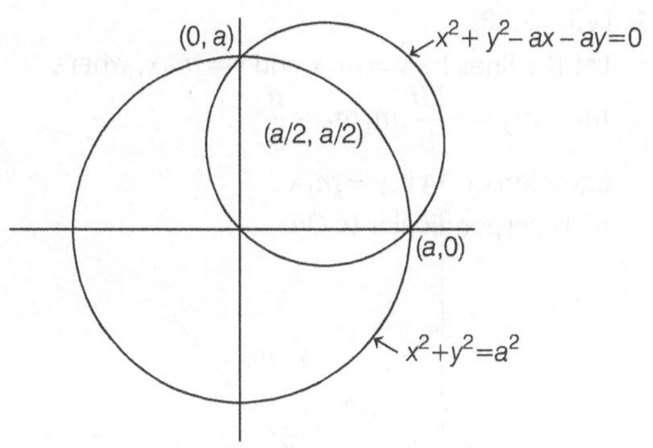

$\therefore$ Combined equation is $xy = 0$.

28. (b) Given line

$$y = 3x + 2$$

Given equation $x^2 + 2xy + 3y^2 + 4x + 8y - 11 = 0$

Now, by homogenising, we get

$$x^2 + 2xy + 3y^2 + 4x\left(\frac{y-3x}{2}\right) + 8y\left(\frac{y-3x}{2}\right)$$

$$-11\left(\frac{y-3x}{2}\right)^2 = 0$$

$$\Rightarrow \quad x^2 + 2xy + 3y^2 + 2xy - 6x^2 + 4y^2$$

$$-12xy - \frac{11}{4}(y^2 + 9x^2 - 6xy) = 0$$

$$\Rightarrow \quad x^2\left(1 - 6 - \frac{99}{4}\right) + y^2\left(3 + 4 - \frac{11}{4}\right)$$

$$+xy\left(-8 + \frac{66}{4}\right) = 0$$

$$\Rightarrow \quad -\frac{119}{4}x^2 + \frac{17}{4}y^2 + \frac{34}{4}xy = 0$$

$$\Rightarrow \quad -7x^2 + 2xy + y^2 = 0$$

$$\therefore \quad \tan\theta = \left|\frac{2\sqrt{h^2 - ab}}{a+b}\right|$$

$$= \left|\frac{2\sqrt{1^2 - (-7)}}{-7+1}\right| = \frac{2\sqrt{2}}{3}$$

$$\Rightarrow \quad \theta = \tan^{-1}\left(\frac{2\sqrt{2}}{3}\right)$$

29. (c) Given line $y = 2\sqrt{2}x + c$

$$\Rightarrow \quad \frac{y - 2\sqrt{2}x}{c} = 1$$

Circle $x^2 + y^2 = 2$

By homogenisation

$$x^2 + y^2 = 2\left(\frac{y - 2\sqrt{2}x}{c}\right)^2$$

$$\Rightarrow \quad c^2(x^2 + y^2) = 2(y^2 + 8x^2 - 4\sqrt{2}xy)$$

$$\Rightarrow \quad x^2(c^2 - 16) + y^2(c^2 - 2) + 8\sqrt{2}xy = 0 \qquad \text{...(i)}$$

$\because$ Eq. (i) represent a pair of straight line and lines are perpendicular to each other

$$\Rightarrow \quad a + b = 0$$

$$\Rightarrow \quad c^2 - 16 + c^2 - 2 = 0$$

$$\Rightarrow \quad 2c^2 - 18 = 0$$

$$\Rightarrow \quad c^2 - 9 = 0$$

$$\Rightarrow \quad c = \pm 3$$

Numerical Type Questions

1. (0.7)

Given lines are $3x^2 + 7xy + 2y^2 = 0$

$$\Rightarrow \quad 3x^2 + 6xy + xy + 2y^2 = 0$$

$$\Rightarrow \quad 3x(x + 2y) + y(x + 2y) = 0$$

$$\Rightarrow \quad (x + 2y)(3x + y) = 0$$

$$\Rightarrow \quad x + 2y = 0 \text{ and } 3x + y = 0$$

Thus, the equation of AL and AM are

$$2x - y + \lambda = 0$$

$$3y - x + \mu = 0$$

which passes through $A\ (1, 1)$

$$\lambda = -1 \quad \Big| \quad \mu = -2$$

$\therefore$ Equation are $2x - y - 1 = 0, 3y - x - 2 = 0$

Solving

$x + 2y = 0$ and $3y - x - 2 = 0$, we get

$$M\left(\frac{2}{5}, \frac{-1}{5}\right)$$

$3x + y = 0$ and $2x - y - 1 = 0$, we get

$$L\left(\frac{-1}{5}, \frac{3}{5}\right)$$

$\therefore$ Area of the quadrilateral $ALMO$

$$= \frac{1}{2}\left|\begin{array}{cc} 1 & 1 \\ -1/5 & 3/5 \\ 0 & 0 \\ 2/5 & -1/5 \\ 1 & 1 \end{array}\right|$$

$$= \frac{1}{2}\left(\frac{3 + 1 + 2 + 1}{5}\right) = \frac{7}{10}$$

$= 0.7$ sq unit

2. (2.6)

Given lines are

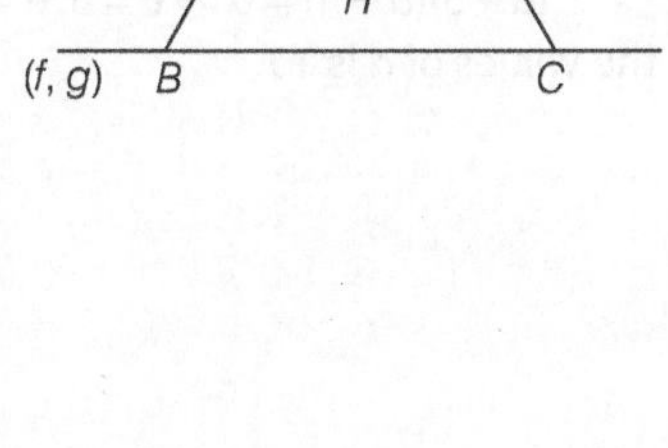

$$y^2 - 8xy - 9x^2 = 0$$

$$\Rightarrow \quad y^2 - 9xy + xy - 9x^2 = 0$$

$$\Rightarrow \quad y(y-9x) + x(y-9x) = 0$$

$$\Rightarrow \quad (y-9x)(y+x) = 0$$

$$\Rightarrow \quad y + x = 0,\ y - 9x = 0$$

Equations of AB and AC are

$$x + 9y = \alpha + 9\beta$$

and $$x - y = \alpha + \beta$$

Here, B and C are the images of A w.r.t. the lines $y = 9x$ and $y = -x$.

So, $C = (-\beta, -\alpha)$

and $$B = \left(\frac{9\beta - 40\alpha}{41}, \frac{40\beta + 9\alpha}{41}\right)$$

Now, P, B and C are collinear, so

$$\begin{vmatrix} f & g & 1 \\ \dfrac{9\beta - 40\alpha}{41} & \dfrac{40\beta + 9\alpha}{41} & 1 \\ -\beta & -\alpha & 1 \end{vmatrix} = 0$$

$$\Rightarrow \quad \begin{vmatrix} f & g & 1 \\ 9\beta - 40\alpha & 40\beta + 9\alpha & 41 \\ \beta & \alpha & -1 \end{vmatrix} = 0$$

$$\Rightarrow \quad \begin{vmatrix} f & g & 1 \\ 9\beta - 40\alpha & 40\beta + 9\alpha & 41 \\ f+\beta & g+\alpha & 0 \end{vmatrix} = 0$$

$$\Rightarrow \quad \begin{vmatrix} 9\beta - 40\alpha & 40\beta + 9\alpha \\ f+\beta & g+\alpha \end{vmatrix}$$

$$-41 \begin{vmatrix} f & g \\ f+\beta & g+\alpha \end{vmatrix} = 0$$

$$\Rightarrow \quad (g+\alpha)(9\beta - 40\alpha) - (f+\beta)(40\beta + 9\alpha) - 41(f\alpha - g\beta) = 0$$

$$\Rightarrow \quad 40(\alpha^2 + \beta^2) + 10(4g + 5f)\alpha + 10(4f - 5g)\beta = 0$$

$$\Rightarrow \quad 4(\alpha^2 + \beta^2) + (4g + 5f)\alpha + (4f - 5g)\beta = 0$$

Hence, the locus $A(\alpha, \beta)$ is

$$4(x^2 + y^2) + (4g + 5f)x + (4f - 5g)y = 0$$

Given, $a(x^2 + y^2) + (bg + cf)x + (cf - bg)y = 0$

$$\therefore \quad a = 4, b = 4, c = 5$$

Now, $$\frac{a+b+c}{5} = \frac{4+4+5}{5} = \frac{13}{5} = 2.6$$

3. (1)

As we know that, if $ax^3 + bx^2y + cxy^2 + dy^3 = 0$ represents two perpendicular lines then

$a^2 + ac + bd + d^2 = 0$

Now, given equation is $ax^3 - 9x^2y - xy^2 + 4y^3 = 0$

$$\therefore \quad a^2 - a - 36 + 16 = 0$$

$$\Rightarrow \quad a^2 - a - 20 = 0$$

$$\Rightarrow \quad (a-5)(a+4) = 0 \Rightarrow a = 5, -4$$

$\therefore$ Sum of the values of a is $+1$.

4. $(\sqrt{5} = 2.23)$

Let the lines be $y = m_1x$ and $y = m_2x$, where

$m_1 + m_2 = -\dfrac{2h}{b}, m_1m_2 = \dfrac{a}{b}$

Equation of OA is $y = m_1x$

BE is perpendicular to OA.

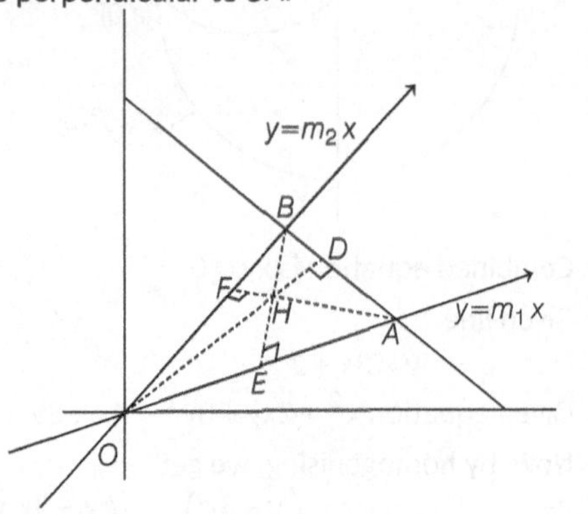

So, BE $x + m_1y + \lambda_1 = 0$ which is passing through $H(l, m)$, we get

$$l + mm_1 + \lambda_1 = 0 \Rightarrow \lambda_1 = -(l + mm)$$

Thus, $BE: x + m_1y - (l + mm_1) = 0$

Similarly $AF: x + m_2y - (l + mm_2) = 0$

Solving OA and AF, we get

$$A = \left(\frac{l + mm_2}{1 + m_1m_2}, \frac{m_1(l + mm_2)}{1 + m_1m_2}\right)$$

Solving OB and BE, we get

$$B = \left(\frac{l + mm_1}{1 + m_1m_2}, \frac{m_2(l + mm_2)}{1 + m_1m_2}\right)$$

Equation of AB is $lx + my + \lambda_3 = 0$ which is passing through A.

So, $$l\left(\frac{l + mm_2}{1 + m_1m_2}\right) + m\left(\frac{m_1(l + mm_2)}{1 + m_1m_2}\right) + \lambda_3 = 0$$

$$\lambda_3 = -\left(\frac{l^2 + lm(m_2 + m_1) + m^2(m_1m_2)}{1 + m_1m_2}\right)$$

$$= -\frac{\left[l^2 - \left(\dfrac{2h}{b}\right)lm + \dfrac{a}{b}m^2\right]}{1 + \dfrac{a}{b}}$$

$$= -\left(\frac{bl^2 - 2hlm + am^2}{a+b}\right)$$

Hence, equation of third side is

$$lx + my = \left(\frac{bl^2 - 2hlm + am^2}{a+b}\right)$$

$$\Rightarrow \quad (lx + my)(a+b) = (bl^2 - 2hlm + am^2)$$

$$\lambda = 1, \mu = 2$$

$$\Rightarrow \quad \sqrt{\lambda^2 + \mu^2} = \sqrt{1+4} = \sqrt{5} = 2.23$$

5. (4)
Given pair of straight lines

$$x^2-3y^2-2xy+8y-4=0$$

$\Rightarrow \quad (x-3y+2)(x+y-2)=0$

Now, $(-5,-1)$ lies on $x-3y+2=0$

In limiting case line passing through $(-5,-1)$ can be parallel to $x+y-2=0$ i.e. $m>-1$

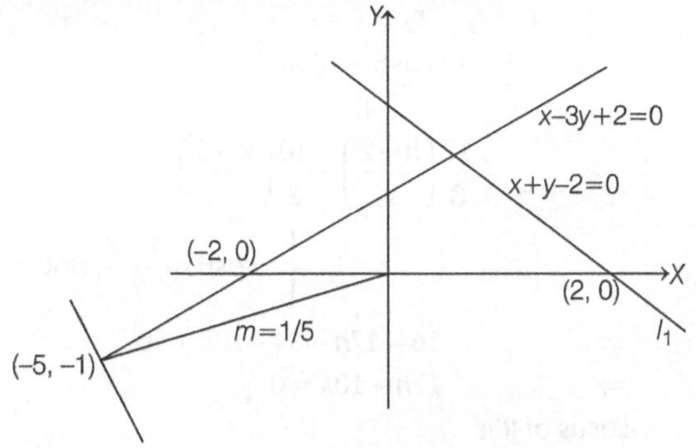

and maximum slope can occur if it passes through (0, 0),

i.e. $m<\frac{1}{5}$

$\Rightarrow \quad m\in\left(-1,\frac{1}{5}\right)$

$\Rightarrow \quad a=-1$ and $b=\frac{1}{5}$

$\therefore \quad a+\frac{1}{b}=-1+5=4$

6. (8)
Given pair of lines $x^2-(y^2-4y+4)=0$

$\Rightarrow \quad x^2-(y-2)^2=0$

$\Rightarrow \quad (x-y+2)(x+y-2)=0$

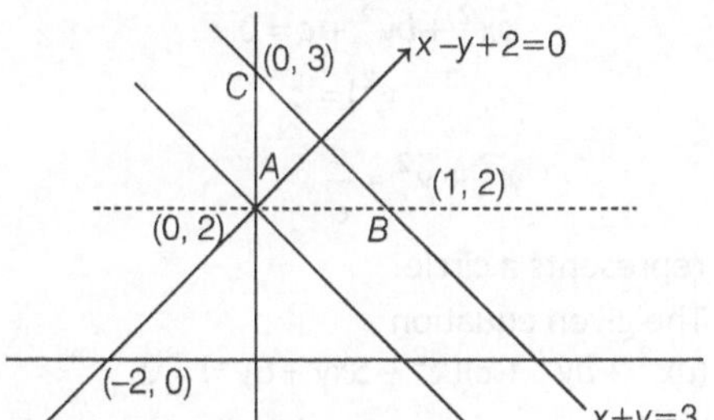

$\therefore$ Equation lines which bisect the angle between the lines

$$x-y+2=0$$

and $\quad x+y-2=0$

is $y=2$ and $x=0$

Now, the vertices of triangle with $y=2, x=0$ and $x+y=3$

$$A=(0,2), B=(1,2), C=(0,3)$$

$\because ABC$ is right angled triangle.

$\therefore \quad A=\frac{1}{2}\times AB\times AC$

$=\frac{1}{2}\times 1\times 1=\frac{1}{2}$

$\therefore \quad 16A=16\times\frac{1}{2}=8$ sq units

7. (0.5)
Let the lines $y=m_1x$ and $y=m_2x$

where, $m_1+m_2=-\frac{2h}{b}$

$$m_1m_2=\frac{a}{b}$$

Thus, the point O, A and B are

$$O(0,0), A=\left(\frac{1}{l+mm_1},\frac{m_1}{l+mm_1}\right)$$

and $\quad B=\left(\frac{1}{l+mm_2},\frac{m_2}{l+mm_2}\right)$

There centroid $(\alpha,\beta)=\left(\frac{1}{3}\left\{\frac{1}{l+mm_1}+\frac{1}{l+mm_2}\right\},\right.$

$\left.\frac{1}{3}\left\{\frac{m_1}{l+mm_1}+\frac{m_2}{l+mm_2}\right\}\right)$

$\therefore \quad \alpha=\frac{1}{3}\left(\frac{2l+(m_1+m_2)m}{(l+mm_1)(l+mm_2)}\right)$

and $\quad \beta=\frac{1}{3}\left(\frac{l(m_1+m_2)+2mm_1m_2}{(l+mm_1)(l+mm_2)}\right)$

$$\alpha=\frac{2}{3}\left(\frac{bl-hm}{bl^2-2hlm+am^2}\right)$$

and $\quad \beta=\frac{2}{3}\left(\frac{am-hl}{bl^2-2hlm+am^2}\right)$

Thus, $\quad \frac{\alpha}{bl-hm}=\frac{\beta}{am-hl}=\frac{2}{3}\left(\frac{1}{bl^2-2hlm+am^2}\right)$

Given, $\quad \frac{\alpha}{bl-hm}=\frac{\beta}{am-hl}=\frac{p}{q}\left(\frac{l}{bl^2-2hlm+am^2}\right)$

$\therefore \quad \frac{p+q}{10}=\frac{2+3}{10}=\frac{5}{10}=0.5$

Level 3

1. (b) Let $y=m_1x$ and $y=m_2x$ be the lines represented by $ax^2+2hxy+by^2=0$

$\therefore \quad m_1+m_2=-\frac{2h}{b} \quad$ ...(i)

$m_1m_2=\frac{a}{b} \quad$ ...(ii)

Syppose $A(\alpha,0)$ and the line $x=\alpha$ meets $y=m_1x$ at B and $y=m_2x$ at C

Thus, $B=(\alpha, m_1\alpha)$ and $C(\alpha, m_2\alpha)$

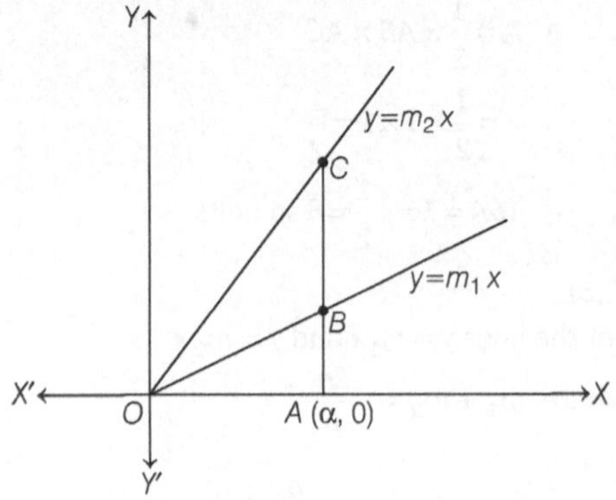

$$AB = BC$$

$$\Rightarrow \quad \frac{m_2\alpha + 0}{2} = m_1\alpha$$

$$\Rightarrow \quad m_2 = 2m_1$$

By Eq. (i), $\quad 3m_1 = -\frac{2h}{b}$

$$\Rightarrow \quad m_1 = -\frac{2h}{3b}$$

and by Eq. (ii), $2m_1^2 = \frac{a}{b}$

$$2\left(\frac{-2h}{3b}\right)^2 = \frac{a}{b}$$

$$\Rightarrow \quad 8h^2 = 9ab$$

2. (a) We have, $2x^2 - 5xy + 2y^2 = 0$

$$\Rightarrow \quad 2x^2 - 4xy - xy + 2y^2 = 0$$

$$\Rightarrow \quad 2x(x-2y) - y(x-2y) = 0$$

$$\Rightarrow \quad (x-2y)(2x-y) = 0$$

$\Rightarrow \quad x - 2y = 0$ and $2x - y = 0$

Let $P=(a,b), Q=(c,d)$ and $R=(h,k)$

and $RB = r, PB = r_1$ and $QB = r_2$

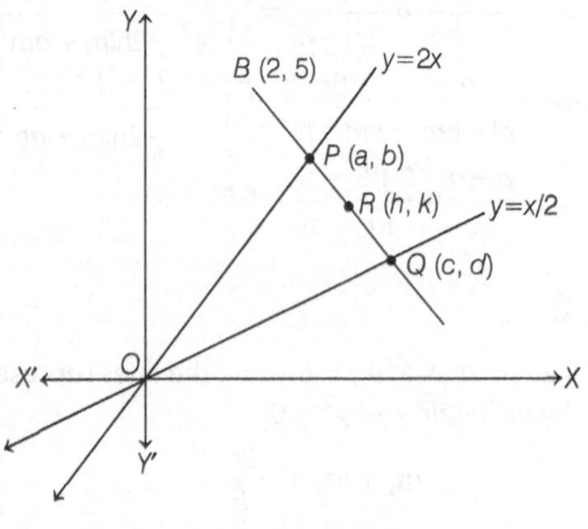

So, $\quad P=(2+r_1\cos\theta, 5+r_1\sin\theta)$

$\quad Q=(2+r_2\cos\theta, 5+r_2\sin\theta)$

and $\quad R=(2+r\cos\theta, 5+r\sin\theta)$

$\because$ P lies on $y=2x$

$$\therefore \quad \frac{1}{r_1} = 2\cos\theta - \sin\theta$$

Also Q lies on $x=2y$

$$\Rightarrow \quad \frac{1}{r_2} = \frac{\cos\theta - 2\sin\theta}{8}$$

Given, r_1, r, r_2 are in HP.

$$\therefore \quad \frac{2}{r} = \frac{1}{r_1} + \frac{1}{r_2}$$

$$\frac{2}{r} = \frac{17\cos\theta - 10\sin\theta}{8}$$

$$= \frac{17}{8}\left(\frac{h-2}{r}\right) - \frac{10}{8}\left(\frac{k-5}{r}\right)$$

$$\left\{\therefore \cos\theta = \frac{h-2}{r}, \sin\theta = \frac{k-5}{r}\right\}$$

$$\Rightarrow \quad 16 = 17h - 34 - 10k + 50$$

$$\Rightarrow \quad 17h - 10k = 0$$

Locus of R is

$$\Rightarrow \quad 17x - 10y = 0$$

3. (b) Let a and b be non-zero real numbers.

Therefore, the given equation

$$(ax^2 + by^2 + c)(x^2 - 5xy + 6y^2) = 0$$

$$\Rightarrow \quad x^2 - 5xy + 6y^2 = 0$$

$$\Rightarrow \quad (x-2y)(x-3y) = 0$$

$\Rightarrow \quad x = 2y$ and $x = 3y$

represents two straight lines passing through origin

or

$$ax^2 + by^2 + c = 0$$

When $c=0$ and a and b are of same sign.

Then, $ax^2 + by^2 + c = 0, x = 0, y = 0$

which is a point specified as the origin.

When $a=b$ and c is sign opposite to that of a

$$ax^2 + by^2 + c = 0$$

$$\Rightarrow \quad a(x^2 + y^2) = c$$

$$\Rightarrow \quad x^2 + y^2 = \frac{c}{a}$$

represents a circle.

$\therefore$ The given equation

$(ax^2 + by^2 + c)(x^2 - 5xy + 6y^2) = 0$

may represents two straight lines and a circle.

4. (b) Let the equation of lines represented by $ax^2 + 2hxy + by^2 = 0$ be $y - m_1x = 0$ and $y - m_2x = 0$ and one diagonal AC be $lx + my = 1$

Therefore, $m_1 + m_2 = -\frac{2h}{b}$

and $\quad m_1m_2 = \frac{a}{b}$

Now, solving the equation of OA and OC with the line AC, we get the coordinates of

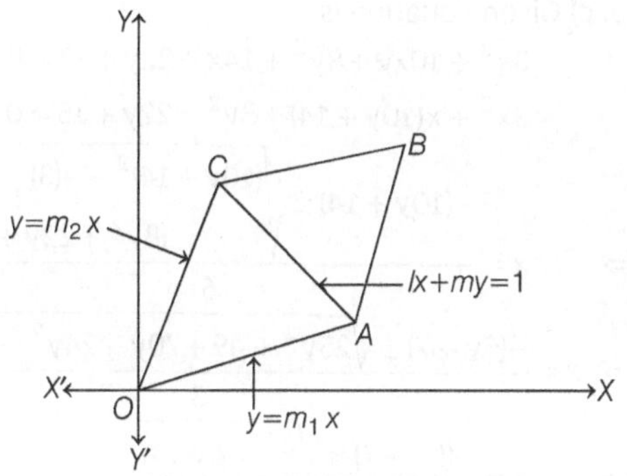

$$A\left(\frac{1}{l+mm_1}, \frac{m_1}{l+mm_1}\right)$$

and $$C\left(\frac{1}{l+mm_2}, \frac{m_2}{l+mm_2}\right)$$

Now, find the coordinates of mid-point of AC and the equation of diagonal through this mid-point and origin.

The required equation is $x(am-hl)=(lb-mh)y$

5. (b) Given equation is

$$(x^2+y^2)\sin^2\alpha=(x\cos\beta-y\sin\beta)^2$$

$$x^2(\sin^2\alpha-\cos^2\beta)+2xy\sin\beta\cos\beta +y^2(\sin^2\alpha-\sin^2\beta)=0 \quad \text{...(i)}$$

The homogeneous equation of second degree is

$$ax^2+2hxy+by^2=0 \quad \text{...(ii)}$$

On comparing Eqs. (i) and (ii),

$$a=\sin^2\alpha-\cos^2\beta$$
$$h=\sin\beta\cos\beta$$
$$b=\sin^2\alpha-\sin^2\beta$$

Let the angle between the lines representing by Eq. (i) is θ.

$$\therefore \quad \tan\theta=2\frac{\sqrt{h^2-ab}}{|a+b|}$$

$$=2\frac{\sqrt{\sin^2\beta\cos^2\beta-(\sin^2\alpha-\cos^2\beta)(\sin^2\alpha-\sin^2\beta)}}{|\sin^2\alpha-\cos^2\beta+\sin^2\alpha-\sin^2\beta|}$$

$$=\frac{2\sqrt{\sin^2\beta\cos^2\beta-\sin^4\alpha+\sin^2\alpha\sin^2\beta+\sin^2\alpha\cos^2\beta-\sin^2\beta\cos^2\beta}}{|2\sin^2\alpha-1|}$$

$$=\frac{2\sqrt{\sin^2\alpha(-\sin^2\alpha+\sin^2\beta+\cos^2\beta)}}{|2\sin^2\alpha-1|}$$

$$=\frac{2\sqrt{\sin^2\alpha(1-\sin^2\alpha)}}{|-\cos 2\alpha|}$$

$$=\frac{2\sin\alpha\cos\alpha}{\cos 2\alpha}=\tan 2\alpha$$

$$\theta=2\alpha$$

6. (d) The combined equation of the bisectors of the angles between the lines of the first pair is

$$\frac{x^2-y^2}{2-1}=\frac{xy}{3}$$

$$\Rightarrow \quad x^2-y^2=\frac{1}{3}xy$$

and the combined equation of the bisectors of the angles between the lines of the second pair is

$$\Rightarrow \quad \frac{x^2-y^2}{4-1}=\frac{xy}{9}$$

$$\Rightarrow \quad x^2-y^2=\frac{1}{3}xy$$

It is clear that the two pairs are equally inclined to each other.

Hence, the angle between l_2 and m_1 is α.

7. (b) Given equation

$$5x^2+12xy-6y^2+4x-2y+3=0$$

and line $x+ky=1$

After homogenising, we get

$$5x^2+12xy-6y^2+4x(x+ky) -2y(x+ky)+3(x+ky)^2=0$$

$$\Rightarrow \quad 9x^2-6y^2-2ky^2+10xy+4kxy +3(x^2+k^2y^2+2kxy)=0$$

$$\Rightarrow \quad 12x^2+y^2(3k^2-2k-6)+xy(10k+10)=0$$

$$\therefore \quad m_1+m_2=-\frac{2h}{b}$$

$$=\frac{-(10k+10)}{3k^2-2k-6}$$

$\therefore$ Equally inclined

$$m_1+m_2=0$$
$$10k+10=0$$
$$k=-1$$

8. (a,b,d)

Equation of the lines given by

$$x^2+2hxy+y^2=0$$

be $y=m_1x$ and $y=m_2x$

$\therefore$ These make an angle α with $y+x=0$ whose slope is -1.

So, $$\frac{m_1+1}{1-m_1}=\tan\alpha=\frac{-1-m_2}{1-m_2}$$

$$m_1+m_2=\frac{(\tan\alpha-1)^2+(\tan\alpha+1)^2}{\tan^2\alpha-1}$$

$$=-\frac{2\sec^2\alpha\times\cos^2\alpha}{\cos^2\alpha}$$

$$\Rightarrow \quad -2\sec 2\alpha=-2h$$

$$\Rightarrow \quad \sec 2\alpha=h$$

$$\Rightarrow \quad \cos 2\alpha = \frac{1}{h}$$

$$\Rightarrow \quad 2\cos^2 \alpha - 1 = \frac{1}{h}$$

$$\Rightarrow \quad \cos\alpha = \sqrt{\frac{1+h}{2h}}$$

$$\Rightarrow \quad \cot\alpha = \sqrt{\frac{h+1}{h-1}}$$

9. (a,d) Combined equations of the sides of the triangle is

$$(x^2 - y^2)(2x + 3y - 6) = 0$$

Separate equations of sides are

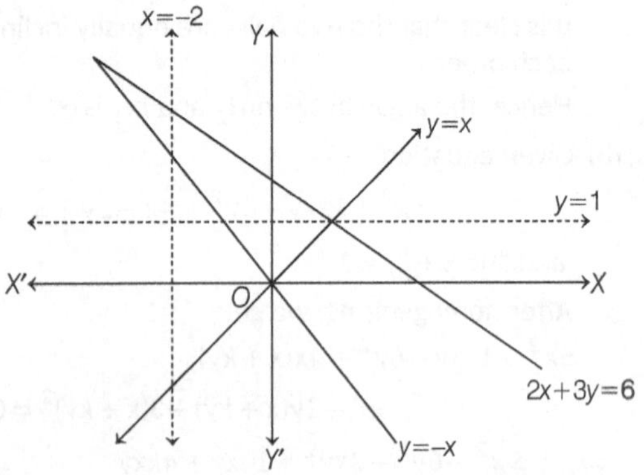

$$x - y = 0,\ x + y = 0 \text{ and } 2x + 3y - 6 = 0$$

The point $(-2, a)$ moves on the line $x = -2$
and $(b, 1)$ moves on the line $y = 1$
From the figure,
the y-coordinate of points of intersection of $x = -2$ with $y = -x$ and $2x + 3y = 6$ give the range of values of a.
The x-coordinates of the points of intersection of $y = 1$ with $y = -x$ and $y = x$ give the range of values of b.

10. (a, b, c, d)

Let $\frac{2}{3}x^2 + \frac{p}{3}xy + y^2 = (y - mx)(y - m'x)$

and $-\frac{2}{3}x^2 + \frac{q}{-3}xy + y^2 = \left(y + \frac{1}{m}x\right)(y - m'x)$

Then, $m + m' = -\frac{p}{3},\ mm' = \frac{2}{3}$ ...(i)

$$\frac{1}{m} - m' = -\frac{q}{3},\ -\frac{m'}{m} = -\frac{2}{3} \qquad \text{...(ii)}$$

From Eqs. (i) and (ii),

$$m^2 = 1 \Rightarrow m = \pm 1$$

If $\quad m = 1,\ m' = \frac{2}{3}$

So, $\quad p = -5,\ q = -1$

and $\quad m = -1,\ m' = -\frac{2}{3}$

So, $\quad p = 5,\ q = 1$

11. (b, c) Given equation is

$$3x^2 + 10xy + 8y^2 + 14x + 22y + 15 = 0$$

$$3x^2 + x(10y + 14) + 8y^2 + 22y + 15 = 0$$

$$\Rightarrow \quad x = \frac{-(10y+14) \pm \sqrt{(10y+14)^2 - 4(3)(8y^2 + 22y + 15)}}{6}$$

$$\Rightarrow \quad x = \frac{-(5y+7) \pm \sqrt{25y^2 + 49 + 70y - 24y^2 - 66y - 45}}{3}$$

$$\Rightarrow \quad x = \frac{-(5y+7) \pm \sqrt{y^2 + 4y + 4}}{3}$$

$$\Rightarrow \quad 3x = -(5y+7) \pm (y+2)$$

$$\Rightarrow \quad 3x = -5y - 7 \pm (y+2)$$

Taking (+)

$$3x = -5y - 7 + y + 2$$
$$3x = -4y - 5$$
$$3x + 4y + 5 = 0$$

Taking (–)

$$3x = -5y - 7 - y - 2$$
$$3x + 6y + 9 = 0$$
$$(x + 2y + 3) = 0$$

$\therefore$ Lines are $3x + 4y + 5 = 0$

and $\quad x + 2y + 3 = 0$

$\therefore \quad m_1 = -\frac{3}{4}$

and $\quad m_2 = -\frac{1}{2}$

$$m_1 + m_2 = -\frac{3}{4} - \frac{1}{2} = -\frac{5}{4}$$

$$m_1 m_2 = \frac{3}{8}$$

$$\Rightarrow \quad \tan\theta = \left|\frac{-\frac{1}{2} + \frac{3}{4}}{1 + \frac{3}{8}}\right| = \frac{2}{11}$$

$$\therefore \quad \sin\theta = \frac{2}{5\sqrt{5}}$$

12. (a,b,c,d)

Equation of the lines joining the origin to the points of intersection of the given lines is

$$3x^2 + mxy - 4x + 1 = 0$$

$$3x^2 + mxy - 4x(2x + y) + 1(2x + 3y)^2 = 0$$

[by homogenisation]

which are perpendicular for all values of m.

Paragraph-1 (13-15)

Given lines $2x^2 - 5xy + 2y^2 = 0$ be the two sides of a parallelogram

$$2x^2 - 5xy + 2y^2 = 0$$

$$2x^2 - 4xy - xy + 2y^2 = 0$$
$$2x(x-2y) - y(x-2y) = 0$$
$$(x-2y)(2x-y) = 0$$
$$x - 2y = 0, \quad 2x - y = 0$$
$$y = \frac{x}{2} \text{ and } y = 2x$$

$\therefore$ Points B is

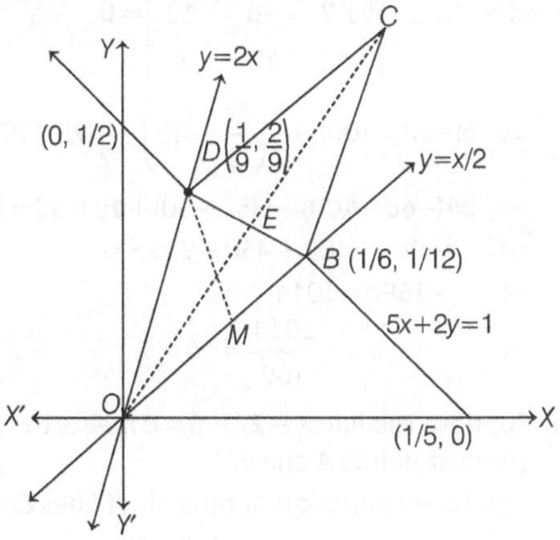

Solving $y = \frac{x}{2}$ and $5x + 2y = 1$

$$5x + x = 1$$
$$\Rightarrow \quad 6x = 1 \Rightarrow x = \frac{1}{6}$$
$$\Rightarrow \quad y = \frac{1}{12}$$
$$\therefore \quad B\left(\frac{1}{6}, \frac{1}{12}\right)$$

and D solving, $y = 2x$ and $5x + 2y = 1$

$$9x = 1$$
$$x = \frac{1}{9} \text{ and } y = \frac{2}{9}$$
$$D\left(\frac{1}{9}, \frac{2}{9}\right)$$

$$\text{Mid-point of } BD = \left(\frac{\frac{1}{6}+\frac{1}{9}}{2}, \frac{\frac{1}{12}+\frac{2}{9}}{2}\right)$$
$$= \left(\frac{5}{36}, \frac{11}{72}\right)$$

13. (a) Equation of one side is $x - 2y = 0$

14. (b) Equation of diagonal OC is

$$y - 0 = \frac{11/72}{5/36}(x - 0)$$
$$y = \frac{11}{10}x$$
$$10y - 11x = 0$$
$$\text{or} \quad 11x - 10y = 0$$

15. (a) Area of parallelogram = Base × Height = $OB \times DM$

$$OB = \sqrt{\left(\frac{1}{6}\right)^2 + \left(\frac{1}{12}\right)^2}$$
$$= \frac{1}{6}\sqrt{1 + \frac{1}{4}} = \frac{\sqrt{5}}{12}$$
$$DM = \left|\frac{\frac{1}{9} - \frac{4}{9}}{\sqrt{1+4}}\right| \quad \left\{\because \text{line } x - 2y = 0, \text{ point}\left(\frac{1}{9}, \frac{2}{9}\right)\right\}$$
$$= \frac{1}{3\sqrt{5}}$$
$$\therefore \text{Area} = \frac{\sqrt{5}}{12} \times \frac{1}{3\sqrt{5}} = \frac{1}{36} \text{ sq units}$$

Paragraph 2 (Q. 16-18)

16. (b) Suppose $lx + my = 1$ is a line meeting the curve $3x^2 - y^2 - 2x + 4y = 0$ at points A and B. Therefore, the combined equation of the pair of lines OA and OB is

$$3x^2 - y^2 - (2x - 4y)(lx + my) = 0$$
$$\because \quad \angle AOB = 90^\circ$$
$$\therefore \quad \text{Coefficient of } x^2 + \text{Coefficient of } y^2 = 0$$
$$3 - 2l + (-1 + 4m) = 0$$
$$\Rightarrow \quad l - 2m - 1 = 0$$

$\therefore$ Line $lx + my - 1 = 0$ passes through the point (1,–2).

17. (a) Suppose the line $lx + my = 1$ intersects $x^2 + y^2 = a^2$ at points A and B.

$\therefore$ The combined equation of pair of lines OA and OB is

$$x^2 + y^2 - a^2(lx + my)^2 = 0$$
$$\angle AOB = 90^\circ$$
$$\therefore \text{Coefficient of } x^2 + \text{Coefficient of } y^2 = 0$$
$$(1 - a^2l^2) + (1 - a^2m^2) = 0$$
$$\Rightarrow \quad a^2(l^2 + m^2) = 2 \Rightarrow l^2 + m^2 = \frac{2}{a^2}$$

18. (d) Line $y = mx + c$

$$y^2 = 4ax$$

At the homogenisation $y^2 - 4ax\left(\frac{y - mx}{c}\right) = 0 \quad \text{...(i)}$

Eq. (i) represent a pair of straight lines which are perpendicular to each other.

$\Rightarrow$ Coefficient of x^2 + Coefficient of $y^2 = 0$

$$\frac{4am}{c} + 1 = 0 \Rightarrow c = -4am$$

Equation of line is

$$y = mx + c$$
$$y = mx - 4am$$
$$y = m(x - 4a)$$

which passes through the fixed point $(4a, 0)$.

19. $a \to (q, s)$, $b \to p$, $c \to q$, $d \to r$

(a) Let m and m^2 be the slope of the lines represented by

$$ax^2 + 2hxy + by^2 = 0, \text{ then}$$

$$m + m^2 = -\frac{2h}{b} \quad \text{...(i)}$$

and $$mm^2 = \frac{a}{b} \quad \text{...(ii)}$$

By Eqs. (i) and (ii), $(m + m^2)^3 = \left(-\frac{2h}{b}\right)^3$

$$\Rightarrow \quad m^3 + m^6 + 3mm^2(m + m^2) = -\frac{8h^3}{b^3}$$

$$\Rightarrow \quad \frac{a}{b} + \frac{a^2}{b^2} + \frac{3a}{b}\left(-\frac{2h}{b}\right) = -\frac{8h^3}{b^3}$$

$$\Rightarrow \quad ab + a^2 - 6abh = -\frac{8h^3}{b} \Rightarrow \frac{a+b}{h} + \frac{8h^2}{ab} = 6$$

(b) Let $y = m_1x$ and $y = m_2x$ lines represented by $x^2 + 4xy + y^2 = 0$, then $m_1 + m_2 = -4$ and $m_1m_2 = 1$

Then, the product of perpendicular drawn from the point (1, 2)

$$= \frac{|(2 - m_1)(2 - m_2)|}{\sqrt{1 + m_1^2}\sqrt{1 + m_2^2}}$$

$$= \left|\frac{4 - 2(m_1 + m_2) + m_1m_2}{\sqrt{(1 + (m_1 + m_2)^2 - 2m_1m_2 + (m_1m_2)^2}}\right|$$

$$= \left|\frac{4 + 8 + 1}{\sqrt{1 + 16 - 2 + 1}}\right| = \frac{13}{4} \text{ units}$$

$$\therefore \lambda = \frac{13}{4} = 3.25$$

$\Rightarrow [\lambda] = 3$, an odd prime number.

(c) Let $y = mx$ be the common line, then

$$b_1m^2 + 2h_1m + a_1 = 0$$

and $$b_2m^2 + 2h_2m + a_2 = 0$$

$$\Rightarrow \quad \begin{vmatrix} b_1 & 2h_1 \\ b_2 & 2h_2 \end{vmatrix} \times \begin{vmatrix} 2h_1 & a_1 \\ 2h_2 & a_2 \end{vmatrix} = \begin{vmatrix} a_1 & b_1 \\ a_2 & b_2 \end{vmatrix}^2$$

$$\Rightarrow \quad 4(b_1h_2 - b_2h_1)(a_2h_1 - a_1h_2) = (a_1b_2 - a_2b_1)^2$$

$$\Rightarrow \quad \frac{(a_1b_2 - a_2b_1)^2}{(b_1h_2 - b_2h_1)(a_2h_1 - a_1h_2)} = 4$$

(d) Given, line pair is $x^2 - 4xy + 4y^2 + x - 2y - 6 = 0$

$$\Rightarrow \quad (x - 2y)^2 + (x - 2y) - 6 = 0$$

$$\Rightarrow \quad (x - 2y + 3)(x - 2y - 2) = 0$$

$$\therefore \quad \text{Distance between lines} = \left|\frac{3 - (-2)}{\sqrt{1 + 4}}\right| = \sqrt{5} \text{ units}$$

$$\therefore \quad \lambda = \sqrt{5}$$

$$\therefore \quad [\lambda] = [\sqrt{5}] = 2$$

20. $a \to s$, $b \to p$, $c \to r, t$ $d \to q$

(a) $6x^2 + 5xy - 6y^2 + 9x + 20y + c = 0$

$$\Rightarrow \quad a = 6, b = -6, h = \frac{5}{2}, g = \frac{9}{2}, f = 10 \quad \text{...(i)}$$

The given equation represents a pair of straight lines.

$$\therefore \quad \begin{vmatrix} 6 & 5/2 & 9/2 \\ 5/2 & -6 & 10 \\ 9/2 & 10 & c \end{vmatrix} = 0$$

$$\Rightarrow \quad 6(-6c - 100) - \frac{5}{2}\left(\frac{5c}{2} - 45\right) + \frac{9}{2}(25 + 27) = 0$$

$$\Rightarrow \quad 24(-6c - 100) - 5(5c - 90) + 18 \times 52 = 0$$

$$\Rightarrow \quad -169c - 2400 + 450 + 936 = 0$$

$$\Rightarrow \quad -169c = 1014$$

$$\Rightarrow \quad c = -\frac{1014}{169} = -6$$

(b) Suppose the line $x - 2y - 6 = 0$ meets the given curve at points A and B.

$\therefore$ Combined equation of the pair of lines OA and OB is

$$2x^2 + 6xy + 3y^2 + (4x + 2y)\left(\frac{x - 2y}{6}\right) - 36\left(\frac{x - 2y}{6}\right)^2 = 0$$

In this equation, the coefficient of x^2 + coefficient of y^2 is

$$\left(2 + \frac{2}{3} - 1\right) + \left(3 - \frac{2}{3} - 4\right) = \left(1 + \frac{2}{3}\right) + \left(-\frac{2}{3} - 1\right) = 0$$

$$\therefore \quad \angle AOB = \frac{\pi}{2}$$

(c) The combined equation of the pair of lines is

$$2x^2 - 2xy + 3y^2 + \frac{(2x - y)(x + 2y)}{k} - \left(\frac{x + 2y}{k}\right)^2 = 0$$

$\therefore$ Given two lines are at right angled.

coefficient of x^2 + coefficient of $y^2 = 0$

$$\Rightarrow \quad \left(2 + \frac{2}{k} - \frac{1}{k^2}\right) + \left(3 - \frac{2}{k} - \frac{4}{k^2}\right) = 0$$

$$\Rightarrow \quad 5k^2 - 5 = 0$$

$$\Rightarrow \quad k = \pm 1$$

(d) If $ax^2 + 2hxy + by^2 + 2gx + 2fy + c = 0$ represents a pair of parallel lines, then the distance between them is

$$2\sqrt{\frac{g^2 - ac}{a(a + b)}}$$

We have, $a = 9, b = 16, g = -6$ and $c = -12$

$$\therefore \quad d = 2\sqrt{\frac{g^2 - ac}{a(a + b)}} = 2\sqrt{\frac{36 + 108}{9 \times 25}} = \frac{2 \times 12}{3 \times 5} = \frac{8}{5}$$

$$\Rightarrow \quad 5d = 8$$

Numerical Type Questions

Question Stem (21 to 23)

$$ax^2+3xy-2y^2-5x+5y+c=0$$

$$a=a, b=-2, h=\frac{3}{2}, g=-\frac{5}{2}, f=\frac{5}{2}, c=c$$

21. (5) As lines are perpendicular.

$\therefore \quad a-2=0 \Rightarrow a=2$

$\{\because$ coefficient of x^2 + coefficient of $y^2=0\}$

Using $\Delta=0$

$$abc+2fgh-af^2-bg^2-ch^2=0$$

$$2(-2)c+2\left(\frac{5}{2}\right)\left(-\frac{5}{2}\right)\left(\frac{3}{2}\right)-2\left(\frac{5}{2}\right)^2-(-2)\left(-\frac{5}{2}\right)^2-c\left(\frac{3}{2}\right)^2=0$$

$\Rightarrow \quad c+3=0$

$\Rightarrow \quad c=-3$

The two lines of

$$2x^2+3xy-3y^2-5x+5y-3=0$$

are $\quad x+2y-3=0$

and $\quad 2x-y+1=0$

x-intercepts, $x_1=3, x_2=-\frac{1}{2}$

y-intercepts, $y_1=\frac{3}{2}, y_2=1$

$\Rightarrow \quad x_1+x_2+y_1+y_2=5$

22. (1.75) $CM^2=\left(\frac{5}{4}-\frac{1}{5}\right)^2+\frac{49}{25}$

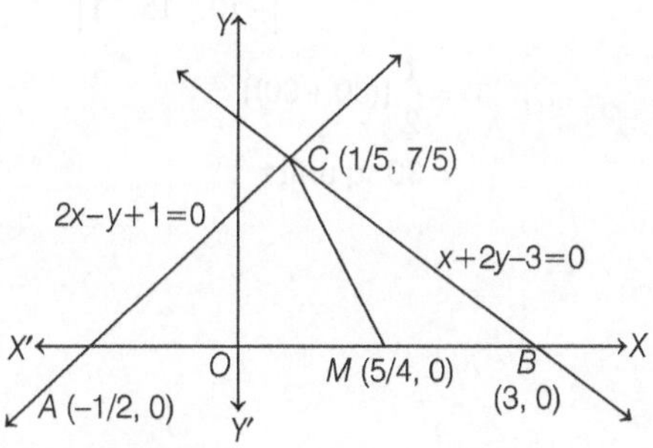

$$=\frac{441}{400}+\frac{49}{25}$$

$$=\frac{441+784}{400}=\frac{1225}{400}=\frac{49}{16}=\frac{7}{4}$$

23. (1.5) Circumcircle of ΔABC

$$\left(x+\frac{1}{2}\right)(x-3)+y^2=0$$

$\Rightarrow \quad (2x+1)(x-3)+2y^2=0$

$\Rightarrow \quad 2(x^2+y^2)-5x-3=0$

$$x^2+y^2-\frac{5}{2}x-\frac{3}{2}=0$$

$\therefore$ Given it is orthogonal to $x^2+y^2-4y+k=0$ ...(i)

which is orthogonal to Eq. (i), we get

$$0+0=k-\frac{3}{2} \Rightarrow k=\frac{3}{2}$$

24. (0.4)

$$\angle ONP=\angle OMP=\frac{\pi}{2}$$

$\therefore O, M, P, N$ are concyclic with diameter OP.

Let $P(H, k)$ be the point.

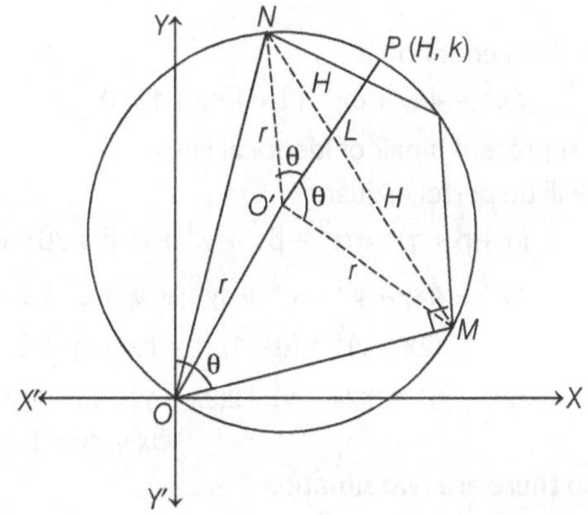

$\therefore$ Coordinates of $O'=\left(\frac{H}{2},\frac{K}{2}\right)$

$\therefore$ Radius $\Rightarrow r=O'N=OM=OO'$

$$\Rightarrow \quad \sqrt{\left(\frac{H}{2}\right)^2+\left(\frac{K}{2}\right)^2}=\sqrt{\frac{H^2+K^2}{4}}$$

$$=\frac{\sqrt{H^2+K^2}}{2}$$

Let $\angle MON=\theta$

$\therefore \quad \angle MO'N=2\theta$

$\Rightarrow \quad \angle NO'L=\angle MO'L=\theta$

and given $MN=2n$ then $NL=LM=n$

Now, in $\Delta NO'L$

$$\sin\theta=\frac{NL}{O'N}=\frac{h}{r}=\frac{h}{\frac{\sqrt{H^2+K^2}}{2}}$$

$$=\frac{2h}{\sqrt{H^2+K^2}} \quad \text{...(i)}$$

But angle between the lines represented by $ax^2+2hxy+by^2=0$ is

$$\tan\theta=\frac{2\sqrt{h^2-ab}}{|a+b|}$$

$$\sin\theta=\frac{2\sqrt{h^2-ab}}{\sqrt{(a-b)^2+4h^2}} \quad \text{...(ii)}$$

By Eqs. (i) and (ii),

$$\frac{2\sqrt{h^2-ab}}{\sqrt{(a-b)^2+4h^2}}=\frac{2h}{\sqrt{H^2+K^2}}$$

Squaring both the sides, we get

$$(h^2-ab)(H^2+K^2)=h^2\{(a-b)^2+4h^2)$$

Locus is

$$(h^2-ab)(x^2+y^2)=h^2((a-b)^2+4h^2)$$

$\therefore \qquad \lambda=4$

and $\qquad \frac{\lambda}{10}=0.4$

25. (2)

The given equation is

$$4x^2-4xy+ay^2+bx+cy+1=0$$

$\therefore$ It represent a pair of identical lines.

It will be perfect square

$$(\alpha+\beta+\gamma)^2=\alpha^2+\beta^2+\gamma^2+2\alpha\beta+2\beta\gamma+2\gamma\alpha$$

$\therefore \qquad 4x^2-4xy+y^2-y^2+ay^2+bx+cy+1=0$

$\Rightarrow \qquad (2x-y)^2+(a-1)y^2+bx+cy+1=0$

$\Rightarrow \qquad (2x-y)^2+2(2x-y)-2(2x-y)+(a-1)y^2+bx+cy+1=0$

So there are two situation.

Case I

$$(2x-y)^2+2(2x-y)+1+(a-1)y^2+x(b-4)+y(c+2)=0$$

$$(2x-y+1)^2+(a-1)y^2+x(b-4)+y(c+2)=0$$

It represents identical pair is straight lines,

$\therefore \qquad a=1, b=4, c=-2$

or $\qquad a_1=1, b_1=4, c_1=-2$

Case II

$$(2x-y)^2-2(2x-y)+1+(a-1)y^2+x(b+4)+y(c-2)=0$$

$\Rightarrow \qquad (2x-y+1)^2+(a-1)y^2+x(b+4)+y(c-2)=0$

$\therefore$ If represents identical pair of straight line

$\therefore \qquad a-1=0, b+4=0, c-2=0$

$\qquad a=1, b=-4, c=2$

$\therefore \qquad a_2=1, b_2=-4\, c_2=2$

Now, the value

$$a_1+b_1+c_1+a_2+b_2+c_2$$
$$=1+4-2+1-4+2=2$$

26. (30 sq units)

Given, $9x^3+9x^2y-45x^2=4y^3+4xy^2-20y^3$

$\Rightarrow \qquad 9x^2(x+y-5)=4y^2(y+x-5)$

$\Rightarrow \qquad (9x^2-4y^2)(x+y-5)=0$

$\Rightarrow \qquad (x+y-5)(3x-2y)(3x+2y)=0$

$\therefore$ Lines are

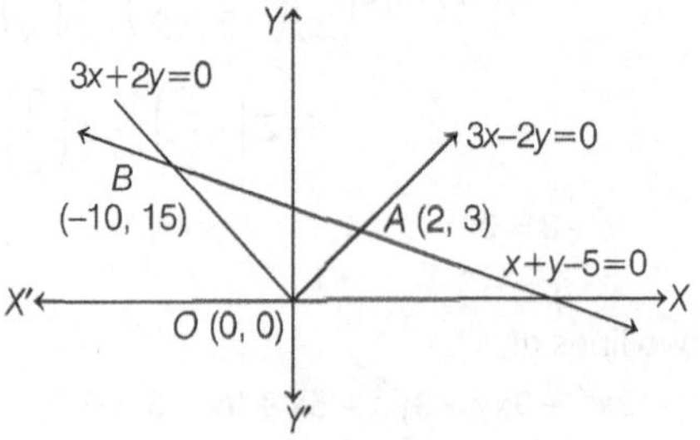

$$x+y-5=0 \qquad \text{...(i)}$$
$$3x-2y=0 \qquad \text{...(ii)}$$
$$3x+2y=0 \qquad \text{...(iii)}$$

Points of intersections of Eqs. (i), (ii) and (iii) are A, B, O.

By solving

$$A(2,3), O(0,0), B(-10,15)$$

$\therefore$ Area of ΔABO

$$=\frac{1}{2}\text{ mod of}\begin{vmatrix}0 & 0 & 1\\ 2 & 3 & 1\\ -10 & 15 & 1\end{vmatrix}$$

$$=\frac{1}{2}\{(30+30)\}$$

$$=30 \text{ sq units}$$

CHAPTER

04

CIRCLE

LEVEL 01 \\ Starter Level

4.1 Equation of Circle

1 Find the centre and radius of the circle

(i) $x^2 + y^2 - 4x - 4y - 8 = 0$

(ii) $2x^2 + 2y^2 + 3x + 4y = 0$

(iii) $x^2 + y^2 + 2x \sin\theta + 2y \cos\theta - 8 = 0$

(iv) $2x^2 + \lambda xy + 2y^2 + (\lambda^2 - 4)x + (3\lambda + 6)y - 5 + \lambda = 0$ for some λ

2 Find the equation of circle

(i) which touches the X-axis at the origin and whose radius is 4.

(ii) which touches the both axes and whose radius is 5.

(iii) passing through the points (4, 1) and (6, 5) and whose centre is on the line $4x + y = 16$.

(iv) whose end points of diameter are (1, –2) and (3, – 4).

(v) whose centre (2, 3) and passes through the point of intersection of the lines $3x - 2y - 1 = 0$ and $4x + y - 27 = 0$.

(vi) the end points of whose diameter are the centres of the circles $x^2 + y^2 + 16x - 14y = 1$ and $x^2 + y^2 - 4x + 10y = 2$.

(vii) whose centre is (4, 3) and touching the circle $x^2 + y^2 = 1$.

(viii) concentric with the circle $x^2 + y^2 + 4x + 6y + 11 = 0$ and passing through the point (5, 4).

(ix) which is circumscribed about the triangle whose vertices (–2, 3), (5, 2) and (6, – 1).

3 Find the length of intercepts on both the axes made by following circle

(i) $x^2 + y^2 + 10x - 8y + 16 = 0$

(ii) $2x^2 + 2y^2 - 14x - 6y + 3 = 0$

4 Find the equation of the circle whose centre is (3, 4) and which touches the line $5x + 12y = 1$.

5 A circle of radius 2 lies in the first quadrant and touches both the axes of coordinates. Find the equation of the circle with centre at (6, 5) and touches the above circle externally.

6 Find the equation of a circle passing through the point (1, 2) and (3, 4) and touching the line $3x + y - 3 = 0$.

7 Find the equation of the circle passing through the points (1, 1), (2, – 1) and (3, 2).

8 Find the equation of the circle whose centre is (3, – 1) and which cut-off an intercept of the length 6 from the line $2x - 5y + 18 = 0$.

9 One of the diameters of the cirlce circumscribing the rectangle *ABCD* is $4y = x + 7$. If *A* and *B* are the points (– 3, 4) and (5, 4), respectively. Find the area of the rectangle.

4.2 Point, Line and Circle

10 Find the position of following points with respect to the circle $x^2 + y^2 - 6x - 10y - 2 = 0$

(i) (3, – 1) (ii) (– 1, 3)

(iii) (0, – 1) (iv) (– 3, 0)

11 Find the coordinate of points of intesection of the following line and circle

(i) Line : $5x - y + 2 = 0$

Circle : $x^2 + y^2 - 13x - 4y - 9 = 0$

(ii) Line : $2x + y - 1 = 0$

Circle : $x^2 + y^2 + 6x - 4y + 8 = 0$

12 Find the equation of tangent(s)

(i) to the circle $x^2 + y^2 = 2$ at (1, 1).

(ii) to the circle $x^2 + y^2 - 12x + 4y - 16 = 0$ which are perpendicular to the line $4x - 3y + 8 = 0$.

(iii) to the circle $x^2 + y^2 = 25$ at points whose abscissa is 4.

(iv) to the circle $x^2 + y^2 = 4$ which are parallel to the straight line $x + 2y + 3 = 0$

13 Find the length of the chord $4x - 3y = 5$ of the circle $x^2 + y^2 + 3x - y - 10 = 0$.

14 If $4l^2 - 5m^2 + 6l + 1 = 0$ and $lx + my + 1 = 0$ touches a fixed circle. Find the centre and radius of the circle.

15 Find the locus of the point (l, m), so that $lx + my = 1$ touches the circle $x^2 + y^2 = a^2$.

16 Circles of radii 36 and 9 touches externally. Find the radius of the circle which touches the two circles externally and also their common tangents.

17 Find the length of chord intercepted by circle $x^2 + y^2 = 9$ and the line $x - y + 2 = 0$.

18 Find the equation of normal

(i) to the circle $x^2 + y^2 - 5x + 2y - 48 = 0$ at the point (5, 6)

(ii) to the cirlce $x^2 + y^2 = 24$, which is parallel to the line $x + 2y = 3$.

19 Find the equation of the pair of tangent drawn to the circle $x^2 + y^2 - 2x + 4y = 0$ from the point (0, 1).

20 Find the angle between the tangents drawn at the point (5, 12) and (12, −5) to the circle $x^2 + y^2 = 169$.

21 Find the number of common tangents to the two circles $x^2 + y^2 - 8x + 2y = 0$ and $x^2 + y^2 - 2x - 16y + 25 = 0$.

22 Find the equation of the chord

(i) to the circle $x^2 + y^2 = 9$ whose middle point is (1, −2).

(ii) to the circle $x^2 + y^2 + 2x + 3y + 1 = 0$ from the point (1, 2).

(iii) that is common to the circles $2x^2 + 2y^2 + 7x - 5y + 2 = 0$ and $x^2 + y^2 - 4x + 8y - 18 = 0$.

4.3 Family of Circle

23 Find the equation of circle passing through the point of intersection of circle $x^2 + y^2 - 4x + 2y - 3 = 0$ and line $3x - 4y + 5 = 0$ and

(i) also passes through the point (1, 1).

(ii) having x-coordinate of centre is −3.

(iii) x-intercept is $2\sqrt{7}$.

(iv) concentric with $2x^2 + 2y^2 - 6x + 4y + 7 = 0$.

(v) touching the cirlce $x^2 + y^2 = 8$ at (2, 2).

24 Find the equation of circle passing through the point of intersection of circles $x^2 + y^2 + x + 2y - 4 = 0$ and $x^2 + y^2 - 2x - 4y - 6 = 0$ and

(i) also passes through (1, −2).

(ii) centre lies on $3x + y = 2$.

(iii) having y-intercept is 0.

25 Find the values of λ if following circles are orthogonal to each other

(i) $C_1 : x^2 + y^2 + 3x - 4y + 7 = 0$

$C_2 : x^2 + y^2 - 4x + 3y - 3\lambda = 0$

(ii) $C_1 : 2x^2 + 2y^2 - 4x - 3y - 7 = 0$

$C_2 : 3x^2 + 3y^2 - 7x + 8y - 19\lambda = 0$

(iii) $C_1 : (x-2)^2 + (y-3)^2 = 4$

$C_2 : (x+1)^2 + (y+\lambda)^2 = 5 + \lambda^2$

26 The straight line $x + y = 1$ meets the circle $x^2 + y^2 - 6x - 8y = 0$ at A and B. Find the equation of circle of which AB is a diameter.

27 Find the equation of the circle which passes through the points of intersection of circles $x^2 + y^2 - 2x - 6y + 6 = 0$ and $x^2 + y^2 + 2x - 6y + 6 = 0$ and intersects the circle $x^2 + y^2 + 4x + 6y + 4 = 0$ orthogonally.

4.4 Miscellaneous

28 Find the radical axis of following circles

(i) $C_1: x^2+y^2-3x+2y+7=0$

$C_2: 3x^2+3y^2-7x+8y+2=0$

(ii) $C_1: 3x^2+3y^2-2x+4y-11=0$

$C_2: 5x^2+5y^2+8x+14=0$

29 Find the radical centre of following circles

(i) $C_1: x^2+y^2+2x+3y-7=0$

$C_2: x^2+y^2-3x+7y+8=0$

$C_3: x^2+y^2+2x+6y-1=0$

(ii) $C_1: 3x^2+3y^2-2x+7y+8=0$

$C_2: 2x^2+2y^2-x+1=0$

$C_3: x^2+y^2-x+3y+2=0$

30 Find the locus of a point which moves, so that the ratio of the length of tangent to the circle $x^2+y^2+4x+3=0$ and $x^2+y^2-6x+5=0$ is 2 : 3.

31 Find the equations of the circle orthogonal to both the circle $x^2+y^2+3x-5y+6=0$ and $4x^2+4y^2-28x+29=0$ and whose centre lie on the line $3x+4y+1=0$.

32 From the point $A(0, 3)$ on the circle $x^2+4x+(y-3)^2=0$ a chord AB is drawn and extended to a point M such that $AM=2AB$. Find the locus of M.

33 Find the locus of the mid-point of the chords of the cirlce $x^2+y^2+2gx+2fy+c=0$, which subtends an angle of 120° at the centre of the circle.

LEVEL 02 \ JEE Main Level

Single Choice Correct Questions

4.1 Equation of Circle

1 In the XY-plane, the length of the shortest path from (0, 0) to (12, 16) that does not go inside the circle $(x-6)^2+(y-8)^2=25$ is

(a) $10\sqrt{3}$ (b) $10\sqrt{5}$ (c) $10\sqrt{3}+\frac{5\pi}{3}$ (d) $10+5\pi$

2 If the radius of the circle $x^2+y^2-18x+12y+\lambda=0$ be 11, then λ is equal to

(a) 347 (b) 4 (c) -4 (d) 49

3 The minimum distance of point $(-3, -6)$ from the circle $x^2+y^2-10x-12y-92=0$ is

(a) $4\sqrt{13}-\sqrt{151}$ (b) $4\sqrt{13}-\sqrt{153}$

(c) $4\sqrt{3}+\sqrt{153}$ (d) None of these

4 The image of the centre of the circle $x^2 + y^2 = a^2$ with respect to the mirror $x + y = 1$ is

(a) $\left(\frac{1}{\sqrt{2}}, \sqrt{2}\right)$ (b) $(\sqrt{2}, \sqrt{2})$ (c) $(\sqrt{2}, 2\sqrt{2})$ (d) None of these

5 The equation of image of circle $x^2 + y^2 + 16x - 24y + 183 = 0$ by the line mirror $4x + 7y + 13 = 0$ is

(a) $x^2 + y^2 + 32x - 4y + 235 = 0$ (b) $x^2 + y^2 + 32x + 4y - 235 = 0$

(c) $x^2 + y^2 + 32x - 4y - 235 = 0$ (d) $x^2 + y^2 + 32x + 4y + 235 = 0$

6 The radius of the circle with the polar equation $r^2 - 8r(\sqrt{3}\cos\theta + \sin\theta) + 15 = 0$ is

(a) 18 (b) 7 (c) 6 (d) 5

7 A circle is inscribed into a rhombus $ABCD$ with one angle 60°. The distance from the centre of the circle to the nearest vertex is equal to 1. If P is any point of the cirlce, then $(PA)^2 + (PB)^2 + (PC)^2 + (PD)^2$ is

(a) 12 (b) 11 (c) 9 (d) 10

8 The locus of the centre of a circle which passes through the origin and cuts-off a length $2b$ from the line $x = c$ is

(a) $y^2 + 2cx = b^2 + c^2$ (b) $x^2 + cx = b^2 + c^2$

(c) $y^2 + 2cy = b^2 + c^2$ (d) None of these

9 Let $A_0A_1A_2A_3A_4A_5$ be a regular hexagon inscribed in a circle of unit radius. Then, the product of the length of the line segments A_0A_1, A_0A_2 and A_0A_4 is

(a) $\frac{3}{4}$ (b) $3\sqrt{3}$ (c) 3 (d) $\frac{3\sqrt{3}}{2}$

10 For which value of a, in equation $x^2 + y^2 + (a^2 - 4)xy + 2x + 2y + a = 0$ represents circle (not point circle)

(a) 3 (b) 2 (c) 4 (d) None of these

11 The area of an equilateral triangle inscribed in the circle $x^2 + y^2 - 2x = 0$, is

(a) $\frac{3\sqrt{3}}{2}$ (b) $\frac{3\sqrt{3}}{4}$ (c) $\frac{3\sqrt{3}}{8}$ (d) None of these

12 The area of square inscribed in a circle $x^2 + y^2 - 6x - 8y = 0$, is

(a) 100 sq units (b) 50 sq units (c) 25 sq units (d) None of these

13 Consider a circle C which touches the Y-axis at (0, 6) and cuts off an intercept $6\sqrt{5}$ on the X-axis, then the radius of circle C is equal to **[JEE Main 2021]**

(a) $\sqrt{53}$ (b) 9 (c) 8 (d) $\sqrt{82}$

14 If the area of an equilateral triangle inscribed in the circle $x^2 + y^2 + 10x + 12y + c = 0$, is $27\sqrt{3}$ sq units, then c is equal to **[JEE Main 2019]**

(a) 20 (b) 25 (c) – 25 (d) 13

4.2 Point, Line and Circle

15 The value of α for which the point $(\alpha, \alpha + 2)$ is an interior point of smaller segment of the curve $x^2 + y^2 - 4 = 0$ made by the chord of the curve, whose equation is $3x + 4y + 12 = 0$, is

(a) $\left(-\infty, -\frac{20}{7}\right)$ (b) $(-2, 0)$ (c) $\left(-\infty, \frac{20}{7}\right)$ (d) $\alpha \in \phi$

16 A tangent at a point on the circle $x^2 + y^2 = a^2$ intersects a concentric circle C at two points A and B. The tangents to the circle C at A and B meet at a point on the circle $x^2 + y^2 = b^2$, then the equation of circle C is

(a) $x^2 + y^2 = ab$
(b) $x^2 + y^2 = (a-b)^2$
(c) $x^2 + y^2 = (a+b)^2$
(d) $x^2 + y^2 = a^2 + b^2$

17 Tangents are drawn from (4, 4) to the circle $x^2 + y^2 - 2x - 2y - 7 = 0$ to meet the circle at A and B. The length of the chord AB is

(a) $2\sqrt{3}$
(b) $3\sqrt{2}$
(c) $2\sqrt{6}$
(d) $6\sqrt{2}$

18 A variable circle C has the equation $x^2 + y^2 - 2(t^2 - 3t + 1)x - 2(t^2 + 2t)y + t = 0$, where t is a parameter. If the power of point $P(a, b)$ with respect to the circle C is constant, then the ordered pair (a, b) is

(a) $\left(\frac{1}{10}, -\frac{1}{10}\right)$
(b) $\left(-\frac{1}{10}, \frac{1}{10}\right)$
(c) $\left(\frac{1}{10}, \frac{1}{10}\right)$
(d) $\left(-\frac{1}{10}, -\frac{1}{10}\right)$

19 Point (3, – 1) is centre of a circle. This circle cuts-off a chord of length 6 units on the line $2x + 5y + 18 = 0$. The equation of the circle, is

(a) $29x^2 + 29y^2 - 174x + 58y - 332 = 0$
(b) $29x^2 + 29y + 174x - 58y + 332 = 0$
(c) $29x^2 + 29y^2 - 174x - 58y + 332 = 0$
(d) $29x^2 + 29y^2 + 74x - 68y + 332 = 0$

20 The line $2x - y + 1 = 0$ is a tangent to the circle at the point (2, 5) and the centre of the circle lies on $x - 2y = 4$. Then, the radius of the circle is **[JEE Main 2021]**

(a) $5\sqrt{3}$
(b) $4\sqrt{5}$
(c) $3\sqrt{5}$
(d) $5\sqrt{4}$

21 The locus of the centre of a circle touching circle $x^2 + y^2 - 4y - 2x = 2\sqrt{3} - 1$ internally and tangents on which from (1, 2) is making a 60° angle each other, is

(a) $(x-1)^2 + (y-2)^2 = 3$
(b) $(x-2)^2 + (y-1)^2 = 1 + 2\sqrt{3}$
(c) $x^2 + y^2 = 1$
(d) None of these

22 The number of points $P(x, y)$ lying inside or on the circle $x^2 + y^2 = 9$ and satisfying the equation $\tan^4 x + \cot^4 x + 2 = 4\sin^2 y$, is

(a) 2
(b) 4
(c) 8
(d) infinite

23 If the tangent at point P on the circle $x^2 + y^2 + 6x + 6y = 2$ meets the straight line $5x - 2y + 6 = 0$ at a point Q on Y-axis, then the length of PQ is

(a) 4
(b) $2\sqrt{5}$
(c) 5
(d) $3\sqrt{5}$

24 The locus of the centre of circle such that the point (2, 3) is the mid-point of the chord $5x + 2y = 16$, is

(a) $2x - 5y - 11 = 0$
(b) $2x + 5y - 11 = 0$
(c) $2x + 5y + 11 = 0$
(d) None of these

25 The range of values of $\lambda (\lambda > 0)$ such that the angle θ between the pair of tangents drawn from $(\lambda, 0)$ to the circle $x^2 + y^2 = 4$ lies in $\left(\frac{\pi}{2}, \frac{2\pi}{3}\right)$, is

(a) $\left(\frac{4}{\sqrt{3}}, 2\sqrt{2}\right)$
(b) $(0, \sqrt{2})$
(c) (1, 2)
(d) None of these

26 Two tangents are drawn from a point P to the cirlce $x^2 + y^2 - 2x - 4y + 4 = 0$ such that the angle between them tangents is $\tan^{-1}\left(\frac{12}{5}\right)$, where $\tan^{-1}\left(\frac{12}{5}\right) \in (0, \pi)$. If the centre of the circle is denoted by C and these tangents touch the circle at points A and B, then the ratio of the area of ΔPAB and ΔCAB is **[JEE Main 2021]**

(a) 3 : 1 (b) 9 : 4 (c) 2 : 1 (d) 11 : 4

27 If a line $y = mx + c$ is a tangent to the circle $(x - 3)^2 + y^2 = 1$ and it is perpendicular to a line L_1, where L_1 is the tangent to the circle $x^2 + y^2 = 1$ at the point $\left(\frac{1}{\sqrt{2}}, \frac{1}{\sqrt{2}}\right)$, then **[JEE Main 2020]**

(a) $c^2 + 6c + 7 = 0$ (b) $c^2 - 7c + 6 = 0$

(c) $c^2 - 6c + 7 = 0$ (d) $c^2 + 7c + 6 = 0$

28 The tangent and the normal lines at the point $(\sqrt{3}, 1)$ to the circle $x^2 + y^2 = 4$ and X-axis form a triangle. The area of this triangle (in sq unit) is **[JEE Main 2019]**

(a) $\frac{4}{\sqrt{3}}$ (b) $\frac{1}{\sqrt{3}}$ (c) $\frac{2}{\sqrt{3}}$ (d) $\frac{1}{3}$

29 If two distinct chord, drawn from the point (p, q) on the circle $x^2 + y^2 = px + qy$ (where, $pq \neq 0$) are bisected by X-axis, then

(a) $p^2 = q^2$ (b) $p^2 = 8q^2$ (c) $p^2 < 8q^2$ (d) $p^2 > 8q^2$

30 Middle point of the chord of the circle $x^2 + y^2 = 25$ intercepted on the line $x - 2y = 2$ is

(a) $\left(\frac{3}{5}, \frac{4}{5}\right)$ (b) $(-2, -2)$ (c) $\left(\frac{2}{5}, -\frac{4}{5}\right)$ (d) $\left(\frac{8}{3}, \frac{1}{3}\right)$

31 The line $ax + by + c = 0$ is normal to the circle $x^2 + y^2 + 2gx + 2fy + d = 0$

(a) $ag + bf + c = 0$ (b) $ag + bf - c = 0$

(c) $ag - bf + c = 0$ (d) $ag - bf - c = 0$

4.3 Family of Circle

32 The circle passing through the point $(-1, 0)$ and touching the Y-axis at $(0, 2)$ also passes through the point

(a) $\left(-\frac{3}{2}, 0\right)$ (b) $\left(-\frac{5}{2}, 2\right)$ (c) $\left(-\frac{3}{2}, \frac{5}{2}\right)$ (d) $(-4, 0)$

33 The locus of the centre of a circle of radius 2 which rolls on the outside of circle $x^2 + y^2 + 3x - 6y - 9 = 0$, is

(a) $x^2 + y^2 + 4x - 6y + 5 = 0$ (b) $x^2 + y^2 + 3x - 6y - 31 = 0$

(c) $x^2 + y^2 + 3x - 6y + \frac{29}{4} = 0$ (d) None of these

34 A circle C_1 of radius 2 touches both X and Y-axes. Another circle C_2 whose radius is greater than 2 touches circle C_1 and both the axes. Then, radius of circle C_2 is

(a) $6 - 4\sqrt{2}$ (b) $6 + 4\sqrt{2}$ (c) $6 - 4\sqrt{3}$ (d) $6 + 4\sqrt{3}$

35 Two circles of radii 4 cm and 1 cm touch each other externally and θ is the angle contained by their direct common tangents, then $\sin\theta$ is equal to

(a) $\frac{24}{25}$ (b) $\frac{12}{25}$ (c) $\frac{3}{4}$ (d) None of these

36 If the circles $x^2+y^2+5kx+2y+k=0$ and $2(x^2+y^2)+2kx+3y-1=0$ $(k \in R)$ intersect at the points P and Q, then the line $4x+5y-k=0$ passes through P and Q for **[JEE Main 2019]**

(a) exactly two values of k
(b) no value of k
(c) exactly one value of k
(d) infinite many values of k

37 The common tangent to the circles $x^2+y^2=4$ and $x^2+y^2+6x+8y-24=0$ also passes through the point **[JEE Main 2019]**

(a) $(6,-2)$
(b) $(4,-2)$
(c) $(-4,6)$
(d) $(-6,4)$

38 If $\left(\alpha, \frac{1}{\alpha}\right), \left(\beta, \frac{1}{\beta}\right), \left(\gamma, \frac{1}{\gamma}\right)$ and $\left(\delta, \frac{1}{\delta}\right)$ are four distinct points on a circle of radius 4 units, then $\alpha\beta\gamma\delta$ is equal to

(a) 4
(b) $\frac{1}{4}$
(c) 1
(d) 16

39 If the radius of the circle $(x-1)^2+(y-2)^2=1$ and $(x-7)^2+(y-10)^2=4$ are increasing uniformly with respect to time as 0.3 and 0.4 unit per sec, then they will touch each other at t is equal to

(a) 45 s
(b) 90 s
(c) 11 s
(d) 135 s

40 Angle of intersection of the curves $r=\sin\theta+\cos\theta$ and $r=2\sin\theta$ is equal to

(a) $\frac{\pi}{2}$
(b) $\frac{\pi}{3}$
(c) $\frac{\pi}{4}$
(d) None of these

41 The circle $x^2+y^2-4x-8y+16=0$ rolls up the tanget to it at $(2+\sqrt{3}, 2)$ by 2 units, the equation of the circle in the new position is

(a) $(x-3)^2+(y-4+\sqrt{3})^2=4$
(b) $(x-3)^2+(y-4-\sqrt{3})^2=4$
(c) $(x+3)^2+(y-4-\sqrt{3})^2=4$
(d) None of these

42 The locus of centre of the circle of the which touches the circle $x^2+(y-1)^2=1$ externally and also touches X-axis, is

(a) $\{(x, y): x^2+(y-1)^2=4\} \cup \{(x, y)=y<0\}$
(b) $\{(x, y): x^2=4y\} \cup \{(0, y): y<0\}$
(c) $\{(x, y): x^2=y\} \cup \{(0, y): y<0\}$
(d) $\{(x, y): x^2=4y\} \cup \{(x, y): y<0\}$

43 The locus of the mid-point of the chord of the circle $x^2+y^2-ax-by=0$ which subtend a right angle at $\left(\frac{a}{2}, \frac{b}{2}\right)$, is

(a) $ax+by=0$
(b) $ax+by=a^2+b^2$
(c) $x^2+y^2-ax-by+\frac{a^2+b^2}{8}=0$
(d) $x^2+y^2-ax-by-\frac{a^2+b^2}{8}=0$

44 Equation of circle passing through intersection of two circles $x^2+y^2+13x-3y=0$ and $2x^2+2y^2+4x-7y-25=0$ and also through (1, 1), is

(a) $4x^2+4y^2-30x-10y-25=0$
(b) $4x^2+4y^2+30x-13y-25=0$
(c) $4x^2-4y^2-17x-10y+25=0$
(d) None of these

45 Equation of a circle that passes through the point (1, 2) which bisects the circumference of the circle $x^2 + y^2 = 9$ and is orthogonal to the circle $x^2 + y^2 - 2x + 8y - 7 = 0$, is

(a) $x^2 + y^2 + 8x - 2y - 9 = 0$
(b) $x^2 + y^2 + 8x + 2y - 17 = 0$
(c) $x^2 + y^2 - 8x + 2y - 1 = 0$
(d) None of these

46 The equation of the circle whose centre lies on the line $x + y - 11 = 0$ and which passes through the intersection of the circle $x^2 + y^2 - 3x + 2y - 4 = 0$ and the line $2x + 5y + 2 = 0$, is

(a) $x^2 + y^2 - 9x - 13y + 10 = 0$
(b) $x^2 + y^2 + 9x - 13y - 10 = 0$
(c) $x^2 + y^2 - 9x - 13y - 10 = 0$
(d) None of these

47 The extremities of a diagonal of a rectangle are (– 4, 4) and (6, – 1). A circle circumscribes the reactange and cuts an intercept *AB* on the *Y*-axis. Then, the area of the triangle formed by *AB* and the two tangents at *A* and *B*, is

(a) $\left(\frac{11}{2}\right)^2$
(b) $\frac{(11)^2}{2}$
(c) $\left(\frac{11}{2}\right)^3$
(d) $\left(\frac{11}{2}\right)^4$

48 The members of a family of circles are given by the equation $2(x^2 + y^2) + \lambda x - (1 + \lambda^2)y - 10 = 0$. The number of circle belonging to the family that are cut orthogonally by the fixed circle $x^2 + y^2 + 4x + 6y + 3 = 0$, is

(a) 2
(b) 1
(c) 0
(d) None of these

49 Three concentric cirlces of which the biggest is $x^2 + y^2 = 1$, have their radii in AP. If the line $y = x + 1$ cuts all the circles in real and distinct points. The interval in which the common difference of the AP will lie, is

(a) $\left(0, \frac{1}{4}\right)$
(b) $\left(0, \frac{1}{2\sqrt{2}}\right)$
(c) $\left(0, \frac{2 - \sqrt{2}}{4}\right)$
(d) None of these

50 A ray of light incident at the point (–2, – 1) gets reflected from the tangent at (0, – 1) to the circle $x^2 + y^2 = 1$. The reflected ray touches the circle. The equation of the line along which the incident ray moves, is

(a) $4x - 3y + 11 = 0$
(b) $4x + 3y + 11 = 0$
(c) $3x + 4y + 11 = 0$
(d) $4x + 3y + 7 = 0$

51 The locus of the point of intersection of the tangent to the circle $x^2 + y^2 = r^2$, which include an angle of 45°, is the curve $(x^2 + y^2)^2 = \lambda r^2 (x^2 + y^2 - r^2)$. The value of λ is

(a) 2
(b) 4
(c) 8
(d) 16

4.4 Miscellaneous

52 The pole of the straight line $9x + y - 28 = 0$ with respect to the circle $2x^2 + 2y^2 - 3x + 5y - 7 = 0$, is

(a) (3, 1)
(b) (1, 3)
(c) (3, – 1)
(d) (– 3, 1)

53 The polar of a point with respect to the circle $x^2 + y^2 - 2\lambda x + c^2 = 0$

(a) passes through $(\lambda, 0)$
(b) passes through a fixed point if $\lambda = 2$
(c) passes through a fixed point for all value of λ
(d) is a tangent to the circle

54 Number of points (x, y) having integral coordinates satisfying the condition $x^2 + y^2 = 25$, is **[JEE Main 2021]**

(a) 69 (b) 80 (c) 81 (d) 77

55 Let Z be the set of all integers,

$$A = \{(x, y) \in Z \times Z : (x-2)^2 + y^2 \le 4\}$$
$$B = \{(x, y) \in Z \times Z : x^2 + y^2 \le 4\}$$
$$C = \{(x, y) \in Z \times Z : (x-2)^2 + (y-2)^2 \le 4\}$$

If the total number of relation from $A \cap B$ to $A \cap C$ is 2^P, then the value of P is **[JEE Main 2021]**

(a) 16 (b) 25 (c) 49 (d) 9

56 The sum of the squares of the lengths of the chords intercepted on the circle, $x^2 + y^2 = 16$ by the lines $x + y = n$, $n \in N$, where N is the set of all natural numbers, is

(a) 210 (b) 160
(c) 320 (d) 105

57 If a variable line $3x + 4y - \lambda = 0$ is such that the two circles $x^2 + y^2 - 2x - 2y + 1 = 0$ and $x^2 + y^2 - 18x - 2y + 78 = 0$ are on its opposite sides, then the set of all values of λ is the interval

(a) (23, 31) (b) (2, 17)
(c) [13, 23] (d) [12, 21]

58 Three circles of radii a, b, c $(a < b < c)$ touch each other externally. If they have X-axis as a common tangent, then **[JEE Main 2019]**

(a) a, b, c are in AP (b) $\sqrt{a}, \sqrt{b}, \sqrt{c}$ are in AP
(c) $\frac{1}{\sqrt{b}} + \frac{1}{\sqrt{c}} = \frac{1}{\sqrt{a}}$ (d) $\frac{1}{\sqrt{b}} = \frac{1}{\sqrt{a}} + \frac{1}{\sqrt{c}}$

Numerical Value Type Questions

59 Two parallel chords of a circle of radius 2 are at a distance $\sqrt{3} + 1$ apart. If the chords subtends at the centre, angle of $\frac{\pi}{k}$ and $\frac{2\pi}{k}$, where $k > 0$, then the value of $[k]$, where $[.]$ represent greatest integer function is ....

60 If the curves $x^2 + y^2 - 6x + 8 = 0$ and $x^2 + y^2 - 8x + 16 - k = 0$ $(k > 0)$ touch each other at a point, then the largest value of k is .... **[JEE Main 2020]**

61 The diameter of the circle, whose centre lies on the line $x + y = 2$ in the first quadrant and which touches both the lines $x = 3$ and $y = 2$ is .... **[JEE Main 2020]**

62 Let the equation $x^2 + y^2 + px + (1-p)y + 5 = 0$ represents circles of varying radius $r \in (0, 5]$. Then, the number of elements in the set $S = \{q : q = p^2$ and q is an integer$\}$ is .... **[JEE Main 2021]**

63 Let O be the centre of the circle $x^2 + y^2 - 2x - 4y - 20 = 0$. Suppose, the tangents at the points $A(1, 7)$ and $C(4, -2)$ on the circle meet at B. The area of the quadrilateral $OABC$ is ....

64 ΔABC is right angled at A. The circle with centre A and radius AB cuts BC and AC internally at D and E, respectively if $BD = 20$ and $DC = 16$ and length $AC = p\sqrt{q}$, where q is not perfect square, then $p + q$ is equal to ....

65 If the circle C_1 touches X-axis and the line $y = x + \tan\theta \left(\theta \in 0, \frac{\pi}{2}\right)$ in the first quadrant and circle C_2 touches the line $y = x \tan\theta$, Y-axis and circle C_1 in such a way that ratio of radius of C_1 to radius of C_2 is 2 : 1, then value of $\tan\frac{\theta}{2} = \frac{\sqrt{a} - b}{c}$, where a, b, c are relatively prime, then $a + b + c$ is equal to ....

66 There exist two circles passing through $(-1, 1)$ and touching the line $x + y = 2$ and $x - y = 2$ whose radii are r_1 and r_2 and $r_1 + r_2 = \lambda\sqrt{2}$, then λ is equal to ....

67 The circle $x^2 + y^2 - 4x - 4y + 4 = 0$ is inscribed in a triangle which has two of its sides along the coordinate axes. The locus of the circumcentre of the triangle is $x + y - xy + k\sqrt{x^2 + y^2} = 0$, then k is equal to ....

68 The value of r such that the area of the triangle formed by the pair of tangents drawn to the circle $x^2 + y^2 = r^2$ from the point $P(6, 8)$ and its chord of contact is maximum is ....

69 The line joining the points $A(3, 4)$ and $B(1, 0)$ cuts the circle $x^2 + y^2 = 4$ at points P and Q. If $\frac{AP}{PQ} = \lambda$ and $\frac{AQ}{QB} = \mu$, then the value of $|\lambda\mu|$ is ....

70 The centre of two circles C_1 and C_2 each of unit radius are at a distance of 6 units from each other. Let P be the mid-point of the line segment joining the centres of C_1 and C_2, C be a circle touching the circle C_1 and C_2 externally. If a common tangent to C_1, which is passing through P, is also a common tangent to C_1 and C_2, then the radius of C is equal to ....

71 Consider the family of circles $x^2 + y^2 - 2x - 2\lambda y - 8 = 0$ passing through two fixed points A and B, then the distance between the points A and B is ....

72 The acute angle between the line $3x - 4y = 5$ and the circle $x^2 + y^2 - 4x + 2y - 4 = 0$ is θ, then the value of $9\cos\theta$ is ....

73 The circle $x^2 + y^2 - 4x - 4y + 4 = 0$ inscribed in a triangle which has two of its sides along the coordinate axis. The locus of the circumcentre of the triangle is $x + y - xy + k(x^2 + y^2)^{\frac{1}{2}} = 0$, then the value of k is ....

74 Two circles C_1 and C_2 both pass through the points $A(1, 2)$ and $E(2, 1)$ and touch the line $4x - 2y = 9$ at B and D, respectively. The possible coordinates of a point C such that the quadrilateral $ABCD$ is a parallelogram, is (a, b), then the value of $|ab|$ is ....

75 Let $ABCD$ be a quadrilateral with area 18, with side AB parallel to the side CD and $AB = 2CD$. Let AD be perpendicular to AB and CD. If a circle is drawn inside the quadrilateral $ABCD$ touching are the sides, then its radius is equal to ....

76 The number of ordered pair of integers (a, b) satisfy all the following inequalities $a^2 + b^2 < 16$, $a^2 + b^2 < 8a$, $a^2 + b^2 < 8b$, is ....

77 The radii of the smallest and the largest circle passing through $(\sqrt{3}, \sqrt{2})$ and touching the circle $x^2 + y^2 - 2\sqrt{2}y - 2 = 0$ are r_1 and r_2, respectively, then the mean of r_1 and r_2 is ....

78 The distance between centres of two circles is equal to 1. Let x be the side of a rhombus two opposite vertices of which lie on one circle and the other two on the other if radii of the circles are 6 and 3, then the value of $[x]$ {where [.] denotes greatest integer function} is ....

79 Two circles of radii 4 and 2 touch each other internally. The side of equilateral trianlge one vertex of which coincides with the point of tangency and other two lying on the two given circles, is ....

80 Two circles of radius 8 are placed inside a semi-circle of radius 25. The two circles are each tangent to the diameter and to the semi-circle. If the distance between the centres of the two circles is λ, then the value of $\frac{60}{\lambda}$ is ....

LEVEL 03 \ JEE Advanced Level

Single Choice Correct Questions

1 From a point P on the line $2x + y + 4 = 0$ which is nearest to the circle $x^2 + y^2 - 12y + 35 = 0$ tangents are drawn to given circle. The area of quadrilateral $PACB$ (where 'C' be the centre of circle and PA and PB are tangents) is

(a) 8 (b) $\sqrt{110}$ (c) $\sqrt{19}$ (d) None of these

2 Two circles are given as $x^2 + y^2 + 14x - 6y + 40 = 0$ and $x^2 + y^2 - 2x + 6y + 7 = 0$ with their centres as C_1 and C_2. If equation of another circle whose centre C_3 lies on the line $3x + 4y - 16 = 0$ and touches the circle C_1 externally and also $C_1C_2 + C_2C_3 + C_3C_1$ is minimum, is $x^2 + y^2 + ax + by + c = 0$, then the value of $a + b + c$ is

(a) 2 (b) 0 (c) 16 (d) None of these

3 Let O be the centre of the circle $x^2 + y^2 = r^2$, where $r = \sqrt{\frac{5}{2}}$. Suppose, PQ is a chord of this circle and the equation of the line passing through P and Q is $2x + 4y = 5$. If the centre of the circumcircle of ΔOPQ lies on the line $x + 2y = 9$, then the value of r is **[JEE Advanced 2020]**

(a) 3 (b) 2

(c) 5 (d) None of these

4 Let T be the line passing through the point $P(-2, 7)$ and $Q(2, -5)$. Let F_1 be the set of all circles (S_1, S_2) such that T is tangents to S_1 at P and tangents to S_2 at Q and also such that S_1 and S_2 touch each other at a point say, M. Let E_1 be the set representing the locus of M as the pair (S_1, S_2) varies in F_1. Let the set of all straight lines segment joining a pair of distinct points of E_1 and passing through the point $R(1, 1)$ be F_2. Let E_2 be the set of the mid-point of the line segment in the set F_2. Then, which of the following statement is true?

(a) The point $(4, -6)$ lies in E_1

(b) The point $\left(\frac{4}{5}, \frac{7}{5}\right)$ does not lie in E_2

(c) The point $\left(\frac{1}{2}, 1\right)$ lies in E_2

(d) The point $\left(0, \frac{3}{2}\right)$ does not lie in E_1

5 The line $lx + my + n = 0$ intersect the curve $ax^2 + 2hxy + by^2 = 1$ at the point P and Q. The circle on PQ as diameter pass through origin, then which of the following statement is true?

(a) $n^2(a + b) = l^2 + m^2$ (b) $m^2(a + b) = l^2 + n^2$

(c) $n^2(a + b) = l^2 + m^2$ (d) None of these

6 Let the base AB of a ΔABC be fixed and the vertices C lie on a fixed circle of radius r. Lines through A and B are drawn to intersect CB and CA respectively, at E and F such that $CE : EB = 1:2$ and $CF : FA = 1:2$. If the point of intersection P of these lines on the median through AE for all positions of AB, then the locus of P is

(a) a circle of radius $\frac{r}{2}$ (b) a circle of radius $2r$

(c) a parabola of latusrectum $4r$ (d) a rectangular hyperbola

7 For each natural number k, let C_k denotes the circle with radius k cm and centre at the origin. On the circle C_k, α-particle moves k cm in the counter-clockwise direction. After completing its motion on C_k the particle moves C_{k+1} in the radial direction. The motion of the particle continues in this manner. The particle starts at (1, 0). If the particle crosses the positive direction of the X-axis for the first time on the circle, then n is

(a) 5 (b) 6 (c) 7 (d) 8

8 Consider a triangle whose two sides lie on the X-axis and the line $x + y + 1 = 0$. If the orthocentre of triangle is (1, 1), then the equation of circle passing through the vertices of the triangle, is **[JEE Advanced 2021]**

(a) $x^2 + y^2 - 3x + y = 0$ (b) $x^2 + y^2 + x + 3y = 0$

(c) $x^2 + y^2 + 2y - 1 = 0$ (d) $x^2 + y^2 + x + y = 0$

9 The range of values of λ for which the variable line $3x + 4y - \lambda = 0$ lies between the circles $x^2 + y^2 - 2x - 2y + 1 = 0$ and $x^2 + y^2 - 18x - 2y + 78 = 0$ without intercepting a chord on other circle is

(a) $\lambda \le 21$ (b) $12 \le \lambda \le 21$ (c) $\lambda \ge 12$ (d) $15 \le \lambda \le 25$

10 Let a circle be given by $2x(x - a) + y(2y - b) = 0, (a \ne 0), b \ne 0)$, which of the following relation is valid between a and b if two chords, each bisected by X-axis, can be drawn to the circle from $\left(a, \frac{b}{2}\right)$?

(a) $a^2 > 2b^2$ (b) $a^2 > 4b^2$ (c) $a^2 < 2b^2$ (d) $a^2 < 4b^2$

11 If the circle $C_1 : x^2 + y^2 = 16$ intersects another circle C_2 of radius 5 in such a manner that the common chord is of maximum length and has a slope equal to $\frac{3}{4}$, then the coordinates of the centre of C_2 are

(a) $\left(\pm\frac{9}{5}, \pm\frac{12}{5}\right)$ (b) $\left(\mp\frac{9}{5}, \pm\frac{12}{5}\right)$ (c) $\left(\pm\frac{12}{5}, \pm\frac{9}{5}\right)$ (d) $\left(\pm\frac{12}{5}, \mp\frac{9}{5}\right)$

12 The complete set of values of a for which atleast one point (x, y) satisfies all the relations $x^2 + y^2 \le 16, (x - a)^2 + (y - a)^2 = 9, (x + a)^2 + (y - a)^2 = 9$

(a) $[-\sqrt{3}, \sqrt{3}]$ (b) $[-3, 3]$ (c) $[-2\sqrt{3}, 2\sqrt{3}]$ (d) None of these

More Than One Correct

1 A straight line through the vertex P of a ΔPQR intersects the side QR at the point S and the circumcircle of the ΔPQR at the point T. If S is not the centre of the circumcircle, then

(a) $\frac{1}{PS} + \frac{1}{ST} < \frac{2}{\sqrt{QS \times SR}}$ (b) $\frac{1}{PS} + \frac{1}{ST} > \frac{2}{\sqrt{QS \times SR}}$

(c) $\frac{1}{PS} + \frac{1}{ST} < \frac{4}{QR}$ (d) $\frac{1}{PS} + \frac{1}{ST} > \frac{4}{QR}$

2 A(2, 0) is a point on the circle $(x+2)^2+(y-3)^2=25$. A line through A(2, 0) making an angle of 45° with the tangent to the circle A is drawn. Then, the equations of the circles on these lines which are at a distance of $5\sqrt{2}$ units from point A and of radius 3, are

(a) $(x-1)^2+(y-7)^2=9$ (b) $(x-3)^2+(y+7)^2=9$

(c) $(x-9)^2+(y-1)^2=9$ (d) $(x+5)^2+(y+1)^2=9$

3 If α and β are roots of the equation $ax^2+bx+c=0$ and α', β' are roots of $a'x^2+b'x+c'=0$ and the circle having $A(\alpha, \alpha')$ and $B(\beta, \beta')$ as diametric end points passes through the origin and the point $\left(\frac{b}{a}, \frac{b'}{a'}\right)$, then

(a) $a'c+ac'=0$ (b) $a'b+ab'=0$

(c) $b'c+bc'=0$ (d) $a^2b'^2+a'^2b^2=0$

4 Point M moved on circle $(x-4)^2+(y-8)^2=20$. Then, it broke away from it and moving along a tangent to the circle, cuts the X-axis at the point $(-2, 0)$. The coordinates of a point on the circle at which the moving point broke away, are

(a) $\left(\frac{-3}{5}, \frac{46}{5}\right)$ (b) $\left(\frac{-2}{5}, \frac{44}{5}\right)$ (c) (6, 4) (d) (3, 5)

5 In a right angled triangle, the length of the sides are a and b $(0<a<b)$. A circle passes through the mid-point of the smaller side and touches the hypotenuse at its mid-point. Then,

(a) centre of circle is $\left(\frac{2a^2-b^2}{4a}, \frac{b}{2}\right)$ (b) centre of circle is $\left(\frac{b^2-a^2}{4a}, \frac{b}{4}\right)$

(c) radius of the circle is $\frac{b}{4a}\sqrt{a^2+b^2}$ (d) radius of circle is $\frac{a}{4b}\sqrt{a^2+b^2}$

6 The circle $C_1: x^2+y^2=3$ with centre O, intersects the parabola $x^2=2y$ at the point P in the first quadrant. Let the tangent to the circle C_1 at P touches other two circles C_2 and C_3 at R_2 and R_3, respectively. Suppose, C_2 and C_3 have equal radii $2\sqrt{3}$ and centres Q_2 and Q_3, respectively. If Q_2 and Q_3 lie on the Y-axis, then **[JEE Advanced 2016]**

(a) $Q_2Q_3=12$

(b) $R_2R_3=4\sqrt{6}$

(c) area of ΔOR_2R_3 is $6\sqrt{2}$

(d) area of ΔPQ_2Q_3 is $4\sqrt{2}$

7 Let x and y be real variables satisfying the $x^2+y^2+8x-10y-40=0$. Let $a=\max\{\sqrt{(x+2)^2+(y-3)^2}\}$ and $b=\min\{\sqrt{(x+2)^2+(y-3)^3}\}$, then

(a) $a+b=18$ (b) $a+b=\sqrt{2}$

(c) $a-b=4\sqrt{2}$ (d) $a\cdot b=73$

8 Three sides of a triangle have the equations $L_i=y-m_ix=0, i=1, 2, 3$, then $L_1L_2+\lambda L_2L_3+\mu L_3L_1=0$, where $\lambda_1\mu\neq 0$, is the equation of the circumcircle of the triangle if

(a) $1+\lambda+\mu=m_1m_2+\lambda m_2m_3+\lambda m_3m_1$

(b) $m_1(1+\mu)+m_2(1+\lambda)+m_3(\mu+\lambda)=0$

(c) $\frac{1}{m_3}+\frac{1}{m_1}+\frac{1}{m_2}=1+\lambda+\mu$

(d) None of the above

9 The circles $x^2+y^2+2x+4y-20=0$ and $x^2+y^2+6x-8y+10=0$

(a) are such that the number of common tangents on them is 2

(b) are orthogonal

(c) are such that the length of common tangent is $5\left(\frac{12}{5}\right)^{1/4}$

(d) are such that the length of their common chord is $5\sqrt{\frac{3}{2}}$

10 AC is diameter of circle and AB is a tangent BC meet the circle again at D. $AC=1$, $AB=a$ and $CD=b$, then

(a) $ab>1$ (b) $ab<1$

(c) $\frac{b}{a}>\frac{1}{a^2+\frac{1}{2}}$ (d) $\frac{b}{a}<\frac{1}{a^2+\frac{1}{2}}$

11 Let RS be the diameter of the circle $x^2+y^2=4$, where S is the point (2, 0). Let P be a variable point (other than R and S) on the circle and tangents to the circle at S and P meet at the point Q. The normal to the circle at P intersects a line drawn through Q parallel to RS at point E. Then, the locus of E through the point(s)

(a) $\left(\frac{1}{3},\frac{2\sqrt{2}}{\sqrt{3}}\right)$ (b) $\left(\frac{1}{3},\frac{2\sqrt{3}}{\sqrt{2}}\right)$

(c) $\left(\frac{1}{3},-\frac{2\sqrt{2}}{\sqrt{3}}\right)$ (d) $\left(\frac{1}{\sqrt{3}},\frac{2\sqrt{2}}{\sqrt{3}}\right)$

12 Which of the following statement(s) is (are) true?

(a) Let C be a circle with centre O and HK is chord of contact of tangents drawn from a point A. OA intersects the circle C at P and Q and B is mid-point of HK, then AB is the harmonic mean of AP and AQ

(b) Let the bisector of $\angle A$ of ΔABC meets BC in D and the circumcircle of ΔABC is E, then AD is less than the GM of AB and AC

(c) The equation of pair of tangents is $SS_1=T^2$

(d) Equation of chord when mid-point is given, is $T=0$

13 If the area of the quadrilateral formed by the tangents from the origin to the circle $x^2+y^2+6x-10y+C=0$ and radii corresponding to the points of contact is 15, then a value of C is

(a) 9 (b) 4 (c) 5 (d) 25

14 Circle touching X-axis at a distance 3 from the origin and having an intercept of length $2\sqrt{7}$ on Y-axis is/are

(a) $x^2+y^2-6x+8y+9=0$ (b) $x^2+y^2-6x+7y+9=0$

(c) $x^2+y^2-6x-8y+9=0$ (d) $x^2+y^2-6x-7y+9=0$

15 The circles $x^2+y^2-8x+2y+8=0$ and $x^2+y^2-2x-6y+10-a^2=0$ have exactly two common tangents, then a can be in

(a) $(-\infty,-2)$ (b) $(2,\infty)$

(c) $(-8,-3]$ (d) $[3,8)$

Linked Comprehension Type Questions

Paragraph I Directions (Q.Nos. 16-18)

ABCD is a square of side length 2 units. C_1 is the circle inscribed in *ABCD* and C_2 is the circumcircle of *ABCD*. *L* is a fixed line in the same plane through the point *A*. Answer the following questions :

16 If *P* is any point on C_1 and *Q* is any point on C_2, then $\frac{(PA)^2+(PB)^2+(PC)^2+(PD)^2}{(QA)^2+(QB)^2+(QC)^2+(QD)^2}=$

(a) 0.75 (b) 1.25 (c) 1 (d) 0.5

17 A circle touches the line *L* and C_1 externally such that both circles are on the same side of *L*. Then, the locus of the centre *C* is

(a) an ellipse (b) a parabola (c) hyperbola (d) a pair of lines

18 A line *M* through *A* is drawn parallel to *BD*. Point *S* moves such that its distance from the line *BD* and vertex are equal. If locus of *S* cuts *M* at T_2 and T_3 and *AC* at T_1. Then, the area of $\Delta T_1T_2T_3$ is

(a) $\frac{1}{2}$ sq unit (b) $\frac{2}{3}$ sq unit (c) 1 sq unit (d) 2 sq units

Paragraph II Directions (Q.Nos. 19-21)

Let *A, B, C* and *D* lies on the same line such that $AB=BC=CD=1$. The points *A* and *C* are also joined by a semicircle with *AC* as diameter and *P* is a variable point on this semicircle such that $\angle PBD=\theta, 0\le\theta\le\pi$, let *R* is the region bounded by arc *AP*, the straight line *PD* and line *AD*.

19 The maximum possible area of region *R* is

(a) $\frac{2\pi-\sqrt{3}}{12}$ (b) $\frac{2\pi+3\sqrt{3}}{6}$ (c) $\frac{2\pi+6\sqrt{3}}{12}$ (d) $\frac{2\pi+3\sqrt{3}}{12}$

20 Let *L* is the perimeter of region *P*, then *L* is equal to

(a) $3-\theta+\sqrt{5-4\cos\theta}$ (b) $3+\pi+\sqrt{5-4\cos\theta}$

(c) $3+\pi-\theta+\sqrt{5-4\cos\theta}$ (d) $3+\pi+\theta+\sqrt{5+4\cos\theta}$

21 The non-negative difference of greatest and least values of *L* is

(a) $3-\sqrt{3}+\frac{\pi}{3}$ (b) $\sqrt{3}-3+\frac{\pi}{3}$ (c) $\sqrt{3}+3+\frac{2\pi}{3}$ (d) $\pi-2$

Paragraph III Directions (Q.Nos. 22-24)

Consider an isosceles triangle *T* with base 10 units and height 12 units, define a sequence $C_1, C_2, C_3, \ldots$ of circles such that C_1 is in circle of *T* and C_{i+1} is tangent to C_2 and both legs of isosceles triangle for $i\ge1$.

22 Radius of circle C_1 is

(a) $\frac{5}{3}$ (b) $\frac{7}{3}$

(c) $\frac{10}{3}$ (d) None of these

23 Ratio of radius C_{i+1} to radius of C_i is

(a) $\frac{2}{9}$ (b) $\frac{4}{9}$ (c) $\frac{5}{9}$ (d) $\frac{6}{9}$

24 The total area of all possible circles is

(a) $\frac{164\pi}{13}$ (b) $\frac{170\pi}{13}$ (c) $\frac{175\pi}{13}$ (d) $\frac{180\pi}{13}$

Paragraph IV Directions (Q.Nos. 25-26)

P is a variable point on the line $L = 0$. Tangents are drawn to the circle $x^2 + y^2 = 4$ from P to touch it at Q and R. The parallelogram $PQSR$ is completed.

25 If $L = 2x + y - 6 = 0$, then the locus of circumcentre of ΔPQR is

(a) $2x - y = 4$ (b) $2x + y = 3$ (c) $x - 2y = 4$ (d) $x + 2y = 3$

26 If $P = (6, 8)$, then the area of ΔQRS is

(a) $\frac{6^{3/2}}{25}$ sq units (b) $\frac{(24)^{3/2}}{25}$ sq units (c) $\frac{48\sqrt{6}}{25}$ sq units (d) $\frac{192\sqrt{6}}{25}$ sq units

Matrix Match Type Questions

27 Let the circle $C_1 : x^2 + y^2 = 9$ and $C_2 : (x-3)^2 + (y-4)^2 = 16$ intersect at the point X and Y. Suppose that another circle $C_3 : (x-h)^2 + (y-k)^2 = r^2$ satisfies the following conditions

(i) Centre of C_3 collinear with the centres of C_1 and C_2

(ii) C_1 and C_2 both lies inside C_3 and

(iii) C_3 touches C_1 at M and C_2 at N

Let the line through X and Y intersects C_3 at Z and W and let a common tangent of C_1 and C_3 be a tangent to the parabola $x^2 = 8\alpha y$.

Match the following columns **[JEE Advanced 2019]**

Column I		Column II	
(i)	$2h + k$	A.	6
(ii)	$\frac{\text{Length of } ZW}{\text{Length of } XY}$	B.	$\sqrt{6}$
(iii)	$\frac{\text{Area of } \Delta MZN}{\text{Area of } \Delta ZMW}$	C.	$\frac{5}{4}$
(iv)	α	D.	$\frac{21}{5}$
		E.	$2\sqrt{6}$
		F.	$\frac{10}{3}$

28 Match the columns

	Column I		Column II
(i)	The length of the common chord of two circles of radii 3 and 4 units which intersect orthogonally is $\frac{K}{5}$, then K is equal to	A.	1
(ii)	The circumference of the circle $x^2 + y^2 + 4x + 12y + p = 0$ is bisected by the circle $x^2 + y^2 - 2x + 8y + q = 0$, then $p + q$ is equal to	B.	24
(iii)	Number of distinct chord o f the circle $2x(x - \sqrt{2}) + y(2y - 1) = 0$, chords are passing through the point $\left(\sqrt{2}, \frac{1}{2}\right)$ and are bisected on X-axis, is	C.	32
(iv)	One of the diameters of the circle circumscribing the rectangle $ABCD$ is $4y = x + 7$. If A and B are the points $(-3, 4)$ and $(5, 4)$ respectively, then the area of rectangle is	D.	36

29 There are two circles in a parallelogram. One of them of radius 3 is inscribed in the parallelogram and the other touches two sides of the parallelogram and the first circle. The distance between the points of tangency which lie on the same side of the parallelogram is equal to 3.

	Column I		Column II
(i)	The radius of the other circle is	A.	$\frac{75}{2}$
(ii)	Area of the parallelogram is equal to	B.	75
(iii)	Let d_1, d_2 denote the lengths of the diagonals of parallelogram than the product $d_1 d_2$ is equal to	C.	$5\sqrt{31}$
(iv)	Let d_1, d_2 the diagonals of the parallelogram, then the value of $d_1 + d_2$ is equal to	D.	$\frac{3}{4}$

30 Match the following columns

	Column I		Column II
(i)	Number of values of a for which the common chord of the circles $x^2 + y^2 = 8$ and $(x - a)^2 + y^2 = 8$	A.	4
(ii)	A chord of the circles $(x - 1)^2 + y^2 = 4$ lies along the line $y = 22\sqrt{3}(x - 1)$. The length of the chord is equal to	B.	2
(iii)	The number of circles touching all the three lines $3x + 7y = 2$, $21x + 49y = 5$ and $9x + 21y = 1$, are	C.	0
(iv)	If radii of the smallest and largest circle passing through the point $(\sqrt{3}, \sqrt{2})$ and touching the circle $x^2 + y^2 - 2\sqrt{2}y - 2 = 0$ are r_1 and r_2 respectively, then the mean of r_1 and r_2 is	D.	1

31 Match the following columns

	Column I		Column II
(i)	If (α, β) is a point on circle whose centre is on X-axis which touches line $y = -x$ at $(2, -2)$, then greatest value of α is	A.	22
(ii)	From point on circle $x^2 + y^2 = 36$ tangents are drawn to hyperbola $x^2 - y^2 = 36$, locus of mid-point of chord of contact is $(x^2 - y^2)^2 = 12l(x^2 + y^2)$, then l is equal to	B.	2
(iii)	The sum of squares of length of chord intercepted by the line $x + y = n, n \in N$ on circle $x^2 + y^2 = 4$	C.	3
(iv)	Three distinct lines are drawn in a plane and there are exactly n circles in the plane tangent to all three lines, then n can be	D.	$4 + 2\sqrt{2}$

32 $A(-2, 0)$ and $B(2, 0)$ are two fixed points and P is a point such that $PA - PB = 2$. Let S be the circle $x^2 + y^2 = r^2$, then match the following columns

	Column I		Column II
(i)	If $r = 2$, then number of points P satisfying $PA - PB = 2$ and lying on $x^2 + y^2 = r^2$, is	A.	2
(ii)	If $r = 1$, then number of points P satisfying $PA - PB = 2$ and lying on $x^2 + y^2 = r^2$, is	B.	4
(iii)	For $r = 2$, number of common tangents is	C.	0
(iv)	For $r = \frac{1}{2}$, number of common tangents is	D.	1

Numerical Value Type Questions

Question Stem Directions (Q.Nos. 33-35) The answer to each question is a **NUMERICAL VALUE.**

Consider two circles S_1 and S_2 (externally touching) having centres at points A and B whose radii are 1 and 2, respectively. A tangent to circle S_1 from point B intersects the circle S_1 at point C. D is chosen on circle S_2, so that AC is parallel to BD and the two segments BC and AD do not intersect. Segment AD intersects the circles S_1 at E. The line through B and E intersects the circle S_1 at another point.

33 The length of segment EF is ....

34 The area of ΔABD is ....

35 The length of segment DE is ....

Question Stem Directions (Q.Nos. 36-38) The answer to each question is a **NUMERICAL VALUE.**

In an acute ΔABC, point H is the intersection point of altitude CE to AB and altitude BD to AC. A circle with DE as its diameter intersects AB and AC at point F and G respectively. If $BC = 25$, $BD = 20$ and $BE = 7$

36 The sum of length of all the sides of ΔABC is ....

37 Area of circle S is ....

38 Let FG and AH intersect at point K, then the length of AK is equal to ....

39 If centres of three circles C_1, C_2, C_3 are collinear such that C_2 passes through the centres O_1, O_3 and areas of C_1, C_2, C_3 are in arithmetic progression, then $\left(\frac{r_1^3 + r_2^3 + r_3^3}{r_1 r_2 r_3}\right)$ is equal to ....

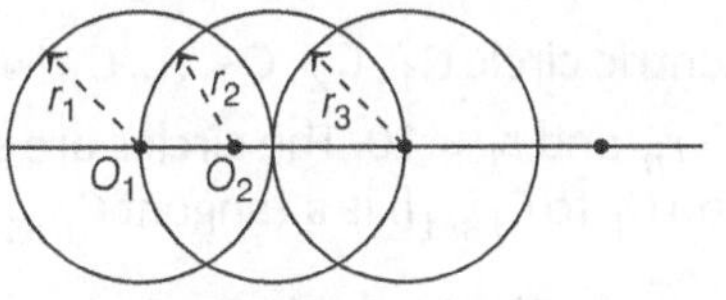

40 The number of straight lines which pass through the intersection of the lines $x - 2y - 5 = 0$, $7x + y = 50$ and divide the circumference of the circle $x^2 + y^2 = 100$ into two arcs whose lengths are in the ratio 2 :1, is ......

41 Through the point of intersection P which has integral coordinates of the circles $x^2 + y^2 = 1$ and $x^2 + y^2 + 2x + 4y + 1 = 0$, a common chord APB is drawn terminating on the two circles such that the chords AP and BP of the given circle. Subtend equal angles at the centre. If the equation of chord, $ax + by + c = 0$, then the value of $a + b + c$, is ......

42 From a point P tangent drawn to the circles $x^2 + y^2 + x - 3 = 0, 3x^2 + 3y^2 - 5x + 3y = 0$ and $4x^2 + 4y^2 + 8x + 7y + 9 = 0$ are of equal length. If the equation of the circle through P which touches the line $x + y = 5$ at the point $(6, -1)$ is $x^2 + y^2 + ax + by + 2c = 0$, then the value of $a + b + c$, is ......

43 A circle passes through three points A, B and C with the line segment AC as its diameter. A line passing through A intersects the chord BC at a point D inside the circle. If angles DAB and CAB are α and β, respectively and the distance between the point A and the mid-point of the line segment DC is d. If the area of the circle in terms of d, α and β

$\frac{k\,\pi d^2}{l\cos^2\alpha + m\cos^2\beta + n\cos\alpha\cos\beta\cos(\beta - \alpha)}$, then the value of $\frac{2k + l + m}{n}$ is ......

44 Let the straight line $L: \tan(\cot^{-1} 2)\, x - y = 4$ be rotated through an angle $\cot^{-1} 3$ about the point $M(0, -4)$ in anti-clockwise sense. After rotation the line become tangent to the circle which lies in fourth quadrant and also touches coordinate axes. The sum of radii of all possible circles, is ......

45 The difference between the radii of the largest and the smallest circles, which have their centre on the which circumference of the circle $x^2 + y^2 + 2x + 4y - 4 = 0$ and passes through the point (a, b) lying outside the given circles, is ......

46 If the radius of the smallest circles which touches the line $3x - y = 6$ at $(1, -3)$ and also touches the line $x = y$ is $\sqrt{a}\,(10 - 4\sqrt{b})$, where a and b are prime numbers, then the value of $(b + a)^3 + (b - a)^3$ is ......

47 Let two parallel lines L_1 and L_2 with positive slope are tangent to the circle $C_1 : x^2 + y^2 - 2x - 16y + 64 = 0$. If L_1 is also tangent to the circle $C_2 : x^2 + y^2 - 2x + 2y - 2 = 0$ and equation of L_2 is $a\sqrt{a}x - by + c - a\sqrt{a} = 0$, where $a, b, c \in N$ then the value of $a + b + c$, is ......

48 A circle is inscribed in a triangle with sides of lengths 3, 4 and 5. A second circle, interior to the triangle, is tangent to the first circle and to both sides of the larger acute angle of the triangle. If the radius of the second circle can be expressed in the form $\frac{\sin k}{\cos w}$, where k and w are in degrees and lie in the interval $(0, 90°)$, then the value of $k + w$ is ......

49 Inside a semi-circle of unit radius, two circle of radii r_1 and r_2 are drawn, touching the diameter and the semi-circle and also touching each other externally. If maximum value of $r_1 + r_2 = P(\sqrt{P} - 1)$, where $P \in Z$, then P is equal to ......

50 Consider series of n concentric circle $C_1, C_2, C_3, \ldots, C_n$ with radii $r_1, r_2, r_3, \ldots, r_n$ respectively satisfying $r_1 > r_2 > r_3 \ldots > r_n$ and $r_1 = 10$. The circles are such that the chord of contact of tangents from any point on C_i to $C_{i+1}h$ is a tangent C_{i+2} $(i = 1, 2, 3, \ldots)$. The value of $\lim_{h \to \infty} \sum_{i=1}^{n} r_i$ if the angle between the tangents from any point of C_1 to C_2 is $\frac{\pi}{3}$, is ......

ANSWER KEY

Level 1

1. (i) Centre = (2, 2) and Radius = 4 (ii) Centre = $\left(-\frac{3}{4}, -1\right)$ and Radius = $\frac{5}{4}$

(iii) Centre = $(-\sin\theta, -\cos\theta)$ and Radius = 3 (iv) Centre = $\left(1, -\frac{3}{2}\right)$ and Radius = $\frac{\sqrt{23}}{2}$

2. (i) $x^2 + y^2 \pm 8y = 0$ (ii) $x^2 + y^2 \pm 10x \pm 10y + 25 = 0$

(iii) $x^2 + y^2 - 6x - 8y + 15 = 0$ (iv) $x^2 + y^2 - 4x + 6y + 11 = 0$

(v) $x^2 + y^2 - 4x - 6y - 12 = 0$ (vi) $x^2 + y^2 + 6x - 2y - 51 = 0$

(vii) $x^2 + y^2 - 8x - 6y + 9 = 0$ (viii) $x^2 + y^2 + 4x + 6y - 85 = 0$

(ix) $x^2 + y^2 - 2x + 2y - 23 = 0$

3. (i) x-intercept = 6 and y-intercept = 0 (ii) x-intercept = $\sqrt{43}$ and y-intercept = $\sqrt{3}$

4. $(x-3)^2 + (y-4)^2 = \left(\frac{62}{13}\right)^2$ 5. $x^2 + y^2 - 12x - 10y + 52 = 0$

6. $x^2 + y^2 - 8x - 2y + 7 = 0$ and $x^2 + y^2 - 3x - 7y + 12 = 0$

7. $x^2 + y^2 - 5x - y + 4 = 0$ 8. $x^2 + y^2 - 6x + 2y - 28 = 0$

9. 32 sq units

10. (i) Lies on the circle (ii) Lies inside the circle

(iii) Lies outside the circle (iv) Lies outside the circle

11. (i) (1, 7) and $\left(\frac{-1}{2}, \frac{-1}{2}\right)$

(ii) Line touches the circle at point (– 1, 3) or intersect at the point (– 1, 3)

12. (i) $x + y = 2$ (ii) $3x + 4y - 10 \mp 5\sqrt{56} = 0$

(iii) $4x - 3y = 25$ (iv) $x + 2y \pm \sqrt{20} = 0$

13. 5 units 14. Centre = (3, 0) and radius = $\sqrt{5}$ units 15. $x^2 + y^2 = \frac{1}{a^2}$ 16. 4 units

17. $2\sqrt{7}$ 18. (i) $14x - 5y - 40 = 0$ (ii) $x + 2y = 1$

19. $2x - y + 1 = 0$ and $x + 2y - 2 = 0$ 20. 90° 21. 2

22. (i) $x - 2y = 5$ (ii) $4x + 7y + 10 = 0$ (iii) $15x - 21y + 38 = 0$

23. (i) $4(x^2 + y^2) - 7x - 4y + 3 = 0$

(ii) $3(x^2 + y^2) + 18x - 34y + 41 = 0$

(iii) $x^2 + y^2 - 4x + 2y - 3 = 0$ and $9(x^2 + y^2) + 96x - 158y + 193 = 0$

(iv) Not possible

(v) $3(x^2 + y^2) - 15x + 10y - 14 = 0$

24. (i) $7(x^2 + y^2) + x + 2y - 32 = 0$ (ii) $5(x^2 + y^2) - 4x - 8y - 26 = 0$ (iii) Not possible

25. (i) $\lambda = \frac{19}{3}$ (ii) $\lambda = \frac{-23}{19}$ (iii) $\lambda = -\frac{3}{2}$

26. $x^2 + y^2 - 2y - 6 = 0$ 27. $x^2 + y^2 + 14x - 6y + 6 = 0$

28. (i) $x + y = \frac{19}{2}$ (ii) $34x - 20y + 97 = 0$

29. (i) $\left(\frac{7}{5}, -2\right)$ (ii) $\left(-\frac{9}{2}, -\frac{5}{4}\right)$

30. $5x^2 + 5y^2 + 60x + 7 = 0$ 31. $4(x^2 + y^2) + 2y - 29 = 0$ 32. $x^2 + y^2 + 8x - 6y + 9 = 0$

33. $4x^2 + 4y^2 + 8gx + 8fy + 3g^2 + 3f^2 + c = 0$

Level 2

1. (c) 2. (c) 3. (b) 4. (d) 5. (d) 6. (b) 7. (b)
8. (a) 9. (c) 10. (d) 11. (b) 12. (b) 13. (b) 14. (b)
15. (d) 16. (a) 17. (b) 18. (b) 19. (a) 20. (c) 21. (a)
22. (c) 23. (c) 24. (a) 25. (a) 26. (b) 27. (a) 28. (c)
29. (d) 30. (c) 31. (b) 32. (d) 33. (b) 34. (b) 35. (a)
36. (b) 37. (a) 38. (c) 39. (b) 40. (c) 41. (b) 42. (b)
43. (c) 44. (b) 45. (a) 46. (c) 47. (c) 48. (a) 49. (c)
50. (b) 51. (c) 52. (c) 53. (c) 54. (a) 55. (b) 56. (a)
57. (d) 58. (c) 59. 3 60. 36 61. 3 62. 61 63. 75
64. 32 65. 22 66. (6) 67. 1 68. 5 69. 7 70. 8
71. 6 72. 3 73. 1 74. 4 75. 2 76. 6 77. 1
78. 6 79. 4 80. 2

Level 3

1. (c) 2. (b) 3. (b) 4. (d) 5. (a) 6. (a) 7. (c)
8. (b) 9. (b) 10. (a) 11. (b) 12. (b)

More than One Correct

1. (b, d) 2. (a, b, c, d) 3. (a, d) 4. (b, c) 5. (a, c) 6. (a, b, c) 7. (a, c, d)
8. (a, b) 9. (a, b, c, d) 10. (b, c) 11. (a, c) 12. (a, b, c) 13. (a, d) 14. (a, c)
15. (a, b, c, d)

Lined Comprehension Type Questions

16. (a) 17. (b) 18. (c) 19. (b) 20. (c) 21. (c) 22. (c)
23. (b) 24. (d) 25. (b) 26. (d)

Match the Column

27. (i)–A, (ii)–B, (iii)–C, (iv)–F
28. (i)–B, (ii)–D, (iii)–A, (iv)–C
29. (i)–D, (ii)–A, (iii)–B, (iv)–C
30. (i)–B, (ii)–A, (iii)–C, (iv)–D
31. (i)–D, (ii)–C, (iii)–A, (iv)–B
32. (i)–A, (ii)–C, (iii)–C, (iv)–A

Numerical Value Type Questions

33. 1.54 34. 2.83 35. 2 36. 80 37. 176.63 38. 8.63 39. 3
40. 2 41. 1 42. 6 43. 2 44. 8 45. 6 46. 370
47. 14 48. 54 49. 2 50. 20

SOLUTIONS

Level 1

1. (i) $x^2+y^2-4x-4y-8=0$

$\because$ Circle of equation $x^2+y^2+2gx+2fy+c=0$ is having centre $(-g,-f)$ and radius $=\sqrt{g^2+f^2-c}$

$\because\ 2g=-4;\quad 2f=-4;\quad c=-8$

$g=-2;\quad f=-2;\quad c=-8$

$\therefore$ Centre (2, 2) and radius $=\sqrt{(-2)^2+(-2)^2+8}$

$=\sqrt{16}=4$

(ii) Circle : $2x^2+2y^2+3x+4y=0$

$\Rightarrow\quad x^2+y^2+\frac{3}{2}x+2y=0$

$2g=\frac{3}{2};\quad 2f=2;\quad c=0$

$g=\frac{3}{4};\quad f=1;$

$\therefore$ Centre $=(-g,-f)=\left(-\frac{3}{4},-1\right)$

$\therefore$ Radius $=\sqrt{g^2+f^2-c}$

$=\sqrt{\frac{9}{16}+1}=\sqrt{\frac{25}{16}}=\frac{5}{4}$

(iii) Circle : $x^2+y^2+2x\sin\theta+2y\cos\theta-8=0$

$2g=2\sin\theta;\quad 2f=2\cos\theta;\quad c=-8$

$g=\sin\theta;\quad f=\cos\theta;\quad c=-8$

$\therefore$ Centre is $(-g,-f)=(-\sin\theta,-\cos\theta)$

and radius $=\sqrt{g^2+f^2-c}$

$=\sqrt{\sin^2\theta+\cos^2\theta-(-8)}$

$=\sqrt{1+8}=\sqrt{9}=3$

(iv) Circle :

$2x^2+2y^2+\lambda xy+(\lambda^2-4)x+(3\lambda+6)y-5+\lambda=0$ for some λ

$\Rightarrow\ x^2+y^2+\frac{\lambda}{2}xy+\left(\frac{\lambda^2-4}{2}\right)x+\left(\frac{3\lambda+6}{2}\right)y-\frac{5}{2}+\frac{\lambda}{2}=0$

$\because$ It represents a circle for some λ.

$\therefore$ Coefficient of $xy=0$

$\frac{\lambda}{2}=0$

$\Rightarrow\quad \lambda=0$

$\therefore$ Equation is $x^2+y^2-2x+3y-\frac{5}{2}=0$

$\therefore\quad 2g=-2;\quad 2f=3;\quad c=-\frac{5}{2}$

$g=-1;\quad f=\frac{3}{2};$

$\therefore$ Centre is $(-g,-f)\Rightarrow\left(1,-\frac{3}{2}\right)$

and radius is $\sqrt{g^2+f^2-c}=\sqrt{1+\frac{9}{4}+\frac{5}{2}}$

$=\sqrt{\frac{4+9+10}{4}}=\sqrt{\frac{23}{4}}$

$=\frac{\sqrt{23}}{2}$

2. (i) $\because$ Circle touches at X-axis at origin and radius is 4 units.

$\Rightarrow$ Centre $(0,\pm 4)$

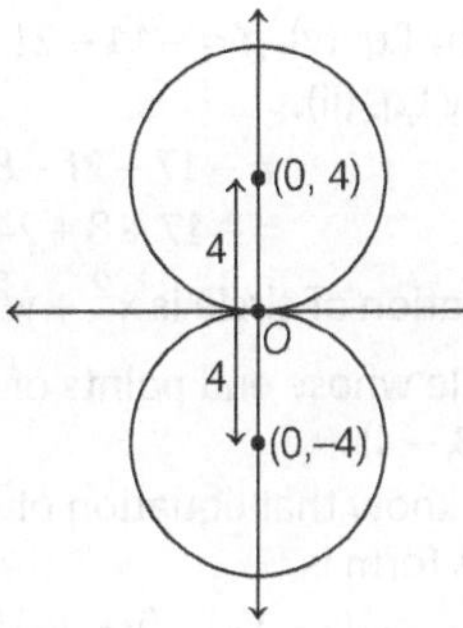

$\therefore$ Equation of circle is

$(x-0)^2+(y\pm 4)^2=4^2$

$\Rightarrow\quad x^2+y^2\pm 8y+16=16$

$\Rightarrow\quad x^2+y^2\pm 8y=0$

(ii) $\because$ Circle touches both the axes with 5 units.

$\therefore$ Centre is $(\pm 5,\pm 5)$

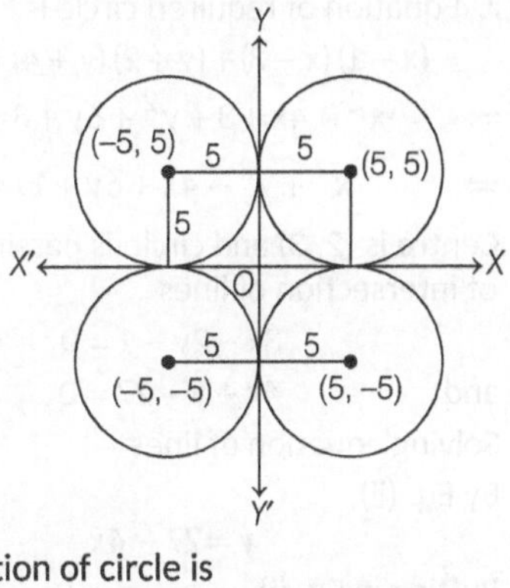

$\therefore$ Equation of circle is

$(x-(\pm 5))^2+(y-(\pm 5))^2=(5)^2$

$\Rightarrow\quad x^2+y^2\pm 10x\pm 10y+25=0$

(iii) ∵ Circle is passing through points (4, 1) and (6, 5).

Let the cirlce be $x^2 + y^2 + 2gx + 2fy + c = 0$ ...(i)

Hence, put (4, 1) and (6, 5),

$P : (4, 1) \Rightarrow 16 + 1 + 8g + 2f + c = 0$

$\Rightarrow \quad 8g + 2f + c = -17$ ...(ii)

$Q : (6, 5)$

$\Rightarrow \quad 36 + 25 + 12g + 10f + c = 0$

$\Rightarrow \quad 12g + 10f + c = -61$ ...(iii)

∵ Centre $(-g, -f)$ lies on $4x + y = 16$

$\Rightarrow \quad -4g - f = 16$ ...(iv)

By Eqs. (ii) and (iii), we get

$$-4g - 8f = 44$$
$$g + 2f = -11$$
$$g = -11 - 2f \quad \text{...(v)}$$

Put it in Eq. (iv),

$$-4(-11 - 2f) - f = 16$$

$\Rightarrow \quad 44 + 8f - f = 16$

$\Rightarrow \quad 7f = 16 - 44$

$\Rightarrow \quad f = -\frac{28}{7} = -4$

Now by Eq. (v), $g = -11 - 2(-4) \Rightarrow g = -3$

and by Eq. (ii),

$$c = -17 - 2f - 8g$$
$$= -17 + 8 + 24 = 15$$

∴ Equation of circle is $x^2 + y^2 - 6x - 8y + 15 = 0$

(iv) ∵ Circle whose end points of diamter are $(1, -2)$ and $(3, -4)$.

∵ We know that equation of circle in diametric end points form is

$$(x - x_1)(x - x_2) + (y - y_1)(y - y_2) = 0$$

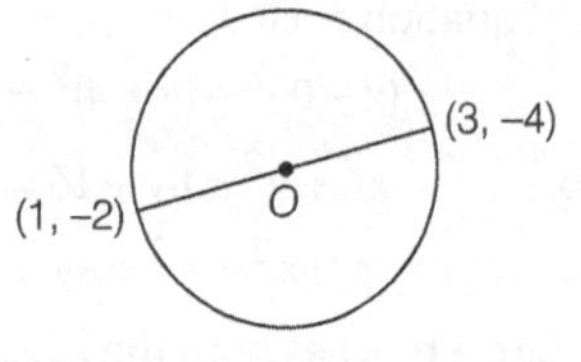

where (x_1, y_1) and (x_2, y_2) are diametric end point

∴ Equation of required circle is

$$(x - 1)(x - 3) + (y + 2)(y + 4) = 0$$

$\Rightarrow \quad x^2 - 4x + 3 + y^2 + 6y + 8 = 0$

$\Rightarrow \quad x^2 + y^2 - 4x + 6y + 11 = 0$

(v) Centre is (2, 3) and circle is passing through point of intersection of lines

$$3x - 2y - 1 = 0 \quad \text{...(i)}$$

and $\quad 4x + y - 27 = 0$ ...(ii)

Solving equation of lines

By Eq. (ii),

$$y = 27 - 4x$$

Putting in Eq. (i),

$$3x - 2(27 - 4x) - 1 = 0$$
$$3x - 54 + 8x - 1 = 0$$
$$11x - 55 = 0, \quad x = 5$$
$$y = 27 - 4 \times 5 = 7$$

∴ Circle is passing through (5, 7).

∴ Radius is $\sqrt{(5-2)^2 + (7-3)^2} = \sqrt{3^2 + 4^2} = 5$

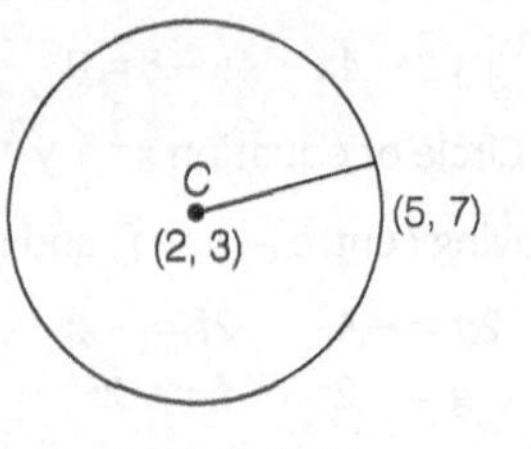

∴ Equation of required circle is

$$(x - 2)^2 + (y - 3)^2 = (5)^2$$
$$x^2 + y^2 - 4x - 6y - 12 = 0$$

(vi) Given

$$C_1 : x^2 + y^2 + 16x - 14y = 1$$

O_1 (centre) $= (-8, 7)$

$$C_2 : x^2 + y^2 - 4x + 10y = 2$$

O_2 (centre) $= (2, -5)$

∵ O_1 and O_2 are diametric end points of required circle.

∴ Equation of required circle is

$$(x - x_1)(x - x_2) + (y - y_1)(y - y_2) = 0$$

$\Rightarrow \quad (x + 8)(x - 2) + (y - 7)(y + 5) = 0$

$\Rightarrow \quad x^2 + 6x - 16 + y^2 - 2y - 35 = 0$

$\Rightarrow \quad x^2 + y^2 + 6x - 2y - 51 = 0$

(vii) Given centre of required circle is (4, 3) and it touches the circle $x^2 + y^2 = 1$

$\Rightarrow O(0, 0), r_1 = 1$

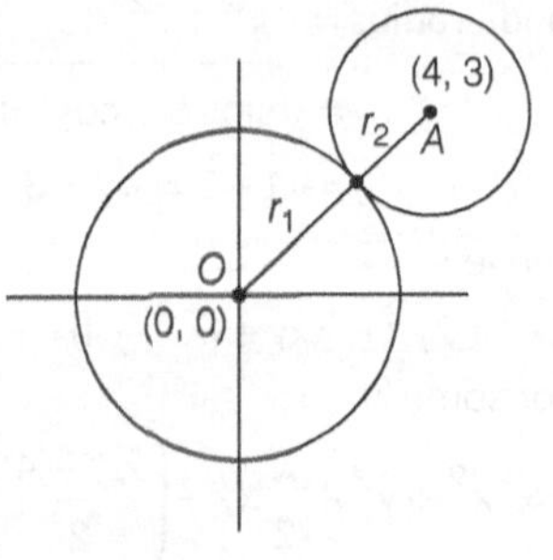

∵ Both circle touches each other.

∴ $\quad OA = r_1 + r_2$

$\Rightarrow \quad \sqrt{(4-0)^2 + (3-0)^2} = 1 + r_2$

$\Rightarrow \quad 5 = 1 + r_2$

$\Rightarrow \quad r_2 = 4$

∴ Equation of required circle is

$$(x - 4)^2 + (y - 3)^2 = 4^2$$

$\Rightarrow \quad x^2 + y^2 - 8x - 6y + 9 = 0$

(viii) $\because$ Required circle is concentric with the circle $x^2+y^2+4x+6y+11=0$ whose centre is $(-2,-3)$ and passing through $(5, 4)$.

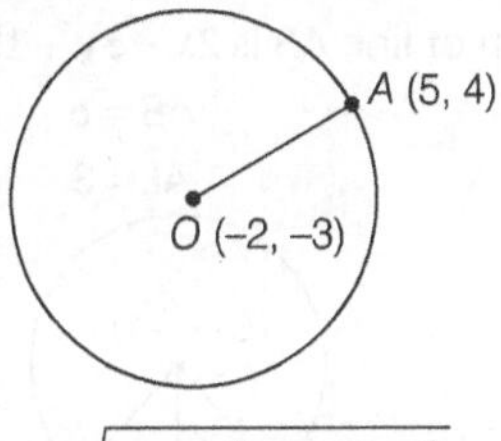

$\therefore$ Radius $= OA = \sqrt{(5+2)^2+(4+3)^2} = 7\sqrt{2}$

$\therefore$ Equation of required circle is

$$(x+2)^2+(y+3)^2=(7\sqrt{2})^2$$

$$x^2+y^2+4x+6y+13=98$$

$$x^2+y^2+4x+6y-85=0$$

(ix) Let (a,b) be the centre and r be the radius of the required circle.

$\because$ Circle passes through the given vertices therefore, their distance from centre is radius

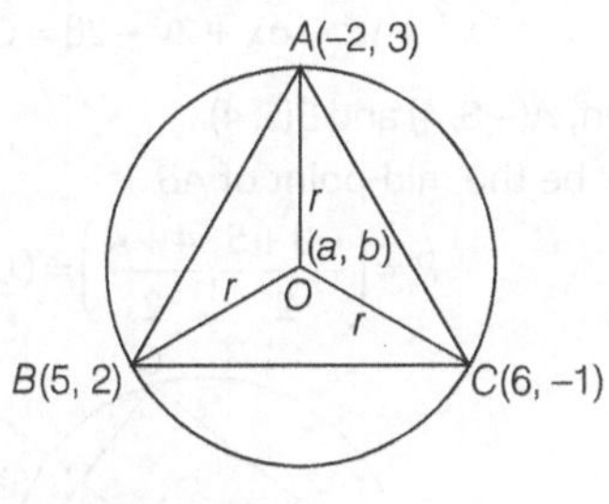

Using distance formula

$OA = r \Rightarrow OA^2 = r^2 \Rightarrow (a+2)^2+(b-3)^2=r^2$

$\Rightarrow \quad a^2+b^2+4a-6b+13=r^2 \quad$ ...(i)

$OB = r \Rightarrow OB^2 = r^2$

$\Rightarrow \quad (a-5)^2+(b-2)^2=r^2$

$\Rightarrow \quad a^2+b^2-10a-4b+29=r^2 \quad$ ...(ii)

$OC = r \Rightarrow OC^2=r^2 \Rightarrow (a-6)^2+(b+1)^2=r^2$

$\Rightarrow \quad a^2+b^2-12a+2b+37=r^2 \quad$ ...(iii)

Subtracting Eq. (ii) from Eq. (i), we get

$$14a-2b-16=0$$

$\Rightarrow \quad 14a-2b=16 \quad$ ...(iv)

Subtracting Eq. (iii) from Eq. (ii), we get

$$2a-6b-8=0 \Rightarrow a-3b=4 \quad \text{...(v)}$$

Solving Eqs. (iv) and (v),

$$14a-2b=16$$

$\Rightarrow \quad 14a-42b=56$

$\Rightarrow \quad 40b=-40$

$\Rightarrow \quad b=-1$ and $a=4-3=1$

$\therefore \quad a=1, b=-1$

By Eq. (i),

$(1)^2+(-1)^2+4(1)-6(-1)+13=r^2$

$\Rightarrow \quad 1+1+4+6+13=r^2$

$\Rightarrow \quad r^2=25 \Rightarrow r=5$

$\therefore$ Equation of required circle

$$(x-1)^2+(y+1)^2=5^2$$

$\Rightarrow \quad x^2+y^2-2x+2y-23=0$

3. (i) Circle : $x^2+y^2+10x-8y+16=0$

$2g=10; \quad 2f=-8; \quad c=16$

$g=5; \quad f=-4; \quad c=16$

$\therefore$ x-intercept $=2\sqrt{g^2-c}=2\sqrt{(5)^2-16}$

$=2\sqrt{25-16}=6$

and y-intercept $=2\sqrt{f^2-c}=2\sqrt{(-4)^2-16}=0$

(ii) Circle : $2x^2+2y^2-14x-6y+3=0$

$$x^2+y^2-7x-3y+\frac{3}{2}=0$$

$2g=-7; \quad 2f=-3; \quad c=\frac{3}{2}$

$g=-\frac{7}{2}; \quad f=-\frac{3}{2};$

$\therefore$ x-intercept $=2\sqrt{g^2-c}=2\sqrt{\frac{49}{4}-\frac{3}{2}}$

$=2\sqrt{\left(\frac{49-6}{4}\right)}=2\sqrt{\frac{43}{4}}=\sqrt{43}$

and y-intercept $=2\sqrt{f^2-c}=2\sqrt{\left(-\frac{3}{2}\right)^2-\frac{3}{2}}$

$=2\left(\sqrt{\frac{9-6}{4}}\right)=2\frac{\sqrt{3}}{2}=\sqrt{3}$

4. Let O be the centre of circle and r be the radius of circle, then

r = length of the perpendicular from (3, 4) to line

$$5x+12y-1=0$$

$$=\left|\frac{5(3)+12(4)-1}{\sqrt{5^2+12^2}}\right|=\left|\frac{15+48-1}{13}\right|=\frac{62}{13}$$

Now, the equation of the required circle will be

$$(x-3)^2+(y-4)^2=\left(\frac{62}{13}\right)^2$$

5. Given, $AC=2$, $A(2,2)$ and $B(6,5)$

$$AB=\sqrt{(6-2)^2+(5-2)^2}$$

$\Rightarrow \quad AB=\sqrt{16+9}=5$

$\therefore \quad BC=AB-AC=5-2=3$

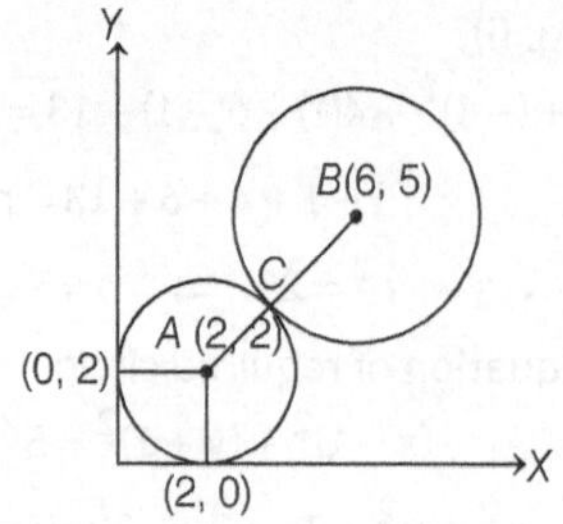

Hence, equation of circle whose centre is (6, 5) and radius is 3 will be $(x-6)^2+(y-5)^2=(3)^2$

$$x^2+y^2-12x-10y+52=0$$

6. Let the circle be $x^2+y^2+2gx+2fy+c=0$ ...(i)

Since it passes through (1, 2) and (3, 4)

$$5+2g+4f+c=0 \quad \text{...(ii)}$$

and $$25+6g+8f+c=0 \quad \text{...(iii)}$$

Subtracting Eqs. (ii) from Eq. (iii), we get

$$g+f+5=0 \quad \text{...(iv)}$$

Again, $3x+y-3=0$ is a tangent to circle (i)

$$\left|\frac{-3g-f-3}{\sqrt{10}}\right|=\sqrt{g^2+f^2-c}$$

$\Rightarrow$ $(3g+f+3)^2=10(g^2+f^2-c)$

$\Rightarrow$ $(3g+f+3)^2=10(g^2+f^2+5+2g+4f)$ [$\because$ from Eq. (ii)]

$\Rightarrow$ $(2g-5+3)^2=10[g^2+(g+5)^2+5+2g-4g-20]$ [$\because$ putting the value of f from Eq. (iv)]

$\Rightarrow$ $(2g-2)^2=10(2g^2+8g+10)$

$\Rightarrow$ $g^2-2g+1=5(g^2+4g+5)$

$\Rightarrow$ $2g^2+11g+12=0$

$\Rightarrow$ $(g+4)(2g+3)=0 \Rightarrow g=-4,-\frac{3}{2}$

and from Eq. (iv), $f=-1,-\frac{7}{2}$ and $c=7, 12$

Required circles are

$$x^2+y^2-8x-2y+7=0$$

and $$x^2+y^2-3x-7y+12=0$$

7. Let the equation of the circle be

$$x^2+y^2+2gx+2fy+c=0$$

Since, it passes through (1, 1), (2, –1) and (3, 2)

$$1+1+2g+2f+c=0$$

$\Rightarrow$ $4+1+4g-2f+c=0$

$\Rightarrow$ $9+4+6g+4f+c=0$

$\Rightarrow$ $2g+2f+c=-2$ ...(i)

$\Rightarrow$ $4g-2f+c=-5$ ...(ii)

$\Rightarrow$ $6g+4f+c=-13$ ...(iii)

Solving Eqs. (i), (ii) and (iii), we get

$$g=-\frac{5}{2}, f=-\frac{1}{2}, c=4$$

Hence, equation of circle is

$$x^2+y^2-5x-y+4=0$$

8. Let O be the centre of the circle, then O (3, – 1)

Equation of line AB is $2x-5y+18=0$

and $AB=6$

$\therefore$ $AL=3$

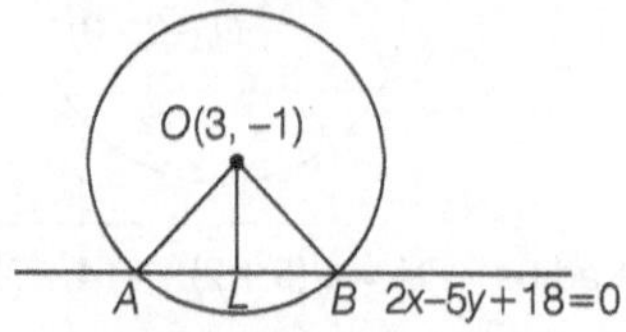

OL = Length of the perpendicular from O to AB

$$=\frac{|2(3)-5(-1)+18|}{\sqrt{2^2+(-5)^2}}=\sqrt{29}$$

$\therefore$ Radius of the circle,

$$OA=\sqrt{AL^2+OL^2}=\sqrt{3^2+29}=\sqrt{38}$$

Now, equation of the required circle is

$$(x-3)^2+(y+1)^2=38$$

$\Rightarrow$ $x^2+y^2-6x+2y-28=0$

9. Given, $A(-3, 4)$ and $B(5, 4)$.

Let P be the mid-point of AB

$\therefore$ $$P=\left(\frac{-3+5}{2},\frac{4+4}{2}\right)=(1, 4)$$

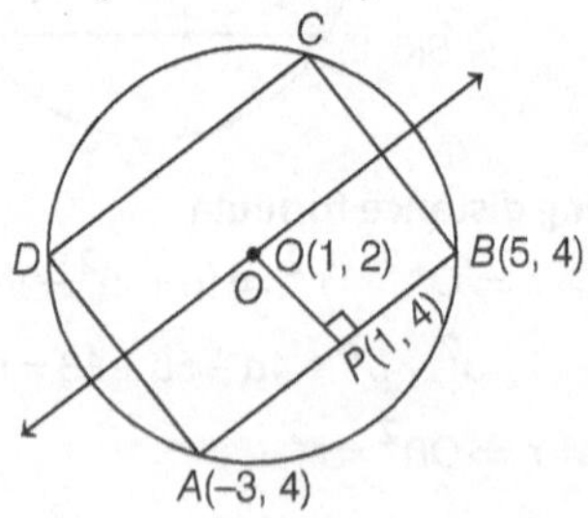

Equation of AB,

$$y-4=\frac{4-4}{5+3}(x+3) \Rightarrow y=4$$

Clearly, AB is parallel to X-axis.

$\therefore$ Perpendicular bisector of AB will be parallel to Y-axis and since it passes through the point $P(1, 4)$.

$\therefore$ Its equation will be $x=1$ ...(i)

Equation of one diameter of circle is $4y=x+7$ ...(ii)

Solving Eqs. (i) and (ii), we get $x=1$ and $y=2$

O is the centre of circle, then O(1, 2).

$\therefore$ Length of perpendicular from O to AB

i.e. $$OP=\sqrt{(1-1)^2+(4-2)^2}=2$$

One side of the rectangle $=2OP=4$

Also, other side of rectangle

$$AB=\sqrt{(5+3)^2+(4-4)^2}=8$$

$\therefore$ Area of rectangle $ABCD=4\times8=32$ sq units

10. Circle : $S = x^2 + y^2 - 6x - 10y - 2 = 0$

(i) Point $P(3, -1)$

$$S_1 = (3)^2 + (-1)^2 - 6(3) - 10(-1) - 2$$
$$= 9 + 1 - 18 + 10 - 2$$
$$= 20 - 20 = 0$$

$\therefore$ P lies on the circle.

(ii) Point $Q(-1, 3)$

$$S_1 = (-1)^2 + (3)^2 - 6(-1) - 10(3) - 2$$
$$= 1 + 9 + 6 - 30 - 2$$
$$= -16 < 0$$

$\therefore$ Point Q lies inside the circle.

(iii) Point $R(0, -1)$

$$S_1 = (0)^2 + (-1)^2 - 6(0) - 10(-1) - 2$$
$$= 1 + 10 - 2$$
$$= 9 > 0$$

$\therefore$ Point R lies outside the circle.

(iv) Point $T = (-3, 0)$

$$S_1 = (-3)^2 + (0)^2 - 6(-3) - 10(0) - 2$$
$$= 9 + 18 - 2$$
$$= 25 > 0$$

$\therefore$ Point T lies outside the circle.

11. (i) Line $5x - y + 2 = 0$

$\Rightarrow \quad y = 5x + 2 \quad$ ...(i)

Circle $x^2 + y^2 - 13x - 4y - 9 = 0 \quad$ ...(ii)

Putting the value of y from Eq. (i) to Eq. (ii), we get

$$x^2 + (5x + 2)^2 - 13x - 4(5x + 2) - 9 = 0$$
$$\Rightarrow \quad 26x^2 - 13x - 13 = 0$$
$$\Rightarrow \quad 2x^2 - x - 1 = 0$$
$$\Rightarrow \quad (2x + 1)(x - 1) = 0$$
$$x = 1 \quad \text{or} \quad x = -\frac{1}{2}$$

When $x = 1 \Rightarrow y = 5(1) + 2 = 7$

$$x = -\frac{1}{2} \Rightarrow y = 5\left(-\frac{1}{2}\right) + 2 = -\frac{1}{2}$$

$\therefore$ Coordinates of points of intersection are $(1, 7)$ and $\left(-\frac{1}{2}, -\frac{1}{2}\right)$.

(ii) Line : $2x + y - 1 = 0$

$\Rightarrow \quad y = 1 - 2x \quad$ ...(i)

Circle $x^2 + y^2 + 6x - 4y + 8 = 0 \quad$ ...(ii)

Putting the value of y from Eq. (i) to Eq. (ii), we get

$$x^2 + (1 - 2x)^2 + 6x - 4(1 - 2x) + 8 = 0$$
$$\Rightarrow \quad x^2 + 1 + 4x^2 - 4x + 6x - 4 + 8x + 8 = 0$$
$$\Rightarrow \quad 5x^2 + 10x + 5 = 0$$
$$\Rightarrow \quad x^2 + 2x + 1 = 0$$
$$\Rightarrow \quad (x + 1)^2 = 0$$
$$x = -1 \text{ and } y = 1 - 2(-1) = 3$$

$\therefore$ Line touches the circle at point $(-1, 3)$ or intersect at the point $(-1, 3)$.

12. (i) Circle $x^2 + y^2 = 2$

Point (1, 1)

$\therefore$ Equation of tangent at point (x_1, y_1) to the circle $x^2 + y^2 + 2gx + 2fy + c = 0$ is

$$xx_1 + yy_1 + 2g\left(\frac{x + x_1}{2}\right) + 2f\left(\frac{y + y_1}{2}\right) + c = 0$$

Equation of required tangent is

$$x(1) + y(1) = 2$$
$$\Rightarrow \quad x + y = 2$$

(ii) Circle $x^2 + y^2 - 12x + 4y - 16 = 0$

$$\Rightarrow \quad (x - 6)^2 + (y + 2)^2 = 56$$

Tangents are perpendicular to the line

$$4x - 3y + 8 = 0$$

Slope is $\frac{4}{3}$

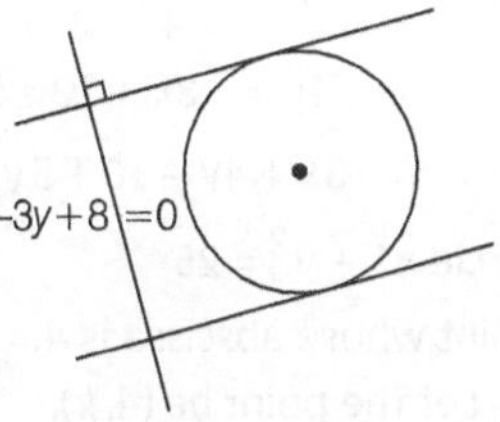

Let equation of tangent is $y = mx + c$

$\therefore$ $m \times$ slope of given line $= -1$

$$m \times \frac{4}{3} = -1$$
$$m = -\frac{3}{4}$$

$\therefore$ Equation becomes $y = -\frac{3x}{4} + c$

Circle $x^2 + y^2 - 12x + 4y - 16 = 0$

Putting $y = -\frac{3x}{4} + c$ in equation of circle

$$x^2 + \left(-\frac{3x}{4} + c\right)^2 - 12x + 4\left(-\frac{3x}{4} + c\right) - 16 = 0$$
$$\Rightarrow x^2 + \frac{9x^2}{16} + c^2 - \frac{3xc}{2} - 12x - 3x + 4c - 16 = 0$$
$$\Rightarrow \quad \frac{25x^2}{16} - 15x - \frac{3xc}{2} + c^2 + 4c - 16 = 0$$
$$\Rightarrow \quad \frac{25}{16}x^2 - x\left(15 + \frac{3c}{2}\right) + c^2 + 4c - 16 = 0$$

$\therefore$ It is tangent $D = 0$

$$\left(15 + \frac{3c}{2}\right)^2 - 4\left(\frac{25}{16}\right)(c^2 + 4c - 16) = 0$$

$\Rightarrow \quad \frac{9}{4}(10+c)^2 - \frac{25}{4}(c^2+4c-16)=0$

$\Rightarrow 9(100+c^2+20c)-25(c^2+4c-16)=0$

$\Rightarrow 900+9c^2+180c-25c^2-100c+400=0$

$\Rightarrow \quad -16c^2+80c+1300=0$

$\Rightarrow \quad -4(4c^2-20c-325)=0$

$\Rightarrow \quad 4c^2-20c-325=0$

$$c=\frac{20\pm\sqrt{400-4(4)(-325)}}{4\times 2}$$

$$=\frac{5\pm\sqrt{25+325}}{2}$$

$$=\frac{5\pm\sqrt{350}}{2}$$

$$=\frac{5}{2}\pm\frac{5\sqrt{14}}{2}$$

$\therefore$ Equation of tangent

$$y=-\frac{3x}{4}+\frac{5}{2}\pm\frac{5\sqrt{14}}{2}$$

$$4y=-3x+10\pm 5\times 2\sqrt{14}$$

$$3x+4y-10\mp 5\sqrt{56}=0$$

(iii) Circle $x^2+y^2=25$

Point whose abscissa is 4.

$\therefore$ Let the point be $(4, k)$.

$\because$ It lies on the circle.

$\Rightarrow \quad (4)^2+k^2=25$

$$k^2=9$$

$$k=\pm 3$$

$\therefore$ Points are $(4, 3)$ and $(4, -3)$.

$\therefore$ Equation of tangent at (4, 3) is

$\Rightarrow \quad x(4)+y(3)=25$

$\Rightarrow \quad 4x+3y=25$

and equation of tangent at $(4, -3)$ is

$$x(4)+y(-3)=25$$

$\Rightarrow \quad 4x-3y=25$

(iv) Circle $x^2+y^2=4$ ...(i)

Tangent is parallel to $x+2y+3=0$

$\therefore$ Let equation of tangent is $x+2y+k=0$

$\Rightarrow \quad x=-k-2y$

$\therefore$ Putting $x=-k-2y$ in Eq. (i),

$$(-k-2y)^2+y^2=4$$

$\Rightarrow \quad k^2+4y^2+4ky+y^2=4$

$\Rightarrow \quad 5y^2+4ky+k^2-4=0$

$D=0$ (condition of tangency)

$\Rightarrow \quad (4k)^2-4(5)(k^2-4)=0$

$\Rightarrow \quad 16k^2-4(5)(k^2-4)=0$

$\Rightarrow \quad 16k^2-20k^2+80=0$

$\Rightarrow \quad -4k^2=-80$

$\Rightarrow \quad k^2=20$

$\Rightarrow \quad k=\pm\sqrt{20}$

$\therefore$ Equation of tangents are $x+2y\pm\sqrt{20}=0$.

13. Given line is $4x-3y-5=0$

and given circle is $x^2+y^2+3x-y-10=0$

Let O be the centre and r be the radius of circle, then

$$O\left(-\frac{3}{2},\frac{1}{2}\right) \text{ and } r^2=\frac{9}{4}+\frac{1}{4}+10=\frac{25}{2}$$

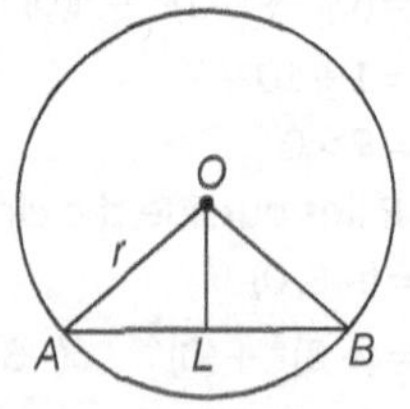

$$OL=\left|\frac{4\left(-\frac{3}{2}\right)-3\left(\frac{1}{2}\right)-5}{\sqrt{16+9}}\right|$$

$$OL=\frac{5}{2}$$

$\therefore \quad AB=2AL=2\sqrt{r^2-OL^2}$

$$=2\sqrt{\frac{25}{2}-\frac{25}{4}}=5 \text{ units}$$

14. Given, $4l^2-5m^2+6l+1=0$ ...(i)

Given line $\quad lx+my+1=0$ ...(ii)

$lx+my+1=0$ touches the circle whose centre is (α, β) and radius is r, then

$$\left|\frac{l\alpha+m\beta+1}{\sqrt{l^2+m^2}}\right|=r$$

$\Rightarrow \quad (l\alpha+m\beta+1)^2=r^2(l^2+m^2)$

$\Rightarrow \quad l^2\alpha^2+m^2\beta^2+1+2l\alpha+2m\beta+2lm\alpha\beta$
$$=r^2l^2+r^2m^2$$

$\Rightarrow \quad (\alpha^2-r^2)l^2+(\beta^2-r^2)m^2+2lm\alpha\beta$
$$+2l\alpha+2m\beta+1=0 \quad \text{...(iii)}$$

Comparing Eqs. (i) and (iii), we get

$$\alpha^2-r^2=4 \quad \text{and} \quad \beta^2-r^2=-5$$

Solving these equations, we get

$$\alpha=3, \beta=0, r=\sqrt{5}$$

$\therefore$ Centre of circle (3, 0) and radius $=\sqrt{5}$ units

15. Since, $y = mx + c$ is tangent to $x^2 + y^2 = a^2$, if

$c = \pm a\sqrt{1+m^2}$

$\therefore y = -\frac{lx}{m} + \frac{1}{m}$ is tangent to $x^2 + y^2 = a^2$, if

$$\frac{1}{m} = \pm \frac{a}{m}\sqrt{l^2 + m^2}$$

$$\Rightarrow \quad l^2 + m^2 = \frac{1}{a^2}$$

Hence, locus of the point (l, m) is $x^2 + y^2 = \frac{1}{a^2}$

16.

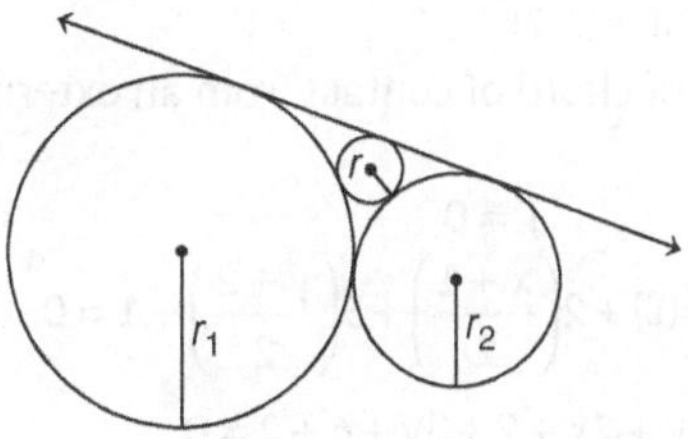

$$2\sqrt{rr_1} + 2\sqrt{rr_2} = 2\sqrt{r_1 r_2}$$

$$\Rightarrow \quad \sqrt{r}(6+3) = 6 \times 3$$

$$\Rightarrow \quad \sqrt{r} = \frac{18}{9}$$

$$\Rightarrow \quad \sqrt{r} = 2$$

$$\Rightarrow \quad r = 4$$

$\therefore$ Radius of circle = 4 units

17. Given equation of circle

$$x^2 + y^2 = 9$$

$$OA = OB = r = 3$$

$$OP = \left(\frac{0 - 0 + 2}{\sqrt{1+1}}\right) = \sqrt{2}$$

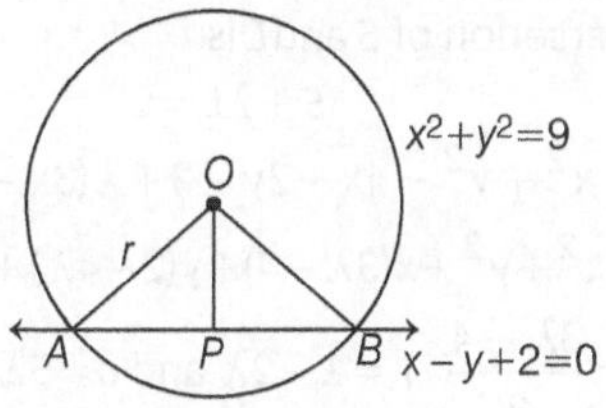

Length of chord $AB = 2AP$

$$= 2\sqrt{OA^2 - OP^2}$$

$$= 2\sqrt{9-2} = 2\sqrt{7}$$

18. (i) Circle $x^2 + y^2 - 5x + 2y - 48 = 0$

$$2g = -5; \quad 2f = 2$$

$$g = -\frac{5}{2}; \quad f = 1$$

Centre $(-g, -f) \Rightarrow \left(\frac{5}{2}, -1\right)$

and given point (5, 6)

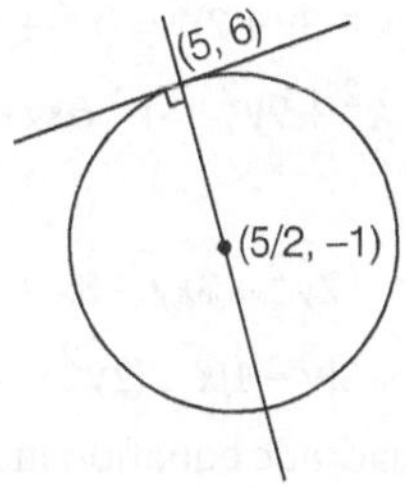

$\therefore$ Normal passes through the centre.

$\therefore$ Equation of normal = Equation of line passing through (5, 6) and $\left(\frac{5}{2}, -1\right)$

$$\Rightarrow \quad (y - 6) = \frac{-1-6}{\frac{5}{2} - 5}(x - 5)$$

$$\Rightarrow \quad (y - 6) = \frac{-7 \times 2}{-5}(x - 5)$$

$$\Rightarrow \quad (y - 6) = \frac{14}{5}(x - 5)$$

$$\Rightarrow \quad 5y - 30 = 14x - 70$$

$$\Rightarrow \quad 14x - 5y - 40 = 0$$

(ii) Circle $x^2 + y^2 = 2x$

$\Rightarrow \quad x^2 + y^2 - 2x = 0$

Centre (1, 0)

Normal is parallel to $x + 2y = 3$

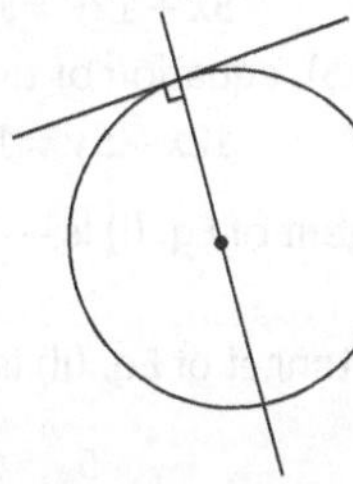

Let equation of normal is $x + 2y = k$

$\because$ It passes through the centre.

$$\therefore \quad 1 + 2(0) = k$$

$$k = 1$$

$\therefore$ Equation of normal is $x + 2y = 1$

19. Circle $x^2 + y^2 - 2x + 4y = 0$

From the point (0, 1),

Pair of tangents $\Rightarrow T^2 = SS_1$

$$S = x^2 + y^2 - 2x + 4y = 0$$

$$S_1 = (0)^2 + (1)^2 - 2(0) + 4(1) = 1 + 4 = 5$$

$$T = x(0) + y(1) - 2\left(\frac{x+0}{2}\right) + 4\left(\frac{y+1}{2}\right)$$

$$= y - x + 2(y+1)$$

$$= -x + 3y + 2 = 0$$

$$\therefore \quad T^2 = SS_1$$

$\Rightarrow \quad (-x+3y+2)^2=(x^2+y^2-2x+4y)(5)$

$\Rightarrow \quad x^2+9y^2+4-6xy-4x+12y$

$=5x^2+5y^2-10x+20y$

$\Rightarrow \quad 2x^2-2y^2+3xy-3x+4y-2=0$

$\Rightarrow \quad 2x^2+3(y-1)x-(2y^2-4y+2)=0$

This is a quadratic equation in x, by using quadratic formula,

$$x=\frac{-3(y-1)\pm\sqrt{9(y-1)^2+8(2y^2-4y+2)}}{2\times 2}$$

$\Rightarrow \quad 4x+3(y-1)=\pm\sqrt{25y^2-50y+25}$

$\Rightarrow \quad 4x+3y-3=\pm 5(y-1)$

$\Rightarrow \quad 4x+3y-3=5y-5$

$\Rightarrow \quad 4x-2y=-2$

$\Rightarrow \quad 2x-y=-1$

$\Rightarrow \quad 4x+3y-3=-5y+5$

$\Rightarrow \quad 4x+8y=8$

$\Rightarrow \quad x+2y=2$

$\therefore$ Separate equation of tangents are

$$2x-y+1=0$$

and $\quad x+2y-2=0$

20. Since, both the points lie on the circle.

At (5, 12), equation of tangent is

$$5x+12y=169 \quad \text{...(i)}$$

and at (12, – 5), equation of tangent is

$$12x-5y=169 \quad \text{...(ii)}$$

Slope of tangent of Eq. (i) is $-\frac{5}{12}$ (say m_1)

and slope of tanget of Eq. (ii) is $\frac{12}{5}$ (say m_2)

$\therefore \quad m_1m_2=-\frac{5}{12}\times\frac{12}{5}=-1$

Tangents are perpendicular to each other.

$\therefore$ Angle between tangents is 90°.

21. Given, equation of circle

$$x^2+y^2-8x+2y=0$$

Centre, $C_1(4,-1), r_1=\sqrt{16+1}=\sqrt{17}$

and equation of circle

$$x^2+y^2-2x-16y+25=0$$

Centre $C_2=(1,8), r_2=\sqrt{1+64-25}=\sqrt{40}$

Now, $\quad C_1C_2=\sqrt{(4-1)^2+(-1-8)^2}$

$=\sqrt{9+81}=\sqrt{90}$

$r_1+r_2=\sqrt{17}+\sqrt{40}$

Clearly, $C_1C_2<r_1+r_2$.

Hence, the number of common tangent are 2.

22. (i) Circle $x^2+y^2=9$

Middle point of chord = (1, – 2)

$\because$ Equation of chord whose middle point is given is

$$T=S_1$$

So w.r.t. (1, –2) and circle $x^2+y^2-9=0$

Equation of chord is

$$x(1)+y(-2)-9=(1)^2+(-2)^2-9$$

$\Rightarrow \quad x-2y-9=1+4-9$

$\Rightarrow \quad x-2y=5$

(ii) Circle $x^2+y^2+2x+3y+1=0$

External point = (1,2)

$\because$ Equation of chord of contact from an external point is

$$T=0$$

$\Rightarrow \quad x(1)+y(2)+2\left(\frac{x+1}{2}\right)+3\left(\frac{y+2}{2}\right)+1=0$

$\Rightarrow \quad 2x+4y+2x+2+3y+6+2=0$

$\Rightarrow \quad 4x+7y+10=0$

(iii) $S_1: 2x^2+2y^2+7x-5y+2=0$

$\Rightarrow \quad x^2+y^2+\frac{7}{2}x-\frac{5}{2}y+1=0$

$S_2: x^2+y^2-4x+8y-18=0$

$\because$ Equation of common chord to the circles is

$$S_1-S_2=0$$

i.e. $\quad \frac{15}{2}x-\frac{21}{2}y+19=0$

$$15x-21y+38=0$$

23. $S: x^2+y^2-4x+2y-3=0$

$L: 3x-4y+5=0$

Equation of circle passing through point of intersetion of S and L is

$\Rightarrow \quad S+\lambda L=0$

$\Rightarrow x^2+y^2-4x+2y-3+\lambda(3x-4y+5)=0$

$\Rightarrow x^2+y^2+x(3\lambda-4)+y(2-4\lambda)+(5\lambda-3)=0 \quad \text{...(i)}$

$g=\frac{3\lambda-4}{2}, f=1-2\lambda$ and $c=5\lambda-3$

(i) Passes through (1, 1)

$\therefore$ It will satisfy the Eq. (i)

$\Rightarrow \quad 1^2+1^2+1(3\lambda-4)+(2-4\lambda)+5\lambda-3=0$

$\Rightarrow \quad 1+1+3\lambda-4+2-4\lambda+5\lambda-3=0$

$\Rightarrow \quad 4\lambda-3=0$

$$\lambda=\frac{3}{4}$$

$\therefore$ Equation of circle is

$$x^2+y^2+\left(\frac{9}{4}-4\right)x+y(2-3)+\frac{15}{4}-3=0$$

$\Rightarrow \quad 4(x^2+y^2)-7x-4y+3=0$

(ii) Having x-cooridinate of centre = – 3

$\therefore \quad -g = -3$

$\Rightarrow \quad \frac{3\lambda - 4}{2} = 3$

$\Rightarrow \quad 3\lambda = 10 \Rightarrow \lambda = \frac{10}{3}$

$\therefore$ Equation of required circle is

$x^2 + y^2 + x(10-4) + y\left(2 - \frac{40}{3}\right) + \frac{50}{3} - 3 = 0$

$\Rightarrow \quad 3(x^2 + y^2) + 18x - 34y + 41 = 0$

(iii) x-intercept $= 2\sqrt{7}$

$2\sqrt{g^2 - c} = 2\sqrt{\left(\frac{3\lambda - 4}{2}\right)^2 - (5\lambda - 3)}$

$= \sqrt{9\lambda^2 - 24\lambda + 16 - 20\lambda + 12}$

$= \sqrt{9\lambda^2 - 44\lambda + 28} = 2\sqrt{7}$ [given]

$9\lambda^2 - 44\lambda + 28 = 28$

$\Rightarrow \quad 9\lambda^2 - 44\lambda = 0$

$\Rightarrow \quad \lambda(9\lambda - 44) = 0$

$\Rightarrow \quad \lambda = 0, \frac{44}{9}$

$\therefore$ Required equation of circles are

$x^2 + y^2 - 4x + 2y - 3 = 0$

and $9(x^2 + y^2) + 96x - 158y + 193 = 0$

(iv) Concentric with $2x^2 + 2y^2 - 6x + 4y + 7 = 0$

$\Rightarrow \quad x^2 + y^2 - 3x + 2y + \frac{7}{2} = 0$

Centre $(-g, -f) = \left(\frac{3}{2}, -1\right)$

$\therefore \quad -\frac{3\lambda - 4}{2} = \frac{3}{2}$

$\Rightarrow \quad 3\lambda - 4 = -3$

$\Rightarrow \quad 3\lambda = 1$

$\Rightarrow \quad \lambda = \frac{1}{3}$

and $\quad -(1 - 2\lambda) = -1$

$\Rightarrow \quad -1 + 2\lambda = -1$

$\Rightarrow \quad \lambda = 0$

$\because$ No unique value of λ.

$\therefore$ No such circle is possible.

(v) Touching the cirlce at (2, 2)

$\Rightarrow$ Circle will passes through (2, 2)

$\therefore \quad 2^2 + 2^2 + 2(3\lambda - 4) + 2(2 - 4\lambda) + 5\lambda - 3 = 0$

$\Rightarrow \quad 8 + 6\lambda - 8 + 4 - 8\lambda + 5\lambda - 3 = 0$

$\Rightarrow \quad 3\lambda = -1$

$\Rightarrow \quad \lambda = -\frac{1}{3}$

$\therefore$ Equation of circle is

$x^2 + y^2 + x(-1-4) + y\left(2 + \frac{4}{3}\right) + 5\left(-\frac{1}{3}\right) - 3 = 0$

$\Rightarrow \quad 3(x^2 + y^2) - 15x + 10y - 14 = 0$

24. $S_1 : x^2 + y^2 + x + 2y - 4 = 0$

$S_2 : x^2 + y^2 - 2x - 4y - 6 = 0$

Equation of circle passing through point of intersection of S_1 and S_2 is

$S_1 + \lambda S_2 = 0$

$\Rightarrow x^2(\lambda + 1) + y^2(\lambda + 1) + x(1 - 2\lambda) + y(2 - 4\lambda) - 4 - 6\lambda = 0$

$\Rightarrow x^2 + y^2 + \left(\frac{1 - 2\lambda}{1 + \lambda}\right)x + \left(\frac{2 - 4\lambda}{1 + \lambda}\right)y + \frac{(-4 - 6\lambda)}{1 + \lambda} = 0$

(i) Passes through (1, – 2)

$\therefore \quad (1 + \lambda)(1 + 4) + 1(1 - 2\lambda) - 2(2 - 4\lambda) - 4 - 6\lambda = 0$

$\Rightarrow (\lambda + 4\lambda - 2\lambda + 8\lambda - 6\lambda) + 1 + 4 + 1 - 4 - 4 = 0$

$\Rightarrow \quad 5\lambda - 2 = 0$

$\Rightarrow \quad \lambda = \frac{2}{5}$

$\therefore$ Required equation is

$\frac{7}{5}(x^2 + y^2) + \frac{1}{5}x + y\left(\frac{2}{5}\right) - \frac{32}{5} = 0$

$\Rightarrow \quad 7(x^2 + y^2) + x + 2y - 32 = 0$

(ii) Centre lies on $3x + y = 2$

$\because$ Centre is $(-g, -f) = \left(\frac{2\lambda - 1}{2(\lambda + 1)}, \frac{4\lambda - 2}{2(\lambda + 1)}\right)$

$\therefore \quad 3\left(\frac{2\lambda - 1}{2(\lambda + 1)}\right) + \frac{4\lambda - 2}{2(\lambda + 1)} = 2$

$\Rightarrow \quad 6\lambda - 3 + 4\lambda - 2 = 4\lambda + 4$

$\Rightarrow \quad 6\lambda = 9$

$\Rightarrow \quad \lambda = \frac{9}{6} = \frac{3}{2}$

$\therefore$ Equation of circle is

$\frac{5}{2}(x^2 + y^2) - 2x - 4y - 13 = 0$

$\Rightarrow \quad 5(x^2 + y^2) - 4x - 8y - 26 = 0$

(iii) $\because$ y-intercept = 0

$2\sqrt{f^2 - c} = 0$

$f^2 = c$

$\left[\frac{4\lambda - 2}{2(\lambda + 1)}\right]^2 = \frac{-4 - 6\lambda}{\lambda + 1}$

$\Rightarrow \quad (2\lambda - 1)^2 = (-4 - 6\lambda)(\lambda + 1)$

$\Rightarrow 4\lambda^2 - 4\lambda + 1 = -4\lambda - 4 - 6\lambda^2 - 6\lambda$

$\Rightarrow 10\lambda^2 + 6\lambda + 5 = 0$

$$\lambda = \frac{-6 \pm \sqrt{36-200}}{2}$$

$$= \frac{-6 \pm \sqrt{-164}}{2} \text{ (imaginary)}$$

$\therefore$ No such circle is possible.

25. (i) $C_1 : x^2 + y^2 + 3x - 4y + 7 = 0$

$$g_1 = \frac{3}{2}; \quad f_1 = -2; \quad c_1 = 7$$

$$C_2 : x^2 + y^2 - 4x + 3y - 3\lambda = 0$$

$$g_2 = -2; \quad f_2 = \frac{3}{2}; \quad c_2 = -3\lambda$$

$\because$ C_1 and C_2 are orthogonal.

$\therefore$ $2(g_1g_2 + f_1f_2) = c_1 + c_2$

$\Rightarrow$ $2\left(\frac{3}{2}\right)(-2) + 2(-2)\left(\frac{3}{2}\right) = 7 - 3\lambda$

$\Rightarrow -6 - 6 - = 7 - 3\lambda \Rightarrow 3\lambda = 7 + 12$

$\Rightarrow$ $\lambda = \frac{19}{3}$

(ii) $C_1 : 2x^2 + 2y^2 - 4x - 3y - 7 = 0$

$$x^2 + y^2 - 2x - \frac{3}{2}y - \frac{7}{2} = 0$$

$$g_1 = -1; \quad f_1 = -\frac{3}{4}; \quad c_1 = -\frac{7}{2}$$

C_2: $3x^2 + 3y^2 - 7x + 8y - 19\lambda = 0$

$\Rightarrow$ $x^2 + y^2 - \frac{7}{3}x + \frac{8}{3}y - \frac{19\lambda}{3} = 0$

$$g_2 = -\frac{7}{6}; \quad f_2 = \frac{8}{6}; \quad c_2 = -\frac{19\lambda}{6}$$

$\because$ C_1 and C_2 are orhogonal.

$$2g_1g_2 + 2f_1f_2 = c_1 + c_2$$

$\Rightarrow 2(-1)\left(-\frac{7}{6}\right) + 2\left(-\frac{3}{4}\right)\left(\frac{8}{6}\right) = -\frac{7}{2} - \frac{19}{6}\lambda$

$\Rightarrow$ $\frac{7}{3} - 2 = \frac{-21 - 19\lambda}{6}$

$\Rightarrow$ $\frac{7-6}{3} = \frac{-21-19\lambda}{6}$

$\Rightarrow$ $2 = -21 - 19\lambda$

$\Rightarrow$ $19\lambda = -23$

$\Rightarrow$ $\lambda = -\frac{23}{19}$

(iii) $C_1 = (2, 3)$; $\quad g_1 = -2$; $\quad f_1 = -3$; $\quad c_1 = 9$

$C_2 : (-1, -\lambda)$; $\quad g_2 = 1$; $\quad f_2 = \lambda$; $\quad c_2 = -4$

$\because$ C_1 and C_2 are orthogonal.

$$2g_1g_2 + 2f_1f_2 = c_1 + c_2$$

$\Rightarrow$ $2(-2)(1) + 2(-3)(\lambda) = 9 - 4$

$\Rightarrow$ $-4 - 6\lambda = 9 - 4$

$\Rightarrow$ $-6\lambda = 9$

$\Rightarrow$ $\lambda = -\frac{3}{2}$

26. Required equation of circle is

$$x^2 + y^2 - 6x - 8y + \lambda(x + y - 1) = 0$$

$\Rightarrow$ $x^2 + y^2 - (6 - \lambda)x - (8 - \lambda)y - \lambda = 0$

whose centre is $\left(3 - \frac{\lambda}{2}, 4 - \frac{\lambda}{2}\right)$ and which lies on the line $x + y - 1 = 0$

$\Rightarrow$ $3 - \frac{\lambda}{2} + 4 - \frac{\lambda}{2} - 1 = 0$

$\Rightarrow$ $\lambda = 6$

Hence, required equation is

$$x^2 + y^2 - 6x - 8y + 6x + 6y - 6 = 0$$

$\Rightarrow$ $x^2 + y^2 - 2y - 6 = 0$

27. Equation of the circle passing through the points of intersection of the circles

$$x^2 + y^2 - 2x - 6y + 6 = 0 \quad \text{...(i)}$$

and $\quad x^2 + y^2 + 2x - 6y + 6 = 0 \quad$ ...(ii)

is $\quad x^2 + y^2 - 2x - 6y + 6 + \lambda(-4x) = 0 \quad$ ...(iii)

Cirlce (iii) is orthogonal to the circle

$$x^2 + y^2 + 4x + 6y + 4 = 0$$

$\therefore$ $2(-1 - 2\lambda)(2) + 2(-3)(3) = 4 + 6$

$\Rightarrow$ $\lambda = -4$

$\therefore$ Required circle is $x^2 + y^2 + 14x - 6y + 6 = 0$

28. (i) $C_1 : x^2 + y^2 - 3x + 2y + 7 = 0 \to S_1$

$C_2 : 3x^2 + 3y^2 - 7x + 8y + 2 = 0$

$\Rightarrow$ $x^2 + y^2 - \frac{7}{3}x + \frac{8}{3}y + \frac{2}{3} = 0 \to S_2$

Equation of radical axis is

$$S_1 - S_2 = 0$$

$\Rightarrow \left(-3x + \frac{7}{3}x\right) + \left(2y - \frac{8}{3}y\right) + 7 - \frac{2}{3} = 0$

$\Rightarrow$ $-9x + 7x + 6y - 8y + 21 - 2 = 0$

$\Rightarrow$ $-2x - 2y + 19 = 0$

$\Rightarrow$ $x + y = \frac{19}{2}$

(ii) $C_1 : x^2 + y^2 - \frac{2}{3}x + \frac{4}{3}y - \frac{11}{3} = 0 \to S_1$

C_2: $\quad x^2 + y^2 + \frac{8}{5}x + \frac{14}{5} = 0 \to S_2$

$\therefore$ Radical axis is

$$S_1 - S_2 = 0$$

$\Rightarrow \left(-\frac{2}{3} - \frac{8}{5}\right)x + \frac{4}{3}y - \frac{11}{3} - \frac{14}{5} = 0$

$\Rightarrow (-10 - 24)x + 20y - 55 - 42 = 0$

$\Rightarrow$ $-34x + 20y - 97 = 0$

$\Rightarrow$ $34x - 20y + 97 = 0$

29. (i) $S_1 : x^2+y^2+2x+3y-7=0$

$S_2 : x^2+y^2-3x+7y+8=0$

$S_3 : \ x^2+y^2+2x+6y-1=0$

For radical centre,

$$S_1 - S_2 = 0$$

and $$S_2 - S_3 = 0$$

$$5x-4y-15=0 \quad ...(i)$$

and $$-5x+y+9=0 \quad ...(ii)$$

By Eqs. (i) and (ii), we get

$$-3y-6=0$$

$$y=-2, \quad x=\frac{7}{5}$$

$\therefore$ Radical centre is $\left(\frac{7}{5},-2\right)$.

(ii) $S_1 = x^2+y^2-\frac{2}{3}x+\frac{7}{3}y+\frac{8}{3}=0$

$S_2 = x^2+y^2-\frac{x}{2}+\frac{1}{2}=0$

$S_3 = x^2+y^2-x+3y+2=0$

For radical centre,

Solve $S_1-S_2=0$ and $S_2-S_3=0$

$$\left(-\frac{2}{3}+\frac{1}{2}\right)x+\frac{7}{3}y+\left(\frac{8}{3}-\frac{1}{2}\right)=0$$

and $$\left(-\frac{x}{2}+x\right)-3y+\frac{1}{2}-2=0$$

$\Rightarrow$ $$(-4+3)x+14y+16-3=0$$

and $$x-6y-3=0$$

$\Rightarrow$ $$-x+14y+13=0 \quad ...(i)$$

and $$x-6y-3=0 \quad ...(ii)$$

By Eqs. (i) and (ii),

$$8y+10=0$$

$$y=-\frac{5}{4}$$

and $$x=3+6y$$

$$=3-\frac{30}{4}$$

$$=-\frac{18}{4}=-\frac{9}{2}$$

$\therefore$ Radical centre is $\left(-\frac{9}{2},-\frac{5}{4}\right)$.

30. Given equations of circles are

$$x^2+y^2+4x+3=0$$

and $$x^2+y^2-6x+5=0$$

Length of tangents are $\sqrt{x_1^2+y_1^2+4x_1+3}$ and $\sqrt{x_1^2+y_1^2-6x_1+5}$, respectively.

Given, $$\frac{\sqrt{x_1^2+y_1^2+4x_1+3}}{\sqrt{x_1^2+y_1^2-6x_1+5}}=\frac{2}{3}$$

$\Rightarrow$ $$\frac{x_1^2+y_1^2+4x_1+3}{x_1^2+y_1^2-6x_1+5}=\frac{4}{9}$$

$\Rightarrow$ $$9x_1^2+9y_1^2+36x_1+27=4x_1^2+4y_1^2-24x_1+20$$

$\Rightarrow$ $$5x_1^2+5y_1^2+60x_1+7=0$$

$\therefore$ Locus of the point is

$$5x^2+5y^2+60x+7=0$$

31. Given circles are

$$x^2+y^2+3x-5y+6=0 \quad ...(i)$$

and $$4x^2+4y^2-28x+29=0$$

$\Rightarrow$ $$x^2+y^2-7x+\frac{29}{4}=0 \quad ...(ii)$$

Let the required circle be

$$x^2+y^2+2gx+2fy+c=0 \quad ...(iii)$$

Since circle (iii) cuts circles (i) and (ii) orthogonally.

$\therefore$ $$2g\left(\frac{3}{2}\right)+2f\left(-\frac{5}{2}\right)=c+6$$

$\Rightarrow$ $$3g-5f=c+6 \quad ...(iv)$$

and $$-7g=c+\frac{29}{4} \quad ...(v)$$

Subtracting Eq. (v) from Eq. (iv), we get

$$40g-20f=-5 \quad ...(vi)$$

Since centre $(-g,-f)$ of circle (iii) lies on the line.

$$3x+4y+1=0$$

$\therefore$ $$-3g-4f=-1 \quad ...(vii)$$

Solving Eqs. (vi) and (vii), we get

$$g=0, f=\frac{1}{4}$$

From Eq. (v), we get $c=-\frac{29}{4}$

Required equation of circle is

$$x^2+y^2+\frac{1}{2}y-\frac{29}{4}=0$$

$\Rightarrow$ $$4(x^2+y^2)+2y-29=0$$

32. Given circle is $x^2+y^2+4x-6y+9=0$

Let $M(\alpha,\beta)$.

Since $AM=2AB$

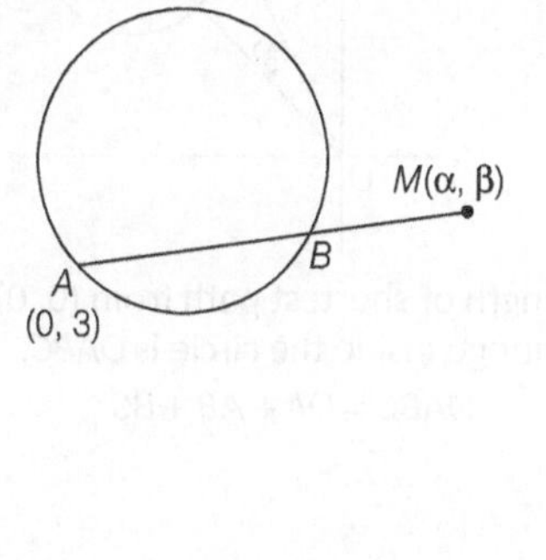

$\therefore$ B is the middle-point of AM.

$$\therefore \quad B = \left(\frac{\alpha}{2}, \frac{\beta+3}{2}\right)$$

$\because B\left(\frac{\alpha}{2}, \frac{\beta+3}{2}\right)$ lie on circle

$$\therefore \quad \left(\frac{\alpha}{2}\right)^2 + \left(\frac{\beta+3}{2}\right)^2 + 4\cdot\frac{\alpha}{2} - 6\left(\frac{\beta+3}{2}\right) + 9 = 0$$

$$\Rightarrow \quad \alpha^2 + (\beta+3)^2 + 8\alpha - 12(\beta+3) + 36 = 0$$

$$\Rightarrow \quad \alpha^2 + \beta^2 + 8\alpha - 6\beta + 9 = 0$$

Hence, locus of $M(\alpha, \beta)$ is

$$x^2 + y^2 + 8x - 6y + 9 = 0$$

33. Let the mid-point of chord AB is $P(\alpha, \beta)$.

Centre of circle $O(-g, -f)$.

$$\therefore \quad OP = \sqrt{(g+\alpha)^2 + (f+\beta)^2}$$

$$OA = r = \sqrt{g^2 + f^2 - c}$$

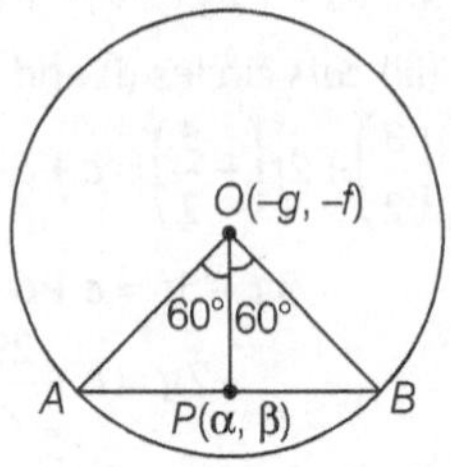

From ΔAOP, $\quad \cos 60° = \frac{OP}{OA}$

$$\Rightarrow \quad \frac{1}{2} = \frac{OP}{OA} \Rightarrow 4OP^2 = OA^2$$

$$\Rightarrow \quad 4[(g+\alpha)^2 + (f+\beta)^2] = g^2 + f^2 - c$$

Hence, locus of P is

$$4x^2 + 4y^2 + 8gx + 8fy + 3g^2 + 3f^2 + c = 0$$

Level 2

1. (c) Given equation of circle $(x-6)^2 + (y-8)^2 = 25$

Centre $(6, 8)$, $r = 5$

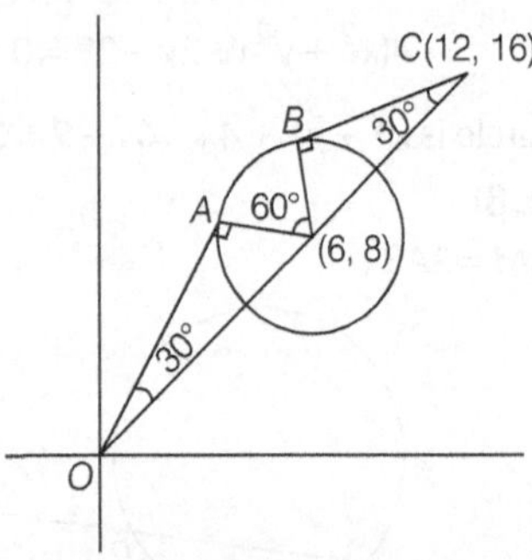

The length of shortest path from (0, 0) to (12, 16) that does not go inside the circle is $OABC$.

$$OABC = OA + AB + BC$$

$$\therefore \quad OA = \sqrt{6^2 + 8^2 - 25} = 5\sqrt{3}$$

$$AB = \text{length of an arc } AB = \frac{\pi}{3} \times 5 = \frac{5\pi}{3}$$

$$BC = \sqrt{(12-6)^2 + (16-8)^2 - 25} = 5\sqrt{3}$$

Hence, shortest path

$$= 5\sqrt{3} + \frac{5\pi}{3} + 5\sqrt{3} = 10\sqrt{3} + \frac{5\pi}{3}$$

2. (c) Circle : $x^2 + y^2 - 18x + 12y + \lambda = 0$

$$2g = -18; \quad 2f = 12; \quad c = \lambda$$

$$g = -9; \quad f = 6; \quad c = \lambda$$

$$\text{Radius} = \sqrt{g^2 + f^2 - c} = \sqrt{81 + 36 - \lambda} = 11$$

$$\Rightarrow \quad 117 - \lambda = 121$$

$$\Rightarrow \quad \lambda = 117 - 121 = -4$$

3. (b) Circle : $x^2 + y^2 - 10x - 12y - 92 = 0$

Point : $(-3, -6)$

$$2g = -10, \quad 2f = -12, \quad c = -92$$

$$g = -5, \quad f = -6, \quad c = -92$$

$\therefore$ Centre $(-g, -f) = (5, 6)$

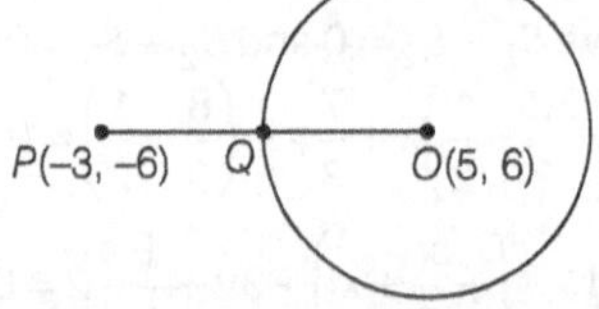

and radius $= \sqrt{g^2 + f^2 - c} = \sqrt{25 + 36 + 92} = \sqrt{153}$

$\therefore$ Minimum distance between point and circle is

$$PQ = OP - OQ \quad \text{...(i)}$$

$$= OP - r$$

Now, $\quad OP = \sqrt{(5+3)^2 + (6+6)^2} = \sqrt{64 + 144}$

$$= \sqrt{208} = 2\sqrt{52} = 4\sqrt{13}$$

$$\therefore \quad PQ = 4\sqrt{13} - \sqrt{153}$$

4. (d) Let P be the image of the origin with respect to the mirror line $x + y = 1$

Since, $OA = OB$

$\because Q$ is mid-point of AB.

So, $Q\left(\frac{1}{2}, \frac{1}{2}\right)$.

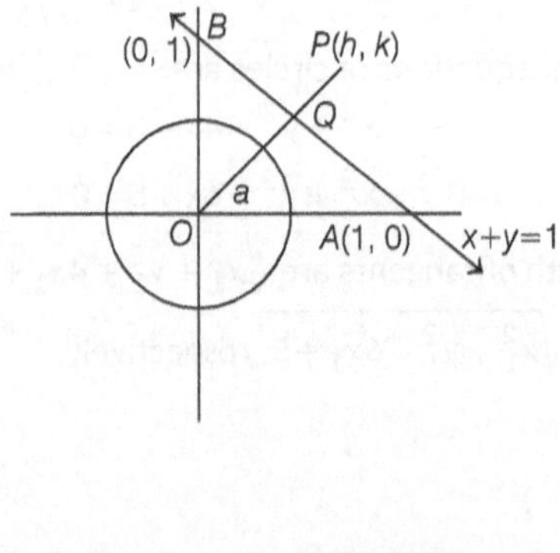

Let the cordinate of $P(h, k)$.
Since Q is mid-point of OP.

$$\therefore \quad \frac{0+h}{2}=\frac{1}{2}, \frac{0+k}{2}=\frac{1}{2} \Rightarrow h=k=1$$

Hence, coordinate of $P(1,1)$.

5. (d) Given circle is

$$x^2+y^2+16x-24y+183=0 \quad \text{...(i)}$$

Centre $=(-8,12)$

$$r=\sqrt{64+144-183}=\sqrt{25}=5$$

Given line $\quad 4x+7y+13=0 \quad$ ...(ii)

Let (α,β) be the image of $(-8,12)$ w.r.t. Eq. (ii)

$$\frac{\alpha+8}{4}=\frac{\beta-12}{7}=-\frac{2(4(-8)+7(12)+13)}{4^2+7^2}$$

$$\frac{\alpha+8}{4}=\frac{\beta-12}{7}=-\frac{2(65)}{65}=-2$$

$$\Rightarrow \quad \alpha=-8-8=-16$$

and $\quad \beta=-14+12=-2$

$\therefore$ Equation of circle $(x+16)^2+(y+2)^2=5^2$

$$\Rightarrow \quad x^2+y^2+32x+4y+235=0$$

6. (b) Given

$$r^2-8(\sqrt{3}r\cos\theta+r\sin\theta)+15=0$$

We know $x=r\cos\theta$ and $y=r\sin\theta$

or $\quad r^2=x^2+y^2$

$\therefore$ Equation may be written as

$$x^2+y^2-8(\sqrt{3}x+y)+15=0$$

$$\Rightarrow \quad x^2+y^2-8\sqrt{3}x-8y+15=0$$

Now, $\quad \text{radius}=\sqrt{(4\sqrt{3})^2+(4)^2-15}=7$

7. (b) In ΔOPD,

$$\sin 30^\circ=\frac{OP}{OD}$$

$$OP=OD\sin 30^\circ$$

$$r=\sqrt{3}\times\frac{1}{2}=\frac{\sqrt{3}}{2}$$

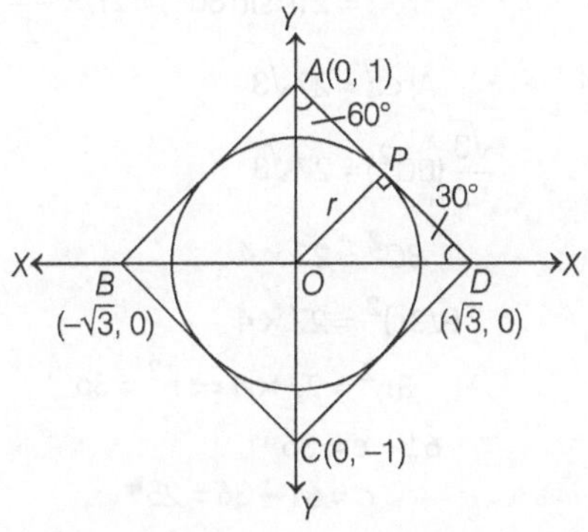

Let cordinate of $P(x,y)$,

$$\therefore \quad |PA|^2+|PB|^2+|PC|^2+|PD|^2$$

$$=x^2+(y-1)^2+(x+\sqrt{3})^2+y^2+x^2+(y+1)^2+(x-\sqrt{3})^2+y^2$$

$$=4(x^2+y^2+2)$$

$$=4\left(\frac{3}{4}+2\right)=11 \qquad \left[\because x^2+y^2=\frac{3}{4}\right]$$

8. (a) Let the centre be $C(h,k)$.

$\because$ Circle passes through origin.

$$\therefore \quad \text{Radius}=OC=\sqrt{h^2+k^2}$$

Let $x=c$ meets the circle in A and B

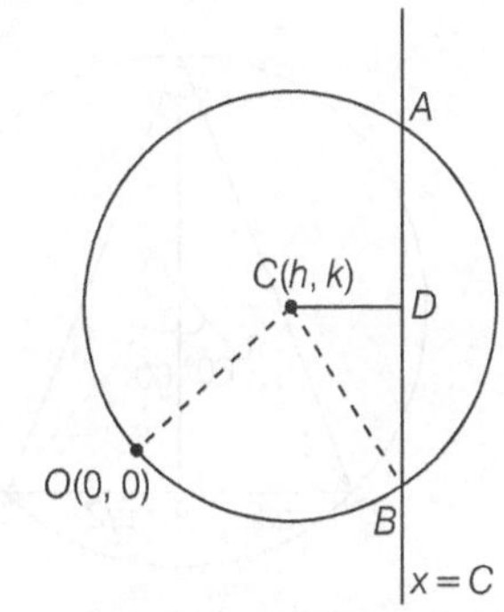

$$\Rightarrow \quad CD \perp AB$$

$$\Rightarrow \quad CB^2=CD^2+BD^2$$

$$\Rightarrow \quad OC^2=CD^2+b^2 \ (\because CB=OC \text{ and } BD=b)$$

$$\Rightarrow \quad h^2+k^2=(h-c)^2+b^2$$

$$\Rightarrow k^2+2ch-c^2-b^2=0$$

Locus is $y^2+2cx=b^2+c^2$

9. (c) $\quad (A_0A_1)^2=\frac{1}{4}+\frac{3}{4}=1$

$$A_0A_1=1$$

$$A_0A_2=\sqrt{3}$$

$$A_0A_4=\sqrt{3}$$

$$\Rightarrow (A_0A_1)(A_0A_2)(A_0A_4)=3$$

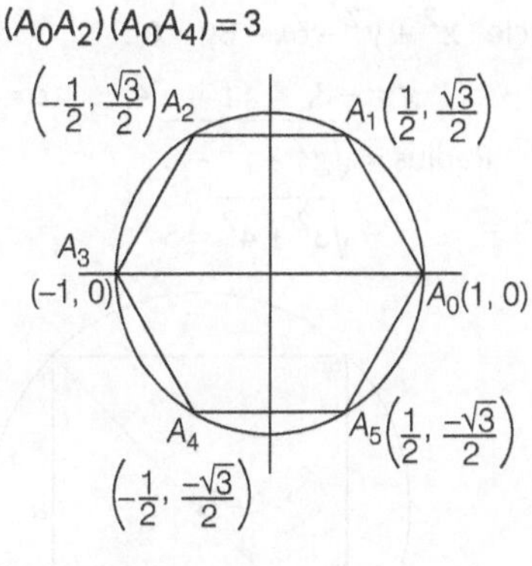

10. (d) $\because$ General equation of circle is

$$x^2+y^2+2gx+2fy+c=0$$

Now, the given equation

$$x^2+y^2+(a^2-4)xy+2x+2y+a=0$$

$\therefore \quad a^2 - 4 = 0$

$\Rightarrow \quad a = \pm 2$

Now, centre is $(-1, -1)$ and radius $= \sqrt{g^2 + f^2 - c}$

$= \sqrt{1 + 1 - a} = \sqrt{2 - a}$

$\because$ At $a = 2$ it is a point circle.
but at $a = -2$ it is a real circle.

11. (b) Circle : $x^2 + y^2 - 2x = 0$

$$(x-1)^2 + y^2 = 1$$

Radius = 1, Centre $\Rightarrow (1, 0)$

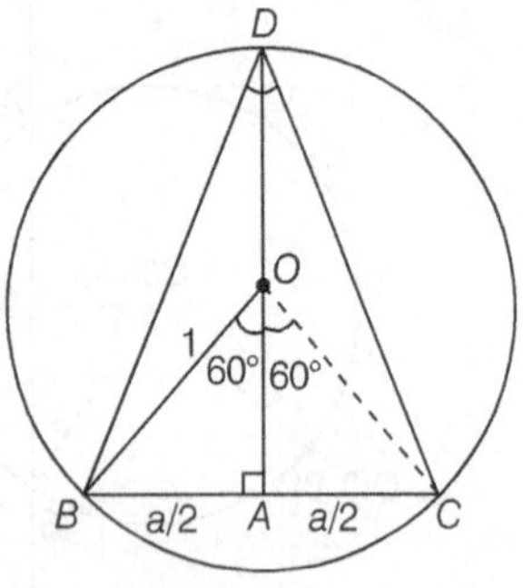

Let a be side is equilateral triangle.
In ΔOAB,

$$\frac{AB}{OB} = \sin 60°$$

$$\frac{a}{2} = \frac{\sqrt{3}}{2}$$

$$\Rightarrow \quad a = \sqrt{3}$$

$\therefore$ Area of equilateral triangle $= \frac{\sqrt{3}}{4} a^2$

$= \frac{\sqrt{3}}{4} (\sqrt{3})^2$

$= \frac{3\sqrt{3}}{4}$ sq units

12. (b) Circle : $x^2 + y^2 - 6x - 8y = 0$

$$g = -3, \quad f = -4, \quad c = 0$$

Radius $= \sqrt{g^2 + f^2 - c}$

$= \sqrt{3^2 + 4^2} = 5$

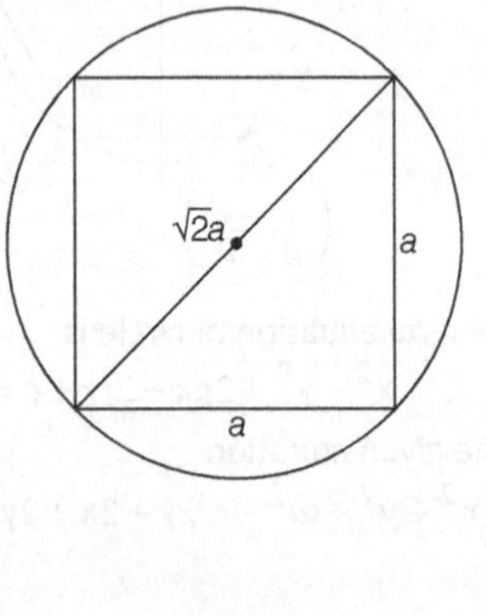

Let a be the side of square.

$\therefore \quad \sqrt{2}a = 2r$

$$a = \frac{2r}{\sqrt{2}} = \frac{2 \times 5}{\sqrt{2}} = 5\sqrt{2}$$

$\therefore$ Area of square $= a^2 = (5\sqrt{2}) = 50$ sq units

13. (b) $\because$ Circle touches the Y-axis at $(0, 6)$ with x-intercept $6\sqrt{5}$.

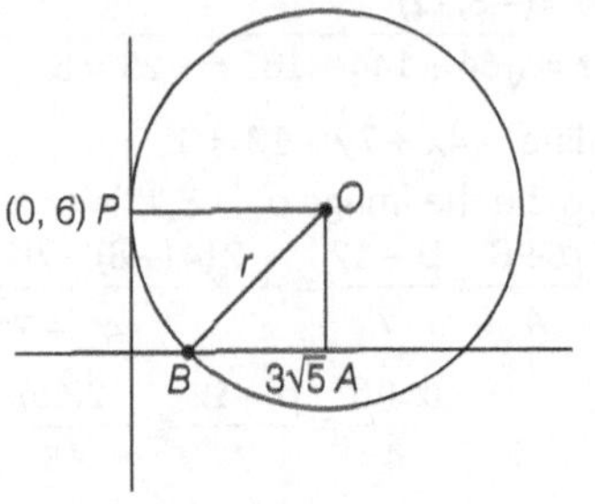

$\therefore$ In ΔOAB, $\quad AB = \frac{1}{2} \times 6\sqrt{5} = 3\sqrt{5}$

$OA = 6$

$\therefore \quad r = \sqrt{AB^2 + OA^2} = \sqrt{45 + 36} = \sqrt{81}$

$r = 9$

14. (b) Centre $(-g, -f) = (-5, -6)$

$$r = \sqrt{g^2 + f^2 - c} = \sqrt{25 + 36 - c} = \sqrt{61 - c}$$

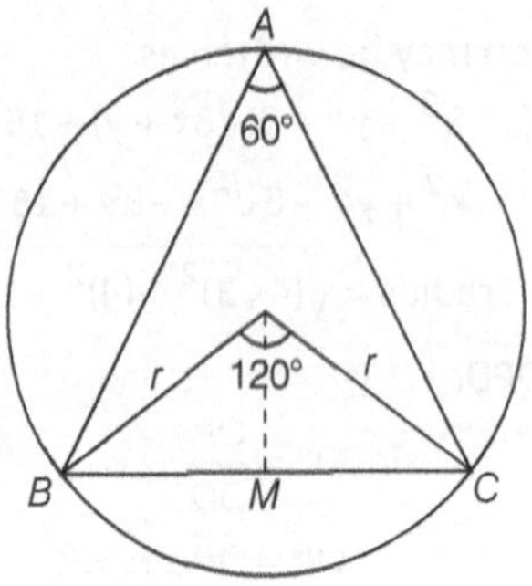

$\because \quad BC = 2BM$

$= 2(r \sin 60°) = 2r \times \frac{\sqrt{3}}{2} = \sqrt{3}r$

and $\quad$ Area $= 27\sqrt{3}$

$$\frac{\sqrt{3}}{4}(BC^2) = 27\sqrt{3}$$

$$BC^2 = 27 \times 4$$

$$(\sqrt{3}r)^2 = 27 \times 4$$

$$3r^2 = 27 \times 4 \Rightarrow r^2 = 36$$

$$\Rightarrow \quad 61 - c = 36$$

$$c = 61 - 36 = 25$$

15. (d) Given equation of circle

$$x^2 + y^2 - 4 = 0 \quad \text{...(i)}$$

Equation of line $3x + 4y + 12 = 0$ ...(ii)

From Eqs. (i) and (ii), we get

$$x^2 + \left(\frac{-3x - 12}{4}\right)^2 - 4 = 0$$

$\Rightarrow$ $16x^2 + 9x^2 + 72x + 144 - 64 = 0$

$\Rightarrow$ $25x^2 + 72x + 80 = 0$

$D = (72)^2 - 4 \times 25 \times 80 < 0$

Hence, line does not intersect the circle.

$\therefore$ $\alpha \in \phi$

16. (a) Equation of AB is $xh + yk = r^2$

AB is tangent to $x^2 + y^2 = a^2$

$\Rightarrow$ $\frac{r^2}{\sqrt{h^2 + k^2}} = a$

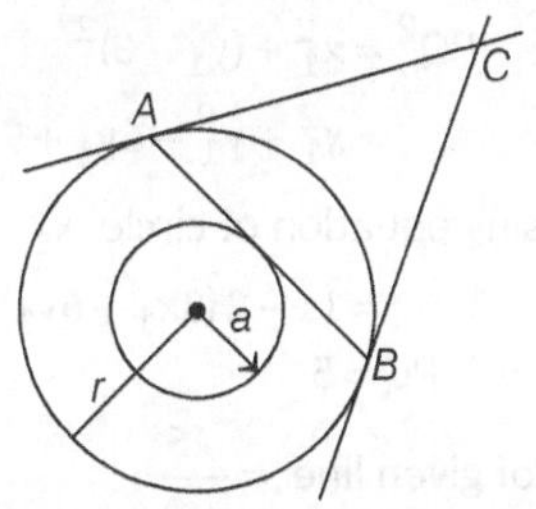

$\Rightarrow$ $r^2 = ab$ $[\because h^2 + k^2 = b^2]$

Equation of circle is

$$x^2 + y^2 = ab$$

17. (b) Equation of circle $x^2 + y^2 - 2x - 2y - 7 = 0$

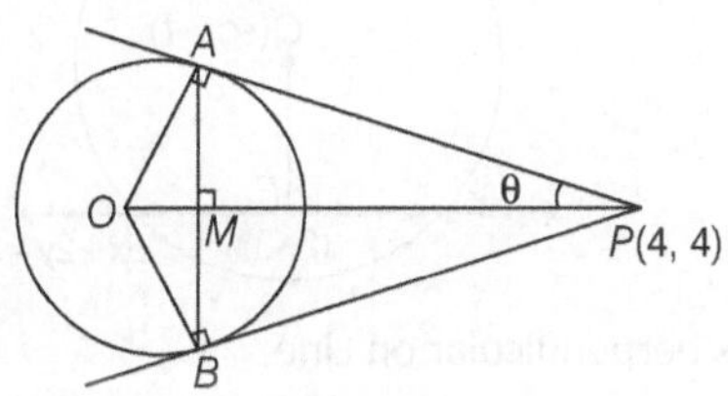

Length of tangent

$PA = \sqrt{4^2 + 4^2 - 2(4) - 2(4) - 7}$

$PA = \sqrt{16 + 16 - 8 - 8 - 7}$

$PA = 3$

Radius of circle $= \sqrt{1 + 1 + 7} = 3$

In ΔOAP,

$$\tan\theta = \frac{OA}{AP} = \frac{3}{3} = 1$$

$$\theta = \frac{\pi}{4}$$

In ΔAMP, $\sin 45° = \frac{AM}{AP}$

$\Rightarrow$ $\frac{1}{\sqrt{2}} = \frac{AM}{3}$

$\Rightarrow$ $AM = \frac{3}{\sqrt{2}}$

Length of chord $AB = 2AM = 2 \times \frac{3}{\sqrt{2}} = 3\sqrt{2}$

18. (b) Power of point $P(a, b)$

$= a^2 + b^2 - 2(t^2 - 3t + 1)a - 2(t^2 + 2t)b + t$

$= -(2a + 2b)t^2 + (6a - 4b + 1)t + a^2 + b^2 - 2a$

$\Rightarrow$ $2a + 2b = 0$

$\Rightarrow$ $a = -b$

Now, $6a - 4b + 1 = 0$

$\Rightarrow$ $6a + 4a + 1 = 0$

$\Rightarrow$ $a = -\frac{1}{10}, b = \frac{1}{10}$

$\therefore$ $P = \left(-\frac{1}{10}, \frac{1}{10}\right)$

19. (a) Centre : $(3, -1)$

Equation of chord $2x + 5y + 18 = 0$ ...(i)

$\because CM \perp AB$

and $AB = 6$

$\Rightarrow$ $AM = \frac{1}{2} \times AB = 3$

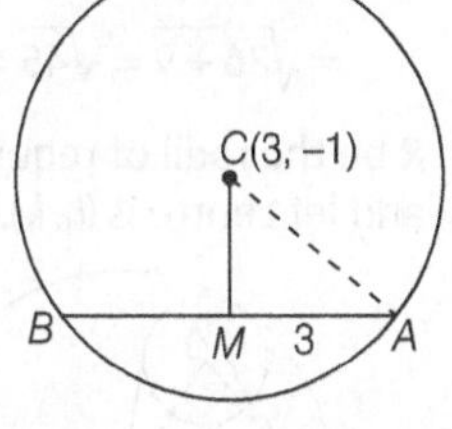

CM = Length of perpendicular drawn from $C(3, -1)$ on the line (i)

$$= \left|\frac{2 \times 3 + 5(-1) + 18}{\sqrt{2^2 + 5^2}}\right| = \frac{19}{\sqrt{29}}$$

Radius $= AC = \sqrt{AM^2 + CM^2}$

$$= \sqrt{3^2 + \frac{(19)^2}{29}}$$

$$= \frac{\sqrt{29 \times 9 + 19 \times 19}}{\sqrt{29}}$$

$$= \frac{\sqrt{622}}{\sqrt{29}}$$

Equation of circle $(x - 3)^2 + (y + 1)^2 = \frac{622}{29}$

$\Rightarrow$ $x^2 - 6x + 9 + y^2 + 2y + 1 = \frac{622}{29}$

$\Rightarrow$ $x^2 + y^2 - 6x + 2y + 10 - \frac{622}{29} = 0$

$\Rightarrow$ $29x^2 + 29y^2 - 174x + 58y - 332 = 0$

20. (c) Line $(L_1): 2x - y + 1 = 0$

$$m_1 = +2$$

Line $(L_2): x - 2y = 4$

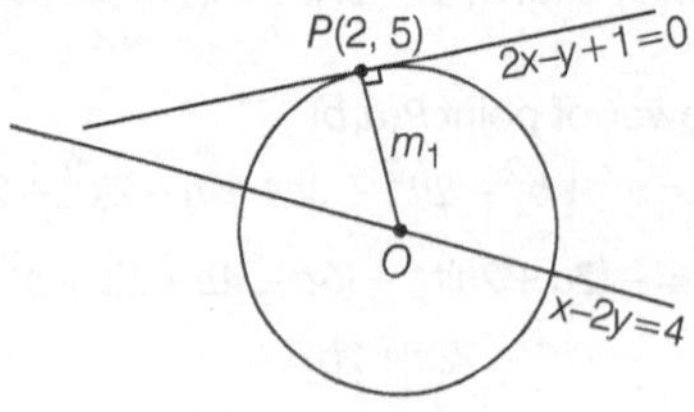

Let $O\left(a, \frac{a-4}{2}\right)$ be the centre of circle.

$\because OP \perp L_1$

$$\Rightarrow \quad m_{OP} \times m_{L_1} = -1$$

$$\Rightarrow \quad \frac{\left(\frac{a-4}{2} - 5\right)}{(a-2)} \times 2 = -1$$

$$\frac{a-14}{a-2} = -1$$

$$\Rightarrow \quad a - 14 = -a + 2$$

$$\Rightarrow \quad 2a = 16 \Rightarrow a = 8$$

$\because$ Centre $\Rightarrow (8, 2)$

$$\text{Radius} = OP = \sqrt{(8-2)^2 + (2-5)^2}$$

$$= \sqrt{36+9} = \sqrt{45} = 3\sqrt{5}$$

21. (a) Let r and R be the radii of required and given circles respectively and let centre is (h, k).

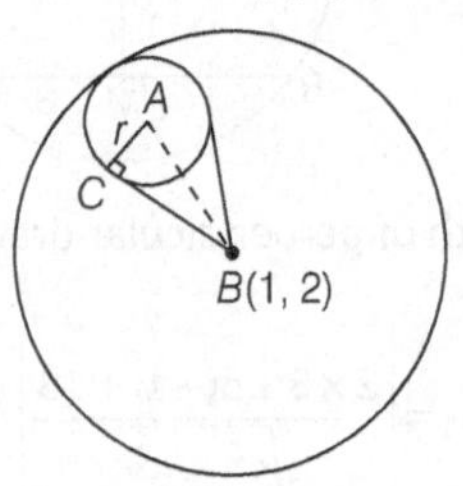

By given condition

$$\sqrt{(h-1)^2 + (k-2)^2} = R - r \quad \text{...(i)}$$

Now, $$\frac{r}{AB} = \tan 30°$$

$$\Rightarrow \quad r = AB \tan 30°$$

$$\Rightarrow \quad r = (R - r)\frac{1}{\sqrt{3}}$$

$$\Rightarrow \quad r = \frac{R}{1+\sqrt{3}} \quad (\because AB = R - r)$$

By Eq. (i),

$$\sqrt{(h-1)^2 + (k-2)^2} = R - \frac{R}{1+\sqrt{3}} = \frac{R(\sqrt{3})}{(\sqrt{3}+1)}$$

Now, $$R = 1 + \sqrt{3}$$

$$\Rightarrow \sqrt{(h-1)^2 + (k-2)^2} = \sqrt{3}$$

$\therefore$ Locus is $(x-1)^2 + (y-2)^2 = 3$

22. (c) $\tan^4 x + \cot^4 x \geq 2$

$$4 \sin^2 y \geq 4 \Rightarrow \sin^2 y \geq 1$$

$$\Rightarrow \quad \sin y = 1, -1 \quad y = \pm\frac{\pi}{2}, \pm\frac{3\pi}{2} \ldots$$

But $$y = \pm\frac{\pi}{2}, \pm\frac{3\pi}{2}$$

and corresponding values to each value of y these are two values of x so total number of points are 8.

23. (c) Circle : $x^2 + y^2 + 6x + 6y = 2$

Let $P(x_1, y_1)$ be a point.

Equation of tangent at point $P(x_1, y_1)$

$$xx_1 + yy_1 + 3(x + x_1) + 3(y + y_1) - 2 = 0 \quad \text{...(i)}$$

Coordinate of Q satisfy Eq. (i),

$$5x - 2y + 6 = 0 \text{ and } x = 0$$

So, $3x_1 + 6y_1 + 7 = 0$ and $Q(0, 3)$

$$\because \quad PQ^2 = x_1^2 + (y_1 - 3)^2$$

$$= x_1^2 + y_1^2 - 6y_1 + 9 = 11 - 6x_1 - 12y_1$$

[using equation of circle, $x_1^2 + y_1^2 + 6x_1 + 6y_1 = 2$]

$$= 11 - 2(3x_1 + 6y_1) = 11 - 2(-7) = 25$$

$$\therefore \quad PQ = 5$$

24. (a) Slope of given line $= -\frac{5}{2}$

$$\text{Slope of } OP = \frac{3+f}{2+g}$$

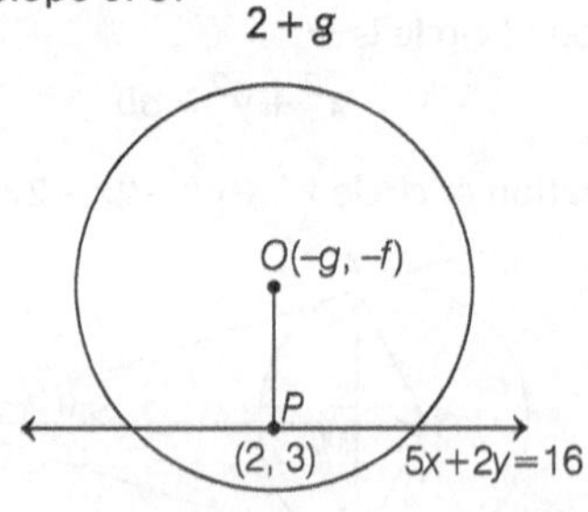

OP is perpendicular on Line.

$$\because \quad m_{\text{line}} \times m_{OP} = -1$$

$$\Rightarrow \quad -\frac{5}{2} \times \frac{3+f}{2+g} = -1$$

$$15 + 5f = 4 + 2g$$

Locus is $2x - 5y - 11 = 0$

25. (a) We have, $$\frac{\pi}{2} < \theta < \frac{2\pi}{3}$$

$$\frac{\pi}{4} < \frac{\theta}{2} < \frac{\pi}{3}$$

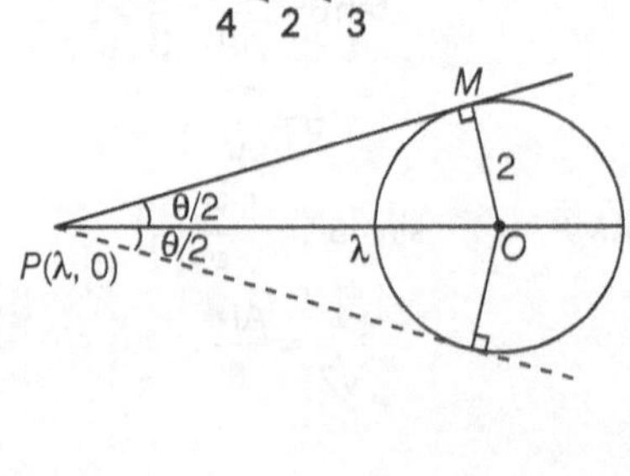

$$\frac{1}{\sqrt{2}}<\sin\frac{\theta}{2}<\frac{\sqrt{3}}{2}$$

But $\sin\frac{\theta}{2}=\frac{2}{\lambda}$

$$\frac{1}{\sqrt{2}}<\frac{2}{\lambda}<\frac{\sqrt{3}}{2}\Rightarrow\frac{4}{\sqrt{3}}<\lambda<2\sqrt{2}$$

26. (b) Let $\theta=\tan^{-1}\frac{12}{5}$

$$\tan\theta=\frac{12}{5}$$

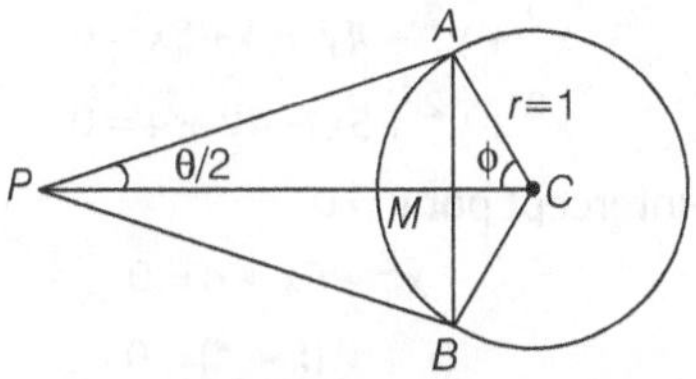

$$\frac{2\tan\frac{\theta}{2}}{1-\tan^2\frac{\theta}{2}}=\frac{12}{5}\Rightarrow\tan\frac{\theta}{2}=\frac{2}{3}$$

$$\sin\frac{\theta}{2}=\frac{2}{\sqrt{13}}\text{ and }\cos\frac{\theta}{2}=\frac{3}{\sqrt{13}}$$

In ΔCAP, $\tan\frac{\theta}{2}=\frac{1}{AP}$

$$AP=\frac{3}{2}$$

In ΔAPM, $\sin\frac{\theta}{2}=\frac{AM}{AP}$ and $\cos\frac{\theta}{2}=\frac{PM}{AP}$

$$AM=\frac{3}{\sqrt{13}}\quad\Rightarrow PM=\frac{9}{2\sqrt{13}}$$

$\therefore$ $AB=\frac{6}{\sqrt{13}}$

$\therefore$ Area of $\Delta PAB=\frac{1}{2}\times AB\times PM$

$$=\frac{1}{2}\times\frac{6}{\sqrt{13}}\times\frac{9}{2\sqrt{13}}=\frac{27}{26}$$

Now $\phi=90°\ \frac{\theta}{2}$

In ΔCAM, $\cos\phi=\frac{CM}{CA}$

$\Rightarrow$ $CM=1\cos\left(\frac{\pi}{2}-\frac{\theta}{2}\right)=\sin\frac{\theta}{2}=\frac{2}{\sqrt{13}}$

$\therefore$ Area of $\Delta CAB=\frac{1}{2}\times AB\times CM$

$$=\frac{1}{2}\times\frac{6}{\sqrt{13}}\times\frac{2}{\sqrt{13}}=\frac{6}{13}$$

$$\therefore\ \frac{\text{Area of }\Delta PAB}{\text{Area of }\Delta CAB}=\frac{\frac{27}{26}}{\frac{6}{13}}=\frac{9}{4}$$

27. (a) For circle $x^2+y^2=1$ tangent at point $P\left(\frac{1}{\sqrt{2}},\frac{1}{\sqrt{2}}\right)$ is

$T=0$

$\Rightarrow$ $\frac{1}{\sqrt{2}}x+\frac{1}{\sqrt{2}}y-1=0$

$$x+y-\sqrt{2}=0 \quad\text{...(i)}$$

Perpendicular line to line Eq. (i) is

$$x-y+c=0$$

This is tangent to the circle $(x-3)^2+y^2=1$

Also, we know, perpendicular distance from centre = Radius

$\therefore$ $\left|\frac{3-0+c}{\sqrt{2}}\right|=1$

$\Rightarrow$ $|3+c|=\sqrt{2}$

$$(3+c)^2=2$$

$\Rightarrow$ $c^2+6c+9-2=0$

$\Rightarrow$ $c^2+6c+7=0$

28. (c) Equation of tangent to circle $x^2+y^2=4$ at point $(\sqrt{3},1)$ is $\sqrt{3}x+y=4$

Slope of this tangent $=-\sqrt{3}$

$\therefore$ Slope of normal at point $(\sqrt{3},1)$ is $\frac{1}{\sqrt{3}}$.

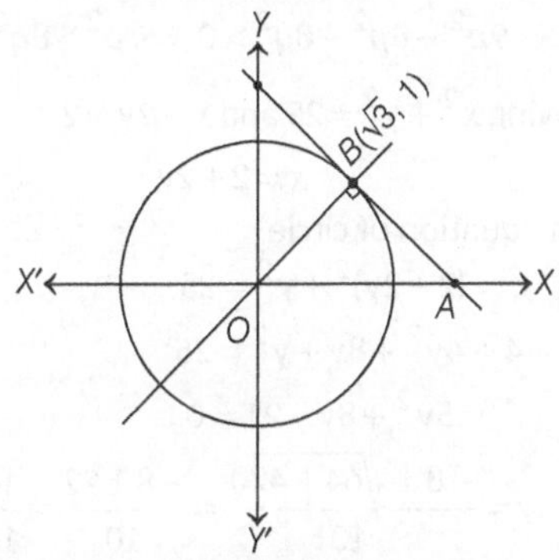

$\therefore$ Equation of normal at point $(\sqrt{3},1)$ is

$$y-1=\frac{1}{\sqrt{3}}(x-\sqrt{3})$$

$\Rightarrow$ $x-\sqrt{3}y=0$

So, normal passes through the centre of circle. Tangent cut the X-axis at point A when, $y=0$

$\therefore$ $x=\frac{4}{\sqrt{3}}\Rightarrow A\left(\frac{4}{\sqrt{3}},0\right)$

Area of ΔOAB

$$A=\frac{1}{2}\begin{vmatrix}0&0&1\\\frac{4}{\sqrt{3}}&0&1\\\sqrt{3}&1&1\end{vmatrix}=\frac{1}{2}\times\frac{4}{\sqrt{3}}=\frac{2}{\sqrt{3}}$$

29. (d) From the equation of circle, it is clear that circle passes through origin. Let AB is chord of circle $A \equiv (p, q)$ C is the mid-point and coordinate of C is $(h, 0)$.

Then, the coordinate of B is $(-p+2h, -q)$.

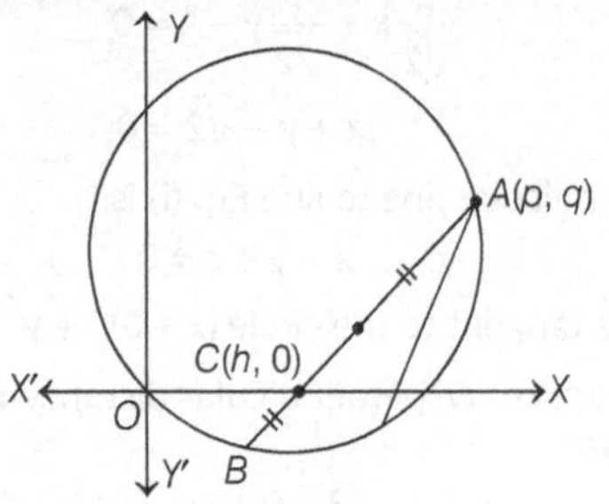

$\because$ B lies on the circle $x^2 + y^2 = px + qy$, then we have

$$(-p+2h)^2 + (-q)^2 = p(-p+2h) + q(-q)$$

$$\Rightarrow \quad p^2 + 4h^2 - 4ph + q^2 = -p^2 + 2hp - q^2$$

$$\Rightarrow 2p^2 + 2q^2 - 6ph + 4h^2 = 0$$

$$\Rightarrow \quad 2h^2 - 3ph + p^2 + q^2 = 0 \qquad \text{...(i)}$$

There are given two distinct chords which are bisected at X-axis, then there will be two distinct values of h satisfying Eq. (i).

So, discriminant of quadratic equation > 0

$$\Rightarrow \quad D > 0$$

$$\Rightarrow (-3p)^2 - 4(2)(p^2 + q^2) > 0$$

$$\Rightarrow \quad 9p^2 - 8p^2 - 8q^2 > 0 \Rightarrow p^2 > 8q^2$$

30. (c) Soving $x^2 + y^2 = 25$ and $x - 2y = 2$

$$x = 2 + 2y$$

Put in equation of circle

$$(2+2y)^2 + y^2 = 25$$

$$4 + 4y^2 + 8y + y^2 = 25$$

$$5y^2 + 8y - 21 = 0$$

$$y = \frac{-8 \pm \sqrt{64+420}}{10} = \frac{-8 \pm 22}{10} = \frac{14}{10}, -\frac{30}{10} = \frac{7}{5}, -3$$

$$\therefore \quad x = 2 + 2\left(\frac{7}{5}\right) \text{ and } 2 + 2(-3)$$

$$\Rightarrow \quad x = \frac{24}{5} \text{ and } -4$$

$\therefore$ Point are $A(-4, -3)$ and $B\left(\frac{24}{5}, \frac{7}{5}\right)$.

$$\therefore \text{Mid-point of } AB = \left(\frac{-4+\frac{24}{5}}{2}, \frac{-3+\frac{7}{5}}{2}\right) = \left(\frac{2}{5}, -\frac{4}{5}\right)$$

31. (b) The centre of given circle is $(-g, -f)$ and the given line $ax + by + c = 0$ is normal to the circle. It will passes through the centre of circle.

$$\Rightarrow \quad a(-g) + b(-f) + c = 0$$

$$\Rightarrow \quad ag + bf - c = 0$$

32. (d) Equation of circle passing through a point (x_1, y_1) and touching the line L is given by

$$(x - x_1)^2 + (y - y_1)^2 + \lambda L = 0$$

$\therefore$ Equation of circle passing through $(0, 2)$ and touching $x = 0$ is

$$(x-0)^2 + (y-2)^2 + \lambda x = 0 \qquad \text{...(i)}$$

Also it passes through $(-1, 0)$

$$\Rightarrow \quad 1 + 4 - \lambda = 0$$

$$\Rightarrow \quad \lambda = 5$$

Eq. (i) becomes

$$x^2 + y^2 - 4y + 4 + 5x = 0$$

$$\Rightarrow \quad x^2 + y^2 + 5x - 4y + 4 = 0$$

For x-intercept put $y = 0$

$$\Rightarrow \quad x^2 + 5x + 4 = 0$$

$$\Rightarrow \quad (x+1)(x+4) = 0$$

$$\Rightarrow \quad x = -1, -4$$

Hence, required point $(-4, 0)$.

33. (b)Let (h, k) be the centre of the circle which rolls on the outside of the given circle. Centre of the given circle $\left(-\frac{3}{2}, 3\right)$ and its radius $= \sqrt{\frac{9}{4} + 9 + 9} = \frac{9}{2}$

Clearly, (h, k) is always at a distance equal to the sum $\left(\frac{9}{2} + 2\right) = \frac{13}{2}$ of the radii of two circle from $\left(-\frac{3}{2}, 3\right)$.

Therefore,

$$\left(h + \frac{3}{2}\right)^2 + (k-3)^2 = \left(\frac{13}{2}\right)^2$$

$$\Rightarrow h^2 + k^2 + 3h - 6k + \frac{9}{4} + 9 - \frac{169}{4} = 0$$

$$\therefore \quad \text{Locus is } x^2 + y^2 + 3x - 6y - 31 = 0$$

34. (b) First circle touches both axes and radius is 2 units.

$\therefore$ Centre is $(2, 2)$.

Let radius of other circle be 'a' and this circle also touches both the axes.

$\therefore$ Centre of circle is (a, a).

$\because$ Circle touches the first circle

$$\therefore \quad C_1C_2 = r_1 + r_2$$

$$\Rightarrow \quad \sqrt{(a-2)^2 + (a-2)^2} = a + 2$$

$$\sqrt{2}(a-2) = a + 2$$

$$a - 2 = \pm \frac{1}{\sqrt{2}}(a+2)$$

$$\left(1 \mp \frac{1}{\sqrt{2}}\right) a = (2 \pm \sqrt{2})$$

$$\Rightarrow \quad a = \frac{\sqrt{2}(\sqrt{2} \pm 1)}{(\sqrt{2} \mp 1)} = 6 \pm 4\sqrt{2}$$

$$\because \quad a > 2;\ a = 6 + 4\sqrt{2}$$

35. (a) Here, $\sin\frac{\theta}{2}=\frac{3}{5}$, $\cos\frac{\theta}{2}=\frac{4}{5}$

$\therefore \quad \sin\theta = 2\sin\frac{\theta}{2}\cos\frac{\theta}{2}=\frac{24}{25}$

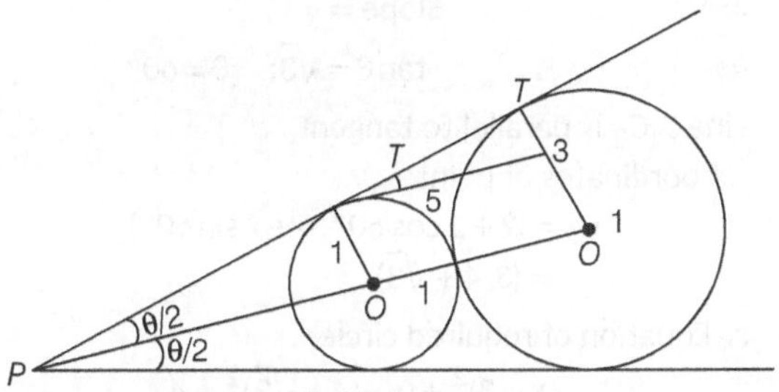

36. (b) $S_1 \equiv x^2+y^2+5kx+2y+k=0$

$S_2 \equiv x^2+y^2+kx+\frac{3}{2}y-\frac{1}{2}=0$

Equation of common chord

$$S_1 - S_2 = 0$$

$\Rightarrow \quad 4kx+\frac{y}{2}+k+\frac{1}{2}=0$ ...(i)

$\Rightarrow \quad 4x+5y-k=0$ ...(ii) (given)

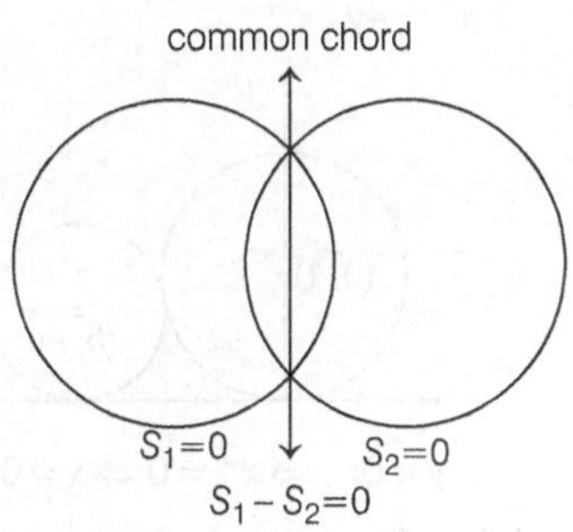

On comparing Eqs. (i) and (ii).

$$\frac{4k}{4}=\frac{1}{10}=\frac{2k+1}{-2k}$$

$$k=\frac{1}{10} \text{ and } -2k=20k+10$$

$\Rightarrow \quad k=\frac{1}{10}$ and $k=-\frac{5}{11}$

$\because \quad \frac{1}{10}\neq-\frac{5}{11}$

So, there is no such value of k, for which lines $4x+5y-k=0$ passes through points P and Q.

$\therefore$ Number of value of k exists.

37. (a) For the cirlce $x^2+y^2=4$ and $C_1(0,0)$ and $r_1=2$

For the circle $x^2+y^2+6x+8y-24=0$

$$C_2=(-3,-4) \text{ and } r_2=7$$

$\therefore \quad C_1C_2=\sqrt{3^2+4^2}=5$

$$r_1+r_2=2, |r_1-r_2|=5$$

$\because \quad C_1C_2=|r_1-r_2|$

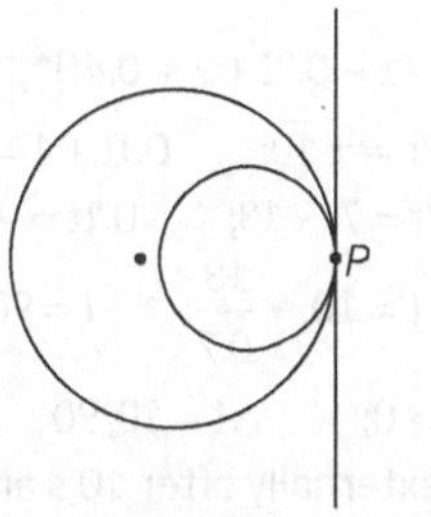

$\therefore$ Circle touches internally here equation is common tangent is same as common chord.

$\Rightarrow \quad S_1-S_2=0$

$\Rightarrow \quad (x^2+y^2-4)-(x^2+y^2+6x+8y-24)=0$

$\Rightarrow \quad -6x-8y+20=0$

$$3x+4y-10=0$$

38. (c) Let the centre (h, k)

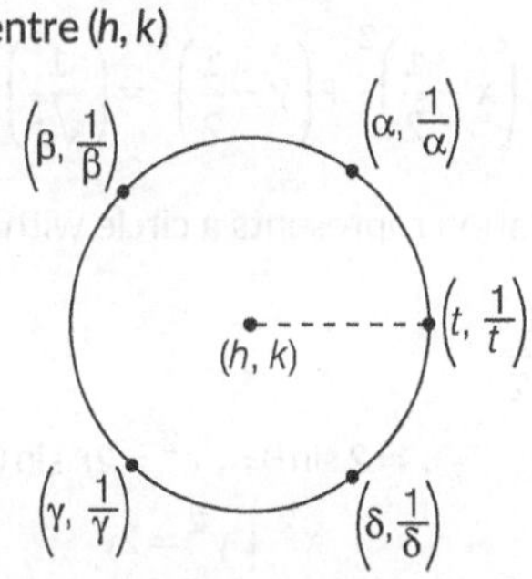

Equation $\quad \left(h-\frac{1}{t}\right)^2+\left(k-\frac{1}{t}\right)^2=16$

$$h^2+k^2-2ht-\frac{2k}{t}+t^2+\frac{1}{t^2}=16$$

$$t^4-2ht^3+t^2(h^2+k^2-16)-2kt+1=0$$

Roots of this equation is $\alpha, \beta, \gamma, \delta$.

$\therefore \quad \alpha\beta\gamma\delta=1$

39. (b) Given circles are

$(x-1)^2+(y-2)^2=1$ ...(i)

and $\quad (x-7)^2+(y-10)^2=4$ ...(ii)

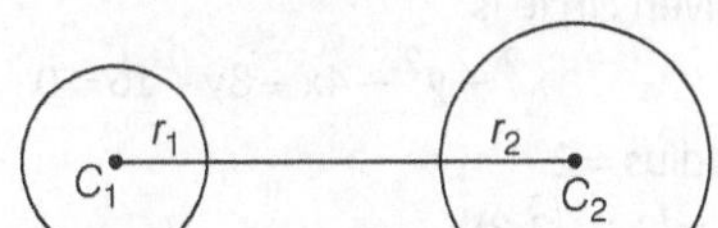

Let $C_1(1,2), C_2(7,10)$ and $r_1=1$ and $r_2=2$

$$C_1C_2=\sqrt{(7-1)^2+(10-2)^2}=10$$

$$r_1+r_2=3$$

$$C_1C_2>r_1+r_2$$

$\therefore$ Circles are separated.

Radius of both the circles at time t are $(1+0.3t)$ and $(2+0.4t)$ for the circle to touch each other.

$$C_1C_2^2=[(r_1+0.3t)\pm(r_2+0.4t)]^2$$

$\Rightarrow \quad 100=[(1+0.3t)\pm(2+0.4t)]^2$

$$\Rightarrow \quad 100 = (1 + 0.3t + 2 + 0.4t)^2, (1 + 0.3t - 2 - 0.4t)^2$$

$$\Rightarrow \quad 3 + 0.7t = \pm 10; \qquad 0.1t + 1 = \pm 10$$

$$\Rightarrow \quad 0.7t = 7, -13; \qquad 0.1t = 9, -11$$

$$\Rightarrow \quad t = 10, -\frac{13}{0.7}; \qquad t = 90, -110$$

$$\because \quad t > 0; \qquad \therefore t = 10, 90$$

$\therefore$ Touch externally after 10 s and internally after 90 s.

40. (c) The equation of the given curves are

$$r = \sin\theta + \cos\theta$$

On multiplying by r both sides

$$r^2 = r\sin\theta + r\cos\theta$$

Let $x = r\cos\theta$, $y = r\sin\theta$

$$\Rightarrow \quad x^2 + y^2 = y + x$$

$$\Rightarrow \quad \left(x - \frac{1}{2}\right)^2 + \left(y - \frac{1}{2}\right)^2 = \left(\frac{1}{\sqrt{2}}\right)^2 \qquad \text{...(i)}$$

This equation represents a circle with centre $\left(\frac{1}{2}, \frac{1}{2}\right)$ and radius $\frac{1}{\sqrt{2}}$

and $\qquad r = 2\sin\theta \Rightarrow r^2 = 2r\sin\theta$

$$\Rightarrow \quad x^2 + y^2 = 2y$$

This equation also represents a circle with centre (0, 1) and radius = 1

$$C_1C_2 = \sqrt{\frac{1}{4} + \frac{1}{4}} = \frac{1}{\sqrt{2}}$$

Hence, angle of intersection of two circles

$$\Rightarrow \cos^{-1}\left(\frac{r_1^2 + r_2^2 - d^2}{2r_1r_2}\right) = \cos^{-1}\left(\frac{\frac{1}{2} + 1 - \frac{1}{2}}{2 \times \frac{1}{\sqrt{2}} \times 1}\right)$$

$$= \cos^{-1}\left(\frac{1}{\sqrt{2}}\right) = \frac{\pi}{4}$$

41. (b) Given circle is

$$x^2 + y^2 - 4x - 8y + 16 = 0$$

$\Rightarrow$ Radius = 2

Let $P \equiv (2 + \sqrt{3}, 3)$

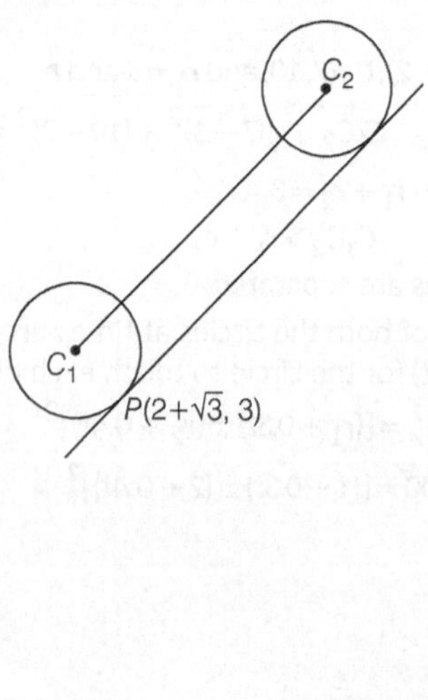

$\therefore$ Equation of tangent at P

$$x(2 + \sqrt{3}) + y(3) - 2(x + 2 + \sqrt{3}) - 4(y + 3) + 16 = 0$$

$$\Rightarrow \quad \sqrt{3}x - y - 2\sqrt{3} = 0$$

$$\Rightarrow \quad \text{Slope} = \sqrt{3}$$

$$\Rightarrow \quad \tan\theta = \sqrt{3}; \quad \theta = 60°$$

Line C_1C_2 is parallel to tangent.

$\therefore$ Coordinates of point

$$C_2 = (2 + 2\cos 60°, 4 + 2\sin 60°)$$

$$= (3, 4 + \sqrt{3})$$

$\therefore$ Equation of required circle

$$(x - 3)^2 + (y - 4 - \sqrt{3})^2 = 4$$

42. (b) According to given question,

$$\sqrt{(h - 0)^2 + (k - 1)^2} = 1 + |k|$$

$$\Rightarrow \quad h^2 + (k - 1)^2 = (1 + |k|)^2$$

$$\Rightarrow \quad h^2 = 2k + 2|k|$$

$\therefore$ Locus is $x^2 = 2y + 2|y|$

$$y > 0 \Rightarrow x^2 = 2y + 2y = 4y$$

$$\Rightarrow \quad x^2 = 4y$$

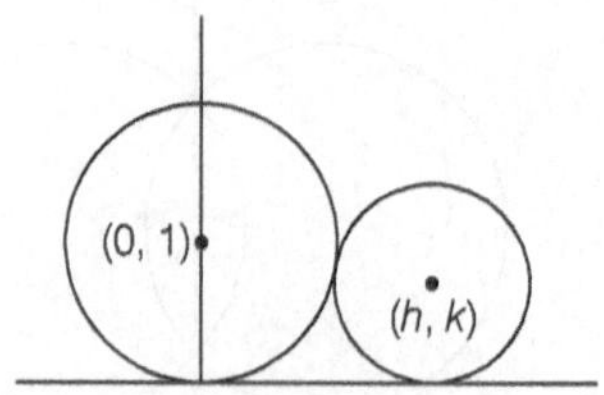

$$y < 0 \quad \Rightarrow x^2 = 0 \Rightarrow x = 0$$

43. (c) Given $x^2 + y^2 - ax - by = 0$

Centre $\left(\frac{a}{2}, \frac{b}{2}\right)$, Radius $= \frac{\sqrt{a^2 + b^2}}{2}$

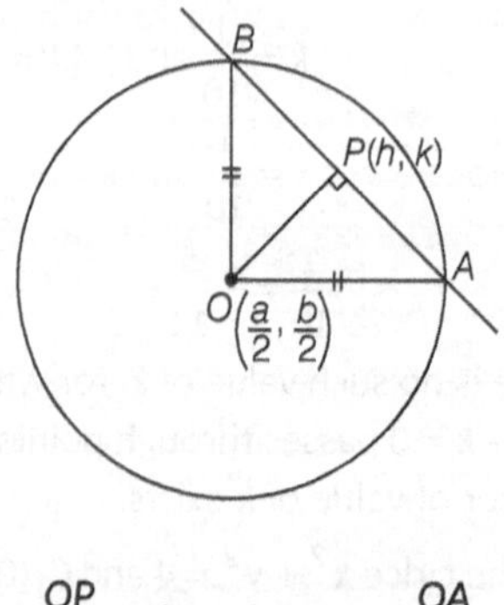

In ΔOPA,

$$\frac{OP}{OA} = \sin 45° \Rightarrow OP = \frac{OA}{\sqrt{2}}$$

$$\Rightarrow \quad \frac{\sqrt{a^2 + b^2}}{2\sqrt{2}} = \sqrt{\left(h - \frac{a}{2}\right)^2 + \left(k - \frac{b}{2}\right)^2}$$

$$\Rightarrow \quad \frac{a^2 + b^2}{8} = h^2 - ah + \frac{a^2}{4} + k^2 - kb + \frac{b^2}{4}$$

$$\Rightarrow \quad h^2+k^2-ah-kb+\frac{a^2+b^2}{8}=0$$

$$\Rightarrow \quad x^2+y^2-ax-by+\frac{a^2+b^2}{8}=0$$

44. (b) Equation of family of circles (equation of circles passing through point of intersections of two given circles)

$$S_1+\lambda S_2=0$$

$$(x^2+y^2+13x-3y)+\lambda(2x^2+2y^2+4x-7y-25)=0$$

$\because$ The required circle passes through (1, 1).

$$\therefore \quad 1^2+1^2+13-3+\lambda(2+2+4-7-25)=0$$

$$\lambda=\frac{1}{2}$$

$\therefore$ Required equation is

$$4x^2+4y^2+30x-13y-25=0$$

45. (a) $S_1 \equiv x^2+y^2-9=0$

$$S_2=x^2+y^2-2x+8y-7=0$$

Let $S \equiv x^2+y^2+2gx+2fy+c=0$ be the required circle.

$\because S=0$ passes through (1, 2).

$\therefore$ We have, $2g+4f+c=-5$ ...(i)

$S=0$ bisect the circumference $S_1=0$

$\Rightarrow S-S'=0$ passes through (0, 0)

$$\Rightarrow \quad c+9=0$$

$$\Rightarrow \quad c=-9 \quad \text{...(ii)}$$

$S=0$ and $S_2=0$ cut orthogonally.

$$\Rightarrow \quad 2g(-1)+2f(4)=c-7$$

$$\Rightarrow \quad -2g+8f=c-7 \quad \text{...(iii)}$$

By Eqs. (i), (ii) and (iii),

$$g=4, f=-1 \text{ and } c=-9$$

$\therefore$ Required circle is $x^2+y^2+8x-2y-9=0$

46. (c) Let $S \equiv x^2+y^2-3x+2y-4=0$

and $\quad L=2x+5y+2=0$

The required circle equation is of the form

$$S+\lambda L \equiv x^2+y^2-3x+2y-4+\lambda(2x+5y+2)=0$$

$$\equiv x^2+y^2-(3-2\lambda)x+(2+5\lambda)y-4+2\lambda=0$$

The centre is $\left[\frac{3-2\lambda}{2}, -\frac{(2+5\lambda)}{2}\right]$ lies on the line

$$x+y-11=0$$

$$\Rightarrow \quad \frac{3-2\lambda}{2}-\frac{2+5\lambda}{2}-11=0$$

$$\Rightarrow \quad -7\lambda-21=0$$

$$\Rightarrow \quad \lambda=-3$$

$\therefore$ Equation of required circle is

$$x^2+y^2-3x+2y-4-3(2x+5y+2)=0$$

$$\Rightarrow \quad x^2+y^2-9x-13y-10=0$$

47. (c) The circumcircle of the rectangle is

$$(x+4)(x-6)+(y-4)(y+1)=0$$

$$\Rightarrow \quad x^2+y^2-2x-3y-28=0 \quad \text{...(i)}$$

Length of y-intercept $=2\sqrt{f^2-c}=2\sqrt{\frac{9}{4}+28}=11$

$$\therefore \quad AB=11$$

$\therefore$ For ordinate of A and B are given by

$$y^2-3y-28=0$$

$$\Rightarrow \quad y=7, -4$$

$\therefore A(0, 7)$ and $B(0, -4)$

Equation of tangent at $A(0, 7)$

$$x(0)+y(7)-(x+0)-\frac{3}{2}(y+7)-28=0$$

$$\Rightarrow \quad 2x-11y+77=0 \quad \text{...(ii)}$$

and at $B(0, -4)$

$$x(0)+y(-4)-(x+0)-\frac{3}{2}(y-4)-28=0$$

$$\Rightarrow \quad 2x+11y+44=0 \quad \text{...(iii)}$$

Solving Eqs. (ii) and (iii), we get point $C \equiv \left(-\frac{121}{4}, \frac{3}{2}\right)$

which is point of intersection of tangents at A and B.

$\therefore$ Area of $\Delta ABC=\frac{1}{2}(AB \times \text{Height})$

$$=\frac{1}{2} \times 11 \times \frac{121}{4}=\left(\frac{11}{2}\right)^3$$

48. (a) $x^2+y^2+\frac{\lambda}{2}x-\frac{1+\lambda^2}{2}y-5=0$

and $x^2+y^2+4x+6y+3=0$ cut orthogonally, i.e.

$$2g_1g_2+2f_1f_2=c_1+c_2$$

$$2 \times \frac{\lambda}{4} \times 2+2\left(\frac{1+\lambda^2}{4}\right)(3)=-5+3$$

$$\Rightarrow \quad \lambda-\frac{3}{2}(1+\lambda^2)=-2$$

$$\Rightarrow \quad 2\lambda-3-3\lambda^2=-4$$

$$\Rightarrow \quad 3\lambda^2-2\lambda-1=0$$

$$\Rightarrow \quad (\lambda-1)(3\lambda+1)=0$$

$$\Rightarrow \quad \lambda=1, -\frac{1}{3}$$

$\therefore$ Two real values of λ.

$\therefore$ There are two circles belongs to this family.

49. (c) For largest common difference the smallest circle just touches the line.

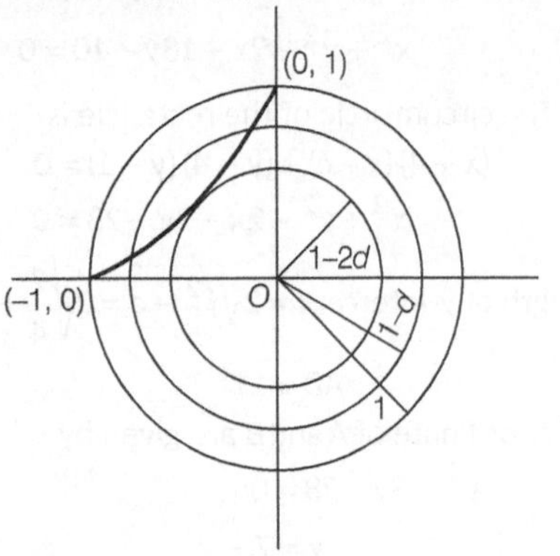

$$\Rightarrow \quad 1-2d=\frac{1}{\sqrt{2}}$$

$$\Rightarrow \quad d=\frac{\sqrt{2}-1}{2\sqrt{2}}$$

$$d\in\left(0,\frac{\sqrt{2}-1}{2\sqrt{2}}\right)$$

$$\Rightarrow \quad d\in\left(0,\frac{2-\sqrt{2}}{4}\right)$$

50. (b) $\tan 2\theta=\dfrac{2\left(\dfrac{1}{2}\right)}{1-\dfrac{1}{4}}=\dfrac{4}{3}$

Slope of $QR=-\dfrac{4}{3}$

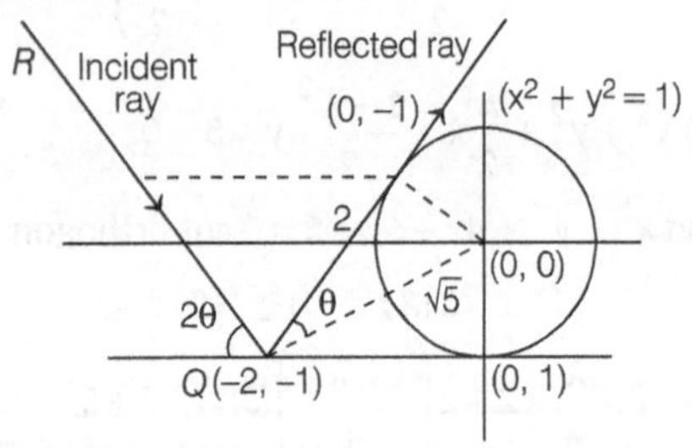

$\therefore$ Equation of incident ray QR is $(y+1)=-\dfrac{4}{3}(x+2)$

$$\Rightarrow \quad 3y+4x+11=0$$

51. (c) Let length of tangent $PA=L$

Radius of circle $OA=r$

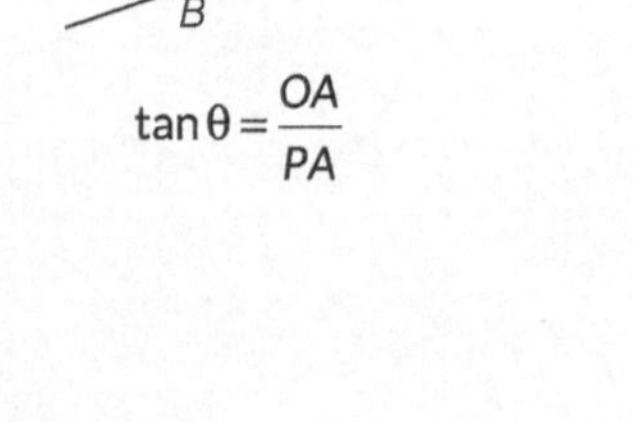

In ΔOAP, $\quad \tan\theta=\dfrac{OA}{PA}$

$$\tan\theta=\frac{r}{L}$$

$$\tan 2\theta=\frac{2\tan\theta}{1-\tan^2\theta}$$

$$\tan 45^\circ=\frac{\dfrac{2r}{L}}{1-\dfrac{r^2}{L^2}} \quad [\because 2\theta=45^\circ]$$

$$\Rightarrow \quad 1=\frac{2rL}{L^2-r^2}$$

$$\Rightarrow \quad 1=\frac{2r\sqrt{h^2+k^2-r^2}}{h^2+k^2-r^2-r^2}$$

$$[\because PA=L=\sqrt{h^2+k^2-r^2}]$$

$$\Rightarrow (h^2+k^2-2r^2)^2=4r^2(h^2+k^2-r^2)$$

$$\Rightarrow (h^2+k^2)^2+4r^4-4r^2(h^2+k^2)=4r^2(h^2+k^2-r^2)$$

$$\Rightarrow (h^2+k^2)^2+8r^4=8r^2(h^2+k^2)$$

$$\Rightarrow (h^2+k^2)^2=8r^2(h^2+k^2-r^2)$$

$\therefore$ Locus of P is

$$(x^2+y^2)^2=8r^2(x^2+y^2-r^2)$$

$$\therefore \quad \lambda=8$$

52. (c) $L: 9x+y-28=0 \quad$ ...(i)

$S: 2x^2+2y^2-3x+5y-7=0$

$$x^2+y^2-\frac{3}{2}x+\frac{5}{2}y-\frac{7}{2}=0 \quad \text{...(ii)}$$

Let pole be (x_1, y_1), then polar w.r.t. circle is

$$xx_1+yy_1+g(x+x_1)+f(y+y_1)+c=0$$

$$\Rightarrow \quad xx_1+yy_1-\frac{3}{4}(x+x_1)+\frac{5}{2}(y+y_1)-\frac{7}{2}=0$$

$$\Rightarrow x\left(x_1-\frac{3}{4}\right)+y\left(y_1+\frac{5}{4}\right)-\frac{3}{4}x_1+\frac{5}{4}y_1-\frac{7}{2}=0$$

$$\Rightarrow 4x\left(x_1-\frac{3}{4}\right)+4y\left(y_1+\frac{5}{4}\right)-(3x_1-5y_1+14)=0$$

Compare with Eq. (i)

$$4\left(x_1-\frac{3}{4}\right)=9;\quad x_1=3$$

and $\quad 4\left(y_1+\dfrac{5}{4}\right)=1;\quad y_1=-1$

$\Rightarrow (3,-1)$

53. (c) Let point be (x_1, y_1) and its polar w.r.t. $x^2+y^2-2\lambda x+c^2=0$ is given by

$$xx_1+yy_1+g(x+x_1)+f(y+y_1)+c^2=0$$

$$\Rightarrow \quad xx_1+yy_1-\lambda(x+x_1)+c^2=0$$

$$x(x_1-\lambda)-\lambda x_1+yy_1+c^2=0$$

Clearly, it passes through fixed point for all values of λ.

54. (a) $x^2+y^2<25$

On X-axis and Y-axis $4\times4+1=17$

$$x=1 \quad y=1,2,3,4$$
$$x=2 \quad y=1,2,3,4$$
$$x=3 \quad y=1,2,3$$
$$x=4 \quad y=1,3$$

In Ist quadrant $=13$

In all quadrant $=13\times4=52$

$\therefore$ Required number of points $=52+17=69$

55. (b) $C_1:(x-2)^2+y^2=4$

$$A=\{(x,y):(x-2)^2+y^2\le 4\}$$

$C_2: x^2+y^2=4 \quad B=\{(x,y):x^2+y^2\le 4\}$

$C_3:(x-2)^2+(y-2)^2=4$

$C=\{(x,y):(x-2)^2+(y-2)^2\le 4\}$

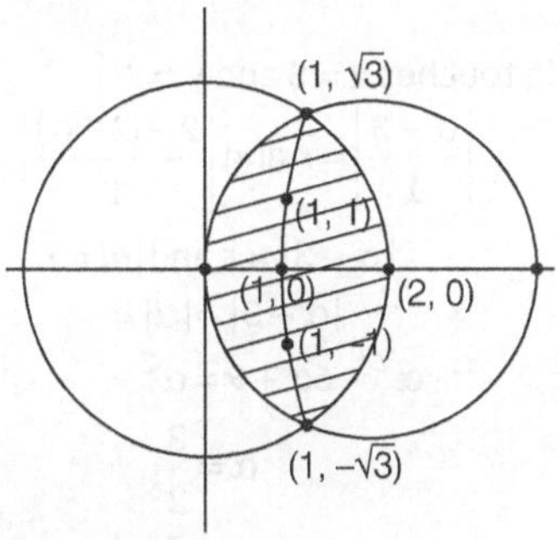

Shaded region is $A\cap B$

Number of common points is $C_1C_2=5$

i.e. (0, 0), (1, 0), (2, 0), (1, 1), (1, –1)

Similarly in C_2 and C_3 is 5.

$\therefore$ Number of relations

$$=2^{\text{number of elements in }(A\cap B)\times(B\cap C)}$$
$$=2^{5\times5}=2^{25}$$

$\Rightarrow \quad P=25$

56. (a) Let the chord $x+y=n$ cuts the circle $x^2+y^2=16$ at A and B.

Let of perpendicular from O on AB

$$OM=\left|\frac{0+0-n}{\sqrt{1^2+1^2}}\right|=\frac{n}{\sqrt2}$$

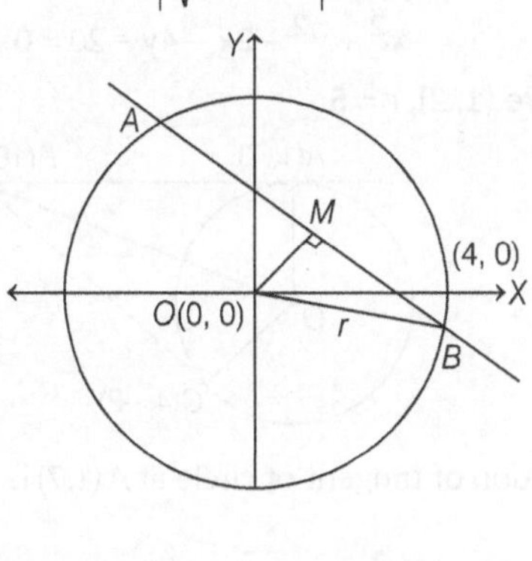

Radius of circle $=4$

$\because \quad OM<r$

$$\frac{n}{\sqrt2}<4$$
$$n<5.65$$

$\therefore$ Possible values of $n=1,2,3,4,5$

$\therefore$ Length of chord $AB, l=2\sqrt{4^2-\left(\dfrac{n}{\sqrt2}\right)^2}$

$$l=2\sqrt{16-\frac{n^2}{2}}\Rightarrow l=\sqrt{64-2n^2}$$

$\therefore$ For $n=1, \quad l^2=62$

$n=2, \quad l^2=56$

$n=3, \quad l^2=46$

$n=4, \quad l^2=32$

$n=5, \quad l^2=14$

$\therefore$ Sum of squares of length of chords

$$=62+56+46+32+14=210$$

57. (d) Centre of circles are opposite sides is line

$$(3+4-\lambda)(27+4-\lambda)<0$$
$$(\lambda-7)(\lambda-31)<0$$
$$\lambda\in(7,31)$$

Distance from $S_1 \Rightarrow \left|\dfrac{3+4-\lambda}{5}\right|\ge1$

$\Rightarrow \quad \lambda\in(-\infty,2]\cup[12,\infty)$

Distance from $S_2 \Rightarrow \left|\dfrac{27+4-\lambda}{5}\right|\ge2$

$\Rightarrow \quad \lambda\in(-\infty,27]\cup(41,\infty)$

So, $\lambda\in[12,21]$

58. (c) According to given information, we have the following figure.

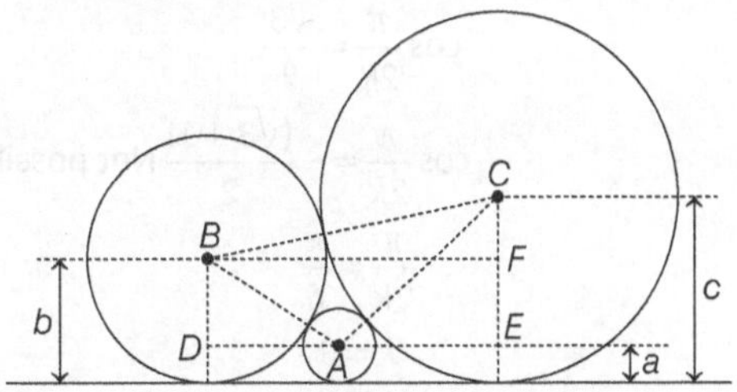

where A,B,C are the centres of the circles

Clearly, $AB=a+b$ (sum of radii) and $BD=b-a$

$\therefore \quad AD=\sqrt{(a+b)^2-(b-a)^2}$

[using Pythagoras theorem in ΔABD]

$$=2\sqrt{ab}$$

Similarly, $AC=a+c$ and $CE=c-a$

$\therefore$ In ΔACE, $AE=\sqrt{(a+c)^2-(c-a)^2}=2\sqrt{ac}$

Similarly, $BC=b+c$ and $CF=c-b$

$\therefore$ In ΔBCF, $BF = \sqrt{(b+c)^2 - (c-b)^2} = 2\sqrt{bc}$

$\because \quad AD + AE = BF$

$\therefore \quad 2\sqrt{ab} + 2\sqrt{ac} = 2\sqrt{bc}$

$\Rightarrow \quad \dfrac{1}{\sqrt{c}} + \dfrac{1}{\sqrt{b}} = \dfrac{1}{\sqrt{a}}$

Numerical value Type Questions

59. (3) Since the distance between parallel chords is greater than radius, therefore both chords lie on opposite side of centre.

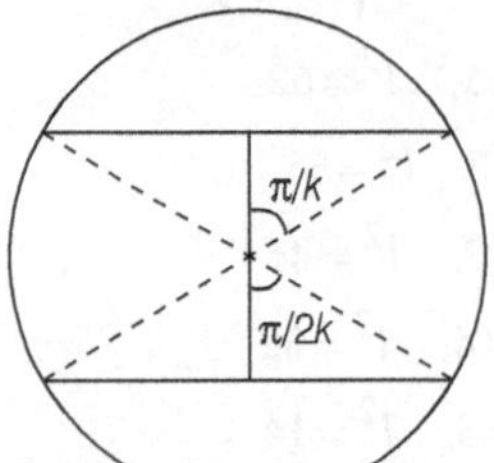

$$2\cos\frac{\pi}{2k} + 2\cos\frac{\pi}{k} = \sqrt{3}+1$$

Let $\quad \dfrac{\pi}{2k} = \theta$

$\therefore \quad 2\cos\theta + 2\cos 2\theta = \sqrt{3}+1$

$\Rightarrow \quad 2\cos\theta + 2(2\cos^2\theta - 1) = \sqrt{3}+1$

$\Rightarrow 4\cos^2\theta + 2\cos\theta - (3+\sqrt{3}) = 0$

$\therefore \quad \cos\theta = \dfrac{-2 \pm \sqrt{4 + 16(3+\sqrt{3})}}{2\times 4}$

$= \dfrac{-1 \pm (2\sqrt{3}+1)}{4}$

$= \dfrac{\sqrt{3}}{2}, \dfrac{-(\sqrt{3}+1)}{2}$

$\because \quad \cos\dfrac{\pi}{2R} = \dfrac{\sqrt{3}}{2}$

$\cos\dfrac{\pi}{2k} = -\dfrac{(\sqrt{3}+1)}{2}$ Not possible

$\dfrac{\pi}{2k} = \dfrac{\pi}{6}$

$k = 3$

$[k] = 3$

60. (36) $C_1 : x^2 + y^2 - 6x + 8 = 0$

$O_1(3, 0)$ and $r_1 = 1$

$C_2 : x^2 + y^2 - 8y + 16 - k = 0$

$O_2(0, 4)$ and $r_2 = \sqrt{k}$

$\because$ C_1 and C_2 touch each other,

$\Rightarrow \quad O_1O_2 = |r_1 \pm r_2|$

$\Rightarrow \quad 5 = |1 \pm \sqrt{k}|$

$\therefore \quad 1 + \sqrt{k} = 5$ or $\sqrt{k} - 1 = 5$

$k = 16$ or $k = 36$

$\therefore$ Largest value of k is 36.

61. (3) $\because$ Centre lies on $x + y = 2$ and in first quadrant, centre $= (\alpha, 2-\alpha)$.

where, $\alpha > 0$ and $2 - \alpha > 0$

$\Rightarrow \quad 0 < \alpha < 2$

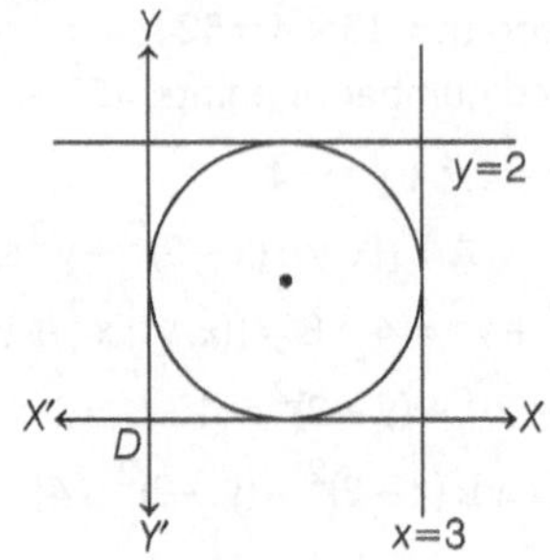

$\because$ Circle touches $x = 3$ and $y = 2$

$\therefore \quad \left|\dfrac{\alpha - 3}{1}\right| = r$ and $\left|\dfrac{2-(2-\alpha)}{1}\right| = r$

$\Rightarrow \quad |\alpha - 3| = r$ and $|\alpha| = r$

$\Rightarrow \quad |\alpha - 3| = |\alpha|$

$\Rightarrow \quad \alpha^2 - 6\alpha + 9 = \alpha^2$

$\Rightarrow \quad \alpha = \dfrac{3}{2}$

$\therefore \quad r = \dfrac{3}{2}$

$\Rightarrow$ diameter $= 2r = 3$

62. (61) $r = \sqrt{\dfrac{p^2}{4} + \dfrac{(1-p)^2}{4} - 5}$

$= \sqrt{\dfrac{2p^2 - 2p - 19}{2}}$

$\because \quad r \in (0, 5]$

$0 < 2p^2 - 2p - 19 \le 10$

$\Rightarrow \quad p \in \left[\dfrac{1-\sqrt{239}}{2}, 1 - \dfrac{\sqrt{39}}{2}\right] \cup \left(1 + \dfrac{\sqrt{39}}{2}, \dfrac{1+\sqrt{239}}{2}\right]$

So number of integral values of P^2 is 61.

63. (75) Given equation of circle

$$x^2 + y^2 - 2x - 4y - 20 = 0$$

Centre (1, 2), $r = 5$

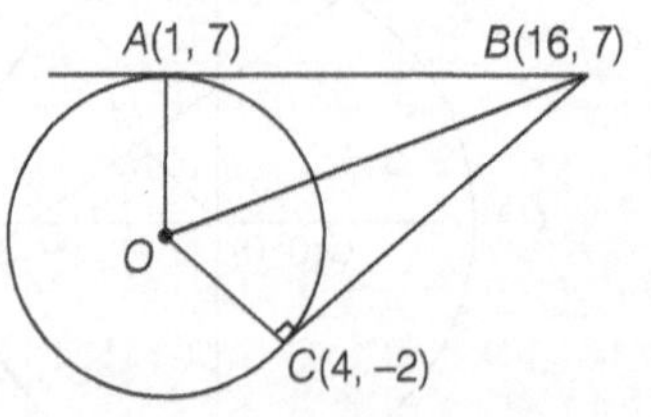

Equation of tangent of circle at $A(1, 7)$ is

$$x+7y-\frac{2(x+1)}{2}-\frac{4(y+7)}{2}-20=0$$

$$\Rightarrow \quad x+7y-x-1-2y-14-20=0$$

$$\Rightarrow \quad y=7 \quad \text{...(i)}$$

Equation of tangent of circle at $C(4,-2)$ is

$$4x-2y-2\left(\frac{x+4}{2}\right)-4\left(\frac{y-2}{2}\right)-20=0$$

$$4x-2y-x-4-2y+4-20=0$$

$$3x-4y-20=0 \quad \text{...(ii)}$$

Solving Eqs. (i) and (ii), we get

$B(16,7)$

$$AB=BC=\sqrt{(16-1)^2+(7-7)^2}=15$$

$\therefore$ Area of quadrilateral $OABC = 2$ area of ΔOAB

$$=2\times\frac{1}{2}(OA\times AB)$$

$$=5\times 15=75$$

64. (32) We have, $BD=20$

$CD=16$

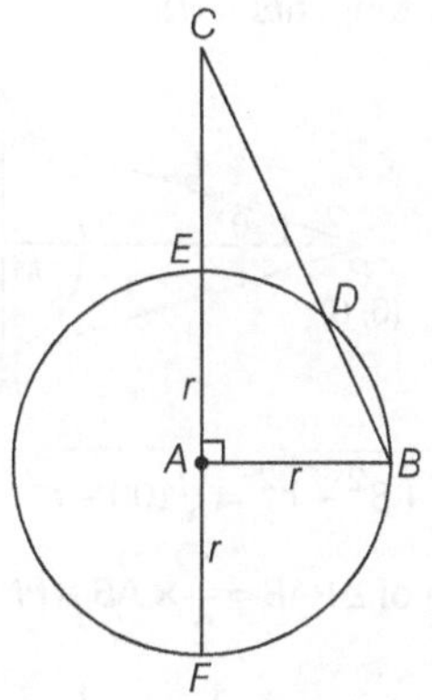

$AB =$ radius of circle $= r$

In ΔABC, $AC^2+AB^2=BC^2$

$$\Rightarrow \quad AC^2+r^2=(36)^2 \quad \text{...(i)}$$

We know

$$CE\times CF=CD\times CB$$

$$(AC-AE)\times(AC+AF)=16\times 36$$

$$(AC-r)(AC+r)=16\times 36$$

$$AC^2-r^2=16\times 36 \quad \text{...(ii)}$$

From Eqs. (i) and (ii), we get

$$2AC^2=36^2+16\times 36$$

$$\Rightarrow \quad 2AC^2=36(36+16)$$

$$\Rightarrow \quad AC^2=18\times 52$$

$$\Rightarrow \quad AC=\sqrt{18\times 52}$$

$$\Rightarrow \quad AC=6\sqrt{26}$$

$\therefore$ $p=6$ and $q=26$

and $p+q=6+26=32$

65. (22) Given, slope of $OA=\tan\theta$

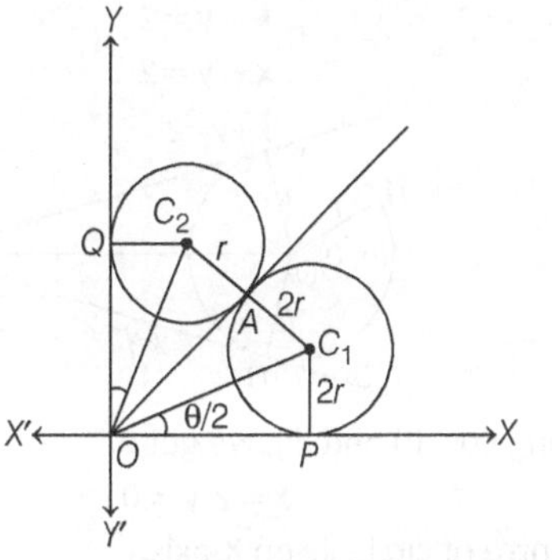

$$\therefore \quad \angle AOP=\theta$$

$$\angle C_1OP=\frac{\theta}{2}$$

$$\angle AOQ=\frac{\pi}{2}-\theta$$

$$\angle C_2OQ=\frac{\pi}{4}-\frac{\theta}{2}$$

In ΔOQC_2,

$$OA=r\cot\left(\frac{\pi}{4}-\frac{\theta}{2}\right)$$

In ΔOAC_1,

$$OA=2r\cot\frac{\theta}{2}$$

$$\therefore \quad r\cot\left(\frac{\pi}{2}-\frac{\theta}{2}\right)=2r\cot\frac{\theta}{2}$$

$$\Rightarrow \quad \cot\left(\frac{\pi}{4}-\frac{\theta}{2}\right)=2\cot\frac{\theta}{2}$$

$$\Rightarrow \quad \tan\frac{\theta}{2}=2\tan\left(\frac{\pi}{4}-\frac{\theta}{2}\right)$$

$$\Rightarrow \quad \tan\frac{\theta}{2}=\frac{2\left(1-\tan\frac{\theta}{2}\right)}{1+\tan\frac{\theta}{2}}$$

$$\Rightarrow \quad \tan\frac{\theta}{2}+\tan^2\frac{\theta}{2}=2-2\tan\frac{\theta}{2}$$

$$\Rightarrow \quad \tan^2\frac{\theta}{2}+3\tan\frac{\theta}{2}-2=0$$

$$\Rightarrow \quad \tan\frac{\theta}{2}=\frac{-3\pm\sqrt{9+8}}{2}$$

$$\Rightarrow \quad \tan\frac{\theta}{2}=\frac{-3+\sqrt{17}}{2} \quad \left[\because \tan\frac{\theta}{2}>0\right]$$

$$\Rightarrow \quad \tan\frac{\theta}{2}=\frac{\sqrt{17}-3}{2}$$

Hence, $a=17, b=3, c=2$

$$\therefore \quad a+b+c=17+3+2=22$$

66. (6) Given equation of tangents are

$$x + y = 2 \quad \text{...(i)}$$
$$x - y = 2 \quad \text{...(ii)}$$

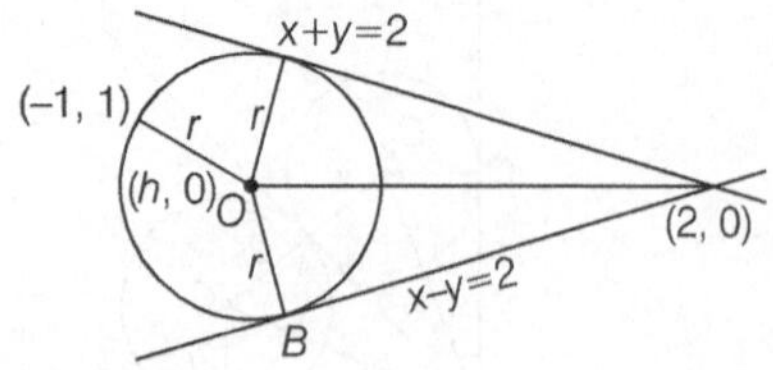

Solving Eqs. (i) and (ii), we get

$$x = 2, y = 0$$

∴ Centre of circle lie on X-axis.

Let centre of circle is $(h, 0)$

$$r = \sqrt{(h+1)^2 + (1)^2}$$

$$OA = r = \left|\frac{h+0-2}{\sqrt{2}}\right|$$

[OA is perpendicular distance to the line $x + y = 1$]

$$r = \frac{h-2}{\sqrt{2}}$$

$$\therefore \quad (h+1)^2 + 1 = \left(\frac{h-2}{\sqrt{2}}\right)^2$$

$$\Rightarrow \quad h^2 + 2h + 1 + 1 = \frac{h^2 - 4h + 4}{2}$$

$$\Rightarrow \quad 2h^2 + 4h + 4 = h^2 - 4h + 4$$

$$\Rightarrow \quad h^2 + 8h = 0$$

$$\Rightarrow \quad h = 0, -8$$

∴ Equation of circle are

$$x^2 + y^2 = 2$$

and $$(x+8)^2 + y^2 = 50$$

$$\therefore \quad r_1 = \sqrt{2} \text{ and } r_2 = 5\sqrt{2}$$

$$\therefore \quad r_1 + r_2 = \sqrt{2} + 5\sqrt{2} = 6\sqrt{2}$$

Hence, $\lambda = 6$

67. (1) Given circle is $x^2 + y^2 - 4x - 4y + 4 = 0$ ...(i)

Centre $(2, 2)$, $r = 2$

Let the hypotenuse of triangle meet OX and OY at A and B respectively.

Let $A(a, 0), B(0, b)$

Area of $\Delta AOB = \Delta OCA + \Delta OCB + \Delta ACB$

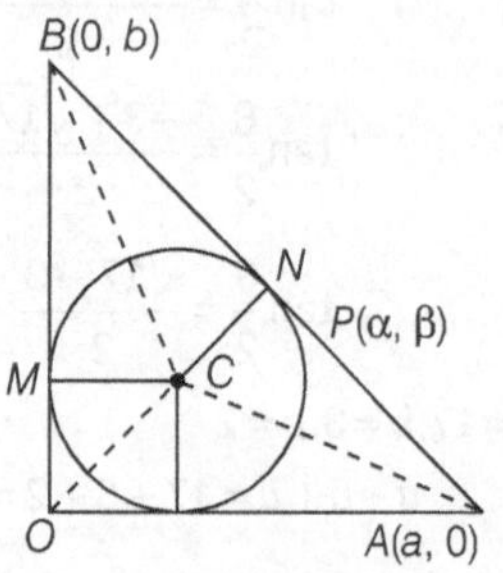

$$\Rightarrow \quad \frac{1}{2}ab = \frac{1}{2}(2a) + \frac{1}{2}(2b) + \frac{1}{2}2\sqrt{a^2+b^2}$$

$$\Rightarrow \quad ab = 2(a+b) + 2\sqrt{a^2+b^2} \quad \text{...(ii)}$$

Let $P(\alpha, \beta)$ be the circumcentre of ΔOAB.

Since ΔOAB is right angle. Its circumcentre is mid-point of hypotenuse.

$$\therefore \quad \alpha = \frac{a}{2} \text{ and } \beta = \frac{b}{2}$$

$$\Rightarrow \quad a = 2\alpha \text{ and } b = 2\beta$$

From Eq. (ii), $4\alpha\beta = 4\alpha + 4\beta + 4\sqrt{\alpha^2 + \beta^2}$

$$\Rightarrow \quad \alpha\beta = \alpha + \beta + \sqrt{\alpha^2 + \beta^2}$$

Hence, equation of the locus $P(\alpha, \beta)$ is

$$xy = x + y + \sqrt{x^2 + y^2}$$

$$\Rightarrow \quad x + y - xy + \sqrt{x^2 + y^2} = 0$$

Hence, $k = 1$

68. (5) $\because PA = PB$ ...(i)

$\because$ Length of tangents $= \sqrt{S_1}$

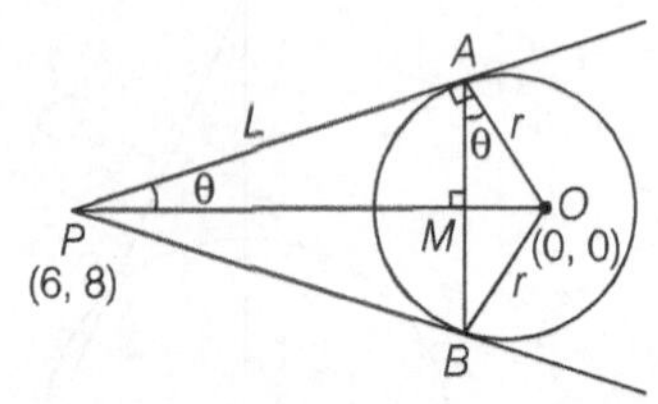

$$\Rightarrow L = \sqrt{6^2 + 8^2 - r^2} = \sqrt{100 - r^2}$$

$$\because \quad \text{Area of } \Delta PAB = \frac{1}{2} \times AB \times PM$$

$$= \frac{1}{2} \times 2r\frac{L}{OP} \times \frac{L^2}{OP}$$

$$\left[\begin{aligned} \because AB = 2AM &= 2r\cos\theta \\ &= 2r \cdot \frac{L}{OP} \\ PM = L\cos\theta &= L \cdot \frac{L}{OP} \end{aligned}\right]$$

$$= \frac{rL^3}{(OP)^2} = \frac{r(100 - r^2)^{3/2}}{6^2 + 8^2}$$

$$A = \frac{r(100 - r^2)^{3/2}}{100}$$

For maximum

$$\frac{dA}{dr} = \frac{1}{100}\left[(100 - r^2)^{3/2} - \frac{3}{2}r(100 - r^2)^{1/2}(2r)\right]$$

$$= \frac{1}{100}\sqrt{100 - r^2}\,[100 - r^2 - 3r^2]$$

$$\frac{dA}{dr} = 0$$

$100 - 4r^2 = 0 \Rightarrow r = 5$

Also, $\frac{dA}{dr} > 0$ for $r < 5$ and is negative for $r > 5$.

$\therefore$ A is maximum at $r = 5$

69. (7) The point dividing the segment AB in the ratio $k:1$, then point is $\left(\frac{k+3}{k+1}, \frac{4}{k+1}\right)$.

This lies on the circle $x^2 + y^2 = 4$

$\Rightarrow \quad (k+3)^2 + 16 = 4(k+1)^2$

$\Rightarrow \quad 3k^2 + 2k - 21 = 0$

For λ and μ are the roots.

Therefore, $|\lambda\mu| = \left|-\frac{21}{3}\right| = |-7| = 7$

70. (8) $\therefore MN = MP + PN$

$= \sqrt{3^2 - 1^2} + \sqrt{3^2 - 1^2} = 4\sqrt{2}$

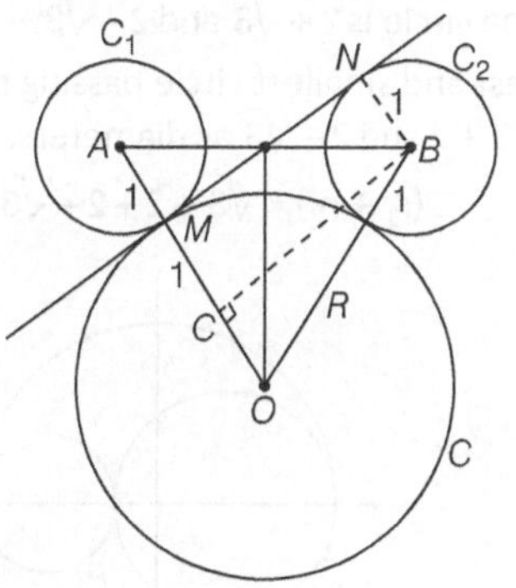

$\because \quad MN = BC = \sqrt{(OB)^2 - (OC)^2}$

$= \sqrt{(R+1)^2 - (R-1)^2}$

(where, R the radius of circle C)

$32 = (R+1+R-1)(R+1-R+1) \Rightarrow 4R = 32$

$R = 8$

71. (6) $x^2 + y^2 - 2x - 2\lambda y - 8 = 0$

$(x^2 + y^2 - 2x - 8) - 2\lambda y = 0$

$\Rightarrow \quad S + \lambda L = 0$

$S : x^2 + y^2 - 2x - 8 = 0$

$L : y = 0$

Solving S and L, we get

$x^2 - 2x - 8 = 0$

$(x-4)(x+2) = 0$

$\Rightarrow \quad x = -2, 4$

$\Rightarrow A(-2, 0)$ and $B = (4, 0)$

$\therefore$ Distance between A and $B = AB = 6$

72. (3) $\because \theta$ is the angle between the tangent and the line.
Circle with centre $(2, -1)$ and radius $= 3$

Perpendicular from centre on $3x - 4y = 5$ is

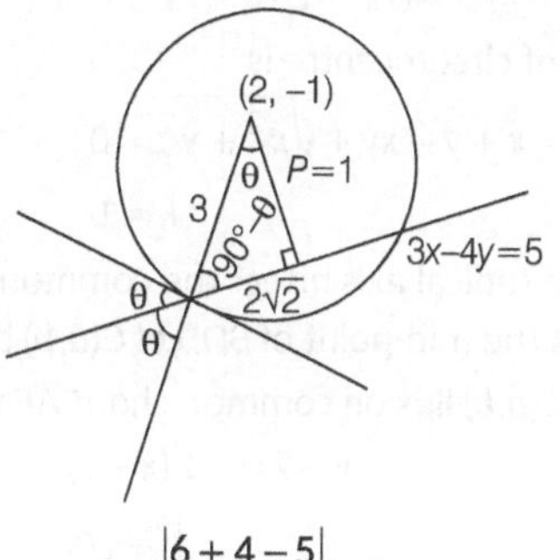

$P = \left|\frac{6+4-5}{5}\right| = 1$

$\Rightarrow \quad \sin(90 - \theta) = \frac{1}{3}$

$\cos\theta = \frac{1}{3}$

$\Rightarrow \quad 9\cos\theta = 9 \times \frac{1}{3} = 3$

73. (1) The given circle is

$x^2 + y^2 - 4x - 4y + 4 = 0$

$(x-2)^2 + (y-2)^2 = 4$

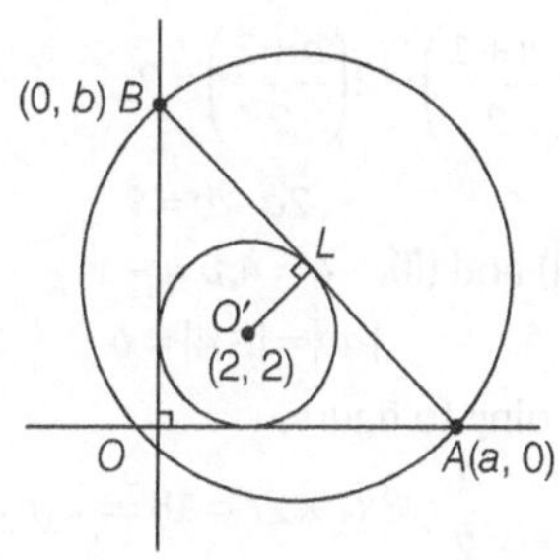

Let the coordinates of A and B are $(a, 0)$ and $(b, 0)$.
AB is diameter of circumcircle.

Hence, circumcentre is $\left(\frac{a}{2}, \frac{b}{2}\right)$.

Let $x_1 = \frac{a}{2}, y_1 = \frac{b}{2}$ ...(i)

Equation of AB, $\quad \frac{x}{a} + \frac{y}{b} = 1$

$\because \quad O'L = 2$

$$\frac{\frac{2}{a} + \frac{2}{b} - 1}{\sqrt{\frac{1}{a^2} + \frac{1}{b^2}}} = 2$$

{(0, 0) and (2, 2) lies on the same side of AB}

$\Rightarrow \quad \frac{2a + 2b - ab}{\sqrt{a^2 + b^2}} = -2$

By Eqs. (i) and (ii), we get

$4x_1 + 4y_1 - 4x_1y_1 = -4\sqrt{x_1^2 + y_1^2}$

$$\Rightarrow x_1 + y_1 - x_1y_1 + \sqrt{x_1^2 + y_1^2} = 0$$

Locus of circumcentre is

$$x + y - xy + \sqrt{x^2 + y^2} = 0$$

$$\therefore \quad k = 1$$

74. (4) The radical axis bisect the common tangent BD.

$\therefore$ M is the mid-point of BD. Let $C(a,b)$ be the point.

Now, $C(a,b)$ lies on common chord AE which is

$$y - 2 = -1(x - 1)$$

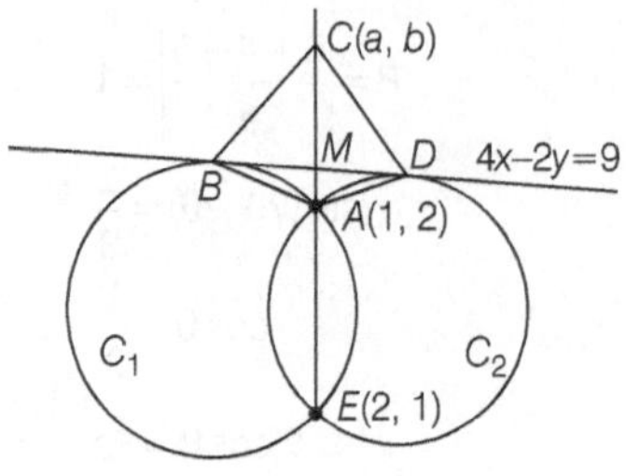

$$\Rightarrow \quad x + y = 3$$

$$\therefore \quad a + b = 3 \quad \text{...(i)}$$

Now, $M\left(\dfrac{a+1}{2}, \dfrac{b+2}{2}\right)$ lies on $4x - 2y = 9$

$$4\left(\frac{a+1}{2}\right) - 2\left(\frac{b+2}{2}\right) = 9$$

$$2a - b = 9 \quad \text{...(ii)}$$

By Eqs. (i) and (ii), $a = 4, b = -1$

$$\therefore \quad |ab| = |-4| = 4$$

75. (2) According to figure,

$$\frac{1}{2} \times 3x_1 \times 2r = 18 \Rightarrow x_1 r = 6 \quad \text{...(i)}$$

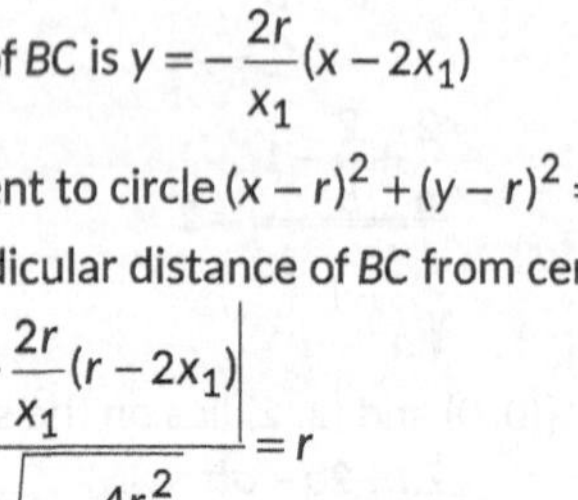

Equation of BC is $y = -\dfrac{2r}{x_1}(x - 2x_1)$

BC is tangent to circle $(x-r)^2 + (y-r)^2 = r^2$

$\therefore$ Perpendicular distance of BC from centre = Radius

$$\frac{\left|r + \dfrac{2r}{x_1}(r - 2x_1)\right|}{\sqrt{1 + \dfrac{4r^2}{x_1^2}}} = r$$

$$\Rightarrow \quad \frac{2r^2}{x_1} - 3r = r\sqrt{1 + \frac{4r^2}{x_1^2}}$$

$$\Rightarrow \quad (2r^2 - 3x_1)^2 = x_1^2 + 4r^2$$

$$r_1 x_1 = \frac{2}{3}x_1^2 \Rightarrow 3r = 2x_1 \quad \text{...(ii)}$$

By Eqs. (i) and (ii), we get

$$r = 2 \text{ units}$$

76. (6) Points satisfying the inequalities are common to all 3 circles given by (1, 1), (1, 2), (2, 1), (2, 2), (2, 3), (3, 2). Number of ordered pairs $(a,b) = 6$

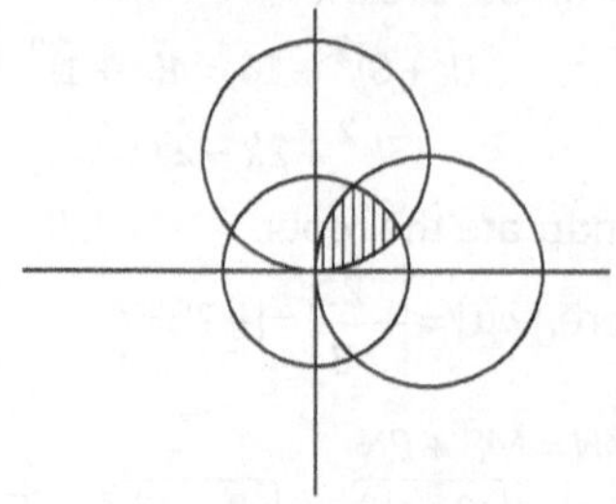

77. (1) Maximum and minimum distance of point $P(\sqrt{3}, \sqrt{2})$ from the circle is $2 + \sqrt{3}$ and $2 - \sqrt{3}$.

$\therefore$ Largest and smallest circle passing through P must have $\sqrt{3} + 2$ and $2 - \sqrt{3}$ as diameters.

$$\Rightarrow \quad 2(r_1 + r_2) = \sqrt{3} + 2 + 2 - \sqrt{3} = 4$$

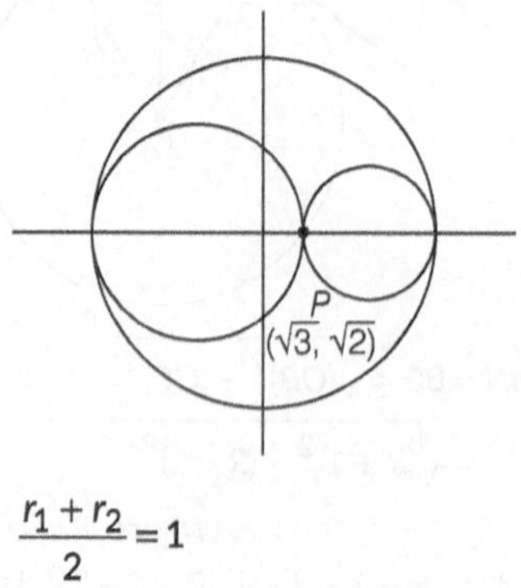

$$\Rightarrow \quad \frac{r_1 + r_2}{2} = 1$$

78. (6) $\because$ Diagonal of rhombus bisect each other at 90°.

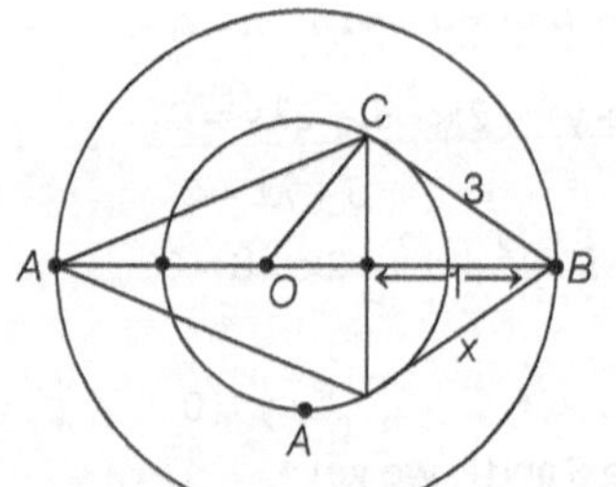

$$O'C = \sqrt{OC^2 - OO'^2}$$

$$O'C = \sqrt{9 - 1} = \sqrt{8}$$

$$x = \sqrt{O'C^2 + O'B^2}$$

$$= \sqrt{8 + 36} = \sqrt{44}$$

$$= [\sqrt{44}] = 6$$

79. (4) $\because PQR$ is an equilateral triangle

$$PQ = QR = PR$$

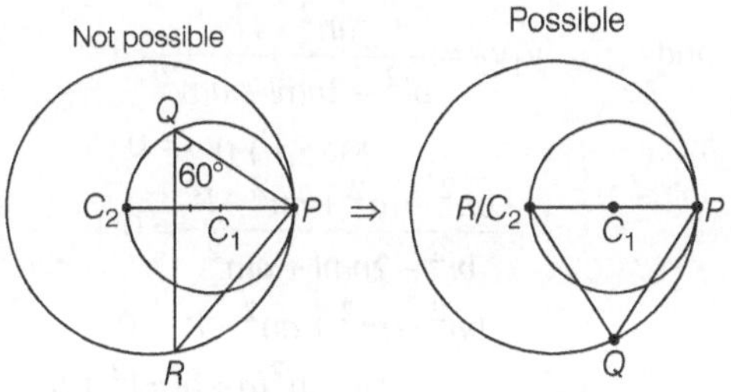

$\because \quad PQ < C_2P$ (diameter of smaller circle)

and $\quad PR > C_2P$

$\because$ So it is possible when,

$$PQ = C_2P$$

$\therefore \quad PQ = 4$

80. (2) Let AE be diameter of semi-circle where $AE = 50$ and two circles are touching according to figure each having radius 8 units.

$\because OC = CM - OM = 25 - 8 = 17$

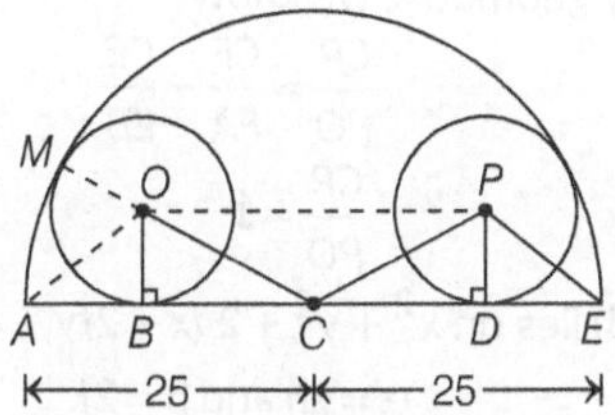

$\therefore$ In ΔOBC, $\quad OB^2 + BC^2 = OC^2$

$$BC^2 = 17^2 - 8^2 = 289 - 64 = 225$$

$\Rightarrow \quad BC = 15$

Similarly $\quad CD = 15$

$\therefore \quad OP = BD = BC + CD$

$$= 15 + 15 = 30 = \lambda$$

$\therefore \quad \frac{60}{\lambda} = \frac{60}{30} = 2$

Level 3

1. (c) Nearest point will be foot of the perpendicular drawn from centre (0, 6)

$$\frac{x-0}{2} = \frac{y-6}{1} = \frac{-(0+6+4)}{4+1} = -2$$

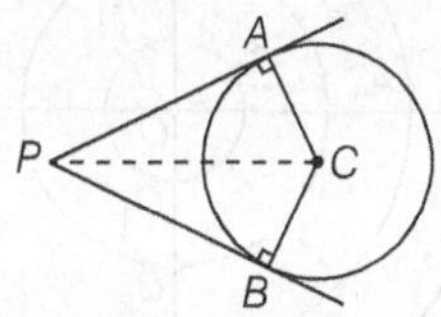

$\Rightarrow \quad (x, y) = (-4, 4)$

$\because \quad PC = \sqrt{(-4)^2 + (2)^2}$

Radius $= CA = CB = \sqrt{36 - 35} = 1$

$$PA = \sqrt{PC^2 - CA^2} = \sqrt{20 - 1} = \sqrt{19}$$

Area of quadrilateral $PACB = 2 \times$ Area of ΔPCA

$$= 2 \times \frac{1}{2} \times PA \times AC$$

$$= 2 \times \frac{1}{2} \times \sqrt{19} \times 1 = \sqrt{19}$$

2. (b) Image of the centre $C_2(1, -3)$ in the line $3x + 4y - 16 = 0$ is $P(7, 5)$

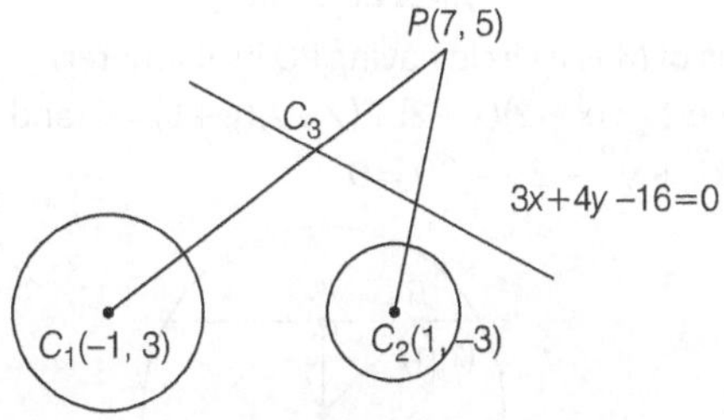

Now, $C_1C_2 + C_2C_3 + C_3C_1$ to be minimum C_1, C_3 and P should be on the same line so,

$$C_3 = (0, 4)$$

Distance between C_3 and $C_1 = \sqrt{50} = 5\sqrt{2}$

So, radius of $C_1 = 3\sqrt{2}$

So, radius of $C_3 = 2\sqrt{2}$

Equation of C_3

$$(x-0)^2 + (y-4)^2 = (2\sqrt{2})^2$$

$$x^2 + y^2 - 8y + 8 = 0$$

$$a = 0, \; b = -8, \; c = 8$$

$$a + b + c = 0$$

3. (b) $OA = \frac{\sqrt{5}}{2}, OC = \frac{4}{\sqrt{5}}$

$$CQ = OC = \frac{4}{\sqrt{5}} \text{ and } CA = \frac{3}{2\sqrt{5}}$$

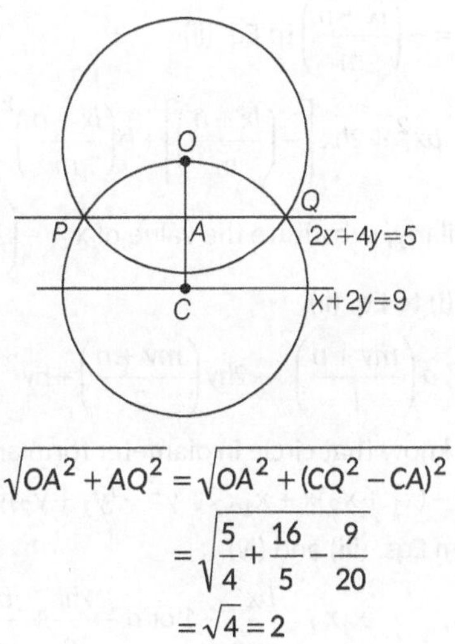

$\therefore \quad OQ = \sqrt{OA^2 + AQ^2} = \sqrt{OA^2 + (CQ^2 - CA)^2}$

$$= \sqrt{\frac{5}{4} + \frac{16}{5} - \frac{9}{20}}$$

$$= \sqrt{4} = 2$$

4. (d)

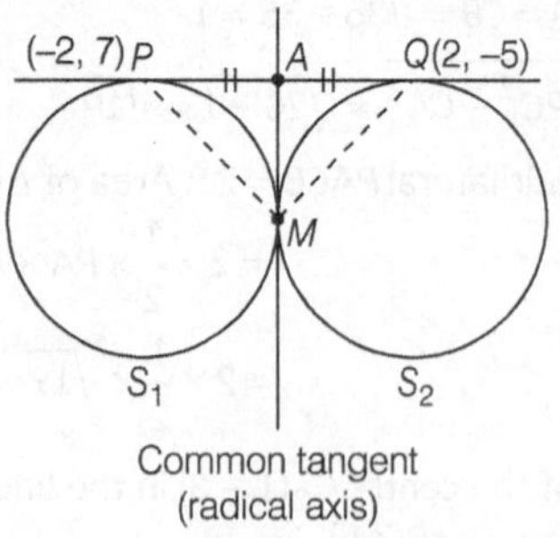

$\Rightarrow \quad AP = AQ = AM$

Locus of M is a circle having PQ its diameter.

Hence, $E_1 : (x-2)(x+2)+(y-7)(y+5)=0$ and $x \neq \pm 2$

$\Rightarrow x^2 + y^2 - 2y - 39 = 0$

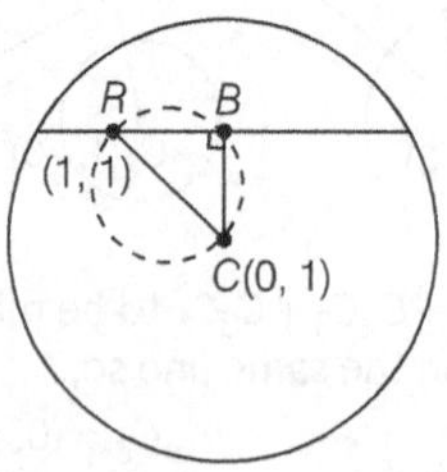

Locus of mid-points of chords passing through (1, 1).

$$\Rightarrow \quad BR^2 + BC^2 = CR^2$$

$$\Rightarrow \quad (h-1)^2 + (k-1)^2 + h^2 + (k-1)^2 = 1$$

$$\Rightarrow \quad h^2 + k^2 - 2h - 2K + 2 + h^2 + k^2 - 2k + 1 = 0$$

$$\Rightarrow \quad 2h^2 + 2k^2 - 2h - 4k + 2 = 0$$

$$\Rightarrow \quad h^2 + k^2 - h - 2k + 1 = 0$$

Locus $\quad x^2 + y^2 - x - 2y + 1 = 0$

5. (a) Line : $\quad lx + my + n = 0 \quad$ ...(i)

Intersect the curve :

$$ax^2 + 2hxy + by^2 - 1 = 0 \quad \text{...(ii)}$$

Substitute value of y from Eq. (i)

$$y = -\left(\frac{lx+n}{m}\right) \text{ to Eq. (ii)}$$

$$ax^2 + 2hx\left[-\left(\frac{lx+n}{m}\right)\right] + b\left(\frac{lx+n}{m}\right)^2 - 1 = 0 \quad \text{...(iii)}$$

Similarly, substitute the value of $x = -\left(\frac{my+n}{l}\right)$ from Eq. (i) to Eq. (ii)

$$a\left(\frac{my+n}{l}\right)^2 - 2hy\left(\frac{my+n}{2}\right) + by^2 - 1 = 0 \quad \text{...(iv)}$$

We know that circle in diameter form are

$$x^2 - (x_1 + x_2)x + x_1x_2 + y^2 - (y_1 + y_2)y + y_1y_2 = 0$$

From Eqs. (iii) and (iv),

$$x_1x_2 = \frac{bx^2}{m^2} - 1 \text{ or } a - \frac{2hl}{m} + \frac{bl}{m^2}$$

$$x_1x_2 = \frac{bn^2 - m^2}{am^2 - 2nlm + bl^2}$$

and $\quad y_1y_2 = \dfrac{an^2 - l^2}{bl^2 - 2nml + am^2}$

$$\because \quad x_1x_2 + y_1y_2 = 0$$

$$\frac{bn^2 - m^2 + an^2 - l^2}{bl^2 - 2nml + am^2} = 0$$

$$bn^2 - m^2 + an^2 - l^2 = 0$$

$$n^2(a+b) = l^2 + m^2$$

6. (a)

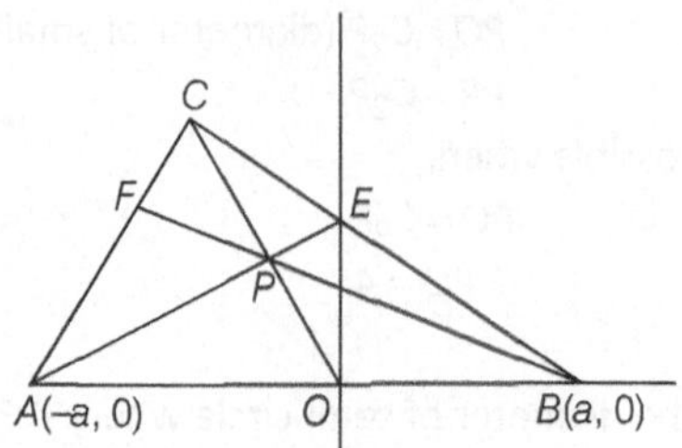

Let A and B be (– a, 0) and (a, 0). Also let P be (h, k). Then by geometry, we know

$$\frac{CP}{PO} = \frac{CF}{FA} + \frac{CE}{EB}$$

$$\therefore \quad \frac{CP}{PO} = 1$$

If C (α, β) lies on $x^2 + y^2 + 2gx + 2fy + c = 0$, then

$$\alpha = 2h \text{ and } \beta = 2k$$

$$\Rightarrow \quad 4(h^2 + k^2 + gh + fk) + c = 0$$

$\therefore$ Locus of P(h, k) is $x^2 + y^2 + gx + fy + \frac{c}{4} = 0$ which is a

$$\text{Radius of circle} = \sqrt{\left(\frac{g}{2}\right)^2 + \left(\frac{f}{2}\right)^2 - \frac{c}{4}}$$

$$= \frac{1}{2}\sqrt{g^2 + f^2 - c} = \frac{r}{2}$$

7. (c) It is given that C_1 has centre (0, 0) and radius equal to 1. Similarly, C_2 has centre (0, 0) and radius 2 and C_k has the centre (0, 0) and radius k.

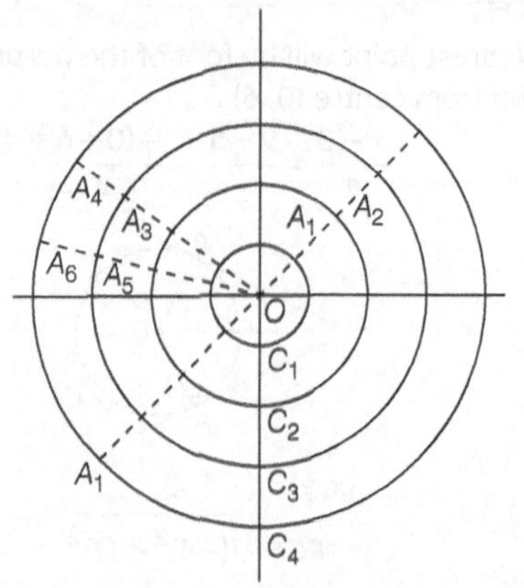

Now particle starts it motion from (1, 0) and moves 1 radian on first circle, then particle shifts from C_1 to C_2.
After that particle moves 1 radian on C_2 and then particle shifts from C_2 to C_3.
Similarly particle moves on 'n' circles.
Now $n > 2\pi$ because particle crosses the X-axis for the first time.
On C_n, then n is least positive integer.

$$\therefore \quad n = 7$$

8. (b) Image of orthocentre about any of the side of the triangle lies on circumcircle.
So image of (1, 1) about $y = 0$ is $(1, -1)$.
So image of (1, 1) about $x + y + 1 = 0$ is $(-2, -2)$
and intersection of $y = 0, x + y + 1 = 0$ is $(-1, 0)$.
Let the equation of circle is

$$x^2 + y^2 + 2gx + 2fy + c = 0$$

$$(1, -1) \rightarrow 2 + 2g - 2f + c = 0 \quad \text{...(i)}$$

$$(-1, 0) \rightarrow 1 - 2g + c = 0 \quad \text{...(ii)}$$

$$(-2, -2) \rightarrow 8 - 4g - 4f + c = 0 \quad \text{...(iii)}$$

By Eqs. (i), (ii) and (iii),

$$g = \frac{1}{2}; f = \frac{3}{2}; c = 0$$

So, equation is $x^2 + y^2 + x + 3y = 0$

9. (b) The given circle

$$S_1 : x^2 + y^2 - 2x - 2y + 1 = 0 \quad \text{...(i)}$$

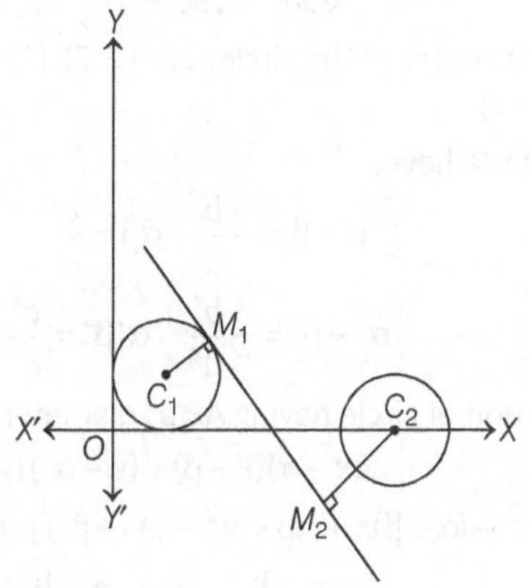

has centre $C_1(1, 1)$ and radius

$$= r_1 = 1$$

The other given circle

$$S_2 : x^2 + y^2 - 18x - 2y + 78 = 0 \quad \text{...(ii)}$$

According to required condition, we have

$$C_1M_1 \geq r_1$$

$$\left| \frac{3 + 4 - \lambda}{\sqrt{3^2 + 4^2}} \right| \geq 1$$

$$\lambda - 7 \geq 5$$

[$\because C_1$ lies below the line, $7 - \lambda$ is a negative quantity]

$$\Rightarrow \quad \lambda \geq 12$$

Also, we have $\quad C_2M_2 \geq r_2$

$$\Rightarrow \quad \left| \frac{27 + 4 - \lambda}{\sqrt{3^2 + 4^2}} \right| \geq 2$$

$$31 - \lambda \geq 10$$

$$\lambda \leq 21$$

[C_2 lies above the line, $31 - \lambda$ is positive quantity]

$$\therefore \quad 12 \leq \lambda \leq 21$$

10. (a) The given circle is

$$2x(x - a) + y(2y - b) = 0$$

or $$x(x - a) + y\left(y - \frac{b}{2}\right) = 0$$

or $$x^2 + y^2 - ax - \frac{by}{2} = 0 \quad \text{...(i)}$$

Let one of the chord through $\left(a, \frac{b}{2}\right)$ be bisected at the point $(h, 0)$, then the equation of the chord having $(h, 0)$ as mid-point is

$$T = S_1$$

$$\Rightarrow \quad hx + 0y - \frac{a}{2}(x + h) - \frac{b}{4}(y + 0) = h^2 - ah$$

$$\Rightarrow \quad \left(h - \frac{a}{2}\right)x - \frac{by}{4} - \frac{ah}{2} = h^2 - ah \quad \text{...(ii)}$$

Now, Eq. (ii) will pass through $\left(a, \frac{b}{2}\right)$ if

$$\left(h - \frac{a}{2}\right)a - \frac{b}{4} \cdot \frac{b}{2} - \frac{a}{2} \cdot h = h^2 - ah$$

$$\Rightarrow \quad h^2 - \frac{3}{2}ah + \frac{a^2}{2} + \frac{b^2}{8} = 0 \quad \text{...(iii)}$$

According to the given conditions Eq. (iii) must have 2 distinct real roots. It is possible if the discriminant of Eq. (iii) is greater than 0.

$$\Rightarrow \quad \frac{9}{4}a^2 - 4\left(\frac{a^2}{2} + \frac{b^2}{8}\right) > 0$$

$$\Rightarrow \quad \frac{a^2}{4} - \frac{b^2}{2} > 0 \Rightarrow a^2 > 2b^2$$

11. (b) Slope of line perpendicular to chord having slope $\frac{3}{4}$ is

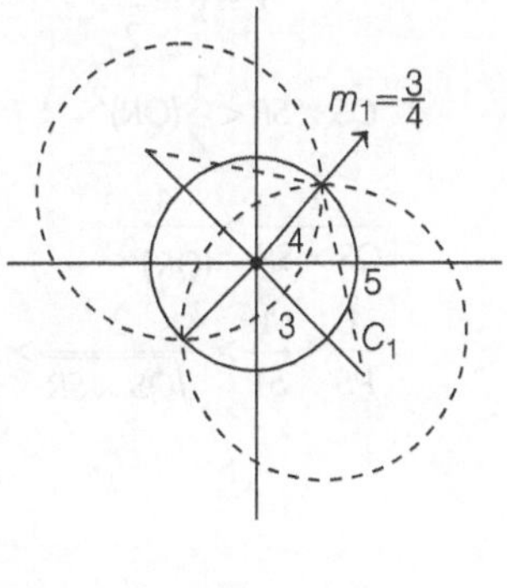

$$m_2 = -\frac{4}{3} = \tan q$$

$$\sin q = \frac{4}{5}, \cos q = -\frac{3}{5}$$

$\therefore$ Required centre is

$$\left(0 \pm 3\left(-\frac{3}{5}\right), 0 \pm -3\left(\frac{4}{5}\right)\right)$$

$$= \left(\mp\frac{9}{5}, \pm\frac{12}{5}\right)$$

12. (b)

$$x^2 + y^2 \le 16 \quad \text{...(i)}$$

$$(x-a)^2 + (y-a)^2 = 9 \quad \text{...(ii)}$$

$$(x+a)^2 + (y-a)^2 = 9 \quad \text{...(iii)}$$

Eqs. (i) and (ii) have a solution if $\sqrt{a^2 + a^2} \le 4 + 3$

$$\Rightarrow \quad |a| = \le \frac{7}{\sqrt{2}} \quad \text{...(iv)}$$

$$\{\because C_1C_2 \le r_1 + r_2\}$$

Eqs. (i) and (iii) have a solution if

$$\sqrt{a^2 + a^2} \le 4 + 3$$

$$\Rightarrow \quad |a| \le \frac{7}{\sqrt{2}} \quad \text{...(v)}$$

Eqs. (ii) and (iii) have a solution if

$$3 - 3 \le |2a| \le 3 + 3$$

$$0 \le |a| \le 3 \quad \text{...(vi)}$$

By Eqs. (iv), (v) and (vi), we get

$$|a| \le 3$$

$$\Rightarrow \quad -3 \le a \le 3$$

Multiple Correct Type Questions

1. (b, d) Since S is not the centre of circumcircle

$$PS \ne ST, QS \ne SR$$

$$\because \quad AM > GM$$

$$\frac{1}{2}\left(\frac{1}{PS} + \frac{1}{ST}\right) > \frac{1}{\sqrt{PS \times ST}}$$

$$\Rightarrow \quad \frac{1}{PS} + \frac{1}{ST} > \frac{2}{\sqrt{QS \times SR}} \quad \{\because PS \times ST = QS \times SR\}$$

We know that if $x + y = k$ a constant, then xy is maximum, if

$$x = y = \frac{k}{2}$$

Thus, $\quad QS \times SR < \frac{1}{4}(QR)^2$

$$\Rightarrow \quad \frac{1}{QS \times SR} > \frac{4}{(QR)^2}$$

$$\Rightarrow \quad \frac{1}{PS} + \frac{1}{ST} > \frac{2}{\sqrt{QS \times SR}} > \frac{4}{QR}$$

2. (a, b, c, d)

The given circle is $x^2 + y^2 + 4x - 6y - 12 = 0$, therefore the equation of the tangent to this circle at $A(2, 0)$ is

$$x(2) + y(0) + 2(x+2) - 3(y+0) - 12 = 0$$

$$\Rightarrow \quad 4x - 3y - 8 = 0 \quad \text{...(i)}$$

Let $y = m(x-2)$ be a line through $A(2, 0)$.

This line makes an angle of 45° with line [Eq. (i)]

So, we have

$$\left|\frac{m - \frac{4}{3}}{1 + \frac{4m}{3}}\right| = 1$$

$$\Rightarrow \quad 3m - 4 = \pm|3 + 4m|$$

$$\Rightarrow \quad m = -7 \text{ or } m = \frac{1}{7}$$

Hence, the lines are

$$x - 7y - 2 = 0 \text{ and } 7x + y - 14 = 0$$

$\because$ The centres of the circles are at a distance of $5\sqrt{2}$ from $A(2, 0)$.

We have, $\quad \dfrac{x-2}{-\frac{1}{\sqrt{50}}} = \dfrac{y}{\frac{7}{\sqrt{50}}} = \pm 5\sqrt{2}$

and $\quad \dfrac{x-2}{\frac{7}{\sqrt{50}}} = \dfrac{y}{\frac{1}{\sqrt{50}}} = \pm 5\sqrt{2}$

$\therefore$ The centre of the circles are $(1, 7), (3, -7), (9, 1)$ and $(-5, -1)$.

3. (a, d) We have,

$$\alpha + \beta = -\frac{b}{a}; \quad \alpha\beta = \frac{c}{a}$$

$$\alpha' + \beta' = -\frac{b'}{a'}; \quad \alpha'\beta' = \frac{c'}{a'}$$

Equation of circle having AB as diameter is

$$(x - \alpha)(x - \beta) + (y - \alpha')(y - \beta') = 0$$

$$\Rightarrow x^2 - (\alpha + \beta)x + \alpha\beta + y^2 - (\alpha' + \beta')y + \alpha'\beta' = 0$$

$$\Rightarrow \quad x^2 + \frac{b}{a}x + \frac{c}{a} + y^2 + \frac{b'}{a'}y + \frac{c'}{a'} = 0$$

$$\Rightarrow \quad aa'(x^2 + y^2) + a'bx + ab'y + a'c + ac' = 0$$

$\because$ It passes through origin is $\left(\frac{b}{a}, \frac{b'}{a}\right)$.

$$\therefore \quad a'c + ac' = 0$$

and $\quad aa'\left(\dfrac{b^2}{a^2} + \dfrac{b'^2}{a'^2}\right) + a'b \times \dfrac{b}{a} + ab' \times \dfrac{b'}{a'} = 0$

$$\Rightarrow \quad a'^2b^2 + a^2b'^2 = 0$$

4. (b, c)

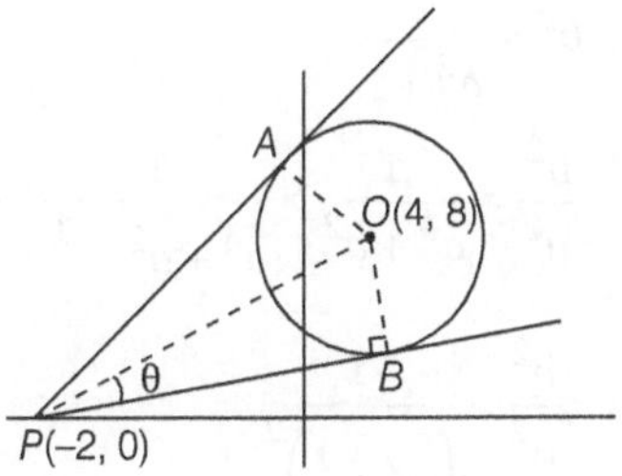

$$\sin\theta = \frac{2\sqrt{5}}{10} = \frac{1}{\sqrt{5}}$$

Slope of PA and PB are $\tan(\alpha \pm \theta)\left\{\text{where } \tan\alpha = \frac{8}{6} = \frac{4}{3}\right\}$

$$\Rightarrow \frac{\frac{4}{3}+\frac{1}{2}}{1-\frac{4}{3}\cdot\frac{1}{2}}, \frac{\frac{4}{3}-\frac{1}{2}}{1+\frac{4}{3}\cdot\frac{1}{2}} = \frac{11}{2}, \frac{5}{10}$$

$$\therefore \quad A, B = \left[4+2\sqrt{5}\left(-\frac{11}{5\sqrt{5}}\right), 8+2\sqrt{5}\left(\frac{2}{5\sqrt{5}}\right)\right]$$

$$\text{and}\left[4-2\sqrt{5}\left(-\frac{1}{\sqrt{5}}\right), 8-2\sqrt{5}\left(\frac{2}{\sqrt{5}}\right)\right] \equiv \left(-\frac{2}{5}, \frac{44}{5}\right), (6, 4)$$

5. (a,c) Let the vertices of the triangle be $O(0, 0)$, $A(a, 0)$ and $B(0, b)$. Let $C(h, k)$ be the centre and r be the radius of the given circle.

$\because$ The circle touches AB at the point $D\left(\frac{a}{2}, \frac{b}{2}\right)$, the centre $C(h, k)$ lies on the line.

$$y - \frac{b}{2} = \frac{a}{b}\left(x - \frac{a}{2}\right) \text{ or } 2ax - 2by = a^2 - b^2$$

$$\therefore\ 2ah - 2bk = a^2 - b^2 \qquad \text{...(i)}$$

$$\text{Now,} \quad r^2 = CD^2 = \left(h - \frac{a}{2}\right)^2 + \left(k - \frac{b}{2}\right)^2 \qquad \text{...(ii)}$$

$$r^2 = \left(h - \frac{a}{2}\right)^2 + k^2 \qquad \text{...(iii)}$$

From Eqs. (ii) and (iii), we get

$$k = \frac{b}{2}$$

From Eq. (i), $\quad h = \dfrac{(2a^2 - b^2)}{4a}$

$$\therefore \text{ We get} \quad r^2 = \left(\frac{2a^2 - b^2}{4a} - \frac{a}{2}\right)^2 + \left(\frac{b}{4}\right)^2$$

$$= \frac{b^4}{16a^2} - \frac{b^2}{16}$$

$$r^2 = \frac{b^4 + a^2b^2}{16a^2} = \frac{b^2}{(4a)^2}(a^2 + b^2)$$

$$r = \frac{b}{4a}\sqrt{a^2 + b^2}$$

6. (a,b,c) On solving $x^2 + y^2 = 3$ and $x^2 = 2y$, we get point $P(\sqrt{2}, 1)$.

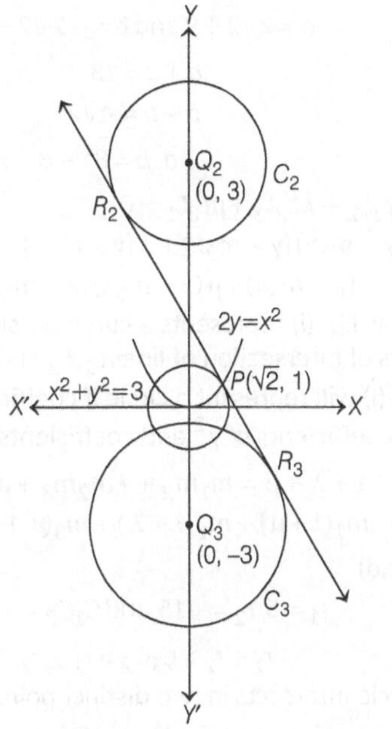

Equation of tangent at $P \Rightarrow \sqrt{2}x + y = 3$

Let Q_2 be $(0, k)$ and radius is $2\sqrt{3}$.

$$\therefore \quad \left|\frac{\sqrt{2}(0) + k - 3}{\sqrt{2+1}}\right| = 2\sqrt{3}$$

$$\therefore \quad k = 9, -3$$

$$Q_2(0, 9) \text{ and } Q_3(0, -3)$$

$$\therefore \quad Q_2Q_3 = 12$$

R_2R_3 is internal common tangent of circle C_2 and C_3.

$$\therefore \quad R_2R_3 = \sqrt{(Q_2Q_3)^2 - (2\sqrt{3} + 2\sqrt{3})^2}$$

$$= \sqrt{12^2 - 48} = \sqrt{96} = 4\sqrt{6}$$

Perpendicular distance of origin O from R_2R_3 is equal to radius of circle $C_1 = \sqrt{3}$

Hence, area of $\Delta OR_2R_3 = \frac{1}{2}(R_2R_3)\sqrt{3} = \frac{1}{2} \times 4\sqrt{6} \times \sqrt{3}$

$$= 6\sqrt{2}$$

Perpendicular distance of P from $Q_2Q_3 = \sqrt{2}$

$\therefore$ Area of $\quad \Delta PQ_2Q_3 = \frac{1}{2} \times 12 \times \sqrt{2} = 6\sqrt{2}$

7. (a,c,d) $x^2 + y^2 + 8x - 10y - 40 = 0$

$\Rightarrow$ centre (–4, 5), radius –9.

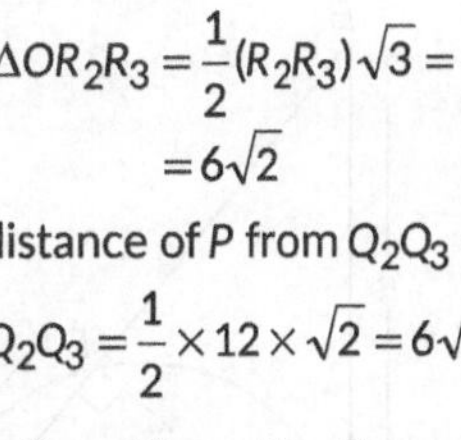

Distance of the centre $(-4, 5)$ from the point $(-2, 3)$ is $\sqrt{4+4}=2\sqrt{2}$

$$\therefore \quad a=2\sqrt{2}+9 \text{ and } b=-2\sqrt{2}+9$$

$$a+b=18$$

$$a-b=4\sqrt{2}$$

$$a \cdot b=81-8=73$$

8. (a,b) $L_1L_2+\lambda L_2L_3+\mu L_3L_1=0$

$$\Rightarrow (y-m_1x)(y-m_2x)+\lambda(y-m_2x)$$

$$(y-m_3x)+\mu(y-m_3x)(y-m_1x)=0 \quad \text{...(i)}$$

Clearly, Eq. (i) represents a curve passing through points of intersection of lines L_1, L_2 and L_3.

$\therefore$ Eq. (i) will represent a circle if coefficient of x^2 = coefficients of y^2 and coefficients of $xy=0$

$$\therefore \quad 1+\lambda+\mu=m_1m_2+\lambda m_2m_3+\mu m_1m_3$$

and $\quad m_1(1+\mu)+m_2(1+\lambda)+m_3(\mu+\lambda)=0$

9. (a,b,c,d)

$$r_1=5, r_2=\sqrt{15} \text{ and } C_1C_2=\sqrt{40}$$

$$\Rightarrow \quad r_1+r_2>C_1C_2>r_1-r_2$$

$\therefore$ Circle intersects in two distinct points.

There are two common tangents.

Also, $2g_1g_2+2f_1f_2=2(1)(3)+2(2)(-4)=-10$

and $\quad C_1+C_2=-20+10=-10$

$\therefore$ Two circles are orthogonal.

$$\text{Length of common chord}=\frac{2r_1r_2}{\sqrt{r_1^2+r_2^2}}=5\sqrt{\frac{3}{2}}$$

$$\text{Length of common tangent}=\sqrt{(C_1C_2)^2-(r_1-r_2)^2}$$

$$=5\left(\frac{12}{5}\right)^{1/4}$$

10. (b,c) $\tan C=\dfrac{AD}{b}=\dfrac{a}{1}$

$$\Rightarrow \quad AD=ab<AC=1 \quad \Rightarrow ab<1$$

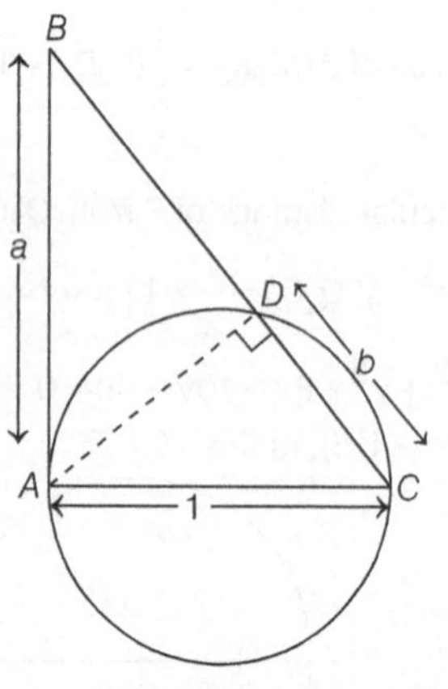

$$(AC)^2=AD^2+CD^2 \quad \Rightarrow \quad 1=b^2a^2+b^2$$

$$\Rightarrow \quad b^2=\frac{1}{a^2+1}$$

$$\Rightarrow \quad \frac{b^2}{a^2}=\frac{1}{a^4+a^2}>\frac{1}{a^4+a^2+\frac{1}{4}}$$

$$\Rightarrow \quad \frac{b^2}{a^2}>\frac{1}{\left(a^2+\frac{1}{2}\right)^2}$$

$$\Rightarrow \quad \frac{b}{a}>\frac{1}{a^2+\frac{1}{2}}$$

11. (a,c)

Circle $\quad x^2+y^2=4$

Equation of tangent at $P(2\cos\theta, 2\sin\theta)$

$$2x\cos\theta+2y\sin\theta=4$$

$$x\cos\theta+y\sin\theta=2$$

Equation of normal at $P \Rightarrow y=x\tan\theta$.

Equation of tangent at $S \Rightarrow x=2$

$$\therefore \quad Q=\left(2, 2\left(\frac{1-\cos\theta}{\sin\theta}\right)\right)\equiv\left(2, 2\tan\frac{\theta}{2}\right)$$

$\therefore$ Equation of line through Q and parallel to RS is

$$y=2\tan\frac{\theta}{2}$$

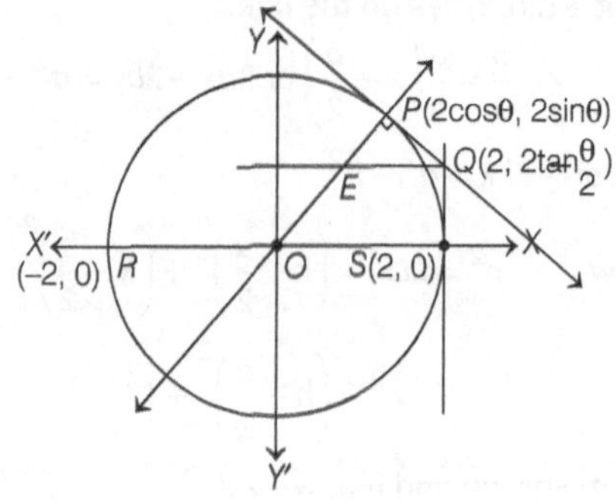

$\because$ Intersection point E of normal and $y=2\tan\dfrac{\theta}{2}$

$$2\tan\frac{\theta}{2}=x\tan\theta$$

$$\Rightarrow \quad x=1-\tan^2\frac{\theta}{2}$$

$$\therefore \text{ Locus of } E \Rightarrow x=1-\left(\frac{y}{2}\right)^2$$

$$y^2=4-4x$$

It satisfied the point $\left(\dfrac{1}{3}, \dfrac{2\sqrt{2}}{\sqrt{3}}\right)$ and $\left(\dfrac{1}{3}, \dfrac{-2\sqrt{2}}{\sqrt{3}}\right)$.

12. (a,b,c) (a) $AP + AQ = \dfrac{(OA - r) + (OA + r)}{2} = OA$

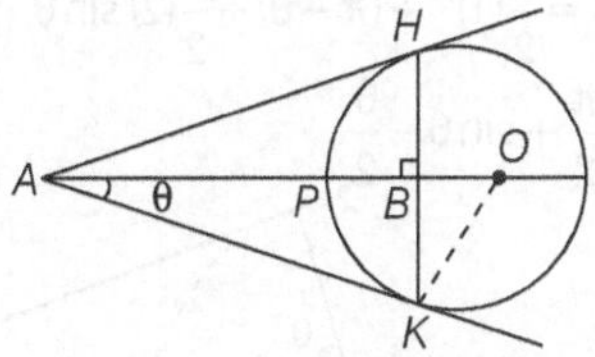

$$\cos\theta = \frac{AK}{OA} = \frac{AB}{AK}$$

$$\Rightarrow \quad AK^2 = (OA)(AB)$$

$$\Rightarrow \quad AK^2 = (AP)(AQ)$$

$$(OA)(AB) = (AP)(AQ)$$

$$\Rightarrow \quad \left(\frac{AP + AQ}{2}\right) AB = (AP)(AQ)$$

$$\Rightarrow \quad AB = \frac{2(AP)(AQ)}{AP + AQ}$$

(b) $\because \; \Delta ABD \sim \Delta AEC$

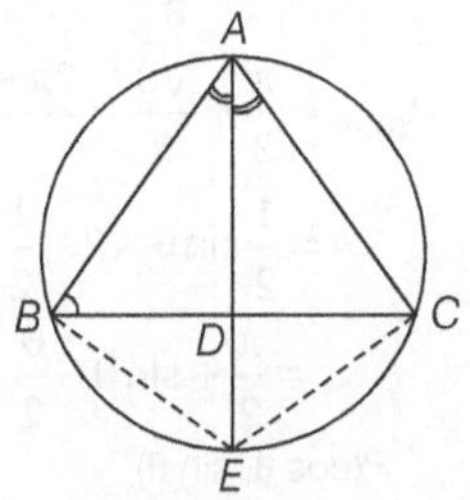

$$\Rightarrow \quad \frac{AB}{AE} = \frac{AD}{AC}$$

$$\Rightarrow \quad AD \times AE = AB \times AC > AD^2$$

$$\therefore \quad AD < \sqrt{(AB)(AC)}$$

(c) $\because$ Equation of pair of tangents is $SS_1 = T^2$

(d) Equation of chord when mid-point is given

$$T = S_1$$

$$\therefore \quad |y_1 - y_2| = 2\sqrt{7}$$

$$\lambda^2 - 36 = 28 \Rightarrow \lambda^2 = 64$$

$$\lambda = \pm 8$$

$\therefore$ So, equation of circles are

$$x^2 + y^2 - 6x \pm 8y + 9 = 0$$

Similarly equation of other circle will be

$$x^2 + y^2 + 6x \pm 8y + 9 = 0$$

13. (a,d) Area of quadrilateral $OACB = 15$

$$= 2\left(\frac{1}{2} \times OA \times AC\right) = OA \times OC$$

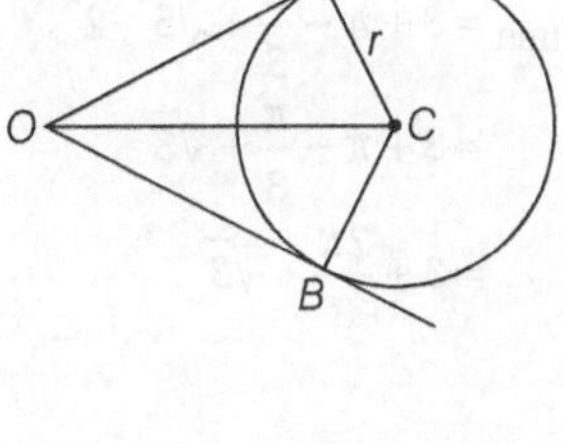

$$\Rightarrow \quad \sqrt{C} \times \sqrt{9 + 25} - C = 15$$

$$\Rightarrow \quad C = 9, 25$$

14. (a,c) Let the equation of circle be

$$(x - 3)^2 + (y - 0)^2 + \lambda y = 0$$

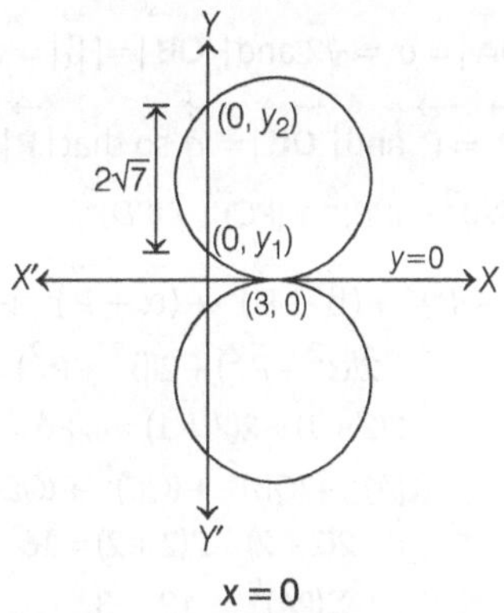

Put $\quad x = 0$

$$\Rightarrow \quad y^2 + \lambda y + 9 = 0$$

Now, $\quad y_1 + y_2 = -\lambda$

$$y_1 y_2 = 9$$

$$|y_1 - y_2| = 2\sqrt{7}$$

$$\Rightarrow \quad (y_1 - y_2)^2 = 28$$

$$y_1^2 + y_2^2 - 2y_1 y_2 = 28$$

$$\Rightarrow \quad (y_1 + y_2)^2 - 4y_1 y_2 = 8$$

$$\Rightarrow \quad (-\lambda)^2 - 4 \times 9 = 28$$

$$\Rightarrow \quad \lambda^2 - 36 = 28$$

$$\lambda^2 = 28 + 36 = 64$$

$$\Rightarrow \quad \lambda = \pm 8$$

$\therefore$ Equation is $x^2 + y^2 - 6x + 8y + 9 = 0$

$$x^2 + y^2 - 6x - 8y + 9 = 0$$

15. (a,b,c,d)

$$x^2 + y^2 - 8x + 2y + 8 = 0$$

$$x^2 + y^2 - 2x - 6y + 10 - a^2 = 0$$

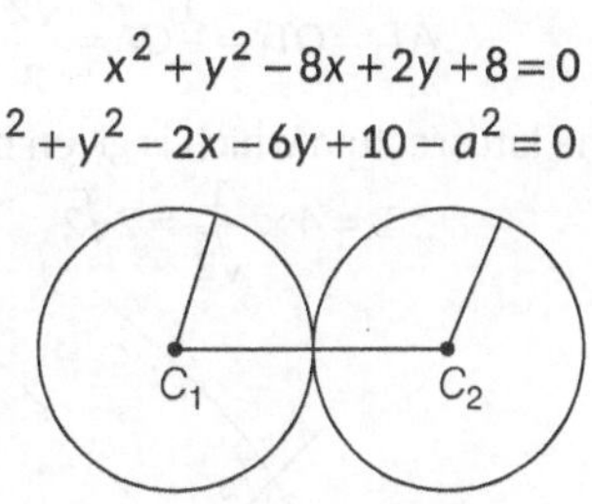

$$|r_1 - r_2| < C_1C_2 < r_1 + r_2$$

$C_1(4, -1)$ and $C_2(1, 3)$

$$r_1 = \sqrt{16 + 1 - 8} = 3$$

$$r_2 = \sqrt{1 + 9 - 10 + a^2} = |a|$$

$$C_1C_2 = \sqrt{9 + 16} = 5$$

$$|3 - |a|| \le 5 < 3 + |a|$$

$$|a| > 2$$

$$a \in (-\infty, 2) \cup (2, \infty)$$

Sol. (Q.Nos. 16-18)

16. (a) Let O be the centre of the square so that O is centre of both C_1 and C_2 (see Fig.). Take O as origin let $\overrightarrow{OA} = \vec{\alpha}$ and $\overrightarrow{OB} = \vec{\beta}$. So that $\overrightarrow{OC} = -\vec{\alpha}$, $\overrightarrow{OD} = -\vec{\beta}$

and $|\overrightarrow{OA}| = \alpha = \sqrt{2}$ and $|\overrightarrow{OB}| = |\vec{\beta}| = \sqrt{2}$

Let $\overrightarrow{OP} = \vec{P}$ and $|\overrightarrow{OC}| = q$ so that $|\vec{P}| = 1$ and $q = \sqrt{2}$

Now, $(PA)^2 + (PB)^2 + (PC)^2 + (PD)^2$

$\Rightarrow \quad (\vec{\alpha} - \vec{P})^2 + (\vec{\beta} - \vec{P})^2 + (\vec{\alpha} + \vec{P})^2 + (\vec{\beta} + \vec{P})^2$

$\Rightarrow \quad 2(\alpha^2 + P^2) + 2(\beta^2 + P^2)$

$= 2(2+1) + 2(2+1) = 6 + 6 = 12$

Now, $(QA)^2 + (QB)^2 + (QC)^2 + (QD)^2$

$\Rightarrow \quad 2(2+2) + 2(2+2) = 16$

$\therefore \quad \dfrac{\Sigma(PA)^2}{\Sigma(QA)^2} = \dfrac{12}{16} = \dfrac{3}{4} = 0.75$

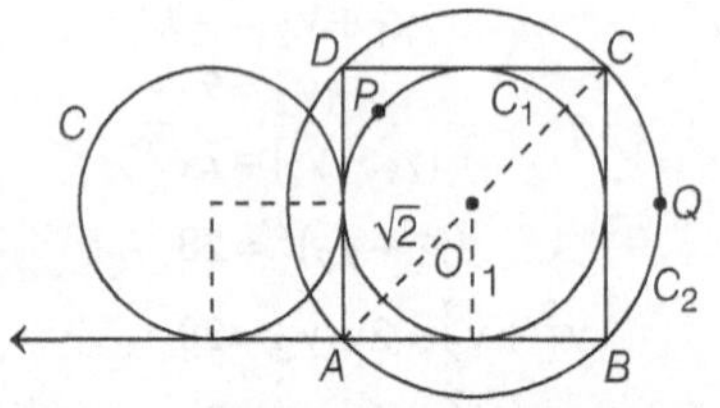

17. (b) Suppose L_1 is a line parallel to L at a unit distance from L such that L_1 and the circle C_1 are on opposite sides to the line L.

Therefore, the centre of the circle C is equidistant from the line L_1 and the centre of the circle C_2. Hence, the centre of C is a parabola with focus at the centre of C_2 and directrix L_1.

18. (c) Clearly, the locus S is a parabola with A as a focus and BD as directrix.

$$AT_1 = OT_1 = \frac{1}{2}OA = \frac{\sqrt{2}}{2} = \frac{1}{\sqrt{2}}$$

T_2T_3 is the latusrectum which is given by

$$= 4 \times \frac{1}{\sqrt{2}} = 2\sqrt{2}$$

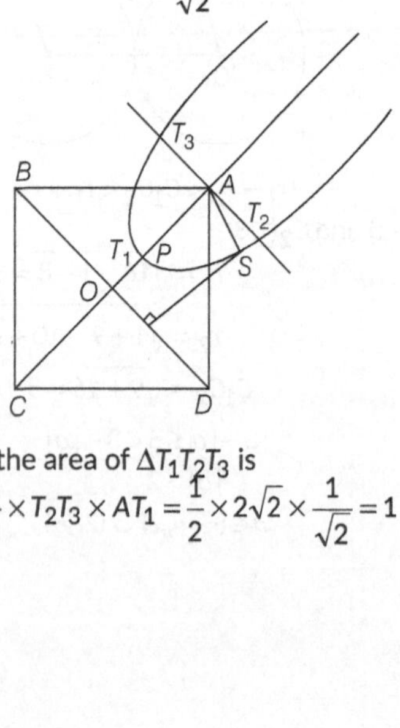

Therefore, the area of $\Delta T_1T_2T_3$ is

$$= \frac{1}{2} \times T_2T_3 \times AT_1 = \frac{1}{2} \times 2\sqrt{2} \times \frac{1}{\sqrt{2}} = 1$$

Sol. (Q.Nos. 19-21)

19. (b) Area $= \dfrac{1}{2}(1)^2 + (\pi - \theta) + \dfrac{1}{2}(2)\sin\theta$

$$A = \frac{\pi}{2} + \sin\theta - \frac{\theta}{2}$$

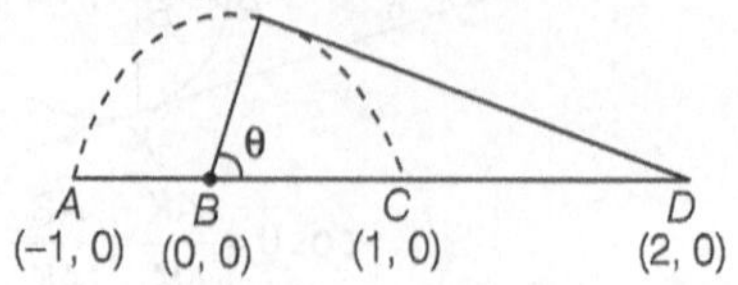

Now for maximum, $\dfrac{dA}{d\theta} = \cos\theta - \dfrac{1}{2}$

$$\frac{dA}{d\theta} = 0$$

$\Rightarrow \quad \cos\theta = \dfrac{1}{2} \Rightarrow \theta = \dfrac{\pi}{3}$

$\oplus \quad \pi/3 \quad \ominus$

$\therefore$ A is maximum for $\theta = \dfrac{\pi}{3}$

$\Rightarrow \quad A_{max} = \dfrac{\pi}{3} + \dfrac{\sqrt{3}}{2} = \dfrac{2\pi + 3\sqrt{3}}{6}$

In this also $\quad A = \dfrac{1}{2}\sin\theta \times 2 + \dfrac{1}{2}(\pi - \theta)$

$$= \frac{\pi}{2} + \sin\theta - \frac{\theta}{2}$$

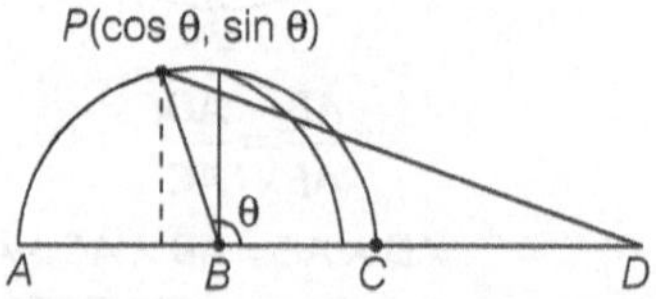

20. (c) $\quad L = 3 + \pi - \theta + \sqrt{(2-\cos\theta)^2 + \sin^2\theta}$

$$= 3 + \pi - \theta + \sqrt{5 - 4\cos\theta}$$

21. (c) $\quad \dfrac{dL}{d\theta} = -1 + \dfrac{4\sin\theta}{2\sqrt{5 - 4\cos\theta}}$

$\Rightarrow \quad -1 + \dfrac{2\sin\theta}{\sqrt{5 - 4\cos\theta}}$

Now, $\quad \dfrac{dL}{d\theta} = 0 \Rightarrow 2\sin\theta = \sqrt{5 - 4\cos\theta}$

$\Rightarrow \quad 4\sin^2\theta = 5 - 4\cos\theta \Rightarrow \cos\theta = \dfrac{1}{2}$

and $\quad \dfrac{d^2L}{d\theta^2} = \dfrac{10\cos\theta - 8\cos^2\theta - 4}{(5 - 4\cos\theta)^{3/2}}$

At $\quad \cos\theta = \dfrac{1}{2}, \dfrac{d^2L}{d\theta^2} = \dfrac{5}{3\sqrt{3}} > 0$

$\therefore \quad L_{min} = 3 + \pi - \dfrac{\pi}{3} + \sqrt{5 - 2}$

$$= 3 + \pi - \frac{\pi}{3} + \sqrt{3}$$

$$= 3 + \frac{2\pi}{3} + \sqrt{3}$$

Sol. (Q.Nos. 22-24)

22. (c) In radius $= \dfrac{A}{S} = \dfrac{\frac{1}{2} \times 10 \times 12}{18} = \dfrac{10}{3}$

23. (b)

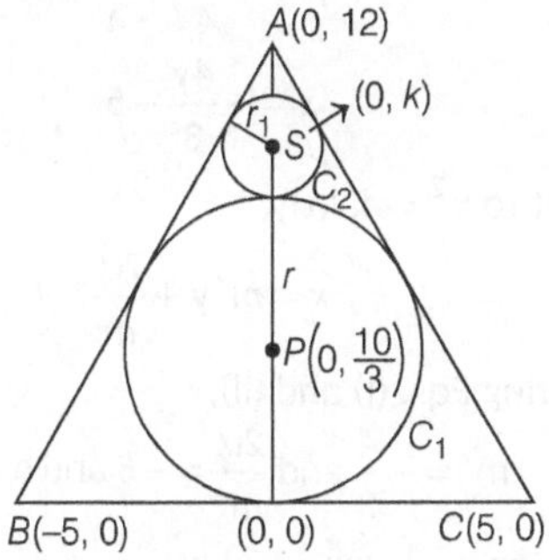

From figure

$$PS = r + r_1$$

$$k - \frac{10}{3} = r + r_1$$

Let r_1 is radius of circle C_2

$$k - \frac{10}{3} = \frac{10}{3} + r_1$$

$$k - \frac{20}{3} = r_1 \quad \text{...(i)}$$

Now equation of AC is $\dfrac{x}{5} + \dfrac{y}{12} = 1$

Now, radius r_1 is

$$r_1 = \left| \frac{\frac{K}{12} - 1}{\sqrt{\left(\frac{1}{5}\right)^2 + \left(\frac{1}{12}\right)^2}} \right| = \left| \frac{5K - 60}{13} \right|$$

$$K = -\frac{13}{5} r_1 + 12 \quad \{\text{as } K > 0, K < 12 \text{ and } r_1 > 0\}$$

$$\Rightarrow \quad r_1 + \frac{20}{3} = \frac{-13}{5} r_1 + 12$$

$$\Rightarrow \quad r_1 = \frac{40}{27}$$

$$\Rightarrow \quad \frac{r_1}{r} = \frac{4}{9}$$

24. (d) Due to symmetry ratio of radius C_{i+1} to C_i is same as that of C_2 to C_1.

Now, radius are in GP with common ratio $= \dfrac{4}{9}$

Area will be GP of with common ratio $= \dfrac{16}{81}$

$$\therefore \text{ Sum of areas} = \frac{\pi\left(\frac{10}{3}\right)^2}{\pi\left(\frac{16}{81}\right)^2} = \frac{180\pi}{13}$$

Sol. (Q.Nos. 25-26)

25. (b) $PQ = PR$, i.e. parallelogram $PQRS$ is a rhombus.

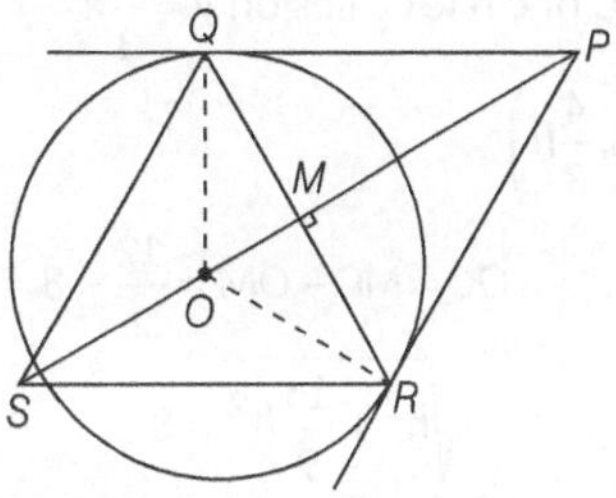

$\therefore$ Mid-point of QR

= Mid-point of PS and QR is perpendicular to PS

$\therefore$ S is the mirror image of P w.r.t. QR.

$$L = 2x + y = 6$$

Let $\quad P = (k, 6 - 2k)$

$$\angle PQO = \angle PRO = \frac{\pi}{2}$$

OP is diameter of circumcircle PQR, then centre $\left(\dfrac{k}{2}, 3 - k\right)$

$$x = \frac{k}{2} \Rightarrow k = 2x$$

$$y = 3 - k \Rightarrow 2x + y = 3$$

26. (d) $P(6, 8)$ Given equation of QR is $6x + 8y = 4$

or $\quad 3x + 4y - 2 = 0$

$$PM = \frac{48}{5} \text{ and } PQ = \sqrt{96}$$

$$QM = \sqrt{96 - \frac{(48)^2}{25}} = \sqrt{\frac{96}{25}}$$

and $\quad QR = 2\sqrt{\dfrac{96}{25}}$

Area of $\Delta PQR = \dfrac{1}{2} \times PM \times QR = \dfrac{192\sqrt{6}}{25}$

$PQSR$ is a rhombus.

Area of ΔQRS = Area of $\Delta PQR = \dfrac{192\sqrt{6}}{25}$ sq units

Match the column type questions

27. (i) → A, (ii) → B, (iii) → C, (iv) → F

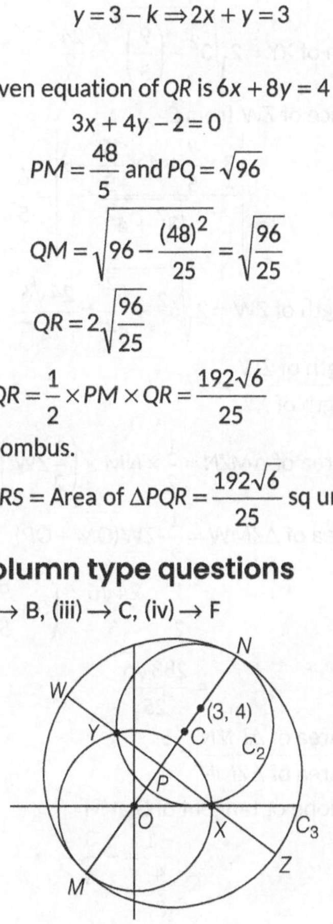

(i) $2r = MN = 3 + \sqrt{3^2 + 4^2} + 4 = 12$

$\therefore \quad r = 6$

Centre C of circle C_3 lies on $y = \frac{4}{3}x$

Let $C\left(h, \frac{4}{3}h\right)$

$$OC = MC - OM = \frac{12}{2} - 3 = 3$$

$$\therefore \quad \sqrt{h^2 + \frac{16}{9}h^2} = 3$$

$$\Rightarrow \quad \frac{5h}{3} = 3 \quad \Rightarrow \quad h = \frac{9}{5}$$

$$k = \frac{4}{3}h = \frac{12}{5}$$

$$\therefore \quad 2h + k = \frac{18}{5} + \frac{12}{5} = 6$$

(ii) Equation of line ZW

$$C_1 = C_2$$

$$\Rightarrow \quad 3x + 4y = 9$$

Distance of ZW from (0, 0)

$$\Rightarrow \quad \left|\frac{-9}{\sqrt{3^2 + 4^2}}\right| = \frac{9}{5}$$

Length of $XY = 2\sqrt{3^2 - \left(\frac{9}{5}\right)^2} = \frac{24}{5}$

Distance of ZW from C

$$\Rightarrow \quad \left|\frac{3 \times \frac{9}{5} + 4 \times \frac{12}{5} - 9}{\sqrt{3^2 + 4^2}}\right| = \frac{6}{5}$$

Length of $ZW = 2\sqrt{6^2 - \frac{6^2}{5^2}} = \frac{24\sqrt{6}}{5}$

$\therefore \frac{\text{Length of } ZW}{\text{Length of } XY} = \sqrt{6}$

(iii) Area of $\Delta MZN = \frac{1}{2} \times NM \times \left(\frac{1}{2}ZW\right) = \frac{72\sqrt{6}}{5}$

Area of $\Delta ZMW = \frac{1}{2} \cdot ZW(OM + OP)$

$$= \frac{1}{2} \times \frac{24\sqrt{6}}{5} \times \left(3 + \frac{9}{5}\right)$$

$$= \frac{288\sqrt{6}}{25}$$

$\therefore \quad \frac{\text{Area of } \Delta MZN}{\text{Area of } \Delta ZMN} = \frac{5}{4}$

(iv) Slope of tangent of G at M

$$\Rightarrow \quad \frac{-1}{\frac{4}{3}} = -\frac{3}{4}$$

$\therefore$ Equation of tangent $y = mx - 3\sqrt{1 + m^2}$

$$\Rightarrow \quad y = -\frac{3}{4}x - 3\sqrt{1 + \frac{9}{16}}$$

$$\Rightarrow \quad y = -\frac{3x}{4} - \frac{15}{4}$$

$$x = -\frac{4y}{3} - 5 \quad \text{...(i)}$$

Tangent to $x^2 = 4(2\alpha)y$

$$\Rightarrow \quad x = m'y + \frac{2\alpha}{m} \quad \text{...(ii)}$$

Comparing Eqs. (i) and (ii),

$$m' = -\frac{4}{3} \text{ and } \frac{2\alpha}{m'} = -5 \text{ and } \alpha = \frac{10}{3}$$

28. (i) $\rightarrow$ B, (ii) $\rightarrow$ D, (iii) $\rightarrow$ A, (iv) $\rightarrow$ C

(i) Let length of common chord be $2a$, then

$$\sqrt{9 - a^2} + \sqrt{16 - a^2} = 5$$

$$\sqrt{16 - a^2} = 5 - \sqrt{9 - a^2}$$

$$16 - a^2 = 25 + 9 - a^2 - 10\sqrt{9 - a^2}$$

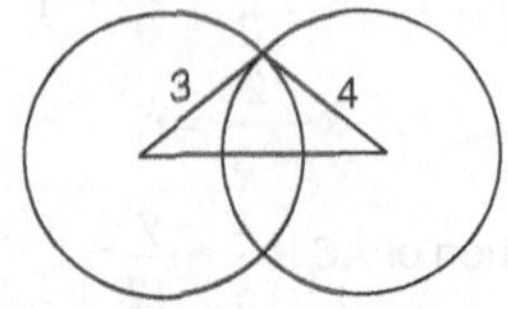

$$\Rightarrow \quad 10\sqrt{9 - a^2} = 18 \quad \Rightarrow \quad 100a^2 = 576$$

$$\Rightarrow \quad a = \frac{24}{10}$$

$$\Rightarrow \quad 2a = \frac{24}{5} = \frac{k}{5} \Rightarrow k = 24$$

(ii) Equation of common chord is $6x + 4y + p + q = 0$

Common chord pass through centre $(-2, -6)$ of circle

$$x^2 + y^2 + 4x + 12y + p = 0$$

$$\therefore \quad p + q = 36$$

(iii) Equation of circle is

$$2x^2 + 2y^2 - 2\sqrt{2}x - y = 0$$

Let $(\alpha, 0)$ be mid-point of a chord, then equation of the chord is

$$2\alpha x - \sqrt{2}(x + \alpha) - \frac{1}{2}(y + 0) = 2\alpha^2 - 2\sqrt{2}\alpha$$

$\because$ It passes through the point $\left(\sqrt{2}, \frac{1}{2}\right)$.

$$\therefore \; 2\sqrt{2}\alpha - \sqrt{2}(\sqrt{2} + \alpha) - \frac{1}{4} = 2\alpha^2 - 2\sqrt{2}\alpha$$

$$\Rightarrow \quad 8\alpha^2 - 12\sqrt{2}\alpha + 9 = 0$$

$$(2\sqrt{2}\alpha - 3)^2 = 0 \Rightarrow \alpha = \frac{3}{2\sqrt{2}}, \frac{3}{2\sqrt{2}}$$

$\therefore$ Number of chords is 1.

(iv) Mid-point of $AB = (1, 4)$

$\therefore$ Equation perpendicular bisector of AB is $x = 1$

A diameter of the circle is $4y = x + 7$

$\therefore$ Centre of the circle is $(1, 2)$.

$\therefore$ Sides of the rectange are 8 and 4.

$\therefore$ Area $= 32$

29. (i) → D, (ii) → A, (iii) → B, (iv) → C

(i) $2\sqrt{Rr} = 3 \Rightarrow r = \frac{9}{4 \times 3} = \frac{3}{4}$

(ii) $\frac{r}{R} = \frac{x}{x+3} = \frac{1}{4} \Rightarrow x = 1$

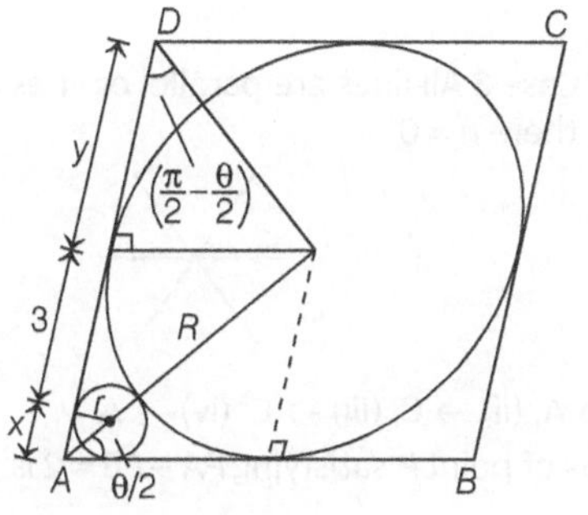

$$\tan\frac{\theta}{2} = \frac{3}{4}$$

$$y = R\cot\left(\frac{\pi}{2} - \frac{\theta}{2}\right)$$

$$= 3\tan\frac{\theta}{2} = \frac{9}{4}$$

$ABCD$ is a rhombus with side $= 4 + \frac{9}{4} = \frac{25}{4}$

Area of $ABCD = \frac{25}{4} \times \frac{25}{4} \sin\theta$

$$= \frac{25}{4} \times \frac{25}{4} \frac{\left(2 \times \frac{3}{4}\right)}{\left(1 + \frac{9}{16}\right)} = \frac{75}{2}$$

(iii) $A = \frac{1}{2} d_1 d_2 = \frac{75}{2}$

$$d_1 d_2 = 75$$

(iv) $\frac{d_1^2}{4} + \frac{d_2^2}{4} = \frac{625}{16}$

$$= \frac{(d_1 + d_2)^2 - 2d_1 d_2}{16}$$

$\Rightarrow$ $d_1 + d_2 = \sqrt{775} = 5\sqrt{31}$

30. (i) → B, (ii) → A, (iii) → C, (iv) → D

(i) Equation of common chord is

$$x = \frac{a}{2}$$

Since it subtend $\frac{\pi}{2}$ at origin

$$x^2 + y^2 = \left(\frac{2x}{a}\right)^2$$

$\Rightarrow$ $\left(1 - \frac{4}{a^2}\right)x^2 + y^2 = 0$

$\therefore$ Coefficient of x^2 + Coefficient of $y^2 = 0$

$\therefore$ $1 - \frac{4}{a^2} + 1 = 0 \Rightarrow a = \pm\sqrt{2}$

(ii) Length of chord $= 2\sqrt{r^2 - P^2}$

$$P = 0$$

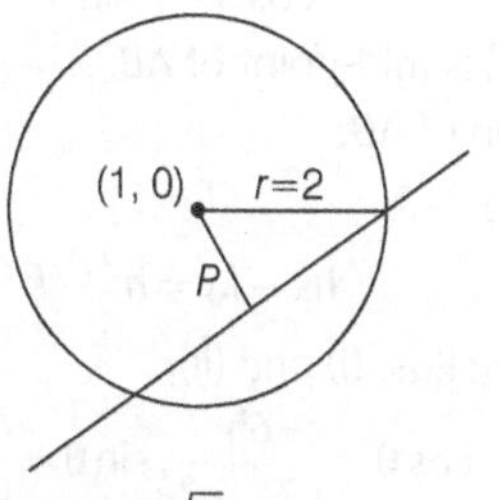

Length of chord $= 2\sqrt{4} = 4$

(iii) $\because$ All three lines are parallel, so, there no circle touching all the lines.

(iv) $(\sqrt{3}, \sqrt{2})$ lies inside the circle

$S: x^2 + y^2 - 2\sqrt{2}y - 2 = 0$

$r_1 = 2 - \sqrt{3}$ and $r_2 = 2 + \sqrt{3}$

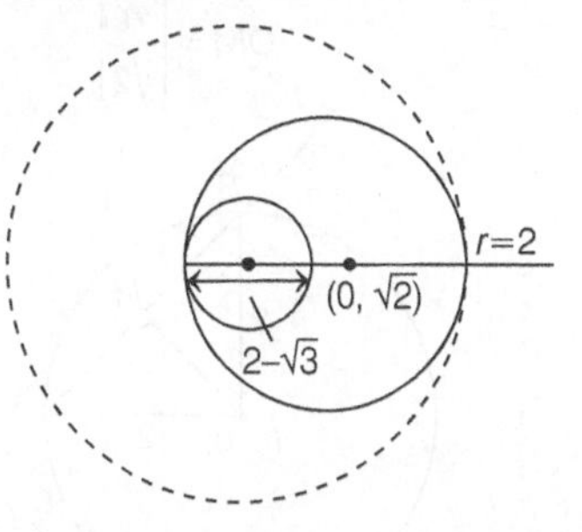

$\therefore$ Mean $= \frac{r_1 + r_2}{2} = 2$

31. (i) → D, (ii) → C, (iii) → A, (iv) → B

(i) Slope of line $= -1$

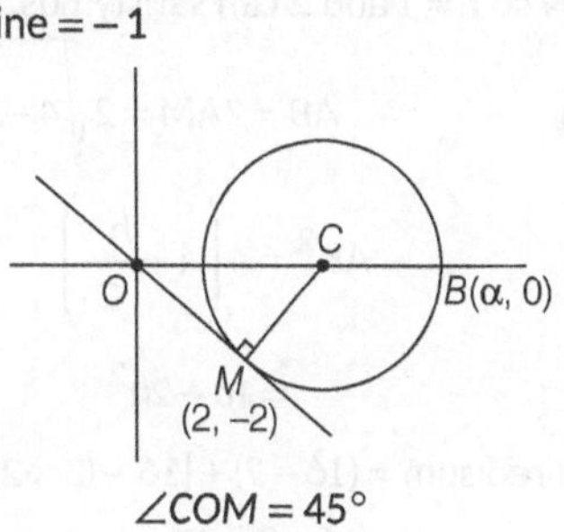

$\therefore$ $\angle COM = 45°$

$\therefore$ $OM = 2\sqrt{2}$

So, $OC = 4$

$\therefore$ So α will be maximum at point B.

$$\alpha = 4 + 2\sqrt{2}$$

(ii) Equation of chord of contact is $T = 0$

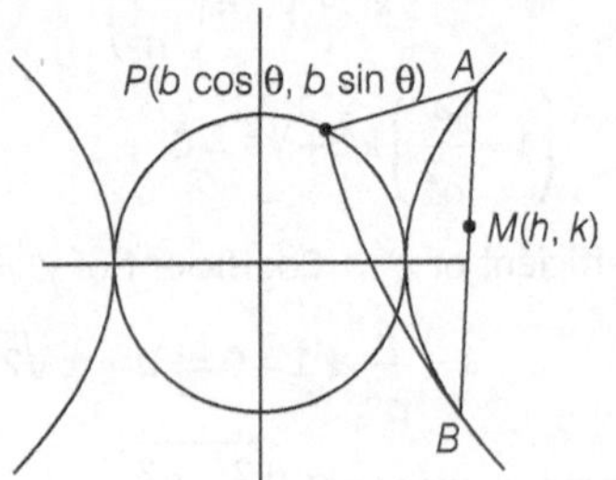

$$x\cos\theta - y\sin\theta = 6 \quad ...(i)$$

Let $M(h, k)$ is mid-point of AB.

$\therefore$ Equation of AB,

$$T = S_1$$

$$\Rightarrow \quad hx - ky = h^2 - k^2 \quad ...(ii)$$

Comparing Eqs. (i) and (ii),

$$\cos\theta = \frac{6h}{h^2 - k^2}, \sin\theta = \frac{6k}{h^2 - k^2}$$

Squaring and add both

$$(x^2 - y^2)^2 = 36(x^2 + y^2)$$

$$\Rightarrow \quad l = 3$$

(iii) Line will cut circle if

$$OM < 2 \quad ...(i)$$

Now, $$OM = \left|\frac{n}{\sqrt{2}}\right| \quad ...(ii)$$

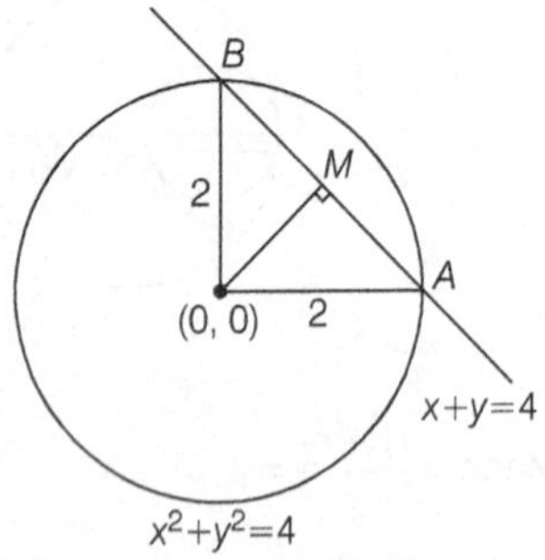

As $x \in N$ so $n = 1$ and 2 can satisfy Eqs. (i) and (ii)

Length $$AB = 2AM = 2\sqrt{4 - \frac{h^2}{2}}$$

$$AB^2 = 4\left(4 - \frac{h^2}{2}\right)$$

$$= 16 - 2h^2$$

$\therefore$ Required sum $= (16 - 2) + [16 - (2 \times 2^2)]$

$= 14 + 8 = 22$

(iv) **Case I** If lines form a triangle, then

$n = 4$

i.e. excircle and 1 circle

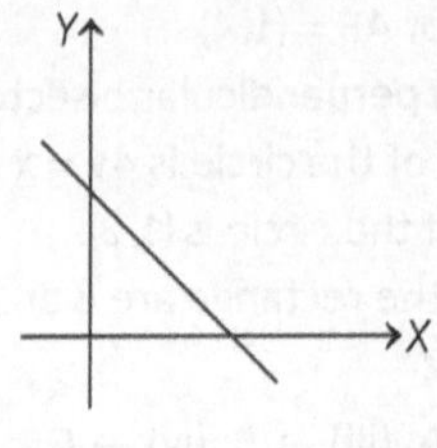

Case 2 Two lines are parallel and third intersect them, then $n = 2$

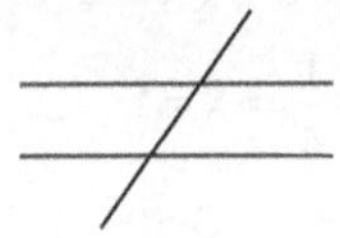

Case 3 All lines are parallel or lines are concurrent there $n = 0$

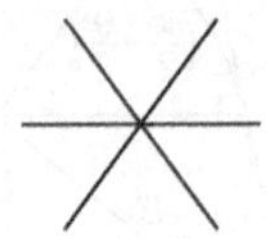

32. (i) → A, (ii) → C, (iii) → C, (iv) → A

Locus of point P satisfying $PA - PB = 2$ is a branch of hyperbola $x^2 - \frac{y^2}{3} = 1$ (right of origin)

For $r = 2$, Circle and hyperbola cuts at two points

For $r = 1$, There is no point of intersection

For $r = 2$, No common tangents

For $r = \frac{1}{2}$, Only two common tangents are possible

Sol. (Q.Nos. 33-35)

Sol. $AB = 3$

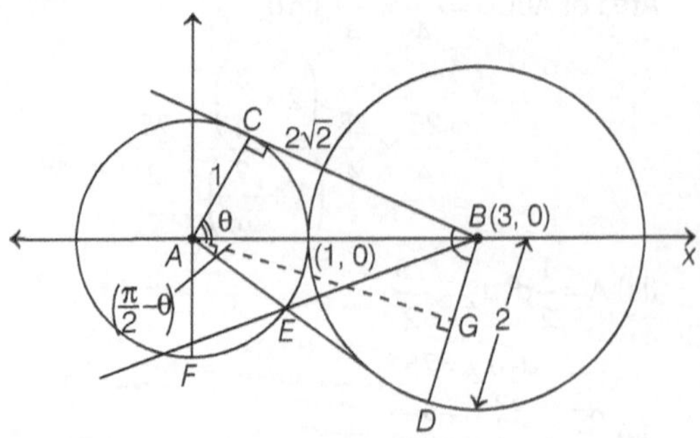

$$\Rightarrow \quad BC = \sqrt{AB^2 - AC^2}$$

$$= \sqrt{9 - 1} = 2\sqrt{2}$$

$\therefore$ Let $\angle BAC = \theta$

$\Rightarrow \quad \tan\theta = 2\sqrt{2}$

$AG = 2\sqrt{2}$

$\{\because AG \perp BD \Rightarrow AGBC$ is rectangle$\}$

$\because \quad m_{BD} = 2\sqrt{2} \quad \{\because AC \| BG \Rightarrow AC \| BD\}$

So, $$D=\left[3-2\left(\frac{1}{3}\right),0-2\left(\frac{2\sqrt{2}}{3}\right)\right]$$

$$=\left(\frac{7}{3},-\frac{4\sqrt{2}}{3}\right)$$

$$m_{AE}=-\tan(\pi-2\theta)=\tan 2\theta$$

$$=\frac{2(2\sqrt{2})}{1-8}=\frac{-4\sqrt{2}}{7}$$

$$E\equiv\left(0-1\left(-\frac{7}{9}\right)\right),0-1\left(\frac{4\sqrt{2}}{9}\right)$$

$$=\left(\frac{7}{9},\frac{-4\sqrt{2}}{9}\right)$$

$\therefore$ Equation of $BC=y=\frac{\sqrt{2}}{5}(x-3)$

solving intersection of line.

BE with circle S_1

$$x^2+\frac{2}{25}(x-3)^2=1$$

$$\Rightarrow \quad 27x^2-12x-9=0$$

$$\Rightarrow \quad (9x-7)(3x+1)=0$$

$$F\equiv\left(-\frac{1}{3},\frac{-2\sqrt{2}}{3}\right)$$

33. (1.54) $EF=\sqrt{\left(\frac{7}{9}+\frac{1}{3}\right)^2+\left(\frac{4\sqrt{2}}{9}-\frac{2\sqrt{2}}{3}\right)^2}=\frac{1}{3}\sqrt{100+8}$

$$\Rightarrow \quad \frac{1}{9}\sqrt{108}=\frac{6\sqrt{3}}{9}=\frac{2\sqrt{3}}{3}=1.54$$

34. (2.83) Area of $\Delta ABD=\frac{1}{2}\times\text{Base}\times\text{Height}$

$$=\frac{1}{2}\times BD\times AG$$

$$=\frac{1}{2}\times 2\times 2\sqrt{2} \quad [AG=BC]$$

$$=2\sqrt{2}=2.83$$

35. (2) $DE=\sqrt{\left(\frac{7}{3}-\frac{7}{9}\right)^2+\left(\frac{-4\sqrt{2}}{3}+\frac{4\sqrt{2}}{9}\right)^2}$

$$=\sqrt{\left(\frac{14}{9}\right)^2+\left(\frac{8\sqrt{2}}{9}\right)^2}$$

$$=\frac{1}{9}\sqrt{196+128}$$

$$=\frac{\sqrt{324}}{9}=\frac{18}{9}=2$$

Sol. (Q.Nos. 36-38)

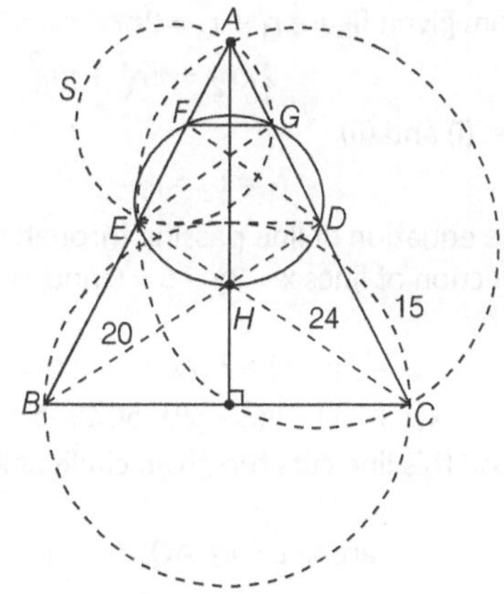

$$\sin C=\frac{4}{5},\sin B=\frac{24}{25}$$

$$\tan A=-\tan(B+C)=-\left(\frac{\frac{24}{7}+\frac{4}{3}}{1-\frac{24}{7}\times\frac{4}{3}}\right)$$

$$=\frac{4}{3}=\tan C$$

36. (80) $\because \tan A=\tan C$

$$\Rightarrow \quad A=C$$

$$\Rightarrow \quad BC=AB=25$$

$$\Rightarrow \quad AC=2(CD)=30$$

$$\Rightarrow \quad AB+BC+CA=80$$

$$DE=AD=DC=15$$

37. (176.63) Area of $S=\frac{\pi(15)^2}{4}=\frac{225\pi}{4}=176.63$

38. (8.64) $\because$ $AEPC$ is cyclic quadrilateral with AC as diameter.

$\because$ $EFGD$ is cyclic.

$$\Rightarrow \quad \angle AFG=\angle GDE$$

$\because$ $BEDC$ is cyclic $\Rightarrow \angle GDE=\angle ABC$

$$\Rightarrow \quad \angle AFG=\angle ABC \Rightarrow FG \parallel BC$$

$$\therefore \quad \Delta AFG\sim\Delta ABC$$

$$\Rightarrow \quad \frac{AK}{AP}=\frac{AF}{AB}$$

$$\Rightarrow \quad AF=\frac{1}{2}AE=9 \quad [\because AD=DE]$$

$$\therefore \quad AK=\frac{9\times 24}{25}$$

$$=\frac{216}{25}=8.64$$

Numerical Value Type Questions

39. (3) From given figure $r_1 + r_3 = 2r_2$ ...(i)

and $2\pi r_2^2 = \pi r_1^2 + \pi r_3^2$ ...(ii)

By Eqs. (i) and (ii)

$$r_1 = r_2 = r_3$$

40. (2) The equation of line passing through the point of intersection of lines $x - 2y - 5 = 0$ and $7x + y = 50$ is given by

$$(x - 2y - 5) + \lambda(7x + y - 50) = 0$$

$\Rightarrow \quad x(1 + 7\lambda) + y(\lambda - 2) - 50\lambda - 5 = 0$ ...(i)

Suppose this line cuts the given circle at A and B such that

$$\text{arc } APB : \text{arc } AQB = 2 : 1$$

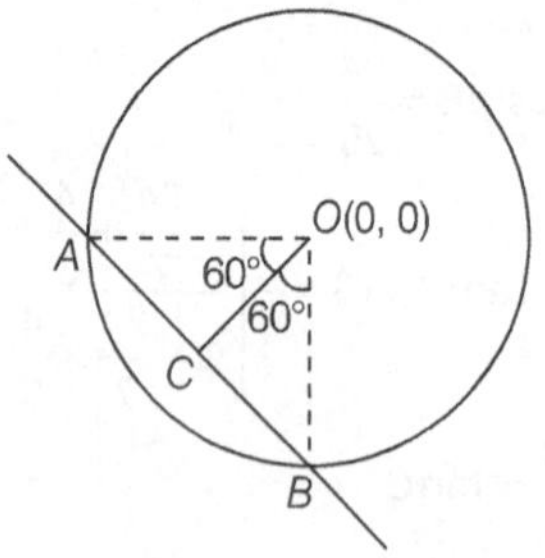

$\therefore \quad \angle AOB = 120°$

$\Rightarrow \quad \angle AOC = \angle BOC = 60°$

Now, $OC = \left|\dfrac{-50\lambda - 5}{\sqrt{(1 + 7\lambda)^2 + (\lambda - 2)^2}}\right| = \dfrac{50\lambda + 5}{\sqrt{50\lambda^2 + 10\lambda + 5}}$

and OA = radius of circle $x^2 + y^2 = 100$ is 10

In ΔAOC, we have

$$\cos 60° = \frac{OC}{OA}$$

$\Rightarrow \quad OC = OA \cos 60° = 10 \times \dfrac{1}{2} = 5$

$\Rightarrow \quad \dfrac{50\lambda + 5}{\sqrt{50\lambda^2 + 10\lambda + 5}} = 5$

$\Rightarrow \quad (10\lambda + 1)^2 = 50\lambda^2 + 10\lambda + 5$

$\Rightarrow \quad 25\lambda^2 + 5\lambda - 2 = 0$

$\Rightarrow \quad (5\lambda + 2)(5\lambda - 1) = 0$

$$\lambda = \frac{1}{5}, -\frac{2}{5}$$

$\therefore$ Required lines $= 4x - 3y = 25$

and $3x + 4y = 25$

$\therefore$ Number of lines = 2

41. (1) The given circle intersect at the point $(-1, 0)$ and $\left(\dfrac{3}{4}, -\dfrac{4}{5}\right)$

$\because P$ has integral coordinates, the coordinates of P are $(-1, 0)$

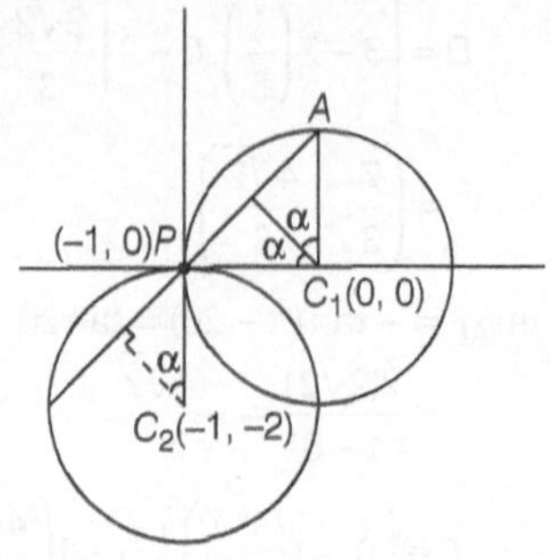

Let an equation of the chord AB through P is $y = m(x + 1)$

Let AP and BP make the same angle $2a$ at the centre $C_1(0, 0)$ and $C_2(-1, -2)$ respectively of the given circles.

$$\angle APC_1 = \angle BPC_2 = \frac{\pi}{2} - \alpha \quad \text{...(i)}$$

Now slope of $C_1P = 0$ and slope of $C_2P = \dfrac{2}{m'}$, where $m' = 0$ from Eq. (i), we have

$$\tan\left(\frac{\pi}{2} - \alpha\right) = \left|\frac{m - 0}{1 + m \cdot 0}\right| = \left|\frac{m - \frac{2}{m'}}{1 + m\frac{2}{m'}}\right|$$

$$|m| = \left|\frac{mm' - 2}{m' + 2m}\right| = \left|\frac{1}{m}\right|$$

$$m^2 = 1 \Rightarrow m = \pm 1$$

So that required equation of the chord is $y = x + 1$. Corresponding to the value of $m = 1$ (for $m = -1$, the equation will represent the chord PAB, A and B lying on the same side of P)

Hence, we do not consider it.

$\because \quad x - y + 1 = 0$

$\Rightarrow \quad a = 1, b = -1, c = 1$

$\Rightarrow \quad a + b + c = 1$

42. (6) Here point P is the radical centre of these circles, on solving like $S_1 - S_2 = 0$ and $S_2 - S_3 = 0$, we get

Point $P(0, -3)$

Given required circle passes through P and touches $x + y = 5$ at point $(6, -1)$

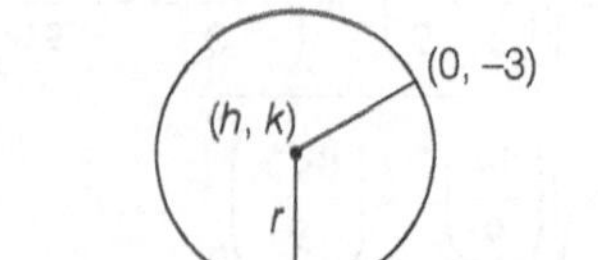

Let the centre of required circle (h, k).

$\therefore \quad \sqrt{h^2 + (k + 3)^2} = \sqrt{(h - 6)^2 + (k + 1)^2}$

$\Rightarrow \quad 3h + k = 7$ ...(i)

Now, equation of line perpendicular $x + y = 5$ is given by

$$x - y = \lambda$$

$\because$ It passes through (6, – 1), then

$$6+1=\lambda \Rightarrow \lambda=7$$

This also passes (h, k).

$$\therefore \quad h-k=\lambda \quad \text{...(ii)}$$

From Eqs. (i) and (ii),

$$h=\frac{7}{2}, k=-\frac{7}{2}$$

$$\therefore \quad (h, k)=\left(\frac{7}{2},-\frac{7}{2}\right)$$

$\therefore$ Radius

$$r=\sqrt{\left(\frac{7}{2}\right)^2+\left(3-\frac{7}{2}\right)^2}=\sqrt{\frac{50}{4}}=\frac{5\sqrt{2}}{2}$$

$\therefore$ Equation of circle

$$\left(x-\frac{7}{2}\right)^2+\left(y+\frac{7}{2}\right)^2=\frac{50}{4}$$

$$\Rightarrow \quad x^2+y^2-7x+7y+12=0$$

Given $\quad x^2+y^2+ax+by+2c=0$

$$\Rightarrow \quad a=-7, b=7$$

$$2c=12 \Rightarrow c=6$$

$$\therefore \quad a+b+c=6$$

43. (2) Let 'r' be the radius of circle, then $AC=2r$

$\because$ AC is the diameter.

$$\therefore \quad \angle ABC=90^\circ$$

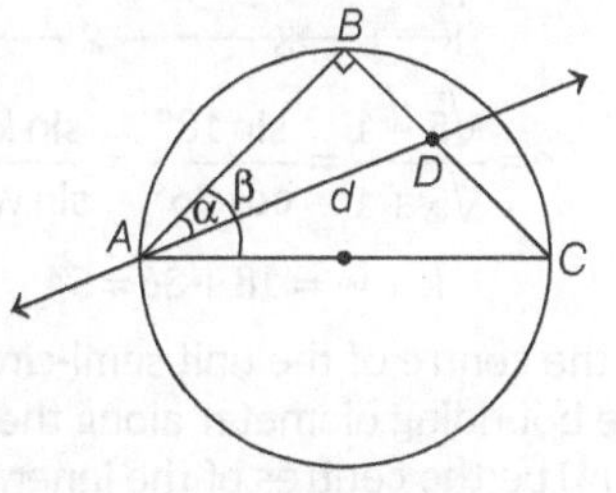

$\because$ In ΔABC,

$$BC=2r\sin\beta, AB=2r\cos\beta$$

In right angled ΔABC,

$$BD=AB\tan\alpha=2r\cos\beta\tan\alpha$$

$$AD=AB\sec\alpha=2r\cos\beta\sec\alpha$$

$$\therefore \quad DC=BC-BD$$

$$=2r\sin\beta-2r\cos\beta\tan\alpha$$

Now, E is the mid-point of DC.

$$\therefore \quad DE=\frac{DC}{2}=\frac{2r\sin\beta-2r\cos\beta\tan\alpha}{2}$$

$$\therefore \quad DE=r\sin\beta-r\cos\beta\tan\alpha$$

$\because$ In ΔADC, AE is median.

$$\therefore \quad 2(AE^2+DE^2)=AD^2+AC^2$$

$$\Rightarrow \quad 2[d^2+r^2(\sin\beta-\cos\beta\tan\alpha)^2]$$

$$=4r^2\cos^2\beta\sec^2\alpha+4r^2$$

$$\Rightarrow \quad r^2=\frac{d^2\cos^2\alpha}{\cos^2\alpha+\cos^2\beta+2\cos\alpha\cos\beta\cos(\beta-\alpha)}$$

$$\therefore \quad \text{Area}=\pi r^2$$

$$=\frac{\pi d^2\cos^2\alpha}{\cos^2\alpha+\cos^2\beta+2\cos\alpha\, 2\cos\beta\cos(\beta-\alpha)}$$

$\because k=1, l=1, m=1$ and $n=2$

Now, $\quad \dfrac{2k+l+m}{n}=\dfrac{2+1+1}{2}=2$

44. (8) $\theta=\alpha+\beta=\tan^{-1}\frac{1}{2}+\tan^{-1}\frac{1}{3}$

$$=\tan^{-1}\left(\frac{\frac{1}{2}+\frac{1}{3}}{1-\frac{1}{2}\cdot\frac{1}{3}}\right)=\tan^{-1}1$$

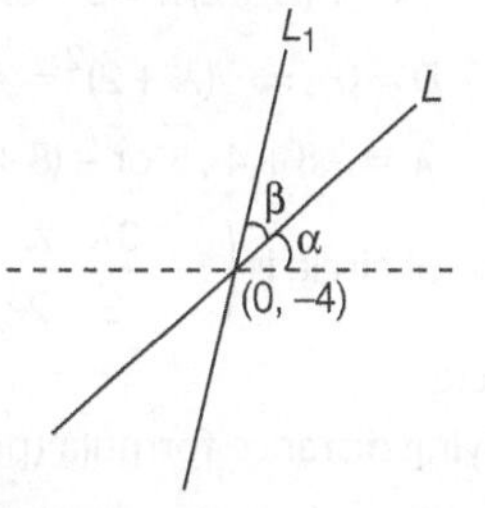

$$\theta=\frac{\pi}{4}$$

Equation of L_1 is $y+4=x \Rightarrow y-x+4=0$

Let the equation of circle is

$$(x-r)^2+(y+r)^2=r^2$$

$$\Rightarrow \quad x^2+y^2-2xr+2yr+r^2=0$$

$$r=\left|\frac{-r-r+4}{\sqrt{2}}\right|$$

$$\Rightarrow \quad 2r^2-16r+16=0 \quad \text{...(i)}$$

$\because$ Sum of radii = Sum of roots of Eq. (i)

$$=-\frac{(-16)}{2}=8$$

45. (6) Given circle is $(x+1)^2+(y+2)^2=9$ has radius = 3

Points on the circle which are nearest and farthest to the point $P(a, b)$ are Q and R respectively.

$\because$ Difference of largest and smallest distance of P to the circle $=PR-PQ=QR=2r=6$

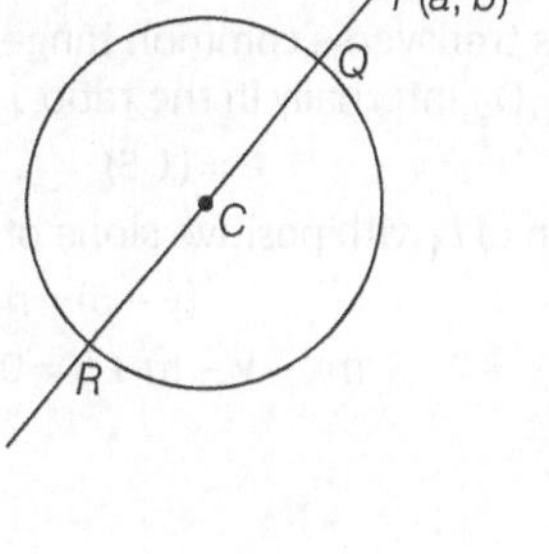

46. (370) The require circle is $S+\lambda L=0$

$$(x-1)^2+(y+3)^2+\lambda(3x-y-6)=0$$

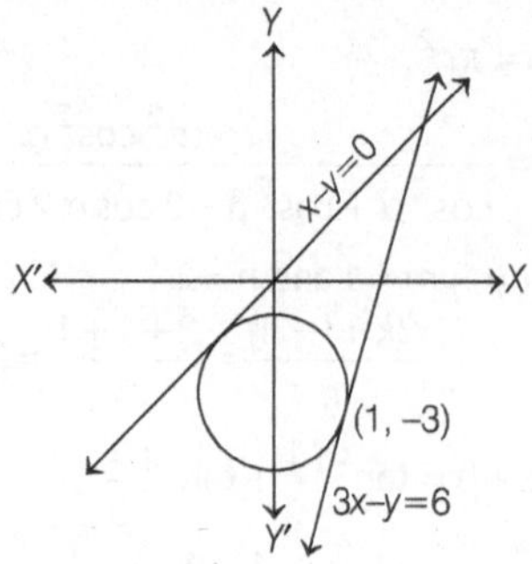

$$\Rightarrow \quad x^2+y^2+(3\lambda-2)x+(6-\lambda)y+(10-6\lambda)=0 \quad \text{...(i)}$$

Put $y=x$ in Eq. (i), we get

$$x^2+(\lambda+2)x+5-3\lambda=0$$

$$D=0 \Rightarrow (\lambda+2)^2=4(5-3\lambda)$$

$$\Rightarrow \quad \lambda=-8+4\sqrt{5} \text{ or } -(8+4\sqrt{5})$$

Also centre of circle is $\left(1-\frac{3\lambda}{2},\frac{\lambda}{2}-3\right)$ and $(1,-3)$ lies on the circle.

$\therefore$ On applying distance formula (put $\lambda=-8+4\sqrt{5}$)

$\therefore$ We get radius of smallest circle $=\sqrt{2}(10-4\sqrt{5})$

$\therefore a=2$ and $b=5$

$$\therefore \quad (b+a)^3+(b-a)^3=343+27=370$$

47. (14) $C_1 \Rightarrow$ Centre (1, 8), radius $=1$

$C_2 \Rightarrow$ Centre (1, −1), radius $=2$

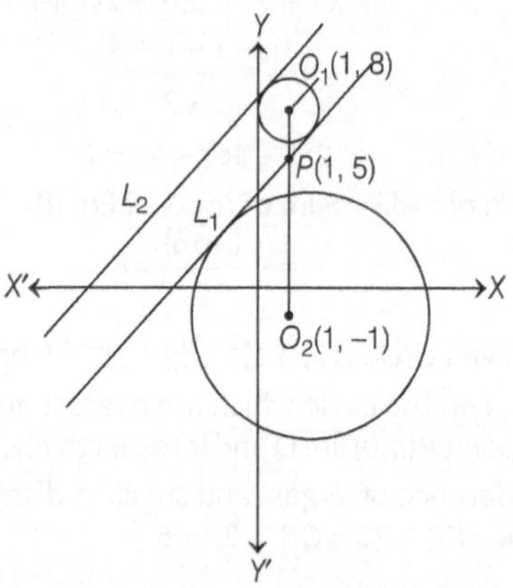

Line L_1 is transverse common tangent of C_1 and C_2, then it cut O_1, O_2 internally in the ratio 1 : 2 at point P, then

$$P=(1,5)$$

Equation of L_1 with positive slope of

$$(y-5)=m(x-1)$$

$$mx-y-m+5=0$$

If it is tangent to circle C_2

$$\left|\frac{m+1-m+5}{\sqrt{m^2+1}}\right|=2 \Rightarrow m=2\sqrt{2}$$

$$\Rightarrow \quad 2\sqrt{2}x-y-2\sqrt{2}+5=0$$

Equation of $L_2 \Rightarrow 2\sqrt{2}x-y+k=0$

and $\left|\frac{2\sqrt{2}-8+k}{3}\right|=1 \Rightarrow k=\pm(3+8-2\sqrt{2})$

$$a=2, b=1 \text{ and } c=11$$

So, $a+b+c=2+1+11=14$

48. (54) Inradius of ΔABC, $r_0=\frac{A}{5}=\frac{6}{6}=1$

$$\sin\frac{C}{2}=\frac{1-r}{1+r}=\frac{1}{\sqrt{5}} \quad \left\{\because \sin C=\frac{4}{5}\right\}$$

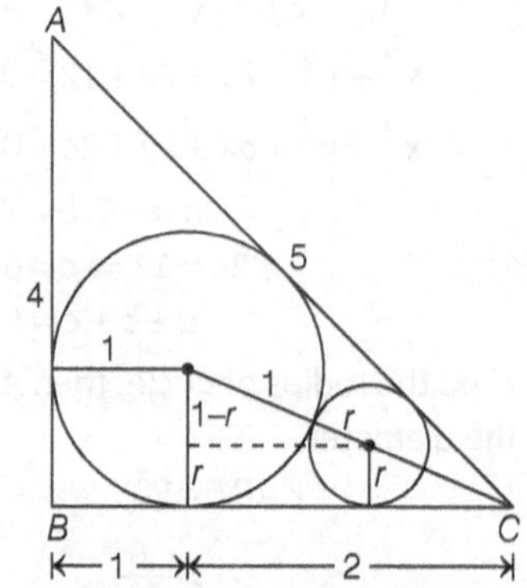

$$r=\frac{\sqrt{5}-1}{\sqrt{5}+1}=\frac{\sin 18^\circ}{\cos 36^\circ}=\frac{\sin k}{\sin w}$$

$$\therefore \quad k+w=18+36=54$$

49. (2) Let the centre of the unit semi-circle be at the origin and the bounding diameter along the X-axis, let (a, r_1) and (b, r_2) be the centres of the inner circles of radii r_1 and r_2

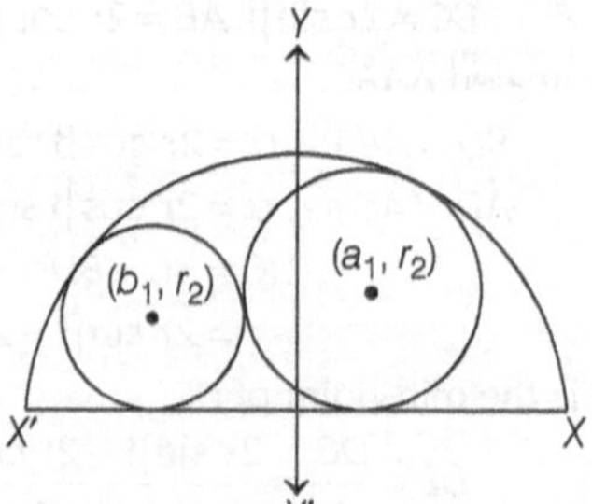

$\because$ These circles touch the semi-circle

$$1-r_1=\sqrt{a^2+r_1^2} \Rightarrow 1-a^2=2r_1 \quad \text{...(i)}$$

and $$1-r_2=\sqrt{b^2+r_2^2} \Rightarrow 1-b^2=2r_2 \quad \text{...(ii)}$$

From Eqs. (i) and (ii), we get

$$r_1 + r_2 = 1 - \frac{a^2 + b^2}{2} \le 1 - |a||b|$$

and the maximum value occurs for the sign of equality, where

$|a| = |b| \Rightarrow r_1 = r_2$ so that $|a| = |b| = r_1$

Hence, Eq. (i), gives $1 - r_1^2 = 2r_1$

or $\quad r_1^2 + 2r_1 - 1 = 0 \Rightarrow r_1 = -1 + \sqrt{2}$

Maximum $r_1 + r_2 = 2(\sqrt{2} - 1)$

$\Rightarrow \quad P = 2$

50. (20) Consider three concentric circles C_1, C_2, C_3 with radii r_1, r_2, r_3.

Let

$$C_1 \equiv x^2 + y^2 = r_1^2$$
$$C_2 = x^2 + y^2 = r_2^2$$
$$C_3 = x^2 + y^2 = r_3^2$$

Equation of chord of contact of tangent from $P(x_1, y_1)$ to C_2 is $xx_1 + yy_1 - r_2^2 = 0$

If this is a tangent to C_3, we get

$$\frac{r_2^2}{r_1} = r_3 \Rightarrow r_2^2 = r_1 r_3$$

$\Rightarrow r_1, r_2, r_3$ all are in GP.

$\Rightarrow r_1, r_2, r_3, \ldots r_n$ are in GP.

Also, angle between the tangents from C_1 to C_2 is $\frac{\pi}{3}$

$$\Rightarrow \quad \sin\frac{\pi}{6} = \frac{1}{2} = \frac{r_2}{r_1} \Rightarrow r_2 = \frac{r_1}{2}$$

$$\Rightarrow \quad \lim_{h \to \infty} \sum_{i=1}^{n} r_i = r_1 + r_2 + \ldots \infty$$

$$= 10 + \frac{10}{2} + \frac{10}{2^2} + \ldots \infty$$

$$= 10\left(1 + \frac{1}{2} + \frac{1}{2^2} + \ldots \infty\right)$$

$$= 10\left(\frac{1}{1 - \frac{1}{2}}\right) = 20$$

CHAPTER

05

PARABOLA

LEVEL 01 \\ Starter Level

5.1 Equation of Parabola

1 (i) Check the following equations represents a parabola or not.

(a) $3x^2 + 4y - 7 = 0$

(b) $2x^2 + 6y^2 - 24xy + 2x + 4y - 7 = 0$

(ii) Find the value of λ for which following conic represent a parabola.

(a) $x^2 + 2y^2 - 4\lambda xy + 4x + 2y - 1 = 0$

(b) $\lambda x^2 + 3y^2 + \lambda xy + 2x + 6y - 8 = 0$

(iii) Find the centre of the following conic.

(a) $14x^2 - 4xy + 11y^2 - 44x - 58y + 71 = 0$

(b) $3x^2 + 6xy + y^2 - 4x + 6y + 8 = 0$

2 Find the equation of parabola for the following.

(a) Vertex (0, 0) and focus (3, 0)

(b) Vertex (0, 0) and focus (– 5, 0)

(c) Vertex (0, 0) and focus (0, 4)

(d) Vertex (0, 0) and focus (0, – 7)

3 Find the equation of parabola.

(a) In parametric form whose equation is $y^2 = 4x$.

(b) Whose focus is at (3, – 4) and directrix is $x + y - 2 = 0$

(c) Whose vertex at (3, 2) and focus at (5, 2)

(d) Whose vertex at (– 3, 0) and directrix is $x + 5 = 0$

(e) Whose axis parallel to Y-axis and passes through the points (0, 0), (10, 12) and (30, 8)

(f) Which has its axis along the X-axis and passes through the points (3, 2) and (– 2, – 1)

(g) With its vertex at the point (2, 3) and axis parallel to Y-axis and which passes through the points (4, 5)

(h) With latus rectum joining the points (3, 6) and (3, – 2)

4 Find the vertex, axis, focus, directrix and latus rectum of following parabola :

(a) $y^2 = 3x + 4y + 2$ (b) $x^2 = y + 4x + 2$

(c) $9y^2 - 16x - 12y + 57 = 0$ (d) $x^2 + 8x + 12y + 4 = 0$

5 (a) M is the foot of the perpendicular from a point P on parabola $y^2 = 8x$ to its directrix and ΔSPM is an equilateral triangle, where S is the focus then find SP.

(b) An equilateral triangle is inscribed in the parabola $y^2 = 4x$ such that one vertex of this triangle coincides with the vertex of the parabola. Then, find the side length of this triangle.

(c) A right angled ΔABC is inscribed in parabola $y^2 = 4x$, where A is vertex of parabola and $\angle BAC = \frac{\pi}{2}$. If $AB = \sqrt{5}$, then find the area of ΔABC.

(d) AP is perpendicular to PB, where A is vertex of parabola $y^2 = 4x$ and P on the parabola. B is on the axis of parabola. Then, find the locus of centroid to ΔPAB.

6 (a) The girder of the bridge is in the form of parabola. Its heights is 10 feet and base is 100 feet wide. How high is it 20 feet from the centre?

(b) An arch is in the form of a parabola with its axis vertical. The arch is 10 cm high and 5 m wide at the base. How wide is it 2 m from the vertex of the parabola?

5.2 Point and Parabola

7 (i) Find the position of following points w.r.t. parabola $y^2 = 8x$

(a) (2, 1) (b) (3, – 1) (c) (– 4, 2) (d) (0, 0)

(ii) The point $(a, 2a)$ is an interior point of the region bounded by the parabola $y^2 = 16x$ and the double ordinate through the focus, then find values of a.

8 (i) If the chord joining $P(at_1^2, 2at_1)$ and $Q(at_2^2, 2at_2)$ is the focal chord, then find t_1t_2.

(ii) Find the length of focal chord of parabola $y^2 = 4x$ makes an angle 45° with the X-axis.

5.3 Line and Parabola

9 (i) For the parabola $y^2 = 8x$, find the values of λ for which the line $y = 2x - \lambda$ is

(a) Intersecting line (b) Tangent

(c) Neither intersecting nor touching

(ii) Find the equation of tangent to the parabola $y^2 = 4x + 5$

(a) at point (1, 3) (b) in parametric form at $P(t)$

(c) having slope 1 (d) which is parallel to the line $y = 2x + 7$

(iii) Find the equation of common tangents of the following.

(a) $y^2 = 4x, x^2 = 4y$ (b) $y^2 = 16x, x^2 + y^2 = 8$

(c) $y^2 = 8x, xy = -1$ (d) $y = x^2$ and $y = -(x-2)^2$

(e) $x^2 + y^2 - 6y + 4 = 0, y^2 = x$

(iv) Find the point of contact.

(a) If a tangent to the parabola $y^2 = 8x$ is $y = x + 2$.

(b) $x \cos \alpha + y \sin \alpha = p$ touches the parabola $y^2 = 4ax$.

(c) If a tangent to the parabola $y^2 = 8x$ makes an angle of 45° with the straight line $y = 3x + 5$

(v) Find the shortest distance between

(a) The line $y = x - 2$ and the parabola $y = x^2 + 3x + 2$

(b) The line $x + y = 4$ and the parabola $y^2 + 4x + 4y = 0$

10 Find the equation of normal

(a) to the parabola $x^2 = 4y$ at (2, 1)

(b) to the parabola $y^2 = 4x$ in parametric form.

(c) to the parabola $y^2 = 16x$ which is parallel to $y = 2x + 1$.

(d) to the parabola $y = x^2 - x - \frac{1}{2}$ which has equal intercept on axis. Also find the point where this normal meets the curve again.

(e) to the parabola $y^2 = 4x$ which pass through (15, 12).

(f) To the parabola $y^2 = 16x$ at ends of latus rectum.

11 (a) If $x + y = k$ is a normal to the parabola $y^2 = 12x$, then find the value of k.

(b) If the normal at $P(18, 12)$ to the parabola $y^2 = 8x$ cuts it again Q, find PQ.

(c) Find the minimum distance between the curves $y^2 = 4x$ and $x^2 + y^2 - 12x + 31 = 0$

(d) Find the angle made by the normal chord at the point whose ordinate is equal to abscissa to the focus of parabola $y^2 = 4ax$

12 (a) Find the locus of point of intersection of the normals at the end of the focal chord of the parabola $y^2 = 4ax$.

(b) Find the locus of the point of intersection of two normal act to a parabola $y^2 = 4ax$. Which are at right angles to one another?

13 (a) Find the number of distinct normals that can be drawn from (– 2, 1) to the parabola $y^2 - 4x - 2y - 3 = 0$

(b) If 3 distinct normals can be drawn to the parabola $y^2 - 2y = 4x - 9$ from the point $(2a, b)$, then the range of the values of a.

(c) If the normals from any point to the parabola $y^2 = 4x$ cut the line $x = 2$ in points whose ordinates are in AP, then find the relation between slopes of tangents at co-normal points.

(d) If the three normals from a point to parabola $y^2 = 8x$ cut the axis in points whose distance from the vertex are in AP, then find the locus of that points.

5.4 Pair and Tangents, Director Circles and Chord of Contact, Pole and Polar

14 (a) Find the equation of pair of tangents to the parabola $y^2 - 4x - 6y + 7 = 0$ from the point (– 2, – 1).

(b) Find the equation of the chord of contact to the tangents from the point (– 1, 2) to the parabola $y^2 = 12x$.

(c) Find the equation of chord of the parabola $y^2 = 8x$ which is bisected at (2, 3).

(d) Find the length of chord of contact drawn from (– 2, 1) to the parabola $y^2 = 4x$.

15 (a) Find the locus of point of intersection of two tangents to a parabola $y^2 = 4x$ which make a given angle $\frac{\pi}{3}$ with one another.

(b) Find the area of triangle formed by the tangents drawn from (– 3, 1) to $y^2 = 8x$ and their chord of contact.

(c) Find the locus of the mid-points of the chords of the parabola $y^2 = 16x$ which subtends 90° at vertex of the parabola.

(d) Find the locus of middle points of normal chords of the parabola $y^2 = 4ax$.

16 (i) Find the equation of director circle of the following parabolas.

(a) $y^2 = x + 2$ (b) $x^2 = 4x + 4y$

(c) $y^2 = 4x + 4y - 8$

(ii) Find the polar of the following points w.r.t. given parabola.

(a) Point (1, 2), Parabola $y^2 = 4x$ (c) Point (2, 3), Parabola $x^2 = 16y$

(iii) Find the length of tangent, subtangent, normal and subnormal to $y^2 = 3x$ at (3, 3).

(iv) Find the diameter of parabola $y^2 = 16x$ w.r.t. chords $y = 2x + 3$.

17 (a) Find the locus of poles of normal chords of $y^2 = 4ax$.

(b) In the parabola $y^2 = 4ax$, find the locus of the poles of all chords of constant length 2 C.

(c) Find the locus of poles of chords of the parabola $y^2 = 4ax$ which are at a consistant distance b from the vertex.

5.5 Miscellaneous

18 (a) If a parabolic reflector is 20 cm in diameter and 5 cm deep. Find focus.

(b) A ray of light moving parallel to the X-axis gets reflected from a parabolic mirror whose equation is $(y - 4)^2 = 8(x + 1)$ after reflection, the ray passes through the point (α, β), then find $\alpha + \beta + 10$.

(c) A ray of light is moving along the line $y = x + 2$ get reflected from a parabolic mirror, whose equation is $y^2 = 4x + 8$. After reflection the ray does not pass through the focus of parabola. Find the equation of the line which containing the reflected ray.

(d) Find the length of latus rectum of the parabola $(p^2 + q^2)(x^2 + y^2) = (qx + py - pq)^2$.

LEVEL 02 \ JEE Main Level

Single Choice Correct Questions

5.1 Equation of Parabola

1 A parabola is drawn with its focus at (3, 4) and vertex at the focus of the parabola $y^2 - 12x - 4y + 4 = 0$. The equation of the parabola is

(a) $x^2 + 6x - 8y - 25 = 0$ (b) $y^2 - 8x - 6y + 25 = 0$

(c) $x^2 - 6x + 8y - 25 = 0$ (d) $x^2 - 6x - 8y + 25 = 0$

2 Axis of a parabola is $y = x$ and vertex and focus are at a distance $\sqrt{2}$ and $2\sqrt{2}$ units respectively from the origin. Then, equation of the parabola is

(a) $(x - y)^2 = 8(x + y - 2)$ (b) $(x + y)^2 = 2(x + y - 2)$

(c) $(x - y)^2 = 4(x + y - 2)$ (d) $(x + y)^2 = 2(x - y + 2)$

3 The axis of a parabola is $3x + 4y + 6 = 0$, its vertex is $(-2, 0)$ and latusrectum is 4 in length, then the equation of parabola is

(a) $9x^2 + 24xy + 16y^2 - 44x + 108y - 124 = 0$

(b) $9x^2 + 24xy + 16y^2 - 22x + 108y - 125 = 0$

(c) $9x^2 - 24xy + 16y^2 + 44x - 108y - 124 = 0$

(d) None of the above

4 The locus of the vertices of the family of parabolas $y = \frac{a^3x^2}{3} + \frac{a^2x}{2} - 2a$ is

(a) $xy = \frac{3}{4}$ (b) $xy = \frac{35}{16}$ (c) $xy = \frac{64}{105}$ (d) $xy = \frac{105}{64}$

5 The line $x - 1 = 0$ is the directrix of the parabola $y^2 - kx + 8 = 0$. Then, one of the values of k is

(a) $\frac{1}{8}$ (b) 8 (c) 4 (d) $\frac{1}{4}$

6 Consider the parabola $y^2 = 4x$, $A \equiv (4, -4)$ and $B = (9, 6)$ be two fixed points on the parabola. Let 'C' be a moving point on the parabola between A and B such that the area of the triangle ABC is maximum the coordinate of 'C' is

(a) $\left(\frac{1}{4}, 1\right)$ (b) $(4, 4)$ (c) $(3, 2\sqrt{3})$ (d) $(3, -2\sqrt{3})$

5.2 Point and Parabola

7 If the point $P(2a, a)$ lies interior region between vertex and latus rectum of the parabola $y = -\frac{x^2}{16}$, then a lies in the interval

(a) $(-16, 0)$ (b) $(-4, 0)$ (c) $(-4, -2)$ (d) ϕ

8 $P(x, y)$ is a variable point on the parabola $y^2 = 4ax$ and $Q(x + c, y + c)$ is another variable point, where c is a constant. The locus of mid-point of PQ is

(a) parabola (b) ellipse (c) hyperbola (d) circle

9 Set of values of a for which the point $(\alpha, 1)$ lies inside the curves $C_1 : x^2 + y^2 = 4$ and $C_2 : y^2 = 4x$ is

(a) $|\alpha| < \sqrt{3}$ (b) $|\alpha| < 2$ (c) $\frac{1}{4} < \alpha < \sqrt{3}$ (d) None of these

10 The locus of the mid-point of the line segment joining the focus of the parabola $y^2 = 4ax$ to a moving point of the parabola, is another parabola whose directrix is **[JEE Main 2021]**

(a) $x = 0$ (b) $x = -\frac{a}{2}$ (c) $x = a$ (d) $x = \frac{a}{2}$

11 Let P be a point on the parabola, $y^2 = 12x$ and N be the foot of the perpendicular drawn from P on the axis of the parabola. A line is now drawn through the mid-point M of PN, parallel to its axis which meets the parabola at Q. If the y-intercept of the line NQ is $\frac{4}{3}$, then **[JEE Main 2021]**

(a) $MQ = \frac{1}{3}$ (b) $PN = 4$ (c) $PN = 3$ (d) $MQ = \frac{1}{4}$

12 Let P be a point on the parpendicular $y^2 = 3(2x - 3)$ and M is the foot of perpendicular drawn from P on the directrix of the parabola, then length of each side of an equilateral ΔSMP, where 'S' is the focus of the parabola is

(a) 2 units (b) 4 units (c) 6 units (d) 8 units

13 The ΔPQR of area A is inscribed in the parabola $y^2 = 4ax$ such that the vertex P lies at the vertex of the parabola and the base QR is a focal chord. The modulus of the difference of the ordinates of the points Q and R is

(a) $\frac{A}{2a}$ (b) $\frac{A}{a}$ (c) $\frac{2A}{a}$ (d) $\frac{4A}{a}$

14 The length of a focal chord of the parabola $y^2 = 4ax$ at a distance b from the vertex is c. Then,

(a) $2a^2 = bc$ (b) $a^3 = b^2c$ (c) $ac = b^2$ (d) $b^2c = 4a^3$

15 Let O be the vertex and Q be any point on the parabola $x^2 = 8y$. If the point P divides the line segment OQ internally in the ratio 1 : 3, then locus of P is

(a) $x^2 = y$ (b) $y^2 = x$ (c) $y^2 = 2x$ (d) $x^2 = 2y$

5.3 Line and Parabola

16 If the line $y - \sqrt{3}x + 3 = 0$ cuts the parabola $y^2 = x + 2$ at P and Q, then $AP \cdot AQ$ is equal to [where $A = (\sqrt{3}, 0)$]

(a) $\frac{2(\sqrt{3}+2)}{3}$ (b) $\frac{4\sqrt{3}}{2}$ (c) $\frac{4(2-\sqrt{2})}{3}$ (d) $\frac{4(\sqrt{3}+2)}{3}$

17 If the line $x + y - 1 = 0$ touches the parabola $y^2 = kx$, then the value of k is

(a) 4 (b) -4 (c) 2 (d) -2

18 A line L passing through the focus of the parabola $y^2 = 4(x-1)$ intersect the parabola in two distinct points. If 'm' is the slope of the line L. Then,

(a) $-1 < m < 1$ (b) $m < -1$ or $m > 1$

(c) $m \in R$ (d) None of these

19 Let C be the locus of the mirror image of a point on the parabola $y^2 = 4x$ w.r.t. the line $y = x$. Then, the equation of tangent to C at $P(2, 1)$ is **[JEE Main 2021]**

(a) $x - y = 1$ (b) $2x + y = 5$ (c) $x + 3y = 5$ (d) $x + 2y = 4$

20 A line is drawn from $A(-2, 0)$ to intersect the curve $y^2 = 4x$ in P and Q in the first quadrant such that $\frac{1}{AP} + \frac{1}{AQ} < \frac{1}{4}$, then slope of the line is always

(a) $> \sqrt{3}$ (b) $> \frac{1}{\sqrt{3}}$ (c) $> \sqrt{2}$ (d) $> \frac{1}{\sqrt{3}}$

21 A tangent to the parabola $y^2 = 8x$ makes an angle of 45° with the straight line $y = 3x + 5$. The equation of the tangent and its point of contact are

(a) $2x + y - 1 = 0$ at $\left(\frac{1}{2}, 2\right)$ and $x - 2y - 8 = 0$ at $(8, 8)$

(b) $2x + y - 1 = 0$ at $(1, -2)$ and $x + 2y + 8 = 0$ at $(8, 8)$

(c) $2x + y + 1 = 0$ at $\left(\frac{1}{2}, -2\right)$ and $x - 2y + 8 = 0$ at $(8, 8)$

(d) None of the above

22 Equation of a common tangent to the circle $x^2 + y^2 - 6x = 0$ and the parabola $y^2 = 4x$ is

(a) $\sqrt{3}y = 3x + 1$ (b) $2\sqrt{3}y = 12x + 1$

(c) $\sqrt{3}y = x + 3$ (d) $2\sqrt{3}y = -x - 12$

23 The tangent at (1, 7) to the curve $x^2 = y - 6x$ touches the circle $x^2 + y^2 + 16x + 12y + c = 0$ at

(a) (0, 7) (b) (− 6, 7) (c) (6, − 7) (d) (− 6, − 7)

24 A circle with the centre lying on the focus of the parabola $y^2 = 2px$ such that it touches the directrix of the parabola. Then, a point of intersection of circle and the parabola is

(a) $\left(\frac{p}{2}, p\right)$ (b) $\left(\frac{p}{2}, 2p\right)$ (c) $\left(-\frac{p}{2}, p\right)$ (d) $\left(-\frac{p}{2}, -p\right)$

25 The parabola $y^2 = 4x$ and the circle $(x-6)^2 + y^2 = r^2$ will have no common tangent, if 'r' is.

(a) greater than $\sqrt{20}$ (b) less than $\sqrt{20}$

(c) greater than $\sqrt{18}$ (d) lies between $\sqrt{20}$ to $\sqrt{28}$

26 L is the normal to the parabola $y^2 = 4x$ and passes through the point (9, 6). If the slope of normal is positive, then its equation is

(a) $y - x + 3 = 0$ (b) $y + 3x - 33 = 0$ (c) $y - 2x + 12 = 0$ (d) $y + x - 15 = 0$

27 If the normal at the point (1, 2) on the parabola $y^2 = 4x$ meets the parabola again at the point $(t^2, 2t)$, then t is equal to

(a) 1 (b) – 1 (c) 3 (d) – 3

28 The normal to the parabola $y^2 = 4ax$ at $(at_1^2, 2at_1)$ meets the curve again at $(at_2^2, 2at_2)$, then the number of point, where $f\{f(x)\}$ is discontinuous, is equal to, where $f(x) = \frac{1}{1-x}$

(a) $t_1^2 + t_1t_2$ (b) $-t_1^2 - t_1t_2$ (c) $t_1^2 - t_1t_2$ (d) None of these

29 The length of normal drawn from the point on the axis of the parabola $y^2 = 8x$ whose distance from the focus is 8 units is equal to.

(a) 8 units (b) 6 units (c) $4\sqrt{3}$ units (d) 10 units

30 If a normal to the parabola $y^2 = 8x$ makes 45° angle with X-axis, then foot of normal will be

(a) (2, 4) (b) (2, – 4) (c) (8, 8) (d) (8, – 8)

31 The equation of the common normal to the parabolas $y^2 = 4ax$ and $x^2 = 4ay$, is

(a) $y = x$ (b) $x + y = a$ (c) $y = x - 3a$ (d) $x + y = 3a$

32 Parabola $(y-\alpha)^2 = 4a(x-\beta)$ and $(y-\alpha)^2 = 4a'(x-\beta')$ will have a common normal (other than the normal passing through vertex of parabola) if

(a) $\frac{2(a-a')}{\beta'-\beta} < 1$ (b) $\frac{2(a-a')}{\beta-\beta'} < 1$ (c) $\frac{2(a'-a)}{\beta+\beta'} > 1$ (d) $\frac{2(a'-a)}{\beta'+\beta} < 1$

33 The condition that the parabola $y^2 = 4c(x-d)$ and $y^2 = 4ax$ have a common normal other than X-axis is $(a > 0, c > 0)$ is

(a) $2a < 2c + d$ (b) $2c < 2a + d$ (c) $2d < 2a + c$ (d) $2d < 2c + a$

34 $\text{Min}\,[(x_1 - x_2)^2 + (3 + \sqrt{1 - x_1^2} - \sqrt{4x_2})^2]\ \forall\, x_1, x_2 \in R$ is

(a) $4\sqrt{5} + 1$ (b) $3 - 2\sqrt{2}$ (c) $\sqrt{5} + 1$ (d) $\sqrt{5} - 1$

35 The line $x - y - 1 = 0$ meets the parabola $y^2 = 4x$ at A and B. Normals at A and B meet at C. If D is a point on the curve such that CD is normal at D, then the coordinates of D are

(a) (4, 4) (b) (4, – 4) (c) (2, – 4) (d) (2, 4)

36 From the point (15, 12), three normals are drawn to the parabola $y^2 = 4x$. The centroid of the triangle formed by feet of normals is

(a) $\left(\frac{26}{3}, 0\right)$ (b) (4, 0) (c) $\left(\frac{16}{3}, 0\right)$ (d) None of these

37 If $c > \frac{1}{2}$, then three normals can be drawn from the point $(c, 0)$ to the parabola $y^2 = x$ out of which, one is the X-axis. If the other two normals are at right angles, then the value of c is

(a) $\frac{4}{3}$ (b) $\frac{3}{4}$ (c) $\frac{2}{3}$ (d) $\frac{3}{2}$

5.4 Pair of Tangents, Chord of Contact, Length of Tangent, Normal, Subtangent and Subnormal

38 The tangents from the origin to the parabola $y^2 + 4 = 4x$ are inclined at

(a) $\frac{\pi}{6}$ (b) $\frac{\pi}{4}$ (c) $\frac{\pi}{3}$ (d) $\frac{\pi}{2}$

39 The point of intersection of tangents drawn at the end points of latusrectum of the parabola $y^2 = 4x$ is

(a) $(-2, 0)$ (b) $(-3, 0)$ (c) $(-1, 0)$ (d) None of these

40 From an external point P a pair of tangent lines are drawn to the parabola $y^2 = 4x$. If θ_1 and θ_2 are the inclination of these tangents with the axis of X such that $\theta_1 + \theta_2 = \frac{\pi}{4}$, then locus of P is

(a) $x - y + 1 = 0$ (b) $x + y - 1 = 0$
(c) $x - y - 1 = 0$ (d) $x + y + 1 = 0$

41 Tangents are drawn from the point $(-1, 2)$ on the parabola $y^2 = 4x$. The length, these tangents will intercept on the line $x = 2$, is

(a) 6 units (b) $6\sqrt{2}$ units (c) $2\sqrt{6}$ units (d) None of these

42 If two tangents drawn from the point (α, β) to the parabola $y^2 = 4x$ be such that the slope of one tangent is double of the other, then

(a) $\beta = \frac{2}{9}\alpha^2$ (b) $\alpha = \frac{2}{9}\beta^2$ (c) $2\alpha = 9\beta^2$ (d) None of these

43 Locus of the point of intersection of the perpendicular tangents of the curve $y^2 + 4y - 6x - 2 = 0$ is

(a) $2x - 1$ (b) $2x + 3 = 0$ (c) $2y + 3 = 0$ (d) $2x + 5 = 0$

44 If P is a point on the line $x + 4a = 0$ and QR is the chord of contact of P w.r.t. $y^2 = 4ax$, then $\angle QOR$ (where O is the vertex) is equal to

(a) 45° (b) 60° (c) 30° (d) 90°

45 The chord of a parabola $y^2 = 6x$ are passing through the point $(9, 5)$. Then, the middle points of these chords lie on the curve represented by the equation

(a) $y^2 - 5y - 3x + 27 = 0$ (b) $y^2 + 5y - 3x - 27 = 0$
(c) $y^2 - 5y - 3x - 27 = 0$ (d) $y^2 + 5y - 3x + 27 = 0$

46 PQ is chord of the parabola $y^2 = 4x$ whose perpendicular bisector meets the axis at M and the ordinate of the mid-point PQ meets the axis at N. Then, length of MN is equal to

(a) 2 units (b) 2.5 units (c) 3 units (d) 4 units

47 If two chords drawn from the point $A(4, 4)$ to the parabola $x^2 = 4y$ are bisected by line $y = mx$, then interval in which m lies is

(a) $(-2\sqrt{2}, 2\sqrt{2})$
(b) $(-\infty, -\sqrt{2}) \cup (\sqrt{2}, \infty)$
(c) $(-\infty, -2\sqrt{2} - 2) \cup (2\sqrt{2} - 2, \infty)$
(d) None of the above

48 The number of common chords of the parabolas $x = y^2 - 6y + 11$ and $y = x^2 - 6x + 11$, are

(a) 1 (b) 2 (c) 4 (d) 6

49 A circle is drawn with centre at the focus of parabola $y^2 = 4x$ and the common chord is equidistant from the focus and the vertex. Then, the equation of circle is

(a) $(x-1)^2 + y^2 = \frac{9}{4}$ (b) $(x-1)^2 = \frac{9}{16} - y^2$ (c) $(y-1)^2 + x^2 = \frac{9}{4}$ (d) $(y-1)^2 + x^2 = \frac{9}{16}$

50 A circle is described whose centres is the vertex and whose diameter is three-quarters of the latusrectum of the parabola $y^2 = 4ax$. If PQ is the common chord of the circle and the parabola and L_1L_2 is the latusrectum, then the area (in sq units) of the trapezium PL_1L_2Q is

(a) $3\sqrt{2}a^2$ (b) $2\sqrt{2}a^2$ (c) $4a^2$ (d) $\left(\frac{2+\sqrt{2}}{2}\right)a^2$

51 Polar of a point w.r.t. $y^2 = 4bx$ touch $x^2 = 4ay$, then locus of this point is

(a) $xy = ab$ (b) $xy = -2ab$ (c) $xy = 2ab$ (d) None of these

52 The length of subtangent w.r.t. point (1, 3) for the parabola $x^2 = y - 2$ is

(a) $\frac{1}{2}$ unit (b) $\frac{5}{2}$ units (c) $\frac{3}{2}$ units (d) None of these

5.5 Miscellaneous

53 A circle of radius r touches the parabola $y^2 = 4ax$, $(a > 0)$ at the vertex and the centre lies on the axis of the parabola. Further, the circle completely lies within the parabola. The largest possible value of r is

(a) $2a$ (b) $3a$ (c) $4a$ (d) a

54 A ray of light travels along a line $y = 4$ and strikes the surface of a curves $y^2 = 4(x + y)$, then equation of the line along which reflected ray travel is

(a) $x = 0$ (b) $x = 2$ (c) $x + y = 4$ (d) $2x + y = 4$

55 The minimum area of circle which touches the parabolas $y = x^2 + 1$ and $x = y^2 + 1$ is

(a) $\frac{9\pi}{16}$ sq units (b) $\frac{9\pi}{32}$ sq units (c) $\frac{9\pi}{8}$ sq units (d) $\frac{9\pi}{4}$ sq units

56 Let us define a region R in XY-plane as a set of points satisfy $[x^2] = y$ {where $[x]$ denotes greatest integer $\le x$} the region R defines

(a) a parabola whose axis is horizontal (b) a parabola whose axis is vertical
(c) a integer point of the parabola $y = x^2$ (d) None of these

Numerical Value Type Questions

57 If the orthocentre of the triangle formed by the points t_1, t_2, t_3 on the parabola $y^2 = 4ax$ is the focus, then the value of $|t_1t_2 + t_2t_3 + t_3t_1|$ is

58 Let the curve C be the mirror image of the parabola $y^2 = 4x$ w.r.t. the line $x + y + 4 = 0$. If A and B are the points of intersection of C with the line $y = -5$, then the distance between A and B is

59 A parabola of latusrectum 8 units touches a fixed equal parabola, the axes of the two curves being parallel. Find the latusrectum of the locus of the vertex of the moving curve.

60 The towers of a bridge, hung in the form of a parabola, have their tops 30 m above the roadway and are 200 m apart. If the cable is 5 m above the roadway at the cente of the bridge, find the length of the vertical supporting cable 30 m from the centre.

61 Let $ABCD$ be a square of side length 2 units. A line M is drawn through A parallel to BD. A point S moves such that its distance from the line BD and the vertex A are equal. If the focus of S cut M at T_2 and T_3 and AC at T_1, then area of $\Delta T_1T_2T_3$ is

62 If the straight lines $y - b = m_1(x + a)$ and $y - b = m_2(x + a)$ are the tangents of $y^2 = 4ax$, then value of $\sqrt{m_1m_2 + 4}$ upto 2 decimal.

63 If the focal chord of $y^2 = 16x$ is a tangent to $(x - 6)^2 + y^2 = 2$, then sum of the possible values of the slope of chord.

64 If the tangent at the point $P(2, 4)$ to the parabola $y^2 = 8x$ meets the parabola $y^2 = 8x + 5$ at Q and R, then find the square of the distance between Q and R.

65 A line is a common tangent to the circle $(x - 3)^2 + y^2 = 9$ and the parabola $y^2 = 4x$. If the two points of contact (a, b) and (c, d) are distinct and lie in the first quadrant, then $2(a + c)$ is equal to .... **[JEE Main 2021]**

66 From the point $O(-1, 2)$ tangents lines are drawn to the parabola $y^2 = 4x$. If the area of the triangle formed by the chord of contact and the tangents is A, then the value of $\frac{A}{\sqrt{2}}$ is

67 Normals are drawn from the point P with slopes m_1, m_2, m_3 to the parabola $y^2 = 4x$. If locus of P with $m_1m_2 = \alpha$ is a part of the parabola itself, then the value of α is

68 A variable chord PQ of parabola $y^2 = 4ax$ is drawn parallel to the line $y = x$, if the locus of point of intersection of normals at P and Q is $ax + by - c$, then value of $\frac{a + b}{100}$ is

LEVEL 03 \ JEE Advanced Level

Single Choice Correct Questions

1 If the line $x - 1 = 0$ is the directrix of the parabola $y^2 - kx + 8 = 0, k \neq 0$ and the parabola intersects the circle $x^2 + y^2 = 4$ in two real distinct points, then the value of k is

(a) -4 (b) -8 (c) 4 (d) None of these.

2 Through the vertex 'O' of the parabola $y^2 = 4ax$ two chords OP and OQ are drawn and the circles on OP and OQ as diameters intersects at R. If θ_1, θ_2 and ϕ are the angles made with the axis by the tangents at P and Q on the parabola and by OR, then the value of $\cot\theta_1 + \cot\theta_2$ is equal to

(a) $-2\tan\phi$ (b) $-2\tan(\pi - \phi)$ (c) 0 (d) $2\cot\phi$

3 Two parabolas C and D intersect at two different points, where C is $y = x^2 - 3$ and D is $y = kx^2$. The intersection at which the x value is positive designated point A and $x = a$ at this intersection the tangent line l at A to curve D intersects curve C at point B, other than A. If x value of point B is 1, then a is equal to

(a) 1 (b) 2 (c) 3 (d) 4

4 P is a point on the directrix of the parabola $y^2 = 4ax$ and Q is the point of contact of a tangent drawn from P to the parabola. Then, the mid point PQ lies on the curve

(a) $y^2(2x + a) = a(3x + a)$

(b) $y^2(3x + a) = a(2x + a)^2$

(c) $y^2(2x + a) = a(3x + 2a)^2$

(d) $y^2(2x + a) = a(3x + a)^2$

5 Let P be a point on the parabola $y = 4 - x^2$ lying in the first quadrant. Tangent to the parabola at P meets the X and Y-axes at A and B, respectively. The minimum possible area of ΔOAB is (where O is origin).

(a) $\frac{64}{3\sqrt{3}}$ sq units (b) $\frac{32}{3\sqrt{3}}$ sq units (c) $96\sqrt{3}$ sq units (d) $192\sqrt{3}$ sq units

6 Length of the intercept on the normal chord at the point $P(at^2, 2at)$ of the parabola $y^2 = 4ax$ made by circle described on the focal distance of the point P as diameter is

(a) $a\sqrt{2+t^2}$ (b) $\frac{a}{2}\sqrt{1+t^2}$ (c) $2a\sqrt{1+t^2}$ (d) $a\sqrt{1+t^2}$

7 Two equal parabolas have the same focus and their axes are at right angle. A normal to the parabola is perpendicular to a normal to other parabola. Then, the locus of the point of intersection of those normals, is

(a) $(x - y)^2 = 2a(x + y)$ (b) $(x + y)^2 = 2a(x + y)$

(c) $(x + y)^2 = 4a(x - y)$ (d) $(x - y)^2 = 4a(x + y)$

8 A circle touches the parabola $y^2 = 4x$ at the point (1, 2) and also the directrix. The y-co-ordinate of the point of contact of the circle and directrix is

(a) $\sqrt{2}$ (b) 2 (c) $2\sqrt{2}$ (d) 4

9 The tangent and normal at $P(t)$, for all real positive t, to the parabola $y^2 = 4ax$ meet the axis of the parabola in T and G respectively, then the angle at which the tangent at P to the parabola is inclined to the tangent at 'P' to the circle through the point P, T and G is

(a) $\cot^{-1} t$ (b) $\cot^{-1} t^2$ (c) $\tan^{-1} t$ (d) $\tan^{-1} t^2$

10 From point A, common tangents are drawn to the circle $x^2 + y^2 = \frac{a^2}{2}$ and parabola $y^2 = 4ax$. The area of the quadrilateral formed by common tangents, the chord of contact of point A w.r.t. circle and the parabola is

(a) $\frac{15a^2}{2}$ sq units (b) $\frac{15a^2}{4}$ sq units (c) $4a^2$ sq units (d) $5a^2$ sq units

11 The side of a triangle touches a parabola $y^2 = 4ax$ and two of its angular points lies on another parabola $y^2 = 4b(x + c)$ with its axis in the same direction, then locus of third angular point is

(a) $a^2y^2 = (2b - a)^2(ax + 4bc)$ (b) $a^2y^2 = (2b - a)^2(ax - 4bc)$

(c) $a^2y^2 = (2b - a)(ax + 4bc)$ (d) None of these

12 The locus of the poles of chords which subtends a constant angle α at the vertex of parabola $y^2 = 4ax$ is

(a) $(x + 4a)^2 = 4\tan^2\alpha(y^2 - 4ah)$ (b) $(x + 4a)^2 = 4\cot^2\alpha(y^2 - 4ah)$

(c) $(x + 4a)^2 = 4\tan\alpha(y^2 - 4ah)$ (d) None of these

More Than One Correct

13 If C is a circle described on the focal chord of the parabola $y^2 = 4x$ as diameter which is inclined at an angle of 45° with the axis, then the

(a) radius of the circle is 2 units

(b) the centre of the circle is (3, 2)

(c) the line $x + 1 = 0$ touches the circle

(d) the circle $x^2 + y^2 + 2x - 6y + 3 = 0$ is orthogonal to C

14 Let A and B be two distinct points on the parabola $y^2 = 4x$. If the axis of the parabola touches a circle of radius r and AB as its diameter, then the slope of line AB is

(a) $-\frac{1}{r}$ (b) $\frac{1}{r}$ (c) $\frac{2}{r}$ (d) $-\frac{2}{r}$

15 If P_1P_2 and Q_1Q_2 are two focal chords of a parabola $y^2 = 4ax$ at right angles, then

(a) area of the quadrilateral $P_1Q_1P_2Q_2$ is minimum when the chords are inclined at an angle $\frac{\pi}{4}$ to the axis of the parabola.

(b) minimum area is twice the area of the square on the latusrectum of the parabola

(c) minimum area of quadrilateral $P_1Q_1P_2Q_2$ cannot be found

(d) minimum area is thrice the area of the square on the latusrectum of the parabola

16 Tangents to the parabola $y^2 = 4ax$ at point Q meets the tangents at vertex A at point B and the axis of parabola at T, P is any point on this tangent and N is the foot of perpendicular from Q on SP, where S is focus, M is the foot of perpendicular from Q on the directrix, then

(a) B bisects QT (b) B trisects QT (c) $QM = SN$ (d) $QM = 2SN$

17 The equation of the line that touches the curves $y = x|x|$ and $x^2 + (y-2)^2 = 4$, where $x \neq 0$, is

(a) $y = 4\sqrt{5}x + 20$ (b) $y = 4\sqrt{3}x - 12$ (c) $y = 0$ (d) $y = 4\sqrt{5}x - 20$

18 Let the equation of a circle and a parabola be $x^2 + y^2 - 4x - 6 = 0$ and $y^2 = 9x$ respectively. Then,

(a) $(1, -3)$ is a point on the common chord

(b) the equation of the common chord is $y + 1 = 0$

(c) the length of common chord is 6 units

(d) the equation of common chord is $x - 1 = 0$

19 Let 'P' be point on the parabola $y^2 = 4ax$ which is at the shortest distance from the centre S of the circle $x^2 + y^2 - 4x - 16y + 64 = 0$. Let Q be the point on the circle dividing the line segment SP internally, then **[JEE Advanced 2016]**

(a) $SP = 2\sqrt{5}$

(b) $SQ : QP = (\sqrt{5} + 1) : 2$

(c) the x-intercept of the normal to the parabola at P is 6

(d) the slope of the tangent to the circle at Q is $\frac{1}{2}$

20 Variable circle is described to pass through point (1, 0) and tangent to the curve $y = \tan(\tan^{-1} x)$. The locus of the centre of the circle is a parabola whose

(a) length of the latusrectum is $2\sqrt{2}$ units

(b) axis of symmetry has the equation $x + y = 1$

(c) vertex has the coordinates $\left(\frac{3}{4}, \frac{1}{4}\right)$

(d) None of the above

21 Through a point $P(-2, 0)$ tangents PQ and PR are drawn to the parabola $y^2 = 8x$. Two circles each passing through the focus of the parabola and one touching at Q and the other at R are drawn. Which of the following point(s) w.r.t. the ΔPQR lie(s) on the common chord of the two circles?

(a) Centroid (b) Orthocentre

(c) Incentre (d) Circumcentre

22 If from the vertex of a parabola $y^2 = 4ax$ a pair of chords be drawn at right angles to one another and with these chords as adjacent sides a rectangle be made, then the locus of the further angle of the triangle is
(a) an equal parabola
(b) a parabola with focus at $(9a, 0)$
(c) a parabola with directrix as $x - 7a = 0$
(d) not a parabola

Linked Comprehension Type

Paragraph I Directions (Q.Nos. 23-24)

Let a, r, s, t be non-zero real numbers. Let $P(at^2, 2at)$, $Q\left(\frac{a}{t^2}, -\frac{2a}{t}\right)$, $R(ar^2, 2ar)$ and $S(as^2, 2as)$ be distinct points on parabola $y^2 = 4ax$. Suppose that PQ is the focal chord and lines QR and PK are parallel, where K is point $(2a, 0)$

23 The value of r is
(a) $-\frac{1}{t}$ (b) $\frac{t^2+1}{t}$ (c) $\frac{1}{t}$ (d) $\frac{t^2-1}{t}$

24 If $st = 1$, then the tangent at P and the normal at S to the parabola meet at a point whose ordinate is
(a) $\frac{(t^2+1)^2}{2t^3}$ (b) $\frac{a(t^2+1)^2}{2t^3}$ (c) $\frac{a(t^2+1)^3}{2t^3}$ (d) $\frac{a(t^2+2)^2}{t^3}$

Paragraph II Direction (Q. Nos. 25-27)

$P(2t^2, 4t)$ is a point on the parabola $y^2 = 8x$ and $Q(h, k)$ is a point on the tangent at P and external to the circle $x^2 + y^2 = 8$. Based on the information, answer the following questions.

25 As Q moves on the tangent at P, the locus of the point of intersection of the chord of contact of Q w.r.t. the circle at the tangent at P is
(a) $y^2 - x^2 = 4$ (b) $y^2 = 2x$ (c) $y^2 - 2x^2 = 4$ (d) $y^2 = -4x$

26 The point in second quadrant from which perpendicular tangents can be drawn to both the parabola and the circle is
(a) $(-2, 2\sqrt{3})$ (b) $(-1, \sqrt{2})$ (c) $(-\sqrt{2}, \sqrt{2})$ (d) $(-3, 2\sqrt{3})$

27 If AB is the chord of contact of $Q(h, k)$ w.r.t. the circle $x^2 + y^2 = 8$, then the circumcentre of ΔAQB lies on the curve (when $t = 2$)
(a) $x + 2y = 4$ (b) $x - 2y = 4$ (c) $2y = x + 4$ (d) $x + 2y + 4 = 0$

Paragraph III Direction (Q. Nos. 28-32)

If P is a point moving on a parabola $y^2 = 4ax$ and Q is moving point on the circle $x^2 + y^2 - 24ay + 128a^2 = 0$. The points P and Q will be closest when they lie along the normal to the parabola $y^2 = 4ax$ passing through the centre of the circle

28 If the normal at $(at^2, 2at)$ of the parabola passes through the centre of the circle, the value of t must be
(a) 1 (b) 2 (c) 3 (d) 4

29 The shortest distance between P and Q must be
(a) $a(\sqrt{2} - 1)$ (b) $2a(\sqrt{5} - 1)$ (c) $4a(\sqrt{5} - 1)$ (d) $4a(\sqrt{5} + 1)$

30 If P and Q are closest, then point P must be

(a) $(1, 2a)$ (b) $(2a, 2\sqrt{2}a)$

(c) $(4a, 4a)$ (d) $(5a, 4a)$

31 When P and Q are closest then point Q must be

(a) $\left(\frac{4a}{\sqrt{3}}, 6a-\frac{8a}{\sqrt{5}}\right)$ (b) $\left(\frac{4a}{\sqrt{5}}, 12a-\frac{8a}{\sqrt{5}}\right)$

(c) $\left(\frac{4a}{\sqrt{5}}, -12a-\frac{8a}{\sqrt{5}}\right)$ (d) $\left(\frac{4a}{\sqrt{5}}, -8a-\frac{12a}{\sqrt{5}}\right)$

32 When P and Q are closest and the diameter QR of the circle is drawn through Q, the x-coordinate of R is

(a) $-\frac{2a}{\sqrt{5}}$ (b) $-\frac{8a}{\sqrt{5}}$

(c) $-\frac{4a}{\sqrt{5}}$ (d) $-\frac{6a}{\sqrt{5}}$

Paragraph IV Direction (Q. Nos. 33-37)

If a source of light is placed at the fixed point of a parabola and if the parabola is reflecting surface, the ray will bounce back in a line parallel to the axis of the parabola.

33 A ray of light is coming along the line $y=2$ from the positive direction of X-axis and strikes a concave mirror whose intersection with the XY-plane is parabola $y^2=8x$, the equation of the reflected ray is

(a) $2x+3y=4$ (b) $3x+2y=6$

(c) $4x+3y=8$ (d) $5x+4y=10$

34 A ray of light moving parallel to the X-axis gets reflected from a parabolic mirror whose equation is $y^2+10y-4x+17=0$. After reflection, the ray must pass through the point

(a) $(-2, -5)$ (b) $(-1, -5)$

(c) $(-3, -5)$ (d) $(-4, -5)$

35 A ray of light is coming along the line $x=2$ from the positive direction of Y-axis and strikes a concave mirror whose intersection with XY-plane is a parabola $x^2=4y$, the equation of the reflected ray after second reflection is

(a) $2x+y=1$ (b) $3x-2y+2=0$

(c) $y=1$ (d) None of these

36 Two ray of light coming along the line $y=1$ and $y=-2$ from the positive direction of X-axis and strikes a concave mirror whose intersection with the XY-plane is a parabola $y^2=x$ at A and B respectively. The reflected ray pass through a fixed point C, then area of ΔABC is

(a) $\frac{21}{8}$ sq units (b) $\frac{19}{2}$ sq units

(c) $\frac{17}{2}$ sq units (d) $\frac{15}{2}$ sq units

37 A ray of light is coming along the line $x=\lambda(\lambda<0)$ from the positive direction of Y-axis and strikes a concave mirror $x^2=4(y-1)$ the ray reflected from it is $x+y=2$, if this reflected ray again strikes, then the equation of new reflected ray is

(a) $y+1=2\sqrt{2}$ (b) $x+1=2\sqrt{2}$

(c) $y+2=2\sqrt{2}$ (d) $x+2=2\sqrt{2}$

Paragraph V Direction (Q. Nos. 38-42)

Let S be the focus be considered as pole (origin) and take SA as the initial line a vertical line ZM be considered as directrix. Let P be any point on the conic and let its coordinates be (r, θ), so that $SP = r$ and $\angle ZSP = \theta$ and further let $SL = l$ (semi latusrectum).

On the basis of given information answer the following question

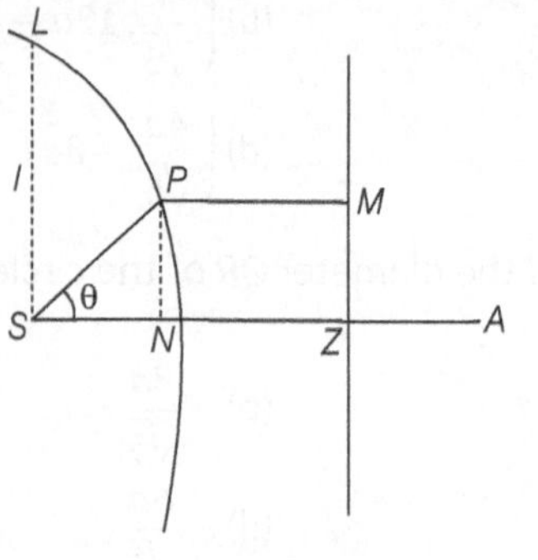

38 The polar equation of conic is

(a) $r = l(1 + e\cos\theta)$ (b) $r = l(l - e\cos\theta)$ (c) $r = \frac{l}{1 + e\cos\theta}$ (d) $r = \frac{l}{1 - e\cos\theta}$

39 If e is the eccentricity of the conic, then SZ is equal to

(a) $\frac{l}{e^2}$ (b) $\frac{l}{e}$ (c) le^2 (d) le

40 The equation of the directrix of the conic corresponding to the focus which has been chosen as the pole is

(a) $\frac{l}{r} = e\cos\theta$ (b) $\frac{r}{l} = e\cos\theta$ (c) $\frac{l}{r} = \frac{\cos\theta}{e}$ (d) $\frac{r}{e} = \frac{e}{\cos\theta}$

41 The polar equation of parabola must be

(a) $r = \frac{l}{2}\sin^2\frac{\theta}{2}$ (b) $r = \frac{l}{2}\operatorname{cosec}^2\frac{\theta}{2}$ (c) $r = \frac{l}{2}\cos^2\frac{\theta}{2}$ (d) $r = \frac{l}{2}\sec^2\frac{\theta}{2}$

42 If the axis SZ of the conic makes an angle α with the initial SA, then the equation of conic will be

(a) $r\{l + e\cos(\theta + a)\}$ (b) $r[l(1 - e\cos(\theta - a)]$

(c) $r = \frac{l}{1 + e\cos(\theta + a)}$ (d) $r = \frac{l}{1 + e\cos(\theta - a)}$

Matrix Match Type Questions

43 Match the following

	Column I		Column II
(i)	If the line $x - 1 = 0$ is the directrix of the parabola $y^2 - kx + 8 = 0$, then the value of k is	A.	2
(ii)	If l is the length of an equilateral triangle inscribed in the parabola $y^2 = 4x$ with one vertex at the origin then, $\frac{l}{2\sqrt{3}}$ is equal to	B.	4
(iii)	The latusrectum of a parabola having $(3, 5)$ and $(3, -3)$ as extremities of the latusrectum is	C.	8
(iv)	If $(2, 0)$ is the vertex and Y-axis as the directrix, then its focus $(a, 0)$ when a equals	D.	-4
		E.	-8

44 Match the following

	Column I		Column II
(i)	If PQ is any focal chord of the parabola $y^2 = 32x$ and length of PQ can never be less than λ units, then λ is divisible by	A.	2
(ii)	A tangent is drawn to the parabola $y^2 = 4x$ at the point 'P' whose abscissa lies in the interval [1, 4]. If maximum possible area of triangle formed by the tangent at P, ordinate of the point 'P' and the X-axis is λ units, then λ is divisible by	B.	3
(iii)	The normal at the ends of the latusrectum of the parabola $y^2 = 4x$ meet the parabola again at A and A', if λ is divisible by	C.	4
		D.	6
		E.	8

45 Match the column

	Column I		Column II
(i)	The point from which perpendicular tangents can be drawn to the parabola $y^2 = 8x$ is	A.	$(-2, 1)$
(ii)	The line $x + y = -3$ touches the parabola $y^2 = 12x$ at the point	B.	$(4, 6)$
(iii)	$4x + 3y - 34 = 0$ is normal to the parabola $y^2 = 9x$ at the point	C.	$(3, -6)$
(iv)	The line parallel to $4y - x + 3 = 0$ touches the parabola $y^2 = 7x$ at the point	D.	$(28, 14)$
		E.	$(-2, 5)$

46 Match the column

	Column I		Column II
(i)	The x-coordinate of points on the axis of the parabola $y^2 - 4x - 2y + 5 = 0$ from which all the three normals to the parabola are real is	A.	4
(ii)	The x-coordinate of points on the axis of the parabola $4y^2 - 32x + 4y + 65 = 0$ from which all the three normal to the parabola are real is	B.	5
(iii)	The x-coordinate of points on the axis of the parabola $4y^2 - 16x - 4y + 41 = 0$ from which all the three normals to the parabolas are real is	C.	6
		D.	7
		E.	8

47 Match the column.

	Column I		Column II
(i)	If the parabola $x^2 = ay$ make an intercept of length $\sqrt{40}$ units on the line $y - 2x = 1$, then the values of a are	A.	-2
(ii)	If the tangents drawn from the point (0, 2) to the parabola $y^2 = 4ax$ are inclined at an angle $\frac{3\pi}{4}$, then the values of a are	B.	1
(iii)	If two distinct chords of a parabola $y^2 = 4ax$ passing through $(a, 2a)$ are bisected on the line $x + y = 1$, then the length of L-R is	C.	2
(iv)	If the focus of the parabola $x^2 - ay + 3 = 0$ is (0, 2) and if two values of a are a_1, a_2 such that $a_1 > a_2$, then the value of $\frac{a_1}{a_2}$ is	D.	3

48 Match column I

	Column I		Column II
(i)	The equation of director circle to the parabola $y^2+4x+4y=0$ is $ax+b=0$, then (a, b) is	A.	(1, 2)
(ii)	Tangents are drawn from the point (2, 3) to the parabola $y^2=4x$, then the points of contact are	B.	(4, 4)
(iii)	From a point P on the circle $x^2+y^2=5$, the equation of chord of contact to the parabola $y^2=4x$ is $y=2(x-2)$. Then, the coordinates of point P will be	C.	$(-2, 1)$
(iv)	$P(4,-4)$ and Q are points on the parabola $y^2=4x$ such that area of ΔPOQ is 6 sq units where O is vertex, then the coordinates of Q may be	D.	$(9,-6)$
		E.	$(1,-2)$

Numerical Value Type Questions

Question Stem Direction (Q. Nos. 49-51) The answer to each question is a **NUMERICAL VALUE.**

Consider the circle $x^2+y^2=9$ and the parabola $y^2=8x$. They intersect at P and Q in the first and the fourth quadrants respectively. The tangents to the circle at P and Q intersect X-axis at R and tangents to the parabola at P and Q intersect X-axis at S. Based on the above information answer the following questions.

49 The ratio of the area of ΔPQS and ΔPQR is ....

50 The radius of the circumcircle of ΔPRS is ....

51 The radius of in circle of ΔPQR is ....

52 Tangent is drawn at any point (x_1, y_1) on the parabola $y^2=4ax$. Now, tangents are drawn from any point on this tangent to the circle $x^2+y^2=a^2$ such that all the chord of contact pass through a fixed point (x_2, y_2). A relation $\lambda x_1 y_2^2+\mu x_2 y_1^2=0$ hold between the points, then the value $\dfrac{\lambda+\mu}{50}$ is

53 A parabola is drawn to pass through A and B, then ends of diameter of a given circle of radius a and to have a directrix a tangent to a concentric circle of radius b. The axes being AB and the perpendicular diameter. If the locus of the focus of the parabola is $\dfrac{x^2}{mb^2}+\dfrac{ny^2}{b^2-a^2}=\dfrac{l}{2}$, then find $\sqrt{m+n+l}$.

54 If the normals of the parabola $y^2=4x$ drawn at the end points of its latusrectum are tangents to the circle $(x-3)^2+(y+2)^2=r^2$, then the value of r^2 is **[2015 Advanced]**

ANSWER KEY

Level 1

1. (i) (a) Parabola (b) not

(ii) (a) $\lambda = \pm \frac{1}{\sqrt{2}}$ (b) $\lambda = 0, 12$

(iii) (a) (2, 3) (b) $\left(-\frac{11}{6}, \frac{5}{2}\right)$

2. (a) $y^2 - 12x = 0$ (b) $y^2 + 20x = 0$ (c) $x^2 - 16y = 0$ (d) $x^2 + 28y = 0$

3. (a) $x = t^2$ and $y = 2t$ (b) $x^2 + y^2 - 2xy - 8x + 20y + 46 = 0$ (c) $(y-2)^2 = 8(x-3)$

(d) $y^2 = 8(x+3)$ (e) $y = -\frac{7}{150}x^2 + \frac{5x}{3}$ (f) $5y^2 = 3x + 11$

(g) $x^2 - 4x - 2y + 10 = 0$ (h) $(y-2)^2 = -8(x-1)$

4. (a) Vertex $(-2, 2)$; axis $(y = 2)$; focus $\left(-\frac{5}{4}, 2\right)$; directrix $(4x + 11 = 0)$; latus rectum $(4a = 3)$

(b) Vertex $(2, -6)$; axis $(x + 2)$; focus $\left(2, -\frac{23}{4}\right)$; directrix $(4y + 25 = 0)$; latus rectum $(4a = 1)$;

(c) vertex $\left(\frac{53}{16}, \frac{2}{3}\right)$; axis $(3y - 2 = 0)$; focus $\left(\frac{541}{144}, \frac{2}{3}\right)$; directrix $(144x - 413) = 0$; latus rectum $\left(4a = \frac{16}{9}\right)$

(d) vertex $(-4, 1)$; axis $(x + 4 = 0)$; focus $(-4, -2)$; directrix $(y = 4)$; latus rectum $(4a = 12)$

5. (a) 8 units (b) $8\sqrt{3}$ units (c) 20 sq units (d) $6x - 8 = 9y^2$

6. (a) 8.4 feet (b) 2.236 m

7. (i) (a) interior (b) outside (c) lies on the parabola (d) lies on the parabola

(ii) focus $(a, 2a)$, $a \in (0, 4)$

8. (i) -1 (ii) 8

9. (i) (a) $\lambda > -1$ (b) $\lambda = -1$ (c) $\lambda < -1$

(ii) (a) $3y - 2x - 7 = 0$ (b) $ty = \left(x + \frac{5}{4}\right) + t^2$

(c) $4x - 4y - 9 = 0$ (d) $y = 2x + 3$

(iii) (a) $x + y + 1 = 0$ (b) $y = -x - 4$ (c) $y = x + 2$

(d) $4x - y - 4 = 0$ (e) $x - 2y = -1$

(iv) (a) (2, 4) (b) $(a\tan^2\alpha, -2a\tan\alpha)$ (c) (8, 8)

(v) (a) $\frac{3}{\sqrt{2}}$ (b) $2\sqrt{2}$

10. (a) $x + y - 3 = 0$ (b) $y + tx = 2t + t^2$ (c) $y = 2x - 48$

(d) $\left(\frac{1}{\sqrt{2}}, -\frac{1}{\sqrt{2}}\right)$ (e) $y = -4x + 72$ (f) $x - y - 12 = 0$

11. (a) $k = 9$ (b) $PQ = \frac{80}{9}\sqrt{10}$ (c) $\sqrt{5}$ (d) $\pi/2$

12. (a) $y^2 = a(x - 3a)$ (b) $y^2 = a(x - 3a)$

13. (a) one (b) $a > 2$ (c) $m_2^2 = m_1m_3$ (d) $27y^2 = (x-4)^3$

14. (a) $2x^2 - 3y^2 + 8xy + 16x + 10y + 21 = 0$ (b) $y = 3x - 3$ (c) $3y - 2x - 5 = 0$ (d) $3\sqrt{5}$

15. (a) $3x^2 + 10x + 3 = y^2$ (b) $\frac{125}{2}$ sq units

(c) $y^2 - 8x + 128 = 0$ (d) $y^4 - 2ak^2(h - 2a) + 8a^4 = 0$

16. (i) (a) $4x + 9 = 0$ (b) $y + 2 = 0$ (c) $x = 0$

(ii) (a) $y = x + 1$ (b) $x = 4y + 12$ (c) tangent = $3\sqrt{5}$; normal = $\frac{3}{2}\sqrt{5}$; subtangent = 6; subnormal = $\frac{3}{2}$

(iv) $y = 4$

17. (a) $(x + 2a)y^2 + 4a^3 = 0$ (b) $(y^2 - 4ax)(y^2 + 4a^2) = 4a^2c^2$

(c) $4a^2x^2 = b^2(4a^2 + y^2)$

18. (a) (5, 0)   (b) 15   (c) $x - 7y + 26 = 0$ (d) $\frac{2pq}{\sqrt{p^2 + q^2}}$

Level 2

1. (d) **2.** (a) **3.** (a) **4.** (d) **5.** (c) **6.** (a) **7.** (b)
8. (a) **9.** (c) **10.** (a) **11.** (d) **12.** (c) **13.** (c) **14.** (d)
15. (d) **16.** (d) **17.** (b) **18.** (d) **19.** (a) **20.** (a) **21.** (c)
22. (c) **23.** (d) **24.** (a) **25.** (b) **26.** (b) **27.** (d) **28.** (b)
29. (a) **30.** (b) **31.** (d) **32.** (a) **33.** (a) **34.** (b) **35.** (b)
36. (a) **37.** (b) **38.** (d) **39.** (c) **40.** (c) **41.** (b) **42.** (b)
43. (d) **44.** (d) **45.** (a) **46.** (a) **47.** (c) **48.** (d) **49.** (a)
50. (d) **51.** (b) **52.** (c) **53.** (a) **54.** (a) **55.** (b) **56.** (d)
57. (5) **58.** (4) **59.** (16) **60.** (7.25) **61.** (1) **62.** (1.73) **63.** (0)
64. (40) **65.** (9) **66.** (8) **67.** (2) **68.** (0.01)

Level 3

1. (b) **2.** (a) **3.** (c) **4.** (d) **5.** (b) **6.** (d) **7.** (a)
8. (c) **9.** (c) **10.** (b) **11.** (a) **12.** (b) **13.** (c) **14.** (c, d)
15. (a, b) **16.** (a, c) **17.** (a, b, c) **18.** (a, c, d) **19.** (a, c, d) **20.** (b, c) **21.** (a, b, c, d)
22. (a, b, c) **23.** (d) **24.** (b) **25.** (d) **26.** (a) **27.** (c) **28.** (b)
29. (c) **30.** (c) **31.** (b) **32.** (c) **33.** (c) **34.** (b) **35.** (d)
36. (a) **37.** (d) **38.** (c) **39.** (b) **40.** (a) **41.** (d) **42.** (d)
43. (i)–(B, E), (ii)–B, (iii)–C, (iv)–B
44. (i)–A, C, E (ii)–A, C, E, (iii)–A, B, C, D
45. (i)–A, E (ii)–C, (iii)–B, (iv)–D
46. (i)–A, B, C, D, E (ii)–D, E, (iii)–B, C, D, E
47. (i)–A, B (ii)–A, C, (iii)–B, C, D (iv)–D
48. (i)–E, (ii)–A, B, (iii)–C, (iv)–A, D
49. (28.28) **50.** (5.2) **51.** (2) **52.** (0.1) **53.** (2) **54.** (2)

SOLUTIONS

Level 1

1. (i) (a) Given equation $3x^2+4y-7=0$

Here, $a=3, b=0, h=0, g=0, f=2, c=-7$

$\because \quad \Delta = abc+2fgh-af^2-bg^2-ch^2$

$= 0+0-3(2)^2-0-0=-12 \neq 0$

$\therefore$ It represents a non-degenerate conic.

Now, $h^2=0$,

$ab=0$

$\because \quad h^2=ab$

$\therefore$ It represents a parabola.

(b) Given equation

$2x^2+6y^2-24xy+2x+4y-7=0$

Here, $a=2, b=6, h=-12, g=1, f=2$ and $c=-7$

$\Delta = abc+2fgh-af^2-bg^2-ch^2$

$=2(6)(-7)+2(2)(1)(-12)-2(2)^2-6(1)^2-(-7)(-12)^2$

$=-84-48-8-6+1008$

$=862 \neq 0$

and $h^2=(-12)^2=144$

$ab=12$

$\because \quad h^2 \neq ab$

$\therefore$ It does not represent a parabola.

(ii) (a) Given equation

$x^2+2y^2-4\lambda xy+4x+2y-1=0$

Here, $a=1, b=2, h=-2\lambda, g=2, f=1$ and $c=-1$

$\therefore$ It represent a parabola.

$\therefore \Delta \neq 0$ and $h^2=ab$

$\Delta = abc+2fgh-af^2-bg^2-ch^2$

$=1(2)(-1)+2(1)(2)(-2\lambda)-1(1)^2-2(2)^2-(-1)(-2\lambda)^2$

$=-2-8\lambda-1-8+4\lambda^2=4\lambda^2-8\lambda-11$

$\because \quad h^2=ab$

$(-2\lambda)^2=2\times 1$

$4\lambda^2=2$

$\lambda^2=\frac{2}{4}=\frac{1}{2} \Rightarrow \lambda=\pm\frac{1}{\sqrt{2}}$

Now, $\quad \Delta = 4\left(\pm\frac{1}{\sqrt{2}}\right)^2-8\left(\pm\frac{1}{\sqrt{2}}\right)-11$

$\Rightarrow \quad 2\mp\frac{8}{\sqrt{2}}-11 \neq 0$

$\therefore \quad \lambda=\pm\frac{1}{\sqrt{2}}$

(b) Given equation

$\lambda x^2+3y^2+\lambda xy+2x+6y-8=0$

Here, $a=\lambda, b=3, h=\frac{\lambda}{2}, g=1, f=3$ and $c=-8$

Now, $\Delta = abc+2fgh-af^2-bg^2-ch^2$

$=\lambda(3)(-8)+2(3)(1)\left(\frac{\lambda}{2}\right)-\lambda(3)^2-3(1)^2-(-8)\left(\frac{\lambda}{2}\right)^2$

$=-24\lambda+3\lambda-9\lambda-3+2\lambda^2$

$=2\lambda^2-30\lambda-3$

Since, it represents a parabola.

$\therefore \quad h^2=ab$

$\Rightarrow \quad \left(\frac{\lambda}{2}\right)^2=3\lambda$

$\Rightarrow \quad \lambda^2-12\lambda=0$

$\Rightarrow \quad \lambda(\lambda-12)=0$

$\Rightarrow \quad \lambda=0,12$

When $\lambda=0$, then $\Delta=-3\neq 0$

When $\lambda=12$, then $\Delta=2(12)^2-30(12)-3\neq 0$

$\therefore \quad \lambda=0,12$

(iii) (a) Let $S: 14x^2-4xy+11y^2-44x-58y+71=0$

Now, for centre of conic

$\frac{\partial S}{\partial x}=0$

$\Rightarrow \quad 28x-4y-44=0$

$\Rightarrow \quad 7x-y-11=0 \quad$ ...(i)

$\frac{\partial S}{\partial y}=0$

$\Rightarrow \quad -4x+22y-58=0$

$\Rightarrow \quad -2x+11y-29=0 \quad$ ...(ii)

On solving Eqs. (i) and (ii), we get

$x=2$ and $y=3$

$\therefore$ Centre is (2, 3).

(b) Let $S: 3x^2+6xy+y^2-4x+6y+8=0$

$\frac{\partial S}{\partial x}=0$

$\Rightarrow \quad 6x+6y-4=0$

$\Rightarrow \quad 3x+3y-2=0 \quad$ ...(i)

$\frac{\partial S}{\partial y}=0$

$\Rightarrow \quad 6x+2y+6=0$

$\Rightarrow \quad 3x+y+3=0 \quad$ ...(ii)

On solving Eqs. (i) and (ii), we get

$$x = -\frac{11}{6}, y = \frac{5}{2}$$

$\therefore$ Centre is $\left(\frac{-11}{6}, \frac{5}{2}\right)$.

2. (a) Given, vertex (0, 0) and focus $(a, 0) = (3, 0)$

$\therefore \quad a = 3$

$\because$ Axis is X-axis.

$\therefore$ Equation of parabola is $y^2 = -4ax$

$\Rightarrow \quad y^2 = 4(3)x \Rightarrow y^2 = 12x$

(b) Given, vertex (0, 0) and focus $(-a, 0) = (-5, 0)$

$\therefore \quad a = 5$

Axis is X-axis.

$\therefore$ Equation of parabola is $y^2 = -4ax$

$$y^2 = -4(5)x \Rightarrow y^2 = -20x$$

(c) Given, vertex (0, 0) and focus $(0, a) = (0, 4)$

$\therefore \quad a = 4$

Axis is Y-axis.

$\therefore$ Equation of parabola is $x^2 = 4ay$

$\Rightarrow \quad x^2 = 16y$

(d) Vertex (0, 0) and focus $(0, -a) = (0, -7)$

$\therefore \quad a = 7$

Axis is Y-axis.

$\therefore$ Equation of parabola is $x^2 = -4ay$

$$x^2 = -4(7)y$$

$$x^2 = -28y$$

3. (a) Given, equation is $y^2 = 4x$

$$4a = 4$$

$$a = 1$$

Parametric form of parabola is

$x = at^2, y = 2at$, where t is a parameter.

$\therefore \quad x = t^2$ and $y = 2t$

This is required form.

(b) Let $S(3, -4)$ be the focus and ZN be the directrix, whose equation is $x + y - 2 = 0$.

Let $P(x, y)$ be any point on the parabola.

Join SP and draw $PM \perp ZN$.

Then, $SP = PM$

$\because \quad PS = \sqrt{(x-3)^2 + (y+4)^2} \quad$ ...(i)

PM = Perpendicular distance from P to line ZN.

$$= \left|\frac{x + y - 2}{\sqrt{1^2 + 1^2}}\right| \quad \text{...(ii)}$$

By Eqs. (i) and (ii),

$$\sqrt{(x-3)^2 + (y+4)^2} = \left|\frac{x + y - 2}{\sqrt{2}}\right|$$

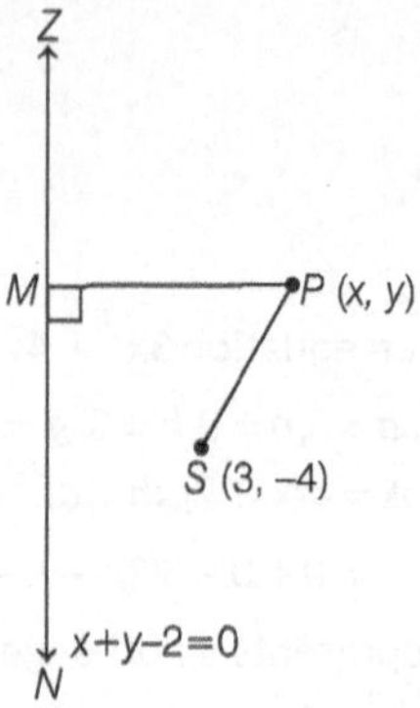

On squaring both sides,

$$x^2 - 6x + 9 + y^2 + 8y + 16 = \frac{x^2 + y^2 + 4 + 2xy - 4y - 4x}{2}$$

$$\Rightarrow 2x^2 - 12x + 2y^2 + 16y + 50 = x^2 + y^2 + 2xy - 4x - 4y + 4$$

$$\Rightarrow x^2 + y^2 - 2xy - 8x + 20y + 46 = 0$$

which is the required euqation of parabola.

(c) The points (3, 2) and (5, 2) lie on the horizontal line $y = 2$.

$\therefore$ The axis of the parabola is the horizontal line $y - 2 = 0$

Distance between the vertex and focus is 2.

$\therefore \quad a = 2$

$\therefore$ Equation of parabola is

$$(y-k)^2 = 4a(x-h)$$

$\Rightarrow \quad (y-2)^2 = 4(2)(x-3)$

$\Rightarrow \quad (y-2)^2 = 8(x-3)$

(d) $\because$ The line passing through the focus and perpendicular to the directrix is X-axis.

$\therefore$ Axis of the required parabola is X-axis.

Let the coordinates of the focus be $S(a, 0)$.

$\because$ The vertex is the mid-point of line joining the focus and the point $(-5, 0)$ where directrix $x + 5 = 0$ meets the axis.

$\therefore \quad -3 = \frac{a-5}{2}$

$\Rightarrow \quad a = -1$

Thus, coordinates of the focus is $(-1, 0)$.

Let $P(x, y)$ be a point on the parabola.

$\therefore$ By definition of parabola

i.e. $\quad PS = PM$

$\Rightarrow \quad \sqrt{(x+1)^2 + y^2} = (x+5)$

$$(x+1)^2 + y^2 = (x+5)^2$$

$\Rightarrow \quad y^2 = (x^2 + 25 + 10x) - (x^2 + 1 + 2x)$

$$y^2 = 8(x+3)$$

(e) $\because$ The axis of the parabola is parallel to Y-axis, the equation of the parabola may be taken as

$y = Ax^2 + Bx + C$ where, A, B, C are constants ...(i)

$\because$ The parabola (i) passes through points (0,0), (10, 12) and (30, 8).

We have,

$$0 = A(0)^2 + B(0) + C \Rightarrow C = 0$$

and $$12 = A(10)^2 + B(10) + C$$

$$\Rightarrow \quad 100A + 10B = 12 \quad ...(ii)$$

and $$8 = A(30)^2 + B(30) + C$$

$$\Rightarrow \quad 900A + 30B = 8 \quad ...(iii)$$

Multiplying Eq. (ii) by 3 and then subtracting from Eq. (iii), we get

$$600A = -28$$

$$A = -\frac{28}{600} = -\frac{7}{150}$$

On substituting $A = -\frac{7}{150}$ in Eq. (ii), we get

$$100 \times \left(-\frac{7}{150}\right) + 10B = 12$$

$$10B = 12 + \frac{14}{3} = \frac{50}{3}$$

$$\Rightarrow \quad B = \frac{5}{3}$$

$\therefore$ The required equation to parabola is

$$y = \frac{-7}{150}x^2 + \frac{5}{3}x$$

(f) $\because$ The axis of parabola is along X-axis.

$\therefore$ Let vertex be $(h, 0)$.

Taking $4a$ as the length of latus rectum, then the equation of parabola is

$$(y - 0)^2 = 4a(x - h) \quad ...(i)$$

$\because$ Eq. (i) passes through (3, 2) and $(-2, -1)$, we have

$$4 = 4a(3 - h) \quad ...(ii)$$

and $$1 = 4a(-2 - h) \quad ...(iii)$$

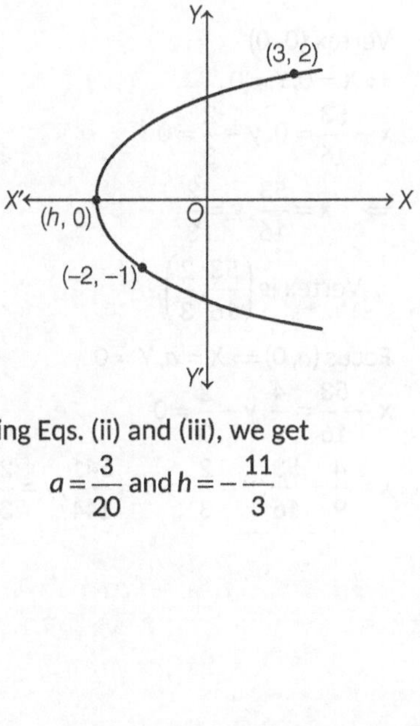

On solving Eqs. (ii) and (iii), we get

$$a = \frac{3}{20} \text{ and } h = -\frac{11}{3}$$

$\therefore$ Required equation is

$$y^2 = \frac{3}{5}\left(x + \frac{11}{3}\right)$$

$$5y^2 = 3x + 11$$

(g) Here, the equation of parabola with vertex (2, 3) and axis parallel to Y-axis is $(x - 2)^2 = 4a(y - 3)$

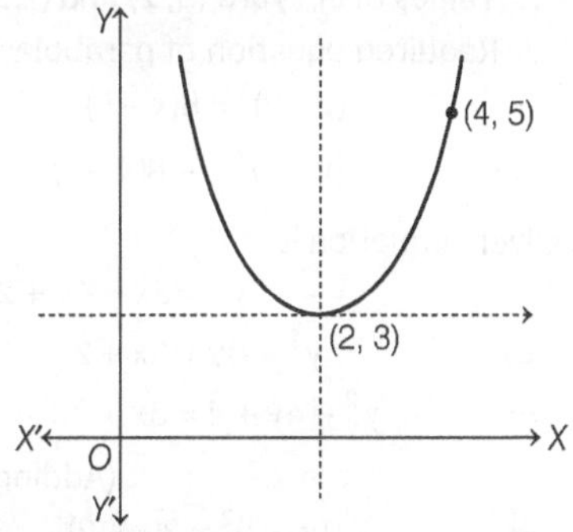

$\because$ (4, 5) lies on the parabola.

$$\therefore \quad (4 - 2)^2 = 4a(5 - 3)$$

$$\Rightarrow \quad a = \frac{1}{2}$$

Hence, the required equation of parabola is

$$(x - 2)^2 = 2(y - 3)$$

$$\Rightarrow \quad x^2 - 4x - 2y + 10 = 0$$

(h) Slope of (3, 6) and $(3, -2)$ is $\frac{-2-6}{3-3} = \infty$

$\because$ Latus rectum is perpendicular to axis.

$\therefore$ Axis parallel to X-axis.

The equation of the two possible parabolas will be of the form

$$(y - k)^2 = \pm 4a(x - h) \quad ...(i)$$

$\because$ Latus rectum $= \sqrt{(3-3)^2 + (6+2)^2} = 8$

$$\therefore \quad 4a = 8 \Rightarrow a = 2$$

From Eq. (i), we have

$$(y - k)^2 = \pm 8(x - h)$$

$\because$ (3, 6) and $(3, -2)$ lie on the parabola, then

$$(6 - k)^2 = \pm 8(3 - h) \quad ...(ii)$$

and $$(-2 - k)^2 = \pm 8(3 - h) \quad ...(iii)$$

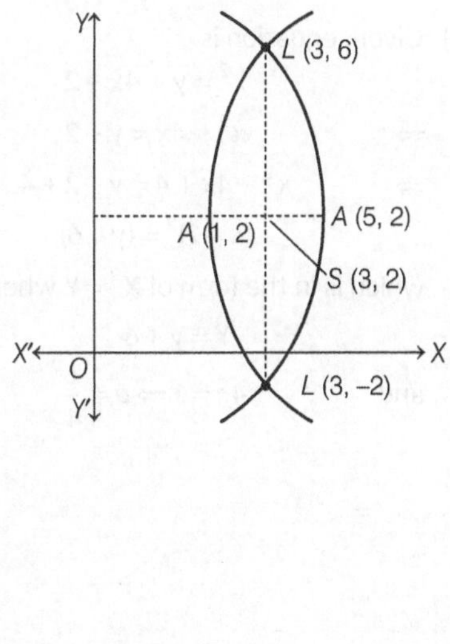

On solving Eqs. (i) and (iii), we get

$$k=2$$

From Eq. (ii),

$$16=\pm 8(3-h)$$
$$h=3\pm 2$$
$$h=5,1$$

$\therefore$ Values of (h, k) are (5, 2) and (1, 2).

$\therefore$ Required equation of parabolas is

$$(y-2)^2=8(x-5)$$

and $$(y-2)^2=-8(x-1)$$

4. (a) Given, equation is

$$y^2=3x+4y+2$$
$$\Rightarrow \quad y^2-4y=3x+2$$
$$\Rightarrow \quad y^2-4y+4=3x+2+4$$

(Adding 4 on both sides)

$$\Rightarrow \quad (y-2)^2=3(x+2)$$
$$\Rightarrow \quad Y^2=3X$$

[where, $Y=y-2,\ X=x+2$]

$$4a=3 \Rightarrow a=\frac{3}{4}$$

Vertex $(0,0) \Rightarrow X=0, Y=0$

$\Rightarrow x+2=0, y-2=0 \Rightarrow x=-2, y=2$

$\therefore$ The vertex is $(-2, 2)$.

Focus $(a, 0)$

$\Rightarrow X=a, Y=0$

$x+2=\frac{3}{4}, y-2=0 \Rightarrow x=-\frac{5}{4}, y=2$

$\therefore$ Focus is $\left(\frac{-5}{4}, 2\right)$

Latus rectum $\Rightarrow 4a=3$

Directrix : $X+a=0$

$$\Rightarrow \quad x+2+\frac{3}{4}=0$$
$$\Rightarrow \quad x=\frac{-11}{4}$$
$$\Rightarrow \quad 4x+11=0$$

Axis, $$Y=0$$
$$y-2=0$$
$$y=2$$

(b) Given, equation is

$$x^2=y+4x+2$$
$$\Rightarrow \quad x^2-4x=y+2$$
$$\Rightarrow \quad x^2-4x+4=y+2+4$$
$$(x-2)^2=(y+6)$$

which is in the form of $X^2=Y$ where, $X=x-2$

$$Y=y+6$$

and $$4a=1 \Rightarrow a=\frac{1}{4}$$

Vertex (0, 0)

$$X=0 \text{ and } Y=0$$
$$x-2=0 \text{ and } y+6=0$$
$$x=2 \text{ and } y=-6$$

$\therefore$ Vertex is $(2, -6)$.

Focus $(0, a)$

$$X=0 \text{ and } Y=a$$
$$x-2=0 \text{ and } y+6=\frac{1}{4}$$
$$x=2 \text{ and } y=-\frac{23}{4}$$

$\therefore$ Focus is $\left(2, -\frac{23}{4}\right)$.

Latus rectum, $4a=1$

Directrix, $$Y+a=0$$
$$y+6+\frac{1}{4}=0$$
$$\Rightarrow \quad 4y+25=0$$

Axis, $$X=0$$
$$x-2=0$$
$$x=2$$

(c) Given, equation is

$$9y^2-16x-12y+57=0$$
$$\Rightarrow \quad 9y^2-12y=16x-57$$
$$\Rightarrow \quad y^2-\frac{12}{9}y=\frac{1}{9}(16x-57)$$
$$\Rightarrow \quad y^2-\frac{4}{3}y+\frac{4}{9}=\frac{1}{9}(16x-57)+\frac{4}{9}$$
$$\Rightarrow \quad \left(y-\frac{2}{3}\right)^2=\frac{1}{9}(16x-53)$$
$$\Rightarrow \quad \left(y-\frac{2}{3}\right)^2=\frac{16}{9}\left(x-\frac{53}{16}\right)$$

$\Rightarrow Y^2=\frac{16}{9}X$ [where, $X=x-\frac{53}{16}, Y=y-\frac{2}{3}$ and $4a=\frac{16}{9} \Rightarrow a=\frac{4}{9}$]

Vertex (0, 0)

$\Rightarrow X=0, Y=0$

$x-\frac{53}{16}=0,\ y-\frac{2}{3}=0$

$\Rightarrow \quad x=\frac{53}{16}, y=\frac{2}{3}$

$\therefore$ Vertex is $\left(\frac{53}{16}, \frac{2}{3}\right)$.

Focus $(a, 0) \Rightarrow X=a, Y=0$

$x-\frac{53}{16}=\frac{4}{9}, y-\frac{2}{3}=0$

$x=\frac{4}{9}+\frac{53}{16}, y=\frac{2}{3} \Rightarrow x=\frac{541}{144}, y=\frac{2}{3}$

$\therefore$ Focus is $\left(\frac{541}{144}, \frac{2}{3}\right)$.

Latus rectum, $4a = \frac{16}{9}$.

Directrix, $\quad X + a = 0$

$$x - \frac{53}{16} + \frac{4}{9} = 0$$

$\Rightarrow \quad 144x - 477 + 64 = 0$

$\Rightarrow \quad 144x - 413 = 0$

Axis $Y = 0 \Rightarrow \quad y - \frac{2}{3} = 0$

$\Rightarrow \quad 3y - 2 = 0$

(d) Given, equation is

$$x^2 + 8x + 12y + 4 = 0$$

$$x^2 + 8x = -12y - 4$$

$$x^2 + 8x + 16 = -12y - 4 + 16$$

$$(x+4)^2 = -12(y-1)$$

$$X^2 = -4aY$$

where, $X = x + 4, Y = y - 1$ and $4a = +12 \Rightarrow a = 3$

Vertex (0, 0)

$\therefore X = 0, Y = 0$

$x + 4 = 0, y - 1 = 0$

$x = -4, y = 1$

$\therefore$ Vertex is $(-4, 1)$.

Focus $(0, -a) \Rightarrow X = 0, Y = -a$

$x + 4 = 0, y - 1 = -3$

$x = -4, y = -2$

$\therefore$ Focus is $(-4, -2)$.

Latus rectum $= 4a = 12$

Directrix $\Rightarrow Y = a$

$\Rightarrow \quad y - 1 = 3$

$\quad y = 4$

Axis, $\quad X = 0$

$\quad x + 4 = 0$

5. (a) Equation of parabola $\Rightarrow y^2 = 8x$

where, $4a = 8 \Rightarrow a = 2$

$\because$ We know that for parabola

$$SP = PM$$

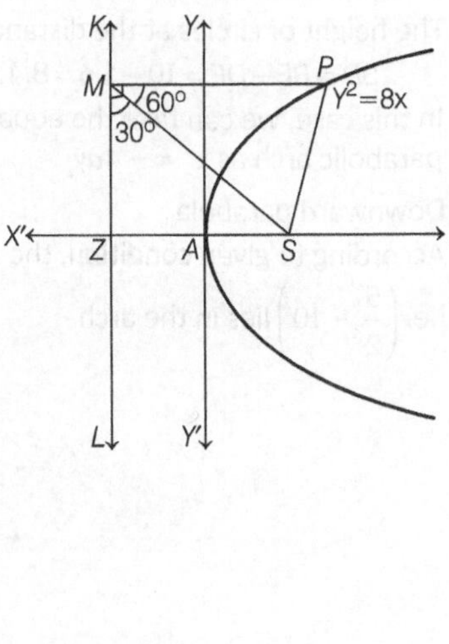

$\because$ SPM is equilateral triangle.

$\therefore \quad SP = PM = SM$

$\Rightarrow \quad \angle PMS = 60°$

$\Rightarrow \quad \angle SMZ = 90° - 60° = 30°$

In ΔSMZ,

$$\sin 30° = \frac{SZ}{SM}$$

$\Rightarrow \quad \frac{1}{2} = \frac{2a}{SM}$

$\Rightarrow \quad SM = 4a$

$\Rightarrow \quad SM = 8$

$\therefore \quad SM = SP = 8$ units

(b) If ΔOAB is equilateral, then $OA = OB = AB = P$.

Thus, AB will be a double ordinate of the parabola.

Thus, $\angle AOM = \angle MOB = \frac{\pi}{6}$

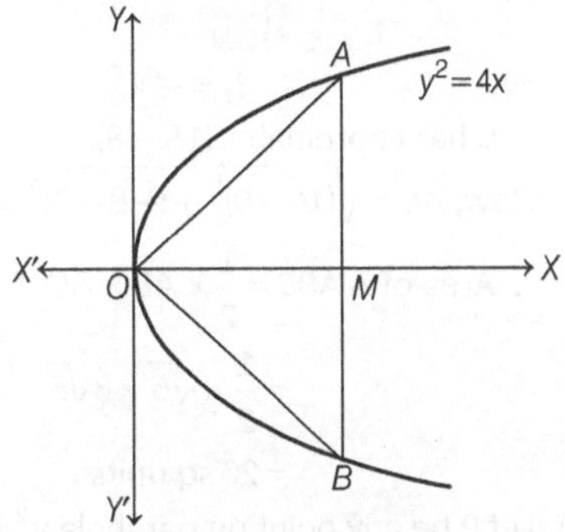

$\Rightarrow OM = P \cos\frac{\pi}{6}$ and $AM = P \sin\frac{\pi}{6}$

Then, A has coordinates $\left(\frac{\sqrt{3}}{2}P, \frac{P}{2}\right)$.

A lies on the parabola $y^2 = 4x$

$$\frac{P^2}{4} = 4\left(\frac{\sqrt{3}}{2}P\right)$$

$$P = 8\sqrt{3}$$

$\therefore$ Length of sides of ΔABO is $8\sqrt{3}$ units.

(c) Given, $y^2 = 4x$ and $\angle BAC = 90°$

$\because \quad AB = \sqrt{5}$

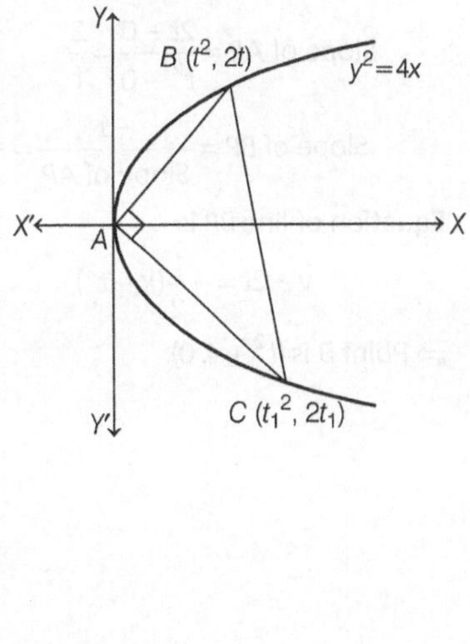

Let us assume B be $(t^2, 2t)$ and C be $(t_1^2, 2t_1)$ in parametric form.

$\because \quad AB = \sqrt{5}$

$$\sqrt{(t^2-0)^2+(2t-0)^2} = \sqrt{5}$$

$$t^4 + 4t^2 - 5 = 0$$

$$(t^2-1)(t^2+5) = 0$$

$t^2 - 1 = 0, \quad t^2 = -5$ [Not possible for real value]

$\therefore \quad t^2 = 1$

$$t = \pm 1$$

Point $B\,(1, 2)$.

Also $\quad m_{AB} \cdot m_{AC} = -1$

$$\Rightarrow \frac{2t-0}{t^2-0} \times \frac{2t_1 - 0}{t_1^2 - 0} = -1$$

$$\Rightarrow \frac{2}{t} \cdot \frac{2}{t_1} = -1$$

$$\Rightarrow t_1 = -4$$

$\therefore$ C has coordinates $(16, -8)$.

Now, $AC = \sqrt{(16-0)^2 + (-8-0)^2} = \sqrt{320} = 8\sqrt{5}$

$$\therefore \text{ Area of } \Delta ABC = \frac{1}{2} \times AB \times AC$$

$$= \frac{1}{2} \times \sqrt{5} \times 8\sqrt{5}$$

$$= 20 \text{ sq units}$$

(d) Let P be any point on parabola $y^2 = 4x$, where $\angle APB = 90°$

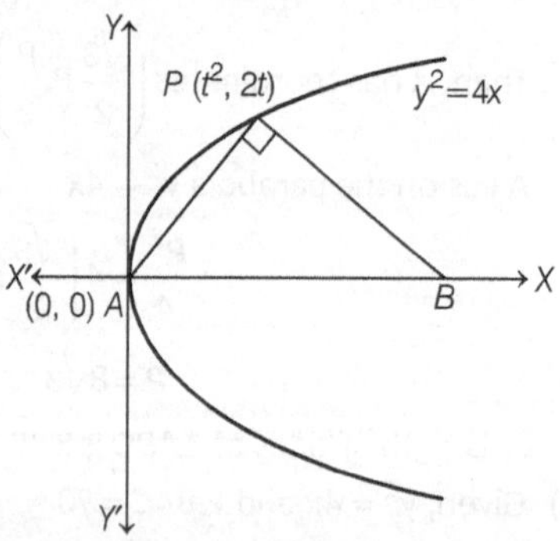

$\Rightarrow$ Let $P(at^2, 2at) = P(t^2, 2t)$

$$\text{Slope of } AP = \frac{2t-0}{t^2-0} = \frac{2}{t}$$

$$\text{Slope of } BP = \frac{-1}{\text{Slope of } AP} = -\frac{t}{2}$$

Equation of line BP is

$$y - 2t = -\frac{t}{2}(x - t^2)$$

$\Rightarrow$ Point B is $(t^2 + 4, 0)$.

Now, let centroid of ΔPAB is (h, k).

$$\therefore \quad h = \frac{t^2 + t^2 + 4}{3}, k = \frac{2t}{3}$$

$$3h - 4 = 2t^2 \text{ and } \frac{3k}{2} = t$$

$$\therefore \quad 3h - 4 = 2\left(\frac{3k}{2}\right)^2.$$

$\therefore$ Required locus is

$6x - 8 = 9y^2$ which is a parabola.

6. (a) Let AX and AY be taken as the axis of X and Y respectively.

Equation of parabola $x^2 = -4ay$ ...(i)

Given, $\quad CA = 10$ feet

$FG = 100$ feet

$CF = CG = 50$ feet

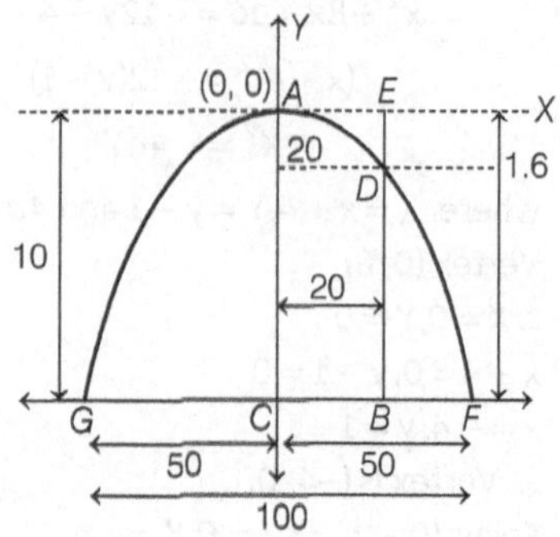

$\therefore$ Coordinates of F is $(50, -10)$.

$\because$ The parabola passes through $(50, -10)$.

$$\therefore \quad (50)^2 = -4a(-10)$$

$$\Rightarrow \quad 4a = \frac{2500}{10} \Rightarrow 250$$

By Eq. (i),

$$x^2 = -250y \quad \text{...(ii)}$$

when, $\quad x = 20$

Then, by Eq. (ii)

$$(20)^2 = -250y$$

$$y = -\frac{400}{250} = -1.6$$

$\therefore$ Coordinates of D be $(20, -1.6)$.

The height of circles at the distance 20 feet

$$BD = BE - DE = 10 - 1.6 = 8.4 \text{ feet}.$$

(b) In this case, we can take the equation of the parabolic arch as $x^2 = -4ay$ ...(i)

Downward parabola

According to given condition, the point $(2.5, -10)$, i.e. $\left(\frac{5}{2}, -10\right)$ lies in the arch

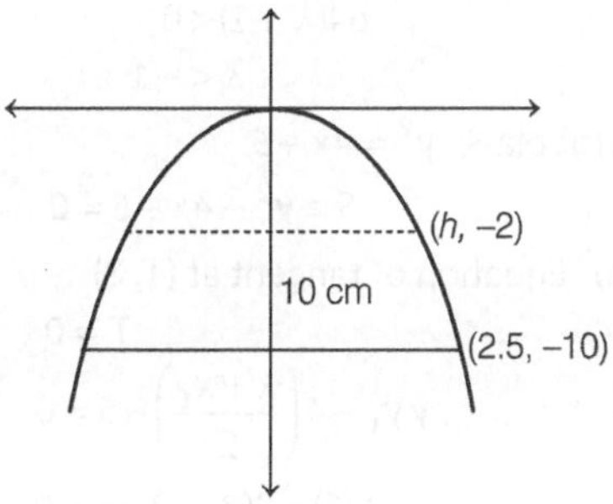

$\therefore \qquad \left(\frac{5}{2}\right)^2 = -4a(-10)$

$a = \frac{25}{4 \times 40} = \frac{5}{32}$

$\therefore$ Eq. (i) becomes

$x^2 = -4\left(\frac{5}{32}\right)y$

$\Rightarrow \qquad x^2 = -\frac{5}{8}y \qquad \text{...(ii)}$

Let $2h$ be the width of the arch 2 m below the vertex, then the point $(h, -2)$ lies on Eq. (ii).

$\Rightarrow \qquad h^2 = -\frac{5}{8}(-2)$

$\Rightarrow \qquad h^2 = \frac{5}{4}$

$\Rightarrow \qquad 2h = \sqrt{5}$

$\therefore$ The required width of the arch (2 m between the vertex)

$= 2h$

$= \sqrt{5}$ m

$= 2.236$ m

7. (i) Given, parabola, $S \equiv y^2 - 8x = 0$

(a) For point $A(2, 1)$,

$S_1 = (1)^2 - 8(2)$

$= 1 - 16 = -15 < 0$

$\therefore$ Point (2, 1) lies in the interior region.

(b) For point $B(-1, 3)$

$S_1 = (3)^2 - 8(-1)$

$= 9 + 8 = 17 > 0$

$\therefore$ Point $(-1, 3)$ lies in the outside region.

(c) For point $C(2, -4)$

$S_1 = (-4)^2 - 8(2)$

$= 16 - 16 = 0$

$\therefore$ Point $(2, -4)$ lies on the parabola.

(d) For point $D(0, 0)$

$S_1 = 0 - 0 = 0$

$\therefore$ Point (0, 0) lies on the parabola.

(ii) Given, equation

$S \equiv y^2 - 16x = 0$

$\because (a, 2a)$ lies in the interior region.

$\therefore \qquad S_1 < 0$

$(2a)^2 - 16a < 0$

$4a^2 - 16a < 0$

$a^2 - 4a < 0$

$a(a-4) < 0$

$\therefore \qquad a \in (0, 4)$

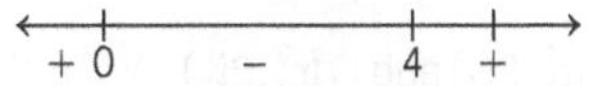

for $0 < a < 4$ point $(a, 2a)$ lies in the interior region.

8. (i) Let $y^2 = 4ax$ be a parabola. If PQ is a focal chord.

Consider any two points on the parabola $(at_1^2, 2at_1)$ and $(at_2^2, 2at_2)$.

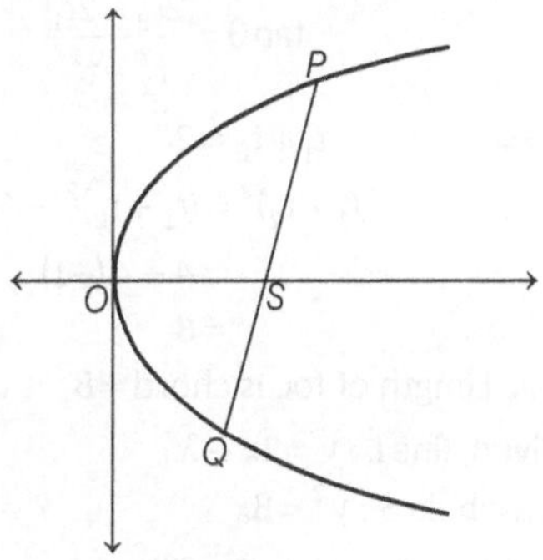

$\because PQ$ passes through the focus $S(a, 0)$.

So, P, S and a are collinear.

$\Rightarrow \qquad m_{PS} = m_{QS}$

$\frac{2at_1 - 0}{at_1^2 - a} = \frac{0 - 2at_2}{a - at_2^2}$

$\Rightarrow \qquad \frac{2at_1}{at_1^2 - a} = \frac{2at_2}{at_2^2 - a}$

$\Rightarrow \qquad \frac{t_1}{t_1^2 - 1} = \frac{t_2}{t_2^2 - 1}$

$\Rightarrow \qquad t_1(t_2^2 - 1) = t_2(t_1^2 - 1)$

$\Rightarrow t_1t_2^2 - t_1 - t_2t_1^2 + t_2 = 0$

$\Rightarrow t_1t_2(t_2 - t_1) + 1(t_2 - t_1) = 0$

$\Rightarrow \qquad (t_2 - t_1)(t_1t_2 + 1) = 0$

$\because \qquad t_2 - t_1 \neq 0$

$\therefore \qquad t_1t_2 = -1$

(ii) Given, $y^2 = 4x$

$4a = 4$

$a = 1$

Let us assume the end points of focal chord are $P(at_1^2, 2at_1)$ and $Q(at_2^2, 2at_2)$

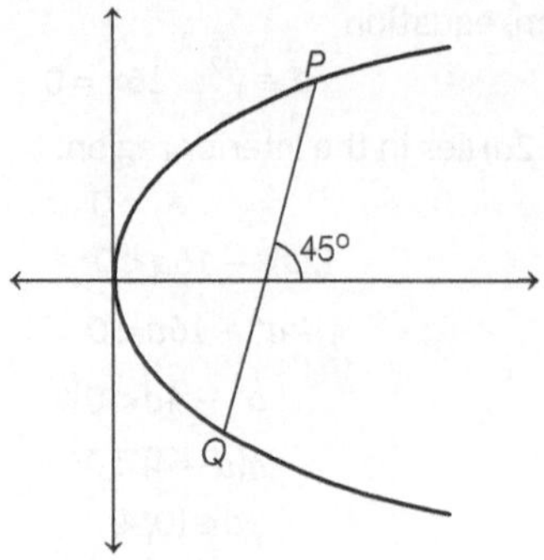

$\because P \Rightarrow (t_1^2, 2t_1)$ and $Q\,(t_2^2, 2t_2)$

$\because$ Length of focal chord for P and Q is

$$= a(t_1 - t_2)^2$$
$$= (t_1 - t_2)^2$$

Now, PQ makes an angle 45° with X-axis.

$$\tan\theta = \frac{2t_2 - 2t_1}{t_2^2 - t_1^2}$$

$$\Rightarrow \quad t_1 + t_2 = 2$$

$$\Rightarrow \quad (t_1 - t_2)^2 = (t_1 + t_2)^2 - 4t_1t_2$$
$$= 4 - 4(-1) \qquad \{\because t_1t_2 = -1\}$$
$$= 8$$

$\therefore$ Length of focus chord $= 8$

9. (i) Given, line $L : y = 2x - \lambda$ ...(i)

Parabola $S : y^2 = 8x$ ...(ii)

Put $y = 2x - \lambda$ in Eq. (ii),

$$(2x - \lambda)^2 = 8x$$
$$4x^2 - 4x\lambda + \lambda^2 - 8x = 0$$
$$4x^2 - 4x(\lambda + 2) + \lambda^2 = 0 \qquad \text{...(iii)}$$

$a = 4, b = -4(\lambda + 2), c = \lambda^2$

$$D = b^2 - 4ac$$
$$= [-4(\lambda + 2)]^2 - 4(4)(\lambda^2)$$
$$= 16[(\lambda + 2)^2 - \lambda^2]$$
$$= 16[\lambda^2 + 4\lambda + 4 - \lambda^2]$$
$$= 16(4\lambda + 4)$$
$$= 64(\lambda + 1)$$

(a) Now, for intersecting lines Eq. (iii) has two real and distinct roots.

$$\therefore \quad D > 0$$
$$64(\lambda + 1) > 0$$
$$\lambda > -1$$

(b) For tangent

$$D = 0$$
$$64(\lambda + 1) = 0$$
$$\lambda = -1$$

(c) For neither intersecting nor touching

$$D < 0$$
$$64(\lambda + 1) < 0$$
$$\lambda < -1$$

(ii) Parabola $S : y^2 = 4x + 5$

$$S \equiv y^2 - 4x - 5 = 0$$

(a) Equation of tangent at (1, 3)

$$T = 0$$
$$yy_1 - 4\left(\frac{x + x_1}{2}\right) - 5 = 0$$
$$y(3) - 2(x + 1) - 5 = 0$$
$$3y - 2x - 2 - 5 = 0$$
$$3y - 2x - 7 = 0$$

(b) Parabola : $y^2 = 4x + 5$

$$y^2 = 4\left(x + \frac{5}{4}\right)$$

$Y^2 = 4AX$ where, $Y = y, A = 1, X = x + \dfrac{5}{4}$.

$$P(t) = \left(-\frac{5}{4} + At^2, 2At\right) \Rightarrow \left(-\frac{5}{4} + t^2, 2t\right)$$

$\Rightarrow$ Equation of tangent is $ty = \left(x + \dfrac{5}{4}\right) + t^2$

Required equation in parametric form.

(c) Let us assume the equation of tangent is $y = mx + c$. According to question, slope is 1.

$$\therefore \quad y = x + c$$

Put $y = x + c$ in parabola $y^2 - 4x - 5 = 0$, we get

$$(x + c)^2 - 4x - 5 = 0$$
$$x^2 + 2xc + c^2 - 4x - 5 = 0$$
$$x^2 + 2x(c + 2) + c^2 - 5 = 0$$

$A = 1, B = 2(c + 2), C = c^2 - 5$

$D = 0$

$$B^2 - 4AC = 0$$
$$\Rightarrow \quad [2(c + 2)]^2 - 4(1)(c^2 - 5) = 0$$
$$\Rightarrow \quad 4(c^2 + 4c + 4) - 4(c^2 - 5) = 0$$
$$\Rightarrow \quad c^2 + 4c + 4 - c^2 + 5 = 0$$
$$\Rightarrow \quad 4c + 9 = 0$$
$$c = -\frac{9}{4}$$

$\therefore$ Equation of tangent is

$$y = x - \frac{9}{4}$$
$$\Rightarrow \quad 4y = 4x - 9$$
$$4x - 4y - 9 = 0$$

(d) Equation of tangent parallel to the line $y = 2x + 7$.

Let us assume the tangent is $y = 2x + \lambda$ ...(i)

If it is a tangent to the parabola, then it will meet it in two coincident points.

Put $y = 2x + \lambda$ in $y^2 - 4x - 5 = 0$

$$(2x+\lambda)^2 - 4x - 5 = 0$$
$$\Rightarrow \quad 4x^2 + 4x\lambda + \lambda^2 - 4x - 5 = 0$$
$$\Rightarrow \quad 4x^2 + 4x(\lambda - 1) + \lambda^2 - 5 = 0$$
$$D = 0$$
$$[4(\lambda - 1)]^2 - 4(4)(\lambda^2 - 5) = 0$$
$$\Rightarrow \quad (\lambda - 1)^2 - (\lambda^2 - 5) = 0$$
$$\Rightarrow \quad \lambda^2 - 2\lambda + 1 - \lambda^2 + 5 = 0$$
$$\Rightarrow \quad 2\lambda = 6$$
$$\lambda = 3$$

$\therefore$ Equation of tangent is $y = 2x + 3$.

9. (iii) (a) Given curve,

$$y^2 = 4x \qquad \text{...(i)}$$

and $$x^2 = 4y \qquad \text{...(ii)}$$

Let the equation of tangent be $y = mx + \dfrac{a}{m}$ for Eq. (i)

$$y = mx + \frac{1}{m} \qquad [\because 4a = 4, a = 1]$$

If it is tangent to $x^2 = 4y$, then $x^2 = 4\left(mx + \dfrac{1}{m}\right)$

$$\Rightarrow \quad mx^2 - 4m^2x - 4 = 0$$
$$\Rightarrow \quad D = 0$$
$$\Rightarrow (-4m^2)^2 - 4(m)(-4) = 0$$
$$\Rightarrow \quad 16m^4 + 16m = 0$$
$$\Rightarrow \quad 16m(m^3 + 1) = 0$$

$m = 0, m = -1$

$\therefore$ **Equation of tangent**

$\because m = 0$ will not satisfy the given tangent.

So, $m = -1$

$\therefore$ Common tangent is $y = -x - 1$

$$\Rightarrow \quad x + y + 1 = 0$$

(b) Given curve, $\quad y^2 = 16x \qquad$ ...(i)

and $\quad x^2 + y^2 = 8 \qquad$ ...(ii)

Let the equation of tangent to the parabola $y^2 = 16x$ is

$$y = mx + \frac{4}{m}$$

$$\Rightarrow m^2x - my + 4 = 0 \qquad \text{...(iii)}$$

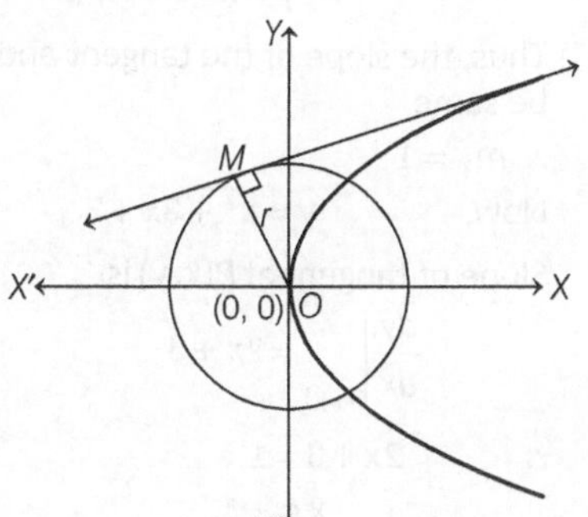

Eq. (iii) be a tangent to the circle $x^2 + y^2 = 8$.

$\therefore$ Length of perpendicular from the centre to tangent = Radius

$$\left|\frac{0 - 0 + 4}{\sqrt{m^4 + m^2}}\right| = (2\sqrt{2})$$

$$\Rightarrow \quad \left|\frac{4}{m\sqrt{m^2 + 1}}\right| = 2\sqrt{2}$$

$$\Rightarrow \quad \frac{16}{m^2(m^2 + 1)} = 8$$

$$\Rightarrow \quad m^2(m^2 + 1) = 2$$
$$\Rightarrow \quad m^4 + m^2 - 2 = 0$$
$$\Rightarrow \quad (m^2 + 2)(m^2 - 1) = 0$$

$\therefore m^2 - 1 = 0, \quad m^2 + 2 = 0$

$m = \pm 1, m^2 = -2$ (Not possible)

$\therefore$ Equation of common tangents are

$y = \pm x \pm 4$ or $y = x + 4$ and $y = -x - 4$

(c) Let the equation of tangent to the parabola $y^2 = 8x$ is

$$y = mx + \frac{2}{m} \qquad \text{...(i)}$$

$\because$ If it is tangent to the curve $xy = -1$

$$\therefore \quad x\left(mx + \frac{2}{m}\right) = -1$$

$$\Rightarrow \quad mx^2 + \frac{2}{m}x + 1 = 0$$
$$\Rightarrow \quad m^2x^2 + 2x + m = 0$$
$$\Rightarrow \quad D = 0$$
$$\Rightarrow \quad (2)^2 - 4(m^2)(m) = 0$$
$$\Rightarrow \quad m^3 - 1 = 0 \Rightarrow m = 1$$

$\therefore$ Equation of tangent is $y = x + 2$.

(d) Any point on the parabola $y = x^2$ is (t, t^2)

Now, the tangent at (t, t^2) is

$$xx_1 = \frac{1}{2}(y + y_1)$$
$$xt = \frac{1}{2}(y + t^2)$$
$$2tx - y - t^2 = 0$$

If it is tangent to the parabola $y = -(x-2)^2$, then

$$2tx - t^2 = -(x-2)^2$$
$$x^2 + 2(t-2)x + 4 - t^2 = 0$$

and $$D = 0$$

$$[2(t-2)]^2 - 4(1)(4 - t^2) = 0$$
$$\Rightarrow \quad 4 + t^2 - 4t - 4 + t^2 = 0$$

$\Rightarrow \quad t^2 - 2t = 0$

$t = 0$ or 2

$t = 0$ is not possible

$\therefore$ Equation of common tangent is

$$4x - y - 4 = 0$$

(e) Any point on the parabola $y^2 = x$ can be considered as (t^2, t).

$\therefore$ Equation of tangent at (t^2, t) is

$$yt = \frac{(x + t^2)}{2}$$

$\Rightarrow \quad 2yt = x + t^2$

$x - 2yt + t^2 = 0$

If it is tangent to circle $x^2 + y^2 - 6y + 4 = 0$, then

$(2yt - t^2)^2 + y^2 - 6(y) + 4 = 0$

$\Rightarrow (4t^2 + 1)y^2 - 2(2t^3 + 3)y + (t^4 + 4) = 0$

$\because \quad D = 0$

$\Rightarrow \quad 4(2t^3 + 3)^2 - 4(4t^2 + 1)(t^4 + 4) = 0$

$\Rightarrow \quad t^4 - 12t^3 + 16t^2 - 5 = 0$

$\Rightarrow \quad (t - 1)(t^3 - 11t^2 + 5t + 5) = 0$

$\therefore \quad t = 1$

$\therefore$ Equation of common tangent is

$$x - 2y = -1$$

9. (iv) (a) Parabola $y^2 = 8x$

Equation of tangent $y = x + 2$

Put $\quad y = x + 2$ in $y^2 = 8x$

$(x + 2)^2 = 8x$

$\Rightarrow \quad x^2 + 4x + 4 - 8x = 0$

$\Rightarrow \quad x^2 - 4x + 4 = 0$

$\Rightarrow \quad (x - 2)^2 = 0$

$\Rightarrow \quad x = 2$

$\therefore \quad y = x + 2 = 2 + 2 = 4$

$\therefore$ Point of contact (2, 4).

(b) Equation of tangent

$x \cos\alpha + y \sin\alpha = p$

$y = -x \cot\alpha + p \operatorname{cosec}\alpha$

$m = -\cot\alpha, c = p \operatorname{cosec}\alpha$

$\because$ The line touches the parabola $y^2 = 4ax$

$\therefore \quad c = \frac{a}{m}$ or $cm = a$

$(p \operatorname{cosec}\alpha)(-\cot\alpha) = a$

$p\cos\alpha + a\sin^2\alpha = 0$

Point of contact $\left(\frac{a}{m^2}, \frac{2a}{m}\right)$

$$= \left(\frac{a}{\cot^2\alpha}, \frac{2a}{-\cot\alpha}\right)$$

$$= (a\tan^2\alpha, -2a\tan\alpha)$$

(c) Parabola $y^2 = 8x$

Equation of tangent to the parabola $y^2 = 4ax$ at $(at^2, 2at)$ is $ty = x + at^2$

Here, $a = 2$.

So, the equation of the tangent at $(2t^2, 4t)$ to the parabola $y^2 = 8x$

is $\quad ty = x + 2t^2 \quad$ ...(i)

Slope of Eq. (i), $\frac{1}{t}$

and given that, tangent and lines $y = 3x + 5$ makes an angle 45°

$$\left|\frac{m_1 - m_2}{1 + m_1 m_2}\right| = \tan 45^\circ$$

$$\Rightarrow \frac{\frac{1}{t} - 3}{1 + \frac{3}{t}} = \pm 1 \Rightarrow \frac{1 - 3t}{t + 3} = 1 \text{ or } \frac{1 - 3t}{t + 3} = -1$$

$1 - 3t = t + 3$ or $1 - 3t = -t - 3$

$4t = -2$ or $2t = 4$

$t = -\frac{1}{2}$ or $t = 2$

For $t = -\frac{1}{2}$ tangent is $2x + y + 1 = 0 \Rightarrow y = -2x - 1$

Point of contact $\left(\frac{a}{m^2}, \frac{+2a}{m}\right)$

$$\Rightarrow \left(\frac{2}{4}, \frac{2 \times 2}{-2}\right) = \left(\frac{1}{2}, -2\right)$$

and $t = 2$, tangent is $2y = x + 8$

Point of contact $\left(\frac{2}{(1/2)^2}, \frac{2 \times 2}{(1/2)}\right)$

$= (8, 8)$

9. (v) (a) The shortest distance between a line and the parabola means the shortest distance between a line and a tangent to the parabola parallel to given line.

Given line, $y = x - 2$

$P: y = x^2 + 3x + 2$

Thus, the slope of the tangent and the line will be same

$\therefore m_T = 1$

Now, $\quad y = x^2 + 3x + 2$

Slope of tangent at $P(x, y)$ is

$$\left.\frac{dy}{dx}\right|_{(x,\, y)} = 2x + 3$$

$\therefore \quad 2x + 3 = 1$

$x = -1$

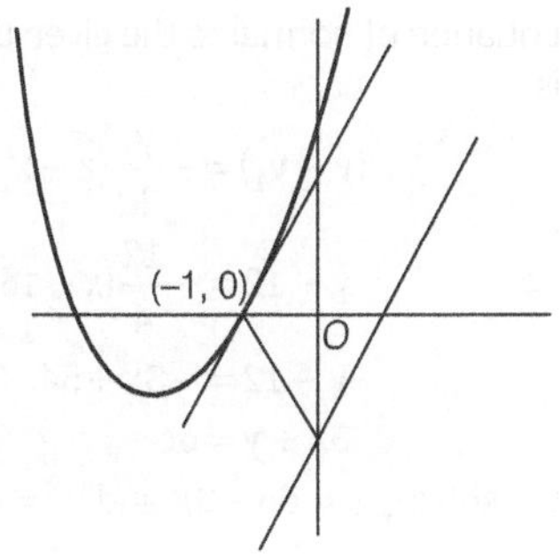

and $\quad y=(-1)^2+3(-1)+2=0$

$\therefore P(-1,0)$.

Required shortest distance

$$=\left|\frac{-1+0-2}{\sqrt{1^2+1^2}}\right|=\frac{3}{\sqrt{2}}$$

(b) Given line, $\quad x+y=4$

$$y=-x+4 \quad \text{...(i)}$$

Parabola $y^2+4x+4y=0 \quad$ ...(ii)

$$2y\frac{dy}{dx}+4+4\frac{dy}{dx}=0$$

$$\frac{dy}{dx}=\frac{-4}{2y+4}=\frac{-2}{y+2}$$

$$m_T=-1$$

$$\frac{-2}{y+2}=-1$$

$$-y-2=-2 \Rightarrow y=0$$

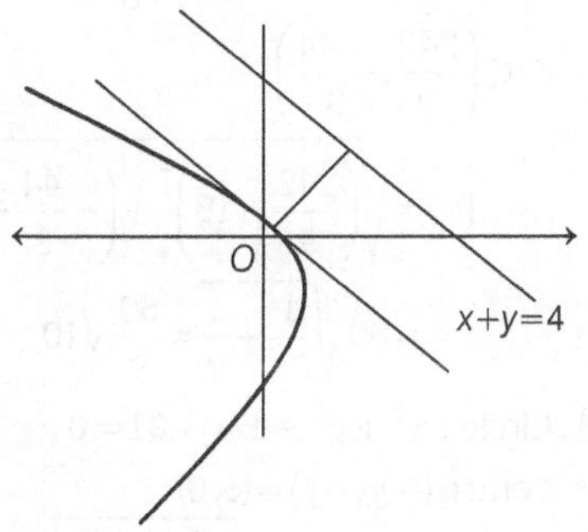

By Eq. (ii),

$$0+4x=0 \Rightarrow x=0$$

Point on parabola is (0, 0).

$$\therefore \text{Shortest distance}=\left|\frac{0+0-4}{\sqrt{1^2+1^2}}\right|=\frac{4}{\sqrt{2}}=2\sqrt{2}$$

10. (a) Parabola : $x^2=4y$

$\Rightarrow \quad 4a=4, \quad a=1$

Equation of normal at (2, 1)

$\because$ Equation of normal at (x_1, y_1) to the parabola $x^2=4ay$ is $y-y_1=\frac{-2a}{x_1}(x-x_1)$.

$\therefore$ Equation is $(y-1)=\frac{-2(1)}{2}(x-2)$

$\Rightarrow \quad y-1=-(x-2)$

$\Rightarrow \quad y=-x+2+1$

$\Rightarrow \quad x+y-3=0$

(b) Parabola $y^2=4x \Rightarrow 4a=4 \Rightarrow a=1$

Equation of normal in parametric form

$$y+tx=2at+at^2$$

$\therefore$ Required equation is

$$y+tx=2t+t^2.$$

(c) Parabola $y^2=16x$

$\Rightarrow \quad 4a=16 \Rightarrow a=4$

normal is parallel to $y=2x+1$

$\therefore$ Slope of normal $m=2$

$\therefore$ Equation of normal is,

$$y=mx-2am-am^3$$

$$y=2x-2(4)(2)-4(2)^3$$

$$y=2x-16-32$$

$$y=2x-48$$

(d) $S: y=x^2-x-\frac{1}{2}$

$$y=x^2-x+\frac{1}{4}-\frac{3}{4}$$

$$\left(y+\frac{3}{4}\right)=\left(x-\frac{1}{2}\right)^2$$

$X^2=4AY$, where $Y=y+\frac{3}{4}, X=x-\frac{1}{2}$

$$4A=1 \Rightarrow A=\frac{1}{4}$$

$\because$ Slope of normal $(m)=-1$ (given)

$\because$ Equation of normal for given slope m is

$Y=mX-2am-am^3$

$$\left(y+\frac{3}{4}\right)=-1\left(x-\frac{1}{2}\right)-2\left(\frac{1}{4}\right)(-1)-\frac{1}{4}(-1)^3$$

$$\Rightarrow y+\frac{3}{4}=-x+\frac{1}{2}+\frac{1}{2}-\frac{1}{4}$$

$\Rightarrow \quad y=-x$

Now, normal meets again to parabola

$$-x=x^2-x-\frac{1}{2}$$

$$x^2=\pm\frac{1}{2}, \quad x=\pm\frac{1}{\sqrt{2}}$$

$\therefore \quad x=\frac{1}{\sqrt{2}}, \quad y=-\frac{1}{\sqrt{2}}$

$\quad x=-\frac{1}{\sqrt{2}}, \quad y=\frac{1}{\sqrt{2}}$

$\therefore$ Normal meets at point $\left(\frac{1}{\sqrt{2}},-\frac{1}{\sqrt{2}}\right)$.

(e) Parabola $S: y^2 = 4x$

Point $P(15, 12)$.

$\because$ We know that, equation of normal to the parabola $y^2 = 4ax$ having slope m is

$$y = mx - 2am - am^3$$

$\because$ Here, $a = 1$

$$\therefore \quad y = mx - 2m - m^3$$

$\therefore$ It passes through (15, 12)

$$12 = 15m - 2m - m^3$$

$$m^3 - 13m + 12 = 0$$

$$(m-1)(m-3)(m+4) = 0$$

$\therefore$ $m = 1, \quad m = 3, \quad m = -4$

Equation of normal is

At $m = 1 \Rightarrow y = x - 2 - 1 \Rightarrow y = x - 3$

At $m = 3 \Rightarrow y = 3x - 6 - 27 \Rightarrow y = 3x - 33$

At

$m = -4 \Rightarrow y = -4x + 8 + 64 \Rightarrow y = -4x + 72$

(f) Parabola : $y^2 = 16x$

$$4a = 16$$

$$a = 4$$

Ends of latus rectum,

$(a, 2a)$ and $(a, -2a)$

$\Rightarrow (4, 8)$ and $(4, -8)$

$\because$ Slope of normal at (x_1, y_1) to the parabola $y^2 = 4ax$ is $-\dfrac{y_1}{2a}$.

$$\therefore \quad m_1 = \frac{-8}{2(4)} = -1, \quad m_2 = \frac{-(-8)}{2(4)} = 1$$

Now, equation of normals are

$$y - y_1 = m_1(x - x_1)$$

$$y - 8 = -1(x - 4)$$

$$y + x = 12$$

$$x + y - 12 = 0$$

$$y - y_2 = m_2(x - x_2)$$

$$y + 8 = 1(x - 4)$$

$$y - x = -12$$

$$x - y - 12 = 0$$

11. (a) Parabola $S: y^2 = 12x$

$$4a = 12 \Rightarrow a = 3$$

Given line, $x + y = k \Rightarrow y = -x + k$

The line will be normal to given parabola if

$$k = -2am - am^3$$

$$= -2(3)(-1) - 3(-1)^3$$

$$= +6 + 3 = 9$$

$$\therefore \quad k = 9$$

(b) Equation of parabola

$S: y^2 = 8x$

$4a = 8 \Rightarrow a = 2$

Equation of normal to the given parabola at $P(18, 12)$ is

$$(y - y_1) = -\frac{y_1}{2a}(x - x_1)$$

$$\Rightarrow \quad y - 12 = -\frac{12}{4}(x - 18)$$

$$y - 12 = -3x + 54$$

$$3x + y = 66$$

On solving $y = 66 - 3x$ and $y^2 = 8x$

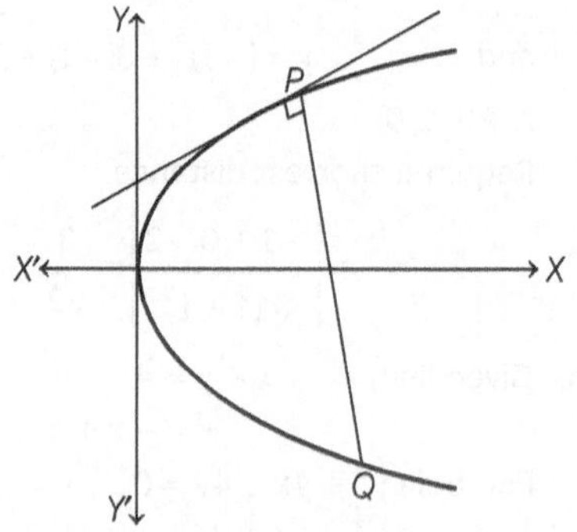

$$(66 - 3x)^2 = 8x$$

$$\Rightarrow 9x^2 - 396x + 4356 - 8x = 0$$

$$\Rightarrow \quad 9x^2 - 404x + 4356 = 0$$

After solving, we get

$$x = 18, \frac{242}{9}$$

$$y = 12, \frac{-44}{3}$$

$$\therefore Q\left(\frac{242}{9}, \frac{-44}{3}\right)$$

$$\therefore \quad PQ = \sqrt{\left(\frac{242}{9} - 18\right)^2 + \left(-\frac{44}{3} - 12\right)^2}$$

$$= 80\sqrt{\frac{1}{81} + \frac{1}{9}} = \frac{80}{9}\sqrt{10}$$

(c) Circle : $x^2 + y^2 - 12x + 31 = 0$

Centre $(-g, -f) = (6, 0)$

$$\text{radius} = \sqrt{g^2 + f^2 - c}$$

$$= \sqrt{36 - 31} = \sqrt{5}$$

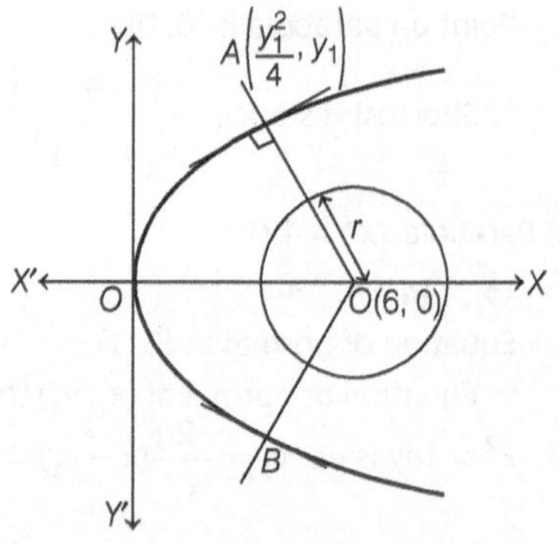

Parabola : $y^2 = 4x$

Now, the shortest distance always occur along common normal.

∴ Normal for parabola

$$\frac{dy}{dx} = \frac{2}{y}$$

Slope of normal at $A\left(\frac{y_1^2}{4}, y_1\right)$ is $-\frac{y_1}{2}$.

Now, slope of $OA = -\frac{y_1}{2}$

$$\frac{y_1 - 0}{\frac{y_1^2}{4} - 6} = -\frac{y_1}{2}$$

$$2y_1 + y_1\left(\frac{y_1^2}{4} - 6\right) = 0$$

$$y_1\left(2 + \frac{y_1^2}{4} - 6\right) = 0$$

$$y_1 = 0 \text{ and } \frac{y_1^2}{4} = 4$$

$$y_1 = \pm 4$$

∴ Points are (0, 0), $A(4, 4)$ and $B = (4, -4)$

Shortest distance $= OA - r$

$$\{\because OA = \sqrt{2^2 + 4^2} = \sqrt{20} = 2\sqrt{5}\}$$

$$= 2\sqrt{5} - \sqrt{5} = \sqrt{5}$$

(d) $S : y^2 = 4ax$

Let the normal at $P(at_1^2, 2at_1)$ meet the curve at $Q(at_2^2, 2at_2)$.

∴ PQ is normal chord.

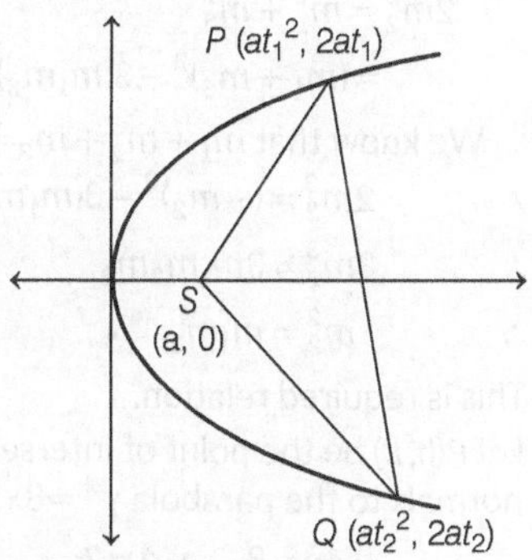

Equation of normal at $(at_1^2, 2at_1)$

$$y = -t_1x + 2at_1 + at_1^3$$

meets at $Q(at_2^2, 2at_2)$.

$\therefore \quad 2at_2 = -at_1t_2^2 + 2at_1 + at_1^3$

$\Rightarrow \quad 2a(t_2 - t_1) + at_1(t_2^2 - t_1^2) = 0$

$\Rightarrow \quad a(t_2 - t_1)(2 + t_1(t_2 + t_1)) = 0$

$$a(t_2 - t_1) \neq 0$$

$\Rightarrow \quad t_1(t_2 + t_1) = -2$

$$t_2 = -t_1 - \frac{2}{t_1} \qquad \text{...(i)}$$

By the given condition,

$$2at_1 = at_1^2 \Rightarrow t_1 = 2$$

$\because \quad t_1 = 2 \Rightarrow t_2 = -2 - \frac{2}{2} = -3$ [by Eq. (i)]

∴ $P(4a, 4a)$ and $Q(9a, -6a)$.

Now, slope of $SP = \frac{4a - 0}{4a - a} = \frac{4}{3}$

Slope of $SQ = \frac{-6a - 0}{9a - a} = \frac{-3}{4}$

∵ Slope of $SP \times$ Slope of $SQ = -1$

$\therefore \quad SP \perp SQ$.

∴ Required angle is $\frac{\pi}{2}$.

12. (a) $S : y^2 = 4ax$

Let $P(at_1^2, 2at_1)$ and $Q(at_2^2, 2at_2)$ are end points of focal chord.

Equation of normal at P and Q are

$$y = -t_1x + 2at_1 + at_1^3 \qquad \text{...(i)}$$

and $\quad y = -t_2x + 2at_2 + at_2^3 \qquad$ ...(ii)

After solving Eqs. (i) and (ii), we get

$$x = 2a + a(t_1^2 + t_2^2 + t_1t_2)$$

$$y = -at_1t_2(t_1 + t_2)$$

Also PQ is focal chord

$$t_1t_2 = -1$$

∴ Locus of point of intersection (h, k)

$$h = 2a + a((t_1 + t_2)^2 - t_1t_2)$$

$$= 2a + a[(t_1 + t_2)^2 + 1]$$

and $\quad k = -a(-1)(t_1 + t_2)$

$$k = a(t_1 + t_2)$$

$$t_1 + t_2 = \frac{k}{a}$$

$\therefore \quad h = 2a + a\left[\left(\frac{k}{a}\right)^2 + 1\right]$

$$h = 2a + \left(\frac{k^2 + a^2}{a}\right)$$

$$ah = 2a^2 + k^2 + a^2$$

$$k^2 = ah - 3a^2$$

∴ Locus is $y^2 = a(x - 3a)$

(b) The equation of the normal to parabola $y^2 = 4ax$ is

$y = mx - 2am - am^3$

∵ It passes through the point (h, k).

If $\quad k = mh - 2am - am^3$

$$am^3 + (2a - h)m + k = 0 \qquad \text{...(i)}$$

Let the roots of above equation is m_1, m_2 and m_3.
Let the perpendicular normals correspond to the values of m_1 and m_2.

$\therefore \quad m_1 m_2 = -1$

By Eq. (i), $\quad m_1 m_2 m_3 = -\frac{k}{a}$

$$-1(m_3) = -\frac{k}{a}$$

$$m_3 = \frac{k}{a}$$

$\because$ m_3 is root of Eq. (i).

$\therefore \quad a\left(\frac{k}{a}\right)^3 + (2a-h)\left(\frac{k}{a}\right) + k = 0$

$\Rightarrow \quad k^3 + a(2a-h)k + a^2 k = 0$

$\Rightarrow \quad k^2 + 2a^2 - ha + a^2 = 0$

$\Rightarrow \quad k^2 = ha - 3a^2$

$\therefore$ Locus is $y^2 = a(x-3a)$

13. (a) Parabola $S: y^2 - 4x - 2y - 3 = 0$

$$(y-1)^2 = 4(x+1)$$

So, the axis is $y - 1 = 0$
$\because$ Vertex is $(-1, 1)$.

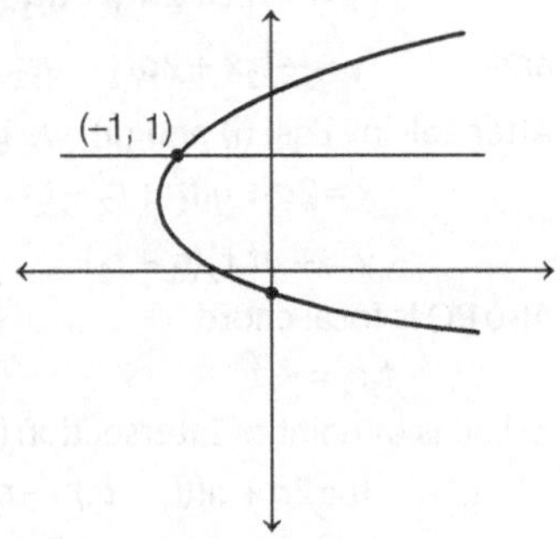

Also $(-2, 1)$ lies on the axis and it is exterior to the parabola.
Because $S_1: 1^2 - 4(-2) - 2(1) - 3$

$$1 + 8 - 2 - 3$$

$$4 > 0$$

$\therefore$ Only one normal is possible.

(b) Given parabola, $y^2 - 2y = 4x - 9$

$$(y-1)^2 = 4(x-2)$$

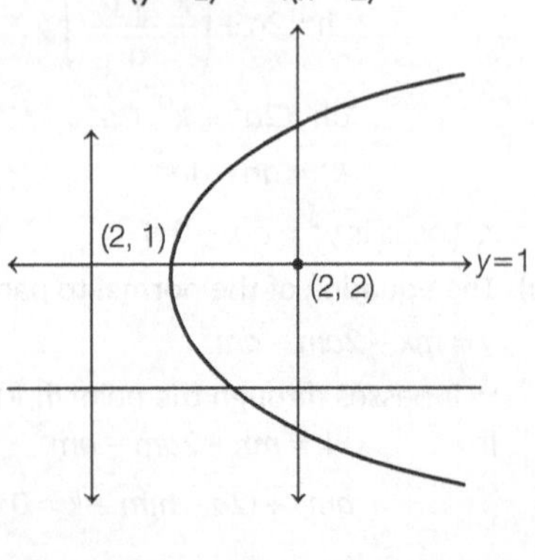

Vertex $\Rightarrow (2, 1)$ Focus $\Rightarrow (1+1, 2) = (2, 2)$
Axis $\Rightarrow y = 1$
$\because$ From point $(2a, b)$ 3 normal can be drawn.
$\because$ Equation is normal at (h, k) for $y^2 = 4ax$.

$$k = mh - 2m - am^3$$

$$am^3 + m(2a-h) + k = 0 \quad \text{...(i)}$$

$$m_1 + m_2 + m_3 = 0$$

$$m_1 m_2 + m_2 m_3 + m_3 m_1 = \frac{2a-h}{a}$$

$$m_1 m_2 m_3 = \frac{k}{a}$$

When 3 normals are real, then all roots of equation are real.

$\therefore \quad m_1^2 + m_2^2 + m_3^2 > 0$

$$(m_1 + m_2 + m_3)^2 - 2(m_1 m_2 + m_2 m_3 + m_3 m_1) > 0$$

$$0 - \frac{2(2a-h)}{a} > 0$$

$$h - 2a > 0$$

For $a > 0$.
$\because$ $h > 2$(x-coordinate of focus).
In given equation,
Points is $(2a, b)$ and focus is (2, 2)

$$2a > 2(2) \Rightarrow a > 2$$

(c) Equation of normal to the parabola $y^2 = 4x$ is given by

$$y = -mx + 2m + m^3$$

$\because$ It intersects at $x = 2$, we get

$$y = -2m + 2m + m^3 \Rightarrow y = m^3$$

Let the three ordinates m_1^3, m_2^3 and m_3^3 are in AP

$$2m_2^3 = m_1^3 + m_3^3$$

$$= (m_1 + m_3)^3 - 3(m_1 m_3)(m_1 + m_3) \quad \text{...(i)}$$

$\because$ We know that $m_1 + m_2 + m_3 = 0$, we get

$$2m_2^3 = (-m_2)^3 - 3(m_1 m_3)(-m_2)$$

$\therefore \quad 3m_2^3 = 3m_1 m_2 m_3$

$\therefore \quad m_2^2 = m_1 m_3$

This is required relation.

(d) Let $P(h, k)$ be the point of intersection of three normals to the parabola $y^2 = 8x$.

$$4a = 8 \Rightarrow a = 2$$

$\therefore$ Equation of normal at $P(h, k)$

$$y = mx - 2am - am^3$$

$\therefore \quad k = mh - 4m - 2m^3 \quad \text{...(i)}$

$$2m^3 + (4-h)m + k = 0$$

Let m_1, m_2 and m_3 are roots.

$\because \quad m_1 + m_2 + m_3 = 0 \quad \text{...(ii)}$

$$m_1 m_2 + m_2 m_3 + m_3 m_1 = \frac{4-h}{2} \quad \text{...(iii)}$$

$$m_1m_2m_3 = -\frac{k}{2} \quad ...(iv)$$

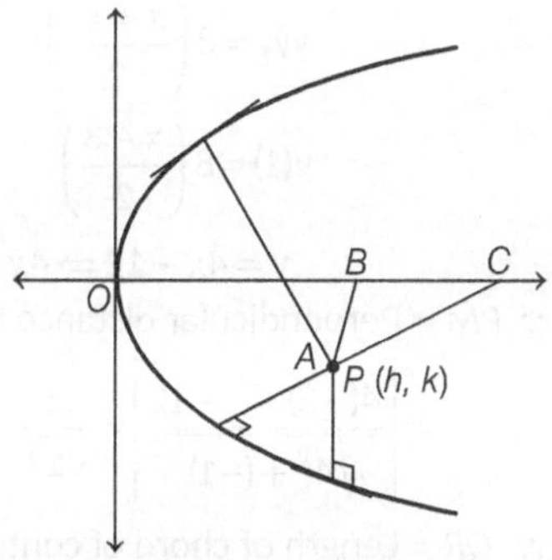

$\because$ Equation cuts the axis on passes, i.e. $y=0$ at $(2m^2+4, 0)$ $\{0 = mh - 4m - 2x, h = 4 + 2m^2\}$

$\therefore$ Normal through (h, k) cut the axis at $A(4+2m_1^2, 0), B = (4+2m_2^2, 0)$ and $C(4+2m_3^2, 0)$

Let $V(0, 0)$ be the vertex of parabola $y^2 = 8x$

$\therefore$ $VA = 4+2m_1^2, VB = 4+2m_2^2, VC = 4+2m_3^2$

$\because$ VA, VB and VC are in AP.

$$2VB = VA + VC$$

$$\Rightarrow \quad 8+4m_2^2 = 8+2m_1^2+2m_3^2$$

$$\Rightarrow \quad 2m_2^2 = m_1^2 + m_3^2$$

$$= (m_1+m_3)^2 - 2m_1m_3.$$

$$\Rightarrow \quad 2m_2^2 = (m_1+m_3)^2 - 2m_1m_3$$

$$[\because m_1+m_2+m_3 = 0]$$

$$\Rightarrow \quad 2m_2^2 = (-m_2)^2 - 2m_1m_3$$

$$\Rightarrow \quad m_2^2 = -2m_1m_3$$

$$\Rightarrow \quad m_2^3 = -2m_1m_2m_3$$

$$\Rightarrow \quad m_2^3 = -2\left(-\frac{k}{2}\right) \quad \text{[by Eq. (iv)]}$$

$$\Rightarrow \quad m_2^3 = k$$

Now, by Eq. (iii),

$$\Rightarrow \quad m_2(m_1+m_3) + m_3m_1 = \frac{4-h}{2}$$

$$\Rightarrow \quad m_2(-m_2) + \frac{m_3m_1m_2}{m_2} = \frac{4-h}{2}$$

$$\Rightarrow \quad -m_2^2 - \left(\frac{k}{2}\right)\frac{1}{m_2} = \frac{4-h}{2}$$

$$\Rightarrow \quad -m_2^3 - \frac{k}{2} = \left(\frac{4-h}{2}\right)m_2$$

$$\Rightarrow \quad -k - \frac{k}{2} = \left(\frac{4-h}{2}\right)m_2$$

$$\Rightarrow \quad -3\frac{k}{2} = \left(\frac{4-h}{2}\right)m_2$$

$$\Rightarrow \quad 3k = (h-4)m_2$$

Taking cube on both sides, we get

$$27k^3 = (h-4)^3 m_2^3$$

$$27k^3 = (h-4)^3 (k)$$

$$27k^2 = (h-4)^3$$

$$\therefore \quad \text{Locus } 27y^2 = (x-4)^3.$$

14. (a) Parabola $S: y^2 - 4x - 6y + 7 = 0$

Point $(-2, -1)$

$$\because \quad S_1 : (-1)^2 - 4(-2) - 6(-1) + 7$$

$$\Rightarrow \quad 1+8+6+7$$

$$\Rightarrow \quad 22 > 0$$

$\therefore$ Point $(-2, -1)$ lies outside the parabola.

$\therefore$ Equation of pair of tangent

$$T^2 = SS_1$$

$$\left[y(-1) - 4\left(\frac{x-2}{2}\right) - 6\left(\frac{y-1}{2}\right) + 7\right]^2$$

$$= (y^2 - 4x - 6y + 7)(22)$$

$$\Rightarrow \quad [-y - 2x + 4 - 3y + 3 + 7]^2$$

$$= 22y^2 - 88x - 132y + 154$$

$$\Rightarrow \quad (14 - 2x - 4y)^2 = 22y^2 - 88x - 132y + 154$$

$$\Rightarrow \quad 196 + 4x^2 + 16y^2 - 56x + 16xy - 112y$$

$$= 22y^2 - 88x - 132y + 154$$

$$\Rightarrow \quad 4x^2 - 6y^2 + 16xy + 32x + 20y + 42 = 0$$

$$\Rightarrow \quad 2x^2 - 3y^2 + 8xy + 16x + 10y + 21 = 0$$

(b) $S: y^2 = 12x$

Point $(-1, 2)$

Chord of contact from an external point

$$T = 0$$

$$yy_1 = 12\left(\frac{x+x_1}{2}\right)$$

$$\Rightarrow \quad y(2) = 12\left(\frac{x-1}{2}\right)$$

$$\Rightarrow \quad y = 3x - 3$$

(c) $S: y^2 = 4x$

Point (2, 3)

$\because$ The equation of chord of the parabola $y^2 = 4x$ which is bisected at (2, 3) is

$$T = S_1$$

$$yy_1 - 4\left(\frac{x+x_1}{2}\right) = y_1^2 - 4(x_1)$$

$$\Rightarrow \quad y(3) - 4\left(\frac{x+2}{2}\right) = (3)^2 - 4(2)$$

$$\Rightarrow \quad 3y - 2x - 4 = 9 - 8$$

$$\Rightarrow \quad 3y - 2x - 5 = 0$$

(d) $S: y^2 - 4x = 0 \Rightarrow 4a = 4 \Rightarrow a = 1$

Point : $(x_1, y_1) = (-2, 1) \Rightarrow x_1 = -2, y_1 = 1$

$\therefore$ Length of chord of contact is

$$= \frac{1}{|a|}\sqrt{(y_1^2 - 4ax_1)(y_1^2 + 4a^2)}$$

$$= \frac{1}{1}\sqrt{\{(1)^2 - 4(-2)\}\{1 + 4(1)^2\}}$$

$$= \sqrt{(1+8)(1+4)} = \sqrt{9 \times 5} = 3\sqrt{5}$$

15. (a) $S: y^2 = 4x$

Let the point of intersection of the tangents be $R(\alpha, \beta)$.

The equation of tangent to the parabola at $(t_1^2, 2t_1)$ and $(t_2^2, 2t_2)$ are $t_1 y = x + t_1^2$ ...(i)

And $t_2 y = x + t_2^2$ ...(ii)

So, $m_1 = \frac{1}{t_1}$ and $m_2 = \frac{1}{t_2}$.

and the point of intersection of Eqs. (i) and (ii) is

$\Rightarrow [t_1 t_2, (t_1 + t_2)]$

Thus, $\alpha = t_1 t_2$ and $\beta = t_1 + t_2$

Let θ be the angle between two tangent given $\theta = 60°$

$$\tan\theta = \left|\frac{\frac{1}{t_1} - \frac{1}{t_2}}{1 + \frac{1}{t_1}\cdot\frac{1}{t_2}}\right|$$

$$\Rightarrow \tan 60° = \left|\frac{t_2 - t_1}{1 + t_1 t_2}\right|$$

$$(\sqrt{3})^2 = \frac{(t_2 - t_1)^2}{(1 + t_1 t_2)^2}$$

$$\Rightarrow 3(1 + t_1 t_2)^2 = (t_2 - t_1)^2$$

$$\Rightarrow 3(1 + \alpha)^2 = (t_1 + t_2)^2 - 4t_1 t_2$$

$$\Rightarrow 3(1 + \alpha)^2 = \beta^2 - 4\alpha$$

Locus $3(x + 1)^2 = y^2 - 4x$

$\Rightarrow 3x^2 + 6x + 3 = y^2 - 4x$

$\Rightarrow 3x^2 + 10x + 3 = y^2$

(b) Given, $S: y^2 = 8x$

$4a = 8 \Rightarrow a = 2$

Point $(-3, 1) \Rightarrow (x_1, y_1)$

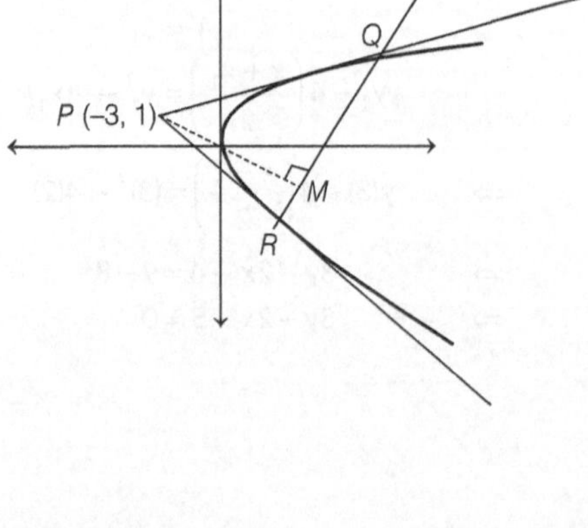

Equation of QR

$$T = 0$$

$$yy_1 = 8\left(\frac{x + x_1}{2}\right)$$

$$y(1) = 8\left(\frac{x - 3}{2}\right)$$

$$y = 4x - 12 \Rightarrow 4x - y - 12 = 0$$

$\because PM$ = Perpendicular distance from P to QR

$$= \left|\frac{4(-3) - 1 - 12}{\sqrt{4^2 + (-1)^2}}\right| = \frac{25}{\sqrt{17}}$$

$\therefore QR$ = Length of chord of contact

$$= \sqrt{(y_1^2 - 4ax_1)\cdot(y_1^2 + 4a^2)}$$

$$= \sqrt{(1)^2 - 8(-3)(1^2 + 4(2)^2]}$$

$$= \sqrt{(1+24)} = \sqrt{(1+24)(1+16)}$$

$$= \sqrt{25 \times 17} = 5\sqrt{17}$$

$\therefore$ Area of $\Delta PQR = \frac{1}{2} \times QR \times PM$

$$= \frac{1}{2} \times 5\sqrt{17} \times \frac{25}{\sqrt{17}} = \frac{125}{2} \text{ sq units}$$

(c) $S: y^2 = 16x$

Let $P(h, k)$ be the mid-point of a chord QR of the parabola $y^2 = 16x$, then equation of chord QR is

$$T = S_1$$

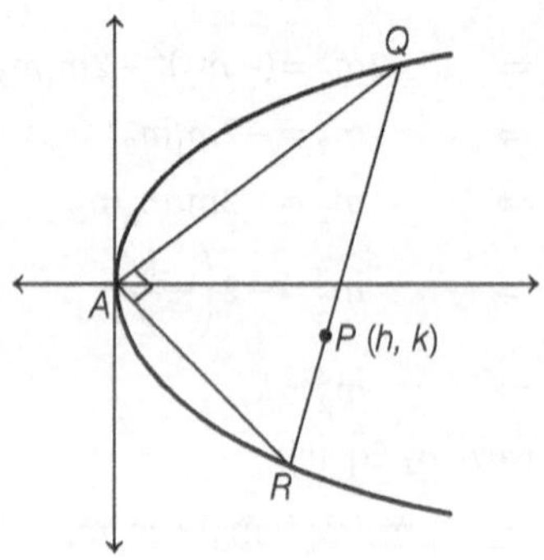

$$ky - 16\left(\frac{x + h}{2}\right) = k^2 - 16h$$

$$ky - 8(x + h) = k^2 - 16h$$

$$\Rightarrow ky - 8x = k^2 - 8h \quad \text{...(i)}$$

If Δ is the vertex of the parabola.

For combined equation of AQ and AR making homogeneous of $y^2 = 16x$ with help of Eq. (i)

$$\Rightarrow y^2 = 16x\left(\frac{ky - 8x}{k^2 - 8h}\right)$$

$$\Rightarrow y^2(k^2 - 8h) = 16x(ky - 8x)$$

$$\Rightarrow y^2(k^2 - 8h) - 16kxy + 128x^2 = 0$$

$\because \quad \angle QAR = 90°$

Coefficient of x^2 + Coefficient of $y^2 = 0$

$$k^2 - 8h + 128 = 0$$

$\therefore$ Locus is $y^2 - 8x + 128 = 0$

(d) $S : y^2 = 4ax$

Equation of normal chord at any point $(at^2, 2at)$ of parabola $y^2 = 4ax$ is

$$y + tx = 2at + at^3 \quad \text{...(i)}$$

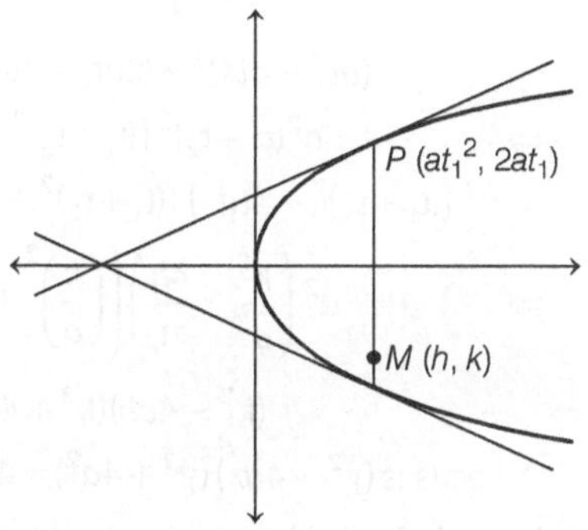

But, if $M(h, k)$ be its middle point, then equation must be

$$T = S_1$$

$$\Rightarrow \quad yk - 2a(x + h) = k^2 - 4ah$$

$$\Rightarrow \quad yk - 2ax = k^2 - 2ah \quad \text{...(ii)}$$

Eqs. (i) and (ii) are identical Eq. (i)

$$\therefore \quad \underset{\text{(i)}}{\frac{1}{k} = \frac{t}{-2a}} = \underset{\text{(ii)}}{\frac{2at + at^3}{k^2 - 2ah}}$$

First relation $t = \frac{-2a}{k}$

Second relation $\frac{t}{-2a} = \frac{2at + at^3}{(k^2 - 2ah)}$

$$k^2 - 2ah = (-2a)(2a + at^2)$$

$$\Rightarrow \quad (k^2 - 2ah) = -2a\left\{2a + a\left(\frac{-2a}{k}\right)^2\right\}$$

$$\Rightarrow \quad (k^2 - 2ah) = -(2a)(2a)\frac{(k^2 + 2a^2)}{k^2}$$

$$\Rightarrow \quad k^4 - 2ahk^2 = -4a^2k^2 - 8a^4$$

$$\Rightarrow \quad k^4 - 2ak^2(h - 2a) + 8a^4 = 0$$

$\therefore$ Locus is $y^4 - 2ay^2(x - 2a) + 8a^4 = 0$

16. (i) (a) Given parabola, $y^2 = x + 2$

$$Y^2 = X$$

where, $X = x + 2$ and $Y = y$

We have, $4a = 1 \Rightarrow a = \frac{1}{4}$

$\therefore$ Equation of director circle is

$$X + a = 0$$

$$x + 2 + \frac{1}{4} = 0$$

$$4x + 9 = 0$$

(b) Parabola $x^2 = 4x + 4y$

$$x^2 - 4x + 4 = 4y + 4$$

$$(x - 2)^2 = 4(y + 1)$$

$$X^2 = 4Y$$

where, $X = x - 2, Y = y + 1$

$$X^2 = 4Y$$

and $\quad 4a = 4 \Rightarrow a = 1$

$\therefore$ Equation of director circle is $Y + a = 0$

$$y + 1 + 1 = 0$$

$$y + 2 = 0$$

(c) Parabola $y^2 = 4x + 4y - 8$

$$y^2 - 4y + 4 = 4x - 4$$

$$(y - 2)^2 = 4(x - 1)$$

$$Y^2 = 4X$$

where, $\quad Y = y - 2$ and $X = x - 1$

$$4a = 4 \Rightarrow a = 1$$

$\therefore$ Equation of director circle is $X + a = 0$

$$x - 1 + 1 = 0$$

$$x = 0$$

(ii) (a) Point (1, 2) and Parabola $y^2 = 4x$

$$4a = 4$$

$$a = 1$$

$\therefore$

$$\text{Polar} = T = 0$$

$$yy_1 = 2a(x + x_1)$$

$$y(2) = 2(1)(x + 1)$$

$$y = x + 1$$

(b) Point (2, 3), Parabola $x^2 = 16y$

Polar w.r.t. point (2, 3)

$$xx_1 = 16\left(\frac{y + y_1}{2}\right)$$

$$x(2) = 16\,\frac{(y + 3)}{2}$$

$$x = 4y + 12$$

(iii) $S : y^2 = 3x$

Point $P(3, 3)$

Equation of tangent at $P(3, 3)$

$$y(3) = 3\left(\frac{x + 3}{2}\right)$$

$$\Rightarrow \quad 2y = x + 3$$

$$\Rightarrow \quad y = \frac{1}{2}x + \frac{3}{2}$$

$$\Rightarrow \quad m = \frac{1}{2} \Rightarrow \tan\theta = \frac{1}{2}$$

Length of tangent $= y_1 \operatorname{cosec}\theta = 3\times\sqrt{1+2^2} = 3\sqrt{5}$

Length of normal $= y_1 \sec\theta = 3\sqrt{1+\left(\frac{1}{2}\right)^2} = \frac{3}{2}\sqrt{5}$

Length of subtangent $= y_1 \cot\theta = 3\times 2 = 6$

Length of subnormal $= y_1 \tan\theta = 3\times\frac{1}{2} = \frac{3}{2}$

(iv) $\because$ Equation of parabola $\Rightarrow y^2 = 16x$

$$4a = 16 \Rightarrow a = 4$$

Equation of chord is $y = 2x + 3$

$$m = 2$$

$\therefore$ Equation of diameter $\Rightarrow y = \frac{2a}{m}$

$$\therefore \quad y = \frac{2(4)}{2}$$

$$y = 4$$

17. (a) $S: y^2 = 4ax$ ...(i)

Equation of normal for Eq. (i)

$$y = mx - 2am - am^3 \quad \text{...(ii)}$$

Let (h, k) be the pole of Eq. (ii) w.r.t. Eq. (i).

$\therefore$ Eq. (ii) is polar of (h, k) w.r.t Eq. (i)

$$\therefore \quad y(k) = 2a(x+h)$$

$$ky = 2ax + 2ah \quad \text{...(iii)}$$

On comparing Eqs. (ii) and (iii),

$$\frac{k}{1} = \frac{2a}{m} = \frac{2ah}{-2am - am^3}$$

(i) (ii)

From Eq. (i),

$$k = \frac{2a}{m} \quad \text{... (iv)}$$

From Eq. (ii),

$$h = -2a - m^2 \quad \text{...(v)}$$

By Eqs. (iv) and (v),

$$h = -2a - a\left(\frac{2a}{k}\right)^2$$

$$\Rightarrow \quad (h+2a)k^2 + 4a^3 = 0$$

$\therefore$ Locus is $(x+2a)y^2 + 4a^3 = 0$

(b) Let PQ be the chord of parabola $y^2 = 4ax$

If (h, k) is the pole of line PQ.

Then, (h, k) be the point of intersection of tangents at P and Q to the parabola $y^2 = 4ax$

$\because$ We know that point of intersection is $(at_1t_2, a(t_1+t_2))$,

$$h = at_1t_2 \text{ and } k = a(t_1+t_2)$$

$$t_1t_2 = \frac{h}{a}$$

$$t_1 + t_2 = \frac{k}{a}$$

Given, $(PQ) = 2c$

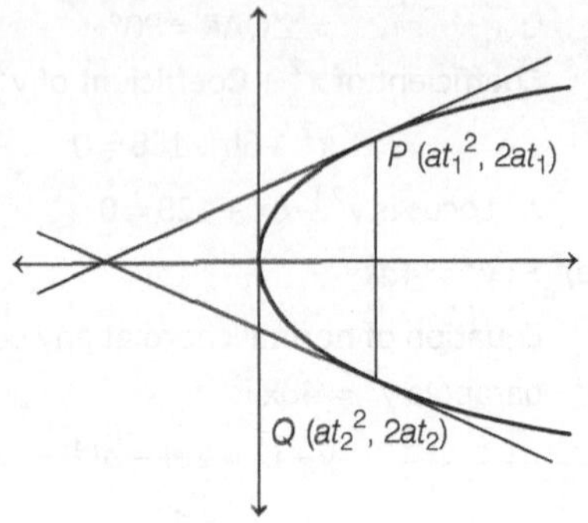

$$\Rightarrow \quad (at_1^2 - at_2^2)^2 + (2at_1 - 2at_2)^2 = 4c^2$$

$$\Rightarrow \quad a^2(t_1-t_2)^2[(t_1+t_2)^2+4] = 4c^2$$

$$\Rightarrow a^2\{(t_1+t_2)^2 - 4t_1t_2\}\{(t_1+t_2)^2+4\} = 4c^2$$

$$\Rightarrow \quad a^2\left(\frac{k^2}{a^2} - \frac{4h}{a}\right)\left(\left(\frac{k}{a}\right)^2 + 4\right) = 4c^2$$

$$\Rightarrow \quad (k^2 - 4ah)(k^2 + 4a^2) = 4a^2c^2$$

$\because$ Locus is $(y^2 - 4ax)(y^2 + 4a^2) = 4a^2c^2$

(c) Let pole be (h, k)

Equation of chord = Equation of polar of (h, k) is

$$ky = 2a(x+h)$$

$$\Rightarrow ky - 2ax - 2ah = 0$$

Its distance from $(0, 0)$ is b.

$$\left|\frac{2ah}{\sqrt{4a^2+k^2}}\right| = b$$

On squaring both sides, we get

$$4a^2h^2 = b^2(4a^2 + k^2)$$

$\because$ Locus is $4a^2x^2 = b^2(4a^2 + y^2)$

18. (a) Let the equation of the parabola be taken as

$$y^2 = 4ax$$

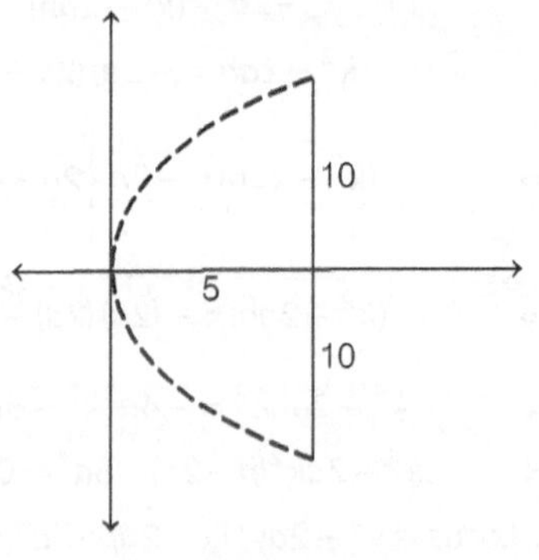

Then, the point $(5, 10)$ must lie on it

$$\Rightarrow \quad (10)^2 = 4a(5)$$

$$4a = \frac{100}{5} = 20 \Rightarrow a = 5$$

$\Rightarrow$ The focus of the reflector is at mid-point of the given diameter.

$\therefore$ Focus is $(5, 0)$.

18. (b) As we know that all rays of light parallel to X-axis of the parabola are reflected through the focus of the parabola.

Given parabola, $(y-4)^2 = 8(x+1)$

$$Y^2 = 8X$$

where, $Y = y-4$, $X = x+1$

Focus of the parabola is $(a, 0)$.

$\therefore \quad X = a$ and $Y = 0$

$x+1 = 2$ and $y-4 = 0$

$x = 1$ and $y = 4$

$\therefore$ Focus is $(1, 4)$.

Thus, $\quad \alpha = 1$ and $\beta = 4$

Now, $\alpha + \beta + 10 = 1+4+10 = 15$

(c) Let the line $y = x+2$ intersects the parabola at P.

On solving $y = x+2$ and the parabola $y^2 = 4(x+2)$, we get

$$(x+2)^2 = 4(x+2)$$

$\Rightarrow \quad (x+2)(x+2-4) = 0$

$\Rightarrow \quad x = 2, -2$

At $x = 2, y = 4$

$\therefore$ Point P is $(2, 4)$.

Now, the equation of tangent to the parabola $y^2 = 4(x+2)$ at $P(2, 4)$ is

$$yy_1 = 4\frac{(x+x_1)}{2} + 8$$

$\Rightarrow \quad y(4) = 2(x+2) + 8$

$\Rightarrow \quad x - 2y + 6 = 0$

Let IP be the incident ray, PM be the reflected ray and PN be the normal.

As we know that the normal is equally inclined with an incident ray and reflected ray.

Now, the slope of $IP = 1$

Slope of $PN = -2$

and let slope of reflected ray $= m$

Then, $\quad \dfrac{1+2}{1-2} = \dfrac{-2-m}{1-2m}$

$\Rightarrow \quad m = \dfrac{1}{7}$

$\therefore$ Equation of the reflected ray is

$$y - 4 = \frac{1}{7}(x-2)$$

$\Rightarrow \quad 7y - 28 = x - 2$

$\Rightarrow \quad x - 7y + 26 = 0$

(d) Given equation

$$(p^2+q^2)(x^2+y^2) = (qx+py-pq)^2$$

$\Rightarrow \quad x^2 + y^2 = \dfrac{(qx+py-pq)^2}{p^2+q^2}$

Taking square root,

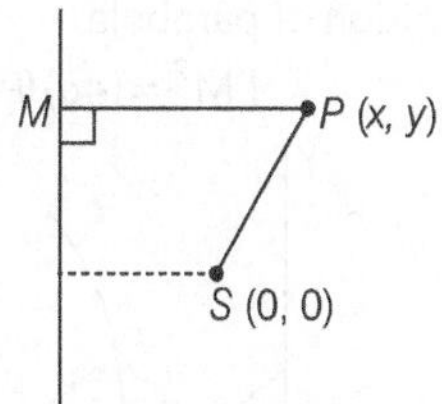

$$\sqrt{x^2+y^2} = \frac{|qx+py-pq|}{\sqrt{p^2+q^2}}$$

$$\Rightarrow \sqrt{(x-0)^2+(y-0)^2} = \frac{|qx+py-pq|}{\sqrt{p^2+q^2}}$$

which is of the form $SP = PM$

$\because$ Distance from focus $S(0, 0)$ to directrix $qx + py - pq$ is

$\Rightarrow \quad \dfrac{1}{2}$ Latusrectum $= \dfrac{pq}{\sqrt{p^2+q^2}}$

$\therefore$ Latusrectum $= \dfrac{2pq}{\sqrt{p^2+q^2}}$.

Level II

1. (d) Given, focus $= (3, 4)$

Vertex of required parabola = Focus of the parabola

$$y^2 - 12x - 4y + 4 = 0$$

Now, focus of parabola

$\Rightarrow \quad y^2 - 12x - 4y + 4 = 0$

$$(y-2)^2 = 12x$$

$$Y^2 = 12X$$

where, $Y = y-2$, $X = x$ and $4a = 12 \Rightarrow a = 3$

Now, focus is $(a, 0)$

$\Rightarrow \quad X = a$ and $Y = 0$

$\Rightarrow \quad x = 3$ and $y - 2 = 0 \Rightarrow y = 2$

$\therefore$ Focus is $(3, 2)$

and vertex of parabola is $(3, 2)$.

$\because$ Focus and vertex of required parabola lies on line $x = 3$.

$\therefore$ Axis of parabola is $x = 3$

Latusrectum $= 4(4-2) = 4(2) = 8$

$\therefore$ Equation of parabola

$$(x-3)^2 = 8(y-2)$$

$\Rightarrow \quad x^2 - 6x + 9 = 8y - 16$

$\Rightarrow \quad x^2 - 6x - 8y + 25 = 0$

2. (a) $\because$ Distance of vertex from origin is $\sqrt{2}$ and focus is $2\sqrt{2}$ units.

$\therefore$ $V(1, 1)$ and $F(2, 2)$ (i.e. lying on $y = x$)

where, length of latusrectum $= 4a = 4\sqrt{2}$ units

$\{\because a = \sqrt{2}\}$

∴ By definition of parabola

$$PM^2 = (4a)(PN) \quad ...(i)$$

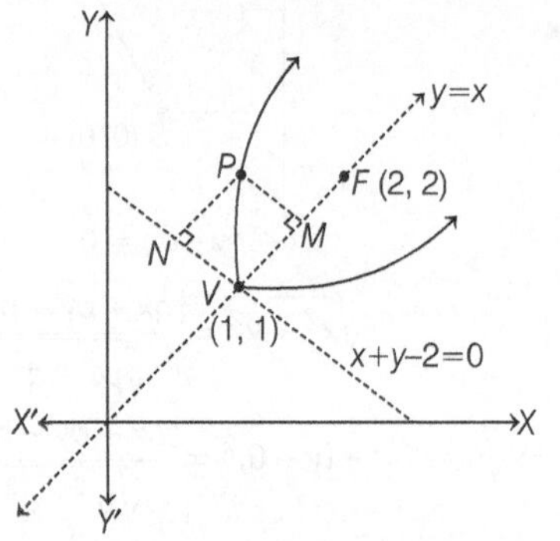

where PN is length of perpendicular upon $x + y - 2 = 0$ i.e. tangent at vertex.

Using Eq. (i),

$$\frac{(x-y)^2}{2} = 4\sqrt{2}\left(\frac{x+y-2}{\sqrt{2}}\right)$$

$$\Rightarrow \quad (x-y)^2 = 8(x+y-2)$$

3. (a) The axis of the parabola is $3x + 4y + 6 = 0$ ...(i)

Its slope is $= -\frac{3}{4}$

∴ Gradient of its perpendicular line $= \frac{4}{3}$

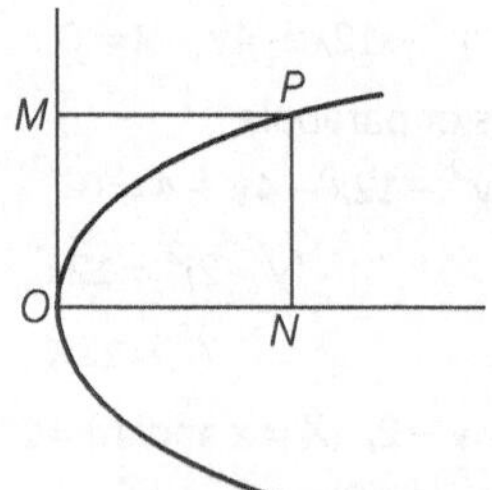

Equation of the line parallel to the line (i) and passing through vertex $A(-2, 0)$ is

$$y - 0 = \frac{4}{3}(x+2)$$

$$\Rightarrow \quad 3y = 4x + 8$$

$$\Rightarrow \quad 4x - 3y + 8 = 0 \quad ...(ii)$$

Let $P(x, y)$ be any point on the parabola.

∴ Length of perpendicular drawn from $P(x, y)$ to line (i) is

$$PN = \left|\frac{3x+4y+6}{\sqrt{3^2+4^2}}\right| = \left|\frac{3x+4y+6}{5}\right|$$

and length of the perpendicular drawn from $P(x, y)$ to the line (ii)

$$PM = \left|\frac{4x-3y+8}{\sqrt{4^2+(-3)^2}}\right| = \left|\frac{4x-3y+8}{5}\right|$$

∴ Equation of parabola is $Y^2 = 4aX$

$$\Rightarrow \quad PN^2 = 4PM \quad [\because \text{here } 4a = 4]$$

$$\Rightarrow \quad \left(\frac{3x+4y+6}{5}\right)^2 = 4\left(\frac{4x-3y+8}{5}\right)$$

$$\Rightarrow \quad (3x+4y+6)^2 = 20(4x-3y+8)$$

$$\Rightarrow 9x^2 + 16y^2 + 24xy + 36x + 48y + 36 = 80x - 60y + 160$$

$$\Rightarrow 9x^2 + 16y^2 + 24xy - 44x + 108y - 124 = 0$$

This is the required equation of the parabola.

4. (d) Given equation is

$$y = \frac{a^3x^2}{3} + \frac{a^2x}{2} - 2a$$

$$\Rightarrow \quad y = \frac{a^3}{3}\left(x^2 + \frac{3x}{2a}\right) - 2a$$

$$\Rightarrow \quad \frac{3}{a^3}(y+2a) = x^2 + \frac{3x}{2a} + \left(\frac{3}{4a}\right)^2 - \left(\frac{3}{4a}\right)^2$$

$$\Rightarrow \quad \frac{3}{a^3}y + \frac{6}{a^2} + \frac{9}{16a^2} = \left(x + \frac{3}{4a}\right)^2$$

$$\Rightarrow \quad \frac{3}{a^3}\left(y + 2a + \frac{3a}{16}\right) = \left(x + \frac{3}{4a}\right)^2$$

$$\Rightarrow \quad \left(x + \frac{3}{4a}\right)^2 = \frac{3}{a^3}\left(y + \frac{35a}{16}\right)$$

If its vertex be (x, y), then

$$x = \frac{-3}{4a} \text{ and } y = \frac{-35a}{16}$$

∴ Eliminating 'a', we get

$$xy = \frac{105}{64}$$

5. (c) The given parabola,

$$y^2 - kx + 8 = 0$$

$$\Rightarrow \quad y^2 = kx - 8$$

$$\Rightarrow \quad y^2 = k\left(x - \frac{8}{k}\right)$$

$$\Rightarrow \quad Y^2 = 4aX$$

where, $Y = y, X = x - \frac{8}{k}$ and $4a = k \Rightarrow a = \frac{k}{4}$

∵ The equation of directrix

$$X + a = 0$$

$$\Rightarrow \quad x - \frac{8}{k} + \frac{k}{4} = 0$$

But given directrix is $x - 1 = 0$

$$\therefore \quad \frac{8}{k} - \frac{k}{4} = 1$$

$$\Rightarrow \quad 32 - k^2 = 4k$$

$$\Rightarrow \quad k^2 + 4k - 32 = 0$$

$$\Rightarrow \quad (k+8)(k-4) = 0$$

$$k = -8, 4$$

6. (a) Area of ΔABC is maximum, if CD is maximum because AB is fixed.

Equation of AB is

$$\Rightarrow \quad y-6=\frac{6+4}{9-4}(x-9)$$

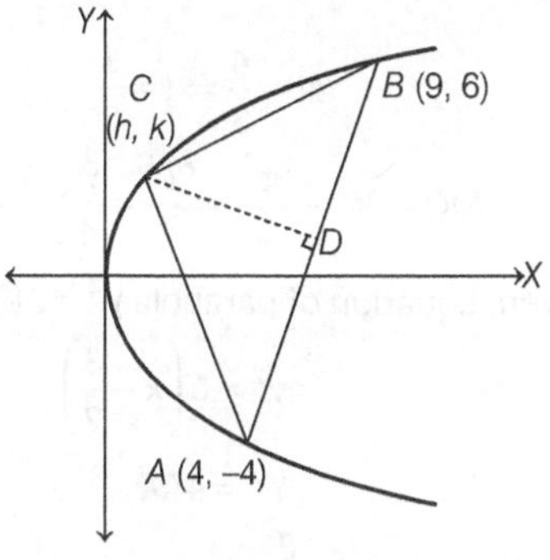

$$\Rightarrow \quad y-6=2(x-9)$$

$$y=2x-12 \Rightarrow 2x-y-12=0$$

$$\therefore \quad CD=\left|\frac{2h-k-12}{\sqrt{2^2+(1)^2}}\right|$$

$$\left[\because C \text{ lies on } y^2=4x,\ k^2=4h \text{ and } h=\frac{k^2}{4}\right]$$

$$\Rightarrow \quad CD=\frac{|2h-k-12|}{\sqrt{5}}$$

$$\Rightarrow \quad CD=\left|\frac{\frac{k^2}{2}-k-12}{\sqrt{5}}\right|$$

$$\Rightarrow \quad \frac{k^2-2k-24}{2\sqrt{5}}$$

$$CD=\frac{|k-1)^2-25|}{20}=\frac{25-(k-1)^2}{20}$$

$\therefore$ CD maximum when, $k-1=0$

$$k=1$$

$$\therefore \quad h=\frac{1}{4}$$

$\therefore$ C is $\left(\frac{1}{4},1\right)$.

7. (b) Given parabola, $S: y=-\frac{x^2}{16}$

$$\Rightarrow \quad x^2=-16y$$

$$S: x^2+16y=0$$

$\because$ $P(2a, a)$ lies inside the parabola.

$$\therefore \quad S_1<0$$

$$\Rightarrow \quad 4a^2+16a<0$$

$$\Rightarrow \quad 4a(a+4)<0$$

$$\Rightarrow \quad a=0,-4$$

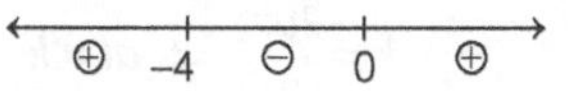

$\therefore$ $a \in (-4, 0)$

8. (a) Let mid-point of PQ be (α, β).

$$\therefore \quad \alpha=\frac{x+x+c}{2}, \quad \beta=\frac{y+y+c}{2}$$

$$\alpha=x+\frac{c}{2}, \quad \beta=y+\frac{c}{2}$$

$$\Rightarrow \quad x=\alpha-\frac{c}{2}, \quad y=\beta-\frac{c}{2}$$

$\because$ (x, y) lies on parabola $y^2=4ax$

$$\therefore \quad \left(\beta-\frac{c}{2}\right)^2=4a\left(\alpha-\frac{c}{2}\right)$$

$\therefore$ Locus is $\left(y-\frac{c}{2}\right)^2=4a\left(x-\frac{c}{2}\right)$

which is parabola.

9. (c) Given, $\quad C_1: x^2+y^2-4=0$

$$C_2: y^2-4x=0$$

$\because$ Point $(\alpha, 1)$ lies inside the curve C_1 and C_2 both

$$\therefore \quad S_1<0 \text{ and } S_2<0$$

$$\Rightarrow \quad \alpha^2+1^2-4<0 \text{ and } 1^2-4\alpha<0$$

$$\alpha^2-3<0 \text{ and } 4\alpha>1$$

$$\Rightarrow \quad (\alpha-\sqrt{3})(\alpha+\sqrt{3})<0 \text{ and } \alpha>\frac{1}{4}$$

Common

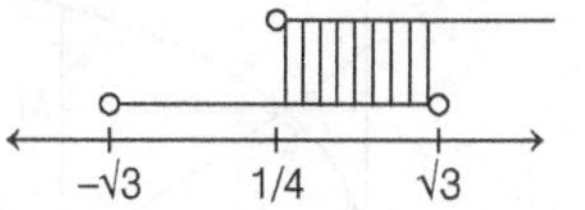

$\therefore$ $\alpha \in \left(\frac{1}{4}, \sqrt{3}\right)$ or $\frac{1}{4}<\alpha<\sqrt{3}$

10. (a) Given equation of parabola is

$$y^2=4ax$$

Focus : $S(a, 0)$

Let any point on the parabola be

$P(at^2, 2at)$ and let the mid-point of PS be $M(h, k)$

$$\therefore \quad h=\frac{a+at^2}{2} \Rightarrow h=\frac{a}{2}(1+t^2) \quad \text{...(i)}$$

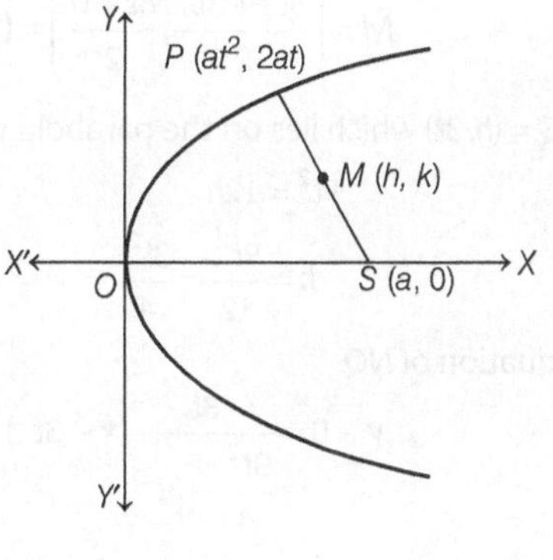

and $\qquad k=\dfrac{2at}{2} \Rightarrow at=k$

$\Rightarrow \qquad t=\dfrac{k}{a}$ ...(ii)

Put $t=\dfrac{k}{a}$ in Eq. (i),

$$h=\frac{a}{2}\left[1+\left(\frac{k}{a}\right)^2\right]$$

$\Rightarrow \qquad h=\dfrac{a}{2}\times\dfrac{1}{a^2}(k^2+a^2)$

$\Rightarrow \qquad 2ah=k^2+a^2$

$\therefore$ Locus is $y^2=2ax-a^2$

$\Rightarrow \qquad y^2=2a\left(x-\dfrac{a}{2}\right)$

which is the form of parabola $Y^2=4AX$

where, $Y=y$, $X=x-\dfrac{a}{2}$ and $4A=2a \Rightarrow A=\dfrac{a}{2}$

$\therefore$ Directrix is $X+A=0$

$$x-\frac{a}{2}+\frac{a}{2}=0$$

$$x=0$$

11. (d) Given, parabola $y^2=12x$

$\because \qquad 4a=12 \Rightarrow a=3$

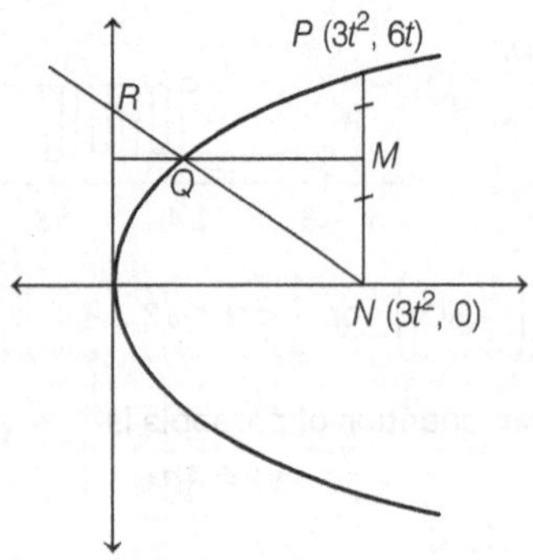

Let P be any point on parabola

$\Rightarrow \qquad P=(at^2, 2at)=(3t^2, 6t)$

$\Rightarrow \qquad N=(3t^2, 0)$

$\therefore \qquad M=\left(\dfrac{3t^2+3t^2}{2}, \dfrac{6t+0}{2}\right)=(3t^2, 3t)$

Let $Q=(h, 3t)$ which lies on the parabola $y^2=12x$

$\therefore \qquad 9t^2=12h$

$\Rightarrow \qquad h=\dfrac{9t^2}{12}=\dfrac{3t^2}{4}$

$\therefore$ Equation of NQ

$$y-0=\frac{3t}{\frac{3t^2}{4}-3t^2}(x-3t^2)$$

$\Rightarrow \qquad y=\dfrac{-4}{3t}(x-3t^2)$

For y-intercept $x=0$

$\therefore \qquad y$-intersept $=4t$

Now, given y-intercept of line NQ is $\dfrac{4}{3}$.

$\therefore \qquad 4t=\dfrac{4}{3} \Rightarrow t=\dfrac{1}{3}$

$\therefore \qquad MQ=3t^2-\dfrac{3t^2}{4}=\dfrac{9t^2}{4}=\dfrac{9}{4}\times\dfrac{1}{9}=\dfrac{1}{4}$

12. (c) Given, equation of parabola $y^2=3(2x-3)$

$$y^2=6\left(x-\frac{3}{2}\right)$$

$\because \qquad Y^2=4AX$

where, $Y=y, X=x-\dfrac{3}{2}$

and $\quad 4A=6 \Rightarrow A=\dfrac{3}{2}$

$\therefore$ Focus $(S)=x-\dfrac{3}{2}=\dfrac{3}{2}$

$\qquad x=3$ and $y=0 \Rightarrow (3, 0)$

Equation of directrix

$$X+A=0$$

$\Rightarrow \qquad x-\dfrac{3}{2}+\dfrac{3}{2}=0$

$\Rightarrow \qquad x=0$

Let the coordinates of point P on parabola $(At^2, 2At)$

$$X=At^2, \quad Y=2At$$

$\Rightarrow \qquad X=\dfrac{3}{2}t^2, \quad Y=3t$

$\Rightarrow \qquad x-\dfrac{3}{2}=\dfrac{3}{2}t^2, \quad y=3t$

$\therefore P\left(\dfrac{3}{2}+\dfrac{3}{2}t^2, 3t\right)$

$\therefore$ Coordinates of $M=(0, 3t)$

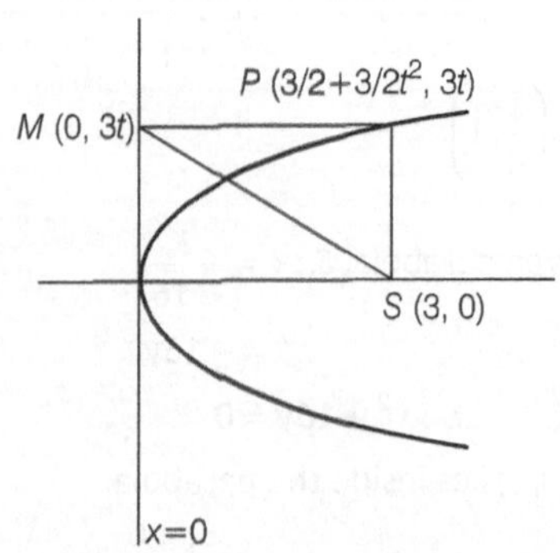

$\therefore \qquad MS=\sqrt{9+9t^2}=3\sqrt{1+t^2}$

and $\qquad MP=\dfrac{3}{2}+\dfrac{3}{2}t^2$

$\therefore \quad MS = PM$

$\therefore \quad 9 + 9t^2 = \left\{\frac{3}{2}(1+t^2)\right\}^2$

$\Rightarrow \quad 1 + t^2 = \frac{1}{4}(1+t^2)^2$

$\Rightarrow \quad 1 = \frac{1}{4}(1+t^2)$

$\Rightarrow \quad 1 + t^2 = 4$

Side of Δ SPM.

$\therefore \quad MS = 3\sqrt{4} = 3 \times 2 = 6$ units

13. (c) Let R be the point on parabola $y^2 = 4ax$.

Then, $R = (at^2, 2at)$

$\because$ RQ is focal chord

$\therefore \quad Q = \left(\frac{a}{t^2}, \frac{2a}{t}\right)$

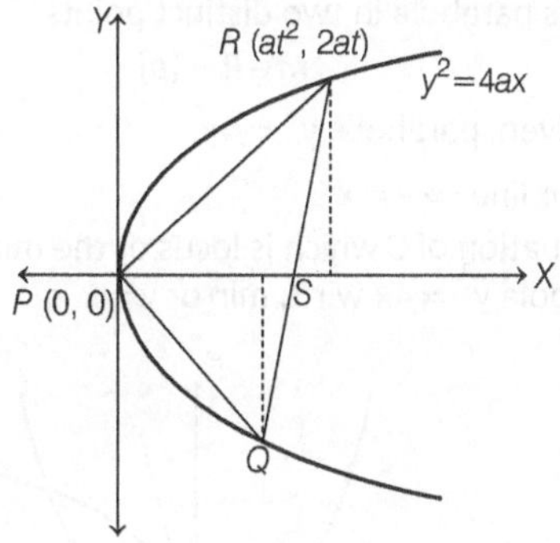

$\because$ Area of $\Delta PQR = \frac{1}{2}$ mod of $\begin{vmatrix} 0 & 0 & 1 \\ at^2 & 2at & 1 \\ \frac{a}{t^2} & \frac{2a}{t} & 1 \end{vmatrix}$

$= \frac{1}{2}\left|(at^2)\left(\frac{2a}{t}\right) - (2at)\left(\frac{a}{t^2}\right)\right|$

$= \frac{1}{2}\left|2a^2t - \frac{2a^2}{t}\right|$

$2A = 2a^2\left|t - \frac{1}{t}\right|$

$\Rightarrow \quad 2a\left|t - \frac{1}{t}\right| = \frac{2A}{a}$

$\because$ Difference of ordinate $\left|2at - \frac{2a}{t}\right|$

$= 2a\left|t - \frac{1}{t}\right| = \frac{2A}{a}$

14. (d) Let the ends of the focal chord be $(at_1^2, 2at_1)$ and $(at_2^2, 2at_2)$, then $t_1t_2 = -1$

$\Rightarrow \quad c^2 = [\{a(t_1^2 - t_2^2)\}^2 + \{2a(t_1 - t_2)\}^2]$

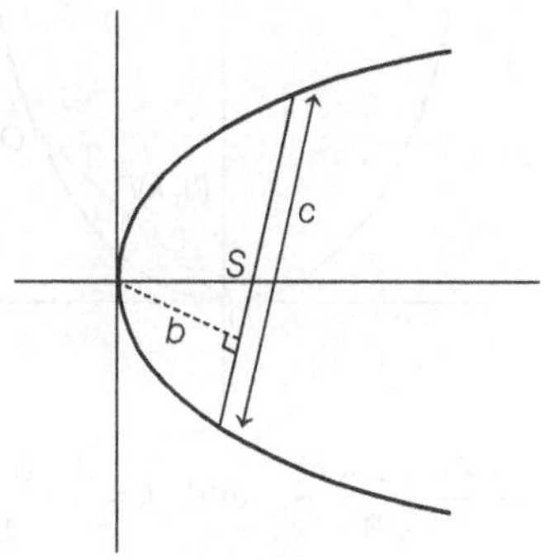

The equation of focal chord is

$$y - 2at_1 = \frac{2at_1 - 2at_2}{at_1^2 - at_2^2}(x - at_1^2)$$

or $\quad y - 2at_1 = \frac{2}{t_1 + t_2}(x - at_1^2) \quad$ ...(i)

$\therefore \quad b =$ Distance between (0, 0) and line (i)

$$= \left|\frac{2at_1 + \frac{2}{t_1+t_2}(-at_1^2)}{\sqrt{1 + \frac{4}{(t_1+t_2)^2}}}\right| = \frac{2at_1t_2}{\sqrt{4 + (t_1+t_2)^2}}$$

$\therefore \quad b = \left|\frac{-2a}{\sqrt{2 + t_1^2 + t_2^2}}\right| \quad$ ...(ii) $\{\because t_1t_2 = -1\}$

Also, $c^2 = a^2(t_1 - t_2)^2(\{t_1 + t_2\}^2 + 4)$

$= a^2(t_1^2 + t_2^2 + 2)(t_1^2 + t_2^2 + 2)$

$c = a(2 + t_1^2 + t_2^2)$

$\therefore \quad \frac{c}{a} = 2 + t_1^2 + t_2^2 \quad$ ...(iii)

Now, by Eq. (ii)

$$b^2 = \frac{4a^2}{2 + t_1^2 + t_2^2} \Rightarrow 2 + t_1^2 + t_2^2 = \frac{4a^2}{b^2}$$

From Eq. (iii), we get

$$\frac{c}{a} = \frac{4a^2}{b^2}$$

$\Rightarrow \quad b^2c = 4a^3$

15. (d) Equation of parabola is $x^2 = 8y$, which is the form of $x^2 = 4ay$, we get

$$4a = 8 \Rightarrow a = 2$$

Let point Q on parabola

$$(2at, at^2) \Rightarrow (4t, 2t^2)$$

Let $P(h, k)$ be the point which divides the line segment joining vertex $O(0, 0)$ and $Q(4t, 2t^2)$ in the ratio 1 : 3.

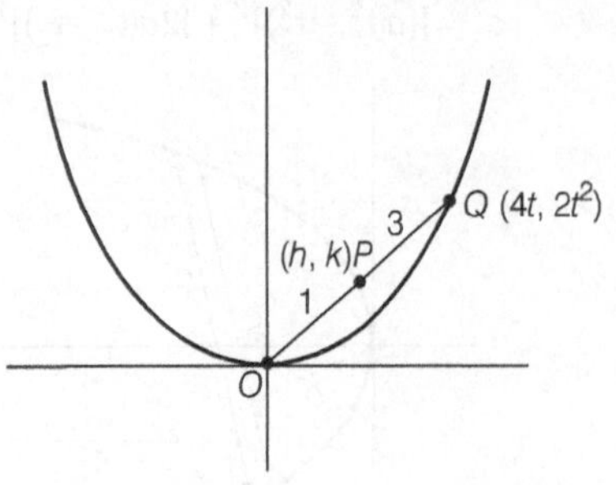

$\therefore \quad h = \dfrac{3 \times 0 + 1 \times 4t}{1+3}$ and $k = \dfrac{3 \times 0 + 2t^2 \times 1}{1+3}$

$$h = \frac{4t}{4} \text{ and } k = \frac{2t^2}{4}$$

$$\therefore \quad k = \frac{2(h^2)}{4}$$

$$k = \frac{h^2}{2}$$

$$\Rightarrow \quad h^2 = 2k$$

$\therefore$ Locus is $x^2 = 2y$

16. (d) Given, $y - \sqrt{3}x + 3 = 0$ can be written as

$$\frac{y-0}{\frac{\sqrt{3}}{2}} = \frac{x-\sqrt{3}}{1/2} = r \quad \text{...(i)}$$

$$\Rightarrow \quad y = \frac{r\sqrt{3}}{2} \text{ and } x = \sqrt{3} + \frac{r}{2}$$

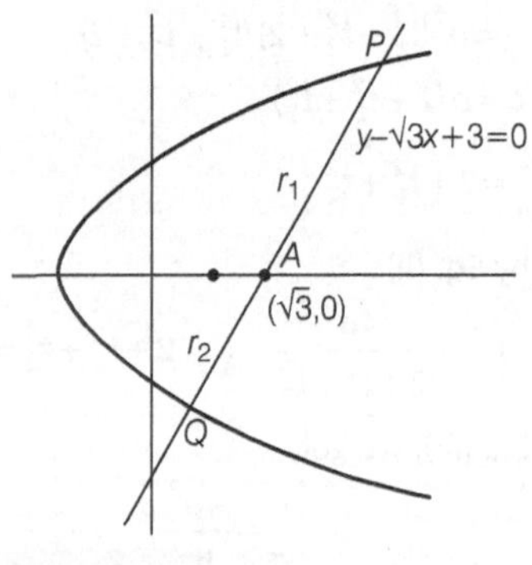

Given, parabola $y^2 = x + 2$ ...(ii)

From Eq. (i) and (ii), we get

$$\frac{3r^2}{4} = \sqrt{3} + \frac{r}{2} + 2$$

$$\Rightarrow \quad 3r^2 - 2r - (4\sqrt{3} + 8) = 0 \quad \text{...(iii)}$$

$\because$ r_1 and r_2 are the roots of Eq. (iii) and distance of AP and AQ.

$$\therefore \quad AP \times AQ = |r_1 r_2| = \left|\frac{c}{a}\right|$$

$$= \left|\frac{-(4\sqrt{3}+8)}{3}\right| = \frac{4(\sqrt{3}+2)}{3}$$

17. (b) $\because$ Line $L : x + y - 1 = 0$ is tangent for parabola $y^2 = kx$

$\because L : y = -x + 1$ and $y^2 = kx$

$\therefore \quad m = -1, c = 1$ and $4a = k$

$$\Rightarrow \quad a = \frac{k}{4}$$

From condition of tangency

$$c = \frac{a}{m}$$

$$\Rightarrow \quad 1 = \frac{\frac{k}{4}}{-1}$$

$$\frac{k}{-4} = 1$$

$$\Rightarrow \quad k = -4$$

18. (d) We know that

Any line passing through focus other than axis always meets parabola in two distinct points

$$\therefore \quad m \in R - \{o\}$$

19. (a) Given, parabola, $y^2 = 4x$

Mirror line $\Rightarrow y = x$

$\therefore$ Equation of C which is locus of the mirror image of parabola $y^2 = 4x$ w.r.t. mirror $y = x$ is

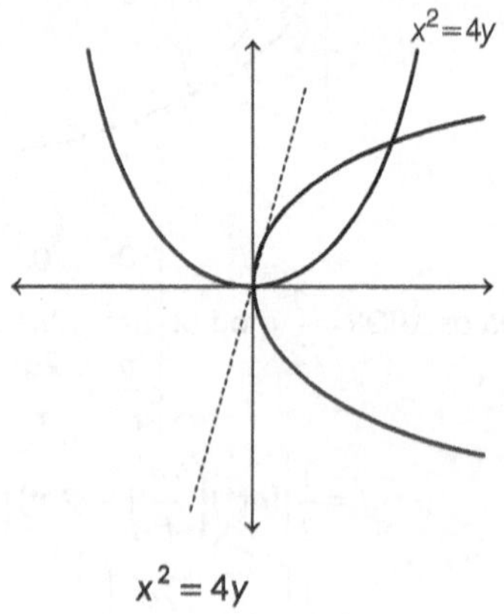

$$\therefore \quad x^2 = 4y$$

Now, tangent of parabola $x^2 = 4y$ at (2, 1) is given by

$$xx_1 = 4\left(\frac{y+y_1}{2}\right)$$

$$\Rightarrow \quad x(2) = 2(y+1)$$

$$\Rightarrow \quad x = y + 1$$

$$\Rightarrow \quad x - y = 1$$

20. (a) $\because$ A line from point $A(-2, 0)$ cut the parabola at P and Q in first quadrant.

Now, let as assume it makes angle θ with X-axis and $\theta \in \left(0, \frac{\pi}{2}\right)$.

$\therefore P(-2 + r\cos\theta, r\sin\theta)$ and Q lies on the parabola $y^2 = 4x$

$$\Rightarrow \quad (r^2 \sin^2\theta) - 4(-2 + r\cos\theta) = 0$$

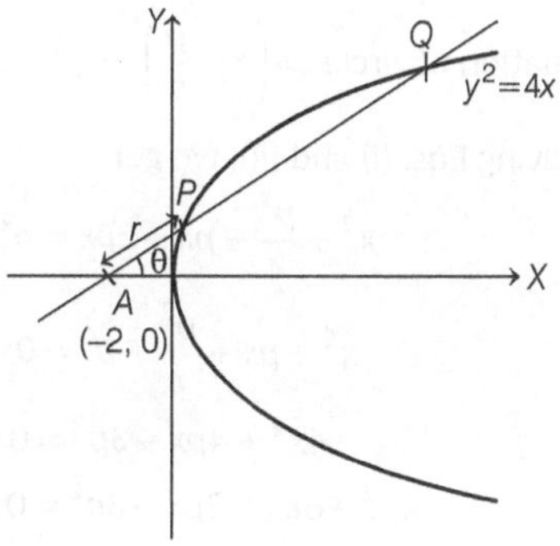

$$\Rightarrow \quad r^2 \sin^2\theta - 4r\cos\theta + 8 = 0$$

$$\Rightarrow \quad r_1 + r_2 = \frac{4\cos\theta}{\sin^2\theta}$$

$$\Rightarrow \quad r_1 \cdot r_2 = \frac{8}{\sin^2\theta}$$

Now, $$\frac{1}{AP} + \frac{1}{AQ} < \frac{1}{4}$$

$$\Rightarrow \quad \frac{1}{r_1} + \frac{1}{r_2} < \frac{1}{4}$$

$$\Rightarrow \quad \frac{r_1 + r_2}{r_1 r_2} < \frac{1}{4}$$

$$\Rightarrow \quad \frac{4\cos\theta / \sin^2\theta}{8 / \sin^2\theta} < \frac{1}{4}$$

$$\Rightarrow \quad \frac{\cos\theta}{2} < \frac{1}{4}$$

$$\Rightarrow \quad \cos\theta < \frac{1}{2}$$

$$\Rightarrow \quad \tan\theta > \sqrt{3} \qquad \left\{\because \theta \in \left(0, \frac{\pi}{2}\right)\right\}$$

$$\Rightarrow \quad m > \sqrt{3}$$

21. (c) Given, $y^2 = 8x \Rightarrow 4a = 8 \Rightarrow a = 2$

$\because$ We know that equation of tangent is slope form for parabola $y^2 = 4ax$ is (let m be slope of tangent)

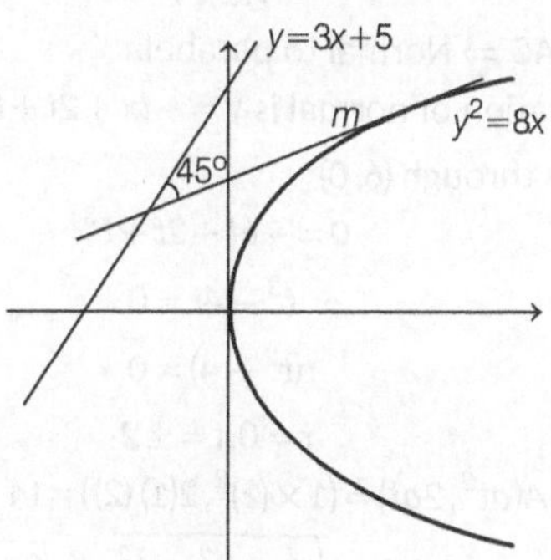

$$y = mx + \frac{a}{m}$$

Here, $a = 2$

$$\therefore \quad y = mx + \frac{2}{m}$$

According to question,

$$\tan\theta = \left|\frac{m_1 - m_2}{1 + m_1 m_2}\right|$$

$$\Rightarrow \quad \tan 45° = \left|\frac{m - 3}{1 + 3m}\right|$$

$$\Rightarrow \quad \pm 1 = \frac{m - 3}{1 + 3m}$$

$$\Rightarrow \quad 1 + 3m = \pm(m - 3)$$

$$\Rightarrow \quad 3m + 1 = m - 3 \qquad 3m + 1 = -m + 3$$

$$\Rightarrow \quad 2m = -4 \qquad 4m = 2$$

$$\Rightarrow \quad m = -2 \qquad m = \frac{1}{2}$$

Equation of tangent	Equation of tangent
$y = -2x + \frac{2}{(-2)}$	$y = \frac{x}{2} + \frac{2}{1/2}$
$\Rightarrow \quad y = -2x - 1$	$2y = x + 8$
$\Rightarrow 2x + y + 1 = 0$	$x - 2y + 8 = 0$
at $POC \Rightarrow \left(\frac{a}{m^2}, \frac{2a}{m}\right)$	at $POC \Rightarrow \left(\frac{a}{m^2}, \frac{2a}{m}\right)$
$\Rightarrow \left(\frac{2}{4}, \frac{4}{-2}\right) = \left(\frac{1}{2}, -2\right)$	$= \left(\frac{2}{1/4}, \frac{4}{1/2}\right) = (8, 8)$

22. (c) $\because$ We know that equation of tangent to parabola

$$y^2 = 4ax \text{ is } y = mx + \frac{a}{m}$$

$$\therefore \text{ For } \quad y^2 = 4x \Rightarrow 4a = 4 \Rightarrow a = 1$$

$$\Rightarrow \text{Equation of tangent} \Rightarrow y = mx + \frac{1}{m}$$

$$m^2 x - my + 1 = 0 \qquad \text{...(i)}$$

Now, line (i) is also tangent of circle

$$x^2 + y^2 - 6x = 0$$

Centre of circle (3, 0) and radius = 3

$\because$ Distance between centre and tangent = radius

$$\therefore \quad \left|\frac{m^2(3) - m(0) + 1}{\sqrt{m^4 + m^2}}\right| = 3$$

$$\Rightarrow \quad \frac{(3m^2 + 1)^2}{(\sqrt{m^4 + m^2})^2} = (3)^2$$

$$\Rightarrow \quad (3m^2 + 1)^2 = 9(m^4 + m^2)$$

$$\Rightarrow \quad 9m^4 + 6m^2 + 1 = 9m^4 + 9m^2$$

$$\Rightarrow \quad 3m^2 = 1$$

or $$m = \infty$$

$$\left\{\because \lim_{m \to \infty} \frac{3m^2 + 1}{\sqrt{m^4 + m^2}} \Rightarrow \lim_{m \to \infty} \frac{m^2\left(3 + \frac{1}{m^2}\right)}{m^2 \sqrt{1 + \frac{1}{m^2}}} = 3\right\}$$

$$\Rightarrow \qquad m = \pm\frac{1}{\sqrt{3}} \text{ or } m = \infty$$

$\because$ Equation of common tangent are $x = 0$

$$y = \frac{x}{\sqrt{3}} + \sqrt{3} \text{ and } y = \frac{-x}{\sqrt{3}} - \sqrt{3}$$

$$\therefore \qquad x = 0, \sqrt{3}y = x + 3 \text{ and } \sqrt{3}y = -x - 3$$

$\Rightarrow$ Equation of tangents

23. (d) The tangent at (1, 7) to the parabola $x^2 = y - 6x$ is

$$x(1) = \frac{1}{2}(y + 7) - 6$$

$\{$Replacing $x^2 \to xx_1, 2y \to y + y_1\}$

$$\Rightarrow \qquad y = 2x + 5$$

which is also tangent to the circle.

$$x^2 + y^2 + 16x + 12y + c = 0$$

$\therefore$ $x^2 + (2x + 5)^2 + 16x + 12(2x + 5) + c = 0$ have equal and real roots

$$\Rightarrow \qquad 5x^2 + 60x + 85 + c = 0$$

$$\Rightarrow \qquad \alpha = \beta$$

$$\Rightarrow \qquad \alpha + \beta = \frac{-b}{a}$$

$$\Rightarrow \qquad 2\alpha = \frac{-60}{5}$$

$$\Rightarrow \qquad \alpha = -6$$

$$\therefore \qquad x = -6$$

and $\qquad y = 2x + 5$

$$= 2(-6) + 5$$

$$y = -7$$

$\therefore$ Point of contact $(-6, -7)$.

24. (a) Given, parabola $y^2 = 2px$ $\qquad$ ...(i)

$$4a = 2p \Rightarrow a = \frac{p}{2}$$

$\therefore$ Focus $= \left(\frac{p}{2}, 0\right)$ and directrix $\Rightarrow x + a = 0$

$$x + \frac{p}{2} = 0$$

Focus is centre of circle, having radius r.

$\therefore$ Equation of circle is

$$(y - 0)^2 + \left(x - \frac{p}{2}\right)^2 = r^2$$

$$\Rightarrow \qquad \left(x - \frac{p}{2}\right)^2 + y^2 = r^2$$

$\because$ It touch the directrix $x + \frac{p}{2} = 0$, then

r = Distance between $x = -\frac{p}{2}$ and $\left(\frac{p}{2}, 0\right) = |p|$

$\because$ Equation of circle $\Rightarrow \left(x - \frac{p}{2}\right)^2 + y^2 = p^2$ $\qquad$ ...(ii)

On solving Eqs. (i) and (ii), we get

$$x^2 + \frac{p^2}{4} - px + 2px = p^2$$

$$\Rightarrow \qquad x^2 + px + \frac{p^2}{4} - p^2 = 0$$

$$\Rightarrow \qquad 4x^2 + 4px - 3p^2 = 0$$

$$\Rightarrow \qquad 4x^2 + 6px - 2px - 3p^2 = 0$$

$$\Rightarrow \qquad 2x(2x + 3p) - p(2x + 3p) = 0$$

$$\Rightarrow \qquad (2x + 3p)(2x - p) = 0$$

$$\Rightarrow \qquad x = \frac{-3p}{2}, x = \frac{p}{2}$$

$\because$ $x = \frac{p}{2}$, we get

$$y = \pm\sqrt{2px} = \pm p.$$

$\because$ Points are $\left(\frac{p}{2}, \pm p\right)$.

25. (b) Parabola $\Rightarrow y^2 = 4x$

Circle : $(x - 6)^2 + y^2 = r^2$

Centre = (6, 0) and radius = r

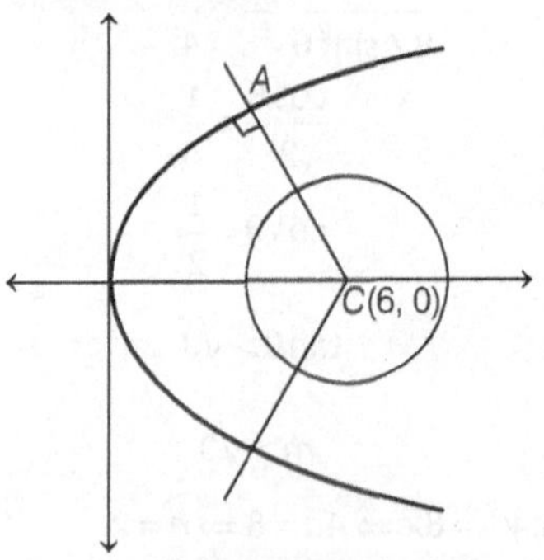

$\because$ For no common tangent circle lies inside the parabola

$$\Rightarrow \qquad AC > r$$

Now, $AC \Rightarrow$ Normal to parabola

$\because$ Equation of normal is $y = -tx + 2t + t^3$.

Passes through (6, 0)

$$\therefore \qquad 0 = -6t + 2t + t^3$$

$$t^3 - 4t = 0$$

$$t(t^2 - 4) = 0$$

$$t = 0, t = \pm 2$$

$$\therefore \qquad A(at^2, 2at) = (1 \times (2)^2, 2(1)(2)) = (4, 4)$$

$$AC = \sqrt{(6-4)^2 + 4^2} = \sqrt{20}$$

$$\therefore \qquad AC > r \Rightarrow \sqrt{20} > r \text{ or } r < \sqrt{20}$$

26. (b) Given parabola, $y^2 = 4x$

$$4a = 4 \Rightarrow a = 1$$

$\because$ Equation of normal at point $P(at^2, 2at)$
$\Rightarrow (t^2, 2t)$ is $tx + y = 2t + t^3$

$\because$ (9, 6) lies on parabola and also passes through normal

$\therefore \quad t^2 = 9$ or $2t = 6 \Rightarrow t = 3$

$\therefore$ Equation of normal is $3x + y = 6 + 27$

$$3x + y - 33 = 0$$

27. (d) Given, parabola, $y^2 = 4x$, $4a = 4 \Rightarrow a = 1$
and point (1, 2).

$\because$ Equation of normal at $P(at_1^2, 2at_1) \Rightarrow (t_1^2, 2t_1)$
is $t_1x + y = 2t_1 + t_1^3$

$\because \quad t_1^2 = 1$ and $2t_1 = 2 \Rightarrow t_1 = 1$

$\therefore$ Equation is $x + y = 2 + 1$

$$x + y = 3 \quad \text{...(i)}$$

Meets again $\quad y^2 = 4x \quad$ ...(ii)

$\therefore$ Solving Eqs. (i) and (ii),

$$(3 - x)^2 = 4x$$

$\Rightarrow \quad x^2 + 9 - 6x = 4x$

$\Rightarrow \quad x^2 - 10x + 9 = 0$

$\Rightarrow \quad (x - 9)(x - 1) = 0$

$\Rightarrow \quad x = 1$ and $x = 9$

$\therefore$ Point $(9, -6)$ and given that $(t^2, 2t)$

$$t = -3$$

28. (b) $\because f(x) = \dfrac{1}{1 - x}$

$$f\{f(x)\} = \frac{1}{1 - \frac{1}{1 - x}} = \frac{1 - x}{-x} = \frac{x - 1}{x}$$

$\because f\{f(x)\}$ is discontinuous at $x = 0, 1$

$\therefore$ Number of discontinuity = 2

$\because$ A normal to the parabola $y^2 = 4ax$ at $(at_1^2, 2at_1)$ meets again at $(at_2^2, 2at_2)$

$\therefore \quad t_2 = -t_1 - \dfrac{2}{t_1}$

$\therefore \quad -(t_1t_2 + t_1^2) = 2$

= Number of point of discontinuity

29. (a) Given parabola, $y^2 = 8x$

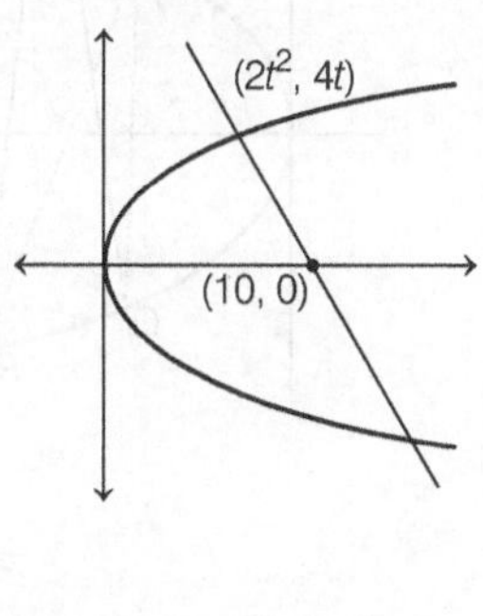

$\because$ Equation of normal at any point $P(at^2, 2at)$ is

$$y + tx = 2at + at^3$$

$\because \quad a = 2$

$$y + tx = 4t + 2t^3$$

$$y + tx - 4t - 2t^3 = 0$$

$\because$ Focus is $(a, 0)$ or (2, 0)

So, the point on axis is (10, 0).

$\because$ Normal passes through (10, 0)

$$0 + 10t = 4t + 2t^3$$

$\Rightarrow \quad 6 = 2t^2$

$\Rightarrow \quad t^2 = 3 \Rightarrow t = \sqrt{3}$

So, the normal is at point $(6, 4\sqrt{3})$

So, required length is $\sqrt{(10 - 6)^2 + (4\sqrt{3} - 0)^2}$

$$= \sqrt{(16 + 48)} = \sqrt{64} = 8 \text{ units}$$

30. (b) $\because$ We know that for parabola $y^2 = 4ax$, the equation of normal having slope m, is
$y = mx - 2am - am^3$, where foot of perpendicular is $(am^2, -2am)$

$\because$ Here, $y^2 = 8x$ and $\theta = 45°$

$\therefore \quad 4a = 8$ and $m = \tan 45°$

$$a = 2 \text{ and } m = 1$$

$\therefore$ Foot of perpendicular is $(2(1)^2, -2(2)(1)) = (2, -4)$.

31. (d) Given parabolas

$$y^2 = 4ax \text{ and } x^2 = 4ay.$$

$\because$ One parabola is the image of the other parabola in the line $x = y$

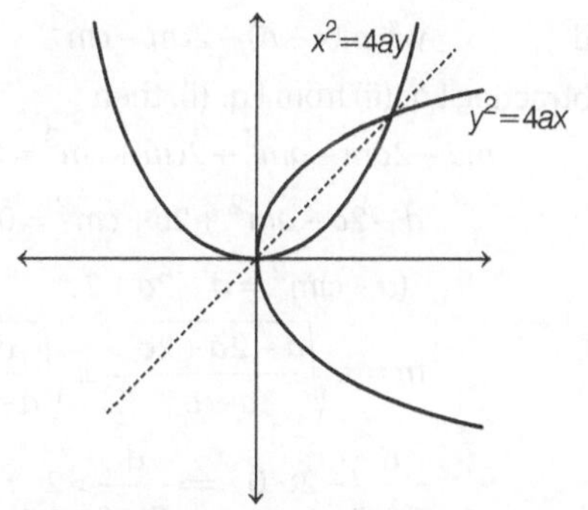

The slope of common normal is -1

$\because$ It is perpendicular to $x - y = 0$

Normal at t is

$$xt + y = at^3 + 2at$$

$$\text{Slope} = -t = -1 \quad [\because t = 1]$$

$\therefore$ Common normal

$$x + y = a + 2a$$

$$x + y = 3a$$

32. (a) $\because$ Equation of normal of parabola

$$(y-\alpha)^2 = 4a(x-\beta)$$

is $\quad (y-\alpha) = m(x-\beta) - 2am - am^3 \quad$ ...(i)

Equation of normal of parabola $(y-\alpha)^2 = 4a'(x-\beta')$ is

$$(y-\alpha) = m(x-\beta') - 2a'm - a'm^3 \quad \text{...(i)}$$

$\{\because$ Normal are same$\}$

$$\therefore \quad 1 = \frac{m\beta + 2am + am^3}{m\beta' + 2a'm + a'm^3}$$

$$\because \quad m(\beta-\beta') + 2m(a-a') + m^3(a-a') = 0$$

$$m = 0 \text{ or } m^2(a-a') + 2(a-a') + (\beta-\beta') = 0$$

$$m = \pm\sqrt{\frac{2(a'-a) + (\beta'-\beta)}{a-a'}}$$

$$\therefore \quad \frac{2(a'-a) + (\beta'-\beta)}{a-a'} > 0$$

or $$\frac{2(a-a') + (\beta-\beta')}{a'-a} > 0$$

$$\Rightarrow \quad -2 + \left(\frac{\beta-\beta'}{a'-a}\right) > 0$$

$$\Rightarrow \quad \frac{\beta-\beta'}{a'-a} > 2$$

$$\Rightarrow \quad \frac{\beta-\beta'}{2(a'-a)} > 1 \Rightarrow \frac{2(a'-a)}{\beta-\beta'} < 1$$

or $$\frac{2(a-a')}{\beta'-\beta} < 1$$

33. (a) Normal of parabolas $y^2 = 4ax$ and $y^2 = 4c(x-d)$ in terms of slope are

$$y = mx - 2am - am^3 \quad \text{...(i)}$$

and $$y = m(x-d) - 2cm - cm^3 \quad \text{...(ii)}$$

Subtracting Eq. (ii) from Eq. (i), then

$$md - 2am - am^3 + 2cm + cm^3 = 0 \quad (\because m \neq 0\}$$

$$\Rightarrow \quad d - 2a - am^2 + 2c + cm^2 = 0$$

$$\Rightarrow \quad (a-c)m^2 = d - 2a + 2c$$

$$\Rightarrow \quad m = \pm\sqrt{\frac{d-2a+2c}{a-c}} = \pm\sqrt{\frac{d}{a-c} - 2}$$

$$\because \quad \frac{d}{a-c} - 2 > 0 \Rightarrow \frac{d}{a-c} > 2$$

$$d > 2a - 2c$$

$$2a < d + 2c$$

34. (b) Let, $y_1 = 3 + \sqrt{1-x_1^2}$ and $y_2 = \sqrt{4x_2}$

$$x_1^2 + (y_1-3)^2 = 1 \text{ and } y_2^2 = 4x_2$$

Thus, (x_1, y_1) lies on the circle $x^2 + (y-3)^2 = 1$ and (x_2, y_2) lies on parabola $y^2 = 4x$

Thus, the given expression is the shortest distance between the curve $x^2 + (y-3)^2 = 1$ and $y^2 = 4x$.

Now, the shortest distance always occurs along the common normal to the curves and normal to circle passes through the centre of the circle.

Normal to parabola $y^2 = 4x$ is $y = mx - 2m - m^3$.

It passes through centre of circle (0, 3).

Therefore, $3 = -2m - m^3 \Rightarrow m^3 + 2m + 3 = 0$

which has only one real value, $m = -1$

Hence, the corresponding point on the parabola is (1, 2).

$$\therefore \quad \sqrt{(x_1-x_2)^2 + (3 + \sqrt{1-x_1^2} - \sqrt{4x_2})^2} = \sqrt{(x_1-x_2)^2 + (y_1-y_2)^2}$$

$\Rightarrow$ Distance between (1, 2) and (0, 3) – radius

$$= \sqrt{1+1} - 1 = \sqrt{2} - 1$$

$$\Rightarrow \quad (x_1-x_2)^2 + (3 + \sqrt{1-x_1^2} - \sqrt{4x_2})^2 = (\sqrt{2}-1)^2$$

$$\therefore \quad \min[(x_1-x_2)^2 + (3 + \sqrt{1-x_1^2} - \sqrt{4x_2})^2] = 3 - 2\sqrt{2}$$

35. (b) Substituting $y = x - 1$ in $y^2 = 4x$, we get

$$x^2 - 6x + 1 = 0$$

whose roots are

$$x = \frac{6 \pm \sqrt{36-4}}{2} = 3 \pm 2\sqrt{2}$$

Now, at $x = 3 + 2\sqrt{2} \Rightarrow y = x - 1 = 2 + 2\sqrt{2}$

and $\quad x = 3 - 2\sqrt{2} \Rightarrow y = x - 1 = 2 - 2\sqrt{2}$

Let $A(3+2\sqrt{2}, 2+\sqrt{2})$ and $B(3-2\sqrt{2}, 2-2\sqrt{2})$.

$\because$ The normals at A, B and D meet at C and the algebric sum of the ordinates of A, B and D is zero.

$$\Rightarrow \quad 2 + 2\sqrt{2} + 2 - 2\sqrt{2} + y_3 = 0 \Rightarrow y_3 = -4$$

$$\therefore \quad D \Rightarrow y_3^2 = 4x_3$$

$$16 = 4x_3 \Rightarrow x_3 = 4$$

$$\therefore \quad D \Rightarrow (4, -4)$$

36. (a) $y^2 = 4x \Rightarrow 4a = 4 \Rightarrow a = 1$

Let $P(at_1^2, 2at_1) \equiv (t_1^2, 2t_1)$, $Q(at_2^2, 2at_2) \equiv (t_2^2, 2t_2)$

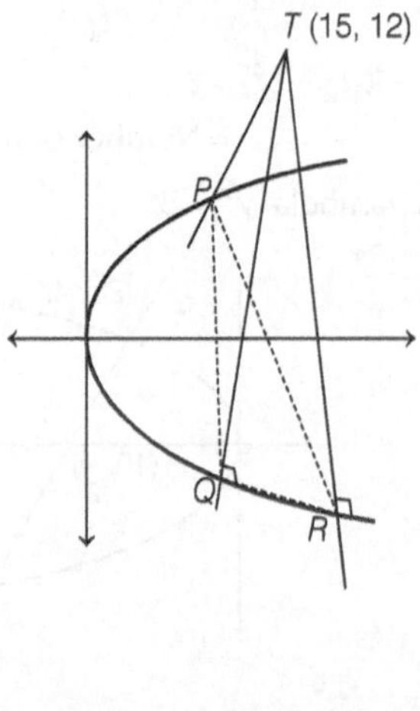

and $R(at_3^2, 2at_3) \equiv (t_3^2, 2t_3)$ be the feet of normals drawn from (15, 12).

We know that, equation of normal is

$$y + tx = 2at + at^3$$

$$y + tx = 2t + t^3$$

$$t^3 + t(2 - x) - y = 0$$

Passes through (15, 12)

$$t^3 + (2 - 15)t - 12 = 0$$

$$t^3 - 13t - 12 = 0$$

$$\because \quad t_1 + t_2 + t_3 = 0$$

$$t_1t_2 + t_2t_3 + t_3t_1 = -13$$

$$t_1t_2t_3 = 12$$

$$\because \text{ Centroid of } \Delta PQR = \left(\frac{t_1^2 + t_2^2 + t_3^2}{3}, \frac{2(t_1 + t_2 + t_3)}{3}\right)$$

$$\Rightarrow \left(\frac{(t_1 + t_2 + t_3)^2 - 2(t_1t_2 + t_2t_3 + t_3t_1)}{3}, \frac{0}{3}\right)$$

$$\Rightarrow \left(\frac{0 - 2(-13)}{3}, 0\right) \Rightarrow \left(\frac{26}{3}, 0\right)$$

37. (b) The normal at $(at^2, 2at)$, where $a = \frac{1}{4}$. Which is drawn to $y^2 = x$ is

$$tx + y = 2\left(\frac{1}{4}\right)t + \frac{1}{4}t^3$$

$$\Rightarrow \quad 4tx + 4y = 2t + t^3$$

This passes through $(c, 0)$, we get

$$4tc = 2t + t^3$$

$$\Rightarrow \quad t(t^2 + 2 - 4c) = 0$$

$$\Rightarrow \quad t = 0 \text{ or } t = \pm\sqrt{4c - 2}$$

$$\because \quad c > \frac{1}{2} \Rightarrow 4c - 2 > 0$$

It is clear that $t = 0$ gives that X-axis is normal at (0, 0). The other two normals are at right angle $\Rightarrow$ Product of their slope

$$\Rightarrow \quad \sqrt{4c - 2} \times (-\sqrt{4c - 2}) = -1$$

$$4c - 2 = 1 \Rightarrow c = \frac{3}{4}$$

38. (d) Given parabola, $S : y^2 + 4 = 4x$

$S \Rightarrow y^2 - 4x + 4 = 0$

Equation of pair of tangents drawn from (0, 0) to the parabola $SS_1 = T^2$

$$\Rightarrow (y^2 - 4x + 4)(0 - 0 + 4) = \left[y(0) - 4\left(\frac{x + 0}{2}\right) + 4\right]^2$$

$$\Rightarrow \quad (y^2 - 4x + 4)(4) = 4(x - 2)^2$$

$$\Rightarrow \quad y^2 - 4x + 4 = x^2 - 4x + 4$$

$$\Rightarrow \quad x^2 - y^2 = 0 \Leftrightarrow (x - y)(x + y) = 0$$

This represents a pair of straight line.

$\therefore$ Angle between the lines $= \frac{\pi}{2}$

$\{\because$ coefficient of x^2 + coefficient of $y^2 = 0\}$

39. (c) Equation of tangent at (x_1, y_1) for parabola $y^2 = 4ax$ is

$$yy_1 = 2a(x + x_1)$$

Here, $\quad y^2 = 4x \Rightarrow 4a = 4 \Rightarrow a = 1$

End points of latusrectum are $L(1, 2)$ and $L'(1, -2)$.

Equation of tangent at $L \Rightarrow 2y = 2(x + 1)$ ...(i)

and at $L' \Rightarrow -2y = 2(x + 1)$ ...(ii)

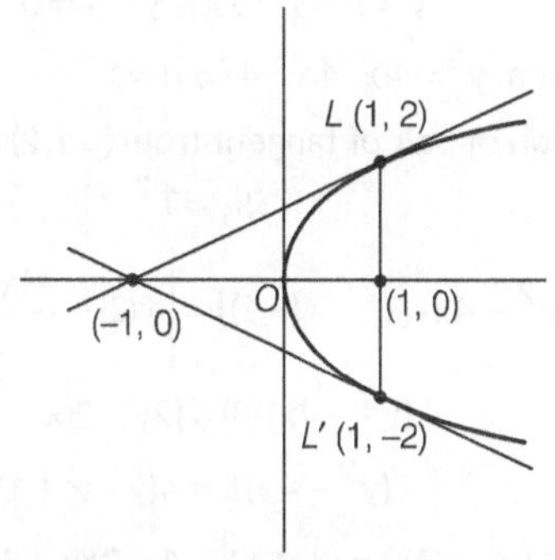

Solving Eqs. (i) and (ii), we get

$$x = -1 \text{ and } y = 0$$

$\therefore$ Required point of intersection is $(-1, 0)$.

40. (c) Given parabola, $y^2 = 4x, 4a = 4 \Rightarrow a = 1$

Let points $A(at_1^2, 2at_1) = (t_1^2, 2t_1)$

and $B(at_2^2, 2at_2) = (t_2^2, 2t_2)$ are 2 points.

Now, the points of intersection of PA and PB is $P = [at_1t_2, a(t_1 + t_2)] = (x, y)$

$$\Rightarrow \quad (t_1t_2, t_1 + t_2) = (x, y)$$

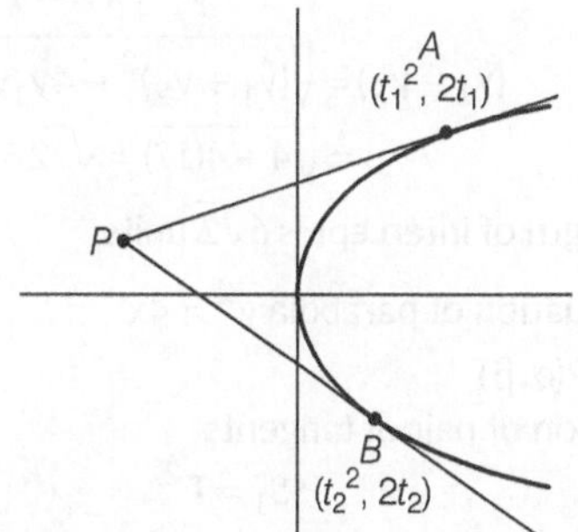

Now, slope of $AP = \frac{1}{t_1}$ and $BP = \frac{1}{t_2}$

$$\therefore \quad \tan\theta_1 = \frac{1}{t_1} \text{ and } \tan\theta_2 = \frac{1}{t_2}$$

Now, according to question

$$\theta_1+\theta_2=\frac{\pi}{4}$$

$$\Rightarrow \quad \tan(\theta_1+\theta_2)=\tan\frac{\pi}{4}$$

$$\Rightarrow \quad \frac{\tan\theta_1+\tan\theta_2}{1-\tan\theta_1\tan\theta_2} \Rightarrow \frac{\frac{1}{t_1}+\frac{1}{t_2}}{1-\frac{1}{t_1}\cdot\frac{1}{t_2}}=1$$

$$\Rightarrow \quad \frac{t_1+t_2}{t_1t_2-1}=1$$

$$\Rightarrow \quad \frac{y}{x-1}=1$$

$$\Rightarrow \quad y=x-1 \Rightarrow x-y-1=0$$

41. (b) Given, $y^2=4x$, $4a=4 \Rightarrow a=1$

Equation of pair of tangent from $(-1,2)$ is

$$SS_1=T^2$$

$$\Rightarrow \quad (y^2-4x)\{2^2-4(-1)\}=\left[y(2)-\frac{4(x-1)}{2}\right]^2$$

$$\Rightarrow \quad (y^2-4x)(8)=[2y-2(x-1)]^2$$

$$\Rightarrow \quad (y^2-4x)8=4(y-x+1)^2$$

$$\Rightarrow \quad 2(y^2-4x)=y^2+x^2+1-2xy-2x+2y$$

$$\Rightarrow \quad 2y^2-8x=y^2+x^2+1-2xy-2x+2y$$

$$\Rightarrow \quad x^2-y^2-2xy+6x+2y+1=0$$

Intercept with $x=2$

$$(2)^2-y^2-2(2)y+6(2)+2y+1=0$$

$$\Rightarrow \quad 4-y^2-4y+12+2y+1=0$$

$$\Rightarrow \quad -y^2-2y+17=0$$

$$\Rightarrow \quad y^2+2y-17=0$$

$$\Rightarrow \quad y_1+y_2=-2$$

$$\Rightarrow \quad y_1\cdot y_2=-17$$

$$\therefore \quad (y_1-y_2)=\sqrt{(y_1+y_2)^2-4y_1y_2}$$

$$=\sqrt{4+4(17)}=\sqrt{72}=6\sqrt{2}$$

$\therefore$ Length of intercept is $6\sqrt{2}$ units.

42. (b) Equation of parabola $y^2=4x$

Point $P(\alpha,\beta)$

Equation of pair of tangents

$$SS_1=T^2$$

$$\Rightarrow \quad (y^2-4x)(\beta^2-4\alpha)=[y\beta-2(x+\alpha)]^2$$

$$\Rightarrow \quad \beta^2y^2-4\alpha y^2-4x\beta^2+16x\alpha$$

$$=y^2\beta^2+4(x^2+\alpha^2+2x\alpha)-4y\beta(x+\alpha)$$

$$\Rightarrow \quad 4x^2+4\alpha^2+8\alpha x-4\beta xy-4\alpha\beta y+4\alpha y^2$$

$$+4\beta^2x-16\alpha x=0$$

$$\Rightarrow 4x^2+4\alpha y^2-4\beta xy+4x(\beta^2-2\alpha)-(4\alpha\beta)y+4\alpha^2=0$$

$$\Rightarrow x^2+\alpha y^2-\beta xy+(\beta^2-2\alpha)x-(\alpha\beta)y+\alpha^2=0$$

$$a=1, b=\alpha, h=-\frac{\beta}{2}, g=\frac{\beta^2-2\alpha}{2}, f=\frac{-\alpha\beta}{2} \text{ and } c=\alpha^2$$

$$m_1+m_2=\frac{2h}{b}=\frac{-\beta}{\alpha} \qquad \text{...(i)}$$

$$m_1m_2=\frac{a}{b}=\frac{1}{\alpha} \qquad \text{...(ii)}$$

$$\because \quad m_1=2m_2$$

From Eqs. (i) and (ii),

$$\therefore \quad 3m_2=-\frac{\beta}{\alpha} \text{ and } 2m_2^2=\frac{1}{\alpha}$$

$$m_2=\frac{-\beta}{3\alpha} \text{ and } 2\left(\frac{-\beta}{3\alpha}\right)^2=\frac{1}{\alpha}$$

$$2\beta^2=9\alpha \quad \Rightarrow \quad \alpha=\frac{2}{9}\beta^2$$

43. (d) Given parabola,

$$y^2+4y-6x-2=0$$

$$\Rightarrow \quad y^2+4y+4=6x+6$$

$$\Rightarrow \quad (y+2)^2=6(x+1)$$

$$\Rightarrow \quad Y^2=6X$$

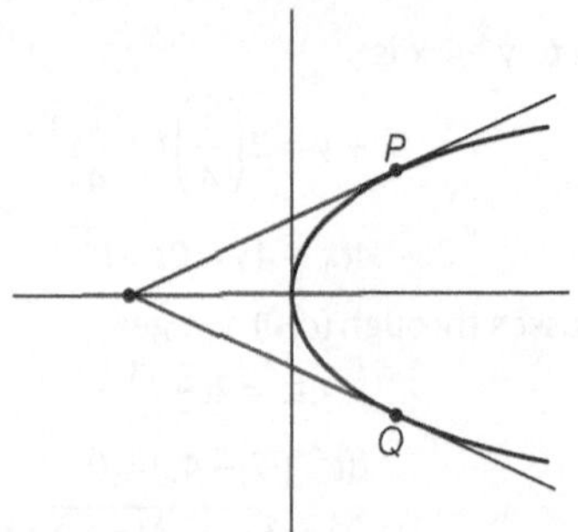

where, $Y=y+2$, $X=x+1$

For vertex $y+2=0$, $x+1=0$

$$y=-2 \text{ and } x=-1 \Rightarrow (-1,-2)$$

General point on parabola

$$X=at_1^2, \quad Y=2at_1$$

$$x+1=at_1^2, \quad y+2=2at_1$$

$$x=at_1^2-1, \quad y=2at_1-2$$

Similarly, second point $(at_2^2-1, 2at_2-2)$.

$\because$ Point of intersect w.r.t. $(at_1^2, 2at_1)$ and $(at_2^2, 2at_2)$ is $\{at_1t_2, 2a(t_1+t_2)\}$

$$\therefore \quad x+1=at_1t_2 \quad y+2=2a(t_1+t_2)$$

$$\because \quad t_1t_2=-1$$

$\because$ Locus is $x+1=-\frac{3}{2}$

$$\Rightarrow \quad 2x+2+3=0$$

$$\Rightarrow \quad 2x+5=0$$

44. (d) Given line, $x+4a=0$ ...(i)

and parabola $y^2=4ax$ ...(ii)

Let a point $P(h,k)$ on line (i).

$\because \quad h+4a=0 \Rightarrow h=-4a$

$\therefore$ Chord of contact at $P(h,k)$

$$T=0$$

$$\Rightarrow yy_1-2a(x+x_1)=0$$

$$\Rightarrow ky-2a(x+h)=0$$

$$\Rightarrow ky-2ax-2ah=0$$

$$\Rightarrow ky-2ax+8a^2=0 \quad [\because \text{put } h=-4a]$$

$$\Rightarrow \frac{2ax-ky}{8a^2}=1$$

$\therefore$ Combined equation of pair of lines OQ and OR is

$$\Rightarrow y^2-4ax\left(\frac{2ax-ky}{8a^2}\right)=0$$

$$\Rightarrow 2ay^2-2ax^2+4akxy=0$$

$\because$ Coeff of x^2 + Coeff of $y^2=2a-2a=0$

$\therefore$ The lines are perpendicular to each other

$$\angle QOR=90°$$

45. (a) Given parabola, $S: y^2=6x \Rightarrow 4a=6 \Rightarrow a=\frac{3}{2}$

Equation of chord with mid-point (h,k)

$$T=S_1$$

$$\Rightarrow yk-6\left(\frac{x+h}{2}\right)=k^2-6h$$

$$\Rightarrow ky-3x-3h=k^2-6h$$

$$\Rightarrow ky-3x+3h-k^2=0$$

$\because$ It passes through (9, 5)

$$5k-27+3h-k^2=0$$

$$\Rightarrow k^2-5k=3(h-9)$$

$\therefore$ Locus is $y^2-5y-3x+27=0$

46. (a) Given parabola, $y^2=4x$

$$4a=4 \Rightarrow a=1$$

Let $P=(at_1^2,2at_1) \Rightarrow (t_1^2,2t_1)$

and $Q=(at_2^2,2at_2) \Rightarrow (t_2^2,2t_2)$

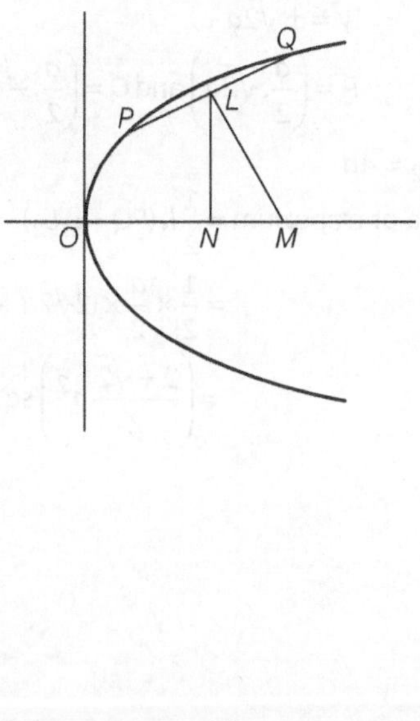

Let L be mid-point of PQ.

$$\therefore \quad L=\left(\frac{1}{2}(t_1^2+t_2^2),(t_1+t_2)\right)$$

Slope of chord PQ is $\Rightarrow \frac{2}{t_1+t_2}$

Hence, the equation of the perpendicular bisector of the chord PQ is

$$y-(t_1+t_2)=-\left(\frac{t_1+t_2}{2}\right)\left(x-\frac{t_1^2+t_2^2}{2}\right)$$

Put $y=0$, we get $x=2+\frac{t_1^2+t_2^2}{2}$

$\therefore$ Point M is $\left(2+\frac{t_1^2+t_2^2}{2},0\right)$ ...(i)

Also, $N=\left(\frac{t_1^2+t_2^2}{2},0\right)$ ...(ii)

By Eqs. (i) and (ii),

$$OM=2+\frac{t_1^2+t_2^2}{2}$$

$$ON=\frac{t_1^2+t_2^2}{2}$$

$$MN=OM-ON=2 \text{ units}$$

47. (c) Point (4, 4) lies on parabola.

Let the point of intersection of the line $y=mx$ with the line chords be $(\alpha, m\alpha)$

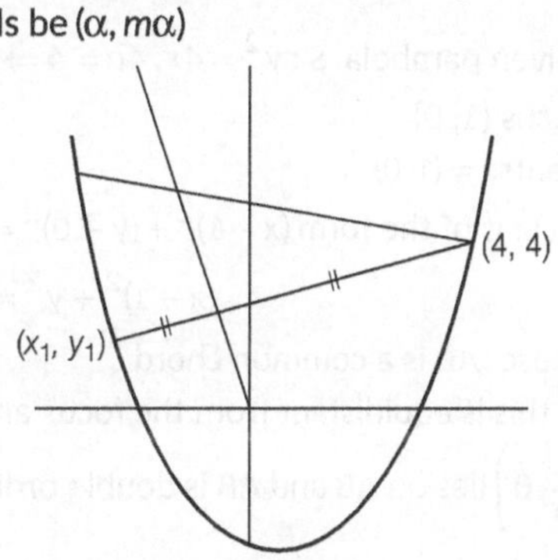

Then, $\alpha=\frac{4+x_1}{2}$

$$\Rightarrow x_1=2\alpha-4$$

and $m\alpha=\frac{4+y_1}{2}$

$$\Rightarrow y_1=2m\alpha-4$$

As (x_1,y_1) lies on the curve

$$\therefore \quad (2\alpha-4)^2=4(2m\alpha-4)$$

$$\Rightarrow 4\alpha^2+16-16\alpha=8(m\alpha-2)$$

$$\Rightarrow 4\alpha^2-8\alpha(2+m)+32=0$$

For two distinct chords

$$\therefore \quad D>0$$

$\Rightarrow \quad \{8(2+m)\}^2 - 4(4)(32) > 0$

$\Rightarrow \quad (2+m)^2 - 8 > 0$

$\Rightarrow \quad 2+m > 2\sqrt{2}$ or $2+m < -2\sqrt{2}$

$\Rightarrow \quad m > 2\sqrt{2}-2$ or $m < -2\sqrt{2}-2$

48. (d) Given parabola,

$$x = y^2 - 6y + 11 \quad \text{...(i)}$$

and $$y = x^2 - 6x + 11 \quad \text{...(ii)}$$

When we put $x \leftrightarrow y$ in Eq. (i), we get $y = x^2 - 6x + 11$ which is identical with parabola written in Eq. (ii).

$\therefore$ The given parabolas are symmetrical about the line $y = x$

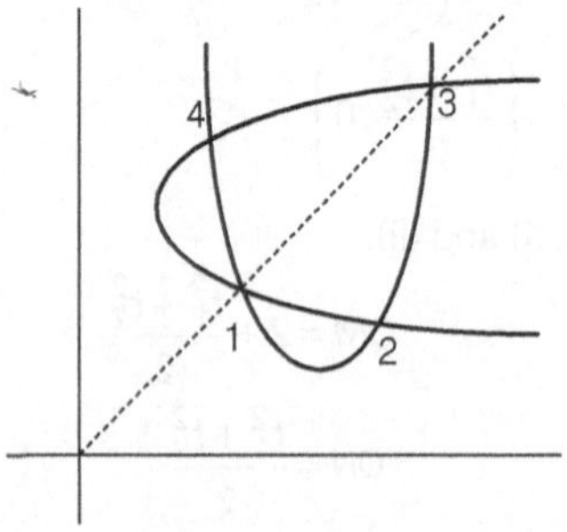

They intersect each other at 4 distinct points

$\therefore$ Number of common chords $= {}^4C_2 = \dfrac{4 \times 3}{2} = 6$

49. (a) Given parabola, $S: y^2 = 4x$, $4a = 4 \Rightarrow a = 1$

$\therefore$ Focus (1, 0)

$\Rightarrow$ Centre $= (1, 0)$

$\because$ Circle is of the form $(x-1)^2 + (y-0)^2 = r^2$

$\Rightarrow \quad (x-1)^2 + y^2 = r^2$

Suppose, AB is a common chord

$\because$ If this is equidistant from the focus and the vertex $M\left(\dfrac{1}{2}, 0\right)$ lies on AB and AB is double ordinate of the parabola.

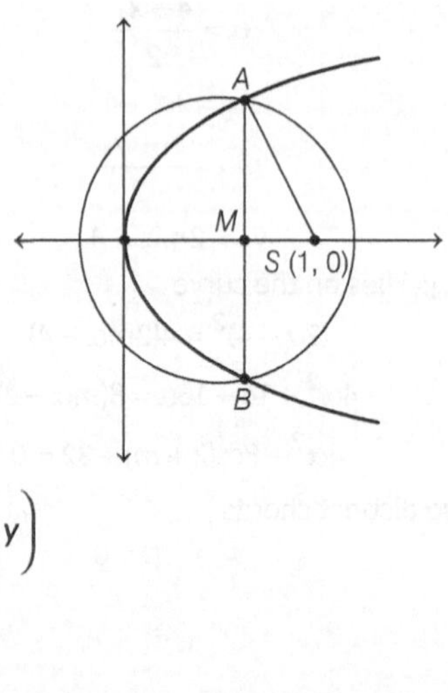

Let $A = \left(\dfrac{1}{2}, y\right)$

So, that $y^2 = 4\left(\dfrac{1}{2}\right) \Rightarrow y = \pm\sqrt{2}$

$\Rightarrow \quad A = \left(\dfrac{1}{2}, \sqrt{2}\right)$ and $B = \left(\dfrac{1}{2}, -\sqrt{2}\right)$

$\because \Delta AMS$ is right angle triangle.

$\therefore \quad SA^2 = SM^2 + MA^2 = \left(\dfrac{1}{2}\right)^2 + 2 = \dfrac{9}{4} = (\text{radius})^2$

$\therefore$ Equation of circle is

$$(x-1)^2 + y^2 = \frac{9}{4}$$

50. (d) Three quarters of the latusrectum $= 3a$

$\therefore$ Equation of circle is $x^2 + y^2 = \left(\dfrac{3a}{2}\right)^2$

and equation of parabola is $y^2 = 4ax$

$\therefore$ Point of intersection of parabola and circle is

$$x^2 + 4ax - \left(\frac{3a}{2}\right)^2 = 0$$

$\Rightarrow \quad 4x^2 + 16ax - 9a^2 = 0$

$\Rightarrow \quad 4x^2 + 18ax - 2ax - 9a^2 = 0$

$\Rightarrow \quad 2x(2x + 9a) - a(2x + 9a) = 0$

$\Rightarrow \quad (2x + 9a)(2x - a) = 0$

$\Rightarrow \quad x = \dfrac{-9a}{2}, x = \dfrac{a}{2}$

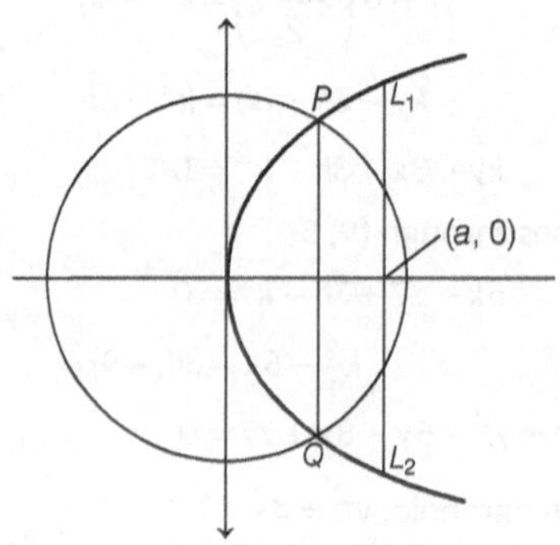

But $y^2 > 0$ and $x > 0$

$\therefore \quad x = \dfrac{a}{2}$ is only possible solution.

$\Rightarrow \quad y = \pm\sqrt{2}a$

$\therefore \quad P = \left(\dfrac{a}{2}, \sqrt{2}a\right)$ and $Q = \left(\dfrac{a}{2}, -\sqrt{2}a\right)$.

$\therefore \quad L_1L_2 = 4a$

$\therefore$ Area of trapezium $= \dfrac{1}{2}h(PQ + L_1L_2)$

$= \dfrac{1}{2} \times \dfrac{a}{2} \times (2\sqrt{2}a + 4a)$

$= \left(\dfrac{2+\sqrt{2}}{2}a^2\right)$ sq units

51. (b) Let the point be (h, k), then the polar of (h, k) w.r.t. $y^2 = 4bx$ is $ky = 2b(x+h)$

$$y = \frac{2b}{k}x + \frac{2bh}{k} \quad \text{...(i)}$$

Comparing Eq. (i) by $y = mx + c$

$$m = \frac{2b}{k} \text{ and } c = \frac{2bh}{k}$$

$\because$ Eq. (i) touches the parabola $x^2 = 4ay$

$\therefore \quad c + am^2 = 0$

$\Rightarrow \quad \frac{2bh}{k} + a\left(\frac{2b}{k}\right)^2 = 0$

$\Rightarrow \quad hk + 2ab = 0$

$hk = -2ab$

$\therefore$ Locus is $xy = -2ab$

52. (c) Parabola $x^2 = y - 2$

Equation of tangent at (1, 3) is

$$x(1) = \frac{y+3}{2} - 2$$

$\Rightarrow \quad 2x = y + 3 - 4$

$\Rightarrow \quad y = 2x + 1$

$\Rightarrow \quad \tan\theta = 2$

$\therefore$ Length of subtangent $= y_1 \cot\theta$

$$= 3\cot\theta = \frac{3}{2} \text{ units}$$

53. (a) Equation of the circle is

$$(x-y)^2 + y^2 = r^2$$

If $P(at^2, 2at)$ is any point on the parabola, then

$$(r - at^2)^2 + 4a^2t^2 \geq r^2$$

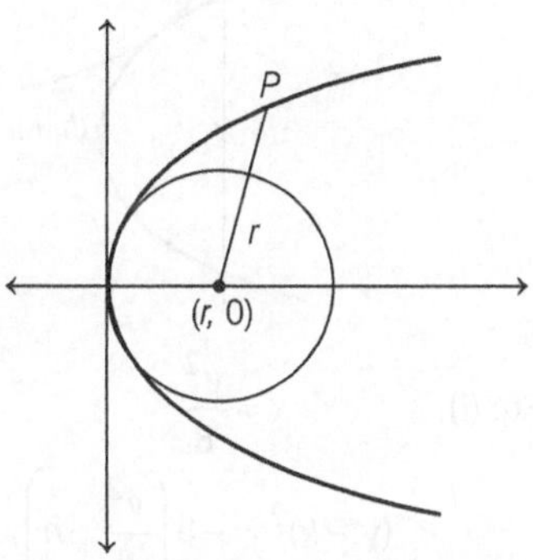

Equality holds, when $t = 0$

$\Rightarrow \quad a^2t^4 - 2art^2 + 4a^2t^2 \geq 0$

$\Rightarrow \quad at^2 - 2r + 4a \geq 0$

$\Rightarrow \quad r \leq \frac{a}{2}(t^2 + 4)$

where the equality holds when $t = 0$

$\therefore$ Maximum value of r is $2a$.

54. (a) $\because$ Curve is $y^2 - 4y = 4x$

$\Rightarrow \quad (y-2)^2 = 4(x+1)$

$4a = 4 \Rightarrow a = 1$

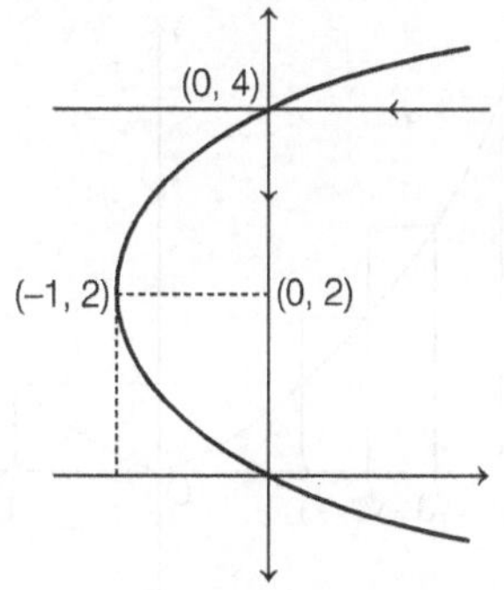

Focus :

$\therefore \quad y - 2 = 0, \quad x + 1 = 1$

$y = 2, \quad x = 0$

$\therefore$ (0, 2)

Point of intersection of the curve and $y = 4$ is (0, 4)

From the reflection property of parabola, reflected ray passes through the focus.

$\therefore$ $x = 0$ is required line.

55. (b) The parabola $y = x^2 + 1$ and $x = y^2 + 1$ are symmetrical about $y = x$

$\therefore$ The tangent at point A is parallel to $y = x$.

Slope = 1, then $\frac{dy}{dx} \Rightarrow 2x = 1$

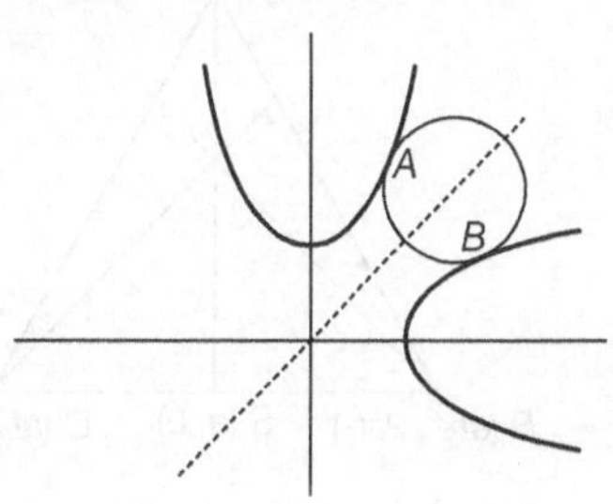

$$x = \frac{1}{2} \Rightarrow y = \frac{5}{4}$$

$\therefore \quad A = \left(\frac{1}{2}, \frac{5}{4}\right)$ and $B = \left(\frac{5}{4}, \frac{1}{2}\right)$

$\therefore \quad$ Radius $= \frac{1}{2}AB$

$$= \frac{1}{2}\sqrt{\left(\frac{1}{2} - \frac{5}{4}\right)^2 + \left(\frac{5}{4} - \frac{1}{2}\right)^2}$$

$$= \frac{3\sqrt{2}}{8} \text{ units}$$

$\therefore \quad$ Area $= \pi r^2$

$$= \pi\left(\frac{3\sqrt{2}}{8}\right)^2 = \frac{9\pi}{32} \text{ sq units}$$

56. (d) $\because [x^2]=[y]$

If $0 \le y < 1 \Rightarrow [y]=0 \Rightarrow [x^2]=0$

$0 < x^2 < 1 \Rightarrow x \in (-1, 1)$

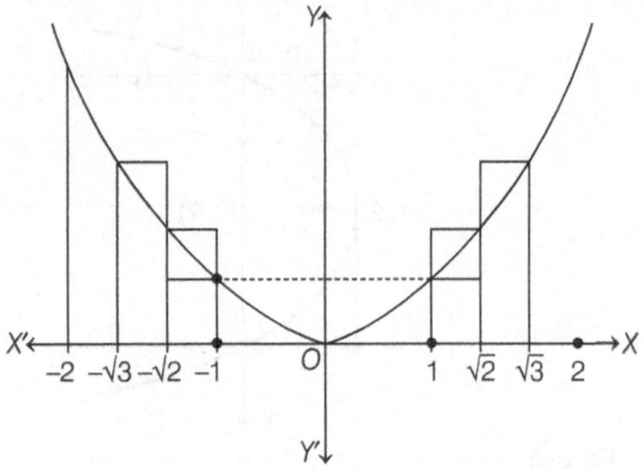

For $\quad 1 \le y < 2 \Rightarrow [y]=1 \Rightarrow [x^2]=1$

$1 \le x^2 < 2 \Rightarrow x \in (-\sqrt{2}, -1] \cup [1, \sqrt{2})$

For $\quad 2 \le y < 3$, when $[y]=2$, then $[x^2]=2$

$2 \le x^2 < 3\, x \in (-\sqrt{3}, -\sqrt{2}] \cup [\sqrt{2}, \sqrt{3})$

The graph of the region will not only contain of the parabola $y=x^2$ but $[x^2]=[y]$ contain only points rectangles of sides 1, 2; 1, $\sqrt{2}-1$; 1, $\sqrt{3}-\sqrt{2}$ etc.

$\therefore$ a, b, c are incorrects.

57. (5) $\because SA$ is perpendicular to BC [S is focus $(a, 0)$]

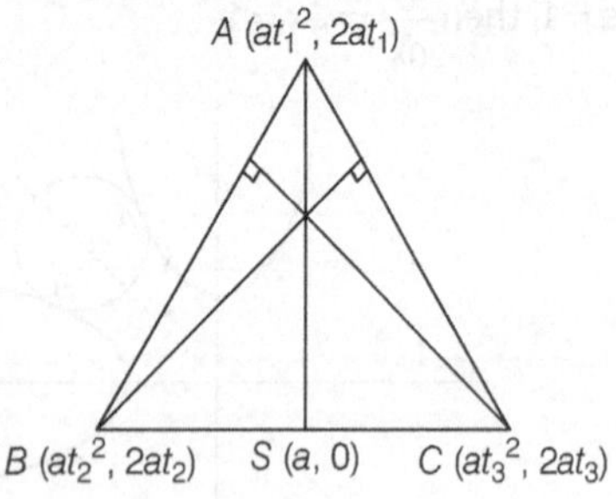

$$\Rightarrow \left(\frac{2at_1 - 0}{at_1^2 - a}\right) \cdot \left(\frac{2at_3 - 2at_2}{at_3^2 - at_2^2}\right) = -1$$

$$\Rightarrow \frac{2t_1}{t_1^2 - 1} \cdot \frac{2}{t_3 + t_2} = -1$$

$$\Rightarrow 4t_1 + t_1^2(t_2 + t_3) = t_2 + t_3 \quad \text{...(i)}$$

Similarly, $4t_2 + t_2^2(t_3 + t_1) = t_3 + t_2$ ...(ii)

On subtracting Eq. (ii) from Eq. (i), then

$$4 + (t_1t_2 + t_2t_3 + t_3t_1) = -1$$

$$|t_1t_2 + t_2t_3 + t_3t_1| = |-5| = 5$$

58. (4) Let $P(t^2, 2t)$ be a point on the curve $y^2 = 4x$, whose image is $Q(x, y)$ on $x + y + 4 = 0$, then

$$\frac{x - t^2}{1} = \frac{y - 2t}{1} = \frac{-2(t^2 + 2t + 4)}{1 + 1^2}$$

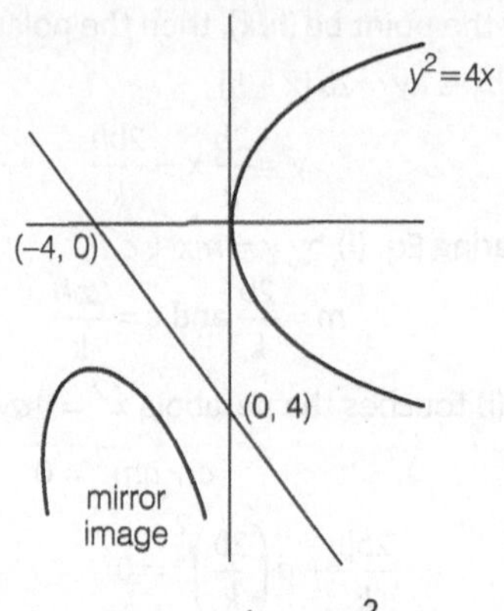

$$\Rightarrow \quad x = -2t - 4 \text{ and } y = -t^2 - 4$$

Now, the straight line $y = -5$ meets the mirror image.

$$\therefore \quad -t^2 - 4 = -5$$

$$t^2 = 1 \Rightarrow t = \pm 1$$

Thus, point of intersections of A and B are $(-6, -5)$ and $(-2, -5)$.

$$\therefore \quad AB = \sqrt{(-2+6)^2 + (-5+5)^2} = 4$$

59. (16) Let the given parabola, $y^2 = 4ax$, where $4a = 8$

$\therefore$ Equation is $y^2 = 8x$ ...(i)

If the vertex of moving parabola is (h, k), then equation of moving parabola

$$(y - k)^2 = -8(x - h) \quad \text{...(ii)}$$

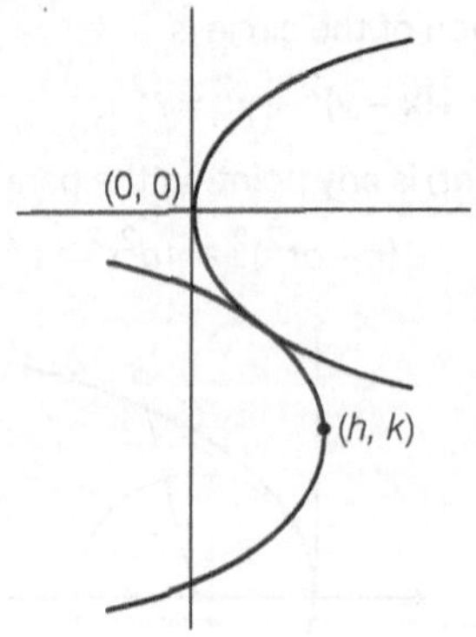

From Eq. (i), $\quad x = \dfrac{y^2}{8}$

$$\Rightarrow (y-k)^2 = -8\left(\frac{y^2}{8} - h\right) \quad \text{[from Eq. (ii)]}$$

$$\Rightarrow y^2 - 2ky + k^2 = -y^2 + 8h$$

$$\Rightarrow 2y^2 - 2ky + k^2 - 8h = 0$$

$\because$ Two parabolas Eqs. (i) and (ii) touch each others.

$$\therefore \quad D = 0$$

$$B^2 - 4AC = 0$$

$$\Rightarrow (-2k)^2 - 4(2)(k^2 - 8h) = 0$$

$\Rightarrow \quad 4k^2 - 8(k^2 - 8h) = 0$

$\Rightarrow \quad k^2 - 2k^2 + 16h = 0 \Rightarrow k^2 = 16h$

$\therefore$ Locus is $y^2 = 16x$

$\because$ Latusrectum $= 4a = 16$

60. (7.25) Let CAB be the bridge and $X'OX$ be the roadway. Let A be the centre of the bridge. Taking $X'OX$ as X-axis and Y-axis along OA, we find that the coordinates of A are (0, 5).

Clearly the bridge is in the shape of a parabola having its vertex at $A(0,5)$.

Let its equation be $x^2 = 4a(y-5)$ ...(i)

It passes through $B(100, 30)$.

$\therefore \quad (100)^2 = 4a(30-5) \Rightarrow a = 100$

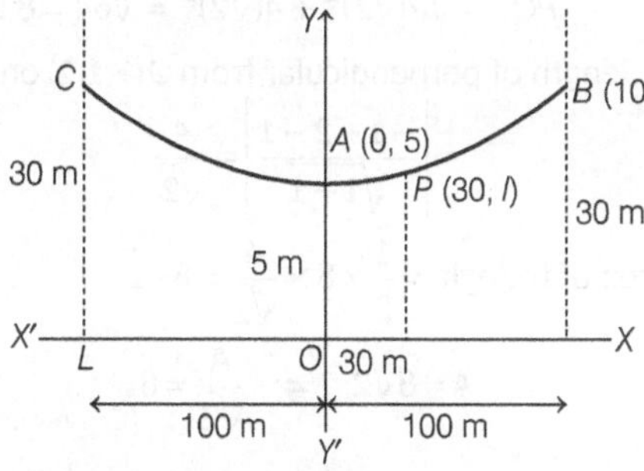

Put $a = 100$ in Eq. (i), we get

$$x^2 = 400(y-5) \quad ...(ii)$$

Let l m be the length of vertical supporting cable 30 m from the centre.

Then, $P(30, l)$ lies on Eq. (ii),

$\therefore \quad 900 = 400(l-5)$

$$\frac{9}{4} = l - 5$$

$$l = \frac{9}{4} + 5 = \frac{29}{4} = 7.25 \text{ m}$$

$\therefore$ Length of vertical supporting cable is 7.25 m.

61. (1) Obviously locus of S is a parabola with focus A and directrix along BD. Obviously AC is the axis of this parabola and its vertex is at T_1 and line M is along its latusrectum.

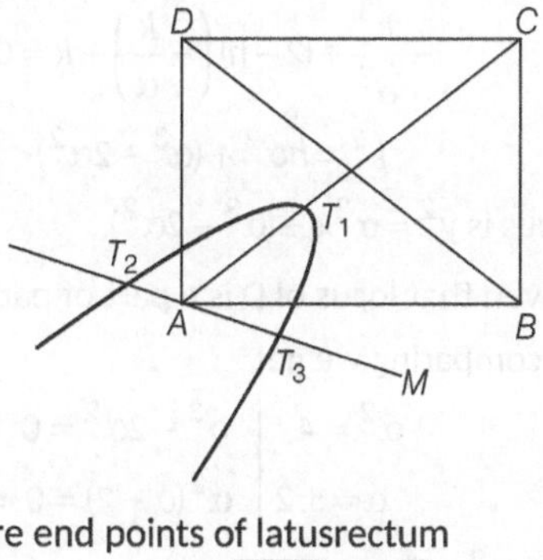

So, T_2, T_3 are end points of latusrectum

$\because \quad AC = \sqrt{4+4} = 2\sqrt{2}$

$$AT_1 = \frac{1}{4}AC = \frac{1}{\sqrt{2}}$$

$\Rightarrow \quad T_2T_3 = 4\left(\frac{1}{\sqrt{2}}\right) = \frac{4}{\sqrt{2}}$

$\therefore$ Area of $\Delta T_1T_2T_3 = \frac{1}{2} \times \frac{4}{\sqrt{2}} \times \frac{1}{\sqrt{2}} = 1$ sq unit

62. (1.73) Given lines, $y - b = m_1(x+a)$ and $y - b = m_2(x+a)$ are lines pass through $(-a, b)$ which a point lying on the directrix of parabola $y^2 = 4ax$.

Thus, we can conclude at both lines are perpendicular to each other.

$\because \quad m_1m_2 = -1$

$\therefore \quad \sqrt{m_1m_2 + 4} = \sqrt{3} = 1.73$

63. (0) Let the focal chord be $y = mx + c$ which is passing through the focus of parabola $y^2 = 16x$

$\Rightarrow \quad 0 = 4m + c \quad$ {$\because$ focus (4, 0)}

$\Rightarrow \quad c = -4m$

Thus, equation of focal chord is $y = mx - 4m$ ...(i)

$\because$ Eq. (i) is tangent to the circle $(x-6)^2 + y^2 = 2$

So, distance between centre and line (i) = radius of circle

$$\left|\frac{6m - 4m}{\sqrt{m^2+1}}\right| = \sqrt{2}$$

[$\because$ Centre (6, 0) and radius $= \sqrt{2}$]

$\Rightarrow \quad \left(\frac{2m}{\sqrt{m^2+1}}\right)^2 = 2$

$\Rightarrow \quad 4m^2 = 2m^2 + 2$

$\Rightarrow \quad 2m^2 = 2 \Rightarrow m^2 = 1 \Rightarrow m = \pm 1$

$\therefore$ Sum of possible values of $m = -1 + 1 = 0$

64. (40) The equation of the tangent at P to the parabola $y^2 = 8x$ is

$$4y = 4(x+2)$$

$$y = x + 2 \quad ...(i)$$

Given, parabola is $y^2 = 8x + 5$ ...(ii)

Solving Eqs. (i) and (ii), we get

$$(x+2)^2 = 8x + 5$$

$$x^2 + 4x + 4 = 8x + 5$$

$$x^2 - 4x - 1 = 0$$

$\Rightarrow \quad (x-2)^2 = (\sqrt{5})^2$

$$x = 2 \pm \sqrt{5}$$

and $\quad y = 4 \pm \sqrt{5} \quad$ [from Eq. (i)]

$\therefore$ $Q(2+\sqrt{5}, 4+\sqrt{5})$ and $R(2-\sqrt{5}, 4-\sqrt{5})$

$\therefore \quad QR = \sqrt{(2\sqrt{5})^2 + (2\sqrt{5})^2} = \sqrt{40}$

$$QR^2 = 40$$

65. (9) Given, circle $(x-3)^2+y^2=9$

and parabola $y^2=4x$, $4a=4 \Rightarrow a=1$

Let tangent of parabola $y=mx+\dfrac{a}{m}$

$$\Rightarrow \quad y=mx+\frac{1}{m} \quad \Rightarrow m^2x-my+1=0$$

The above line also a tangent to circle $(x-3)^2+y^2=9$

$\therefore$ Perpendicular distance from centre (3, 0) = radius

$$\left|\frac{3m^2-0+1}{\sqrt{m^2+m^4}}\right|=3$$

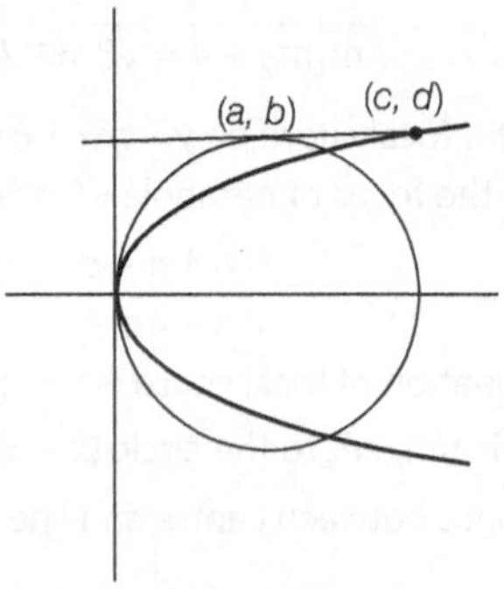

$$\Rightarrow \quad (3m^2+1)^2=9(m^2+m^4)$$

$$\Rightarrow \quad 9m^4+1+6m^2=9m^2+9m^4$$

$$3m^2=1 \Rightarrow m=\pm\frac{1}{\sqrt{3}}$$

$\therefore$ Tangent is $y=\dfrac{1}{\sqrt{3}}x+\sqrt{3}$. It will be used $m=\dfrac{1}{\sqrt{3}}$

or $\quad y=-\dfrac{1}{\sqrt{3}}x-\sqrt{3}$ (rejected)

$\therefore$ For parabola point of contact

$$=\left(\frac{a}{m^2},\frac{2a}{m}\right)=(3,2\sqrt{3})=(c,d)$$

For circle

Solving $y=\dfrac{x}{\sqrt{3}}+\sqrt{3}$ and $(x-3)^2+y^2=9$

$$\Rightarrow \quad x^2-6x+9+\frac{x^2}{3}+3+2x=9$$

$$\Rightarrow \quad 4x^2-12x+9=0$$

$$\Rightarrow \quad (2x-3)^2=0 \Rightarrow x=\frac{3}{2}$$

$$\therefore \quad y=\frac{1}{\sqrt{3}}\times\frac{3}{2}+\sqrt{3} \Rightarrow \frac{3\sqrt{3}}{2}$$

$$\therefore \quad (a,b)=\left(\frac{3}{2},\frac{3\sqrt{3}}{2}\right)$$

$$\therefore \quad 2(a+c)=2\left(\frac{3}{2}+3\right)=9$$

66. (8) Chord of contact w.r.t. point $O(-1,2)$ for parabola $y^2=4x$ is

$$\Rightarrow \quad y(2)=2(x-1) \qquad [\because yy_1=2a(x+x_1)]$$

$$\Rightarrow \quad y=x-1$$

Solving $y=x-1$ with parabola

$$(x-1)^2=4x$$

$$\Rightarrow \quad x^2-2x+1-4x=0$$

$$\Rightarrow \quad x^2-6x+1=0$$

$$\Rightarrow \quad x=\frac{6\pm\sqrt{36-4}}{2}=3\pm2\sqrt{2}$$

$$\because \quad y=2\pm2\sqrt{2}$$

$\therefore$ $P(3+2\sqrt{2},2+2\sqrt{2})$ and $Q=(3-2\sqrt{2},2-2\sqrt{2})$

$$PQ^2=\sqrt{(4\sqrt{2})^2+4(\sqrt{2})^2}=\sqrt{64}=8 \text{ units}$$

Also, length of perpendicular from $O(-1,2)$ on PQ is

$$=\left|\frac{-1-2-1}{\sqrt{1+1}}\right|=\frac{4}{\sqrt{2}}$$

$$\therefore \text{ Area of triangle}=\frac{1}{2}\times8\times\frac{4}{\sqrt{2}}=8\sqrt{2}$$

$$A=8\sqrt{2} \Rightarrow \frac{A}{\sqrt{2}}=8$$

67. (2) Equation of normal to the parabola $y^2=4x$ having slope m is $y=mx-2m-m^3$.

$\because$ It passes through the point $P(h,k)$

$$mh-k-2m-m^3=0$$

$$\Rightarrow \quad m^3+(2-h)m+k=0 \begin{cases} m_1 \\ m_2 \\ m_3 \end{cases} \quad \text{...(i)}$$

$$\therefore \quad m_1+m_2+m_3=0$$

$$\Sigma m_1m_2=2-h$$

$$m_1m_2m_3=-k$$

But given, $m_1\cdot m_2=\alpha \Rightarrow m_3=-\dfrac{k}{\alpha}$

Put $m_3=\dfrac{-k}{\alpha}$ in Eq. (i) because m_3 is root of that equation

$$-\frac{k^3}{\alpha^3}+(2-h)\left(-\frac{k}{\alpha}\right)+k=0$$

$$\Rightarrow \quad k^2=h\alpha^2+(\alpha^3-2\alpha^2)$$

$\therefore$ Locus is $y^2=\alpha^2x+(\alpha^3-2\alpha^2)$.

But given that locus of D is a part of parabola $y^2=4x$

$\therefore$ By comparing we get

$$\alpha^2=4 \;\Big|\; \alpha^3-2\alpha^2=0$$

$$\Rightarrow \quad \alpha=\pm2 \;\Big|\; \alpha^2(\alpha-2)=0 \Rightarrow \alpha=0,2$$

$$\therefore \quad \alpha=2$$

68. (0.01) $\because$ Chord is parallel to the line $y = x$.

$\therefore$ Let equation of chord is

$$x - y + \lambda = 0 \quad ...(i)$$

And end points of chord on parabola $y^2 = 4ax$ are $P(at_1^2, 2at_1)$ and $Q(at_2^2, 2at_2)$.

$\therefore$ Equation of variable chord

$$y - 2at_1 = \frac{2a(t_2 - t_1)}{a(t_2^2 - t_1^2)}(x - at_1^2)$$

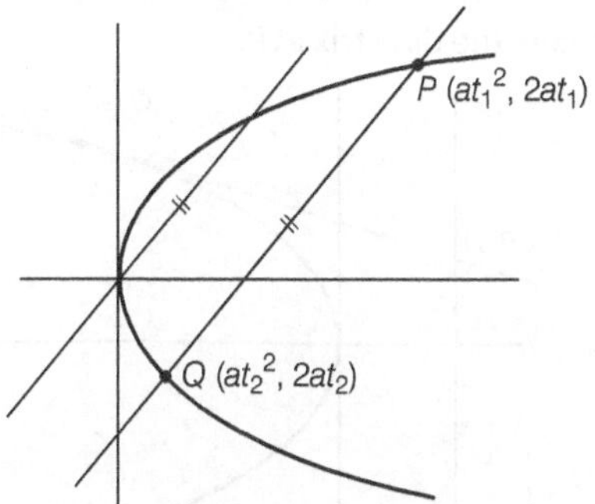

$$\Rightarrow \quad (t_1 + t_2)y - 2at_1(t_1 + t_2) = 2(x - at_1^2)$$

$$\Rightarrow \quad 2x - y(t_1 + t_2) + 2at_1t_2 = 0 \quad ...(ii)$$

Comparing Eqs. (i) and (ii),

$$\frac{2}{1} = \frac{(t_1 + t_2)}{1} = \frac{2at_1t_2}{\lambda}$$

$$\therefore \quad t_1 + t_2 = 2 \quad ...(iii)$$

$$at_1t_2 = \frac{k}{\lambda} \quad ...(iv)$$

$\therefore$ Equation of normal from point P and Q

$$y = t_1x + 2at_1 + at_1^3 \quad ...(v)$$

and

$$y = t_2x + 2at_2 + at_2^3 \quad ...(vi)$$

Solving above Eqs. (v) and (vi) and assume point of intersection is (h, k).

$$\therefore \quad h = 2a + a(t_1^2 + t_2^2 + t_1t_2)$$

By Eq. (iii), $h = 2a + a\{(2)^2 - t_1t_2\}$

$$h = 2a + 4a - at_1t_2$$

$$h = 6a - \frac{k}{\lambda} \quad \text{[from Eq. (iv)]}$$

and

$$k = -at_1t_2(t_1 + t_2)$$

$$k = \frac{-k}{\lambda}(2) \Rightarrow \lambda = -2$$

Put $\lambda = -2$

$$h = 6a + \frac{k}{2} \Rightarrow 2h = 12a + k \Rightarrow 2h - k = 12a$$

$\therefore$ Locus is $2x - y - 12a = 0$

$\therefore$ $ax + by - c = 0$

$\therefore$ $a = 2$ and $b = -1$

$$\frac{a+b}{100} = \frac{2-1}{100} = \frac{1}{100} = 0.01$$

Level 3

1. (b) Given parabola, $y^2 - kx + 8 = 0$

$$y^2 = k\left(x - \frac{8}{k}\right) \Rightarrow Y^2 = 4AX$$

where, $Y = y, X = x - \frac{8}{k}$ and $4A = k \Rightarrow A = \frac{k}{4}$

$\because$ Equation of directrix is

$$X + A = 0$$

$$\Rightarrow \quad x - \frac{8}{k} + \frac{k}{4} = 0 \quad ...(i)$$

and the given directrix $x - 1 = 0$ ...(ii)

Comparing Eqs. (i) and (ii),

$$-\frac{8}{k} + \frac{k}{4} = -1$$

$$\Rightarrow \quad k^2 - 32 = -4k$$

$$\Rightarrow \quad k^2 + 4k - 32 = 0$$

$$\Rightarrow \quad k^2 + 8k - 4k - 32 = 0$$

$$\Rightarrow \quad k(k+8) - 4(k+8) = 0$$

$$\Rightarrow \quad (k+8)(k-4) = 0$$

$$\Rightarrow \quad k = 4, -8$$

For $k = 4$ the parabola is $y^2 = 4(x-2)$ whose vertex is $(2, 0)$ and touches the circle $x^2 + y^2 = 4$ at the vertex

$\therefore k \neq 4$

For $k = -8$ the parabola is $y^2 = -8(x+1)$ with intersects the circle $x^2 + y^2 = 4$ at two real and distinct points.

2. (a) $\because$ Angle made by diameter at any point is $\frac{\pi}{2}$

$$\therefore \quad \angle ORP = \angle ORQ = \frac{\pi}{2}$$

Let $P(at_1^2, 2at_1)$ and $Q(at_2^2, 2at_2)$ are two points on parabola $y^2 = 4ax$

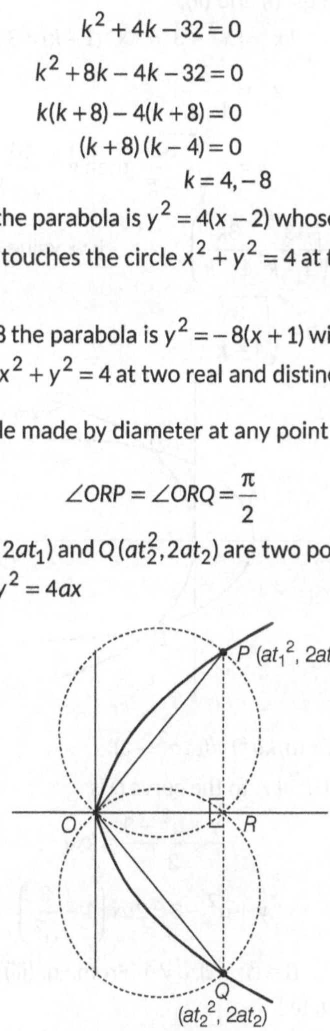

$\because$ Slope of OP and OQ are $\frac{1}{t_1}$ and $\frac{1}{t_2}$.

$\therefore$ Slope of $PQ = \dfrac{2at_2 - 2at_1}{at_2^2 - at_1^2} = \dfrac{2}{t_1 + t_2}$

$\because PQ \perp OR.$

$\therefore$ Slope of $OR = -\left(\dfrac{t_1+t_2}{2}\right) = \tan\phi$ ...(i)

Now,

$\Rightarrow \quad \dfrac{1}{t_1} = \tan\theta_1$ ...(ii)

and $\quad \dfrac{1}{t_2} = \tan\theta_2$ ...(iii)

Using Eqs. (i) and (iii), we get

$$\cot\theta_1 + \cot\theta_2 = t_1 + t_2 = -2\tan\phi \quad \text{[By Eq. (i)]}$$

3. (c) Given parabola $C: y = x^2 - 3$ ...(i)

and $\quad D: y = kx^2$ ...(ii)

Solving Eqs. (i) and (ii),

$$kx^2 = x^2 - 3 \Rightarrow x^2(1-k) = 3$$

$\Rightarrow \quad x^2 = \dfrac{3}{1-k}$

$$x = \pm\sqrt{\frac{3}{1-k}}, \text{ then } y = \frac{3k}{1-k}$$

$\therefore \quad A\left(\sqrt{\dfrac{3}{1-k}}, \dfrac{3k}{1-k}\right)$ [$\because$ x value of A is positive]

$\therefore \quad a = \sqrt{\dfrac{3}{1-k}}$

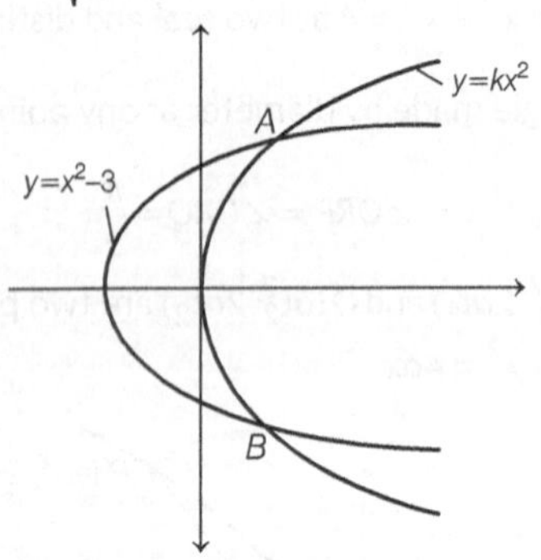

Then, $A = (a, ka^2) \equiv (a, a^2 - 3)$ ...(iii)

Tangent 'l' at A to the curve D is

$$\frac{y + a^2 - 3}{2} = kxa$$

$\Rightarrow \quad y + a^2 - 3 = 2ax\left(1 - \dfrac{3}{a^2}\right)$

$\because \quad B \equiv (1, -2), a \neq 1$ {From Eq. (iii)} ...(iv)

From Eq. (iv),

$$-2 + a^2 - 3 = 2a\left(1 - \frac{3}{a^2}\right)$$

$\Rightarrow \quad a^3 - 5a = 2a^2 - 6$

$\Rightarrow \quad a^3 - 2a^2 - 5a + 6 = 0$

$\Rightarrow \quad (a-1)(a+2)(a-3) = 0$

$\Rightarrow \quad a = 3 \quad \{\because a \neq 1, a \neq -2\}$

4. (d) Given parabola $y^2 = 4ax$

Let $P = (-a, y)$ be a point on the directrix and $Q(at^2, 2at)$ be the point of contact of tangent from P.

The tangent at Q is

$$ty = x + at^2$$

This cuts the directrix at P.

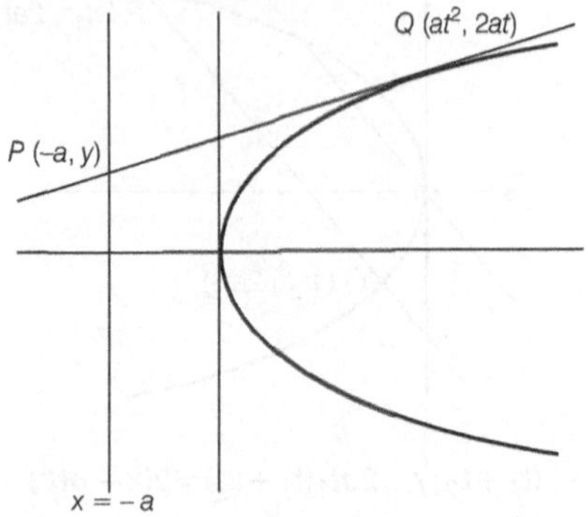

So, $\quad y = \dfrac{a(t^2 - 1)}{t}$

$\therefore \quad P = \left(-a, \dfrac{a(t^2-1)}{t}\right)$ ...(i)

Suppose in (x_1, y_1) is the mid-point of PQ

$$2x_1 = a(t^2 - 1)$$

$\Rightarrow \quad 2x_1 + a = at^2$ ...(ii)

and $\quad 2y_1 = 2at + \dfrac{a(t^2-1)}{t}$

$\Rightarrow \quad 2y_1 t = 3at^2 - a$ ...(iii)

By Eq. (ii),

$$t = \sqrt{\frac{2x_1 + a}{a}}$$

Put the value of 't' in Eq. (iii),

$$2y_1\left(\sqrt{\frac{2x_1+a}{a}}\right) = 3a\left(\frac{2x_1+a}{a}\right) - a$$

$\Rightarrow \quad 2y_1\sqrt{\dfrac{2x_1+a}{a}} = 6x_1 + 2a$

$\Rightarrow \quad y_1\sqrt{\dfrac{2x_1+a}{a}} = 3x_1 + a$

On squaring both sides, we get

$$y_1^2\left(\frac{2x_1+a}{a}\right) = (3x_1 + a)^2$$

$\Rightarrow \quad y_1^2(2x_1 + a) = a(3x_1 + a)^2$

$\therefore$ Point (x, y) lies on curve $y^2(2x + a) = a(3x + a)^2$.

5. (b) The given equation of parabola

$$x^2 = -(y-4) \quad \text{...(i)}$$

Now, slope of tangent at point $P(x_1, y_1)$ is $= -2x_1$

$\therefore$ Equation of tangent at P

$$2xx_1 + y = y_1 + 2x_1^2$$

$\therefore$ Point $A \Rightarrow \left(\dfrac{y_1 + 2x_1^2}{2x_1}, 0\right)$

and point $B \Rightarrow (0, y_1 + 2x_1^2)$

$\therefore$ Area of $\Delta OAB = \dfrac{1}{2} \times OA \times OB$

$$\Delta = \frac{(y_1 + 2x_1^2)^2}{4x_1} \Rightarrow \frac{(4 - x_1^2 + 2x_1^2)^2}{4x_1} \Rightarrow \frac{(x_1^2+4)^2}{4x_1}$$

For minimum value of Δ

$$\frac{d\Delta}{dx_1} = \frac{1}{4}\left\{\frac{4x_1^2(x_1^2+4) - (x_1^2+4)^2}{x_1^2}\right\}$$

$$= \frac{(x_1^2+4)}{4x_1^2}(3x_1^2 - 4)$$

$$\frac{d\Delta}{dx_1} = 0 \Rightarrow x_1^2 = \frac{4}{3} \text{ or } x_1 = \frac{2}{\sqrt{3}}$$

$\because \dfrac{d\Delta}{dx_1}$ changes sign from (–) to (+), when $x_1 = \dfrac{2}{\sqrt{3}}$

$\therefore$ Δ is minimum at $x_1 = \dfrac{2}{\sqrt{3}}$

$$\therefore \quad \Delta = \frac{\left(\frac{4}{3}+4\right)^2}{4\left(\frac{2}{\sqrt{3}}\right)} = \frac{256}{9} \times \frac{\sqrt{3}}{8} = \frac{32}{3\sqrt{3}} \text{ sq units}$$

6. (d) Given parabola is, $y^2 = 4ax$

Focus $S(a, 0)$.

Focal distance SP is diameter of circle.

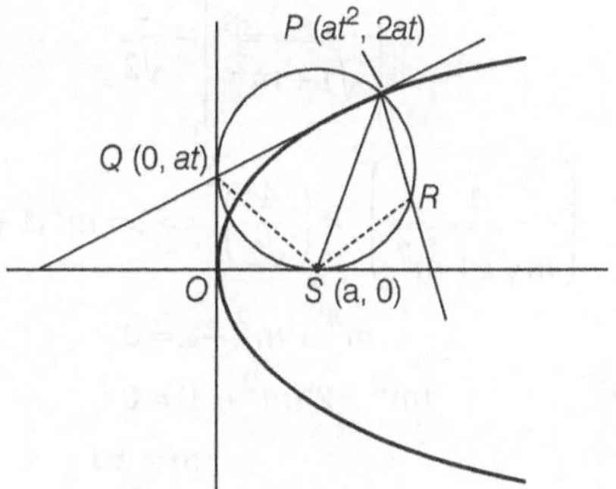

Equation of tangent at $P(at^2, 2at)$

$$\Rightarrow \quad ty = x + at^2$$

For point Q, put $x = 0$

$$y = at$$

$\because$ PR is normal chord of parabola

$$\therefore \quad \angle QPR = 90^\circ$$

and $\angle SQP = \angle SRP = 90^\circ$ (angle made by diameter)

$$\because \quad \angle QSR = 90^\circ$$

$\because$ All angles are equal.

$\therefore$ $PQSR$ is a rectangle.

Now, $\quad QS = PR$

$$\Rightarrow \quad PR = \sqrt{a^2 + a^2t^2} = a\sqrt{1+t^2}$$

7. (a) Consider the common focus as the origin and their axes.

Let us assume the equation of the parabolas for given conditions are

$$y^2 = 4a(x+a) \text{ and } x^2 = 4a(y+a)$$

$\therefore$ Equation of normals for both the parabolas having slopes m and m', respectively are given by

$$y = m(x+a) - 2am - am^3 \quad \text{...(i)}$$

and $\quad x = m'(y+a) - 2am' - am'^3 \quad$ ...(ii)

$\because$ The normal are perpendicular to each other.

$$\therefore \quad m\left(\frac{1}{m'}\right) = -1$$

$$\therefore \quad m' = -m$$

Put $m' = -m$ in Eq. (ii), we get

$$x = -m(y+a) + 2am + am^3 \quad \text{...(iii)}$$

Adding Eqs. (i) and (iii), we get

$$m = \frac{x+y}{x-y} \quad \text{...(iv)}$$

Put $m = \dfrac{x+y}{x-y}$ in Eq. (i), we get

$$y = \frac{x+y}{x-y}(x+a) - 2a\left(\frac{x+y}{x-y}\right) - a\left(\frac{x+y}{x-y}\right)^3$$

$\Rightarrow y(x-y)^3$

$$= (x+y)\{(x+a)(x-y)^2 - 2a(x-y)^2 - a(x+y)^2\}$$

$$= (x+y)\{(x-y)^2\{x+a-2a\} - a(x+y)^2\}$$

$$= (x+y)\{(x-y)^2(x-a) - a(x+y)^2$$

$$\Rightarrow y(x-y)^3 = (x+y)\{x(x-y)^2 - a[(x-y)^2 + (x+y)^2]\}$$

$$= (x+y)[x(x-y)^2 - 2a(x^2+y^2)]$$

$$y(x-y)^3 = (x+y)(x)(x-y)^2 - 2a(x+y)(x^2+y^2)$$

$$\Rightarrow 2a(x+y)(x^2+y^2) = x(x+y)(x-y)^2 - y(x-y)^3$$

$$= (x-y)^2[x(x+y) - y(x-y)]$$

$$= (x-y)^2[x^2 + xy - xy + y^2]$$

$$\Rightarrow 2a(x+y)(x^2+y^2) = (x-y)^2(x^2+y^2)$$

$$2a(x+y) = (x-y)^2$$

8. (c) Given parabola, $y^2 = 4x$

$$4a = 4 \Rightarrow a = 1$$

Focus $(a, 0) \Rightarrow (1, 0)$

and (1, 2) is a point that is upper end of latus rectum.

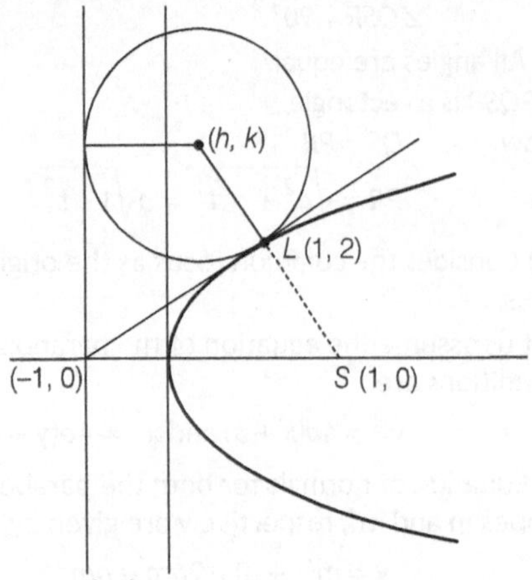

Tangent at L for parabola

$$y(2)-2(x+1)=0$$

$$\Rightarrow \quad y-x-1=0 \Rightarrow y=x+1$$

Let (h, k) be the centre of the circle that lies on the normal to the parabola at (1, 2) whose equation is $x+y=3$.

$$\because \quad h+k=3 \quad \text{...(i)}$$

$\because$ The circle touches the directrix $x+1=0$

We have, $(h+1)^2=(h-1)^2+(k-2)^2=(\text{radius})^2$ ...(ii)

By Eq. (i), $4h=(k-2)^2=2 \Rightarrow 4h=(3-h-2)^2$

$$\Rightarrow \quad h^2-6h+1=0 \Rightarrow h=3\pm 2\sqrt{2}$$

$$\therefore \quad k=\pm 2\sqrt{2}$$

$\because$ Circle above the axis.

$\therefore$ $k=2\sqrt{2}$ which is y-coordinate of the point of contact of the circle with the directrix.

9. (c) Equation of parabola $y^2=4ax$

$$P(t)=(at^2, 2at)$$

Equation of tangent at $P(t) \Rightarrow y=\dfrac{x}{t}+at$ ...(i)

Equation of normal at $P(t) \Rightarrow y=-tx+2at+at^3$ ...(ii)

For point $T \Rightarrow$ Put $y=0$ in Eq. (i), we get

$$x=-at^2$$

$$\therefore \quad T=(-at^2, 0)$$

and for point $G \Rightarrow$ Put $y=0$ in Eq. (ii), we get

$$x=2a+at^2$$

$$\therefore \quad G \Rightarrow (2a+at^2, 0)$$

$\because$ $PT \perp PG$.

$\therefore$ The circle passing through P, T, G will have its centre at mid-point of TG.

$\therefore$ $C=(a, 0)$

$\therefore$ Slope of $PC=\dfrac{2at}{a(t^2-1)}=\dfrac{2t}{t^2-1}$

$\therefore$ Slope of tangent at $P=\dfrac{1-t^2}{2t}$

$\therefore$ Slope of tangent to parabola $=\dfrac{1}{t}$

$$\therefore \tan(\theta_1-\theta_2)=\left|\frac{\tan\theta_1-\tan\theta_2}{1+\tan\theta_1\tan\theta_2}\right|$$

$$=\left|\frac{\frac{1}{t}-\frac{1-t^2}{2t}}{1+\frac{1}{t}\cdot\frac{1-t^2}{2t}}\right|=\left|\frac{(1+t^2)t}{1+t^2}\right|=|t|$$

$$\therefore \quad |\theta_1-\theta_2|=\tan^{-1}t$$

10. (b) Given parabola is $y^2=4ax$

and circle $x^2+y^2=\dfrac{a^2}{2}$

Equation of tangent w.r.t. parabola having slope m

$$y=mx+\frac{a}{m} \quad \text{...(i)}$$

Eq. (i) also tangent for $x^2+y^2=\dfrac{a^2}{2}$.

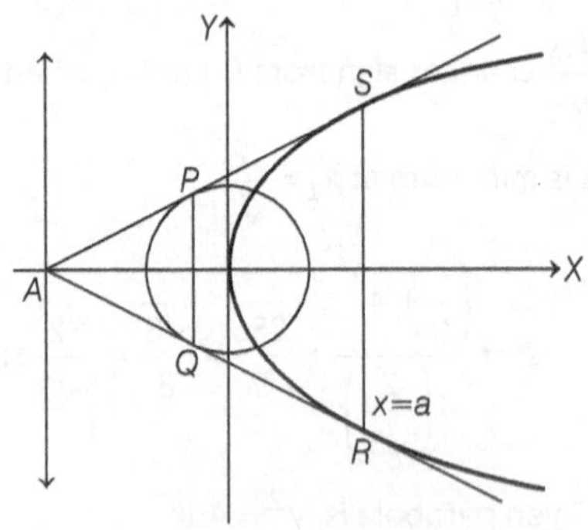

$\therefore$ Distance between centre and line (i) = radius.

$$\left|\frac{0-0-\frac{a}{m}}{\sqrt{1+m^2}}\right|=\frac{a}{\sqrt{2}}$$

$$\left(\frac{1}{m\sqrt{1+m^2}}\right)^2=\left(\frac{1}{\sqrt{2}}\right)^2 \Rightarrow 2=m^2(1+m^2)$$

$$\Rightarrow \quad m^4+m^2-2=0$$

$$(m^2+2)(m^2-1)=0$$

$$\therefore \quad m=\pm 1$$

$\therefore$ Equation of pair of tangents are

$$y=x+a \quad \text{...(ii)}$$

$$\text{and} \quad y=-x-a \quad \text{...(iii)}$$

And point of intersection of Eqs. (ii) and (iii) is $A(-a, 0)$ which lies on directrix of parabola.

The chord of contact for $A(-a, 0)$ w.r.t. to circle

$$x^2 + y^2 = \frac{a^2}{2}$$

$$T = 0$$

$$\Rightarrow \quad -ax = \frac{a^2}{2} \Rightarrow x = -\frac{a}{2}$$

and w.r.t. parabola $x = a$

Let PQ be $x = -\frac{a}{2}$ and RS be $x = a$

So that $PQRS$ is a trapezium.

$\therefore$ Area $= \frac{1}{2}(PQ + RS) \times$ (distance between PQ and RS)

$$= \frac{1}{2}(a + 4a) \times \left(a + \frac{a}{2}\right) = \frac{15a^2}{4} \text{ sq units}$$

11. (a) Consider the equation obtained by putting the coordinates of A and B in $y^2 = 4b(x + c)$

$$a^2(t_1 + t_2)^2 = 4b(at_1t_2 + c) \quad \text{...(i)}$$

$$a^2(t_2 + t_3)^2 = 4b(at_2t_3 + c) \quad \text{...(ii)}$$

This implies that, t_1 and t_3 are roots of the equation.

$$a^2(t + t_2)^2 = 4b(att_2 + c) \begin{matrix} t_1 \\ t_2 \end{matrix}$$

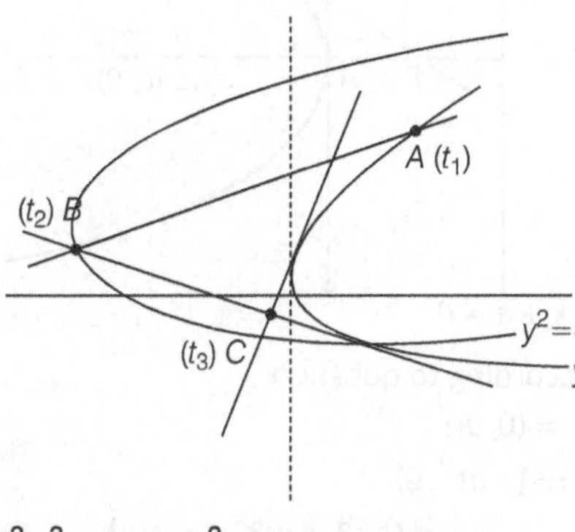

i.e. $a^2(t^2 + 2tt_2 + t_2^2) = 4batt_2 + 4bc$

$$\Rightarrow a^2t^2 + 2t(a^2t_2 - 2abt_2) + a^2t_2^2 - 4bc = 0$$

$$\Rightarrow \quad t_1 + t_3 = -\frac{a^2t_2 - 2abt_2}{a^2} = t_2\left(\frac{2b - a}{a}\right)$$

and $\quad t_1t_3 = \frac{a^2t_2^2 - 4bc}{a^2}$

$$\therefore \quad \frac{k}{a} = t_2\left(\frac{2b - a}{a}\right) \Rightarrow t_2 = \frac{k}{2b - a}$$

and $\quad \frac{h}{a} = \frac{a^2t_2^2 - 4bc}{a^2} \Rightarrow \frac{ah + 4bc}{a^2} = t_2^2$

$$\Rightarrow \quad \frac{k^2}{(2b - a)^2} = \frac{ah + 4bc}{a^2}$$

$\therefore$ Locus $a^2y^2 = (2b - a)^2(ax + 4bc)$.

12. (b) Let (h, k) be the coordinate of the polar of the variable chord.

The equation of polar (h, k) w.r.t. parabola $y^2 = 4ax$ is

$$ky = 2a(x + h)$$

$$\Rightarrow \quad \frac{ky - 2ax}{2ah} = 1 \quad \text{...(i)}$$

Now, making the equation of parabola homogeneous with the help of Eq. (i), we get

$$y^2 = 4ax\left(\frac{ky - 2ax}{2ah}\right)$$

$$\Rightarrow \quad 4ax^2 - 2kxy + hy^2 = 0$$

$A = 4a, B = h$ and $H = -k$

$\therefore \alpha$ is the angle between the lines.

$$\therefore \quad \tan\alpha = \left|\frac{2\sqrt{H^2 - AB}}{A + B}\right| = \frac{2\sqrt{k^2 - 4ah}}{4a + h}$$

$$\Rightarrow \quad \tan^2\alpha = \frac{4(k^2 - 4ah)}{(4a + h)^2}$$

$$\Rightarrow \quad (h + 4a)^2 = 4\cot^2\alpha\,(k^2 - 4ah)$$

$\therefore$ Locus is $(x + 4a)^2 = 4\cot^2\alpha\,(y^2 - 4ah)$

13. (c) Parabola : $y^2 = 4x$, $4a = 4 \Rightarrow a = 1$

Let $P(at_1^2, 2at_1) = (t_1^2, 2t_1)$

$Q \Rightarrow \left(\frac{1}{t_1^2}, -\frac{2}{t_1}\right)$ are end points of the focal chord.

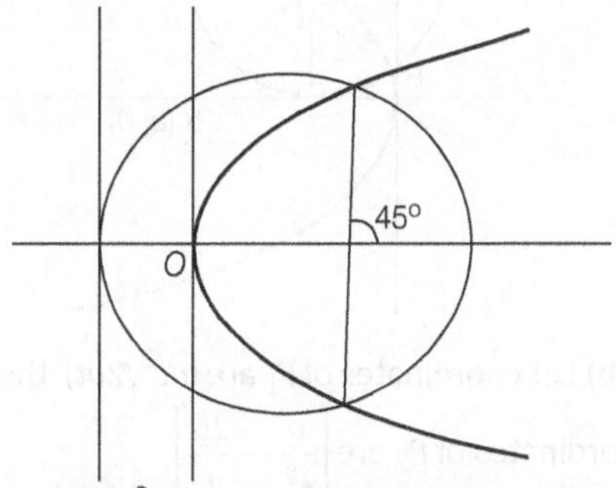

$x+a=0$

$\because$ We know that the circle, focal chord as diameter touches the directrix of parabola.

Directrix $x + a = 0 \Rightarrow x + 1 = 0$

$\therefore$ Equation of circle

$$(x - t_1^2)\left(x - \frac{1}{t_1^2}\right) + (y - 2t_1)\left(y + \frac{2}{t_1}\right) = 0$$

$$\Rightarrow x^2 - \left(t_1^2 + \frac{1}{t_1^2}\right)x + 1 + y^2 - 2\left(t_1 - \frac{1}{t_1}\right)^2 y - 4 = 0$$

The slope of focal chord PQ is equal to

So, we have $\dfrac{2\left(t_1 + \frac{1}{t_1}\right)}{t_1^2 - \frac{1}{t_1^2}} = 1$

$$\Rightarrow \quad \frac{2}{t_1 - \frac{1}{t_1}} = 1 \Rightarrow t_1 - \frac{1}{t_1} = 2$$

$\Rightarrow t_1^2 - 2t_1 - 1 = 0 \Rightarrow t_1 = 1 \pm \sqrt{2}$

Here, we neglect $(1+\sqrt{2})$ because on substituting $t_1 = 1+\sqrt{2}$ to find the radius of the circle, we get a rather big equation which does not fit to the option provided in the question.

$\therefore$ Option (c) the line $x+1=0$ touches the circle, is correct. Parametric coordinates of the parabola are $(t^2, 2t)$, ($\because a = 1$).

14. (c, d) Let $A = (t_1^2, 2t_1)$ and $B = (t_2^2, 2t_2)$, $(t_1 \neq t_2)$

Therefore, slope of AB is $= \dfrac{2}{t_1+t_2}$

Now, C is the centre of the circle described on AB as diameter which is given by $\left(\dfrac{t_1^2+t_2^2}{2}, t_1+t_2\right)$

$\therefore$ The radius is $|t_1+t_2| = r$ [$\because$ the circle touches the axis]

So, the slope of $AB = \pm \dfrac{2}{r}$

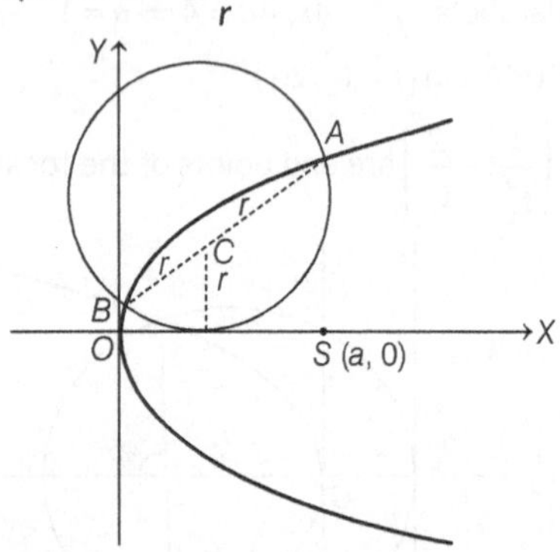

15. (a,b) Let coordinates of P_1 are $(at^2, 2at)$, then coordinates of P_2 are $\left[\dfrac{a}{t^2}, -\dfrac{2a}{t}\right]$

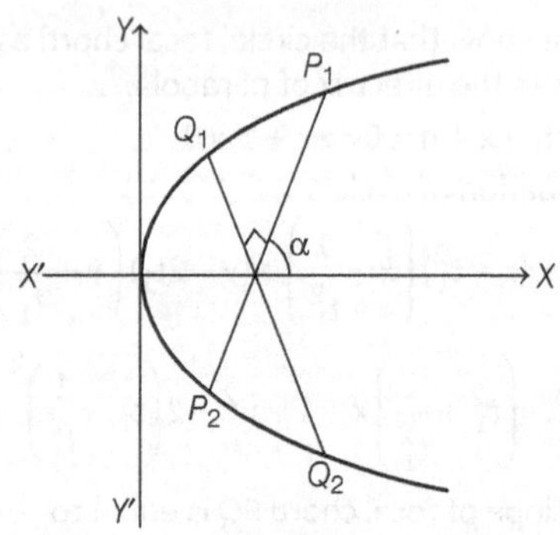

Let focal chord P_1P_2 makes an angle α with X-axis, then

$$\tan\alpha = \frac{2at - \left(-\dfrac{2a}{t}\right)}{at^2 - \dfrac{a}{t^2}} = \frac{2}{t - \dfrac{1}{t}}$$

$$\Rightarrow \quad t - \frac{1}{t} = 2\cot\alpha$$

Now, $P_1P_2 = a\left[t + \dfrac{1}{t}\right]^2 = a\left[\left(t - \dfrac{1}{t}\right)^2 + 4\right]$

$= a\{4\cot^2\alpha + 4\} = 4a\,\text{cosec}^2\,\alpha$ [from Eq. (i)]

Similarly, $Q_1Q_2 = 4a\,\text{cosec}^2(90^\circ - \alpha) = 4a\sec^2\alpha$

$\therefore$ Area of quadrilateral $P_1Q_1P_2Q_2 = \dfrac{1}{2}(P_1P_2)(Q_1Q_2)$

$= 8a^2\sec^2\alpha\,\text{cosec}^2\,\alpha = 32a^2$

$\therefore$ Minimum area $= 32a^2 = 2\,(\text{Latusrectum})^2$ and is inclined at

$\alpha = \dfrac{\pi}{4}$ [$\because \text{cosec}^2\alpha = 1, \sec^2\alpha = 1$]

16. (a,c) Given parabola, $y^2 = 4ax$

Tangent at point $Q(at^2, 2at)$

$$ty = x + at^2$$

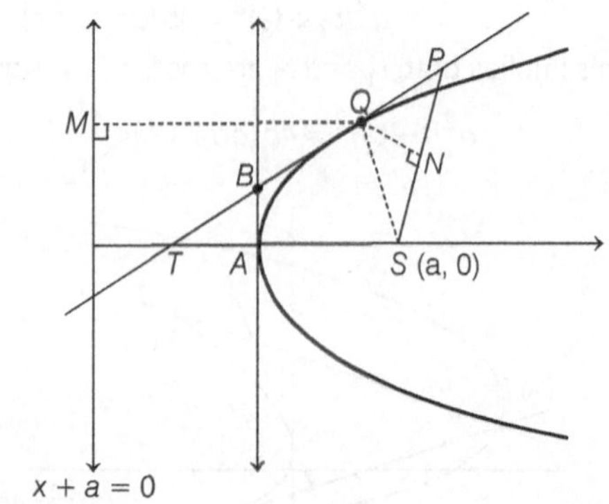

Now, according to question

Point $B \Rightarrow (0, at)$

Point $T \Rightarrow (-at^2, 0)$

Mid-point of $QT = \left(\dfrac{at^2 - at^2}{2}, \dfrac{0+2at}{2}\right)$

$= (0, at) = B$

$\therefore$ B is mid-point QT.

17. (a,b,c) $\because$ $y = x|x| = \begin{bmatrix} x^2, & x \geq 0 \\ -x^2, & x < 0 \end{bmatrix}$

$\because$ Equation of tangent in terms of slope (m) of $x^2 = 4ay$ is

$$y = mx - am^2 \qquad \text{...(i)}$$

Also, line (i) touches the circle $x^2 + (y-2)^2 = 4$, then

$$\left|\frac{2 + am^2}{\sqrt{1+m^2}}\right| = 2$$

$$\Rightarrow \quad 4 + a^2m^4 + 4am^2 = 4 + 4m^2$$

$$\Rightarrow \quad m^2 = \frac{4-4a}{a^2} \text{ and } m^2 = 0$$

Put $4a = 1$ for $y = x^2$ and $x \geq 0$

$\Rightarrow m^2 = 48 \Rightarrow m = 4\sqrt{3}$

Equation of tangent is $y = 4\sqrt{3}x - 12, y = 0$

Put $4 - 4a = 5$ for $y = -x^2$ and $x < 0$

$\Rightarrow m^2 = 80 \Rightarrow m = 4\sqrt{5}$

tangent is $y = 4\sqrt{5}x + 20$.

$\because$ Common tangents are

$$y = 0, y = 4\sqrt{3}x - 12 \text{ and } y = 4\sqrt{5}\,x + 20$$

18. (a,c,d)

Given parabola : $y^2 = 9x$ ...(i)

Circle : $x^2 + y^2 - 4x - 6 = 0$ ...(ii)

Solving Eqs. (i) and (ii),

$$x^2 + 9x - 4x - 6 = 0$$

$$\Rightarrow \quad x^2 + 5x - 6 = 0$$

$$\Rightarrow \quad (x+6)(x-1) = 0$$

$$x = -6 \quad \text{or} \quad x = 1$$

$$y^2 = -54 \quad \text{or} \quad y^2 = 9$$

$$y \in \phi \quad \text{or} \quad y = \pm 3$$

$\therefore$ When $x = 1 \Rightarrow y = 3$

$x = 1 \Rightarrow y = -3$

$\therefore$ $A(1, 3)$ and $B(1, -3)$ are common points

equation of common chord $\Rightarrow x = 1 \Rightarrow x - 1 = 0$

length of common chord $= AB = \sqrt{(6)^2} = 6$

19. (a,c,d) Tangent to $y^2 = 4ax$ at $(t^2, 2t)$ is.

$$y(2t) = 2(x + t^2)$$

$$yt = x + t^2$$

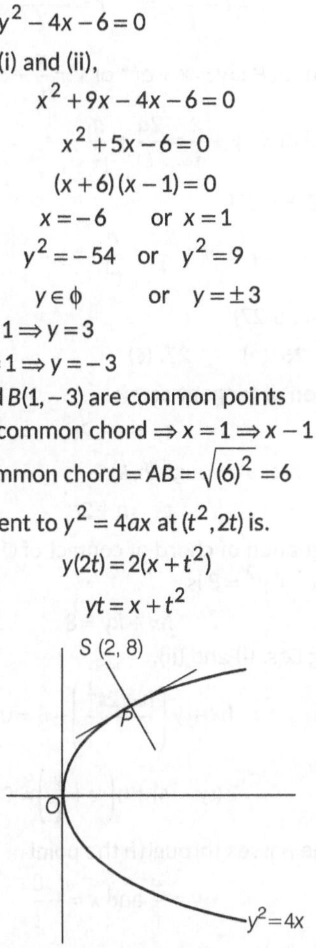

Equation of normal at $P(t^2, 2t)$ is

$$y + tx = 2t + t^3$$

$\because$ Normal at P passes through centre of circle $S(2,8)$

$$\therefore \quad 8 + 2t = 2t + t^3$$

$$t = 2$$

i.e., $\quad P = (4, 4)$

$\because$ Shortest distance between two curves lie along their common normal and common normal will pass through the centre of circle.

$$\therefore \quad SP = \sqrt{(4-2)^2 + (4-8)^2} = 2\sqrt{5}$$

$\therefore$ Option (a) is correct.

Also $\quad SQ = 2 \Rightarrow PQ = SP - SQ = 2\sqrt{5} - 2$

Thus, $\quad \dfrac{SQ}{QP} = \dfrac{1}{\sqrt{5}-1} = \dfrac{\sqrt{5}+1}{4}$

$\therefore$ Option (b) is incorrect.

Now, x-intercept of normal is $x = 2 + 2^2 = 6$

Option (c) is correct.

$\therefore$ Slope of tangent $= \dfrac{1}{t} = \dfrac{1}{2}$

$\therefore$ Option (d) is correct.

20. (b,c) $y = \tan \tan^{-1} x = x$

Let (h, k) be the centre, then $\left|\dfrac{h-k}{\sqrt{2}}\right| = \sqrt{(h-1)^2 + k^2}$

$$\Rightarrow \quad (h-k)^2 = 2(h^2 + k^2 - 2h + 1)$$

$$\Rightarrow h^2 + k^2 + 2hk - 4h + 2 = 0 \Rightarrow (h+k)^2 = 4h - 2$$

$$\Rightarrow \quad (h+k)^2 - 2(h+k) + 1 = 4h - 2 - 2(h+k) + 1$$

$$\Rightarrow \quad (h+k-1)^2 = 2\left(h - k - \frac{1}{2}\right)$$

$\therefore$ Locus is $(x + y - 1)^2 = 2\left(x - y - \dfrac{1}{2}\right)$

$$\left(\frac{x+y-1}{\sqrt{2}}\right)^2 = \frac{2}{\sqrt{2}}\left(\frac{x - y - \frac{1}{2}}{\sqrt{2}}\right)$$

$\therefore$ Axis of symmetry is $x + y = 1$

Length of latusrectum $= \sqrt{2}$

Vertex is the point of intersection of

$$x + y = 1 \text{ and } x - y = \frac{1}{2} \Rightarrow \left(\frac{3}{4}, \frac{1}{4}\right)$$

21. (a,b,c,d) Given, parabola $y^2 = 8x$

$$4a = 8 \Rightarrow a = 2$$

$\because$ $P(-2, 0)$ lies on the directrix and the axis.

The tangent will have slope $m = \pm 1$

$\therefore$ Equation are $y = x + 2$ and $y = -x - 2$

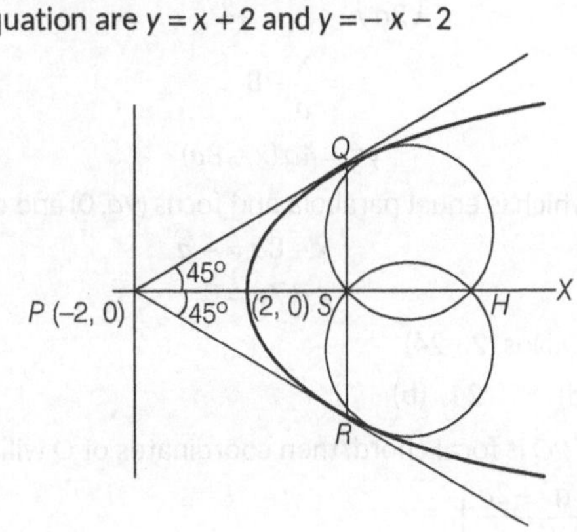

The chord of contact of tangents is QR as $x = 2$ (i.e., L-R)

$\because$ Common chord of two circles is X-axis.

$\because$ $\angle QPR = 90°$

$\therefore$ Circumcentre is (2, 0) on X-axis.

and orthocentre is $P(-2, 0)$ on X-axis

$\Rightarrow$ Centroid and incentre also lies on X-axis because circumcentre, orthocentre, centroid and incentre are collinear.

22. (a,b,c) $P \equiv (at^2, 2at)$ and $R = (at_1^2, 2at_1)$

$\because$ $OP \perp OR$.

$\therefore$ Slope of $OP \times$ Slope of $OR = -1$

$$\frac{2}{t} \times \frac{2}{t_1} = -1$$

$$\Rightarrow \quad t_1 = -\frac{4}{t}$$

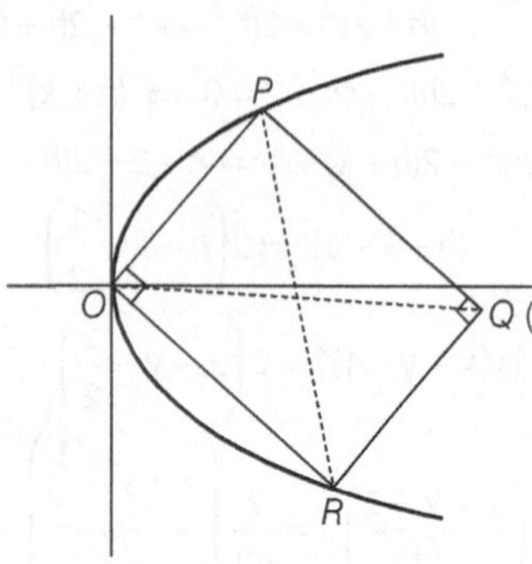

Now, coordinates of R is $\left(\frac{16a}{t^2}, \frac{-8a}{t}\right)$

$\because$ $OPQR$ is a rectangle.

$\therefore$ Mid-point of OQ = Mid-point of PR

$$\Rightarrow \quad x = at^2 + \frac{16a}{t^2} \Rightarrow \frac{x}{a} = t^2 + \frac{16}{t^2} \quad \text{...(i)}$$

$$\text{and} \quad y = 2at - \frac{8a}{t} \Rightarrow \frac{y}{2a} = t - \frac{4}{t} \quad \text{...(ii)}$$

From Eqs. (i) and (ii), we get

$$\left(\frac{y}{2a}\right)^2 = t^2 + \frac{16}{t^2} - 8$$

$$= \frac{x}{a} - 8$$

$$y^2 = 4a(x - 8a)$$

which is equal parabola and focus $(9a, 0)$ and directrix

$$x - 8a = -a$$

$$x - 7a = 0$$

Sol. (Q.Nos. 23-24)

23. (d) **24.** (b)

If PQ is focal chord, then coordinates of Q will be $\left(\frac{a}{t^2}, \frac{-2a}{t}\right)$

Now, slope of QR = slope of PK

$$\frac{2ar + \frac{2a}{t}}{ar^2 - \frac{a}{t^2}} = \frac{2at}{at^2 - 2a}$$

$$\Rightarrow \quad \frac{r + \frac{1}{t}}{r^2 - \frac{1}{t^2}} = \frac{t}{t^2 - 2} \Rightarrow \frac{1}{r - \frac{1}{t}} = \frac{t}{t^2 - 2} \Rightarrow r - \frac{1}{t} = \frac{t^2 - 2}{t}$$

$$\Rightarrow \quad r = \frac{t^2 - 2}{t} + \frac{1}{t}$$

$$r = t - \frac{1}{t} \quad \Rightarrow \quad r = \frac{t^2 - 1}{t}$$

24. Tangent at $P: ty = x + at^2$ or $y = \frac{x}{t} + at$

Normal at $S: y + \frac{x}{t} = \frac{2a}{t} + \frac{a}{t^3}$

Solving, we get

$$2y = at + \frac{2a}{t} + \frac{a}{t^3} \Rightarrow y = \frac{a(t^2 + 1)^2}{2t^3}$$

Sol. (Q.Nos. 25-27)

25. (d) **26.** (a) **27.** (c)

25. Equation of tangent at P is

$$ty = x + 2t^2$$

which passes through (h, k)

$$\Rightarrow \quad tk = h + 2t^2 \quad \text{...(i)}$$

The equation of chord of contact of $Q(h, k)$ w.r.t. the circle $x^2 + y^2 = 8$ is

$$hx + ky = 8 \quad \text{...(ii)}$$

Solving Eqs. (i) and (ii),

$$hx + y\left(\frac{h + 2t^2}{t}\right) - 8 = 0$$

$$\Rightarrow \quad 2(ty - 4) + h\left(x + \frac{y}{t}\right) = 0$$

The line passes through h the point

$$y = \frac{4}{t} \text{ and } x = -\frac{8}{t}$$

$\therefore$ Locus is $\frac{y}{4} = \frac{-x}{y} \Rightarrow y^2 = -4x$

26. The required point lies on the director circle of the circle and directrix of parabola. The directrix of the parabola is $x + 2 = 0$ and director circle of the given circle is $x^2 + y^2 = (2\sqrt{2} \times \sqrt{2})^2 = 16$

Now, $x = -2 \Rightarrow (-2)^2 + y^2 = 16$

and $y^2 = 12 \Rightarrow y = \pm 2\sqrt{3}$

$\therefore$ Therefore, the required point $(-2, 2\sqrt{3})$.

27. The equation of the circumcircle of ΔAQB is

$$(x^2+y^2-8)+\lambda(hx+ky-8)=0$$

This should pass through (0, 0) (centre of the circle).

$\therefore \quad -8+\lambda(-8)=0 \Rightarrow \lambda=-1$

$\therefore$ Circumcircle is

$$x^2+y^2-hx-ky=0 \quad \text{...(i)}$$

So that, $\left(\frac{h}{2},\frac{k}{2}\right)$ its centre.

If (x, y) is the circumcentre of ΔAQB.

Then, $x=\frac{h}{2}$ and $y=\frac{k}{2} \Rightarrow h=2x$ and $k=2y$

Put h and k in equation of tangent at point $P(h, k)$ and $t=2$, we get

$$2(2y)=2x+8$$
$$x-2y+4=0$$

Sol. (Q.Nos. 28-32)

28. (b) **29.** (c) **30.** (c) **31.** (b) **32.** (c)

Given parabola, $y^2=4ax$

Let P be the point on parabola $\Rightarrow (at^2, 2at)$

Equation of normal at P is

$$y=-tx+2at+at^3 \quad \text{...(i)}$$

Circle : $x^2+y^2-24ay+128a^2=0$

and centre $(0, 12a)$

$\therefore$ Radius $=\sqrt{144a^2-128a^2}=4a$

28. Normal at $P(at^2, 2at)$ is in Eq. (i) which passes through the centre.

$\therefore \quad 12a=0+2at+at^3$

$\Rightarrow \quad 12=2t+t^3$

$\Rightarrow \quad t^3+2t-12=0$

$\Rightarrow \quad (t-2)(t^2+2t-6)=0$

$\Rightarrow \quad t-2=0 \Rightarrow t=2$

29. Equation of tangent at P

$\because t=2$

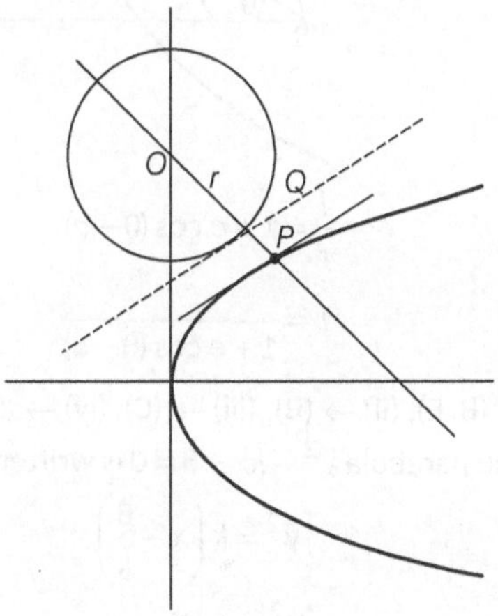

$P=(4a, 4a)$

$\Rightarrow \quad y(4a)=\frac{4a(x+4a)}{2}$

$\Rightarrow \quad 2y=x+4a \Rightarrow x-2y+4a=0$

Distance between centre of circle and tangent

$$d=\left|\frac{0-24a+4a}{\sqrt{5}}\right|=\frac{20a}{\sqrt{5}}=4\sqrt{5}a$$

$\therefore$ Shortest distance $=d-r=4\sqrt{5}a-4a=4a(\sqrt{5}-1)$

30. P is $(at^2, 2at)=(4a, 4a)$

31. Equation of normal $y=-2x+4a+8a=-2x+12a$

Solving with circle, $\quad x^2+y^2-24ay+128a^2=0$

$\Rightarrow \quad x^2+(12a-2x)^2-24a(-2x+12a)+128a^2=0$

$\Rightarrow \quad x^2+144a^2+4x^2-48ax+48ax-288a^2+128a^2=0$

$\Rightarrow \quad 5x^2-16a^2=0$

$\Rightarrow \quad x^2=\frac{16a^2}{5} \Rightarrow x=\pm\frac{4a}{\sqrt{5}}$

$\Rightarrow \quad x=\frac{4a}{\sqrt{5}}$ and $y=\frac{-8a}{\sqrt{5}}+12a$

32. $Q=\left(\frac{4a}{\sqrt{5}}, \frac{-8a}{\sqrt{5}}+12a\right)$ and $C=(0, 12a)$

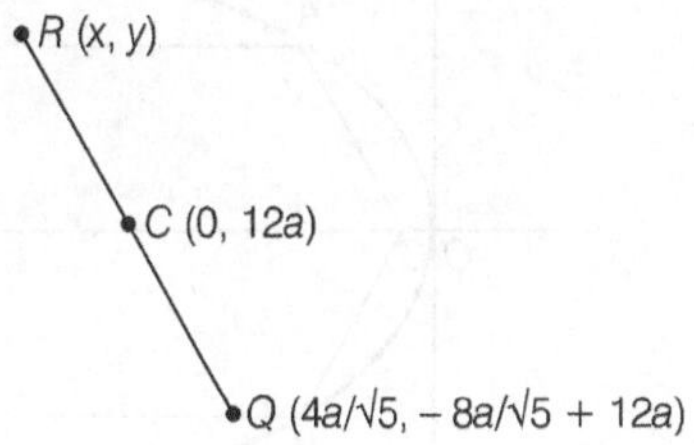

$\therefore x=-\frac{4a}{\sqrt{5}}$

Sol. (Q.Nos. 33-37)

33. (c) **34.** (b) **35.** (d) **36.** (a) **37.** (d)

33. Point of intersection of $y=2$ and $y^2=8x$ is $P\left(\frac{1}{2}, 2\right)$ and focus of the parabola is $S(2, 0)$.

$\therefore$ Equation of the reflected ray is

$$y-0=\frac{2-0}{\frac{1}{2}-2}(x-2)$$

$\Rightarrow \quad y=-\frac{4}{3}(x-2) \Rightarrow 4x+3y=8$

34. $\because \; y^2+10y-4x+17=0$

$\Rightarrow (y+5)^2-25-4x+17=0$

$\Rightarrow \quad (y+5)^2=4x+8 \Rightarrow (y+5)^2=4(x+2)$

Let $y+5=Y$ and $x+2=X$, then $Y^2=4X$

Focus is $X=1$ and $Y=0$ i.e. $(-1, -5)$.

After reflection, the ray must pass through focus $(-1, -5)$.

35. Solving $x = 2$ and $x^2 = 4y$, i.e. (2, 1)

Focus is (0, 1)

Equation of first reflection is $y = 1$ which is perpendicular to incident ray.

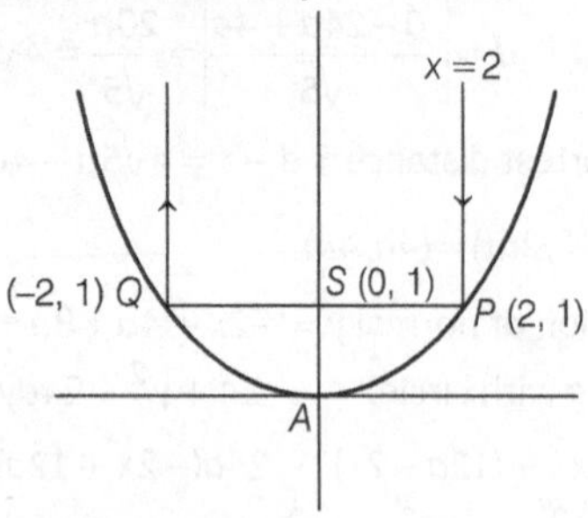

Hence, second reflected ray is parallel to incident ray and pass through $(-2, 1)$ is $x = -2$ or $x + 2 = 0$

36. Solving $y = 1$ and $y^2 = x$

Then, $A = (1, 1)$ and solving, $y = -2$ and $y^2 = x$

Then, $B = (4, -2)$

$\because$ After reflection both reflected rays pass through focus of the parabola $y^2 = x$ i.e. $C = \left(\frac{1}{4}, 0\right)$

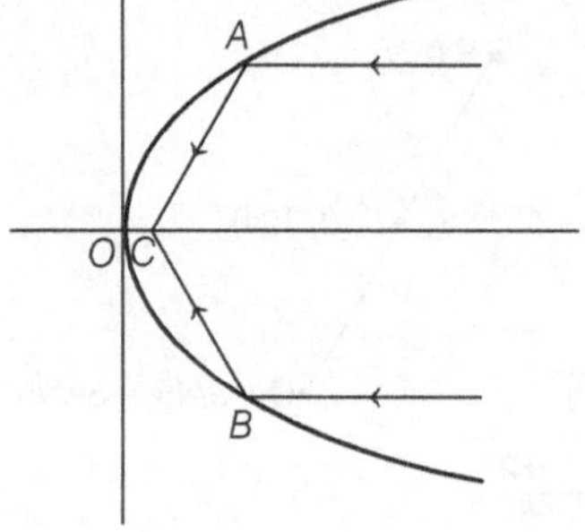

$$\therefore \text{ Required area} = \frac{1}{2}\begin{Vmatrix} 1 & 1 & 1 \\ 4 & -2 & 1 \\ \frac{1}{4} & 0 & 1 \end{Vmatrix} = \frac{21}{8} \text{ sq units}$$

37. Focus of the parabola $x^2 = 4(y - 1)$ is $S = (0, 2)$

$\because$ First reflected ray pass through S.

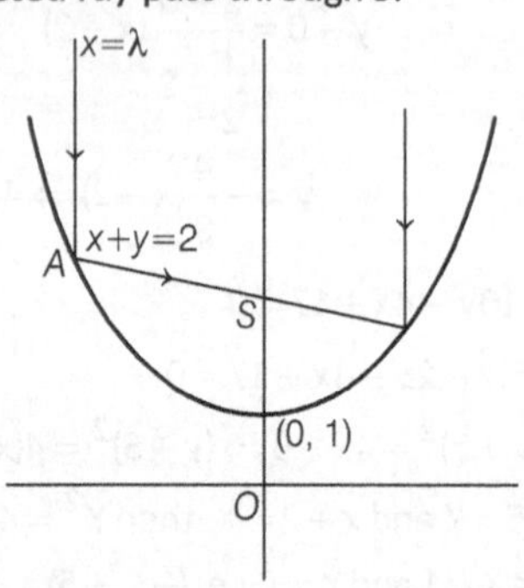

Now, solving $x + y = 2$ and $x^2 = 4(y - 1)$

$$\Rightarrow \quad x^2 = 4(2 - x - 1)$$

$$\Rightarrow \quad x^2 + 4x - 4 = 0$$

$$\therefore \quad x = \frac{-4 \pm \sqrt{16 + 16}}{2} = -2 \pm 2\sqrt{2}$$

Then, equation of new reflected ray parallel to $x = \lambda$ is $x = -2 + 2\sqrt{2}$

or $\quad x + 2 = 2\sqrt{2}$

Sol. (Q.Nos. 38-42)

38. (c) **39.** (b) **40.** (a) **41.** (d) **42.** (d)

38. $\because$ $SP = e\,PM = e\,NZ = e\,(SZ - SN) = e \cdot SZ - e \cdot SN$

$$= l - e \cdot (r \cos \theta)$$

or $\quad r = l - er \cos \theta$

$$\Rightarrow \quad r\,(1 + e \cos \theta) = l$$

$$\therefore \quad \frac{l}{r} = 1 + e \cos \theta \qquad \text{...(i)}$$

39. $\because$ $e \cdot SZ =$ The semi latusrectum $SL = l$ (given)

$$\therefore \quad SZ = \frac{l}{e}$$

40. $\because$ $r \cos \theta = SZ = \frac{l}{e}$ or $\frac{l}{r} = e \cos \theta$

41. Put $e = 1$ in Eq. (i), then the equation of parabola is

$$\frac{l}{r} = 1 + \cos \theta = 2 \cos^2 \left(\frac{\theta}{2}\right)$$

$$\therefore \quad r = \frac{l}{2} \sec^2 \left(\frac{\theta}{2}\right)$$

42. $\because$ The axis SZ of the conic makes any angle a with the initial line SA, the equation of the conic will be

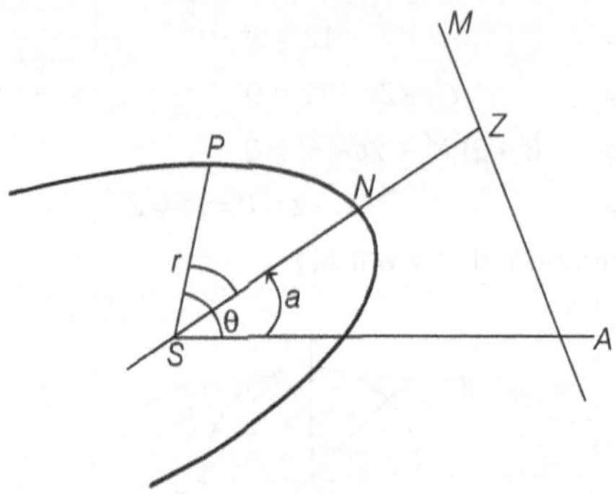

$$\frac{l}{r} = 1 + e \cos (\theta - a)$$

$$\therefore \quad r = \frac{l}{1 + e \cos (\theta - a)}$$

43. (i) $\rightarrow$ (B, E), (ii) $\rightarrow$ (B), (iii) $\rightarrow$ (C), (iv) $\rightarrow$ (B)

(i) The parabola $y^2 - kx + 8 = 0$ is written as

$$y^2 = k\left(x - \frac{8}{k}\right)$$

$$\therefore \quad Y^2 = 4AX$$

$$\Rightarrow \quad 4A = k \mid X = x - \frac{8}{k}$$

$$\Rightarrow \quad A = \frac{k}{4}$$

$\therefore$ Now, directrix is $X + A = 0$

$$\Rightarrow \quad x - \frac{8}{k} = \frac{-k}{4}$$

$$\Rightarrow \quad x + \frac{k^2 - 32}{4k} = 0$$

$$\Rightarrow \quad x = -\left(\frac{k^2 - 32}{4k}\right)$$

Given, $x - 1 = 0$ is directrix

$$\Rightarrow \quad \frac{k^2 - 32}{4k} = -1 \Rightarrow k^2 - 32 = -4k$$

$$\Rightarrow \quad k^2 + 4k - 32 = 0 \Rightarrow (k+8)(k-4) = 0$$

$$\Rightarrow \quad k = -8, 4$$

(ii) ΔOPM is equilateral triangle,

$$OP = OM = PM = l$$

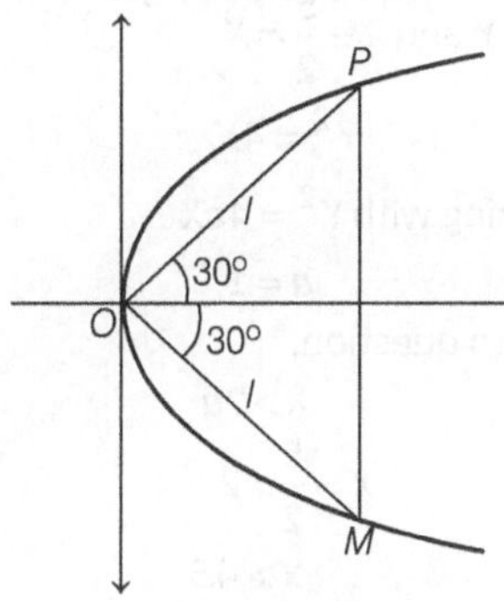

$$\therefore P \Rightarrow (l \cos 30°, l \sin 30°) = \left(\frac{l\sqrt{3}}{2}, \frac{l}{2}\right)$$

$\because$ P lies on parabola $y^2 = 4x$

$$\therefore \quad \left(\frac{l}{2}\right)^2 = 4\left(\frac{l\sqrt{3}}{2}\right)$$

$$\Rightarrow \quad l = 8\sqrt{3}$$

$$\frac{l}{2\sqrt{3}} = \frac{8\sqrt{3}}{2\sqrt{3}} = 4$$

(iii) Length of latusrectum $= \sqrt{(3-3)^2 + (5+3)^2} = 8$ units

(iv) Distance of the vertex from the directrix (i.e. Y-axis) is 2 and, if is equal to half the distance of the focus from the directrix so that the focus is (4, 0). Therefore, $a = 4$.

44. (i) $\rightarrow$ (A, C, E), (ii) $\rightarrow$ (A, C, E), (iii) $\rightarrow$ (A, B, C, D)

(i) Let $P(at^2, 2at)$ be any point on the parabola $y^2 = 4ax$ and $S(a, 0)$ be the focus, then the other end of focal chord through P will be $Q\left(\frac{a}{t^2}, \frac{-2a}{t}\right)$.

Then, length of focal chord $PQ = a\left(t + \frac{1}{t}\right)^2$

$$t + \frac{1}{t} \geq 2 \quad \Rightarrow a\left(t + \frac{1}{t}\right)^2 \geq 4a$$

or $\quad PQ \geq 32 \quad \{\because 4a = 32\}$

$$\lambda = 32$$

(ii) Equation of tangent to parabola $y^2 = 4x$ at $P(t^2, 2t)$ is

$$ty = x + t^2$$

Given, $\quad t^2 \in [1, 4]$

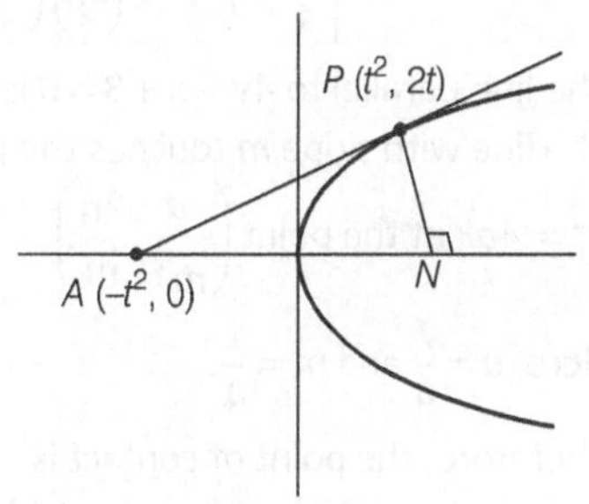

$$\therefore \text{ Area of } \Delta APN = \frac{1}{2}(AN)(PN) = \frac{1}{2}(2t^2)(2t)$$

$$\Rightarrow \quad \Delta = 2t^3 = 2(t^2)^{3/2}$$

$$\because \quad t \in [1, 4]$$

Δ_{max} occurs when $t^2 = 4$

$$\Delta_{max} = 2(4)^{3/2} = 16 \text{ sq units}$$

$$\therefore \quad \lambda = 16$$

(iii) Given parabola, $y^2 = 4x$. Now, the ends of latusrectum are $P(1, 2)$ and $Q(1, -2)$ or $P(t^2, 2t)$ and $Q(t_1^2, 2t_1)$, where $t = 1$ and $t_1 = -1$

We know that the other end of normal is given by

$$t_2 = -t - \frac{2}{t}$$

$$\Rightarrow \quad A(t_2^2, 2t_2) \text{ and } A'(t_3^2, 2t_3)$$

where, $\quad t_2 = -3$ and $t_3 = 3$

or $\quad A(9, -6)$ and $A' = (9, 6)$

$$\therefore \quad AA' = 12 \text{ units}$$

$$\lambda = 12$$

45. (i) $\rightarrow$ (A, E), (ii) $\rightarrow$ (C), (iii) $\rightarrow$ (B), (iv) $\rightarrow$ (D)

(i) It is known that the locus of the point from which perpendicular tangents is drawn to a parabola is the directrix i.e. $x + 2 = 0$ on which the points $(-2, 1)$ and $(-2, 5)$ lie.

(ii) The line

$$y = mx + \frac{a}{m}$$

Touches $y^2 = 4ax$ at $\left(\frac{a}{m^2}, \frac{2a}{m}\right)$

The given lines is $y=-x-3\,(a=3, m=-1)$ and hence it touches the parabola $y^2=12x$ at

$$\left(\frac{a}{m^2}, \frac{2a}{m}\right)=(3,-6)$$

(iii) It is known that, the line $y=mx-2am-am^3$ is a normal to the parabola $y^2=4ax$ at the point $(am^2, -2am)$. In the present case, $m=-\frac{4}{3}$ and $a=\frac{9}{4}$. Therefore,

$$(am^2, -2am)=\left[\frac{9}{4}\times\frac{16}{9}, -2\left(\frac{9}{4}\right)\left(\frac{-4}{3}\right)\right]=(4,6)$$

(iv) The line parallel to $4y-x+3=0$ is $4y-x+c=0$. The line with slope m touches the parabola $y^2=4ax$ at the point $\left(\frac{a}{m^2}, \frac{2a}{m}\right)$

Here, $a=\frac{7}{4}$ and $m=\frac{1}{4}$.

Therefore, the point of contact is

$$\left(\frac{a}{m^2}, \frac{2a}{m}\right)=\left(\frac{7}{4}\times 16, 2\left(\frac{7}{4}\right)\times 4\right)$$

$$=(28, 14)$$

46. (i) $\to$ (A, B, C, D, E), (ii) $\to$ (D, E), (iii) $\to$ (B, C, D, E)

If three normals drawn to any parabola $y^2=4ax$ from a given point (h, k) be real, then $h>2a$.

(i) $\because y^2-4x-2y+5=0$

$$\Rightarrow \quad (y-1)^2=4(x-1)$$

Let $y-1=Y$ and $x-1=X$

$$\therefore \quad Y^2=4X$$

On comparing with $Y^2=4aX$

$\therefore\ a=1$

According to question,

$$X>2a \Rightarrow x-1>2 \text{ or } x>3$$

$$\therefore \quad x=4,5,6,7,8$$

(ii) $\because 4y^2-32x+4y+65=0$

$$\Rightarrow \quad 4(y^2+y)=32x-65$$

$$\Rightarrow \quad 4\left[\left(y+\frac{1}{2}\right)^2-\frac{1}{4}\right]=32x-65$$

$$\Rightarrow \quad 4\left(y+\frac{1}{2}\right)^2=32x-64$$

$$\text{or} \quad \left(y+\frac{1}{2}\right)^2=8(x-2)$$

Let $y+\frac{1}{2}=Y$ and $x-2=X$

$$\therefore \quad Y^2=8X$$

On comparing with $Y^2=4aX$

$$\therefore \quad a=2$$

According to question,

$$X>2a$$

$$\Rightarrow \quad x-2>4$$

$$\therefore \quad x>6$$

$$\therefore \quad x=7,8$$

(iii) $\because 4y^2-16x-4y+41=0$

$$\Rightarrow \quad 4(y^2-y)=16x-41$$

$$\Rightarrow \quad 4\left[\left(y-\frac{1}{2}\right)^2-\frac{1}{4}\right]=16x-41$$

$$\Rightarrow \quad 4\left(y-\frac{1}{2}\right)^2=16x-40$$

$$\text{or} \quad \left(y-\frac{1}{2}\right)^2=4\left(x-\frac{5}{2}\right)$$

Let $y-\frac{1}{2}=Y$ and $x-\frac{5}{2}=X$

$$\therefore \quad Y^2=4X$$

On comparing with $Y^2=4aX$

$$\therefore \quad a=1$$

According to question,

$$X>2a$$

$$\Rightarrow \quad x-\frac{5}{2}>2$$

$$x>4.5$$

$\therefore\ x=5,6,7,8$

47. (i) $\to$ (A, B), (ii) $\to$ (A, C), (iii) $\to$ (B, C, D), (iv) $\to$ (D)

(i) Given parabola is, $x^2=ay$ ...(i)

and the given lines is $y-2x=1$ ...(ii)

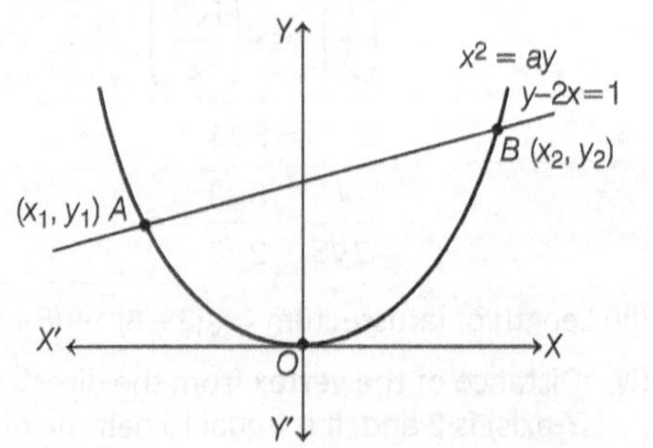

On solving Eqs. (i) and (ii), we get

$$x^2=a(2x+1) \quad \Rightarrow x^2-2ax-a=0$$

Let coordinates of A and B are (x_1, y_1) and (x_2, y_2) respectively, then

$$|x_1-x_2|=\frac{\sqrt{D}}{a}=\frac{\sqrt{4a^2+4a}}{1}=2\sqrt{a^2+a} \quad \text{...(iii)}$$

Also, AB parallel on $y-2x=1$

$\therefore \quad y_1-2x_1=1$ and $y_2-2x_2=1$

or $\quad y_2-y_1=2(x_2-x_1) \quad ...(iv)$

$\therefore$ Length of $AB=\sqrt{(x_2-x_1)^2+(y_2-y_1)^2}$

$=\sqrt{5}|x_1-x_2| \quad$ [from Eq. (iv)]

$=2\sqrt{5}\sqrt{a^2+a} \quad$ [from Eq. (iii)]

Given, $2\sqrt{5}\sqrt{a^2+a}=\sqrt{40}$

$$a^2+a=2 \Rightarrow a^2+a-2=0$$

$\Rightarrow \quad (a+2)(a-1)=0 \Rightarrow a=-2,1$

(ii) If tangents are drawn from (x_1, y_1) to the parabola $y^2=4ax$ and angle between tangents is θ, then

$$|\tan\theta|=\frac{\sqrt{y_1^2-4ax_1}}{|x_1+a|}$$

Here, $x_1=0$, $y_1=2$ and $\theta=\frac{3\pi}{4}$, then $|-1|=\frac{2}{|0+a|}$

$\Rightarrow \quad |a|=2 \Rightarrow a=\pm 2$

(iii) Let the other end be at $(at^2, 2at)$.

So, the mid-point is $\left(\frac{at^2+a}{2}, \frac{2at+2a}{2}\right)$.

Which satisfy $x+y=1$

$$\frac{at^2+a}{2}+\frac{2at+2a}{2}=1$$

$\Rightarrow \quad at^2+2at+3a-2=0$

Since, two distinct chords are possible, so $D>0$

$$4a^2-4a(3a-2)>0$$

$\Rightarrow \quad -8a^2+8a>0 \Rightarrow -8a(a-1)>0$

$\Rightarrow \quad 8a(a-1)<0 \Rightarrow a(a-1)<0$

$\Rightarrow \quad 0<a<1 \Rightarrow 0<4a<4 \Rightarrow L-R \in (0,4)$

Length of latusrectum can be 1, 2, or 3.

(iv) The given parabola, $x^2-ay+3=0$

$$x^2=a\left(y-\frac{3}{a}\right) \Rightarrow X^2=4AY$$

where, $X=x, 4A=a \Rightarrow A=\frac{a}{4}$ and $Y=y-\frac{3}{a}$

For focus $X=0, Y=A$

$$x=0 \text{ and } y-\frac{3}{a}=\frac{a}{4}$$

Focus is $\left(0, \frac{3}{a}+\frac{a}{4}\right)$ given focus is $(0, 2)$

$$\frac{3}{a}+\frac{a}{4}=2$$

$\Rightarrow \quad 12+a^2-8a=0$

$\Rightarrow \quad (a-6)(a-2)=0$

$\Rightarrow \quad a=2,6 \Rightarrow a_1=6$ and $a_2=2$

$$\frac{a_1}{a_2}=\frac{6}{2}=3$$

48. (i) $\rightarrow$ (E), (ii) $\rightarrow$ (A, B), (iii) $\rightarrow$ (C), (iv) $\rightarrow$ (A, D)

(i) Equation of parabola $y^2+4x+4y=0$

$$(y+2)^2=-4(x-1)$$

$$Y^2=-4AX$$

where, $Y=y+2, \quad X=x-1$ and $4A=4 \Rightarrow A=1$

$\therefore$ Equation of directrix $X-A=0$

$\Rightarrow \quad x-1-1=0$

$\Rightarrow \quad x-2=0$

$\Rightarrow \quad ax+b=0$

$\Rightarrow \quad a=1 \; b=-2$

(ii) Equation of tangent at $(t^2, 2t)$ on $y^2=4x$

$$ty=x+t^2$$

It passes through the point (2, 3), then

$$3t=2+t^2 \Rightarrow t^2-3t+2=0$$

$\Rightarrow \quad (t-1)(t-2)=0 \Rightarrow t=1,2$

The point of contact is (1, 2) or (4, 4).

(iii) Let $P(\sqrt{5}\cos\theta, \sqrt{5}\sin\theta)$, then the chord of contact of the parabola $y^2=4x$ w.r.t. P is

$$y\cdot(\sqrt{5}\sin\theta)=2(x+\sqrt{5}\cos\theta)$$

$$y=\frac{2x}{\sqrt{5}\sin\theta}+2\cot\theta$$

On comparing with $y=2(x-2)$, then

$$\sqrt{5}\sin\theta=1 \Rightarrow \cot\theta=-2$$

$$\sqrt{5}\sin\theta=1 \Rightarrow \sqrt{5}\cos\theta=-2$$

$\therefore$ P is $(-2, 1)$.

(iv) Let coordinates of Q be $(t^2, 2t)$

Now, area of ΔOPQ is

$$\frac{1}{2}\begin{vmatrix} t^2 & 2t \\ 4 & -4 \end{vmatrix}=\pm 6$$

$\Rightarrow \quad 2t^2+4t=\pm 6$

$\Rightarrow \quad t^2+2t\pm 3=0$

$\Rightarrow \quad t^2+2t+3=0$ (not possible),

$\Rightarrow \quad t^2+2t-3=0 \Rightarrow (t+3)(t-1)=0$

$$t=1,-3$$

$\therefore$ Q are (1, 2) or (9, −6).

Sol. (Q.Nos. 49-51)

Given equation, $\quad x^2+y^2=9 \quad ...(i)$

$$y^2=8x \quad ...(ii)$$

Solving Eqs. (i) and (ii),

$$x^2+8x-9=0$$

$$(x+9)(x-1)=0$$

$$x=-9, \quad x=1$$

$$y^2=-72, \quad y^2=8$$

$\Rightarrow \quad y = \pm 2\sqrt{2}$

$\because$ Point $P(1, 2\sqrt{2})$ and $Q(1, -2\sqrt{2})$.

Tangents to the parabola at P and Q, respectively are

$y(2\sqrt{2}) = 4(x+1)$ and $y(-2\sqrt{2}) = 4(x+1)$

$\therefore S = (-1, 0)$

The tangent to circle at P is

$$x(1) + y(2\sqrt{2}) = 9$$

$\therefore R = (9, 0)$

Therefore, $P(1, 2\sqrt{2})$, $Q(1, -2\sqrt{2})$, $R = (9, 0)$ and $S = (-1, 0)$

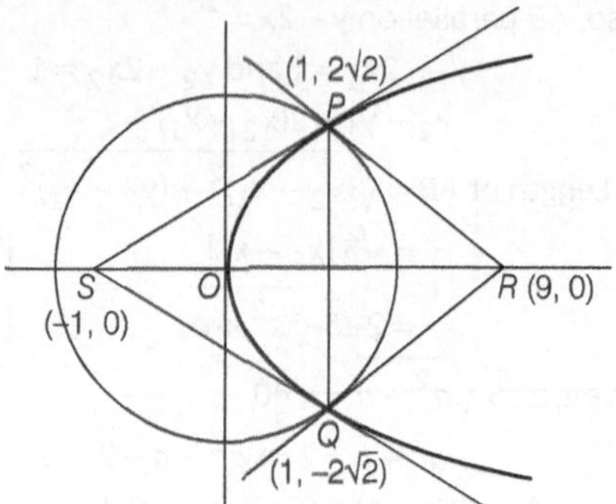

49. (28.28) The area of $\Delta PQR = \frac{1}{2} \text{mod} \begin{vmatrix} 1 & 2\sqrt{2} & 1 \\ 1 & -2\sqrt{2} & 1 \\ 9 & 0 & 1 \end{vmatrix}$

$$= \frac{1}{2} |1(-2\sqrt{2} - 0) - 2\sqrt{2}(1-9) + 1(0 + 18\sqrt{2})|$$

$$= 16\sqrt{2} \text{ sq units}$$

The area of ΔPQS is given $= \frac{1}{2} \text{mod} \begin{vmatrix} 1 & 2\sqrt{2} & 1 \\ 1 & -2\sqrt{2} & 1 \\ -1 & 0 & 1 \end{vmatrix}$

$$= \frac{1}{2} |1(-2\sqrt{2} - 0) - 2\sqrt{2}(1+1) + 1(0 - 2\sqrt{2})|$$

$$= 4\sqrt{2} \text{ sq units}$$

$\therefore$ The sum of the area of ΔPQS and ΔPQS is

$$16\sqrt{2} + 4\sqrt{2} = 20\sqrt{2} = 28.28 \text{ sq units}$$

50. (5.2) The area of ΔPRS is given by

$$= \frac{1}{2} \text{mod} \begin{vmatrix} 1 & 2\sqrt{2} & 1 \\ 9 & 0 & 1 \\ -1 & 0 & 1 \end{vmatrix}$$

$$= \frac{1}{2} |-2\sqrt{2}\ (9+1)|$$

$$= 10\sqrt{2}$$

Now, $PS = 2\sqrt{3}$, $RS = 10$ and $PR = 6\sqrt{2}$

Therefore, the circum radius of ΔPQS is

$$= \frac{abc}{4\Delta} = \frac{(PR)(PS)(RS)}{4 ar (\Delta PRS)}$$

$$= \frac{(6\sqrt{2})(2\sqrt{3})(10)}{4(10\sqrt{2})}$$

$$= 3\sqrt{3} = 5.1961$$

51. (2) For ΔPQR,

The area $= 16\sqrt{2}$

Now, $PQ = 4\sqrt{2}$, $PR = 6\sqrt{2}$ and $QR = 6\sqrt{2}$

$\therefore$ The inradius of ΔPQR is given by

$$= \frac{\Delta}{3} = \frac{\text{Area}}{\text{Semi-perimeter}} = \frac{16\sqrt{2}}{8\sqrt{2}} = 2$$

52. (0.1) Tangent at (x_1, y_1) on the parabola $y^2 = 4ax$

$$\therefore \quad yy_1 = 2a(x + x_1)$$

Any point on this tangent will be $\left(x', \frac{2a(x' + x_1)}{y_1}\right)$

Equation of chord of contact of the point $\left(x', \frac{2a(x' + x_1)}{y_1}\right)$ w.r.t. circle $x^2 + y^2 = a^2$ is

$$xx' + y\left\{\frac{(2a(x' + x_1)}{y_1}\right\} = a^2$$

$$\Rightarrow \quad xx'y_1 + 2ayx' + 2ayx_1 = a^2 y_1$$

$$\Rightarrow \quad (2ayx_1 - a^2 y_1) + x'(xy_1 + 2ay) = 0$$

which is family of straight lines passing through point of intersection of

$$2ayx_1 - a^2 y_1 = 0 \text{ and } xy_1 + 2ay = 0$$

$$\therefore \quad y = \frac{ay_1}{2x_1}$$

and $\quad x = \frac{-2ay}{y_1} = \left(\frac{-2a}{y_1}\right)\left(\frac{ay_1}{2x_1}\right)$

$$x = -\frac{a^2}{x_1}$$

The fixed point is (x_2, y_2).

$$\therefore \quad x_2 = -\frac{a^2}{x_1} \text{ and } y_2 = \frac{ay_1}{2x_1}$$

Now, $\quad \frac{x_1}{x_2} = \left(\frac{x_1}{-a^2 / x_1}\right) = \frac{x_1^2}{-a^2}$

and $\quad \frac{y_1}{y_2} = \frac{y_1}{\frac{ay_1}{2x_1}} = \frac{2x_1}{a}$

Now, $\quad 4\left(\frac{x_1}{x_2}\right) + \left(\frac{y_1}{y_2}\right)^2 = -\frac{4x_1^2}{a^2} + \frac{4x_1^2}{a^2} = 0$

$\therefore \lambda = 4$ and $\mu = 1$

$$\therefore \quad \frac{\lambda + \mu}{50} = \frac{5}{50} = 0.1$$

53. (2) Let AB taken along X-axis, its mid-point O as origin and a line through O perpendicular to AB as Y-axis.

Equation of circle on AB as diameter is

$$x^2 + y^2 = a^2 \quad \text{...(i)}$$

The equation of a circle of radius b and concentric to circle (i) is

$$x^2 + y^2 = b^2 \quad \text{...(ii)}$$

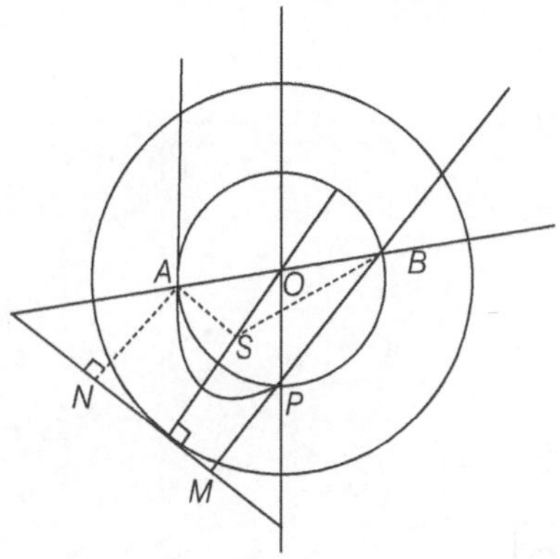

Let $S(h, k)$ be the focus of the parabola which passes through $A(-a, 0)$ and $B(a, 0)$

Let equation of tangent to circle (ii) be

$$x\cos\theta + y\sin\theta = 6 \quad \text{...(iii)}$$

This is taken as directrix, then $SP = PM$

$$\Rightarrow BS = BM \Rightarrow (h-a)^2 + (k-0)^2 = (a\cos\theta - b)^2 \quad \text{...(iv)}$$

and $AS = AN$

$$(h+a)^2 + (k-0)^2 = (a\cos\theta + b)^2 \quad \text{...(v)}$$

On adding Eqs. (iv) and (v)

$$a^2\cos^2\theta + b^2 = h^2 + k^2 + a^2 \quad \text{...(vi)}$$

On subtracting Eqs. (iv) and (v)

$$4ab\cos\theta = 4ah$$

$$\Rightarrow \cos\theta = \frac{h}{b} \quad \text{...(vii)}$$

From Eqs. (vi) and (vii), we get

$$\frac{a^2h^2}{b^2} + b^2 = h^2 + k^2 + a^2$$

$$\Rightarrow \frac{h^2(b^2-a^2)}{a^2} + k^2 = (b^2 - a^2)$$

$$\Rightarrow \frac{h^2}{a^2} + \frac{k^2}{b^2 - a^2} = 1$$

$\therefore$ Locus is $\dfrac{x^2}{a^2} + \dfrac{y^2}{b^2 - a^2} = 1$

By compare $\dfrac{x^2}{ma^2} + \dfrac{ny^2}{b^2 - a^2} = \dfrac{l}{2}$

$\therefore m = 1, n = 1$ and $\dfrac{l}{2} = 1 \Rightarrow l = 2$

$\therefore \sqrt{m + n + l} = \sqrt{1 + 1 + 2} = \sqrt{4} = 2$

54. (2) End points of latusrectum are $(a, \pm 2a)$ i.e. $(1, \pm 2)$.

Equation of normal at (x_1, y_1) is

$$\frac{y - y_1}{x - x_1} = -\frac{y_1}{2a}$$

i.e. $\dfrac{y-2}{x-1} = -\dfrac{2}{2}$ and $\dfrac{y+2}{x-1} = \dfrac{2}{2}$

$\Rightarrow \quad x + y = 3$

and $\quad x - y = 3$

which is tangent to $(x-3)^2 + (y+2)^2 = r^2$

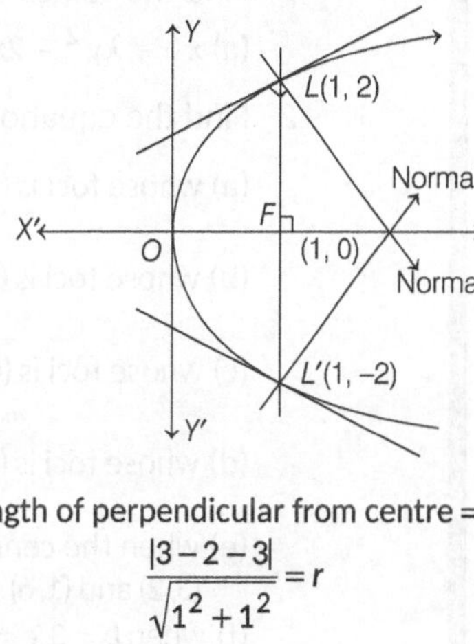

$\therefore$ Length of perpendicular from centre = Radius

$$\Rightarrow \frac{|3 - 2 - 3|}{\sqrt{1^2 + 1^2}} = r$$

$$\therefore \quad r^2 = 2$$

CHAPTER

06

ELLIPSE

LEVEL 01 \ Starter Level

6.1 Equation of Ellipse

1 Find the values of λ for which following equations represents an ellipse

(a) $x^2 + \lambda y^2 - 2xy + 4x + 6y - 3 = 0$ (b) $x^2 - y^2 + 5xy + 2x - 7y + \lambda = 0$

2 Find the equation of ellipse

(a) whose foci is $(\pm 4, 0)$ and eccentricity $= \frac{1}{3}$.

(b) whose foci is $(\pm 5, 0)$ and $x = \frac{36}{5}$ as one directrix.

(c) whose foci is $(0, \pm 3)$ and eccentricity $= \frac{1}{2}$.

(d) whose foci is $(0, \pm 2)$ and $y = \frac{25}{4}$ as one directrix.

(e) when the centre is at (0, 0), major axis is on the Y-axis and passes through the points (3, 2) and (1, 6).

(f) when $b = 3, c = 4$, centre is at the origin and foci is on X-axis.

(g) with $e = \frac{3}{4}$ foci on Y-axis, centre at the origin and passing through the point (6, 4).

3 Find the equation of ellipse

(a) whose eccentricity is $\frac{1}{2}$, focus is the point $(-1, 1)$ and directrix is the line $x - y + 3 = 0$.

(b) whose focii are $(-1, 5)$ and $(5, 5)$ and whose major axis is 10 units.

(c) whose axes are parallel to the coordinate axes having its centre at the point $(2, -3)$, one focus at $(3, -3)$ and one vertex at $(4, -3)$.

(d) whose focus is the centre of circle $x^2 + y^2 - 4y + 1 = 0$ and centre is vertex of $y^2 = 8x$ with $e = \frac{1}{4}$.

(e) whose one focus is same as focus of $x^2 + 4y - 2x + 7 = 0$ and eccentricity is radius of circle $4x^2 + 4y^2 = 1$ with $x = 1$ as a major axis with length is 10 units.

4 Find the centre, vertex, focus, directrix, equation of major axis, equation of minor axis, eccentricity, length of major axis, length of minor axis and latusrectum for the following ellipse.

(a) $4x^2 + 9y^2 = 36$

(b) $2x^2 + y^2 = 2$

(c) $4x^2 + y^2 - 8x + 2y + 1 = 0$

(d) $2x^2 + 3y^2 - 4x - 12y + 13 = 0$

5 (a) Find the equation of locus of all points, the sum of whose distances from (3, 0) and (9, 0) is 12 units and check locus represents an ellipse or not.

(b) A man running a race-course notes that the sum of the distances from the two flag posts from him is always 10 m and the distance between the flag posts is 8 m. Find the equation of the locus traced by the man in standard form.

(c) A rod *AB* of length 15 cm rests in between two coordiantes axes in such a way that the end-point *A* lies on *X*-axis and end point *B* lies on *Y*-axis. A point $P(x, y)$ is taken on the rod in such a way that $AP = 6$ cm. Find the locus of point *P*.

(d) An archway is in the shape of a semi-ellipse, the road level being the major axis. If breadth of the archway is 30 ft and a man 6 ft tall touches the top when 2 ft from the side, find the greatest height of the arch.

6.2 Point and Ellipse

6 Find the position of given points w.r.t. the given ellipse

(a) point $(4, -3)$, ellipse $5x^2 + 7y^2 = 140$

(b) point $(1, 2)$, ellipse $x^2 + 2y^2 = 50$

(c) point $(-1, 3)$, ellipse $x^2 + 4y^2 = 37$

7 Find the values of α, if

(a) $P(\alpha, \alpha + 1)$ lies outside the ellipse $x^2 + 2y^2 = 1$

(b) $P(2\alpha, 3\alpha)$ lies on the ellipse $3x^2 + 4y^2 = 12$

(c) $P\left(7 - \frac{5\alpha}{4}, -\alpha\right)$ lies inside the ellipse $\frac{x^2}{25} + \frac{y^2}{16} = 1$

8 Find the parametric form of the following ellipse

(a) $\frac{(x-3)^2}{4} + \frac{(y-4)^2}{25} = 1$

(b) $4x^2 + y^2 - 8x + 2y + 1 = 0$

9 (a) Find the area of the greatest isosceles triangle that can be inscribed in the ellipse $16x^2 + 25y^2 = 400$ having its vertex coincident with one extremity of major axis.

(b) Find the area of the greatest rectangle that can be inscribed in the ellipse $4x^2 + 9y^2 = 36$.

6.3 Line and Ellipse

10 (a) Find the point of interseciton of a line $x + y = 1$ and ellipse $x^2 + 2y^2 = 1$ and length of intercept made by the line.

(b) Find the value of *k* for which the line $y = 2x + k$, is

(i) intersecting,

(ii) tangent

(iii) neither intersecting nor tangent w.r.t. ellipse $2x^2 + y^2 = 1$.

11 (i) Find the equation of tangent

(a) at a point $(-1, 2)$ to the ellipse $4x^2+9y^2=40$.

(b) in parametric form for the ellipse $\frac{x^2}{25}+\frac{y^2}{16}=1$.

(c) which makes an angle $\theta=\frac{\pi}{6}$ to the ellipse $3x^2+4y^2=1$.

(d) having slope 2 for the ellipse $3x^2+4y^2=12$.

(e) which is perpendicular to the line $4x-3y=1$ to the ellipse $9x^2+y^2=1$.

(f) which is parallel to $x+y=1$ to the ellipse $4x^2+3y^2=12$.

(ii) (a) If the line $x+9y=19$ touches the ellipse $2x^2+9y^2=38$, then find the point of contact.

(b) Find the point of contact, if a tangent to the ellipse $3x^2+4y^2=12$ makes an angle of 45° with the straight line $y=2x+1$.

(iii) (a) Find the maximum area of the ellipse $\frac{x^2}{a^2}+\frac{y^2}{b^2}=1$ which touches the line $y=3x+2$.

(b) Find the point of intersections of tangents at end points of latusrectum for the ellipse $\frac{x^2}{25}+\frac{y^2}{16}=1$.

(iv) (a) For the ellipse $\frac{x^2}{a^2}+\frac{y^2}{b^2}=1$, tangent at any point meets the tangents drawn at vertices of ellipse at two points P and Q. Find the product of PA and QB, where A and B are vertices.

(b) Find the locus of the foot of perpendicular drawn from the centre upon any tangent to the ellipse $\frac{x^2}{a^2}+\frac{y^2}{b^2}=1$.

(c) Find the locus of the mid-point of the portion of the tangent to the tangent to ellipse $\frac{x^2}{a^2}+\frac{y^2}{b^2}=1$.

12 Find the equation of normal

(a) at point $(2, 3)$ to the ellipse $2x^2+5y^2=53$

(b) which makes an angle $\frac{\pi}{3}$ to the axis of the ellipse $x^2+2y^2=1$.

(c) having slope of normal -2 to the ellipse $\frac{x^2}{16}+\frac{y^2}{9}=1$

(d) to the ellipse at the negative end of latusrectum of ellipse $\frac{x^2}{9}+\frac{y^2}{4}=1$

(e) to the ellipse $\frac{x^2}{9}+\frac{y^2}{25}=1$ which is parallel to line $3x+2y-1=0$. Also, find the point of contact.

13 (a) If the normal at the point $P(\theta)$ to the ellipse $5x^2+14y^2=70$ intersects it again at the point $Q(2\theta)$, find the value of $\cos\theta$.

(b) The normal at any point P on the ellipse $\frac{x^2}{a^2}+\frac{y^2}{b^2}=1$ meets the major axis and minor axis at G and G', respectively and CF is perpendicular upon the normal from the centre C of the ellipse, then find the value of $PF\cdot PG$ and $PF\cdot PG'$.

(c) If the normal at end of latusrectum passes through the opposite end of minor axis of ellipse $\frac{x^2}{a^2}+\frac{y^2}{b^2}=1$, find eccentricity.

(d) Find the maximum value of $\tan\theta$, where θ is angle between CP and normal at P of ellipse $\frac{x^2}{a^2}+\frac{y^2}{b^2}=1$, where C is centre of ellipse.

14. (a) How many maximum normal can be drawn to the ellipse $2x^2+y^2=3$ from the point

(i) (1, 1) (ii) (0, 1) (iii) (2, 1)

(b) Find the eccentric angles of the following points w.r.t. to given ellipse, is

(i) (1, 2) w.r.t. the ellipse $2x^2+3y^2=14$.

(ii) (–1, 3) w.r.t. the ellipse $3x^2+y^2=12$.

(c) Find the condition of eccentric angles of conormal points of an ellipse $\frac{x^2}{a^2}+\frac{y^2}{b^2}=1$.

(d) Find the equation of fixed curve through which conormal points of an ellipse $\frac{x^2}{a^2}+\frac{y^2}{b^2}=1$ are passed.

6.4 Pair of tangents, Chord of Contact, Conjugate Diameter, Director circle, Auxiliary circles, Subtangent and Normal

15 (a) Find the equation of pair of tangents from the point (1, 3) to the ellipse $4x^2+9y^2=36$.

(b) Find the angle between the pair of tangents from the point (1, 2) to the ellipse $3x^2+2y^2=5$

(c) Find the value λ, if pair of tangents from $(\lambda, 1)$ to the ellipse $x^2+2y^2=10$ are mutually perpendicular to each other.

16 Find the locus of point of intersection of tangents to the ellipse $x^2+2y^2=1$, if the difference of the eccentric angle of the points is $\frac{2\pi}{3}$.

17 (a) Find the equation of chord of contact to ellipse $x^2+4y^2=1$ with respect to point (2, 1)

(b) If from a point P tangents PQ and PR are drawn to the ellipse $\frac{x^2}{2}+y^2=1$ so that equation of QR is $x+3y=1$, then find the coordinates of P.

(c) Tangents are drawn from the points on the line $x-y-5=0$ to $x^2+4y^2=4$. Then, all the chords of contact pass through a fixed point, find its coordinates.

18 (a) Find the locus of the point, the chord of contact of tangents from which to the ellipse $\frac{x^2}{a^2}+\frac{y^2}{b^2}=1$ subtends a right angle at the centre of the ellipse.

(b) If two sides of a triangle inscribed in the ellipse $\frac{x^2}{a^2}+\frac{y^2}{b^2}=1$ are parallel to two given lines, then find the equation of third side.

19 (a) Find the equation of chord of the ellipse $x^2+2y^2=16$, whose mid-point is (1, 2).

(b) Tangents are drawn from the point $P(3, 2)$ to the ellipse $x^2+4y^2=9$. Find the middle point of chord of contact w.r.t. the given point P.

(c) Find the length of the chord of the ellipse $\frac{x^2}{25}+\frac{y^2}{16}=1$, whose middle point is $\left(\frac{1}{2}, \frac{2}{5}\right)$

20 (a) Find the locus of mid-point of a focal chord of the ellipse $\frac{x^2}{a^2}+\frac{y^2}{b^2}=1$.

(b) Find the locus of mid-point of the normal chords of the ellipse $\frac{x^2}{a^2}+\frac{y^2}{b^2}=1$

(c) Find the locus of mid-point of chord of ellipse $\frac{x^2}{a^2}+\frac{y^2}{b^2}=1$ which subtends a right angle at the centre of ellipse.

21 (a) Find the equation of the diameter of the ellipse $\frac{x^2}{a^2}+\frac{y^2}{b^2}=1$, where slope of chord is m.

(b) Find the coordinates of four extremities of two conjugate diameters of an ellipse $\frac{x^2}{a^2}+\frac{y^2}{b^2}=1$

(c) For the ellipse $\frac{x^2}{a^2}+\frac{y^2}{b^2}=1$, find the equation of the diameter conjugate to $ax-by=0$

(d) Find the eccentricity of the ellipse, if $y=x$ and $2x+3y=0$ are the equations of a pair of its conjugate diameters.

22 (a) Find the sum of squares of any two conjugate semi-diameters of an ellipse $\frac{x^2}{a^2}+\frac{y^2}{b^2}=1$

(b) If P and D be the ends of the conjugate diameters of an ellipse, find the locus of mid-point of PD.

(c) If CP and CD are the conjugate diameters of the ellipse $\frac{x^2}{a^2}+\frac{y^2}{b^2}=1$, then find the locus of orthocentre of ΔPCD.

23 (a) Find the equation of director circle for the ellipse $5x^2+9y^2=45$.

(b) Find the locus of the mid-points of the chords of contact on the director circle of the ellipse $\frac{x^2}{a^2}+\frac{y^2}{b^2}=1$.

24 (a) Find the equation of auxiliary circle of ellipse $\frac{x^2}{16}+\frac{y^2}{9}=1$.

(b) If equation of auxiliary circle of an ellipse is $x^2+y^2=4$ and eccentricity of ellipse $\frac{1}{\sqrt{2}}$, then find the equations of possible ellipse.

(c) Two points P and Q are corresponding points on an ellipse $\frac{x^2}{a^2}+\frac{y^2}{b^2}=1$ and its auxiliary circle. Then, find the locus of point of intersection of normal at P and normal at Q.

25 (a) Find the pole of the straight line $x + 4y = 4$ w.r.t. ellipse $x^2 + 4y^2 = 4$.

(b) Find the locus of poles of normal chords of an ellipse $\frac{x^2}{a^2} + \frac{y^2}{b^2} = 1$.

(c) Find the polar of the focus of an ellipse $\frac{x^2}{25} + \frac{y^2}{16} = 1$.

(d) The perpendicular from the centre of an ellipse $\frac{x^2}{a^2} + \frac{y^2}{b^2} = 1$ on the polar of a point w.r.t. the ellipse is equal to C (constant), find the locus of the point.

26 Find the length of tangent, normal, subtangent and subnormal to ellipse $\frac{x^2}{25} + \frac{y^2}{9} = 2$, at point $(5, -3)$.

6.5 Miscellaneous

27 (a) Find the equation of ellipse passing through the point of intersections of two ellipse $3x^2 + 2y^2 - 6x - 12y + 63 = 0$ and $4x^2 + 2y^2 - 19x - 120y + 35 = 0$, also passes through $(1, 1)$.

(b) Find the equation of common tangent in the first quadrant to circle $x^2 + y^2 = 16$ and the ellipse $\frac{x^2}{25} + \frac{y^2}{4} = 1$. Also, find the length of intercept of tangent between the coordinate axes.

(c) Tangents are drawn form any point P on the parabola $(y - 2)^2 = 4(x - 1)$ to the ellipse $\frac{x^2}{4} + \frac{y^2}{1} = 1$, which are mutually perpendicular to each other. Find the locus of the point P.

(d) If the tangent drawn at a point $(t^2, 2t), t \neq 0$ on the parabola $y^2 = 4x$ is the same as the normal drawn at a point $(\sqrt{5} \cos \phi, 2 \sin \phi)$ on the ellipse $4x^2 + 5y^2 = 20$. Find the value of t and ϕ.

(e) Find the locus of the poles of the tangents to $\frac{x^2}{a^2} + \frac{y^2}{b^2} = 1$ w.r.t the concentric ellipse $c^2x^2 + d^2y^2 = 1$.

(f) If the point of intersection of the ellipse $\frac{x^2}{a^2} + \frac{y^2}{b^2} = 1$ and $\frac{x^2}{c^2} + \frac{y^2}{d^2} = 1$ be at the extremities of the conjugate diameters of the former, find the relation between, a, b, c and d.

28 (a) A ray is emanating from the point $(-3, 0)$ is incident on the ellipse $16x^2 + 25y^2 = 400$ at the point P with ordinate 4. Find the equation of reflected ray after first reflection.

(b) A ray is coming along the line $x - y + 2 = 0$ on the ellipse $3x^2 + 4y^2 = 12$. After striking the ellipse mirror it is then reflected. Find the equation of line containing the reflected ray.

29. (a) Find the equation of the ellipse whose axes are the length of 6 and $2\sqrt{6}$ units and their equations of axis are $x - 3y + 3 = 0$ and $3x + y - 1 = 0$, respectively.

(b) Find the equation of major axis and minor axis, centre, foci, directrix, eccentricity and length of latusrectum of the ellipse $4(x - 2y + 1)^2 + 9(2x + y + 2)^2 = 25$.

LEVEL 02 \ JEE Main Level

Single Choice Correct Questions

6.1 Equation of Ellipse

1 The equation $\frac{x^2}{2-r}+\frac{y^2}{r-5}+1=0$ represents an ellipse, if

(a) $r>2$ (b) $2<r<5$ (c) $r>5$ (d) None of these

2 An ellipse $x^2+4y^2=4$ is inscribed in a rectangle aligned with the coordinate axes, which in turn is inscribed in another ellipse that passes through the point (4, 0). Then, equation of ellipse is

(a) $x^2+16y^2=16$ (b) $x^2+12y^2=16$

(c) $x^2+12y^2=12$ (d) $x^2+16y^2=12$

3 The equation $(5x-1)^2+(5y-2)^2=(\lambda^2-2\lambda+1)(3x+4y-1)^2$ represents an ellipse, then λ

(a) $\notin (0, 1)$ (b) $\notin (2, \infty)$ (c) $\notin (1, 2)$ (d) $\notin \left(\frac{1}{2}, 2\right)$

4 An ellipse has OB as semi-minor axis, F and F' is focii and $\angle FBF'$ is a right angle. Then, the eccentricity of the ellipse is

(a) $\frac{1}{4}$ (b) $\frac{1}{\sqrt{3}}$ (c) $\frac{1}{\sqrt{2}}$ (d) $\frac{1}{2}$

5 There are exactly two points on the ellipse $\frac{x^2}{a^2}+\frac{y^2}{b^2}=1$ whose distance from its centre is the same and is equal to $\frac{\sqrt{a^2+2b^2}}{2}$. Then, the eccentricity of ellipse is

(a) $\frac{1}{2}$ (b) $\frac{1}{\sqrt{2}}$ (c) $\frac{1}{3}$ (d) $\frac{1}{3\sqrt{2}}$

6 A point P on the ellipse $x^2+2y^2=2$ is a distance $\sqrt{2}$ units from its focus S. The ratio of its distances from the directrix of the ellipse, is

(a) 1 : 3 (b) 1 : 2 (c) 1 : 1 (d) None of these

7 If $\frac{x^2}{f(4a)}+\frac{y^2}{f(a^2-5)}=1$ represents an ellipse with major axis as Y-axis and f is a decreasing function such that $f(x)>0\ \forall\ x \in R$, then complete set of values of a, is

(a) $(-\infty, 1)$ (b) $(-\infty, -1) \cup (5, \infty)$ (c) $(-1, 4)$ (d) None of these

8 An ellipse is described by using an endless string where is passed over two pins. If the axes are 6 cm and 4 cm, the necessary length of the string and the distance between the pins respectively (in cm(s)), are

(a) 6 and $2\sqrt{5}$ (b) 6 and $\sqrt{5}$ (c) 4 and $2\sqrt{5}$ (d) $6+2\sqrt{5}$ and $2\sqrt{5}$

9 In an ellipse, with centre at the origin, if the difference of the length of major axis and minor axis is 10 units and one of the focii is at $(0, 5\sqrt{3})$, then the length of its latusrectum is **[JEE Mains 2019]**

(a) 5 units (b) 10 units (c) 8 units (d) 6 units

6.2 Point and Ellipse

10 Locus of the point which divides double ordinates of ellipse $\frac{x^2}{a^2}+\frac{y^2}{b^2}=1$ in the ratio 1 : 2 internally is

(a) $\frac{x^2}{a^2}+\frac{9y^2}{b^2}=1$
(b) $\frac{x^2}{a^2}+\frac{9y^2}{b^2}=\frac{1}{9}$
(c) $\frac{9x^2}{a^2}+\frac{9y^2}{b^2}=1$
(d) None of these

11 Let F be the focus on positive X-axis of ellipse $\frac{x^2}{16}+\frac{y^2}{9}=1$ and A and B are two points such that $\angle AFB=90°$, if $A(\theta)=\frac{\pi}{6}$, then AF^2+BF^2 is equal to

(a) $274-16\sqrt{7}(\sqrt{3}-1)$
(b) $39-4\sqrt{7}(\sqrt{3}-1)$
(c) $274-48\sqrt{21}+32\sqrt{7}$
(d) None of these

12 If S and S' are the foci of the ellipse $\frac{x^2}{25}+\frac{y^2}{16}=1$ and P is any point on it, then difference of maximum and minimum of $SP\cdot S'P$, is equal to

(a) 16 (b) 9 (c) 15 (d) 25

13 The base of a triangle is the major axis of the ellipse $\frac{x^2}{16}+\frac{y^2}{9}=1$ and third vertex moves on the ellipse, then the maximum area of the triangle will be

(a) 6 sq units (b) 72 sq units (c) 12 sq units (d) None of these

14 A circle has the same centre as an ellipse and passes through the foci F_1 and F_2 of the ellipse, such that the two curves intersect in 4 points. Let 'P' be any one of their point of intersection. If the major axis of the ellipse is 17 units and the area of ΔPF_1F_2 is 30 sq units, then the distance between the foci is

(a) 11 units (b) 12 units (c) 13 units (d) 15 units

6.3 Line and Ellipse

15 Find the complete set of values of λ for which the line $3x+y=k$ cuts ellipse $x^2+2y^2=4$ at two real and distinct points

(a) $(-\infty, 2)$ (b) $(-\sqrt{38}, \sqrt{38})$ (c) $(2, \infty)$ (d) None of these

16 The locus of the middle point of the intercept of the tangents drawn from an external point to the ellipse $x^2+2y^2=2$ between the coordinate axes is

(a) $\frac{1}{x^2}+\frac{1}{2y^2}=1$
(b) $\frac{1}{4x^2}+\frac{1}{2y^2}=1$
(c) $\frac{1}{2x^2}+\frac{1}{4y^2}=1$
(d) $\frac{1}{2x^2}+\frac{1}{y^2}=1$

17 A tangent to the ellipse $\frac{x^2}{25}+\frac{y^2}{16}=1$ at any point P meets the line $x=0$ at a point Q. Let R be the image of Q in the line $y=x$, then the circle whose extremities of a diameter are Q and R passes through a fixed point. The fixed point, is

(a) (3, 0) (b) (5, 0) (c) (0, 0) (d) (4, 0)

18 If $\sqrt{3}bx + ay = 2ab$ touches the ellipse $\frac{x^2}{a^2} + \frac{y^2}{b^2} = 1$, then the eccentric angle of the point of contact is

(a) $\frac{\pi}{6}$ (b) $\frac{\pi}{4}$ (c) $\frac{\pi}{3}$ (d) $\frac{\pi}{2}$

19 The sum of squares of the perpendiculars on any tangents to the ellipse $\frac{x^2}{a^2} + \frac{y^2}{b^2} = 1$ from two points on the minor axis each at *ae* distance from the centre is

(a) $2a^2$ units (b) $2b^2$ units (c) $a^2 + b^2$ units (d) $a^2 - b^2$ units

20 Tangent is drawn at ellipse $\frac{x^2}{27} + y^2 = 1$ at $(3\sqrt{3}\cos\theta, \sin\theta)$, where $\theta \in \left(0, \frac{\pi}{2}\right)$. Then, the value of θ such that sum of intercepts on axes made by this tangent is minimum, is

(a) $\frac{\pi}{3}$ (b) $\frac{\pi}{6}$ (c) $\frac{\pi}{8}$ (d) $\frac{\pi}{4}$

21 The point, at shortest distance from the line $x + y = 7$ and lying on the ellipse $x^2 + 2y^2 = 6$ has coordinates is

(a) $(\sqrt{2}, \sqrt{2})$ (b) $(0, \sqrt{3})$ (c) $(2, 1)$ (d) $\left(\sqrt{5}, \frac{1}{\sqrt{2}}\right)$

22 The area of the rectangle formed by the perpendiculars from the centre of the ellipse $\frac{x^2}{9} + \frac{y^2}{4} = 1$ to the tangent and normal at the point, whose eccentric angle is $\frac{\pi}{4}$, is

(a) $\frac{30}{17}$ sq units (b) $\frac{30}{13}$ sq units (c) $\frac{27}{17}$ sq units (d) $\frac{27}{13}$ sq units

23 Number of distinct normal lines that can be drawn to the ellipse $\frac{x^2}{169} + \frac{y^2}{25} = 1$ from the point $P(0, 6)$ is

(a) 1 (b) 2 (c) 3 (d) 4

24 Consider the particle travelling clockwise on the elliptical path $\frac{x^2}{100} + \frac{y^2}{25} = 1$. The partical leaves the orbit at the point (8, −3) and travels in a straight line normal to ellipse. At what point will the particle cross the *Y*-axis

(a) (0, 9) (b) (0, −9) (c) no such point (d) None of these

25 The point *P* on the ellipse $\frac{x^2}{4} + y^2 = 1$ for which area of ΔPON is maximum, where *O* is origin and *N* is the foot of perpendicular to the normal at *P*, is

(a) $\left(\frac{4}{\sqrt{5}}, \frac{1}{\sqrt{5}}\right)$ (b) $\left(-\frac{4}{\sqrt{5}}, \frac{1}{\sqrt{5}}\right)$

(c) $\left(\frac{-4}{\sqrt{5}}, \frac{-1}{\sqrt{5}}\right)$ (d) All of these

26 The normal at a point *P* on the ellipse $x^2 + 4y^2 = 16$ meets the *X*-axis at *Q*. If *M* is the mid-point of the line segment *PQ*, then the locus of *M* intersects the latusrectum of the given ellipse at the points

(a) $\left(\pm\frac{3\sqrt{5}}{2}, \pm\frac{2}{7}\right)$ (b) $\left(\pm\frac{3\sqrt{5}}{2}, \pm\frac{\sqrt{12}}{4}\right)$ (c) $\left(\pm 2\sqrt{3}, \pm\frac{1}{7}\right)$ (d) $\left(\pm 2\sqrt{3}, \pm\frac{4\sqrt{3}}{7}\right)$

27 If ϕ is the angle between the normals at θ and $\frac{\pi}{2}+\theta$ on the ellipse $\frac{x^2}{a^2}+\frac{y^2}{b^2}=1$, then $\frac{2\sqrt{1-e^2}}{e^2}$ is equal to

(a) $\sin 2\theta \tan\phi$ (b) $\cos 2\theta \tan\phi$ (c) $\tan 2\theta \sin\phi$ (d) None of these

28 If α, β, γ are the eccentric angles of three points on the ellipse $\frac{x^2}{a^2}+\frac{y^2}{b^2}=1$ at which normals are concurrent, then

(a) $\sin(\alpha+\beta)+\sin(\beta+\gamma)+\sin(\gamma+\alpha)=0$ (b) $\sin(\alpha-\beta)+\sin(\beta-\gamma)+\sin(\gamma-\alpha)=0$

(c) $\sin 2\gamma+\sin 2\beta+\sin 2\gamma=0$ (d) None of these

29 If PT and PN are tangent and normal to ellipse $\frac{x^2}{a^2}+\frac{y^2}{b^2}=1$ at point P, then the value of $\angle SPN$ is (where S and S' aare focii of ellipse)

(a) $\angle S'PN$ (b) $\angle S'NP$ (c) $\angle SNP$ (d) None of these

6.4 Pair of tangents, Chord of Contact, Conjugate Diameter, Director circle, Auxiliary circles, Subtangent and Subnormal

30 Angle between the tangents drawn from point (8, 5) to the ellipse $\frac{x^2}{16}+\frac{y^2}{25}=1$ is l, then $[l]$ equal to

(a) 1 (b) 0 (c) 2 (d) None of these

31 From point $P(8, 27)$ tangents PQ and PR are drawn to the ellipse $\frac{x^2}{4}+\frac{y^2}{9}=1$. Then, the angle subtended by QR at origin is

(a) $\tan^{-1}\left(\frac{\sqrt{6}}{65}\right)$ (b) $\tan^{-1}\left(\frac{4\sqrt{6}}{65}\right)$ (c) $\tan^{-1}\left(\frac{8\sqrt{2}}{65}\right)$ (d) $\tan^{-1}\left(\frac{48\sqrt{6}}{455}\right)$

32 Tangents are drawn from the point $P(3, 4)$ to the ellipse $\frac{x^2}{9}+\frac{y^2}{4}=1$ touching the ellipse at point A and B. The equation of the locus of the point whose distances from the point P and the line AB are equal to

(a) $9x^2+y^2-6xy-54x-62y+241=0$ (b) $x^2+9y^2+6xy-54x+62y-241=0$

(c) $9x^2+9y^2-6xy-54x-62y-241=0$ (d) $x^2+y^2-2xy+27x+31y+20=0$

33 Equation of the chord of contact of pair of tangents drawn to the ellipse $4x^2+9y^2=36$. From the point (m, n), where $m \cdot n = m + n$, m, n being non-zero positive integers is

(a) $2x+9y=18$ (b) $2x+2y=1$ (c) $4x+9y=18$ (d) None of these

34 If PQ be chord of the ellipse $\frac{x^2}{a^2}+\frac{y^2}{b^2}=1$ which subtends right angle at the centre, then the distance from the centre is

(a) $\frac{ab}{\sqrt{a^2+b^2}}$ (b) $\sqrt{a^2+b^2}$

(c) $\sqrt{ab}$ (d) it depends on the slope of the chord

35 The equation of line passing through the centre and bisecting the chord $7x + y - 1 = 0$ of the ellipse $\frac{x^2}{1} + \frac{y^2}{7} = 1$ is

(a) $x = y$ (b) $2x = y$ (c) $x = 2y$ (d) $x + y = 0$

36 If (α, β) are the eccentric angle of the extremities of a focal chord of the ellipse $\frac{x^2}{16} + \frac{y^2}{9} = 1$, then $\tan\frac{\alpha}{2} \cdot \tan\frac{\beta}{2}$ equals to

(a) $\frac{8\sqrt{7} - 23}{9}$ (b) $\frac{23 - 8\sqrt{7}}{9}$ (c) $\frac{\sqrt{7} + 4}{\sqrt{7} - 4}$ (d) None of these

37 If ΔPQR is an equilateral triangle inscribed in the auxiliary circle of the ellipse $\frac{x^2}{a^2} + \frac{y^2}{b^2} = 1$, $(a > b)$ and $\Delta P'Q'R'$ is corresponding triangle inscribed within the ellipse, then centroid of the $\Delta P'Q'R'$ lies at

(a) centre of ellipse (b) focus of ellipse
(c) between focus and centre on major axis (d) None of the above

38 The line passing through the extremity A of the major axis and extremity B of the minor axis of the ellipse $x^2 + 9y^2 = 9$ meets the auxiliary circle at point M. Then, the area of the triangle with vertices at A, M and the origin O, is

(a) $\frac{31}{10}$ sq units (b) $\frac{29}{10}$ sq units (c) $\frac{21}{10}$ sq units (d) $\frac{27}{10}$ sq units

39 Any ordinate MP of the ellipse $\frac{x^2}{25} + \frac{y^2}{9} = 1$ meets the auxiliary circle at Q, then locus of the point of intersection of normals at P and Q to the respective curves is

(a) $x^2 + y^2 = 8$ (b) $x^2 + y^2 = 34$
(c) $x^2 + y^2 = 64$ (d) $x^2 + y^2 = 15$

40 Find the equation of line joining the point of intersection of director circle of the ellipse $\frac{x^2}{4} + \frac{y^2}{3} = 1$, $y = x$ and focus of given ellipse is

(a) $(\sqrt{7} - \sqrt{2})\, y - \sqrt{7}x + \sqrt{7} = 0$ (b) $(\sqrt{7} + \sqrt{2})\, y - \sqrt{7}x - \sqrt{7} = 0$
(c) $(\sqrt{7} + \sqrt{2})\, y - \sqrt{7}x + \sqrt{7} = 0$ (d) All of these

41 If the polar of ellipse $\frac{x^2}{16} + \frac{y^2}{9} = 1$ touches the another ellipse $x^2 + 2y^2 = 2$, then the locus of its pole is

(a) $\frac{x^2}{64} + \frac{y^2}{128} = 1$ (b) $\frac{x^2}{128} + \frac{y^2}{81} = 1$
(c) $\frac{x^2}{81} + \frac{y^2}{128} = 1$ (d) None of these

42 The eccentricity of an ellipse whose pair of conjugate diameter are $y = 3x$ and $2y = -5x$ is

(a) $\sqrt{\frac{12}{13}}$ (b) $\sqrt{\frac{14}{15}}$ (c) $\sqrt{\frac{13}{15}}$ (d) None of these

43 For the ellipse $3x^2 + 4y^2 = 12$, then the conjugate diameter of line joining (6, 5) to origin is

(a) $10y + 5x = 0$ (b) $11y + 9x = 0$ (c) $10y - 9x = 0$ (d) $10y + 9x = 0$

44 The relation between the length of subtangent and subnormal to the ellipse $\frac{x^2}{16}+\frac{y^2}{9}=1$ at $P\left(\frac{\pi}{4}\right)$ is

(a) $\frac{\text{length of subtangent}}{\text{length of subnormal}}=16$ (b) length of subtangent $=36$

(c) length of subnormal $=\frac{9}{4}$ (d) All of these

6.5 Miscellaneous

45 If the ellipse $\frac{x^2}{4}+\frac{y^2}{1}=1$ meets the ellipse $\frac{x^2}{1}+\frac{y^2}{a^2}=1$ in four distinct points and $a^2=b^2-5b+7$, then b does not lie in

(a) [4, 5] (b) $(-\infty,-2)\cup(3,\infty)$ (c) $(-\infty, 0)$ (d) [2, 3]

46 a and b are positive real numbers such that $a>b$. If the area of the ellipse $ax^2+by^2=3$ equal area of the ellipse $(a+b)x^2+(a-b)y^2=3$, then $\frac{a}{b}$ is equal to

(a) $\frac{\sqrt{5}+1}{4}$ (b) $\frac{\sqrt{5}+1}{2}$ (c) $\frac{\sqrt{6}-1}{2}$ (d) $\frac{\sqrt{5}-1}{4}$

47 The angle subtended by the common tangent of the two ellipse $\frac{(x-4)^2}{25}+\frac{y^2}{4}=1$ and $\frac{(x+1)^2}{1}+\frac{y^2}{4}=1$ at the origin is

(a) $\frac{\pi}{2}$ (b) $\frac{\pi}{4}$ (c) $\frac{\pi}{3}$ (d) $\frac{\pi}{6}$

48 Point 'O' is the centre of an ellipse with major axis AB and minor axis CD. Point F is one of the focii. If $OF=6$ and the diameter of the inscribed circle of ΔOCF is 2 units, then $(AB)(CD)$ is equal to

(a) 35 (b) 45 (c) 55 (d) 65

49 The product of the slopes of the common tangents of the ellipse $\frac{x^2}{4}+y^2=1$ and the circle $x^2+y^2=3$ is

(a) 1 (b) -2 (c) -3 (d) -1

50 If a tangent of slope 2 of the ellipse $\frac{x^2}{a^2}+\frac{y^2}{b^2}=1$ is normal to the circle $x^2+y^2+4x+1=0$, then the maximum value of ab is equal to

(a) 2 (b) 1 (c) 4 (d) 3

51 The equation of the circle passing through the point of intersection of ellipse $\frac{x^2}{a^2}+\frac{y^2}{b^2}=1$ and $\frac{x^2}{b^2}+\frac{y^2}{a^2}=1$, is

(a) $x^2+y^2=a^2$ (b) $x^2+y^2=b^2$

(c) $x^2+y^2=\frac{a^2b^2}{a^2+b^2}$ (d) $x^2+y^2=\frac{2a^2b^2}{a^2+b^2}$

52 A ray emanating from the point (–4, 0) is incident on the ellipse $9x^2 + 25y^2 = 225$ at the point P with ordinate 3. The equation of the reflected ray after first reflection is

(a) $3x + 4y = 10$ (b) $3x + 4y = 12$ (c) $4x + 3y = 12$ (d) None of these

53 The centre of the ellipse $12(2x - 3y + 1)^2 + 16(3x + 2y - 2)^2 = 48$ is

(a) $\left(\frac{4}{13}, \frac{7}{13}\right)$ (b) $\left(\frac{1}{13}, \frac{4}{13}\right)$ (c) $\left(\frac{\pm 2}{13}, \frac{5}{13}\right)$ (d) None of these

54 Let θ be the acute angle between the tangents to the ellipse $\frac{x^2}{9} + \frac{y^2}{1} = 1$ and the circle $x^2 + y^2 = 3$ at their point of intersection in the first quadrant. Then, $\tan\theta$ is equal to **[JEE Main 2021]**

(a) $\frac{5}{2\sqrt{3}}$ (b) $\frac{2}{\sqrt{3}}$ (c) $\frac{4}{\sqrt{3}}$ (d) 2

Numerical Value Type Questions

55 Find the radius of the circle passing through the focii of the ellipse $\frac{x^2}{16} + \frac{y^2}{9} = 1$ and having its centre at (0, 3).

56 If the maximum distance of any point on the ellipse $x^2 + 2xy + 2y^2 = 1$ from its centre be r. If r is in the form of $\sqrt{\frac{a}{b - \sqrt{c}}}$, then find $\frac{a + b + c}{30}$ upto 3 decimal.

57 The number of following statements which are true in the case of the ellipse $x^2 + 4y^2 - 2x - 16y + 13 = 0$, is

Statement I The latusrectum of the ellipse is 1.

Statement II The distance between the foci is $x\sqrt{3}$.

Statement III The sum of focal distance of a point $P(x, y)$ on the ellipse is 4 units.

Statement IV The line $y = 3$ meets the tangent drawn at the vertices of the ellipse at points P and Q. Then, PQ subtends a right angle at either of the focii.

58 P and Q are two points on the ellipse $\frac{x^2}{25} + \frac{y^2}{9} = 1$ such that sum of their ordinates is 3. If locus of point of intersection of tangents at P and Q is $ax^2 + by^2 + 2gx + 2fy + c = 0$, then find $\frac{a + b + c + g - f}{1000}$.

59 An ellipse $\frac{x^2}{25} + \frac{y^2}{9} = 1$ is rotated through a right angle in its own plane about its centre, which is fixed. If the locus of the point of intersection of a tangent to the ellipse in its original position with the tangent at the same point of the curve in its new position $ax^4 + by^4 + cx^2 + dy^2 + ex^2y^2 + fxy + g = 0, (a > 0)$, then find $a + b + c + d + e + f + g$.

60 Find the length of the sides of a square which can be made by four perpendicular tangents to the ellipse $\frac{x^2}{16} + \frac{y^2}{9} = 1$.

61 If two vertices of a rectangle lie on $y = 2x + c$ and other two vertices are (0, 4) and (–1, 2). Find the sum of c and coordinates of other two vertices such that area of ellipse inscribed in the rectangle is $\frac{5\pi}{2}$ sq units.

62 A tangent to the ellipse $x^2 + 4y^2 = 4$ meets the ellipse $x^2 + 2y^2 = 6$ at P and Q. Let angle between tangents at P and Q of the ellipse $x^2 + 2y^2 = 6$ is $\frac{\pi}{k}$, find the value of k.

63 A normal is drawn to ellipse $\frac{x^2}{(a^2+2a+2)^2} + \frac{y^2}{(a^2+1)^2} = 1$, whose centre is at O. If maximum radius of circle, centered at origin and touching the normal is 5, then find the positive value of a.

64 Tangents drawn from the point $P(2, 3)$ to the circle $x^2 + y^2 - 8x + 6y + 1 = 0$ touch the circle at the points A and B. The circumcircle of the ΔPAB cuts the director circle of ellipse $\frac{(x+5)^2}{9} + \frac{(y-3)^2}{b^2} = 1$ orthogonally, then the value of $\frac{b^2}{6}$, is

65 ΔABC is an isoceles triangle with its base BC twice its altitude. A point P moves within the triangle such that square of its distance from BC is half the rectangle contained by its distances from the two side. If 'e' is the eccentricity of ellipse of locus of point P. Then, value of $[5e^2]$ {[·] denotes greatest integers less than equal to x}, is

66 Suppose x and y are real numbers such that $x^2 + 9y^2 - 4x + 6y + 4 = 0$, then maximum value of $\frac{4x-9y}{20}$, is

67 If the equation of the curve on reflection of the ellipse $\frac{(x-4)^2}{16} + \frac{(y-3)^2}{9} = 1$ about the line $x - y - 2 = 0$ is $16x^2 + 9y^2 + k_1 x - 36y + k_2 = 0$, then $\frac{k_1 + k_2}{22}$ is

LEVEL 03 \ JEE Advanced Level

Single Choice Correct Type Questions

1 A focus of an ellipse is at the origin. The directrix is the line $x = 4$ and the eccentricity is $\frac{1}{2}$. Then, the length of the semi-major axis is

(a) $\frac{8}{3}$ units (b) $\frac{2}{3}$ units (c) $\frac{4}{3}$ units (d) $\frac{5}{3}$ units

2 A series of ellipse $E_1, E_2, \ldots, E_n$ are drawn such that E_n touches E_{n-1} at the extremities of the major axis of E_{n-1} and focii of E_n coincide with the extremities of minor axis of E_{n-1}. If eccentricity of the ellipses is independent of n, then the value of eccentricity is

(a) $\frac{\sqrt{5}}{3}$ (b) $\frac{\sqrt{5}+1}{4}$ (c) $\frac{\sqrt{5}-1}{2}$ (d) $\frac{\sqrt{5}-1}{3}$

3 If $a = [t^2 - 3t + 4]$ and $b = [3 + 5t]$, where [·] denotes the greatest integer function, then the latusrectum of the ellipse $\frac{x^2}{a^2} + \frac{y^2}{b^2} = 1$ at $t = \frac{3}{2}$, is

(a) 20 units (b) 10 units (c) $\frac{1}{5}$ units (d) $\frac{1}{10}$ units

4 If base of a triangle is a and sum of its sides is $2s$, then locus of its incircle is (where a, s are constant)
(a) a circle (b) an ellipse (c) a pair of straight line (d) None of these

5 Any point P of an ellipse is joined to the extremities of the major axis. Then, the angle made by the portion of a directrix interecepted by extremities at the corresponding focus is
(a) $\frac{\pi}{6}$ (b) $\frac{\pi}{3}$ (c) $\frac{\pi}{2}$ (d) $\frac{2\pi}{7}$

6 The point of intersection of the perpendicular from a focus upon any tangent and the line joining the centre of the ellipse to point of contact of the ellipse, lies on
(a) directrix (b) major axis (c) minor axis (d) None of these

7 The sum of the squares of the perpendiculars on any tangent to $\frac{x^2}{a^2}+\frac{y^2}{b^2}=1$ from two points on the minor axis, each of which is at a distance $\sqrt{a^2-b^2}$ from the centre, is
(a) $2b^2$ (b) $2a^2$ (c) a^2 (d) b^2

8 If the tangent at any point of an ellipse $\frac{x^2}{a^2}+\frac{y^2}{b^2}=1$ makes an angle α with the major axis and an angle β with the focal radius of the point of contact, then eccentricity of ellipse is
(a) $\frac{\cos\beta}{\cos\alpha}$ (b) $\frac{\sin\beta}{\sin\alpha}$ (c) $\frac{\cos\alpha}{\cos\beta}$ (d) $\frac{\sin\alpha}{\sin\beta}$

9 Let P_i and $P_i{}'$ be the feet of perpendiculars drawn from focii S, S' on a tangent T_i to an ellipse whose length of semi-major axis is 20 units. If $\sum_{i=1}^{10}(SP_i)(S'P_i{}')=2560$, then the value of eccentricity of ellipse is
(a) $\frac{1}{5}$ (b) $\frac{2}{5}$ (c) $\frac{3}{5}$ (d) $\frac{4}{5}$

10. The normal GP is produced to Q so that $GQ=hGP$, for the ellipse $\frac{x^2}{a^2}+\frac{y^2}{b^2}=1$, then the locus at point Q is {where P is point of contact}
(a) $\frac{x^2}{a^2(x+e^2-xe^2)^2}+\frac{y^2}{xb^2}=1$ (b) $\frac{x^2}{a^2\,(x+e^2-xe^2)^2}+\frac{y^2}{x^2b^2}=1$
(c) $\frac{x^2}{x^2b^2}+\frac{y^2}{a^2\,(x+e^2-xe^2)}=1$ (d) None of these

11 Let ΔABC be an equilateral triangle inscribed in the circle $x^2+y^2=a^2$. Suppose perpendiculars from A, B, C to the major axis of the ellipse $\frac{x^2}{a^2}+\frac{y^2}{b^2}=1$, $(a>b)$ meets the ellipse at P, Q, R respectively, so that P, Q, R lie on the same side of the major axis as are the corresponding points A, B and C, then the normal to the ellipse drawn at points P, Q, R
(a) are concurrent (b) are not concurrent
(c) are collinear (d) None of these

12 If the chord through point θ_1 and θ_2 on an ellipse $\frac{x^2}{a^2}+\frac{y^2}{b^2}=1$ intersects the major axis at $(d, 0)$, then $\tan\frac{\theta_1}{2}\cdot\tan\frac{\theta_2}{2}$ is

(a) $\frac{d-a}{d+a}$ (b) $\frac{d-a}{d+2a}$ (c) $\frac{d-2a}{d+2a}$ (d) $\frac{d-a}{2d+a}$

13 Find the quadratic equation whose one root is abscissa of focus and other roots is $\frac{25}{41}$ times the square of lengh of chord of the ellipse $\frac{x^2}{25}+\frac{y^2}{16}=1$, whose middle point is $\left(\frac{1}{2}, \frac{2}{5}\right)$ is

(a) $x^2-52x+141=0$ (b) $x^2-52x+147=0$

(c) $x^2+52x+147=0$ (d) None of these

14 In an ellipse referred to its centre, the length of the subtangent corresponding to the point $\left(3, \frac{12}{5}\right)$ is $\frac{16}{3}$, then eccentricity is

(a) $\frac{14}{15}$ (b) $\frac{3}{5}$ (c) $\frac{4}{5}$ (d) $\frac{2}{5}$

15 A pair of conjugate diameters is produced to meet the directrix, then the orthocentre of triangle so formed is at the

(a) centre (b) focus

(c) vertex (d) None of these

16 If circle be described on two semi-conjugate diameter of the ellipse as diameters, then the locus of their seocnd points of intersection is

(a) $2(x^2+y^2)^2=(a^2x^2+b^2y^2)$ (b) $2(x^2+y^2)=(a^2x^2+b^2y^2)^2$

(c) $2(x^2+y^2)=(b^2x^2+a^2y^2)$ (d) $2(ax^2+y^2)=(x^2+b^2y^2)^2$

17 A circle of radius r is concentric with the ellipse $\frac{x^2}{a^2}+\frac{y^2}{b^2}=1$, then the angle between common tangent and the major axis is

(a) $\tan^{-1}\sqrt{\frac{r^2-b^2}{r^2-a^2}}$ (b) $\tan^{-1}\sqrt{\frac{r^2-b^2}{a^2-r^2}}$

(c) $\tan^{-1}\sqrt{\frac{b^2-r^2}{r^2-a^2}}$ (d) None of these

18 Common tangents are drawn to the parabola $y^2=4x$ and the ellipse $3x^2+8y^2=48$ touching the parabola at A and B and the ellipse at C and D. Then, area of the quadrilateral $ABCD$ is

(a) $55\sqrt{3}$ sq units (b) $55\sqrt{2}$ sq units

(c) $55\sqrt{6}$ sq units (d) $55\sqrt{5}$ sq units

19 Chords of the ellipse $\frac{x^2}{a^2}+\frac{y^2}{b^2}=1$ touches the parabola $ay^2=-2b^2x$, then locus of their poles is

(a) $by^2=2a^2x$ (b) $ay^2=b^2x$

(c) $2ay^2=b^2x$ (d) $ay^2=2b^2x$

More Than One Correct

20 Consider the ellipse $\dfrac{x^2}{f(k^2+2k+5)}+\dfrac{y^2}{f(k+11)}=1$ and $f(x)$ is a positive decreasing function, then

(a) The set of values of k, for which the major axis is X-axis is $(-3, 2)$
(b) The set of values of k, for which the major axis is Y-axis is $(-\infty, 2)$
(c) The set of values of k, for which the major axis is Y-axis is $(-\infty, -3) \cup (2, \infty)$
(d) The set of values of k, for which the major axis is Y-axis is $(-3, \infty)$

21 A parallelogram circumscribes the ellipse $\dfrac{x^2}{9}+\dfrac{y^2}{4}=1$ and two of its opposite angular points lie on straight lines $x^2=\lambda^2$, $\lambda \neq 0$ the locus of the other two vertices is

(a) ellipse, if $\lambda > 3$
(b) circle, if $\lambda \in (0, 3)$
(c) circle, if $\lambda = \dfrac{9}{\sqrt{5}}$
(d) ellipse, if $\lambda \in (0, 3)$

22 Consider two straight lines, each of which is tangent to both the circle $x^2+y^2=\dfrac{1}{2}$ and the parabola $y^2=4x$. Let these lines intersect at the point Q. Consider the ellipse whose centre is at the origin $O(0, 0)$ and whose semi-major axis is OQ. If the length of the minor axis of this ellipse is $\sqrt{2}$ units, then which of the following statement(s) is are true **[JEE Advanced 2018]**

(a) For the ellipse, the eccentricity is $\dfrac{1}{\sqrt{2}}$ and the length of the latusrectum is 1 unit.
(b) For the ellipse, the eccentricity is $\dfrac{1}{2}$ and length of the latusrectum is $\dfrac{1}{2}$ unit.
(c) The area of region bounded by the ellipse between $x=\dfrac{1}{\sqrt{2}}$ and $x=1$ is $\dfrac{1}{4\sqrt{2}}(\pi-2)$ sq units.
(d) The area of the region bonded by the ellipse between the lines $x=\dfrac{1}{\sqrt{2}}$ and $x=1$ is $\dfrac{1}{16}(\pi-2)$ sq units.

23 Let E_1 and E_2 be two ellipses whose centres are at the origin. The major axes of E_1 and E_2 lie along the X-axis and Y-axis, respectively. Let 'S' be the circle $x^2+(y-1)^2=2$. The straight line $x+y=3$ touches the curves S_1, E_1 and E_2 at P, Q and R respectively. Suppose $RQ=PR=\dfrac{2\sqrt{2}}{3}$. If e_1 and e_2 are the eccentricities of E_1 and E_2 respectively, then the correct expression (s) is/are **[JEE Advanced 2015]**

(a) $e_1^2+e_2^2=\dfrac{43}{40}$
(b) $e_1e_2=\dfrac{\sqrt{7}}{2\sqrt{10}}$
(c) $|e_1^2-e_2^2|=\dfrac{5}{8}$
(d) $e_1e_2=\dfrac{\sqrt{3}}{4}$.

24 If a pair of variable straight lines $x^2+4y^2+\alpha xy=0$. (where α is a real number) cut the ellipse $x^2+4y^2=4$ at two points A and B, then the locus of the point of intersection of tangent at A and B is

(a) $x-2y=0$
(b) $2x-y=0$
(c) $x+2y=0$
(d) $2x+y=0$

25 Let $S_1: 3x^2 + 4y^2 - 1 = 0$, $S_2: |x + y| - 1 = 0$ and $S_3: |y - x| - 1 = 0$ are the given curves. Then, which of the following is/are not true.

(a) Area enclosed by the exterior to the auxiliary circle of $S_1 = 0, S_2 \le 0$ and $S_3 \le 0$ is $\left(2 - \frac{\pi}{3}\right)$ sq units.

(b) Area enclosed by the exterior to the director circle of $S_1 = 0, S_2 \le 0$ and $S_3 \le 0$ is $\left(2 - \frac{\pi}{3}\right)$ sq units.

(c) Area enclosed by $S_2 \le 0$ and $S_3 \le 0$ is 1 sq unit.

(d) Area enclosed by $S_1 \ge 0, S_2 \le 0$ and $S_3 \le 0$ is $\left(2 - \frac{\pi}{3\sqrt{2}}\right)$ sq units.

26 For a point P on ellipse the circle with PS and PS', where S, S' represent focii of ellipse as diameter intersect the auxiliary circle of ellipse at A, A_1 and B, B_1, respectively. Then, which of the following is/are correct?

(a) A and A_1 coincide, B and B_1 coincide

(b) Segment AB is tangent to ellipse at P

(c) Tangents at A and B on auxiliary circles are perpendicular

(d) SA and $S'B$ are parallel.

27 The locus of incentre of ΔPF_1F_2, where P is a variable point lying on ellipse $\frac{x^2}{a^2} + \frac{y^2}{b^2} = 1$ with F_1, F_2 as foci and eccentricity 'e' is a conic, whose

(a) equation is $\frac{x^2}{a^2e^2} + \frac{y^2(1+e)^2}{b^2e^2} = 1$

(b) equation is $\frac{x^2}{a^2e^2} + \frac{y^2(1-e)^2}{b^2e^2} = 1$

(c) eccentricity is $\sqrt{\frac{2e}{1+e}}$

(d) eccentricity is $\sqrt{\frac{2e}{1-e}}$

28 Which of the following is/are true?

(a) There are infinite positive integral values of a for which

$(13x - 1)^2 + (13y - 2)^2 = \left\{\frac{5x + 12y - 1}{a}\right\}^2$ represent an ellipse

(b) The minimum distance of a point (1, 2) from the ellipse $4x^2 + 9y^2 + 8x - 36y + 4 = 0$ is 1 unit.

(c) If from a point $P(0, \alpha)$ two normal other than axes are drawn to the ellipse $\frac{x^2}{25} + \frac{y^2}{16} = 1$, then $|\alpha| < \frac{9}{4}$.

(d) If the length of latusrectum of an ellipse is one-third of its major axis, then its eccentricity is equal to $\frac{1}{\sqrt{3}}$.

Linked Comprehension Type

Paragraph I Directions (Q.Nos. 29-31)

An ellipse has major axis of length 4 units and minor axis of length 2 units. It is slipping between the coordinate axes in the first quadrant while maintaining contact with both X-axis and Y-axis.

29 The locus of the centre of the ellipse is

(a) $x + y = 3$

(b) $(y - 1)^2 = 4(x - 2)$

(c) $x^2 + y^2 = 5$

(d) $x^2 + y^2 = 20$

30 The locus of focii of the ellipse is

(a) $x^2+y^2=10$ (b) $x^4+y^4-x^2y^2=1$

(c) $x^2+y^2(x^2+y^2)=4$ (d) $x^2+y^2+\frac{1}{x^2}+\frac{1}{y^2}=16$

31 When the ordinate of centre of ellipse is 1, a light ray passing through origin and centre of ellipse strike the tangent at vertex of the parabola $y^2+4x-10y-15=0$. The reflected ray cuts Y-axis at the point

(a) (0, 8) (b) (0, 10) (c) (0, 12) (d) (0, 16)

Paragraph II Directions (Q.Nos. 32-36)

The standard equation of an ellipse and the general equation of a circle are respectively given by the equations.

$E:\frac{x^2}{a^2}+\frac{y^2}{b^2}=1$ and $C: x^2+y^2+2gx+2fy+c=0$, then the equation $E+\lambda C=0, \lambda \neq 0$ i.e.

$\left\{\left(\frac{x^2}{a^2}+\frac{y^2}{b^2}-1\right)+\lambda(x^2+y^2+2gx+2fy+c)=0\right\}$ represents a curve which passes through the common points at the ellipse and the circle. On the basis of above information, answer the following questions.

32 If $\alpha, \beta, \gamma, \delta$ be the eccentric angles of the four concyclic points on the ellipse E. Then, $\alpha+\beta+\gamma+\delta$ is equal to

(a) $n\pi, n \in I$ (b) $(n+1)\pi, n \in I$

(c) $2n\pi, n \in I$ (d) $(2n+1)\pi, n \in I$

33 If common chords of E and C are inclined at angles θ and ϕ with the axis of E, then

(a) $\theta=\phi$ (b) $\theta=\phi=2\tan^{-1}\frac{b}{a}$

(c) $\theta+\phi=\frac{\pi}{2}$ (d) $\theta+\phi=\pi$

34 The radius of the circle passing through the points of intersection E and bisectors of the quadrant is

(a) $\frac{ab}{\sqrt{a^2+b^2}}$ (b) $\frac{ab\sqrt{2}}{\sqrt{a^2+b^2}}$ (c) $\frac{2ab}{\sqrt{a^2+b^2}}$ (d) $\frac{a^2-b^2}{\sqrt{a^2+b^2}}$

35 Let the eccentric angles of three points P, Q and R on the ellipse E are α, $\frac{\pi}{2}+\alpha$ and $\pi+\alpha$. A circle C pass through P, Q and R cuts the ellipse E again at S_1, then the eccentric angle of 'S', is

(a) $\frac{\pi}{2}-3\alpha$ (b) $\pi-3\alpha$ (c) $\frac{3\pi}{2}-3\alpha$ (d) $2\pi-3\alpha$

36 The centre of circle C which pass through the 3 points on E whose eccentric angles are α, β and γ, are

(a) $\left\{\left(\frac{a^2-b^2}{4a}\right)(\cos\alpha+\cos\beta+\cos\gamma+\cos(\alpha+\beta+\gamma)),\right.$

$\left.\left(\frac{a^2-b^2}{4b}\right)(\sin\alpha+\sin\beta+\sin\gamma+\sin(\alpha+\beta+\gamma))\right\}$

(b) $\left\{\left(\frac{a^2 - b^2}{4a}\right)(\cos\alpha + \cos\beta + \cos\gamma - \cos(\alpha+\beta+\gamma)),\right.$

$\left.\left(\frac{b^2 - a^2}{4b}\right)(\sin\alpha + \sin\beta + \sin\gamma + \sin(\alpha+\beta+\gamma))\right\}$

(c) $\left\{\left(\frac{a^2 - b^2}{4a}\right)(\cos\alpha + \cos\beta + \cos\gamma - \cos(\alpha+\beta+\gamma)),\right.$

$\left.\left(\frac{a^2 - b^2}{4b}\right)(\sin\alpha + \sin\beta + \sin\gamma - \sin(\alpha+\beta+\gamma))\right\}$

(d) None of the above

Paragraph III Directions (Q.Nos. 37-39)

A coplanar beam of light emerging from a point source has the equation $\lambda x - y + 2(1+\lambda) = 0$, $\lambda \in R$, the rays of the beam strike an elliptical surface and gets reflected. The reflected rays form another convergent beam having equation $\mu x - y + 2(1-\mu) = 0, \mu \in R$. Further it is found that the foot of the perpendicular from the point (2, 2) upon any tangent to the ellipse lies on the circle $x^2 + y^2 - 4y - 5 = 0$

37 The eccentricity of the ellipse is equal to

(a) $\frac{1}{3}$ (b) $\frac{1}{\sqrt{2}}$ (c) $\frac{2}{3}$ (d) $\frac{1}{2}$

38 The area of the largest triangle that an incident ray and the corresponding reflected ray can enclose with axis of an ellipse is equal to

(a) $4\sqrt{5}$ sq units (b) $2\sqrt{5}$ sq units (c) $\sqrt{5}$ sq units (d) None of these

39 Total distance travelled by an incident ray and the corresponding reflected ray is the least is the point of incidence coincides with

(a) an end of the minor with (b) an end of the major axis

(c) an end of this latusrectum (d) None of these

Matrix Match Type Questions

40 Match the following

	Column I		Column II
(i)	The number of rational points on the ellipse $\frac{x^2}{9} + \frac{y^2}{4} = 1$ is	(A)	Infinite
(ii)	The number of integral points on the ellipse $\frac{x^2}{9} + \frac{y^2}{4} = 1$ is	(B)	4
(iii)	The number of rational points on the ellipse $\frac{x^2}{3} + y^2 = 1$, is	(C)	2
(iv)	The number of integral point on the ellipse $\frac{x^2}{3} + y^2 = 1$ is	(D)	0

41 Match the following

	Column I		Column II
(i)	The minimum and maximum distance of a point (2, 6) from the ellipse $9x^2+8y^2-36x-16y-28=0$ are L and G, then	(A)	$L+G=10$
(ii)	The minimum and maximum distance of a point (1, 2) from the ellipse $4x^2+9y^2+8x-36y+4=0$ are L and G, then	(B)	$L+G=6$
(iii)	The minmum and maximum distance of a point $\left(\frac{9}{5},\frac{12}{5}\right)$ from the ellipse $4(3x+4y)^2+9(4x-3y)^2=900$ are L and G, then	(C)	$G-L=2\sqrt{7}$
(iv)	The minimum and maximum distance of a point $(0,\sqrt{7})$ to the ellipse $16x^2+9y^2=144$ are L and G, then	(D)	$G-L=6$
		(E)	$L^G+G^L=6$

42 Match the following

	Column I		Column II
(i)	If the tangent to the ellipse $x^2+4y^2=16$ at the point $P(\phi)$ is a normal to the circle $x^2+y^2-8x-4y=0$, then $\frac{\phi}{2}$ may	(A)	$\frac{\pi}{2}$
(ii)	The eccentric angle of a point on the ellipse $x^2+3y^2=6$ at a distance 2 units from the centre of the ellipse is/are	(B)	$\cos^{-1}\left(-\frac{2}{3}\right)$
(iii)	The eccentric angles of intersection of the ellipse $x^2+4y^2=4$ and the parabola $x^2+1=y$ is	(C)	$\frac{\pi}{4}$
(iv)	If the normal at the point $P(\theta)$ to the ellipse $\frac{x^2}{14}+\frac{y^2}{5}=1$ intersects it again at the point $Q(2\theta)$, then θ is	(D)	$\frac{5\pi}{4}$

43 Match the column.

	Column I		Column II
(i)	If $E: 2x^2+y^2=2$ and directror circle of E is C_1, director circle of C_1 is C_2, director circle of C_2 is C_3, director circle of C_3 is C_4 and so on. If $r_1, r_2, r_3, \ldots$ are the radii of $C_1, C_2, C_3, \ldots$ respectively, then	(A)	$r_1^2, r_2^2, r_3^2, \ldots$ are in GP with common ratio 2
(ii)	If $E: 3x^2+2y^2=6$ and director circle of E is C_1, director circle of C_1 is C_2, director circle of C_2 is C_3 and so on. If $r_1, r_2, r_3, \ldots, r_a$ are radii of $C_1, C_2, C_3, \ldots$ respectively, then	(B)	GM of r_1^2, r_2^2, r_3^2 is 6.
(iii)	If $E: 3x^2+4y^2=12$ and director circle of E is C_1, director circle of C_1 is C_2, director circle of C_2 is C_3 and so on. If $r_1, r_2, r_3, \ldots$ are radii of $C_1, C_2, \ldots$ respectiely, then	(C)	$r_1^2+r_2^2+\ldots r_n^2=r_1^2(2^n-1)$
		(D)	GM of r_1^2, r_2^2, r_3^2 is 10
		(E)	GM of r_1^2, r_2^2, r_3^2 is 14.

44 Match the following

	Column I		Column II
(i)	If the angle between the straight lines joining the foci and one of the ends of the minor axis of the ellipse $\frac{x^2}{a^2}+\frac{y^2}{b^2}=1$ is 90°. Find its eccentricity.	(A)	4
(ii)	For an ellipse $\frac{x^2}{9}+\frac{y^2}{4}=1$ with vertices A and A', tangent drawn at the point P in the first quadrant meets the Y-axis in Q and chord $A'P$ meets the Y-axis in M. If O is the origin, then OQ^2-MQ^2 equals to	(B)	2
(iii)	The x-coordinate of points on the axis of the parabola $4y^2-32x+4y+65=0$ from which all the three normals to the parabola are real is	(C)	$\frac{1}{\sqrt{2}}$
(iv)	The area of parallelogram inscribed in the ellipse $\frac{x^2}{2^2}+\frac{y^2}{\left(\frac{1}{2}\right)^2}=1$, whose diagonals are conjugate diameters of the ellipse are given by	(D)	1
		(E)	8

Numerical Value Type Questions

Questions Stem Directions (Q.Nos. 45-46) The answer to each question is a **NUMERICAL VALUE.**

Let an ellipse having major axis and minor axis parallel to X-axis and Y-axis, respectively. Its two focii S and S' are (2, 1), (4, 1) and a line $x+y=9$ is a tangent to this ellipse at point P

45 $\sqrt{13}$ times eccentricity of the ellipse is ....

46 $\frac{1}{\sqrt{13}}$ times length of major axis is ....

Questions Stem Directions (Q.Nos. 47-48) The answer to each question is a **NUMERICAL VALUE.**

Let $F_1(x_1, 0)$ and $F_2(x_2, 0)$ for $x_1<0$ and $x_2>0$ be the focii of the ellipse $\frac{x^2}{9}+\frac{y^2}{8}=1$. Suppose a parabola having vertex at the origin and focus at F_2 intersects the ellipse at point M in the first quadrant and at point N in the fourth quadrant. **[JEE Advanced 2016]**

47 The orthocentre of the ΔF_1MN is (a, b), then $b-a$ is ....

48 If the tangents to the ellipse at M and N meet at R and the normal to the parabola at M meets the X-axis at Q and the ratio of area of the ΔMQR to area of the quadrilateral MF_1NF_2 is $c=d$ then $c+d$ is ....

49 Consider the ellipse, $E:\frac{x^2}{a^2}+\frac{y^2}{b^2}=1$, centred at point '$O$' and having AB and CD as its major and minor axes, respectively. If S_1 be one of the focii of the ellipse, radius of incircle of ΔOCS_1 be 1 unit and $OS_1=6$ units, then the value of $\frac{a-b}{20}$ is

50 Origin O is the centre of two concentric circles whose radii are a and b. A line OPQ is drawn to cut the inner circle in P and outer circle in Q. PR is drawn parallel to the Y-axis and QR is drawn parallel to the X-axis. The locus of R is an ellipse touching the two circles, if the foci of this ellipse lie on the inner circle, if eccenricity is $\sqrt{2}\lambda$, then $\frac{\lambda}{25}$ is

51 An isoscles ΔABC is inscribed in the circle $x^2+y^2=16$, $A=(0,4)$. From points A, B and C three ordinates are drawn which cut the ellipse $\frac{x^2}{16}+\frac{y^2}{9}=1$ out the points P, Q and R, respectively such that A and P are on the opposite sides, Q and B are on the same side and C and D are on the same side of the major axis. If ΔPQR is right angled at P and θ is the smallest angle of the ΔABC, then the value of $16\tan^2\theta+8\tan\theta+9$.

52 Variable pairs of chords at right angles and drawn through any point P (with eccentric angle $\frac{\pi}{4}$) on the ellipse $\frac{x^2}{4}+y^2=1$, to meet the ellipse at two points say A and B. If the line joining A and B passes through a fixed point $Q(a,b)$ such that a^2+b^2 has the value equal to $\frac{m}{n}$, where m, n are relatively prime positive integers, then find $\left[\frac{m+n}{3}\right]$, where $[\cdot]$ greatest integer function.

53 Let E be the ellipse $\frac{x^2}{16}+\frac{y^2}{9}=1$. For any three distinct point P, Q and Q' on E. Let $M(P,Q)$ be the mid-point of the line segment joining P and Q and $M(P,Q')$ be the mid-point of the line segment joining P and Q'. Then, the maximum possible value of the distance between $M(P,Q)$ and $M(P,Q')$ as P, Q and Q' vary on E, is ...... **[JEE Advanced 2021]**

54 Suppose that the foci of the ellipse $\frac{x^2}{9}+\frac{y^2}{5}=1$ are $(F_1,0)$ and $(F_2,0)$, where $F_1>0$ and $F_2<0$. Let P_1 and P_2 be two parabolas with a common vertices at $(0,0)$ with foci at $(F_1,0)$ and $(2F_2,0)$, respectively. Let T_1 be a tangent to P_1 which passes through $(F_1,0)$. If m_1 is the slope of T_1 and m_2 is the slope of T_2, then the value of $\left(\frac{1}{m_1^2}+m_2^2\right)$ is **[JEE Advanced 2015]**

55 Let P be a point on the ellipse $\frac{x^2}{a^2}+\frac{y^2}{b^2}=1$, $0<b<a$. Let the line parallel to Y-axis passing through P meet the circle $x^2+y^2=a^2$ at the point Q such that P and Q are on the same side of X-axis. For two positive real numbers r and s. If the locus of the point R on PQ such that $\frac{PR}{RQ}=\frac{r}{s}$ as P varies over the ellipse, is $\frac{x^2}{a^2}+\frac{y^2(r+s)^2}{\lambda(bs+ar)^\mu}=1$, then $\frac{\lambda+\mu}{100}$ is

56 'O' is the origin and also the centre of two concentric circles having radii of the inner and outer circle as 'a' and 'b', respectively. A line OPQ is drawn to cut the inner circle in P and the outer circle in Q. PR is drawn parallel to the Y-axis and QR is drawn parallel to the X-axis. If the foci of this ellipse lie on the inner circle, find the square of eccentricity.

57 Consider the family of circles $x^2+y^2=r^2$, where $2<r<5$. If in the first quadrant, the common tangent to a circle of this family and the ellipse $\frac{x^2}{25}+\frac{y^2}{4}=1$ meets the coordinate axes at A and B. If the equation of locus of the mid-point of AB is $\lambda x^2+\mu y^2=\lambda x^2 y^2$, then value of $\sqrt{\lambda\mu}$ is

58 If the normals at (x_r, y_r), where $r=1, 2, 3, 4$ to the ellipse $\frac{x^2}{a^2}+\frac{y^2}{b^2}=1$ are concurrent, then value of λ is, if $(x_1+x_2+x_3+x_4)\left(\frac{1}{x_1}+\frac{1}{x_2}+\frac{1}{x_3}+\frac{1}{x_4}\right)=\frac{100}{\lambda}$.

59 Two tangents to an ellipse $\frac{x^2}{a^2}+\frac{y^2}{b^2}=1$ are drawn which intersect at right angles. Let l_1 and l_2 be the intercepts on these tangents made by the auxiliary circle $x^2+y^2=a^2$. If $l_1^m+l_2^n=(SS')^p$, then find $\frac{(mnp)^{1/3}}{50}$.

ANSWER KEY

Level 1

1. (a) $\lambda > 1$ (b) no real values

2. (a) $\frac{x^2}{144}+\frac{y^2}{128}=1$ (b) $\frac{x^2}{36}+\frac{y^2}{11}=1$ (c) $\frac{x^2}{27}+\frac{y^2}{36}=1$ (d) $\frac{4x^2}{21}+\frac{4y^2}{25}=1$

(e) $\frac{x^2}{10}+\frac{y^2}{40}=1$ (f) $\frac{x^2}{25}+\frac{y^2}{9}=1$ (g) $16x^2+7y^2=688$

3. (a) $7x^2+7y^2+2xy+10x-10y+7=0$ (b) $\frac{(x-2)^2}{25}+\frac{(y-5)^2}{16}=1$

(c) $\frac{(x-2)^2}{4}+\frac{(y+3)^2}{3}=1$ (d) $\frac{x^2}{60}+\frac{y^2}{64}=1$ (e) $\frac{4(x-1)^2}{75}+\frac{(y-2)^2}{25}=1$

4. (a) Center O (0, 0); Vertex ($\pm$ 3, 0); Focus ($\pm\sqrt{5}$, 0); Directrix $\left(x=\pm\frac{9}{\sqrt{5}}\right)$;

equation of minor axis = Y-axis;

equation of major axis = X-axis;

length of minor axis ($2b=4$);

length of major axis ($2a=6$);

latus rectum $\left(\frac{8}{3}\right)$;

Eccentricity $=\frac{\sqrt{5}}{3}$

(b) Centre (0, 0); Vertex $(0,\pm\sqrt{2})$; Focus $(0,\pm 1)$; Directrix $(y=\pm 2)$;

equation of minor axis = X-axis;

equation of major axis = Y-axis;

length of minor axis = ($2a=2$);

length of major axis = ($2b=2\sqrt{2}$);

latus rectum $=\sqrt{2}$;

Eccentricity $=\left(\frac{1}{\sqrt{2}}\right)$

(c) Centre (1, 1); Vertex (1, 1) and (1, − 3); Focus $(1,-1\pm\sqrt{3})$; Directrix $\left(y=-1\pm\frac{4}{\sqrt{3}}\right)$;

equation of minor axis = ($x=1$);

equation of major axis = ($y=-1$);

length of minor axis = ($2a=2$);

length of major axis= ($2b=4$) ;

latus rectum = 1;

Eccentricity $=\frac{\sqrt{3}}{2}$

(d) Centre (1, 2); Vertex $\left(1\pm\frac{1}{\sqrt{2}},2\right)$; Focus $\left(1\pm\frac{1}{\sqrt{6}},2\right)$; Directrix $\left(x=1\pm\frac{\sqrt{3}}{\sqrt{2}}\right)$;

equation of minor axis = ($x=2$);

equation of major axis = ($y=2$);

length of minor axis = ($2a=\sqrt{2}$);

length of major axis = $\left(2b=\frac{2}{\sqrt{3}}\right)$;

latus rectum $=\left(\frac{2\sqrt{2}}{3}\right)$;

Eccentricity $=\left(\frac{1}{\sqrt{3}}\right)$

5. (a) Ellipse (b) $\frac{x^2}{25}+\frac{y^2}{9}=1$ (c) $\frac{x^2}{81}+\frac{y^2}{36}=1$ (d) $\frac{45}{\sqrt{14}}ft$

6. (a) Outside (b) inside (c) lies on the ellipse

7. (a) $\alpha \in (-\infty, -1) \cup \left(-\frac{1}{3}, \infty\right)$ (b) $\alpha = \pm\frac{1}{2}$ (c) $\alpha \in \left(\frac{12}{5}, \frac{16}{5}\right)$

8. (a) $x = 3 + 2\cos\theta$ and $y = 4 + 5\sin\theta$ (b) $x = \cos\theta + 1$ and $y = 2\sin\theta - 1$

9. (a) $15\sqrt{3}$ (b) 12 sq units

10. (a) $\frac{2\sqrt{2}}{3}$ units (b) (i) $k \in (-\sqrt{3}, \sqrt{3})$ (ii) $k = \pm\sqrt{3}$ (iii) $k \in (-\infty, -\sqrt{3}) \cup (\sqrt{3}, \infty)$

11. (i) (a) $9y - 2x - 20 = 0$ (b) $\frac{x}{5}\cos\theta + \frac{y}{4}\sin\theta = 1$ (c) $3x + 2y = 2$

(d) $y = 2x \pm \sqrt{19}$ (e) $4y + 3x = \pm\sqrt{17}$ (f) $y + x = \pm\sqrt{7}$

(ii) (a) (1, 2) (b) $\left(\mp\frac{4}{\sqrt{31}}, \pm\frac{9}{\sqrt{31}}\right)$

(iii) (a) $\frac{2\pi}{3}$ (b) $\left(-\frac{25}{3}, 0\right)$

(iv) (a) b^2 (b) $(x^2 + y^2)^2 = a^2x^2 + b^2y^2$ (c) $\frac{a^2}{x^2} + \frac{b^2}{y^2} = 4$

12. (a) $15x - 4y - 18 = 0$ (b) $4\sqrt{3} - 2\sqrt{2}y - \sqrt{3} = 0$ (c) $y = -2x \pm \frac{7}{\sqrt{13}}$

(d) $-3x + 3\sqrt{5}y = 5\sqrt{5}$ (e) $\left(\frac{6}{\sqrt{29}}, \frac{-25}{\sqrt{29}}\right)$ and $\left(\frac{-6}{\sqrt{29}}, \frac{25}{\sqrt{29}}\right)$

13. (a) $\cos\theta = -\frac{2}{3}$ (b) $PF \cdot PG = b^2$ and $PF \cdot PG' = a^2$ (c) $\sqrt{\frac{\sqrt{5}-1}{2}}$ (d) $\frac{a^2 - b^2}{2ab}$

14. (a) (i) 3 (ii) 4 (iii) 2

(b) (i) $\theta = \sin^{-1}\left(\sqrt{\frac{6}{7}}\right)$ and $\cos^{-1}\left(\frac{1}{\sqrt{7}}\right)$ (ii) $\theta = \frac{2\pi}{3}$ or $\theta = \frac{2\pi}{3}$

(c) $\alpha + \beta + \gamma + \delta = (2n + 1)\pi, n \in I$

(d) $(a^2 - b^2)xy + b^2kx - a^2hy = 0$

15. (a) $45x^2 - 72y^2 - 54xy + 72x + 486y - 765 = 0$ (b) $\theta = \tan^{-1}\left(\frac{12}{\sqrt{5}}\right)$ (c) $\lambda = \pm\sqrt{14}$

16. $x^2 + 2y^2 = 4$

17. (a) $2x + 4y - 1 = 0$ (b) (2, 3) (c) $\left(\frac{4}{5}, \frac{-1}{5}\right)$

18. (a) $\frac{x^2}{a^4} + \frac{y^2}{b^4} = \frac{1}{a^2} + \frac{1}{b^2}$ (b) $\frac{x^2}{a^2\lambda^2} + \frac{y^2}{b^2\lambda^2} = 1$

19. (a) $x + 4y = 9$ (b) $\left(\frac{27}{25}, \frac{18}{25}\right)$ (c) $\frac{7}{5}\sqrt{41}$ units

20. (a) $\frac{x^2}{a^2} + \frac{y^2}{b^2} = \frac{he}{a}$ (b) $(a^2 - b^2)^{-2}$ (c) $\left(\frac{x^2}{a^4} + \frac{y^2}{b^4}\right)$

21. (a) $y = \left(\frac{-b^2}{a^2m}\right)x$ (b) $P(a\cos\theta, b\sin\theta), Q(-a\cos\theta, b\sin\theta), S(-a\sin\theta, b\cos\theta)$ and $R(a\sin\theta, b\cos\theta)$

(c) $a^3y + b^3x = 0$ (d) $\frac{1}{\sqrt{3}}$

22. (a) $a^2 + b^2$ (b) $\frac{x^2}{a^2} + \frac{y^2}{b^2} = \frac{1}{2}$ (c) $2(a^2x^2 + b^2y^2)^3 = (a^2 - b^2)^2(a^2x^2 - b^2y^2)^2$

23. (a) $x^2 + y^2 = 14$ (b) $x^2 + y^2 = (a^2 + b^2)\left(\frac{x^2}{a^2} + \frac{y^2}{b^2}\right)^2$

24. (a) $x^2 + y^2 = 16$ (b) $\frac{x^2}{2} + \frac{y^2}{4} = 1$ (c) $x^2 + y^2 = (a + b)^2$

25. (a) (1, 1) (b) $\left(\frac{a^6}{x^2} + \frac{b^6}{y^2}\right) = (a^2 - b^2)^2$ (c) $\frac{3x}{25} + \frac{y(0)}{16} = 1$ (d) $\frac{x^2}{a^4} + \frac{y^2}{b^4} = \frac{1}{c^2}$

26. Length of tangent $= \sqrt{34}$ units

Length of normal $= \frac{3}{5}\sqrt{34}$ units

Length of subtangent = 5 units

Length of subnormal $= \frac{9}{5}$ units

27. (a) $247x^2 + 148y^2 - 769x - 3588y + 3962 = 0$

(b) Equation of common tangent= $(2x + 3\sqrt{y} = 4\sqrt{7})$

Axes point = $A(2\sqrt{7}, 0)$ and $B(0, \sqrt{7})$

Length of intercept $= \frac{14}{\sqrt{3}}$

(c) $x^2 + y^2 = 5$

(d) $t = \pm \frac{1}{\sqrt{5}}$ and $\phi\ 2n\pi \pm \cos^{-1}\left(\frac{-1}{\sqrt{5}}\right), 0 \le \phi\ 2\pi$

(e) $a^2c^4x^2 + b^2d^2y^2 = 1$

(f) $\frac{a^2}{c^2} + \frac{b^2}{d^2} = 2$

28. (a) $4x + 3y = 12$ (b) $x + y = 2$

29. (a) $21x^2 - 6xy + 29y^2 + 6x - 58y - 151 = 0$

(b) Equation of major axis = $(2x + y + 2 = 0)$;

Equation of minor axis = $(x - 2y + 1 = 0)$;

Centre = (– 1, 0);

foci = $\left(-1 \pm \frac{\sqrt{5}}{6}; \mp \frac{\sqrt{5}}{2} = 0\right)$;

Directrix = $\left(x - 2y + 1 \mp \frac{3\sqrt{5}}{2} = 0\right)$;

Eccentricity= $\left(\frac{\sqrt{5}}{3}\right)$;

and length of latus rectum $= \frac{20}{9\sqrt{5}}$

Level 2

1. (b)	**2.** (b)	**3.** (b)	**4.** (c)	**5.** (b)	**6.** (c)	**7.** (d)
8. (d)	**9.** (a)	**10.** (a)	**11.** (b)	**12.** (b)	**13.** (c)	**14.** (c)
15. (b)	**16.** (c)	**17.** (c)	**18.** (a)	**19.** (a)	**20.** (b)	**21.** (c)
22. (b)	**23.** (d)	**24.** (c)	**25.** (d)	**26.** (c)	**27.** (a)	**28.** (a)
29. (a)	**30.** (a)	**31.** (d)	**32.** (a)	**33.** (c)	**34.** (a)	**35.** (a)
36. (a)	**37.** (a)	**38.** (d)	**39.** (c)	**40.** (d)	**41.** (b)	**42.** (c)
43. (d)	**44.** (a)	**45.** (d)	**46.** (b)	**47.** (a)	**48.** (d)	**49.** (b)
50. (c)	**51.** (d)	**52.** (b)	**53.** (a)	**54.** (b)		
55. (4)	**56.** (0.333)	**57.** (3)	**58.** (0.109)	**59.** (707.2)	**60.** (7.070)	**61.** (17)
62. (2)	**63.** (2)	**64.** (9)	**65.** (3)	**66.** (0.8)	**67.** (6)	

Level 3

1. (a)	**2.** (c)	**3.** (c)	**4.** (b)	**5.** (c)	**6.** (a)	**7.** (b)
8. (a)	**9.** (c)	**10.** (b)	**11.** (a)	**12.** (a)	**13.** (b)	**14.** (c)
15. (b)	**16.** (a)	**17.** (b)	**18.** (b)	**19.** (d)	**20.** (a, c)	**21.** (a, c)
22. (a, c)	**23.** (a, b)	**24.** (a, c)	**25.** (b, c, d)	**26.** (a, b, d)	**27.** (a, c)	**28.** (a, b, c)
29. (c)	**30.** (d)	**31.** (b)	**32.** (c)	**33.** (d)	**34.** (b)	**35.** (a)
36. (d)	**37.** (c)	**38.** (b)	**39.** (d)			

40. (i)–A, (ii)–B, (iii)–A, (iv)–C

41. (i)–A, D, (ii)–B, E (iii)–B, D, (iv)–C

42. (i)–A, C, (ii)–C, D, (iii)–A, (iv)–B

43. (i)–A, B, C, (ii)–A, C, D, (iii)–A, C, E

44. (i)–C, (ii)–A, (iii)–E, (iv)–B

45. (1)	**46.** (2)	**47.** (0.9)	**48.** (13)	**49.** (0.2)	**50.** (0.02)	**51.** (24)
52. (6)	**53.** (4)	**54.** (4)	**55.** (0.03)	**56.** (0.5)	**57.** (10)	**58.** (25)
59. (0.04)						

SOLUTIONS

Level 1

1. (a) Given equation,

$$x^2+\lambda y^2-2xy+4x+6y-3=0$$

$\because a=1, b=\lambda, h=-1, g=2, f=3$ and $c=-3$

$\because$ We know that any second degree equation represents an ellipse, when $\Delta \neq 0$ and $h^2<ab$.

$$\therefore \quad \Delta = abc+2fgh-af^2-bg^2-ch^2$$

$$\Delta=(1)(\lambda)(-3)+2(3)(2)(-1)-(1)(3)^2-(\lambda)(2)^2-(-3)(-1)^2$$

$$=-3\lambda-12-9-4\lambda+3$$

$$=-7\lambda-18$$

Now, $h^2<ab$

$$\Rightarrow \quad (-1)^2<\lambda \Rightarrow \lambda>1$$

(b) Given equation

$$x^2-y^2+5xy+2x-7y+\lambda=0$$

$$a=1, b=-1, h=\frac{5}{2}, g=1, f=-\frac{7}{2} \text{ and } c=\lambda$$

$$\therefore \quad \Delta=abc+2fgh-af^2-bg^2-ch^2$$

$$=(1)(-1)(\lambda)+2\left(-\frac{7}{2}\right)(1)\left(\frac{5}{2}\right)-1\left(-\frac{7}{2}\right)^2-(-1)(1)^2-(\lambda)\left(\frac{5}{2}\right)^2$$

$$=-\lambda-\frac{35}{2}-\frac{49}{4}+1-\frac{25}{4}\lambda$$

$$=\frac{-29\lambda}{4}-\frac{119}{4}+1$$

$$=\frac{-29\lambda-115}{4}$$

$\because$ For ellipse $h^2<ab$

$$\Rightarrow \quad \left(\frac{5}{2}\right)^2<(1)(-1)$$

$$\Rightarrow \quad \frac{25}{4}<-1$$

$\because \frac{25}{4}$ is not less than -1

$\therefore$ For no real values of λ the given equation represents an ellipse.

2. (a) Given, foci $(\pm 4, 0)$ and eccentricity $(e)=\frac{1}{3}$.

$\because$ Clearly see that the given condition represents an standard form of ellipse $\frac{x^2}{a^2}+\frac{y^2}{b^2}=1$,

where focus is $(\pm ae, 0) \Rightarrow ae=4$ and $e=\frac{1}{3}$

$$\Rightarrow \quad a=\frac{4}{e}=\frac{4}{\frac{1}{3}}=12$$

$$\therefore \quad b^2=a^2(1-e^2)$$

$$=(12)^2\left\{1-\left(\frac{1}{3}\right)^2\right\}$$

$$=(12)^2\times\frac{8}{9}=\frac{32}{3}\times 12=128$$

$\therefore$ Equation of ellipse

$$\frac{x^2}{a^2}+\frac{y^2}{b^2}=1 \Rightarrow \frac{x^2}{144}+\frac{y^2}{128}=1$$

(b) Given, foci $(\pm 5, 0)$ and one directrix $\Rightarrow x=\frac{36}{5}$

So, major axis is the X-axis and if $2a=$ length of major axis. Then, $2ae=$ distance between focus $=10$ units

$$\Rightarrow \quad ae=5 \quad \text{...(i)}$$

and

$$x=\frac{a}{e}$$

$$\therefore \quad \frac{a}{e}=\frac{36}{5} \quad \text{...(ii)}$$

Multiplying Eqs. (i) and (ii), we get

$$a^2=36 \Rightarrow a=6 \Rightarrow e=\frac{5}{6}$$

$$\because \quad b^2=a^2(1-e^2) \Rightarrow b^2=36\left(1-\frac{25}{36}\right)=11$$

$\therefore$ Required equation

$$\Rightarrow \quad \frac{x^2}{a^2}+\frac{y^2}{b^2}=1 \Rightarrow \frac{x^2}{36}+\frac{y^2}{11}=1$$

(c) Given, focii $(0, \pm 3)$ and eccentricity $=\frac{1}{2}$

$\because$ Focii lies on Y-axis.

$\therefore$ The equation of ellipse in standard form is $\frac{x^2}{a^2}+\frac{y^2}{b^2}=1, a<b$

$$\Rightarrow \quad 2be=6 \text{ and } e=\frac{1}{2}$$

$$\Rightarrow \quad be=3$$

$$\Rightarrow \quad b=\frac{3}{e}=\frac{3}{\frac{1}{2}}=6$$

$$\Rightarrow \quad b^2=36$$

Now, $a^2 = b^2(1-e^2)$

$$= 36\left(1-\frac{1}{4}\right) = 36 \times \frac{3}{4}$$

$\Rightarrow \quad a^2 = 27$

$\therefore$ Equation of ellipse.

$$\frac{x^2}{a^2} + \frac{y^2}{b^2} = 1$$

$$\Rightarrow \quad \frac{x^2}{27} + \frac{y^2}{36} = 1$$

(d) Given, focii $(0, \pm 2)$ and $y = \frac{25}{4}$ one directirx. So, major axis is Y-axis and $2be = 2$

$$be = 1 \quad \text{...(i)}$$

$$\Rightarrow \quad \frac{b}{e} = \frac{25}{4} \quad \text{...(ii)}$$

On multiplying Eqs (i) and (ii), we get

$$\Rightarrow \quad b^2 = \frac{25}{4}$$

$$\therefore \quad e = \frac{2}{5}$$

$$\Rightarrow \quad a^2 = \frac{25}{4}\left(1 - \frac{4}{25}\right) = \frac{21}{4}.$$

$[\because a^2 = b^2(1-e^2)]$

$\therefore$ Equation of ellipse

$$\frac{4x^2}{21} + \frac{4y^2}{25} = 1$$

(e) Given, centre is at $(0, 0)$ and major axis is along Y-axis.
So, the equation of the ellipse is of the form.

$$\frac{x^2}{a^2} + \frac{y^2}{b^2} = 1, \quad b > a.$$

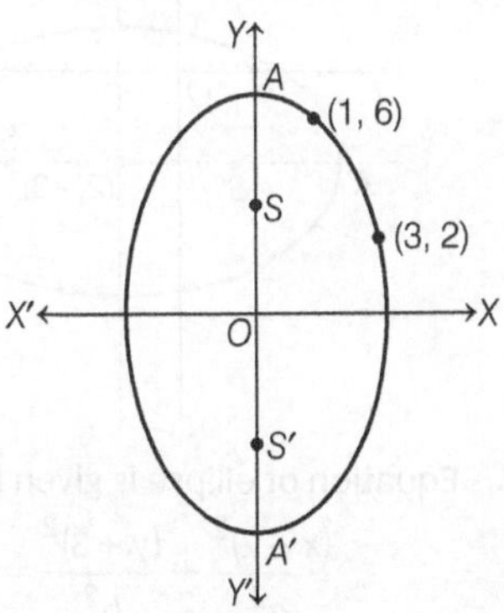

$\because$ $(3,2)$ and $(1,6)$ lies on ellipse

$$\therefore \quad \frac{9}{a^2} + \frac{4}{b^2} = 1 \quad \text{...(i)}$$

$$\text{and} \quad \frac{1}{a^2} + \frac{36}{b^2} = 1 \quad \text{...(ii)}$$

On multiplying Eq. (i) by 9 and subtract Eq. (ii) from it, we get

$$\Rightarrow \quad \frac{80}{a^2} = 8 \Rightarrow a^2 = 10$$

$$\therefore \quad \frac{1}{b^2} = \frac{1 - \frac{9}{a^2}}{4}$$

$$\Rightarrow \quad b^2 = 40$$

$\therefore$ Equation of ellipse is

$$\frac{x^2}{10} + \frac{y^2}{40} = 1$$

(f) Given centre is at origin, focii are on the X-axis, therefore the equation of the ellipse is of the form.

$$\frac{x^2}{a^2} + \frac{y^2}{b^2} = 1 \text{ and } c^2 = a^2 - b^2 \text{ or } c = ae$$

It is given that $b = 3$ and $c = 4$ i.e. $ae = 4$

Also, $\quad b^2 = a^2(1-e^2)$

$$\Rightarrow \quad a^2 = b^2 + a^2e^2 = 3^2 + 4^2 = 5^2$$

$$\Rightarrow \quad a^2 = 25$$

$\therefore$ Equation of ellipse is

$$\frac{x^2}{25} + \frac{y^2}{9} = 1$$

(g) Given, $e = \frac{3}{4}$ centre is $(0, 0)$

Let the equation of the ellipse is

$$\frac{x^2}{a^2} + \frac{y^2}{b^2} = 1, \; b > a$$

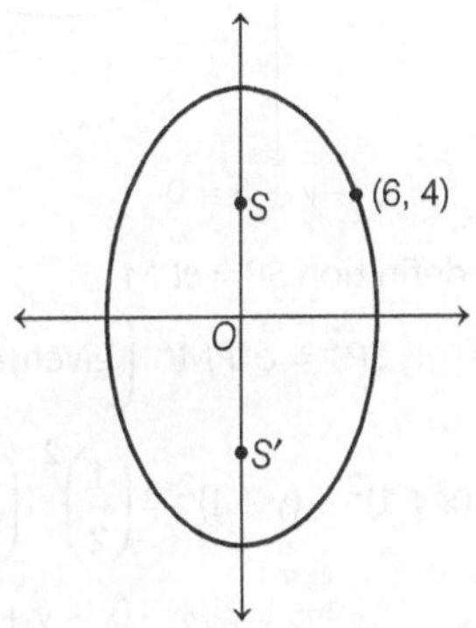

$$\Rightarrow \quad e^2 = 1 - \frac{a^2}{b^2}$$

$$\Rightarrow \quad \frac{9}{16} = 1 - \frac{a^2}{b^2}$$

$$\Rightarrow \quad \frac{a^2}{b^2} = \frac{7}{16}$$

$$\Rightarrow \quad a^2 = \frac{7}{16}b^2.$$

$\therefore$ Equation of ellipse

$$\frac{16x^2}{7b^2}+\frac{y^2}{b^2}=1 \Rightarrow \frac{16x^2+7y^2}{7b^2}=1$$

$$\Rightarrow \quad 16x^2+7y^2=7b^2$$

So, (6, 4) lies on ellipse

$$\Rightarrow \quad 16(6)^2+7(4)^2=7b^2$$

$$\Rightarrow \quad 16(36+7)=7b^2$$

$$\Rightarrow \quad 16\times\frac{43}{7}=b^2$$

$$\therefore \quad a^2=\frac{7}{16}\times 16\times\frac{43}{7}$$

$$\Rightarrow \quad a^2=43$$

$\therefore$ Equation of ellipse is

$$16x^2+7y^2=7\times 16\times\frac{43}{7}$$

$$\Rightarrow \quad 16x^2+7y^2=688$$

3. (a) Let $P(x, y)$ be any point on the ellipse.
Let focus be $S(-1, 1)$ and equation of line be $x-y+3=0$
Join S and P.
Drawn $PM \perp ZM$.

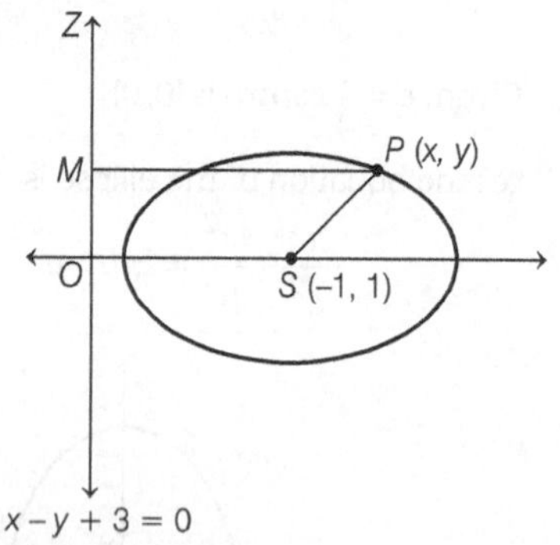

By definition $SP=ePM$

$$SP^2=e^2PM^2 \left(\text{given}, e=\frac{1}{2}\right)$$

$$\Rightarrow (x+1)^2+(y-1)^2=\left(\frac{1}{2}\right)^2\cdot\left(\frac{x-y+3}{\sqrt{2}}\right)^2$$

$$(x+1)^2+(y-1)^2=\frac{(x-y+3)^2}{8}$$

$$\Rightarrow 8(x^2+y^2+2x-2y+2)=x^2+y^2+9 -2xy+6x-6y$$

$$\Rightarrow 7x^2+7y^2+2xy+10x-10y+7=0$$

(b) $\because$ The y-coordinates of two focii S and S' are same, the major axis of the ellipse will be parallel to X-axis.

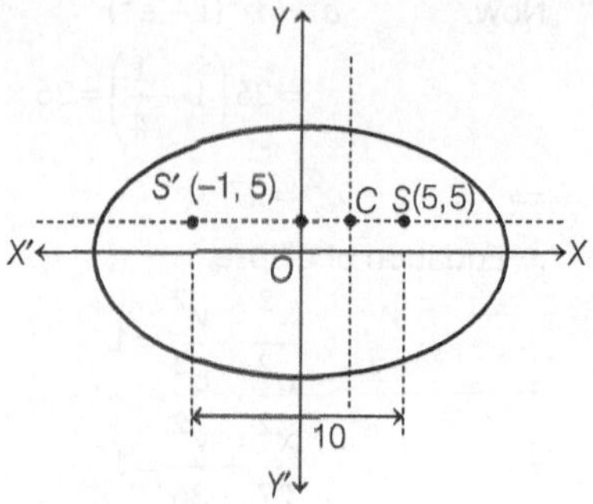

So, its equation is the form

$$\Rightarrow \quad \frac{(x-h)^2}{a^2}+\frac{(y-k)^2}{b^2}=1$$

Mid-point of focii

$$\therefore \quad (h, k)=\left(\frac{-1+5}{2}, \frac{5+5}{2}\right)=(2, 5)$$

Given, $\quad 2a=10 \Rightarrow a=5$

$$\therefore \quad SS'=\sqrt{(-1-5)^2+(5-5)^2}=6$$

$$\Rightarrow \quad 2ae=6$$

$$\Rightarrow \quad 10e=6$$

$$\Rightarrow \quad e=\frac{3}{5}$$

$$\therefore \quad b^2=a^2(1-e^2)=25\left(1-\frac{9}{25}\right)=16$$

$\therefore$ Equation of ellipse is $\dfrac{(x-2)^2}{25}+\dfrac{(y-5)^2}{16}=1$

(c) Let $C(2, -3)$ be the centre of ellipse.
Let $2a$ and $2b$ be the length of major axis and minor axis of the ellipse.

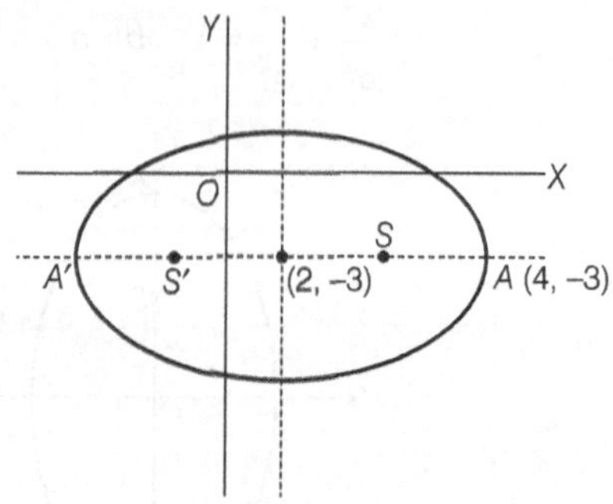

$\therefore$ Equation of ellipse is given by

$$\frac{(x-2)^2}{a^2}+\frac{(y+3)^2}{b^2}=1$$

Let $A(4, -3)$ be the given vertex of ellipse
$\therefore CA=a$ i.e. the length of semi-major axis

$$\Rightarrow \quad a=\sqrt{(4-2)^2+(-3+3)^2}=2$$

As we know that, the distance between the focus and centre of the ellipse is equal to ae

$\therefore \quad CS = ae$

$\Rightarrow \quad \sqrt{(3-2)^2 + (-3+3)^2} = 1$

$1 = 2e \qquad \{\because a = 2\}$

$\Rightarrow \quad e = \frac{1}{2}$

Now, $b^2 = a^2(1-e^2)$

$= 4\left(1 - \frac{1}{4}\right) = 3$

$\therefore$ Equation of ellipse $\frac{(x-2)^2}{4} + \frac{(y+3)^2}{3} = 1$

(d) **For ellipse**

Focus = Centre of circle $x^2 + y^2 - 4y + 1 = 0$

$\equiv (-g, -f) = (0, 2)$

Centre = vertex of parabola $y^2 = 8x$

$= (0, 0)$

$\because$ Focus and centre both lies on Y-axis

$\therefore$ Y-axis is major axis of ellipse.

$\therefore$ Equation of ellipse is of the form $\frac{x^2}{a^2} + \frac{y^2}{b^2} = 1$, $b > a$

Now, $be = 2$

$\because \quad e = \frac{1}{4}$

$\Rightarrow \quad b = \frac{2}{\frac{1}{4}} = 8$

Now, $\quad a^2 = b^2(1-e^2)$

$= (8)^2\left[1 - \left(\frac{1}{4}\right)^2\right]$

$= 64 \times \frac{15}{16} = 4 \times 15 = 60$

$\therefore$ Equation of ellipse $\frac{x^2}{60} + \frac{y^2}{64} = 1$

(e) **For ellipse**

Focus = Focus of parabola $x^2 + 4y - 2x + 7 = 0$

$\Rightarrow \quad (x-1)^2 = -4\left(y + \frac{6}{4}\right)$

$\Rightarrow \quad (x-1)^2 = -4\left(y + \frac{3}{2}\right)$

On comparing $X^2 = -4AY$, we get

$X = x - 1, A = 1$ and $Y = y + \frac{3}{2}$

$\therefore$ Focus $(0, A)$

$\Rightarrow \quad x - 1 = 0, \quad y + \frac{3}{2} = 1$

$\Rightarrow \quad x = 1, \quad y = \frac{-1}{2}.$

$\therefore$ Focus is $\left(1, -\frac{1}{2}\right)$

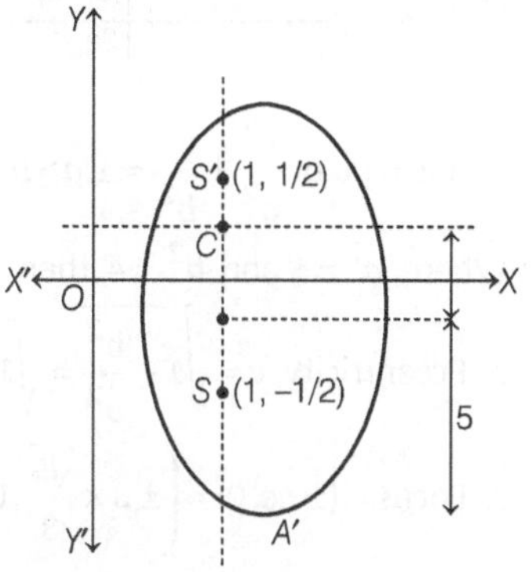

Given, e = radius of circle $4x^2 + 4y^2 = 1$

$\Rightarrow \quad x^2 + y^2 = \frac{1}{4}$

$\therefore \quad r = \frac{1}{2}$

$\therefore \quad e = \frac{1}{2}$

$\because x = 1$ is major axis

$\Rightarrow \quad 2b = 10 \Rightarrow b = 5$

Now, $\quad a^2 = b^2(1-e^2)$

$= 5^2\left[1 - \left(\frac{1}{2}\right)^2\right]$

$\Rightarrow \quad 25 \times \frac{3}{4} = \frac{75}{4}$

Now, second focus is $2be = 2 \times 5 \times \frac{1}{2} = 5$

$\therefore \quad$ Focus $\left(1, 5 - \frac{1}{2}\right) = \left(1, \frac{9}{2}\right)$

Now, centre is (1, 2)

Now, equation of ellipse $\frac{4(x-1)^2}{75} + \frac{(y-2)^2}{25} = 1$

4. (a) Given, equation

$4x^2 + 9y^2 = 36$

$\Rightarrow \quad \frac{4x^2}{36} + \frac{9y^2}{36} = 1$

$\Rightarrow \quad \frac{x^2}{9} + \frac{y^2}{4} = 1$

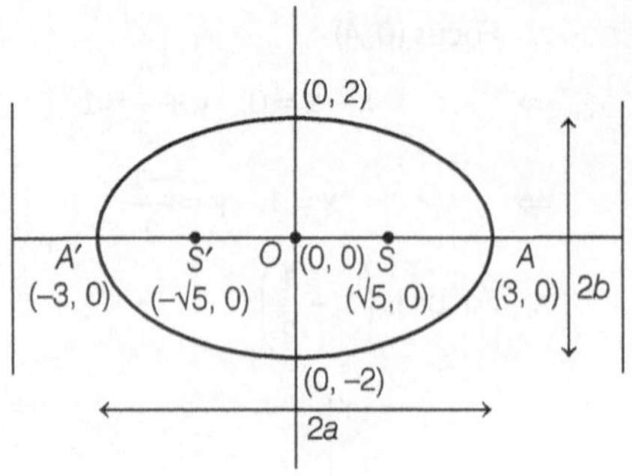

$\therefore$ It is type of $\frac{x^2}{a^2}+\frac{y^2}{b^2}=1, a>b$

Where, $a^2=9$ and $b^2=4$, then $a=3$ and $b=2$

$\therefore$ Eccentricity, $e=\sqrt{1-\frac{b^2}{a^2}}=\sqrt{1-\frac{4}{9}}=\frac{\sqrt{5}}{3}$

$\therefore$ Focus $=(\pm ae, 0)=\left(\pm 3\times\frac{\sqrt{5}}{3}, 0\right)=(\pm\sqrt{5}, 0)$

Now, vertex is $\Rightarrow (\pm a, 0)=(\pm 3, 0)$

Centre O (0, 0).

Major axis = X-axis

Length of major axis $=2a=6$

Minor axis = Y-axis

$\therefore$ Length of minor axis $=2b=4$

$\therefore$ Equation of directrix are, $x=\pm\frac{a}{e}=\pm\frac{3}{\sqrt{5}}\times 3$

$\Rightarrow \quad x=\pm\frac{9}{\sqrt{5}}.$

$\therefore$ Length of latusrectum $\Rightarrow \frac{2b^2}{a}=2\times\frac{4}{3}=\frac{8}{3}$

(b) Given, equation $2x^2+y^2=2$

$\Rightarrow \quad \frac{x^2}{1}+\frac{y^2}{2}=1$

$\Rightarrow \quad a^2=1$ and $b^2=2$

$\because b>a$, hence major axis is Y-axis.

$\therefore$ Eccentricity, $e=\sqrt{1-\frac{a^2}{b^2}}=\sqrt{1-\frac{1}{2}}=\frac{1}{\sqrt{2}}$

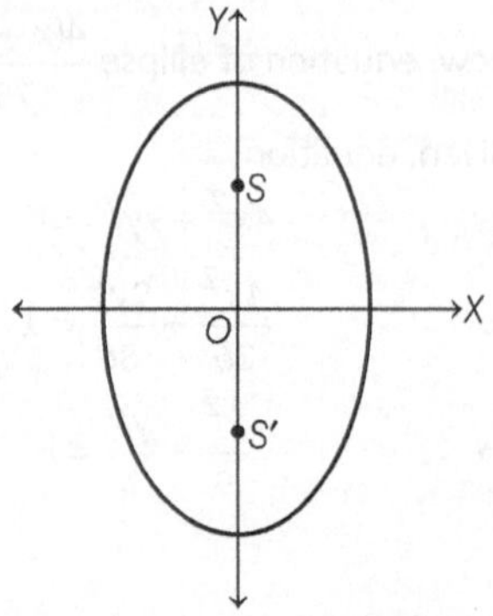

$\therefore$ Focus is $(0, \pm be)=(0, \pm 1)$

Vertex $\Rightarrow (0, \pm b)=(0, \pm\sqrt{2})$

Centre (0, 0)

Major axis $\Rightarrow$ Y-axis $\Rightarrow x=0$

Minor axis $\Rightarrow$ X-axis $\Rightarrow y=0$

Length of major axis $\Rightarrow 2b=2\sqrt{2}$

Length of minor axis $\Rightarrow 2a=2$

Directrix, $y=\pm\frac{b}{e}=\pm\frac{\sqrt{2}}{\frac{1}{\sqrt{2}}}=\pm 2 \Rightarrow y=\pm 2$

$\therefore$ Length of latusrectum $\Rightarrow \frac{2a^2}{b}=\frac{2}{\sqrt{2}}=\sqrt{2}$

(c) Given, equation is $4x^2+y^2-8x+2y+1=0$

$\Rightarrow \quad 4(x^2-2x)+y^2+2y+1=0$

$\Rightarrow 4(x^2-2x+1-1)+y^2+2y+1=0$

$\Rightarrow \quad 4(x-1)^2-4+(y+1)^2=0$

$\Rightarrow \quad 4(x-1)^2+(y+1)^2=4$

$\Rightarrow \quad \frac{(x-1)^2}{1}+\frac{(y+1)^2}{4}=1$

$\Rightarrow \frac{X^2}{1}+\frac{Y^2}{4}=1$, where $X=x-1$ and $Y=y+1$

Here, $a^2=1\Rightarrow a=1$ and $b^2=4\Rightarrow b=2$

$\because b>a$

$\therefore$ Eccentricity, $e=\sqrt{1-\frac{a^2}{b^2}}=\sqrt{1-\frac{1}{4}}=\frac{\sqrt{3}}{2}$

Focus $(0, \pm be)\Rightarrow X=0, Y=\pm be$

$\Rightarrow \quad x-1=0, y+1=\pm\sqrt{3}$

$\Rightarrow \quad x=1, y=-1\pm\sqrt{3}$

$\therefore$ Focus $\Rightarrow (1, -1\pm\sqrt{3})$

Vertex $(0, \pm b)\Rightarrow X=0, Y=\pm be$

$\Rightarrow \quad x-1=0, y+1=\pm\sqrt{3}$

$\Rightarrow \quad x=1, y=1, \pm\sqrt{3}$

$\therefore$ Vertex is (1, 1) and (1, –3)

Centre $\Rightarrow$ (1, 1)

Major axis $\Rightarrow Y=0\Rightarrow y+1=0\Rightarrow y=-1$

Minor axis $\Rightarrow X=0\Rightarrow x-1=0\Rightarrow x=1$

Length of Major axis $=2b=4$

Length of Minor axis $=2a=2$

Directrix, $Y=\pm\frac{b}{e}\Rightarrow y+1=\pm\frac{2\times 2}{\sqrt{3}}$

$$y=-1\pm\frac{4}{\sqrt{3}}$$

$\therefore$ Directrix $\Rightarrow y=-1\pm\frac{4}{\sqrt{3}}$

$\therefore$ Length of latusrectum $=\frac{2a^2}{b}=\frac{2\times 1}{2}=1$

(d) The given equation can be rewritten as

$$2(x^2 - 2x) + 3(y^2 - 4y) + 13 = 0$$

$$\Rightarrow \quad 2(x-1)^2 + 3(y-2)^2 = 1$$

$$\Rightarrow \quad \frac{(x-1)^2}{(1/\sqrt{2})^2} + \frac{(y-2)^2}{(1/\sqrt{3})^2} = 1$$

which is the form $\frac{x^2}{a^2} + \frac{y^2}{b^2} = 1$, where $X = x - 1$ and $Y = y - 2$

$$a = \frac{1}{\sqrt{2}} \text{ and } b = \frac{1}{\sqrt{3}}$$

$\Rightarrow$ Centre $X = 0, Y = 0$, i.e. $(1, 2)$

$\therefore$ Length of major axis $= 2a = \sqrt{2}$

$\therefore$ Length of minor axis $= 2b = \frac{2}{\sqrt{3}}$

and $e = \sqrt{1 - \frac{b^2}{a^2}} = \frac{1}{\sqrt{3}}$, then foci are $\left(1 \pm \frac{1}{\sqrt{6}}, 2\right)$

Vertex $\Rightarrow (\pm a, 0) \Rightarrow X = \pm\frac{1}{\sqrt{2}}$ and $Y = 0$

$$x - 1 = \pm\frac{1}{\sqrt{2}} \text{ and } y - 2 = 0$$

$$\Rightarrow \quad x = 1 \pm \frac{1}{\sqrt{2}} \text{ and } y = 2$$

$$\Rightarrow \quad \left(1 \pm \frac{1}{\sqrt{2}}, 2\right)$$

$\therefore$ Directrix, $X = \pm\frac{a}{e}$

$$\Rightarrow \quad x - 1 = \pm\frac{\frac{1}{\sqrt{2}}}{\frac{1}{\sqrt{3}}} \Rightarrow x = 1 \pm \frac{\sqrt{3}}{\sqrt{2}}$$

$\therefore$ Length of Latusrectum $= \frac{2b^2}{a} = \frac{2 \times \frac{1}{3}}{\frac{1}{\sqrt{2}}} = \frac{2\sqrt{2}}{3}$

5. (a) Let the variable point be $P(h, k)$ and given points are $A(3, 0)$ and $B(9, 0)$.

According to the question,

$$PA + PB = 12$$

$$\Rightarrow \sqrt{(h-3)^2 + (k-0)^2} + \sqrt{(h-9)^2 + (k-0)^2} = 12$$

$$= \sqrt{(h-3)^2 + k^2} = 12 - \sqrt{(h-9)^2 + k^2}$$

$$\Rightarrow \quad (h-3)^2 + k^2 = 144 + (h-9)^2 + k^2 - 2(12)\sqrt{(h-9)^2 + k^2}$$

$$\Rightarrow h^2 - 6h + 9 + k^2 - 144 - h^2 + 18h - 81 - k^2 = -24\sqrt{h^2 - 18h + 81 + k^2}$$

$$\Rightarrow \quad 12h - 216 = -24(\sqrt{h^2 - 18h + 81 + k^2})$$

$$\Rightarrow \quad -12(18 - h) = -24(\sqrt{h^2 - 18h + 81 + k^2})$$

$$\Rightarrow \quad (18 - h) = 2(\sqrt{h^2 - 18h + 81 + k^2})$$

On squaring both sides, we get

$$324 + h^2 - 36h = 4h^2 - 72h + 324 + 4k^2$$

$$\Rightarrow \quad 3h^2 + 4k^2 - 36h = 0$$

$\therefore$ Locus is $3x^2 + y^2 - 36x = 0$

$3(x^2 - 12x + 36) + y^2 - 108 = 0$

$$\Rightarrow \quad \frac{(x-6)^2}{36} + \frac{y^2}{108} = 0$$

$\therefore$ It represents an ellipse.

(b) According to given condition, the path traced by the man is an ellipse having its focii at two flag posts. Let the equation of the ellipse is

$$\frac{x^2}{a^2} + \frac{y^2}{b^2} = 1 \quad \text{...(i)}$$

As the sum of distance of the man from two flag posts is 10 m.

$\Rightarrow$ The sum of the focal distance of a point on the ellipse is 10 m.

$$\therefore \quad SP + S'P = 10$$

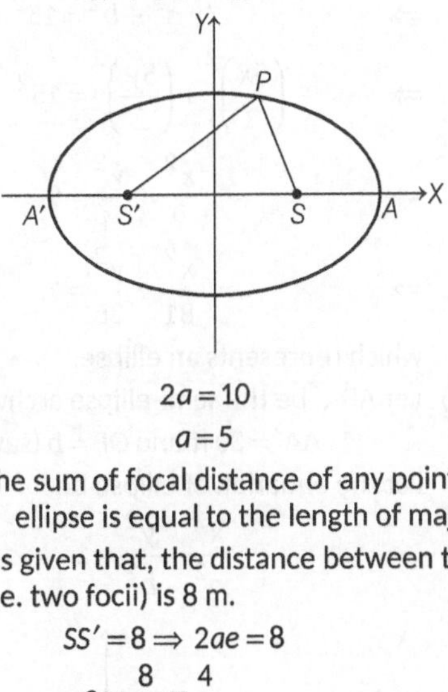

$$\Rightarrow \quad 2a = 10$$

$$\Rightarrow \quad a = 5$$

{$\because$ the sum of focal distance of any point on the ellipse is equal to the length of major axis}

Also, it is given that, the distance between the flag posts (i.e. two focii) is 8 m.

$$\therefore \quad SS' = 8 \Rightarrow 2ae = 8$$

$$e = \frac{8}{10} = \frac{4}{5}$$

and $\quad b^2 = a^2(1 - e^2) = 25\left(1 - \frac{16}{25}\right) = 9$

$\therefore$ Equation of ellipse or locus of path traced by man.

$$\frac{x^2}{a^2} + \frac{y^2}{b^2} = 1$$

$$\Rightarrow \quad \frac{x^2}{25} + \frac{y^2}{9} = 1$$

(c) The position of rod as per figure.

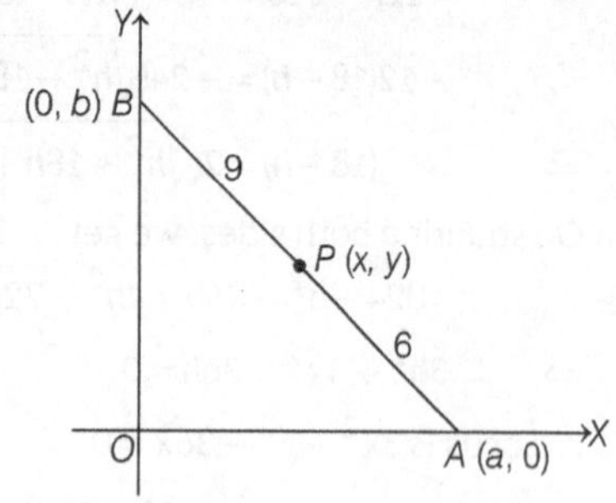

Let $A(a,0), B(0,b)$, and $P(x,y)$ be the coordinates of A, B and P, respectively.

Given, $AP = 6$ cm, so $BP = 9$ cm.

$\therefore$ Point P divides AB in the ratio $6:9 \Rightarrow 2:3$

$\therefore$ By section formula, we have,

$$x = \frac{2(0)+3(a)}{2+3} \text{ and } y = \frac{2(b)+3(0)}{2+3}$$

$$x = \frac{3a}{5} \text{ and } y = \frac{2b}{5}$$

$$\Rightarrow \quad a = \frac{5x}{3} \text{ and } b = \frac{5y}{2}$$

In ΔOAB,

$$OA^2 + OB^2 = AB^2$$

$$\Rightarrow \quad a^2 + b^2 = 15^2$$

$$\Rightarrow \quad \left(\frac{5x}{3}\right)^2 + \left(\frac{5y}{2}\right)^2 = 15^2$$

$$\Rightarrow \quad \frac{x^2}{9} + \frac{y^2}{4} = 9$$

$$\Rightarrow \quad \frac{x^2}{81} + \frac{y^2}{36} = 1,$$

which represents an ellipse.

(d) Let ABA' be the semi-ellipse archway such that $AA' = 30$ ft and $OB = b$ (say)

Let the equation of ellipse be

$$\frac{x^2}{a^2} + \frac{y^2}{b^2} = 1 \quad \text{...(i)}$$

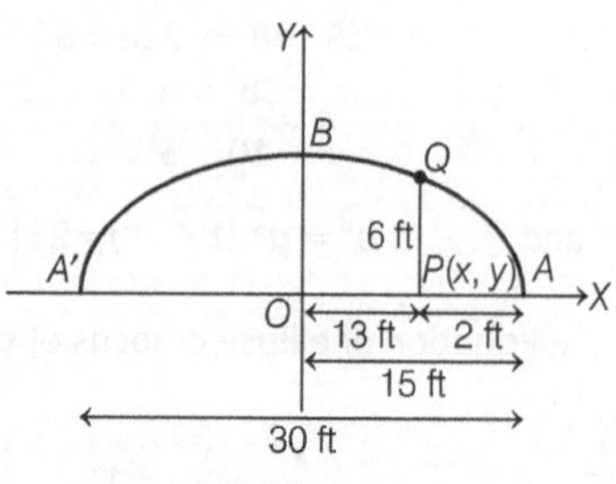

Let $P(x,y)$ be a point at a distance 2 ft from point A and man stands at P touches the ellipse at Q.

$\therefore$ $PQ = 6, AA' = 30 \Rightarrow 2a = 30 \Rightarrow a = 15$

$\therefore$ $OP = 15 - 2 = 13$

$\therefore$ The point Q is (13,6) lies on ellipse (i)

$\because$ $a = 15$ and Q lies on the ellipse

$$\therefore \quad \frac{(13)^2}{(15)^2} + \frac{6^2}{b^2} = 1$$

$$\Rightarrow \quad \frac{36}{b^2} = 1 - \frac{169}{225}$$

$$\Rightarrow \quad b^2 = \frac{225}{56} \times 36 = \frac{225 \times 9}{14}$$

$$\Rightarrow \quad b = \frac{15 \times 3}{\sqrt{14}} = \frac{45}{\sqrt{14}}$$

Hence, greatest height, $b = \frac{45}{\sqrt{14}}$ ft

6. (a) Point $(4, -3)$

Ellipse $S: 5x^2 + 7y^2 - 140 = 0$

$\because$ Now, $S_1 \Rightarrow 5(4)^2 + 7(-3)^2 - 140$

$\Rightarrow \quad 80 + 63 - 140 \Rightarrow 3 > 0$

$\therefore$ Point $(4, -3)$ lies outside the ellipse.

(b) Point $(1, 2)$

Ellipse $S: x^2 + 2y^2 - 50 = 0$

Now, $\quad S_1 = (1)^2 + 2(2)^2 - 50$

$\Rightarrow \quad 1 + 8 - 50 = -41 < 0$

$\therefore$ Point (1, 2) lies inside the ellipse

(c) Point (-1, 3)

Ellipse $S: x^2 + 4y^2 - 37 = 0$

$$S_1 = (-1)^2 + 4(3)^2 - 37 = 1 + 36 - 37 = 0$$

$\therefore$ Point lies on ellipse.

7. (a) Point $P(\alpha, \alpha+1)$

Ellipse $S: x^2 + 2y^2 - 1 = 0$

$\because$ Point P lies outside the ellipse

$$\therefore \quad S_1 > 0$$

$$\Rightarrow \quad (\alpha)^2 + 2(\alpha+1)^2 - 1 > 0$$

$$\Rightarrow \alpha^2 + 2\alpha^2 + 4\alpha + 2 - 1 > 0$$

$$\Rightarrow \quad 3\alpha^2 + 4\alpha + 1 > 0$$

$$\Rightarrow \quad 3\alpha^2 + 3\alpha + \alpha + 1 > 0$$

$$\Rightarrow \quad 3\alpha(\alpha+1) + 1(\alpha+1) > 0$$

$$\Rightarrow \quad (\alpha+1)(3\alpha+1) > 0$$

$$\Rightarrow \quad \alpha = -1, -\frac{1}{3}$$

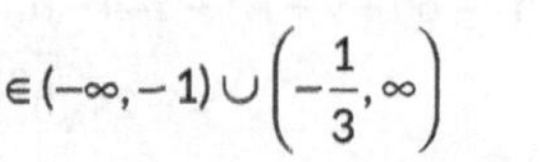

$$\therefore \quad \alpha \in (-\infty, -1) \cup \left(-\frac{1}{3}, \infty\right)$$

(b) Point $P(2\alpha, 3\alpha)$

Ellipse $S: 3x^2 + 4y^2 - 12 = 0$

$\because$ Point P lies on ellipse

$\therefore \quad S_1 = 0$

$\Rightarrow 3(2\alpha)^2 + 4(3\alpha)^2 - 12 = 0$

$\Rightarrow \quad 12\alpha^2 + 36\alpha^2 = 12 \Rightarrow 48\alpha^2 = 12$

$\Rightarrow \quad \alpha^2 = \frac{12}{48}$

$\Rightarrow \quad \alpha^2 = \frac{1}{4} \Rightarrow \alpha = \pm\frac{1}{2}$

(c) Point $P\left(7 - \frac{5\alpha}{4}, -\alpha\right)$

Ellipse $S: \frac{x^2}{25} + \frac{y^2}{16} - 1 = 0$

$\because$ P lies inside the ellipse.

$\therefore S_1 < 0$

$$\frac{\left(7 - \frac{5\alpha}{4}\right)^2}{25} + \frac{(-\alpha)^2}{16} - 1 < 0$$

$$(28 - 5\alpha)^2 + 25\alpha^2 - 400 < 0$$

$\Rightarrow 25\alpha^2 + 784 - 280\alpha + 25\alpha^2 - 400 < 0$

$\Rightarrow \quad 50\alpha^2 - 280\alpha + 384 < 0$

$\Rightarrow \quad 25\alpha^2 - 140\alpha + 192 < 0$

$\Rightarrow \quad 25\alpha^2 - 80\alpha - 60\alpha + 192 < 0$

$\Rightarrow \quad 5\alpha(5\alpha - 16) - 12(5\alpha - 16) < 0$

$\Rightarrow \quad (5\alpha - 16)(5\alpha - 12) < 0$

$\oplus$ 12/5 $\ominus$ 16/5 $\oplus$

$\therefore \quad \alpha \in \left(\frac{12}{5}, \frac{16}{5}\right)$

8. (a) Given, ellipse $S: \frac{(x-3)^2}{4} + \frac{(y-4)^2}{25} = 1$

$\Rightarrow \quad a^2 = 4$ and $b^2 = 25$

$\Rightarrow \quad a = 2$ and $b = 5$

$\because$ Parametric form of ellipse $\frac{X^2}{a^2} + \frac{Y^2}{b^2} = 1$, is

$X = a\cos\theta$ and $Y = b\sin\theta$

$\therefore$ For given ellipse

$x - 3 = 2\cos\theta$

$x = 3 + 2\cos\theta$

and $\quad y - 4 = 5\sin\theta$

$y = 4 + 5\sin\theta$

$\therefore$ Parametric form is

$x = 3 + 2\cos\theta$ and $y = 4 + 5\sin\theta$

(b) Given ellipse,

$4x^2 + y^2 - 8x + 2y + 1 = 0$

$4(x^2 - 2x + 1) + (y^2 + 2y + 1) = 4$

$\Rightarrow \quad 4(x-1)^2 + (y+1)^2 = 4$

$\Rightarrow \quad \frac{(x-1)^2}{1} + \frac{(y+1)^2}{4} = 1$

$\Rightarrow \quad a^2 = 1$ and $b^2 = 4$

$a = 1$ and $b = 2$

$\therefore$ Parametric form of ellipse $\frac{X^2}{a^2} + \frac{Y^2}{b^2} = 1$, is

$x - 1 = a\cos\theta$ and $y + 1 = b\sin\theta$

$x = 1 + \cos\theta$ and $y = 2\sin\theta - 1$

$\therefore$ Parametric form is

$x = \cos\theta + 1$ and $y = 2\sin\theta - 1$

9. (a) Given, ellipse $16x^2 + 25y^2 = 400$

$$\frac{x^2}{25} + \frac{y^2}{16} = 1$$

$\Rightarrow \quad a^2 = 25$ and $b^2 = 16$

$\Rightarrow \quad a = 5$ and $b = 4$

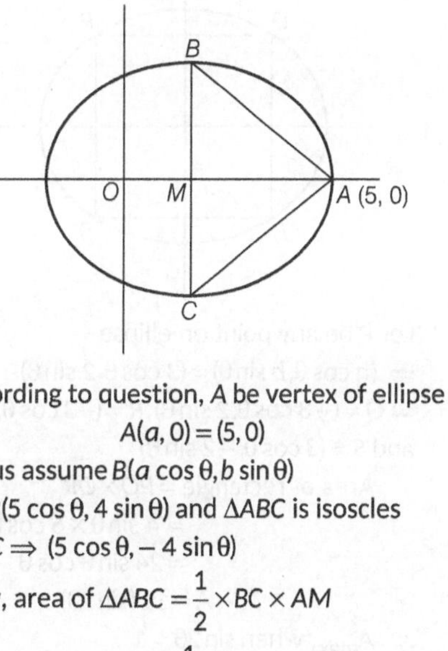

According to question, A be vertex of ellipse

$\therefore \quad A(a, 0) = (5, 0)$

Let us assume $B(a\cos\theta, b\sin\theta)$

$\Rightarrow B(5\cos\theta, 4\sin\theta)$ and ΔABC is isoscles

$\therefore \quad C \Rightarrow (5\cos\theta, -4\sin\theta)$

Now, area of $\Delta ABC = \frac{1}{2} \times BC \times AM$

$= \frac{1}{2} \times (8\sin\theta) \times (5 - 5\cos\theta)$

$= \frac{1}{2} \times 8 \times 5(\sin\theta)(1 - \cos\theta)$

$= 20(\sin\theta - \sin\theta\cos\theta)$

Now, maximum value of A

$A = 20(\sin\theta - \sin\theta\cos\theta)$

$$\frac{dA}{d\theta} = 20(\cos\theta + \sin\theta(\sin\theta) - \cos\theta\cos\theta) = 0$$

$\cos\theta + \sin^2\theta - \cos^2\theta = 0$

$\Rightarrow \quad \cos\theta + 1 - \cos^2\theta - \cos^2\theta = 0$

$$\Rightarrow \quad 2\cos^2\theta - \cos\theta - 1 = 0$$

$$\Rightarrow \quad (2\cos\theta + 1)(\cos\theta - 1) = 0$$

$$\Rightarrow \quad \cos\theta = -\frac{1}{2}, \cos\theta = 1 \quad \text{(Not possible)}$$

$$\therefore \quad \theta \in (0, \pi)$$

$$\Rightarrow \quad \theta = \frac{2\pi}{3}$$

$$\text{At } \theta = \frac{2\pi}{3} \Rightarrow \frac{d^2A}{d\theta^2} < 0$$

$\therefore$ A is maximum

$\therefore$ Maximum area

$$\Rightarrow 20\left(\frac{\sqrt{3}}{2} - \frac{\sqrt{3}}{2} \times \left(-\frac{1}{2}\right)\right) = 20\left(\frac{3\sqrt{3}}{4}\right) = 15\sqrt{3}$$

(b) Given, ellipse

$$4x^2 + 9y^2 = 36$$

$$\Rightarrow \quad \frac{x^2}{9} + \frac{y^2}{4} = 1$$

$$\Rightarrow \quad a^2 = 9 \text{ and } b^2 = 4$$

$$\Rightarrow \quad a = 3 \text{ and } b = 2$$

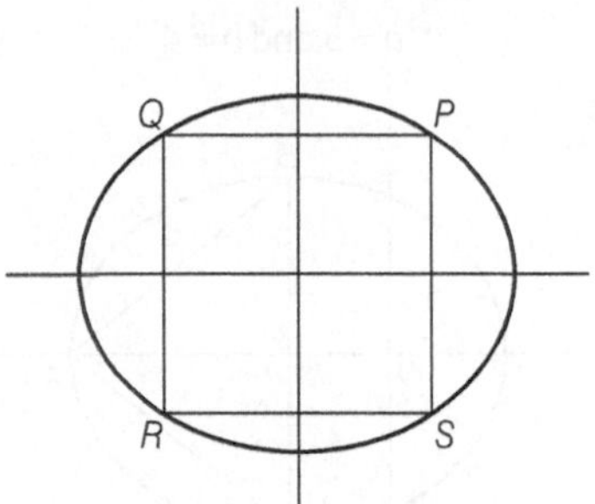

Let P be any point on ellipse

$\Rightarrow (a\cos\theta, b\sin\theta) = (3\cos\theta, 2\sin\theta)$

$\Rightarrow Q = (-3\cos\theta, 2\sin\theta), R = (-3\cos\theta, -2\sin\theta)$

and $S = (3\cos\theta, -2\sin\theta)$

$$\therefore \quad \text{Area of rectangle} = PQ \times QR$$

$$= 4\sin\theta \times 6\cos\theta$$

$$= 24\sin\theta\cos\theta$$

$$A = 12(\sin 2\theta)$$

$\therefore$ A_{max}, when $\sin 2\theta = 1$

$$\Rightarrow \quad 2\theta = \frac{\pi}{2} \Rightarrow \theta = \frac{\pi}{4}$$

Now, maximum area A is 12 sq units

10. (a) Line $x + y = 1$ ...(i)

Ellipse $x^2 + 2y^2 = 1$ ...(ii)

Solving Eqs. (i) and (ii),

Put $y = 1 - x$

$$\Rightarrow \quad x^2 + 2(1-x)^2 - 1 = 0$$

$$\Rightarrow \quad x^2 + 2x^2 - 4x + 2 - 1 = 0$$

$$\Rightarrow \quad 3x^2 - 4x + 1 = 0$$

$$3x^2 - 3x - x + 1 = 0$$

$$3x(x-1) - 1(x-1) = 0$$

$$(x-1)(3x-1) = 0$$

$x = 1$	$x = \frac{1}{3}$
$y = 1 - 1$	$y = 1 - \frac{1}{3}$
$= 0$	$= \frac{2}{3}$

$\therefore$ Points are $A(1, 0)$ and $B\left(\frac{1}{3}, \frac{2}{3}\right)$.

Now, length of intercept made by line $= AB$

$$= \sqrt{\left(1 - \frac{1}{3}\right)^2 + \left(0 - \frac{2}{3}\right)^2}$$

$$= \sqrt{\frac{4}{9} + \frac{4}{9}}$$

$$= \sqrt{\frac{8}{9}} = \frac{2\sqrt{2}}{3} \text{ units}$$

(b) Line : $y = 2x + k$ ...(i)

Ellipse : $2x^2 + y^2 = 1$ ...(ii)

Now, solving Eqs. (i) and (ii), we get

$$2x^2 + (2x + k)^2 - 1 = 0$$

$$\Rightarrow 2x^2 + 4x^2 + 4kx + k^2 - 1 = 0$$

$$\Rightarrow \quad 6x^2 + 4kx + 1(k^2 - 1) = 0 \quad \text{...(iii)}$$

$$\Rightarrow A = 6, B = 4k \text{ and } C = k^2 - 1$$

$$\therefore \quad D = B^2 - 4AC$$

$$= (4k)^2 - 4(6)(k^2 - 1)$$

$$= 16k^2 - 24(k^2 - 1)$$

$$= 8(2k^2 - 3k^2 + 3)$$

$$= 8(3 - k^2)$$

Now, (i) line intersect the ellipse

$\Rightarrow$ two real and distinct roots of Eq. (iii)

$$\Rightarrow \quad D > 0$$

$$\Rightarrow \quad 8(3 - 1k^2) > 0$$

$$\Rightarrow \quad (\sqrt{3} - k)(\sqrt{3} + k) > 0$$

$$\Rightarrow \quad k \in (-\sqrt{3}, \sqrt{3})$$

(ii) For tangent $\Rightarrow$ two real and equal roots

$$\Rightarrow \quad D = 0$$

$$\Rightarrow \quad 8(3 - k^2) = 0$$

$$k = \pm\sqrt{3}$$

(iii) For neither intersecting nor touching

$\Rightarrow$ no real roots.

$\Rightarrow D < 0 \Rightarrow (\sqrt{3} - k)(\sqrt{3} + k) < 0$

$\Rightarrow k \in (-\infty, -\sqrt{3}) \cup (\sqrt{3}, \infty)$

11. (i) (a) Ellipse : $4x^2+9y^2=40$

$$\Rightarrow \quad \frac{x^2}{10}+\frac{9y^2}{40}=1$$

Point $(-1,2)$

$\therefore$ Equation of tangent at point (x_1, y_1) to the ellipse $\frac{x^2}{a^2}+\frac{y^2}{b^2}=1$ is given by

$$\Rightarrow \quad \frac{xx_1}{a^2}+\frac{yy_1}{b^2}=1$$

$\therefore$ Required equation of tangent

$$\Rightarrow \quad \frac{x(-1)}{10}+\frac{9y(2)}{40}=1$$

$$\Rightarrow \quad \frac{-x}{10}+\frac{9y}{20}=1$$

$$\Rightarrow \quad -2x+9y=20$$

$$\Rightarrow \quad 9y-2x-20=0$$

(b) Ellipse : $\frac{x^2}{25}+\frac{y^2}{16}=1$

$$\Rightarrow \quad a^2=25 \text{ and } b^2=16$$

$$\Rightarrow \quad a=5 \text{ and } b=4$$

Parametric form is $x=a\cos\theta, y=b\sin\theta$

$$x=5\cos\theta, y=4\sin\theta$$

$\therefore$ Equation of tangent at $(5\cos\theta, 4\sin\theta)$, is

$$\frac{x(5\cos\theta)}{25}+\frac{y(4\sin\theta)}{16}=1$$

$$\Rightarrow \quad \frac{x}{5}\cos\theta+\frac{y}{4}\sin\theta=1$$

(c) Ellipse : $3x^2+4y^2=1$

$$\Rightarrow \quad \frac{x^2}{\frac{1}{3}}+\frac{y^2}{\frac{1}{4}}=1$$

Point $(a\cos\theta, b\sin\theta)$

$$\Rightarrow \quad a=\frac{1}{\sqrt{3}}, b=\frac{1}{2} \text{ and } \theta=\frac{\pi}{6}$$

$$\therefore \text{ Point is } \left(\frac{1}{\sqrt{3}}\cos\frac{\pi}{6}, \frac{1}{2}\sin\frac{\pi}{6}\right)$$

$$=\left(\frac{1}{\sqrt{3}}\times\frac{\sqrt{3}}{2}, \frac{1}{2}\times\frac{1}{2}\right)=\left(\frac{1}{2},\frac{1}{4}\right)$$

$\therefore$ Equation of tangent is

$$\Rightarrow 3x\left(\frac{1}{2}\right)+4y\left(\frac{1}{4}\right)=1$$

$\{\because$ ellipse $3x^2+4y^2=1$ and tangent $3xx_1+4yy_1=1\}$

$$\Rightarrow 3x+2y=2$$

(d) Given, slope $(m)=2$

Ellipse : $3x^2+4y^2=12$

$$\Rightarrow \quad \frac{x^2}{4}+\frac{y^2}{3}=1$$

$$a^2=4 \text{ and } b^2=3$$

$\because$ Equation of tangent having slope (m) for ellipse $\frac{x^2}{a^2}+\frac{y^2}{b^2}=1$ is given by $y=mx\pm\sqrt{a^2m^2+b^2}$, where $m=2$.

$\therefore$ Equation is $y=2x\pm\sqrt{4(2)^2+3}$

$$y=2x\pm\sqrt{19}$$

(e) $\because$ Tangent is perpendicular to the line $4x-3y=1$

$$\therefore \text{ Slope of tangent}=\frac{-1}{\frac{4}{3}}=\frac{-3}{4}.$$

$$\therefore \quad m=-\frac{3}{4}$$

Equation of ellipse : $9x^2+y^2=1$

$$\frac{x^2}{\frac{1}{9}}+\frac{y^2}{1}=1$$

$$a^2=\frac{1}{9} \text{ and } b^2=1$$

$\therefore$ Equation of tangent is

$$y=mx\pm\sqrt{a^2m^2+b^2}$$

$$=-\frac{3}{4}x\pm\sqrt{\frac{1}{9}\times\frac{9}{16}+1}$$

$$=-\frac{3}{4}x\pm\frac{1}{4}\sqrt{17}$$

$$\Rightarrow \quad 4y+3x=\pm\sqrt{17}$$

(f) $\because$ Tangent is parallel to the line $y+x=1$

$\therefore$ Slope of tangent $(m)=-1$

Given, equation of ellipse is $4x^2+3y^2=12$

$$\Rightarrow \quad \frac{x^2}{3}+\frac{y^2}{4}=1$$

Now, $m=-1, a^2=3$ and $b^2=4$

$\therefore$ Equation of tangent is

$$y=mx\pm\sqrt{a^2m^2+b^2}$$

$$=-x\pm\sqrt{3(1)+4}$$

$$\Rightarrow \quad y+x=\pm\sqrt{7}$$

(ii) (a) Given, equation of tangent is $x+9y=19$ and Ellipse : $2x^2+9y^2=38$

$$\frac{x^2}{19}+\frac{9y^2}{38}=1$$

$$\Rightarrow \quad \frac{x^2}{19}+\frac{y^2}{\frac{38}{9}}=1$$

Let a point $P\left(\sqrt{19}\cos\theta, \frac{\sqrt{38}}{3}\sin\theta\right)$ be on ellipse,

where tangent is $x+9y=19$

$\therefore$ Equation of tangent at P

$$\left(\frac{\sqrt{19}\cos\theta}{19}\right)x+\left(\frac{\frac{\sqrt{38}}{3}\sin\theta}{\frac{38}{9}}\right)y=1$$

$$\Rightarrow \quad \frac{(\cos\theta)\,x}{\sqrt{19}}+\frac{\sin\theta}{\frac{\sqrt{38}}{3}}y=1$$

Given, line $x+9y=19$ both are identical

$$\frac{\cos\theta}{\sqrt{19}}=\frac{\frac{\sin\theta}{\frac{\sqrt{38}}{3}}}{9}=\frac{1}{19}$$

$$\Rightarrow \quad \cos\theta=\frac{\sqrt{19}}{19}$$

and $\quad \sin\theta=\frac{9}{19}\times\frac{\sqrt{38}}{3}$

$$\Rightarrow \quad \cos\theta=\frac{1}{\sqrt{19}}$$

and $\quad \sin\theta=\frac{3\sqrt{2}}{\sqrt{19}}$

$\therefore$ Point of contact is $\left(\sqrt{19}\times\frac{1}{\sqrt{19}}, \frac{\sqrt{38}}{3}\times\frac{3\sqrt{2}}{\sqrt{19}}\right)$

$$=(1,2)$$

(b) Equation of ellipse : $3x^2+4y^2=12$

$$\Rightarrow \quad \frac{x^2}{4}+\frac{y^2}{3}=1$$

$$a^2=4 \text{ and } b^2=3$$

Let the slope of tangent is m.

$\therefore$ Tangent makes an angle 45° with the line

$$y=2x+1$$

$\therefore \quad m_1=m, m_2=2$ and $\theta=45°$

$$\therefore \quad \tan\theta=\left|\frac{m_1-m_2}{1+m_1m_2}\right|$$

$$\Rightarrow \tan 45°=\left|\frac{m-2}{1+2m}\right|$$

$$\Rightarrow \frac{m-2}{1+2m}=\pm 1$$

$\Rightarrow \quad m-2=\pm 1(1+2m)$

$\Rightarrow \quad m+2=1+2m$

$\Rightarrow \quad m=1$

or $\quad m-2=-1-2m$

$\Rightarrow \quad 3m=1 \Rightarrow m=\frac{1}{3}$

$\therefore$ Point of contact

$$\left(\mp\frac{a^2m}{\sqrt{a^2m^2+b^2}}, \pm\frac{b^2}{\sqrt{a^2m^2+b^2}}\right)$$

At $m=1$,

$$\left(\mp\frac{4(1)}{\sqrt{4+3}}, \pm\frac{3}{\sqrt{4+3}}\right)$$

$$\left(\mp\frac{4}{\sqrt{7}}, \pm\frac{3}{\sqrt{7}}\right)$$

At $m=\frac{1}{3}$,

$$\left(\mp\frac{\frac{4}{3}}{\sqrt{\frac{4}{9}+3}}, \pm\frac{3}{\sqrt{\frac{4}{9}+3}}\right)$$

$$\left(\mp\frac{4}{\sqrt{31}}, \pm\frac{9}{\sqrt{31}}\right)$$

(iii) (a) $\because$ Line $y=3x+2$ touches ellipse $\frac{x^2}{a^2}+\frac{y^2}{b^2}=1$

Here, $\quad m=3$ and $c=2$

$$\because \quad c^2=a^2m^2+b^2$$

$$4=9a^2+b^2 \quad \text{...(i)}$$

Now, we know that for two numbers a and b

AM≥ GM

For $9a^2, b^2$

$$\Rightarrow \quad \frac{9a^2+b^2}{2}\geq\sqrt{9a^2b^2}$$

$$\Rightarrow \quad \frac{4}{2}\geq 3ab \quad \text{[from Eq. (i)]}$$

$$\Rightarrow \quad 3ab\leq 2$$

$$\Rightarrow \quad ab\leq\frac{2}{3} \quad \text{...(ii)}$$

Now, area of ellipse $=\pi ab$

Multiplying Eqs. (ii) by π, we get

$$\therefore \quad \pi ab\leq\frac{2}{3}\pi$$

$\therefore$ Area of ellipse $\leq\frac{2\pi}{3}$

$\therefore$ Maximum area $=\frac{2\pi}{3}$

(b) Ellipse : $\frac{x^2}{25}+\frac{y^2}{16}=1$

$$\Rightarrow \quad a^2=25 \text{ and } b^2=16$$

$\Rightarrow \quad a=5$ and $b=4$

$\Rightarrow \quad e=\sqrt{1-\frac{16}{25}}=\frac{3}{5}$

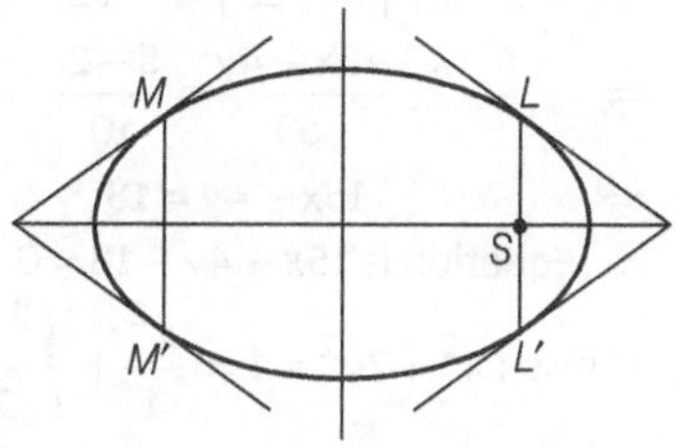

End points of latusrectum $\left(ae, \frac{b^2}{a}\right)$ and $\left(ae, -\frac{b^2}{a}\right)$

$L\left(3, \frac{16}{5}\right)$ and $L'\left(3, -\frac{16}{5}\right)$

$\therefore$ Equation of tangents at L and L'

$$\frac{3x}{25}+\frac{y}{5}=1 \quad \text{...(i)}$$

$$\frac{3x}{25}-\frac{y}{5}=1 \quad \text{...(ii)}$$

Adding Eqs. (i) and (ii), we get

$$\frac{3x}{25}=1 \Rightarrow x=\frac{25}{3} \quad \Rightarrow y=0$$

$\therefore$ Point of intersection is $\left(\frac{25}{3}, 0\right)$

Similarly, the point of intersection of tangent at M and M' is $\left(-\frac{25}{3}, 0\right)$.

(iv) (a) Ellipse : $\frac{x^2}{a^2}+\frac{y^2}{b^2}=1$

Any point M on ellipse is $(a\cos\theta, b\sin\theta)$

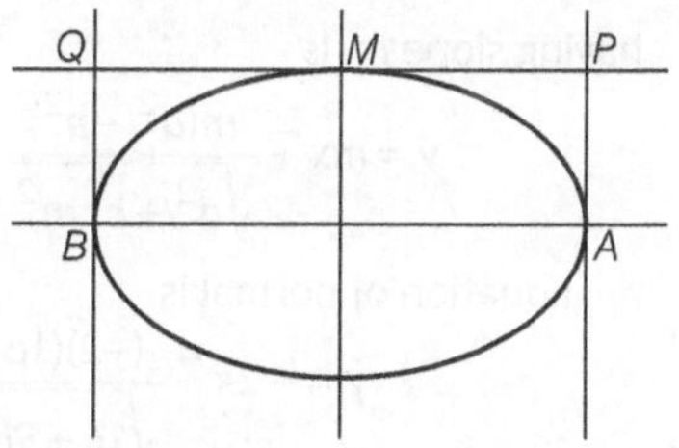

$\therefore$ Equation of tangent at M.

$$\Rightarrow \quad \frac{x\cos\theta}{a}+\frac{y\sin\theta}{b}=1$$

$$\Rightarrow \quad y=\frac{b}{\sin\theta}\left(1-\frac{x\cos\theta}{a}\right)$$

$\because$ Equation of tangent A and B are $x=a, x=-a$.

$\therefore$ Point $A\left(a, \frac{b}{\sin\theta}(1-\cos\theta)\right)$

Point $B\left(-a, \frac{b}{\sin\theta}(1+\cos\theta)\right)$

Now, $\quad PA=\frac{b}{\sin\theta}(1-\cos\theta)$

and $\quad PB=\frac{b}{\sin\theta}(1+\cos\theta)$

$$\therefore \quad PA \times PB=\frac{b}{\sin\theta}\times(1-\cos\theta)\times\frac{b}{\sin\theta}(1+\cos\theta)$$

$$\Rightarrow \frac{b^2}{\sin^2\theta}(1-\cos^2\theta)=\frac{b^2}{\sin^2\theta}\times\sin^2\theta=b^2$$

(b) Ellipse : $\frac{x^2}{a^2}+\frac{y^2}{b^2}=1$,

Centre $O(0, 0)$

Let foot of perpendicular is $P(h, k)$

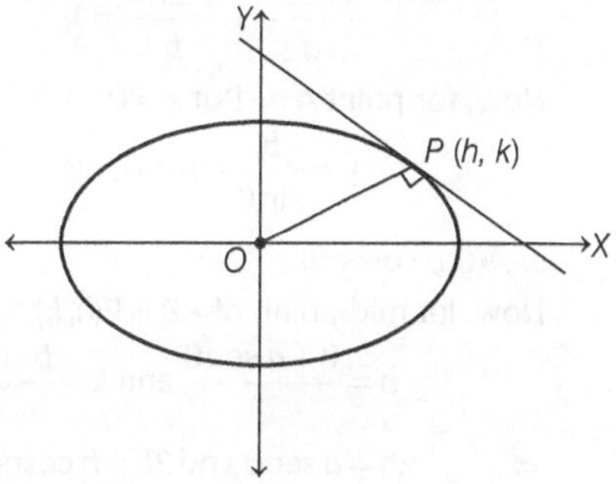

Then, the slope of OP is $=\frac{k}{h}$

(where O is centre)

$\Rightarrow$ Then, slope of tangent $=-\frac{h}{k}$.

$\therefore$ Equation of tangent passing through (h, k)

$$y-k=-\frac{h}{k}(x-h)$$

$$\Rightarrow \quad y=-\frac{hx}{k}+\frac{h^2}{k}+k$$

$$y=-\frac{h}{k}x+\frac{h^2+k^2}{k} \quad \text{...(i)}$$

$$y=mx+c$$

$\because$ Equation (i) touches the ellipse

$$c^2=a^2m^2+b^2$$

$$\left(\frac{h^2+k^2}{k}\right)^2=a^2\left(-\frac{h}{k}\right)^2+b^2$$

$$\Rightarrow \quad (h^2+k^2)^2=a^2h^2+b^2k^2$$

$\therefore$ Locus is $(x^2+y^2)^2=a^2x^2+b^2y^2$

(c) Equation of ellipse : $\frac{x^2}{a^2}+\frac{y^2}{b^2}=1.$

Let P be any point on ellipse

$\therefore\ P=(a\cos\theta, b\sin\theta)$

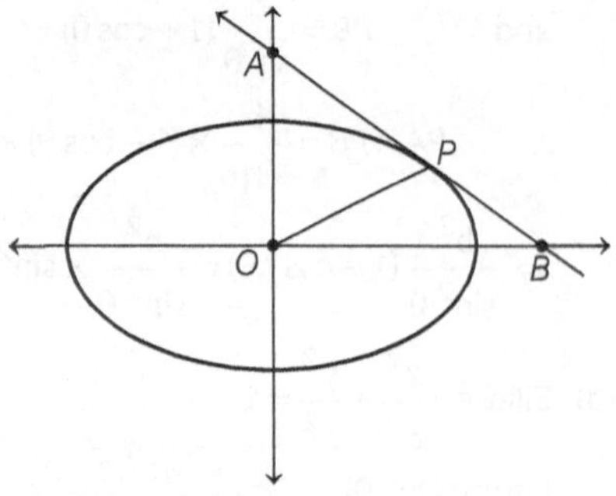

$\therefore$ Equation of tangent at P

$$\frac{x\cos\theta}{a}+\frac{y\sin\theta}{b}=1$$

Now, for point $A \Rightarrow$ Put $x=0$

$$y=\frac{b}{\sin\theta}=b\operatorname{cosec}\theta$$

$\therefore A(0, b\operatorname{cosec}\theta)$

Now, let mid-point of AB is $P(h,k)$

$$h=\frac{0+a\sec\theta}{2} \text{ and } k=\frac{b\operatorname{cosec}\theta+0}{2}$$

$$\Rightarrow \quad 2h=a\sec\theta \text{ and } 2k=b\operatorname{cosec}\theta$$

$$\Rightarrow \quad \cos\theta=\frac{a}{2h} \text{ and } \sin\theta=\frac{b}{2k}$$

$$\because \quad \sin^2\theta+\cos^2\theta=1$$

$$\Rightarrow \quad \frac{b^2}{4k^2}+\frac{a^2}{4h^2}=1$$

$\therefore$ Locus is $\frac{b^2}{y^2}+\frac{a^2}{x^2}=4$

or $\quad \frac{a^2}{x^2}+\frac{b^2}{y^2}=4$

12. (a) Ellipse : $2x^2+5y^2=53$

$$\frac{x^2}{\frac{53}{2}}+\frac{y^2}{\frac{53}{5}}=1$$

$$a^2=\frac{53}{2} \text{ and } b^2=\frac{53}{5}$$

Point $=(2,3)$

$\because$ Equation of normal to the ellipse $\frac{x^2}{a^2}+\frac{y^2}{b^2}=1$ at point $P(x_1, y_1)$ is

$$\frac{a^2x}{x_1}-\frac{b^2y}{y_1}=a^2-b^2$$

$\therefore$ Equation is $\frac{53}{2}\frac{x}{2}-\frac{53}{5}\frac{y}{3}=\frac{53}{2}-\frac{53}{5}$

$$\Rightarrow \quad 53\left[\frac{x}{4}-\frac{y}{15}\right]=53\left(\frac{1}{2}-\frac{1}{5}\right)$$

$$\Rightarrow \quad \frac{15x-4y}{60}=\frac{5-2}{10}$$

$$\Rightarrow \quad 15x-4y=18$$

$\therefore$ Equation is $15x-4y-18=0$

(b) Ellipse : $x^2+2y^2=1 \Rightarrow \frac{x^2}{1}+\frac{y^2}{1/2}=1$

$$a^2=1 \text{ and } b^2=\frac{1}{2}$$

$\because$ Let the point is $(a\cos\theta, b\sin\theta)$

$\because\ a=1, b=\frac{1}{\sqrt{2}}$ and $\theta=\frac{\pi}{3}$

$\therefore$ Equation of normal is

$$\frac{ax}{\cos\theta}-\frac{by}{\sin\theta}=a^2-b^2$$

$$\Rightarrow \quad \frac{x}{\cos\frac{\pi}{3}}-\frac{\frac{1}{\sqrt{2}}y}{\sin\frac{\pi}{3}}=1-\frac{1}{2}$$

$$\Rightarrow \quad 2x-\frac{\sqrt{2}}{\sqrt{3}}y=\frac{1}{2}$$

$$\Rightarrow \quad 4\sqrt{3}x-2\sqrt{2}\,y-\sqrt{3}=0$$

(c) Ellipse : $\frac{x^2}{16}+\frac{y^2}{9}=1$

$\because\ a^2=16$ and $b^2=9$

Slope $\Rightarrow m=-2$

$\therefore$ Equation of normal to the ellipse $\frac{x^2}{a^2}+\frac{y^2}{b^2}=1$ having slope m, is

$$y=mx\mp\frac{m(a^2-b^2)}{\sqrt{a^2+b^2m^2}}$$

$\therefore$ Equation of normal is

$$y=-2x\mp\frac{(-2)(16-9)}{\sqrt{16+9(-2)^2}}$$

$$\Rightarrow \quad y=-2x\mp\frac{-14}{\sqrt{52}}$$

$$\Rightarrow \quad y=-2x\pm\frac{7}{\sqrt{13}}$$

(d) Equation of ellipse : $\frac{x^2}{9}+\frac{y^2}{4}=1$

$\therefore\ a^2=9, b^2=4$ and $e=\sqrt{1-\frac{b^2}{a^2}}=\frac{\sqrt{5}}{3}$

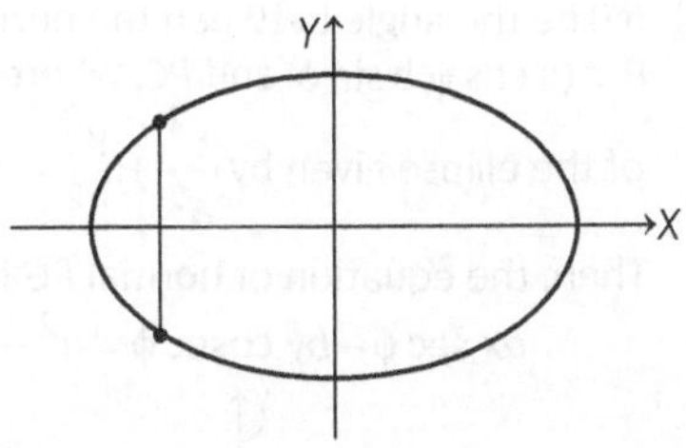

$\therefore$ Negative end of latusrectum $\left(-ae, -\frac{b^2}{a}\right)$

$$\left(-9\times\frac{\sqrt{5}}{3}, -\frac{4}{3}\right)$$

$\Rightarrow$ $(-3\sqrt{5}, -4/3)$

$\therefore$ Equation of normal $\frac{a^2x}{x_1} - \frac{b^2y}{y_1} = a^2 - b^2$

$$\Rightarrow \frac{9x}{-3\sqrt{5}} - \frac{4y}{-4/3} = 9 - 4$$

$$\Rightarrow \frac{-3x}{\sqrt{5}} + 3y = 5 \Rightarrow -3x + 3\sqrt{5}y = 5\sqrt{5}$$

(e) Ellipse : $\frac{x^2}{9} + \frac{y^2}{25} = 1$

Normal parallel to line $3x + 2y - 1 = 0$

$\therefore$ Slope of normal $= -\frac{3}{2}$

$$\Rightarrow m = -\frac{3}{2}$$

$$a^2 = 9 \text{ and } b^2 = 25$$

$\therefore$ Equation of normal is

$$y = mx \mp \frac{m(a^2 - b^2)}{\sqrt{a^2 + b^2m^2}}$$

$$\Rightarrow y = -\frac{3}{2}x \mp \frac{\left(-\frac{3}{2}\right)(9-25)}{\sqrt{9 + 25\times\left(-\frac{3}{2}\right)^2}}$$

$$\Rightarrow y = -\frac{3}{2}x \mp \frac{24\times 2}{\sqrt{36+225}}$$

$$\Rightarrow y = \frac{-3}{2}x \mp \frac{48}{\sqrt{261}}$$

$$\Rightarrow y = -\frac{3}{2}x \mp \frac{48}{3\sqrt{29}}$$

$$\therefore y = -\frac{3}{2}x \mp \frac{16}{\sqrt{29}}$$

Now, point of contacts are

$$\left(\pm\frac{a^2}{\sqrt{a^2+b^2m^2}}, \pm\frac{mb^2}{\sqrt{a^2+b^2m^2}}\right)$$

$$= \left(\pm\frac{9}{\frac{3}{2}\sqrt{29}}, \pm\frac{\left(-\frac{3}{2}\right)25}{\frac{3}{2}\sqrt{29}}\right)$$

$$\Rightarrow \left(\pm\frac{6}{\sqrt{29}}, \mp\frac{25}{\sqrt{29}}\right)$$

$$\Rightarrow \left(\frac{6}{\sqrt{29}}, \frac{-25}{\sqrt{29}}\right) \text{ and } \left(\frac{-6}{\sqrt{29}}, \frac{25}{\sqrt{29}}\right)$$

13. (a) Given equation, $5x^2 + 14y^2 = 70$

$$\Rightarrow \frac{x^2}{14} + \frac{y^2}{5} = 1$$

$\because$ $a^2 = 14$ and $b^2 = 5$

$\because$ The equation of normal at $P(\theta)$ is

$$\frac{ax}{\cos\theta} - \frac{by}{\sin\theta} = a^2 - b^2 \quad \text{...(i)}$$

$\because Q(2\theta)$ lies on normal

$$\therefore \frac{a}{\cos\theta}(a\cos 2\theta) - \frac{b}{\sin\theta}(b\sin 2\theta) = a^2 - b^2$$

$$\Rightarrow a^2\frac{(2\cos^2\theta - 1)}{\cos\theta} - 2b^2\cos\theta = a^2 - b^2$$

$$\Rightarrow 2a^2\cos^2\theta - a^2 - 2b^2\cos^2\theta = (a^2 - b^2)\cos\theta$$

$$\Rightarrow 28\cos^2\theta - 14 - 10\cos^2\theta = 9\cos\theta$$

$$\Rightarrow 18\cos^2\theta - 9\cos\theta - 14 = 0$$

$$\Rightarrow 18\cos^2\theta + 12\cos\theta - 21\cos\theta - 14 = 0$$

$$\Rightarrow 6\cos\theta(3\cos\theta + 2) - 7(3\cos\theta + 2) = 0$$

$$\Rightarrow (6\cos\theta - 7)(3\cos\theta + 2) = 0$$

$$\Rightarrow \cos\theta = \frac{7}{6}, -\frac{2}{3}$$

$\because$ $\cos\theta$ cannot be $\frac{7}{6}$ $\quad (\because \cos\theta \in [-1, 1])$

$\therefore$ $\cos\theta = -\frac{2}{3}$

(b) Let the ellipse be $\frac{x^2}{a^2} + \frac{y^2}{b^2} = 1$ and $P(a\cos\theta, b\sin\theta)$ be any point on the ellipse

The equation of normal at P

$$ax\sec\theta - by\,\text{cosec}\,\theta = a^2 - b^2 \quad \text{...(i)}$$

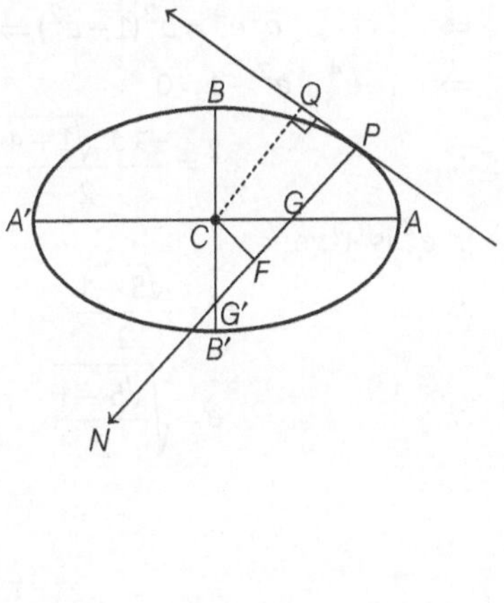

$\because$ The line (i) meets the major axis at G and minor axis at G', respectively, then

$$G=\left(\frac{a^2-b^2}{a}\cos\theta, 0\right) \Rightarrow G'=\left(0, -\frac{a^2-b^2}{b}\sin\theta\right)$$

$PF = CQ =$ length of perpendicular from $C(0, 0)$ on the tangent

$\frac{x}{a}\cos\theta+\frac{y}{b}\sin\theta=1$ at P

$$=\left|\frac{ab}{\sqrt{a^2\sin^2\theta+b^2\cos^2\theta}}\right|$$

Also, $PG=\frac{b}{a}\sqrt{b^2\cos^2\theta+a^2\sin^2\theta}$

and $PG'=\frac{a}{b}\sqrt{b^2\cos^2\theta+a^2\sin^2\theta}$

Then, $PF\cdot PG=b^2$ and $PF\cdot PG'=a^2$.

(c) The equation of normal at $L=\left(ae, \frac{b^2}{a}\right)$ is given by

$$\frac{a^2x^2}{ae}-\frac{b^2y^2}{\frac{b^2}{a}}=a^2-b^2$$

$$\Rightarrow \quad \frac{x^2}{e}-y^2=\frac{a^2-b^2}{a} \quad \text{...(i)}$$

According to question, normal passes through opposite of latusrectum i.e.
It passes through $(0, -b)$

$$\Rightarrow \quad \frac{0}{e}+b^2=\frac{(a^2-b^2)}{a}$$

$$\Rightarrow \quad a^2-b^2-ab^2=0 \quad \text{...(ii)}$$

Using $\quad b^2=a^2(1-e^2)$

$$\Rightarrow \quad b^2=a^2-a^2e^2$$

$$\Rightarrow \quad a^2e^2=a^2-b^2$$

$$\Rightarrow \quad a^2e^2=ab^2 \quad \text{[from Eq. (ii)]}$$

On squaring both sides, we get

$$a^2e^4=b^2$$

$$\Rightarrow \quad a^2e^4=a^2(1-e^2) \Rightarrow e^4=1-e^2$$

$$\Rightarrow \quad e^4+e^2-1=0$$

$$\therefore \quad e^2=\frac{-1\pm\sqrt{1+4}}{2}$$

$\because$ e^2 is '+' ve

$$\therefore \quad e^2=\frac{\sqrt{5}-1}{2}$$

$$\therefore \quad e=\sqrt{\frac{\sqrt{5}-1}{2}}$$

(d) If θ be the angle between the normal PG at $P\equiv(a\cos\phi, b\sin\phi)$ and PC, where C is the centre of the ellipse given by $\frac{x^2}{a^2}+\frac{y^2}{b^2}=1$

Then, the equation of normal PG is

$$ax\sec\phi-by\,\text{cosec}\,\phi=a^2-b^2$$

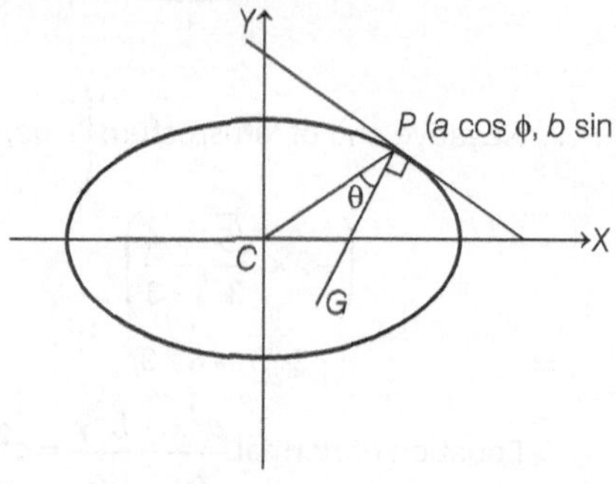

Its slope $\frac{-a\sec\phi}{-b\,\text{cosec}\,\phi}=\frac{a}{b}\tan\phi=m_1$ (say)

The slope of $PC=\frac{b\sin\phi}{a\cos\phi}=\frac{b}{a}\tan\phi=m_2$ (say)

$$\because \text{ Now, } \tan\theta=\frac{m_1-m_2}{1+m_1m_2}=\frac{\frac{a}{b}\tan\phi-\frac{b}{a}\tan\phi}{1+\frac{a}{b}\tan\phi\cdot\frac{b}{a}\tan\phi}$$

$$=\frac{\left(\frac{a}{b}-\frac{b}{a}\right)\tan\phi}{1+\tan^2\phi}$$

$$=\frac{a^2-b^2}{ab}\times\frac{\tan\phi}{\sec^2\phi}$$

$$=\frac{a^2-b^2}{2(ab)}\times 2\sin\phi\cos\phi$$

$$=\frac{a^2-b^2}{2ab}\cdot(\sin 2\phi)$$

$\because$ $\tan\theta$ is maximum, when $\sin 2\phi=1$

$\therefore$ Maximum value of $\tan\theta=\frac{a^2-b^2}{2ab}$

14. (a) (i) Equation of ellipse $E: 2x^2+y^2-3=0$

Now, for point $A(1, 1)$

$$S_1=2(1)^2+(1)^2-3=0$$

$\therefore$ Point A lies on ellipse

$\therefore$ 3 normals can be drawn

(ii) For point $B(0, 1)$

$$S_1=2(0)^2+(1)^2-3=-2<0$$

$\therefore$ Point B lies inside the ellipse

$\therefore$ 4 normals can be drawn.

(iii) For point $C(2, 1)$, $S_1: 2(2^2)+(1)^2-3=6>0$

$\therefore$ Point C lies outside the ellipse

$\therefore$ 2 normals can be drawn.

(b) (i) Point (1, 2)

Ellipse : $2x^2+3y^2=14 \Rightarrow \frac{x^2}{7}+\frac{y^2}{\frac{14}{3}}=1$

$\because$ 'θ' be the eccentric angle of any point in ellipse

$\therefore$ Point is $(a\cos\theta, b\sin\theta)$

Now, $a^2=7$ and $b^2=\frac{14}{3}$

$\therefore$ $a\cos\theta=1$ and $b\sin\theta=2$

$\Rightarrow \quad \sqrt{7}\cos\theta=1, \sqrt{\frac{14}{3}}\sin\theta=2$

$\Rightarrow \quad \cos\theta=\frac{1}{\sqrt{7}}, \sin\theta=\sqrt{\frac{6}{7}}$

$\Rightarrow \quad \theta=\sin^{-1}\left(\sqrt{\frac{6}{7}}\right)$

and $\quad \cos^{-1}\left(\frac{1}{\sqrt{7}}\right)$

(ii) Point $(-1, 3)$

Ellipse : $3x^2+y^2=12$

$\Rightarrow \quad \frac{x^2}{4}+\frac{y^2}{12}=1$

$\Rightarrow \quad a^2=4$ and $b^2=12$

$\Rightarrow \quad a=2$ and $b=2\sqrt{3}$

Now, let 'θ' be the eccentric angle

$\therefore$ $a\cos\theta=-1, \; b\sin\theta=3$

$\Rightarrow \quad 2\cos\theta=-1, \; 2\sqrt{3}\sin\theta=3$

$\Rightarrow \quad \cos\theta=\frac{-1}{2}, \; \sin\theta=\frac{\sqrt{3}}{2}.$

$\Rightarrow \quad \theta=\frac{2\pi}{3}$ or $\theta=\frac{2\pi}{3}.$

(c) Given equation, $\frac{x^2}{a^2}+\frac{y^2}{b^2}=1$

Let $\theta(h, k)$ be any given point and let $P(a\cos\phi, b\sin\phi)$ be any point on the ellipse

$$\frac{x^2}{a^2}+\frac{y^2}{b^2}=1$$

$\therefore$ Equation of normal at $P(a\cos\phi, b\sin\phi)$ is

$ax\sec\phi-by\operatorname{cosec}\phi=a^2-b^2$

It passes through $Q(h, k)$

$\therefore \quad ah\sec\phi-bk\operatorname{cosec}\phi=a^2-b^2$

$$\Rightarrow \quad \frac{ah}{\cos\phi}-\frac{bk}{\sin\phi}=a^2-b^2 \quad \text{...(i)}$$

$$\Rightarrow \quad \frac{\frac{ah}{1-\tan^2\frac{\phi}{2}}}{1+\tan^2\frac{\phi}{2}}-\frac{\frac{bk}{2\tan\frac{\phi}{2}}}{1+\tan^2\frac{\phi}{2}}=a^2-b^2$$

Let $\tan\frac{\phi}{2}=t$

$$\therefore \quad \frac{ah(1+t^2)}{1-t^2}-\frac{bk(1+t^2)}{2t}=(a^2-b^2)$$

$2tah(1+t^2)-bk(1+t^2)(1-t^2)=2t(1-t^2)(a^2-b^2)$

$\Rightarrow \quad 2aht+2aht^3-bk+bkt^4=2ta^2-2tb^2-2a^2t^3+2b^2t^3$

$\Rightarrow bkt^4+2\{ah+(a^2-b^2)\}t^3+2\{ah-(a^2-b^2)\}t-bk=0$ ...(ii)

with is fourth degree equation in t.

$\therefore$ 4 normals can be drawn.

$\therefore$ Has four different values say $\alpha, \beta, \gamma, \delta$

$\therefore \quad t=\frac{\tan\phi}{2}.$

$\therefore \quad t_1=\tan\frac{\alpha}{2}$

$t_2=\tan\frac{\beta}{2}$

$t_3=\tan\frac{\gamma}{2}$

$t_4=\tan\frac{\delta}{2}$

$$\therefore \quad \tan\left(\frac{\alpha}{2}+\frac{\beta}{2}+\frac{\gamma}{2}+\frac{\delta}{2}\right)=\frac{S_1-S_3}{1-S_2+S_4}$$

$\because$ By Eq. (ii)

$$\Rightarrow \quad \Sigma\tan\frac{\alpha}{2}=\frac{-2\{ah+(a^2-b^2)\}}{bk}$$

$$\therefore \quad \Sigma\tan\frac{\alpha}{2}\tan\frac{\beta}{2}=0$$

$$\therefore \; \Sigma\tan\frac{\alpha}{2}\tan\frac{\beta}{2}\tan\frac{\gamma}{2}=\frac{-2\{ah-(a^2-b^2)\}}{bk}$$

$$\therefore \quad \tan\frac{\alpha}{2}\tan\frac{\beta}{2}\tan\frac{\gamma}{2}=\frac{-bk}{bk}=-1$$

$$\therefore \quad \tan\left(\frac{\alpha+\beta+\gamma+\delta}{2}\right)$$

$$=\frac{\frac{-2\{ah+(a^2-b^2)\}}{bk}+\frac{2\{ah-(a^2-b^2)\}}{bk}}{1-0-1}=\infty$$

$$\therefore \quad \tan\left(\frac{\alpha+\beta+\gamma+\delta}{2}\right)=\infty$$

$$\therefore \quad \frac{\alpha+\beta+\gamma+\delta}{2}=\text{odd multiple of }\frac{\pi}{2}$$

$$\Rightarrow \quad \frac{\alpha+\beta+\gamma+\delta}{2}=\frac{(2n+1)\pi}{2}$$

$\Rightarrow \quad \alpha+\beta+\gamma+\delta=(2n+1)\pi, n\in I.$

(d) Let $P(x_1, y_1), Q(x_2, y_2), R(x_3, y_3)$ and $T(x_4, y_4)$ be conormal points so that normal drawn from them meet in $T(h, k)$

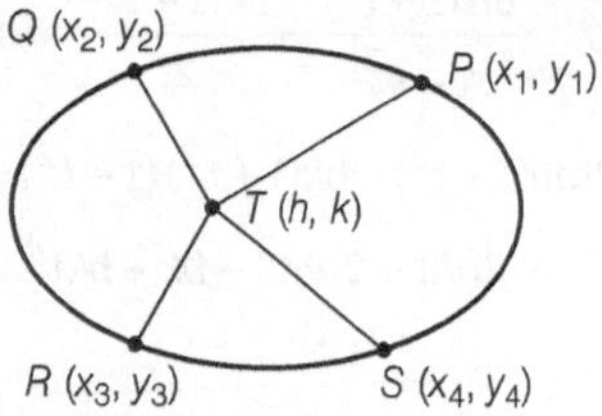

The equation of normal at $P(x_1, y_1)$ is

$$\frac{a^2x}{x_1} - \frac{b^2y}{y_1} = a^2 - b^2$$

$\Rightarrow \quad (a^2 - b^2)\, x_1y_1 + b^2yx_1 - a^2xy_1 = 0$

But the point $T(h, k)$ lies on it.

$\therefore \quad (a^2 - b^2)x_1y_1 + b^2kx_1 - a^2hy_1 = 0$

Similarly, for point Q, R and S are

$$(a^2 - b^2)x_2y_2 + b^2kx_2 - a^2hy_2 = 0$$

$$(a^2 - b^2)x_3y_3 + b^2kx_3 - a^2hy_3 = 0$$

and $(a^2 - b^2)x_4y_4 + b^2kx_4 - a^2hy_4 = 0$

$\therefore$ P, Q, R, S, are lie on the curve

$$(a^2 - b^2)\, xy + b^2kx - a^2hy = 0.$$

15. (a) Point (1, 3)

Ellipse : $4x^2 + 9y^2 = 36$

$\Rightarrow S : 4x^2 + 9y^2 - 36 = 0$

$S_1 : 4(1)^2 + 9(3)^2 - 36$

$\Rightarrow \quad 4 + 81 - 36 = 49$

Equation of Pair of tangents

$\therefore \quad SS_1 = T^2$

$$T = 4x(1) + 9y(3) - 36$$

$$= 4x + 27y - 36$$

$\Rightarrow \quad (4x^2 + 9y^2 - 36)(49) = (4x + 27y - 36)^2$

$\Rightarrow \quad 196x^2 + 441y^2 - 1764 = 16x^2 + 729y^2 + 1296 + 216xy - 1944y - 288x$

$\Rightarrow \quad 180x^2 - 288y^2 - 216xy + 288x + 1944y - 3060 = 0$

$\Rightarrow 90x^2 - 144y^2 - 108xy + 144x + 972y - 1530 = 0$

$\Rightarrow \quad 45x^2 - 72y^2 - 54xy + 72x + 486y - 765 = 0$

(b) Point (1, 2)

Ellipse $S : 3x^2 + 2y^2 - 5 = 0$

$$S_1 = 3(1)^2 + 2(2)^2 - 5 = 3 + 8 - 5 = 6.$$

$$T = 3x(1) + 2y(2) - 5 = 3x + 4y - 5$$

$\therefore$ Equation of pair of tangents

$$SS_1 = T^2$$

$$6(3x^2 + 2y^2 - 5) = (3x + 4y - 5)^2$$

$\Rightarrow \quad 18x^2 + 12y^2 - 30 = 9x^2 + 16y^2 + 24xy - 40y - 30x + 25$

$\Rightarrow 9x^2 - 4y^2 - 24xy + 30x + 40y - 55 = 0$

$$a = 9, b = -4 \text{ and } h = -12$$

Let θ be the angle between tangents

$$\therefore \quad \tan\theta = \left|\frac{2\sqrt{h^2 - ab}}{a + b}\right|$$

$$\tan\theta = 2\left|\frac{\sqrt{144 + 36}}{9 - 4}\right| = \frac{2 \times \sqrt{180}}{5} \times \frac{\sqrt{5}}{\sqrt{5}}$$

$$= \frac{2 \times 30}{5\sqrt{5}} = \frac{12}{\sqrt{5}}$$

$$\therefore \quad \theta = \tan^{-1}\left(\frac{12}{\sqrt{5}}\right)$$

(c) Point $(\lambda, 1)$

Ellipse $S : x^2 + 2y^2 - 10 = 0$

$\Rightarrow \quad S_1 = \lambda^2 + 2(1) - 10 = \lambda^2 - 8$

$\Rightarrow \quad T = x(\lambda) + 2y(1) - 10$

$\Rightarrow \quad x\lambda + 2y - 10 = 0$

Equation of pair of tangents

$\therefore \quad SS_1 = T^2$

$\Rightarrow (x^2 + 2y^2 - 10)(\lambda^2 - 8) = (x\lambda + 2y - 10)^2$

$\Rightarrow \lambda^2x^2 + 2\lambda^2y^2 - 10\lambda^2 - 8x^2 - 16y^2 + 80$

$= x^2\lambda^2 + 4y^2 + 100 + 4xy\lambda - 40y - 20x\lambda$

$\Rightarrow 8x^2 + y^2(20 - 2\lambda^2) + 4xy\lambda + 10\lambda^2 + 8x^2 - 80 + 100 - 40y - 20x\lambda = 0$

$\therefore \quad a = 8$ and $b = 20 - 2\lambda^2$

$\therefore$ Tangents are mutually perpendicular

$$a + b = 0$$

$\Rightarrow \quad 8 + 20 - 2\lambda^2 = 0$

$\Rightarrow \quad 2\lambda^2 = 28$

$\Rightarrow \quad \lambda^2 = 14$

$\Rightarrow \quad \lambda = \pm\sqrt{14}$

16. $\because$ We know that $P(\alpha)$ and $P(\beta)$ are two point of contacts where tangents are drawn and intersect at $R(h, k)$.

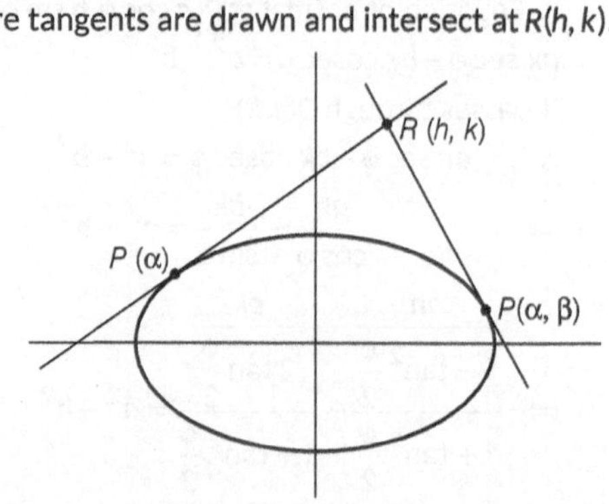

Then, $R(h,k)=\left\{\dfrac{a\cos\left(\dfrac{\alpha+\beta}{2}\right)}{\cos\left(\dfrac{\alpha-\beta}{2}\right)},\dfrac{b\sin\left(\dfrac{\alpha+\beta}{2}\right)}{\cos\left(\dfrac{\alpha-\beta}{2}\right)}\right\}$

$\because$ Given, ellipse $x^2+2y^2=1 \Rightarrow \dfrac{x^2}{1}+\dfrac{y^2}{\frac{1}{2}}=1.$

$\Rightarrow \quad a^2=1,\ b^2=\dfrac{1}{2}$

$\Rightarrow \quad a=1,\ b=\dfrac{1}{\sqrt{2}}$

and given $|\alpha-\beta|=\left(\dfrac{2\pi}{3}\right)$

$\therefore \quad h=\dfrac{\cos\left(\dfrac{\alpha+\beta}{2}\right)}{\cos\dfrac{2\pi}{3}}$

$\Rightarrow \quad h=\dfrac{\cos\left(\dfrac{\alpha+\beta}{2}\right)}{-\dfrac{1}{2}}$

$\Rightarrow \quad -\dfrac{h}{2}=\cos\left(\dfrac{\alpha+\beta}{2}\right)$

and $\quad k=-2\left(\dfrac{1}{\sqrt{2}}\right)\sin\left(\dfrac{\alpha+\beta}{2}\right)$

$\Rightarrow \quad \sin\left(\dfrac{\alpha+\beta}{2}\right)=\dfrac{-k}{\sqrt{2}}$

$\therefore \quad \sin^2\left(\dfrac{\alpha+\beta}{2}\right)+\cos^2\left(\dfrac{\alpha+\beta}{2}\right)=1$

$\Rightarrow \quad \left(-\dfrac{h}{2}\right)^2+\left(\dfrac{-k}{\sqrt{2}}\right)^2=1$

$\Rightarrow \quad \dfrac{h^2}{4}+\dfrac{k^2}{2}=1$

$\Rightarrow \quad h^2+2k^2=4$

$\therefore$ Locus is $x^2+2y^2=4$

17. (a) Ellipse, $S: x^2+4y^2-1=0$

Point $P\ (2,1)$

$\because S_1: 2^2+4(1)^2-1$

$\Rightarrow 4+4-1=7>0$

$\therefore$ P lies outside the ellipse.

Now, equation of chord of contact from an external point

$T=0$

$\Rightarrow \quad x(2)+4y(1)-1=0$

$\Rightarrow \quad 2x+4y-1=0$

(b) Ellipse : $\dfrac{x^2}{2}+y^2=1$

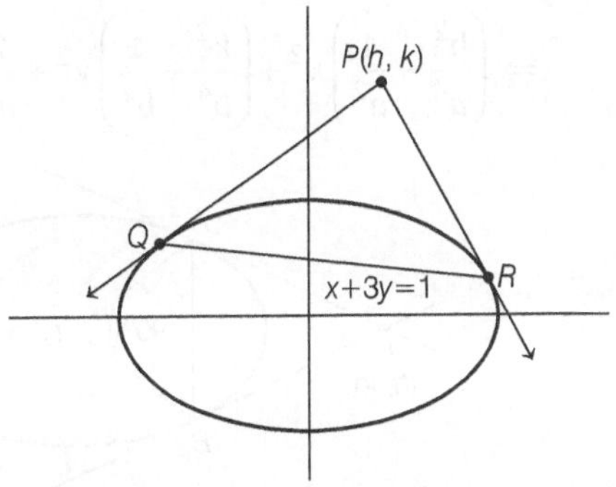

Let coordinates of $P(h,k)$ the equation of QR (i.e. chord of contact)

$T=0$

$\dfrac{hx}{2}+ky=1 \quad$ ...(i)

But it given as $x+3y=1 \quad$ ...(ii)

Eqs. (i) and (ii) are identical

$\Rightarrow \quad \dfrac{\frac{h}{2}}{1}=\dfrac{k}{3}=1$

$\Rightarrow \quad h=2$ and $k=3$

$\therefore$ Coordinates of P are $(2,3)$

(c) Let any pont on line $x-y-5=0$ will be of the form $(t,t-5)$, where $t\in R$.

Chord of contact of point w.r.t. ellipse $x^2+4y^2=4$

$\Rightarrow \quad tx+4(t-5)y-4=0$

$\Rightarrow \quad -20y-4+t(x+4y)=0$

that represents a family of straight lines, each member of this family passes through the point of intersection of straight lines

$-20y-4=0,\quad x+4y=0$

$\Rightarrow \quad y=-\dfrac{4}{20},\ x=-4y$

$\Rightarrow \quad y=-\dfrac{1}{5}\ \Rightarrow x=\dfrac{4}{5}$

$\therefore$ Required point is $\left(\dfrac{4}{5},\dfrac{-1}{5}\right)$.

18. (a) $\because$ The equation of chord of contact of tangents drawn from $P(h,k)$ to the ellipse $\dfrac{x^2}{a^2}+\dfrac{y^2}{b^2}=1$ is

$\dfrac{hx}{a^2}+\dfrac{ky}{b^2}=1 \quad$ ...(i)

The equation of the straight lines CA and CB is obtained by making homogeneous ellipse $\dfrac{x^2}{a^2}+\dfrac{y^2}{b^2}=1$ with the help of Eq. (i).

$$\therefore \quad \frac{x^2}{a^2}+\frac{y^2}{b^2}=\left(\frac{hx}{a^2}+\frac{ky}{b^2}\right)^2$$

$$\Rightarrow \left(\frac{h^2}{a^4}-\frac{1}{a^2}\right)x^2+\left(\frac{k^2}{b^4}-\frac{1}{b^2}\right)y^2+\frac{2hk}{a^2b^2}xy=0$$

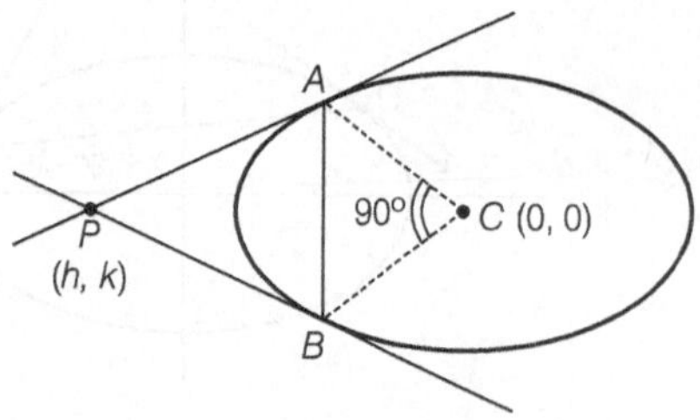

$$\Rightarrow \quad \frac{h^2}{a^4}-\frac{1}{a^2}+\frac{k^2}{b^4}-\frac{1}{b^2}=0$$

$$\Rightarrow \quad \frac{h^2}{a^4}+\frac{k^2}{b^4}=\frac{1}{a^2}+\frac{1}{b^2}$$

$\therefore$ Locus is $\frac{x^2}{a^4}+\frac{y^2}{b^4}=\frac{1}{a^2}+\frac{1}{b^2}$

(b) Let verticies of ΔABC be points on the ellipse whose eccentric angles are $\theta_1, \theta_2, \theta_3$ respectively.

Chord AB is

$$\frac{x}{a}\cos\left(\frac{\theta_1+\theta_2}{2}\right)+\frac{y}{b}\sin\left(\frac{\theta_1+\theta_2}{2}\right)$$

$$=\cos\left(\frac{\theta_1-\theta_2}{2}\right) \quad \text{...(i)}$$

$\because$ Slope of chord $=-\frac{b}{a}\cot\left(\frac{\theta_1+\theta_2}{2}\right)=$ constant

Similarly, $\theta_2+\theta_3=2\beta$

$\Rightarrow \quad \theta_1+\theta_2=2\alpha$ (assume)

$\therefore \quad \theta_1-\theta_3=2(\alpha-\beta)$.

Chord CA (third side)

$$\Rightarrow \frac{x}{a}\cos\left(\frac{\theta_1+\theta_3}{2}\right)+\frac{y}{b}\sin\left(\frac{\theta_1+\theta_3}{2}\right)$$

$$=\cos\left(\frac{\theta_1-\theta_3}{2}\right) \quad \text{...(ii)}$$

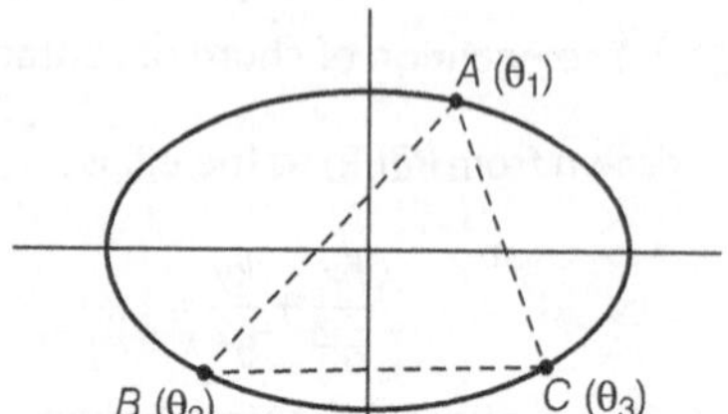

By Eqs. (i) or (ii) we can say

$$\Rightarrow \frac{x}{a}\cos\theta+\frac{y}{b}\sin\theta=\cos(\alpha-\beta)=\lambda$$

$$\Rightarrow \frac{x}{a\lambda}\cos\theta+\frac{y}{b\lambda}\sin\theta=1$$

That represents as tangent to ellipse

$$\frac{x^2}{a^2\lambda^2}+\frac{y^2}{b^2\lambda^2}=1$$

19. (a) Given, ellipse $S: x^2+2y^2-16=0$

Middle point of chord = (1, 2)

$\therefore$ Equation of chord $T=S_1$

$\Rightarrow \quad x(1)+2y(2)-16=(1)^2+2(2)^2-16$

$\Rightarrow \quad x+4y-16=1+8-16 \Rightarrow x+4y=9$

(b) Given, ellipse $S: x^2+4y^2-9=0$ and given external point (3, 2)

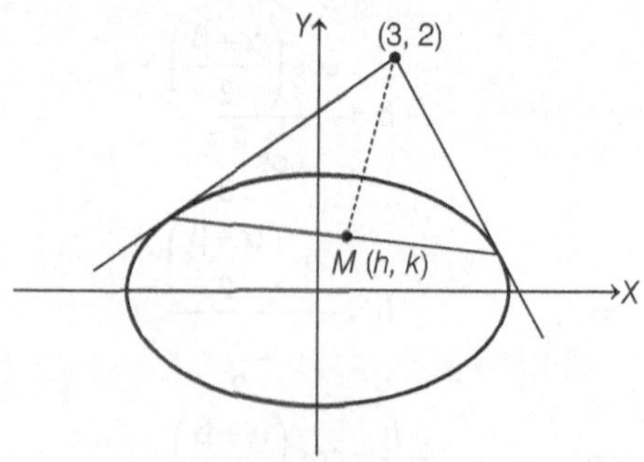

$\therefore$ Equation of chord of contact $T=0$

$$x(3)+4y(2)-9=0$$

$$3x+8y-9=0 \quad \text{...(i)}$$

Let mid-point of chord is $M(h, k)$

$\therefore$ Equation of same chord is $T=S_1$

$\Rightarrow \quad xh+4yk-9=h^2+4k^2-9$

$\Rightarrow \quad xh+4yk-(h^2+4k^2)=0 \quad \text{...(ii)}$

Eqs. (i) and (ii) are identical

$$\therefore \quad \frac{3}{h}=\frac{8}{4k}=\frac{-9}{-(h^2+4k^2)}$$

$$2h=3k \quad \text{and} \quad 3h=h^2+4k^2$$

$$\Rightarrow \quad 3h=h^2+4\times\frac{4h^2}{9} \Rightarrow \frac{25h^2}{9}=3h$$

$$\Rightarrow \quad h=\frac{27}{25} \text{ and } k=\frac{18}{25}$$

$\therefore$ Mid-point is $\left(\frac{27}{25}, \frac{18}{25}\right)$

(c) $\because$ The gives ellipse, $S: \frac{x^2}{25}+\frac{y^2}{16}-1=0$

Middle-point of chord $=\left(\frac{1}{2}, \frac{2}{5}\right)$

$\therefore$ Equation of chord having $\left(\frac{1}{2}, \frac{2}{5}\right)$ as mid-point is

$$T=S_1$$

$$\Rightarrow \frac{x\left(\frac{1}{2}\right)}{25}+\frac{y\left(\frac{2}{5}\right)}{16}-1=\frac{\left(\frac{1}{2}\right)^2}{25}+\frac{\left(\frac{2}{5}\right)^2}{16}-1$$

$$\Rightarrow \frac{x}{50}+\frac{y}{40}=\frac{1}{100}+\frac{1}{100}$$

$$\Rightarrow \frac{4x+5y}{200}=\frac{2}{100}$$

$$\Rightarrow 4x+5y=4$$

$$\Rightarrow 5y=4(1-x) \quad \text{...(i)}$$

Solving with ellipse

$$16x^2+25y^2-400=0$$

$$\Rightarrow 16x^2+[4(1-x)]^2-400=0 \quad \text{[from Eq. (i)]}$$

$$\Rightarrow 16x^2+16(x^2+1-2x)-400=0$$

$$\Rightarrow x^2+x^2+1-2x-25=0$$

$$\Rightarrow 2x^2-2x-24=0$$

$$\Rightarrow x^2-x-12=0$$

$$\Rightarrow (x-4)(x+3)=0$$

For $x-4=0$ and $x+3=0$

$$x=4 \text{ or } x=-3$$

$$\therefore \quad y=\frac{4(1-4)}{5}=\frac{-12}{5}$$

and

$$y=\frac{4(1+3)}{5}=\frac{16}{5}$$

$\therefore$ Point $\left(4, \frac{-12}{5}\right)$ or $\left(-3, \frac{16}{5}\right)$

$$\therefore \text{ Length of chord}=\sqrt{(4+3)^2+\left(\frac{16}{5}+\frac{12}{5}\right)^2}$$

$$=7\sqrt{\frac{41}{25}}=\frac{7}{5}\sqrt{41} \text{ units}$$

20. (a) Ellipse, $S: \frac{x^2}{a^2}+\frac{y^2}{b^2}-1=0$

Let the mid-point of the focal chord of the given ellipse is (h, k)

Then, equation of chord is $T=S_1$

$$\Rightarrow \frac{x(h)}{a^2}+\frac{y(k)}{b^2}-1=\frac{h^2}{a^2}+\frac{k^2}{b^2}-1$$

$$\Rightarrow \frac{hx}{a^2}+\frac{ky}{b^2}=\frac{h^2}{a^2}+\frac{k^2}{b^2} \quad \text{...(i)}$$

$\because$ It passes through focus $(ae, 0)$.

$$\therefore \frac{h(ae)}{a^2}+\frac{k(0)}{b^2}=\frac{h^2}{a^2}+\frac{k^2}{b^2}$$

$$\Rightarrow \frac{he}{a}=\frac{h^2}{a^2}+\frac{k^2}{b^2}$$

$\therefore$ Locus is $\frac{x^2}{a^2}+\frac{y^2}{b^2}=\frac{he}{a}$.

(b) Let the point P be $(a\cos\theta, b\sin\theta)$

$\therefore$ The equation of normal at P to the ellipse is

$$ax\sec\theta-by\,\text{cosec}\,\theta=a^2-b^2 \quad \text{...(i)}$$

Let the mid-point be $M(h, k)$

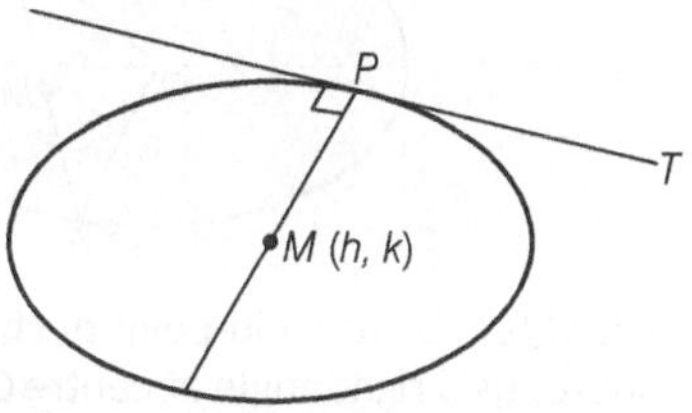

$\therefore$ Equation of chord bisected at (h, k) is

$$T=S_1$$

$$\frac{xh}{a^2}+\frac{yk}{b^2}=\frac{h^2}{a^2}+\frac{k^2}{b^2} \quad \text{...(ii)}$$

$\because$ Eqs. (i) and (ii) are identical

$$\therefore \frac{a\sec\theta}{\frac{h}{a^2}}=\frac{-b\,\text{cosec}\,\theta}{\frac{k}{b^2}}=\frac{a^2-b^2}{\frac{h^2}{a^2}+\frac{k^2}{b^2}}$$

From first and last

$$\therefore \cos\theta=\frac{\left(\frac{h^2}{a^2}+\frac{k^2}{b^2}\right)}{\frac{h}{a^3(a^2-b^2)}}$$

From second and last

$$\sin\theta=\frac{\left(\frac{h^2}{a^2}+\frac{k^2}{b^2}\right)}{-\frac{k}{b^3(a^2-b^2)}}$$

$\because \sin^2\theta+\cos^2\theta=1$

$$\Rightarrow \frac{\left(\frac{h^2}{a^2}+\frac{k^2}{b^2}\right)^2}{\frac{k^2}{b^6(a^2-b^2)^2}}+\frac{\left(\frac{h^2}{a^2}+\frac{k^2}{b^2}\right)^2}{\frac{h^2}{a^6(a^2-b^2)^2}}=1$$

$$\Rightarrow \left(\frac{h^2}{a^2}+\frac{k^2}{b^2}\right)^2 \cdot \left\{\frac{b^6}{k^2}+\frac{a^6}{h^2}\right\}=\frac{1}{(a^2-b^2)^2}$$

$$\Rightarrow \therefore \text{ Locus is } \left(\frac{a^6}{x^2}+\frac{b^6}{y^2}\right)\left(\frac{x^2}{a^2}+\frac{y^2}{b^2}\right)^2=(a^2-b^2)^{-2}$$

(c) Given, ellipse $\frac{x^2}{a^2}+\frac{y^2}{b^2}=1$ ...(i)

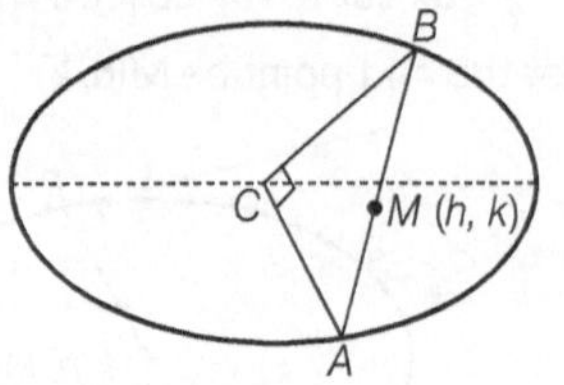

Let $M(h, k)$ be the mid-point of chord AB which subtends a right angle at centre $C(0, 0)$.
Now, equation of chord bisected at $M(h, k)$

$$\Rightarrow \quad T=S_1$$

$$\Rightarrow \quad \frac{hx}{a^2}+\frac{ky}{b^2}=\frac{h^2}{a^2}+\frac{k^2}{b^2}$$

$$\Rightarrow \quad \frac{\left(\frac{hx}{a^2}+\frac{ky}{b^2}\right)}{\frac{h^2}{a^2}+\frac{k^2}{b^2}}=1 \quad \text{...(ii)}$$

∵ Centre is (0, 0)
So, we have to use homogenisation to get pair of straight line passing through origin.
∴ Eq. (i) becomes

$$\Rightarrow \quad \left(\frac{x^2}{a^2}+\frac{y^2}{b^2}\right)=\left(\frac{\frac{hx}{a^2}+\frac{ky}{b^2}}{\frac{h^2}{a^2}+\frac{k^2}{b^2}}\right)^2$$

$$\Rightarrow \quad \left(\frac{x^2}{a^2}+\frac{y^2}{b^2}\right)\left(\frac{h^2}{a^2}+\frac{k^2}{b^2}\right)^2=\left(\frac{hx}{a^2}+\frac{ky}{b^2}\right)^2$$

$$=\frac{h^2x^2}{a^4}+\frac{k^2y^2}{b^4}+\frac{2hxky}{a^2b^2}$$

∵ Chord subtends right angle at (0, 0)
∴ Coefficient of x^2 + Coefficient of $y^2=0$

$$\therefore \quad \left(\frac{h^2}{a^2}+\frac{k^2}{b^2}\right)^2\cdot\left(\frac{1}{a^2}+\frac{1}{b^2}\right)=\frac{h^2}{a^4}+\frac{k^2}{b^4}$$

$$\therefore \text{ Locus is } \left(\frac{x^2}{a^2}+\frac{y^2}{b^2}\right)^2\left(\frac{1}{a^2}+\frac{1}{b^2}\right)=\frac{x^2}{a^4}+\frac{y^2}{b^4}$$

21. (a) ∵ We know that the locus of the mid-point of all parallel chord of an ellipse is diameter of ellipse
Let $P(h, k)$ be the mid-point of chord $y=mx+c$ of the ellipse

$$\frac{x^2}{a^2}+\frac{y^2}{b^2}=1$$

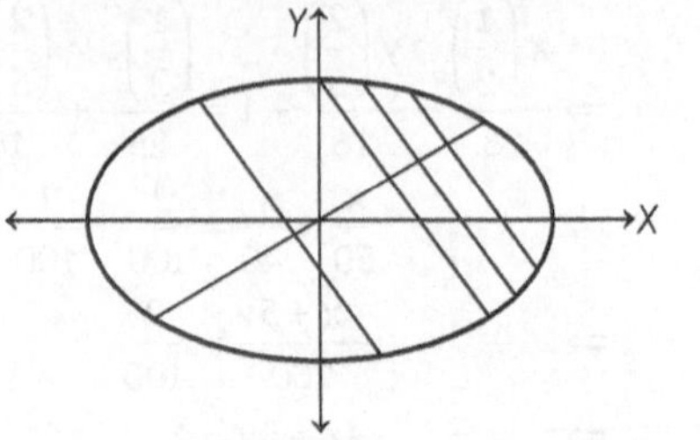

The equation of chord when it bisected at $P(h, k)$

$$T=S_1$$

$$\Rightarrow \quad \frac{xh}{a^2}+\frac{ky}{b^2}=\frac{h^2}{a^2}+\frac{k^2}{b^2}$$

$$\therefore \quad \text{Slope is}=\left(\frac{-h}{a^2}\right)\left(\frac{b^2}{k}\right)$$

$$\text{given,} \quad m=\frac{-h}{a^2}\cdot\frac{b^2}{k}$$

$$\Rightarrow \quad k=-\frac{b^2}{a^2m}\cdot h$$

$$\therefore \text{ Locus is } y=\left(\frac{-b^2}{a^2m}\right)x$$

i.e. equation of diameter.

(b) Two diameters are said to be conjugate when each bisects all chords parallel to the other.
It $y=m_1x$ and $y=m_2x$ be two conjugate diameters of an ellipse, then $m_1m_2=\frac{-b^2}{a^2}$.

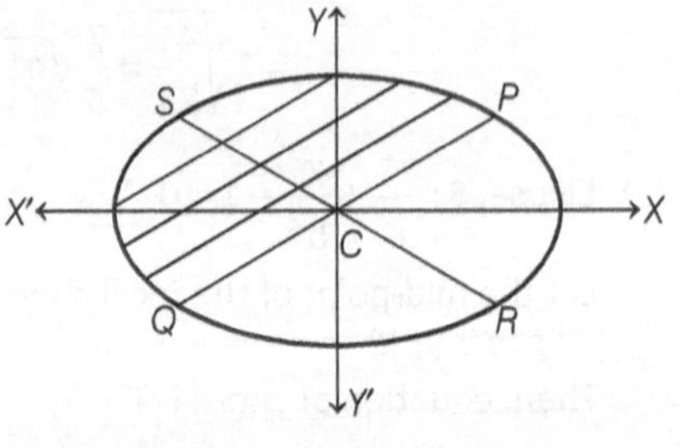

Let PQ and RS be two conjugate diameters. Then, the coordiantes of four exetremities of two conjugate diameters, are
$P(a\cos\theta, b\sin\theta)$, $Q(-a\cos\theta, b\sin\theta)$,
$S(-a\sin\theta, b\cos\theta)$
and $R(a\sin\theta, -b\cos\theta)$

(c) The equation of given diameter, is

$$ax-by=0$$

$$\Rightarrow \quad y=\frac{a}{b}x \quad \text{...(i)}$$

$$\text{Thus,} \quad m_1=\frac{a}{b}$$

Let the diameter conjugate to (i) be $y = m_2 x$.

$\because$ We know that $m_1 m_2 = \dfrac{-b^2}{a}$

$\Rightarrow \quad \dfrac{a}{b} m_2 = \dfrac{-b^2}{a^2} \Rightarrow m_2 = \dfrac{-b^3}{a^3}$

$\because$ The required equation is $y = \dfrac{-b^3}{a^3} x$

$\Rightarrow \quad a^3 y + b^3 x = 0$

(d) Given, equation of diameter are

$$y = x \text{ and } 2x + 3y = 0$$

$$\Rightarrow \quad y = \frac{-2}{3} x$$

$$m_1 = 1 \text{ and } m_2 = -\frac{2}{3}$$

$$\because \quad m_1 m_2 = \frac{-2}{3}$$

$$\Rightarrow \quad \frac{-b^2}{a^2} = \frac{-2}{3}$$

$$\Rightarrow \quad \frac{b^2}{a^2} = \frac{2}{3}$$

$$\therefore \quad e = \sqrt{1 - \frac{b^2}{a^2}} = \sqrt{1 - \frac{2}{3}} = \frac{1}{\sqrt{3}}$$

22. (a) Let CP and CD be two conjugate semi-diameters of an ellipse $\dfrac{x^2}{a^2} + \dfrac{y^2}{b^2} = 1$ and let eccentric angle of P is θ, then eccentric angle of D is $\left(\dfrac{\pi}{2} + \theta\right)$.

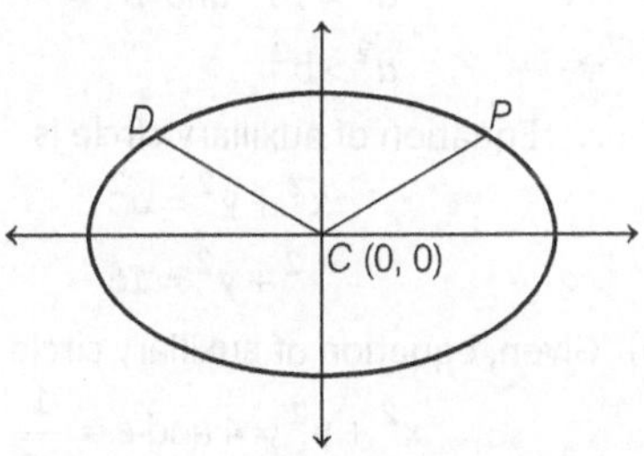

Therefore, coordiantes of P and D are

$$P = (a \cos \theta, b \sin \theta)$$

and $\quad D = \left\{ a \cos\left(\dfrac{\pi}{2} + \theta\right), b \sin\left(\dfrac{\pi}{2} + \theta\right) \right\}$

$$\Rightarrow \quad D = (-a \sin \theta, b \cos \theta)$$

Now, $CP^2 + CD^2$

$$= a^2 \cos^2 \theta + b^2 \sin^2 \theta + a^2 \sin^2 \theta + b^2 \cos^2 \theta$$

$$= a^2 (\cos^2 \theta + \sin^2 \theta) + b^2 (\sin^2 \theta + \cos^2 \theta)$$

$$= a^2 + b^2$$

(b) Given, equation of ellipse $\dfrac{x^2}{a^2} + \dfrac{y^2}{b^2} = 1$

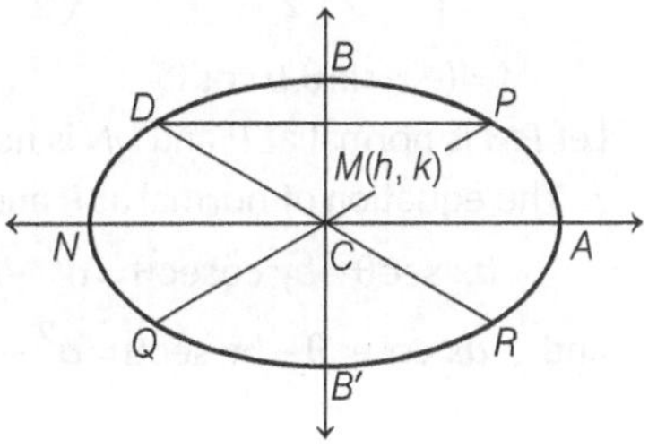

Let the eccentric angle of P is θ

$\Rightarrow P (a \cos \theta, b \sin \theta)$

$\therefore$ Eccentric angle of Q is $\left(\dfrac{\pi}{2} + \theta\right)$

$$\therefore \quad D = \left\{ a \cos\left(\frac{\pi}{2} + \theta\right), b \sin\left(\frac{\pi}{2} + \theta\right) \right\}$$

$$= (-a \sin \theta, b \cos \theta)$$

Let $M = (h, k)$ be the mid-point of PD

$$\therefore \quad h = \frac{a \cos \theta - a \sin \theta}{2}$$

$$\text{and} \quad k = \frac{b \sin \theta + b \cos \theta}{2}$$

$$\frac{2h}{a} = \cos \theta - \sin \theta \qquad \text{...(i)}$$

$$\text{and} \quad \frac{2k}{b} = \sin \theta + \cos \theta \qquad \text{...(ii)}$$

Squaring Eqs. (i) and (ii) and adding, we get

$$\Rightarrow \quad \left(\frac{2h}{a}\right)^2 + \left(\frac{2k}{b}\right)^2 = \cos^2 \theta + \sin^2 \theta$$

$$-2 \sin \theta \cos \theta + \sin^2 \theta + \cos^2 \theta + 2 \sin \theta \cos \theta$$

$$\Rightarrow \quad \frac{4h^2}{a^2} + \frac{4k^2}{b^2} = 2$$

$$\Rightarrow \quad \frac{h^2}{a^2} + \frac{k^2}{b^2} = \frac{1}{2}$$

$\therefore$ Locus is $\dfrac{x^2}{a^2} + \dfrac{y^2}{b^2} = \dfrac{1}{2}$.

(c) Ellipse : $\dfrac{x^2}{a^2} + \dfrac{y^2}{b^2} = 1$

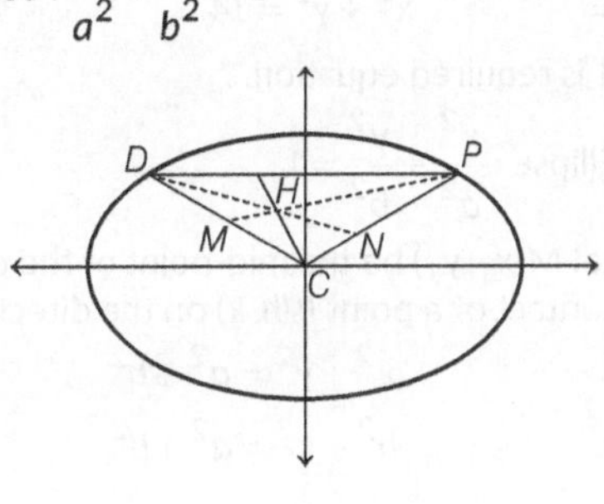

Let $P(a\cos\theta, b\sin\theta)$

$$\Rightarrow \quad D=\left\{a\cos\left(\frac{\pi}{2}+\theta\right), b\sin\left(\frac{\pi}{2}+\theta\right)\right\}$$

$$=(-a\sin\theta, b\cos\theta)$$

Let PM is normal at P and DN is normal at D.

$\therefore$ The equation of normal at P and D are

$$ax\sec\theta - by\,\text{cosec}\,\theta = a^2-b^2 \quad \text{...(i)}$$

and $$ax\,\text{cosec}\,\theta - by\sec\theta = a^2-b^2 \quad \text{...(ii)}$$

$\{\theta$ replace $\frac{\pi}{2}+\theta\}$

$\because H(\alpha,\beta)$ be point of intersection of normals i.e. orthocentre

$$a\alpha\sec\theta - b\beta\,\text{cosec}\,\theta = a^2-b^2$$

$$a\alpha\,\text{cosec}\,\theta - b\beta\sec\theta = a^2-b^2$$

$$\Rightarrow \frac{\sec\theta}{a\alpha-b\beta}=\frac{\text{cosec}\,\theta}{-(a\alpha+b\beta)}=\frac{a^2-b^2}{(a^2\alpha^2+b^2\beta^2)}$$

$$\Rightarrow \cos\theta=\frac{a^2\alpha^2+b^2\beta^2}{(a^2-b^2)(a\alpha-b\beta)},$$

and $$\sin\theta=\frac{-(a^2\alpha^2+b^2\beta^2)}{(a^2-b^2)(a\alpha+b\beta)}$$

$$\because \quad \sin^2\theta+\cos^2\theta=1$$

$$\Rightarrow \frac{(a^2\alpha^2+b^2\beta^2)^2}{(a^2-b^2)^2}\left\{\frac{1}{(a\alpha-b\beta)^2}+\frac{1}{(a\alpha+b\beta)^2}\right\}=1$$

$$\Rightarrow (a^2\alpha^2+b^2\beta^2)^2(2a^2\alpha^2+2b^2\beta^2)$$
$$=(a^2-b^2)^2(a\alpha-b\beta)^2(a\alpha+b\beta)^2$$

$\therefore$ Locus is

$$2(a^2x^2+b^2y^2)^3=(a^2-b^2)^2(a^2x^2-b^2y^2)^2$$

23. (a) Given, ellipse $5x^2+9y^2=45$

$$\Rightarrow \quad \frac{x^2}{9}+\frac{y^2}{5}=1$$

$$\because \quad a^2=9 \text{ and } b^2=5$$

$\because$ Equation of director circle is

$$x^2+y^2=a^2+b^2$$

$$\therefore \quad x^2+y^2=9+5$$

$$\Rightarrow \quad x^2+y^2=14$$

It is required equation.

(b) Ellipse : $\frac{x^2}{a^2}+\frac{y^2}{b^2}=1.$

Let $M(x_1, y_1)$ be the mid-point of the chord of contact of a point $P(h, k)$ on the director circle

$$x^2+y^2=a^2+b^2$$

$$\therefore \quad h^2+k^2=a^2+b^2 \quad \text{...(i)}$$

and $$\frac{xx_1}{a^2}+\frac{yy_1}{b^2}=\frac{x_1^2}{a^2}+\frac{y_1^2}{b^2} \quad \text{...(ii)}$$

{equation of chord}

Also, the chord of contact of $P(h, k)$ w.r.t. ellipse is

$$\frac{hx}{a^2}+\frac{ky}{b^2}=1 \quad \text{...(iii)}$$

By Eqs. (ii) and (iii), we get

$$\frac{\frac{h}{a^2}}{\frac{x_1}{a^2}}=\frac{\frac{k}{b^2}}{\frac{y_1}{b^2}}=\frac{1}{\frac{x_1^2}{a^2}+\frac{y_1^2}{b^2}}$$

$$\Rightarrow \quad \frac{h}{x_1}=\frac{k}{y_1}=\frac{1}{\left(\frac{x_1}{a}\right)^2+\left(\frac{y_1}{b}\right)^2}$$

$$\therefore \quad h=\frac{x_1}{\frac{x_1^2}{a^2}+\frac{y_1^2}{b^2}} \text{ and } k=\frac{y_1}{\frac{x_1^2}{a^2}+\frac{y_1^2}{b^2}}$$

Substituting the values of h, k in Eq. (i), we get

$$\frac{x_1^2}{\left(\frac{x_1^2}{a^2}+\frac{y_1^2}{b^2}\right)^2}+\frac{y_1^2}{\left(\frac{x_1^2}{a^2}+\frac{y_1^2}{b^2}\right)^2}=a^2+b^2$$

$\therefore$ Locus at (x, y) is

$$x^2+y^2=(a^2+b^2)\left(\frac{x^2}{a^2}+\frac{y^2}{b^2}\right)^2$$

24. (a) Given, equation of ellipse

$$\frac{x^2}{16}+\frac{y^2}{9}=1$$

$$\because \quad a^2=16 \text{ and } b^2=9$$

$$\because \quad a^2>b^2$$

$\therefore$ Equation of auxiliary circle is

$$x^2+y^2=a^2$$

$$\Rightarrow \quad x^2+y^2=16$$

(b) Given, equation of auxiliary circle

$$x^2+y^2=4 \text{ and } e=\frac{1}{\sqrt{2}}$$

Case I When $a>b$

$$\therefore \quad a^2=4$$

$$\therefore \quad e=\sqrt{1-\frac{b^2}{a^2}} \Rightarrow e^2=1-\frac{b^2}{a^2}$$

$$\Rightarrow \quad b^2=a^2(1-e^2)$$

$$\Rightarrow \quad b^2=4\left(1-\frac{1}{2}\right)=2$$

$\therefore$ Equation of ellipse is $\frac{x^2}{4}+\frac{y^2}{2}=1$

Case II When $a<b$

$\therefore \quad b^2 = 4$

$\therefore \quad a^2 = b^2(1-e^2) = 4\left(1-\frac{1}{2}\right) = 2$

$\therefore$ Equation is $\frac{x^2}{2}+\frac{y^2}{4}=1$

(c) Given, equation of ellipse is $\frac{x^2}{a^2}+\frac{y^2}{b^2}=1$

$\therefore$ Its auxiliary circle $\Rightarrow x^2+y^2=a^2$

Let $P(a\cos\alpha, b\sin\alpha)$ be the point on ellipse

$\therefore \quad Q \Rightarrow (a\cos\alpha, a\sin\alpha)$

Now, equation of normal at P

$$ax\sec\alpha - by\,\text{cosec}\,\alpha = a^2-b^2 \quad \text{...(i)}$$

$\therefore$ Equation of normal at Q

$$y = (\tan\alpha)x \quad \text{...(ii)}$$

Let (h, k) be the point of intersection of Eqs. (i) and (ii), we get

$\therefore \quad k = \tan\alpha h$

$\Rightarrow \quad \tan\alpha = \frac{k}{h}$

$\sin\alpha = \frac{k}{\sqrt{h^2+k^2}}$

and $\quad \cos\alpha = \frac{h}{\sqrt{h^2+k^2}}$

$\because$ (h, k) also lies on Eq. (i)

$\therefore \quad ah\left(\frac{\sqrt{h^2+k^2}}{h}\right) - bk\left(\frac{\sqrt{h^2+k^2}}{k}\right) = a^2-b^2$

$\Rightarrow \quad (a-b)(\sqrt{h^2+k^2}) = a^2-b^2$

$\Rightarrow \quad \sqrt{h^2+k^2} = a+b$

$\Rightarrow \quad h^2+k^2 = (a+b)^2$

$\therefore$ Locus is $x^2+y^2=(a+b)^2$

25. (a) Let the pole be (h, k)

The equation of polar w.r.t. the ellipse $x^2+4y^2=4$ is

$$hx+4ky=4$$

which is identical with

$$x+4y=4$$

$\therefore \quad \frac{h}{1}=\frac{4k}{4}=\frac{4}{4}$

$h=1$ and $k=1$

Hence, the pole is $(1, 1)$.

(b) Given, equation of ellipse $\frac{x^2}{a^2}+\frac{y^2}{b^2}=1$

Let (h, k) be the pole.

$\therefore$ Equation of polar is

$$\frac{hx}{a^2}+\frac{ky}{b^2}=1$$

It is normal to ellipse

$\therefore$ It is identical with $ax\sec\theta - by\,\text{cosec}\,\theta = a^2-b^2$

$\therefore \quad \frac{\frac{h}{a^2}}{a\sec\theta} = \frac{\frac{k}{b^2}}{-b\,\text{cosec}\,\theta} = \frac{1}{a^2-b^2}$

$\Rightarrow \quad \cos\theta = \frac{a^3}{h(a^2-b^2)}$ and $\sin\theta = \frac{-b^3}{k(a^2-b^2)}$

$\because \sin^2\theta+\cos^2\theta=1$

$\Rightarrow \quad \frac{b^6}{k^2(a^2-b^2)^2}+\frac{a^6}{h^2(a^2-b^2)^2}=1$

$\therefore$ Locus is $\left(\frac{a^6}{x^2}+\frac{b^6}{y^2}\right)=(a^2-b^2)^2$

(c) Given, ellipse is $\frac{x^2}{25}+\frac{y^2}{16}=1$

$a^2=25$ and $b^2=16$

$\therefore \quad e=\sqrt{1-\frac{16}{25}}=\frac{3}{5} \quad \left[\because e=\sqrt{1-\frac{b^2}{a^2}}\right]$

$\therefore$ Focus is $(ae, 0)=(3, 0)$

$\therefore$ Equation of polar is

$$\frac{xx_1}{a^2}+\frac{yy_1}{b^2}=1 \Rightarrow \frac{x(3)}{25}+\frac{y(0)}{16}=1$$

i.e. directrix.

(d) Given, $\frac{x^2}{a^2}+\frac{y^2}{b^2}=1$

Let pole be (h, k)

$\therefore$ Equation of polar $\Rightarrow \frac{hx}{a^2}+\frac{ky}{b^2}=1$

$\because$ Given that distance between centre and polar $=C$

$\therefore \quad \left|\frac{0+0-1}{\sqrt{\frac{h^2}{a^4}+\frac{k^2}{b^4}}}\right| = C$

Squaring both sides, we get

$\Rightarrow \quad \frac{1}{\frac{h^2}{a^4}+\frac{k^2}{b^4}}=C^2$

$\therefore \quad \frac{h^2}{a^4}+\frac{k^2}{b^4}=\frac{1}{C^2}$

$\therefore$ Locus is $\frac{x^2}{a^4}+\frac{y^2}{b^4}=\frac{1}{C^2}$.

26. Given, equation of ellipse $\frac{x^2}{25}+\frac{y^2}{9}=2$

$$\Rightarrow \quad \frac{x^2}{50}+\frac{y^2}{18}=1$$

$$a^2=50 \text{ and } b^2=18$$

Now, at point $(5,-3)$ equation of tangent is

$$\frac{x(5)}{50}+\frac{y(-3)}{18}=1$$

$$\Rightarrow \quad \frac{x}{10}+\frac{y}{-6}=1$$

$$\Rightarrow \quad \frac{y}{6}=\frac{x}{10}-1$$

$$\Rightarrow \quad y=\frac{6}{10}x-6$$

$$\therefore \quad \tan\theta=\frac{6}{10}=\frac{3}{5}$$

$$\because \quad x_1=5 \text{ and } y_1=-3$$

$$\therefore \quad \sin\theta=\frac{3}{\sqrt{34}}$$

and $\quad \cos\theta=\frac{5}{\sqrt{34}}$

Length of tangent $=|y_1 \operatorname{cosec}\theta|$

$$=\left|-3\times\frac{\sqrt{34}}{3}\right|=\sqrt{34} \text{ units}$$

Length of normal $=|y_1 \sec\theta|$

$$=\left|-3\times\frac{\sqrt{34}}{5}\right|=\frac{3}{5}\sqrt{34} \text{ units}$$

Length of subtangent $=|y_1\cot\theta|=\left|-3\times\frac{5}{3}\right|=5$ units

Length of subnormal $=|y_1\tan\theta|=\left|-3\times\frac{3}{5}\right|=\frac{9}{5}$ units

27. (a) Given, ellipse $S: 3x^2+2y^2-6x-12y+63=0$ ...(i)

and $\quad S'=4x^2+2y^2-19x-120y+35=0$ ...(ii)

Family of ellipse passing through point of intersection of given ellipse

$$S+\lambda S'=0$$

$$\Rightarrow (3x^2+2y^2-6x-12y+63)+\lambda(4x^2+2y^2-19x-120y+35)=0 \quad \text{...(iii)}$$

$\because$ The required ellipse also passes $(1,1)$

$$\{3(1)^2+2(1)^2-6-12+63\}+\lambda(4+2-19-120+35)=0$$

$$(5-6-12+63)+\lambda(98)=0$$

$$\Rightarrow \quad 50=98\lambda$$

$$\Rightarrow \quad \lambda=\frac{50}{98}=\frac{25}{49}$$

By Eq. (iii) required ellipse

$$3x^2+2y^2-6x-12y+63+\frac{25}{49}(4x^2+2y^2-19x-120y+35)=0$$

$$\Rightarrow 147x^2+98y^2-294x-588y+3087+100x^2+50y^2-475x-3000y+875=0$$

$$\Rightarrow \quad 247x^2+148y^2-769x-3588y+3962=0$$

(b) Given, equation of ellipse $\frac{x^2}{25}+\frac{y^2}{4}=1$ and circle $x^2+y^2=16$ equation of tangent to ellipse at $(5\cos\theta, 2\sin\theta)$, is

$$\frac{x\cos\theta}{5}+\frac{y\sin\theta}{2}=1 \quad \text{...(i)}$$

$\because$ It also touches the circle.

$\therefore$ Distance between centre and line (i) = radius

$$\left|\frac{0+0-1}{\sqrt{\left(\frac{\cos\theta}{5}\right)^2+\left(\frac{\sin\theta}{2}\right)^2}}\right|=4$$

$$\Rightarrow \quad \left|\frac{-10}{\sqrt{4\cos^2\theta+25\sin^2\theta}}\right|=4$$

$$\Rightarrow \quad \frac{100}{4\cos^2\theta+25\sin^2\theta}=16$$

$$\Rightarrow \quad 4(4\cos^2\theta+25\sin^2\theta)=25$$

$$\Rightarrow \quad 84\cos^2\theta=75$$

$$\Rightarrow \quad \cos^2\theta=\frac{75}{84}$$

$$\therefore \quad \sin^2\theta=\frac{9}{84} \text{ and } \tan^2\theta=\frac{9}{75}$$

$\therefore$ Tangent is in first quadrant

$\therefore$ Slope must be negative

$$y=\left(-\frac{2}{5}\cot\theta\right)x+\frac{2}{\sin\theta}$$

$$\Rightarrow \quad y=-\frac{2}{5}\sqrt{\frac{75}{9}}x+\frac{2}{3}\sqrt{84}$$

$$\Rightarrow \quad y=-\frac{2}{5}\times\frac{5}{\sqrt{3}}x+\frac{2\times2\times\sqrt{21}}{3}$$

$$\Rightarrow 2x+\sqrt{3}y=4\sqrt{7}$$

Therefore, the common tangent is $2x+\sqrt{3}y=4\sqrt{7}$

The common tangent meets the axes in the points $A(2\sqrt{7},0)$ and $B\left(0,\frac{4\sqrt{7}}{\sqrt{3}}\right)$.

$$\therefore \quad AB=\sqrt{28+\frac{112}{3}}=\frac{\sqrt{196}}{\sqrt{3}}=\frac{14}{\sqrt{3}}$$

(c) As we know that the locus of point of intersection of two perpendicular tangents to an ellipse is the director circle

Now, ellipse $\Rightarrow \frac{x^2}{4}+\frac{y^2}{1}=1.$

$\therefore$ Equation of locus of point is = Equation of director circle

$$x^2+y^2=a^2+b^2$$

$$\Rightarrow \quad x^2+y^2=4+1 \Rightarrow x^2+y^2=5$$

(d) Parabola : $y^2=4x$

Ellipse : $4x^2+5y^2=20 \Rightarrow \frac{x^2}{5}+\frac{y^2}{4}=1$

$\therefore$ Equation of tangent at $P(t^2, 2t)$ to $y^2=4x$, is

$$\Rightarrow \quad yt=x+t^2 \quad \text{...(i)}$$

$\therefore$ Equation of normal at $Q(\sqrt{5}\cos\phi, 2\sin\phi)$

$$2y\cos\phi = x\sqrt{5}\sin\phi - \sin\phi\cos\phi \quad \text{...(ii)}$$

$\because$ Line (i) and (ii) are identical

$$\therefore \quad \frac{t}{2\cos\phi}=\frac{1}{\sqrt{5}\sin\phi}=\frac{t^2}{-\sin\phi\cos\phi}$$

$$\Rightarrow \quad t=\frac{2}{\sqrt{5}}\cot\phi \text{ and } t^2=-\frac{\cos\phi}{\sqrt{5}}$$

$$\therefore \quad \frac{4}{5}\cot^2\phi=\frac{-\cos\phi}{\sqrt{5}}$$

$$\Rightarrow \quad \frac{\cos\phi}{5}\left(\frac{4\cos\phi}{\sin^2\phi}+\sqrt{5}\right)=0$$

$\cos\phi \neq 0, t \neq 0$

$\therefore \sqrt{5}(1-\cos^2\phi)+4\cos\phi=0$

$$\Rightarrow \quad \cos\phi=-\frac{1}{\sqrt{5}}$$

$$\phi=2n\pi \pm \cos^{-1}\left(-\frac{1}{\sqrt{5}}\right), 0 \le \phi < 2\pi$$

$\therefore$ Corresponding value of t

$$t^2=-\frac{\cos\phi}{\sqrt{5}}=\frac{1}{5}$$

$$\Rightarrow \quad t=\pm\frac{1}{\sqrt{5}}$$

(e) Let the pole be (h, k)

The equation of polar w.r.t. the ellipse $c^2x^2+d^2y^2=1$, is

$$\Rightarrow \quad c^2hx+d^2ky=1$$

$$\Rightarrow \quad d^2ky=-c^2hx+1$$

$$\Rightarrow \quad y=\left(-\frac{c^2h}{d^2k}\right)x+\frac{1}{d^2k}$$

which is tangent to the ellipse $\frac{x^2}{a^2}+\frac{y^2}{b^2}=1.$

So, $\quad c^2=a^2m^2+b^2$

$$\Rightarrow \quad \left(\frac{1}{d^2/k^2}\right)^2=a^2\left(\frac{-c^2h}{d^2k^2}\right)^2+b^2$$

$$\Rightarrow a^2c^4h^2+b^2d^2k^2=1$$

$\therefore$ Locus is $a^2c^4x^2+b^2d^2y^2=1$

(f) The given ellipse are

$$\frac{x^2}{a^2}+\frac{y^2}{b^2}=1 \quad \text{...(i)}$$

and

$$\frac{x^2}{c^2}+\frac{y^2}{d^2}=1 \quad \text{...(ii)}$$

The equation of lines are

$$x^2\left(\frac{1}{a^2}-\frac{1}{c^2}\right)+y^2\left(\frac{1}{b^2}-\frac{1}{d^2}\right)=0 \quad \text{...(iii)}$$

which represents a pair of lines through origin

If $y=mx$ be one of the lines, then $y=mx$ satisfy the Eq. (iii)

$$\Rightarrow \quad \left(\frac{1}{a^2}-\frac{1}{c^2}\right)+m^2\left(\frac{1}{b^2}-\frac{1}{d^2}\right)=0$$

$$\Rightarrow \quad \left(\frac{1}{b^2}-\frac{1}{d^2}\right)m^2+\left(\frac{1}{a^2}-\frac{1}{c^2}\right)=0 \quad \text{...(iv)}$$

Let m_1, m_2 are roots of Eq. (iv)

$$m_1m_2=\frac{\frac{1}{a^2}-\frac{1}{c^2}}{\frac{1}{b^2}-\frac{1}{d^2}}$$

$\because$ For conjugate diameter

$$m_1m_2=\frac{-b^2}{a^2}$$

$$\therefore \quad -\frac{b^2}{a^2}=\frac{\frac{1}{a^2}-\frac{1}{c^2}}{\frac{1}{b^2}-\frac{1}{d^2}}$$

$$\Rightarrow \quad a^2\left(\frac{1}{a^2}-\frac{1}{c^2}\right)=-b^2\left(\frac{1}{b^2}-\frac{1}{d^2}\right)$$

$$\Rightarrow \quad \left(1-\frac{a^2}{c^2}\right)=\left(-1+\frac{b^2}{d^2}\right) \Rightarrow \frac{a^2}{c^2}+\frac{b^2}{d^2}=2$$

28. (a) The given equation of an ellipse is

$$16x^2+25y^2=400$$

$$\Rightarrow \quad \frac{x^2}{25}+\frac{y^2}{16}=1$$

$$\Rightarrow a=5, b=4 \text{ and } e=\sqrt{1-\frac{16}{25}}=\frac{3}{5}$$

From the reflection property of an ellipse the reflection ray passes through the focus.

Thus, the coordiantes of focus is $(ae, 0)=(3, 0)$

When $y=4 \Rightarrow x=0$

So, the point is (0, 4)

So, the equation of the reflection ray is

$$\frac{x}{3}+\frac{y}{4}=1$$

$$\Rightarrow \quad 4x+3y=12$$

(b) The equation of the incident ray is $x-y+2=0$

$$\Rightarrow \quad \frac{x}{-2}+\frac{y}{2}=1$$

The equation of ellipse

$$\frac{x^2}{4}+\frac{y^2}{3}=1$$

Clearly the coordinates of the focii are (-2, 0) and (2, 0)

$\because$ The incident ray $x-y+2=0$ intersect the ellipse at (0,2)

So, the equation of the reflection ray = the equation of the joining (0, 2) and (2, 0)

$$\Rightarrow \quad \frac{x}{2}+\frac{y}{2}=1 \Rightarrow x+y=2$$

29. (a) Axis are $x-3y+3=0$ and $3x+y-1=0$

Length are 6 and $2\sqrt{6}$ i.e. $a=3$ and $\sqrt{6}$

$\therefore$ Equation of ellipse is

$$\Rightarrow \left[\frac{\left(\frac{x-3y+3}{\sqrt{1+9}}\right)}{\sqrt{6}}\right]^2+\left[\frac{\frac{3x+y-1}{\sqrt{1+9}}}{3}\right]^2=1$$

$$\Rightarrow \quad \frac{(x-3y+3)^2}{60}+\frac{(3x+y-1)^2}{90}=1$$

$$\Rightarrow 3(x-3y+3)^2+2(3x+y-1)^2=180$$

$$\Rightarrow 21x^2-6xy+29y^2+6x-58y-151=0$$

(b) The equation of ellipse

$$4(x-2y+1)^2+9(2x+y+2)^2=25$$

$$\Rightarrow 4\times5\left(\frac{x-2y+1}{\sqrt{5}}\right)^2+9(5)\left(\frac{2x+y+2}{\sqrt{5}}\right)^2=25$$

$$\Rightarrow \quad \frac{\left(\frac{x-2y+1}{\sqrt{5}}\right)^2}{\frac{5}{4}}+\frac{\left(\frac{2x+y+2}{\sqrt{5}}\right)^2}{\frac{5}{9}}=1$$

or

$$\frac{X^2}{a^2}+\frac{Y^2}{b^2}=1$$

$$a>b$$

Equation of major axis $\Rightarrow Y=0$

$$\left(\frac{2x+y+2}{\sqrt{5}}\right)=0$$

$$\Rightarrow \quad 2x+y+2=0$$

Equation of minor axis $\Rightarrow X=0$

$$\left(\frac{x-2y+1}{\sqrt{5}}\right)=0$$

$$x-2y+1=0$$

$\therefore$ Centre $\Rightarrow X=0$ and $Y=0$

$x-2y+1=0$ and $2x+y+2=0$, we get

$$x=-1 \text{ and } y=0$$

$\therefore$ Centre $(-1,0)$

$$\therefore \text{ Length of latusrectum}=\frac{2b^2}{a}=\frac{2\times\frac{5}{9}}{\frac{\sqrt{5}}{2}}=\frac{20}{9\sqrt{5}}$$

$$\therefore \quad \text{Eccentricity, } e=\sqrt{1-\frac{b^2}{a^2}}$$

$$\Rightarrow \quad b^2=a^2(1-e^2)$$

$$\Rightarrow \quad \frac{5}{9}=\frac{5}{4}(1-e^2)$$

$$\Rightarrow \quad \frac{4}{9}=1-e^2$$

$$\Rightarrow \quad e^2=1-\frac{4}{9}$$

$$\Rightarrow \quad e=\frac{\sqrt{5}}{3}$$

For focii

$X=\pm ae$	$Y=0 \qquad \because a=\sqrt{\frac{5}{4}}, e=\frac{\sqrt{5}}{3}$
$\frac{x-2y+1}{\sqrt{5}}=\pm\frac{5}{6}$	$2x+y+2=0$ ...(ii) $\therefore ae=\frac{5}{6}$
$x-2y+1=\pm\frac{5\sqrt{5}}{6}$ ...(i)	$2x+y+2=0\times2$
	$\Rightarrow 4x+2y+y=0$ ...(iii)

Solving Eqs. (i) and (iii), we get

$$5x+5=\pm\frac{5\sqrt{5}}{6}$$

$$x=-1\pm\frac{\sqrt{5}}{6}$$

and

$$y=-2-2x$$

$$=-2+2\mp\frac{\sqrt{5}}{3}=\mp\frac{\sqrt{5}}{3}$$

$$\therefore \text{ Focii} \Rightarrow \left(-1\pm\frac{\sqrt{5}}{6}, \mp\frac{\sqrt{5}}{3}\right)$$

For directrix

$$X=\pm\frac{a}{e} \Rightarrow \frac{x-2y+1}{\sqrt{5}}=\pm\frac{\sqrt{5}/2}{\sqrt{5}/3}$$

$$\Rightarrow \quad x-2y+1=\pm\frac{3\sqrt{5}}{2}$$

$$\Rightarrow \quad x-2y+1\mp\frac{3\sqrt{5}}{2}=0$$

Level 2

1. (b) Given equation

$$\frac{x^2}{2-r}+\frac{y^2}{r-5}=-1$$

$$\Rightarrow \quad \frac{x^2}{r-2}+\frac{y^2}{5-r}=1$$

$\because$ It represents an ellipse

$$r-2>0, \quad 5-r>0$$

$$\Rightarrow \quad r-2\neq 5-r$$

$$r>2 \text{ or } r<5$$

$$\Rightarrow \quad 2r=7\Rightarrow r=\frac{3}{2}$$

$\therefore$ For ellipse $2<r<5$

2. (b) Given that, $x^2+4y^2=4$

$$\Rightarrow \quad \frac{x^2}{4}+\frac{y^2}{1}=1$$

$$\therefore \quad a^2=4 \text{ and } b^2=1$$

$$a=2 \text{ and } b=1$$

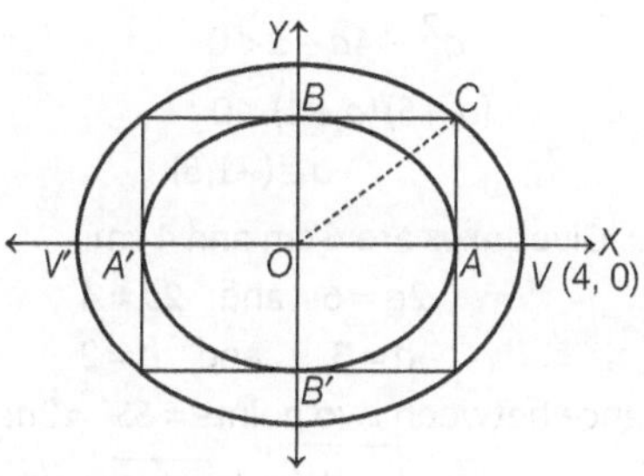

Length of minor axes is 1 unit

$\because \quad B=(0,1)$ and $A=(2,0)$

$\therefore$ Point C is (2, 1)

$\because$ Given that vertex (4, 0) is required ellipse

$\therefore$ Equation of ellipse is $\frac{x^2}{4^2}+\frac{y^2}{b^2}=1$

$\because$ It passes through C (2, 1), we get

$$\frac{2^2}{16}+\frac{1}{b^2}=1$$

$$\Rightarrow \quad b^2=\frac{1}{1-\frac{1}{4}}$$

$$\Rightarrow \quad b^2=\frac{4}{3}$$

$\therefore$ Equation is $\frac{x^2}{16}+\frac{3y^2}{4}=1$

$$\Rightarrow \quad x^2+12y^2=16$$

3. (b) We know that equation of ellipse is

$$SP=ePM$$

Now, SP = distance between two points.

PM = perpendicular distance between a point on line

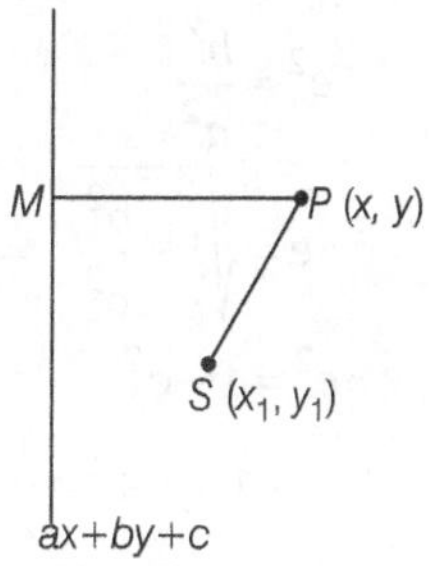

$$\therefore \quad SP=\sqrt{(x-x_1)^2+(y-y_1)^2}$$

$$PM=\left|\frac{ax+by+c}{\sqrt{a^2+b^2}}\right|$$

Now, given form

$(5x-1)^2+(5y-2)^2=(\lambda^2-2\lambda+1)(3x+4y-1)^2$

Convert it in general form

$$25\left\{\left(x-\frac{1}{5}\right)^2+\left(y-\frac{2}{5}\right)^2\right\}$$

$$=(\lambda^2-2\lambda+1)\left(\frac{3x+4y-1}{\sqrt{3^2+4^2}}\right)^2\times 5^2$$

$$\Rightarrow \left(x-\frac{1}{5}\right)^2+\left(y-\frac{2}{5}\right)^2=(\lambda^2-2\lambda+1)\left(\frac{3x+4y-1}{\sqrt{25}}\right)^2$$

$$\because \quad e=\lambda^2-2\lambda+1$$

$$\because \quad 0<e<1$$

$$\Rightarrow \quad 0<(\lambda^2-2\lambda+1)<1$$

$$\Rightarrow \quad (\lambda-1)^2<1$$

$$\Rightarrow \quad -1<\lambda-1<1$$

$$\Rightarrow \quad 0<\lambda<2$$

4. (c) $\because$ Given,

$$\angle FBF'=90^\circ \Rightarrow FB\perp F'B$$

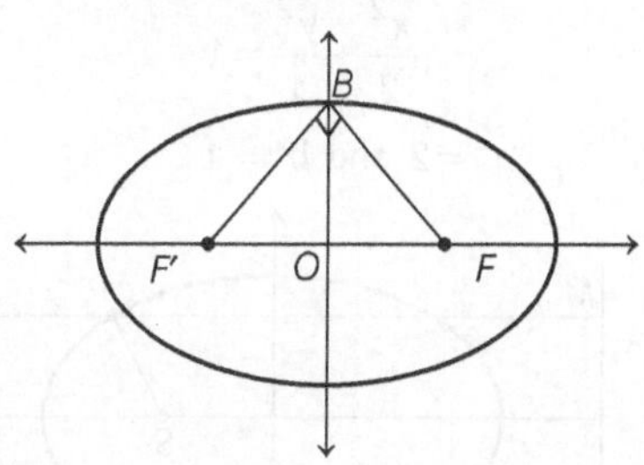

$\because$ Slope of $F'B\times$ Slope of $FB=-1$

$\because$ F and F' are $(ae,0)$ and $(-ae,0)$

and $B=(0,b)$

$$\therefore \quad \frac{b-0}{0+ae}\times\frac{b-0}{0-ae}=-1$$

$\Rightarrow \quad \dfrac{b^2}{a^2e^2}=1 \Rightarrow b^2=a^2e^2$

$\Rightarrow \quad e^2=\dfrac{b^2}{a^2}$

$\Rightarrow \quad e=\sqrt{1-\dfrac{b^2}{a^2}} \Rightarrow e=\sqrt{1-e^2}$

$\Rightarrow \quad e^2=1-e^2$

$\Rightarrow \quad 2e^2=1 \Rightarrow e^2=\dfrac{1}{2}$

$\Rightarrow \quad e=\pm\dfrac{1}{\sqrt{2}}$

$\because \quad e>0 \Rightarrow e=\dfrac{1}{\sqrt{2}}$

5. (b) $\because$ Given, ellipse $\dfrac{x^2}{a^2}+\dfrac{y^2}{b^2}=1$

$\because$ There are exactly two points in the ellipse $\dfrac{x^2}{a^2}+\dfrac{y^2}{b^2}=1$, whose distance from centre is same, the points would be either end points of the major axis or of the minor axis.

But

$$\frac{\sqrt{a^2+2b^2}}{2}<a$$

So, the points are the vertices of minor axis.

Hence, $\quad b=\dfrac{\sqrt{a^2+2b^2}}{2}$

$a^2=2b^2$

$\therefore \quad \dfrac{b^2}{a^2}=\dfrac{1}{2}$

$\because \quad e=\sqrt{1-\dfrac{b^2}{a^2}}=\sqrt{1-\dfrac{1}{2}}=\dfrac{1}{\sqrt{2}}$

6. (c) Given, ellipse $x^2+2y^2=2$

$\Rightarrow \quad \dfrac{x^2}{2}+\dfrac{y^2}{1}=1$

$a^2=2$ and $b^2=1$

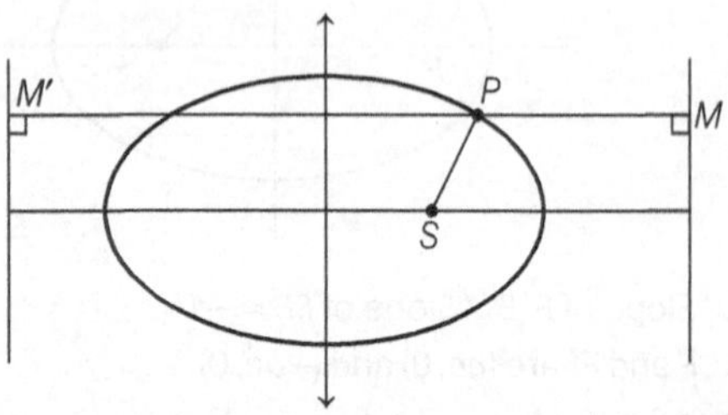

$\therefore \quad e=\sqrt{1-\dfrac{b^2}{a^2}}=\sqrt{1-\dfrac{1}{2}}=\dfrac{1}{\sqrt{2}}$

Given, $PS=\sqrt{2}=ePM$

$\Rightarrow \quad \sqrt{2}=\dfrac{1}{\sqrt{2}}PM \Rightarrow PM=2$

But $\quad MM'=\dfrac{2a}{e}=\dfrac{2\sqrt{2}}{\frac{1}{\sqrt{2}}}=4$

$\therefore \quad PM'=2$

$\therefore \quad PM:PM'=2:2=1:1$

7. (d) $a^2=f(4a)$ and $b^2=f(a^2-5)$

$\because f(x)$ is decreasing function

$f(x)>0$

$\because$ Y-axis as major axis

$\therefore \quad b>a$

$b^2>a^2$

$f(a^2-5)>f(4a)$

$\because \quad 0<f(4a)<f(a^2-5)$

$\because f(x)$ is decreasing

$\Rightarrow \quad 4a>a^2-5$

$\Rightarrow \quad a^2-4a-5<0$

$\Rightarrow \quad (a-5)(a+1)<0$

$a\in(-1,5)$

8. (d) $\because$ Gives axes are 6 cm and 4 cm.

$\therefore \quad 2a=6$ and $2b=4$

$a=3$ and $b=2$

Distance between two points $=SS'=2ae$

$\therefore \quad e=\sqrt{1-\dfrac{b^2}{a^2}}=\sqrt{1-\dfrac{4}{9}}=\dfrac{\sqrt{5}}{3}$

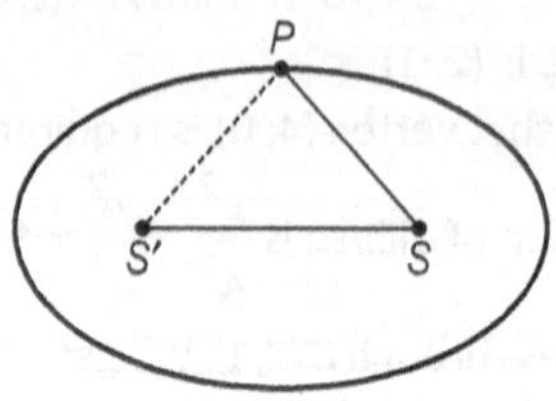

$\therefore \quad 2ae=2\times3\times\dfrac{\sqrt{5}}{3}=2\sqrt{5}$

$\therefore$ Length of string $=SPSS'=SP+PS+SS'$

$=2a+2ae$

$=(6+2\sqrt{5})$ units

9. (a) One of the focus of the ellipse $\dfrac{x^2}{a^2}+\dfrac{y^2}{b^2}=1$ is on Y-axis $(0,5\sqrt{3})$

$\Rightarrow \quad be=5\sqrt{3} \quad$ ...(i)

{where e is eccentricity of ellipse}

According to the question,

$2b-2a=10$

$$\Rightarrow \qquad b-a=5 \qquad \text{...(ii)}$$

$$\Rightarrow \qquad b>a$$

On squaring Eq. (i) both the sides, we get

$$b^2e^2=75$$

$$\Rightarrow \qquad b^2\left(1-\frac{a^2}{b^2}\right)=75$$

$$\Rightarrow \qquad b^2-a^2=75$$

$$\Rightarrow \qquad (b-a)(b+a)=75$$

$$\Rightarrow \qquad 5(b+a)=75 \qquad \text{[from Eq. (ii)]}$$

$$\Rightarrow \qquad b+a=15 \qquad \text{...(iii)}$$

By solving Eqs. (ii) and (iii), we get

$$b=10 \text{ and } a=5$$

$$\therefore \text{ Length of latusrectum} = \frac{2a^2}{b}=\frac{2\times 25}{10}=5 \text{ units}$$

10. (a) Given ellipse,

$$\frac{x^2}{a^2}+\frac{y^2}{b^2}=1$$

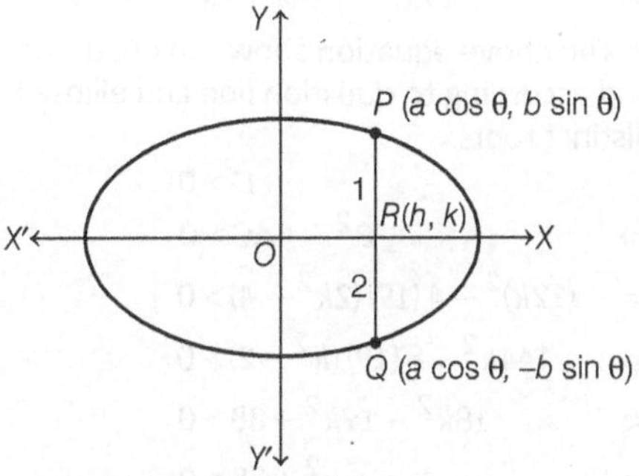

Let $P(a\cos\theta, b\sin\theta)$ be one end of double ordinate of given ellipse second end point $Q(a\cos\theta, -b\sin\theta)$

$\because$ Point $R(h, k)$ divides PQ in $1:2$ internally.

$\therefore$ By section formula,

$$h=\frac{2a\cos\theta+a\cos\theta}{3},\ k=\frac{2b\sin\theta-b\sin\theta}{3}$$

$$\Rightarrow \quad 3h=3a\cos\theta,\ 3k=b\sin\theta$$

$$\Rightarrow \quad \frac{h}{a}=\cos\theta,\ \frac{3k}{b}=\sin\theta$$

$$\because \qquad \sin^2\theta+\cos^2\theta=1$$

$$\Rightarrow \qquad \left(\frac{h}{a}\right)^2+\left(\frac{3k}{b}\right)^2=1$$

$$\Rightarrow \qquad \frac{h^2}{a^2}+\frac{9k^2}{b^2}=1$$

$$\therefore \quad \text{Locus is } \frac{x^2}{a^2}+\frac{9y^2}{b^2}=1$$

11. (b) Given ellipse, $\frac{x^2}{16}+\frac{y^2}{9}=1$

$$a^2=16 \text{ and } b^2=9$$

$$\Rightarrow \qquad a=4 \text{ and } b=3$$

$$\therefore \qquad e=\sqrt{1-\frac{b^2}{a^2}}=\sqrt{1-\frac{9}{16}}=\frac{\sqrt{7}}{4}$$

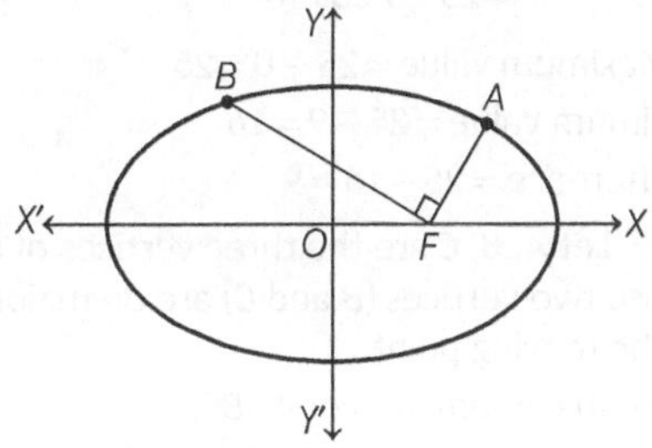

$$F=(ae,0)=\left(4\times\frac{\sqrt{7}}{4},0\right)=(\sqrt{7},0)$$

$$AF^2=(4\cos\theta-\sqrt{7})^2+(3\sin\theta)^2$$

$$=16\cos^2\theta+7-8\sqrt{7}\cos\theta+9\sin^2\theta$$

$$BF^2=(-4\sin\theta-\sqrt{7})^2+(3\cos\theta-0)^2$$

$$=16\sin^2\theta+7+8\sqrt{7}\sin\theta+9\cos^2\theta$$

$$\therefore\ AF^2+BF^2=16\cos^2\theta+16\sin^2\theta+14$$

$$+8\sqrt{7}(\sin\theta-\cos\theta)+9\sin^2\theta+9\cos^2\theta$$

$$=16+14+9+8\sqrt{7}(\sin\theta-\cos\theta)$$

$$=39+8\sqrt{7}(\sin\theta-\cos\theta)$$

$$\text{At} \qquad \theta=\frac{\pi}{6}$$

$$\Rightarrow \quad 39+8\sqrt{7}\left(\frac{1}{2}-\frac{\sqrt{3}}{2}\right)$$

$$\Rightarrow \quad 39-4\sqrt{7}(\sqrt{3}-1)$$

12. (b) $\because$ It is given P be any point on the ellipse

$$\frac{x^2}{25}+\frac{y^2}{16}=1$$

$$\because \qquad a=5 \text{ and } b=4$$

$$\therefore\ P(a\cos\theta, b\sin\theta)=P(5\cos\theta, 4\sin\theta)$$

Now, S and S' are foci of ellipse

$$\because \qquad S=(ae,0) \text{ and } S'=(-ae,0)$$

$$\therefore \qquad e=\sqrt{1-\frac{b^2}{a^2}}=\sqrt{1-\frac{16}{25}}=\frac{3}{5}$$

$$\therefore \qquad S=(3,0) \text{ and } S'=(-3,0)$$

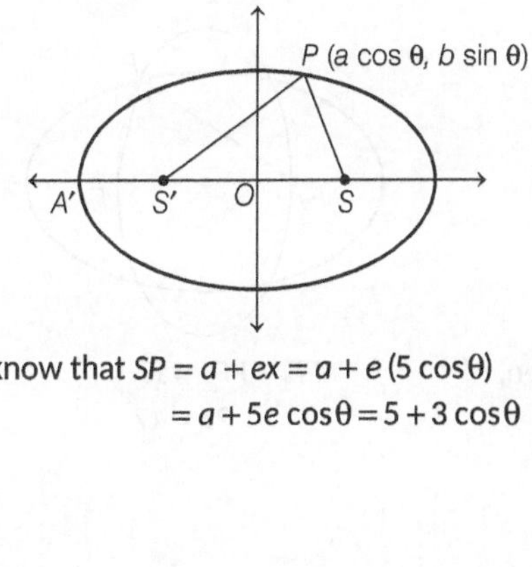

$\because$ We know that $SP=a+ex=a+e(5\cos\theta)$

$$=a+5e\cos\theta=5+3\cos\theta$$

and $\quad S'P = a - ex = a - 5e\cos\theta = 5 - 3\cos\theta$

$\therefore \quad SP \cdot S'P = (5 + 3\cos\theta)(5 - 3\cos\theta)$

$= 25 - 9\cos^2\theta$

$\therefore$ Maximum value $= 25 - 0 = 25$

Minimum value $= 25 - 9 = 16$

$\because$ Difference $= 25 - 16 = 9$

13. (c) $\because$ Let A, B, C are the three vertices of a triangle whose two vertices (B and C) are on major axis and A be the moving point.

$\therefore$ For maximum area of ΔABC.

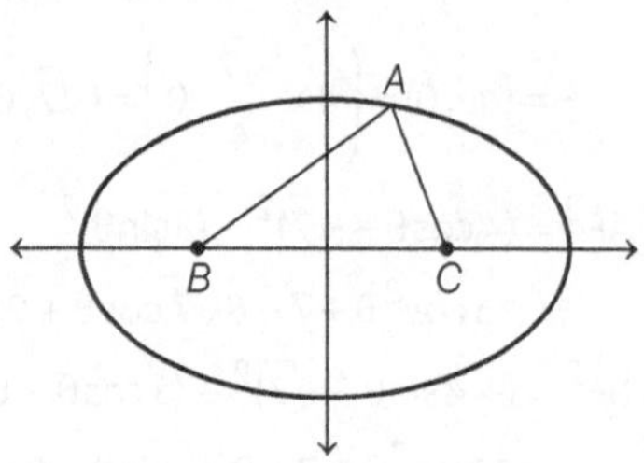

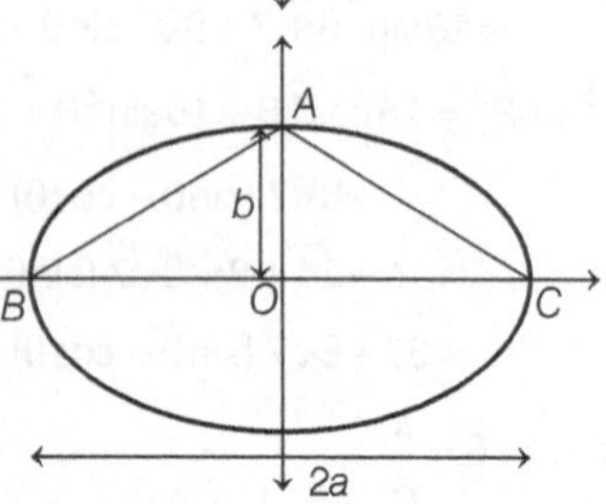

First base will be maximum and altitude should also be maximum

So, it clearly see that axes is maximum when B and C are vertices and A be on minor axis as per given in figure.

$\therefore \quad (\Delta ABC)_{max} = \frac{1}{2} \times BC \times AO$

$$= \frac{1}{2} \times 8 \times 3 = 12 \qquad \left[\begin{array}{l} \because \frac{x^2}{16} + \frac{y^2}{9} = 1 \\ a^2 = 16 \text{ and } b^2 = 9 \\ a = 4 \text{ and } b = 3 \\ 2a = 8 \end{array}\right.$$

14. (c) $\because$ It is given that F_1 and F_2 are foci of the ellipse.

$\because$ We know that sum distance of any point from the focus of the ellipse is length of major axis

$$PF_1 + PF_2 = 2a$$

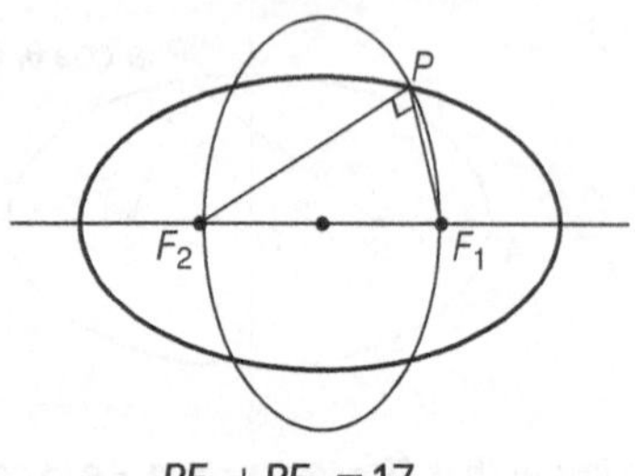

Given, $\quad PF_1 + PF_2 = 17$

$\therefore \quad 2a = 17$

$\because \quad$ Area of triangle $= 30$

$\Rightarrow \quad \frac{1}{2} \times (PF_1)(PF_2) = 30$

$\Rightarrow \quad PF_1(17 - PF_1) = 60 \Rightarrow 17PF_1 - PF_1^2 = 60$

$\Rightarrow \quad PF_1^2 - 17PF_1 + 60 = 0$

$\Rightarrow \quad (PF_1 - 12)(PF_1 - 5) = 0$

$\Rightarrow \quad PF_1 = 12, 5$

$\because \quad PF_1 < PF_2$

$\therefore \quad PF_1 = 5 \;\Rightarrow PF_2 = 12$

$\therefore \quad F_1F_2 = \sqrt{5^2 + 12^2} = 13$ units

15. (b) Line : $3x + y = k$

$\Rightarrow \quad y = k - 3x \qquad$ ...(i)

Ellipse : $\quad x^2 + 2y^2 = 4 \qquad$ ...(ii)

Solving Eqs (i) and (ii), we get

$\Rightarrow \quad x^2 + 2(k - 3x)^2 = 4$

$\Rightarrow \quad x^2 + 2k^2 + 18x^2 - 12kx = 4$

$\Rightarrow \quad 19x^2 - 12kx + 2k^2 - 4 = 0$

$\because$ The above equation shows an quadratic equation and according to question line and ellipse have real and distinct roots.

$D > 0$

$\Rightarrow \quad B^2 - 4AC > 0$

$\Rightarrow \quad (12k)^2 - 4(19)(2k^2 - 4) > 0$

$\Rightarrow \quad 144k^2 - 8(19)(k^2 - 2) > 0$

$\Rightarrow \quad 18k^2 - 19k^2 + 38 > 0$

$\Rightarrow \quad -k^2 + 38 > 0$

$\Rightarrow \quad k^2 < 38$

$\Rightarrow \quad (k - \sqrt{38})(k + \sqrt{38}) < 0$

$\therefore \quad k \in (-\sqrt{38}, \sqrt{38})$

16. (c) Let point R on the ellipse is $R(a\cos\theta, b\sin\theta)$

$\because$ Ellipse is $x^2 + 2y^2 = 2$

$$\frac{x^2}{2} + \frac{y^2}{1} = 1$$

$\Rightarrow \quad a^2 = 2$ and $b^2 = 1$

$\Rightarrow \quad a = \sqrt{2}$ and $b = 1$

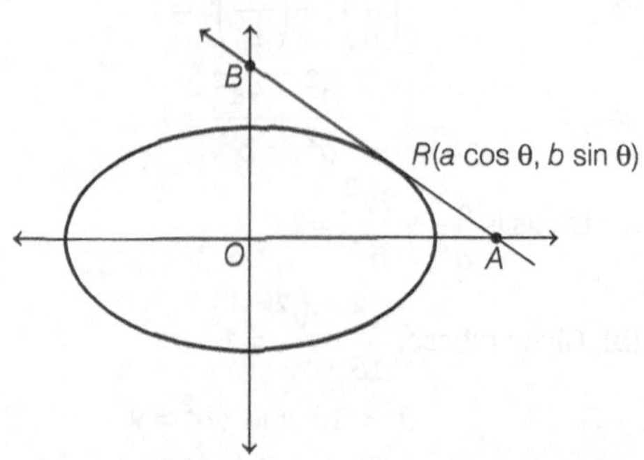

$\therefore R(\sqrt{2}\cos\theta, \sin\theta)$

$\therefore$ Equation of tangent at R.

$$\Rightarrow \quad \frac{x(\sqrt{2}\cos\theta)}{2} + y(\sin\theta) = 1$$

$\therefore$ For A put $y = 0 \Rightarrow x = \dfrac{\sqrt{2}}{\cos\theta}$

$A \Rightarrow (\sqrt{2}\sec\theta, 0)$

For B put $x = 0 \Rightarrow y = \dfrac{1}{\sin\theta}$

$B \Rightarrow (0, \operatorname{cosec}\theta)$

$\because Q(h, k)$ be the middle point of AB

$$\therefore \quad h = \frac{\sqrt{2}\sec\theta + 0}{2}$$

and $$k = \frac{\operatorname{cosec}\theta + 0}{2}$$

$$\Rightarrow \quad \sqrt{2}h = \sec\theta, \ \sin\theta = \frac{1}{2k}$$

and $$\cos\theta = \frac{1}{\sqrt{2}h}$$

$$\because \quad \sin^2\theta + \cos^2\theta = 1$$

$$\Rightarrow \quad \frac{1}{2h^2} + \frac{1}{4k^2} = 1$$

$\therefore$ Locus is $\dfrac{1}{2x^2} + \dfrac{1}{4y^2} = 1$

17. (c) Given, ellipse $\dfrac{x^2}{25} + \dfrac{y^2}{16} = 1$

$$\because \quad a^2 = 25 \text{ and } b^2 = 16$$

Let any point P on the ellipse be $(a\cos\theta, b\sin\theta) \Rightarrow (5\cos\theta, 4\sin\theta)$

$\therefore$ Equation of tangent at P

$$\frac{x\cos\theta}{5} + \frac{y\sin\theta}{4} = 1$$

It meets the line $x = 0$ at $Q(0, 4\operatorname{cosec}\theta)$

Image of Q in the line $y = x$ is $R(4\operatorname{cosec}\theta, 0)$

$\because$ Equation of circle is

$$x(x - 4\operatorname{cosec}\theta) + y(y - \operatorname{cosec}\theta) = 0$$

$$\Rightarrow \quad x^2 - 4x\operatorname{cosec}\theta + y^2 - y\operatorname{cosec}\theta = 0$$

$$\Rightarrow \quad (x^2 + y^2) - \operatorname{cosec}\theta(4x + y) = 0$$

It represents a family of circles passing through the intersection of $x^2 + y^2 = 0$ and $4x + y = 0$

Now, solving them, we get

$$x = 0 \text{ and } y = 0$$

$\therefore$ Fixed point is $(0, 0)$.

18. (a) Given, ellipse $\dfrac{x^2}{a^2} + \dfrac{y^2}{b^2} = 1$

and Tangent $\sqrt{3}bx + ay = 2ab$

$$\Rightarrow \quad \frac{\sqrt{3}bx}{2ab} + \frac{ay}{2ab} = 1$$

$$\Rightarrow \quad \frac{x}{\left(\frac{2}{\sqrt{3}}\right)a} + \frac{y}{2b} = 1 \quad \text{...(i)}$$

Let $P(\theta)$ be the point of contact and whose eccentric angle is θ

$\Rightarrow P = (a\cos\theta, b\sin\theta)$

$\therefore$ Equation of tangent at P

$$\frac{x}{a}\cos\theta + \frac{y}{b}\sin\theta = 1 \quad \text{...(ii)}$$

Eqs (i) and (ii) are identical

$$\because \quad \frac{\frac{2}{\sqrt{3}}a}{\frac{a}{\cos\theta}} = \frac{2b}{\frac{b}{\sin\theta}} = 1$$

$$\frac{2\cos\theta}{\sqrt{3}} = 2\sin\theta = 1$$

$$\Rightarrow \quad \sin\theta = \frac{1}{2} \text{ and } \cos\theta = \frac{\sqrt{3}}{2}$$

$$\Rightarrow \quad \theta = \frac{\pi}{6} \text{ and } \theta = \frac{\pi}{6}$$

$\therefore$ Required eccentric angle is $\dfrac{\pi}{6}$

19. (a) Any tangent to the ellipse $\dfrac{x^2}{a^2} + \dfrac{y^2}{b^2} = 1$ having slope m is

$$y = mx + \sqrt{a^2m^2 + b^2}$$

Points on minor axis are $A(0, ae)$ and $B(0, -ae)$

Now, length of perpendicular from

$$A = \left| \frac{\sqrt{a^2m^2 + b^2} - ae}{\sqrt{1 + m^2}} \right|$$

Length of perpendicular from

$$B = \left| \frac{\sqrt{a^2m^2 + b^2} + ae}{\sqrt{1 + m^2}} \right|$$

Now, sum of squares of the lengths is

$$\frac{(\sqrt{a^2m^2 + b^2} - ae)^2}{1 + m^2} + \frac{(\sqrt{a^2m^2 + b^2} + ae)^2}{1 + m^2}$$

$$\Rightarrow \quad \frac{1}{1 + m^2} 2(a^2m^2 + b^2 + a^2e^2)$$

$$\Rightarrow \quad \frac{2}{1 + m^2}(a^2m^2 + a^2 - a^2e^2 + a^2e^2)$$

$$[\because b^2 = a^2(1 - e^2) = a^2 - a^2e^2]$$

$$= \frac{2a^2(1 + m^2)}{1 + m^2} = 2a^2$$

20. (b) Given, point $(3\sqrt{3}\cos\theta, \sin\theta)$ and ellipse $\frac{x^2}{27}+y^2=1$

$\therefore$ Equation of tangent is $\frac{x}{3\sqrt{3}}\cos\theta + y\sin\theta = 1$

$\therefore$ Intercepts are $\left(0, \frac{1}{\sin\theta}\right)\left(\frac{3\sqrt{3}}{\cos\theta}, 0\right)$

Sum of intercept $(S) = \frac{1}{\sin\theta}+\frac{3\sqrt{3}}{\cos\theta}$

$= 3\sqrt{3}\sec\theta + \operatorname{cosec}\theta$

For minimum of (S)

$\frac{dS}{d\theta} \Rightarrow 3\sqrt{3}\sec\theta\tan\theta - \operatorname{cosec}\theta\cot\theta = 0$

$\Rightarrow \quad \tan^3\theta = \frac{1}{3\sqrt{3}}$

$\Rightarrow \quad \tan\theta = \frac{1}{\sqrt{3}} \Rightarrow \theta = \frac{\pi}{6}$

At $\theta = \frac{\pi}{6} \quad \Rightarrow \frac{d^2S}{d\theta^2} > 0$

$\therefore$ S will be minimum at $\theta = \frac{\pi}{6}$

21. (c) $\because$ The tangent at the point of shortest distance from the line $x+y=7$ is parallel to the given line

Any point on the ellipse $\frac{x^2}{6}+\frac{y^2}{3}=1$ is $(\sqrt{6}\cos\theta, \sqrt{3}\sin\theta)$

Equation of tangent $\Rightarrow \frac{x\cos\theta}{\sqrt{6}}+\frac{y\sin\theta}{\sqrt{3}}=1$

It is parallel to $x+y=7$

$\frac{\cos\theta}{\sqrt{6}} = \frac{\sin\theta}{\sqrt{3}} \Rightarrow \tan\theta = \frac{1}{\sqrt{2}}$

$\Rightarrow \quad \sin\theta = \frac{1}{\sqrt{3}}$ and $\cos\theta = \frac{\sqrt{2}}{\sqrt{3}}$

$\therefore$ Point is (2, 1)

22. (b) Ellipse : $\frac{x^2}{9}+\frac{y^2}{4}=1$

$\Rightarrow \quad a^2 = 9$ and $b^2 = 4$

$\Rightarrow \quad a = 3$ and $b = 2$

Let $P(a\cos\theta, b\sin\theta) = (3\cos\theta, 2\sin\theta)$ be any point on ellipse.

$\Rightarrow \quad \theta = \frac{\pi}{4}$ (given)

$\therefore$ Equation of tangent at

$P\left(3\cos\frac{\pi}{4}, 2\sin\frac{\pi}{4}\right) = P\left(\frac{3}{\sqrt{2}}, \frac{2}{\sqrt{2}}\right)$ is $\frac{x}{9}\cdot\frac{3}{\sqrt{2}}+\frac{y}{4}\cdot\frac{2}{\sqrt{2}}=1$

$\Rightarrow \quad \frac{x}{3\sqrt{2}}+\frac{y}{2\sqrt{2}}=1$

$\Rightarrow 2x+3y = 6\sqrt{2} \Rightarrow$ Slope $= \frac{-2}{3}$

$\therefore$ Equation of normal

$\Rightarrow \quad \left(y-\frac{2}{\sqrt{2}}\right) = \frac{3}{2}\left(x-\frac{3}{\sqrt{2}}\right)$

$\left(\text{slope of normal} = \frac{-1}{\frac{-2}{3}} = \frac{3}{2}\right)$

$\Rightarrow \quad 2y - \frac{4}{\sqrt{2}} = 3x - \frac{9}{\sqrt{2}}$

$\Rightarrow \quad 3x - 2y = \frac{5}{\sqrt{2}}$

$\because$ Perpendicular from centre to tangent and normals are $OM \perp ON$

$\therefore \quad OM = \left|\frac{6\sqrt{2}}{\sqrt{2^2+3^2}}\right| = \frac{6\sqrt{2}}{\sqrt{13}}$

$ON = \left|\frac{\frac{5}{\sqrt{2}}}{\sqrt{2^2+3^2}}\right| = \frac{5}{\sqrt{2}}\cdot\left(\frac{1}{\sqrt{13}}\right)$

Thus all $\Rightarrow OM \times ON = \frac{6\sqrt{2}}{\sqrt{13}} \times \frac{5}{\sqrt{2}\sqrt{13}} = \frac{30}{13}$ sq units

23. (d) Given, $\frac{x^2}{169}+\frac{y^2}{25}=1$

Equation of normal at the point $(13\cos\theta, 5\sin\theta)$ is

$\frac{13x}{\cos\theta}-\frac{5y}{\sin\theta}=144$

$\because$ It passes through (0, 6)

$\Rightarrow \quad 15 + 72\sin\theta = 0$

$\sin\theta = -\frac{15}{72} = \frac{-5}{24}$

$\because \quad \theta = 2\pi - \sin^{-1}\left(\frac{5}{24}\right)$

and $\pi + \sin^{-1}\left(\frac{5}{24}\right)$

Also, Y-axis is one of the normals.

Four normals can be drawn.

24. (c) $\because$ Given, equation of ellipse is $\frac{x^2}{100}+\frac{y^2}{25}=1$

$\because$ Particle left the orbit from (8, –3) and normal to ellipse.

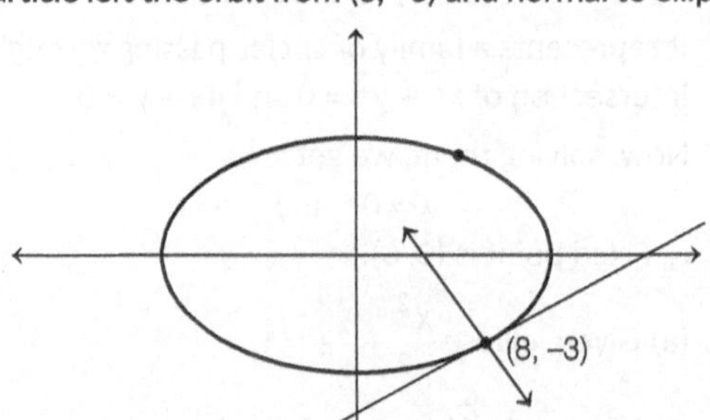

$\because$ Equation of normal at (8, -3)

$$\frac{a^2x}{x_1} - \frac{b^2y}{y_1} = a^2 - b^2$$

$$\Rightarrow \quad \frac{a^2x}{8} - \frac{b^2y}{(-3)} = 75$$

$$\Rightarrow \quad \frac{100}{8}x + \frac{25}{3}y = 75$$

For Y-axis put $x = 0$

$$\Rightarrow \quad y = \frac{75 \times 3}{25}$$

$\therefore$ Point is (0, 9)

But it moves outwards direction i.e. no such point

25. (d) Let $P(2\cos\theta, \sin\theta)$

Equation of normal at P

$$2x\sec\theta - y\,\text{cosec}\,\theta = 3$$

$\therefore \quad OM \perp PM$

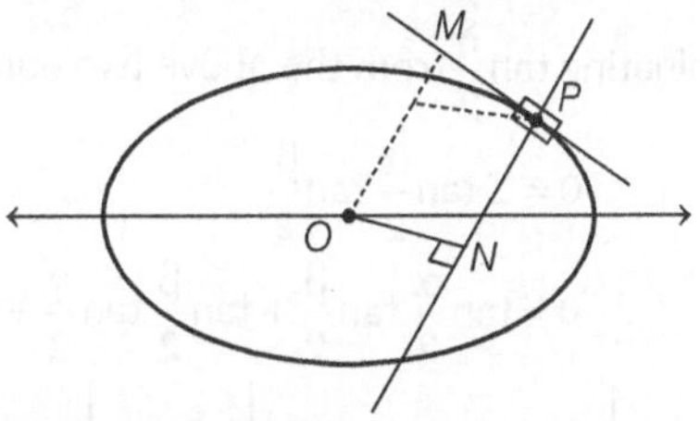

$\therefore$ Equation of tangent at P

$$\frac{x\cos\theta}{2} + y\sin\theta = 1$$

$$\because \quad ON = \left|\frac{3}{\sqrt{4\sec^2\theta + \text{cosec}^2\theta}}\right|$$

$$OM = \left|\frac{1}{\sqrt{\left(\frac{\cos\theta}{2}\right)^2 + (\sin\theta)^2}}\right|$$

$\because$ $ONPM$ is rectangle.

$\therefore$ Area of $\Delta OPN = \frac{1}{2}$ area of rectangle $ONPQ$

$$= \frac{1}{2}\left[\frac{3}{\sqrt{4\sec^2\theta + \text{cosec}^2\theta}} \times \frac{1}{\sqrt{\frac{\cos^2\theta}{4} + \sin^2\theta}}\right]$$

$$= \frac{3}{2\sqrt{1 + \frac{1}{4}\cot^2\theta + 4\tan^2\theta + 1}}$$

$$= \frac{3}{2\sqrt{\left(4\tan^2\theta + 2 + \frac{1}{4}\cot^2\theta\right)}}$$

$$= \frac{3}{2\sqrt{\left(2\tan\theta + \frac{1}{2}\cot\theta\right)^2}}$$

$$= \frac{3}{2\left|2\tan\theta + \frac{1}{2}\cot\theta\right|}$$

$\because$ Area will be maximum, when $2\tan\theta = \frac{1}{2}\cot\theta$ using application of derivatives

$$\Rightarrow \quad \tan^2\theta = \frac{1}{4} \quad \Rightarrow \tan\theta = \pm\frac{1}{2}$$

$$\tan\theta = \frac{1}{2} \Rightarrow P = \left(\frac{4}{\sqrt{5}}, \frac{1}{\sqrt{5}}\right) \text{ or } \left(\frac{-4}{\sqrt{5}}, \frac{-1}{\sqrt{5}}\right)$$

$$\tan\theta = \frac{-1}{2} \quad \Rightarrow P = \left(\frac{-4}{\sqrt{5}}, \frac{1}{\sqrt{5}}\right) \text{ or } \left(\frac{4}{\sqrt{5}}, \frac{-1}{\sqrt{5}}\right)$$

26. (c) Given, ellipse is $\frac{x^2}{16} + \frac{y^2}{4} = 1$

$\because$ Here, $a^2 = 16$ and $b^2 = 4$

$$a = 4 \quad \text{and} \quad b = 2$$

$\therefore$ Equation of normal

$$ax\sec\theta - by\,\text{cosec}\,\theta = a^2 - b^2$$

$$\Rightarrow \quad 4x\sec\theta - 2y\,\text{cosec}\,\theta = 16 - 4$$

$$\Rightarrow \quad 2x\sec\theta - y\,\text{cosec}\,\theta = 6$$

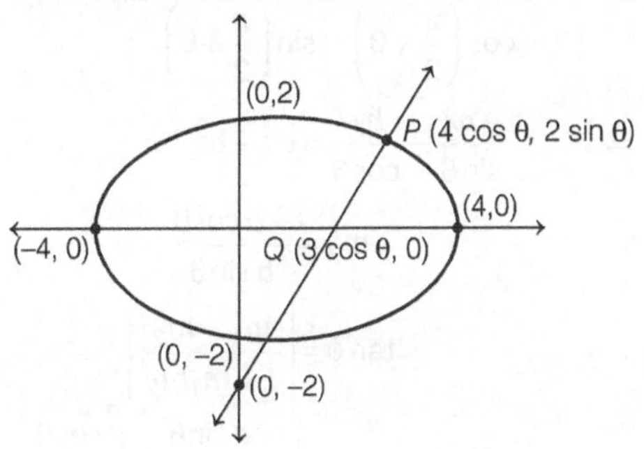

$\therefore$ Q (Put $y = 0$)

$$x = 3\cos\theta$$

$\because$ $Q(3\cos\theta, 0)$

$$\therefore \quad M = \text{Mid-point of } PQ$$

$$= \left(\frac{7\cos\theta}{2}, \frac{2\sin\theta}{2}\right)$$

$$= \left(\frac{7\cos\theta}{2}, \sin\theta\right) = (h, k)$$

$$h = \frac{7\cos\theta}{2} \quad \Rightarrow \cos\theta = \frac{2h}{7} \quad \text{...(i)}$$

$$k = \sin\theta \quad \text{...(ii)}$$

On squaring and adding Eqs. (i) and (ii), we get

$$\frac{4h^2}{49} + k^2 = 1$$

$\because$ Locus is $\dfrac{4x^2}{49}+y^2=1$ ...(iii)

For given ellipse, $e^2=1-\dfrac{b^2}{a^2}=1-\dfrac{4}{16}=\dfrac{3}{4}\Rightarrow e=\dfrac{\sqrt{3}}{2}$

$\because \quad x=\pm 4\times\dfrac{\sqrt{3}}{2}=\pm 2\sqrt{3} \quad (\because x=\pm ae)$ ...(iv)

By Eq. (iii),

$$\frac{4}{49}\times 12+y^2=1$$

$$\Rightarrow \quad y^2=1-\frac{49}{48}=\frac{1}{49}$$

$$\Rightarrow \quad y=\pm\frac{1}{7}$$

$\because$ Required points $\left(\pm 2\sqrt{3},\pm\dfrac{1}{7}\right)$

27. (a) Given, ellipse is $\dfrac{x^2}{a^2}+\dfrac{y^2}{b^2}=1$

Normal at $\theta \Rightarrow \dfrac{ax}{\cos\theta}-\dfrac{by}{\sin\theta}=a^2-b^2$

$$\therefore \quad m_1=\frac{a\sin\theta}{b\cos\theta}$$

Normal at $\left(\dfrac{\pi}{2}+\theta\right)$

$$\Rightarrow \quad \frac{ax}{\cos\left(\frac{\pi}{2}+\theta\right)}-\frac{by}{\sin\left(\frac{\pi}{2}+\theta\right)}=a^2-b^2$$

$$\Rightarrow \quad \frac{-ax}{\sin\theta}-\frac{by}{\cos\theta}=a^2-b^2$$

$$\because \quad m_2=\frac{-a\cos\theta}{b\sin\theta}$$

$$\because \quad \tan\phi=\left|\frac{m_1-m_2}{1+m_1m_2}\right|$$

$$=\left|\frac{\frac{a\sin\theta}{b\cos\theta}+\frac{a\cos\theta}{b\sin\theta}}{1+\frac{a\sin\theta}{b\cos\theta}\cdot\frac{-a\cos\theta}{b\sin\theta}}\right|$$

$$=\left|\frac{\frac{a}{b}\left(\frac{\sin^2\theta+\cos^2\theta}{\sin\theta\cdot\cos\theta}\right)}{1-\frac{a^2}{b^2}}\right|$$

$$=\left|\frac{\frac{a}{b}\left(\frac{2}{2\sin\theta\cos\theta}\right)}{\frac{b^2-a^2}{b^2}}\right|$$

$$\Rightarrow \quad \frac{2ab}{a^2e^2}\cdot\frac{1}{\sin 2\theta}$$

$$\Rightarrow \quad a^2e^2=a^2-b^2$$

$$\therefore \quad \tan\phi=\frac{2ab}{a^2e^2}\cdot\frac{1}{\sin 2\theta}$$

$$\therefore \quad \tan\phi\cdot\sin 2\theta=\frac{2ab}{a^2e^2}=\frac{2a\cdot a\sqrt{1-e^2}}{a^2e^2}$$

$$\Rightarrow \quad \frac{2\sqrt{1-e^2}}{e^2}=\tan\phi\sin 2\theta$$

28. (a) Suppose the normals at α,β and γ are concurrent at (h,k) and let δ be the foci of the fourth normal from (h,k)

$\because$ We know that (using property of concurrency of normal)

$$\Sigma\tan\frac{\alpha}{2}\tan\frac{\beta}{2}=0 \quad \text{...(i)}$$

and $\quad \tan\dfrac{\alpha}{2}\tan\dfrac{\beta}{2}\tan\dfrac{\gamma}{2}\tan\dfrac{\delta}{2}=-1$ ...(ii)

Eliminating $\tan\dfrac{\delta}{2}$ from the above two equation gives

$$0=\Sigma\tan\frac{\alpha}{2}\tan\frac{\beta}{2}$$

$$0=\tan\frac{\alpha}{2}\tan\frac{\beta}{2}+\tan\frac{\beta}{2}\tan\frac{\gamma}{2}+\tan\frac{\gamma}{2}\tan\frac{\alpha}{2}$$

$$+\tan\frac{\alpha}{2}\left[\frac{-1}{\tan\frac{\alpha}{2}\tan\frac{\beta}{2}\tan\frac{\gamma}{2}}\right]+\tan\frac{\beta}{2}\left[\frac{-1}{\tan\frac{\alpha}{2}\tan\frac{\beta}{2}\tan\frac{\gamma}{2}}\right]$$

$$+\tan\frac{\gamma}{2}\left[\frac{-1}{\tan\frac{\alpha}{2}\tan\frac{\beta}{2}\tan\frac{\gamma}{2}}\right]$$

$$0=\left[\tan\frac{\alpha}{2}\tan\frac{\beta}{2}-\cot\frac{\alpha}{2}\cot\frac{\beta}{2}\right]$$

$$+\left[\tan\frac{\beta}{2}\tan\frac{\gamma}{2}-\cot\frac{\beta}{2}\cot\frac{\gamma}{2}\right]$$

$$+\left[\tan\frac{\gamma}{2}\tan\frac{\alpha}{2}-\cot\frac{\gamma}{2}\cot\frac{\alpha}{2}\right]$$

$$0=-\frac{\left[\cos^2\frac{\alpha}{2}\cos\frac{\beta}{2}\right]-\left[\sin^2\frac{\alpha}{2}\sin\frac{\beta}{2}\right]}{\left[\cos\frac{\alpha}{2}\cos\frac{\beta}{2}\sin\frac{\alpha}{2}\sin\frac{\beta}{2}\right]}$$

$$-\frac{\left[\cos^2\frac{\beta}{2}\cos\frac{\gamma}{2}\right]-\left[\sin^2\frac{\beta}{2}\sin\frac{\gamma}{2}\right]}{\left[\cos\frac{\beta}{2}\cos\frac{\gamma}{2}\sin\frac{\beta}{2}\sin\frac{\gamma}{2}\right]}$$

$$-\frac{\left[\cos^2\frac{\gamma}{2}\cos\frac{\alpha}{2}\right]-\left[\sin^2\frac{\gamma}{2}\sin\frac{\alpha}{2}\right]}{\left[\cos\frac{\gamma}{2}\cos\frac{\alpha}{2}\right]\left[\sin\frac{\gamma}{2}\sin\frac{\alpha}{2}\right]}$$

$$\text{Therefore,}\quad \Sigma\frac{\dfrac{4\cos(\alpha+\beta)}{2\cos\left(\alpha-\dfrac{\beta}{2}\right)}}{\sin\alpha\cdot\sin\beta}=0$$

$$\Rightarrow \quad \Sigma\, 2(\cos\alpha+\cos\beta)=0$$

$$\Rightarrow \sin\gamma(\cos\alpha+\cos\beta)+\sin\alpha(\cos\beta+\cos\gamma)+\sin\beta(\cos\gamma+\cos\alpha)=0$$

$$\Rightarrow \sin(\beta+\gamma)+\sin(\gamma+\alpha)+\sin(\alpha+\beta)=0$$

29. (a)

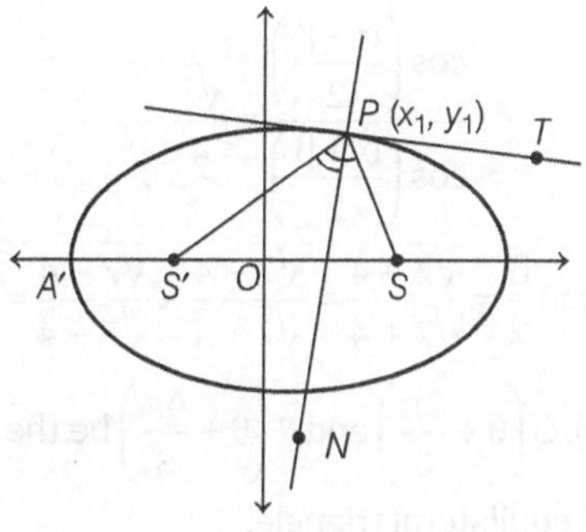

$\because$ PT and PN be the tangent and normal at $P(x_1, y_1)$ to the ellipse $\dfrac{x^2}{a^2}+\dfrac{y^2}{b^2}=1$

$\because$ Equation of tangent is $\Rightarrow \dfrac{xx_1}{a^2}+\dfrac{yy_1}{b^2}=1$

$\because$ Slope of tangent is $m_1=-\dfrac{b^2}{a^2}\dfrac{x_1}{y_1}$ and slope of normal is $m_2=\dfrac{a^2y_1}{b^2x_1}$

If $S(ae, 0)$ and $S'(-ae, 0)$ are the focii, then slope of SP is

$$m_3=\frac{y_1}{x_1-ae}$$

and slope of $S'P$ is $m_4=\dfrac{y_1}{x_1+ae}$

$$\Rightarrow \quad \tan\angle SPN=\frac{m_3-m_2}{1+m_2m_3}=\frac{aey_1}{b^2}$$

$\{$using $a^2-b^2=a^2e^2\}$

Similarly, $\quad \tan\angle S'PN=\dfrac{aey_1}{b^2}$

$$\Rightarrow \quad \angle SPN=\angle S'PN$$

30. (a) Given, ellipse is $S:\dfrac{x^2}{16}+\dfrac{y^2}{25}-1=0$

Equation of pair of tangents from point (8,5).

$$SS_1=T^2$$

$$\Rightarrow \left(\frac{x^2}{16}+\frac{y^2}{25}-1\right)\left(\frac{8^2}{16}+\frac{5^2}{25}-1\right)=\left(\frac{x(8)}{16}+\frac{y(5)}{25}-1\right)^2$$

$$\Rightarrow \quad \left(\frac{x^2}{16}+\frac{y^2}{25}-1\right)(4)=\left(\frac{x}{2}+\frac{y}{5}-1\right)^2$$

$$\Rightarrow \quad \frac{x^2}{4}+\frac{4y^2}{25}-4=\frac{x^2}{4}+\frac{y^2}{25}+\frac{2xy}{2\times5}-x-\frac{2y}{5}+1$$

$$\Rightarrow \quad \frac{x^2}{4}-\frac{x^2}{4}+\frac{4y^2}{25}-\frac{y^2}{25}-\frac{xy}{5}+x+\frac{2y}{5}-5=0$$

$$\Rightarrow \quad 4y^2-y^2-5xy+25x+10y-125=0$$

$$\Rightarrow \quad 3y^2-5xy+25x+10y-125=0$$

$$\therefore \quad a=0,\quad b=3\quad\text{and}\quad h=\frac{-5}{2}$$

$\because$ Let θ be the angle between both the tangent

$$\tan\theta=\left|\frac{2\sqrt{h^2-ab}}{a+b}\right|=\left|\frac{2\times\frac{5}{2}}{3}\right|$$

$$\Rightarrow \quad \tan\theta=\frac{5}{3}\quad\Rightarrow \theta=\tan^{-1}\frac{5}{3}$$

$$\Rightarrow \quad \theta\simeq 59^\circ$$

$$\Rightarrow \quad \theta=59^\circ\times\frac{\pi}{180}$$

$$\Rightarrow \quad l\simeq\frac{59\pi}{180}\Rightarrow [l]=1$$

31. (d) Equation of QR is $\Rightarrow T=0$ (chord of contact).

$$\frac{8x}{4}+\frac{27y}{9}=1$$

$$\Rightarrow \quad 2x+3y=1$$

Now, equation of the pair of lines passing through origin and points Q, R is given by

$$\left(\frac{x^2}{4}+\frac{y^2}{9}\right)=(2x+3y)^2$$

(using homogenisation)

$$\Rightarrow \quad 9x^2+4y^2=36(4x^2+9y^2+12xy)$$

$$\Rightarrow \quad 135x^2+432xy+320y^2=0$$

$$h=216,\quad a=135\quad\text{and}\quad b=320$$

$$\Rightarrow \quad \tan\theta=\left|\frac{2\sqrt{h^2-ab}}{a+b}\right|$$

$$\Rightarrow \quad \tan\theta=\frac{2\sqrt{(216)^2-135\times320}}{455}$$

$$\Rightarrow \quad \theta=\tan^{-1}\frac{8\sqrt{2916-2700}}{455}=\tan^{-1}\frac{48\sqrt{6}}{455}$$

32. (a) AB being the chord of contact of the ellipse from $P(3, 4)$ has its equation $T=0$

$$\frac{3x}{9}+\frac{4y}{4}=1$$

$$\Rightarrow \quad x+3y=3$$

If (h, k) is any point on the locus, then

$$\sqrt{(h-3)^2+(k-4)^2}=\left|\frac{h+3k-3}{\sqrt{1+9}}\right|$$

On squaring both sides, we get

$$10(h^2+k^2-6h-8k+25)=(h+3k-3)^2$$

Locus of (h, k) is $9x^2+y^2-6xy-54x-62y+241=0$

33. (c) Given, $m\cdot n=m+n$ and ellipse $\frac{x^2}{9}+\frac{y^2}{4}=1$

$$\Rightarrow \quad mn-m=n \Rightarrow m(n-1)=n$$

$\because$ m and n are positive integer

$\therefore$ n is divisible by $n-1$

$\Rightarrow n=2$ and $m=\frac{2}{1}=2$

Now, point is (2, 2)

$\therefore$ Equation of chord of contact w.r.t. (2, 2) is $T=0$

$$\Rightarrow \quad \frac{2x}{9}+\frac{2y}{4}=1 \Rightarrow 4x+9y=18$$

34. (a) Let $x\cos\alpha+y\sin\alpha=P$ ...(i)

be the chord PQ, then P is the desired distance

Homogenising the equations of ellipse with the help of Eq. (i), we get combined equations of OP and OQ

$$\Rightarrow \quad \frac{x^2}{a^2}+\frac{y^2}{b^2}=\left(\frac{x\cos\alpha+y\sin\alpha}{P}\right)^2$$

$$\Rightarrow \left(\frac{1}{a^2}-\frac{\cos^2\alpha}{P^2}\right)x^2+\left(\frac{1}{b^2}-\frac{\sin^2\alpha}{P^2}\right)y^2-\frac{2xy\sin\alpha\cos\alpha}{P^2}=0$$

$\because$ $OP\perp OQ$, we get

$$\frac{1}{a^2}-\frac{\cos^2\alpha}{P^2}+\frac{1}{b^2}-\frac{\sin^2\alpha}{P^2}=0$$

$$\Rightarrow \quad P=\frac{ab}{\sqrt{a^2+b^2}}$$

35. (a) Let (h, k) be the mid-point of the chord $7x+y-1=0$

$\therefore$ Equation of chord w.r.t. ellipse $\frac{x^2}{1}+\frac{y^2}{7}-1=0$ is

$$T=S_1$$

$$\Rightarrow \quad \frac{hx}{1}+\frac{ky}{7}-1=\frac{h^2}{1}+\frac{k^2}{7}-1$$

$$\Rightarrow \quad hx+\frac{ky}{7}=h^2+\frac{k^2}{7} \quad \text{...(i)}$$

and $\quad 7x+y=1$ ...(ii)

Eqs. (i) and (ii) are identical

$$\therefore \quad \frac{h}{7}=\frac{k}{7} \Rightarrow h=k$$

$\therefore$ Locus is $y=x \Rightarrow x-y=0$

36. (a) Given, ellipse is $\frac{x^2}{16}+\frac{y^2}{9}=1$

$$a^2=16 \text{ and } b^2=9$$

$$\therefore \quad e=\sqrt{1-\frac{b^2}{a^2}}=\sqrt{1-\frac{9}{16}}=\frac{\sqrt{7}}{4}$$

Let $P(4\cos\alpha, 3\sin\alpha)$ and $Q(4\cos\beta, 3\sin\beta)$ be a focal chord of the ellipse passing through the focus at $(\sqrt{7}, 0)$,

then $\quad \frac{3\sin\beta}{4\cos\beta-\sqrt{7}}=\frac{3\sin\alpha}{4\cos\alpha-\sqrt{7}}$

$$\Rightarrow \quad \frac{\sin(\alpha-\beta)}{\sin\alpha-\sin\beta}=\frac{\sqrt{7}}{4}$$

$$\Rightarrow \quad \frac{\cos\left(\frac{\alpha-\beta}{2}\right)}{\cos\left(\frac{\alpha+\beta}{2}\right)}=\frac{\sqrt{7}}{4}$$

$$\Rightarrow \tan\frac{\alpha}{2}\cdot\tan\frac{\beta}{2}=\frac{\sqrt{7}-4}{\sqrt{7}+4}=\frac{\sqrt{7}-4}{\sqrt{7}+4}\times\frac{\sqrt{7}-4}{\sqrt{7}-4}=\frac{8\sqrt{7}-23}{9}$$

37. (a) Let $P(\theta)$, $Q\left(\theta+\frac{2\pi}{3}\right)$ and $R\left(\theta+\frac{4\pi}{3}\right)$ be the three vertices of equilateral triangle.

Then, $P'=(a\cos\theta, b\sin\theta)$

$$Q'=\left\{a\cos\left(\theta+\frac{2\pi}{3}\right), b\sin\left(\theta+\frac{2\pi}{3}\right)\right\}$$

$$R'=\left\{a\cos\left(\theta+\frac{4\pi}{3}\right), b\sin\left(\theta+\frac{4\pi}{3}\right)\right\}$$

Let centroid of $\Delta P'Q'R'=(x', y')$

$$\therefore \quad x'=a\left[\frac{\cos\theta+\cos\left(\theta+\frac{2\pi}{3}\right)+\cos\left(\theta+\frac{4\pi}{3}\right)}{3}\right]$$

$$=\frac{a}{3}\left[\cos\theta+2\cos(\theta+\pi)\cos\frac{\pi}{3}\right]=0$$

$$y'=\frac{a}{3}\left[\sin\theta+\sin\left(\theta+\frac{2\pi}{3}\right)+\sin\left(\theta+\frac{4\pi}{3}\right)\right]$$

$$=\frac{a}{3}\left[\sin\theta+2\sin(\theta+\pi)\sin\frac{\pi}{3}\right]=0$$

$\therefore$ $(x', y')=(0, 0) \Rightarrow$ centre of ellipse

38. (d) Given, ellipse $x^2+9y^2=9$

$$\Rightarrow \quad \frac{x^2}{9}+\frac{y^2}{1}=1$$

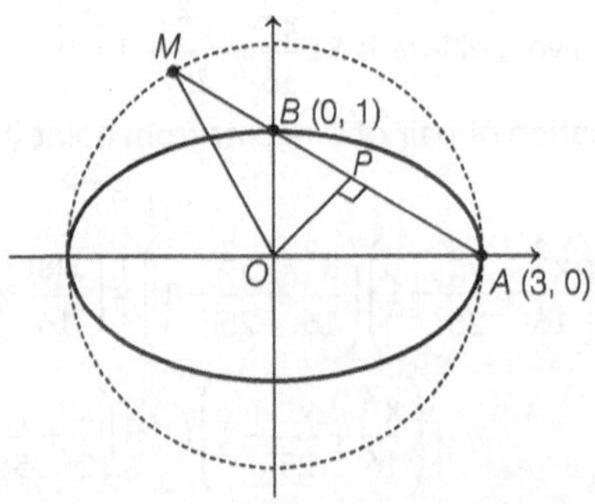

Now, equation of line AM is

$$y-0=\frac{1-0}{0-3}(x-3)$$

$$\Rightarrow \quad x+3y-3=0 \quad \text{...(i)}$$

Perpendicular distance from O (0, 0)

$$OP=\left|\frac{-3}{\sqrt{1^2+3^2}}\right|=\frac{3}{\sqrt{10}} \text{ units}$$

Now, AM

Equation of auxiliary circle at M is $x^2+y^2=a^2$

$$\Rightarrow \quad x^2+y^2=9 \quad \text{...(ii)}$$

Now, from Eq. (i), $x=3(1-y)$

then, $9(1-y)^2+y^2=9$ [from Eq. (ii)]

$$\Rightarrow \quad 9+9y^2-18y+y^2=9$$

$$\Rightarrow \quad 10y^2=18y$$

$$\Rightarrow \quad y=\frac{18}{10}=\frac{9}{5} \text{ and } x=\frac{-12}{5}$$

$$\Rightarrow \quad M=\left(\frac{-12}{5},\frac{9}{5}\right) \text{ and } A=(3,0)$$

$$\therefore \quad AM=\sqrt{\left(\frac{27}{5}\right)^2+\left(\frac{9}{5}\right)^2}$$

$$=\frac{9}{5}\sqrt{9+1}=\frac{9\sqrt{10}}{5}$$

$$\therefore \text{Area of triangle}=\frac{1}{2}\times AM\times OP$$

$$=\frac{1}{2}\times\frac{9\sqrt{10}}{5}\times\frac{3}{\sqrt{10}}=\frac{27}{10} \text{ sq units}$$

39. (c) Ellipse : $\frac{x^2}{25}+\frac{y^2}{9}=1$

Equation of normal at point $P\,(5\cos\theta, 3\sin\theta)$ on the ellipse is $5x\sec\theta-3y\operatorname{cosec}\theta=25-9=16$

Equation of normal of the auxiliary circle i.e.

$$\Rightarrow \quad x^2+y^2=a^2 \Rightarrow x^2+y^2=25$$

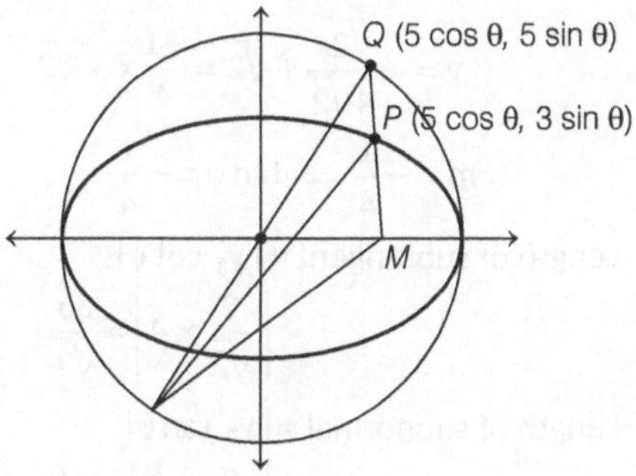

At point Q is $y=(\tan\theta)x$

$$\Rightarrow \quad \tan\theta=\frac{y}{x}$$

$$\Rightarrow \quad \sec\theta=\sqrt{1+\frac{y^2}{x^2}} \Rightarrow \operatorname{cosec}\theta=\sqrt{1+\frac{x^2}{y^2}}$$

Eliminating θ

$$\Rightarrow \quad 5x\sqrt{\frac{x^2+y^2}{x^2}}-3y\sqrt{\frac{y^2+x^2}{y^2}}=16$$

$$\Rightarrow \quad (5-3)\sqrt{x^2+y^2}=16$$

$$\Rightarrow \quad x^2+y^2=64$$

40. (d) Equation of ellipse

$$\frac{x^2}{4}+\frac{y^2}{3}=1 \qquad \because a^2=4, b^2=3$$

$\therefore$ Equation of director circle is

$$x^2+y^2=a^2+b^2$$

$$x^2+y^2=7$$

$\because$ Point of intersection of director circle and the line $y=x$ is

$$2x^2=7$$

$$\Rightarrow \quad x=\pm\sqrt{\frac{7}{2}}, \quad y=\pm\sqrt{\frac{7}{2}}$$

$\because$ Points $\left(\sqrt{\frac{7}{2}},\sqrt{\frac{7}{2}}\right)$ and $\left(-\sqrt{\frac{7}{2}},-\sqrt{\frac{7}{2}}\right)$

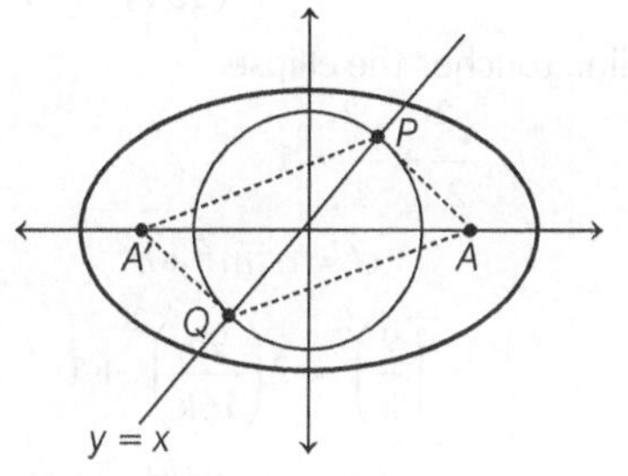

Focus of ellipse

$$(\pm ae, 0)=\left(\pm 2\left(\frac{1}{2}\right),0\right) \qquad \because e=\sqrt{1-\frac{3}{4}}=\sqrt{\frac{1}{4}}=\frac{1}{2}$$

$$=(\pm 1, 0)$$

and points are

$$P\left(\sqrt{\frac{7}{2}},\sqrt{\frac{7}{2}}\right) \text{ and } Q\left(-\sqrt{\frac{7}{2}}-\sqrt{\frac{7}{2}}\right)$$

$$A(1,0) \text{ and } A'(-1,0)$$

$$\therefore \quad PA=(y-0)=\frac{\sqrt{\frac{7}{2}}}{\sqrt{\frac{7}{2}}-1}(x-1)$$

$$y=\frac{\sqrt{7}}{\sqrt{7}-\sqrt{2}}(x-1)$$

$$\Rightarrow \quad (\sqrt{7}-\sqrt{2})y-\sqrt{7}x+\sqrt{7}=0$$

$$PA' \Rightarrow y-0=\frac{\sqrt{\frac{7}{2}}}{\sqrt{\frac{7}{2}}+1}(x+1)$$

$\Rightarrow \quad (\sqrt{7}+\sqrt{2})y-\sqrt{7}x-\sqrt{7}=0$

$$QA = y-0=\frac{-\sqrt{\frac{7}{2}}}{-\sqrt{\frac{7}{2}}-1}(x-1)$$

$\Rightarrow \quad (\sqrt{7}+\sqrt{2})y-\sqrt{7}x+\sqrt{7}=0$

$$QA' \Rightarrow y-0=\frac{-\sqrt{\frac{7}{2}}}{-\sqrt{\frac{7}{2}}+1}(x+1)$$

$\Rightarrow \quad (\sqrt{7}-\sqrt{2})y-\sqrt{7}x-\sqrt{7}=0$

41. (b) Given, equation of ellipse

$$\frac{x^2}{16}+\frac{y^2}{9}=1$$

Let $P(h, k)$ be the pole, then the equation of the polar is

$$T=0$$

$$\Rightarrow \quad \frac{hx}{16}+\frac{ky}{9}=1 \Rightarrow \frac{ky}{9}=-\frac{hx}{16}+1$$

$$\Rightarrow \quad y=-\left(\frac{9h}{16k}\right)x+\frac{9}{k}$$

The line touches the ellipse

$$\frac{x^2}{2}+\frac{y^2}{1}=1$$

$$\therefore \quad c^2=a^2m^2+b^2$$

$$\Rightarrow \quad \left(\frac{9}{k}\right)^2=2\left(\frac{9h}{16k}\right)^2+1$$

$$\Rightarrow \quad \frac{81}{k^2}=2\times\frac{81h^2}{256k^2}+1$$

$$\Rightarrow \quad 81=\frac{81h^2}{128}+k^2$$

$$\Rightarrow \quad 1=\frac{81h^2}{81\times128}+\frac{k^2}{81}$$

$$\therefore \quad \frac{h^2}{128}+\frac{k^2}{81}=1$$

$\therefore$ Locus is $\frac{x^2}{128}+\frac{y^2}{81}=1$

42. (c) Let the ellipse is $\frac{x^2}{a^2}+\frac{y^2}{b^2}=1$

$\because$ $y=3x$ and $2y=-5x$ is a pair of conjugate diameter, therefore

$$\Rightarrow \quad m_1m_2=-\frac{b^2}{a^2}\left\{m_1=3 \text{ and } m_2=\frac{-5}{2}\right\}$$

$$\therefore \quad 3\times\left(\frac{-5}{2}\right)=\frac{-b^2}{a^2}\Rightarrow\frac{b^2}{a^2}=\frac{15}{2}$$

$\therefore$ 'e' of the ellipse is lies between (0, 1)

$$\therefore \quad e=\sqrt{1-\frac{a^2}{b^2}}=\sqrt{1-\frac{2}{15}}=\sqrt{\frac{13}{15}}$$

43. (d) Ellipse : $3x^2+4y^2=12 \Rightarrow \frac{x^2}{4}+\frac{y^2}{3}=1$

$$\therefore \quad a^2=4 \text{ and } b^2=3$$

Now, the line is $y=\frac{5}{6}x$

$$\therefore \quad m_1=\frac{5}{6}$$

Let $y=m_2x$ be the conjugate diameter

Now, we know that

$$m_1m_2=\frac{-b^2}{a^2}$$

$$\Rightarrow \quad \frac{5}{6}\times m_2=\frac{-3}{4}$$

$$\Rightarrow \quad m_2=\frac{-3\times6}{4\times5}=\frac{-18}{20}=-\frac{9}{10}$$

$\therefore$ Equation of conjugate diameter

$$y=\frac{-9}{10}x \Rightarrow 10y+9x=0$$

44. (a) Given, ellipse is $\frac{x^2}{16}+\frac{y^2}{9}=1$

Now, point $P\left(\frac{\pi}{4}\right)=\left(a\cos\frac{\pi}{4}, b\sin\frac{\pi}{4}\right)$

$$=\left(4\times\frac{1}{\sqrt{2}}, 9\times\frac{1}{\sqrt{2}}\right)=\left(\frac{4}{\sqrt{2}}, \frac{9}{\sqrt{2}}\right)$$

Now, equation of tangent at $P\left(\frac{\pi}{4}\right)$

$$\Rightarrow \quad \frac{x}{16}\times\frac{4}{\sqrt{2}}+\frac{y}{9}\left(\frac{9}{\sqrt{2}}\right)=1$$

$$\Rightarrow \quad \frac{x}{4\sqrt{2}}+\frac{y}{\sqrt{2}}=1$$

$$\Rightarrow \quad y=-\frac{\sqrt{2}x}{4\sqrt{2}}+\sqrt{2}=\frac{-1}{4}x+\sqrt{2}$$

$$\therefore \quad m=-\frac{1}{4} \Rightarrow \tan\phi=-\frac{1}{4}$$

$\therefore$ Length of subtangent $=|y_1 \cot\phi|$

$$=\left|\frac{9}{\sqrt{2}}\times4\right|=\frac{36}{\sqrt{2}}$$

$\therefore$ Length of subnormal $=|y_1\tan\phi|$

$$=\left|\frac{9}{\sqrt{2}}\times\frac{1}{4}\right|=\frac{9}{4\sqrt{2}}$$

Now, $\frac{\text{length of subtangent}}{\text{length of subnormal}}=\frac{36}{\frac{9}{4}}=16$

45. (d) $\because$ Given, ellipse are

$\Rightarrow \quad \frac{x^2}{4}+\frac{y^2}{1}=1$ and $\frac{x^2}{1}+\frac{y^2}{a^2}=1$

$\Rightarrow \quad a_1^2=4,\ b_1^2=1$ and $a_2^2=1, b_2^2=a^2$

$a^2=b^2-5b+7$ for the two ellipse to intersect at four distinct point

$$b_2>1$$
$$b_2^2>1$$
$$\Rightarrow \quad b^2-5b+7>1$$
$$\Rightarrow \quad b^2-5b+6>0$$
$$\Rightarrow \quad (b-2)(b-3)>0$$
$$\therefore \quad b\in(-\infty,2)\cup(3,\infty)$$

$\therefore$ b does not lies in [2, 3]

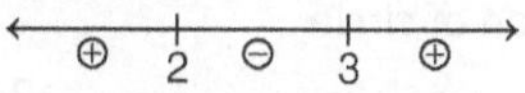

46. (b) Given, ellipse : $S_1: ax^2+by^2=3$

$$\Rightarrow \quad \frac{x^2}{\frac{3}{a}}+\frac{y^2}{\frac{3}{b}}=1$$
$$\Rightarrow \quad A_1^2=\frac{3}{a}$$
$$\text{and} \quad B_1^2=\frac{3}{b}$$
$$S_2:(a+b)x^2+(a-b)y^2=3$$
$$\Rightarrow \quad \frac{x^2}{\frac{3}{a+b}}+\frac{y^2}{\frac{3}{a-b}}=1$$
$$\Rightarrow \quad A_2^2=\frac{3}{a+b}$$
$$\text{and} \quad B_2^2=\frac{3}{a-b}$$

$\because$ Area of ellipse is $=\pi AB$

According to the question,

$$\pi A_1B_1=\pi A_2B_2$$
$$\Rightarrow \quad \sqrt{\frac{3}{a}}\sqrt{\frac{3}{b}}=\sqrt{\frac{3}{a+b}}\sqrt{\frac{3}{a-b}}$$
$$\Rightarrow \quad a^2-b^2=ab$$
$$\Rightarrow \quad a^2-ab-b^2=0$$
$$\Rightarrow \quad \left(\frac{a}{b}\right)^2-\left(\frac{a}{b}\right)-1=0$$
$$\Rightarrow \quad t^2-t-1=0 \qquad \left\{\because \frac{a}{b}=t\right\}$$
$$\therefore \quad t=\frac{1\pm\sqrt{1+4}}{2}=\frac{\sqrt{5}+1}{2} \qquad \{\because a, b \text{ are +ve}\}$$
$$\Rightarrow \quad \frac{a}{b}=\frac{\sqrt{5}+1}{2}$$

47. (a) Given, ellipse

$$\frac{(x-4)^2}{25}+\frac{y^2}{4}=1 \text{ and } \frac{(x+1)^2}{1}+\frac{y^2}{4}=1$$

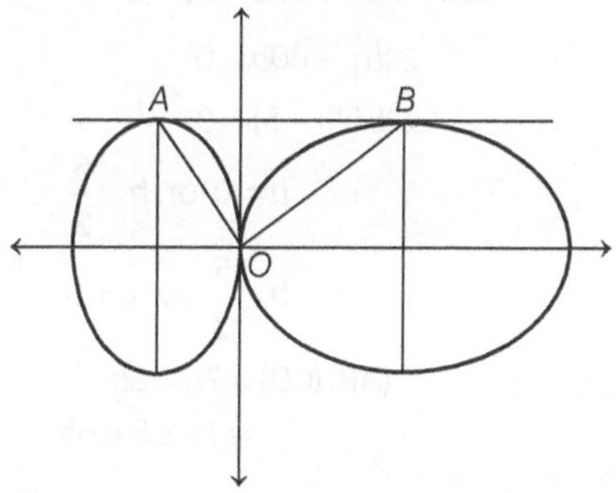

$\because$ $B(4, 2)$ is one end of the minor axis of the ellipse

$$\frac{(x-4)^2}{25}+\frac{y^2}{4}=1$$

and $A(-1, 2)$ is one end of the major axis of the second ellipse, therefore

$$AB=5, OB=\sqrt{16+4}=\sqrt{20}$$
$$\text{and} \quad OA=\sqrt{1+4}=\sqrt{5}$$
$$\because \quad OA^2+OB^2=20+5=25=AB^2$$

$\because$ OAB is right angled triangle.

$$\because \quad \angle AOB=\frac{\pi}{2}$$

48. (d) Let $OC=b, OF=6$ (given) and $CF=a$

$\because$ Given, Inradius $=1$

$$r=\frac{\Delta}{S}=1$$
$$\Rightarrow \quad \frac{\frac{1}{2}\times b\times 6}{\frac{a+b+6}{2}}=1$$

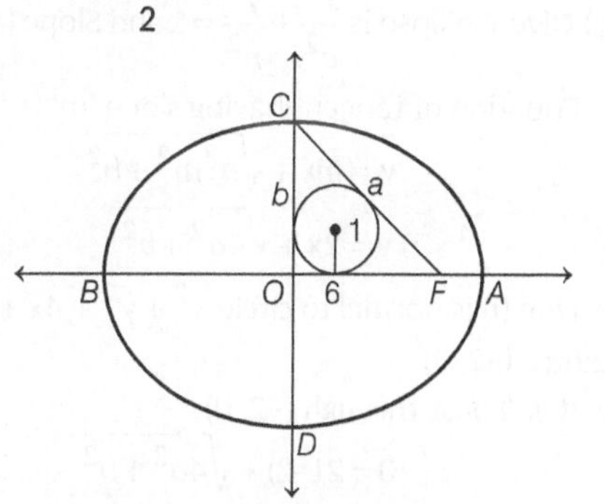

$$\Rightarrow \quad 6b=a+b+6$$
$$\Rightarrow \quad 5b-a=6 \qquad \text{...(i)}$$
$$\Rightarrow \quad a=5b-6$$
$$\text{and} \quad OF=6$$
$$\Rightarrow \quad OF^2=36$$
$$\Rightarrow \quad a^2e^2=36$$
$$\Rightarrow \quad a^2-b^2=36$$

$$(5b-6)^2 - b^2 = 36 \quad \text{[from Eq. (i)]}$$

$$(5b-6)^2 = 36 + b^2$$

$\Rightarrow$ $25b^2 + 36 - 60b = 36 + b^2$

$\Rightarrow$ $24b^2 - 60b = 0$

$\Rightarrow$ $12b(2b-5) = 0$

$\Rightarrow$ $b = 0$ or $b = \frac{5}{2}$

$\because$ $b = \frac{5}{2} \Rightarrow a = 5 \times \frac{5}{2} - 6 = \frac{13}{2}$

Thus, $(AB)(CD) = 2a \times 2b$

$= 13 \times 5 = 65$

49. (b) $\because$ Equation of tangent to the ellipse $\frac{x^2}{a^2} + \frac{y^2}{b^2} = 1$ having slope 'm' is $y = mx + \sqrt{a^2m^2 + b^2}$

$\therefore$ The line $y = mx + \sqrt{4m^2+1}$ $\left\{\because S: \frac{x^2}{4} + y^2 = 1\right\}$ ...(i)

$\because$ The line touches the circle $x^2 + y^2 = 3$

$\therefore$ Radius = distance between centre (0, 0) to line (i)

$$\sqrt{3} = \left|\frac{\sqrt{4m^2+1}}{\sqrt{1+m^2}}\right|$$

On squaring both sides, we get

$$3 = \frac{4m^2+1}{m^2+1}$$

$\Rightarrow$ $3m^2 + 3 = 4m^2 + 1$

$\Rightarrow$ $m^2 = 2 \Rightarrow m = \pm\sqrt{2}$

$\therefore$ Product of slope is -2.

50. (c) Given, ellipse is $\frac{x^2}{a^2} + \frac{y^2}{b^2} = 1$ and Slope $(m) = 2$

$\because$ Equation of tangent having slope 'm' is

$$y = mx + \sqrt{a^2m^2 + b^2}$$

$$y = 2x + \sqrt{4a^2 + b^2} \quad \text{...(i)}$$

$\because$ Line (i) is normal to circle $x^2 + y^2 + 4x + 1 = 0$ having centre $(-2, 0)$

$\because$ It will pass through $(-2, 0)$

$$0 = 2(-2) + \sqrt{4a^2 + b^2}$$

$\Rightarrow$ $4 = \sqrt{4a^2 + b^2} \Rightarrow 4a^2 + b^2 = 16$

$\because$ $AM \geq GM$

$\Rightarrow$ $\frac{4a^2 + b^2}{2} \geq \sqrt{4a^2b^2}$

$\Rightarrow$ $\frac{16}{2} \geq 2ab \Rightarrow ab \leq 4$

51. (d) $\because$ Equation of ellipse passing through points of intersection of the two given ellipse is $S_1 + \lambda S_2 = 0$

$$\left(\frac{x^2}{a^2} + \frac{y^2}{b^2} - 1\right) + \lambda\left(\frac{x^2}{b^2} + \frac{y^2}{a^2} - 1\right) = 0$$

$\Rightarrow$ $x^2\left(\frac{1}{a^2} + \frac{\lambda}{b^2}\right) + y^2\left(\frac{1}{b^2} + \frac{\lambda}{a^2}\right) + (-1-\lambda) = 0$ ...(i)

$\because$ Eq. (i) represents circle

Coefficient of x^2 = Coefficient of y^2

$\because$ $\frac{1}{a^2} + \frac{\lambda}{b^2} = \frac{1}{b^2} + \frac{\lambda}{a^2}$

$\Rightarrow$ $\lambda\left(\frac{1}{b^2} - \frac{1}{a^2}\right) = \left(\frac{1}{b^2} - \frac{1}{a^2}\right)$

$\Rightarrow$ $\lambda = 1$

$\therefore$ Equation of circle

$$x^2 + y^2 = \frac{2}{\frac{1}{a^2} + \frac{1}{b^2}} = \frac{2a^2b^2}{a^2 + b^2}$$

52. (b) For point P, y-coordinate $= 3$

Given, ellipse is $9x^2 + 25y^2 = 225$

$\Rightarrow$ $\frac{9x^2}{225} + \frac{25y^2}{225} = 1$

$\Rightarrow$ $\frac{x^2}{25} + \frac{y^2}{9} = 1$

$\Rightarrow$ $a^2 = 25$ and $b^2 = 9$

$\therefore$ $e = \sqrt{1 - \frac{b^2}{a^2}} = \sqrt{1 - \frac{9}{25}} = \frac{4}{5}$

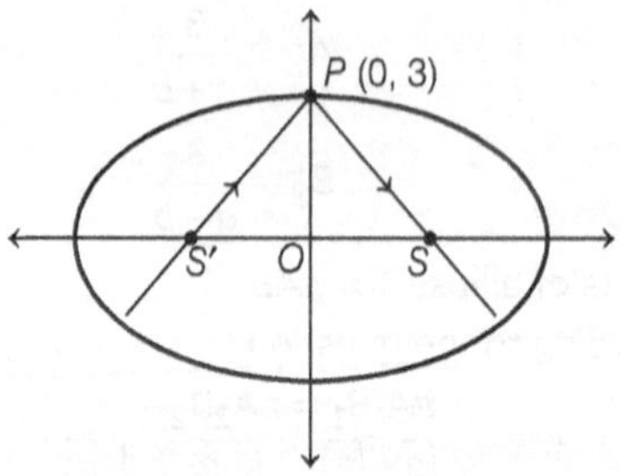

Therefore, coordinate of P

Put $y = 3$

$$\frac{x^2}{25} + \frac{9}{9} = 1$$

$$\frac{x^2}{25} = 0$$

$\Rightarrow$ $x = 0$

$\because$ $P(0, 3)$

Now, focii $\Rightarrow (\pm ae, 0) = (\pm 4, 0)$

Therefore, reflected ray PS is = line joining $P(0, 3)$, $S(4, 0)$

$$\Rightarrow \quad y-0=\frac{3-0}{0-4}(x-4)$$

$$\Rightarrow \quad -4y=3x-12$$

$$\Rightarrow \quad 3x+4y-12=0 \Rightarrow 3x+4y=12$$

53. (a) Given, ellipse

$$12(2x-3y+1)^2+16(3x+2y-2)^2=48$$

Ellipse can be written as

$$12\left(\frac{2x-3y+1}{\sqrt{13}}\times\sqrt{13}\right)^2+16\left(\frac{3x+2y-2}{\sqrt{13}}\times\sqrt{13}\right)^2=48$$

$$\Rightarrow 12\times 13\left(\frac{2x-3y+1}{\sqrt{13}}\right)^2+16\times 13\left(\frac{3x+2y-2}{\sqrt{13}}\right)^2=48$$

$$\Rightarrow \quad \frac{\left(\frac{2x-3y+1}{\sqrt{13}}\right)^2}{\frac{48}{12\times 13}}+\frac{\left(\frac{3x+2y-2}{\sqrt{13}}\right)^2}{\frac{48}{13\times 16}}=1$$

$$\Rightarrow \quad \frac{X^2}{\frac{4}{13}}+\frac{Y^2}{\frac{3}{13}}=1$$

$\therefore$ Centre of conic $2x-3y+1=0$

$$2x=3y-1$$

$$x=\frac{3y-1}{2} \quad \text{...(i)}$$

and $\quad 3x+2y-2=0 \quad$ ...(ii)

Put $x=\frac{3y-1}{2}$ in Eq. (ii)

$$\Rightarrow \quad 3\left(\frac{3y-1}{2}\right)+2y-2=0$$

$$\Rightarrow \quad 9y-3+4y-4=0 \Rightarrow 13y=7$$

$$\Rightarrow \quad y=\frac{7}{13} \quad \text{and} \quad x=\frac{\frac{21}{13}-1}{2}=\frac{4}{13}$$

$\because$ Centre is $\left(\frac{4}{13},\frac{7}{13}\right)$

54. (b) Given, ellipse $\frac{x^2}{9}+\frac{y^2}{1}=1 \quad$ ... (i)

and circle $x^2+y^2=3 \quad$ ... (iii)

The point of intersection by solving Eqs. (i) and (ii) in first quadrant $(3/2, \sqrt{3}/2)$.

Differentiating Eqs. (i) and (ii) w.r.t. x, we have

Let $m_1=\frac{dy}{dx}=\frac{-x}{9y}$ and $m_2=\frac{dy}{dx}=\frac{-x}{y}$

At $\left(\frac{3}{2},\frac{\sqrt{3}}{2}\right)$

$$m_1=-\frac{1}{3\sqrt{3}},\ m_2=-\sqrt{3}$$

If angle between both curves is θ, then

$$\tan\theta=\left|\frac{m_1-m_2}{1+m_1m_2}\right|$$

$$=\left|\frac{\frac{-1}{3\sqrt{3}}+\sqrt{3}}{1+\left(\frac{-1}{3\sqrt{3}}\right)(-\sqrt{3})}\right|=\frac{2}{\sqrt{3}}$$

55. (4) Given, ellipse is

$$\frac{x^2}{16}+\frac{y^2}{9}=1$$

$$\Rightarrow \quad a^2=16$$

and $\quad b^2=9$

$$\therefore \quad e=\sqrt{1-\frac{b^2}{a^2}}=\sqrt{1-\frac{9}{16}}=\frac{\sqrt{7}}{4}$$

$\therefore \quad$ Foci $=(\pm ae, 0)=(\pm\sqrt{7}, 0)$

and radius of circle is $=\sqrt{7+9}=\sqrt{16}=4$ units

56. (0.333)

Any point on the ellipse be $(a\cos\theta, a\sin\theta)$

Clearly the centre is (0, 0)

Now, distance $=r$

$$\Rightarrow \sqrt{a^2\cos^2\theta+a^2\sin^2\theta}=a$$

Also, $a^2\cos^2\theta+2a^2\sin\theta\cos\theta+2a^2\sin^2\theta=1$

$$a^2=\frac{1}{1+\sin 2\theta+\sin^2\theta}$$

$$\therefore \quad r=a=\frac{1}{\sqrt{1+\sin 2\theta+\sin^2\theta}}$$

$$r=\frac{1}{\sqrt{\frac{3}{2}+\sin 2\theta-\frac{1}{2}\cos 2\theta}}$$

r will be maximum when $P=\frac{3}{2}+\sin 2\theta-\frac{1}{2}\cos 2\theta$ is minimum

$$P_{min}=\frac{3}{2}-\sqrt{1^2+\left(\frac{1}{2}\right)^2}$$

$\{\because a\sin\theta+b\cos\in[-\sqrt{a^2+b^2}, \sqrt{a^2+b^2}]\}$

$$=\frac{3}{2}-\sqrt{1+\frac{1}{4}}$$

$$=\frac{3}{2}-\frac{\sqrt{5}}{2}=\frac{3-\sqrt{5}}{2}$$

$$\therefore \quad r_{max}=\frac{1}{\sqrt{\frac{3-\sqrt{5}}{2}}}=\frac{\sqrt{2}}{\sqrt{3-\sqrt{5}}}=\sqrt{\frac{a}{b-\sqrt{c}}}$$

$\therefore a=2,\ b=3$ and $c=5$

$$\because \quad \frac{a+b+c}{30}=\frac{2+3+5}{30}=\frac{10}{30}=\frac{1}{3}=0.333$$

57. (3) Given, equation is

$$(x-1)^2+4(y-2)^2=-13+1+16=4$$

$$\Rightarrow \quad \frac{(x-1)^2}{4}+\frac{(y-2)^2}{1}=1$$

Here, $a^2=4$ and $b^2=1$

$$\therefore \quad e=\sqrt{1-\frac{1}{4}}=\frac{\sqrt{3}}{2}$$

$\therefore$ Length of latusrectum $=\frac{2b^2}{a}=\frac{2(1)}{2}=1$

Statement I is true.

$\frac{X^2}{4}+\frac{Y^2}{1}=1$, where $X=x-1, Y=y-2$

The focii $\Rightarrow (1\pm\sqrt{3},2)$

Distance between the focii is $=(1+\sqrt{3})-(1-\sqrt{3})=2\sqrt{3}$

Statement II is false

$\because$ Sum of focal distance is

$$SP+S'P=2a=2(2)=4$$

Statement III is true.

Tangents at vertices

Vertices $X=0$ and $Y=0$

$\Rightarrow \quad x-1=\pm 2$ and $y-2=0$

$\Rightarrow \quad x=1\pm 2$ and $y=2$

$\Rightarrow \quad x=3,-1$

Vertices $\Rightarrow$ (3, 2) and (-1, 2)

$\because$ Tangent at vertices is $x=3$ or $x=-1$

which meets the line $y=3$ at (3, 3) at (-1, 3)

$\therefore P=(3,3)$ and $Q=(-1,3)$

$$m_{SP}\times m_{SQ}=\left(\frac{3-2}{3-1-\sqrt{3}}\right)\left(\frac{3-2}{-1-1-\sqrt{3}}\right)=-1$$

$$\Rightarrow \quad m_{SP}\perp m_{SQ}$$

$\therefore$ Statement IV is true.

58. (0.109) Let (h, k) be the point of intersection of tangents at θ_1 and θ_2.

$$\therefore \quad h=\frac{a\cos\left(\frac{\theta_1+\theta_2}{2}\right)}{\cos\left(\frac{\theta_1-\theta_2}{2}\right)}$$

and

$$k=\frac{b\sin\left(\frac{\theta_1+\theta_2}{2}\right)}{\cos\left(\frac{\theta_1-\theta_2}{2}\right)}$$

$$\frac{h}{a}=\frac{\cos\left(\frac{\theta_1+\theta_2}{2}\right)}{\cos\left(\frac{\theta_1-\theta_2}{2}\right)} \quad \text{...(i)}$$

$$\frac{k}{b}=\frac{\sin\left(\frac{\theta_1+\theta_2}{2}\right)}{\cos\left(\frac{\theta_1-\theta_2}{2}\right)} \quad \text{...(ii)}$$

Squaring and adding both equations,we get

$$\frac{h^2}{a^2}+\frac{k^2}{b^2}=\frac{1}{\cos^2\left(\frac{\theta_1-\theta_2}{2}\right)} \quad \text{...(iii)}$$

It is given that

$$b(\sin\theta_1+\sin\theta_2)=3$$

$$\Rightarrow \quad \sin\theta_1+\sin\theta_2=1 \quad \{\because b=3\}$$

$$\Rightarrow 2\sin\left(\frac{\theta_1+\theta_2}{2}\right)\cos\left(\frac{\theta_1-\theta_2}{2}\right)=1$$

Now, $$\Rightarrow \frac{k}{b}=\frac{\sin\left(\frac{\theta_1+\theta_2}{2}\right)}{\cos\left(\frac{\theta_1-\theta_2}{2}\right)}=\frac{1}{2\cos^2\left(\frac{\theta_1-\theta_2}{2}\right)} \quad \text{...(iv)}$$

From Eqs. (iii) and (iv), we get

$$\frac{h^2}{a^2}+\frac{k^2}{b^2}=\frac{2k}{b}$$

$$\Rightarrow \quad \frac{h^2}{25}+\frac{k^2}{9}=\frac{2k}{3}$$

$\therefore$ Locus is $\frac{x^2}{25}+\frac{y^2}{9}=\frac{2y}{3}$

$$\Rightarrow \quad 9x^2+25y^2=150y$$

$$\Rightarrow \quad 9x^2+25y^2-150y=0$$

$\therefore$ Compare with $ax^2+by^2+2gx+2fy+c=0$, we get

$$a=9, b=25, g=0, f=-75 \text{ and } c=0$$

$\because$ Required value $=\frac{a+b+c+g-f}{1000}$

$$=\frac{9+25+75}{1000}=\frac{109}{1000}=0.109$$

59. (707.2) Given, equation is $\frac{x^2}{25}+\frac{y^2}{9}=1$

$$\Rightarrow \quad a^2=25 \text{ and } b^2=9$$

$$\Rightarrow \quad a=5 \text{ and } b=3$$

The equation of tangent at $(5\cos\theta, 3\sin\theta)$

$$x\frac{5\cos\theta}{25}+y\frac{3\sin\theta}{9}=1$$

$$\frac{x\cos\theta}{5}+\frac{y\sin\theta}{3}=1 \quad \text{...(i)}$$

After rotation equation of tangent is

$$\Rightarrow \quad \frac{x}{5}\cos(90°+\theta)+\frac{y}{3}\sin(90°+\theta)=1$$

$$\Rightarrow \quad \frac{-x}{5}\sin\theta+\frac{y}{3}\cos\theta=1 \quad \text{...(ii)}$$

On subtracting Eqs. (i) and (ii), we get

$$\frac{x}{5}(\sin\theta+\cos\theta)+\frac{y}{3}(\sin\theta-\cos\theta)=0$$

$$\Rightarrow \quad \sin\theta\left(\frac{x}{5}+\frac{y}{3}\right)+\cos\theta\left(\frac{x}{5}-\frac{y}{3}\right)=0$$

$$\Rightarrow \quad \sin\theta(3x+5y)+\cos\theta(3x-5y)=0$$

$$\Rightarrow \quad \sin\theta(3x+5y)=(5y-3x)\cos\theta$$

$$\Rightarrow \quad \frac{\sin\theta}{5y-3x}=\frac{\cos\theta}{3x+5y}$$

$$\tan\theta=\frac{5y-3x}{5y+3x}$$

$$\therefore \quad \sin\theta=\frac{(5y-3x)}{\sqrt{(5y+3x)^2+(5y-3x)^2}}=\frac{5y-3x}{\sqrt{50y^2+18x^2}}$$

$$\therefore \quad \cos\theta=\frac{5y+3x}{\sqrt{(5y+3x)^2+(5y-3x)^2}}=\frac{5y+3x}{\sqrt{50y^2+18x^2}}$$

By Eq. (i),

$$\frac{x}{5}\cdot\frac{(5y+3x)}{\sqrt{50y^2+18x^2}}+\frac{y}{3}\cdot\frac{(5y-3x)}{\sqrt{50y^2+18x^2}}=1$$

$$\Rightarrow \quad 3x(5y+3x)+5y(5y-3x)=15\sqrt{50y^2+18x^2}$$

$$\Rightarrow \quad 15xy+9x^2+25y^2-15xy=15\sqrt{50y^2+18x^2}$$

$$9x^2+25y^2=15(\sqrt{50y^2+18x^2})$$

Solving both sides,

$$(9x^2+25y^2)^2=225(50y^2+18x^2)$$

$$\Rightarrow 81x^4+625y^4+450x^2y^2=11250y^2+4050x^2$$

$$\Rightarrow 81x^4+625y^4+450x^2y^2-11250y^2-4050x^2=0$$

$$\Rightarrow \quad ax^4+by^4+ex^2y^2+dy^2+cx^2+fxy+g=0$$

$a=81,\ b=625,\ e=450,\ c=-4050,\ d=-11250$

$$\therefore \quad \frac{a+b+c+d+e+f+g}{20}=\frac{81+625+450-4050-11250}{20}=\frac{-14144}{20}=-70702$$

Numerical value $=707.2$

60. (7.070) Given, ellipse is

$$\frac{x^2}{16}+\frac{y^2}{9}=1$$

Clearly the vertices of the square parallel on the director circle of the given ellipse

$$\frac{x^2}{16}+\frac{y^2}{9}=1$$

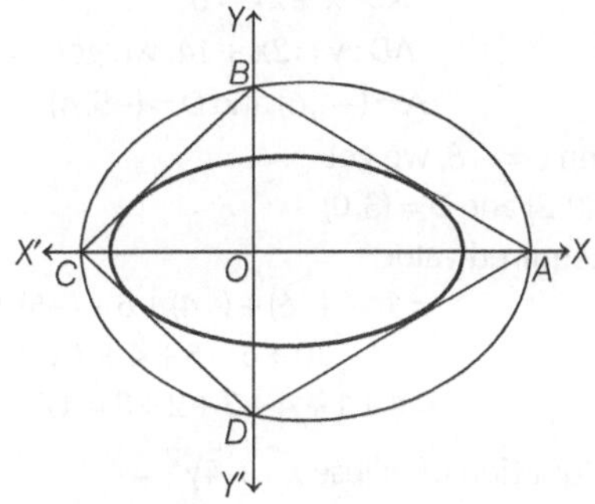

$\because$ The equation of director circle is

$$x^2+y^2=a^2+b^2$$

$$x^2+y^2=16+9=25=r^2$$

$$r=5$$

Thus, the length of $AC=$ (diameter)

$$=2\times5=10$$

$\because\ a\sqrt{2}=10$ (where a is side of square)

$$\Rightarrow \quad a=\frac{10}{\sqrt{2}}=5\sqrt{2}=5\times1.414=7.070$$

61. (17) Let $AD \Rightarrow y=2x+c$

So, $BC \Rightarrow y=2x+4$

$$AB=x+2y=8$$

and $\quad DC=x+2y-3=0$

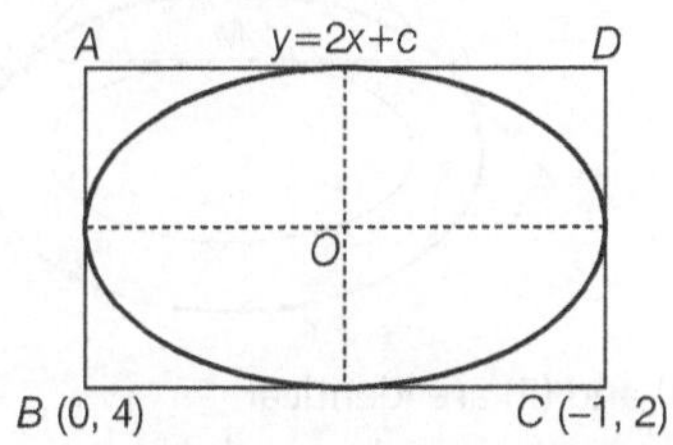

Let $\quad BC=\sqrt{5}$ and $AB=2a$

Clearly $\quad 2a=\sqrt{5} \Rightarrow a=\frac{\sqrt{5}}{2}$

It is given that area of ellipse $=\frac{5\pi}{2}$

$$\Rightarrow \quad \pi ab=\frac{5\pi}{2}$$

$$\Rightarrow \quad \pi\frac{\sqrt{5}}{2}b=\frac{5\pi}{2}\Rightarrow b=\sqrt{5}$$

Also, $\quad b=\left|\frac{c-4}{2\sqrt{5}}\right|=\sqrt{5}$

$$\Rightarrow \quad \frac{c-4}{2\sqrt{5}}=\pm\sqrt{5}$$

$$\Rightarrow \quad c-4=\pm10$$

$$\Rightarrow \quad c=4\pm10$$

$$c=14,-6$$

When $c=14$

On solving the equations

$$AB: x+2y=8$$

and $$AD: y=2x+14, \text{ we get}$$

$$A=(-4,6) \text{ and } D=(-5,4)$$

When $c=-6$, we get

$A=(3,2)$ and $D=(3,0)$

$\therefore$ Required value

$$=14+(-6)+(-4)+6+(-5)+4+(3)+2+3$$
$$=14-10+6-5+4+3+2+3$$
$$=4+1+4+3+2+3=17$$

62. (2) Equation of ellipse $x^2+4y^2=4$

$$\frac{x^2}{4}+y^2=1$$

Equation of tangent at M $(2\cos\theta, \sin\theta)$ is

$$\frac{x\,(2\cos\theta)}{4}+y\sin\theta=1$$

$$\frac{x\cos\theta}{2}+y\sin\theta=1$$

Equation of chord of contact of $R\,(h,k)$ w.r.t.

$$\frac{x^2}{6}+\frac{y^2}{3}=1 \text{ is } \frac{hx}{6}+\frac{ky}{3}=1$$

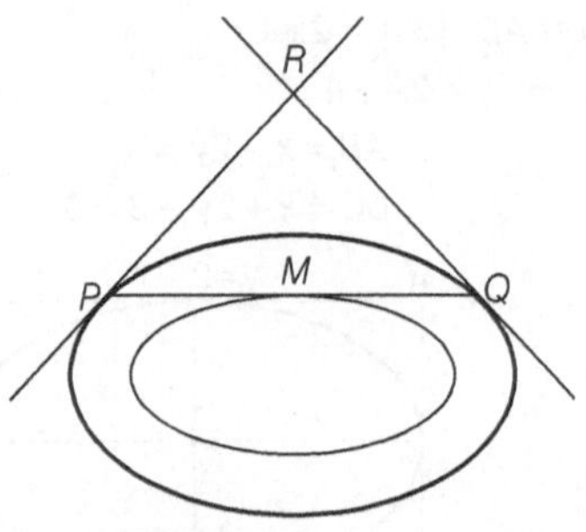

Eqs. (i) and (ii) are identical

$$\frac{\frac{h}{3}}{\cos\theta}=\frac{\frac{k}{3}}{\sin\theta}=1$$

$$\Rightarrow \qquad h^2+k^2=9$$

$\Rightarrow$ (h,k) lies on director circle of ellipse

$$\frac{x^2}{6}+\frac{y^2}{3}=1$$

$\therefore$ Tangent will intersect at right angle $=\frac{\pi}{2}$

$$\therefore \qquad k=2$$

63. (2) Equation of ellipse is

$$\frac{x^2}{(a^2+2a+2)^2}+\frac{y^2}{(a^2+1)^2}=1$$

$$\Rightarrow \qquad \frac{x^2}{A^2}+\frac{y^2}{B^2}=1$$

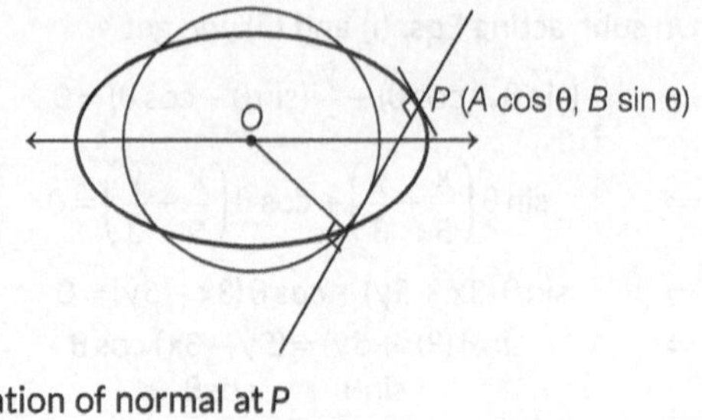

Equation of normal at P

$$Ax\sec\theta-By\,\text{cosec}\,\theta=A^2-B^2$$

$\because$ Radius of circle = perpendicular distance from centre to normal at P

$$=\left|\frac{A^2-B^2}{\sqrt{A^2\sec^2\theta+B^2\text{cosec}^2\theta}}\right|$$

$$=\left|\frac{A^2-B^2}{\sqrt{A^2+B^2+A^2\tan^2\theta+B^2\cot^2\theta}}\right|$$

$$\le\frac{|A^2-B^2|}{\sqrt{A^2+B^2+2AB}}$$

$$\therefore \qquad r_{\max}=|A-B|=a^2+2a+2-(a^2+1)$$
$$=2a+1$$
$$5=2a+1\Rightarrow a=2$$

64. (9) Given, circle is $x^2+y^2-8x+6y+1=0$

Centre O $(4,-3)$

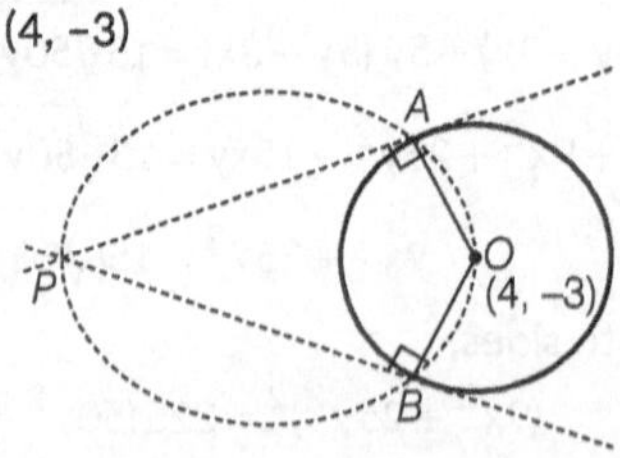

The circumcircle of ΔPAB will circumscribe the quadrilateral $PBOA$. Also, hence one of the diameters must be OP.

$\therefore$ Equation of circumcircle of ΔPAB will be -

$$(x-2)(x-4)+(y-3)(y+3)=0$$

$$\Rightarrow \qquad x^2+y^2-6x-1=0 \qquad \text{...(i)}$$

Director circle of given ellipse will be

$$(x+5)^2+(y-3)^2=9+b^2$$

$$\Rightarrow \qquad x^2+y^2+10x-6y+25-b^2=0 \qquad \text{...(ii)}$$

$\because$ From Eqs (i) and (ii) by applying condition of orthogonality, we get

$$2(g_1g_2+f_1f_2)=c_1+c_2$$

$$\Rightarrow \qquad 2[-3(5)+0(-3)]=-1+25-b^2$$

$$\Rightarrow \qquad -30=24-b^2$$

$$\because \qquad b^2=54$$

$$\therefore \qquad \frac{b^2}{6}=\frac{54}{6}=9$$

65. (3) Let $B=(a,0), C=(-a,0)$ and $A=(0,a)$

Equation of line BA is

$$x+y=a \text{ and that of } CA \text{ is } y=x+a$$

Let $P=(x,y)$ distance from $BC=y$

and area of $PRAS=PR\times PS$

$$=\left(\frac{x+y-a}{\sqrt{2}}\right)\times\left(\frac{x-y+a}{\sqrt{2}}\right)$$

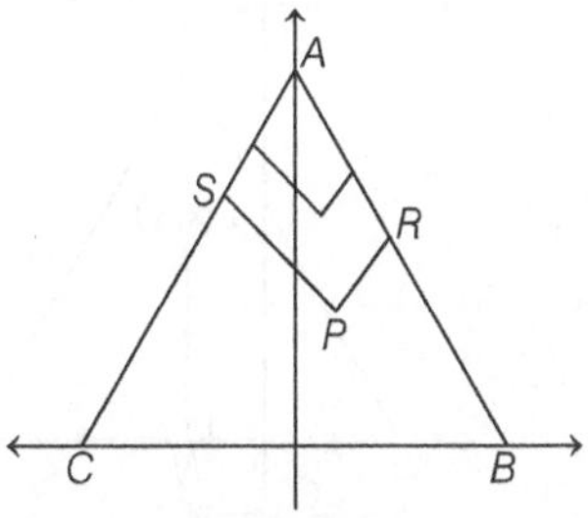

According to the question,

$$y^2=\pm\frac{1}{2}\left\{\frac{x^2-(y-a)^2}{2}\right\}$$

$\because$ $$\pm 4y^2=x^2-y^2-2ay-a^2$$

When it is $+4y^2 \Rightarrow$ It forms a Hyperbola

When it is $-4y^2 \Rightarrow$ It forms an ellipse

$\because$ $$-4y^2=x^2-y^2-2ay-a^2=0$$

$\Rightarrow$ $$x^2+3y^2-2ay-a^2=0$$

$\Rightarrow$ $$x^2+3\left(y-\frac{1}{3}\right)^2=\frac{4a^2}{3}$$

$\Rightarrow$ $$\frac{x^2}{\frac{4a^2}{3}}+\frac{\left(y-\frac{1}{3}\right)^2}{\frac{4a^2}{9}}=1$$

$\therefore$ $$e^2=1-\frac{b^2}{a^2}=1-\frac{\frac{4a^2}{9}}{\frac{4a^2}{3}}=1-\frac{1}{3}=\frac{2}{3}$$

$\Rightarrow$ $$5e^2=\frac{10}{3}\Rightarrow[5e^2]=\left[\frac{10}{3}\right]=3$$

66. (0.8) Given, $x^2+9y^2-4x+6y+4=0$

$\Rightarrow$ $$x^2-4x+9y^2+6y+4=0$$

$\Rightarrow$ $$(x-2)^2+(3y+1)^2=1$$

$\Rightarrow$ $$(x-2)^2+9\left(y+\frac{1}{3}\right)^2=1$$

$\Rightarrow$ $$\frac{(x-2)^2}{1}+\frac{\left(y+\frac{1}{3}\right)^2}{\frac{1}{9}}=1$$

which is equation of ellipse having centre at $\left(2,\frac{-1}{3}\right)$.

General point on ellipse

$$P(4,9)=\left(2+a\cos\theta,\frac{-1}{3}+b\sin\theta\right)\left(2+\cos\theta,\frac{-1}{3}+\frac{1}{3}\sin\theta\right)$$

$\therefore$ $$x=2+\cos\theta \quad\text{and}\quad y=\frac{-1}{3}(1-\sin\theta)$$

$\because$ $$4x-9y=8+4\cos\theta+3-3\sin\theta$$
$$=11+4\cos\theta-3\sin\theta$$
$$f(x,y)=4x-9y$$

$\therefore$ $$f_{max}=11+\sqrt{4^2+3^2}=11+5=16$$

$\{\because$ $a\sin\theta\pm b\cos\theta$ lies between $-\sqrt{a^2+b^2}$ to $\sqrt{a^2+b^2}\}$

$\therefore$ Required answer $=\frac{16}{20}=\frac{4}{5}=0.8$

67. (6) Given line is $x-y-2=0$

Image of (h,k) on ellipse about $x-y-2=0$ is say (h',k')

$\Rightarrow$ $$\frac{h'-h}{1}=\frac{k'-k}{-1}=-2\frac{(h-k-2)}{1^2+1^2}$$

$\Rightarrow$ $$h'-h=k'-k=-h+k+2$$

$\because$ $$h'=h-h+k+2$$

and $$k'=h-k+k-2$$

$\Rightarrow$ $$h'=k+2 \quad\text{and}\quad k'=h-2$$

$\therefore$ $$(h,k)\equiv(k'+2,h'-2)$$

$\therefore$ Equation of ellipse

$$\frac{(x-4)^2}{16}+\frac{(y-3)^2}{9}=1$$

$\Rightarrow$ $$\frac{(h-4)^2}{16}+\frac{(k-3)^2}{9}=1$$

$\Rightarrow$ $$\frac{(k'+2-4)^2}{16}+\frac{(h'-2-3)^2}{9}=1$$

$\Rightarrow$ $$\frac{(k'-2)^2}{16}+\frac{(h'-5)^2}{9}=1$$

$\Rightarrow$ $$16h'^2+9k'^2-36k'-160h'+292=0$$

$\therefore$ Locus is $16x^2+9y^2-36y-160x+292=0$

$\because$ $$\frac{k_1+k_2}{22}=\frac{292-160}{22}=6$$

Level 3

1. (a) $\because$ We know that the distance between focus and directrix $=4$ $\{\because$ focus $=(0,0)$ and directrix $\Rightarrow x=4\}$

$\therefore$ $$\frac{a}{e}-ae=4$$

$$a\left(\frac{1}{e}-e\right)=4 \qquad \left\{\because e=\frac{1}{2}\right\}$$

$\therefore$ $$a\left(2-\frac{1}{2}\right)=4$$

$$\Rightarrow \quad a\left(\frac{3}{2}\right)=4 \Rightarrow a=\frac{8}{3}$$

$\therefore$ Length of semi major axis is $\frac{8}{3}$ units.

2. (c) For E_n, equation of ellipse $\Rightarrow \dfrac{x^2}{a^2}+\dfrac{y^2}{b^2}=1$

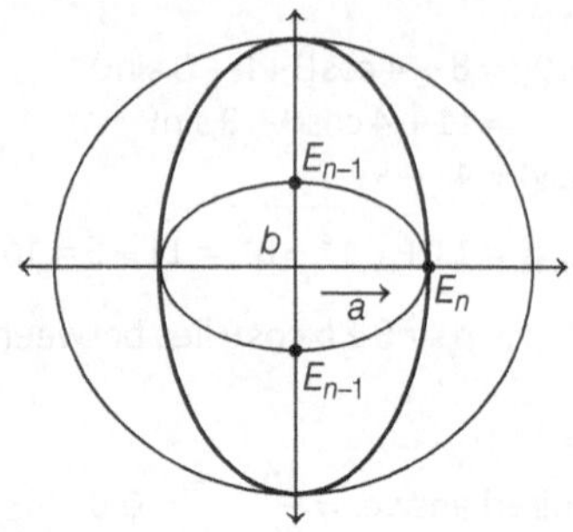

For E_{n-1}, $\dfrac{x^2}{(a')^2}+\dfrac{y^2}{(b')^2}=1$

$$b'=a \quad \text{and} \quad a'e=b \qquad \text{...(i)}$$

$$b'^2=a'^2-a'^2e^2$$

By Eq. (i)

$$a^2=\frac{b^2}{e^2}-b^2 \qquad \left[\because e^2=1-\frac{b^2}{a^2}\right]$$

$$\therefore \quad e^2=\frac{b^2}{a^2}-\frac{b^2e^2}{a^2}=\frac{b^2}{a^2}(1-e^2)$$

$$e^2=(1-e^2)(1-e^2)$$

$$\Rightarrow \quad e^2=(1-e^2)^2$$

$$\Rightarrow \quad e=\pm(1-e^2)$$

$$\because \quad e^2+e-1=0 \text{ and } e^2-e-1=0$$

$$\Rightarrow \quad e=\frac{-1\pm\sqrt{1+4}}{2},\ e=\frac{1\pm\sqrt{1+4}}{2}$$

$$\Rightarrow \quad e=\frac{-1\pm\sqrt{5}}{2},\ e=\frac{1\pm\sqrt{5}}{2}$$

$$\therefore \quad e=\frac{\sqrt{5}-1}{2} \qquad \{\because 0<e<1\}$$

3. (c) $\because$ At $t=\frac{3}{2}$ perpendicular $b>a$

$\therefore$ Length of latusrectum $=\dfrac{2a^2}{b}$

$$=\frac{2[t^2-3t+4]^2}{[3+5t]}=\frac{2\left[\frac{9}{4}-\frac{9}{2}+4\right]^2}{\left[3+\frac{15}{2}\right]}$$

$$\Rightarrow \quad \frac{2[1.75]^2}{[10.5]}=\frac{2}{10}=\frac{1}{5}$$

4. (b) Given that base $BC=a$ and sum of sides

$$=AB+BC+CA$$

$$a+b+c=2s$$

Let ABC is any triangle

Taking BC as axis of X and a perpendicular line through O the mid-point of BC as Y-axis

Let coordinates of the incentre I of the circle be (x, y).

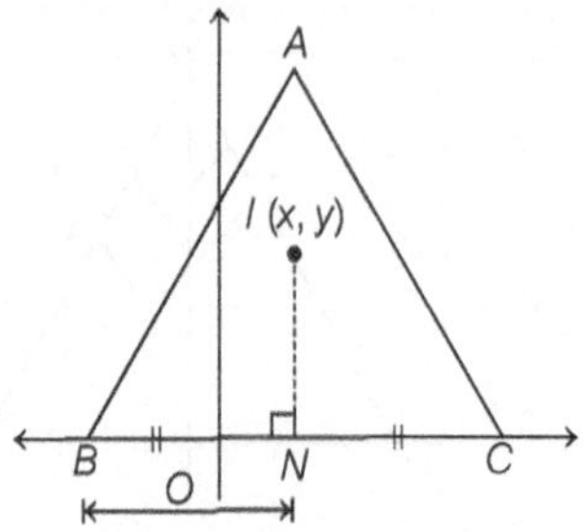

Drop perpendicular IN on BC

$$x=ON=BN-BO$$

$$=(s-b)-\frac{a}{2} \qquad \left\{\therefore OB=OC=\frac{a}{2}\right\}$$

$$=\frac{2s-2b-a}{2}$$

$$=\frac{a+b+c-2b-a}{2}$$

$$x=\frac{c-b}{2}$$

$$c-b=2x \qquad \text{...(i)}$$

$\therefore$ Area of $\Delta ABC=\sqrt{s(s-a)(s-b)(s-c)}$

$$=\sqrt{s(s-a)\left(\frac{a+c-b}{2}\right)\left(\frac{a+b-c}{2}\right)}$$

$$=\frac{1}{2}\sqrt{s(s-a)\{a^2-(c-b)^2\}}$$

$$=\frac{1}{2}\sqrt{s(s-a)\{(a^2)-4x^2\}} \text{ ...[from Eq. (i)]}$$

If radius of incircle be r.

Clearly r is equal the ordinate of the centre I.

$$\because \quad r=y=\frac{\Delta}{s}=\frac{1}{2s}\sqrt{s(s-a)\{a^2-4x^2\}}$$

Put $\quad \lambda=\dfrac{1}{25}\sqrt{s(s-a)}$ constant

$$\{\because a \text{ and } s \text{ are constant}\}$$

$$\therefore \quad y=\lambda\sqrt{a^2-4x^2}$$

$$\Rightarrow \quad y^2=\lambda^2(a^2-4x^2)$$

$$\Rightarrow \quad 4\lambda^2x^2+y^2=a^2\lambda^2$$

$$\Rightarrow \quad \frac{4x^2}{a^2}+\frac{y^2}{a^2\lambda^2}=1$$

$\therefore$ Locus is ellipse.

5. (c) Let the ellipse be $\frac{x^2}{a^2}+\frac{y^2}{b^2}=1$

$$\left\{\begin{array}{l}\because A(a,0) \text{ or } A'(-a,0)\\ S(ae,0) \text{ or } S'(-ae,0)\end{array}\right\}$$

and $P(a\cos\phi, b\sin\phi)$ be a point on it

$\therefore$ Equation of $PA \Rightarrow y-0=\frac{b\sin\phi-0}{a\cos\phi}(x-a)$

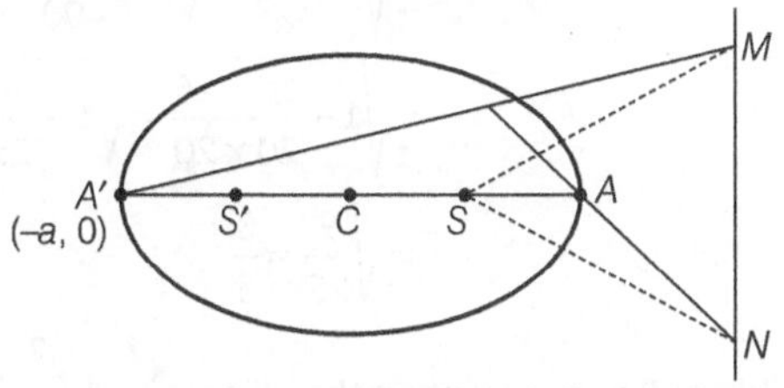

It meets directrix $x=\frac{a}{e}$ in N whose coordinates are

$$\left[\frac{a}{e}, b\sin\phi\cdot\frac{(1-e)}{e(\cos\phi-1)}\right]$$

Equation of PA' is $y=\frac{b\sin\phi}{a\cos\phi+a}(x+a)$

It meets directrix $x=\frac{a}{e}$ in M whose coordinates are

$$\left[\frac{a}{e}, \frac{b\sin\phi(1+e)}{e(\cos\phi+1)}\right)$$

Now, product of slope of SN and SM

$$=\left(\frac{\frac{b\sin\phi(1-e)}{e(\cos\phi-1)}}{\frac{a}{e}-ae}\right)\left(\frac{\frac{b\sin\phi(1+e)}{e(\cos\phi+1)}}{\frac{a}{e}-ae}\right)$$

$$\Rightarrow \frac{b\sin\phi(1-e)e}{e(\cos\phi-1)(a-ae^2)}\times\frac{b\sin\phi(1+e)e}{e(\cos\phi+1)(a-ae^2)}$$

$$\Rightarrow \frac{b^2\sin^2\phi(1-e^2)}{(\cos^2\phi-1)a^2(1-e^2)^2}=\frac{-b^2}{a^2(1-e^2)}\left[\because e^2=1-\frac{b^2}{a^2}\right]$$

$$\Rightarrow \frac{-b^2}{a^2-a^2e^2}=\frac{-b^2}{a^2-(a^2-b^2)}$$

$$=\frac{-b^2}{b^2}=-1$$

$\because$ $SN \perp SM$

$\therefore$ Angle between them is $\frac{\pi}{2}$

6. (a) Any tangent to the ellipse is

$$\frac{x}{a}\cos\theta+\frac{y}{b}\sin\theta=1 \quad \text{...(i)}$$

Its point of contact is $(a\cos\theta, b\sin\theta)$ and slope $=\frac{-b}{a}\cot\theta$ and focus is $S(ae,0)$.

Any line through focus S and perpendicular to tangent (i) is

$$y-0=\frac{a}{b}\tan\theta(x-ae) \quad \text{...(ii)}$$

Also, the equation of CP is

$$y-0=\frac{a}{b}\tan\theta(x-0) \quad \{\because C(0,0)\}\text{...(iii)}$$

Eliminating θ between Eqs. (ii) and (iii), we get

$$\frac{a^2}{b^2}\left(\frac{x-ae}{x}\right)=1$$

$$\Rightarrow \frac{x-ae}{x}=\frac{b^2}{a^2}$$

$$\Rightarrow 1-\frac{ae}{x}=\frac{b^2}{a^2}$$

$$\Rightarrow \frac{ae}{x}=1-\frac{b^2}{a^2}$$

$$\Rightarrow \frac{ae}{x}=e^2$$

$$\Rightarrow x=\frac{ae}{e^2}$$

$$\because x=\frac{a}{e}$$

$\therefore$ It intersect at directrix.

7. (b) The general equation of a tangent to the ellipse is

$$y=mx\pm\sqrt{a^2m^2+b^2} \quad \text{...(i)}$$

Let the points on the minor axis be $P(0, ae)$ and $Q(0,-ae)$ as $b^2=a^2(1-e^2)$

$\therefore$ Length of perpendicular from P on (i) is

$$P_1=\left|\frac{ae\pm\sqrt{a^2m^2+b^2}}{\sqrt{1+m^2}}\right|$$

Similarly, $$P_2=\left|\frac{-ae\pm\sqrt{a^2m^2+b^2}}{\sqrt{1+m^2}}\right|$$

$$\therefore P_1^2+P_2^2=\frac{2}{1+m^2}(a^2e^2+a^2m^2+b^2)$$

$$=\frac{2}{1+m^2}\{a^2-b^2+a^2m^2+b^2\}$$

$$=\frac{2}{1+m^2}(a^2)(1+m^2)=2a^2$$

8. (a) $\because$ $\angle SPN=90^\circ-\beta$

$\because$ $\angle SPN=\angle S'PN$

{$\because$ normal bisect the angle between $S'P$ and SP}

$\therefore$ $\angle SPS'=180^\circ-2\beta$

$\angle PSS'=180^\circ-\alpha+\beta$

$\therefore$ $\angle PS'S=\alpha+\beta-180^\circ$

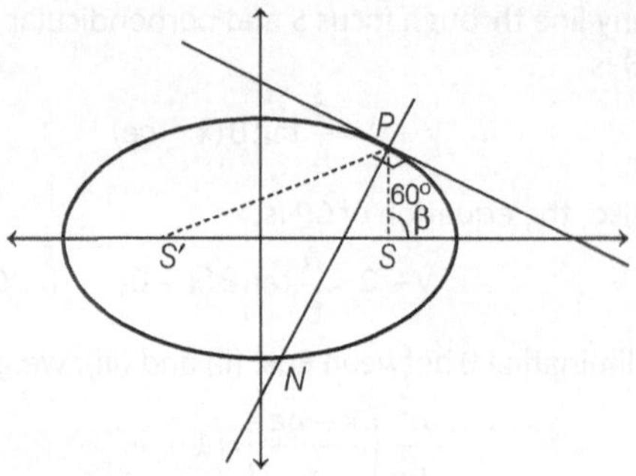

Applying sine rule on $\Delta SPS'$

$$\frac{\sin \angle PS'S}{PS} = \frac{\sin \angle PSS'}{PS'} = \frac{\sin \angle SPS'}{SS'}$$

$$\therefore \quad \frac{\sin(\alpha+\beta-180^\circ)}{PS} = \frac{\sin(180^\circ-(\alpha-\beta))}{PS'}$$

$$= \frac{\sin(180^\circ-2\beta)}{2ae}$$

$$\because \quad PS + PS' = 2a$$

$$\Rightarrow \quad \frac{\sin(\alpha-\beta)-\sin(\alpha+\beta)}{2a} = \frac{\sin 2\beta}{2ae}$$

$$\therefore \quad e = \left|\frac{2\sin\beta\cos\beta}{2\sin\beta\cos\alpha}\right| = \frac{\cos\beta}{\cos\alpha}$$

9. (c)

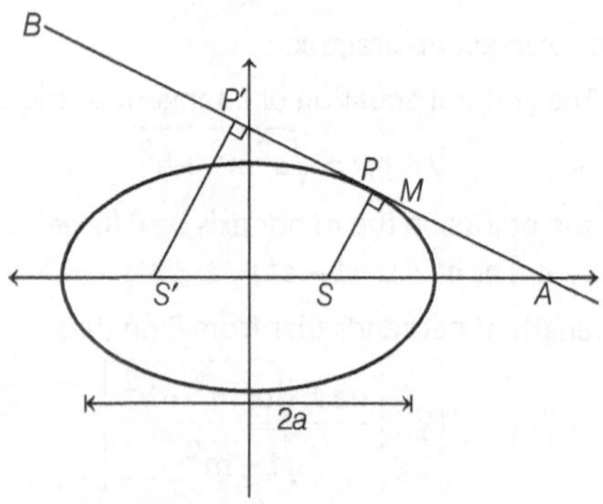

$\because a = 20 \Rightarrow 2a = 40$

Equation of tangent is $y = mx \pm \sqrt{a^2m^2+b^2}$

$SP \perp AB$, $S'P' \perp AB$ and M is point of contact

$\because$ We know that $SP \cdot S'P'$

$$= \left|\frac{mae \pm \sqrt{a^2m^2+b^2}}{\sqrt{1+m^2}}\right| \left|\frac{-mae \pm \sqrt{a^2m^2+b^2}}{\sqrt{1+m^2}}\right|$$

$$= \frac{a^2m^2+b^2-ma^2e^2}{1+m^2}$$

$$= \frac{a^2m^2+b^2-m^2(a^2-b^2)}{1+m^2} = b^2$$

$$\left[\because \begin{array}{l} e^2 = 1-\dfrac{b^2}{a^2} \\ a^2e^2 = a^2-b^2 \end{array}\right]$$

$$\therefore \quad \sum_{i=1}^{10}(SP)(S'P') = b^2 + b^2 + \dots 10 \text{ times}$$

$$= 10b^2 = 2560$$

$$\Rightarrow \quad b^2 = 256$$

$$\Rightarrow \quad b = 16$$

$$\because \quad e = \sqrt{1-\frac{b^2}{a^2}} = \sqrt{1-\frac{256}{400}}$$

$$= \sqrt{1-\frac{16\times16}{20\times20}} = \sqrt{1-\frac{16}{25}}$$

$$= \sqrt{\frac{9}{25}} = \frac{3}{5}$$

10. (b) Let P is any point on the ellipse $\frac{x^2}{a^2}+\frac{y^2}{b^2}=1$, whose coordinates are $(a\cos\phi, b\sin\phi)$.

$\therefore$ Equation of normal

$$PG = ax\sec\phi - by\,\text{cosec}\,\phi = a^2-b^2$$

Solving with $y = 0$, we get

$$x = \frac{a^2-b^2}{a}\cos\phi$$

$$\text{or} \quad x = \frac{a^2\times e^2}{a}\cos\phi = ae^2\cos\phi$$

$\therefore$ abscissa of $G = ae^2\cos\phi$

$$CG = ae^2\cos\phi$$

Draw PL and QM perpendicular from P and Q on X-axis, respectively.

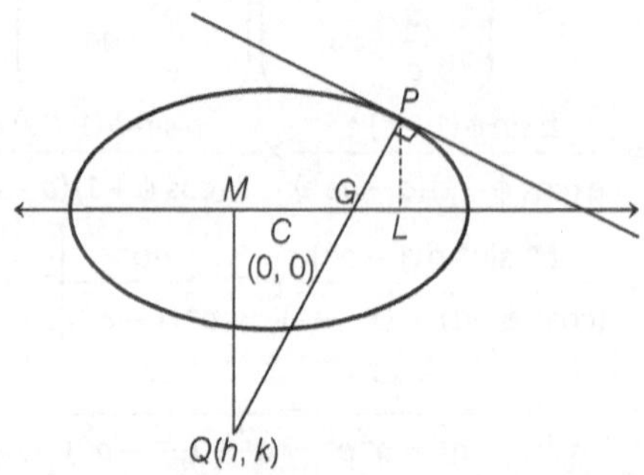

Then, the ΔGPL and ΔGQM are similar.

$$\frac{QM}{PL} = \frac{GM}{GL} = \frac{GQ}{GP} = h\{\because \text{ given } GQ = hGP\}$$

If the coordinates of Q be (h, k), then

$$h = CM = CG + GM$$

$$\Rightarrow \quad h = CG + hGL = CG + h(CL - CG)$$

$$\Rightarrow \quad h = (1-h)CG + hCL$$

$$= (1-h)ae^2\cos\phi + ha\cos\phi$$

$$= (e^2 - he^2 + h)a\cos\phi$$

$$\therefore \quad k = QM = hPL = hb\sin\phi$$

$\because \quad \cos^2\phi + \sin^2\phi = 1$

$\Rightarrow \quad \dfrac{h^2}{(e^2 - he^2 + h)^2 a^2} + \dfrac{k^2}{h^2 b^2} = 1$

$\therefore$ Locus is $\dfrac{x^2}{(e^2 - xe^2 + x)^2 a^2} + \dfrac{y^2}{x^2 b^2} = 1$

11. (a) $\because$ A, B, C are on ellipse and vertical of equilateral triangle.

Let A, B, C are have coordinates

$(a\cos\theta, b\sin\theta)$, $\left[a\cos\left(\theta + \dfrac{2\pi}{3}\right), b\sin\left(\theta + \dfrac{2\pi}{3}\right)\right]$

and $\left[a\cos\left(\theta + \dfrac{4\pi}{3}\right), b\sin\left(\theta + \dfrac{4\pi}{3}\right)\right]$

$\therefore$ We know that normal at three point from same side are concurrent

$\therefore$ A, B, C are concurrent.

12. (a) Equation of chord joining $P(\theta_1)$ and $Q(\theta_2)$

$$\frac{x}{a}\cos\left(\frac{\theta_1 + \theta_2}{2}\right) + \frac{y}{b}\sin\left(\frac{\theta_1 + \theta_2}{2}\right) = \cos\frac{\theta_1 - \theta_2}{2}$$

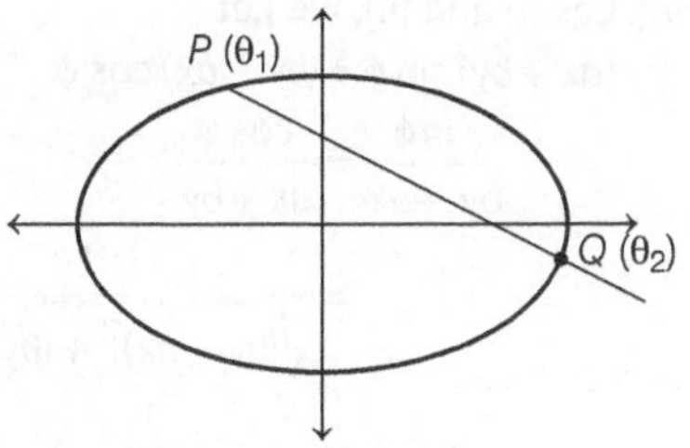

$\because$ $(d, 0)$ lies in it

$\therefore \quad \dfrac{d}{a}\cos\left(\dfrac{\theta_1 + \theta_2}{2}\right) = \cos\left(\dfrac{\theta_1 - \theta_2}{2}\right)$

$\Rightarrow \quad \dfrac{d}{a} = \dfrac{\cos\left(\dfrac{\theta_1 - \theta_2}{2}\right)}{\cos\left(\dfrac{\theta_1 + \theta_2}{2}\right)}$

$\Rightarrow \quad \dfrac{d - a}{d + a} = \dfrac{\cos\left(\dfrac{\theta_1 - \theta_2}{2}\right) - \cos\left(\dfrac{\theta_1 + \theta_2}{2}\right)}{\cos\left(\dfrac{\theta_1 - \theta_2}{2}\right) + \cos\left(\dfrac{\theta_1 + \theta_2}{2}\right)}$

$= \dfrac{2\sin\dfrac{\theta_1}{2}\cdot\sin\dfrac{\theta_2}{2}}{2\cos\dfrac{\theta_1}{2}\cdot\cos\dfrac{\theta_2}{2}}$

$\Rightarrow \quad \dfrac{d - a}{d + a} = \tan\left(\dfrac{\theta_1}{2}\right)\tan\left(\dfrac{\theta_2}{2}\right)$

13. (b) Given, equations is $\dfrac{x^2}{25} + \dfrac{y^2}{16} = 1$ ...(i)

Let roots be (α, β)

$a = 5$ and $b = 4$

$\therefore \quad e = \sqrt{1 - \dfrac{16}{25}} = \dfrac{3}{5}$

$\therefore$ Focus $(\pm ae, 0) = (\pm 3, 0) \Rightarrow \alpha = 3$

Length of chord whose middle part is $\left(\dfrac{1}{2}, \dfrac{2}{5}\right)$

$\therefore$ Equation of chord $T = S_1$

$\Rightarrow \quad \dfrac{x}{25}\left(\dfrac{1}{2}\right) + \dfrac{y}{16}\left(\dfrac{2}{5}\right) = \dfrac{\left(\dfrac{1}{2}\right)^2}{25} + \dfrac{\left(\dfrac{2}{5}\right)^2}{16}$

$\Rightarrow \quad \dfrac{x}{50} + \dfrac{y}{40} = \dfrac{1}{50}$

$\Rightarrow \quad \dfrac{x}{5} + \dfrac{y}{4} = \dfrac{1}{5}$

$\Rightarrow \quad y = \dfrac{-4}{5}(x - 1)$

By Eq. (i), $\dfrac{x^2}{25} + \dfrac{\left(-\dfrac{4}{5}\right)^2 (x-1)^2}{16} = 1$

$\Rightarrow \quad x^2 - x - 12 = 0$

$x_1 = 4$ and $x_2 = -3$

$\therefore$ Length of chord $= \sqrt{(x_1 - x_2)^2 + (y_1 - y_2)^2}$

$= |(x_1 - x_2)\sqrt{1 + m^2}|$

$\Rightarrow \quad l^2 = (7)^2\left(1 + \dfrac{16}{25}\right) = \dfrac{49}{25} \times 41$

$\Rightarrow \quad \dfrac{25l^2}{41} = 49$

$\therefore \quad \alpha = 3$ and $\beta = 49$

$\therefore$ Required equation is

$x^2 - (3 + 49)x + 147 = 0$

$x^2 - 52x + 147 = 0$

14. (c) Let the tangent at point $\left(3, \dfrac{12}{5}\right)$ (say P) meets the axis of the ellipse $\dfrac{x^2}{a^2} + \dfrac{y^2}{b^2} = 1$ in T.

Now, tangent at P will be

$\dfrac{3x}{a^2} + \dfrac{12y}{5b^2} = 1$

Solving with X-axis $y = 0$, we get $x = \dfrac{a^2}{3}$.

$\therefore \quad CT = \dfrac{a^2}{3}$

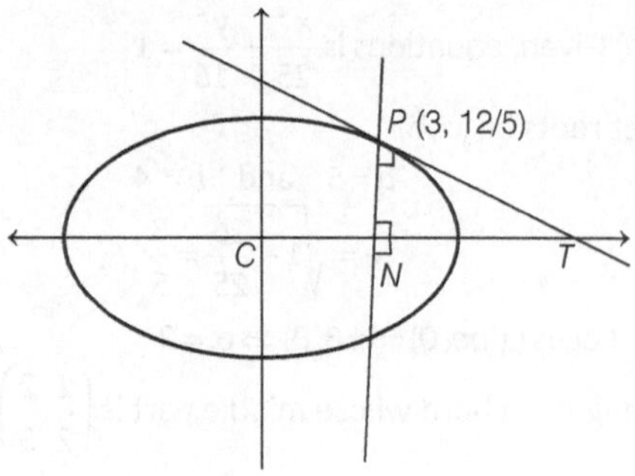

If PN perpendicular to X-axis

$\therefore$ Subnormal (NT) is

$\therefore \quad NT = CT - CN$

$\Rightarrow \quad \dfrac{16}{3} = \dfrac{a^2}{3} - 3$

$\{\because CN = \text{abscissa of } N = \text{abscissa of } P\}$

$a^2 = 25$

$\because \left(3, \dfrac{12}{5}\right)$ lies on ellipse

$\Rightarrow \quad \dfrac{9}{25} + \dfrac{144}{5b^2} = 1$

$\Rightarrow \quad b^2 = 9$

$\because \quad e = \sqrt{1 - \dfrac{9}{25}} = \dfrac{4}{5}$

15. (b) If the coordinates of P are taken as $(a\cos\phi, b\sin\phi)$, then the coordinates of Q' will be

$$\left\{a\cos\left(\frac{\pi}{2}+\phi\right), b\sin\left(\frac{\pi}{2}+\phi\right)\right\} = (-a\sin\phi, b\cos\phi)$$

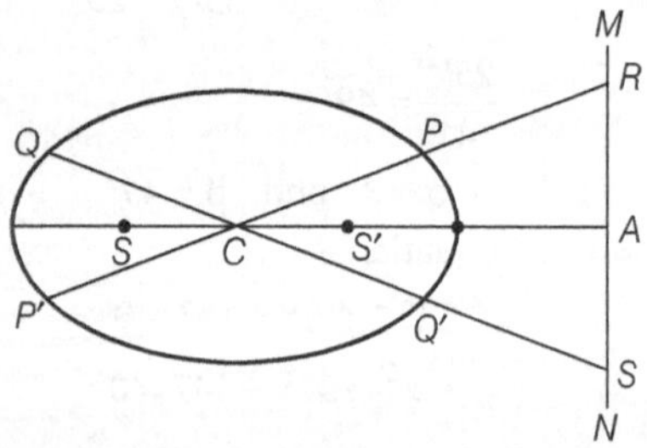

as these all the ends of conjugate diameters CP and $Q'C$ of ellipse $\dfrac{x^2}{a^2} + \dfrac{y^2}{b^2} = 1$

If these diameters are produced let them meet the directrix in points R and S

Then, we have to find the orthocentre of ΔCRS

$\because$ Equation of $P'CP = y = \dfrac{bx}{a}\tan\phi$ ...(i)

and equation of $Q'CQ = y = \dfrac{-bx}{a}\cot\phi$ ...(ii)

By Eqs. (i) and (ii) with directrix, we get

$$R\left(\frac{a}{e}, \frac{b}{e}\tan\phi\right) \text{ and } S\left(\frac{a}{e}, \frac{-b}{e}\cot\phi\right)$$

$\because$ Equation of perpendicular from R on CS

$$y - \frac{b}{e}\tan\phi = \frac{a}{b}\tan\phi\left(x - \frac{a}{e}\right) \quad \text{...(iii)}$$

The line is perpendicular from centre C upon directrix MN will be X-axis i.e. $y = 0$ ...(iv)

Solving Eqs (iii) and (iv), we get

$$x = \frac{a}{e} - \frac{b^2}{ae} \Rightarrow \frac{a^2 - b^2}{ae} = \frac{a^2e^2}{ae} = ae$$

$\therefore$ Point is $(ae, 0)$ that is focus.

16. (a) If CP and CQ be the two conjugate diameters of the ellipse $\dfrac{x^2}{a^2} + \dfrac{y^2}{b^2} = 1$ having the centre $C\,(0, 0)$

$\therefore$ Let $P = (a\cos\phi, b\sin\phi)$, then $Q = (-a\sin\phi, b\cos\phi)$

Equation of circle drawn CP and CQ as diameters

$$(x - 0)(x - a\cos\phi) + (y - 0)(y - b\sin\phi) = 0$$

$\Rightarrow \quad x^2 + y^2 - ax\cos\phi - by\sin\phi = 0$ ...(i)

and $(x - 0)(x + a\sin\phi) + (y - 0)(y - b\cos\phi) = 0$

$\Rightarrow \quad x^2 + y^2 - ax\sin\phi - by\cos\phi = 0$ ...(ii)

Solving Eqs (i) and (ii), we get

$$(ax + by)\sin\phi = (by - ax)\cos\phi$$

$$\Rightarrow \quad \frac{\sin\phi}{by - ax} = \frac{\cos\phi}{ax + by}$$

$$= \frac{1}{\sqrt{(by - ax)^2 + (by + ax)^2}}$$

$$= \frac{1}{\sqrt{2(b^2y^2 + a^2x^2)}}$$

$$\because \quad \sin\phi = \frac{by - ax}{\sqrt{2(b^2y^2 + a^2x^2)}}$$

$$\cos\phi = \frac{ax + by}{\sqrt{2(b^2y^2 + a^2x^2)}}$$

By Eq. (i), we get

$$x^2 + y^2 = \frac{ax(by + ax)}{\sqrt{2(b^2y^2 + a^2x^2)}} + \frac{by(by - ax)}{\sqrt{2(b^2y^2 + a^2x^2)}}$$

$$= \frac{a^2x^2 + b^2y^2}{\sqrt{2}(\sqrt{a^2x^2 + b^2y^2})}$$

$$\Rightarrow x^2 + y^2 = \sqrt{\frac{a^2x^2 + b^2y^2}{2}}$$

Squaring on both sides, we get

$$2(x^2 + y^2)^2 = (a^2x^2 + b^2y^2)$$

17. (b) Given, ellipse $\frac{x^2}{a^2}+\frac{y^2}{b^2}=1$

and circle $x^2+y^2=r^2$

Equation of tangent on circle

$\Rightarrow \quad x\cos\theta+y\sin\theta=r$

and equation of tangent on ellipse

$\Rightarrow \quad y=mx+\sqrt{a^2m^2+b^2}$

If it is tangent to circle, then perpendicular from (0, 0) is r

$\therefore \quad \frac{\sqrt{a^2m^2+b^2}}{\sqrt{1+m^2}}=r$

$\Rightarrow \quad a^2m^2+b^2=r^2(1+m^2)$

$\Rightarrow \quad (a^2-r^2)m^2=r^2-b^2$

$\Rightarrow \quad m=\sqrt{\frac{r^2-b^2}{a^2-r^2}}$

$\Rightarrow \quad \tan\phi=\sqrt{\frac{r^2-b^2}{a^2-r^2}}$

$\Rightarrow \quad \phi=\tan^{-1}\sqrt{\frac{r^2-b^2}{a^2-r^2}}$

18. (b) Let $y=mx+\frac{1}{m}$ be a tangent to the parabola $y^2=4x$

It will touch the ellipse $\frac{x^2}{4^2}+\frac{y^2}{(\sqrt{6})^2}=1$

$$\frac{1}{m^2}=16m^2+6 \quad (\because c^2=a^2m^2+b^2)$$

$\therefore \quad 16m^4+6m^2-1=0$

$\Rightarrow \quad (8m^2-1)(2m^2+1)=0$

$\Rightarrow \quad m=\pm\frac{1}{\sqrt{8}}$

We know that a tangent of slope m touches the parabola $y^2=4ax$ at $\left(\frac{a}{m^2},\frac{2a}{m}\right)$. So, the coordinates of the points of contact of the common tangent of slope $m=\pm\frac{1}{2\sqrt{2}}$ to the parabola $y^2=4x$ are $A(8,4\sqrt{2})$ and $B(8,-4\sqrt{2})$

We also know that of slope m touches the ellipse at

$$\left(\mp\frac{a^2m}{\sqrt{a^2m^2+b^2}},\pm\frac{b^2}{\sqrt{a^2m^2+b^2}}\right)$$

i.e. $m=\pm\frac{1}{2\sqrt{2}}\rightarrow C\left(-2,\frac{3}{\sqrt{2}}\right)$ and $D\left(-2,\frac{-3}{\sqrt{2}}\right)$

Clearly $AB\,||\,CD \Rightarrow ABCD$ is trapezium

We have $AB=8\sqrt{2}$ and $CD=3\sqrt{2}$

Distance between AB and CD is $PQ=8+2=10$ units

$\because$ Area of $ABCD=\frac{1}{2}(AB+CD)\times PQ$

$=\frac{1}{2}(8\sqrt{2}+3\sqrt{2})\times 10$

$=55\sqrt{2}$ sq units

19. (d) Let the coordinates of the pole of any chord of the ellipse $\frac{x^2}{a^2}+\frac{y^2}{b^2}=1$ be (h,k)

Then, equation of the chord i.e. polar of (h,k) is

$$\frac{xh}{a^2}+\frac{yk}{b^2}=1$$

$$b^2xh+a^2yk=a^2b^2 \quad \text{...(i)}$$

It will touch the parabola

$$ay^2=-2b^2x$$

By Eq. (i), $\quad \frac{-ay^2}{2}h+a^2yk=a^2b^2$

$\Rightarrow \quad hy^2-2ayk+2ab^2=0 \quad$ ...(ii)

If the line (i) tangent to the parabola, then the roots of (ii) are equal.

$\therefore \quad D=0$

$\Rightarrow \quad 4a^2k^2-8ab^2h=0$

$\Rightarrow \quad ak^2=2b^2h$

$\therefore$ Locus is $ay^2=2b^2x$

20. (a, c) Given, $\frac{x^2}{f(k^2+2k+5)}+\frac{y^2}{f(k+11)}=1$ and $f(x)$ is decreasing function.

$\therefore \quad f(x_1)<f(x_2)\Rightarrow x_1>x_2$

Now, major axis as X-axis

$\because \quad f(k^2+2k+5)>f(k+11)$

$\Rightarrow \quad k^2+2k+5<k+11$

$\Rightarrow \quad k^2+k-6<0$

$\Rightarrow \quad (k+3)(k-2)<0$

$\because \; k\in(-3,2)$

Now, major axis as Y-axis

$f(k+11)>f(k^2+2k+5)$

$\Rightarrow \quad k+11<k^2+2k+5$

$\Rightarrow \quad (k+3)(k-2)>0$

$\because \quad k\in(-\infty,-3)\cup(2,\infty)$

21. (a, c) Given, ellipse $\frac{x^2}{9}+\frac{y^2}{4}=1$

$\Rightarrow \quad x^2=\lambda^2 \Rightarrow x=\pm\lambda$

$\because$ A be the point of intersection of tangents at $P(\theta_1)$ and $Q(\theta_2)$

$$\because \qquad h=\frac{3\cos\left(\frac{\theta_1+\theta_2}{2}\right)}{\cos\left(\frac{\theta_1-\theta_2}{2}\right)} \qquad ...(i)$$

and
$$k=\frac{2\sin\left(\frac{\theta_1+\theta_2}{2}\right)}{\cos\left(\frac{\theta_1-\theta_2}{2}\right)} \qquad ...(ii)$$

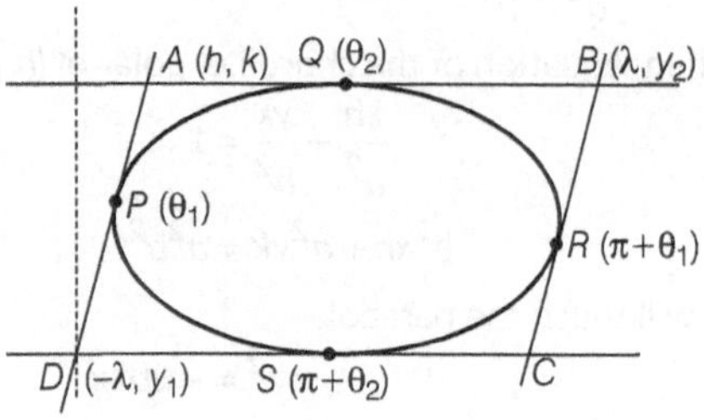

$$\Rightarrow \qquad \frac{h^2}{9}+\frac{k^2}{4}=\frac{1}{\cos^2\left(\frac{\theta_1-\theta_2}{2}\right)}$$

Also,
$$-\lambda=\frac{3\cos\left(\frac{\pi}{2}+\frac{\theta_1+\theta_2}{2}\right)}{\cos\left(\frac{\pi}{2}+\frac{\theta_2-\theta_1}{2}\right)}$$

$$=3\frac{\sin\left(\frac{\theta_1+\theta_2}{2}\right)}{\sin\left(\frac{\theta_2-\theta_1}{2}\right)} \qquad ...(iii)$$

Dividing Eq. (ii) by Eq. (iii), we get

$$\Rightarrow \qquad \tan\left(\frac{\theta_2-\theta_1}{2}\right)=\frac{3k}{-2\lambda}$$

$$\Rightarrow \qquad \frac{h^2}{9}+\frac{k^2}{4}=\sec^2\left(\frac{\theta_1-\theta_2}{2}\right)$$

$$\Rightarrow \qquad \frac{h^2}{9}+\frac{k^2}{4}=1+\frac{9k^2}{4\lambda^2}$$

$\therefore$ Locus is $\frac{x^2}{9}+\frac{y^2}{4}-\frac{9y^2}{4\lambda^2}=1$

$$\because \qquad \frac{x^2}{9}+\frac{y^2}{\left(\frac{4\lambda^2}{\lambda^2-9}\right)}=1$$

$$\Rightarrow \qquad a^2=9 \text{ and } b^2=\frac{4\lambda^2}{\lambda^2-9}$$

For ellipse

$$\therefore \qquad h^2<ab$$

$$\Rightarrow \qquad 0<9\left(\frac{4\lambda^2}{\lambda^2-9}\right)$$

$$\Rightarrow \qquad \frac{\lambda^2}{\lambda^2-9}>0$$

$$\lambda>3$$

It represents an ellipse

$$\oplus \quad -3 \quad \ominus \quad 3 \quad \oplus$$

$\therefore$ For circle $a=b$

$$\Rightarrow \qquad 9=\frac{4\lambda^2}{\lambda^2-9}$$

$$\Rightarrow \qquad 9\lambda^2-81=4\lambda^2$$

$$\Rightarrow \qquad 5\lambda^2=81$$

$$\Rightarrow \qquad \lambda=\frac{9}{\sqrt{5}}$$

22. (a, c) We have,

Equation of circle $x^2+y^2=\frac{1}{2}$

and equation of parabola $y^2=4x$

Let the equation of common tangent of parabola and circle, is

$$y=mx+\frac{1}{m}$$

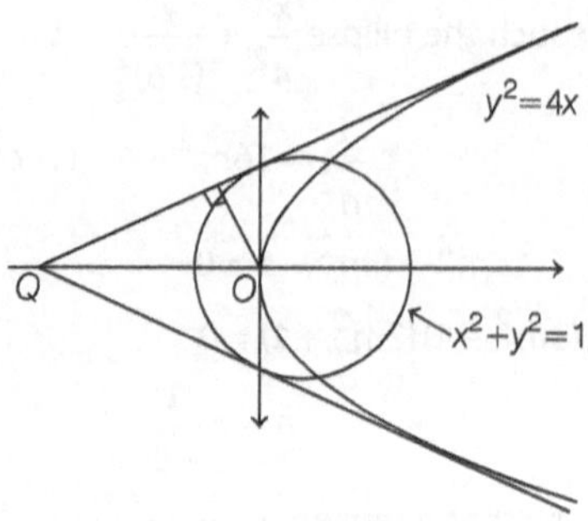

$\because$ Radius of circle $=\frac{1}{\sqrt{2}}$

$$\because \qquad \frac{1}{\sqrt{2}}=\left|\frac{0+0+\frac{1}{m}}{\sqrt{1+m^2}}\right|$$

$$\Rightarrow \quad m^4+m^2-2=0 \Rightarrow m=\pm 1$$

$\because$ Equation of common tangents

$$\Rightarrow \qquad y=x+1 \text{ and } y=-x-1$$

$\therefore$ Intersection point $Q(-1, 0)$

$\therefore$ Equation of ellipse $\frac{x^2}{1}+\frac{y^2}{\frac{1}{2}}=1$

$$\Rightarrow \quad a^2 = 1 \text{ and } b^2 = \frac{1}{2}$$

$$\therefore \quad e = \sqrt{1 - \frac{b^2}{a^2}} = \sqrt{1 - \frac{1}{2}} = \frac{1}{\sqrt{2}}$$

$$\therefore \text{ Length of latusrectum} = \frac{2b^2}{a} = \frac{2\left(\frac{1}{2}\right)}{1} = 1$$

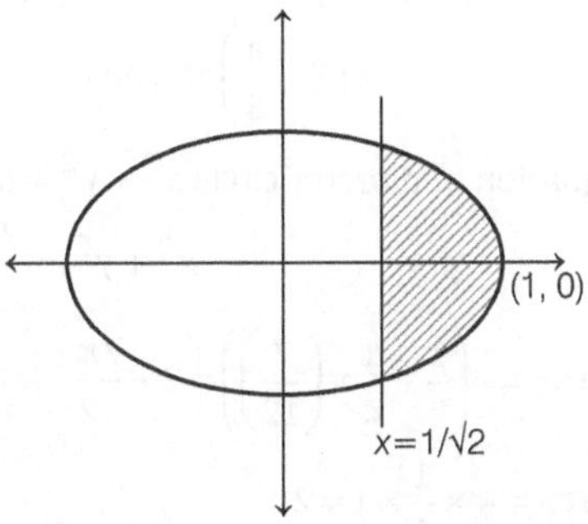

$$\because \text{ Area of shaded region} = 2\int_{\frac{1}{\sqrt{2}}}^{1} \frac{1}{\sqrt{2}}\sqrt{1-x^2}\,dx$$

$$= \sqrt{2}\left[\frac{x}{2}\sqrt{1-x^2} + \frac{1}{2}\sin^{-1}x\right]_{\frac{1}{\sqrt{2}}}^{1}$$

$$= \sqrt{2}\left[\left(0 + \frac{\pi}{4}\right) - \left(\frac{1}{4} + \frac{\pi}{8}\right)\right]$$

$$\Rightarrow \sqrt{2}\left(\frac{\pi}{8} - \frac{1}{4}\right) = \frac{\pi - 2}{4\sqrt{2}} \text{ sq units}$$

23. (a, b)

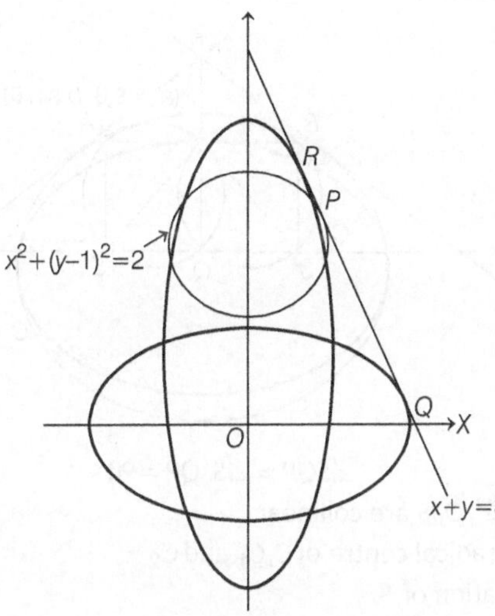

$$E_1 : \frac{x^2}{a^2} + \frac{y^2}{b^2} = 1, \quad a > b$$

$$E_2 : \frac{x^2}{c^2} + \frac{y^2}{d^2} = 1, \quad d > c$$

and $\quad S : x^2 + (y-1)^2 = 2$

as tangent to E_1, E_2 and S is $x + y = 3$

Let the point of contact of tangent be (x_1, y_1) to S

$\because xx_1 + yy_1 - (y + y_1) + 1 = 2$

or $xx_1 + yy_1 - y = 1 + y_1$ same as $x + y = 3$

$$\Rightarrow \quad \frac{x_1}{1} = \frac{y_1 - 1}{1} = \frac{1 + y_1}{3}$$

$$\because \quad x_1 = 1 \text{ and } y_1 = 2$$

$\because P\,(1, 2)$

$$\because \quad PR = PQ = \frac{2\sqrt{2}}{3}$$

Thus, by parametric form

$$\frac{x-1}{\frac{-1}{\sqrt{2}}} = \frac{y-2}{\frac{1}{\sqrt{2}}} = \pm\frac{2\sqrt{2}}{3}$$

$$\therefore \quad x = \frac{5}{3}, y = \frac{4}{3} \text{ and } x = \frac{1}{3}, y = \frac{8}{3}$$

$$\therefore \quad Q = \left(\frac{5}{3}, \frac{4}{3}\right) \text{ and } R = \left(\frac{1}{3}, \frac{8}{3}\right)$$

Now, equation of tangent at Q on ellipse E_1, is

$$\frac{x(5)}{a^2(3)} + \frac{y(4)}{b^2(3)} = 1$$

On comparing with $x + y = 3$

$$a^2 = 5 \text{ and } b^2 = 4$$

$$\because \quad e_1^2 = 1 - \frac{b^2}{a^2} = 1 - \frac{4}{5}$$

$$\Rightarrow \quad e_1^2 = \frac{1}{5}$$

Also, equation of tangent at R on ellipse is

$$\frac{x(1)}{a^2(3)} + \frac{y(8)}{b^2(3)} = 1.$$

On comparing with $x + y = 3$

$$\Rightarrow \quad a^2 = 1 \text{ and } b^2 = 8$$

$$\therefore \quad e_2^2 = 1 - \frac{a^2}{b^2} = 1 - \frac{1}{8} = \frac{7}{8}$$

$$\text{Now,} \quad e_1^2 e_2^2 = \frac{7}{40}$$

$$\Rightarrow \quad e_1 e_2 = \frac{\sqrt{7}}{2\sqrt{10}} \text{ and } e_1^2 + e_2^2 = \frac{43}{40}$$

$$\text{and} \quad (e_1^2 - e_2^2) = \frac{27}{40}$$

24. (a, c) Ellipse : $\frac{x^2}{4} + \frac{y^2}{1} = 1$

Let the point of intersection of tangents at A and B be $P\,(h, k)$, then equation of AB is

$$\frac{xh}{4} + \frac{yk}{1} = 1 \qquad \text{...(i)}$$

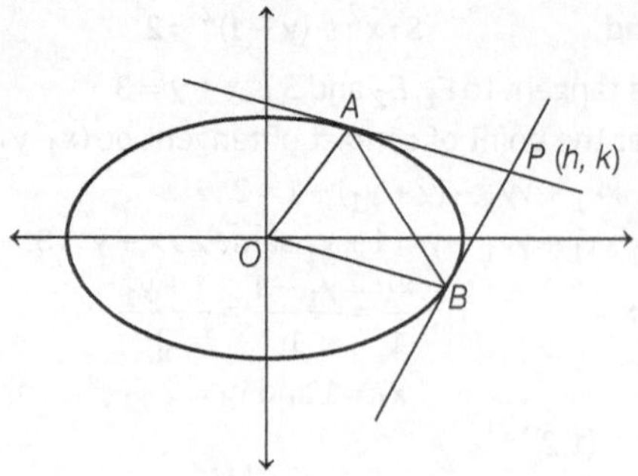

Homogenizing the equation of ellipse using (i)

$$\frac{x^2}{4}+\frac{y^2}{1}=\left(\frac{xh}{4}+\frac{yk}{1}\right)^2$$

$$\Rightarrow \quad x^2\left(\frac{h^2-4}{16}\right)+y^2(k^2-1)+\frac{2hk}{4}xy=0 \quad \text{...(ii)}$$

Given, equation of OA and OB is

$$x^2+4y^2+\alpha xy=0 \quad \text{...(iii)}$$

Eqs. (ii) and (iii) represents same line

$$\therefore \quad \frac{h^2-4}{16}=\frac{k^2-1}{4}=\frac{hk}{2\alpha}$$

$$\Rightarrow \quad h^2-4=4(k^2-1)\Rightarrow h^2-4k^2=0$$

$$\Rightarrow \quad (h-2k)(h+2k)=0$$

$\because$ Locus is $(x-2y)(x+2y)=0$

$$x-2y=0 \text{ or } x+2y=0$$

25. (b, c, d) $S_1: 3x^2+4y^2=1$

$$\Rightarrow \quad \frac{x^2}{\frac{1}{3}}+\frac{y^2}{\frac{1}{4}}=1$$

$$S_2: |x+y|-1=0$$

$$|x+y|=1 \Rightarrow x+y=\pm 1$$

and $\quad S_3: |y-x|-1=0$

$$|y-x|=1 \Rightarrow y-x=\pm 1$$

$\because$ Auxiliary circle of ellipse, $S_1: x^2+y^2=a^2$

$$\Rightarrow \quad x^2+y^2=\frac{1}{3}\Rightarrow 3x^2+3y^2=1$$

Now,

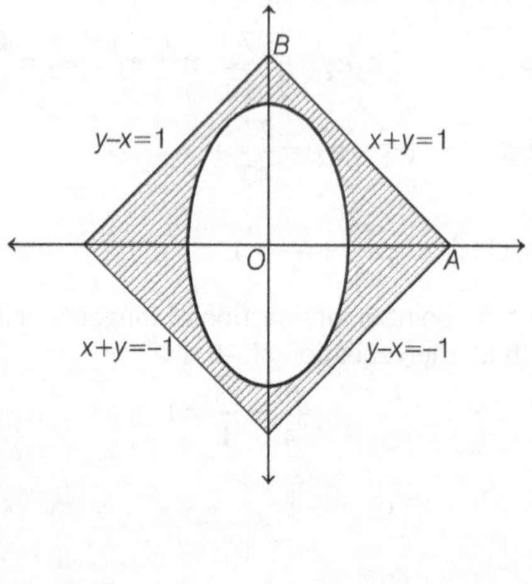

$\because \quad S_1 \geq 0 \Rightarrow 3x^2+3y^2 \geq 1$

$S_2 \leq 0 \quad \Rightarrow |x+y| \leq 1$

$S_3 \leq 0 \quad \Rightarrow |x-y| \leq 1$

$\because$ Required area $=4\left[\text{Area of } \Delta OAB-\frac{1}{4}\pi r^2\right]$

$$=4\left[\frac{1}{2}-\frac{1}{4}\pi\left(\frac{1}{\sqrt{3}}\right)^2\right]$$

$$=\left(2-\frac{\pi}{3}\right) \text{ sq units}$$

$\therefore$ Equation of director circle $x^2+y^2=a^2+b^2$

$$x^2+y^2=\frac{7}{12}$$

(b) Area $=4\left(\frac{1}{2}-\frac{1}{4}\pi\left(\frac{7}{12}\right)\right)=2-\frac{7\pi}{12}$

(c) Area $=4\times\frac{1}{2}\times 1=2$

(d) Area of region $=4\left(\frac{1}{2}-\frac{\pi}{4}\times\frac{1}{\sqrt{3}}\times\frac{1}{2}\right)=2-\frac{\pi}{2\sqrt{3}}$

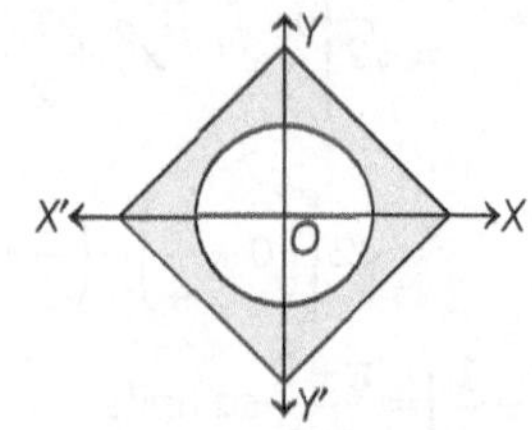

26. (a, b, d)

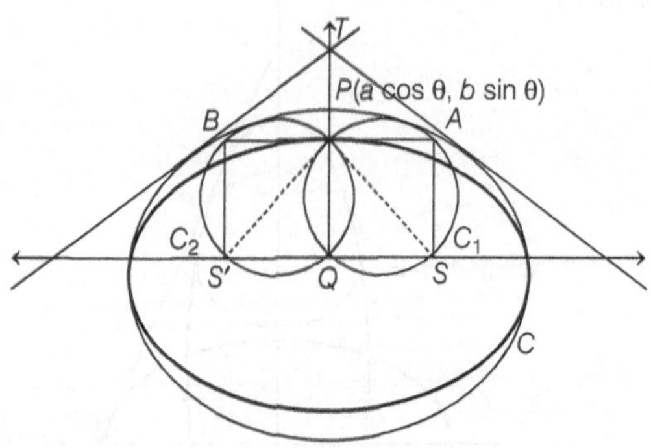

$$\angle SQP=\angle S'QP=90°$$

$\Rightarrow S', Q, S$ are collinear

T is radical centre of C, C_1 and C_2

Equation of

$$C_2: (x-a\cos\theta)(x\mp ae)+(y-b\sin\theta)\,y=0$$

$$C: x^2+y^2=a^2$$

Equation of tangent to C_2 and C_1, is

$$a(\cos\theta-e)\,x+(b\sin\theta)y\mp a^2(e\cos\theta-1)=0$$

Put $x=a\cos\theta$

$\Rightarrow a^2(\cos^2\theta - e\cos\theta) + by\sin\theta \mp a^2(e\cos\theta - 1) = 0$

$\Rightarrow \quad b\sin\theta = a^2\sin^2\theta$

$\Rightarrow \quad y = \frac{a^2}{b}\sin\theta$

Point $y = \left(a\cos\theta, \frac{a^2}{b}\sin\theta\right)$

$\therefore$ Equation of chord w.r.t $x^2 + y^2 = a^2$

$$x\cos\theta + y\frac{a}{b}\sin\theta = a$$

$\Rightarrow \frac{x\cos\theta}{a} + \frac{y\sin\theta}{b} = 1$ which is same as equation of tangent at P for ellipse

$\Rightarrow \quad SA$ and $S'B \perp AB \Rightarrow S'B \| SA$

27. (a, c)

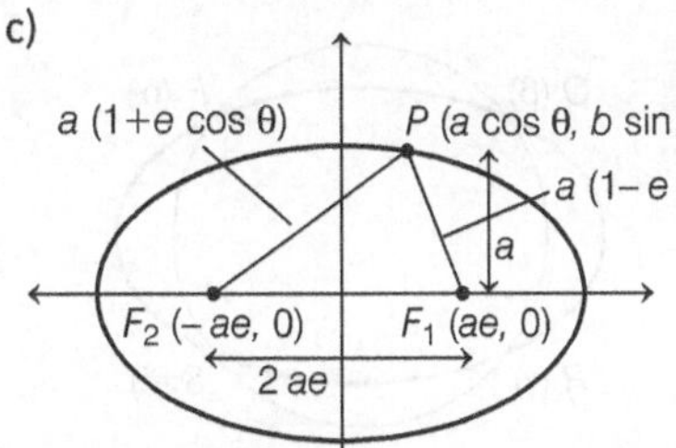

$\because$ We know P be any point on ellipse $\frac{x^2}{a^2} + \frac{y^2}{b^2} = 1$ i.e., $(a\cos\theta, b\sin\theta)$

$\therefore \quad F_1 = (ae, 0), F_2 = (-ae, 0), P(a\cos\theta, b\sin\theta)$,

$PF_2 = a(1 + e\cos\theta), PF_1 = a(1 - e\cos\theta)$ and

$F_1F_2 = 2ae$

$\because \quad h = \frac{ax_1 + bx_2 + cx_3}{a + b + c}$

$$= \frac{a^2e(1 + e\cos\theta) - a^2e(1 - e\cos\theta) + 2a^2e\cos\theta}{2a + 2ae}$$

$\Rightarrow \quad h = ae\cos\theta$ ...(i)

and $\quad k = \frac{2ae\,b\sin\theta}{2a + 2ae} = \frac{eb\sin\theta}{1 + e}$ ...(ii)

From Eqs. (i) and (ii),

$$\frac{h^2}{(ae)^2} + \frac{k^2}{\left(\frac{eb}{1+e}\right)^2} = 1 \quad [\because \cos^2\theta + \sin^2\theta = 1]$$

$\therefore$ Locus of incentre is $\frac{x^2}{a^2e^2} + \frac{y^2}{\left(\frac{be}{1+e}\right)^2} = 1$

$\therefore \quad A^2 = a^2e^2$ and $B^2 = \left(\frac{be}{1+e}\right)^2$

$\therefore \quad (e')^2 = 1 - \frac{B^2}{A^2}$

$$= 1 - \frac{\frac{(be)^2}{(1+e)^2}}{(ae)^2} = 1 - \frac{b^2}{a^2(1+e)^2} = 1 - \frac{1 - e^2}{(1+e)^2}$$

$\left[\because e^2 = 1 - \frac{b^2}{a^2}\right]$

$\Rightarrow \quad (e')^2 = 1 - \frac{(1-e)}{1+e}$

$\Rightarrow \quad (e')^2 = \frac{2e}{1+e}$

$\Rightarrow \quad e' = \sqrt{\frac{2e}{1+e}}$

28. (a, b, c)

(a) Given equation is,

$$\left(x - \frac{1}{13}\right)^2 + \left(y - \frac{2}{13}\right)^2 = \frac{1}{a^2}\left(\frac{5x + 12y - 1}{13}\right)^2$$

It represent an ellipse

$$\frac{1}{a^2} < 1 \Rightarrow a^2 > 1 \Rightarrow a > 1$$

(b) $4x^2 + 9y^2 + 8x - 36y = -4$

$\Rightarrow 4(x^2 + 2x + 1) + 9(y^2 - 4y + 4) = 36$

$\Rightarrow \quad 4(x+1)^2 + 9(y-2)^2 = 36$

$\Rightarrow \quad \frac{(x+1)^2}{9} + \frac{(y-2)^2}{4} = 1$

$\because \quad a = 3, \quad b = 4$ and $e = \sqrt{1 - \frac{4}{9}} = \frac{\sqrt{5}}{3}$

$\because$ (−1, 2) is centre and (1, 2) lies on major axis

$\therefore$ Minimum distance = 1

(c) Equation of normal at $P(\theta)$ is

$5x\sec\theta - 4y\,\mathrm{cosec}\,\theta = 25 - 16$ and it passes through $P(0, \alpha)$.

$$\alpha = \frac{-9}{4\,\mathrm{cosec}\,\theta}$$

$\Rightarrow \quad \alpha = \frac{-9}{4}\sin\theta$

$\Rightarrow \quad |\alpha| < \frac{9}{4}$

(d) $\therefore$ Length of latusrectum

$$= \frac{2b^2}{a} = \frac{2a}{3} \Rightarrow 3b^2 = a^2$$

$\Rightarrow \quad \frac{b^2}{a^2} = \frac{1}{3} \Rightarrow e = \sqrt{1 - \frac{1}{3}} = \sqrt{\frac{2}{3}}$

Sol. (Q.Nos. 29-31)

Given, $2a = 4 \Rightarrow a = 2$

$2b = 2 \Rightarrow b = 1$

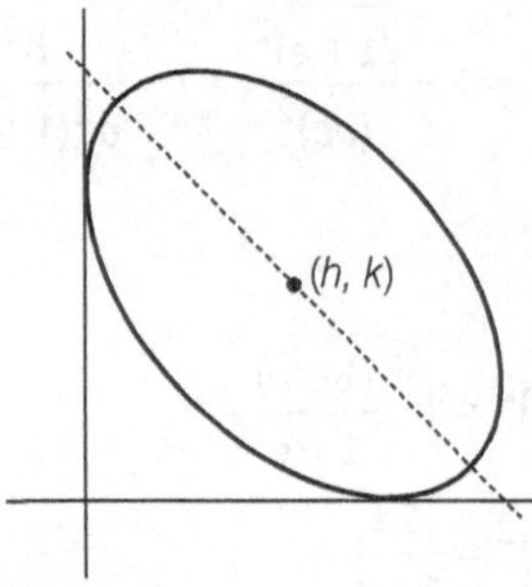

29. (c) $\because$ Origin lies on director circle of ellipse.

$$\because \quad (0-h)^2 + (0-k)^2 = a^2 + b^2$$

$$\Rightarrow \quad h^2 + k^2 = 4 + 1$$

$$\Rightarrow \quad h^2 + k^2 = 5$$

$\because$ Locus is $x^2 + y^2 = 5$

30. (d) $\because$ $x_1 x_2 = b^2 = 1$

and $y_1 y_2 = b^2 = 1$

$\therefore$ $F_1(x_1, y_1)$ and $F_2\left(\frac{1}{x_1}, \frac{1}{y_1}\right)$

i.e. $F_1(h, k)$ and $F_2\left(\frac{1}{h}, \frac{1}{k}\right)$

$$\because \quad \left(h - \frac{1}{h}\right)^2 + \left(k - \frac{1}{k}\right)^2 = (2ae)^2 = (2\sqrt{3})^2 = 12$$

$$\Rightarrow \quad h^2 + \frac{1}{h^2} - 2 + k^2 + \frac{1}{k^2} - 2 = 12$$

$$\Rightarrow \quad \left(h^2 + \frac{1}{h^2}\right) + \left(k^2 + \frac{1}{k^2}\right) = 16$$

$$\Rightarrow \quad x^2 + y^2 + \frac{1}{x^2} + \frac{1}{y^2} = 16$$

31. (b) Ordinate $= 1 \Rightarrow y = 1$

$$\because \quad x^2 + y^2 = 5$$

$$\Rightarrow \quad x^2 + 1^2 = 5 \Rightarrow x = 2$$

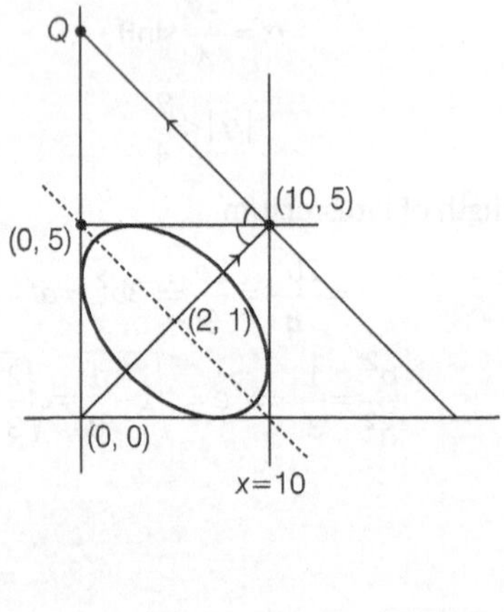

$\because$ Parabola $y^2 + 4x - 10y - 15 = 0$

$$\Rightarrow \quad (y-5)^2 = -4x + 40$$

$$\Rightarrow \quad (y-5)^2 = -4(x-10)$$

On compare $Y^2 = -4AX$, we get

$$A = 1, Y = y - 5 \text{ and } X = x - 10$$

Focus $\Rightarrow$ $(0, A)$

$$\Rightarrow \quad x - 10 = 0$$

$$x = 10 \text{ and } y - 5 = 0$$

$$y = 5$$

$\therefore$ By symmetric we can say that

$$Q = (0, 10)$$

Sol. (Q.Nos. 32-36)

32. (c) Let P, Q, R, S be four concylic points on E, whose eccentric angles are $\alpha, \beta, \gamma, \delta$, respectively.

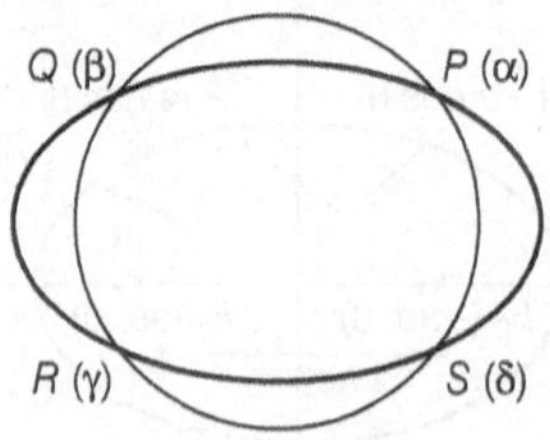

There, equation of chord PQ and RS (taken any two chords

$$\frac{x}{a}\cos\left(\frac{\alpha+\beta}{2}\right) + \frac{y}{b}\sin\left(\frac{\alpha+\beta}{2}\right) - \cos\left(\frac{\alpha-\beta}{2}\right) = 0$$

$$\text{and } \frac{x}{a}\cos\left(\frac{\gamma+\delta}{2}\right) + \frac{y}{b}\sin\left(\frac{\gamma+\delta}{2}\right) - \cos\left(\frac{\gamma-\delta}{2}\right) = 0$$

$\therefore$ The equation of any curve passing through P, Q, R and S is given by

$$\left(\frac{x^2}{a^2} + \frac{y^2}{b^2} - 1\right) + \lambda\left(\frac{x}{a}\cos\left(\frac{\alpha+\beta}{2}\right) + \frac{y}{b}\sin\left(\frac{\alpha+\beta}{2}\right) - \cos\left(\frac{\alpha-\beta}{2}\right)\right) \cdot \left(\frac{x}{a}\cos\left(\frac{\gamma+\delta}{2}\right) + \frac{y}{b}\sin\left(\frac{\gamma+\delta}{2}\right) - \cos\left(\frac{\gamma-\delta}{2}\right)\right)$$

But the given points are concyclic.

$\therefore$ It represents a circle.

$\Rightarrow$ Coeff of x^2 = Coeff of y^2

$$\therefore \cos\left(\frac{\alpha+\beta}{2}\right)\sin\left(\frac{\gamma+\delta}{2}\right) + \sin\left(\frac{\alpha+\beta}{2}\right)\cos\left(\frac{\gamma+\delta}{2}\right) = 0$$

$$\Rightarrow \quad \sin\left(\frac{\alpha+\beta+\gamma+\delta}{2}\right) = 0 = \sin\pi$$

$$\Rightarrow \quad \alpha + \beta + \gamma + \delta = 2n\pi, \quad n \in I$$

33. (d) If the point intersection of chords PR and QS is T

Then, slope of $PR = \frac{-b}{a}\cos\left(\frac{\alpha+\gamma}{2}\right) = \tan\theta$

Given, $\frac{-b}{a}\cos\left\{n\pi-\left(\frac{\beta+\delta}{2}\right)\right\}=0$

$\because \quad \alpha+\beta+\gamma+\delta=2n\pi$

$\Rightarrow \quad \frac{b}{a}\cot\left(\frac{\beta+\delta}{2}\right)=-$ Slope of QS

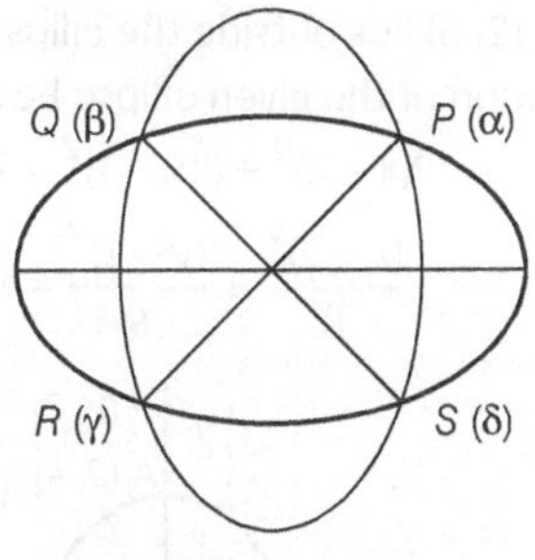

$\Rightarrow \quad \tan\phi=\tan(\pi-\phi)$

$0=\pi-\phi$ or $\theta+\phi=\pi$

34. (b) $\because$ Bisector of quadrants are $y=\pm x$ ...(i)

and ellipse E is $\frac{x^2}{a^2}+\frac{y^2}{b^2}=1$ ...(ii)

Solving Eqs. (i) and (ii), we get

$$x=\pm\frac{ab}{\sqrt{a^2+b^2}},\quad y=\pm\frac{ab}{\sqrt{a^2+b^2}}$$

$$P=\left(\frac{ab}{\sqrt{a^2+b^2}},\frac{ab}{\sqrt{a^2+b^2}}\right),$$

$$Q=\left(\frac{-ab}{\sqrt{a^2+b^2}},\frac{ab}{\sqrt{a^2+b^2}}\right),$$

$$R=\left(\frac{-ab}{\sqrt{a^2+b^2}},\frac{-ab}{\sqrt{a^2+b^2}}\right)$$

and $$S=\left(\frac{ab}{\sqrt{a^2+b^2}},\frac{-ab}{\sqrt{a^2+b^2}}\right)$$

$\because$ P, Q, R, S lie on a circle $x^2+y^2=r^2$,

where $r=OP=OQ=OR=OS$

$=\frac{ab\sqrt{2}}{\sqrt{a^2+b^2}}$, where O is centre of ellipse E.

35. (a) $\because$ Eccentric angles of P, Q and R are α, $\frac{\pi}{2}+\alpha$ and $\pi+\alpha$, respectively.

Let β be the eccentric angle of the fourth point S.

Then, $\quad \alpha+\frac{\pi}{2}+\alpha+\pi+\alpha+\beta=2\pi$

or $\quad \beta=2\pi-\frac{3\pi}{2}-3\alpha$

For $\quad \beta=\frac{\pi}{2}-3\alpha$

36. (d) Let the point of intersection of ellipse

$$\frac{x^2}{a^2}+\frac{y^2}{b^2}=1 \quad \text{...(i)}$$

and circle $x^2+y^2+2gx+2fy+c=0$ ...(ii)

be $\alpha,\beta,\gamma,\delta$

$\because \quad \alpha+\beta+\gamma+\delta=2\pi$

Let ϕ be any point on Eq (i)

$\because \quad x=a\cos\phi$ and $y=b\sin\phi$

This point also lie on Eq. (ii)

$\because a^2\cos^2\phi+b^2\sin^2\phi+2ga\cos\phi$

$+2fb\sin\phi+c=0$ ...(iii)

$\Rightarrow \{(a^2-b^2)\cos^2\phi+2ga\cos\phi+(b^2+c)\}^2$

$=4f^2b^2(1-\cos^2\phi)$

$\Rightarrow \quad (a^2-b^2)^2\cos^4\phi+4ga(a^2-b)^2\cos^3\phi$

$+\{2(a^2-b^2)(b^2+c)+4g^2a^2+4f^2b^2\}$

$\cos^2\phi+4ga(b^2+c)\cos\phi+\{(b^2+c^2)^2-4f^2b^2\}=0$

This is the fourth degree equation in $\cos\phi$

It has four roots (i.e. $\cos\alpha$, $\cos\beta$, $\cos\gamma$, $\cos\delta$)

$\therefore \quad \cos\alpha+\cos\beta+\cos\gamma+\cos\delta=\frac{-4ga}{(a^2-b^2)}$ ...(iv)

Similarly changing Eq. (ii) in $\sin\phi$, we get

$$\sin\alpha+\sin\beta+\sin\gamma+\sin\delta=\frac{-4fb}{b^2-a^2} \quad \text{...(v)}$$

$\because \quad \alpha+\beta+\gamma+\delta=2n\pi$

$\because \quad \delta=2n\pi-(\alpha+\beta+\gamma)$

$\because \quad \sin\delta=-\{\sin(\alpha+\beta+\gamma)\}$

and $\quad \cos\delta=\cos(\alpha+\beta+\gamma)$

Then, from Eqs. (iv) and (v), we get

$$-g=\frac{a^2-b^2}{4a}\times\{\cos\alpha+\cos\beta+\cos\gamma+\cos(\alpha+\beta+\gamma)\}$$

and $$-f=\frac{b^2-a^2}{4b}\{\sin\alpha+\sin\beta+\sin\gamma-\sin(\alpha+\beta+\gamma)\}$$

$\therefore$ Centre of circle $\left\{\frac{a^2-b^2}{4a}(\cos\alpha+\cos\beta+\cos\gamma)\right.$

$+\cos(\alpha+\beta+\gamma),\left(\frac{b^2-a^2}{4b}\right)(\sin\alpha+\sin\beta$

$\left.+\sin\gamma-\sin(\alpha+\beta+\gamma))\right\}$

Sol. (Q.Nos. 37-39)

37. (c) $\lambda x-y+2(1+\lambda)=0$

$\Rightarrow \lambda(x+2)-(y-2)=0$

$\Rightarrow$ This line passes through (-2, 2)

$\mu x-y+2(1-\mu)=0$

$\Rightarrow \quad \mu(x-2)-(y-2)=0$

This lines passes through (2, 2)

Clearly these represent the foci of the ellipse.

So, $\quad 2ae = 4$

The circle $x^2 + y^2 - 4y - 5 = 0$

$\Rightarrow x^2 + (y-2)^2 = 9$ represents the auxiliary circle.

Thus, $a^2 = 9, e = \frac{2}{3}$ and $b^2 = 5$

38. (b) $\therefore$ Area $= \frac{1}{2}(2ae)b = 3 \times \frac{2}{3} \times \sqrt{5} = 2\sqrt{5}$

39. (d) Light ray from one focus will pass through another focus

$\therefore$ (-2, 2) and (2, 2) are the focus of ellipse with $y = 2$ as major axis and corners at (0, 2) and also foot of perpendicular from (2, 2) focus upon any tangent to the ellipse on $x^2 + (y-2)^2 = 9$ auxiliary circle.

$\therefore \quad ae = 2 \Rightarrow a = 3$

Now, $a^2 - b^2 = 4 \Rightarrow a = 3$ and $b = \sqrt{5}$

Equation of ellipse is $\frac{x^2}{9} + \frac{(y-2)^2}{5} = 1$

$P(3\cos\theta, 2 + 5\sin\theta)$ is any point on ellipse light from focus (-2, 2) touches P and gets reflected and passes through (2, 2)

$\Rightarrow S'P + SP = 2a$ is always irrespective of the location at P

40. (i) $\rightarrow$ A, (ii) $\rightarrow$ B, (iii) $\rightarrow$ A, (iv) $\rightarrow$ C

(i) Any point on ellipse $\frac{x^2}{9} + \frac{y^2}{4} = 1$

$P = (3\cos\theta, 2\sin\theta)$

$$= \left\{3\left(\frac{1-\tan^2\frac{\theta}{2}}{1+\tan^2\frac{\theta}{2}}\right), \left(\frac{4\tan\frac{\theta}{2}}{1+\tan^2\frac{\theta}{2}}\right)\right\}$$

$\tan\frac{\theta}{2}$ can take infinite rational value.

(ii) $\frac{x^2}{9} + \frac{y^2}{4} = 1$

$\therefore P(\pm 3, 0), (0, \pm 2) \rightarrow$ only 4 points.

(iii) $P(\sqrt{3}\cos\theta, \sin\theta)$

$$\Rightarrow \frac{\tan\theta}{\sqrt{3}} = k \in Q$$

$$P = \left(\frac{\sqrt{3}}{\sqrt{1+3k^2}}, \frac{\sqrt{3}k}{\sqrt{1+3k^2}}\right)$$

If $\frac{\sqrt{1+3k^2}}{\sqrt{3}} = \lambda \Rightarrow \lambda^2 - k^2 = \frac{1}{3}$

$$(\lambda - k)(\lambda + k) = \frac{1}{3}$$

which can be expressed as product of two rational in infinite ways

(iv) $P \equiv (0, \pm 1) \rightarrow$ 2 points.

41. (i) $\rightarrow$ (A, D), (ii) $\rightarrow$ (B, E), (iii) $\rightarrow$ (B, D), (iv) $\rightarrow$ (C)

(i) Let $S: 9x^2 + 8y^2 - 36x - 16y - 28 = 0$

$\therefore$ Value of S at (2, 6)

$$S_1 = 9(2)^2 + 8(6)^2 - 36(2) - 16(6) - 28$$
$$= 36 + 288 - 72 - 96 - 28 = 128 > 0$$

$\therefore$ Point (2, 6) lies outside the ellipse

The equation of the given ellipse be rewritten as

$$9(x-2)^2 + 8(y-1)^2 = 72$$

$$\Rightarrow \quad \frac{(x-2)^2}{8} + \frac{(y-1)^2}{9} = 1$$

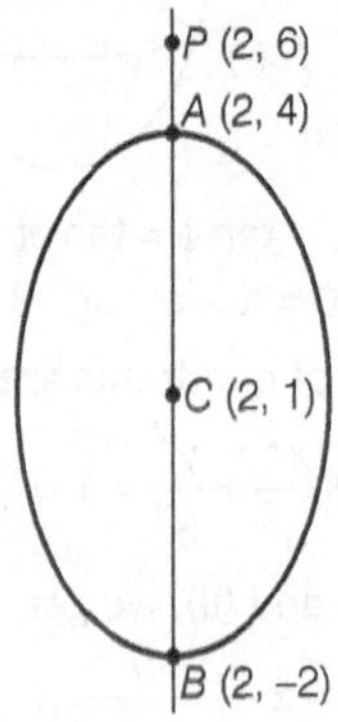

Centre of ellipse (2, 1) and axis perpendicular to Y-axis.

$\therefore$ Vertices are $(x-2) = 0$

and $y - 1 = \pm 3$ or $(2, -2)$ and $(2, 4)$

Minimum distance, $L = PA = 2$

and maximum distance, $G = PB = 8$

$$L + G = 2 + 8 = 10 \text{ and } G - L = 8 - 2 = 6$$

(ii) Let $S: 4x^2 + 9y^2 + 8x - 36y + 4 = 0$

$\therefore$ Value of S at (1, 2) is

$$S_1 = 4(1)^2 + 9(2)^2 + 8(1) - 36(2) + 4$$
$$= 4 + 36 + 8 - 72 + 4 = -20 < 0$$

$\therefore$ Point (1, 2) lies inside the ellipse.

The equation of the given ellipse be written as

$$4(x+1)^2 + 9(y-2)^2 = 36$$

$$\Rightarrow \quad \frac{(x+1)^2}{9} + \frac{(y-2)^2}{4} = 1$$

$\therefore$ Centre of ellipse is (-1, 2) and axis parallel to X-axis

$\therefore$ Vertices are $x + 1 = \pm 3$ and $y - 2 = 0$

$(-4, -2)$ and $(2, 2)$

Minimum distance, $L = PA = 1$

and maximum distance $G = PA' = AA' - PA = 6 - 1$

$$G = 5$$

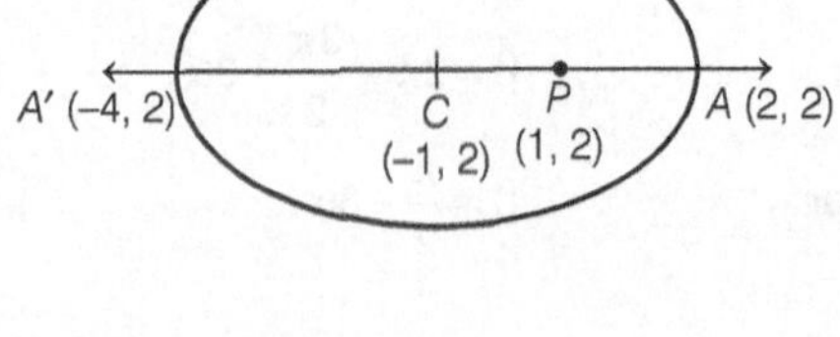

$\therefore \quad L+G=1+5=6$

$\Rightarrow \quad G-L=5-1=4$

$\therefore \quad L^G+G^L=(1)^5+(5)^1=6$

(iii) Here, $3x+4y=0$ and $4x-3y=0$ are mutually perpendicular lines

$\therefore$ Put $X=\dfrac{3x+4y}{\sqrt{3^2+4^2}}$ and $Y=\dfrac{4x-3y}{\sqrt{4^2+3^2}}$

Then, $\quad 4X^2+9Y^2=36$

$$\frac{X^2}{9}+\frac{Y^2}{4}=1$$

$\therefore$ Vertices $X=\pm 3$ and $Y=0$

$\Rightarrow \quad \dfrac{3x+4y}{5}=\pm 3$ and $\dfrac{4x-3y}{5}=0$

$\Rightarrow \quad 3x+4y=\pm 15$ and $y=\dfrac{4x}{3}$.

Vertices are $\left(\dfrac{9}{5},\dfrac{12}{5}\right)$ and $\left(\dfrac{-9}{5},\dfrac{-12}{5}\right)$

$\because$ Given point is a vertex

$\therefore$ Minimum distance $L=0$

and minimum distance $G=$ length of major axis

$=2\times 3=6$

$\therefore \quad L+G=6$ and $G-L=6$

(iv) Given, ellipse : $16x^2+9y^2=144$

$\Rightarrow \quad \dfrac{x^2}{9}+\dfrac{y^2}{16}=1$

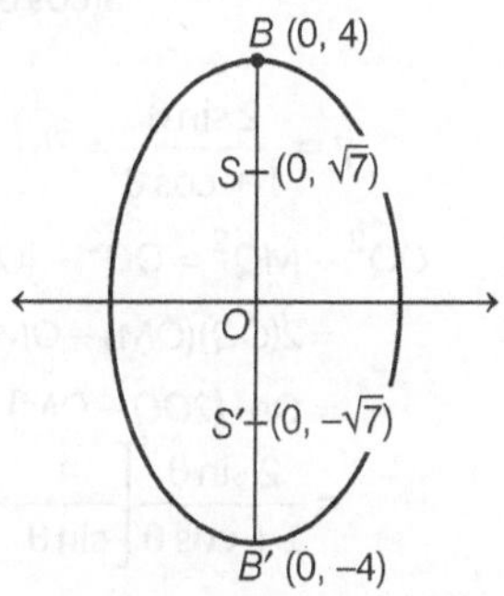

$\Rightarrow \quad a^2=9$ and $b^2=16$

$\therefore \quad e=\sqrt{1-\dfrac{9}{16}}=\dfrac{\sqrt{7}}{4}$

Focus $=(0,\pm be)=(0,\pm\sqrt{7})$

Vertices $=(0,\pm b)=(0,\pm 4)$

$\therefore$ Point $(0,+\sqrt{7})$ is focus

$\therefore$ Minimum distance,

$L=SB$

$\Rightarrow \quad =4-\sqrt{7}$

Maximum distance, $G=SB'=4+\sqrt{7}$

$\therefore \quad L+G=8$ and $G-L=2\sqrt{7}$

42. (i) $\rightarrow$ (A, C), (ii) $\rightarrow$ (C, D), (iii) $\rightarrow$ (A), (iv) $\rightarrow$ (B)

(i) Tangent to ellipse as $P(\phi)$ is

$$\frac{x}{4}\cos\phi+\frac{y}{2}\sin\phi=1$$

It must pass through the centre of the circle. Hence,

$$\frac{4}{4}\cos\phi+\frac{2}{2}\sin\phi=1$$

$\Rightarrow \quad \cos\phi+\sin\phi=1$

$\Rightarrow \quad 1+\sin 2\phi=1$

$\Rightarrow \quad \sin 2\phi=0$

$\Rightarrow \quad 2\phi=0,\pi$

$\Rightarrow \quad \phi=0,\dfrac{\pi}{2} \Rightarrow \dfrac{\phi}{2}=0,\dfrac{\pi}{4}$

(ii) Consider any point $P(\sqrt{6}\cos\theta,\sqrt{2}\sin\theta)$ on ellipse

$$\frac{x^2}{6}+\frac{y^2}{2}=1$$

Given that $OP=2$

$\Rightarrow \quad 6\cos^2\theta+2\sin^2\theta=4$

$\Rightarrow \quad 4\cos^2\theta=2$

$\Rightarrow \quad \cos\theta=\pm\sqrt{\dfrac{2}{4}}=\pm\dfrac{1}{\sqrt{2}}$

$\Rightarrow \quad \theta=\dfrac{\pi}{4}$ or $\dfrac{5\pi}{4}$

(iii) Solving the equation of ellipse and parabola, we have

$$y-1+4y^2=4$$

$\Rightarrow \quad 4y^2+y-5=0$

$\Rightarrow \quad (4y+5)(y-1)=0$

$\Rightarrow \quad y=1$ and $x=0$

The curve touches at $(0,1)$.

$\therefore 2\cos\theta=0$ and $\sin\theta=1 \Rightarrow \theta=\pi/2$

Hence, eccentric angle is $\pi/2$.

(iv) The normal at $P(a\cos\theta, b\sin\theta)$ is

$$\frac{ax}{\cos\theta}-\frac{by}{\sin\theta}=a^2-b^2,$$

where $a^2=14$ and $b^2=5$

It meets the curve again at $Q(2\theta)$ i.e. $(a\cos 2\theta, b\sin 2\theta)$

$\therefore \quad \dfrac{a}{\cos\theta}\cdot a\cos 2\theta-\dfrac{b}{\sin\theta}b\sin 2\theta=a^2-b^2$

$\Rightarrow \quad \dfrac{14}{\cos\theta}\cos 2\theta-\dfrac{5}{\sin\theta}\sin 2\theta=14-5$

$\Rightarrow \quad 28\cos^2\theta-14-10\cos^2\theta=9\cos\theta$

$\Rightarrow \quad 18\cos^2\theta-9\cos\theta-14=0$

$\Rightarrow \quad (6\cos\theta-7)(3\cos\theta+2)=0$

$$\cos\theta=-\frac{2}{3},+\frac{7}{6}$$

$$\theta=\cos^{-1}\left(-\frac{2}{3}\right)$$

43. (i) → (A, B, C), (ii) → (A, C, D), (iii) → (A, C, E)

(i) $E: \frac{x^2}{1} + \frac{y^2}{2} = 1$

Now, $C_1 : x^2 + y^2 = 3$

$C_2 : x^2 + y^2 = 6 \Rightarrow C_3 : x^2 + y^2 = 12$

$\therefore \quad r_1^2 = 3, \quad r_2^2 = 6 \quad \text{and} \quad r_3^2 = 12$

$\therefore \quad r_1^2, r_2^2, r_3^2 \ldots$ are in GP with common ratio 2

$\therefore \quad r_1^2 + r_2^2 + r_3^2 \ldots + r_n^2 = 3 + 6 + 12 + \ldots$

$$= \frac{3(2^n - 1)}{2 - 1} = 3(2^n - 1) = r_1^2(2^n - 1)$$

GM of $r_1^2 r_2^2 r_3^2 \Rightarrow (r_1 r_2 r_3)^{2/3} = (3 \cdot 6 \cdot 12)^{2/3} = 6$

(ii) Given, $E: \frac{x^2}{2} + \frac{y^2}{3} = 1$

Now, $C_1 : x^2 + y^2 = 5$

$C_2 : x^2 + y^2 = 10 \Rightarrow C_3 : x^2 + y^2 = 20$

$\Rightarrow r_1^2 = 5, \quad r_2^2 = 10 \quad \text{and} \quad r_3^2 = 20$

$\therefore \quad r_1^2, r_2^2, r_3^2, \ldots,$ are in GP with common ratio 2

Now, $r_1^2 + r_2^2 + r_3^2 \ldots + r_n^2 = 5 + 10 + 20 + \ldots$

$$= \frac{5(2^n - 1)}{2 - 1} = r_1^2 (2^n - 1)$$

GM of $r_1^2, r_2^2, r_3^2 = (r_1 r_2 r_3)^{2/3} = (5 \cdot 10 \cdot 20)^{2/3} = 10$

(iii) Given, $E: \frac{x^2}{4} + \frac{y^2}{3} = 1$

Now, $C_1 : x^2 + y^2 = 7$

$C_2 : x^2 + y^2 = 14$

$C_3 : x^2 + y^2 = 28$

$r_1^2 = 7, \; r_2^2 = 14 \; \text{and} \; r_3^2 = 28$

$r_1^2, r_2^2, r_3^2 \ldots$ are in GP with common ratio 2.

$r_1^2 + r_2^2 + r_3^2 + \ldots + r_n^2 = 7 + 14 + 28 + \ldots$

$$= 7\frac{(2^n - 1)}{2 - 1} = r_1^2(2^n - 1)$$

GM of $r_1^2, r_2^2, r_3^2 = (r_1 r_2 r_3)^{2/3} = (7 \cdot 14 \cdot 28)^{2/3} = 14$

44. (i) → (C), (ii) → (A), (iii) → (E), (iv) → (B)

(a) $\because$ Given, $S'B \perp SB$

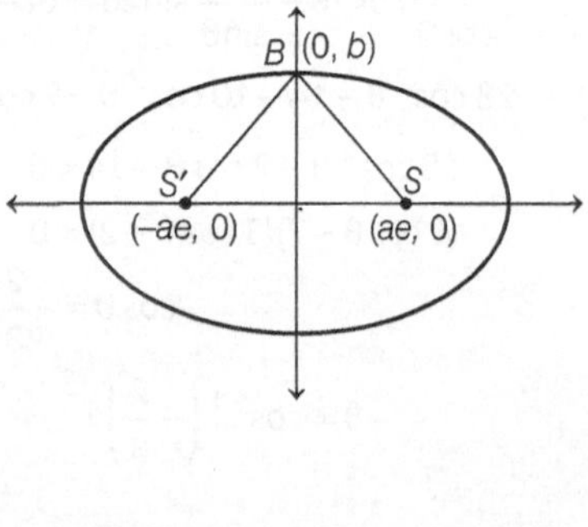

$\Rightarrow \quad m_{S'B} \times m_{SB} = -1$

$\Rightarrow \quad \left(\frac{b - 0}{0 + ae}\right)\left(\frac{b - 0}{0 - ae}\right) = -1$

$\Rightarrow \quad \frac{b^2}{a^2 e^2} = 1$

$\Rightarrow \quad \frac{a^2(1 - e^2)}{a^2 e^2} = 1$

$\Rightarrow \quad e^2 = 1 - e^2$

$\Rightarrow \quad 2e^2 = 1 \Rightarrow e = \frac{1}{\sqrt{2}}$

(ii) $a = 3$ and $b = 2$

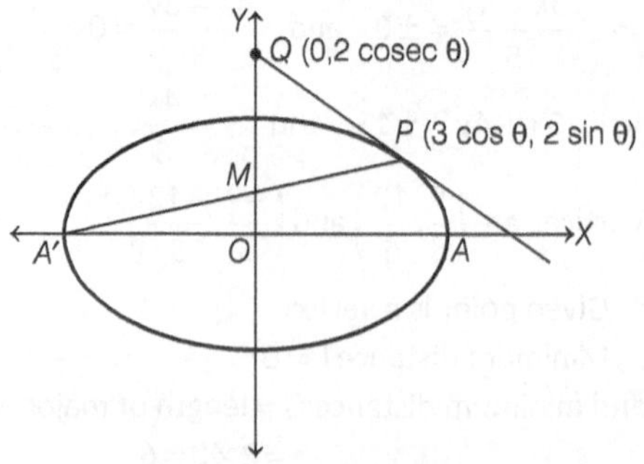

$$T: \frac{x \cos\theta}{3} + \frac{y \sin\theta}{2} = 1$$

$\therefore \quad x = 0 \quad \text{and} \quad y = 2 \operatorname{cosec}\theta$

Equation of chord $A'P \Rightarrow y = \frac{2 \sin\theta}{3(\cos\theta + 1)}(x + 3)$

Put $x = 0$

$$y = \frac{2 \sin\theta}{1 + \cos\theta} = PM$$

$\therefore \quad OQ^2 - MQ^2 = OQ^2 - [OQ - OM]^2$

$= 2(OQ)(OM) - OM^2$

$= OM\,(2OQ - OM)$

$$= \frac{2 \sin\theta}{1 + \cos\theta}\left[\frac{4}{\sin\theta} - \frac{2 \sin\theta}{1 + \cos\theta}\right] = 4$$

(iii) $\left(y + \frac{1}{2}\right)^2 = 8(x - 2)$

On compare $Y^2 = 8X$ where, $Y = y + \frac{1}{2}$ and $X = x - 2$

For 3 normals

$X > 2a, \; x - 2 > 4$

$\Rightarrow \quad x > 6$

(iv) Area of parallelogram $= 2ab = 2(2) \times \left(\frac{1}{2}\right) = 2$ sq units

Sol. (Q.Nos. 45-46)

$\because$ We know that distance between focus $= 2ae$

$2ae = 2 \Rightarrow ae = 1$ and $P_1 P_2 = b^2$

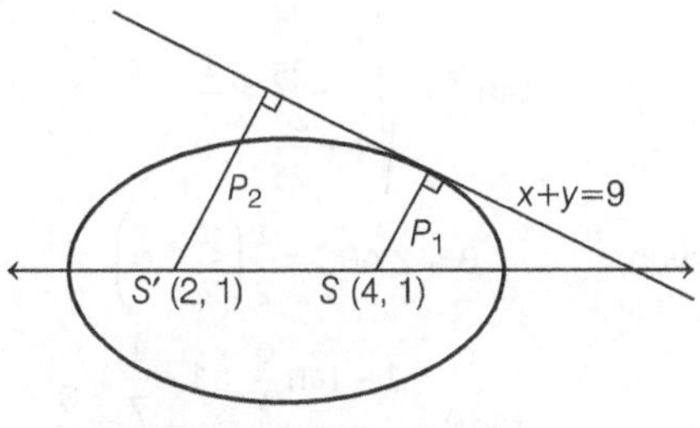

$$\left|\frac{4+1-9}{\sqrt{2}}\right|\left|\frac{2+1-9}{\sqrt{2}}\right|=b^2$$

$\Rightarrow$ $$\frac{4}{\sqrt{2}}\times\frac{6}{\sqrt{2}}=b^2$$

$\Rightarrow$ $$b^2=12$$

$\Rightarrow$ $$a^2-a^2e^2=12$$

$\Rightarrow$ $$a^2-1=12$$

$\Rightarrow$ $$a=\sqrt{13}$$

45. (1) $e=\sqrt{1-\frac{b^2}{a^2}}=\sqrt{1-\frac{12}{13}}=\frac{1}{\sqrt{13}}$

46. (2) $\therefore$ Length of major axis $=2a=2\sqrt{13}$ units

Sol. (Q.Nos. 47-48)

Ellipse : $\frac{x^2}{9}+\frac{y^2}{8}=1$

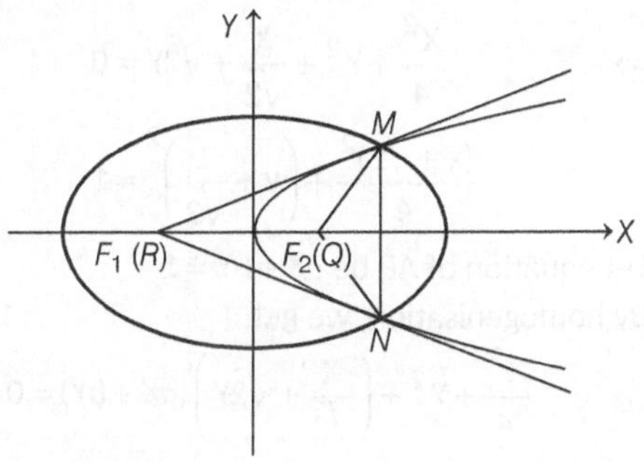

Here, $e=\sqrt{1-\frac{8}{9}}=\frac{1}{3}$

$\therefore$ Focus $(\pm 1, 0)$

$\Rightarrow$ $F_1(-1, 0)$ and $F_2(1, 0)$

The given parabola $y^2=4x$

Thus, $M=\left(\frac{3}{2},\sqrt{6}\right)$ and $N=\left(\frac{3}{2},-\sqrt{6}\right)$

47. (0.9) For orthocentre one altitude is $y=0$ (*MN* is perpendicular)

Other altitude is $(y-\sqrt{6})=\frac{5}{2\sqrt{6}}\left(x-\frac{3}{2}\right)$

$\because$ Orthocentre is $\left(\frac{-9}{10},0\right)$

48. (13) Equation of tangent at *M* and *N* are

$$\frac{x}{6}\pm\frac{y\sqrt{6}}{8}=1$$

$\therefore$ *R* is (6, 0)

$\therefore$ Equation of normal at *M*, is

$$(y-\sqrt{6})=\frac{-\sqrt{6}}{2}\left(x-\frac{3}{2}\right)$$

Thus, *Q* is $\left(\frac{7}{2},0\right)$

$\therefore$ Area $(\Delta MQR)=\frac{1}{2}\times\sqrt{6}\times\frac{5}{2}=\frac{5\sqrt{6}}{4}$

$\therefore$ Area $(MF_1NF_2)=\frac{\sqrt{6}}{2}+\frac{3\sqrt{6}}{2}=\frac{4\sqrt{6}}{2}$

$\therefore$ Required ratio $=\frac{5\sqrt{6}}{4}:\frac{4\sqrt{6}}{2}=\frac{5}{8}$

49. (0.2) $\because$ $OS_1=ae=6$ and $OC=b$ let also $CS_1=a$.

$\because$ Area of $\Delta OCS_1=\frac{1}{2}(OS_1)(OC)=3b$

$\because$ Semi-perimeter of $\Delta OCS_1=\frac{1}{2}(OS_1+OC+CS_1)$

$$\frac{1}{2}(6+a+b) \qquad \text{... (i)}$$

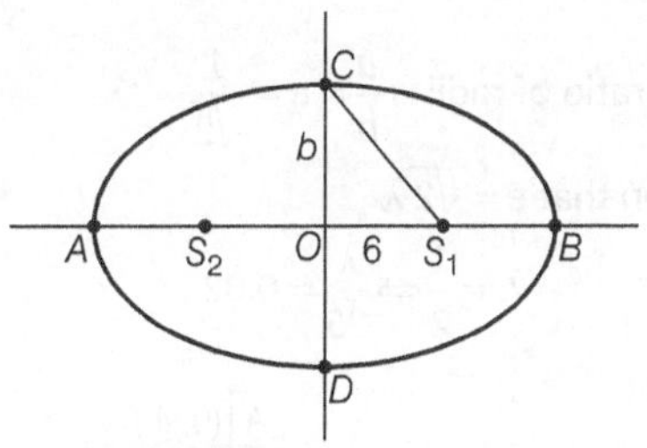

$\because$ Inradius of $\Delta OCS_1=1$

$$\frac{3b}{\frac{1}{2}(6+a+b)}=1 \Rightarrow 5b=6+a \qquad \text{...(ii)}$$

$\because$ $$b^2=a^2-a^2e^2=a^2-36$$

From Eq. (ii),

$$25b^2=36+12a+a^2$$

$\because$ $$25(a^2-36)=36+a^2+12a$$

$\Rightarrow$ $$2a^2-a-78=0$$

$\Rightarrow$ $$a=\frac{13}{2},-6$$

When $a=\frac{13}{2}$, then $b=\frac{5}{2}$

$\therefore$ $$\frac{a-b}{20}=\frac{\frac{13}{2}-\frac{5}{2}}{20}=\frac{4}{20}=0.2.$$

50. (0.02) Let line OPQ makes angle θ with X-axis. So,
$P=(a\cos\theta, a\sin\theta)$, $Q=(b\cos\theta, b\sin\theta)$ and let $R\,(x, y)$
So, $X=a\cos\theta$, $Y=b\sin\theta$
Eliminating θ, we get

$$\frac{x^2}{a^2}+\frac{y^2}{b^2}=1$$

Locus of R is an ellipse

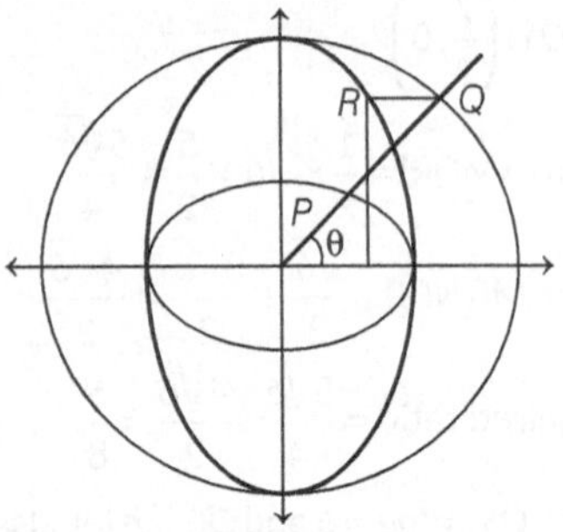

Also $a<b$, so vertices are $(0,b)$ and $(0,-b)$ and extremities of minor axis are $(\pm a, 0)$
So, ellipse touches both inner circle and outer circle.
If focii are $(0, \pm a)$, so $a=be$, then $e=\frac{a}{b}$

$\therefore\quad e=\sqrt{1-e^2}\Rightarrow e^2=1-e^2\Rightarrow e=\frac{1}{\sqrt{2}}$

and ratio of radii is $\frac{a}{b}=e=\frac{1}{\sqrt{2}}$ ...(i)

Given that $e=\sqrt{2}\lambda$

$$\lambda=\frac{1}{2}\Rightarrow\frac{\lambda}{25}=0.02$$

51. (24)

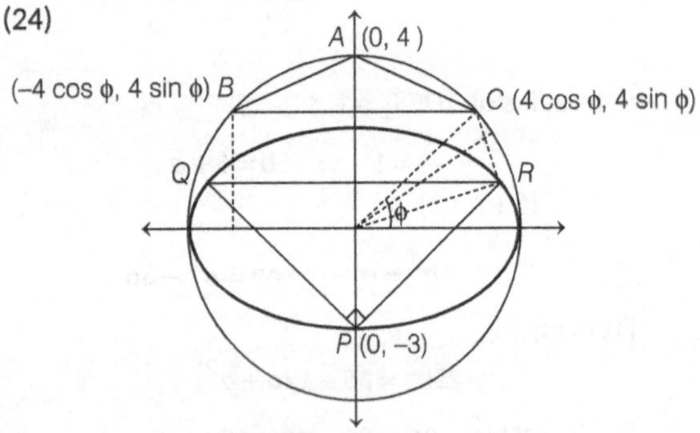

$$\therefore\quad Q=(-4\cos\phi, 3\sin\phi)$$
$$R=(4\cos\phi, 3\sin\phi)$$
$$P=(0,-3)$$

$\therefore PQ\perp PR$

$$\Rightarrow\quad \frac{3(\sin\phi+1)}{4\cos\phi}\left(\frac{3(\sin\phi+1)}{-4\cos\phi}\right)=-1$$

$$\Rightarrow\quad 25\sin^2\phi+18\sin\phi-7=0$$

$$\Rightarrow\quad \sin\phi=\frac{7}{25}\ \{\sin\phi=-1\text{ rejected}\}$$

$$\therefore\quad \tan\frac{\phi}{2}=\sqrt{\frac{1-\frac{24}{25}}{1+\frac{24}{25}}}=\frac{1}{7}$$

Given, $\quad \theta=\angle ABC=\frac{1}{2}\left(\frac{\pi}{2}-\phi\right)$

$$\therefore\quad \tan\theta=\frac{1-\tan\frac{\phi}{2}}{1+\tan\frac{\phi}{2}}=\frac{1-\frac{1}{7}}{1+\frac{1}{7}}=\frac{3}{4}$$

Then, $16\tan^2\theta+8\tan\theta+9$

$$=16\left(\frac{9}{16}\right)+8\left(\frac{3}{4}\right)+9=9+6+9=24$$

52. (6) Let $X=x-\sqrt{2}$ and $Y=y-\frac{1}{\sqrt{2}}$

Equation of ellipse becomes

$$\frac{x^2}{4}+y^2=1$$

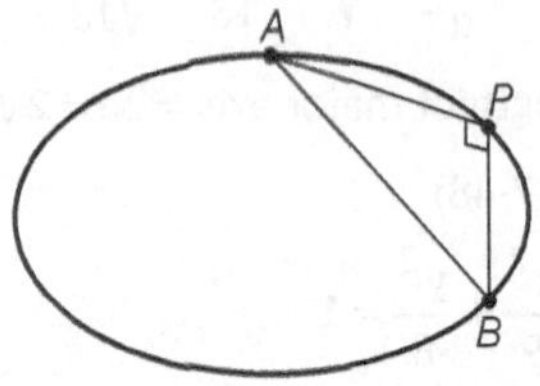

$$\Rightarrow\quad \frac{X^2}{4}+Y^2+\frac{X}{\sqrt{2}}+\sqrt{2}Y=0$$

$$\frac{(x+\sqrt{2})^2}{4}+\left(y+\frac{1}{\sqrt{2}}\right)^2=1$$

Let equation of AB be $aX+bY=1$
By homogenisation, we get

$$\frac{X^2}{4}+Y^2+\left(\frac{X}{\sqrt{2}}+\sqrt{2}Y\right)(aX+bY)=0$$

$\because AP\perp AB$

$$\text{Coeff of }X^2+\text{Coeff of }Y^2=0$$

$$\Rightarrow\quad \frac{a}{\sqrt{2}}+\sqrt{2}b+\frac{5}{4}=0$$

$$\Rightarrow\quad \frac{-4a}{5\sqrt{2}}-\frac{4\sqrt{2}b}{5}=1$$

$\Rightarrow AB$ pass through fix point.

$$\left(\overset{X}{\frac{-4}{5\sqrt{2}}},\overset{Y}{\frac{-4\sqrt{2}}{5}}\right)=\left(\overset{x}{\frac{-4}{5\sqrt{2}}+\sqrt{2}},\overset{y}{\frac{-4\sqrt{2}}{5}+\frac{1}{\sqrt{2}}}\right)$$

$$=\left(\frac{6}{5\sqrt{2}},\frac{-3}{5\sqrt{2}}\right)$$

$$\therefore\quad a=\frac{6}{5\sqrt{2}}$$

and $\qquad b = \dfrac{-3}{5\sqrt{2}}$

$\therefore \qquad a^2 + b^2 = \dfrac{36+9}{50} = \dfrac{9}{10} = \dfrac{m}{n}$

$\Rightarrow \qquad \left[\dfrac{m+n}{3}\right] = \left[\dfrac{19}{3}\right] = 6$

53. (4)

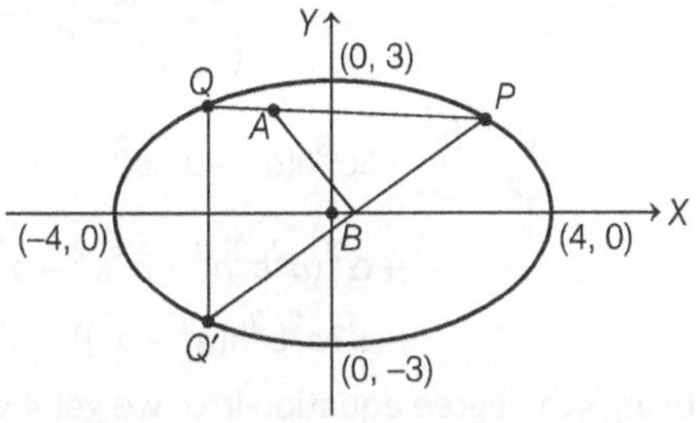

A and B be mid-points of segment PQ and PQ', respectively.

AB = distance between $M(P,Q)$ and $M(P,Q') = \dfrac{1}{2}(QQ')$

Since, Q, Q' must be on E, so maximum of $QQ' = 8$

$\therefore$ Maximum of $AB = \dfrac{8}{2} = 4$

54. (4)

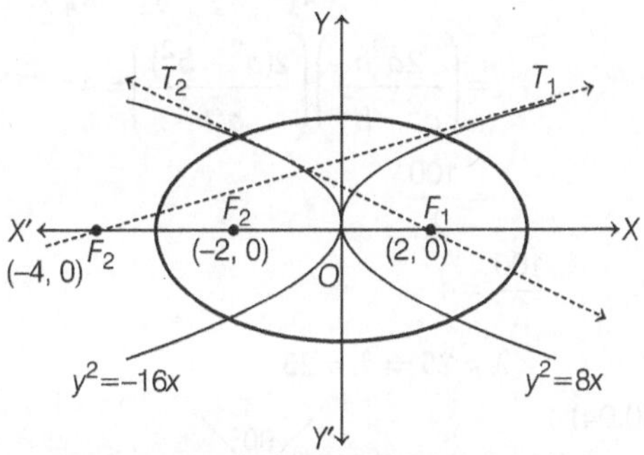

Tangent to P_1 passes throguh $(2F_2, 0) \Rightarrow (-4, 0)$

$$T_1 : y = m_1 x + \frac{2}{m_1}$$

$$0 = -4m_1 + \frac{2}{m_1}$$

$\Rightarrow \qquad m_1^2 = \dfrac{1}{2}$

Also, tangent to P_2 passes through $(F_1, 0)$ i.e., $(2, 0)$

$$T_2 : y = m_2 x + \left(\frac{-4}{m_2}\right)$$

$\Rightarrow \qquad 0 = 2m_2 - \dfrac{4}{m_2}$

$\Rightarrow \qquad m_2^2 = 2$

$\Rightarrow \qquad \dfrac{1}{m_1^2} + m_2^2 = 2 + 2 = 4$

55. (0.03) Let the coordinates of P be $(a\cos\theta, b\sin\theta)$. Then, coordiantes of Q be $(a\cos\theta, a\sin\theta)$.

Let $R(h,k)$ divides PQ in the ratio $r:s$

Then, $\qquad h = \dfrac{s(a\cos\theta) + r(a\cos\theta)}{(r+s)} = a\cos\theta$

$\Rightarrow \qquad \cos\theta = \dfrac{h}{a}$

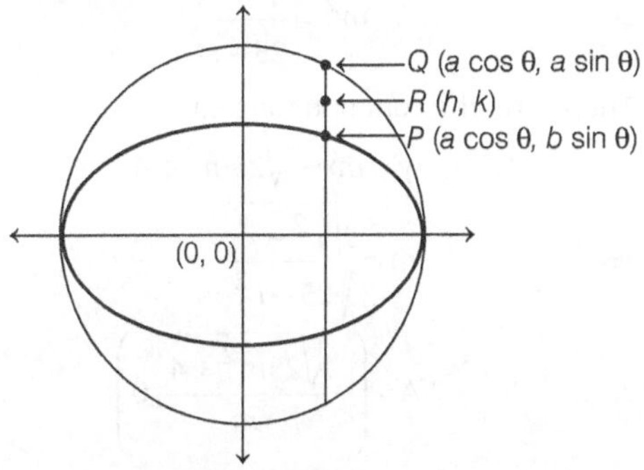

and $\qquad k = \dfrac{s(b\sin\theta) + r(a\sin\theta)}{r+s} = \sin\theta\dfrac{(bs+ar)}{r+s}$

$\Rightarrow \qquad \sin\theta = \dfrac{k(r+s)}{bs+ar}$

$\because \qquad \sin^2\theta + \cos^2\theta = 1$

$\Rightarrow \qquad \dfrac{h^2}{a^2} + k^2\dfrac{(r+s)^2}{(bs+ar)^2} = 1$

$\therefore$ Locus is $\dfrac{x^2}{a^2} + y^2\left\{\dfrac{(r+s)^2}{(bs+ar)^2}\right\} = 1$

On compare $\quad \dfrac{x^2}{a^2} + \dfrac{y^2(r+s)^2}{\lambda(bs+ar)^\mu} = 1$

$\therefore \qquad \lambda = 1$ and $\mu = 2$

$\therefore \qquad \dfrac{\lambda+\mu}{100} = \dfrac{3}{100} = 0.03$

56. (0.5) $Q : (b\cos\theta, b\sin\theta)$

$P : (a\cos\theta, a\sin\theta)$

$R : (a\cos\theta, b\sin\theta)$

$\dfrac{x^2}{a^2} + \dfrac{y^2}{b^2} = 1$ is locus of R which is an ellipse, since focus lie on inner circle.

$\therefore \qquad b = ae$

$\Rightarrow \qquad b^2 = a^2\left(1 - \dfrac{b^2}{a^2}\right)$

$\Rightarrow \qquad \dfrac{b^2}{a^2} = \dfrac{1}{2} \Rightarrow \dfrac{b}{a} = \dfrac{1}{\sqrt{2}}$

$\therefore \qquad e = \sqrt{1 - \dfrac{1}{2}} = \dfrac{1}{\sqrt{2}} \Rightarrow e^2 = \dfrac{1}{2}$

57. (10) The tangent to the ellipse is of the form

$$y = mx + \sqrt{25m^2 + 4}$$

This line touches the circle $x^2 + y^2 = r^2$

$$\therefore \qquad \frac{25\,m^2 + 4}{1 + m^2} = r^2$$

$$\Rightarrow \qquad m^2 = \frac{r^2 - 4}{25 - r^2} \qquad \text{...(i)}$$

Therefore, the common tangent is

$$y = mx + \sqrt{25m^2 + 4}$$

$$\Rightarrow \qquad m = \sqrt{\frac{r^2 - 4}{25 - r^2}}$$

$$\therefore \qquad A = \left(-\frac{\sqrt{25m^2 + 4}}{m}, 0 \right)$$

and $\qquad B = (0, \sqrt{25m^2 + 4})$

Let $M(x_1, y_1)$ be the mid-point of AB so that

$$2x_1 = \frac{-\sqrt{25m^2 + 4}}{m}$$

and $\qquad 2y_1 = \sqrt{25m^2 + 4}$

$$\Rightarrow \qquad 4m^2x_1^2 = 4y_1^2$$

$$\Rightarrow \qquad m^2 = \frac{y_1^2}{x_1^2}$$

Therefore, $2y_1 = \sqrt{25m^2 + 4}$

and $\qquad m^2 = \dfrac{y_1^2}{x_1^2}$

$$\Rightarrow \qquad 4y_1^2 = \frac{25y_1^2 + 4x_1^2}{x_1^2}$$

$$\Rightarrow \qquad 4x_1^2 + 25y_1^2 = 4x_1^2y_1^2$$

$\therefore$ Locus is $4x^2 + 25y^2 = 4x^2y^2 \Rightarrow \lambda = 4$ and $\mu = 25$

$$\therefore \qquad \sqrt{\lambda\mu} = \sqrt{25 \times 4} = 10$$

58. (25) $\because$ Normal at $P(\alpha, \beta)$ to the ellipse $\dfrac{x^2}{a^2} + \dfrac{y^2}{b^2} = 1$

$$\frac{a^2(x - \alpha)}{\alpha} = \frac{b^2(y - \beta)}{\beta} \qquad \text{...(i)}$$

This passes through (h, k)

$$\Rightarrow \qquad \frac{a^2(h - \alpha)}{\alpha} = \frac{b^2(k - \beta)}{\beta}$$

$$\Rightarrow \qquad a^2\beta\,(h - \alpha) = b^2\alpha(k - \beta)$$

$$\Rightarrow \qquad [(h - \alpha)\,a^2 + b^2\alpha]\beta = b^2k\alpha$$

$$\Rightarrow \qquad [a^2h - (a^2 - b^2)\alpha]^2\beta^2 = b^4k^2\alpha^2 \qquad \text{...(ii)}$$

Also, (α, β) lies on the ellipse, which means

$$\frac{\alpha^2}{a^2} + \frac{\beta^2}{b^2} = 1 \qquad \text{...(iii)}$$

Eliminating β from Eqs. (ii) and (iii), we get

$$\Rightarrow \qquad (a^4h^2 + \alpha^2(a^2 - b^2)^2 - 2a^2h\alpha(a^2 - b^2)$$

$$\left(\frac{a^2 - \alpha^2}{a^2} \right) b^2 = b^4k^2\alpha^2$$

$$\Rightarrow \qquad \alpha^4 \frac{(a^2 - b^2)^2 b^2}{a^2} - 2\alpha^3h(a^2 - b^2)b^2$$

$$+ \alpha^2(a^2b^2h^2 + b^4k^2 - b^2(a^2 - b^2)$$

$$+ \alpha(2a^2b^2h(a^2 - b^2)) - a^4b^2h^2 = 0.$$

This being 4th degree equation in α, we get 4 values of α say, x_1, x_2, x_3, x_4.

$$\therefore \quad x_1 + x_2 + x_3 + x_4 = \frac{2b^2h(a^2 - b^2)a^2}{b^2(a^2 - b^2)^2} = \frac{2a^2h}{a^2 - b^2}$$

$$\frac{1}{x_1} + \frac{1}{x_2} + \frac{1}{x_3} + \frac{1}{x^4} = \frac{2a^2b^2h(a^2 - b^2)}{a^4b^2h^2} = \frac{2(a^2 - b^2)}{a^2h}$$

$$\therefore \quad (x_1 + x_2 + x_3 + x_4)\left(\frac{1}{x_1} + \frac{1}{x_2} + \frac{1}{x_3} + \frac{1}{x_4} \right)$$

$$= \left(\frac{2a^2h}{a^2 - b^2} \right)\left(\frac{2(a^2 - b^2)}{a^2h} \right) = 4$$

$$= \frac{100}{\lambda}$$

$$\therefore \qquad \frac{100}{\lambda} = 4$$

$$\lambda = 25 \Rightarrow \lambda = 25$$

59. (0.04)

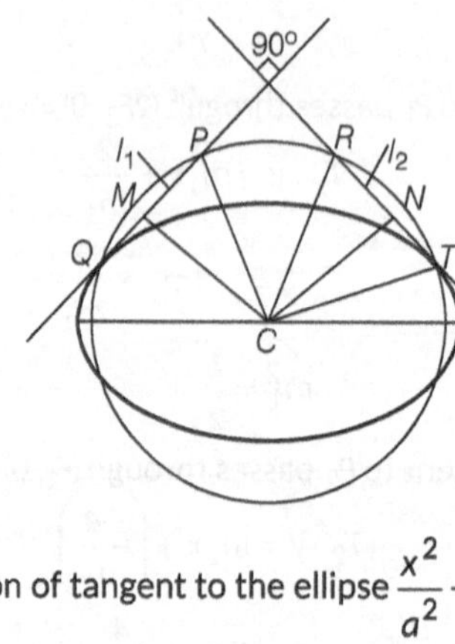

$\because$ Equation of tangent to the ellipse $\dfrac{x^2}{a^2} + \dfrac{y^2}{b^2} = 1$ having slope m is

$$y = mx + \sqrt{a^2m^2 + b^2} \qquad \text{...(i)}$$

and $\qquad y = \dfrac{-x}{m} + \sqrt{\dfrac{a^2}{m^2} + b^2}$

{$\because$ tangents are perpendicular to each other}

$$\Rightarrow \qquad x + my = \sqrt{a^2 + b^2m^2} \qquad \text{...(ii)}$$

Let the intercept of Eqs. (i) and (ii) on auxiliary circle be $l_1 = PQ$ and $l_2 = RT$

Draw CM and CN perpendicular to PQ and RT, respectively where C is (0, 0)

Now, $(CM)^2 = \dfrac{a^2m^2 + b^2}{1+m^2}$ and $(CN)^2 = \dfrac{a^2 + b^2m^2}{1+m^2}$

$$\therefore \qquad l_1 = PQ = 2PM \Rightarrow l_1^2 = 4(PM)^2$$

$$= 4(CP^2 - CM^2)$$

$$= 4\left(a^2 - \frac{a^2m^2 + b^2}{1+m^2}\right)$$

$$= \frac{4}{1+m^2}(a^2 - b^2) \qquad \text{...(iii)}$$

Also, $$l_2^2 = 4\left[a^2 - \frac{a^2 + b^2m^2}{1+m^2}\right]$$

$$= \frac{4m^2}{1+m^2}(a^2 - b^2)$$

Now, $$l_1^2 + l_2^2 = 4(a^2 - b^2)\left(\frac{1}{1+m^2} + \frac{m^2}{1+m^2}\right)$$

$$= 4(a^2 - b^2) = 4a^2e^2$$

$$= (2ae)^2 = (SS')^2.$$

$$\therefore \qquad l_1^2 + l_2^2 = (SS')^2$$

$$\Rightarrow \qquad l_1^m + l_2^n = (SS')^p$$

$$\therefore \qquad m = n = p = 2$$

$$\therefore \qquad \frac{(mnp)^{1/3}}{50} = \frac{(2^3)^{1/3}}{50}$$

$$= \frac{2}{50} = 0.04$$

CHAPTER

07

HYPERBOLA

LEVEL 01 \\ Starter Level

7.1 Equation of Hyperbola

1 Find the value of λ for which the given equation represent the hyperbola

(a) $3x^2 + \lambda y^2 + 4xy - 7x + 8y + 3 = 0$

(b) $2x^2 + 3y^2 + 2\lambda xy + 2x - 6y + 7 = 0$

(c) $\lambda x^2 + 3\lambda y^2 + 4xy - 8x + 5y - 8 = 0$

2 Find the equation of hyperbola

(a) whose vertices are at $(\pm 2, 0)$ and foci at $(\pm 3, 0)$.

(b) whose focii are at $(0, \pm 12)$ and vertices at $\left(0, \pm \frac{\sqrt{11}}{2}\right)$.

(c) in the standard form which passes through the points (1, 1) and (2, – 3).

(d) when vertices are $(\pm 7, 0)$ and $e = 4/3$.

(e) whose foci are at $(0, \pm \sqrt{10})$ and which passes through the point (2, 3).

(f) referred to its axes as the axes of coordinate, if length of its conjugate axis is 6 units and distance between foci is 10 units.

(g) in the standard form having distance between directrices $\frac{4}{3}\sqrt{3}$ units and passing through the point (2, 1).

(h) whose foci coincides with the foci of the ellipse $\frac{x^2}{25} + \frac{y^2}{9} = 1$ and having eccentricity 2.

(i) whose foci $(\pm 3\sqrt{5}, 0)$ and length of latusrectum is 8 units.

3 Find the equation of hyperbola

(a) whose focus is (– 1, 2), eccentricity is 3 and directrix is $x + y - 7 = 0$.

(b) whose foci are (8, 3) and (0, 3) with eccentricity $\frac{4}{3}$.

(c) whose coordinates of vertices are (2, 3) and (8, 3) and distance between foci is 12 units.

(d) whose vertices are at (– 8, – 1) and (16, – 1) one focus is at (17, – 1).

(e) which is the locus of point of intersection of lines $\sqrt{3}x - y - 4\sqrt{3}k = 0$ and $\sqrt{3}kx + ky - 4\sqrt{3} = 0$ for different value of k.

4 Find the vertices, foci, centre, eccentricity, length of axes, length of latusrectum and equation of directrix of the following hyperbola

(a) $\frac{x^2}{9} - \frac{y^2}{25} = 1$ (b) $\frac{y^2}{25} - \frac{x^2}{16} = 1$

(c) $\frac{(x-1)^2}{16} - \frac{(y+2)^2}{9} = 1$ (d) $16x^2 - 3y^2 - 32x - 12y - 44 = 0$

(e) $x^2 - 2y^2 - 2x + 8y - 1 = 0$

5 (a) Find the locus of a point such that the difference of its distances from (4, 0) and (–4, 0) is always equal to 2 units.

(b) $A(-1, 0)$ and $B(2, 0)$ are two given points. A point M is moving in such a way that the $\angle B$ in the triangle remains twice as large as the $\angle A$. Then, find its locus.

(c) Two rods are rotating about two fixed points in opposite directions. If they start from their position of coincidence and one rotates at the rate double that of the other, then find the locus of point of intersection of the two rods.

7.2 Point and Hyperbola

6 Find the position of following points w.r.t. given hyperbola

(a) Point (4, 3) and hyperbola $\frac{x^2}{16} - \frac{y^2}{9} = 1$.

(b) Point (– 1, – 3) and hyperbola $18x^2 - y^2 = 1$.

(c) Point (2, 3) and hyperbola $4y^2 - 9x^2 + 2x - 4 = 0$.

(d) Point (– 1, – 2) and hyperbola $3x^2 - y^2 + 7x + 2y - 11 = 0$.

7 Find the values of λ,

(a) if point $P(1, 2)$ lies outside $x^2 - \lambda y^2 + 2x + 5 = 0$.

(b) if point $P(-1, 3)$ lies inside $x^2 - 4y^2 - 2\lambda x + 11 = 0$.

(c) if point $P(\lambda, -3)$ lies on hyperbola $x^2 - y^2 + 4\lambda y + 7 = 0$.

8 Find the parametric form of the following hyperbola

(a) $\frac{x^2}{16} - \frac{y^2}{9} = 1$

(b) $\frac{(y+2)^2}{49} - \frac{(x-1)^2}{36} = 1$

(c) $4x^2 - 3y^2 + 8x + 6y - 9 = 0$

7.3 Line and Hyperbola

9 (a) Find the value of λ for which the line $y = 2x + \lambda$ intersect the hyperbola $3x^2 - 4y^2 = 12$ at (i) two points (ii) one point (iii) no point.

(b) Find the point of intersections of line $x + 2y - 1 = 0$ with the hyperbola $x^2 - 2y^2 = 1$, also find the length of intercept made by it.

10 Find the equation of tangent

(a) at a point (1, 2) to the hyperbola $27x^2 - 2y^2 = 19$.

(b) in parametric form for hyperbola $\frac{x^2}{4} - \frac{y^2}{36} = 1$ for any angle θ.

(c) which makes an angle $\theta = \pi/4$ with X-axis, to the hyperbola $3x^2 - 4y^2 = 12$.

(d) having slope -1 to the hyperbola $3x^2 - 8y^2 = 24$.

(e) to the hyperbola $4x^2 - 9y^2 = 36$ which is parallel to $5x - 4y + 7 = 0$.

(f) to the hyperbola $9x^2 - 16y^2 = 144$ which is perpendicular to the straight line $3x + 4y + 10 = 0$.

11 (a) If the line $y = mx + \sqrt{a^2m^2 - b^2}$ touches the hyperbola $\frac{x^2}{a^2} - \frac{y^2}{b^2} = 1$ at the point $(a\sec\theta, b\tan\theta)$, then find $\sin\theta$ in terms of a, b, m.

(b) If the line $5x + 12y = 9$ touches the hyperbola $x^2 - 9y^2 = 9$, then find point of contact.

12 (a) Find the product of the foot of perpendiculars from foci upon any tangent to the hyperbola $\frac{x^2}{a^2} - \frac{y^2}{b^2} = 1$.

(b) A variable tangent to the hyperbola $\frac{x^2}{a^2} - \frac{y^2}{b^2} = 1$ meets the tangent at vertex $(a, 0)$ at A and the transverse axis at Q. Find the locus of mid-point of QR.

13 Find the equation of normal

(a) to the hyperbola $5x^2 - 3y^2 - 2 = 0$ at (1, 1).

(b) to the hyperbola $\frac{x^2}{2} - \frac{y^2}{3} = 1$ and makes an angle $\pi/3$ to the X-axis, also find point of contact.

(c) to the hyperbola $x^2 - 2y^2 = 1$ having slope $-2/3$.

(d) to the hyperbola $3x^2 - 4y^2 = 12$ which is parallel to the line $x + y - 1 = 0$

(e) to the positive end point of the hyperbola $\frac{x^2}{16} - \frac{y^2}{9} = 1$.

14 (a) If the normal at $P(\theta)$ on the hyperbola $\frac{x^2}{a^2} - \frac{y^2}{2a^2} = 1$ meets the transverse axis at G, then find $AG : A'G$, where A and A' are the vertices of hyperbola.

(b) Normals are drawn to the hyperbola $\frac{x^2}{a^2} - \frac{y^2}{b^2} = 1$ at points θ_1 and θ_2 meets the conjugate axis at G_1 and G_2 respectively, if $\theta_1 + \theta_2 = \pi/2$, then find $CG_1 \cdot CG_2$ in terms of a, where C is centre, e is eccentricity of hyperbola.

(c) A normal to the hyperbola $\frac{x^2}{a^2} - \frac{y^2}{b^2} = 1$ meets the axes at Q and R. Lines QL and RL are drawn at right angles to the axes and meet at L, find the locus of L.

(d) Find the locus of the foot of perpendicular from the centre upon any normal to the hyperbola $\frac{x^2}{a^2} - \frac{y^2}{b^2} = 1$

15 How many normals can be drawn from the following points to the given hyperbola.

(a) Point (1, 2) and hyperbola $2x^2 - y^2 = 1$

(b) Point $(-7, -3)$ and hyperbola $x^2 - 2y^2 = 1$

(c) Point $(-2, 1)$ and hyperbola $2x^2 - y^2 - 7 = 0$

7.4 Pair of Tangent, Chord of Contact, Conjugate Diameter, Director Circle, Pole, Polar, Asymptotes and Rectangular Hyperbola.

16 (a) Find the number of tangents to the curve $2x^2 - 3y^2 = 12$ from the point $(1, -2)$.

(b) Find the equation of tangents to the curve $4x^2 - 9y^2 = 36$ from the point $(3, 2)$.

(c) Find the angle between the tangent drawn from the point $(4, 3)$ to the hyperbola $\frac{x^2}{16} - \frac{y^2}{9} = 1$.

(d) Find the value of λ for which the pair of tangents drawn from $(2, -\lambda)$ to the hyperbola $3x^2 - 4y^2 = 12$ are mutually perpendicular to each other.

17 (a) Find the locus of point of intersection of perpendicular tangents to the hyperbola $\frac{x^2}{16} - \frac{y^2}{9} = 1$.

(b) Find the locus of the point of intersection of tangents to the hyperbola $\frac{x^2}{a^2} - \frac{y^2}{b^2} = 1$, whose sum of slopes is constant P.

18 (a) Find the equation of the chord of contact of tangents from the point $(2, 3)$ to the hyperbola $\frac{x^2}{9} - \frac{y^2}{4} = 1$.

(b) If from a point P tangent PQ and PR are drawn to the hyperbola $\frac{x^2}{16} - \frac{y^2}{9} = 1$, so that the equation of QR is $x - 2y = 1$, then find the coordinates of P.

(c) Tangents are drawn from the points on the line $x + y + 3 = 0$ to hyperbola $x^2 - 2y^2 = 1$, then all the chord of contact pass through a fixed point. Find its coordinates.

19 (a) Find the equation of the chord of the hyperbola $25x^2 - 16y^2 = 400$ which is bisected at the point $(5, 3)$.

(b) Tangents are drawn from the point $P(3, 2)$ to the hyperbola $\frac{x^2}{25} - \frac{y^2}{16} = 1$, find the middle point of chord of contact w.r.t. given point P.

(c) Find the length of the chord of the hyperbola $x^2 - 4y^2 = 1$ whose middle point is $(3, 1)$.

20 (a) Find the locus of mid-points of normal chord of the hyperbola $\frac{x^2}{a^2} - \frac{y^2}{b^2} = 1$.

(b) Find the locus of mid-point of a chord of the hyperbola $ax^2 - by^2 = 1$ which subtends right angle at the centre of the hyperbola.

(c) Find the locus of the mid-point of focal chord of the hyperbola $\frac{x^2}{a^2} - \frac{y^2}{b^2} = 1$.

(d) Find the condition on 'a' and 'b' for which two distinct chords of the hyperbola $\frac{x^2}{2a^2} - \frac{y^2}{2b^2}$ passing through (a, b) are bisected by the line $x + y = b$.

21 (a) Find the equation of diameter of the hyperbola $\frac{x^2}{9} - \frac{y^2}{4} = 1$, which corresponds the line $3x + 4y + 10 = 0$.

(b) Find the equation of the diameter of the hyperbola $\frac{x^2}{9} - \frac{y^2}{4} = 1$ parallel to the chord $2x + 3y + 5 = 0$.

(c) Find the equation of the diameter which bisect the chord $7x + y - 2 = 0$ of the hyperbola $\frac{x^2}{3} - \frac{y^2}{7} = 1$.

(d) In the hyperbola $16x^2 - 9y^2 = 144$, find the equation of the diameter which is conjugate to the diameter whose equation is $x = 2y$.

22 (a) Find the equation of director circle of hyperbola $9x^2 - 16y^2 = 144$.

(b) If any tangent to the hyperbola $\frac{x^2}{a^2} - \frac{y^2}{b^2} = 1$ with centre C, meets the director circle in P and Q, find the product of slopes of CP and CQ.

23 (a) Find the equation of auxiliary circle of hyperbola $3x^2 - 4y^2 = 12$.

(b) If the tangent at (α, β) of hyperbola $\frac{x^2}{a^2} - \frac{y^2}{b^2} = 1$ meets the auxiliary circle at (x_1, y_1) and (x_2, y_2), then find the relation between β, y_1 and y_2.

24 (a) Find the polar of the focus $(-ae, 0)$ w.r.t. the hyperbola $\frac{x^2}{a^2} - \frac{y^2}{b^2} = 1$.

(b) Find the pole of the line $x - y = 3$ w.r.t. the hyperbola $x^2 - 3y^2 = 3$.

(c) If the polars of (x_1, y_1) and (x_2, y_2) w.r.t. the hyperbola $\frac{x^2}{a^2} - \frac{y^2}{b^2} = 1$ are at right angle. Find the relation between x_1, x_2, y_1, y_2, a and b.

(d) Find the locus of the poles of the normal chords w.r.t. the hyperbola $\frac{x^2}{a^2} - \frac{y^2}{b^2} = 1$.

25 Find the length of tangent, normal, subtangent and subnormal to hyperbola $\frac{x^2}{16} - \frac{y^2}{9} = 1$ at point (4, 3).

26 (a) Find the asymptotes of the curve $xy - 3y - 2x = 0$.

(b) Find the equation of asymptotes of the curve $3x^2 + 10xy + 8y^2 + 14x + 22y + \lambda = 0$.

(c) Find the eccentricity of the hyperbola whose asymptotes are $3x + 4y = 10$ and $4x - 3y = 5$.

(d) Find the equation of hyperbola whose asymptotes are $2x - y = 3$ and $3x + y = 7$ and which pass through the point (1, 1).

(e) The asymptotes of a hyperbola having centre at the point (1, 2) are parallel to the lines $2x + 3y = 0$ and $3x + 2y = 0$. If the hyperbola passes through the point (5, 3), then find the equation.

(f) Find the product of the lengths of the perpendiculars from any point on the hyperbola $x^2 - 2y^2 = 2$ to its asymptotes.

(g) Find the area of triangle formed by any tangent to the hyperbola $\frac{x^2}{9} - \frac{y^2}{4} = 1$ and its asymptotes.

(h) Let P be a variable point on the hyperbola $\frac{x^2}{a^2} - \frac{y^2}{b^2} = 1$ such that its distance from the transverse axis is equal to its distance from an asymptote to the given hyperbola. Find locus of point P.

(i) If P_1 and P_2 are the perpendiculars from any point on the hyperbola $\frac{x^2}{a^2} - \frac{y^2}{b^2} = 1$ on its asymptotes, then find the relation between P_1, P_2, a and b.

27 (a) Find the length of the transverse axis of the rectangular hyperbola $xy = 18$.

(b) If e_1 and e_2 be the eccentricities of the hyperbola, $xy = c^2$ and $x^2 - y^2 = a^2$, find the value of $(e_1 + e_2)^2$.

(c) If A, B and C be three points on the rectangular hyperbola $xy = c^2$, find (i) area of ΔABC (ii) area of triangle formed by tangent at A, B and C.

(d) A triangle has its vertices on a rectangular hyperbola, find the position of orthocentre of triangle w.r.t. the same hyperbola.

(e) If the tangent and normal to a rectangular hyperbola $xy = c^2$ at a point cuts-off intercept a_1 and a_2 on one axis and b_1 and b_2 on the other axis. Find the value of $a_1a_2 + b_1b_2$.

(f) A circle cuts the rectangular hyperbola $xy = 1$ in points (x_r, y_r), $r = 1, 2, 3, 4$, find the value of $\dfrac{x_1 \cdot x_2 \cdot x_3 \cdot x_4}{y_1 \cdot y_2 \cdot y_3 \cdot y_4}$.

(g) Let P_1, P_2, P_3 and P_4 be four points on $xy = c^2$ such that the chord P_1P_2 is perpendicular to the chord P_3P_4. If C is the centre and the lines CP_1, CP_2, CP_3 and CP_4 make angles α, β, γ and δ, respectively with an asymptote, then find $\tan\gamma \cdot \tan\beta \cdot \tan\alpha \cdot \tan\delta$.

(h) A rectangular hyperbola with centre 'C' is cut by a circle of radius r at four points P_1, P_2, P_3 and P_4, then find $CP_1^2 + CP_2^2 + CP_3^2 + CP_4^2$.

7.5 Miscellaneous

28 (a) Find the equations of the common tangent to the curves $\dfrac{x^2}{a^2} - \dfrac{y^2}{b^2} = 1$ and $\dfrac{x^2}{b^2} - \dfrac{y^2}{a^2} = -1$.

(b) Find the equation of common tangents to the curves $\dfrac{x^2}{9} - \dfrac{y^2}{4} = 1$ and $\dfrac{x^2}{9} + \dfrac{y^2}{4} = 1$.

(c) Find the equation of the common tangents to the curves $\dfrac{x^2}{16} - \dfrac{y^2}{9} = 1$ and $x^2 + y^2 = 9$.

(d) Find the equation of common tangents to the curves $y^2 = 8x$ and $\dfrac{x^2}{9} - \dfrac{y^2}{5} = 1$.

(e) From the points on the circle $x^2 + y^2 = a^2$, tangents are drawn to the hyperbola $x^2 - y^2 = a^2$, find the locus of the mid-points of the chord of contact.

(f) Tangents are drawn from a point P to the parabola $y^2 = 4ax$. If the chords of contact of the parabola be a tangent to the hyperbola $\dfrac{x^2}{a^2} - \dfrac{y^2}{b^2} = 1$, find the locus of the point P.

(g) Chords of the hyperbola $x^2 - y^2 = a^2$ touch the parabola $y^2 = 4ax$, find the locus of mid-points of chords.

(h) A variable chord of the circle $x^2 + y^2 = a^2$ touches the hyperbola $\dfrac{x^2}{a^2} - \dfrac{y^2}{b^2} = 1$, find the locus of their mid-points.

(i) Find the locus of the poles w.r.t. parabola $y^2 = 4ax$ of the tangent to the hyperbola $x^2 - y^2 = a^2$.

29 A ray is emanating from the point (5, 0) is incident on the hyperbola $9x^2 - 16y^2 = 144$ at point P with abscissa 8. Find the equation of the reflected ray after first reflection and point P lies in first quadrant.

30 (a) The equations of the transverse and conjugate axis of a hyperbola are $3x+4y-7=0$ and $4x-3y+8=0$, respectively and their respective lengths are 4 units and 6 units. Find the equation of the hyperbola.

(b) Find the coordinates of the foci and the centre of hyperbola

$$\frac{(3x-4y-12)^2}{100}-\frac{(4x+3y-12)^2}{225}=1.$$

(c) Find the eccentricity of the conic $4(2y-x-3)^2-9(2x+y-1)^2=80$.

LEVEL 02 \ JEE Main Level

Single Choice Correct Questions

7.1 Equation of Hyperbola

1 The length of the transverse axis of a hyperbola is 7 units and it passes through the point $(5,-2)$, then equation of the hyperbola is

(a) $\frac{4}{49}x^2-\frac{196}{51}y^2=1$ (b) $\frac{49}{4}x^2-\frac{51}{196}y^2=1$

(c) $\frac{4}{49}x^2-\frac{51}{196}y^2=1$ (d) None of these

2 The distance between the foci of a hyperbola is double the distance between its vertices and the length of its conjugate axis is 6 units. The equation of the hyperbola referred to its axes as axes of coordinates is

(a) $3x^2-y^2=3$ (b) $x^2-3y^2=3$ (c) $3x^2-y^2=9$ (d) $x^2-3y^2=9$

3 A hyperbola having the transverse axis of the length $2\sin\theta$ is confocal (foci of both conic coincide) with the ellipse $3x^2+4y^2=12$. Then, its equation is

(a) $x^2\operatorname{cosec}^2\theta-y^2\sec^2\theta=1$ (b) $x^2\sec^2\theta-y^2\operatorname{cosec}^2\theta=1$

(c) $x^2\sin^2\theta-y^2\cos^2\theta=1$ (d) $x^2\sin^2\theta-y^2\sin^2\theta=1$

4 The equation of transverse axis of hyperbola $(x-3)^2+(y+1)^2=(4x+3y)^2$ is

(a) $x+3y=0$ (b) $4x+3y=9$ (c) $3x-4y=13$ (d) $4x+3y=0$

5 The eccentricity of the conjugate hyperbola of the hyperbola $x^2-3y^2=1$ is

(a) 2 (b) $\frac{2}{\sqrt{3}}$ (c) 4 (d) $\frac{4}{3}$

6 If P is a point on the hyperbola $16x^2-9y^2=144$ whose foci are S_1 and S_2, then PS_1-PS_2 is equal to

(a) 4 (b) 6 (c) 8 (d) 12

7 If e and e' are the eccentricities of the ellipse $5x^2+9y^2=45$ and the hyperbola $5x^2-4y^2=45$ respectively, then ee' is equal to

(a) 9 (b) 4 (c) 5 (d) 1

8 For the hyperbola $\frac{x^2}{\cos^2\alpha}-\frac{y^2}{\sin^2\alpha}=1$, which of the following remains constant, when α varies

(a) abscissae of vertices (b) abscissae of foci

(c) eccentricities (d) directrix

9 The latusrectum of the hyperbola $9x^2-16y^2+72x-32y-16=0$ is

(a) $\frac{9}{2}$ units (b) $\frac{49}{2}$ units (c) $\frac{32}{2}$ units (d) None of these

10 If the foci of the ellipse $\frac{x^2}{16}+\frac{y^2}{b^2}=1$ and the hyperbola $\frac{x^2}{144}-\frac{y^2}{81}=\frac{1}{25}$ coincide, then the value of b^2 is

(a) 1 (b) 5 (c) 7 (d) 9

11 Let LL' be the latusrectum through the focus of hyperbola $\frac{x^2}{a^2}-\frac{y^2}{b^2}=1$ and A' be the farther vertex. If $\Delta A'LL'$ is equilateral, then the eccentricity of the hyperbola is (axes are coordinate axes)

(a) $\sqrt{3}$ (b) $\sqrt{3}+1$ (c) $\frac{\sqrt{3}+1}{\sqrt{2}}$ (d) $\frac{\sqrt{3}+1}{\sqrt{3}}$

12 If the eccentricity of the hyperbola $x^2-y^2\sec^2\alpha=5$ is $\sqrt{3}$ times the eccentricity of the ellipse $x^2\sec^2\alpha+y^2=25$, then the value of α is

(a) $\frac{\pi}{6}$ (b) $\frac{\pi}{4}$ (c) $\frac{\pi}{3}$ (d) $\frac{\pi}{2}$

13 Let e_1 and e_2 be the eccentricities of the ellipse $\frac{x^2}{25}+\frac{y^2}{b^2}=1, b<5$ and the hyperbola $\frac{x^2}{16}-\frac{y^2}{b^2}=1$, respectively satisfying $e_1e_2=1$. If α and β are the distance between the foci of the ellipse and foci of the hyperbola respectively, then the ordered pair (α, β) is **[JEE Main 2020]**

(a) (8, 10) (b) (8, 12) (c) $\left(\frac{24}{5}, 10\right)$ (d) $\left(\frac{20}{3}, 12\right)$

14 Let $0<\theta<\frac{\pi}{2}$, if the eccentricity of the hyperbola $\frac{x^2}{\cos^2\theta}-\frac{y^2}{\sin^2\theta}=1$ is greater than 2, then the length of its latusrectum lies in the interval **[JEE Main 2019]**

(a) $(3, \infty)$ (b) $\left(\frac{3}{2}, 2\right]$ (c) $\left(1, \frac{3}{2}\right]$ (d) (2, 3]

15 Let $S=\{(x,y)\in R^2 : \frac{(x-2)^2}{(29-p)}+\frac{(y+3)^2}{(4-p)}$, $p\neq 4, 29$. Then, S represents

(a) an ellipse, if p is any constant greater than 4.
(b) a hyperbola, if p is any constant between 4 and 29.
(c) a rectangular hyperbola, if p is any constant greater than 29.
(d) no real curve, if p is less than 29.

16 If the hyperbola $x^2+3xy+2y^2+2x+3y+2=0$ and $x^2+3xy+2y^2+2x+3y+c=0$ are conjugate to each other, then value of 'c' is equal to

(a) -2 (b) 4 (c) 0 (d) 1

17 Consider a branch of the hyperbola $x^2-2y^2-2\sqrt{2}x-4\sqrt{2}y-6=0$ with A as one vertex. B is one of the end points of the latusrectum and C is the focus nearest to A. Then, the area (in sq units) of ΔABC is

(a) $1-\sqrt{\frac{2}{3}}$ (b) $\sqrt{\frac{3}{2}}-1$ (c) $1+\sqrt{\frac{2}{3}}$ (d) $\sqrt{\frac{2}{3}}+1$

7.2 Point and Hyperbola

18 If $(\lambda, -1)$ is an exterior point of the curve $4x^2 - 3y^2 = 1$ such that the length of the interval, where λ lies in m, then value of $m + 10$ is

(a) 13 (b) 10 (c) 14 (d) 12

19 The locus of the point (x, y) is {where, $x = \frac{1}{2}(e^t + e^{-t})$ and $y = \frac{1}{2}(e^t - e^{-t})$}

(a) parabola (b) ellipse (c) hyperbola (d) None of these

20 PQ is double ordinate of $\frac{x^2}{a^2} - \frac{y^2}{b^2} = 1$ such that ΔOPQ is an equilateral triangle. If 'e' is the eccentricity of hyperbola, then

(a) $e = \frac{2}{\sqrt{3}}$ (b) $e > \frac{2}{\sqrt{3}}$ (c) $1 < e < \frac{2}{\sqrt{3}}$ (d) $e > \frac{4}{\sqrt{3}}$

21 The locus of centroid of the triangle formed by any point P on the hyperbola $16x^2 - 9y^2 + 32x + 36y - 164 = 0$ and its foci is **[JEE Main 2021]**

(a) $16x^2 - 9y^2 + 32x + 36y - 36 = 0$ (b) $9x^2 - 16y^2 + 36x + 32y - 144 = 0$

(c) $16x^2 - 9y^2 + 32x + 36y - 144 = 0$ (d) $9x^2 - 16y^2 + 36x + 32y - 36 = 0$

7.3 Line and Hyperbola

22 If $ax + by = 1$ is tangent to the hyperbola $\frac{x^2}{a^2} - \frac{y^2}{b^2} = 1$, then $a^2 - b^2$ is equal to

(a) $\frac{1}{a^2e^2}$ (b) a^2e^2 (c) b^2e^2 (d) None of these

23 If values of m for which the line $y = mx + 2\sqrt{5}$ touches the hyperbola $16x^2 - 9y^2 = 144$ are roots of the equation $x^2 - (a + b)x - 4 = 0$, then value of $(a + b)$ is equal to

(a) 2 (b) 4 (c) 0 (d) None of these

24 If $2x - y + 1 = 0$ is a tangent to the hyperbola $\frac{x^2}{a^2} - \frac{y^2}{16} = 1$, then which of the following cannot be sides of a right angled triangle ?

(a) a, 4, 1 (b) a, 4, 2 (c) $2a$, 8, 1 (d) All of these

25 The locus of a point $P(\alpha, \beta)$ which is moving under the condition that the line $y = \alpha x + \beta$ is a tangent to the hyperbola $\frac{x^2}{a^2} - \frac{y^2}{b^2} = 1$ is

(a) a parabola (b) a hyperbola (c) an ellipse (d) a circle

26 The number of possible tangents which can be drawn to the curve $4x^2 - 9y^2 = 36$, which are perpendicular to the straight line $5x + 2y - 10 = 0$, is

(a) 0 (b) 1 (c) 2 (d) 4

27 The coordinates of a point on the hyperbola $\frac{x^2}{24} - \frac{y^2}{18} = 1$, which is nearest to the line $3x + 2y + 1 = 0$, are

(a) (6, 3) (b) (− 6, − 3) (c) (6, − 3) (d) (−6, 3)

28 Let $P(6,3)$ be a point on the hyperbola $\frac{x^2}{a^2}-\frac{y^2}{b^2}=1$. If the normal at the point P intersects the X-axis at $(9, 0)$, then the eccentricity of the hyperbola is

(a) $\sqrt{\frac{5}{2}}$ (b) $\sqrt{\frac{3}{2}}$ (c) $\sqrt{2}$ (d) $\sqrt{3}$

29 The tangent at a point P on the hyperbola $\frac{x^2}{a^2}-\frac{y^2}{b^2}=1$ passes through the point $(0, -b)$ and the normal at P passes through the point $(2a\sqrt{2}, 0)$, then eccentricity of the hyperbola is

(a) 2 (b) $\sqrt{2}$ (c) 3 (d) $\sqrt{2}$

30 Let $P(a \sec\alpha, b\tan\alpha)$ and $Q(a\sec\beta, b\tan\beta)$, where $\alpha+\beta=\frac{\pi}{2}$, be two points on the hyperbola $\frac{x^2}{a^2}-\frac{y^2}{b^2}=1$. If (h,k) is the point of intersection of the normals at P and Q, then k is equal to

(a) $\frac{a^2+b^2}{a}$ (b) $-\left(\frac{a^2+b^2}{a}\right)$ (c) $\frac{a^2+b^2}{b}$ (d) $-\left(\frac{a^2+b^2}{b}\right)$

31 The point $P(-2\sqrt{6}, \sqrt{3})$ lies on the hyperbola $\frac{x^2}{a^2}-\frac{y^2}{b^2}=1$ having eccentricity $\frac{\sqrt{5}}{2}$. If the tangent and normal at P to the hyperbola intersect its conjugate axis at the point Q and R respectively, then QR is equal to

(a) $4\sqrt{3}$ (b) 6 (c) $6\sqrt{3}$ (d) $3\sqrt{6}$

32 Tangent at any point on the hyperbola $\frac{x^2}{a^2}-\frac{y^2}{b^2}=1$ cut the axes at A and B, respectively. If the rectangle $OAPB$ (where O is origin) is completed, then locus of point P is given by

(a) $\frac{a^2}{x^2}-\frac{b^2}{y^2}=1$ (b) $\frac{a^2}{x^2}+\frac{b^2}{y^2}=1$

(c) $\frac{a^2}{y^2}-\frac{b^2}{x^2}=1$ (d) None of these

7.4 Pair of Tangent, Chord of Contact, Conjugate Diameter, Director Circle, Pole, Polar, Asymptotes and Rectangular Hyperbola.

33 The tangents from $(1, 2\sqrt{2})$ to the hyperbola $16x^2-25y^2=400$ include between them an angle is equal to

(a) $\frac{\pi}{6}$ (b) $\frac{\pi}{4}$ (c) $\frac{\pi}{3}$ (d) $\frac{\pi}{2}$

34 The number of points from where a pair of perpendicular tangents can be drawn to the hyperbola $x^2\sec^2\alpha-y^2\operatorname{cosec}^2\alpha=1, \alpha\in\left(0,\frac{\pi}{4}\right)$ are

(a) 0 (b) 1 (c) 2 (d) infinite

35 If $x=9$ is a chord of contact of the hyperbola $x^2-y^2=9$, then the equation of the corresponding pair of tangents is

(a) $9x^2-8y^2+18x-9=0$ (b) $9x^2-8y^2-18x-9=0$

(c) $9x^2-8y^2+18x+9=0$ (d) $9x^2-8y^2-18x+9=0$

36 The locus of the mid-point of the chords of the hyperbola $\frac{x^2}{a^2}-\frac{y^2}{b^2}=1$, passing through a fixed point (α, β) is a hyperbola with centre at $\left(\frac{\alpha}{2}, \frac{\beta}{2}\right)$, is

(a) $\frac{\left(x-\frac{\alpha}{2}\right)^2}{a^2}-\frac{\left(y-\frac{\beta}{2}\right)^2}{b^2}=\frac{\alpha^2}{4a^2}-\frac{\beta^2}{4b^2}$
(b) $\frac{\left(x+\frac{\alpha}{2}\right)^2}{a^2}-\frac{\left(y-\frac{\beta}{2}\right)^2}{b^2}=\frac{\alpha^2}{4a^2}+\frac{\beta^2}{4b^2}$
(c) $\frac{\left(x-\frac{\alpha}{2}\right)^2}{a^2}-\frac{\left(y-\frac{\beta}{2}\right)^2}{b^2}=\frac{\alpha^2}{4a^2}+\frac{\beta^2}{4b^2}$
(d) None of these

37 The locus of the middle points of chords of hyperbola $3x^2-2y^2+4x-6y=0$ parallel to $y=2x$ is

(a) $3x-4y=4$
(b) $3y-4x+4=0$
(c) $4x-4y=3$
(d) $3x-4y=2$

38 The locus of the mid-points of the chords passing through a fixed point (α, β) of the hyperbola $\frac{x^2}{a^2}-\frac{y^2}{b^2}=1$ is

(a) a circle with centre $\left(\frac{\alpha}{2}, \frac{\beta}{2}\right)$
(b) an ellipse with centre $\left(\frac{\alpha}{2}, \frac{\beta}{2}\right)$
(c) a hyperbola with centre $\left(\frac{\alpha}{2}, \frac{\beta}{2}\right)$
(d) straight line through $\left(\frac{\alpha}{2}, \frac{\beta}{2}\right)$

39 If the line $lx+my+n=0$ passes through the extremities of a pair of conjugate diameters of the hyperbola $\frac{x^2}{a^2}-\frac{y^2}{b^2}=1$, then

(a) $a^2l^2-b^2m^2=1$
(b) $a^2l^2-b^2m^2=0$
(c) $a^2l^2+b^2m^2=1$
(d) None of these

40 Let the major axis of a standard ellipse equal to the transverse axis of a standard hyperbola and their director circles have radii equal to $2R$ and R, respectively. If e_1 and e_2 are the eccentricities of the ellipse and hyperbola, then the correct relation is

(a) $4e_1^2-e_2^2=16$
(b) $e_1^2-4e_2^2=2$
(c) $4e_2^2-e_1^2=6$
(d) $e_2^2-4e_1^2=2$

41 A hyperbola passes through (2, 3) and has asymptotes $3x-4y+5=0$ and $12x+5y-40=0$, then the equation of its transverse axis is

(a) $77x-21y-265=0$
(b) $21x-77y+265=0$
(c) $21x-77y-265=0$
(d) $21x+77y-265=0$

42 The combined equation of the asymptotes of the hyperbola $2x^2+5xy+2y^2+4x+5y=0$ is

(a) $2x^2+5xy+2y^2+4x+5y+2=0$
(b) $2x^2+5xy+2y^2+4x+5y-2=0$
(c) $2x^2+5xy+2y^2=0$
(d) None of these

43 From a point $P(1, 2)$ two tangents are drawn to a hyperbola 'H' in which one tangent is drawn to each arm of hyperbola. If the equations of asymptotes of hyperbola H are $\sqrt{3}x-y+5=0$ and $\sqrt{3}x+y-1=0$, then eccentricity of 'H' is

(a) 2
(b) $\frac{2}{\sqrt{3}}$
(c) 2
(d) $\sqrt{3}$

44 If $S=0$ be the equation of the hyperbola $x^2+4xy+3y^2-4x+2y+1=0$, then the value of K for which $S+K=0$ represents its asymptotes, is

(a) 20 (b) – 16 (c) – 22 (d) 18

45 If angle between asymptotes of hyperbola $\frac{x^2}{a^2}-\frac{y^2}{b^2}=1$ is 120° and product of perpendiculars drawn from foci upon its any tangent is 9, then locus of point of intersection of perpendicular tangents of the hyperbola can be

(a) $x^2+y^2=3$ (b) $x^2+y^2=16$

(c) $x^2+y^2=9$ (d) $x^2+y^2=18$

46 The asymptotes of the hyperbola $\frac{x^2}{a^2}-\frac{y^2}{b^2}=1$ from with any tangent to the hyperbola a triangle whose area is $a^2\tan\lambda$ in magnitude, then its eccentricity is

(a) $\sec\lambda$ (b) $\text{cosec}\,\lambda$ (c) $\sec^2\lambda$ (d) $\text{cosec}^2\lambda$

47 C is the centre of the hyperbola $\frac{x^2}{a^2}-\frac{y^2}{b^2}=1$ and the tangent at any point P meets the asymptotes in the points Q and R. The equation of the locus of the centre of the circle circumscribing the ΔCQR is

(a) $4(a^2x^2-b^2y^2)=(a^2+b^2)^2$ (b) $4(a^2x^2+b^2y^2)=(a^2+b^2)^2$

(c) $4(a^2x^2-b^2y^2)=(a^2-b^2)$ (d) None of these

48 If tangents OQ and OR are drawn to variable circles having radius r and the centre lying on the rectangular hyperbola $xy=1$, then locus of circumcentre of ΔOQR is, (O is origin)

(a) $xy=4$ (b) $xy=\frac{1}{4}$ (c) $xy=1$ (d) None of these

49 If foci of hyperbola lie on $y=x$ and one of the asymptotes is $y=2x$, then equation of the hyperbola, given that it passes through (3, 4) is

(a) $x^2-y^2-\frac{5}{2}xy+5=0$ (b) $2x^2-2y^2+5xy+5=0$

(c) $2x^2+2y^2-5xy+10=0$ (d) None of these

50 If normal to the rectangular hyperbola $xy=c^2$ at the point 't' meets the curve again at 't_1', then t^3t_1 has the value, is equal to

(a) 1 (b) – 1 (c) 0 (d) None of these

51 Area of triangle formed by tangent to the hyperbola $xy=16$ at (16, 1) and coordinate axes is equal to

(a) 8 (b) 16 (c) 32 (d) 64

52 Locus of middle points of the parallel chords with gradient 'm' of the rectangular hyperbola $xy=c^2$ is

(a) $y+mx=0$ (b) $y-mx=0$

(c) $my-mx=0$ (d) $my+x=0$

53 The locus of the poles of the normal chords of the rectangular hyperbola $xy=c^2$ is

(a) $(x^2-y^2)=4c^2xy$ (b) $(x^2-y^2)^2+4c^2xy=0$

(c) $(x^2-y^2)^2-4c^2xy=0$ (d) None of these

7.5 Miscellaneous

54 Let 'P' be the point of intersection of the common tangents to the parabola $y^2 = 12x$ and the hyperbola $8x^2 - y^2 = 8$. If S and S' denotes the foci of the hyperbola, where 'S' lies on the positive X-axis, then 'P' divides SS' in a ratio **[JEE Main 2019]**

(a) 14 : 13 (b) 13 : 11 (c) 5 : 4 (d) 2 : 1

55 If the line $y = mx + c$ is a common tangent to the hyperbola $\frac{x^2}{100} - \frac{y^2}{64} = 1$ and the circle $x^2 + y^2 = 36$, then which one of the following is true? **[JEE Main 2020]**

(a) $5m = 4$ (b) $8m + 5 = 0$ (c) $c^2 = 369$ (d) $4c^2 = 369$

56 The locus of the mid-points of the chords of the hyperbola $x^2 - y^2 = 4$, which touch the parabola $y^2 = 8x$, is

(a) $y^3(x-2) = x^2$ (b) $x^3(x-2) = y^2$ (c) $y^2(x-2) = x^3$ (d) $x^2(x-2) = y^3$

57 A hyperbola passes through the foci of the ellipse $\frac{x^2}{25} + \frac{y^2}{16} = 1$ and its transverse and conjugate axes coincide with major and minor axes of the ellipse, respectively. If the product of their eccentricities is one, then the equation of the hyperbola is **[JEE Main 2021]**

(a) $\frac{x^2}{9} - \frac{y^2}{16} = 1$ (b) $\frac{x^2}{9} - \frac{y^2}{4} = 1$

(c) $\frac{x^2}{9} - \frac{y^2}{25} = 1$ (d) $x^2 - y^2 = 9$

58 Suppose, the circle having equation $x^2 + y^2 = 3$ intersects the rectangular hyperbola $xy = 1$ at the points A, B, C and D. The equation $x^2 + y^2 - 3 + \lambda(xy - 1) = 0, \lambda \in R$ represents

(a) a pair of lines through origin for $\lambda = -3$

(b) an ellipse through A, B, C and D for $\lambda = -3$

(c) a parabola through A, B, C and D for $\lambda = -3$

(d) a circle for any $\lambda \in R$

59 $(x-1)(y-2) = 5$ and $(x-1)^2 + (y+2)^2 = r^2$ intersects at four points A, B, C, D and if centroid of ΔABC lies on line $y = 3x - 4$, then locus of D is

(a) $y = 3x$ (b) $x^2 + y^2 + 3x + 1 = 0$

(c) $3y = x + 1$ (d) $y = 3x + 1$

60 If the ellipse $x^2 + \lambda^2 y^2 = \lambda^2 a^2, \lambda^2 > 1$ is confocal with the hyperbola $x^2 - y^2 = a^2$, then which of the following is false?

(a) Ratio of eccentricities of ellipse and hyperbola is $1 : \sqrt{3}$.

(b) Ratio of major axis of ellipse and transverse axis of hyperbola is $\sqrt{3} : 1$.

(c) The ellipse and hyperbola cuts each other orthogonally.

(d) Ratio of the latusrectum of ellipse and hyperbola is 1 : 3.

61 Let 'C' be a curve which is locus of the point of the intersection of lines $x = 2 + m$ and $my = 4 - m$. A circle $S \equiv (x-2)^2 + (y+1)^2 = 25$ intersects the curve C at four points P, Q, R and S. If O is centre of the curve 'C', then $OP^2 + OQ^2 + OR^2 + OS^2$ is

(a) 50 (b) 100 (c) 25 (d) $\frac{25}{2}$

62 The exhaustive set of values of α^2 such that there exists a tangent to ellipse $x^2 + \alpha^2 y^2 = \alpha^2$ such that the portion of the tangent intercepted by the hyperbola $\alpha^2 x^2 - y^2 = 1$, subtends a right angle at centre of the curves is

(a) $\left[\frac{\sqrt{5}+1}{2}, 2\right]$ (b) $(1, 2]$ (c) $\left[\frac{\sqrt{5}-1}{2}, 1\right)$ (d) $\left[\frac{\sqrt{5}-1}{2}, 1\right) \cup \left(1, \frac{\sqrt{5}+1}{2}\right]$

63 If a ray of light incident along the line $3x + (5 - 4\sqrt{2})y = 15$ gets reflected from the hyperbola $\frac{x^2}{16} - \frac{y^2}{9} = 1$, then its reflected ray goes along the line

(a) $x\sqrt{2} - y + 5 = 0$ (b) $\sqrt{2}y - x + 5 = 0$ (c) $\sqrt{2}y - x - 5 = 0$ (d) None of these

64 The equations of the transverse and conjugate axes of a hyperbola are $x + 2y - 3 = 0$ and $2x - y + 4 = 0$, respectively and their respective lengths are $\sqrt{2}$ and $\frac{2}{\sqrt{3}}$ units. The equation of the hyperbola is

(a) $2(x + 2y - 3)^2 - 3(2x - y + 4)^2 = 5$
(b) $2(2x - y + 4)^2 - 3(x + 2y - 3)^2 = 5$
(c) $2(x + 2y - 3)^2 - 3(2x - y + 4)^2 = 1$
(d) $2(2x - y + 4)^2 - 3(x + 2y - 3)^2 = 1$

Numerical Value Type Questions

65 Eccentricity of the hyperbola $[\sqrt{(x-3)^2 + (y-2)^2} - \sqrt{(x+1)^2 + (y+1)^2}] = 1$ is ....

66 If the normal at ϕ on the hyperbola $\frac{x^2}{a^2} - \frac{y^2}{b^2} = 1$ meets the transverse axis at G such that $AG \cdot A'G = a^m (e^n \sec^p \theta - 1)$, where A, A' are the vertices of the hyperbola and m, n and p are positive integers, the value of $\frac{(m+n+p)^2 + 36}{1000}$ is ....

67 The shortest distance between the curves $\frac{x^2}{a^2} - \frac{y^2}{b^2} = 1$ and $4x^2 + 4y^2 = a^2$, $(b > a)$ is $f(a, b)$, then the value of $f(4, 6) + f(2, 3)$, is ....

68 The area of triangle formed by the tangents from the point (3, 2) to hyperbola $x^2 - 9y^2 = 9$ and the chord of contact w.r.t. point (3, 2), is ....

69 From a point P, tangents are drawn to the circle $x^2 + y^2 = a^2$. If the chord of contact of the circle is normal chord of a hyperbola $\frac{x^2}{a^2} - \frac{y^2}{b^2} = 1$ and locus of point P is $\left(\frac{\lambda a^2}{x^2} - \frac{\mu b^2}{y^2}\right) = \left(\frac{a^2 + b^2}{a^2}\right)^r$, then the value of $\frac{\sqrt{\lambda + \mu + r}}{50}$ is ....

70 If the normals at (x_i, y_i), $i = 1, 2, 3, 4$ on the hyperbola $\frac{x^2}{a^2} - \frac{y^2}{b^2} = 1$ are concurrent, then the value of $\frac{1}{100}\left[\frac{(x_1 + x_2 + x_3 + x_4)\left(\frac{1}{x_1} + \frac{1}{x_2} + \frac{1}{x_3} + \frac{1}{x_4}\right)}{(y_1 + y_2 + y_3 + y_4)\left(\frac{1}{y_1} + \frac{1}{y_2} + \frac{1}{y_3} + \frac{1}{y_4}\right)}\right]$ is ....

71 Two tangents to the hyperbola $\frac{x^2}{a^2} - \frac{y^2}{b^2} = 1$ having slopes m_1 and m_2 cut the axes in four concyclic points. The value of $\frac{m_1 \cdot m_2}{10}$ is ....

72 If $P(a \sec \theta, b \tan \theta)$ and $Q(a \sec \phi, b \tan \phi)$ are the ends of a focal chord of $\frac{x^2}{a^2} - \frac{y^2}{b^2} = 1$, if $\tan \frac{\theta}{2} \tan \frac{\phi}{2}$ is $\frac{P+Qe}{R+Se}$, then value of $\frac{P+Q+R+S}{25}$ is ....

73 If the hyperbola $x^2 - y^2 = 4$ is rotated by 45° in anti-clockwise direction about its centre keeping the axis intact, then equation of hyperbola is $xy = -a^2$, where a^2 is equal to ....

74 If a variable line has its intercepts on the coordinates axes e, e', where $\frac{e}{2}, \frac{e'}{2}$ are the eccentricities of a hyperbola and its conjugate hyperbola, then the line always touches the circle $x^2 + y^2 = r^2$, where r is equal to ....

75 If the product of perpendicular distances from any point on the hyperbola $\frac{x^2}{a^2} - \frac{y^2}{b^2} = 1$ of eccentricity $e = \sqrt{3}$ from its asymptotes is equal to 6, the length of the transverse axis of the hyperbola is ....

76 The tangent to the hyperbola $xy = c^2$ at the point P intersects the X-axis at T and the Y-axis at T'. The normal to the hyperbola at P intersects the X-axis and Y-axis at N and N', respectively. The area of ΔPNT and $\Delta PN'T'$ are Δ and Δ', respectively, the value of $\left(\frac{c^2}{\Delta} + \frac{c^2}{\Delta'} + 4\right)$ is ....

77 If P, Q and R be three points on the rectangular hyperbola $xy = c^2$, whose abscissae are x_1, x_2, x_3, if area of ΔPQR is $\left[\frac{(x_1 - x_2)(x_2 - x_3)(x_3 - x_1)}{x_1 \cdot x_2 \cdot x_3}\right] \times \frac{c^\lambda}{\mu}$, then value of $\frac{\lambda + \mu}{100}$ is

78 A circle with centre $(3\alpha, 3\beta)$ and of variable radius cut the rectangular hyperbola $x^2 - y^2 = 9a^2$ at the points P, Q, R, S. If the locus of the centroid of the ΔPQR is $(x - \lambda\alpha)^\mu - (y - \gamma\beta)^2 = a^2$, then the value of $\frac{\lambda + \mu + \gamma}{100}$ is ....

79 If point P lies on $2xy = 1$. A triangle is constructed by P, S and S' (where S and S' are foci). The locus of ex-centre opposite S (S and P lie in the first quadrant) is $(x + py)^2 = (\sqrt{2} - 1)^2 (x - y)^2 + q$, then the value of $p + q$ is ....

80 Chords of circle $x^2 + y^2 = 4$ touch the hyperbola $\frac{x^2}{4} - \frac{y^2}{16} = 1$. The locus of their middle points is the curve $(x^2 + y^2)^2 = \lambda x^2 - 16y^2$, then the value of λ is ....

81 The locus of the point of intersection of the lines $(\sqrt{3})kx + ky - 4\sqrt{3} = 0$ and $\sqrt{3}x - y - 4(\sqrt{3})k = 0$ is a conic, whose eccentricity is ....

[JEE Main 2021]

LEVEL 03 \ JEE Advanced Level

Single Choice Correct Questions

1 Each of the four inequalities given below defines a region the XY-plane. One of these four regions does not have the following property. For any two points (x_1, y_1) and (x_2, y_2) in the region, the point $\left(\frac{x_1+x_2}{2}, \frac{y_1+y_2}{2}\right)$ is also in the region. The inequality defining the region is

(a) $x^2 + 2y^2 \leq 1$ (b) $\max\{|x|, |y|\} \leq 1$

(c) $x^2 - y^2 \leq 1$ (d) $y^2 - x \leq 0$

2 Area of quadrilateral formed with the foci of the hyperbola $\frac{x^2}{a^2} - \frac{y^2}{b^2} = 1$, and $\frac{x^2}{a^2} - \frac{y^2}{b^2} = -1$ is

(a) $4(a^2 + b^2)$ (b) $2(a^2 + b^2)$ (c) $(a^2 + b^2)$ (d) $\frac{1}{2}(a^2 + b^2)$

3 If $y = mx + c$ be a tangent to hyperbola $\frac{x^2}{\lambda^2} - \frac{y^2}{(\lambda^3 + \lambda^2 + \lambda)^2} = 1$, then least value of $16\,m^2$ is equal to

(a) 0 (b) 1 (c) 4 (d) 9

4 **Statement I** Directed circle of hyperbola $\frac{x^2}{a^2} - \frac{y^2}{b^2} + 1 = 0$ is defined only when $b \geq a$.

Statement II Directer circle of hyperbola $\frac{x^2}{25} - \frac{y^2}{9} = 1$ is $x^2 + y^2 = 16$.

(a) Statement I is true, Statement II is true; Statement II is a correct explanation for Statement I.
(b) Statement I is true, Statement II is true; Statement II is not a correct explanation for Statement I.
(c) Statement I is true, Statement II is false.
(d) Statement I is false, Statement II is true.

5 Let $PQ : 2x + y + 6 = 0$ is a chord of the curve $x^2 - 4y^2 = 4$, coordinates of the point $R(\alpha, \beta)$ that satisfy $\alpha^2 + \beta^2 - 1 \leq 0$, such that area of ΔPQR is minimum, are given by

(a) $\left(\frac{-2}{\sqrt{5}}, \frac{1}{\sqrt{5}}\right)$ (b) $\left(\frac{-2}{\sqrt{5}}, \frac{-1}{\sqrt{5}}\right)$ (c) $\left(\frac{2}{\sqrt{5}}, \frac{1}{\sqrt{5}}\right)$ (d) $\left(\frac{2}{\sqrt{5}}, \frac{-1}{\sqrt{5}}\right)$

6 Let the chord $x \cos a + y \sin a = p$ of the hyperbola $\frac{x^2}{16} - \frac{y^2}{18} = 1$ subtends a right angle at the centre, let diameter of the circle, concentric with the hyperbola, to which the given chord is a tangent is d, then $\frac{d}{4}$ is equal to

(a) 4 (b) 5 (c) 6 (d) 7

7 How many number of triangles can be inscribed in the rectangular hyperbola $xy = c^2$, whose sides all touch the parabola $y^2 = 4ax$?

(a) 4 (b) 3 (c) 2 (d) infinte.

8 Let the double ordinate PP' of the hyperbola $\frac{x^2}{4} - \frac{y^2}{3} = 1$ is produced both sides to meet asymptotes of hyperbola in Q and Q': the product $(PQ)(PQ')$ is equal to

(a) 3 (b) 4 (c) 1 (d) 5

9 A series of hyperbola is drawn having a common transverse axis of length $2a$, then the locus of a point p on each hyperbola. Such that its distance from the transverse axis is equal to its distance from an asymptote, is

(a) $(x^2 - y^2)^2 = 4x^2(x^2 - a^2)$ (b) $(x^2 - y^2)^2 = x^2(x^2 - a^2)$

(c) $(x^2 - y^2) = 4y^2(x^2 - a^2)$ (d) $(x^2 - y^2)^2 = y^2(x^2 - a^2)$

10 If $f(x) = x^3 + \alpha x^2 + \beta x + \gamma$, where α, β, γ are rational numbers and two roots of $f(x) = 0$ are eccentricities of a parabola and a rectangular hyperbola, then $\alpha + \beta + \gamma$ is equal to

(a) -1 (b) 0 (c) 1 (d) 2

11 An ellipse has eccentricity $\frac{1}{2}$ and focus at the point $P\left(\frac{1}{2}, 1\right)$ its are directrix is the common tangent nearer to the point P, to the circle $x^2 + y^2 = 1$ and the hyperbola $x^2 - y^2 = 1$. The equation of the ellipse is standard form, is

(a) $9x^2 + 12y^2 = 108$ (b) $9\left(x - \frac{1}{3}\right)^2 + 12(y - 1)^2 = 1$

(c) $9\left(x - \frac{1}{3}\right)^2 + 4(y - 1)^2 = 36$ (d) None of these

More than One Correct

12 The equation $16x^2 - 3y^2 - 32x + 12y - 44 = 0$ represents a hyperbola,

(a) the length of whose transverse axis is $4\sqrt{3}$

(b) the length of whose conjugate axis is 8

(c) whose centre is (1, 2)

(d) whose eccentricity is $\sqrt{\frac{19}{3}}$

13 Which of the following equations in parametric form can represent a hyperbola, where 't' is a parameter?

(a) $x = \frac{a}{2}\left(t + \frac{1}{t}\right)$ and $y = \frac{b}{2}\left(t - \frac{1}{t}\right)$ (b) $\frac{tx}{a} - \frac{y}{b} + t = 0$ and $\frac{x}{a} + \frac{ty}{b} - 1 = 0$

(c) $x = e^t + e^{-t}$ and $y = e^t - e^{-t}$ (d) $x^2 - 6 = 2\cos t$ and $y^2 + 2 = 4\cos^2\left(\frac{t}{2}\right)$

14 The values of 'm' for which a line with slope m is common tangent to the hyperbola $\frac{x^2}{a^2} - \frac{y^2}{b^2} = 1 (a + b)$ and parabola $y^2 = 4a$ can lie in interval

(a) (0, 1) (b) $(\infty, -1) \cup (1, \infty) - \left(\pm\frac{\sqrt{1} + \sqrt{5}}{2}\right)$

(c) $(-1, 0)$ (d) None of these

15 For hyperbola $\frac{x^2}{a^2} - \frac{y^2}{b^2} = 1$. Let n be the number of points on the plane through which perpendicular tangents are drawn.

(a) If $n = 1$, then $e = \sqrt{2}$ (b) If $n > 1$, then $0 < e < \sqrt{2}$

(c) If $n = 0$, then $e > \sqrt{2}$ (d) None of these

16 From any point on the hyperbola

$h_1 : \left(\frac{x^2}{a^2}\right) - \left(\frac{y^2}{b^2}\right) = 1$ tangents are drawn to the hyperbola, $h_2 : \left(\frac{x^2}{a^2}\right) - \left(\frac{y^2}{b^2}\right) = 2$, the area cut-off by the chord of contact on the asymptotes of h_2 is equal to

(a) $\frac{ab}{2}$ (b) $4a^2\sqrt{e^2 - 1}$ (c) $\frac{4b^2}{\sqrt{e^2 - 1}}$ (d) $4ab$

17 The tangent to the hyperbola $xy = c^2$ at the point P intersects the X-axis at T and the Y-axis at T'. The normal to the hyperbola at P intersects the X-axis at N and the Y-axis at N'; the areas of the ΔPNT and $\Delta PN'T'$ are Δ and Δ' respectively, then $\frac{1}{\Delta} + \frac{1}{\Delta'}$

(a) is equal to $\frac{2}{c^2}$ (b) does not depends on t

(c) depends on c (d) is equal to 2

18 Let a and b be positive real numbers such that $a > 1$ and $b < a$. Let P be a point in the first quadrant that lies on the hyperbola $\frac{x^2}{a^2} - \frac{y^2}{b^2} = 1$. Suppose, the tangent to the hyperbola at P passes through the point (1, 0) and suppose, the normal to the hyperbola at P cuts-off equal intercepts on the coordinate axes. Let Δ denote the area of the triangle formed by the tangent at P, the normal at P and the X-axis, if e denotes the eccentricity of the hyperbola, then which of the following statements is/are true?

(a) $1 < e < \sqrt{2}$ (b) $\sqrt{2} < e < 2$ (c) $\Delta = a^4$ (d) $\Delta = b^4$

19 If P is a point on a hyperbola, then

(a) locus of excentre of the circle described opposite to $\angle P$ for $\Delta PSS'$ (S, S' are foci) is tangents at vertex

(b) locus of excentre of the circle described opposite to $\angle S'$ is hyperbola

(c) locus of excentre of the circle described opposite to $\angle P$ for $\Delta PSS'$ (S, S' are foci), is hyperbola

(d) locus of excentre of the circle described opposite to $\angle S'$, is tangent at vertex.

20 If the normal at P to the rectangular hyperbola $x^2 - y^2 = 4$ meets the axes in G, g and C is the centre of the hyperbola, then

(a) $PG = PC$ (b) $Pg = PC$ (c) $PG = Pg$ (d) $Gg = PC$

Linked Comprehension Type

Paragraph I Directions (Q.Nos. 21-23)

Let $P(x, y)$ is a variable point such that $\left|\sqrt{(x-1)^2 + (y-2)^2} - \sqrt{(x-5)^2 + (y-5)^2}\right| = 3$ which represents hyperbola.

21 The eccentricity e' of the corresponding conjugate hyperbola is

(a) $\frac{5}{3}$ (b) $\frac{4}{3}$ (c) $\frac{5}{4}$ (d) $\frac{3}{\sqrt{7}}$

22 Locus of intersection of two perpendicular tangents to the hyperbola is

(a) $(x-3)^2 + \left(y - \frac{7}{2}\right)^2 = \frac{55}{4}$ (b) $(x-3)^2 + \left(y - \frac{7}{2}\right)^2 = \frac{25}{4}$

(c) $(x-3)^2 + \left(y - \frac{7}{2}\right)^2 = \frac{7}{4}$ (d) None of these

23 If origin is shifted to point $\left(3, \frac{7}{2}\right)$ and the axes are rotated through an angle θ in clockwise sense so that equation of given hyperbola changes to the standard form $\frac{x^2}{a^2} - \frac{y^2}{b^2} = 1$, then θ is

(a) $\tan^{-1}\left(\frac{4}{3}\right)$ (b) $\tan^{-1}\left(\frac{3}{4}\right)$

(c) $\tan^{-1}\left(\frac{5}{3}\right)$ (d) $\tan^{-1}\left(\frac{3}{5}\right)$

Paragraph II Directions (Q.Nos. 24-26)

In hyperbola portion of tangent intercepted between asymptotes is bisected at the point of contact. Consider a hyperbola whose centre is at origin. A line $x + y = 2$ touches this hyperbola at $P(1, 1)$ and intersects the asymptotes at A and B such that $AB = 6\sqrt{2}$ units.

24 Equation of asymptotes are

(a) $5xy + 2x^2 + 2y^2 = 0$ (b) $3x^2 + 4xy^2 + 6xy = 0$

(c) $2x^2 + 2y^2 - 5xy = 0$ (d) None of these

25 Angle subtended by AB at centre of the hyperbola is

(a) $\sin^{-1}\frac{4}{5}$ (b) $\sin^{-1}\frac{2}{5}$ (c) $\sin^{-1}\frac{3}{5}$ (d) None of these

26 Equations of the tangent to the hyperbola at $\left(-1, \frac{7}{2}\right)$ is

(a) $5x + 2y = 2$ (b) $3x + 2y = 4$ (c) $3x + 4y = 11$ (d) None of these

Paragraph III Directions (Q.Nos. 27-31)

If a pair of diameters be conjugate to a hyperbola. One of them melt the hyperbola in real points. Say $P(a \sec\theta, b \tan\theta)$ and $P'(-a \sec\theta, -b \tan\theta)$ and the other meets the conjugate hyperbola in real points $Q(a \tan\theta, b \sec\theta)$ and $Q'(-a \tan\theta, -b \sec\theta)$.

Now, answer the following questions:

27 If a pair of conjugate diameters meet the hyperbola and its conjugate in P and Q and O be the centre of the hyperbola, then $OP^2 - OQ^2$ is equal to

(a) $a^2 - b^2$ (b) $\frac{a^2 - b^2}{a^2 + b^2}$ (c) $\frac{a^2 b^2}{a^2 + b^2}$ (d) $a^2 + b^2$

28 If P and Q is the points as referred to the question 27, then the tangents at P and Q to respective hyperbola

(a) are perpendicular

(b) are parallel to diameters

(c) intersect at asymptotes

(d) intersect at circle $x^2 + y^2 = a^2 - b^2$.

29 Suppose, the tangents at P and Q intersect in a point R, then the area of parallelogram $OPRQ$ is

(a) $4ab$ (b) $2ab$ (c) ab (d) None of these

30 Let OP and OQ be two conjugate diameters with respect to a rectangular hyperbola inclined an angled α and β, respectively to the axes. Then,

(a) $OP = OQ$ (b) $\alpha + \beta = \frac{\pi}{2}$

(c) OP and OQ are perpendicular (d) $OP^2 + OQ^2 = 2a^2$, $2a$ being the transverse axis

31 If $y = mx$ ($m \neq 1$) be a diameter with respect to the hyperbola $xy = c^2$, then the diameter conjugate to it is

(a) $y = -\left(\frac{1}{m}\right)x$ (b) $y = -mx$ (c) $y = c^2 mx$ (d) $y = \left(\frac{c^2}{m}\right)x$

Paragraph IV Directions (Q.Nos. 32-34)

A point P moves such that sum of the slopes of the normals drawn from it to the hyperbola $xy = 16$ is equal to the sum of ordinates of feet of normals. The locus of P is a curve C.

32 The equation of the curve C is

(a) $x^2 = 4y$ (b) $x^2 = 16y$ (c) $x^2 = 12y$ (d) $y^2 = 8x$

33 If the tangent to the curve C cuts the coordinate axes at A and B, then the locus of the middle point of AB is

(a) $x^2 = 4y$ (b) $x^2 = 2y$ (c) $x^2 + 2y = 0$ (d) $x^2 + 4y = 0$

34 Area of the equilateral triangle, inscribed in the curve C, having one vertex as the vertex of curve C is

(a) $772\sqrt{3}$ sq units (b) $776\sqrt{3}$ sq units

(c) $760\sqrt{3}$ sq units (d) $768\sqrt{3}$ sq units

Paragraph V Directions (Q.Nos. 35-37)

By appropriately matching the information given in the three columns of the following table. Column 1, 2 and 3 contain conics, equations of tangents to the conics and points of contact respectively.

	Column 1		Column 2		Column 3
(I)	$x^2 + y^2 = a^2$	(i)	$my = m^2 x + a$	(P)	$\left(\frac{a}{m^2}, \frac{2a}{m}\right)$
(II)	$x^2 + a^2 y^2 = a^2$	(ii)	$y = mx + a\sqrt{m^2 + 1}$	(Q)	$\left(-\frac{ma}{\sqrt{m^2+1}}, \frac{a}{\sqrt{m^2+1}}\right)$
(III)	$y^2 = 4ax$	(iii)	$y = mx + \sqrt{a^2 m^2 - 1}$	(R)	$\left(\frac{-a^2 m}{a^2 m^2 + 1}, \frac{1}{\sqrt{a^2 m^2 + 1}}\right)$
(IV)	$x^2 - a^2 y^2 = a^2$	(iv)	$y = mx + \sqrt{a^2 m^2 + 1}$	(S)	$\left(\frac{-a^2 m}{\sqrt{a^2 m^2 - 1}}, \frac{-1}{\sqrt{a^2 m^2 - 1}}\right)$

35 The tangent to a suitable conic (column 1) at $\left(\sqrt{3}, \frac{1}{2}\right)$ is found to be $\sqrt{3}x + 2y = 4$, then which of the following options is the only correct combination? **[JEE Advanced 2018]**

(a) (IV) (iii) (S) (b) (IV) (iv) (S) (c) (II) (iii) (R) (d) (II) (iv) (R)

36 If a tangent to a suitable conic (column 1) is found to be $y = x + 8$ and its point of contact is (8, 16), then which of the following options is the only correct combination?

(a) (I) (ii) (Q) (b) (II) (iv) (R) (c) (III) (i) (P) (d) (III) (ii) (Q)

37 For $a = \sqrt{2}$, if a tangent is drawn to a suitable conic (column 1) at the point of contact (−1, 1), then which of the following options is the only correct combination for obtaining its equations?

(a) (I) (i) (P) (b) (I) (ii) (Q) (c) (II) (ii) (Q) (d) (III) (i) (P)

Matrix Match Type Question

38 Let $H: \frac{x^2}{a^2} - \frac{y^2}{b^2} = 1$, where $a > b$, be a hyperbola in the XY-plane whose conjugate axis LM subtends an angle of 60° at one of its vertices N. Let the area of the ΔLMN be $4\sqrt{3}$.

	Column I		Column II
(i)	The length of the conjugate axis of H is	A.	8
(ii)	The eccentricity of H is	B.	$\frac{4}{\sqrt{3}}$
(iii)	The distance between the foci of H is	C.	$\frac{2}{\sqrt{3}}$
(iv)	The length of the latusrectum of H is	D.	4

[JEE Advanced 2018]

39 Let the foci of the hyperbola $\frac{x^2}{A^2} - \frac{y^2}{B^2} = 1$ be the vertices of the ellipse $\frac{x^2}{a^2} + \frac{y^2}{b^2} = 1$ and the foci of the ellipse be the vertices to the hyperbola. Let the eccentricities of the ellipse and hyperbola be e_E and e_H respectively, then match the following

	Column I		Column II
(i)	$\frac{b}{B}$ is equal to	A.	1
(ii)	$e_H + e_E$ is always greater then	B.	2
(iii)	If angle between the asymptotes of hyperbola is $\frac{2a}{3}$, then $4e_E$ is equal to	C.	3
(iv)	If $e_E^2 = \frac{1}{2}$ and (x, y) is point of intersection of ellipse and the hyperbola, then $\frac{9x^2}{2y^2}$ is	D.	4

40 The locus of a variable point P, whose coordinates are given by

	Column I		Column II
(i)	$x = 3\left(\frac{1-t^2}{1+t^2}\right)$ and $y = \left(\frac{8t}{1+t^2}\right)$ is	A.	an ellipse
(ii)	$x = \frac{1}{2}(e^t + e^{-t})$ and $y = \frac{1}{2}(e^t - e^{-t})$ is	B.	a hyperbola
(iii)	$x = \frac{2}{(e^{i\theta} + e^{-i\theta})}$ and $y = i \cdot \left(\frac{e^{i\theta} - e^{-i\theta}}{e^{i\theta} + e^{-i\theta}}\right)$ is	C.	a circle
(iv)	$x = \frac{i}{2}(e^{-i\theta} - e^{i\theta})$ and $y = \left(\frac{e^{i\theta} + e^{-i\theta}}{\sin 2\theta}\right)$ is	D.	a parabola

41 Match the item of column I with those of column II.

	Column I		Column II
(i)	The locus of the point whose chord of contact with respect to the hyperbola $\frac{x^2}{16}-\frac{y^2}{9}=1$ touches the circle describe on the line joinning the foci is	A.	$(x^2+y^2)^2=4x^2-3y^2$
(ii)	The chords of the circle $x^2+y^2=4$ touches the hyperbola $\frac{x^2}{4}-\frac{y^2}{3}=1$ Then, the locus of the mid-points of these chords is	B.	$x^2+y^2=9$
(iii)	The director circle of the hyperbola $\frac{x^2}{25}-\frac{y^2}{16}=1$ is	C.	$x^2-y^2=32$
(iv)	The distance between the foci of a hyperbola is 16 and its eccentricity is $\sqrt{2}$. Then, the equations of the hyperbola is	D.	$\frac{x^2}{16^2}+\frac{y^2}{9^2}=\frac{1}{25}$

Numerical Value Type Questions

Question Stem Directions (Q.Nos. 42-44) The answer to each question is a **NUMERICAL VALUE.**

Tangents are drawn to the hyperbola $x^2-9y^2=9$ from (3, 2). Answer the following questions.

42 The area of the triangle formed by the tangents and chord contact of (3, 2) is ....

43 The area of the triangle formed by the tangent to the hyperbola at (3, 0) and the two asymptotes, is ....

44 If the mid-point of the intercept of the tangent at (3, 0) between the asymptotes is (α, β) then value of $|\alpha+\beta|$ is ....

Question Stem Directions (Q.Nos. 45-47) The answer to each question is a **NUMERICAL VALUE.**

The locus of the foot of the perpendicular drawn from a focus onto a tangent to the hyperbola is the auxiliary circle. Consider the hyperbola whose foci are (5, 6) and (– 3, –2). The foot of the perpendicular from the focus upon a tangent is the point (2, 5). Answer the following questions.

45 Length of the conjugate axis of the hyperbola is ....

46 If the directrix of the hyperbola corresponding to focus (5, 6) is $ax+by+c=0$, then the value of $|a+b+c|$ is ....

47 Length of the latusrectum of the hyperbola is ....

48 The graphs of $x^2+4y^2+6x-24y+72=0$ and $x^2-y^2+6x+16y-46=0$ intersect at four points. The sum of distance of the four points from points (– 3, 2) is ....

49 The equation of transverse axis of hyperbola (passing through origin) having asymptotes $3x-4y-1=0$ and $4x-3y-6=0$ is $ax+by-c=0$, $a, b, c \in N$ and $gcd\,(a, b, c)=1$, then the value of $\frac{a+b+c}{100}$ is ....

50 A straight line is drawn parallel to the conjugate axis of a hyperbola to meet it and the conjugate hyperbola in the points P and Q. If the tangents at P and Q meet the curve $\frac{y^4}{b^4}\left(\frac{y^2}{b^2}+\frac{x^2}{a^2}\right)=\frac{\lambda x^2}{a^2}$, then the value of $\frac{\lambda}{10}$ is ....

51 A normal to the hyperbola $\frac{x^2}{4}-\frac{y^2}{1}=1$, has equal intercepts on positive X and Y-axes. If this normal touches the ellipse $\frac{x^2}{a^2}+\frac{y^2}{b^2}=1$, then $\frac{9}{25}(a^2+b^2)$ is equal to ....

52 Tangents are drawn from the point (α, β) to the hyperbola $3x^2-2y^2=6$ and are inclined at angles θ and ϕ to the X-axis, if $\tan\theta \cdot \tan\phi = 2$, then the value of $2\alpha^2-\beta^2$ is ....

53 A transversal cut the same branch of a hyperbola $\frac{x^2}{a^2}-\frac{y^2}{b^2}=1$ in P, P' and the asymptotes in Q, Q'; then $(PQ+PQ')-(P'Q'+P'Q)$ is equal to ....

54 Normal drawn to the hyperbola $xy=2$ at the point $P(t)$, meets the hyperbola again at $Q(t_2)$, then minimum distance between the points P and Q is ....

55 A variable straight line of slope 4 intersects the hyperbola $xy=1$ at two points. If the locus of the point which divides the line segment between these two points in the ratio 1 : 2 is $ax^2+by^2+cxy+d=0$, then $\frac{a+b+c+d}{1000}$ is equal to ....

ANSWER KEY

Level 1

1. (a) $\left(-\infty, \frac{4}{3}\right) - \{-3.47\}$ (b) $(-\infty, -\sqrt{6}) \cup (\sqrt{6}, \infty)$ (c) $\left(-\frac{2}{\sqrt{3}}, \frac{2}{\sqrt{3}}\right)\{-0.24\}$

2. (a) $\frac{x^2}{4} - \frac{y^2}{5} = 1$ (b) $\frac{4x^2}{565} - \frac{4y^2}{11} = -1$ (c) $8x^2 - 3y^2 = 5$ (d) $8x^2 - 9y^2 = 343$ (e) $y^2 - x^2 = 5$

(f) $\frac{x^2}{16} - \frac{y^2}{9} = 1$ (g) $3x^2 - 2y^2 = 10$ and $x^2 - 2y^2 = 2$ (h) $3x^2 - y^2 = 12$ (i) $\frac{x^2}{25} - \frac{y^2}{20} = 1$

3. (a) $7x^2 + 7y^2 + 18xy - 130x - 118y + 431 = 0$ (b) $7x^2 - 9y^2 - 56x + 54y - 32 = 0$

(c) $\frac{(x-5)^2}{9} - \frac{(y-3)^2}{27} = 1$ (d) $25x^2 - 144y^2 - 288y - 200x - 3344 = 0$

(e) $3x^2 - y^2 = 48$

4. (a) Vertices $(\pm 3, 0)$, foci $(\pm\sqrt{34}, 0)$, centre $(0, 0)$, eccentricity $\left(\frac{\sqrt{34}}{3}\right)$, length of transverse and conjugate axes (6 and 10), length of latusrectum $\left(\frac{50}{3}\right)$, equation of directrix $\left(x = \pm\frac{9}{\sqrt{34}}\right)$.

(b) Vertices $(0, \pm 5)$, foci $(0, \pm\sqrt{41})$, centre $(0, 0)$, eccentricity $\left(\frac{\sqrt{41}}{5}\right)$, length of transverse and conjugate axes (10 and 8), length of latusrectum $\left(\frac{32}{5}\right)$, equation of directrix $(\sqrt{41}y = \pm 25)$.

(c) Vertices $\{(5, -2), (-3, -2)\}$, foci $\{(6, -2)$ and $(-4, -2)\}$, centre $(1, -2)$, eccentricity $\left(\frac{5}{4}\right)$, length of transverse and conjugate axes (8 and 6), length of latusrectum $\left(\frac{9}{2}\right)$, equation of directrix $\left(x = \frac{21}{5}, \frac{-11}{5}\right)$

(d) Vertices $(1 \pm \sqrt{3}, -2)$, foci $(1 \pm \sqrt{9}, -2)$, centre $(1, -2)$, eccentricity $\left(\sqrt{\frac{19}{3}}\right)$, length of transverse and conjugate axes ($2\sqrt{3}$ and 8), length of latusrectum $\left(\frac{32}{\sqrt{3}}\right)$, equation of directrix $\left(x = 1 \pm \frac{3}{\sqrt{19}}\right)$.

(e) Vertices $(1, 2 \pm \sqrt{3})$, foci $\{(1, 5), (1, -1)\}$, centre $(1, 2)$, eccentricity $(\sqrt{3})$, length of transverse and conjugate axes ($2\sqrt{3}$ and $2\sqrt{6}$), length of latusrectum $(4\sqrt{3})$, equation of directrix $(y = 3, 1)$.

5. (a) $\frac{x^2}{1} - \frac{y^2}{15} = 1$ (b) $\frac{x^2}{1} - \frac{y^2}{3} = 1$ (c) $3x^2 - y^2 - 2ax = 0$

6. (a) Point lies outside the hyperbola. (b) Point lies inside the hyperbola.

(c) Point lies on the hyperbola (d) Point lies outside the hyperbola.

7. (a) $(2, \infty)$ (b) $(12, \infty)$ (c) $6 \pm \sqrt{38}$

8. (a) $(4\sec\theta, 3\tan\theta)$ (b) $(6\tan\theta + 1, 7\sec\theta - 2)$ (c) $\left(\sqrt{\frac{5}{2}}\sec\theta - 1, \sqrt{\frac{10}{3}}\tan\theta + 1\right)$

9. (a) (i) ϕ (ii) ϕ (iii) $\left(-\infty, -\sqrt{\frac{13}{3}}\right) \cup \left(\sqrt{\frac{13}{3}}, \infty\right)$

(b) $(1, 0)$ and $(-3, 2)$; $2\sqrt{5}$ units

10. (a) $27x - 4y = 19$ (b) $\frac{x\sec\theta}{2} - \frac{y\tan\theta}{6} = 1$ (c) $\frac{x}{\sqrt{2}} - \frac{y}{\sqrt{3}} = 1$

(d) $x + y \pm \sqrt{5} = 0$ (e) $5x - 14y \pm \sqrt{161} = 0$ (f) $4x - 3y \pm 5\sqrt{7} = 0$

11. (a) $\sin\theta = \frac{b}{am}$ (b) $\left(5, -\frac{4}{3}\right)$ and $\left(-5, \frac{4}{3}\right)$

12. (a) $P_1P_2 = b^2$ (b) $\frac{4y^2}{b^2} = 1 + \frac{a}{h}$

13. (a) $3x + 5y - 8 = 0$ (b) $x + \sqrt{2}y = 5\sqrt{2}$ and $\left(\pm\frac{\sqrt{2}}{5}, \pm\frac{3}{10}\right)$ (c) $3\sqrt{11}y + 2\sqrt{11}x \pm 9 = 0$

(d) $x + y \pm \sqrt{7} = 0$ (e) $16x + 20y - 125 = 0$

14. (a) $a^2(e^4\sec^2\theta - 1)$ (b) $\frac{a^2e^4}{e^2 - 1}$

(c) $a^2x^2 - b^2y^2 = (a^2 + b^2)^2$ (d) $(x^2 + y^2)^2(a^2y^2 - b^2x^2) = (a^2 + b^2)^2x^2y^2$

15. (a) 4 (b) 4 (c) 4

16. (a) 2 (b) $x = 3, 2x = 3y$ (c) $\tan^{-1}\left(\frac{4}{3}\right)$ (d) No real value

17. (a) $x^2 + y^2 = 7$ (b) $2xy = P(x^2 - a^2)$

18. (a) $8x - 27y = 36$ (b) (16, 18) (c) $\left(-\frac{1}{3}, \frac{1}{6}\right)$

19. (a) $125x - 48y = 481$ (b) (0, 0) and $\left(\frac{300}{11}, \frac{200}{11}\right)$ (c) $\frac{1}{3}\sqrt{1950 + 40\sqrt{78}}$ units

20. (a) $a^4b^4(a^2 + b^2)^2x^2y^2 = (b^2x^2 - a^2y^2)^2(a^6y^2 - b^6y^2)$ (b) $(a - b)(ax^2 - by^2)^2 = a^2x^2 + b^2y^2$

(c) $\frac{x^2}{a^2} - y^2b^2 = \frac{ex}{a}$ (d) $a \neq -b$

21. (a) $16x + 27y = 0$ (b) $3y + 2x = 0$ (c) $x + 3y = 0$ (d) $32x - 9y = 0$

22. (a) $x^2 + y^2 = 7$ (b) $\frac{b^2}{a^2}$

23. (a) $x^2 + y^2 = 4$ (b) $\frac{1}{y_1} + \frac{1}{y_2} = \frac{2}{\beta}$

24. (a) $x = \frac{-a}{e}$ (b) $\left(1, \frac{1}{3}\right)$ (c) $\frac{x_1x_2}{y_1y_2} + \frac{a^4}{b^4} = 0$ (d) $a^6y^2 - b^6x^2 = (a^2 + b^2)^2(x^2y^2)$

25. Length of tangent = 5, length of normal = $\frac{15}{4}$, length of subtangent = 4, length of subnormal = $\frac{9}{4}$

26. (a) $x = 3$ and $y = 2$ (b) $3x^2 + 10xy + 8y^2 + 14x + 22y + 15 = 0$ (c) $\sqrt{2}$

(d) $6x^2 - xy - y^2 - 23x + 4y + 15 = 0$ (e) $(2x + 3y - 8)(3x + 2y - 7) - 154 = 0$ (f) $\frac{2}{3}$

(g) 6 sq units (h) $\tan\theta = a\left|\frac{\sec\theta - \tan\theta}{\sqrt{b^2 + a^2}}\right|$ (i) $\frac{1}{P_1P_2} = \frac{1}{a^2} + \frac{1}{b^2}$

27. (a) 12 units (b) 8 (c) (i) $\frac{c^2}{2} \cdot \frac{(t_1 - t_2)(t_2 - t_3)(t_3 - t_1)}{t_1t_2t_3}$ (ii) $\frac{2c^2(t_1 - t_2)(t_2 - t_3)(t_3 - t_1)}{(t_1 + t_2)(t_2 + t_3)(t_3 + t_1)}$

(d) $\left(\frac{-c}{t_1t_2t_3}, ct_1t_2t_3\right)$ lies on the rectangular hyperbola $xy = c^2$.

(e) 0 (f) 1 (g) 1 (h) $4r^2$

28. (a) $y = \pm x + \sqrt{a^2 - b^2}$ (b) $x = \pm 3$ and $y = \pm 2$ (c) $y = \pm\frac{18}{7}x + \frac{15}{\sqrt{7}}$

(d) $y = \pm x \pm 2$ (e) $a^2(x^2 + y^2) = (x^2 - y^2)^2$ (f) $4a^2x^2 = 4a^4 - y^2b^2$

(g) $x(x^2 - y^2) = ay^2$ (h) $(x^2 + y^2)^2 = (a^2x^2 - b^2y^2)$ (i) $4x^2 + y^2 = 4a^2$

29. $3\sqrt{3}x - 13y + 15\sqrt{3} = 0$

30. (a) $17x^2 + 312xy + 108y^2 - 634x - 312y - 715 = 0$

(b) Focus $\left(\frac{84 + 15\sqrt{3}}{25}, \frac{-12 + 14\sqrt{13}}{25}\right)$, centre $\left(\frac{84}{25}, \frac{-12}{25}\right)$

(c) $\frac{2}{\sqrt{3}}$

Level 2

1. (c)	**2.** (c)	**3.** (a)	**4.** (c)	**5.** (a)	**6.** (b)	**7.** (d)
8. (b)	**9.** (a)	**10.** (c)	**11.** (d)	**12.** (b)	**13.** (a)	**14.** (a)
15. (b)	**16.** (c)	**17.** (b)	**18.** (d)	**19.** (c)	**20.** (b)	**21.** (a)
22. (a)	**23.** (c)	**24.** (d)	**25.** (b)	**26.** (a)	**27.** (c)	**28.** (b)
29. (b)	**30.** (d)	**31.** (c)	**32.** (a)	**33.** (d)	**34.** (d)	**35.** (d)
36. (a)	**37.** (a)	**38.** (c)	**39.** (b)	**40.** (c)	**41.** (d)	**42.** (a)
43. (b)	**44.** (c)	**45.** (d)	**46.** (a)	**47.** (a)	**48.** (b)	**49.** (c)
50. (b)	**51.** (c)	**52.** (a)	**53.** (b)	**54.** (c)	**55.** (d)	**56.** (c)
57. (a)	**58.** (a)	**59.** (a)	**60.** (d)	**61.** (b)	**62.** (a)	**63.** (d)
64. (b)	**65.** (5)	**66.** (0.072)	**67.** (3)	**68.** (8)	**69.** (0.04)	**70.** (1)
71. (0.1)	**72.** (0.08)	**73.** (2)	**74.** (2)	**75.** (6)	**76.** (6)	**77.** (0.04)
78. (0.06)	**79.** (5)	**80.** (4)	**81.** (2)			

Level 3

1. (c)	**2.** (b)	**3.** (d)	**4.** (b)	**5.** (b)	**6.** (c)	**7.** (d)
8. (a)	**9.** (a)	**10.** (a)	**11.** (b)	**12.** (b, c, d)	**13.** (a, c,d)	**14.** (b)
15. (a, b, c)	**16.** (b, c, d)	**17.** (a, b, c)	**18.** (a, d)	**19.** (a, b)	**20.** (a, b, c)	**21.** (c)
22. (d)	**23.** (b)	**24.** (a)	**25.** (c)	**26.** (b)	**27.** (a)	**28.** (c)
29. (c)	**30.** (b)	**31.** (a)	**32.** (b)	**33.** (c)	**34.** (d)	**35.** (d)
36. (c)	**37.** (b)					

38. (i)–D, (ii)–C, (iii)–A, (iv)–B

39. (i)–A, (ii)–B, (iii)–B, (iv)–D

40. (i)–A, (ii)–B, (iii)–B, (iv)–B

41. (i)–D, (ii)–A, (iii)–B, (iv)–C

42. (8)	**43.** (3)	**44.** (3)	**45.** (9.38)	**46.** (7)	**47.** (13.91)	**48.** (40)
49. (0.07)	**50.** (0.4)	**51.** (3)	**52.** (7)	**53.** (0)	**54.** (4)	**55.** (0.025)

SOLUTIONS

Level 1

1. (a) Given, equation is $3x^2+\lambda y^2+4xy-7x+8y+3=0$

$\because$ $a=3, b=\lambda, h=2, g=-7/2, f=4$ and $c=3$

We know that any second degree equation represents a hyperbola, when $\Delta \neq 0$ and $h^2 > ab$

$\therefore$ $\Delta \neq 0$

$$abc+2fgh-af^2-bg^2-ch^2 \neq 0$$

$$3(\lambda)(3)+2(4)(-7/2)(2)-3(4)^2$$

$$-\lambda(-7/2)^2-3(2)^2 \neq 0$$

$$\Rightarrow 9\lambda-56-48+\frac{49\lambda^2}{4}-12 \neq 0$$

$h^2 > ab$

Now, $2^2 > 3(\lambda)$

$\Rightarrow$ $3\lambda < 4$

$\Rightarrow$ $\lambda < \frac{4}{3}$

$$\Rightarrow 36\lambda+49\lambda^2-464 \neq 0$$

$$\Rightarrow 49\lambda^2+36\lambda-464 \neq 0$$

$$\Rightarrow \lambda \neq \frac{-36 \pm \sqrt{92240}}{98}$$

$$= \frac{-36 \pm 303\cdot7}{98}$$

$$= \frac{-36-303\cdot7}{98}, \frac{-36+303\cdot7}{98}$$

$$= \simeq -3.47, \simeq 2.73$$

$\therefore$ It represents and hyperbola, when

$$\lambda \in (-\infty, 4/3) - \left\{\frac{-36-\sqrt{99240}}{98}\right\}$$

or $\lambda \in \left(-\infty, \frac{4}{3}\right) - \{-3.47\}$

(b) Given equation,

$$2x^2+3y^2+2\lambda xy+2x-6y+7=0$$

$\therefore$ $a=2, b=3, h=\lambda, g=1, f=-3$ and $c=7$

We know that any second degree equation represents a hyperbola, when $\Delta \neq 0$ and $h^2 > ab$

$\Delta \neq 0$

$$abc+2fgh-af^2-bg^2$$

$$-ch^2 \neq 0$$

$$\Rightarrow 2(3)(7)+2(-3)(1)(\lambda)$$

$$-2(-3)^2-3(1)^2-7(\lambda^2) \neq 0$$

$$\Rightarrow 42-6\lambda-18-3-7\lambda^2 \neq 0$$

$$\Rightarrow 7\lambda^2+6\lambda+21 \neq 0$$

$$\Rightarrow 7\left(\lambda^2+\frac{6}{7}\lambda+3\right) \neq 0$$

$$\Rightarrow \left(\lambda+\frac{3}{7}\right)^2+3-\frac{9}{49} \neq 0$$

$h^2 > ab$

$\lambda^2 > 2 \times 3$

$\lambda^2 > 6$

$(\lambda-\sqrt{6})(\lambda+\sqrt{6}) > 0$ ... (i)

$\Rightarrow$ $\lambda \in R$ (Never becomes zero).

$\because$ Solving inequity

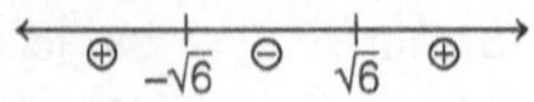

$\therefore$ Required value of λ is

$$(-\infty, -\sqrt{6}) \cup (\sqrt{6}, \infty)$$

(c) Given equation,

$$\lambda x^2+3\lambda y^2+4xy-8x+5y-8=0$$

$\therefore$ $a=\lambda, b=3\lambda, h=2, g=-4, f=5/2$ and $c=-8$

$\Delta \neq 0$

$$\Rightarrow abc+2fgh$$

$$-af^2-bg^2-ch^2 \neq 0$$

$$\Rightarrow \lambda(3\lambda)(-8)+2(5/2)$$

$$(-4)(2)-\lambda(5/2)^2$$

$$-3\lambda(-4)^2-(-8)(2)^2 \neq 0$$

$$\Rightarrow -24\lambda^2-40-\frac{25\lambda}{4}$$

$$-48\lambda+32 \neq 0$$

$$\Rightarrow -24\lambda^2-\frac{217\lambda}{4}-8 \neq 0$$

$$\Rightarrow 24\lambda^2+\frac{217\lambda}{4}+12 \neq 0$$

$$\Rightarrow 96\lambda^2+217\lambda+48 \neq 0$$

$h^2 > ab$

$(2)^2 > \lambda(3\lambda)$

$3\lambda^2 < 4$

$\lambda^2 < 4/3$

$\left(\lambda-\frac{2}{\sqrt{3}}\right)\left(\lambda+\frac{2}{\sqrt{3}}\right) < 0$

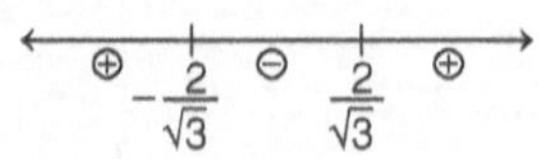

$$\lambda \in \left(\frac{-2}{\sqrt{3}}, \frac{2}{\sqrt{3}}\right)$$

$$\Rightarrow \lambda \neq \frac{-217 \pm \sqrt{(217)^2-4\times96\times48}}{2(96)}$$

$$\neq \frac{-217 \pm \sqrt{47089-18432}}{192}$$

$$\neq \frac{-217 \pm 169.28}{192}$$

$$\Rightarrow \frac{-217-169.28}{192} \simeq -2.01 \quad \Big| \quad \frac{-217+169.28}{192}$$

$$\simeq -0.24$$

$\therefore$ Required value $\lambda \in \left(-\frac{2}{\sqrt{3}}, \frac{2}{\sqrt{3}}\right)\{-0.24\}$

2. (a) $\because$ The vertices and focii lie on the X-axis.

So, let the equation of the required hyperbola be

$$\frac{x^2}{a^2}-\frac{y^2}{b^2}=1$$

The coordinates of its vertices and foci are $(\pm a, 0)$ and $(\pm ae, 0)$, respectively.

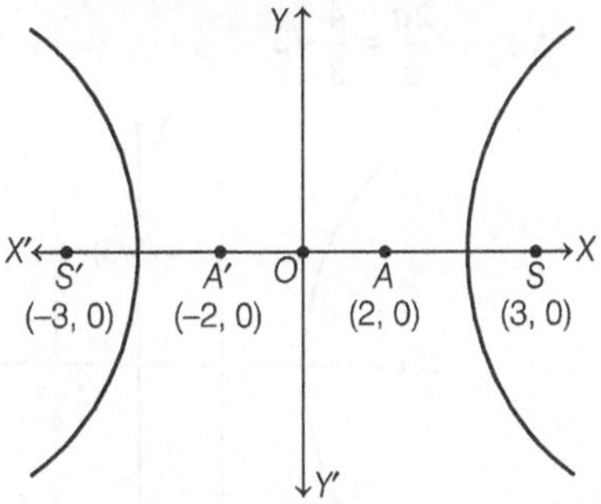

$\because$ The vertices and foci are $(\pm 2, 0)$ and $(\pm 3, 0)$ respectively.

$\therefore \quad a=2$ and $ae=3$

$$e=\frac{3}{2}$$

$$\because \quad b^2=a^2(e^2-1)$$

$$=2^2\left(\left(\frac{3}{2}\right)^2-1\right)$$

$$=9-4=5$$

$\therefore$ Equation of hyperbola is $\frac{x^2}{4}-\frac{y^2}{5}=1$

(b) Given, foci $=(0, \pm 12)$ and vertices $\left(0, \pm\frac{\sqrt{11}}{2}\right)$ which are lies along Y-axis.

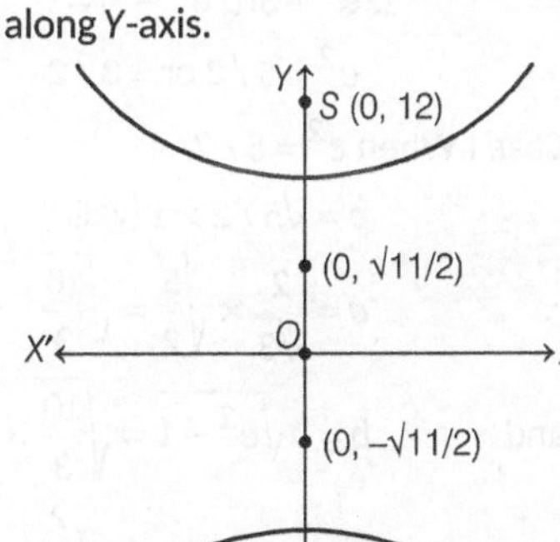

$\therefore$ Transverse axis is along Y-axis.

So, let the equation of hyperbola

$$\frac{x^2}{a^2}-\frac{y^2}{b^2}=-1 \qquad \text{... (i)}$$

$\because$ Focii are $(0, \pm be)$ and vertices $(0, \pm b)$

$\therefore \quad be=12$ and $b=\frac{\sqrt{11}}{2}$

$$\Rightarrow \quad e=\frac{12}{\sqrt{11}}\times 2=\frac{24}{\sqrt{11}}$$

$$\because \quad a^2=b^2(e^2-1)$$

$$\Rightarrow \quad a^2=\frac{11}{4}\left(\frac{24\times 24}{11}-1\right)=\frac{565}{4}$$

$\therefore$ Equation of hyperbola is $\frac{4x^2}{565}-\frac{4y^2}{11}=-1$

(c) Let the equation of hyperbola be $\frac{x^2}{a^2}-\frac{y^2}{b^2}=1$

It passes through (1, 1) and (2, – 3)

$$\therefore \quad \frac{1}{a^2}-\frac{1}{b^2}=1 \qquad \text{... (i)}$$

$$\text{and} \quad \frac{4}{a^2}-\frac{9}{b^2}=1 \qquad \text{... (ii)}$$

Solving Eqs. (i) and (ii) we get $a^2=\frac{5}{8}$ and $b^2=\frac{5}{3}$

Thus, the equation of hyperbola is $\frac{x^2}{5/8}-\frac{y^2}{5/3}=1$

$$\Rightarrow \quad 8x^2-3y^2=5$$

(d) Given, vertices are $(\pm 7, 0)$ which are lie along X-axis.

$\therefore$ Transverse axis is along X-axis, so let the equation of hyperbola is $\frac{x^2}{a^2}-\frac{y^2}{b^2}=1$

The vertices is $(\pm a, 0)$.

$\Rightarrow \quad a=7$

Also, $e=\frac{4}{3}$

$\therefore \quad b^2=a^2(e^2-1)$

$$=49\left[\left(\frac{4}{3}\right)^2-1\right]$$

$$=49\times\left(\frac{16-9}{9}\right)$$

$$=\frac{343}{9}$$

$\therefore$ Equation of hyperbola is $\frac{x^2}{49}-\frac{9y^2}{343}=1$

$$\Rightarrow \quad 7x^2-9y^2=343$$

(e) Given, foci are $(0, \pm\sqrt{10})$, which lie along Y-axis the transverse axis is along Y-axis.

So, let the equation of the hyperbola be

$$\frac{x^2}{a^2}-\frac{y^2}{b^2}=-1$$

$\because$ Coordinate of foci $\Rightarrow (0, \pm be)$

$\therefore \quad be=\sqrt{10}$

Also, hyperbola passes through (2, 3)

$$\frac{4}{a^2}-\frac{9}{b^2}=-1$$

$$\Rightarrow \quad \frac{4}{b^2(e^2-1)}-\frac{9}{b^2}=-1$$

$$\Rightarrow \quad \frac{4}{10-b^2}-\frac{9}{b^2}=-1$$

$$\Rightarrow \quad \frac{4b^2-90+9b^2}{b^2(10-b^2)}=-1$$

$$\Rightarrow \quad 13b^2-90=-10b^2+b^4$$

$$\Rightarrow \quad b^4-23b^2+90=0$$

$$\Rightarrow \quad b^2=\frac{23\pm\sqrt{529-360}}{2}=\frac{23\pm 13}{2}=18, 5$$

$$\therefore \quad b^2=18 \Rightarrow e=\frac{\sqrt{10}}{\sqrt{18}}<1 \text{ (not possible)}$$

$$b^2=5 \Rightarrow e=\frac{\sqrt{10}}{\sqrt{5}}=\sqrt{2} \text{ (possible)}$$

$$\therefore \quad a^2=5(2-1)=5$$

$\therefore$ Equation of hyperbola is $\frac{x^2}{5}-\frac{y^2}{5}=-1$

$$\Rightarrow \quad x^2-y^2=-5$$

or $\quad y^2-x^2=5$

(f) Let the equation of hyperbola be $\frac{x^2}{a^2}-\frac{y^2}{b^2}=1$

Given, length of conjugate axis is

$$2b=6 \Rightarrow b=3$$

and distance between foci = 10

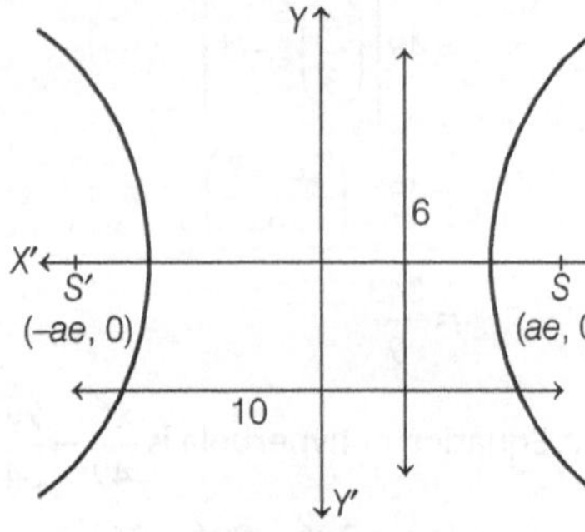

$$2ae=10$$

$$ae=5$$

$$\because \quad b^2=a^2(e^2-1)$$

$$9=25-a^2$$

$$a^2=25-9=16$$

$\therefore$ Equation of hyperbola is $\frac{x^2}{16}-\frac{y^2}{9}=1$

(g) Let the equation of the hyperbola be $\frac{x^2}{a^2}-\frac{y^2}{b^2}=1$

$\because$ Equation of directrices are $x=\pm a/e$

$\therefore$ Distance between the directrices

$$=a/e-(-a/e)=\frac{2a}{e}$$

$$\therefore \quad \frac{2a}{e}=\frac{4}{3}\sqrt{3}$$

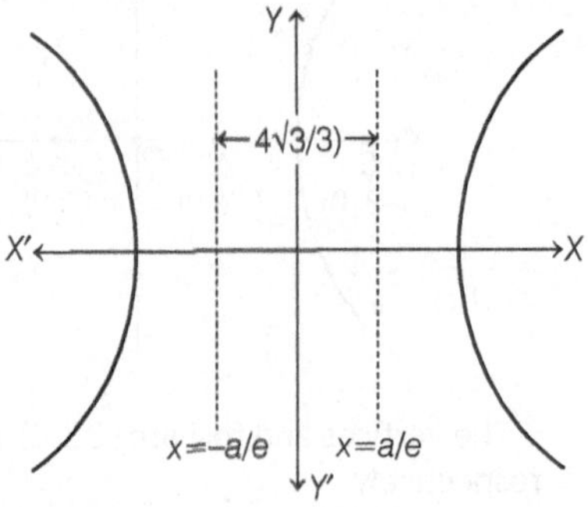

$$\Rightarrow \quad a=\frac{2}{\sqrt{3}}e$$

$\because$ Point (2, 1) lies on it

$$\therefore \quad \frac{4}{a^2}-\frac{1}{b^2}=1$$

$$\Rightarrow \quad \frac{4}{a^2}-\frac{1}{a^2(e^2-1)}=1$$

$$\Rightarrow \quad \frac{4}{\frac{4}{3}e^2}-\frac{1}{\frac{4}{3}e^2(e^2-1)}=1$$

$$\Rightarrow \quad 4e^4-16e^2+15=0$$

$$\Rightarrow \quad (2e^2-3)(2e^2-5)=0$$

$$\Rightarrow \quad e^2=5/2 \text{ or }=3/2$$

Case I When $e^2=5/2$,

$$e=\sqrt{5/2}>1 \text{ (valid)}$$

$$\therefore \quad a=\frac{2}{\sqrt{3}}\times\sqrt{\frac{5}{2}}=\sqrt{\frac{10}{3}}$$

and $\quad b=a\sqrt{e^2-1}=\sqrt{\frac{10}{3}}\times\sqrt{\frac{3}{2}}=\sqrt{5}$

$\therefore$ Equation of hyperbola is $\frac{x^2}{10/3}-\frac{y^2}{5}=1$

$$\Rightarrow \quad 3x^2-2y^2=10$$

Case II When $e^2=3/2$,

$$e=\sqrt{3/2}>1 \text{ (valid)}$$

$$\therefore \quad a=\frac{2}{\sqrt{3}}\sqrt{\frac{3}{2}}=\sqrt{2}$$

and $\quad b=a\sqrt{e^2-1}=\sqrt{2}\sqrt{\frac{3}{2}-1}=1$

$\therefore$ Equation of hyperbola is $\frac{x^2}{2} - \frac{y^2}{1} = 1$

$\Rightarrow \quad x^2 - 2y^2 = 2$

$\therefore$ The required hyperbola are

$3x^2 - 2y^2 = 10$ and $x^2 - 2y^2 = 2$

(h) The given ellipse is $\frac{x^2}{25} + \frac{y^2}{9} = 1$

$\therefore \quad a = 5$ and $b = 3\,(a > b)$

$\therefore \quad e = \sqrt{1 - \frac{b^2}{a^2}} = \sqrt{1 - \frac{9}{25}} = \frac{4}{5}$

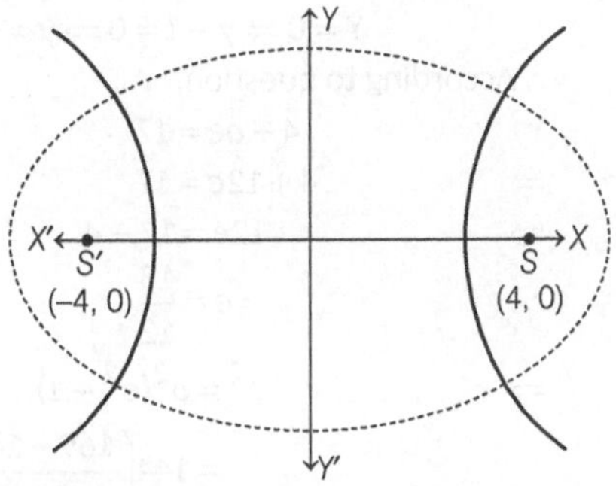

$\therefore$ Foci $\Rightarrow (\pm 4, 0)$

According to question, foci of hyperbola $= (\pm 4, 0)$ and these are on X-axis

$\therefore$ Equation of hyperbola is $\frac{x^2}{A^2} - \frac{y^2}{B^2} = 1$

$\therefore \quad Ae = 4$ and $e = 2$ (given)

$\Rightarrow \quad A = 2$ and $B = A\sqrt{e^2 - 1} = 2(\sqrt{3})$

$\therefore$ Equation of hyperbola is $\frac{x^2}{4} - \frac{y^2}{12} = 1$

$\Rightarrow \quad 3x^2 - y^2 = 12$

(i) Given focii, $(\pm 3\sqrt{5}, 0)$

$\because$ It lies on X-axis.

$\therefore$ Let equation of hyperbola is $\frac{x^2}{a^2} - \frac{y^2}{b^2} = 1$

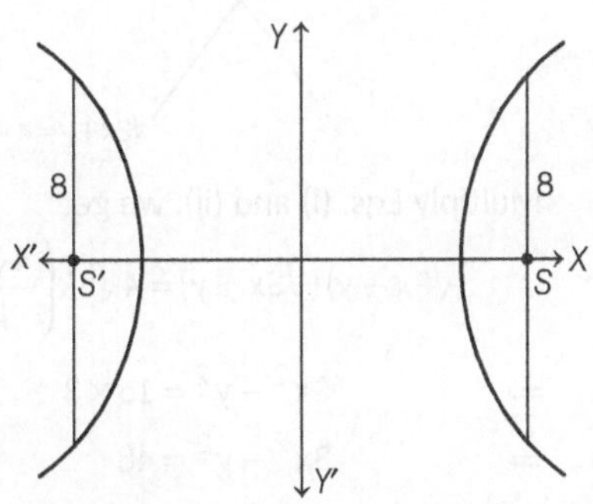

$\therefore$ Foci $\Rightarrow (\pm ae, 0)$

$\Rightarrow \quad ae = 3\sqrt{5}$

Now, length of latusrectum $= \frac{2b^2}{a}$

$8 = \frac{2b^2}{a}$

$\Rightarrow \quad b^2 = 4a$

$\therefore \quad a^2(e^2 - 1) = 4a$

$\Rightarrow \quad 45 - a^2 = 4a$

$\Rightarrow \quad a^2 + 4a - 45 = 0$

$\Rightarrow \quad (a + 9)(a - 5) = 0$

$\Rightarrow \quad a = -9, 5 \quad [\because a > 0]$

$\therefore \quad a = 5$ and $b^2 = 20 \Rightarrow b = \sqrt{20}$

$\therefore$ Equation of hyperbola $\frac{x^2}{25} - \frac{y^2}{20} = 1$

3. (a) Let $P(x, y)$ be any point on the hyperbola, S be the focus and M be the foot of the perpendicular drawn from P on the directrix.

Therefore, from the definition of the hyperbola $PS = ePM$

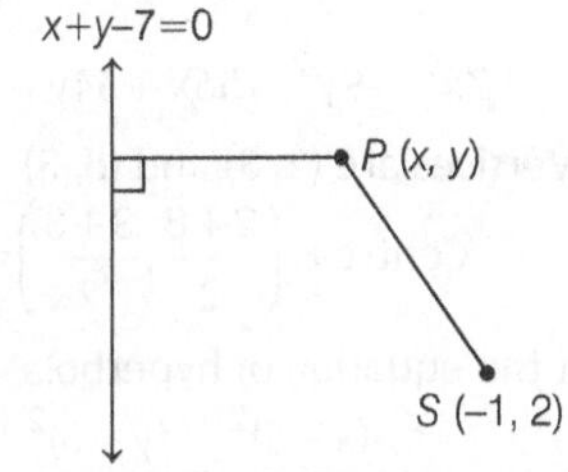

$\Rightarrow \quad PS^2 = e^2(PM)^2$

$\Rightarrow \quad (x + 1)^2 + (y - 2)^2 = 9\left[\frac{|x + y - 7|}{\sqrt{1^2 + 1^2}}\right]^2$

$\Rightarrow x^2 + y^2 + 2x - 4y + 5 = \frac{9(x + y - 7)^2}{2}$

$\Rightarrow \quad 2x^2 + 2y^2 + 4x - 8y + 10$

$= 9x^2 + 9y^2 + 441 + 18xy - 126y - 126x$

$\Rightarrow 7x^2 + 7y^2 + 18xy - 130x - 118y + 431 = 0$

which is required equation of hyperbola.

(b) It is given that foci of hyperbola are $S(8, 3)$ and $S'(0, 3)$

$\because$ Centre of the hyperbola is the mid-point of the line joining of the foci.

$\therefore$ Coordinates of centre are $\left(\frac{8 + 0}{2}, \frac{3 + 3}{2}\right) = (4, 3)$

Let $2a$ and $2b$ be the length of transverse and conjugate axis of hyperbola and e be the eccentricity.

$\therefore$ The required equation of hyperbola is

$\frac{(x - 4)^2}{a^2} - \frac{(y - 3)^2}{b^2} = 1$

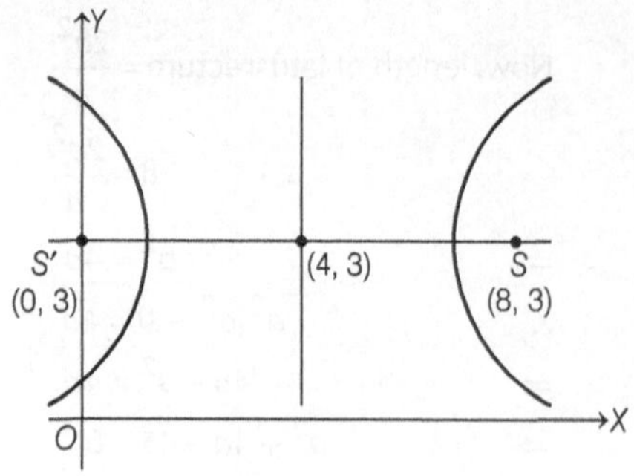

$\because$ $SS' = 2ae$ (S, S' are foci)

$$\Rightarrow \quad \sqrt{8^2 + 0^2} = 2ae$$

$$\Rightarrow \quad 8 = 2ae \Rightarrow ae = 4$$

$$\Rightarrow \quad a = 3 \quad [\because e = 4/3]$$

$$\therefore \quad b^2 = a^2(e^2 - 1)$$

$$= 9\left(\frac{16 - 9}{9}\right) = 7$$

$\therefore$ Equation of hyperbola is $\frac{(x-4)^2}{9} - \frac{(y-3)^2}{7} = 1$

$$\Rightarrow \quad 7x^2 - 9y^2 - 56x + 54y - 32 = 0$$

(c) $\because$ Vertices are (2, 3) and (8, 3)

$\therefore$ Centre is $\left(\frac{2+8}{2}, \frac{3+3}{2}\right) = (5, 3)$

Let the equation of hyperbola

$$\Rightarrow \quad \frac{(x-5)^2}{a^2} - \frac{(y-3)^2}{b^2} = 1$$

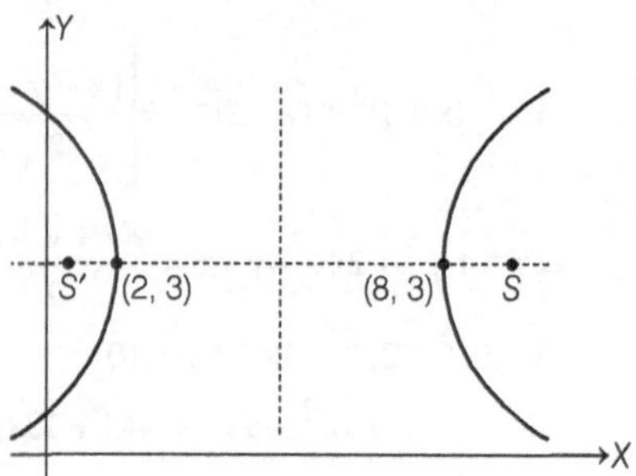

$\because$ Distance between two vertices is $2a$.

$$\Rightarrow \quad 2a = 6 \Rightarrow a = 3$$

and given distance between foci is 12.

$$\therefore \quad 2ae = 12$$

$$\Rightarrow \quad e = \frac{12}{6} = 2$$

$$\therefore \quad b^2 = a^2(e^2 - 1) = 9(3) = 27$$

$\therefore$ Equation of hyperbola is $\frac{(x-5)^2}{9} - \frac{(y-3)^2}{27} = 1$

(d) Vertices are $(-8, -1)$ and $(16, -1)$.

$\therefore$ Centre $= \left(\frac{-8+16}{2}, \frac{-1-1}{2}\right) = (4, -1)$

$\therefore$ Equation of hyperbola is $\frac{(x-4)^2}{a^2} - \frac{(y+1)^2}{b^2} = 1$

which becomes $\frac{X^2}{a^2} - \frac{Y^2}{b^2} = 1$

Now, distance between two vertices $\Rightarrow 2a = 24$

$$\Rightarrow \quad a = 12$$

$\therefore$ Equation becomes $\Rightarrow \frac{X^2}{12^2} - \frac{Y^2}{b^2} = 1$

$\because$ Foci $\Rightarrow (X, Y) \equiv (\pm ae, 0)$

$$\Rightarrow \quad X = \pm ae \Rightarrow x - 4 = \pm ae$$

$$X = 4 \pm ae$$

$$Y = 0 \Rightarrow y + 1 = 0 \Rightarrow y = -1$$

$\therefore$ According to question,

$$4 + ae = 17$$

$$\Rightarrow \quad 4 + 12e = 17$$

$$\Rightarrow \quad 12e = 17 - 4$$

$$\therefore \quad e = \frac{13}{12}$$

$$\Rightarrow \quad b^2 = a^2(e^2 - 1)$$

$$= 144\left(\frac{169 - 144}{144}\right)$$

$$= 25$$

$\therefore$ Equation of hyperbola is

$$\frac{(x-4)^2}{144} - \frac{(y+1)^2}{25} = 1$$

or $25x^2 - 144y^2 - 288y - 200x - 3344 = 0$

(e) Given lines are

$$\sqrt{3}x - y - 4\sqrt{3}\,k = 0$$

$$\Rightarrow \quad \sqrt{3}x - y = 4\sqrt{3}k \quad \text{... (i)}$$

and $\quad \sqrt{3}kx + ky - 4\sqrt{3} = 0$

$$\Rightarrow \quad \sqrt{3}x + y = \frac{4\sqrt{3}}{k} \quad \text{... (ii)}$$

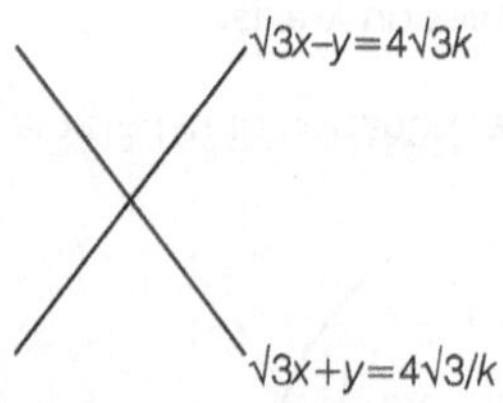

Multiply Eqs. (i) and (ii), we get

$$(\sqrt{3}x - y)(\sqrt{3}x + y) = 4\sqrt{3}k\left(\frac{4\sqrt{3}}{k}\right)$$

$$\Rightarrow \quad 3x^2 - y^2 = 16 \times 3$$

$$\Rightarrow \quad 3x^2 - y^2 = 48$$

which is equation of hyperbola for different values of k.

4. (a) Given, equation $\frac{x^2}{9}-\frac{y^2}{25}=1$

On compare $\frac{x^2}{a^2}-\frac{y^2}{b^2}=1$

$\therefore \quad a^2=9 \Rightarrow a=3$

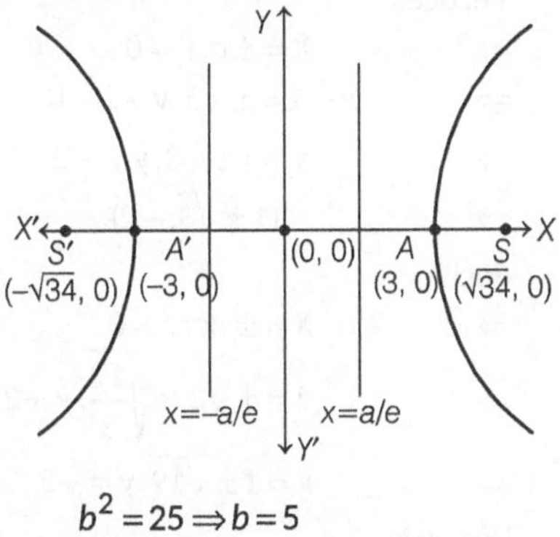

$b^2=25 \Rightarrow b=5$

$\because$ Coefficient of x^2 is positive.

X-axis is transverse axis $\Rightarrow y=0$

Equation of conjugate axis $\Rightarrow x=0$

Length of transverse axis $=2a=6$

Length of conjugate axis $=2b=10$

$\therefore$ Length of latusrectum $=\frac{2b^2}{a}=\frac{50}{3}$

Now, $e=\frac{\sqrt{a^2+b^2}}{a}=\frac{\sqrt{9+25}}{3}=\frac{\sqrt{34}}{3}$

Vertices $(\pm a, 0)=(\pm 3, 0)$

$\Rightarrow$ Centre $=(0, 0)$

Foci $\Rightarrow (\pm ae, 0)=\left(\pm 3\times\frac{\sqrt{34}}{3}, 0\right)$

$=(\pm\sqrt{34}, 0)$

$\therefore$ Equation of directrices are $\Rightarrow x=\pm a/e$

$\Rightarrow \quad x=\pm\frac{3\times 3}{\sqrt{34}}$

$x=\pm\frac{9}{\sqrt{34}}$

(b) Given, equation $\frac{y^2}{25}-\frac{x^2}{16}=1$ ... (i)

$\Rightarrow \quad \frac{x^2}{16}-\frac{y^2}{25}=-1$

On compare $\frac{x^2}{a^2}-\frac{y^2}{b^2}=-1$

$\therefore \quad a^2=16, b^2=25$

$\Rightarrow \quad a=4, b=5$

$\because$ Here coefficient of y^2 in Eq. (i) is positive

$\therefore$ Transverse axis $\Rightarrow Y$-axis $\Rightarrow x=0$

and conjugate axis $\Rightarrow X$-axis $\Rightarrow y=0$

Now, length of transverse axis $=2b=10$

Length of conjugate axis $=2a=8$

Length of latusrectum $=\frac{2a^2}{b}=\frac{2\times 16}{5}=\frac{32}{5}$

$\therefore \quad e=\frac{\sqrt{a^2+b^2}}{b}=\frac{\sqrt{16+25}}{5}=\frac{\sqrt{41}}{5}$

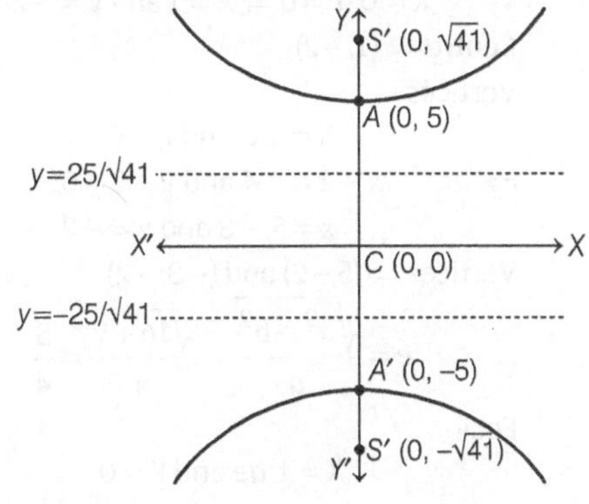

Vertices $=(0, \pm b)=(0, \pm 5)$

Centre $(0, 0)$

Foci $=(0, \pm be)=(0, \pm\sqrt{41})$

$\therefore$ Equation of directrices are $y=\pm b/e$

$\Rightarrow \quad y=\pm 5/(\sqrt{41}/5)=\pm\frac{25}{\sqrt{41}}$

$\Rightarrow \quad \sqrt{41}y=\pm 25$

(c) Given, equation $\frac{(x-1)^2}{16}-\frac{(y+2)^2}{9}=1$

$\Rightarrow \quad \frac{(x-1)^2}{4^2}-\frac{(y+2)^2}{3^2}=1$

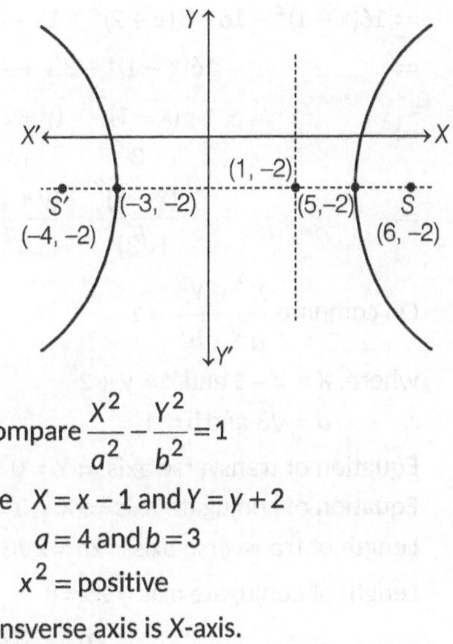

On compare $\frac{X^2}{a^2}-\frac{Y^2}{b^2}=1$

where $X=x-1$ and $Y=y+2$

$\therefore \quad a=4$ and $b=3$

$\therefore \quad x^2=$ positive

$\therefore$ Transverse axis is X-axis.

Equation is $Y=0$

$\Rightarrow \quad y+2=0$

$\Rightarrow \quad y=-2$

and conjugate axis is Y-axis.

Equation is $X=0 \Rightarrow x-1=0 \Rightarrow x=1$

Length of transverse axis $=2a=8$

and length of conjugate axis $=2b=6$

Length of latusrectum $=2b^2/a=9/2$

Centre

$$X=0, Y=0 \Rightarrow x=1 \text{ and } y=-2$$

Centre $\Rightarrow (1,-2)$

Vertices

$$X=\pm a \text{ and } Y=0$$

$$\Rightarrow \quad x-1=\pm 4 \text{ and } y+2=0$$

$$x=5,-3 \text{ and } y=-2$$

Vertices $\Rightarrow (5,-2)$ and $(-3,-2)$

$$\therefore \quad e=\frac{\sqrt{a^2+b^2}}{a}=\frac{\sqrt{16+9}}{4}=\frac{5}{4}$$

Focii

$$X=\pm ae \text{ and } Y=0$$

$$\Rightarrow \quad x-1=\pm 5 \text{ and } y+2=0$$

$$\Rightarrow \quad x=6,-4 \text{ and } y=-2$$

Foci $\Rightarrow (6,-2)$ and $(-4,-2)$.

Directrix $\Rightarrow X=\pm\frac{a}{e}$

$$\Rightarrow \quad x-1=\pm\frac{16}{5}$$

$$\Rightarrow \quad x=\frac{21}{5},\frac{-11}{5}$$

(d) Given equation,

$$16x^2-3y^2-32x-12y-44=0$$

$$\Rightarrow \quad 16(x^2-2x)-3(y^2+4y)-44=0$$

$$\Rightarrow 16(x-1)^2-16-3(y+2)^2+12-44=0$$

$$\Rightarrow \quad 16(x-1)^2-3(y+2)^2=48$$

$$\Rightarrow \quad \frac{(x-1)^2}{3}-\frac{(y+2)^2}{16}=1$$

$$\Rightarrow \quad \frac{(x-1)^2}{(\sqrt{3})^2}-\frac{(y+2)^2}{4^2}=1$$

On compare $\frac{X^2}{a^2}-\frac{Y^2}{b^2}=1$

where, $X=x-1$ and $Y=y+2$

$\therefore \quad a=\sqrt{3}$ and $b=4$

Equation of transverse axis $\Rightarrow Y=0 \Rightarrow y+2=0$

Equation of conjugate axis $\Rightarrow X=0 \Rightarrow x-1=0$

Length of transverse axis $=2a=2\sqrt{3}$

Length of conjugate axis $=2b=8$

$\therefore$ Length of latusrectum $=\frac{2b^2}{a}=\frac{32}{\sqrt{3}}$

Centre

$$\Rightarrow \quad X=0, Y=0$$

$$\Rightarrow \quad x-1=0, y+2=0$$

$$\Rightarrow \quad x=1, y=-2$$

Centre $=(1,-2)$

$$\therefore \quad e=\frac{\sqrt{a^2+b^2}}{a}=\frac{\sqrt{19}}{\sqrt{3}}$$

Vertices

$$\Rightarrow \quad X=\pm a, Y=0$$

$$\Rightarrow \quad x-1=\pm\sqrt{3}, y+2=0$$

$$\Rightarrow \quad x=1\pm\sqrt{3}, y=-2$$

$$\Rightarrow \quad (1\pm\sqrt{3},-2)$$

Foci

$$\Rightarrow \quad X=\pm ae, Y=0$$

$$\Rightarrow \quad x-1=\pm\sqrt{3}\times\sqrt{\frac{19}{3}}, y+2=0$$

$$\Rightarrow \quad x=1\pm\sqrt{19}, y=-2$$

Directrix

$$X=\pm a/e$$

$$\Rightarrow \quad x-1=\pm\frac{\sqrt{3}\times\sqrt{3}}{\sqrt{19}} \Rightarrow x=1\pm\frac{3}{\sqrt{19}}$$

(e) Given, $x^2-2y^2-2x+8y-1=0$

$$(x^2-2x)-2(y^2-4y)-1=0$$

$$\Rightarrow (x-1)^2-2(y-2)^2-1+8-1=0$$

$$\Rightarrow \quad (x-1)^2-2(y-2)^2=-6$$

$$\Rightarrow \quad \frac{(x-1)^2}{6}-\frac{(y-2)^2}{3}=-1$$

On compare $\frac{X^2}{a^2}-\frac{Y^2}{b^2}=-1$

where, $X=x-1$ and $Y=y-2$

$\therefore \quad a=\sqrt{6}$ and $b=\sqrt{3}$

Equation of transverse axis

$$\Rightarrow \quad X=0 \Rightarrow x-1=0 \Rightarrow x=1$$

Equation of conjugate axis

$$Y=0 \Rightarrow y-2=0 \Rightarrow y=2$$

$\therefore$ Centre (1, 2)

Length of transverse axis $=2b=2\sqrt{3}$

Length of conjugate axis $=2a=2\sqrt{6}$

$\therefore$ Length of latusrectum $=\frac{2a^2}{b}=\frac{12}{\sqrt{3}}=4\sqrt{3}$

$$\therefore \quad e=\frac{\sqrt{a^2+b^2}}{b}=\frac{\sqrt{9}}{\sqrt{3}}=\sqrt{3}$$

Vertices

$$\Rightarrow \quad X=0 \text{ and } Y=\pm b$$

$$\Rightarrow \quad x-1=0 \text{ and } y-2=\pm\sqrt{3}$$

$$\Rightarrow \quad x=1 \text{ and } y=2\pm\sqrt{3}$$

Vertices $(1, 2\pm\sqrt{3})$

Foci

$\Rightarrow \quad X = 0$ and $Y = \pm be$

$\Rightarrow \quad x - 1 = 0$ and $y - 2 = \pm(\sqrt{3} \times \sqrt{3})$

$\Rightarrow \quad x = 1$ and $y = 2 \pm 3$

Foci (1, 5) and (1, – 1).

Directrix

$$\Rightarrow \quad Y = \pm b/e \Rightarrow y - 2 = \pm\frac{\sqrt{3}}{3}$$

$$y = 2 \pm 1 \Rightarrow y = 3, 1$$

5. (a) Let (x, y) be any point

$\therefore$ According to question,

$$|\sqrt{(x+4)^2 + (y-0)^2} - \sqrt{(x-4)^2 + (y-0)^2}| = 2$$

$$\Rightarrow \quad \sqrt{(x+4)^2 + y^2} = \pm 2 + \sqrt{(x-4)^2 + y^2}$$

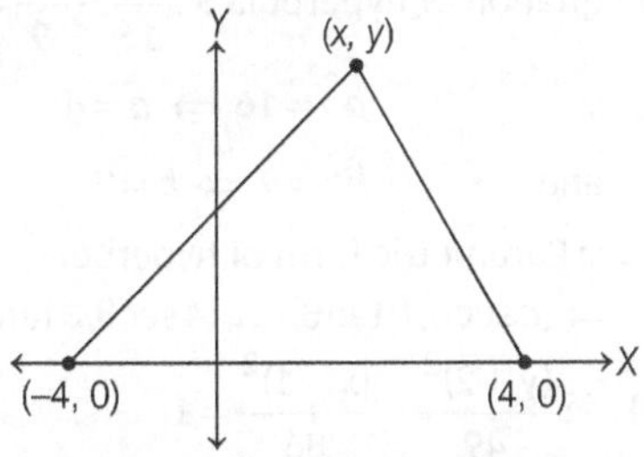

Squaring on both sides, we get

$$(x+4)^2 + y^2 = 4 + (x-4)^2 + y^2 \pm 4\sqrt{(x-4)^2 + y^2}$$

$$\Rightarrow \quad 16x - 4 = \pm 4\sqrt{(x-4)^2 + y^2}$$

$$\Rightarrow \quad 4x - 1 = \pm\sqrt{(x-4)^2 + y^2}$$

$$\Rightarrow \quad (4x-1)^2 = (x-4)^2 + y^2$$

$$\Rightarrow \quad 16x^2 + 1 - 8x = x^2 + 16 - 8x + y^2$$

$$\Rightarrow \quad 15x^2 + 1 - y^2 - 16 = 0$$

$$\Rightarrow \quad 15x^2 - y^2 = 15$$

$$\Rightarrow \quad \frac{x^2}{1} - \frac{y^2}{15} = 1$$

(b) Let m_1 and m_2 be the slope of AM and BM.

$\therefore \quad \angle A = \theta \Rightarrow \angle B = 2\theta$

$$\therefore \quad \tan\theta = m_1 = \frac{k}{h+1}$$

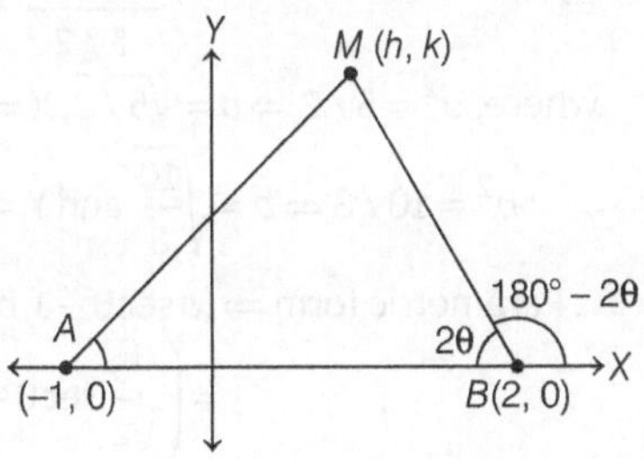

$$\therefore \quad \tan(180° - 2\theta) = m_2 = \frac{k}{h-2}$$

$$\Rightarrow \quad -\tan 2\theta = \frac{k}{h-2}$$

$$\Rightarrow \quad \frac{-2\tan\theta}{1 - \tan^2\theta} = \frac{k}{h-2}$$

$$\Rightarrow \quad \frac{-2(m_1)}{1 - m_1^2} = \frac{k}{h-2}$$

$$\Rightarrow \quad \frac{-2\left(\frac{k}{h+1}\right)}{1 - \left(\frac{k}{h+1}\right)^2} = \frac{k}{h-2}$$

$$\Rightarrow \quad \frac{-2(h+1)}{(h+1)^2 - k^2} = \frac{1}{(h-2)}$$

$$\Rightarrow \quad -2(h+1)(h-2) = (h+1)^2 - k^2$$

$$\Rightarrow \quad -2h^2 - 2h + 4h + 4 = h^2 + 1 + 2h - k^2$$

$$\Rightarrow \quad 3h^2 - k^2 = 3$$

$\therefore$ Required locus is $3x^2 - y^2 = 3$

$$\Rightarrow \quad \frac{x^2}{1} - \frac{y^2}{3} = 1$$

(c) Suppose, two rods are coincident on the X-axis. One rotates about point O and the other about point $A(a, 0)$.

If they rotate according to question, then at some time t, they are in position (given in figure)

From figure, $\tan\theta = \dfrac{k}{h}$

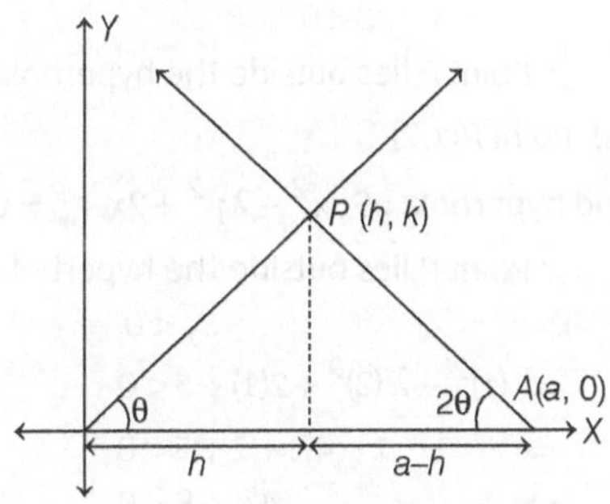

$$\Rightarrow \quad \tan 2\theta = \frac{k}{a-h}$$

$$\therefore \quad \frac{2\tan\theta}{1 - \tan^2\theta} = \frac{k}{a-h}$$

$$\Rightarrow \quad \frac{2\frac{k}{h}}{1 - \frac{k^2}{h^2}} = \frac{k}{a-h}$$

$$\Rightarrow \quad 2h = \frac{(h^2 - k^2)}{a-h}$$

$\Rightarrow \quad 2ah - 2h^2 = h^2 - k^2$

$\Rightarrow \quad 3h^2 - k^2 - 2ah = 0$

$\therefore$ Locus is $3x^2 - y^2 - 2ax = 0$

Locus is hyperbola.

6. (a) Point $P = (4,3)$ and hyperbola $S: \frac{x^2}{16} - \frac{y^2}{9} - 1 = 0$

Now, position of point P

$$S_1 = \frac{(4)^2}{16} - \frac{(3)^2}{9} - 1$$
$$= 1 - 1 - 1$$
$$= -1 < 0$$

$\therefore$ Point P lies outside the hyperbola.

(b) Point $P: (-1,-3)$ and hyperbola $S: 18x^2 - y^2 = 1$

$\therefore \quad S_1 = 18(-1)^2 - (-3)^2 - 1$
$= 18 - 9 - 1$
$= 8 > 0$

$\therefore$ Point P lies inside the hyperbola.

(c) Point $P(2,3)$

and hyperbola $S: 4y^2 - 9x^2 + 2x - 4 = 0$

$\because \quad S_1 = 4(3)^2 - 9(2)^2 + 2(2) - 4$
$= 36 - 36 + 4 - 4 = 0$

$\therefore$ Point P lies on the hyperbola.

(d) Point $P(-1,-2)$

and hyperbola $S: 3x^2 - y^2 + 7x + 2y - 11 = 0$

$\because \quad S_1 = 3(-1)^2 - (-2)^2 + 7(-1) + 2(-2) - 11$
$= 3 - 4 - 7 - 4 - 11$
$= -23 < 0$

$\therefore$ Point P lies outside the hyperbola.

7. (a) Point $P(1,2)$

and hyperbola : $S: x^2 - \lambda y^2 + 2x + 5 = 0$

$\because$ Point P lies outside the hyperbola

$\therefore \quad S_1 < 0$

$\Rightarrow (1)^2 - \lambda(2)^2 + 2(1) + 5 < 0$

$\Rightarrow \quad 1 - 4\lambda + 2 + 5 < 0$

$\Rightarrow \quad -4\lambda + 8 < 0$

$\Rightarrow \quad 4\lambda > 8$

and $\quad \lambda > 2$

$\therefore \quad \lambda \in (2, \infty)$.

(b) Point $P(-1,3)$

and hyperbola $S: x^2 - 4y^2 - 2\lambda x + 11 = 0$

$\because$ Point P lies inside the hyperbola

$\therefore \quad S_1 > 0$

$\Rightarrow (-1)^2 - 4(3)^2 - 2\lambda(-1) + 11 > 0$

$\Rightarrow \quad 1 - 36 + 2\lambda + 11 > 0$

$\Rightarrow \quad 2\lambda - 24 > 0$

and $\quad \lambda > 12$

$\therefore \lambda \in (12, \infty)$

(c) Point $P(\lambda, -3)$

and hyperbola $S: x^2 - y^2 + 4\lambda y + 7 = 0$

$\because$ Point P lies on the hyperbola.

$\therefore \quad S_1 = 0$

$\Rightarrow \lambda^2 - (-3)^2 + 4\lambda(-3) + 7 = 0$

$\Rightarrow \quad \lambda^2 - 9 - 12\lambda + 7 = 0$

$\Rightarrow \quad \lambda^2 - 12\lambda - 2 = 0$

$\Rightarrow \quad \lambda = \frac{12 \pm \sqrt{144 + 8}}{2}$

$\Rightarrow \quad \lambda = 6 \pm \sqrt{38}$

8. (a) Equation of hyperbola $S: \frac{x^2}{16} - \frac{y^2}{9} = 1$

$\because \quad a^2 = 16 \Rightarrow a = 4$

and $\quad b^2 = 9 \Rightarrow b = 3$

$\because$ Parametric form of hyperbola

$\Rightarrow (a\sec\theta, b\tan\theta)$; $\therefore$ $(4\sec\theta, 3\tan\theta)$

(b) $S: \frac{(y+2)^2}{49} - \frac{(x-1)^2}{36} = 1$

$\Rightarrow \quad \frac{(x-1)^2}{36} - \frac{(y+2)^2}{49} = -1$

$\therefore \quad a^2 = 36, b^2 = 49, X = x - 1$ and $Y = y + 2$

$: \quad a = 6, b = 7, x = X + 1, y = Y - 2$

$\therefore$ Parametric form

$= (a\tan\theta + 1, b\sec\theta - 2)$

$\Rightarrow (6\tan\theta + 1, 7\sec\theta - 2)$

(c) Hyperbola : $4x^2 - 3y^2 + 8x + 6y - 9 = 0$

$\Rightarrow \quad 4x^2 + 8x - 3y^2 + 6y - 9 = 0$

$\Rightarrow \quad 4(x^2 + 2x) - 3(y^2 - 2y) - 9 = 0$

$\Rightarrow 4(x^2 + 2x + 1 - 1) - 3(y^2 - 2y + 1 - 1) - 9 = 0$

$\Rightarrow \quad 4(x+1)^2 - 4 - 3(y-1)^2 + 3 - 9 = 0$

$\Rightarrow \quad 4(x+1)^2 - 3(y-1)^2 - 1 - 9 = 0$

$\Rightarrow \quad 4(x+1)^2 - 3(y-1)^2 = 10$

$\Rightarrow \quad \frac{4(x+1)^2}{10} - \frac{3(y-1)^2}{10} = 1$

$\Rightarrow \quad \frac{(x+1)^2}{5/2} - \frac{(y-1)^2}{10/3} = 1$

where, $a^2 = 5/2 \Rightarrow a = \sqrt{5/2}, X = x + 1, x = X - 1$

$b^2 = 10/3 \Rightarrow b = \sqrt{\frac{10}{3}}$ and $Y = y - 1 \Rightarrow y = Y + 1$

$\therefore$ Parametric form $\Rightarrow (a\sec\theta - 1, b\tan\theta + 1)$

$$= \left(\sqrt{\frac{5}{2}}\sec\theta - 1, \sqrt{\frac{10}{3}}\tan\theta + 1\right)$$

9. (a) Line $y : 2x + \lambda$ ... (i)

and hyperbola $S : 3x^2 - 4y^2 = 12$ ... (ii)

Solving Eqs. (i) and (ii), we get

$$3x^2 - 4(2x + \lambda)^2 - 12 = 0$$

$\Rightarrow \quad 3x^2 - 4(4x^2 + \lambda^2 + 2x\lambda) - 12 = 0$

$\Rightarrow \quad 3x^2 - 16x^2 - 4\lambda^2 - 8x\lambda - 12 = 0$

$\Rightarrow \quad -13x^2 - 8x\lambda - 4\lambda^2 - 12 = 0$

$\Rightarrow \quad 13x^2 + (8\lambda)x + (4\lambda^2 + 12) = 0$ ... (iii)

On compare $AX^2 + BX + C = 0$, we get

$$A = 13, B = 8\lambda \text{ and } C = 4\lambda^2 + 12$$

(i) $\because$ Line intersects hyperbola at two points

$\therefore$ Eq. (iii) has two real and distinct roots.

$\Rightarrow \quad D > 0$

$B^2 - 4AC > 0$

$\Rightarrow \quad (8\lambda)^2 - 4(13)(4\lambda^2 + 12) > 0$

$\Rightarrow \quad 64\lambda^2 - 16(13)(\lambda^2 + 3) > 0$

$\Rightarrow \quad 4\lambda^2 - 13(\lambda^2 + 3) > 0$

$\Rightarrow \quad 4\lambda^2 - 13\lambda^2 - 39 > 0$

$\Rightarrow \quad -9\lambda^2 - 39 > 0$

$\Rightarrow \quad 9\lambda^2 + 39 < 0$

$\Rightarrow \quad \lambda^2 + \frac{13}{3} < 0$

$\Rightarrow \quad \lambda \in \phi$

(ii) $\because$ Line intersects at one point

$\therefore$ Eq. (iii) has two real and equal roots.

$\Rightarrow \quad D = 0 \Rightarrow \lambda^2 + \frac{13}{3} = 0$

$\Rightarrow \quad \lambda^2 = \frac{-13}{3} \Rightarrow \lambda \in \phi$

(iii) $\because$ Line does not intersect the hyperbola.

$\therefore$ Eq. (iii) has no real roots.

$\Rightarrow \quad D < 0$

$\Rightarrow \quad \lambda^2 - \frac{13}{3} > 0$

$\Rightarrow \quad \lambda \in \left(-\infty, -\sqrt{\frac{13}{3}}\right) \cup \left(\sqrt{\frac{13}{3}}, \infty\right)$

(b) Line $L : x + 2y - 1 = 0$ ...(i)

and hyperbola $S : x^2 - 2y^2 = 1$ ... (ii)

Solving Eqs. (i) and (ii), we get

$$(1 - 2y)^2 - 2y^2 = 1$$

$\Rightarrow \quad 1 + 4y^2 - 4y - 2y^2 = 1$

$\Rightarrow \quad 2y^2 - 4y = 0$

$\Rightarrow \quad 2y(y - 2) = 0$

$\Rightarrow \quad y = 0 \text{ or } 2$

When $y = 0$, then $x = 1$

When $y = 2$, then $x = -3$ [from Eq. (i)]

$\therefore$ Point of intersections are $A(1, 0)$ and $B(-3, 2)$.

and length of intercept $AB = \sqrt{(-3-1)^2 + (2-0)^2}$

$= \sqrt{16 + 4}$

$= \sqrt{20} = 2\sqrt{5}$ units

10. (a) Given, point (1, 2) and equation of hyperbola :

$$27x^2 - 2y^2 = 19$$

$\because$ Equation of tangent at point $P(x_1, y_1)$ to the hyperbola $\frac{x^2}{a^2} - \frac{y^2}{b^2} = 1$ is

$$\frac{xx_1}{a^2} - \frac{yy_1}{b^2} = 1$$

$\therefore$ Required equation of tangent

$\Rightarrow \quad 27x(1) - 2y(2) = 19$

$27x - 4y = 19$

(b) Given, equation of hyperbola : $\frac{x^2}{4} - \frac{y^2}{36} = 1$

$\therefore \quad a^2 = 4 \text{ and } b^2 = 36$

$\Rightarrow \quad a = 2 \text{ and } b = 6$

$\therefore$ Parametric form of ellipse $\Rightarrow (a\sec\theta, b\tan\theta)$

$= (2\sec\theta, 6\tan\theta)$

$\therefore$ Equation of tangent is

$\Rightarrow \quad \frac{x(2\sec\theta)}{4} - \frac{y(6\tan\theta)}{36} = 1$

$\frac{x\sec\theta}{2} - \frac{y\tan\theta}{6} = 1$

(c) Given, equation of hyperbola : $3x^2 - 4y^2 = 12$

$\Rightarrow \quad \frac{x^2}{4} - \frac{y^2}{3} = 1$

$\because \quad a^2 = 4 \text{ and } b^2 = 3$

$\Rightarrow \quad a = 2 \text{ and } b = \sqrt{3}$

$\because$ Point on hyperbola can be $\Rightarrow (a\sec\theta, b\tan\theta)$

$\Rightarrow \quad (2\sec\theta, \sqrt{3}\tan\theta)$

$\because \quad \theta = \pi/4$

$\Rightarrow$ Point is $(2\sec\pi/4, \sqrt{3}\tan\pi/4)$

$P = (2\sqrt{2}, \sqrt{3})$

$\therefore$ Equation of tangent at point $P(2\sqrt{2}, \sqrt{3})$ is

$$\frac{x(2\sqrt{2})}{4} - \frac{y(\sqrt{3})}{3} = 1$$

$\Rightarrow \quad \frac{x}{\sqrt{2}} - \frac{y}{\sqrt{3}} = 1$

(d) Hyperbola : $3x^2 - 8y^2 = 24$

$$\Rightarrow \quad \frac{3x^2}{24} - \frac{8y^2}{24} = 1$$

$$\Rightarrow \quad \frac{x^2}{8} - \frac{y^2}{3} = 1$$

$\therefore \quad a^2 = 8$ and $b^2 = 3$

$\because$ Slope $(m) = -1$

$\because$ Equation of tangent having slope (m) for the hyperbola is

$$y = mx \pm \sqrt{a^2m^2 - b^2}$$

$\therefore$ Equation is $y = -x \pm \sqrt{8(-1)^2 - (3)}$

$$\Rightarrow \quad y = -x \pm \sqrt{8-3}$$

$$\Rightarrow \quad y = -x \pm \sqrt{5}$$

$$\Rightarrow \quad x + y \pm \sqrt{5} = 0$$

(e) Hyperbola : $4x^2 - 9y^2 = 36$

$$\Rightarrow \quad \frac{x^2}{9} - \frac{y^2}{4} = 1$$

$\because$ Tangent is parallel to the line $5x - 4y + 7 = 0$

$\therefore$ Slope of tangent $(m) = \frac{5}{4}$

$\because \quad a^2 = 9, b^2 = 4$

and $\quad m = 5/4$

$\therefore$ Equation of tangent having slope (m) for the hyperbola is

$$y = mx \pm \sqrt{a^2m^2 - b^2}$$

$$\Rightarrow \quad y = \frac{5x}{4} \pm \sqrt{9(5/4)^2 - 4}$$

$$\Rightarrow \quad y = \frac{5x}{4} \pm \frac{\sqrt{225-64}}{4}$$

$$\Rightarrow \quad 4y = 5x \pm \sqrt{161}$$

$\therefore$ Equation is $5x - 4y \pm \sqrt{161} = 0$

(f) Hyperbola : $9x^2 - 16y^2 = 144$

$$\Rightarrow \quad \frac{x^2}{16} - \frac{y^2}{9} = 1$$

$\therefore \quad a^2 = 16$ and $b^2 = 9$

$\because$ Tangent is perpendicular to the line $3x + 4y + 10 = 0$

$\therefore$ Slope of tangent is $m = \frac{-1}{-3/4} = \frac{4}{3}$

$\therefore$ Equation of tangent is

$$y = mx \pm \sqrt{a^2m^2 - b^2}$$

$$\therefore \quad y = \frac{4}{3}x \pm \sqrt{16\left(\frac{4}{3}\right)^2 - 9}$$

$$= \frac{4}{3}x \pm \sqrt{\frac{256-81}{9}}$$

$$\Rightarrow \quad 3y = 4x \pm \sqrt{175}$$

$$\Rightarrow \quad 4x - 3y \pm 5\sqrt{7} = 0$$

11. (a) $\because P(a\sec\theta, b\tan\theta)$ lies on tangent $y = mx + \sqrt{a^2m^2 - b^2}$ to the hyperbola $\frac{x^2}{a^2} - \frac{y^2}{b^2} = 1$

$\therefore b\tan\theta = am\sec\theta + \sqrt{a^2m^2 - b^2}$

$\Rightarrow b\tan\theta - am\sec\theta = \sqrt{a^2m^2 - b^2}$

Squaring both the sides, we get

$(b\tan\theta - am\sec\theta)^2 = (a^2m^2 - b^2)$

$$\Rightarrow b^2\tan^2\theta + a^2m^2\sec^2\theta - 2abm\sec\theta\tan\theta - a^2m^2 + b^2 = 0$$

$$\Rightarrow \quad b^2(1 + \tan^2\theta) + a^2m^2(\sec^2\theta - 1) - 2abm\sec\theta\tan\theta = 0$$

$$\Rightarrow b^2\sec^2\theta + a^2m^2\tan^2\theta - 2abm\sec\theta\tan\theta = 0$$

$$\Rightarrow \quad (b\sec\theta - am\tan\theta)^2 = 0$$

$$\Rightarrow \quad b\sec\theta - am\tan\theta = 0$$

$$\Rightarrow \quad b\sec\theta = am\tan\theta$$

$$\Rightarrow \quad b = am\sin\theta$$

$$\Rightarrow \quad \sin\theta = \frac{b}{am}$$

(b) Line : $5x + 12y = 9$

Line touch the hyperbola

Hyperbola : $x^2 - 9y^2 = 9$

$$\Rightarrow \quad \frac{x^2}{9} - \frac{y^2}{1} = 1$$

$\therefore \quad a^2 = 9$ and $b^2 = 1$

$\Rightarrow$ (slope of tangent) $(m) = -\frac{5}{12}$

$$\Rightarrow \quad c = \frac{9}{12} = \frac{3}{4}$$

$\because$ If line touches the hyperbola, then point of contact $\left(\pm\frac{a^2m}{c}, \pm\frac{b^2}{c}\right)$

$$= \left(\pm\frac{9(-5/12)}{3/4}, \pm\frac{1}{3/4}\right)$$

$$\Rightarrow \left(\mp 5, \pm\frac{4}{3}\right)$$

$\therefore$ Point of contacts are $(5, -4/3)$ and $(-5, 4/3)$.

12. (a) Let F_1 and F_2 be two foci of the hyperbola

$$\frac{x^2}{a^2}-\frac{y^2}{b^2}=1$$

Then, $F_1=(ae,0)$ and $F_2=(-ae,0)$

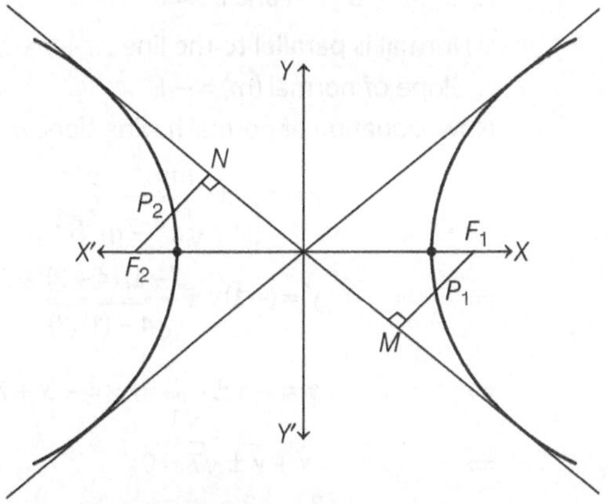

$\because$ Equation of tangent at any point of $P(a\sec\theta, b\tan\theta)$ is

$$\frac{x}{a}\sec\theta-\frac{y}{b}\tan\theta=1 \Rightarrow \frac{x}{a}\sec\theta-\frac{y}{b}\tan\theta-1=0$$

Let $F_1M=P_1$ and $F_2N=P_2$ are perpendicular drawn from foci to tangent.

$$\therefore \quad P_1=\left|\frac{e\sec\theta-1}{\sqrt{\frac{\sec^2\theta}{a^2}+\frac{\tan^2\theta}{b^2}}}\right|$$

and $$P_2=\left|\frac{-e\sec\theta-1}{\sqrt{\frac{\sec^2\theta}{a^2}+\frac{\tan^2\theta}{b^2}}}\right|$$

$$\therefore \quad P_1P_2=\frac{e^2\sec^2\theta-1}{\frac{b^2\sec^2\theta+a^2\tan^2\theta}{a^2b^2}}$$

$$=\frac{b^2(e^2\sec^2\theta-1)}{\left(\frac{b^2}{a^2}\right)\sec^2\theta+\tan^2\theta}$$

$$=\frac{a^2(e^2-1)(e^2\sec^2\theta-1)}{(e^2-1)\sec^2\theta+\tan^2\theta}$$

$$\{\because b^2=a^2(e^2-1)\}$$

$$=\frac{a^2(e^2-1)(e^2\sec^2\theta-1)}{e^2\sec^2\theta-\sec^2\theta+\tan^2\theta}$$

$$=\frac{a^2(e^2-1)(e^2\sec^2\theta-1)}{(e^2\sec^2\theta-1)}$$

$$=a^2(e^2-1)$$

$$\Rightarrow \quad P_1P_2=b^2$$

(b) Let, $P(a\sec\theta, b\tan\theta)$ so that the tangent at P is

$$\frac{x\sec\theta}{a}-\frac{y\tan\theta}{b}=1 \quad \text{...(i)}$$

Transverse axis is $y=0$ ...(ii)

The tangent at $(a,0)$ is $x=a$ ...(iii)

Therefore, $Q(a\cos\theta, 0)$ and $R=\left(a, \frac{b(1-\cos\theta)}{\sin\theta}\right)$

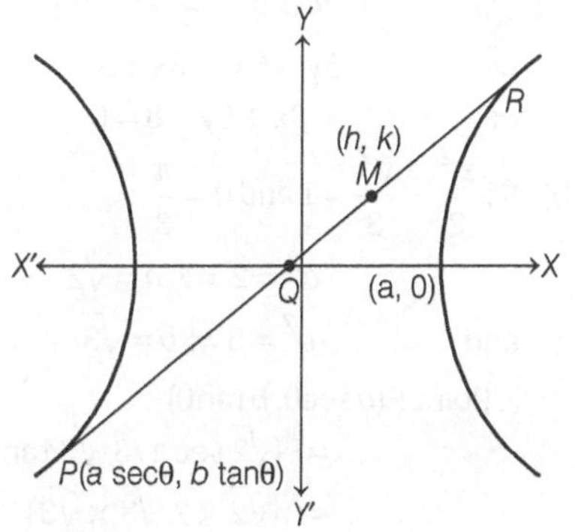

Let $M(h,k)$ be the mid-point of QR, therefore

$$h=\frac{a+a\cos\theta}{2}=\frac{a}{2}(1+\cos\theta)$$

$$=\frac{a}{2}\left(2\cos^2\frac{\theta}{2}\right)$$

$$\Rightarrow \quad h=a\cos^2\frac{\theta}{2} \quad \text{...(i)}$$

$$\Rightarrow \sec^2\frac{\theta}{2}=\frac{a}{h}$$

and $$k=\frac{0+\frac{b(1-\cos\theta)}{\sin\theta}}{2}$$

$$=\frac{b}{2}\left(\frac{2\sin^2\theta/2}{2\sin\theta/2\cos\theta/2}\right)$$

$$\Rightarrow \quad k=\frac{b}{2}\tan\theta/2$$

$$\Rightarrow \quad k^2=\frac{b^2}{4}\left(\tan^2\frac{\theta}{2}\right)$$

$$\Rightarrow \quad \frac{4k^2}{b^2}=1+\sec^2\frac{\theta}{2}$$

$$\Rightarrow \quad \frac{4k^2}{b^2}=1+\left(\frac{a}{h}\right)$$

$$\therefore \text{Locus is } \frac{4y^2}{b^2}=1+\frac{a}{h}$$

13. (a) Point (1, 1)

and hyperbola $S: 5x^2-3y^2-2=0$

$$\Rightarrow \quad 5x^2-3y^2=2$$

$$\Rightarrow \quad \frac{x^2}{2/5}-\frac{y^2}{2/3}=1$$

$\because$ Equation of normal at point (x_1, y_1) is

$$y - y_1 = \frac{-a^2 y_1}{b^2 x_1}(x - x_1)$$

$$\Rightarrow \quad (y-1) = \frac{-2/5(1)}{\frac{2}{3}(1)}(x-1)$$

$$\Rightarrow \quad y - 1 = \frac{-3}{5}(x-1)$$

$$\Rightarrow \quad 5y - 5 = -3x + 3$$

$$\Rightarrow \quad 3x + 5y - 8 = 0$$

(b) $S: \frac{x^2}{2} - \frac{y^2}{3} = 1$ and $\theta = \frac{\pi}{3}$

$\therefore \quad a^2 = 2 \Rightarrow a = \sqrt{2}$

and $\quad b^2 = 3 \Rightarrow b = \sqrt{3}$

$\therefore$ Point $P(a\sec\theta, b\tan\theta)$

$$= P(\sqrt{2}\sec\pi/3, \sqrt{3}\tan\pi/3)$$

$$= P(\sqrt{2}\times 2, \sqrt{3}\times\sqrt{3})$$

$$= P(2\sqrt{2}, 3)$$

$\therefore$ Equation of normal

$$\frac{a^2x}{x_1} + \frac{b^2y}{y_1} = a^2 + b^2$$

$$\Rightarrow \quad \frac{2x}{2\sqrt{2}} + \frac{3y}{3} = 2 + 3 \Rightarrow \frac{x}{\sqrt{2}} + y = 5$$

$$\Rightarrow \quad x + \sqrt{2}y = 5\sqrt{2}$$

$\therefore$ Slope of normal $= -1/\sqrt{2}, c = 5\sqrt{2}$

$\therefore$ Point of contact $\Rightarrow \left(\pm\frac{a^2}{c}, \mp\frac{mb^2}{c}\right)$

$$= \left(\pm\frac{2}{5\sqrt{2}} \mp \frac{3\cdot(-1/\sqrt{2})}{5\sqrt{2}}\right) = \left(\pm\frac{\sqrt{2}}{5} \pm \frac{3}{10}\right)$$

(c) Hyperbola : $x^2 - 2y^2 = 1$ and $m = -2/3$

$$\frac{x^2}{1} - \frac{y^2}{1/2} = 1$$

$\because \quad a^2 = 1$ and $b^2 = 1/2$

$\therefore$ Equation of normal having slope m, is

$$y = mx \pm \frac{m(a^2 + b^2)}{\sqrt{a^2 - m^2b^2}}$$

$$\Rightarrow \quad y = \frac{-2}{3}x \pm \frac{-2/3(1 + 1/2)}{\sqrt{1 - (-2/3)^2(1/2)}}$$

$$\Rightarrow \quad y = \frac{-2}{3}x \pm \frac{-1}{\sqrt{1 + \frac{2}{9}}}$$

$$\Rightarrow \quad y = \frac{-2}{3}x \mp \frac{3}{\sqrt{11}}$$

$$\Rightarrow \quad 3\sqrt{11}y = -2\sqrt{11}x \mp 9$$

$$\Rightarrow \quad 3\sqrt{11}y + 2\sqrt{11}x \pm 9 = 0$$

(d) Hyperbola : $3x^2 - 4y^2 = 12$

$$\Rightarrow \quad \frac{x^2}{4} - \frac{y^2}{3} = 1$$

$\therefore \quad a^2 = 4$ and $b^2 = 3$

$\because$ Normal is parallel to the line : $x + y - 1 = 0$

$\therefore$ Slope of normal $(m) = -1$

Now, equation of normal having slope m is given by

$$y = mx \pm \frac{m(a^2 + b^2)}{\sqrt{a^2 - m^2b^2}}$$

$$\Rightarrow \quad y = (-1)x \pm \frac{(-1)(4+3)}{\sqrt{4 - (1)(3)}}$$

$$\Rightarrow \quad y = -x \pm \frac{-7}{\sqrt{1}} \Rightarrow y = -x \mp 7$$

$$\Rightarrow \quad x + y \pm \sqrt{7} = 0$$

(e) Hyperbola : $\frac{x^2}{16} - \frac{y^2}{9} = 1$

$\therefore \quad a^2 = 16$ and $b^2 = 9$

$\Rightarrow \quad a = 4$ and $b = 3$

$$\therefore \quad e = \sqrt{1 + \frac{b^2}{a^2}}$$

$$= \sqrt{1 + \frac{9}{16}} = \frac{5}{4}$$

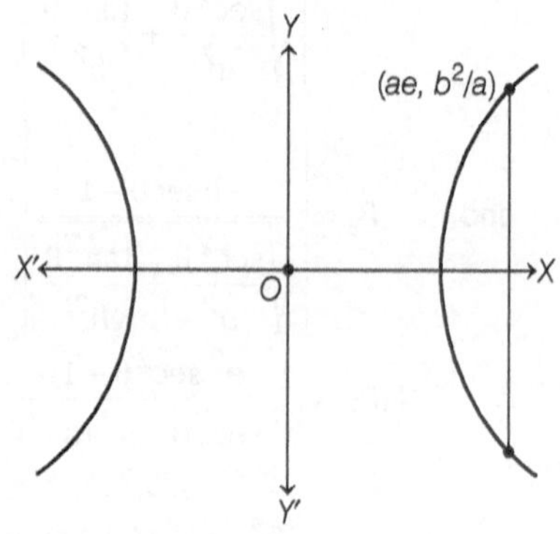

$\therefore$ Positive end of latusrectum $= (ae, b^2/a)$

$$= \left(4 \times 5/4, \frac{9}{4}\right)$$

$$= (5, 9/4)$$

$\therefore$ Equation of normal at (x_1, y_1)

$$\frac{a^2x}{x_1} + \frac{b^2y}{y_1} = a^2 + b^2$$

$$\Rightarrow \quad \frac{16x}{5} + \frac{9y}{9/4} = 16 + 9$$

$$\Rightarrow \quad \frac{16x}{5} + 4y = 25$$

$$\Rightarrow \quad 16x + 20y = 125$$

$\therefore$ Required equation is $16x + 20y - 125 = 0$

14. (a) The equation of the normal at $P(a\sec\theta, b\tan\theta)$ to the given hyperbola is

$$ax\cos\theta + by\cot\theta = a^2 + b^2$$

This meets the transverse axis i.e. X-axis at G

So, the coordinates of G are $\left(\dfrac{a^2+b^2}{a}\sec\theta, 0\right)$

The coordinates of vertices A and A′ are $(a, 0)$ and $(-a, 0)$, respectively.

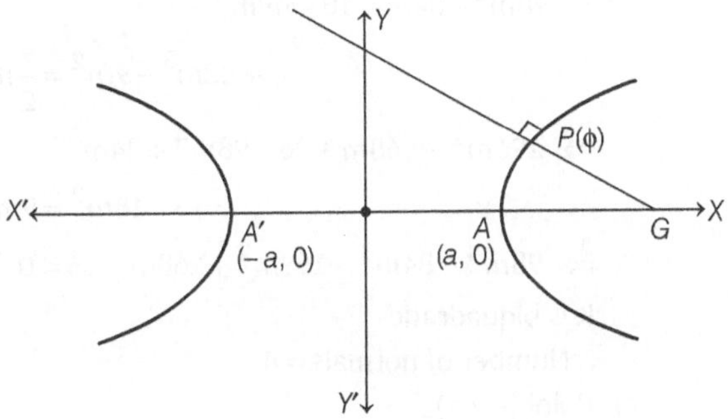

$$\therefore \quad AG : A'G = \left(-a + \frac{a^2+b^2}{a}\sec\theta\right) : \left(a + \frac{b^2+a^2}{a}\sec\theta\right)$$

$$= (-a + ae^2\sec\theta) : (a + ae^2\sec\theta)$$

$$= a^2(e^4\sec^2\theta - 1)$$

(b) Normal at point $P(a\sec\theta_1, b\tan\theta_1)$ is

$$ax\cos\theta_1 + by\cot\theta_1 = a^2 + b^2$$

It meets the conjugate axis at $G_1\left(0, \dfrac{a^2+b^2}{b}\tan\theta_1\right)$

Normal at point $Q(a\sec\theta_2, b\tan\theta_2)$ is

$$ax\cos\theta_2 + by\cot\theta_2 = a^2 + b^2$$

It meets conjugate axis at $G_2\left(0, \dfrac{a^2+b^2}{b}\tan\theta_2\right)$

$$\Rightarrow \quad CG_1 \cdot CG_2 = \frac{(a^2+b^2)^2}{b^2}\tan\theta_1\tan\theta_2$$

$$= \left(\frac{a^2+b^2}{b}\right)^2 \qquad [\because \theta_1 + \theta_2 = \pi/2]$$

$$\Rightarrow \frac{a^4[1+(b^2/a^2)]^2}{b^2} = \frac{a^4e^4}{a^2(e^2-1)}$$

$$= \frac{a^2e^4}{e^2-1}$$

(c) The normal at $P(a\sec\theta, b\tan\theta)$ to the given hyperbola $\dfrac{x^2}{a^2} - \dfrac{y^2}{b^2} = 1$ is

$$ax\cos\theta + by\cot\theta = a^2 + b^2$$

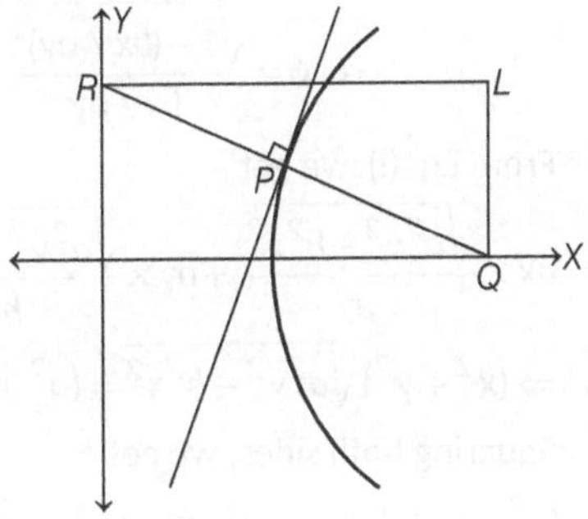

$$\therefore \quad Q = \left(\frac{a^2+b^2}{a}\sec\theta, 0\right)$$

$$R = \left(0, \frac{a^2+b^2}{b}\tan\theta\right)$$

If $L(h, k)$, then

$$\therefore \quad h = \frac{a^2+b^2}{a}\sec\theta \text{ and } k = \frac{a^2+b^2}{b}\tan\theta$$

$$\Rightarrow \sec\theta = \frac{ah}{a^2+b^2} \text{ and } \tan\theta = \frac{kb}{a^2+b^2}$$

$$\because \quad \sec^2\theta - \tan^2\theta = 1$$

$$\Rightarrow \quad \left(\frac{ah}{a^2+b^2}\right)^2 - \left(\frac{kb}{a^2+b^2}\right)^2 = 1$$

$$\Rightarrow \quad a^2h^2 - k^2b^2 = (a^2+b^2)^2$$

$\therefore$ Locus is $a^2x^2 - b^2y^2 = (a^2+b^2)^2$.

(d) The equation of normat at $(a\sec\phi, b\tan\phi)$ to the hyperbola $\dfrac{x^2}{a^2} - \dfrac{y^2}{b^2} = 1$ is

$$ax\cos\phi + by\cot\phi = a^2 + b^2 \qquad \text{... (i)}$$

The equation of any line perpendicular to Eq. (i) and passing through the origin (which is centre) is

$$(b\cot\phi)x - (a\cos\phi)y = 0$$

$$\Rightarrow \quad bx - a\sin\phi y = 0 \qquad \text{... (ii)}$$

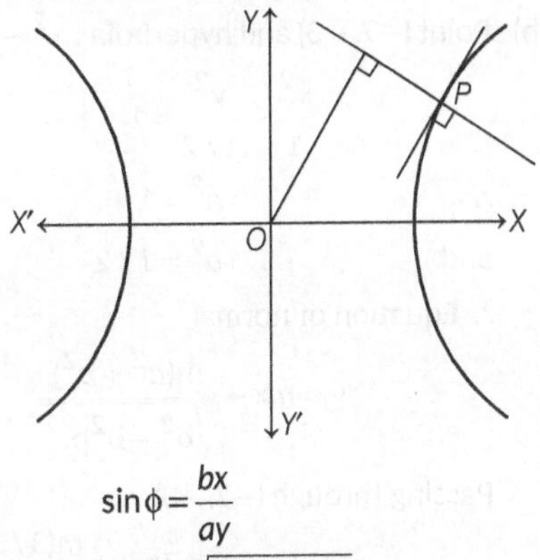

$$\Rightarrow \quad \sin\phi = \frac{bx}{ay}$$

$$\Rightarrow \quad \cos\phi = \sqrt{1 - \left(\frac{bx}{ay}\right)^2}$$

$$\cot\phi = \frac{\sqrt{1-(bx/ay)^2}}{bx/ay}$$

From Eq. (i), we get

$$ax \times \frac{\sqrt{a^2y^2 - b^2x^2}}{ay} + by \times \frac{\sqrt{a^2y^2 - b^2x^2}}{bx} = a^2 + b^2$$

$$\Rightarrow (x^2 + y^2)\sqrt{a^2y^2 - b^2x^2} = (a^2 + b^2)xy$$

Squaring both sides, we get

$$(x^2 + y^2)^2(a^2y^2 - b^2x^2) = (a^2 + b^2)^2x^2y^2$$

which is required locus of the foot of perpendicular.

15. (a) Point (1, 2) and hyperbola : $2x^2 - y^2 = 1$

$$\frac{x^2}{1/2} - \frac{y^2}{1} = 1$$

$\therefore \quad a^2 = 1/2$ and $b^2 = 1$

$\because$ Equation of normal having slope m is given by

$$y = mx - \frac{m(a^2 + b^2)}{\sqrt{a^2 - b^2m^2}}$$

$\because$ It passes through (1, 2)

$$\therefore \quad 2 = m - \frac{m(1/2 + 1)}{\sqrt{1/2 - m^2}}$$

$$\Rightarrow \quad \frac{\frac{3}{2}m}{\frac{\sqrt{1-2m^2}}{\sqrt{2}}} = m - 2 \Rightarrow \frac{3\sqrt{2}\,m}{2\sqrt{1-2m^2}} = m - 2$$

Squaring both the sides, we get

$$18m^2 = 4(m^2 - 4m + 4)(1 - 2m^2)$$

$$\Rightarrow \quad 18m^2 = 4(m^2 - 4m + 4 - 2m^4 + 8m^3 - 8m^2)$$

$$\Rightarrow \quad 8m^4 - 32m^3 + 46m^2 + 16m - 16 = 0$$

$\because$ It is biquadratic.

$\therefore$ Number of normals = 4

(b) Point $(-7, -3)$ and hyperbola : $x^2 - 2y^2 = 1$

$$\frac{x^2}{1} - \frac{y^2}{1/2} = 1$$

$\therefore \quad a^2 = 1$

and $\quad b^2 = 1/2$

$\therefore$ Equation of normal

$$y = mx - \frac{m(a^2 + b^2)}{\sqrt{a^2 - b^2m^2}}$$

Passing through $(-7, -3)$

$$-3 = -7m - \frac{m(3/2)}{\sqrt{1 - \frac{m^2}{2}}}$$

$$\Rightarrow \quad 7m - 3 = \frac{-3m \times \sqrt{2}}{2\sqrt{2 - m^2}}$$

Squaring on both sides, we get

$$(7m - 3)^2 = \frac{18m^2}{4(2 - m^2)}$$

$$\Rightarrow \quad (49m^2 - 42m + 9)(2 - m^2) = \frac{9}{2}m^2$$

$$\Rightarrow 98m^2 - 84m + 18 - 49m^4 + 42m^3 - 9m^2 = \frac{9}{2}m^2$$

$$\Rightarrow \quad 196m^2 - 168m + 36 - 98m^4 + 84m^3 - 18m^2 = 9m^2$$

$$\Rightarrow \quad 98m^4 - 84m^3 - 169m^2 + 168m - 36 = 0$$

It is biquadratic.

$\therefore$ Number of normals = 4

(c) Point $(-2, 1)$

and hyperbola : $2x^2 - y^2 - 7 = 0$

$$\Rightarrow \quad \frac{2x^2}{7} - \frac{y^2}{7} = 1$$

$$\Rightarrow \quad \frac{x^2}{7/2} - \frac{y^2}{7} = 1$$

$\Rightarrow \quad a^2 = 7/2$ and $b^2 = 7$

$\therefore$ Equation of normal

$$y = mx - \frac{m(a^2 + b^2)}{\sqrt{a^2 - m^2b^2}}$$

If passes through $(-2, 1)$.

$$\therefore \quad 1 = -2m - \frac{m(21/2)}{\sqrt{\frac{7}{2} - m^2(7)}}$$

$$\Rightarrow \quad 1 = -2m - \frac{m\left(\frac{21}{2}\right)}{\sqrt{\frac{7}{2}} \cdot \sqrt{1 - 2m^2}}$$

$$\Rightarrow \sqrt{1 - 2m^2}\,(1 + 2m) = -m \times \frac{21}{2} \times \frac{\sqrt{2}}{\sqrt{7}}$$

$$\Rightarrow \sqrt{1 - 2m^2}(1 + 2m) = -m \times \frac{3\sqrt{7}}{\sqrt{2}}$$

Squaring on both sides, we get

$$(1 - 2m^2)(1 + 2m)^2 = m^2 \times \frac{63}{2}$$

$$\Rightarrow \quad (1 + 4m^2 + 4m)(1 - 2m^2) = \frac{63m^2}{2}$$

$$\Rightarrow \quad 18m^4 + 16m^3 + 59m^2 - 8m - 2 = 0$$

$\therefore$ It is biquadratic.

$\therefore$ Number of normals = 4

16. (a) Point $P(1,-2)$ and hyperbola $S: 2x^2-3y^2-12=0$

Now, $S_1 = 2(1)^2 - 3(-2)^2 - 12$

$= 2-12-12 = -22 < 0$

$\therefore$ Point $P(1,-2)$ lies outside the hyperbola.

$\Rightarrow$ Number of tangents $= 2$

(b) Hyperbola : $\frac{x^2}{9}-\frac{y^2}{4}-1=0$

Point (3, 2)

Equation of pair of tangents

$$SS_1 = T^2$$

$$\Rightarrow \left(\frac{x^2}{9}-\frac{y^2}{4}-1\right)\left(\frac{9}{9}-\frac{4}{4}-1\right)=\left(\frac{x(3)}{9}-\frac{y(2)}{4}-1\right)^2$$

$$\Rightarrow \left(\frac{x^2}{9}-\frac{y^2}{4}-1\right)(-1)=\left(\frac{x}{3}-\frac{y}{2}-1\right)^2$$

$$\Rightarrow -\frac{x^2}{9}+\frac{y^2}{4}+1=\frac{x^2}{9}+\frac{y^2}{4}+1-\frac{xy}{3}+y-\frac{2x}{3}$$

$$\Rightarrow \frac{2x^2}{9}-\frac{xy}{3}+y-\frac{2x}{3}=0$$

$$\Rightarrow \frac{x}{3}\left(\frac{2x}{3}-y\right)-1\left(\frac{2x}{3}-y\right)=0$$

$$\Rightarrow \left(\frac{x}{3}-1\right)\left(\frac{2x}{3}-y\right)=0$$

$\therefore$ Tangents are

$$\frac{x}{3}-1=0 \;\Big|\; \frac{2x}{3}-y=0$$

$$x=3 \;\Big|\; \Rightarrow 2x=3y$$

(c) Given, point $P=(4,3)$

Hyperbola $S:=\frac{x^2}{16}-\frac{y^2}{9}-1=0$

$$S_1: \frac{16}{16}-\frac{9}{9}-1=-1<0$$

Point P lies outside the hyperbola.

$\therefore$ Two tangent can be drawn.

Equation of tangents $\Rightarrow SS_1 = T^2$

$$\Rightarrow \left(\frac{x^2}{16}-\frac{y^2}{9}-1\right)(-1)=\left(\frac{4x}{16}-\frac{3y}{9}-1\right)^2$$

$$\Rightarrow -\frac{x^2}{16}+\frac{y^2}{9}+1=\left(\frac{x}{4}-\frac{y}{3}-1\right)^2$$

$$\Rightarrow -\frac{x^2}{16}+\frac{y^2}{9}+1=\frac{x^2}{16}+\frac{y^2}{9}+1-\frac{xy}{6}+\frac{2y}{3}-\frac{x}{2}$$

$$\Rightarrow \frac{2x^2}{16}-\frac{xy}{6}+\frac{2y}{3}-\frac{x}{2}=0$$

$$\Rightarrow \frac{x^2}{8}-\frac{xy}{6}+\frac{2y}{3}-\frac{x}{2}=0$$

$$\Rightarrow 3x^2-4xy-12x+16y=0$$

$\therefore$ $a=3, h=-2$ and $b=0$

$$\therefore \tan\theta = \left|\frac{2\sqrt{h^2-ab}}{a+b}\right| = \frac{4}{3} \Rightarrow \theta = \tan^{-1}\left(\frac{4}{3}\right)$$

(d) Point $(2,-\lambda)$ and hyperbola : $3x^2-4y^2=12$

$$S: \frac{x^2}{4}-\frac{y^2}{3}-1=0$$

Now, point of tangent $SS_1 = T^2$

$$\Rightarrow \left(\frac{x^2}{4}-\frac{y^2}{3}-1\right)\left(1-\frac{\lambda^2}{3}-1\right)=\left(\frac{2x}{4}+\frac{\lambda y}{3}-1\right)^2$$

$$\Rightarrow \left(\frac{x^2}{4}-\frac{y^2}{3}-1\right)\left(\frac{-\lambda^2}{3}\right)$$

$$=\frac{x^2}{4}+\frac{\lambda^2y^2}{9}+1+\frac{\lambda xy}{3}-\frac{2\lambda y}{3}-x$$

$$\Rightarrow \frac{-\lambda^2x^2}{12}+\frac{\lambda^2y^2}{9}+\frac{\lambda^2}{3}=\frac{x^2}{4}+\frac{\lambda^2y^2}{9}+1+\frac{\lambda xy}{3}-\frac{2\lambda y}{3}-x$$

$$\Rightarrow \frac{x^2}{4}+\frac{\lambda^2y^2}{12}+\frac{\lambda xy}{3}-\frac{2\lambda y}{3}-x+1-\frac{\lambda^2}{3}=0$$

$$\therefore \quad a=\left(\frac{1}{4}+\frac{\lambda^2}{12}\right)$$

$\therefore$ Tangents are mutually perpendicular.

$$\therefore \quad a+b=0$$

$$\Rightarrow \frac{1}{4}+\frac{\lambda^2}{12}=0 \Rightarrow \lambda^2=-48$$

No real values of λ.

17. (a) Hyperbola $S: \frac{x^2}{16}-\frac{y^2}{9}-1=0$

Let $P(h,k)$ be the point of intersection of two perpendicular tangents equation of pair of tangents is

$$SS_1 = T^2$$

$$\Rightarrow \left(\frac{x^2}{16}-\frac{y^2}{9}-1\right)\left(\frac{h^2}{16}-\frac{k^2}{9}-1\right)=\left(\frac{hx}{16}-\frac{ky}{9}-1\right)^2$$

$\because$ This represents a pair of straight lines which are mutually perpendicular

$\Rightarrow$ Coefficient of x^2 + Coefficient of $y^2 = 0$

$$\Rightarrow \frac{1}{16}\left(\frac{-k^2}{9}-1\right)-\frac{1}{9}\left(\frac{h^2}{16}-1\right)=0$$

$$\Rightarrow -k^2-9-h^2+16=0$$

$$\Rightarrow h^2+k^2=16-9$$

$\therefore$ Locus is $x^2+y^2=7$

(b) Hyperbola : $\frac{x^2}{a^2}-\frac{y^2}{b^2}-1=0$

Let the point of intersection is $P(h, k)$

$\therefore$ Equation of pair of tangents $SS_1 = T^2$

$$\Rightarrow \left(\frac{x^2}{a^2}-\frac{y^2}{b^2}-1\right)\left(\frac{h^2}{a^2}-\frac{k^2}{b^2}-1\right)=\left(\frac{xh}{a^2}-\frac{ky}{b^2}-1\right)^2$$

$$\Rightarrow \frac{x^2h^2}{a^4}-\frac{x^2k^2}{a^2b^2}-\frac{x^2}{a^2}$$

$$-\frac{y^2h^2}{a^2b^2}+\frac{k^2y^2}{b^4}+\frac{y^2}{b^2}-\frac{h^2}{a^2}+\frac{k^2}{b^2}+1$$

$$=\frac{x^2h^2}{a^4}+\frac{k^2y^2}{b^4}+1-\frac{2hkxy}{a^2b^2}+\frac{2ky}{b^2}-\frac{2hx}{a^2}$$

$$\Rightarrow \frac{x^2k^2}{a^2b^2}+\frac{x^2}{a^2}+\frac{y^2h^2}{a^2b^2}-\frac{y^2}{b^2}$$

$$+\frac{h^2}{a^2}-\frac{k^2}{b^2}-\frac{2hkxy}{a^2b^2}+\frac{2ky}{b^2}-\frac{2hx}{a^2}=0$$

$$\Rightarrow \frac{x^2}{a^2}\left(\frac{k^2}{b^2}+1\right)+\frac{y^2}{b^2}\left(\frac{h^2}{a^2}-1\right)$$

$$+xy\left(\frac{-2kh}{a^2b^2}\right)+y\left(\frac{2k}{b^2}\right)+x\left(\frac{-2h}{a^2}\right)+\frac{h^2}{a^2}-\frac{k^2}{b^2}=1$$

$\because$ Sum of slopes $= P$

$$\therefore \quad \frac{-2H}{B}=P \quad \left\{\because H=\frac{-kh}{a^2b^2} \text{ and } B=\frac{1}{b^2}\left(\frac{h^2}{a^2}-1\right)\right\}$$

$$\Rightarrow \frac{\frac{2kh}{a^2b^2}}{\frac{1}{b^2}\left(\frac{h^2-a^2}{a^2}\right)}=P \Rightarrow 2kh=P(h^2-a^2)$$

$\therefore$ Locus is $2xy = P(x^2-a^2)$

18. (a) Hyperbola : $\frac{x^2}{9}-\frac{y^2}{4}-1=0$

Point $P(2, 3)$.

$\because$ Equation of chord of contact is

$$T=0$$

$$\Rightarrow \frac{x(2)}{9}-\frac{y(3)}{4}-1=0$$

$$\Rightarrow \frac{2x}{9}-\frac{3y}{4}-1=0$$

$$\Rightarrow 8x-27y=36$$

(b) Hyperbola : $\frac{x^2}{16}-\frac{y^2}{9}-1=0$

$\because$ Given, PQ and PR are tangents, QR will be chord of contact and equation of QR

$$\Rightarrow x-2y=1 \quad \text{(given)}$$

$$\Rightarrow x-2y-1=0 \quad \text{...(i)}$$

Let $P(h, k)$ be the point

$\therefore$ Equation of chord of contact w.r.t. P is

$$T=0$$

$$\Rightarrow \frac{hx}{16}-\frac{ky}{9}-1=0 \quad \text{... (ii)}$$

$\because$ Eqs. (i) and (ii) are identical.

$$\therefore \quad \frac{1}{h/16}=\frac{-2}{-k/9}=\frac{-1}{-1}$$

$$\Rightarrow \frac{16}{h}=1 \Rightarrow h=16 \quad \Bigg| \quad \frac{18}{k}=1 \Rightarrow k=18$$

(c) Given, line : $x+y+3=0$

and hyperbola : $x^2-2y^2-1=0$

Let $P(h, -3-h)$ be the any point on the given line

Now, according to question,

Equation of chord of contact at P

$$T=0$$

$$\Rightarrow x(h)-2y(-3-h)-1=0$$

$$\Rightarrow xh+6y+2hy-1=0$$

$$\Rightarrow h(x+2y)+(6y-1)=0$$

$\therefore$ Fixed point is $\Rightarrow x+2y=0$ and $6y-1=0$

$$x=-2y \text{ and } y=1/6$$

$\therefore$ Point is $\left(-\frac{1}{3}, \frac{1}{6}\right)$

19. (a) Hyperbola : $25x^2-16y^2-400=0$

Equation of chord whose middle point is (5, 3)

$$T=S_1 \Rightarrow 0$$

$$\because \quad S_1=25(5^2)-16(3)^2-400$$

$$=625-144-400=81$$

$$T=25(5)x-16(3)y-400$$

$$=125x-48y-400$$

$\therefore$ Required chord

$$125x-48y-400=81$$

$$\Rightarrow 125x-48y=481$$

(b) Chord of contact from external point (3, 2) w.r.t. hyperbola $\frac{x^2}{25}-\frac{y^2}{16}=1$ is given by

$$T=0$$

$$\Rightarrow \frac{3(x)}{25}-\frac{2y}{16}=1 \Rightarrow \frac{3x}{25}-\frac{y}{8}=1 \quad \text{... (i)}$$

Let $P(h, k)$ be the middle point of chord.

$\therefore$ Equation of chord $\Rightarrow T=S_1$

$$\Rightarrow \frac{hx}{25}-\frac{ky}{16}-1=\frac{h^2}{25}-\frac{k^2}{16}-1$$

$$\Rightarrow \frac{hx}{25}-\frac{ky}{16}=\frac{h^2}{25}-\frac{k^2}{16} \quad \text{... (ii)}$$

$\because$ Eqs. (i) and (ii) are identical

$$\frac{3/25}{h/25}=\frac{-1/8}{-k/16}=\frac{1}{\frac{h^2}{25}-\frac{k^2}{16}}$$

$$\Rightarrow \quad \frac{3}{h}=\frac{2}{k}=\frac{1}{\frac{h^2}{25}-\frac{k^2}{16}}$$

$$\Rightarrow \frac{2}{k}=\frac{1}{\frac{h^2}{25}-\frac{k^2}{16}}$$

$$\Rightarrow \quad \frac{3}{h}=\frac{2}{k} \quad \Bigg| \quad \Rightarrow \frac{2h^2}{25}-\frac{k^2}{8}=k$$

$$\Rightarrow 3k=2h \quad \Bigg| \quad \Rightarrow \frac{2}{25}(3k/2)^2-\frac{k^2}{8}=k$$

$$\Rightarrow h=\frac{3k}{2}$$

$$\Rightarrow \quad \frac{9k^2}{50}-\frac{k^2}{8}=k$$

$$\Rightarrow \quad \frac{36k^2-25k^2}{200}=k$$

$$\Rightarrow \quad 11k^2=200k$$

$$\Rightarrow \quad k(11k-200)=0$$

$$k=0 \text{ or } k=\frac{200}{11}$$

i.e. $\quad h=0 \text{ or } h=\frac{200}{11}\times\frac{3}{2}=\frac{300}{11}$

$\therefore$ Points are (0, 0) and $\left(\frac{300}{11},\frac{200}{11}\right)$.

(c) Hyperbola : $x^2-4y^2=1$ ...(i)

Mid-point $P(3,1)$

$\therefore$ Equation of chord

$$T=S_1$$

$$\Rightarrow \quad x(3)-4y(1)-1=3^2-4(1)^2-1$$

$$\Rightarrow \quad 3x-4y=5$$

$$\Rightarrow \quad x=\frac{4y+5}{3} \qquad \text{...(ii)}$$

Put $x=\frac{4y+5}{3}$ in Eq. (i), we get

$$\left(\frac{4y+5}{3}\right)^2-4y^2=1$$

$$\Rightarrow \quad 16y^2+25+40y-12y^2=3$$

$$\Rightarrow \quad 4y^2+40y+22=0$$

$$\Rightarrow \quad 2y^2+20y+11=0$$

$$\Rightarrow \quad y_1+y_2=-\frac{20}{2}=-10$$

$$y_1\cdot y_2=\frac{11}{2}$$

$$y_1\cdot y_2=\sqrt{(y_1+y_2)^2=4y_1y_2}$$

$$=\sqrt{(-10)^2-4\left(\frac{11}{2}\right)}$$

$$=\sqrt{100-22}=\sqrt{78}$$

$$x_1-x_2=\frac{1}{3}(4(y_1-y_2)+5)$$

$$=\frac{1}{3}(4\sqrt{78}+5)$$

$$\text{Length of chord}=\sqrt{(x_1-x_2)^2+(y_1-y_2)^2}$$

$$=\sqrt{\frac{1}{9}(4\sqrt{78}+5)^2+78}$$

$$=\frac{1}{3}\sqrt{1248+25+40\sqrt{78}+702}$$

$$=\frac{1}{3}\sqrt{1950+40\sqrt{78}}$$

20. (a) $\because$ The equation of chord with mid-point (x_1, y_1) is

$$\frac{xx_1}{a^2}-\frac{yy_1}{b^2}=\frac{x_1^2}{a^2}-\frac{y_1^2}{b^2} \qquad \text{... (i)}$$

Let the equation of the chord provided in Eq. (i) is normal to the hyperbola at $P(a\sec\theta, b\tan\theta)$ whose equation is

$$ax\cos\theta+by\cot\theta=a^2+b^2 \qquad \text{... (ii)}$$

$\because$ Eqs. (i) and (ii) are identical.

$$\therefore \frac{x_1/a^2}{a\cos\theta}=\frac{-y_1/b^2}{b\cot\theta}=\frac{(x_1^2/a^2)-(y_1^2/b^2)}{a^2+b^2}$$

$$\Rightarrow \quad \sec\theta=\frac{a^3}{x_1(a^2+b^2)}\{(x_1^2/a^2)-(y_1^2/b^2)\}$$

$$\text{and} \quad \tan\theta=-\frac{b^3}{y_1(a^2+b^2)}\left(\frac{x_1^2}{a^2}-\frac{y_1^2}{b^2}\right)$$

$$\because \quad 1+\tan^2\theta=\sec^2\theta$$

$$\Rightarrow \quad 1=\sec^2\theta-\tan^2\theta$$

$$\Rightarrow \quad 1=\frac{a^6}{x_1^2(a^2+b^2)^2}\left(\frac{x_1^2}{a^2}-\frac{y_1^2}{b^2}\right)^2 -\frac{b^6}{y_1^2(a^2+b^2)^2}\{(x_1^2/a^2)-(y_1^2/b^2)\}^2$$

$$\Rightarrow \quad 1=\frac{(b^2x_1^2-a^2y_1^2)^2(a^6y_1^2-b^6x_1^2)}{a^4b^4(a^2+b^2)^2x_1^2y_1^2}$$

$\therefore$ Locus is $a^4b^4(a^2+b^2)^2x^2y^2$

$$=(b^2x^2-a^2y^2)^2(a^6y^2-b^6x^2)$$

(b) Given, curve is $ax^2-by^2=1$

$$\Rightarrow \quad \frac{x^2}{1/a}-\frac{y^2}{1/b}=1$$

The equation of chord PQ in terms of mid-point (h, k) is

$$ahx - bky = ah^2 - bk^2 \qquad [\because T = S_1]$$

$$\Rightarrow \quad \frac{ahx - bky}{ah^2 - bk^2} = 1$$

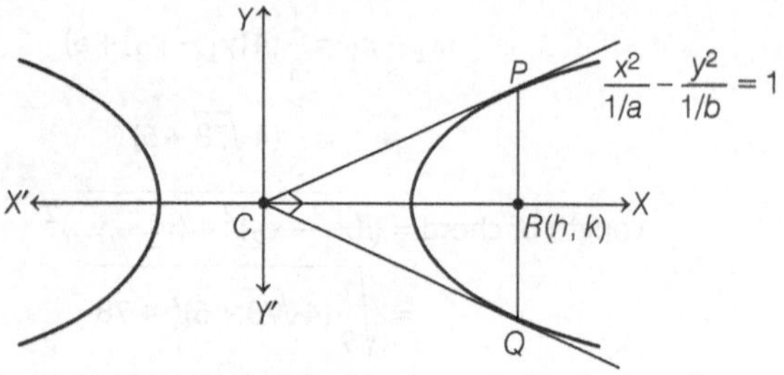

Now, the combined equation of the pair of line CP and CQ (where C is origin) is

$$ax^2 - by^2 - 1\left[\frac{ahx - bky}{ah^2 - bk^2}\right]^2 = 0$$

$$\because \quad \angle PCQ = 90^\circ$$

$\therefore$ Coefficient of x^2 + Coefficient of $y^2 = 0$

$$a - \frac{a^2h^2}{(ah^2 - bk^2)^2} + \left(-b - \frac{b^2k^2}{(ah^2 - bk^2)^2}\right] = 0$$

$$\Rightarrow \quad a - b = \frac{a^2h^2 + b^2k^2}{(ah^2 - bk^2)^2}$$

$$\Rightarrow \quad (a - b)(ah^2 - bk^2)^2 = a^2h^2 + b^2k^2$$

$\therefore$ Locus is $(a - b)(ax^2 - by^2)^2 = a^2x^2 + b^2y^2$

(c) If (h, k) be the mid-point of the chord of the hyperbola $\frac{x^2}{a^2} - \frac{y^2}{b^2} = 1$, then $T = S_1$

$$\Rightarrow \quad \frac{hx}{a^2} - \frac{ky}{b^2} = \frac{h^2}{a^2} - \frac{k^2}{b^2}$$

which is passing through the focus $(ae, 0)$ of the given hyperbola

$$\Rightarrow \quad \frac{aeh}{a^2} = \frac{h^2}{a^2} - \frac{k^2}{b^2} \Rightarrow \frac{h^2}{a^2} - \frac{k^2}{b^2} = \frac{eh}{a}$$

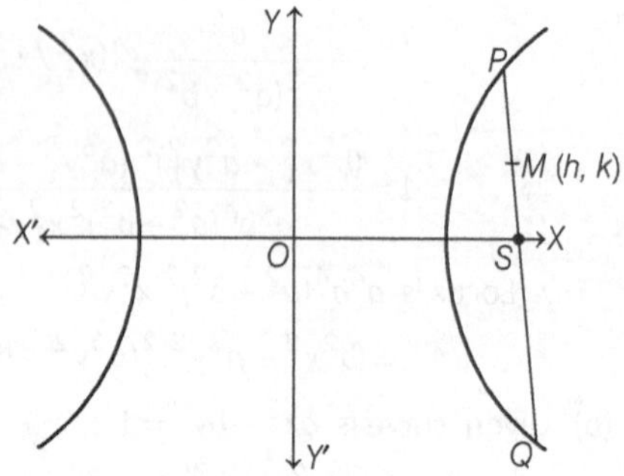

$\therefore$ Locus of (h, k) is

$$\frac{x^2}{a^2} - \frac{y^2}{b^2} = \frac{ex}{a}$$

(d) Let the line $x + y = b$ bisect the chord at $P(\alpha, b - \alpha)$. Therefore, the equation of chord whose mid-point is $P(\alpha, b - \alpha)$ is given by

$$\frac{x\alpha}{2a^2} - \frac{y(b - \alpha)}{2b^2} = \frac{\alpha^2}{2a^2} - \frac{(b - \alpha)^2}{2b^2}$$

$\because$ It passes through (a, b)

$$\Rightarrow \quad \frac{a\alpha}{2a^2} - \frac{b(b - \alpha)}{2b^2} = \frac{\alpha^2}{2a^2} - \frac{(b - \alpha)^2}{2b^2}$$

$$\Rightarrow \quad \alpha^2\left(\frac{1}{a^2} - \frac{1}{b^2}\right) + \alpha\left(\frac{1}{b} - \frac{1}{a}\right) = 0$$

$$\alpha = 0$$

and
$$\alpha = \frac{1}{1/a + 1/b}$$

$\therefore$ The required condition is $a \neq -b$

21. (a) The given line is $3x + 4y + 10 = 0$

$$y = \frac{-3}{4}x + \left(\frac{-5}{2}\right) \qquad \text{... (i)}$$

Hence, the equation of the diameter corresponds to the line (i) is

$$y = \frac{b^2x}{a^2m} = \frac{4x}{9(-3/4)} = \frac{-16}{27}x$$

$$\Rightarrow \quad 16x + 27y = 0$$

(b) The slope of given chord is $m = \frac{-2}{3}$

$\therefore$ The equation of the diameter parallel to the given chord is

$$y = \frac{-b^2x}{a^2m} = \frac{-4x}{9(-2/3)} = \frac{-2}{3}x$$

$$\Rightarrow \quad 3y + 2x = 0$$

(c) $\because$ The equation of the chord is

$$7x + y = 2$$

$$\Rightarrow \quad y = -7x + 2$$

$\therefore$ The equation of the diameters is

$$y = \frac{b^2x}{a^2m} = \frac{7x}{3(-7)} = \frac{-x}{3} \Rightarrow x + 3y = 0$$

(d) Let the equation of the diameter, which is conjugate to $x = 2y$ is

$$y = m_1x$$

As we know that two diameters $y = \left(\frac{1}{2}\right)x$ and $y = m_1x$ are conjugates, if

$$m_1m_2 = \frac{b^2}{a^2}$$

$$\Rightarrow \quad m_1 \times \frac{1}{2} = \frac{16}{9} \Rightarrow m_1 = \frac{32}{9}$$

$\therefore$ The equation of the conjugate diameter is

$$y = \frac{32}{9}x \Rightarrow 32x - 9y = 0$$

22. (a) Hyperbola : $9x^2-16y^2=144$

$$\Rightarrow \quad \frac{x^2}{16}-\frac{y^2}{9}=1$$

$$\therefore \quad a^2=16 \text{ and } b^2=9$$

$\therefore$ Equation of director circle is

$$x^2+y^2=a^2-b^2$$

$$\Rightarrow \quad x^2+y^2=16-9 \Rightarrow x^2+y^2=7$$

(b) $\because$ Equation of hyperbola is

$$\frac{x^2}{a^2}-\frac{y^2}{b^2}=1 \quad \text{... (i)}$$

$\therefore$ Equation of tangent at $(a\sec\theta, b\tan\theta)$ is

$$\frac{x}{a}\sec\theta-\frac{y}{b}\tan\theta=1 \quad \text{... (ii)}$$

$\therefore$ Equation of director circle of Eq. (i) is

$$x^2+y^2=a^2-b^2 \quad \text{... (iii)}$$

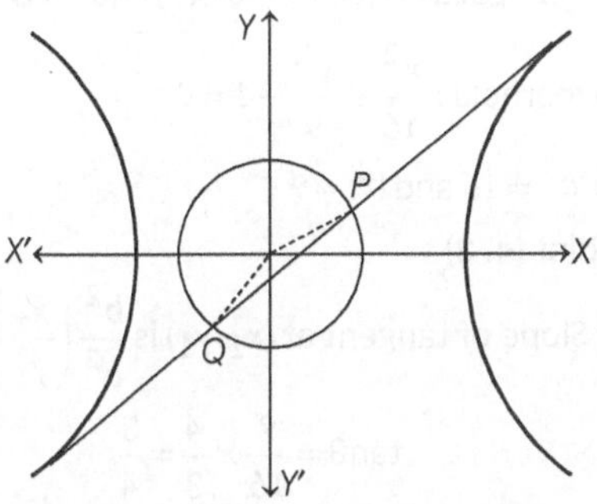

Equation of lines joining the point of intersection of Eqs. (i) and (ii) to the origin is obtained by making Eq. (iii) homogeneous with the help of Eq. (ii).

$$\therefore x^2+y^2=(a^2-b^2)\left(\frac{x}{a}\sec\theta-\frac{y}{b}\tan\theta\right)^2$$

$$\Rightarrow x^2\left(1-\frac{(a^2-b^2)\sec^2\theta}{a^2}\right)+y^2\left(1-\frac{(a^2-b^2)\tan^2\theta}{b^2}\right)+\frac{2(a^2-b^2)\sec\theta\tan\theta}{ab}xy=0 \quad \text{... (iv)}$$

Let m_1 and m_2 are the slope of lines given by Eq. (iv).

$$\therefore \quad m_1m_2=\frac{a}{b}=\frac{1-\frac{(a^2-b^2)\sec^2\theta}{a^2}}{1-\frac{(a^2-b^2)\tan^2\theta}{b^2}}$$

$$=\frac{b^2}{a^2}\left\{\frac{a^2-(a^2-b^2)\sec^2\theta}{b^2-(a^2-b^2)\tan^2\theta}\right\}$$

$$=\frac{b^2}{a^2}\left[\frac{b^2\sec^2\theta-a^2\tan^2\theta}{b^2\sec^2\theta-a^2\tan^2\theta}\right]=\frac{b^2}{a^2}$$

23. (a) Hyperbola : $\frac{x^2}{4}-\frac{y^2}{3}=1$

$$\therefore \quad a^2=4 \text{ and } b^2=3$$

$\because$ Equation of auxiliary circle is

$$x^2+y^2=a^2$$

$$\Rightarrow \quad x^2+y^2=4$$

(b) The tangent at (α, β) is

$$\frac{x\alpha}{a^2}-\frac{y\beta}{b^2}=1 \quad \text{... (i)}$$

and auxiliary circle is

$$x^2+y^2=a^2 \quad \text{... (ii)}$$

From Eq. (i), $x=\frac{a^2}{\alpha}\left(1+\frac{y\beta}{b^2}\right)$ put in Eq. (ii),

we get

$$\frac{a^4}{\alpha^2}\left(1+\frac{y\beta}{b^2}\right)^2+y^2=a^2$$

$$\Rightarrow \quad \frac{a^4[b^4+2\beta b^2y+\beta^2y^2]}{\alpha^2b^4}+y^2=a^2$$

$$\Rightarrow \quad (a^4\beta^2+\alpha^2b^4)y^2+2\beta a^4b^2y-a^4b^4+a^2\alpha^2b^4=0$$

Let y_1 and y_2 are roots of this quadratic equation.

Then, $\quad y_1+y_2=\frac{-2\beta a^4b^2}{a^4\beta^2+\alpha^2b^4}$

and $\quad y_1y_2=\frac{b^4a^2(a^2-\alpha^2)}{a^4\beta^2+\alpha^2b^4}$

$$\therefore \quad \frac{1}{y_1}+\frac{1}{y_2}=\frac{y_1+y_2}{y_1y_2}$$

$$=\frac{-2\beta a^4b^2}{b^4a^2(a^2-\alpha^2)}$$

$$=\frac{-2\beta a^2}{b^2(a^2-\alpha^2)}$$

$$=\frac{-2\beta}{\frac{b^2}{a^2}(a^2-\alpha^2)}$$

$$=\frac{-2\beta}{-\beta^2}$$

$$\left[\because \frac{\alpha^2}{a^2}-\frac{\beta^2}{b^2}=1, \frac{-\beta^2}{b^2}=1-\frac{\alpha^2}{a^2}=\frac{(a^2-\alpha^2)}{a^2}\right]$$

$$=\frac{2}{\beta}$$

$\therefore$ β is harmonic mean between y_1 and y_2.

24. (a) The equation of polar w.r.t. the hyperbola

$\frac{x^2}{a^2}-\frac{y^2}{b^2}=1$ is $\frac{xx_1}{a^2}-\frac{yy_1}{b^2}=1$

$\therefore$ w.r.t. pole (focus) $(-ae, 0)$ is

$$\frac{x(-ae)}{a^2}-\frac{y(0)}{b^2}=1$$

$$\Rightarrow \quad x=-a/e$$

(b) Let (h, k) be the pole of the hyperbola. Then, the equation of the polar from a point (h, k) w.r.t. the hyperbola $x^2-3y^2=3$ is

$$x(h)-3y(k)=3$$

$$\Rightarrow \quad hx-3ky=3 \quad \text{... (i)}$$

Given, equation of polar is

$$x-y=3 \quad \text{... (ii)}$$

$\therefore$ By Eqs. (i) and (ii).

$$\frac{h}{1}=\frac{-3k}{-1}=\frac{3}{3}$$

$$\therefore \quad h=1 \text{ and } k=1/3$$

$\therefore$ Pole is $(1, 1/3)$.

(c) The equation of polar of points (x_1, y_1) and (x_2, y_2) to the hyperbola $\frac{x^2}{a^2}-\frac{y^2}{b^2}=1$ are

$$\frac{xx_1}{a^2}-\frac{yy_1}{b^2}=1 \quad \text{... (i)}$$

and

$$\frac{xx_2}{a^2}-\frac{yy_2}{b^2}=1 \quad \text{... (ii)}$$

$\therefore$ Slope of Eqs. (i) and Eq. (ii) are

$$m_1=\frac{b^2x_1}{a^2y_1} \text{ and } m_2=\frac{b^2x_2}{a^2y_2}$$

$\because$ Polars of the given points are perpendicular

$$\therefore \quad m_1 \times m_2=-1$$

$$\Rightarrow \quad \left(\frac{b^2x_1}{a^2y_1}\right)\left(\frac{b^2x_2}{a^2y_2}\right)=-1$$

$$\Rightarrow \quad \frac{x_1x_2}{y_1y_2}=\frac{-a^4}{b^4}$$

$$\Rightarrow \quad \frac{x_1x_2}{y_1y_2}+\frac{a^4}{b^4}=0$$

(d) Let the pole be (h, k).

The equation of the polar from the point (h, k) w.r.t. the hyperbola $\frac{x^2}{a^2}-\frac{y^2}{b^2}=1$ is

$$\Rightarrow \quad \frac{hx}{a^2}-\frac{ky}{b^2}=1 \quad \text{... (i)}$$

The equation of the normal chord of the given hyperbola at θ is

$$ax\cos\theta-by\cot\theta=a^2+b^2 \quad \text{... (ii)}$$

$\because$ Eqs. (i) and (ii) are identical.

$$\frac{a\cos\theta}{h/a^2}=\frac{b\cot\theta}{k/b^2}=\frac{a^2+b^2}{1}$$

$$\cos\theta=(a^2+b^2)\frac{h}{a^3}$$

$$\Rightarrow \quad \sec\theta=\frac{a^3}{h(a^2+b^2)}$$

$$\cot\theta=\frac{k}{b^3}(a^2+b^2)$$

$$\Rightarrow \quad \tan\theta=\frac{b^3}{k(a^2+b^2)}$$

$\because \sec^2\theta-\tan^2\theta=1$

$$\Rightarrow \quad \frac{a^6}{h^2(a^2+b^2)^2}-\frac{b^6}{k^2(a^2+b^2)^2}=1$$

$\therefore$ Locus is $(a^6y^2-b^6x^2)=(a^2+b^2)^2(x^2y^2)$

25. Hyperbola : $\frac{x^2}{16}-\frac{y^2}{9}-1=0$

$\therefore a^2=16$ and $b^2=9$

Point (4, 3)

$\because$ Slope of tangent at (x_1, y_1) is $\frac{b^2}{a^2}\left(\frac{x_1}{y_1}\right)$

$$\Rightarrow \quad \tan\theta=\frac{9}{16}\times\frac{4}{3}=\frac{3}{4}$$

$$\therefore \quad \sin\theta=3/5 \text{ and } \cos\theta=4/5$$

Length of tangent $=|y_1\operatorname{cosec}\theta|=|3\times5/3|=5$

Length of normal $=|y_1\sec\theta|=|3\times5/4|=15/4$

Length of subtangent $=|y_1\cot\theta|=\left|3\times\frac{4}{3}\right|=4$

Length of subnormal $=|y_1\tan\theta|=|3\times3/4|=9/4$

26. (a) Equations of the asymptotes to the curve $xy-2x-3y=0$ is

$$xy-2x-3y+\lambda=0$$

where, λ is any constant such that it represents two straight lines.

Therefore, $abc+2fgh-af^2-bg^2-ch^2=0$

$$\Rightarrow 0+2\times\left(\frac{-3}{2}\right)\times(-1)\times\left(\frac{1}{2}\right)-0-0-\lambda\left(\frac{1}{2}\right)^2=0$$

$$\Rightarrow \quad \lambda=6$$

Hence, the required asymptotes are

$$xy-2x-3y+6=0$$

$$\Rightarrow \quad (y-2)(x-3)=0$$

$$\Rightarrow \quad y=2 \text{ and } x=3$$

(b) The equations of the asymptotes of the given curve is

$$3x^2+10xy+8y^2+14x+22y+\lambda=0$$

Where λ is any constant such that it represents two straight lines,

Therefore,

$$abc+2fgh-af^2-bg^2-ch^2=0$$

$\Rightarrow 3\cdot 8\cdot \lambda + 2\cdot 7\cdot 11\cdot 5 - 3\cdot 121 - 8\cdot 49 - \lambda\cdot 25 = 0$

$\Rightarrow \quad 24\lambda + 770 - 363 - 392 - 25\lambda = 0$

$\Rightarrow \quad \lambda = 15$

Hence, the combined equation of the given asymptotes is $3x^2+10xy+8y^2+14x+22y+15=0$

(c) Since, the asymptotes are perpendicular to each other, so the hyperbola is rectangular, hence, its eccentricity is $\sqrt{2}$.

(d) The combined equation of the asymptotes is

$$(2x-y-3)(3x+y-7)=0$$

$\Rightarrow 6x^2-xy-y^2-23x+4y+21=0$

Let the equation of the hyperbola be

$$6x^2-xy-y^2-23x+4y+\lambda=0$$

where, λ is any constant such that it represents two straight line which passes through (1, 1), so $\lambda=15$.

Hence, the equation of the hyperbola becomes

$$6x^2-xy-y^2-23x+4y+15=0$$

(e) The combined equation of the asymptotes parallel to the lines

$2x+3y=0$ and $3x+2y=0$ is

$(2x+3y+\lambda)(3x+2y+\mu)=0$

which is passing through (1, 2).

Therefore, $(2\cdot 1+3\cdot 2+\lambda)(3\cdot 1+2\cdot 2+\mu)=0$

$\Rightarrow \quad (8+\lambda)(7+\mu)=0$

$\Rightarrow \quad \lambda=-8$ and $\mu=-7$

Thus, the combined equation of the asymptotes is

$$(2x+3y-8)(3x+2y-7)=0$$

Let the equation of the hyperbola be

$$(2x+3y-8)(3x+2y-7)+v=0$$

which is passing through (5, 3). So,

$$(2\cdot 5+3\cdot 3-8)(3\cdot 5+2\cdot 3-7)+v=0$$

$\Rightarrow \quad 11\times 14+v=0$

$\Rightarrow \quad v=-154$

Hence, the equation of the hyperbola is

$$(2x+3y-8)(3x+2y-7)-154=0$$

(f) The equation of the given hyperbola is

$$x^2-2y^2=2$$

So, the equation of its asymptotes are

$$x-\sqrt{2}\,y=0 \text{ and } x+\sqrt{2}\,y=0.$$

Let any point on the hyperbola be $P(\sqrt{2}\sec\theta, \tan\theta)$.

Let PM and PN are two perpendiculars from the point P to the asymptotes.

Then, $PM\cdot PN$

$$=\left|\frac{\sqrt{2}\sec\theta-\sqrt{2}\tan\theta}{\sqrt{1+2}}\right|\times\left(\frac{\sqrt{2}\sec\theta+\sqrt{2}\tan\theta}{\sqrt{1+2}}\right)$$

$$=\frac{2}{3}\times(\sec\theta-\tan\theta)\times(\sec\theta+\tan\theta)$$

$$=\frac{2}{3}\times(\sec^2\theta-\tan^2\theta)$$

$$=\frac{2}{3}$$

(g) The equation of the given hyperbola is $\frac{x^2}{9}-\frac{y^2}{4}=1$, thus, the equation of the asymptotes are

$$\left(\frac{x}{3}+\frac{y}{2}\right)\left(\frac{x}{3}-\frac{y}{2}\right)=0$$

$\Rightarrow \quad \left(\frac{x}{3}+\frac{y}{2}\right)=0$ and $\left(\frac{x}{3}-\frac{y}{2}\right)=0$

$\Rightarrow \quad 2x+3y=0$ and $2x-3y=0$

The equation of any tangent to the hyperbola $\frac{x^2}{9}-\frac{y^2}{4}=1$ is

$$\frac{x}{3}\sec\theta-\frac{y}{2}\tan\theta=1$$

Let the points of intersection of $2x+3y=0$, $2x-3y=0$ and $\frac{x}{3}\sec\theta-\frac{y}{2}\tan\theta=1$ are, O, P and Q, respectively.

Therefore, $O=(0,0)$

$$P=\left(\frac{3}{\sec\theta+\tan\theta}, \frac{-2}{\sec\theta+\tan\theta}\right)$$

and $Q=\left(\frac{3}{\sec\theta-\tan\theta}, \frac{2}{\sec\theta-\tan\theta}\right)$

Hence, the area of ΔOPQ,

$$=\frac{1}{2}\begin{vmatrix} 0 & 0 & 1 \\ \frac{3}{\sec\theta+\tan\theta} & \frac{2}{\sec\theta+\tan\theta} & 1 \\ \frac{3}{\sec\theta-\tan\theta} & \frac{2}{\sec\theta-\tan\theta} & 1 \end{vmatrix}$$

$$=\frac{1}{2}(6+6)=6 \text{ sq units}$$

(h) Let the point P be $(a\sec\theta, b\tan\theta)$.

The equations of the asymptotes of the given hyperbola $\frac{x^2}{a^2}-\frac{y^2}{b^2}=1$ are $bx-ay=0$ and $bx+ay=0$

Let PM and PN be two perpendiculars from the point P to the transverse axis and the asymptote.

$$bx-ay=0$$

Thus, $\quad PM=b\tan\theta$

and $$PN = \left| \frac{ab\sec\theta - ab\tan\theta}{\sqrt{b^2 + a^2}} \right|$$

It is given that, $PM = PN$

$$\Rightarrow \quad b\tan\theta = \left| \frac{ab\sec\theta - ab\tan\theta}{\sqrt{b^2 + a^2}} \right|$$

$$\Rightarrow \quad \tan\theta = a\left| \frac{\sec\theta - \tan\theta}{\sqrt{b^2 + a^2}} \right|$$

(i) The equations of the asymptotes of the given hyperbola

$$\frac{x^2}{a^2} - \frac{y^2}{b^2} = 1 \text{ are}$$

$$bx - ay = 0 \text{ and } bx + ay = 0$$

Let P be any point on the given hyperbola be $(a\sec\theta, b\tan\theta)$.

It is given that P_1 and P_2 be the lengths of perpendiculars from the point P to the asymptotes

$$bx - ay = 0 \text{ and } bx + ay = 0$$

Thus, $$P_1 = \left| \frac{ab\sec\theta - ab\tan\theta}{\sqrt{b^2 + a^2}} \right|$$

and $$P_2 = \left| \frac{ab\sec\theta + ab\tan\theta}{\sqrt{b^2 + a^2}} \right|$$

Therefore, $\frac{1}{P_1P_2}$

$$= \left(\frac{a^2 + b^2}{(ab\sec\theta - ab\tan\theta)(ab\sec\theta + ab\tan\theta)} \right)$$

$$= \left(\frac{a^2 + b^2}{a^2b^2} \right) = \frac{1}{a^2} + \frac{1}{b^2}$$

27. (a) The given rectangular hyperbola is

$$xy = 18 \quad \text{... (i)}$$

Replacing x by $x\cos(45°) + y\sin(45°)$ and y by $-x\sin(45°) + y\cos(45°)$ in Eq. (i), we get

$$\left(\frac{x}{\sqrt{2}} + \frac{y}{\sqrt{2}} \right)\left(-\frac{x}{\sqrt{2}} + \frac{y}{\sqrt{2}} \right) = 18$$

$$\Rightarrow \quad x^2 - y^2 = -36$$

Hence, the length of the transverse axis

$$= 2a = 2 \cdot 6 = 12 \text{ units}$$

(b) We have, $e_1 = \sqrt{2}$ and $e_2 = \sqrt{2}$

Now, $(e_1 + e_2)^2 = (\sqrt{2} + \sqrt{2})^2 \Rightarrow (2\sqrt{2})^2 = 8$

(c) Let the points A, B, C be $\left(ct_1, \frac{c}{t_1} \right), \left(ct_2, \frac{c}{t_2} \right), \left(ct_3, \frac{c}{t_3} \right)$, respectively, then the area of the ΔABC

$$= \frac{1}{2} \begin{vmatrix} ct_1 & \frac{c}{t_1} & 1 \\ ct_2 & \frac{c}{t_2} & 1 \\ ct_3 & \frac{c}{t_3} & 1 \end{vmatrix}$$

$$= \frac{1}{2} \frac{c^2}{t_1t_2t_3} \begin{vmatrix} t_1^2 & 1 & t_1 \\ t_2^2 & 1 & t_2 \\ t_3^2 & 1 & t_3 \end{vmatrix}$$

$$= \frac{c^2}{2} \times \frac{(t_1 - t_2)(t_2 - t_3)(t_3 - t_1)}{t_1t_2t_3}$$

(ii) The equations of the tangents at A, B and C are

$$\frac{x}{t_1} + yt_1 = 2c \quad \text{...(i)}$$

$$\frac{x}{t_2} + yt_2 = 2c \quad \text{... (ii)}$$

and $$\frac{x}{t_3} + yt_3 = 2c \quad \text{... (iii)}$$

Thus, the points of intersections of Eqs. (i) and (ii), Eqs. (i) and (iii), Eqs. (ii) and (iii), meet at P, Q and R, respectively.

Thus, $$P = \left(\frac{2ct_1t_2}{t_1 + t_2}, \frac{2c}{t_1 + t_2} \right)$$

$$Q = \left(\frac{2ct_1t_3}{t_1 + t_3}, \frac{2c}{t_1 + t_3} \right)$$

and $R = \left(\frac{2ct_2t_3}{t_2 + t_3}, \frac{2c}{t_2 + t_3} \right)$, respectively.

Hence, the area of the ΔPQR

$$= \frac{1}{2} \begin{vmatrix} \frac{2ct_1t_2}{t_1 + t_2} & \frac{2c}{t_1 + t_2} & 1 \\ \frac{2ct_1t_3}{t_1 + t_3} & \frac{2c}{t_1 + t_3} & 1 \\ \frac{2ct_2t_3}{t_2 + t_3} & \frac{2c}{t_2 + t_3} & 1 \end{vmatrix}$$

$$= \frac{2c^2(t_1 - t_2)(t_2 - t_3)(t_3 - t_1)}{(t_1 + t_2)(t_2 + t_3)(t_3 + t_1)}$$

(d) Let P, Q and R are the vertices of a triangle such that

$$P = \left(ct_1, \frac{c}{t_1} \right), Q = \left(ct_2, \frac{c}{t_2} \right) \text{ and } R = \left(ct_3, \frac{c}{t_3} \right)$$

Now, slope of $QR = \frac{c(t_2 - t_3)}{t_2t_3 \cdot c(t_3 - t_2)} = -\frac{1}{t_2t_3}$

Therefore, slope of PM is t_2t_3.

The equations of the perpendicular PM on QR is

$$y - \frac{c}{t_1} = t_2t_3(x - ct_1) \quad \text{... (i)}$$

Similarly, the equations of the perpendicular QN on PR is

$$y-\frac{c}{t_2}=t_1t_3(x-ct_2) \quad \text{... (ii)}$$

Solving Eqs. (i) and (ii), we get

$$x=-\frac{c}{t_1t_2t_3} \text{ and } y=-ct_1t_2t_3$$

Thus, the point $\left(-\frac{c}{t_1t_2t_3},-ct_1t_2t_3\right)$ lies on the rectangular hyperbola $xy=c^2$

Hence, the result.

(e) Any point on the rectangular hyperbola

$$xy=c^2 \text{ is } P\left(ct,\frac{c}{t}\right)$$

The equation of any tangent to the rectangular hyperbola $xy=c^2$ at t is

$$\frac{x}{t}+yt=2c \quad \text{... (i)}$$

The equations of any normal to the rectangular hyperbola $xy=c^2$ at t is

$$\Rightarrow \quad xt^3-yt+xt^4+c=0 \quad \text{... (ii)}$$

Therefore, $a_1=2ct$ and $b_1=\frac{2c}{t}$

and $\quad a_2=c\left(t-\frac{1}{t^3}\right), b_2=c\left(\frac{1}{t}-t^3\right)$

Now, $a_1a_2+b_1b_2=2c^2t\left(t-\frac{1}{t^3}\right)+\frac{2c^2}{t}\left(\frac{1}{t}-t^3\right)$

$$=2c^2\left(t^2-\frac{1}{t^2}+\frac{1}{t^2}-t^2\right)=0$$

(f) Let the equation of the circle be

$$x^2+y^2+2gx+2fy+c=0 \quad \text{... (i)}$$

and the equation of the given hyperbola be

$$xy=1 \quad \text{... (ii)}$$

Solving, we get

$$x^2+\frac{1}{x^2}+2gx+\frac{2f}{x}+c=0$$

$$\Rightarrow \quad x^4+2gx^3+cx^2+2fx+1=0$$

Let its roots are x_1,x_2,x_3 and x_4 then, $x_1x_2x_3x_4=1$

Similarly, can easily prove that

$$y_1y_2y_3y_4=1$$

$$\therefore \quad \frac{x_1\cdot x_2\cdot x_3\cdot x_4}{y_1\cdot y_2\cdot y_3\cdot y_4}=1$$

(g) The line through the centre is $y=x\tan\theta$, so that it meets the curve $xy=c^2$ at point $(c\sqrt{\cot\theta}, c\sqrt{\tan\theta})$.

Hence, the four point are $(c\sqrt{\cot\theta}, c\sqrt{\tan\theta})$ (where $\theta=\alpha,\beta,\gamma$ and δ) the lines P_1P_2 and P_3P_4 as at right angles, this means

$$\left(\frac{c\sqrt{\tan\beta}-c\sqrt{\tan\alpha}}{c\sqrt{\cot\beta}-c\sqrt{\cot\alpha}}\right)\left(\frac{c\sqrt{\tan\delta}-c\sqrt{\tan\gamma}}{c\sqrt{\cot\delta}-c\sqrt{\cot\gamma}}\right)=-1$$

$$\Rightarrow \quad (-\sqrt{\tan\beta}\sqrt{\tan\alpha})(-\sqrt{\tan\delta}\sqrt{\tan\gamma})=-1$$

$$\Rightarrow \quad \tan\alpha\tan\beta\tan\gamma\tan\delta=(-1)^2=1$$

(h) Let the hyperbola be $xy=c^2$

and the circle be

$$S: x^2+y^2+2gx+2fy+k=0$$

Substituting $x=ct$ and $y=c/t$ in equation, we get

$$c^2t^4+2gct^3+kt^2+2fct+c^2=0$$

whose roots are t_1,t_2,t_3, and t_4 where

$$P_r=\left(ct_r,\frac{c}{t_r}\right)$$

(where $r=1,2,3$ and 4). Therefore,

$$t_1+t_2+t_3+t_4=\frac{-2g}{c}$$

$$\therefore \quad \Sigma t_1t_2=\frac{k}{c^2}$$

$$\therefore \quad \Sigma t_1t_2t_3=\frac{-2f}{c}$$

and $\quad t_1t_2t_3=\frac{c^2}{c^2}=1$

Now, $\sum_{i=1}^{4}CP_i^2=c^2\left[t_1^2+t_2^2+t_3^2+\frac{1}{t_1^2}+\frac{1}{t_2^2}+\frac{1}{t_3^2}+\frac{1}{t_4^2}\right]$

$$=c^2\left[(\Sigma t_1)^2-2\Sigma t_1t_2+\left(\Sigma\frac{1}{t_1}\right)^2-2\Sigma\frac{1}{t_1t_2}\right]$$

$$=c^2\left[\frac{4g^2}{c^2}-\frac{2k}{c^2}+\frac{4f^2}{c^2}-\frac{2k}{c^2}\right](\because t_1t_2t_3t_4=1)$$

$$=4(g^2+f^2-k)=4r^2$$

28. (a) The equation of any tangent to the hyperbola $\frac{x^2}{a^2}-\frac{y^2}{b^2}=1$ is

$$y=m_1x+\sqrt{a^2m_1^2-b^2} \quad \text{... (i)}$$

The equation of any tangent to the hyperbola

$$\frac{x^2}{(-b^2)}-\frac{y^2}{(-a^2)}=1 \text{ is}$$

$$y=m_2x+\sqrt{(-b^2)m_2^2-(-a^2)} \quad \text{... (ii)}$$

Since, the Eqs. (i) and (ii) are identical, so

$$a^2m_1^2-b^2=(-b^2)m_2^2-(-a^2)$$

$$\Rightarrow \quad m_1^2=1 \text{ and } m_2^2=1$$

Thus, $\quad m_1=\pm1=m_2$

Hence, the equation of the common tangent are

$$y=\pm x+\sqrt{a^2-b^2}$$

(b) The equations of the given curves are

$$\frac{x^2}{9}-\frac{y^2}{4}=1 \text{ and } \frac{x^2}{9}+\frac{y^2}{4}=1$$

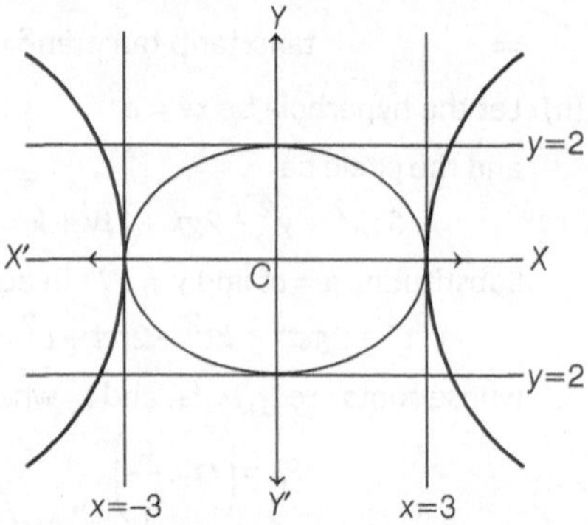

Here, the length of the major axis of the ellipse is equal to the length of the transverse axis and also the length of the minor axis is equal to the length of the conjugate axis, thus the equations of the common tangents are $x=\pm 3$ and $y=\pm 2$.

(c) The equation of any tangent to the hyperbola $\frac{x^2}{16}-\frac{y^2}{9}=1$ is $y=mx+\sqrt{16m^2-9}$

$$\Rightarrow \quad mx-y+\sqrt{16m^2-9}=0 \quad \text{... (i)}$$

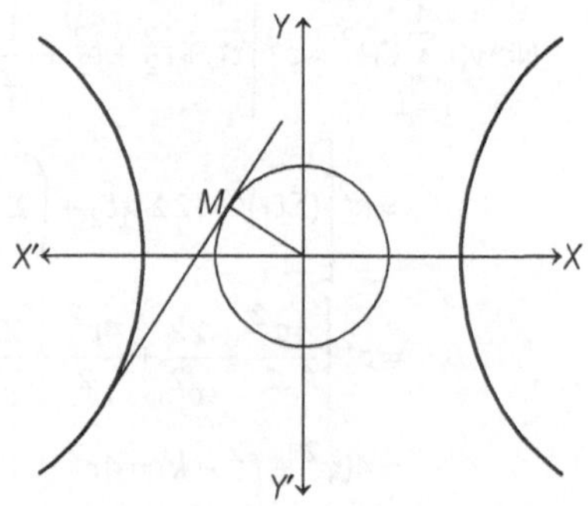

If the tangent (i) is also the tangent to the circle $x^2+y^2=9$, the length of the perpendicular from the centre of the circle is equal to the radius of the circle, so

$$\left|\frac{0+\sqrt{16m^2-9}}{\sqrt{m^2+1}}\right|=3$$

$$\Rightarrow \quad 16m^2-9=9(m^2+1)$$

$$\Rightarrow \quad 16m^2-9m^2=9+9=18$$

$$\Rightarrow \quad 7m^2=18$$

$$m=\pm\frac{18}{7}$$

Hence, the equation of tangents are

$$y=\pm\frac{18}{7}x+\frac{15}{\sqrt{7}}$$

(d) The equation of any tangent to the parabola $y^2=8x$ is

$$y=mx+\frac{2}{m} \quad \text{... (i)}$$

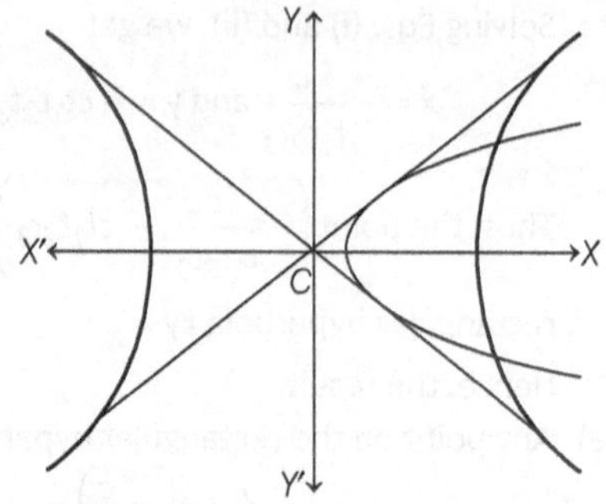

Since, the tangent to the parabola is also a tangent to the hyperbola $\frac{x^2}{9}-\frac{y^2}{5}=1$, so

$$5x^2-9\left(mx+\frac{2}{m}\right)^2=45$$

$$\Rightarrow \quad 5x^2-9\left(m^2x^2+\frac{4}{m^2}+4x\right)=45$$

$$\Rightarrow (5-9m^2)x^2-36x-9\left(\frac{4}{m^2}+5\right)=0$$

Now, $D=0$

$$\Rightarrow \quad (36)^2+36(5-9m^2)\left(\frac{4}{m^2}+5\right)=0$$

$$\Rightarrow \quad 36+\frac{20}{m^2}+25-36-45m^2=0$$

$$\Rightarrow \quad \frac{4}{m^2}+5-9m^2=0$$

$$\Rightarrow \quad 9m^4-5m^2-4=0$$

$$\Rightarrow \quad (m^2-1)(9m^2+4)=0$$

$$\Rightarrow \quad (m^2-1)=0$$

$$\Rightarrow \quad m=\pm 1$$

Hence, the equations of the common tangents are

$$y=\pm x\pm 2.$$

(e) Any point on the circle $x^2+y^2=a^2$ be $(a\cos\theta, a\sin\theta)$.

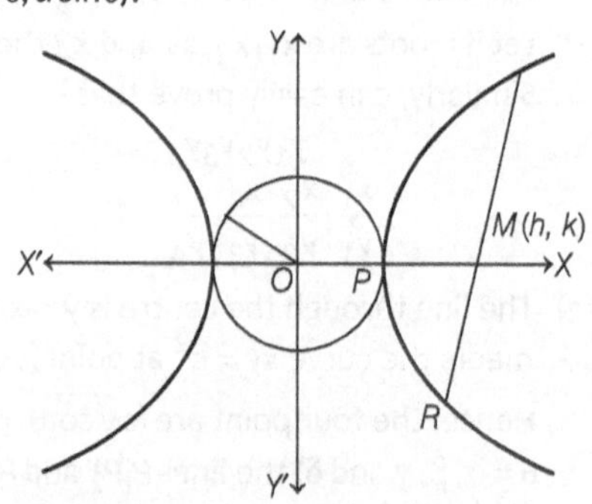

The equation of the chord of contact from the point $(a\cos\theta, a\sin\theta)$ to the hyperbola $x^2 - y^2 = a^2$ is

$$(a\cos\theta)\,x - (a\sin\theta)y = a^2$$

$$\Rightarrow \quad x\cos\theta - y\sin\theta = a \quad \text{... (i)}$$

Let the mid-point be (h, k).

The equation of the chord bisected at (h, k) to the hyperbola $x^2 - y^2 = a^2$, is

$$hx - ky = h^2 - k^2 \quad \text{... (ii)}$$

Since, the Eqs. (i) and (ii) are identical, so

$$\frac{h}{\cos\theta} = \frac{-k}{-\sin\theta} = \frac{h^2 - k^2}{a}$$

$$\Rightarrow \quad \cos\theta = \frac{ah}{h^2 - k^2} \text{ and } \sin\theta = \frac{ak}{h^2 - k^2}$$

Squaring and adding, we get

$$\left(\frac{ah}{h^2 - k^2}\right)^2 + \left(\frac{ak}{h^2 - k^2}\right)^2 = 1$$

$$\Rightarrow \quad a^2(h^2 + k^2) = (h^2 - k^2)^2$$

Hence, the locus of (h, k) is $a^2(x^2 + y^2) = (x^2 - y^2)^2$.

(f) Let the point on the parabola be (h, k)

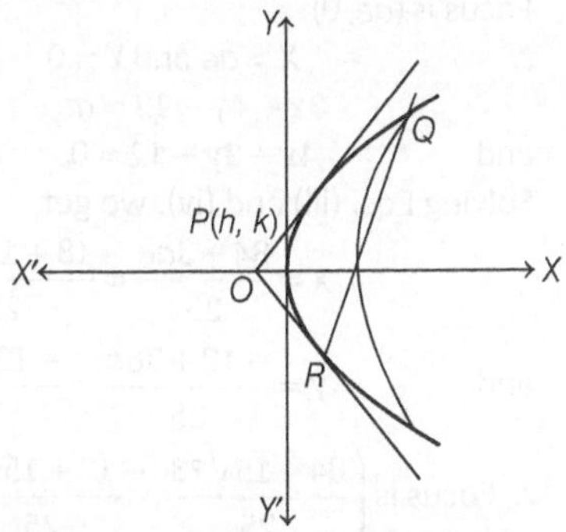

The equation of the chord of contact of the parabola is

$$yk = 2a(x + h) \quad \text{... (i)}$$

$$\Rightarrow \quad y = \frac{2a}{k}x + \frac{2ah}{k} \quad \text{... (ii)}$$

Since, the line (ii) is a tangent to the hyperbola $\frac{x^2}{a^2} - \frac{y^2}{b^2} = 1$, so $c^2 = a^2m^2 - b^2$

$$\Rightarrow \quad \left(\frac{2ah}{k}\right)^2 = a^2\left(\frac{2a}{k}\right)^2 - b^2$$

$$\Rightarrow \quad 4a^2h^2 = 4a^4 - k^2b^2$$

Hence, the locus of (h, k) is

$$4a^2x^2 = 4a^4 - y^2b^2$$

(g) If (h, k) be the mid-point of the chord of the hyperbola $x^2 - y^2 = a^2$, then

$$T = S_1$$

$$\Rightarrow \quad hx - ky = h^2 - k^2$$

$$\Rightarrow \quad ky = hx + (k^2 - h^2)$$

$$\Rightarrow \quad y = \left(\frac{h}{k}\right)x + \left(\frac{k^2 - h^2}{k}\right) \quad \text{... (i)}$$

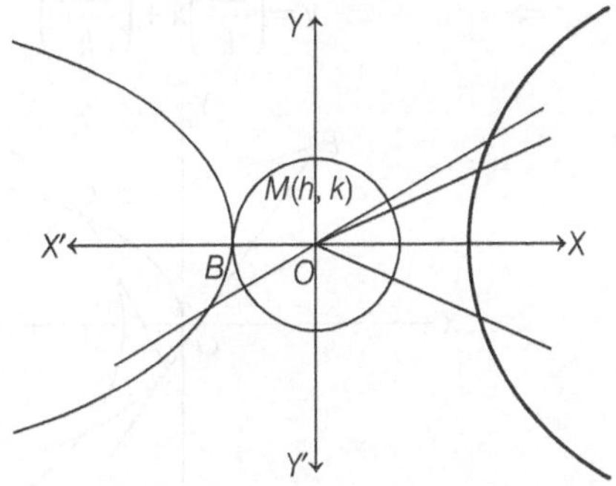

If the line (i) touches the parabola $y^2 = 4ax$, so

$$c = \frac{a}{m}$$

$$\Rightarrow \quad \left(\frac{h^2 - k^2}{k}\right) = \frac{a}{(h/k)} = \frac{ak}{h}$$

$$\Rightarrow \quad h(h^2 - k^2) = ak^2$$

Hence, the locus of (h, k) is $x(x^2 - y^2) = ay^2$

(h) If (h, k) be the mid-point of the chord to the circle $x^2 + y^2 = a^2$, then $T = S$

$$\Rightarrow \quad hx + ky = h^2 + k^2$$

$$\Rightarrow \quad y = \left(-\frac{h}{k}\right)x + \left(\frac{h^2 + k^2}{k}\right) \quad \text{... (i)}$$

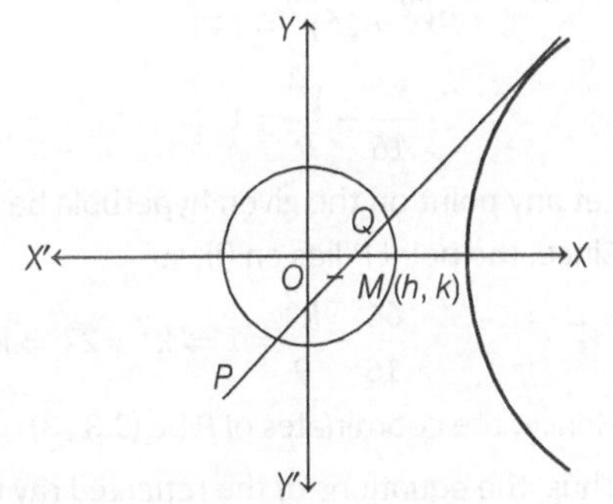

If the line (i) be a tangent to the hyperbola

$$\frac{x^2}{a^2} - \frac{y^2}{b^2} = 1$$

So, $$c^2 = a^2m^2 - b^2$$

$$\Rightarrow \quad \left(\frac{h^2 + k^2}{k}\right)^2 = a^2\left(\frac{h^2}{k^2}\right) - b^2$$

$$\Rightarrow \quad (h^2 + k^2)^2 = (a^2h^2 - b^2k^2)$$

Hence, the locus of (h, k) is

$$(x^2 + y^2)^2 = (a^2x^2 - b^2y^2)$$

(i) Let (h, k) be the pole.

The equation of the polar from the point (h, k) w.r.t. the parabola $y^2 = 4ax$ is

$$yk = 2a(x+h) = 2ax + 2ah$$

$$\Rightarrow \quad y = \left(\frac{2a}{k}\right)x + \left(\frac{2ah}{k}\right) \quad \text{... (i)}$$

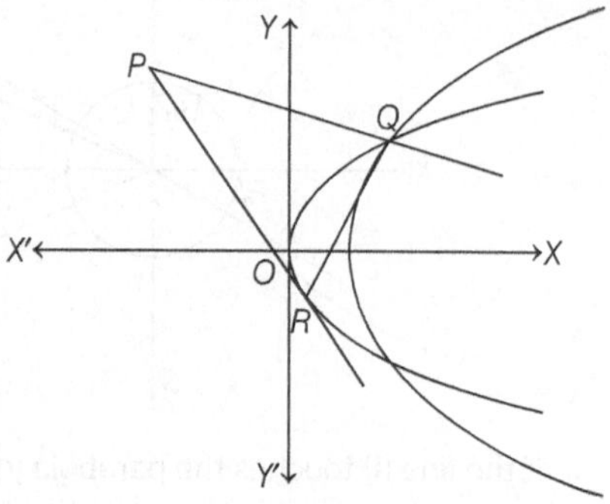

If the line (i) be a tangent to the hyperbola $x^2 - y^2 = a^2$, then $c^2 = a^2m^2 - a^2$

$$\Rightarrow \quad \left(\frac{2ah}{k}\right)^2 = a^2\left(\frac{2a}{k}\right)^2 - a^2$$

$$\Rightarrow \quad 4h^2 = 4a^2 - k^2.$$

$$\Rightarrow \quad h^2 + \frac{k^2}{4} = a^2$$

Hence, the locus of (h, k) is

$$x^2 + \frac{y^2}{4} = a^2 \Rightarrow 4x^2 + y^2 = 4a^2$$

29. The equation of the given hyperbola is

$$9x^2 - 16y^2 = 144$$

$$\Rightarrow \quad \frac{x^2}{16} - \frac{y^2}{9} = 1 \quad \text{... (i)}$$

Let any point on the given hyperbola be $P(8, k)$

Since, the point P lies on (i), so

$$\frac{64}{16} - \frac{k^2}{9} = 1 \Rightarrow k^2 = 27 \Rightarrow k = 3\sqrt{3}$$

Hence, the coordinates of P be $(8, 3\sqrt{3})$

Thus, the equations of the reflected ray is

$$y - 3\sqrt{3} = \frac{0 - 3\sqrt{3}}{-5-8}(x-8)$$

$$\Rightarrow \quad 3\sqrt{3}x - 13y + 15\sqrt{3} = 0$$

30. (a) The equation of required parabola is

$$\frac{\left(\dfrac{3x+4y-7}{\sqrt{3^2+4^2}}\right)^2}{\left(\dfrac{4}{2}\right)^2} - \frac{\left(\dfrac{4x-3y+8}{\sqrt{4^2+(-3)^2}}\right)^2}{\left(\dfrac{6}{2}\right)^2} = 1$$

$$\Rightarrow \quad \frac{1}{100}(3x+4y-7)^2 - \frac{1}{225}(4x-3y+8)^2 = 1$$

$$\Rightarrow 9(3x+4y-7)^2 - 4(16x^2 + 9y^2 + 64 - 24xy + 64x - 48y) = 900$$

$$\Rightarrow 17x^2 + 312xy + 108y^2 - 634x - 312y - 715 = 0$$

(b) Let $3x - 4y - 12 = X$

and $4x + 3y - 12 = Y$

$$\therefore \quad \frac{X^2}{10^2} - \frac{Y^2}{15^2} = 1$$

[$\because$ X and Y are two perpendicular lines]

Centre is point of intersection of $X = 0$ and $Y = 0$

$$\therefore \quad 3x - 4y = 12 \quad \text{... (i)}$$

$$4x + 3y = 12 \quad \text{... (ii)}$$

By solving Eqs. (i) and (ii), we get

$$x = \frac{84}{25} \text{ and } y = \frac{-12}{25}$$

Now, $$e^2 = 1 + \frac{b^2}{a^2}$$

$$= 1 + \frac{225}{100} = \frac{325}{100} \Rightarrow e = \frac{\sqrt{13}}{2}$$

Also, $a = 10, b = 15$

Focus is $(ae, 0)$

$$\therefore \quad X = ae \text{ and } Y = 0$$

$$3x - 4y - 12 = ae \quad \text{...(iii)}$$

and $$4x + 3y - 12 = 0 \quad \text{...(iv)}$$

Solving Eqs. (iii) and (iv), we get

$$x = \frac{84 + 3ae}{25} = \frac{18 + 15\sqrt{13}}{25}$$

and $$y = \frac{-12 + 3ae}{25} = \frac{-12 + 15\sqrt{13}}{25}$$

$$\therefore \text{ Focus is } \left(\frac{84 + 15\sqrt{13}}{25}, \frac{-12 + 15\sqrt{13}}{25}\right).$$

(c) Here, $2y - x - 3 = 0$ and $2x + y - 1 = 0$ are perpendicular to each other

Therefore, the equation of conic, can be written as

$$\Rightarrow \quad 4 \times 5\left(\frac{2y - x - 3}{\sqrt{2^2+1^2}}\right)^2 - 9 \times 5\left[\frac{2x + y - 1}{\sqrt{2^2+1^2}}\right]^2 = 80$$

$$\Rightarrow \quad 4\left[\frac{2y - x - 3}{\sqrt{5}}\right]^2 - 9\left[\frac{2x + y - 1}{\sqrt{5}}\right]^2 = 16$$

On putting $\dfrac{2y - x - 3}{\sqrt{5}} = X$ and $\dfrac{2x + y - 1}{\sqrt{5}} = Y$, then given equation can be written as

$$4X^2 - 9Y^2 = 16 \Rightarrow \frac{X^2}{4} - \frac{Y^2}{(4/3)^2} = 1$$

The eccentricity is given by

$$e = \sqrt{1 + \frac{4/3}{4}} = \frac{2}{\sqrt{3}}$$

Level 2

1. (c) $\because$ Given, length of transverse axis is $=7$

$\therefore \quad 2a = 7$

$\Rightarrow \quad a = \frac{7}{2}$

$\therefore$ Equation of hyperbola

$\Rightarrow \quad \frac{x^2}{a^2} - \frac{y^2}{b^2} = 1$

$\Rightarrow \quad \frac{x^2}{\left(\frac{7}{2}\right)^2} - \frac{y^2}{b^2} = 1$

$\because$ It passes through $(5, -2)$

$\because \quad \frac{(5)^2}{\frac{49}{4}} - \frac{(-2)^2}{b^2} = 1$

$\Rightarrow \quad \frac{4(25)}{49} - \frac{4}{b^2} = 1$

$\Rightarrow \quad \frac{4}{b^2} = \frac{100}{49} - 1$

$\Rightarrow \quad \frac{4}{b^2} = \frac{100-49}{49}$

$\Rightarrow \quad \frac{4}{b^2} = \frac{51}{49}$

$\Rightarrow \quad b^2 = \frac{49 \times 4}{51}$

$= \frac{196}{51}$

$\therefore$ Equation is $\frac{4}{49}x^2 - \frac{51}{196}y^2 = 1.$

2. (c) According to given condition,

$2ae = 2(2a)$

$\Rightarrow \quad e = 2$

$\therefore \quad a = \frac{3}{\sqrt{3}} = \sqrt{3}$

Therefore, the equation of hyperbola is

$\frac{x^2}{3} - \frac{y^2}{9} = 1$

$\Rightarrow \quad 3x^2 - y^2 = 9$

3. (a) Given, $3x^2 + 4y^2 = 12$

$\Rightarrow \quad \frac{x^2}{4} + \frac{y^2}{3} = 1$

$\therefore \quad a^2 = 4 \Rightarrow a = 2$ and $b^2 = 3 \Rightarrow b = \sqrt{3}$

$\therefore \quad e = \sqrt{1 - \frac{b^2}{a^2}} = \sqrt{1 - \frac{3}{4}} = \frac{1}{2}$

$\because \quad ae = 2 \times \frac{1}{2} = 1$

Given, $\quad a = \sin\theta$

$\Rightarrow (\sin\theta)e = 1$

$\Rightarrow \quad e = \operatorname{cosec}\theta$

and $\quad b = a\sqrt{e^2 - 1}$

$= \sin\theta\sqrt{\operatorname{cosec}^2\theta - 1}$

$= \sin\theta \times \cot\theta = \cos\theta$

$\therefore$ Equation of hyperbola is

$\frac{x^2}{\sin^2\theta} - \frac{y^2}{\cos^2\theta} = 1$

$\Rightarrow \quad x^2 \operatorname{cosec}^2\theta - y^2 \sec^2\theta = 1$

4. (c) Given, $(x-3)^2 + (y+1)^2 = (4x+3y)^2$ ...(i)

We know general form of hyperbola

$SP = ePM$

$\Rightarrow \quad SP^2 = e^2 PM^2$

$\Rightarrow \quad (x - x_1)^2 + (y - y_1)^2 = e^2 \left| \frac{ax_1 + by_1 + c}{\sqrt{a^2 + b^2}} \right|^2$ ...(ii)

Now, convert Eq. (i) in the form of Eq. (ii) so, we get

$(x-3)^2 + (y+1)^2 = 5^2\left(\frac{4x+3y}{5}\right)^2$

$\therefore \quad$ Directrix $= 4x + 3y = 0$ and focus $= (3, -1)$

Transverse axis perpendicular to directrix and passing through focus.

$\therefore \quad 3x - 4y = \lambda$

$\Rightarrow \quad 3(3) - 4(-1) = \lambda$

$\Rightarrow \quad \lambda = 9 + 4 = 13$

$\therefore$ Required equation is $3x - 4y = 13$.

5. (a) Given, parabola is $x^2 - 3y^2 = 1$

$\Rightarrow \quad \frac{x^2}{1} - \frac{y^2}{\frac{1}{3}} = 1$

$\because \quad a^2 = 1$ and $b^2 = \frac{1}{3}$

Now, $\quad b^2 = a^2(e^2 - 1)$

$\frac{1}{3} = 1(e^2 - 1)$

Let e' be eccentricity of conjugate hyperbola

$\therefore \quad 1 = \frac{1}{3}(e'^2 - 1)$

$\Rightarrow \quad 3 = e'^2 - 1$

$e'^2 = 4$

$e' = 2$

6. (b) Given, equation is $16x^2 - 9y^2 = 144$

$$\Rightarrow \quad \frac{x^2}{9} - \frac{y^2}{16} = 1$$

$$a^2 = 9 \Rightarrow a = 3$$

and $\quad b^2 = 16 \Rightarrow b = 4$

∴ We known that the difference of length of distances of any point on hyperbola to foci of the hyperbola is double of length of transverse axis

$$\because \quad PS_1 - PS_2 = 2a$$

$$\Rightarrow \quad PS_1 - PS_2 = 2(3) = 6$$

7. (d) Ellipse : $5x^2 + 9y^2 = 45$

$$\Rightarrow \quad \frac{x^2}{9} + \frac{y^2}{5} = 1$$

$$\because \quad a^2 = 9 \text{ and } b^2 = 5$$

$$\because \quad e = \sqrt{1 - \frac{b^2}{a^2}}$$

$$= \sqrt{1 - \frac{5}{9}} = \sqrt{\frac{4}{9}} = \frac{2}{3}$$

Hyperbola : $5x^2 - 4y^2 = 45$

$$\frac{x^2}{9} - \frac{4y^2}{45} = 1$$

$$\therefore \quad a'^2 = 9 \text{ and } b'^2 = \frac{45}{4}$$

$$\because \quad e' = \sqrt{1 + \frac{\frac{45}{4}}{9}}$$

$$= \sqrt{1 + \frac{5}{4}} = \frac{3}{2}$$

$$\therefore \quad ee'^2 = \frac{2}{3} \times \frac{3}{2} = 1$$

8. (b) Hyperbola : $\dfrac{x^2}{\cos^2 \alpha} - \dfrac{y^2}{\sin^2 \alpha} = 1$

$$\because \quad a^2 = \cos^2 \alpha \text{ and } b^2 = \sin^2 \alpha$$

$$\because \quad b^2 = a^2(e^2 - 1)$$

$$\Rightarrow \quad \sin^2 \alpha = \cos^2 \alpha (e^2 - 1)$$

$$\Rightarrow \quad \tan^2 \alpha = e^2 - 1$$

$$\Rightarrow \quad 1 + \tan^2 \alpha = e^2$$

$$\Rightarrow \quad e = \sec \alpha$$

$$\therefore \quad \text{Foci} \Rightarrow (ae, 0) = (\pm 1, 0)$$

⇒ If α varies the abscissae of foci remains constant.

9. (a) Given, equation of hyperbola is

$$9x^2 - 16y^2 + 72x - 32y - 16 = 0$$

$$\Rightarrow \quad 9(x^2 + 8x) - 16(y^2 + 2y) - 16 = 0$$

$$\Rightarrow 9(x^2 + 8x + 16 - 16) - 16(y^2 + 2y + 1 - 1) - 16 = 0$$

$$\Rightarrow \quad 9(x + 4)^2 - 144 - 16(y + 1)^2 + 16 - 16 = 0$$

$$\Rightarrow \quad 9(x + 4)^2 - 16(y + 1)^2 = 144$$

$$\Rightarrow \quad \frac{(x + 4)^2}{16} - \frac{(y + 1)^2}{9} = 1$$

$$\Rightarrow \quad a^2 = 16 \text{ and } b^2 = 9$$

∵ Length of latus rectum $= \dfrac{2b^2}{a} = 2 \times \dfrac{9}{4} = \dfrac{9}{2}$ units

10. (c) Given, equation of hyperbola

$$\frac{x^2}{144} - \frac{y^2}{81} = \frac{1}{25}$$

$$\Rightarrow \quad \frac{x^2}{\frac{144}{25}} - \frac{y^2}{\frac{81}{25}} = 1$$

$$\Rightarrow \quad a^2 = \frac{144}{25} \text{ and } b^2 = \frac{81}{25}$$

$$\Rightarrow \quad a = \frac{12}{5} \text{ and } b = \frac{9}{5}$$

$$\therefore \quad e_1 = \sqrt{1 + \frac{b^2}{a^2}} = \sqrt{1 + \frac{\frac{81}{25}}{\frac{144}{25}}}$$

$$= \frac{\sqrt{225}}{\sqrt{144}} = \frac{15}{12} = \frac{5}{4}$$

Therefore, focii is $= (ae, 0) = \left(\dfrac{12}{5} \times \dfrac{5}{4}, 0\right) = (3, 0)$

∴ Focus of ellipse i.e. $(ae, 0) = (3, 0)$

∵ Ellipse : $\dfrac{x^2}{16} + \dfrac{y^2}{b^2} = 1 \qquad \{\because a^2 = 16 \Rightarrow a = 4\}$

$$\therefore \quad ae = 3 \Rightarrow e = \frac{3}{4}$$

$$\therefore \quad b^2 = a^2(1 - e^2) = 16 \times \left(1 - \frac{9}{16}\right) = 7$$

$$\Rightarrow \quad b^2 = 7$$

11. (d)

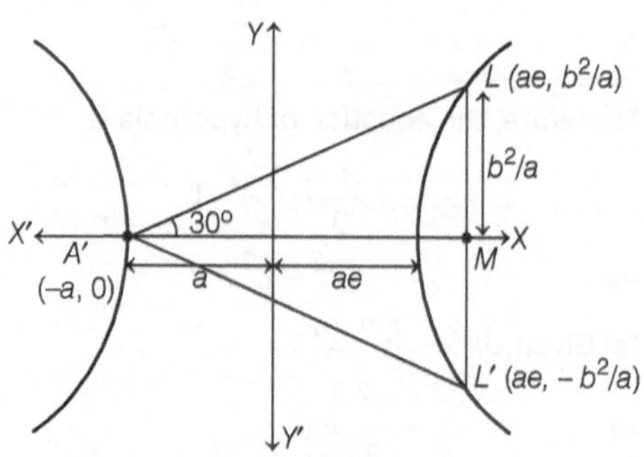

∵ Given, $\Delta A'LL'$ is equilateral triangle.

$$\therefore \quad \angle LA'M = 30°$$

In $\Delta A'ML$,

$$\tan 30° = \frac{LM}{A'M}$$

$$\Rightarrow \quad \tan 30^\circ = \frac{\frac{b^2}{a}}{a + ae}$$

$$\Rightarrow \quad \frac{1}{\sqrt{3}} = \frac{b^2}{a^2(1+e)}$$

$$\Rightarrow \quad \frac{1+e}{\sqrt{3}} = \frac{b^2}{a^2}$$

$$\Rightarrow \quad \frac{1+e}{\sqrt{3}} = e^2 - 1 \qquad \left[\because e^2 = 1 + \frac{b^2}{a^2}\right]$$

$$\Rightarrow \quad \frac{(e+1)}{\sqrt{3}} = (e+1)(e-1)$$

$$\Rightarrow \quad e - 1 = \frac{1}{\sqrt{3}}$$

$$\Rightarrow \quad e = 1 + \frac{1}{\sqrt{3}} = \frac{\sqrt{3}+1}{\sqrt{3}}$$

12. (b) Hyperbola :

$$\frac{x^2}{5} - \frac{y^2}{5\cos^2\alpha} = 1$$

$$\because \quad e_1^2 = 1 + \frac{b^2}{a^2}$$

$$= 1 + \frac{5\cos^2\alpha}{5} = 1 + \cos^2\alpha$$

For ellipse

$$\frac{x^2}{25\cos^2\alpha} + \frac{y^2}{25} = 1$$

$$\Rightarrow \quad a^2 = 25\cos^2\alpha \text{ and } b^2 = 25$$

$$\therefore \quad e_2^2 = 1 - \frac{a^2}{b^2} \qquad [\because b > a]$$

$$= 1 - \frac{25\cos^2\alpha}{25}$$

$$= 1 - \cos^2\alpha$$

$$\Rightarrow \quad e_2^2 = \sin^2\alpha$$

$$\because \quad e_1 = \sqrt{3}e_2$$

$$\Rightarrow \quad e_1^2 = 3e_2^2$$

$$\Rightarrow \quad 1 + \cos^2\alpha = 3\sin^2\alpha$$

$$\Rightarrow \quad 1 + 1 - \sin^2\alpha = 3\sin^2\alpha$$

$$\Rightarrow \quad 2 = 4\sin^2\alpha$$

$$\Rightarrow \quad \sin^2\alpha = \frac{1}{2}$$

$$\Rightarrow \quad \sin\alpha = \frac{1}{\sqrt{2}} = \sin\frac{\pi}{4}$$

$$\Rightarrow \quad \alpha = \frac{\pi}{4}$$

13. (a) For ellipse

$$\frac{x^2}{25} + \frac{y^2}{b^2} = 1 \qquad [\because b < 5]$$

Let e_1 is eccentricity of ellipse

$$\therefore \quad b^2 = a^2(1 - e_1^2)$$

$$\Rightarrow \quad b^2 = 25(1 - e_1^2) \qquad \text{...(i)}$$

Again for hyperbola,

$$\frac{x^2}{16} - \frac{y^2}{b^2} = 1$$

Let e_2 is eccentricity of hyperbola.

$$\therefore \quad b^2 = 16(e_2^2 - 1) \qquad \text{...(ii)}$$

By Eqs. (i) and (ii)

$$25(1 - e_1^2) = 16(e_2^2 - 1)$$

Now, $\quad e_1 e_2 = 1$

$$\therefore \quad 25(1 - e_1^2) = 16\left(\frac{1 - e_1^2}{e_1^2}\right)$$

$$\Rightarrow \quad 25 = 16 \times \frac{1}{e_1^2}$$

$$\Rightarrow \quad e_1^2 = \frac{16}{25}$$

$$\Rightarrow \quad e_1 = \frac{4}{5}$$

$$\therefore \quad e_2 = \frac{5}{4}$$

Now, distance between focii for ellipse $= 2ae_1$

$$= 2 \times 5 \times \frac{4}{5} = 8 = \alpha$$

Distance for hyperbola $= 2ae_2$

$$= 2 \times 4 \times \frac{5}{4} = 10 = \beta$$

$$\therefore \quad (\alpha, \beta) = (8, 10)$$

14. (a) Given hyperbola, $\dfrac{x^2}{\cos^2\theta} - \dfrac{y^2}{\sin^2\theta} = 1$

Here, $a = \cos\theta$ and $b = \sin\theta$

We know that eccentricity of hyperbola is

$$e = \sqrt{1 + \frac{b^2}{a^2}} = \sqrt{1 + \tan^2\theta}$$

$$\because \quad e > 2 \qquad \text{[given]}$$

$$\Rightarrow \quad \sqrt{1 + \tan^2\theta} > 2$$

$$\Rightarrow \quad 1 + \tan^2\theta > 4$$

$$\Rightarrow \quad \tan^2\theta > 3$$

As given $\theta \in \left(0, \dfrac{\pi}{2}\right)$.

Possible value of $\tan\theta > \sqrt{3}$

So, θ can be in the range of $\frac{\pi}{3}<\theta<\frac{\pi}{2}$.

As we know that latusrectum (LR) $=\frac{2b^2}{a}$

$$\therefore \quad LR=\frac{2\sin^2\theta}{\cos\theta}$$

$$=2\tan\theta\sin\theta$$

We know that in the range $\frac{\pi}{3}<\theta<\frac{\pi}{2}$, $\tan\theta$ and $\sin\theta$ both are increasing.

$\therefore$ So, at $\frac{\pi}{3}$ value of latusrectum is minimum and at $\frac{\pi}{2}$ latusrectum will be maximum.

$$\therefore \quad (LR)_{min}=2\tan\frac{\pi}{3}\sin\frac{\pi}{3}=2\times\sqrt{3}\times\frac{\sqrt{3}}{2}=3$$

$$(LR)_{max}=2\tan\frac{\pi}{2}\sin\frac{\pi}{2}=\infty$$

$$\therefore \quad LR\in(3,\infty)$$

15. (b) Given, equation $\frac{x^2}{29-p}+\frac{y^2}{4-p}=1$

For ellipse, $29-p>0, 4-p>0$

$$p<29, p<4$$

$$\therefore \quad p<4$$

From hyperbola, $(29-p)(4-p)<0$

$$(p-29)(p-4)<0$$

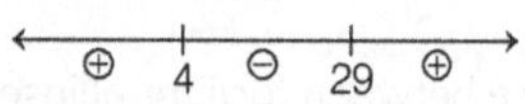

$$\therefore \quad p\in(4,29)$$

16. (c) Hyperbola : $x^2+3xy+2y^2+2x+3y+2=0$

and $\quad x^2+3xy+2y^2+2x+3y+c=0$

$$\therefore \quad \Delta=abc+2fgh-af^2-bg^2-ch^2=0$$

$$A=x^2+3xy+2y^2+2x+3y+\lambda=0$$

$$\therefore \quad \Delta=0$$

$$\Rightarrow \quad 2\lambda+\frac{9}{2}-\frac{9}{4}-2-\frac{9\lambda}{4}=0$$

$$\Rightarrow \quad \lambda=1$$

$$\Rightarrow \quad c+x=2A$$

$$\Rightarrow \quad c+2=2$$

$$\Rightarrow \quad c=0$$

17. (b) Given, equation

$$x^2-2y^2-2\sqrt{2}x-4\sqrt{2}y-6=0$$

This can be written as

$$(x-\sqrt{2})^2-(\sqrt{2}y+2)^2=6+2-4=4$$

$$\Rightarrow \quad (x-\sqrt{2})^2-2(y+\sqrt{2})^2=4$$

$$\Rightarrow \quad \frac{(x-\sqrt{2})^2}{4}-\frac{(y+\sqrt{2})^2}{2}=1$$

$$\Rightarrow \quad \frac{X^2}{4}-\frac{Y^2}{2}=1$$

where, $X=x-\sqrt{2}$ and $Y=y+\sqrt{2}$

$\therefore$ $A=(X=2, Y=0)\Rightarrow x-\sqrt{2}=2, y+\sqrt{2}=0$

$$x=2+\sqrt{2}, y=-\sqrt{2}$$

$$A=(2+\sqrt{2},-\sqrt{2})$$

The centre is $(\sqrt{2},-\sqrt{2})$.

Let 'e' be the eccentricity. Therefore, $b^2=a^2(e^2-1)$

$$2=4(e^2-1)\Rightarrow e=\sqrt{\frac{3}{2}}$$

Hence, $C\Rightarrow(X=ae, Y=0)$

$$\Rightarrow \quad x-\sqrt{2}=2\times\sqrt{\frac{3}{2}}=\sqrt{6}$$

$$y+\sqrt{2}=0\Rightarrow y=-\sqrt{2}$$

$$\because \quad C=(\sqrt{6}+\sqrt{2},-\sqrt{2})$$

Also, $\quad BC=\frac{b^2}{a}=\frac{2}{2}=1$

$$B=\left(X=ae, Y=\frac{be}{4}\right)$$

$$\Rightarrow (\sqrt{2}+\sqrt{2}+\sqrt{6}, 1-\sqrt{2})=(2\sqrt{2}+\sqrt{6}, 1-\sqrt{2})$$

Therefore, the area is $\frac{1}{2}\begin{vmatrix}2+\sqrt{2} & -\sqrt{2} & 1\\ 2\sqrt{2}+\sqrt{6} & 1-\sqrt{2} & 1\\ \sqrt{2}+\sqrt{6} & -\sqrt{2} & 1\end{vmatrix}$

$$=\frac{1}{2}\begin{vmatrix}2+\sqrt{2} & -\sqrt{2} & 1\\ -2+\sqrt{2}+\sqrt{6} & 1 & 0\\ \sqrt{6}-2 & 0 & 0\end{vmatrix}$$

$$[\because R_2\to R_2-R_1, R_3\to R_3-R_1]$$

$$=\frac{1}{2}(\sqrt{6}-2)$$

$$=\sqrt{\frac{3}{2}}-1$$

18. (d) $\because$ The point $(\lambda,-1)$ is an exterior point of the curve

$$4x^2-3y^2-1=0$$

So, $\quad 4(\lambda)^2-3(-1)^2-1<0$

$$\Rightarrow \quad 4\lambda^2-4<0$$

$$\lambda^2-1<0$$

$$(\lambda+1)(\lambda-1)<0$$

$$\Rightarrow \quad -1<\lambda<1$$

$$\lambda\in(-1,1)$$

Thus, the length of interval where λ lies is 2.

$\therefore$ $\quad m=2$ and $m+10=2+10=12$

19. (c) Given, $x=\frac{1}{2}(e^{t}+e^{-t}) \Rightarrow y=\frac{1}{2}(e^{t}-e^{-t})$

Squaring both the sides

$$x^2=\frac{1}{4}(e^{2t}+e^{-2t}+2)$$

and $$y^2=\frac{1}{4}(e^{2t}+e^{-2t}-2)$$

$\therefore$ $$x^2-y^2=\frac{1}{4}$$

$\{e^{2t}+e^{-2t}+2-e^{2t}-e^{-2t}+2\}$

$$=\frac{1}{4}(4)=1$$

$\therefore$ $$x^2-y^2=1$$

which is a hyperbola.

20. (b) Let $P(a\sec\theta, b\tan\theta)$ and $Q(a\sec\theta, -b\tan\theta)$ are two points on hyperbola.

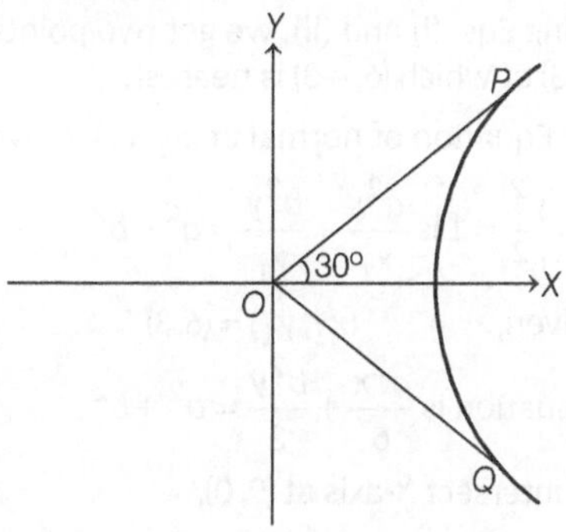

$\because OPQ$ is equilateral triangle (given)

Now, $$\tan 30°=\frac{b\tan\theta}{a\sec\theta}$$

$$\frac{1}{\sqrt{3}}=\frac{b}{a}\sin\theta$$

$$\frac{3b^2}{a^2}=\text{cosec}^2\theta>1$$

$\Rightarrow$ $$3(e^2-1)>1$$

$\Rightarrow$ $$e^2>\frac{1}{3}+1$$

$\Rightarrow$ $$e^2>\frac{4}{3}\Rightarrow e>\frac{2}{\sqrt{3}}$$

21. (a) Given, $16x^2-9y^2+32x+36y-164=0$

$$16(x^2+2x)-9(y^2-4y)-164=0$$

$\Rightarrow 16(x+1)^2-16-9(y-2)^2+36-164=0$

$\Rightarrow$ $$16(x+1)^2-9(y-2)^2=164+16-36$$

$\Rightarrow$ $$16(x+1)^2-9(y-2)^2=144$$

$\Rightarrow$ $$\frac{(x+1)^2}{9}-\frac{(y-2)^6}{16}=1$$

$\Rightarrow$ $$a^2=9,\ b^2=16$$

$$e=\sqrt{1+\frac{16}{9}}=\frac{5}{3}$$

Foci $(-1\pm ae, 2)=(-1\pm 5, 2)$

$=(-1+5, 2)$ and $(-1-5, 2)$

$(4, 2)$ and $(-6, 2)$

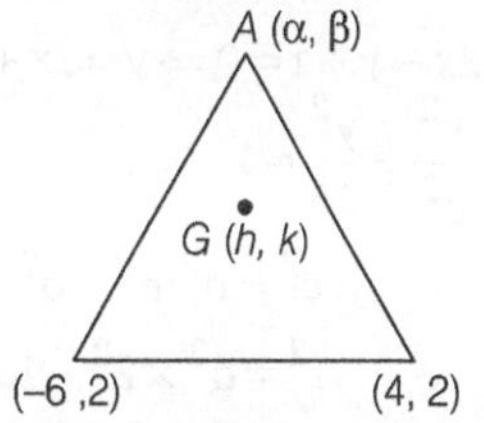

Let the centroid be (h, k) and $A(\alpha, \beta)$ be point on hyperbola.

So, $$h=\frac{\alpha-6+4}{3},\ k=\frac{\beta+2+2}{3}$$

$$\alpha=3h+2, \beta=3k-4$$

$\because (\alpha, \beta)$ lies on on hyperbola, so $S_1=0$

$\Rightarrow$ $16(3h+2+1)^2-9(3k-4-2)^2=144$

$\Rightarrow$ $144(h+1)^2-81(k-2)^2=144$

$\Rightarrow$ $16(h^2+2h+1)-9(k^2-4k+4)=16$

Locus is $16x^2-9y^2+32x+36y-36=0$

22. (a) Tangent at point $P(a\sec\theta, b\tan\theta)$ for the ellipse is

$$\frac{x}{a}\sec\theta-\frac{y}{b}\tan\theta=1 \quad \text{...(i)}$$

Given, tangent is $ax+by=1$ ...(ii)

Comparing Eqs. (i) and (ii), we have

$$\sec\theta=a^2 \text{ and } \tan\theta=-b^2$$

$\because$ $$1+\tan^2\theta=\sec^2\theta$$

$$1+b^4=a^4$$

$$a^4-b^4=1$$

$$(a^2-b^2)(a^2+b^2)=1$$

$\because$ $$a^2+b^2=a^2e^2$$

$\therefore$ $$(a^2-b^2)(a^2e^2)=1$$

$\Rightarrow$ $$a^2-b^2=\frac{1}{a^2e^2}$$

23. (c) $\because$ Equation of hyperbola is

$$\frac{x^2}{9}-\frac{y^2}{16}=1$$

$\because$ Equation of tangent having slope (m) is

$$y=mx+\sqrt{9m^2-16}$$

$$[\because y=mx+\sqrt{a^2m^2-b^2}]$$

Given, $y=mx+2\sqrt{5}$

$\therefore$ $$\sqrt{9m^2-16}=2\sqrt{5}$$

$\Rightarrow \quad 9m^2 - 16 = 20$

$\Rightarrow \quad 9m^2 = 36$

$\Rightarrow \quad m^2 = 4$

$\Rightarrow \quad m = \pm 2$

$\therefore$ Sum of roots $= +2 - 2 = 0$

24. (d) Given, $2x - y + 1 = 0 \Rightarrow y = 2x + 1$ is a tangent to hyperbola $\frac{x^2}{a^2} - \frac{y^2}{16} = 1$

$$\therefore \quad c^2 = a^2m^2 - b^2$$

$$1^2 = a^2 \times 2^2 - 16$$

$$a^2 = \frac{17}{4}$$

$$\Rightarrow \quad a = \frac{\sqrt{17}}{2}$$

$\therefore$ a, 4, 1 and a, 4, 2 and $2a$, 8, 1

i.e. $\frac{\sqrt{17}}{2}$, 4, 1 and $\frac{\sqrt{17}}{2}$, 4, 2 and $\sqrt{17}$, 8, 1

all can not be sides of a right triangle.

25. (b) $\because$ If $y = mx + c$ is the tangent to the hyperbola, then

$$c^2 = a^2m^2 - b^2$$

$$\therefore \quad \beta^2 = a^2\alpha^2 - b^2$$

Hence, the locus of $P(\alpha, \beta)$ is

$$a^2x^2 - y^2 = b^2$$

which is a hyperbola.

26. (a) Tangent to $\frac{x^2}{9} - \frac{y^2}{4} = 1$ at $P(3\sec\theta, 2\tan\theta)$ is

$$\frac{x}{3}\sec\theta - \frac{y}{2}\tan\theta = 1$$

This is perpendicular to

$$5x + 2y - 10 = 0$$

$$\therefore \quad \frac{2\sec\theta}{3\tan\theta} = \frac{2}{5}$$

$$\Rightarrow \quad \sin\theta = \frac{5}{3}$$

which is not possible.

$\therefore$ Number of possible tangents $= 0$

27. (c)

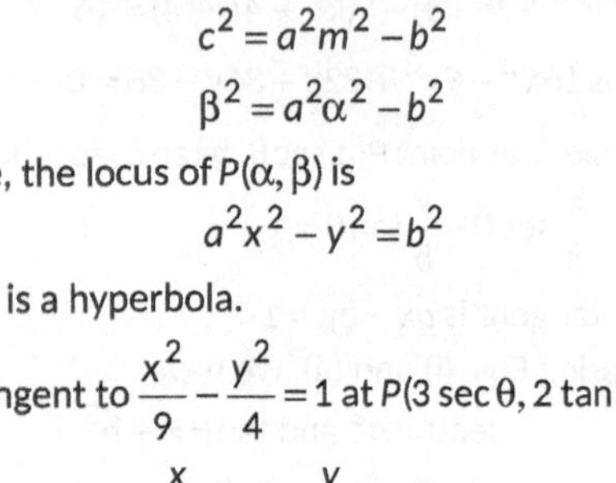

Point P is nearest to the given line, if tangent at P is parallel to the given line.

Now, equation of tangent at $P(x_1, y_1)$ to the hyperbola

$$\frac{x^2}{24} - \frac{y^2}{18} = 1$$

$$\Rightarrow \quad \frac{xx_1}{24} - \frac{yy_1}{18} = 1$$

$\therefore$ Slope of tangent is $\frac{18x_1}{24y_1} = \frac{3x_1}{4y_1}$

Which must be equal to $-\frac{3}{2}$

$$\therefore \quad \frac{3x_1}{4y_1} = -\frac{3}{2} \Rightarrow y_1 = -\frac{x_1}{2} \quad \text{...(i)}$$

Also (x_1, y_1) lies on the curve,

$$\text{Hence,} \quad \frac{x_1^2}{24} - \frac{y_1^2}{18} = 1 \quad \text{...(ii)}$$

Solving Eqs. (i) and (ii), we get two points $(6, -3)$ and $(-6, 3)$ of which $(6, -3)$ is nearest.

28. (b) $\because$ Equation of normal at (x_1, y_1) to hyperbola $\frac{x^2}{a^2} - \frac{y^2}{b^2} = 1$ is $\frac{a^2x}{x_1} + \frac{b^2y}{y_1} = a^2 + b^2$

$\because$ Given, $(x_1, y_1) = (6, 3)$

$\therefore$ Equation is $\frac{a^2x}{6} + \frac{b^2y}{3} = a^2 + b^2$

$\because$ It intersect X-axis at $(9, 0)$.

$$\therefore \quad \frac{a^2(9)}{6} + \frac{b^2(0)}{3} = a^2 + b^2$$

$$\Rightarrow \quad 3a^2 = 2a^2 + 2b^2$$

$$\Rightarrow \quad a^2 = 2b^2 \Rightarrow \frac{a^2}{b^2} = 2$$

$\because$ For hyperbola $\quad b^2 = a^2(e^2 - 1)$

$$\frac{1}{2} = e^2 - 1$$

$$e^2 = 1 + \frac{1}{2} \Rightarrow e = \sqrt{\frac{3}{2}}$$

29. (b) $\because$ Equation of tangent at (x_1, y_1) is $\frac{xx_1}{a^2} - \frac{yy_1}{b^2} = 1$

It passes through $(0, -b)$.

$$\text{So,} \quad 0 + \frac{y_1}{b} = 1 \Rightarrow y_1 = b$$

Equation of normal is $\frac{a^2x}{x_1} + \frac{b^2y}{y_1} = a^2e^2$

which passes through $(2a\sqrt{2}, 0)$.

$$\text{Hence,} \quad \frac{a^2(2a\sqrt{2})}{x_1} = a^2e^2$$

$$x_1 = \frac{2a\sqrt{2}}{e^2}$$

Now, (x_1, y_1) lies on hyperbola

$$\frac{x_1^2}{a^2} - \frac{y_1^2}{b^2} = 1$$

$$\Rightarrow \quad \frac{8}{e^4} - 1 = 1$$

$$\Rightarrow \quad 8/e^4 = 2 \Rightarrow e = \sqrt{2}$$

30. (d) The normal at $P(a \sec\alpha, b \tan\alpha)$ is

$$ax \cos\alpha + by \cot\alpha = a^2 + b^2 \quad \text{...(i)}$$

and the normal at $Q\left(a \sec\left(\frac{\pi}{2} - \alpha\right), b \tan\left(\frac{\pi}{2} - \alpha\right)\right)$

$= (a \operatorname{cosec}\alpha, b \cot\alpha)\left[\because \alpha + \beta = \frac{\pi}{2}\right]$ is

$$ax \sin\alpha + by \tan\alpha = a^2 + b^2 \quad \text{...(ii)}$$

Multiply Eq. (i) by $\sin\alpha$ and Eq. (ii) by $\cos\alpha$ and subtract them, we get

$$by(\cos\alpha - \sin\alpha) = (a^2 + b^2)(\sin\alpha - \cos\alpha)$$

$$y = -\left(\frac{a^2 + b^2}{b}\right)$$

$$\therefore \quad k = -\left(\frac{a^2 + b^2}{b}\right)$$

31. (c) $P(-2\sqrt{6}, \sqrt{3})$ lies on hyperbola $\frac{x^2}{a^2} - \frac{y^2}{b^2} = 1$

$$\Rightarrow \quad \frac{24}{a^2} - \frac{3}{b^2} = 1 \quad \text{...(i)}$$

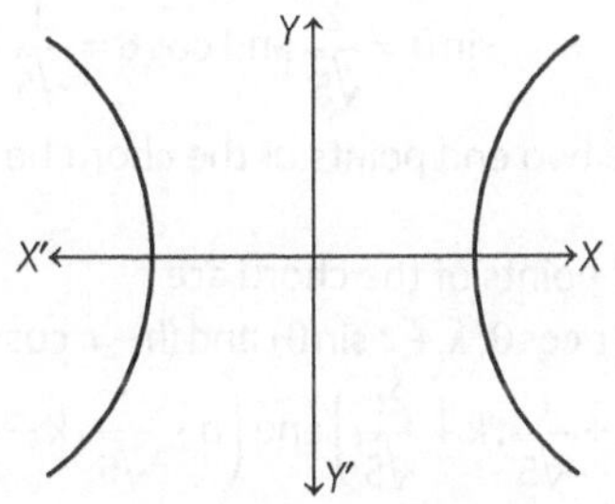

$\because$ Given, $\quad e = \frac{\sqrt{5}}{2}$

and $\quad b^2 = a^2(e^2 - 1)$

$$b^2 = a^2\left(\frac{5}{4} - 1\right)$$

$$4b^2 = a^2$$

Put it in Eq. (i)

$$\frac{6}{b^2} - \frac{3}{b^2} = 1,\ b^2 = 3,\ b = \sqrt{3} \Rightarrow a = 2\sqrt{3}$$

$\therefore$ Equation is $\frac{x^2}{12} - \frac{y^2}{3} = 1$

Tangent at P, $\frac{x}{12}(-2\sqrt{6}) - \frac{y}{3}(\sqrt{3}) = 1$

$$\Rightarrow \quad -\frac{x}{\sqrt{6}} - \frac{y}{\sqrt{3}} = 1$$

For Q put $x = 0$, we get

$$y = -\sqrt{3}$$

$$\therefore \quad Q(0, -\sqrt{3})$$

Normal at P, $(y - \sqrt{3}) = \sqrt{2}(x + 2\sqrt{6})$

For R put $x = 0$ $\quad y = \sqrt{3} + 4\sqrt{3} = 5\sqrt{3}$

$$\therefore \quad R = (0, 5\sqrt{3})$$

$$\therefore \quad QR = 6\sqrt{3}$$

32. (a) Equation of tangents at any point is

$$\frac{x \sec\theta}{a} - \frac{y \tan\theta}{b} = 1$$

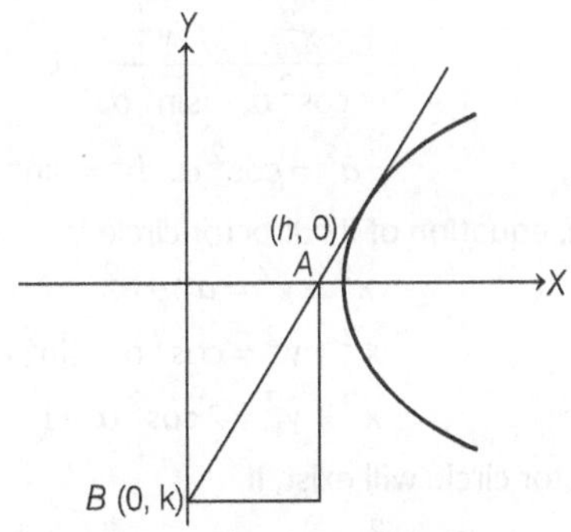

So, for h put $y = 0$

$$h = a \cos\theta \Rightarrow \sec\theta = \frac{a}{h}$$

of B put $x = 0$, $k = -b \cot\theta$, $\tan\theta = -\frac{b}{k}$

$$\because \quad 1 + \tan^2\theta = \sec^2\theta$$

$$1 + \frac{b^2}{k^2} = \frac{a^2}{h^2}$$

$$\frac{a^2}{h^2} - \frac{b^2}{k^2} = 1$$

Locus is $\frac{a^2}{x^2} - \frac{b^2}{y^2} = 1$

33. (d) Given, equation $16x^2 - 25y^2 = 400$

$$\Rightarrow \quad \frac{x^2}{25} - \frac{y^2}{16} = 1$$

$$S \Rightarrow 16x^2 - 25y^2 - 400 = 0$$

Point : $(1, 2\sqrt{2})$

$\because$ The equation of pair of tangents is

$$SS_1 = T^2$$

$$(16x^2 - 25y^2 - 400)(16(1)^2 - 25(2\sqrt{2})^2 - 400)$$
$$= [16x(1) - 25y(2\sqrt{2}) - 400)]^2$$

$$\Rightarrow (16x^2 - 25y^2 - 400)(16 - 200 - 400)$$
$$= (16x - 50\sqrt{2}y - 400)^2$$

$\Rightarrow -584(16x^2-25y^2-400)=(16x-50\sqrt{2}y-400)^2$

$\Rightarrow (16x-50\sqrt{2}y-400)^2+584(16x^2-25y^2-400)=0$

Let m_1 and m_2 are root and slope of tangent, then

$m_1 m_2=\frac{a}{b}$.

$$=\frac{\text{Coefficient of } x^2}{\text{Coefficient of } y^2}$$

$$=\frac{(16)^2+(584)\,16}{(50\sqrt{2})^2-(584)\,(25)}=\frac{9600}{-9600}=-1$$

$\therefore \quad m_1 m_2=-1$

34. (d) Given, $x^2\sec^2\alpha-y^2\operatorname{cosec}^2\alpha=1$

$$\Rightarrow \quad \frac{x^2}{\cos^2\alpha}-\frac{y^2}{\sin^2\alpha}=1$$

Here, $\quad a^2=\cos^2\alpha,\ b^2=\sin^2\alpha$

Then, equation of its director circle is

$$x^2+y^2=a^2-b^2$$

$$\Rightarrow \quad x^2+y^2=\cos^2\alpha-\sin^2\alpha$$

$$x^2+y^2=2\cos^2\alpha-1$$

Director circle will exist, if

$$2\cos^2\alpha-1>0 \Rightarrow \cos^2\alpha>\frac{1}{2}$$

When $\alpha\in\left(0,\frac{\pi}{4}\right)$

$\therefore$ There will exists infinite points.

35. (d) Substitute $x=9$ in the curve equation $x^2-y^2=9$.

So, that $y^2=72$ or $y=\pm 6\sqrt{2}$

Hence, the extremities of the chord of contact are $(9,\pm 6\sqrt{2})$.

The tangent at $(9,6\sqrt{2})$ is $9x-6\sqrt{2}y-9=0$

and the tangent at $(9,-6\sqrt{2})$ is $9x+6\sqrt{2}y-9=0$

Hence, the combined equation of tangents drawn at the ends of chord of contact is

$$(9x-6\sqrt{2}y-9)(9x+6\sqrt{2}y-9)=0$$

$$\Rightarrow 81x^2-72y^2-9(9x-6\sqrt{2}y+9x+6\sqrt{2}y)+81=0$$

$$\Rightarrow \quad 81x^2-72y^2-9(18x)+81=0$$

$$\Rightarrow \quad 9x^2-8y^2-18x+9=0$$

36. (a) Given, equation of hyperbola, $\frac{x^2}{a^2}-\frac{y^2}{b^2}=1$

Let the coordinate of the middle point of the chord be (h,k).

Applying $T=S_1$ will give the equation of chord with mid-point (h,k).

So, $\left(\frac{xh}{a^2}-\frac{yk}{b^2}-1\right)=\left(\frac{h^2}{a^2}-\frac{k^2}{b^2}-1\right)$

This line passes through (α,β), so we get

$$\frac{\alpha h}{a^2}-\frac{\beta k}{b^2}-1=\frac{h^2}{a^2}-\frac{k^2}{b^2}-1$$

$$\Rightarrow \quad \frac{h^2}{a^2}-\frac{\alpha h}{a^2}-\frac{k^2}{b^2}+\frac{\beta k}{b^2}=1$$

$$\Rightarrow \left[\frac{h^2}{a^2}-\frac{2h}{a}\cdot\frac{\alpha}{2a}+\left(\frac{\alpha}{2a}\right)^2\right]-\left[\left(\frac{k}{b}\right)^2-2\left(\frac{k}{b}\right)\left(\frac{\beta}{2b}\right)+\left(\frac{\beta}{2b}\right)^2\right]$$

$$=\left(\frac{\alpha}{2a}\right)^2-\left(\frac{\beta}{2b}\right)^2$$

$$\Rightarrow \left(\frac{h}{a}-\frac{\alpha}{2a}\right)^2-\left(\frac{k}{b}-\frac{\beta}{2b}\right)^2=\frac{\alpha^2}{4a^2}-\frac{\beta^2}{4b^2}$$

$$\Rightarrow \frac{\left(h-\frac{\alpha}{2}\right)^2}{a^2}-\frac{\left(k-\frac{\beta}{2}\right)^2}{b^2}=\frac{\alpha^2}{4a^2}-\frac{\beta^2}{4b^2}$$

$\therefore$ Locus is $\dfrac{\left(x-\frac{\alpha}{2}\right)^2}{a^2}-\dfrac{\left(y-\frac{\beta}{2}\right)^2}{b^2}=\dfrac{\alpha^2}{4a^2}-\dfrac{\beta^2}{4b^2}$

with centre $\left(\frac{\alpha}{2},\frac{\beta}{2}\right)$.

37. (a) Let (h,K) be the mid-points of chords having slope P.

$$\Rightarrow \quad \tan\theta=2$$

$$\Rightarrow \quad \sin\theta=\frac{2}{\sqrt{5}} \text{ and } \cos\theta=\frac{1}{\sqrt{5}}$$

Let the two end points of the chord be distance r from (h,k).

$\Rightarrow$ End points of the chord are

$(h+r\cos\theta,\ k+r\sin\theta)$ and $(h-r\cos\theta,\ k-r\sin\theta)$

$$\Rightarrow \left(h+\frac{r}{\sqrt{5}},k+\frac{2r}{\sqrt{5}}\right) \text{ and } \left(h-\frac{r}{\sqrt{5}},k-\frac{2r}{\sqrt{5}}\right)$$

Putting in the equation of hyperbola

$$3\left(h+\frac{r}{\sqrt{5}}\right)^2-2\left(k+\frac{2r}{\sqrt{5}}\right)^2+4\left(h+\frac{r}{\sqrt{5}}\right)-6\left(k+\frac{2r}{\sqrt{5}}\right)=0 \quad \text{...(i)}$$

$$\text{and } 3\left(h-\frac{r}{\sqrt{5}}\right)^2-2\left(k-\frac{2r}{\sqrt{5}}\right)^2+4\left(h-\frac{r}{\sqrt{5}}\right)-6\left(k-\frac{2r}{\sqrt{5}}\right)=0 \quad \text{...(ii)}$$

Subtracting Eq. (ii) from Eq. (i), we get

$$\frac{12hr}{\sqrt{5}}-\frac{16kr}{\sqrt{5}}+\frac{8r}{\sqrt{5}}-\frac{24r}{\sqrt{5}}=0$$

$$3h-4k-4=0$$

$\therefore$ Locus is $3x-4y-4=0$

38. (c) Equation of hyperbola $\frac{x^2}{a^2}-\frac{y^2}{b^2}=1$

$\because$ Let mid-point of chord is (h, k).

$\therefore$ Equation of chord is

$$T=S_1$$

$$\frac{xh}{a^2}-\frac{ky}{b^2}=\frac{h^2}{a^2}-\frac{k^2}{b^2} \quad \text{...(i)}$$

$\because$ Point (α, β) passes through Eq. (i).

$$\therefore \quad \frac{\alpha h}{a^2}-\frac{\beta y}{b^2}=\frac{h^2}{a^2}-\frac{k^2}{b^2}$$

$$\Rightarrow \left(\frac{h}{a}-\frac{\alpha}{2a}\right)^2-\left(\frac{k}{b}-\frac{\beta}{2b^2}\right)=\frac{\alpha^2}{4a^2}-\frac{\beta^2}{4a^2}$$

$\therefore$ Centre $\frac{h}{a}=\frac{\alpha}{2a}$ and $\frac{k}{b}=\frac{\beta}{2b^2}; h=\frac{\alpha}{2};\ k=\frac{\beta}{2}$

Centre is $\left(\frac{\alpha}{2}, \frac{\beta}{2}\right)$.

39. (b) The extremities of a pair of conjugate diameter of $\frac{x^2}{a^2}-\frac{y^2}{b^2}=1$ are $(a\sec\phi, b\tan\phi)$ and $(a\tan\phi, b\sec\phi)$, respectively.

According to the question, since extremities of a pair of conjugate diameters lie on $lx+my+n=0$

$$\therefore \quad l(a\sec\phi)+m(b\tan\phi)+n=0 \quad \text{...(i)}$$

$$\Rightarrow \quad l(a\tan\phi)+m(b\sec\phi)+n=0 \quad \text{...(ii)}$$

Then, from Eqs. (i) and (ii),

$$al\sec\phi+bm\tan\phi=-n$$

or

$$a^2l^2\sec^2\phi+b^2m^2\tan^2\phi+2ablm\sec\phi\tan\phi=n^2 \quad \text{...(iii)}$$

And from Eq. (ii) $al\tan\phi+bm\sec\phi=-n$

$$a^2l^2\tan^2\phi+b^2m^2\sec^2\phi+2ablm\sec\phi\tan\phi=n^2 \quad \text{...(iv)}$$

Then subtracting Eq. (iv) from Eq. (iii),

$$a^2l^2(\sec^2\phi-\tan^2\phi)+b^2m^2(\tan^2\phi-\sec^2\phi)=0$$

$$\Rightarrow \quad a^2l^2-b^2m^2=0$$

40. (c) Let equation of ellipse is

$$\frac{x^2}{a^2}+\frac{y^2}{b^2}=1 \quad (a>b)$$

$\therefore$ Director circle is $x^2+y^2=a^2+b^2$

And $\quad b^2=a^2(1-e_1^2)$

Given, $\quad a^2+b^2=(2R)^2$

$$a^2+a^2(1-e_1^2)=(2R)^2$$

$$2-e_1^2=\frac{4R^2}{a^2} \quad \text{...(i)}$$

And equation of hyperbola is

$$\frac{x^2}{a^2}-\frac{y^2}{b_1^2}=1$$

And $\quad b_1^2=a^2(e_2^2-1)$

And given, $\quad a^2-b_1^2=R^2$

$$a^2-a^2(e_2^2-1)=R^2$$

$$2-e_2^2=\frac{R^2}{a^2} \quad \text{...(ii)}$$

Dividing Eq. (i) by Eq. (ii), then

$$\frac{2-e_1^2}{2-e_2^2}=4 \Rightarrow 4e_2^2-e_1^2=6$$

41. (d) Transverse axis is the equation of the angle bisector containing point (2, 3) which is given by

$$\frac{3x-4y+5}{5}=\frac{12x+5y-40}{13}$$

$$\Rightarrow \quad 13(3x-4y+5)=5(12x+5y-40)$$

$$\Rightarrow \quad 21x+77y=265$$

42. (a) Let the equation of asymptotes are

$$2x^2+5xy+2y^2+4x+5y+\lambda=0 \quad \text{...(i)}$$

This equation represents a pair of straight lines.

Therefore, $abc+2fgh-af^2-bg^2-ch^2=0$

We have, $\quad 4\lambda+25-\frac{25}{2}-8-\lambda\left(\frac{25}{4}\right)=0$

$$-\frac{9\lambda}{4}+\frac{9}{2}=0$$

$$\lambda=2$$

$\therefore$ Required equation of asymptotes is

$$2x^2+5xy+2y^2+4x+5y+2=0$$

43. (b) Equation of asymptotes are

$$\sqrt{3}x-y+5=0 \text{ and } \sqrt{3}x+y-1=0$$

$a_1=\sqrt{3}, b_1=-1, c_1=5$ and $a_2=\sqrt{3}, b_2=1, c_2=-1$

$\because \ c_1c_2(a_1a_2+b_1b_2)=5(-1)[\sqrt{3}(\sqrt{3})+(1)(-1)]$

$$=-5(3-1)=-10<0$$

Therefore, origin lies in acute angle.

$P(1, 2)$ lies in obtuse angle.

Acute angle between asymptotes

$$\Rightarrow \quad \tan\theta=\left|\frac{m_1+m_2}{1-m_1m_2}\right|$$

$$=\left|\frac{\sqrt{3}+\sqrt{3}}{1-3}\right|$$

$$=|\sqrt{3}|$$

$$\theta=\frac{\pi}{3}$$

$$\therefore \quad e=\sec\frac{\theta}{2}=\sec\frac{\pi}{6}=\frac{2}{\sqrt{3}}$$

44. (c) For equation

$\because K+S=0$ to represents a pair of line $\Rightarrow \Delta=0$

$$\begin{vmatrix} a & h & g \\ h & b & f \\ g & f & c \end{vmatrix}=0$$

$$=\begin{vmatrix} 1 & 2 & -2 \\ 2 & 3 & 1 \\ -2 & 1 & 1+k \end{vmatrix}=0$$

$$=\{3(1+k)-1\}-2(2+2k+2)-2(2+6)=0$$

$$k=-22$$

45. (d) Angle between asymptotes $=2+a_n-1\left(\frac{b}{a}\right)=120°$

$$\tan^{-1}\left(\frac{b}{a}\right)=60°$$

$$\frac{b}{a}=\sqrt{3}$$

$$b=a\sqrt{3}\Rightarrow b^2=3a^2$$

$$\because \quad b^2=9 \Rightarrow a^2=\frac{9}{3}=3$$

Required locus is director circle of the hyperbola and which is

$$x^2+y^2=a^2-b^2=3-9=-6$$

which is not possible.

Now, angle between asymptotes $\Rightarrow 2\tan^{-1}\frac{b}{a}=60°$

$$\Rightarrow \quad \tan^{-1}\frac{b}{a}=30°$$

$$\Rightarrow \quad \frac{b}{a}=\frac{1}{\sqrt{3}}$$

$$\Rightarrow \quad a^2=3b^2$$

$$\because \quad b^2=9$$

$$a^2=27$$

Required locus is $x^2+y^2=a^2-b^2$

$$x^2+y^2=27-9$$

$$x^2+y^2=18$$

46. (a) Any tangent to hyperbola formed a triangle with the asymptotes which has constant area a is

Given, $\quad ab=a^2\tan\lambda$

$$\frac{b}{a}=\tan\lambda$$

$$\Rightarrow \quad e^2-1=\tan^2\lambda$$

$$\Rightarrow \quad e^2=1+\tan^2\lambda$$

$$\Rightarrow \quad e^2=\sec^2\lambda$$

$$\Rightarrow \quad e=\sec\lambda$$

47. (a) Let tangent at $P(a\sec\theta, b\tan\theta)$ is

$$\frac{x}{a}\sec\theta-\frac{y}{b}\tan\theta=1 \quad \text{...(i)}$$

Then, asymptotes $\quad y=\frac{b}{a}x \quad$ ...(ii)

$$y=-\frac{b}{a}x \quad \text{...(iii)}$$

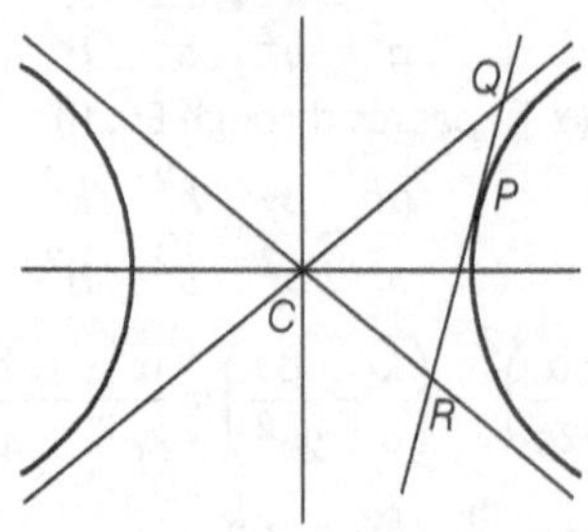

By solving Eqs. (i) and (ii), we get

$$Q=(a(\sec\theta+\tan\theta), b(\sec\theta+\tan\theta))$$

By solving Eqs. (i) and (iii),we get

$$R=(a(\sec\theta-\tan\theta), b(\tan\theta-\sec\theta))$$

Now, let circumcentre of ΔCQR be $O(h,k)$.

Then, $\quad OC^2=OQ^2$

$$\Rightarrow \quad h^2+k^2=\{h-a(\sec\theta+\tan\theta)\}^2+\{k-b(\sec\theta+\tan\theta)\}^2$$

$$\Rightarrow \quad 2(ha+kb)=(\sec\theta+\tan\theta)(a^2+b^2) \quad \text{...(iii)}$$

and $OC^2=OR^2$, we get

$$2(ha-kb)=(\sec\theta-\tan\theta)(a^2+b^2) \quad \text{...(iv)}$$

Multiply Eqs. (iii) and (iv),

$$4\{(ha)^2-(kb)^2\}=(\sec^2\theta-\tan^2\theta)(a^2+b^2)^2$$

$$\Rightarrow \quad 4(h^2a^2-k^2b^2)=(a^2+b^2)^2$$

$$\Rightarrow \quad 4(x^2a^2-y^2b^2)=(a^2+b^2)^2$$

48. (b) Let S be a point on the rectangular hyperbola say $\left(t,\frac{1}{t}\right)$

Now, circumcircle of ΔOQR also passes through S. Therefore, circumcentre is the mid-point of OS.

Hence, $x=\frac{t}{2}, y=\frac{1}{2t}$

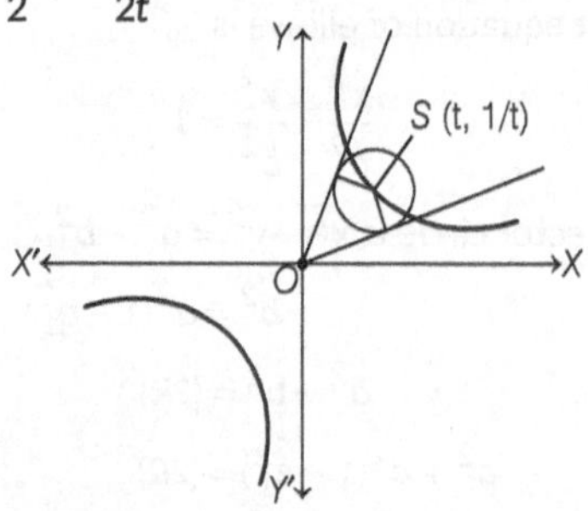

So, the locus of the circumcentre is $xy=\frac{1}{4}$.

49. (c) Foci of hyperbola lie on $y = x$

So, the major axis is $y = x$

Major axis of hyperbola bisects the asymptote.

$\Rightarrow$ Equation of other asymptote is $x = 2y$

$\Rightarrow$ Equation of hyperbola is $(y - 2x)(x - 2y) + k = 0$

Given that, it passes through $(3, 4) \Rightarrow k = -10$

$\therefore$ Required equation is

$$2x^2 + 2y^2 - 5xy + 10 = 0$$

50. (b) Slope of tangent at $P = -\dfrac{1}{t^2}$

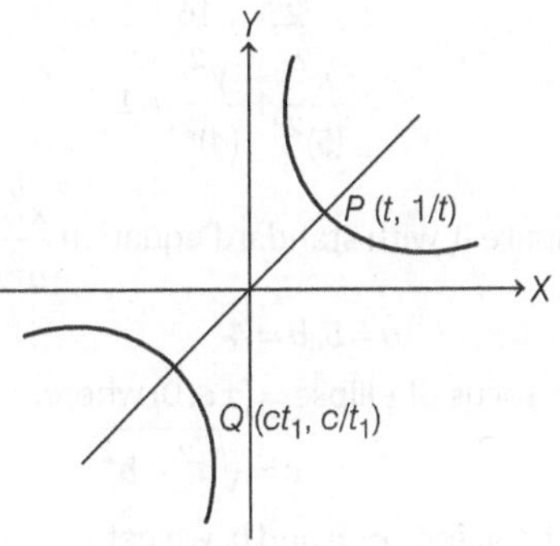

So, slope of normal $= t^2$

$$t^2 = \frac{\dfrac{c}{t_1} - \dfrac{c}{t}}{ct_1 - ct}$$

$$t^2 = \frac{-1}{t_1 t}$$

$$t^3 t_1 = -1$$

51. (c) Tangent to hyperbola $xy = c^2$.

$$\frac{x}{x_1} + \frac{y}{y_1} = 2$$

$\because$ Points is $(16, 1) \Rightarrow x_1 = 16,\ y_1 = 1$

Equation is $\dfrac{x}{16} + \dfrac{y}{1} = 2$

$$x + 16y = 32$$

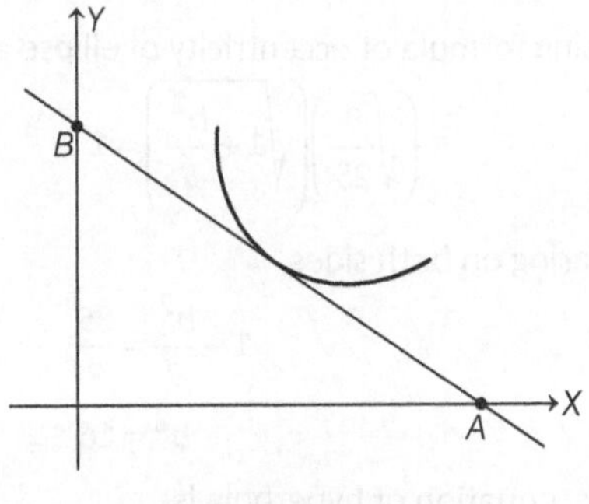

$\therefore \quad A = (32, 0)$ and $B(0, 2)$

$\therefore$ Area of $\Delta OAB = \dfrac{1}{2} \times 32 \times 2 = 32$

52. (a) Let the middle point $M(h, k)$.

Chord $\Rightarrow T = S_1$, where $S = xy - c^2 = 0$

$$\frac{x}{h} + \frac{y}{k} = 2$$

$\therefore \quad$ Slope of line is $-\dfrac{k}{h}$

According to the question,

$$\left(-\frac{k}{h}\right) = m \Rightarrow mh + k = 0$$

Locus $mx + y = 0$

53. (b) The equation of the normal to the rectangular hyperbola is

$$xt^3 - yt = c(t^4 - 1) \qquad \text{...(i)}$$

Let the pole be $M(h, k)$.

Then, equation of polar at M is

$$xk + yh = 2c^2 \qquad \text{...(ii)}$$

$\because$ The lines (i) and (ii) are same line. So,

$$\frac{t^3}{k} = -\frac{t}{h} = \frac{c(t^4 - 1)}{2c^2}$$

$$\Rightarrow \quad \frac{t^3}{k} = -\frac{t}{h} = \frac{t^4 - 1}{2c}$$

Solving, we get

$$-\frac{k}{h^3} = \frac{(t^4 - 1)^2}{4c^2}$$

$$\Rightarrow \quad -\frac{k}{h^3} = \frac{\left(\dfrac{k^2}{h^2} - 1\right)^2}{4c^2}$$

$$\Rightarrow \quad -4c^2 kh = (k^2 - h^2)^2$$

$$\Rightarrow \quad (h^2 - y^2)^2 + 4c^2 kh = 0$$

$\therefore$ Locus is $(x^2 - y^2) + 4c^2 xy = 0$

54. (c) Let equation of common tangent w.r.t. parabola is

$$y = mx + \frac{3}{m}$$

$$\left\{y^2 = 12x, a = 3 \because \text{Equation of tangent, } y = mx + \frac{a}{m}\right\}$$

For hyperbola

$$\therefore \quad c^2 = am^2 - b^2$$

$$\left(\frac{3}{m}\right)^2 = 1(m^2) - 8 \qquad \left\{S : \frac{x^2}{1} - \frac{y^2}{8} = 1\right\}$$

$$\frac{9}{m^2} = m^2 - 8$$

$$m^4 - 8m^2 - 9 = 0$$

$$(m^2 - 9)(m^2 + 1) = 0$$

$$m^2 = 9$$

$$m = \pm 3$$

$\therefore$ Equation are

$$y = 3x + 1 \quad \text{...(i)}$$

and $$y = -3x - 1 \quad \text{...(ii)}$$

$P \Rightarrow$ Solving Eqs. (i) and (ii),

$$y = 0 \text{ and } x = -\frac{1}{3}, P = \left(-\frac{1}{3}, 0\right)$$

$$S = (a, 0) = (3, 0)$$

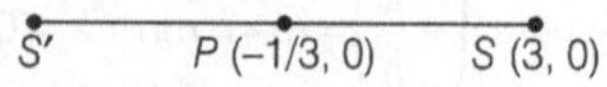

$$S' = \text{Focus of hyperbola}$$
$$= (-ae, 0) = (-3, 0) \quad [\because e = \sqrt{1+8} = 3, -ae = -3]$$

$$\therefore \quad \frac{PS}{PS'} = \frac{\left|3 + \frac{1}{3}\right|}{\left|3 - \frac{1}{3}\right|} = \frac{5}{4}$$

55. (d) $\dfrac{x^2}{100} - \dfrac{y^2}{64} = 1$

$\because$ Condition of tendency

$$c^2 = a^2m^2 - b^2$$
$$c = \pm\sqrt{a^2m^2 - b^2}$$
$$c = \pm\sqrt{100m^2 - 64}$$

Equation of tangent

$$y = mx \pm \sqrt{100m^2 - 64}$$

This tangent is also tangent to circle $x^2 + y^2 = 36$ whose centre (0, 0) and radius = 6

Distance of the tangent from centre is radius,

$$\left|\frac{\sqrt{100m^2 - 64}}{\sqrt{m^2 + 1}}\right| = 6$$

$$\Rightarrow \quad 100m^2 - 64 = 36m^2 + 36$$
$$\Rightarrow \quad 64m^2 = 100$$
$$\Rightarrow \quad m = \frac{10}{8}$$
$$\Rightarrow \quad c^2 = 100 \times \frac{100}{64} - 64$$
$$\Rightarrow \quad c^2 = \frac{(100)^2 - (64)^2}{64} = \frac{164 \times 36}{64}$$
$$\Rightarrow \quad 4c^2 = 369$$

56. (c) $\because$ Chord of contact when mid-point is given,

$$T = S_1$$
$$xh - yk = h^2 - k^2$$

[Let (h, k) be the mid-point of the hyperbola]

$$y = \frac{xh}{k} - \frac{(h^2 - k^2)}{k}$$

This touch the parabola $y^2 = 8x$, then $c = \dfrac{a}{m}$

$$\frac{k^2 - h^2}{k} = \frac{2k}{h}$$

$$\Rightarrow \quad (k^2 - h^2)h = 2k^2$$

$$\therefore \quad \text{Locus is } 2y^2 = x(y^2 - x^2)$$

$$\Rightarrow \quad y^2(x - 2) = x^3$$

57. (a) Given, ellipse is

$$\frac{x^2}{25} + \frac{y^2}{16} = 1$$

or $$\frac{x^2}{(5)^2} + \frac{y^2}{(4)^2} = 1$$

Compare it with standard equation $\dfrac{x^2}{a^2} + \dfrac{y^2}{b^2} = 1$, we get

$$a = 5, b = 4$$

Now, focus of ellipse $= (\pm e, 0)$ where

$$c = \sqrt{a^2 - b^2}$$

Put the values of a and b, we get

$$c = \sqrt{5^2 - 4^2} = \sqrt{25 - 16} = \sqrt{9}$$

$\therefore \quad$ Focus $= (\pm 3, 0)$

According to question, hyperbola passes through the focus of ellipse.

Let equation of hyperbola be $\dfrac{x^2}{a^2} - \dfrac{y^2}{b^2} = 1$

Since, it passes through $(\pm 3, 0)$, we get

$$\frac{(\pm 3)^2}{a^2} - \frac{0}{b^2} = 1, \text{ gives } a = \pm 3 \text{ or } a^2 = 9$$

Also, given that product of eccentrcikes is 1.

Now, (eccentricity of ellipse) (eccentricity of hyperbola) = 1

$$\Rightarrow \quad \left(\sqrt{1 - \frac{16}{25}}\right)\left(\sqrt{1 + \frac{b^2}{9}}\right) = 1$$

[using formula of eccentricity of ellipse and hyperbola]

$$\Rightarrow \quad \left(\sqrt{\frac{9}{25}}\right)\left(\sqrt{1 + \frac{b^2}{9}}\right) = 1$$

Squaring on both sides,

$$1 + \frac{b^2}{9} = \frac{25}{9}$$

$$\Rightarrow \quad b^2 = 16$$

Thus, equation of hyperbola is

$$\frac{x^2}{9} - \frac{y^2}{16} = 1.$$

58. (a) Equation of circle

$$x^2+y^2=3$$

$$\Rightarrow \quad C: x^2+y^2-3=0$$

Rectangular hyperbola $\Rightarrow xy=1$

$$\Rightarrow \quad S=xy-1=0$$

and equation $x^2+y^2-3+\lambda(xy-1)=0$

$\because$ This equation of points of intersection of circle and rectangular hyperbola

$$C+\lambda S=0$$

Now, for $\lambda=-3$

We get, $x^2+y^2-3-3xy+3=0$

$$x^2+y^2-3xy=0$$

That is a pair of straight lines passing through origin.

59. (a) If (x_i, y_i) is the point of intersection of the curves $(x-1)(y-2)=5$ and $(x-1)^2+(y+1)^2=r^2$

$$\therefore \quad \frac{\sum_{i=1}^{4} x_i}{4}=\frac{1+1}{2} \text{ and } \frac{\sum_{i=1}^{4} y_i}{4}=0$$

$$\frac{\sum_{i=1}^{3} x_i}{3}=\frac{4-x_4}{3} \text{ and } \frac{\sum_{i=1}^{3} y_i}{3}=-\frac{y_4}{4}$$

Centroid $\left(\dfrac{\sum_{i=1}^{3} x_i}{3}, \dfrac{\sum_{i=1}^{3} y_i}{3}\right)$ lies on the line $y=3x-4$

$$\therefore \quad -\frac{y_4}{3}=\frac{3(4-x_4)}{3}-4$$

$$y_4=3x_4$$

$\therefore$ Locus of D is $y=3x$

60. (d) Given, ellipse is $\dfrac{x^2}{\lambda^2 a^2}+\dfrac{y^2}{a^2}=1, \lambda^2 a^2>a^2$ and

Let e_1 and e_2 be the eccentricities of ellipse and hyperbola, then

$$a^2=\lambda^2(a^2)(1-e_1^2)$$

or $$e_1=\sqrt{1-\frac{1}{\lambda^2}} \text{ and } e_2=\sqrt{2}$$

Now, $$\lambda a e_1=a e_2$$

$$\Rightarrow \quad \lambda\sqrt{1-\frac{1}{\lambda^2}}=\sqrt{2} \Rightarrow \lambda^2-1=2 \Rightarrow \lambda=\pm\sqrt{3}$$

$$\lambda=\sqrt{3} \qquad [\because \lambda>0]$$

(a) $\dfrac{e_1}{e_2}=\dfrac{1}{\lambda}=\dfrac{1}{\sqrt{3}}$

(b) $\dfrac{\text{Major axis of ellipse}}{\text{Transverse axis of ellipse}}=\dfrac{2\lambda a}{2a}=\lambda=\sqrt{3}$

(c) Equation of tangents of ellipse and hyperbola at (x_1, y_1) are

$$\frac{xx_1}{\lambda^2 a^2}+\frac{yy_1}{a^2}=1 \text{ and } xx_1-yy_1=a^2$$

$$\text{Slopes}=-\frac{x_1}{\lambda^2 y_1} \text{ and } \frac{x_1}{y_1}$$

$$\therefore \quad \text{Product}=\frac{-x_1}{\lambda^2 y_1}\times\frac{x_1}{y_1}=\frac{-x_1^2}{\lambda^2 y_1^2}$$

$$=-\frac{\dfrac{2a^2}{\lambda^2+1}}{\dfrac{a^2(\lambda^2-1)}{\lambda^2+1}}=-\frac{2}{\lambda^2-1}=\frac{-2}{3-1}=-1$$

(d) $\dfrac{\text{Latus rectum of ellipse}}{\text{Latus rectum of hyperbola}}=\dfrac{\dfrac{2a^2}{\lambda a}}{\dfrac{2a^2}{a}} \Rightarrow \dfrac{1}{\lambda}=\dfrac{1}{\sqrt{3}}$

61. (b) $x-2=m$

$$y+1=\frac{4}{m}$$

$$\therefore \quad (x-2)(y+1)=4$$

$$XY=4$$

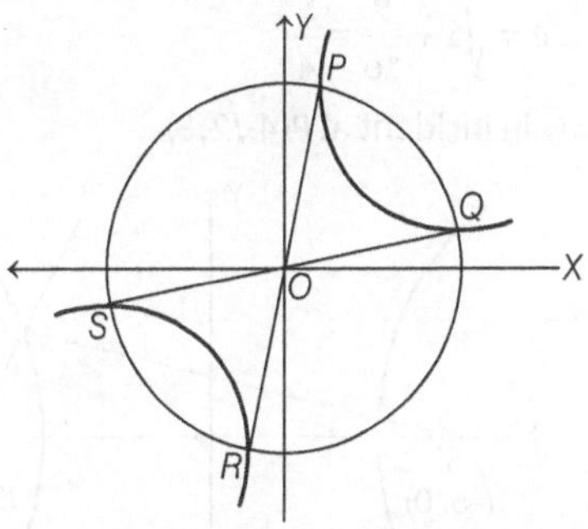

where, $$X=x-2 \text{ and } Y=y+1$$

$$S\equiv(x-2)^2+(y+1)^2=25$$

$$X^2+Y^2=25$$

Curve C and circle S both are concentric.

$$\therefore \quad OP^2+OQ^2+OR^2+OS^2=4r^2$$

$$=4\times 25=100$$

62. (a) Equation of tangent at point $P(a\cos\theta, b\sin\theta)$ is

$$\frac{x}{\alpha}\cos\theta+\frac{y}{1}\sin\theta=1 \qquad \text{...(i)}$$

Let it cut the hyperbola at points P and Q.

Homogenising the hyperbola $\alpha^2x^2-y^2=1$ with the help of Eq. (i), we get

$$\alpha^2x^2-y^2=\left(\frac{x}{\alpha}\cos\theta+y\sin\theta\right)^2$$

This is a pair of straight lines OP and OQ.

Given, $\angle POQ = \frac{\pi}{2}$

$\Rightarrow$ Coefficient of x^2 + Coefficient of $y^2 = 0$

$$\Rightarrow \quad \alpha^2 - \frac{\cos^2\theta}{\alpha^2} - 1 - \sin^2\theta = 0$$

$$\Rightarrow \quad \alpha^2 - \frac{\cos^2\theta}{\alpha^2} - 1 - 1 + \cos^2\theta = 0$$

$$\cos^2\theta = \frac{\alpha^2(2-\alpha^2)}{\alpha^2 - 1}$$

Now, $0 \le \cos^2\theta \le 1$

$$0 \le \frac{\alpha^2(2-\alpha^2)}{\alpha^2-1} \le 1$$

$$\alpha^2 \in \left[\frac{\sqrt{5}+1}{2}, 2\right]$$

63. (d) The given hyperbola is

$$\frac{x^2}{16} - \frac{y^2}{9} = 1$$

Here, $a^2 = 16, b^2 = 9$

$$\therefore \quad e = \sqrt{1 + \frac{9}{16}} = \frac{5}{4}$$

The ray in incident at $P(4\sqrt{2}, 3)$.

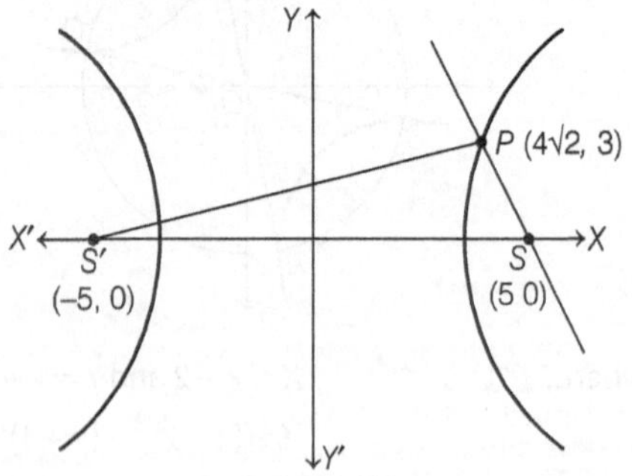

The incident ray passes through (5, 0), so the reflected ray must pass through (– 5, 0).

$\therefore$ The equation of reflected ray = Equation of PS'

$$\Rightarrow \quad (y - 0) = \frac{3-0}{4\sqrt{2}+5}(x+5)$$

$$\Rightarrow \quad y(4\sqrt{2}+5) = 3x + 15$$

$$\Rightarrow \quad 3x - y(4\sqrt{2}+5) + 15 = 0$$

64. (b) The given equation of transverse and conjugate axis are $x + 2y - 3$ and $2x - y + 4 = 0$, respectively.

and length $= \sqrt{2}, \frac{2}{\sqrt{3}}$

$$2a = \sqrt{2},\ 2b = \frac{2}{\sqrt{3}}$$

$$a = \frac{1}{\sqrt{2}},\ b = \frac{1}{\sqrt{3}}$$

$\therefore$ Equation of hyperbola

$$\frac{\left(\frac{2x-y+4}{\sqrt{5}}\right)^2}{\frac{1}{2}} - \frac{\left(\frac{x+2y-3}{\sqrt{5}}\right)^2}{\frac{1}{3}} = 1$$

$$\Rightarrow \quad \frac{2}{5}(2x-y+4)^2 - \frac{3}{5}(x+2y-3)^2 = 1$$

$$2(2x-y+4)^2 - 3(x+2y-3)^2 = 5$$

65. (5) Given, equation

$$[\sqrt{(x-3)^2+(y-2)^2} - \sqrt{(x+1)^2+(y+1)^2}] = 1 \quad \text{...(i)}$$

Let $P(x, y), S(3, 2)$ and $S'(-1, -1)$ are three point.

So, Eq. (i) represents $SP - S'P = 1$

which represents a hyperbola, where S and S' are foci

$$SP - S'P = 2a = 1 \quad \text{...(ii)}$$

$$\therefore \quad SS' = \sqrt{4^2 + 3^2} = 5$$

$\Rightarrow \quad 2ae = 5$ [$\because$ Distance between foci $= 2ae$]

$\Rightarrow \quad (1)\, e = 5$ [by Eq. (ii)]

$\Rightarrow \quad e = 5$

66. (0.072) The equation of any normal to the hyperbola $\frac{x^2}{a^2} - \frac{y^2}{b^2} = 1$ at ϕ is

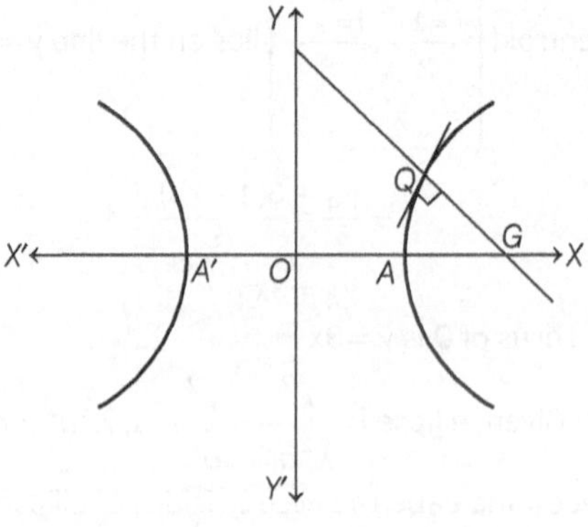

$$ax\cos\phi + by\cot\phi = a^2 + b^2$$

Thus, the coordinates of $G = \left(\frac{a^2+b^2}{a\cos\phi}, 0\right)$

Clearly, the vertices $A = (a, 0)$ and $A' = (-a, 0)$

Now, $AG = \left(\frac{a^2+b^2}{a\cos\phi} - a\right)$

$$A'G = \left(\frac{a^2+b^2}{a\cos\phi} + a\right)$$

Therefore, $AG \cdot A'G = \left(\frac{a^2+b^2}{a\cos\phi} - a\right)\left(\frac{a^2+b^2}{a\cos\phi} + a\right)$

$$= \frac{(a^2+b^2)^2}{a^2}\sec^2(\phi) - a^2$$
$$= (a^2e^2\sec^2\phi - a^2) = a^2(e^2\sec^2\phi - 1)$$
$\therefore$ $m=2, n=2, p=2$

Hence, $\dfrac{(m+n+p)^2+36}{1000} = \dfrac{36+36}{1000} = 0.072$

67. (3) $\because \dfrac{x^2}{a^2} - \dfrac{y^2}{b^2} = 1$ and $x^2+y^2 = \dfrac{a^2}{4}$ having origin as their common centre.

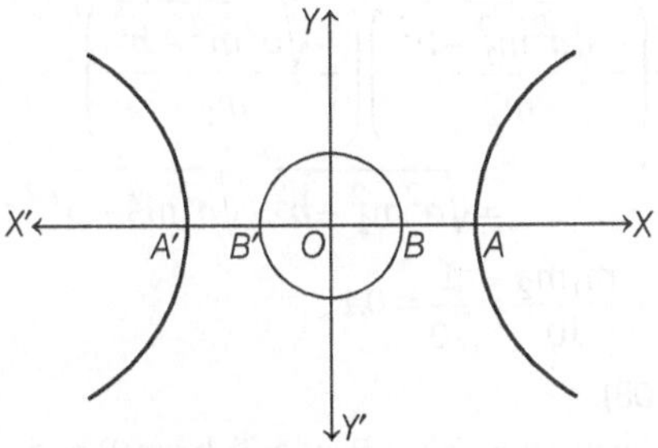

So, vertex is the nearest point.
Hence, shortest distance $= BA$
$$= a - \frac{a}{2} = \frac{a}{2}$$
$\therefore$ $f(a,b) = \dfrac{a}{2}$

$\therefore$ $f(4,6) + f(2,3) = \dfrac{4}{2} + \dfrac{2}{2} = 3$

68. (8) Hyperbola is $x^2 - 9y^2 = 9$ or $\dfrac{x^2}{9} - \dfrac{y^2}{1} = 1$

Equation of tangent is $y = mx \pm \sqrt{a^2m^2 - b^2}$ ...(i)

It passes through (3, 2).

$\Rightarrow$ $2 = 3m \pm \sqrt{9m^2 - 1}$

or $4 + 9m^2 - 12m = 9m^2 - 1 \Rightarrow 0m^2 - 12m + 5 = 0$

Solving, we get values of m as $m_1 = -\dfrac{5}{12}$ and $m_2 = \infty$

Equation of tangent (i) for $m_1 = \dfrac{5}{12}$
$$y = \frac{5}{12}x \pm \sqrt{9\left(\frac{5}{12}\right)^2 - 1}$$
$$y = \frac{5}{12} \pm \frac{3}{4}$$

On taking negative sign point $P(3,2)$ does not satisfy the equation of tangent therefore rejecting negative sign.

$\therefore$ Equation of tangent is $y = \dfrac{5x}{12} + \dfrac{3}{4}$ ...(ii)

How equation of tangent (i) for $m_2 = \infty$ is $x \pm 3 = 0$

Rejective positive sign.

[$\because$ taking positive sign point $P(3,2)$ does not satisfy this equation]

Hence, equation of tangent is $x - 3 = 0$ ...(iii)

Now, equation of chord of contact w.r.t. point $P(3,2)$ is $T = 0$.

or $3x - 18y = 9$

$x - 6y = 3$ ...(iv)

Solving Eqs. (ii) and (iv), $x = -5,\ y = -\dfrac{4}{3}$

Solving Eqs. (iii) and (iv), $x = 3,\ y = 0$

Now, vertices of triangle are (3, 2) (3, 0), $\left(-5, -\dfrac{4}{3}\right)$

$$\therefore \quad \text{Area} = \frac{1}{2}\text{mod}\begin{vmatrix} 3 & 2 & 1 \\ 3 & 0 & 1 \\ -5 & -\frac{4}{3} & 1 \end{vmatrix}$$
$$= \frac{1}{2}\left|3\left(\frac{4}{3}\right) - 2(3+5) + 1(-4)\right|$$
$$= \frac{1}{2}|4 - 16 - 4| = 8 \text{ sq units}$$

69. (0.04) The equation of the chord of contact of the curve $x^2 + y^2 = a^2$ is $hx + ky = a^2$ ...(i)

and the equation of the normal chord of hyperbola is
$$ax\cos\theta + by\cot\theta = a^2 + b^2 \quad \text{...(ii)}$$

Eqs. (i) and (ii) are identical.
$$\because \quad \frac{a\cos\theta}{h} = \frac{b\cot\theta}{k} = \frac{a^2+b^2}{a^2}$$
$$\Rightarrow \quad \cos\theta = \frac{(a^2+b^2)}{a^2}h,\ \cot\theta = \left(\frac{a^2+b^2}{a^2b}\right)k$$
$$\cos\theta = \frac{(a^2+b^2)}{a^3}h \quad \sin\theta = \frac{bh}{ak}$$

We know that $\cos^2\theta + \sin^2\theta = 1$
$$\Rightarrow \quad \left(\frac{a^2+b^2}{a^3}\right)^2 h^2 + \left(\frac{b^2h^2}{a^2k^2}\right) = 1$$
$$\Rightarrow \quad \frac{h^2}{a^2}\left[\frac{b^2}{k^2} + \left(\frac{a^2+b^2}{a^2}\right)^2\right] = 1$$
$$\Rightarrow \quad \frac{b^2}{k^2} + \left(\frac{a^2+b^2}{a^2}\right)^2 = \frac{a^2}{h^2}$$
$$\Rightarrow \quad \frac{a^2}{h^2} - \frac{b^2}{k^2} = \left(\frac{a^2+b^2}{a^2}\right)^2$$

Locus is $\dfrac{a^2}{x^2} - \dfrac{b^2}{y^2} = \left(\dfrac{a^2+b^2}{a^2}\right)^2$

$\because$ $\lambda = 1, \mu = 1$ and $r = 2$
$$\frac{\sqrt{\lambda + \mu + r}}{50} = \frac{\sqrt{4}}{50} = \frac{2}{50} = 0.04$$

70. (1) The equation of any normal to the given hyperbola at (x, y) is

$$\frac{x}{a^2}(k-y)=\frac{y}{b^2}(x-b)$$

$$\Rightarrow \left(\frac{1}{a^2}+\frac{1}{b^2}\right)xy-\frac{y}{b}-\frac{kx}{a^2}=0 \Rightarrow y=\frac{b^2kx}{a^2(e^2x-b)} \quad \text{...(i)}$$

The equation of the hyperbola is

$$\frac{x^2}{a^2}-\frac{y^2}{b^2}=1 \quad \text{...(ii)}$$

Solving Eqs. (i) and (ii), we get

$$\frac{x^2}{a^2}-\frac{b^4k^2x^2}{b^2a^4(e^2x-b)^2}=1$$

$$\Rightarrow a^2e^4x^4-2ba^2e^2x^3-(a^2b^2+b^2k^2+a^4e^4)x^2 + 2ba^4e^2x+a^4b^2=0 \quad \text{...(iii)}$$

Let x_1, x_2, x_3, x_4 all roots of the equation

$$\Sigma x_1=\frac{2b}{e^2};\ \Sigma x_1x_2=-\left(\frac{a^2b^2+b^2k^2+a^4e^2}{a^2e^4}\right)$$

$$\Sigma x_1x_2x_3=-\frac{2ba^2}{e^2};\ x_1x_2x_3x_4=\frac{a^2b^2}{e^4}$$

$$\Sigma\frac{1}{x}=\frac{\Sigma x_1x_2x_3}{x_1x_2x_3x_4}=\frac{2e^2}{b}$$

$$\Rightarrow (1)\ (x_1+x_2+x_3+x_4)\left(\frac{1}{x_1}+\frac{1}{x_2}+\frac{1}{x_3}+\frac{1}{x_4}\right)$$

$$=\frac{2b}{e^2}\times\frac{2e^2}{b}=4$$

Similarly, $(y_1+y_2+y_3+y_4)\left(\frac{1}{y_1}+\frac{1}{y_2}+\frac{1}{y_3}+\frac{1}{y_4}\right)$

$$=(\Sigma y_1)\left(\Sigma\frac{1}{y_1}\right)=4$$

$\therefore$ Required answer $=\frac{4}{4}=1$

71. (0.1) Let the tangent of hyperbola is

$$y=m_1x+\sqrt{a^2m_1^2-b^2}$$

$$y=m_2x+\sqrt{a^2m_2^2-b^2}$$

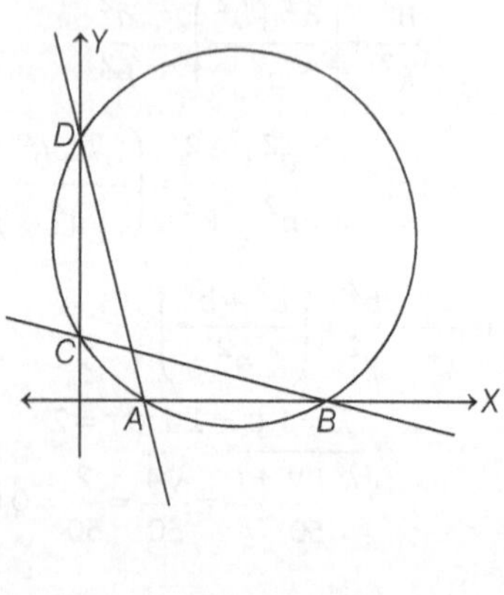

Points of intersection of these tangents with axes are

$$A=\left(-\frac{\sqrt{a^2m_1^2-b^2}}{m_1},0\right), C(0,\sqrt{a^2m_1^2-b^2})$$

$$B=\left(-\frac{\sqrt{a^2m_2^2-b^2}}{m_2},0\right), D(0,\sqrt{a^2m_2^2-b^2})$$

Now, as the four points are concyclic.

$OA\cdot OB=OC\cdot OD$

$$\Rightarrow \left(\frac{-\sqrt{a^2m_1^2-b^2}}{m_1}\right)\left(\frac{-\sqrt{a^2m_2^2-b^2}}{m_2}\right)$$

$$=\sqrt{a^2m_1^2-b^2}\cdot\sqrt{a^2m_2^2-b^2}$$

$$\therefore \quad \frac{m_1m_2}{10}=\frac{1}{10}=0.1$$

72. (0.08)

$\because$ Given, two points $P(a\sec\theta, b\tan\theta)$ and $B(a\sec\phi, b\tan\phi)$.

$\therefore$ Equation of focal chord

$$\frac{x}{a}\cos\left(\frac{\theta-\phi}{2}\right)-\frac{y}{b}\sin\left(\frac{\theta+\phi}{2}\right)=\cos\left(\frac{\theta+\phi}{2}\right)$$

It passes through $(ae, 0)$.

$$\therefore \quad e\cos\left(\frac{\theta-\phi}{2}\right)-0=\cos\left(\frac{\theta+\phi}{2}\right)$$

$$e=-\frac{\cos\left(\frac{\theta+\phi}{2}\right)}{\cos\left(\frac{\theta-\phi}{2}\right)}$$

$$\frac{1}{e}=\frac{\cos\left(\frac{\theta-\phi}{2}\right)}{\cos\left(\frac{\theta+\phi}{2}\right)}$$

Using componendo, we get

$$\frac{1-e}{1+e}=\frac{\cos\left(\frac{\theta+\phi}{2}\right)-\cos\left(\frac{\theta+\phi}{2}\right)}{\cos\left(\frac{\theta-\phi}{2}\right)+\cos\left(\frac{\theta+\phi}{2}\right)}$$

$$\Rightarrow \quad \frac{1-e}{1+e}=\tan\left(\frac{\theta}{2}\right)\tan\frac{\phi}{2}$$

$$\frac{1-e}{1+e}=\frac{P+Qe}{R+Se}$$

$$\Rightarrow \quad P=1, Q=-1$$

$$R=1, S=1$$

$$\therefore \quad \frac{P+Q+R+S}{25}=\frac{2}{25}=0.08$$

73. (2) Given, $x^2 - y^2 = 4$

$$(x-y)(x+y) = 4$$

$\Rightarrow$ $$-\sqrt{2}X'\sqrt{2}Y' = 4$$

$\Rightarrow$ $$X'Y' = \frac{4}{-2} = -2$$

$\therefore$ $$xy = -2$$

$\therefore$ $$a^2 = 2$$

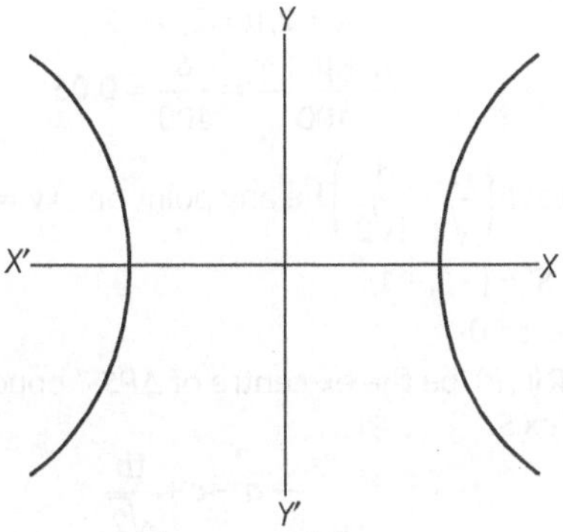

74. (2) Since, $\frac{e}{2}$ and $\frac{e'}{2}$ are eccentricities of a hyperbola and its conjugate.

$\therefore$ $$\frac{4}{e^2} + \frac{4}{e'^2} = 1$$

$\therefore$ $$4 = \frac{e^2 e'^2}{e'^2 + e^2}$$

Line passing through $(e, 0)$ and $(0, e')$.

$\therefore$ $$e'x + ey - ee' = 0 \quad \text{...(i)}$$

It is tangent to the circle $x^2 + y^2 = r^2$

$\therefore$ Distance between line (i) and centre $(0, 0) = r$

$$\left|\frac{ee'}{\sqrt{e^2 + e'^2}}\right| = r$$

$\Rightarrow$ $$r^2 = \frac{ee'^2}{(e^2 + e'^2)} \Rightarrow r^2 = 4$$

$\therefore$ $$r = 2$$

75. (6) The equation of the asymptotes of $\frac{x^2}{a^2} - \frac{y^2}{b^2} = 1$ are

$$\frac{x}{a} + \frac{y}{b} = 0 \text{ and } \frac{x}{a} - \frac{y}{b} = 0$$

i.e. $bx + ay = 0$ and $bx - ay = 0$

Let any position the hyperbola be $P(a\sec\phi, b\tan\phi)$.

It is given that $P_1P_2 = 6$

$$= \left|\frac{ab\sec\theta + ab\tan\theta}{\sqrt{a^2 + b^2}}\right| \left|\frac{ab\sec\theta - ab\tan\theta}{\sqrt{a^2 + b^2}}\right| = 6$$

$\Rightarrow$ $$\frac{a^2b^2\sec^2\theta - a^2b^2\tan^2\theta}{a^2 + b^2} = 6$$

$\Rightarrow$ $$\frac{a^2b^2}{a^2 + b^2} = 4$$

$\Rightarrow$ $$\frac{a^2(2a^2)}{a^2 + 2a^2} = 6$$

$\Rightarrow$ $$2a^2 = 18$$

$\Rightarrow$ $$a^2 = 9 \Rightarrow 2a = 6$$

$\therefore$ Length of transverse axis $= 6$

76. (6) The equation of the tangent to the hyperbola $xy = c^2$ at $\left(ct, \frac{c}{t}\right)$ is $x + yt^2 = 2ct$

Thus, $$T = (2ct, 0) \text{ and } T' = \left(0, \frac{2c}{t}\right)$$

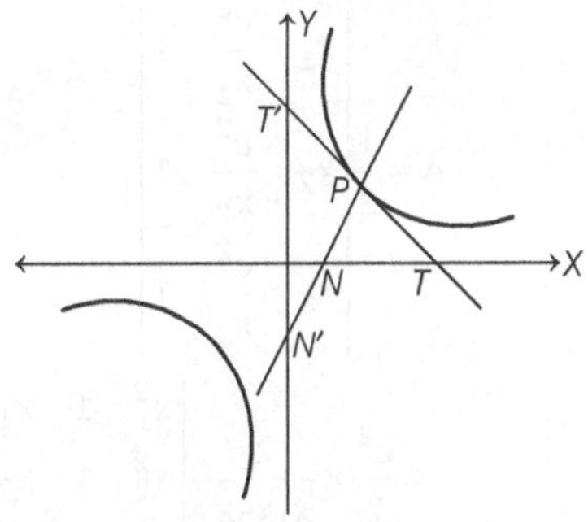

The equation of normal to the hyperbola

$$xy = c^2 \text{ at } \left(ct, \frac{c}{t}\right) \text{ is}$$

$$xt^3 - yt - ct^4 + c = 0$$

Thus, $$N = \left(ct - \frac{c}{t^3}, 0\right)$$

and $$N' = \left(0, \frac{c}{t} - ct^3\right)$$

Now, $$\Delta = \text{Area of } (\Delta PNT)$$

$$= \frac{1}{2}\begin{vmatrix} ct & \frac{c}{t} & 1 \\ ct - \frac{c}{t^3} & 0 & 1 \\ 2ct & 0 & 1 \end{vmatrix}$$

$$= \frac{1}{2}\frac{c^2}{t}\left(t + \frac{1}{t^3}\right) = \frac{c^2}{2}\left(1 + \frac{1}{t^4}\right)$$

and $$\Delta = \text{Area of } (\Delta PN'T') = \frac{1}{2}\begin{vmatrix} ct & \frac{c}{t} & 1 \\ 0 & \frac{c}{t} - ct^3 & 1 \\ 0 & \frac{2c}{t} & 1 \end{vmatrix}$$

$$\Delta' = \left|\frac{ct}{2}\left(\frac{c}{t} - ct^3 - \frac{2c}{t}\right)\right| = \frac{c^2}{2}(t^4 + 1)$$

Now, $$\frac{1}{\Delta}+\frac{1}{\Delta'}=\frac{2}{c^2}\left(\frac{t^4}{t^4+1}\right)+\frac{2}{c^2}\left(\frac{1}{t^4+1}\right)$$

$$=\frac{2}{c^2}\left(\frac{t^4+1}{t^4+1}\right)=\frac{2}{c^2}$$

$$\therefore \quad \frac{c^2}{\Delta}+\frac{c^2}{\Delta'}=2$$

$$\therefore \quad \frac{c^2}{\Delta}+\frac{c^2}{\Delta'}+4=2+4=6$$

77. (0.04) Area of ΔPQR is Δ where ($P(x_1, y_1), Q(x_2, y_2)$ and $R(x_3, y_3)$.

$$y_1=\frac{c^2}{x_1},\quad y_2=\frac{c^2}{x_2},\quad y_3=\frac{c^2}{x_3}$$

$$\Delta=\frac{1}{2}\begin{vmatrix} x_1 & \frac{c^2}{x_1} & 1 \\ x_2 & \frac{c^2}{x_2} & 1 \\ x_3 & \frac{c^2}{x_3} & 1 \end{vmatrix}$$

$$=\frac{c^2}{2}\times\frac{1}{x_1x_2x_3}\begin{vmatrix} x_1^2 & 1 & x_1 \\ x_2^2 & 1 & x_2 \\ x_3^2 & 1 & x_3 \end{vmatrix}$$

$$=\frac{c^2}{2}\frac{(x_1-x_2)(x_2-x_3)(x_3-x_1)}{x_1x_2x_3}$$

$$=\frac{(x_1-x_2)(x_2-x_3)(x_3-x_1)}{(x_1x_2x_3)}\cdot\frac{c^\lambda}{\mu}$$

$$\lambda=2, \mu=2$$

$$\therefore \quad \frac{\lambda+\mu}{100}=\frac{4}{100}=0.04$$

78. (0.06) The equation of the circle with centre $(3\alpha, 3\beta)$ is

$$x^2+y^2-6\alpha x-6\beta y+k=0 \quad \text{...(i)}$$

where, k is variable.

Given, hyperbola is $x^2-y^2=9a^2$ ...(ii)

Eliminating y from Eq. (i) with the help of Eq. (ii), we get

$$(x^2+x^2-9a^2-6\alpha x+k^2)^2=36\beta^2(x^2-9a^2)$$

$$\Rightarrow \quad 4x^4-24\alpha x^2+\ldots=0$$

It is biquadratic equation.

Let the abscissae of four points P, Q, R and S are x_1, x_2, x_3 and x_4, then

$$x_1+x_2+x_3+x_4=6\alpha \quad \text{...(iii)}$$

Similarly, $$y_1+y_2+y_3+y_4=6\beta \quad \text{...(iv)}$$

If (x', y') is the centroid of ΔPQR, then

$$x'=\frac{x_1+x_2+x_3}{3}=\frac{6\alpha-x_4}{3}$$

$$\Rightarrow \quad -3(x'-2\alpha)=x_4$$

and $$y'=\frac{y_1+y_2+y_3}{3}=\frac{6\beta-y_4}{3}$$

$$-3(y'-2\beta)=y_4$$

But, $$x_4^2-y_4^2=9\alpha^2$$

$$\Rightarrow \quad 9(x'-2\alpha)^2-9(y'-2\beta)^2=9a^2$$

$$(x'-2\alpha)^2-(y^2-2\beta)^2=a^2$$

Locus is $(x-2\alpha)^2-(y-2\beta)^2=a^2$

$$\lambda=2, \mu=2, \gamma=2$$

$$\Rightarrow \quad \frac{\lambda+\mu+\gamma}{100}=\frac{6}{100}=0.06$$

79. (5) Let $P\left(\frac{t}{\sqrt{2}}, \frac{1}{t\sqrt{2}}\right)$ be any point on $2xy=1$ and $S=(1,1)$ and $S'=(-1,-1)$.

Here, $t>0$

Let $R(h, k)$ be the ex-centre of $\Delta PSS'$ opposite to vertex S.

$$\Rightarrow \quad h=\frac{-a-c+\frac{tb}{\sqrt{2}}}{a+b-c}$$

and $$k=\frac{-a-c+\frac{b}{\sqrt{2}}}{a+b-c}$$

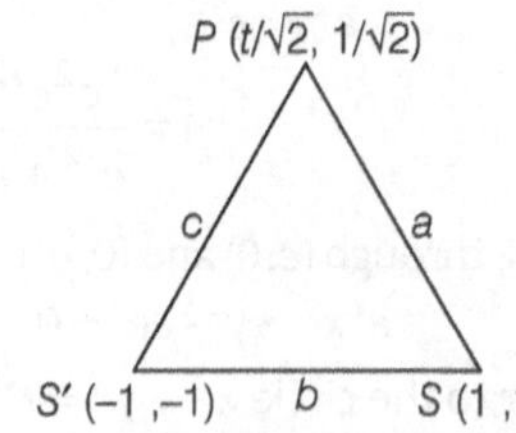

$$\because \quad b=2\sqrt{2}, a=ePM$$

$$=\frac{\sqrt{2}\left(\frac{t}{\sqrt{2}}+\frac{1}{t\sqrt{2}}-1\right)}{\sqrt{2}}$$

$$=\frac{t}{\sqrt{2}}+\frac{1}{t\sqrt{2}}-1$$

and $$c=\sqrt{2}\frac{\left(\frac{t}{\sqrt{2}}+\frac{1}{t\sqrt{2}}+1\right)}{\sqrt{2}}$$

$$=\frac{t}{\sqrt{2}}+\frac{1}{t\sqrt{2}}+1$$

[$\because$ Directrix as $x+y=\pm 1$]

$$\therefore \quad h=\frac{-\sqrt{2}\left(t+\frac{1}{t}\right)+2t}{2\sqrt{2}-2}$$

and $$k=\frac{-\sqrt{2}\left(t+\frac{1}{t}\right)+\frac{2}{t}}{2\sqrt{2}-2}$$

$\therefore \quad h+k=-\left(t+\frac{1}{t}\right)$

and $\quad h-k=\frac{t-\frac{1}{t}}{\sqrt{2}-1}$

$\because \quad \left(t+\frac{1}{t}\right)^2=\left(t-\frac{1}{t}\right)^2+4$

$(h+k)^2=(\sqrt{2}-1)^2(h-k)^2+4$

$\because$ Locus of R is $(x+y)^2=(\sqrt{2}-1)^2(x-y)^2+4$

On comparing, we get

$p=1$ and $q=4$

$p+q=5$

80. (4) Let $P(h, k)$ be the middle point, then the equation of chord whose mid-point $P(h, k)$ is

$$T=S_1$$

$\Rightarrow \quad hx+ky-4=h^2+k^2-4$

$\Rightarrow \quad y=-\frac{h}{k}x+\frac{h^2+k^2}{k}$

It will touch the hyperbola $\frac{x^2}{4}-\frac{y^2}{16}=1$, then

$$\left(\frac{h^2+k^2}{k}\right)^2=4\times\frac{h^2}{k^2}-16$$

$\Rightarrow \quad (h^2+k^2)^2=4h^2-16k^2$

$\because$ Locus of $P(h, k)$ is

$$(x^2+y^2)^2=4x^2-16y^2$$

$\therefore \quad \lambda=4$

81. (2) Given, lines are $\sqrt{3}kx+ky-4\sqrt{3}=0$ ... (i)

$\sqrt{3}x-y-4\sqrt{3}k=0$ ... (ii)

Multiplying Eq. (ii) $\times k$ and then adding Eqs. (i) and (ii),

$$\sqrt{3}kx+ky-4\sqrt{3}=0$$
$$\sqrt{3}kx-ky-4\sqrt{3}k^2=0$$
$$(2\sqrt{3}x)k=4\sqrt{3}+4\sqrt{3}k^2$$

$\therefore \quad x=\frac{4\sqrt{3}(1+k^2)}{2\sqrt{3}k}=2\left(k+\frac{1}{k}\right)$

Subtracting Eq. (i) from Eq. (ii),

$$\sqrt{3}kx+ky-4\sqrt{3}=0$$
$$\sqrt{3}kx-ky-4\sqrt{3}k^2=0$$
$$-\quad+\quad+$$
$$2ky=4\sqrt{3}-4\sqrt{3}k^2$$

$$y=\frac{4\sqrt{3}(1+k^2)}{2k}=2\sqrt{3}\left(k+\frac{1}{k}\right)$$

We have, $x=2\left(k+\frac{1}{k}\right)$ and $y=2\sqrt{3}\left(\frac{1}{k}-k\right)$

$\frac{x}{2}=\left(k+\frac{1}{k}\right)$ ... (iii)

$\frac{y}{2\sqrt{3}}=\left(\frac{1}{k}-k\right)$ ... (iv)

Squaring and subtracting Eq. (iii) from Eq. (iv),

$$\frac{x^2}{4}-\frac{y^2}{12}=\left(k^2+\frac{1}{k^2}+2\right)-\left(\frac{1}{k^2}+k^2-2\right)$$

$$\frac{x^2}{4}-\frac{y^2}{12}=4 \text{ or } \frac{x^2}{16}-\frac{y^2}{48}=1$$

Clearly, this is a hyperbola

$\therefore \quad e^2=1+\frac{b^2}{a^2}=1+\frac{48}{16}=1+3$

$e^2=4$

$\therefore \quad e=\sqrt{4}=2 \quad$ [$\because e$ is positive]

Level 3

1. (c) (a) $x^2+2y^2\leq 1$ represents interior region of an ellipse where on taking any two points the mid-point of that segment will also lie inside that ellipse.

(b) Max$\{|x|,|y|\}\leq 1$

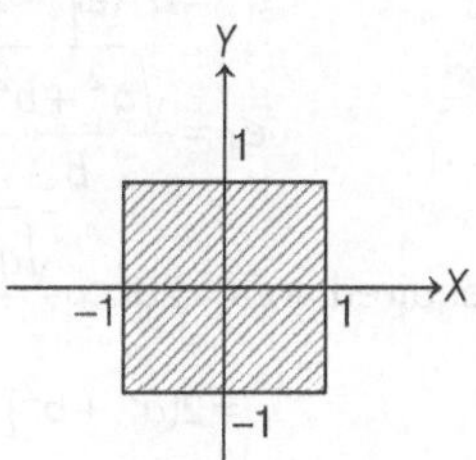

$\Rightarrow |x|\leq 1, |y|\leq 1$

$-1\leq x\leq 1, -1\leq y\leq 1$

which represents the interior region of a square with its side $x=\pm 1$ and $y=\pm 1$ in which for any two points their mid-point also lies inside the region.

(c) $x^2-y^2\leq 1$ represents the exterior region of hyperbola in which if we take 2 points (2, 0) and (– 2, 0), then their mid-point (0, 0) does not lie in the same region.

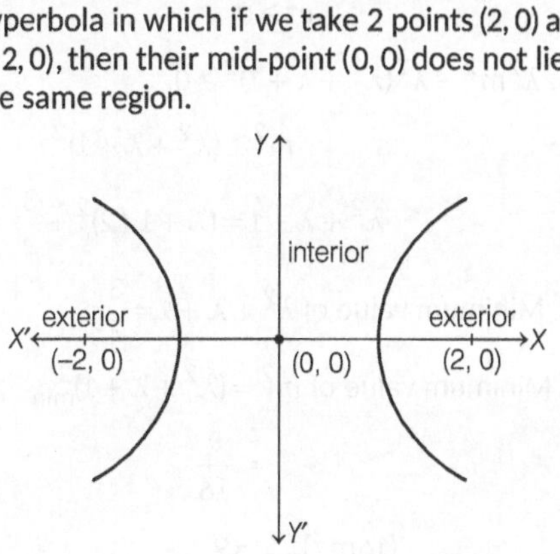

(d) $y^2 \le x$ represents interior region of parabola in which for any two points, their mid-point also lie inside the region.

2. (b) Required area $= 4 \times$ Area of $\Delta S_1 OS_3$

$$= 4 \times \frac{1}{2} \times ae \times be_1$$

$$= 2ab\, ee_1 \quad \dots \text{(i)}$$

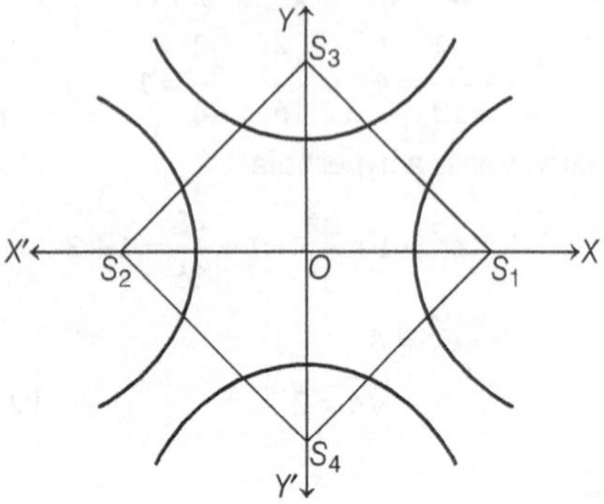

$\because \quad b^2 = a^2(e^2 - 1)$

$\Rightarrow \quad e = \frac{\sqrt{a^2+b^2}}{a} \quad \dots \text{(ii)}$

and $\quad a^2 = b^2(e_1^2 - 1)$

$\Rightarrow \quad e_1 = \frac{\sqrt{a^2+b^2}}{b} \quad \dots \text{(iii)}$

Required area $= 2ab \times \frac{\sqrt{a^2+b^2}}{a} \cdot \frac{\sqrt{a^2+b^2}}{b}$

$$= 2(a^2+b^2)$$

3. (d) $\because y = mx + c$ be tangent to hyperbola

$$\frac{x^2}{\lambda^2} - \frac{y^2}{(\lambda^3+\lambda^2+\lambda)^2} = 1$$

$\therefore$ As per condition of tangency

$$c^2 = a^2m^2 - b^2$$

$$c^2 = \lambda^2 m^2 - (\lambda^3+\lambda^2+\lambda)^2$$

$$c^2 \ge 0$$

$\Rightarrow \lambda^2 m^2 - \lambda^2(\lambda^2+\lambda+1)^2 \ge 0$

$\Rightarrow \quad m^2 \ge (\lambda^2+\lambda+1)^2$

$\because \quad \lambda^2+\lambda+1 = (\lambda+1/2)^2 + \frac{3}{4}$

$\therefore$ Minimum value of $\lambda^2+\lambda+1 = \frac{3}{4}$

$\therefore$ Minimum value of $m^2 = (\lambda^2+\lambda+1)^2_{\min}$

$$= \frac{9}{16}$$

$\therefore \quad (16m^2)_{\min} = 9$

4. (b) Hyperbola $\frac{x^2}{a^2} - \frac{y^2}{b^2} + 1 = 0$ can be rewritten as

$$\frac{x^2}{-a^2} - \frac{y^2}{-b^2} = 1$$

So, the director circle will be

$$x^2 + y^2 = -a^2 - (-b^2) = b^2 - a^2$$

which will be defined only where $b \ge a$ (ie. $b^2 - a^2 \ge 0$)

$\therefore$ Statement I is true and director circle of hyperbola $\frac{x^2}{25} - \frac{y^2}{9} = 1$ is $x^2 + y^2 = 25 - 9 = 16$

$\therefore$ Statement II is true.

Hence, both statement are true but statement II is not a correct explanation of statement I.

5. (b) Curve $x^2 - 4y^2 = 4 \quad \dots \text{(i)}$

Chord of contact $2x + y + 6 = 0 \Rightarrow y = -6 - 2x \quad \dots \text{(ii)}$

Slope of chord $= -2$

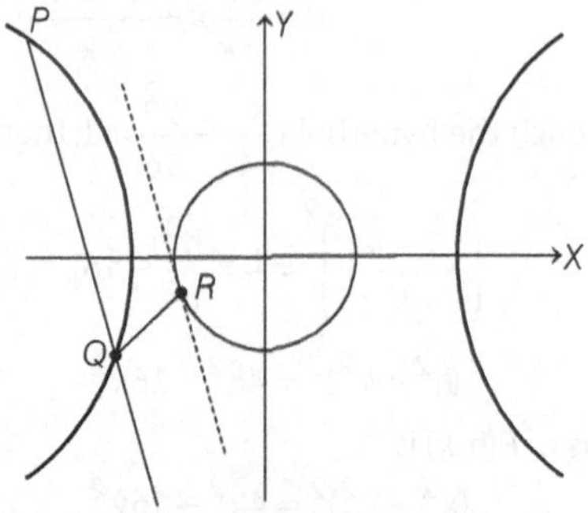

$\because R(\alpha, \beta)$ be the point satisfy $\alpha^2 + \beta^2 - 1 \le 0$

[i.e. interior region of circle $x^2 + y^2 = 1$]

$\because$ It is clearly see that area of ΔPQR is minimum. When tangent at R is parallel to PQ.

Now, tangent at in (w.r.t. circle)

$$x\alpha + y\beta = 1$$

$$\text{Slope} = -\frac{\alpha}{\beta}$$

Slope of chord $= -2$

$\because \quad \frac{\alpha}{\beta} = 2 \Rightarrow \alpha = 2\beta$

$\therefore (\alpha, \beta)$ lies in circle $x^2 + y^2 = 1$

$$\alpha^2 + \beta^2 = 1$$

$$4\beta^2 + \beta^2 = 1,\ 5\beta^2 = 1$$

$$\beta = \pm\frac{1}{\sqrt{5}}$$

$$\alpha = \pm 2/\sqrt{5}$$

$\because (\alpha, \beta)$ lies on third quadrant.

$\therefore \quad (\alpha, \beta) = \left(\frac{-2}{\sqrt{5}}, \frac{-1}{\sqrt{5}}\right)$

6. (c) $\because$ Chord $x\cos\alpha + y\sin\alpha = P$ of the hyperbola

$$\frac{x^2}{16} - \frac{y^2}{18} = 1 \quad \text{... (i)}$$

Subtends a right angle at the centre P.

$$\because \quad \frac{x\cos\alpha + y\sin\alpha}{P} = 1 \quad \text{... (ii)}$$

Homogenising Eq. (i) with help of Eq. (ii)

$$\frac{x^2}{16} - \frac{y^2}{18} = \left(\frac{x\cos\alpha + y\sin\alpha}{P}\right)^2 \quad \text{...(iii)}$$

$\because$ Eq. (iii) represents a pair of lines which are perpendicular.

$\therefore$ Coefficient of x^2 + coefficient of $y^2 = 0$

$$\left(\frac{1}{16} - \frac{\cos^2\alpha}{P^2}\right) + \left(\frac{-1}{18} - \frac{\sin^2\alpha}{P^2}\right) = 0$$

$$\Rightarrow \quad \frac{1}{16} - \frac{1}{18} - \frac{1}{P^2}(\cos^2\alpha + \sin^2\alpha) = 0$$

$$\Rightarrow \quad \frac{1}{P^2} = \frac{18-16}{18 \times 16}$$

$$\Rightarrow \quad \frac{1}{P^2} = \frac{1}{9 \times 16} \Rightarrow P = 3 \times 4 = 12$$

$\therefore x\cos\alpha + y\sin\alpha = \pm 12$ is tangent to circle

$$x^2 + y^2 = r^2 = \left(\frac{d}{2}\right)^2$$

$$\therefore \quad d = 24 \Rightarrow \frac{d}{4} = 6$$

7. (d) From any point $A\left(ct_1, \frac{c}{t_1}\right)$ on the rectangular hyperbola.

$$xy = c^2 \quad \text{... (i)}$$

Tangent AB and AC are drawn to the parabola $y^2 = 4ax$.

Then, $B = \left(ct_2, \frac{c}{t_2}\right)$ and $c = \left(ct_3, \frac{c}{t_3}\right)$

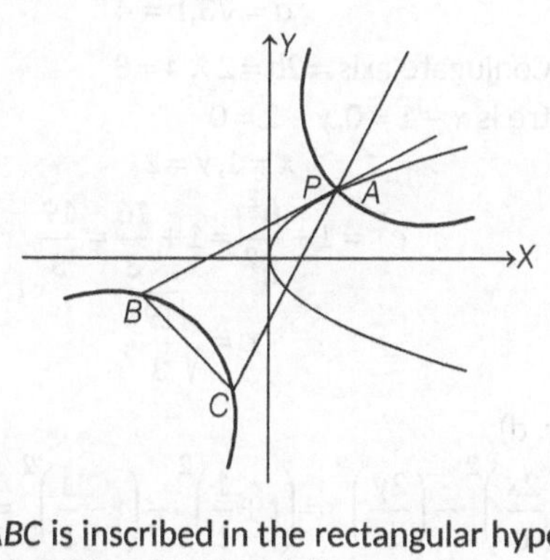

Then, ΔABC is inscribed in the rectangular hyperbola and its sides AB and AC touch the parabola.

$\because BC$ is also a tangent to it.

Equations of chord AB, BC and CA are

$$AB = x + yt_1t_2 = c(t_1 + t_2)$$
$$BC = x + yt_2t_3 = c(t_2 + t_3)$$
$$CA = x + yt_3t_1 = c(t_3 + t_1)$$

$\because AB$ is tangent to a parabola $y^2 = 4ax$, its equation.

$$y = \frac{-1}{t_1t_2} + \frac{c(t_1+t_2)}{t_1t_2}$$

Touching condition $c = \frac{a}{m}$ or $cm = a$

$$\frac{c(t_1+t_2)}{t_1 t_2}\left(\frac{-1}{t_1t_2}\right) = a$$

$$\Rightarrow \quad -at_1^2t_2^2 = c(t_1 + t_2)$$

$$\Rightarrow \quad at_1^2t_2^2 + ct_2 + ct_1 = 0$$

Similarly, $\quad (at_2^2)t_3^2 + ct_3 + ct_1 = 0$

$$\therefore \quad (at_1^2)t^2 + ct + ct_1 = 0 \quad \text{... (ii)}$$

and t_2 and t_3 are the roots of Eq. (ii).

$$\therefore \quad t_2 + t_3 = \frac{-c}{at_1^2} \text{ and } t_2t_3 = \frac{c}{at_1}$$

$$\Rightarrow \quad c = at_1t_2t_3$$

$$\therefore \quad c(t_2 + t_3) = \frac{c^2}{at_1^2}$$

$$= \frac{-a^2t_1^2t_2^2t_3^2}{at_1^2}$$

$$= -at_2^2t_3^2$$

$$\Rightarrow \quad -at_2t_3 = \frac{c(t_2+t_3)}{t_2t_3} \text{ is the condition.}$$

This proves BC to be also a tangent to the parabola. But any point A can be taken in infinite ways anywhere on the curve representing the rectangular hyperbola. Hence, such triangles exist in infinite number.

8. (a) Given hyperbola $\frac{x^2}{4} - \frac{y^2}{3} = 1$

$$\Rightarrow \quad y = \frac{\sqrt{3}}{2}\sqrt{x^2 - 4}$$

$$a = 2, b = \sqrt{3}$$

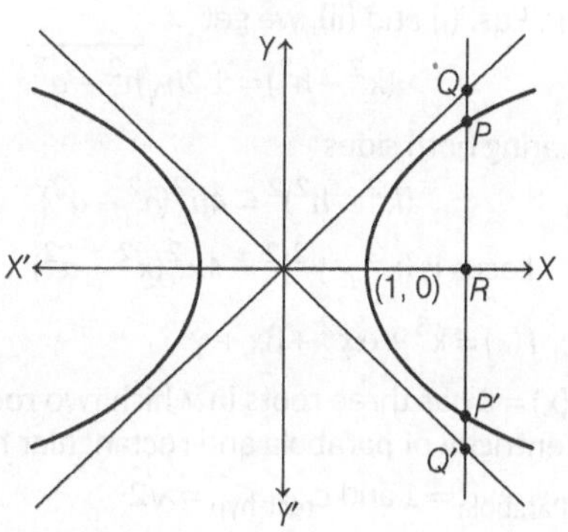

$\because$ Asymptotes $\Rightarrow y = \pm\dfrac{b}{a}x$

$$y = \pm\frac{\sqrt{3}}{2}x$$

And the ends of double ordinate be $P\left(h, \dfrac{\sqrt{3}}{2}\sqrt{h^2-4}\right)$

and $P'\left(h, -\dfrac{\sqrt{3}}{2}\sqrt{h^2-4}\right)$.

Since, mid-point of PP' and QQ' are sum say R.

$$\because \quad Q = \left(h, \frac{\sqrt{3}}{2}h\right), Q'\left(h, \frac{-\sqrt{3}}{2}h\right)$$

$$PQ \cdot PQ' = \left(\frac{\sqrt{3}}{2}\right)^2 [h^2 - (h^2 - 4)]$$

$$= \frac{3}{4} \times 4 = 3$$

9. (a) Let $P(h, k)$ be any point on any one member of hyperbola family having equation $\dfrac{x^2}{a^2} - \dfrac{y^2}{b^2} = 1$, b is any arbitrary constant, then its asymptotes are given by $y = \pm\dfrac{b}{a}x$, then according to question,

$$|k| = \frac{\left|\pm\dfrac{bh}{a} - k\right|}{\sqrt{\dfrac{b^2}{a^2} + 1}}$$

$$\Rightarrow \quad k^2 = \left(\pm\frac{b}{a}h - k\right)^2 \left(\frac{a^2}{a^2 + b^2}\right)$$

$$\text{or} \quad k^2 = \frac{(\pm bh - ak)^2}{a^2 + b^2} \qquad \text{... (i)}$$

Further (h, k) lies on hyperbola

$$\frac{h^2}{a^2} - \frac{k^2}{b^2} = 1$$

$$\Rightarrow \quad b = \sqrt{\frac{a^2k^2}{h^2 - a^2}} \qquad \text{... (ii)}$$

From Eqs. (i) and (ii), we get

$$(k^2 - h^2) = \pm 2h\sqrt{h^2 - a^2}$$

Squaring both sides

$$(k^2 - h^2)^2 = 4h^2(h^2 - a^2)$$

$\because \quad$ Locus is $(x^2 - y^2)^2 = 4x^2(x^2 - a^2)$

10. (a) $\because f(x) = x^3 + \alpha x^2 + \beta x + \gamma$

$\because f(x) = 0$ has three roots in which two roots are eccentricity of parabola and rectangular hyperbola.

$\because e_{\text{Parabola}} = 1$ and $e_{\text{rect-hyp}} = \sqrt{2}$

$$f(1) = 0, f(\sqrt{2}) = 0$$

$\because \alpha, \beta, \gamma$ are rational number.

$\because$ It means $-\sqrt{2}$ is also a root of the equation $f(x) = 0$

$$\therefore \quad f(x) = (x - \sqrt{2})(x + \sqrt{2})(x - 1)$$

$$= (x^2 - 2)(x - 1)$$

$$= x^3 - x^2 - 2x + 2$$

$$\therefore \quad \alpha = -1, \beta = -2, \gamma = 2$$

$$\therefore \quad \alpha + \beta + \gamma = -1 - 2 + 2 = -1$$

11. (b) It is clear from the figure the common tangent to the circle $x^2 + y^2 = 1$ and hyperbola $x^2 - y^2 = 1$ is $x = 1$ (which is nearer to $P(1/2, 1)$ and given one focus at $P(1/2, 1)$, so the equation of the directrix is $x = 1$

Hence, the equation of the ellipse is

$$\sqrt{(x - 1/2)^2 + (y - 1)^2} = \frac{1}{2}|x - 1|$$

$$\Rightarrow 9\left(x - \frac{1}{3}\right)^2 + 12(y - 1)^2 = 1$$

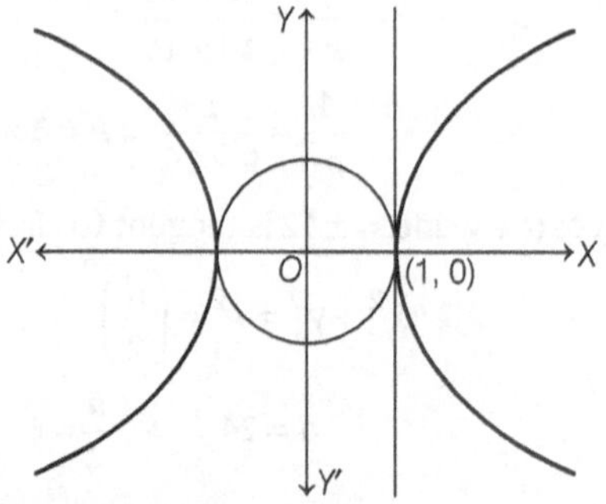

12. (b, c, d) Given equation,

$$16x^2 - 3y^2 - 32x + 12y - 44 = 0$$

$$16(x^2 - 2x) - 3(y^2 - 4y) - 44 = 0$$

$$16(x - 1)^2 - 3(y - 2)^2 = 44 + 16 - 12 = 48$$

$$\Rightarrow \quad \frac{(x - 1)^2}{3} - \frac{(y - 2)^2}{16} = 1$$

$$\Rightarrow \quad a^2 = 3, b^2 = 16$$

$$\Rightarrow \quad a = \sqrt{3}, b = 4$$

$\because$ Conjugate axis $= 2b = 2 \times 4 = 8$

Centre is $x - 1 = 0, y - 2 = 0$

$$x = 1, y = 2$$

$$\text{and} \quad e^2 = 1 + \frac{b^2}{a^2} = 1 + \frac{16}{3} = \frac{19}{3}$$

$$e = \sqrt{\frac{19}{3}}$$

13. (a, c, d)

$$\text{(a)} \left(\frac{2x}{a}\right)^2 - \left(\frac{2y}{b}\right)^2 = \left(t + \frac{1}{t}\right)^2 - \left(t - \frac{1}{t}\right)^2 = 4$$

$$\Rightarrow \quad \frac{x^2}{a^2} - \frac{y^2}{b^2} = 1$$

(b) $\because \quad \frac{tx}{a}-\frac{y}{b}+t=0$

$\Rightarrow \quad t=\frac{ay}{b(x+a)}$ ... (i)

and $\quad \frac{x}{a}+\frac{ty}{b}-1=0$

$\Rightarrow \quad t=\frac{(a-x)b}{ay}$ ... (ii)

By Eqs. (i) and (ii),

$$\frac{ay}{b(x+a)}=\frac{(a-x)b}{ay}$$

$$a^2y^2=b^2(a^2-x^2)$$

$\Rightarrow \quad b^2x^2+a^2y^2=a^2b^2$

$\Rightarrow \quad \frac{x^2}{a^2}+\frac{y^2}{b^2}=1$

(c) $x^2-y^2=(e^t+e^{-t})^2-(e^t-e^{-t})^2=4$

(d) $\because x^2-6=2\cos t$

$\Rightarrow \quad x^2-4=2(1+\cos t)$

$\Rightarrow \quad x^2-4=4\cos^2 t/2$ ... (i)

and $\quad y^2+2=4\cos^2 t/2$ ...(ii)

By Eqs. (i) and (ii) $x^2-4=y^2+2$

or $\quad x^2-y^2=6$

14. (b) Equation of tangent to hyperbola is

$$y=mx\pm\sqrt{a^2m^2-b^2}$$

$\because$ It touches parabola also

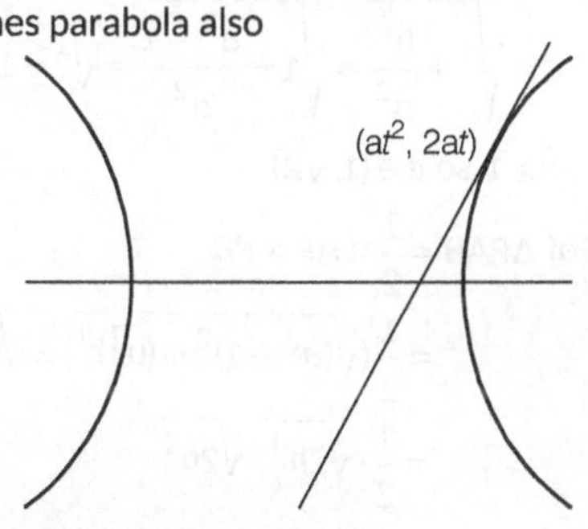

Equation of tangent to parabola

$$c=\frac{a}{m}$$

$\Rightarrow \quad c^2=\frac{a^2}{m^2}$

$\Rightarrow \quad a^2m^2-b^2=\frac{a^2}{m^2}$

$\Rightarrow \quad a^2m^4-b^2m^2-a^2=0$

$\Rightarrow \quad m^2=\frac{b^2\pm\sqrt{b^4+4a^4}}{2a^2}$

$$=\frac{1}{2}\left[\frac{b^2}{a^2}\pm\sqrt{\frac{b^4}{a^4}+4}\right]$$

$\because$ Clearly, $\frac{b^2}{a^2}\neq 1\ m^2>1$

$\Rightarrow \quad m\in(-\infty,-1)\cup(1,\infty)$

and $\quad m^2\neq\frac{1+\sqrt{5}}{2}$

15. (a, b, c) Hyperbola $\frac{x^2}{a^2}-\frac{y^2}{b^2}=1$

$\because$ Locus of point of intersection of perpendicular tangents is director circle $x^2+y^2=a^2-b^2$

Now, $\quad e^2=1+\frac{b^2}{a^2}$

If $a^2>b^2$, then there are infinite (or more than 1) Points on the circle $\Rightarrow e^2<2\Rightarrow e<\sqrt{2}$

If $a^2<b^2$, there does not exist any point on the plane

$\Rightarrow \quad e^2>2\Rightarrow e>\sqrt{2}$

If $a^2=b^2$, there is exactly one point (centre of hyperbola)

$\Rightarrow \quad e=\sqrt{2}$

16. (b, c, d) $\quad h_1=\frac{x^2}{a^2}-\frac{y^2}{b^2}=1$

$$h_2=\frac{x^2}{a^2}-\frac{y^2}{b^2}=2$$

$\Rightarrow \quad \frac{x^2}{2a^2}-\frac{y^2}{2b^2}=1$

Let any point on hyperbola h_1 is $P(a\sec\theta, b\tan\theta)$.

$\therefore$ Then, chord of contact for h_2 w.r.t. P is given by

$$T=0$$

$\frac{x}{a}\sec\theta-\frac{y}{b}\tan\theta=2$ ... (i)

Equation of asymptotes for h_2

$$y=\pm\frac{b}{a}x\Rightarrow y=\frac{b}{a}x$$

Points of asymptotes and Eq. (i) is

$A[2a(\sec\theta+\tan\theta), 2b(\sec\theta+\tan\theta]$

and $\quad B=[2a(\sec\theta-\tan\theta, -2b(\sec\theta-\tan\theta)]$

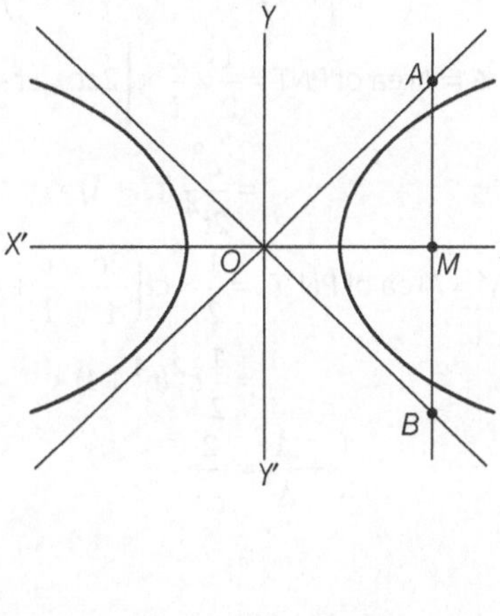

Area of ΔOAB

$$= \frac{1}{2} \text{mod} \begin{vmatrix} 0 & 0 & 1 \\ 2a(\sec\theta + \tan\theta) & 2b(\sec\theta + \tan\theta) & 1 \\ 2a(\sec\theta - \tan\theta) & -2b(\sec\theta - \tan\theta) & 1 \end{vmatrix}$$

$$\Rightarrow \frac{1}{2}|\{-4ab(\sec^2\theta - \tan^2\theta) - 4ab(\sec^2\theta - \tan^2\theta)\}|$$

$$= \frac{1}{2} \times 8ab = 4ab$$

$\because \quad e^2 = 1 + b^2/a^2 = a^2e^2 = a^2 + b^2$

$\Rightarrow \quad b = \sqrt{a^2e^2 - a^2} = a\sqrt{e^2 - 1}$

$$a = \frac{b}{\sqrt{e^2 - 1}}$$

and $\quad 4ab = 4a^2\sqrt{e^2 - 1} = \dfrac{4b^2}{\sqrt{e^2 - 1}}$

17. (a, b, c) Hyperbola $\Rightarrow xy = c^2$

let point $P(ct, c/t)$ on the hyperbola

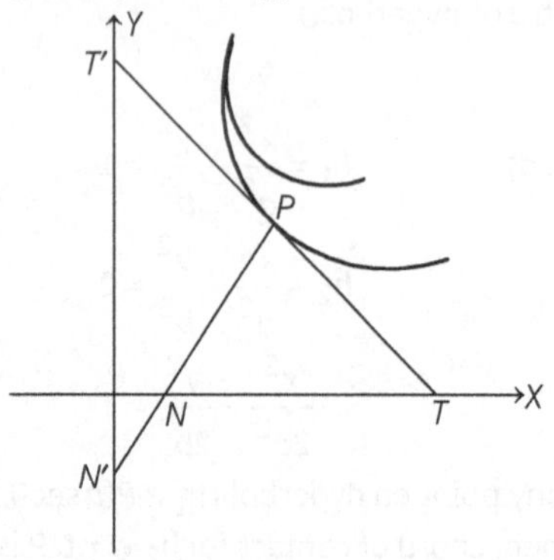

$\therefore$ Equation of tangent at P

$$\frac{x}{t} + ty = 2c \quad \text{...(i)}$$

Normal at P

$$y - \frac{c}{t} = xt^2 - ct^3 \quad \text{...(ii)}$$

Now, for point T and T' put $y = 0$ and $x = 0$ respectively,

We get, $T = (2ct, 0)$ and $T'(0, 2c/t)$

Similarly, $N\left(ct - \dfrac{c}{t^3}, 0\right)$ and $N'\left(0, \dfrac{c}{t} - ct^3\right)$

$$\Delta = \text{Area of } PNT = \frac{1}{2} \times \frac{c}{t} \times \left[2ct - ct + \frac{c}{t^3}\right)$$

$$= \frac{c^2}{2t^4}(t^4 + 1)$$

$$\Delta' = \text{Area of } PN'T' = \frac{1}{2} \times ct\left(\frac{2c}{t} - \frac{c}{t} + ct^3\right)$$

$$= \frac{1}{2}c^2(t^4 + 1)$$

$$\frac{1}{\Delta} + \frac{1}{\Delta'} = \frac{2}{c^2}$$

18. (a, d) Let $P(a\sec\theta, b\tan\theta)$ be a point on hyperbola.

$\therefore$ Equation of tangent at P is

$$\frac{x}{a}\sec\theta - \frac{y}{b}\tan\theta = 1$$

$\because$ (1, 0) lies on the tangent, we get

$$a = \sec\theta$$

Equation of normal at P

$$\frac{ax}{\sec\theta} + \frac{by}{\tan\theta} = a^2 + b^2$$

$\because$ Normal at P makes equal intercept on coordinate axis.

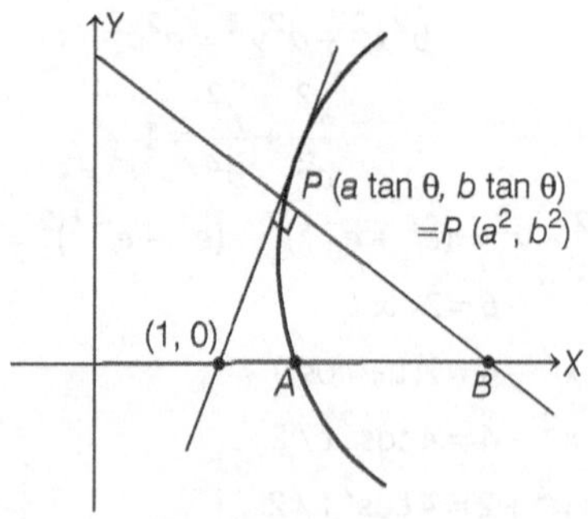

$\therefore$ Slope of normal is -1

So, $\quad \dfrac{-a}{b}\sin\theta = -1$

$$b = \tan\theta$$

$\therefore \quad a^2 - b^2 = 1 \quad \text{... (i)}$

$\because \quad e = \sqrt{1 + \dfrac{b^2}{a^2}} = \sqrt{1 + \dfrac{a^2 - 1}{a^2}} = \sqrt{2 - 1/a^2}$

$\because \quad a > 1$, so $e \in (1, \sqrt{2})$

Area of $\Delta PAB = \dfrac{1}{2} \times AP \times PB$

$$= \frac{1}{2}(\sqrt{(a^2-1)^2 + (b^2)^2}) \times \sqrt{2b^4}$$

$$= \frac{1}{2} \cdot \sqrt{2b^4} \cdot \sqrt{2b^4}$$

$$= \frac{2b^4}{2} = b^4$$

19. (a, b) $\because$ S and S' are foci of hyperbola.

$\therefore \quad S = (ae, 0); \forall S' = (-ae, 0)$

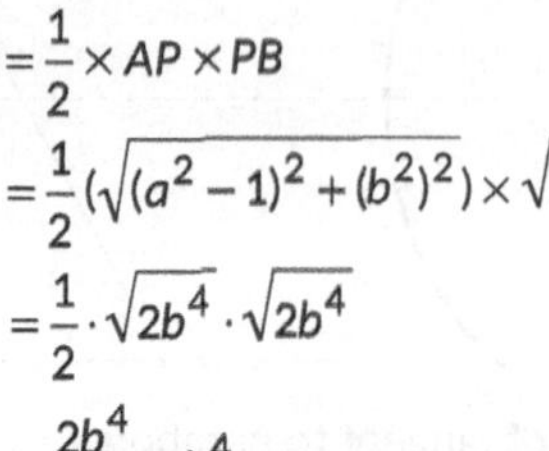

Let $P(a\sec\theta, b\tan\theta)$ be any point on hyperbola.

Let (h, k) be excentre, then

$$h=\frac{a(ae\sec\theta+a)-ae(ae\sec\theta-a)-2ae(a\sec\theta)}{2ae(\sec\theta-1)}$$

$h=-a \Rightarrow x=-a$ (for $a\sec\theta>0$)

Similarly, $x=a$ (for $(a\sec\theta<0)$)

So, locus is $x^2=a^2$

Again let (h,k) excentre opposite $\angle S'$.

$$\therefore h=\frac{2a^2e\sec\theta+a^2e^2\sec\theta+a^2e^2+a^2e^2\sec\theta-a^2e}{2a+2ae},$$

$$k=\frac{2aeb\tan\theta}{2a+2ae}$$

$$h=ae\sec\theta,\ k=\frac{2aeb\tan\theta}{2a+2ae}$$

$\because \quad \sec^2\theta-\tan^2\theta=1$

$$\Rightarrow \left(\frac{h}{ae}\right)^2-k^2\left(\frac{2a+2ae}{2aeb}\right)^2=1 \text{ which is a hyperbola.}$$

20. (a, b, c) Hyperbola $x^2-y^2=4$

$$\frac{x^2}{4}-\frac{y^2}{4}=1$$

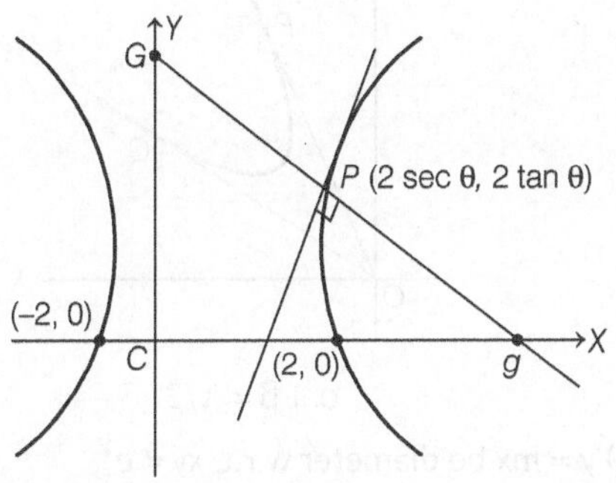

Let $P(2\sec\theta, 2\tan\theta)$ be any point on hyperbola.

Equation of normal

$$\frac{a^2x}{x_1}+\frac{b^2y}{y_1}=a^2+b^2$$

$$\frac{4x}{2\sec\theta}+\frac{4y}{2\tan\theta}=4+4$$

$$2x\cos\theta+2y\cot\theta=8$$

$$x\cos\theta+y\cot\theta=4$$

For g Put $y=0, x=4\sec\theta, g(4\sec\theta,0)$

For G Put $x=0, y=4\tan\theta, G(0,4\tan\theta)$

$$PC=2\sqrt{\sec^2\theta+\tan^2\theta}$$

$$Gg=4\sqrt{\sec^2\theta+\tan^2\theta}$$

$$Pg=\sqrt{(4\sec\theta-2\sec\theta)^2+(2\tan\theta)^2}$$

$$=2\sqrt{\sec^2\theta+\tan^2\theta}$$

$$PG=\sqrt{(2\sec\theta)^2+(2\tan\theta)^2}$$

$$=2\sqrt{\sec^2\theta+\tan^2\theta}$$

$\because \quad PG=Pg=PC$

Sol. (Q.Nos. 21-23)

21. (c) $2a=3$

Distance between the foci (1, 2) and (5, 5) is 5.

$\therefore \quad 2ae=5$

$\therefore \quad e=\frac{5}{3}$

Now, if 'e' is eccentricity of the corresponding conjugate hyperbola, then

$$\frac{1}{e^2}+\frac{1}{e'^2}=1$$

$$\Rightarrow \quad e'=\frac{5}{4}$$

22. (d) Director circle is given by

$$(x-h)^2+(y-k)^2=a^2-b^2$$

where (h,k) is centre, i.e. the

$$\text{mid-point of foci}\left(\frac{1+5}{2},\frac{2+5}{2}\right)=\left(3,\frac{7}{2}\right)$$

$$Y-y=m(X-x)$$

$$b^2=a^2(e^2-1)$$

$$=\left(\frac{3}{2}\right)^2\left(\left(\frac{5}{3}\right)^2-1\right)$$

Therefore, the director circle is

$$(x-3)^2+\left(y-\frac{7}{2}\right)^2=\frac{9}{4}-4$$

$$\Rightarrow \quad (x-3)^2+\left(\frac{y-7}{2}\right)^2=-\frac{7}{4}$$

This does not represents any real point.

23. (b) Slope of transverse axis is $\frac{3}{4}$.

Therefore, the angle of relation is $Q=\tan^{-1}\left(\frac{3}{4}\right)$

Sol. (Q.Nos. 24-26)

24. (a) Equation of tangent in parametric form is given by

$$\frac{x-1}{-1/\sqrt{2}}=\frac{y-1}{1/\sqrt{2}}=\pm 3\sqrt{2}$$

$$\Rightarrow \quad A(4,-2), B(-2,4)$$

$\because$ Equation of asymptotes (OA and OB) are given by

$$y+2=\frac{-2}{4}(x-4) \Rightarrow 2y+x=0$$

and $$y-4=\frac{4}{-2}(x+2) \Rightarrow 2x+y=0$$

Hence, combined equation of asymptotes is

$$(2x+y)(2y+x)=0$$

$$2x^2+2y^2+5xy=0$$

25. (c) $\because m_{OA} = -1/2\ m_{OB} = -2$

$\because \quad \tan\theta = \left|\dfrac{m_{OA} - m_{OB}}{1 + m_{OA}m_{OB}}\right|$

[$\because \theta$ is angle between OA and OB]

$$= \left|\frac{-1/2+2}{1+1}\right|$$

$$\tan\theta = \frac{3}{4}$$

$$\sin\theta = \frac{3}{5} \Rightarrow \theta = \sin^{-1}\frac{3}{5}$$

26. (b) Let the equation of the hyperbola be

$$2x^2 + 2y^2 + 5xy + \lambda = 0$$

It passes through (1, 1).

Therefore, $\quad 2+2+5+\lambda = 0 \Rightarrow \lambda = -9$

So, the hyperbola is

$$2x^2 + 2y^2 + 5xy - 9 = 0$$

Equation of the tangent at $(-1, 7/2)$ is given by

$$2x(-1) + 2y\left(\frac{7}{2}\right) + 5\frac{[x(7/2)] + (-1)(y)}{2} = 9$$

$$\Rightarrow \qquad 3x + 2y = 4$$

Sol. (Q.Nos. 27-31)

27. (a) $\because$ Coordinate of P and Q are $(a\sec\theta, b\tan\theta)$ and $(a\tan\theta, b\sec\theta)$, respectively.

The $\quad OP^2 - OQ^2 = a^2\sec^2\theta + b\tan^2\theta$

$$- a^2\tan^2\theta - b^2\sec^2\theta$$

$$= a^2(\sec^2\theta - \tan^2\theta) + b^2(\tan^2\theta - \sec^2\theta)$$

$$= a^2(1) + b(-1) = a^2 - b^2$$

28. (c) Equation of tangent at P

$$\frac{x}{a}\sec\theta - \frac{y}{b}\tan\theta = 1 \qquad \text{... (i)}$$

$$m = \frac{b}{a}\cdot\frac{\sec\theta}{\tan\theta}$$

Equation of tangent at Q

$$\frac{x}{a}\tan\theta - \frac{y}{b}\sec\theta = 1 \qquad \text{... (ii)}$$

$$m = \frac{b\tan\theta}{a\sec\theta}$$

Equation of asymptotes, $y = \pm\dfrac{b}{a}x$

Solving Eqs. (i) and (ii), we have

$$\frac{x}{a}(\sec\theta - \tan\theta) - \frac{y}{b}(\sec\theta - \tan\theta) = 0$$

$$\frac{x}{a} = \frac{y}{b} \Rightarrow y = \frac{b}{a}x$$

which is an equation of asymptotes.

$\therefore$ Both intersect at asymptotes.

29. (c) $P = (a\sec\theta, b\tan\theta), Q(a\tan\theta, b\sec\theta)$ and $O(0, 0)$

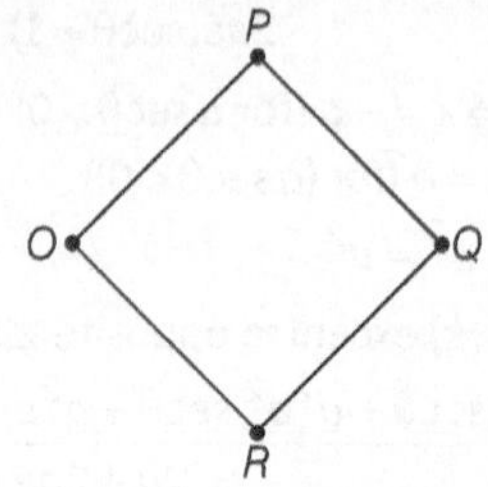

Area of parallel to $OPRQ = 2$, Area of ΔOPQ

$$= 2\times\frac{1}{2}\begin{vmatrix} 0 & 0 & 1 \\ a\sec\theta & b\tan\theta & 1 \\ a\tan\theta & b\sec\theta & 1 \end{vmatrix}$$

$$= ab\sec^2\theta - ab\tan^2\theta$$

$$= ab(\sec^2\theta - \tan^2\theta)$$

$$= ab$$

30. (b) $\because$ We know that conjugate diameter for rectangular asymptote is perpendicular to each other.

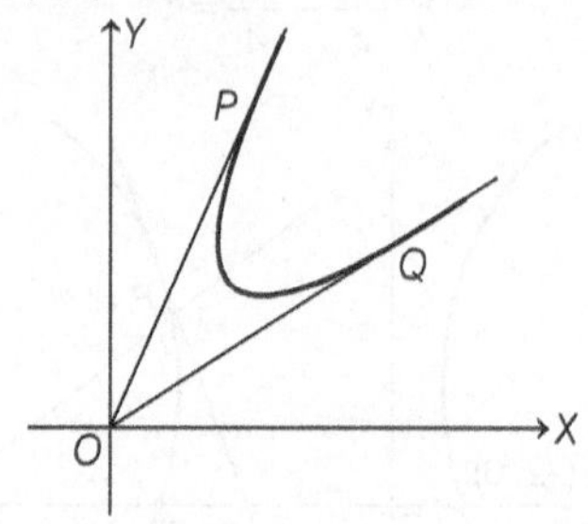

$$\therefore \qquad \alpha + \beta = \pi/2$$

31. (a) $y = mx$ be diameter w.r.t. $xy = e^2$

Hence, $\quad m_1 m_2 = -1$

$$\Rightarrow \qquad m(m_2) = -1 \Rightarrow m_2 = -\frac{1}{m}$$

Equation of diameter is $y = \left(-\dfrac{1}{m}\right)x$

Sol. (Q.Nos. 32-34)

32. (b) Any point on the hyperbola $xy = 16$ is $(4t, 4/t)$.

Normal at this point is $\left(y - \dfrac{4}{t}\right) = t^2(x - 4t)$

If the normal passes through $P(h, k)$, then

$$k - 4/t = t^2(h - 4t)$$

$$\Rightarrow \qquad 4t^4 - t^3h + tk - 4 = 0$$

This equation has roots t_1, t_2, t_3, t_4 which are parameters of the four feet of normals on the hyperbola. Therefore,

$$\Sigma t_1 = \frac{h}{4}, \Sigma t_1 t_2 = 0$$

$$\Sigma t_1 t_2 t_3 = \frac{-k}{4}, t_1 t_2 t_3 t_4 = -1$$

$\because \quad \frac{1}{t_1}+\frac{1}{t_2}+\frac{1}{t_3}+\frac{1}{t_4}=\frac{\Sigma t_1t_2t_3}{t_1t_2t_3t_4}=\frac{k}{4}$

$\Rightarrow \quad y_1+y_2+y_3+y_4=k$

According to question,

$$t_1^2+t_2^2+t_3^2+t_4^2=\frac{h^2}{16}=k$$

$\therefore$ Locus is $x^2=16y$

33. (c) $x^2=16y$

Equation of tangent at P is $x(4t)=\frac{16(y+t^2)}{2}$

$$4tx=8y+8t^2$$
$$tx=2y+2t^2$$
$$A=(2t,0)$$
$$B=(0,-t^2)$$

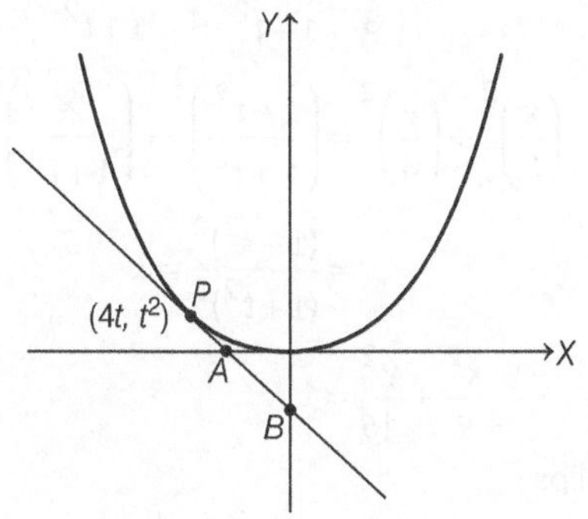

M is mid-point of AB (h,k)

$$h=\frac{2t}{2}, k=\frac{-t^2}{2}$$
$$t=h\Rightarrow k=\frac{-h^2}{2}$$

Locus $x^2+2y=0$

34. (d) $\tan 30^\circ=\frac{4t_1}{t_1^2}=\frac{4}{t_1}$

$\Rightarrow \quad \frac{1}{\sqrt{3}}=\frac{4}{t_1}\Rightarrow t_1=4\sqrt{3}$

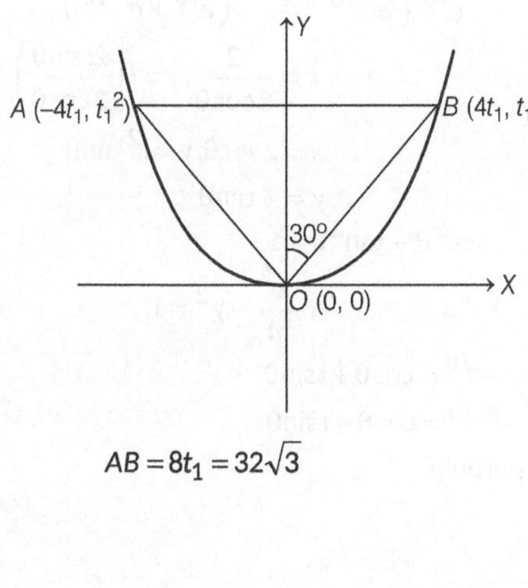

$$AB=8t_1=32\sqrt{3}$$

Area of $(\Delta OAB)=\frac{\sqrt{3}}{4}(AB)^2$

$$=\frac{\sqrt{3}}{4}\times 32\sqrt{3}\times 32\sqrt{3}$$
$$=768\sqrt{3}\text{ sq units}$$

Sol. (Q.Nos. 35-37)

35. (d) Point of contact $(\sqrt{3},1/2)$

and tangent $\sqrt{3}x+2y=4$

$$m=\frac{-\sqrt{3}}{2}$$

$\therefore$ Both the coordinates are positive and m is negative

The possibilities for points are

$$Q\left(-\frac{ma}{\sqrt{m^2+1}},\frac{a}{\sqrt{m^2+1}}\right)\text{ or } R\left(\frac{-a^2m}{\sqrt{a^2m^2+1}},\frac{1}{\sqrt{a^2m^2+1}}\right)$$

For point $Q\left(\frac{\sqrt{3}a}{\sqrt{7}},\frac{2a}{\sqrt{7}}\right)=(\sqrt{3},1/2)$

We get $a=\sqrt{7}$ and $a=\sqrt{7}/4$ which is not possible

For point $R\left(\frac{a^2\sqrt{3}}{\sqrt{3a^2+4}},\frac{2}{\sqrt{3a^2+4}}\right)=(\sqrt{3},1/2)$

$$\frac{a^2}{\sqrt{3a^2+4}}=1\text{ and }\frac{2}{\sqrt{3a^2+4}}=1/2$$
$$a^4-3a^2-4=0\text{ and }3a^2=12$$

$\therefore \quad a^2=4$

Also, for $a^2=4$ equation of ellipse

$x^2+a^2y^2=a^2$ is satisfied for point $(\sqrt{3},1/2)$.

36. (c) Tangent $y=x+8$

$\Rightarrow \quad m=1$

Point (8, 16)

$\therefore$ Both the coordinates as well as m are positive.

The only possibility of points is $\left(\frac{a}{m^2},\frac{2a}{m}\right)=(8,16)$

$\therefore a=8$

Also it satisfies the equation of curve $y^2=4ax$ for the point (8, 16).

And equation of tangent $my=m^2x+a$ is satisfied by

$$m=1\text{ and }a=8$$

$\therefore$ III, (I) P is correct combination.

37. (b) For $a=\sqrt{2}$ and point of contact $(-1,1)$

Equation of circle is satisfied

$$x^2+y^2=2$$

The equation of tangent is

$$-x+y=2$$

$\Rightarrow \quad m=1$

and point of contact $\left(\frac{-ma}{\sqrt{m^2+1}}, \frac{a}{\sqrt{m^2+1}}\right)$

$$= \left(\frac{-\sqrt{2}}{\sqrt{2}}, \frac{\sqrt{2}}{\sqrt{2}}\right) = (-1, 1)$$

∴ I, (II), Q is correct combination.

38. (i) → D, (ii) → C, (iii) → A, (iv) → B

Area of $\Delta LMN = 4\sqrt{3}$ given

$$\Rightarrow \quad \frac{1}{2} \times LM \times ON = 4\sqrt{3}$$

$$\Rightarrow \quad \frac{1}{2} \times 2b \times \sqrt{3}b = 4\sqrt{3}$$

$$\therefore \quad b^2 = 4 \Rightarrow b = 2$$

So, length of conjugate axis of hyperbola $= 2b = 4$

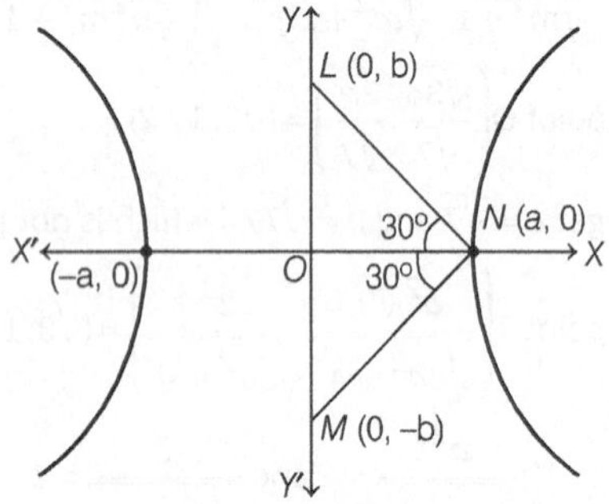

Now, $\quad \tan 30° = \frac{OL}{ON} = \frac{b}{a}$

$$\Rightarrow \quad a = \sqrt{3}b \Rightarrow a = 2\sqrt{3}$$

$$\therefore \quad b^2 = a^2(e^2 - 1)$$

$$\Rightarrow \quad 4 = 12(e^2 - 1)$$

$$e = \sqrt{\frac{4}{3}} = \frac{2}{\sqrt{3}}$$

The distance between the foci of hyperbola $= 2ae$

$$= 2 \times 2\sqrt{3} \times \frac{2}{\sqrt{3}} = 8$$

and length of latusrectum $= \frac{2b^2}{a} = \frac{2 \times 4}{2\sqrt{3}} = \frac{4}{\sqrt{3}}$

39. (i) → A, (ii) → B, (iii) → B, (iv) → D

We have $A = ae_E$ and $a = Ae_H$

$$\Rightarrow \quad e_E \cdot e_H = 1$$

$$\Rightarrow \quad e_E + e_H > 2$$

$$B^2 = A^2(e_H^2 - 1) = a^2(1 - e_E^2) = b^2$$

$$\Rightarrow \quad \frac{b}{B} = 1$$

Also, angle between the asymptotes is

$$2\tan^{-1}\frac{B}{A} = \frac{2\pi}{3}$$

$$\frac{B}{A} = \sqrt{3}$$

$$\Rightarrow \quad \frac{b}{ae_E} = \sqrt{3}$$

$$\Rightarrow \quad e_E^2 = \frac{1}{4}$$

Solving, $\quad \frac{x^2}{a^2} + \frac{y^2}{b^2} = 1$

and $\quad \frac{x^2}{a^2 e_E^2} + \frac{y^2}{b^2} = 1$

We get $\quad x^2 = \frac{2a^2 e_E^2}{b^2(1 - e_E^2)} = 4$

40. (i) → A, (ii) → B, (iii) → B, (iv) → B

(i) $\quad x = 3\left(\frac{1-t^2}{1+t^2}\right), y = \frac{8t}{1+t^2}$

$$\frac{x}{3} = \frac{1-t^2}{1+t^2}, \frac{y}{4} = \frac{2t}{1+t^2}$$

$$\left(\frac{x}{3}\right)^2 + \left(\frac{y}{4}\right)^2 = \left(\frac{1-t^2}{1+t^2}\right)^2 + \left(\frac{2t}{1+t^2}\right)^2$$

$$= \frac{(1+t^2)^2}{(1+t^2)^2} = 1$$

$$\frac{x^2}{9} + \frac{y^2}{16} = 1$$

∴ Ellipse

(ii) $\quad x = \frac{1}{2}(e^t + e^{-t}), y = \frac{1}{2}(e^t - e^{-t})$

$$x^2 = \left(\frac{e^{2t} + e^{-2t} + 2}{4}\right)$$

$$y^2 = \frac{1}{4}(e^{2t} + e^{-2t} - 2)$$

$$x^2 - y^2 = \frac{2}{4} + \frac{2}{4}$$

$$x^2 - y^2 = 1$$

∴ Hyperbola

(iii) $x = \frac{2}{e^{i\theta} + e^{-i\theta}}, y = i \times \left(\frac{e^{i\theta} - e^{-i\theta}}{e^{i\theta} + e^{-i\theta}}\right)$

$$x = \frac{2}{2\cos\theta}, y = i\left(\frac{2i\sin\theta}{2\cos\theta}\right)$$

$$x = 2\sec\theta, y = i^2\tan\theta$$

$$y = -\tan\theta$$

$$\sec^2\theta - \tan^2\theta = 1$$

$$\frac{x^2}{4} - y^2 = 1,$$

$$\because \quad e^{i\theta} = \cos\theta + i\sin\theta$$

$$e^{-i\theta} = \cos\theta - i\sin\theta$$

Hyperbola

(iv) $x = \frac{i}{2}(e^{-i\theta} - e^{i\theta})$ | $y = \frac{e^{i\theta} - e^{-i\theta}}{\sin 2\theta}$

$= -i^2 \frac{2\sin\theta}{2}$ | $= \frac{2\cos\theta}{2\sin\theta\cos\theta}$

$x = \sin\theta$ | $y = \text{cosec}^{i\theta}$

$$xy = 1$$

Hyperbola

41. (i) → D, (ii) → A, (iii) → B, (iv) → C

(i) $P(x_1, y_1)$ be a point whose chord of contact

$$\frac{xx_1}{16} - \frac{yy_1}{9} = 1$$

touches the circle described on the line joining the foci $S(5, 0)$ and $S'(-5, 0)$ whose equation is

$$x^2 + y^2 = 25$$

$$\Rightarrow \quad \left| \frac{0 - 0 - 1}{\sqrt{x_1^2/16^2 + y_1^2/9^2}} \right| = 5$$

$$\Rightarrow \quad \frac{x_1^2}{16^2} + \frac{y_1^2}{9^2} = \frac{1}{25}$$

Locus is $\frac{x^2}{16^2} + \frac{y^2}{9^2} = \frac{1}{25}$

(ii) (x_1, y_1) is mid-point of a chord $x^2 + y^2 = 4$ touching the hyperbola

$$\frac{x^2}{4} - \frac{y^2}{3} = 1$$

That is line $xx_1 + yy_1 = x_1^2 + y_1^2$ touches the hyperbola

$$\Rightarrow \quad \left(\frac{x_1^2 + y_1^2}{y_1} \right)^2 = 4\left(\frac{-x_1}{y_1} \right)^2 = -3$$

$$\Rightarrow \quad (x_1^2 + y_1^2)^2 = 4x_1^2 - 3y_1^2$$

∵ Locus is $(x^2 + y^2)^2 = 4x^2 - 3y^2$

(iii) Director circle of $\frac{x^2}{a^2} - \frac{y^2}{b^2} = 1$ is

$$x^2 + y^2 = a^2 - b^2$$

Hence, $a^2 = 25$ and $b^2 = 16$

∴ Director circle is $x^2 + y^2 = 9$

(iv) Since, $\sqrt{2}$ is the eccentricity of the hyperbola, it must be a rectangular hyperbola.

∴ It is of the form $x^2 - y^2 = a^2$ by hypothesis

$$2ae = 16$$

$$\Rightarrow \quad 2a(\sqrt{2}) = 16$$

$$\Rightarrow \quad a = 4\sqrt{2}$$

∴ Hyperbola is $x^2 - y^2 = 32$

Sol. (Q.Nos. 42-44)

42. (8) $y = mx + \sqrt{9m^2 - 1}$ is a tangent to the hyperbola

$$\frac{x^2}{9} - \frac{y^2}{1} = 1$$

This passes through the point (3, 2) this implies

$$(3m - 2)^2 = 9m^2 - 1$$

$$\Rightarrow \quad 12m = -5$$

$$\Rightarrow \quad m = -\frac{5}{12}$$

Therefore, one tangent is $5x - 12y + 9 = 0$. Also the tangent at the vertex (3, 0) passes through (3, 2). Hence, the other tangent through (3, 2) is $x = 3$, the chord of contact is

$$\frac{3x}{9} - \frac{2y}{1} = 1$$

$$\Rightarrow \quad x - 6y = 3$$

Therefore, the sides of the triangle are

$$5x - 12y + 9 = 0$$

$$x = 3$$

and $\quad x - 6y = 3$

Solving these question, the vertices of the triangle are (3, 2), (3, 0) and $(-5, -4/3)$.

Hence, the area of the triangle is $\frac{1}{2}\begin{vmatrix} 3 & 2 & 1 \\ 3 & 0 & 1 \\ -5 & \frac{-4}{3} & 1 \end{vmatrix}$

$$= \frac{1}{2}\left| 3\left(0 + \frac{4}{3}\right) - 2(3 + 5) + 1(-4 - 0) \right|$$

$$= \frac{1}{2}|4 - 16 - 4|$$

$$= 8 \text{ sq units}$$

43. (3) The area of the triangle formed by two asymptotes and a tangent to the hyperbola

$$\frac{x^2}{a^2} - \frac{y^2}{b^2} = 1$$

is always constant which is equal to ab.

Since, $a = 3, b = 1$

∴ $\quad ab = 3 \times 1 = 3$

44. (3) We know that the portion of the tangent is intercepted between the asymptote is bisected at the point of contact infact the asymptotes are $x = \pm 3y$ and the tangent at (3, 0) is $x = 3$.

Hence, the tangent at (3, 0) intersects the asymptotes at points (3, 1) and (3, −1) so that the mid-point of the segment is (3, 0).

∴ $\alpha + \beta = 3$

Sol. (Q.Nos. 45-47)

45. (9.38) The centre of the hyperbola is

$$\left(\frac{-3+5}{2}, \frac{6-2}{2}\right) = (1, 2)$$

Since, (2, 5) lies on the auxiliary circle, its radius is

$$a = \sqrt{(1-2)^2 + (2-5)^2} = \sqrt{10}$$

The distance between the foci is $2ae$ from the coordinates of the foci, the distance between them is

$$\sqrt{(5+3)^2 + (6+2)^2} = \sqrt{128} = 8\sqrt{2}$$

Now, $\quad 2\sqrt{10}e = 8\sqrt{2} \Rightarrow e = \frac{4}{\sqrt{5}}$

We have, $b^2 = a^2(e^2 - 1) = 10\left(\frac{16}{5} - 1\right)$

$\therefore \quad 2b = 2\sqrt{22} = 9.38$

46. (7) Let the corresponding directrix be $x + y = \lambda$ (because the directrix is perpendicular to transverse axis), the distance of this from the centre is a/e

$$\left|\frac{1+2-\lambda}{\sqrt{2}}\right| = \sqrt{10} \times \frac{\sqrt{5}}{4} = \frac{5}{2\sqrt{2}}$$

$\Rightarrow \quad \lambda = 3 + \frac{5}{2} = \frac{11}{2}$

Therefore, the directrix equation is $2x + 2y - 11 = 0$

$\therefore \ |a + b + c| = 2 + 2 - 11| = 7$

47. (13.91) The length of the latusrectum is

$$\frac{2b^2}{a} = \frac{2(22)}{\sqrt{10}} = \frac{44}{\sqrt{10}} = 13.91$$

48. (40) $\quad (x+3)^2 + (y-12)^2 = 81$

$$(x+3)^2 - (y-8)^2 = -9$$

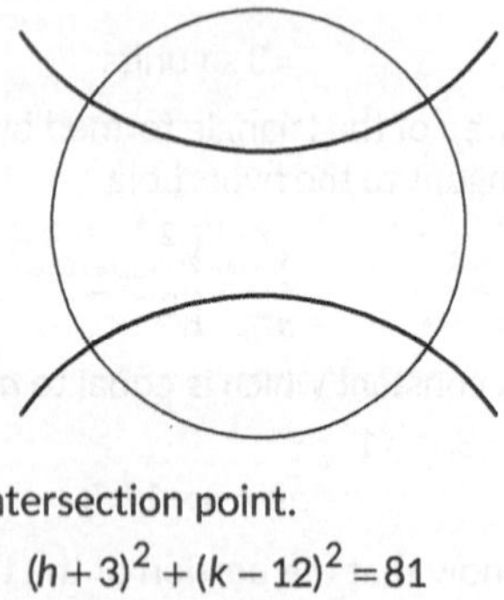

Let (h, k) be intersection point.

Then, $\quad (h+3)^2 + (k-12)^2 = 81 \quad$ ... (i)

$(h+3)^2 - (k-8)^2 = -9 \quad$ ... (ii)

On adding Eqs. (i) and (ii),

$$2h^2 + 12h - 8k + 26 = 0$$

$h^2 + 6x - 4k + 13 = 0 \quad$ ... (iii)

On subtracting Eq. (i) from Eq. (ii),

$$2k^2 - 40k + 118 = 0$$

$k^2 - 20k + 59 = 0 \ \langle \ k_1, k_2 \quad$ ...(iv)

$\therefore \quad k_1 + k_2 = 20, k_1k_2 = 59$

Distance of (h, k) from $(-3, 2)$.

$$d = \sqrt{(h+3)^2 + (k-2)^2} = k$$

$\therefore$ Sum of distance of all intersection points

$$2(k_1 + k_2) = 2(20) = 40$$

49. (0.07) Since, the equation of asymptotes are

$$3x - 4y - 1 = 0$$

and $\quad 4x - 3y - 6 = 0$

$\therefore$ Equation of transverse axis is given by

$$\frac{|3x - 4y - 1|}{\sqrt{3^2 + (-4)^2}} = \frac{|4x - 3y - 6|}{\sqrt{(4)^2 + (-3)^2}}$$

or $\quad |3x - 4y - 1| = |4x - 3y - 6|$

$$3x - 4y - 1 = \pm(4x - 3y - 6)$$

$\Rightarrow \quad x + y - 5 = 0$

and $\quad x - y - 7 = 0$

$\therefore$ Transverse axis is given,

$$ax + by - c = 0 \ a, b, c \in N$$

$\Rightarrow \quad ax + by - c = x + y - 5 = 0$

$$a = 1, b = 1, c = 5$$

$\therefore \quad \frac{a+b+c}{100} = \frac{1+1+5}{100} = 0.07$

50. (0.4) Let the equation of the hyperbola be

$$\frac{x^2}{a^2} - \frac{y^2}{b^2} = 1$$

and its conjugate is

$$\frac{x^2}{a^2} - \frac{y^2}{b^2} = -1$$

The equation of any line parallel to conjugate axis is $x = k$

Let P be $\left(k, \frac{b}{a}\sqrt{k^2 - a^2}\right)$ and Q be $\left(k, \frac{b}{a}\sqrt{k^2 + a^2}\right)$.

The equation of the tangent at P to the hyperbola is

$\frac{xk}{a^2} - \frac{y}{ab}\sqrt{k^2 - a^2} = 1 \quad$ ...(i)

and the equation of the tangent at Q to the conjugate hyperbola is

$\frac{xk}{a^2} - \frac{y}{ab}\sqrt{k^2 - a^2} = -1 \quad$ ... (ii)

Squaring and adding Eqs. (i) and (ii), we get

$$k^2 = \frac{1}{\frac{x^2}{a^4} + \frac{y^2}{a^2b^2}}$$

Solving, we get $x = \frac{1}{2k}\left(\frac{ay}{b}\right)^2$

$$x^2 = \frac{1}{4k^2}\left(\frac{ay}{b}\right)^2$$

$\Rightarrow$ $$x^2 = \frac{1}{4}\left(\frac{ay}{b}\right)^4 \times \left(\frac{x^2}{a^4} + \frac{y^2}{a^2b^2}\right)$$

$\Rightarrow$ $$\frac{4x^2}{a^2} = \frac{y^4}{b^4}\left(\frac{x^2}{a^2} + \frac{y^2}{b^2}\right)$$

$\therefore$ $$\lambda = 4$$

$\therefore$ $$\frac{\lambda}{10} = 0.4$$

51. (3) Equation of normal at $(2\sec\theta, \tan\theta)$ is
$2x\cos\theta + y\cot\theta = 5$

$$\text{Slope} = -1 \Rightarrow \frac{-2\tan\theta}{\sec\theta} = -1$$

$$\theta = \pi/6$$

$\therefore$ Normal becomes $y = -x + 5/\sqrt{3}$

$\because$ It is tangent to ellipse $\frac{x^2}{a^2} + \frac{y^2}{b^2} = 1$

$\therefore$ $$c^2 = a^2m^2 + b^2$$

$$\frac{25}{3} = a^2 + b^2$$

$\therefore$ $$\frac{9}{25}(a^2 + b^2) = \frac{9}{25} \times \frac{25}{3} = 3$$

52. (7) Given hyperbola, $3x^2 - 2y^2 = 6$

$\Rightarrow$ $$\frac{x^2}{2} - \frac{y^2}{3} = 1$$

$\therefore$ $$a^2 = 2, b^2 = 3$$

Tangents from the point (α, β)

$$y = mx \pm \sqrt{a^2m^2 - b^2}$$

$\Rightarrow$ $$(y - mx)^2 = a^2m^2 - b^2$$

$\Rightarrow$ $$(\beta - m\alpha)^2 = a^2m^2 - b^2$$

$\Rightarrow$ $$m^2(\alpha^2 - 2) - 2\alpha\beta m + \beta^2 + 3 = 0$$

$\therefore$ $$m_1m_2 = \frac{\beta^2 + 3}{\alpha^2 - 2} = 2 = \tan\theta \cdot \tan\phi$$

$$\beta^2 + 3 = 2(\alpha^2 - 2)$$

$\Rightarrow$ $$2\alpha^2 - \beta^2 = 7$$

53. (0) Equation of pair of asymptotes is

$$\frac{x^2}{a^2} - \frac{y^2}{b^2} = 0$$

For intersection points of transversal and pair of asymptotes

$$\frac{x^2}{a^2} - \frac{(mx + c)^2}{b^2} = 0 \begin{matrix} x_1 \\ x_2 \end{matrix} \quad \text{...(i)}$$

$$\frac{x^2}{a^2} - \frac{(mx + c)^2}{b^2} = 1 \begin{matrix} x_3 \\ x_4 \end{matrix}$$

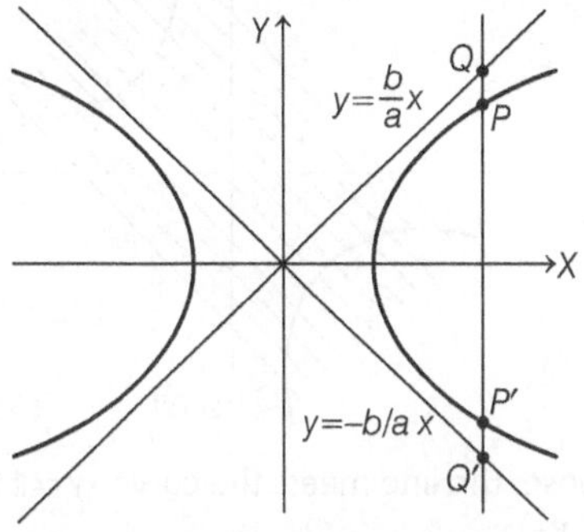

$\Rightarrow$ $$x_1 + x_2 = x_3 + x_4$$

$$y_1 + y_2 = y_3 + y_4$$

$\therefore$ Mid-points of PP' and QQ' are same (say R)

$\therefore$ $$RQ = RQ' \quad \text{... (i)}$$

$$RP = RP' \quad \text{... (ii)}$$

On subtracting Eq. (i) from Eq. (ii),

$$PQ = P'Q'$$

$\Rightarrow$ $$PQ + PP' = P'Q' + PP'$$

$\Rightarrow$ $$P'Q = PQ'$$

$(PQ - P'Q') + (PQ' - P'Q) = 0$

54. (4) $\because$ The normal drawn at $P(t_1)$ meets the hyperbola $xy = 2$ again at $Q(t_2)$, then

$$t_1^3t_2 = -1 \quad \text{... (i)}$$

$\because$ $$P = \left(\sqrt{2}t_1, \frac{\sqrt{2}}{t_1}\right)$$

and $$Q = \left(\sqrt{2}t_2, \frac{\sqrt{2}}{t_2}\right)$$

$\therefore$ Distance $PQ = \sqrt{(\sqrt{2}t_1 - \sqrt{2}t_2)^2 + \left(\frac{\sqrt{2}}{t_1} - \frac{\sqrt{2}}{t_2}\right)^2}$

$$= \sqrt{2}|t_1 - t_2|\sqrt{1 + \frac{1}{t_1^2t_2^2}}$$

$$= \sqrt{2}\,\frac{(t_1^4 + 1)^{3/2}}{t_1^3} \quad \text{[from Eq. (i)]}$$

$\Rightarrow$ $$\sqrt{2}\left(t_1^2 + \frac{1}{t_2^2}\right) \geq \sqrt{2}\,(2)^{3/2}$$

$$= 4 \quad [\because \text{AM} \geq \text{GM}]$$

$\therefore$ Distance $PQ \geq 4$

$\therefore$ Minimum distance between P and Q is 4.

55. (0.025) Let $M(h, k)$ be the point.

The equation of any line through $M(h, k)$ having slope 4 is

$$y - k = 4(x - h)$$

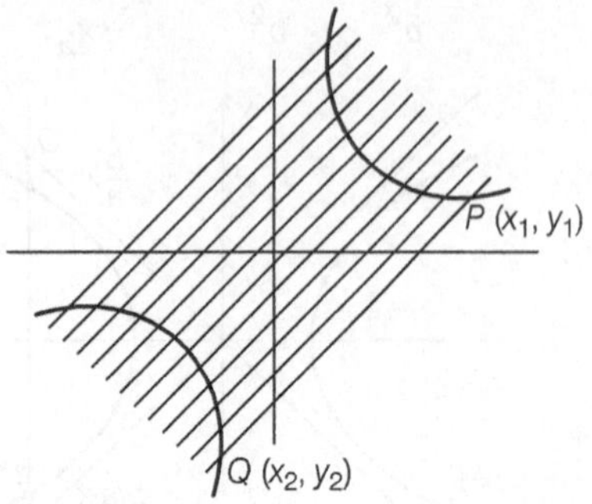

Suppose, the line meets the curve $xy = 1$ at $P(x_1, y_1)$ and $Q(x_2, y_2)$.

Now $\quad y - k = (x - h)$

$\Rightarrow \quad \frac{1}{x} - k = 4(x - h)$

$\Rightarrow \quad 4x^2 - (4h - k)x - 1 = 0$

Let its roots are x_1 and x_2.

$\therefore \quad x_1 + x_2 = \frac{4h - k}{4}$ ... (i)

$x_1 x_2 = -\frac{1}{4}$ ... (ii)

Also $\quad h = \frac{2x_1 + x_2}{3}$

$\Rightarrow \quad 2x_1 + x_2 = 3h$ ... (iii)

From Eqs. (i) and (iii), we get

$$x_1 = 3h - \frac{4h - k}{4} = \frac{8h + k}{4}$$

$$x_2 = 3h - 2\left(\frac{8h + k}{4}\right) = -\left(\frac{2h + k}{2}\right)$$

Putting the values of x_1 and x_2 in Eq. (ii), we get

$$-\left(\frac{8h + k}{4}\right)\frac{(2h + k)}{2} = -\frac{1}{4}$$

$\Rightarrow \quad 16h^2 + 10hk + k^2 - 2 = 0$

Locus is $\quad 16x^2 + 10xy + y^2 - 2 = 0$

$a = 16, b = 1, c = 10, d = -2$

$\Rightarrow \quad \frac{a + b + c + d}{1000} = \frac{25}{1000} = 0.025$